HAESE MATHEMATICS

Specialists in mathematic

Mathematics 9

MYP 4

third edition

Michael Haese
Mark Humphries
Ngoc Vo

for use with
IB Middle Years
Programme

MATHEMATICS 9 MYP 4 third edition

Michael Haese	B.Sc.(Hons.), Ph.D.
Mark Humphries	B.Sc.(Hons.)
Ngoc Vo	B.Ma.Sc.

Published by Haese Mathematics
152 Richmond Road, Marleston, SA 5033, AUSTRALIA
Telephone: +61 8 8210 4666
Email: info@haesemathematics.com
Web: www.haesemathematics.com

National Library of Australia Card Number & ISBN 978-1-922416-34-6

First Edition	2008
Reprinted	2009 (twice), 2010, 2011, 2012
Second Edition	2014
Reprinted	2015, 2016 (twice), 2017, 2018, 2019, 2021
Third Edition	2022
Reprinted	2023, 2024

Cartoon artwork by James Hobbs, Yi-Tung Huang, John Martin, and Ngoc Vo.

Artwork by Hannah Coleman and Brian Houston.

Computer software by Patrick French, Ben Hensley, Huda Kharrufa, Rachel Lee, Linden May, and Han Zong Ng.

Audio recorded by Eloise Quinn-Valentine.

Production work by Hannah Coleman and Michael Mampusti.

Typeset in Australia by Charlotte Frost and Deanne Gallasch. Typeset in Times Roman 10.

Printed in China by Prolong Press Limited.

Acknowledgements: While every attempt has been made to trace and acknowledge copyright, the authors and publishers apologise for any accidental infringement where copyright has proved untraceable. They would be pleased to come to a suitable agreement with the rightful owner.

FOREWORD

Mathematics 9 MYP 4 third edition has been designed and written for the International Baccalaureate Middle Years Programme (IB MYP) Mathematics framework, providing complete coverage of the content and expectations outlined.

This book is suitable for students completing the Standard or Extended content of the MYP Mathematics framework. Some of the material is only intended for students completing the Extended content. This material is marked in dark blue.

Discussions, Activities, Investigations, and Research exercises are used throughout the chapters to develop conceptual understanding. Material is presented in a clear, easy-to-follow style to aid comprehension and retention, especially for English Language Learners. Each chapter ends with extensive review sets and an online multiple choice quiz.

The associated digital Snowflake subscription supports the textbook content with interactive and engaging resources for students and educators.

The Global Context projects highlight the use of mathematics in understanding history, culture, science, society, and the environment. We have aimed to provide a diverse range of topics and styles to create interest for all students and illustrate the real-world application of mathematics.

We have developed this book in consultation with experienced teachers of IB Mathematics internationally but independent of the International Baccalaureate Organisation (IBO). It is not endorsed by the IBO.

We have endeavoured to publish a stimulating and thorough textbook and digital resource to develop and encourage student understanding and nurture an appreciation of mathematics.

We welcome your feedback.

Email: info@haesemathematics.com

Web: www.haesemathematics.com

ONLINE FEATURES

Each textbook comes with a 12 month subscription to the online edition and its range of interactive features. This can be accessed through the **SNOWFLAKE** online learning platform via a web browser or our offline viewer.

To activate your electronic textbook, please contact Haese Mathematics by emailing info@haesemathematics.com with your proof of purchase such as a copy of your receipt.

For general queries regarding **SNOWFLAKE** and online subscriptions:

- Visit our help page: snowflake.haesemathematics.com.au/help
- Contact Haese Mathematics: info@haesemathematics.com

SELF TUTOR

Self Tutor is an engaging feature that supports students learning independently and in classrooms.

Click on any example box to access a step-by-step animation with teacher voice support providing explanation and understanding.

INTERACTIVE LINKS

The **SNOWFLAKE** icons direct you to interactive tools to enhance learning and teaching.

These features include:

- demonstrations to illustrate and animate concepts
- multiple choice quizzes to test understanding
- games to practise and build skills
- tools for graphing and statistics
- printable pages for use in class.

GRAPHICS CALCULATOR INSTRUCTIONS

Graphics calculator instruction booklets are available for the **Casio fx-CG50**, **TI-84 Plus CE**, and the **HP Prime**. Click on the relevant icon below.

CASIO fx-CG50 | TI-84 Plus CE | HP Prime

When additional calculator help may be needed, specific instructions are available from icons within the text.

GRAPHICS CALCULATOR INSTRUCTIONS

ASSUMED KNOWLEDGE

Click on the icon to view the Assumed knowledge chapters.

ASSUMED KNOWLEDGE

TABLE OF CONTENTS

GLOBAL CONTEXTS

The International Baccalaureate Middle Years Programme focuses teaching and learning through six Global Contexts:

- Identities and relationships
- Orientation in space and time
- Personal and cultural expression
- Scientific and technical innovation
- Globalisation and sustainability
- Fairness and development

The Global Contexts help students to develop connections between different subject areas in the curriculum.

GLOBAL CONTEXT — **FAST FASHION**

Global context: Fairness and development
Statement of inquiry: Performing measurements and calculations can help us to responsibly manage natural resources.
Criterion: Applying mathematics in real-life contexts

GLOBAL CONTEXT

Each project contains a series of questions, divided into:

- Factual questions (in green)
- Conceptual questions (in blue)
- Debatable questions (in red).

The projects are also accompanied by the general descriptor and a task-specific descriptor for one of the four assessment criteria.

Chapter 1

Number

Contents:

OPENING PROBLEM

An office building has 15 floors above ground level and 5 floors below ground level. The building has four lifts which start each day at ground level. In the first 10 minutes of the day the lifts perform these movements:

Lift A: up 2 floors, then down 6 floors

Lift B: up 3 floors, down 5 floors, then up 10 floors

Lift C: up 3 floors twice, then down 8 floors

Lift D: down 2 floors, then up 4 floors 3 times in a row.

Things to think about:

a Can you write an *expression* for the position of each lift after 10 minutes?

b What is the position of each lift after 10 minutes?

c The next person wants to enter a lift from ground level.

  i How can you find the *distance* a lift is from ground level using its position?

  ii Which lift is closest to ground level?

The **real numbers** are the set of numbers which we can place on the number line. They include the **integers** or **whole numbers** as well as the numbers in between the integers including decimal numbers and fractions.

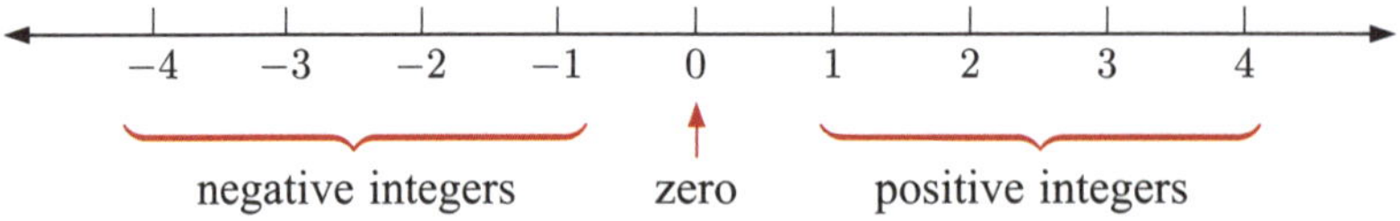

We also define:

- the **counting numbers** as the positive integers 1, 2, 3, 4, 5, ...
- the **natural numbers** as the non-negative integers 0, 1, 2, 3, 4, 5, ...

A EXPONENT NOTATION

We often deal with numbers that are repeatedly multiplied together. Mathematicians use **exponents** or **indices** to represent such products more easily.

Rather than writing $3 \times 3 \times 3 \times 3$, we can write 3^4. We call this **exponent notation**.

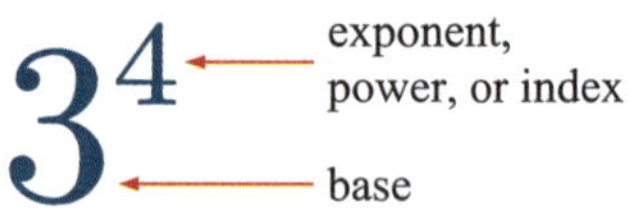

We say that 3 is the **base** and that 4 is the **exponent**, **power**, or **index**.

If n is a positive integer, then a^n is the product of n factors of a.

$$a^n = \underbrace{a \times a \times a \times a \times \dots \times a}_{n \text{ factors}}$$

Example 1 Self Tutor

Evaluate:

a 3^4 **b** $2^4 \times 3^2 \times 7$

a 3^4
$= 3 \times 3 \times 3 \times 3$
$= 81$

b $2^4 \times 3^2 \times 7$
$= 2 \times 2 \times 2 \times 2 \times 3 \times 3 \times 7$
$= 1008$

GRAPHICS CALCULATOR INSTRUCTIONS

NEGATIVE BASES

Consider the following patterns:

$(-1)^1 = -1$
$(-1)^2 = -1 \times -1 = 1$
$(-1)^3 = -1 \times -1 \times -1 = -1$
$(-1)^4 = -1 \times -1 \times -1 \times -1 = 1$

$(-2)^1 = -2$
$(-2)^2 = -2 \times -2 = 4$
$(-2)^3 = -2 \times -2 \times -2 = -8$
$(-2)^4 = -2 \times -2 \times -2 \times -2 = 16$

These patterns lead us to conclude that:

- A **negative** base raised to an **odd** power is **negative**.
- A **negative** base raised to an **even** power is **positive**.

Example 2 Self Tutor

Evaluate:

a -4^2 **b** $(-4)^2$ **c** -2^3 **d** $(-2)^3$

a -4^2
$= -4 \times 4$
$= -16$

b $(-4)^2$
$= -4 \times -4$
$= 16$

c -2^3
$= -2 \times 2 \times 2$
$= -8$

d $(-2)^3$
$= -2 \times -2 \times -2$
$= -8$

GRAPHICS CALCULATOR INSTRUCTIONS

EXERCISE 1A

1 Evaluate:

a 2^3 **b** 2^4 **c** 3^3 **d** 2×5^2

e $2^2 \times 5^2$ **f** $3^3 \times 5$ **g** $3^2 \times 5 \times 7$ **h** $2^2 \times 3^3 \times 11$

i $2^3 \times 3^2 \times 5$ **j** $5^2 \times 7^2 \times 11$ **k** $2^2 \times 5^4 \times 7$ **l** $2^2 \times 3 \times 5^2 \times 7$

2 Write using exponent notation:

a $3 \times 3 \times 3 \times 5$ **b** $2 \times 2 \times 3 \times 3$ **c** $2 \times 3 \times 5 \times 5 \times 5$

3 Evaluate:

a -3^2 **b** $(-3)^2$ **c** -1^3 **d** $(-1)^3$

e $(-5)^2$ **f** -5^2 **g** $(-2)^4$ **h** -2^4

i $(-3)^3$ **j** -3^3 **k** $-(-3)^3$ **l** $-(-4)^3$

4 Evaluate:

a $2^3 \times 3^2 \times (-1)^5$ **b** $(-1)^4 \times 3^3 \times 2^2$ **c** $(-2)^3 \times (-3)^4$

5 Evaluate using a calculator:

a 2^{12} **b** $(-5)^7$ **c** 8^5 **d** $(-9)^5$

e 1.5^4 **f** 1.22^{11} **g** -0.972^{15} **h** $(-1.08)^{23}$

6 Write as a power of 2:

a 16 **b** 64 **c** 256 **d** 512 **e** 1024

7 Write as a power of 3:

a 9 **b** 81 **c** 243 **d** 2187 **e** 59 049

8 By considering $2^1, 2^2, 2^3, 2^4, 2^5, ...$ and looking for a pattern, find the last digit of 2^{33}.

9 Find the last digit of:

a 3^{38} **b** 9^{43} **c** 7^{77}

B THE FUNDAMENTAL THEOREM OF ARITHMETIC

In previous years we have seen that:

A **prime number** is a natural number which has exactly two distinct factors, 1 and itself.

There are infinitely many prime numbers, starting with 2, 3, 5, 7, 11, 13, 17, 19, 23, 29, 31, ...

The **Fundamental Theorem of Arithmetic** states that:

Apart from order, every composite number can be written as a **product of prime factors** in **one and only one way**.

For example, $72 = 2 \times 2 \times 2 \times 3 \times 3$ is the only way of writing 72 as the product of prime factors.

Using exponent notation, we write $72 = 2^3 \times 3^2$.

We call this the **prime factorisation** of 72, and say that 72 is written in **prime factored form**.

There are two methods we can use for writing a composite number in prime factored form:

- In **repeated division**, we systematically divide the number by prime numbers which are its factors, starting with the smallest.
- In a **factor tree**, we find a factor pair for the number and use these factors as branches of the tree. We continue finding factor pairs for each branch until we are left only with prime numbers.

Example 3 — Self Tutor

Write 600 in prime factored form.

Repeated division *or* **Factor tree**

$$\begin{array}{r|l} 2 & 600 \\ \hline 2 & 300 \\ \hline 2 & 150 \\ \hline 3 & 75 \\ \hline 5 & 25 \\ \hline 5 & 5 \\ \hline & 1 \end{array}$$

$\therefore \; 600 = 2 \times 2 \times 2 \times 3 \times 5 \times 5$
$= 2^3 \times 3 \times 5^2$

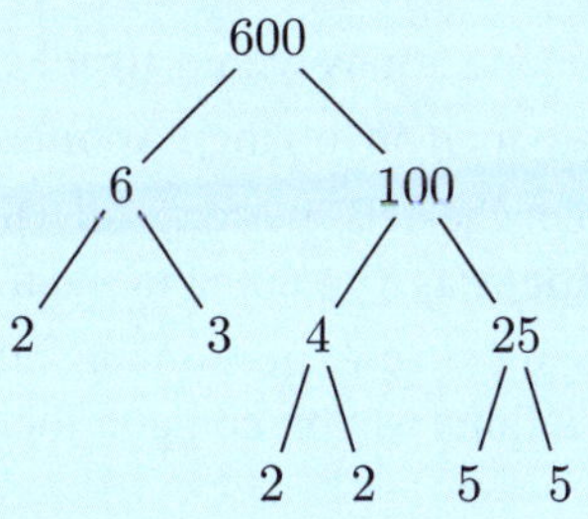

$\therefore \; 600 = 2 \times 3 \times 2 \times 2 \times 5 \times 5$
$= 2^3 \times 3 \times 5^2$

EXERCISE 1B

1 Write as a power of a prime:

a 8 **b** 27 **c** 125 **d** 128
e 343 **f** 729 **g** 361 **h** 1331

2 Write in prime factored form:

a 54 **b** 108 **c** 360 **d** 228
e 196 **f** 756 **g** 936 **h** 1225
i 588 **j** 945 **k** 910 **l** 1274

3 By writing each number in prime factored form, find the highest common factor of:

a 90 and 112 **b** 216 and 336 **c** 345 and 465
d 570 and 524 **e** 720 and 875 **f** 1131 and 1218

4 Find integers a and b greater than 1 such that $a^b \times b^a = 5184$.

5 **a** The prime factorisation of 90 is $2 \times 3^2 \times 5$ or $2^1 \times 3^2 \times 5^1$.
If we define $a^0 = 1$ for all positive integers a, then we can write the factors of 90 as follows:

$1 = 2^0 \times 3^0 \times 5^0$ $\quad 10 = 2^1 \times 3^0 \times 5^1$
$2 = 2^1 \times 3^0 \times 5^0$ $\quad 15 = 2^0 \times 3^1 \times 5^1$
$3 = 2^0 \times 3^1 \times 5^0$ $\quad 18 = 2^1 \times 3^2 \times 5^0$
$5 = 2^0 \times 3^0 \times 5^1$ $\quad 30 = 2^1 \times 3^1 \times 5^1$
$6 = 2^1 \times 3^1 \times 5^0$ $\quad 45 = 2^0 \times 3^2 \times 5^1$
$9 = 2^0 \times 3^2 \times 5^0$ $\quad 90 = 2^1 \times 3^2 \times 5^1$

Explain why the total number of factors of 90 is $2 \times 3 \times 2 = 12$.

b **i** Write 540 in prime factored form.
ii *Hence* write down the total number of factors of 540.

c A number has prime factored form $2^a \times 3^b \times 5^c$. How many factors does it have?

C ORDER OF OPERATIONS

Some expressions contain more than one operation. To evaluate these expressions correctly, we use the following rules:

- Perform the operations within **B**rackets first.
- Calculate any part involving **E**xponents.
- Starting from the left, perform all **D**ivisions and **M**ultiplications as you come to them.
- Restart from the left, performing all **A**dditions and **S**ubtractions as you come to them.

Brackets are **grouping symbols** which indicate a part of an expression which should be evaluated first.

- If an expression contains *one set* of brackets, evaluate that part first.
- If an expression contains *two or more sets* of brackets, one inside the other, evaluate the *innermost set* first.
- In a fraction, we divide *all of* the numerator by *all of* the denominator. So, we treat the numerator and denominator as though they each have brackets around them.

Example 4 **Self Tutor**

Evaluate: $18 - 10 \div 2 + 3^3$

$18 - 10 \div 2 + 3^3$ {exponent first}
$= 18 - 10 \div 2 + 27$ {$\div$ before $-$ and $+$}
$= 18 - 5 + 27$ {$+$ and $-$ from left to right}
$= 13 + 27$
$= 40$

EXERCISE 1C

1 Evaluate:

a $3 + 4 - 9$
b $5 - 12 \div 3$
c $12 + 4 \times 3$
d $-2 \times 5 + 6$
e $3^2 + 6 \div -2$
f $15 \div 3 \times -2$
g $3 - 2 \times 7 + 8$
h $3 \times -2 + 4 \times 6$
i $3 + 18 \div 6 \times 5$
j $24 \div -4 - 3 \times -3$
k $4^2 - 8 \div 2 + 7$
l $18 \div 3^2 - 5$
m $19 - 2^3 \times 3 + 17$
n $3^3 + 18 \div 9 \times 2$
o $3 \times 4^2 - 28 \div 7$

2 Insert an operation $+$, $-$, $\times$, or $\div$ in each space to make the equation true:

a $6 \square 32 \square 8 \square 12 = 22$
b $6 \square 32 \square 8 \square 12 = 70$
c $6 \square 32 \square 8 \square 12 = 96$
d $15 \square 3 \square 12 \square 4 = -3$
e $15 \square 3 \square 12 \square 4 = 159$
f $15 \square 3 \square 12 \square 4 = 48$

Example 5 Self Tutor

Evaluate: $3 \times [12 + (9 \div 3)]$

$$\begin{aligned} & 3 \times [12 + (9 \div 3)] && \{\text{innermost brackets first}\} \\ &= 3 \times [12 + 3] && \{\text{remaining brackets}\} \\ &= 3 \times 15 \\ &= 45 \end{aligned}$$

3 Evaluate:

a $9 \div (7 - 4)$
b $(-2 + 4) \times 3$
c $-2 \times (3 - 6)$
d $(2 - 4) \times 3 + 20$
e $36 \div (-7 + 5^2)$
f $(5 - 8)^2 + 6 \times 2$
g $(6 - 3 \times -4) + 12$
h $(7 - 4) \times (8 - 13)$
i $(12 \div -6)^2 - (7 + 3)$

4 Evaluate:

a $9 - [(4 - 3) + 2 \times 5]$
b $[21 - (4 + 5)] \times 2$
c $12 - [(7 + 4) + 3]$
d $[18 - (15 \div 5)] + 6$
e $108 \div [2 \times (18 \div 3)]$
f $[(4 \times 2) \div (4 \div 2)] \times 3$
g $22 \div [(8 - 6) + 3^2]$
h $[18 - (1 + 6)^2] \times 2$
i $[(14 \div 2) \times (14 \div 7)] \div 2$

Example 6 Self Tutor

Evaluate: $\dfrac{(16 - 7) \div 3}{4 + 2 \times (7 - 3)}$

$$\begin{aligned} & \frac{(16 - 7) \div 3}{4 + 2 \times (7 - 3)} && \{\text{brackets first}\} \\ &= \frac{9 \div 3}{4 + 2 \times 4} && \{\text{evaluate numerator and denominator, } \times \text{ before } +\} \\ &= \frac{3}{4 + 8} && \{\text{evaluate denominator}\} \\ &= \frac{3}{12} \\ &= \frac{1}{4} && \{\text{lowest terms}\} \end{aligned}$$

5 Evaluate:

a $\dfrac{240}{8 \times 6}$
b $\dfrac{27}{17 - 8}$
c $\dfrac{39 \div 3}{14 + 12}$
d $\dfrac{18 + 7}{7 - 2}$
e $\dfrac{58 - 16}{11 - 5}$
f $\dfrac{6 \times 7 + 7}{7}$
g $\dfrac{54}{11 - (2 \times 4)}$
h $\dfrac{(6 + 9) - 5}{7 + (9 - 6)}$
i $\dfrac{2 - 2 \times 2}{2 \times (2 + 2)}$

6 Evaluate:

a $24 \div \dfrac{6^2}{7 - 4}$
b $6 + \dfrac{1 - 8^2}{7}$
c $\dfrac{5 - 3 \times 7}{2^2} + \dfrac{8 \times 9}{12}$

D ABSOLUTE VALUE

The **absolute value** of a number is its *size*, ignoring its sign.

The absolute value of a number is its *distance* from 0 on the number line.

We use the symbol $|$ on either side of a number to indicate its absolute value.

For example:

- The number 5 is 5 units to the right of 0, so $|5| = 5$.
- The number -4 is 4 units to the left of 0, so $|-4| = 4$.

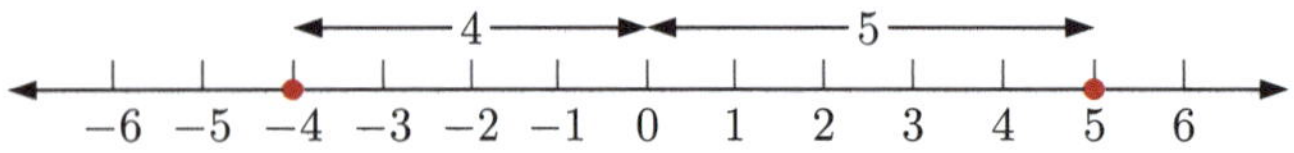

Absolute value signs act as grouping symbols. We need to evaluate the expression within the absolute value signs first, *then* find the absolute value.

Example 7 Self Tutor

Find:

a $|-7 \times -3|$ **b** $|3 - 2 \times 8|$ **c** $\left|\frac{5+9}{4-6}\right|$

a $|-7 \times -3|$
$= |21|$
$= 21$

b $|3 - 2 \times 8|$
$= |3 - 16|$
$= |-13|$
$= 13$

c $\left|\frac{5+9}{4-6}\right|$
$= \left|\frac{14}{-2}\right|$
$= |-7|$
$= 7$

EXERCISE 1D

1 Find:

a $|6|$ **b** $|-1|$ **c** $|-9|$ **d** $|11|$
e $|-13|$ **f** $|4|$ **g** $|-23|$ **h** $|0|$

2 Replace □ with $>$, $<$, or $=$ to make each statement true:

a $|7| \square |5|$ **b** $|-4| \square |-9|$ **c** $|-3| \square |3|$ **d** $|9| \square |-10|$

3 Find:

a $|11 - 4|$ **b** $|5 - 16|$ **c** $|3 \times -8|$
d $|-5 \times -9|$ **e** $|54 \div 9|$ **f** $|64 \div -8|$

4 **a** Find:

i $|9 - 3|$ **ii** $|3 - 3|$ **iii** $|0 - 3|$ **iv** $|-2 - 3|$

b Interpret $|x - 3|$ as a distance on the number line.

5 Find:

a $|4+3\times 6|$ **b** $|8-7\times 4|$ **c** $|10\div 5-4|$

d $|6\times -8+10|$ **e** $|7^2-8|$ **f** $|5\times 4-3^3|$

6 Find:

a $|5\times(2+4)|$ **b** $|42\div(2-5)|$ **c** $|(7-10)\times(1-5)|$

d $\left|\dfrac{6+12}{7-1}\right|$ **e** $\left|\dfrac{8\times 5}{28\div -7}\right|$ **f** $\left|\dfrac{(4+2)^2-1}{13-10\times 2}\right|$

7 Find:

a $3\times|6-8|$ **b** $|4-7|+8$ **c** $\dfrac{|3-12|}{3}$

QUICK QUIZ

MULTIPLE CHOICE QUIZ

REVIEW SET 1A

1 Evaluate:

a 5^3 **b** $2^2\times 5\times 7$ **c** $-(-2)^4$

2 Evaluate using a calculator:

a 3^9 **b** $(-4)^8$ **c** 2.3^5

3 By considering $4^1, 4^2, 4^3, 4^4, 4^5, \ldots$ and looking for a pattern, find the last digit of 4^{99}.

4 Write as a power of a prime:

a 64 **b** 169 **c** 625

5 Write in prime factored form:

a 60 **b** 84 **c** 750

6 By writing each number in prime factored form, find the highest common factor of 1056 and 1848.

7 Evaluate:

a $8-20\div 2^2$ **b** $5\times(2-4)^3$ **c** $\dfrac{6\times 8}{15-(3^2+2)}$

8 Replace □ with $>$, $<$, or $=$ to make each statement true:

a $|8|\ \square\ |-10|$ **b** $|-6|\ \square\ |6|$ **c** $|-11|\ \square\ |-7|$

9 Find:

a $|77\div -7|$ **b** $|12-4\div 2|$ **c** $\left|\dfrac{8\times 12}{10\div -5}\right|$

10 Consider the expression $36 \div 3 - 1 \times 2 + 6 \div 2$.

a Evaluate the expression.

b Insert one set of brackets into the expression to make its value 7.

c Insert two sets of brackets into the expression to make its value 2.

REVIEW SET 1B

1 Write $3 \times 3 \times 5 \times 7 \times 7 \times 7$ using exponent notation.

2 Evaluate:

a $2^3 \times 3$ **b** $(-1)^3 \times 2 \times (-5)^2$

3 Write 32 as a power of 2.

4 Find integers a and b such that $2^a \times 3^b = 144$.

5 Write in prime factored form:

a 135 **b** 1050

6 Evaluate:

a $12 \div -2 \times 3$ **b** $3 \times [10 \div (6-1) + 4]$ **c** $\dfrac{12-(1-9)}{15 \div (1-4)}$

7 Find:

a $|6 \times -3|$ **b** $|5^2 - 6^2|$ **c** $\left|\dfrac{2 - 8 \times 4}{-1-6}\right|$

8 Find:

a $-24 \div |3-6|$ **b** $7 - |9 \times -3| + 2$

9 **a** Write in prime factored form:

i 9 **ii** 20 **iii** 40 **iv** 100 **v** 108 **vi** 144

b Which of the numbers in **a** are perfect squares?

c When perfect squares are written in prime factored form, what do you notice about the *exponent* of each prime factor?

10 Answer the **Opening Problem** on page **12**.

Chapter 2

Algebra: Expressions

Contents:

OPENING PROBLEM

Holly is buying shares in three companies. The prices for each share are shown in the table.

Company	*Share price* ($)
CGX	3.40
XBC	2.19
NGL	5.30

Things to think about:

a Suppose Holly buys 300 CGX shares, 600 XBC shares, and 150 NGL shares. What is the total cost of the shares?

b Now suppose Holly buys x CGX shares. Her financial advisor has recommended that she buy twice as many XBC shares as CGX shares, and half as many NGL shares as CGX shares. Assuming she follows this recommendation, can you write an expression for:

i the number of XBC shares Holly buys **ii** the number of NGL shares Holly buys

iii the total cost of the shares?

c Can you use your expression in **b iii** to check your answer to **a**?

In **algebra** we use letters or symbols to represent unknown values or **variables**.

Together with numbers and operation signs, we use variables to construct mathematical **expressions**.

A ALGEBRAIC NOTATION

In algebra the variables are included in expressions as though they are numbers.

However, we obey some rules which help to simplify the appearance of algebraic expressions.

PRODUCT NOTATION

Product notation is used to represent the *sum* of identical terms.

For example, just as $6+6+6=3\times 6=18$ {3 lots of 6}
we can write $x+x+x=3\times x=3x$ {3 lots of x}

In algebra we agree:

- to **leave out** the $\times$ **signs** between multiplied quantities
- to write **numbers first** in any product
- where products contain two or more letters, we write them in **alphabetical order**.

For example:

- $2a$ is used rather than $2\times a$ or $a2$
- $2ab$ is used rather than $2ba$.

Example 1 Self Tutor

Write using product notation:

a $c+c+c-d\times 7$ **b** $n\times 4\times m-3\times n$ **c** $4\times(x+y)$

a $c+c+c-d\times 7$
$=3\times c-d\times 7$
$=3c-7d$

b $n\times 4\times m-3\times n$
$=4mn-3n$

c $4\times(x+y)$
$=4(x+y)$

EXPONENT NOTATION

Exponent notation is used to represent the *product* of identical factors.

For example, just as $6 \times 6 \times 6 = 6^3$,
we can write $x \times x \times x = x^3$.

Example 2 — Self Tutor

Write using product and exponent notation:

a $p \times p \times p \times p \times 3$ **b** $5 \times z \times z \times z \times y + y \times y$

a $p \times p \times p \times p \times 3$
$= 3p^4$

b $5 \times z \times z \times z \times y + y \times y$
$= 5yz^3 + y^2$

EXERCISE 2A

1 Write using product notation:

a $9 \times x$ **b** $f \times 5$ **c** $q \times p$ **d** $6d \times c$
e $b \times 4 \times a$ **f** $z \times 2y$ **g** $u \times t \times s$ **h** $l \times 3 \times n \times m$
i $a \times b + c$ **j** $d - y \times x$ **k** $2 \times a + 3 \times b$ **l** $4 + m \times 2 + p$

2 Write using product notation:

a $k + k + k + k$ **b** $a + a + b + b + b$ **c** $t + t + t - (u + u)$
d $5 + m + m + m$ **e** $8 - (c + c)$ **f** $9 + g + h + h + h$
g $a + a - b \times 2$ **h** $a \times c + b + b$ **i** $x - y \times 3 + x$

3 Write using product notation:

a $8 \times (a - b)$ **b** $(c + d) \times 6$ **c** $x \times y \times (z + 5)$

4 Write using product and exponent notation:

a $y \times y$ **b** $b \times b \times b \times b$ **c** $4 \times x \times x$
d $a \times 2 \times a$ **e** $y \times y \times y \times 7$ **f** $x \times 3 \times x \times y$
g $m \times n \times n \times m$ **h** $c \times c \times d \times c \times 2$ **i** $6 \times q \times q \times q \times p \times p$

5 Write in expanded form:

a y^3 **b** $4x^4$ **c** $3a^2b$
d $(6p)^2$ **e** $2g^2 + h^3$ **f** $m^5 - 7n^2$

6 Simplify:

a $9 \times a - a \times b \times 4 \times b$ **b** $t \times t + s \times t \times s$
c $2 \times x \times y \times y \times z - z \times z \times 3$ **d** $a \times a \times a - (a + a)$
e $5 \times m \times m + n \times n \times n \times 6$ **f** $m \times n \times m \times n \times m - m \times 2$
g $x + x + x - 8 \times y \times y$ **h** $d + d + d - d \times c \times 6$
i $s + s + s \times t \times t \times 5 \times t$ **j** $g \times g \times g \times 4 - (g + g + g)$
k $x + x + 5 - x \times y \times y$ **l** $a \times 3 \times b \times b + b \times a + b$
m $v \times v \times 10 \times w \times v - 6 \times w \times v \times w + 7 \times w$

B WRITING EXPRESSIONS

It is important to be able to convert between words and algebraic expressions because:

- we need to understand what an expression *means*
- we need to be able to *read* expressions out loud
- worded problems often contain key words such as sum, total, less, and product which tell us which operations we need in the expression.

Example 3 Self Tutor

Write in words:

a $x-5$ **b** $\frac{a+b}{3}$ **c** $3x^2+7$

a $x-5$ is "x minus 5" or "x subtract 5" or "5 less than x".

b $\frac{a+b}{3}$ is "a plus b, all divided by 3" or "the sum of a and b, all divided by 3".

c $3x^2+7$ is "7 more than 3 times the square of x".

EXERCISE 2B

1 Write in words:

a $2a$ **b** pq **c** $\sqrt{m}$ **d** a^2 **e** $a-3$ **f** $b+c$

g $2x+c$ **h** $(2a)^2$ **i** $2a^2$ **j** a^2-c **k** $a-b^2$ **l** $(a-b)^2$

Example 4 Self Tutor

Write as an algebraic expression:

a the sum of p and the square of q **b** the square of the sum of p and q

c the square root of all of 2 times a minus b

a $p+q^2$ **b** $(p+q)^2$ **c** $\sqrt{2a-b}$

2 Write as an algebraic expression:

a the sum of a and c
b the sum of p, q, and r
c the product of a and b
d the sum of r and the square of s
e the square of the sum of r and s
f the sum of the squares of r and s
g the sum of twice a and b
h 5 more than the cube of p
i a less than the square of b
j half the sum of a and b
k the sum of a and a quarter of b
l the square root of the sum of m and n
m the sum of x and its reciprocal
n a quarter of the sum of a and b
o the square root of the sum of the squares of x and y

3 I think of a number, square it, subtract 13, then find the square root of the result. If x is the original number, write an expression for the final result.

DISCUSSION

When we talk about "the difference between 5 and 9", we know to take the smaller number from the larger.

What difficulty might you have when talking about "the difference between 5 and x"?

Can you avoid this difficulty by using absolute value in an expression for "the difference between 5 and x"?

C ALGEBRAIC SUBSTITUTION

If we know the values of the variables in an expression, we can **substitute** the values for the variables. This allows us to **evaluate** the expression.

When we **evaluate** a mathematical expression, we calculate its value for particular numerical values of the variables or unknowns.

If we need to substitute a negative value, we write it in brackets. This helps us obtain the correct sign of each term, and ensures that we do not perform the wrong operation.

Example 5 Self Tutor

If $p = 4$, $q = -2$, and $r = 3$, find the value of:

a $3q^2 - 2r$

b $\dfrac{p - 2q}{2pq + r^3}$

a $3q^2 - 2r = 3 \times (-2)^2 - 2 \times 3$
$= 3 \times 4 - 2 \times 3$
$= 12 - 6$
$= 6$

b $\dfrac{p - 2q}{2pq + r^3} = \dfrac{4 - 2 \times (-2)}{2 \times 4 \times (-2) + 3^3}$
$= \dfrac{4 + 4}{-16 + 27}$
$= \frac{8}{11}$

EXERCISE 2C

1 If $k = 3$, $l = 5$, and $m = -2$, find the value of:

a $4k$
b $9l$
c $2kl$
d klm
e $6m + k$
f $2k + 4m$
g $3m - l$
h $8km - 2l$

2 If $p = 4$, $q = -3$, and $r = 2$, evaluate:

a $\dfrac{p}{q}$
b $\dfrac{p + r}{q}$
c $\dfrac{p + q}{r}$
d $\dfrac{p + 4r}{2q}$
e $\dfrac{p + q + r}{p - r}$
f $\dfrac{8q}{2(p + r)}$
g $\dfrac{3(p + r)}{p + r - 2q}$
h $\dfrac{q + 5p - r}{r - q}$

3 If $x = -1$, $y = -5$, and $z = 4$, evaluate:

a x^2
b z^3
c $x^2 + y^2$
d $(x + y)^2$
e $x^3 + y^3$
f $y^2 - yz$
g $(y + z)^3$
h $2y^2$

Example 6 **Self Tutor**

If $p = 4$, $q = -3$, and $r = 2$, evaluate:

a $\sqrt{p - q + r}$ **b** $\sqrt{p + q^2}$

a $\sqrt{p - q + r}$
$= \sqrt{4 - (-3) + 2}$
$= \sqrt{4 + 3 + 2}$
$= \sqrt{9}$
$= 3$

b $\sqrt{p + q^2}$
$= \sqrt{4 + (-3)^2}$
$= \sqrt{4 + 9}$
$= \sqrt{13}$
≈ 3.61

4 If $a = 4$, $b = -1$, and $c = 3$, evaluate:

a $\sqrt{a} + b$ **b** $\sqrt{a + b}$ **c** $\sqrt{a + c}$ **d** $\sqrt{a + bc}$
e $\sqrt{a^2 + c^2}$ **f** $\sqrt{b^2 + c}$ **g** $\sqrt{a + c - 2b}$ **h** $\sqrt{8c + 2ab}$

5 Evaluate $\dfrac{\sqrt{w - 2y}}{x^2 z}$ for:

a $w = 13$, $x = 2$, $y = 2$, $z = 4$ **b** $w = -1.5$, $x = -2$, $y = -2.75$, $z = 2.5$

6 Evaluate $|x - y\sqrt{z}|$ for:

a $x = 7$, $y = 3$, $z = 4$ **b** $x = -10$, $y = -2$, $z = 9$

Example 7 **Self Tutor**

A vehicle salesman shows Nigel a red bus and a green bus. The red bus has one more than twice the number of seats in the green bus.

a If the green bus has g seats, write an expression for the number of seats in the red bus.
b *Hence* find the number of seats in the red bus, if the green bus has 7 seats.

a The red bus has one more than twice the number of seats in the green bus, so the number of seats in the red bus is $2g + 1$.

b If $g = 7$ then $2g + 1 = 2 \times 7 + 1$
$= 15$

So, if the green bus has 7 seats, then the red bus has 15 seats.

7 Reg has an orange tree, an elm tree, and a peach tree in his backyard. The elm tree is twice as tall as the orange tree, and the peach tree is 3 metres shorter than the elm tree.

a Let the height of the orange tree be h m. Write an expression for the height of the:
i elm tree **ii** peach tree.

b If the orange tree is 5 metres tall, find the height of the:
i elm tree **ii** peach tree.

8 Answer the **Opening Problem** on page **22**.

9 There are 24 chocolates in a box. Aislin eats some of the chocolates, then Brindi eats half of the chocolates that are left.

- **a** If Aislin ate x chocolates, find an expression for the number of chocolates that Brindi ate.
- **b** Find the number of chocolates eaten by Brindi if Aislin ate:
  - **i** 4 chocolates
  - **ii** 8 chocolates
  - **iii** 14 chocolates.

10 The *geometric mean* of two numbers is the square root of their product.

- **a** Write an expression for the geometric mean of x and one more than twice x.
- **b** Evaluate this geometric mean if $x = 4$.

11 The *arithmetic mean* of three numbers is a third of their sum.

- **a** Write an expression for the arithmetic mean of x, y, and z.
- **b** Evaluate the arithmetic mean for $x = 6.2$, $y = 4.8$, and $z = 5.2$.

D THE LANGUAGE OF ALGEBRA

The table below defines some **key words** which we use in algebra:

Word	*Meaning*	*Example(s)*
variable	an unknown value that is represented by a letter or symbol	In $A = \pi r^2$: • A and r are variables • π is *not* a variable because its value is a known constant.
expression	numbers and variables joined by operations	$3x^2 - 2xy$, $\frac{4}{x} + 7x$
equation	two expressions joined by an = sign	$x^2 = x + 4$, $3a - 2 = \frac{1}{a}$
terms	algebraic forms which are separated by + or − signs, the signs being included	$x^2 - 2x + xy - 5$ has four terms: x^2, $-2x$, xy, and -5.
like terms	terms with the same variables to the same powers	In $2x + y^2 + xy - 3x$: • $2x$ and $-3x$ are like terms • $2x$ and y^2 are not like terms • xy and y^2 are not like terms.
constant	a term which does not contain a variable	In $x^2 - 2x + xy - 5$, the term -5 is a constant.
coefficient	the number factor of an algebraic term	In $3x^2 - 2x - xy$: • 3 is the coefficient of x^2 • -2 is the coefficient of x • -1 is the coefficient of xy.

Example 8 **Self Tutor**

Consider $3xy - x^2 + 4 + 2y - xy$.

a Is this an equation or an expression?
b How many terms does it contain?
c State the coefficient of: i y ii x^2
d State the term which is a constant.
e Identify the like terms.

a This is an expression, as it does not contain an = sign.
b There are five terms: $3xy$, $-x^2$, 4, $2y$, and $-xy$.
c i The coefficient of y is 2. ii The coefficient of x^2 is -1.
d The constant is 4.
e The like terms are $3xy$ and $-xy$.

EXERCISE 2D

1 Decide whether each of the following is an expression or an equation:

a $3x - 2y = 4$ b $x^2 + \frac{3}{x} - 2$ c $-1 + 5x = \frac{3}{x}$
d $5 - 2xy + x^2$ e $5 = \frac{1}{2}x^2 - y$ f $3 - \frac{5}{2}x + x^2$

2 State the number of terms in:

a $3 - 2x^2 + 4$ b $5 + \frac{6}{x}$ c $-3xy + 2x^2 - x + 1$

3 State the coefficient of x in:

a $6x$ b $-3x$ c $7 + 2x$ d $8 - x$
e $x + x^2$ f $3 - \frac{x}{2}$ g $xy + 5x$ h $-6x + 3x^2$

4 State the coefficient of x^2 in:

a $3x^2$ b $x - x^2$ c $2xy - 3x^2$ d $\frac{2}{3}x^2 + \frac{1}{x}$

5 State the constant(s) in:

a $2 + 3x$ b $4x^2 - 5$ c $3x^2 - 2x + \frac{1}{2}$ d $5 + x^2 - x - 1$

6 Consider $2x^2 - x + 7 + 5x - 4y$.

a Is this an equation or an expression?
b How many terms does it contain?
c State the coefficient of: i x^2 ii y
d State the term which is a constant.
e Identify the like terms.

7 Identify the like terms in:

a $3x - 2y - x^2 - x$ b $3x - 2xy + y + xy$ c $2x^2 - 3 + x^2 - 4x$
d $5x^2 - 2 + 3x + 7$ e $2x + 2xy + x^2y - xy - 2x^2$

E COLLECTING LIKE TERMS

We have seen that like terms contain the same variables to the same powers.

In an algebraic expression we can add like terms. We call this process **collecting like terms**.

When we add like terms, we add the coefficients of the terms.

For example, since $2 - 3 = -1$,

$$2a - 3a = -1a \text{ or just } -a.$$

We can see this by expanding the product notation:

$$2a - 3a = \underbrace{a + a}_{2a} \underbrace{- a - a - a}_{-3a} = -a$$

Example 9 Self Tutor

Where possible, simplify by collecting like terms:

a $4x + 3x$ **b** $5y - 2y$ **c** $2a - 1 + a$ **d** $mn - 2mn$ **e** $a^2 - 4a$

a $4x + 3x = 7x$ $\{4x$ and $3x$ are like terms$\}$

b $5y - 2y = 3y$ $\{5y$ and $-2y$ are like terms$\}$

c $2a - 1 + a = 3a - 1$ $\{2a$ and a are like terms$\}$

d $mn - 2mn = -mn$ $\{mn$ and $-2mn$ are like terms$\}$

e $a^2 - 4a$ cannot be simplified since a^2 and $-4a$ are not like terms.

EXERCISE 2E

1 Where possible, simplify by collecting like terms:

a $3 + a + 5$ **b** $b + b + 5$ **c** $c - 4 + 7$ **d** $x - 3 + x$

e $4m + m$ **f** $4m - m$ **g** $t - 3t$ **h** $-t + 3t$

i $a^2 + a$ **j** $x^2 + x^2 + 2x$ **k** $2l + 2$ **l** $a^2 - 3a^2 + 3$

m $2x + 5x - 4$ **n** $2x + 5x - 4x$ **o** $3ab + ab$ **p** $9x^2y - x^2y$

2 Simplify, where possible:

a $6x - 6x$ **b** $6x - x$ **c** $6x - 6$

d $2cd + cd$ **e** $pq - 3pq$ **f** $3m^2 - m^2$

g $7b^2 - 6 + 2b^2$ **h** $a + 4 + 3a + 2$ **i** $a + 4 - 3a - 2$

j $2c + 3d - c + 4d$ **k** $3a + 8 - a - 2$ **l** $4x^2 + x - x^2 + 3x$

m $xy + 5 - 6 + 2xy$ **n** $b^2 - 4b - b^2 - 3$ **o** $a^2 + 6a + 2a^2 - 2a$

p $4x - y + 5xy$ **q** $p + q + pq + 2$ **r** $3x^2 - 8x - x^2 - 3x$

3 Simplify, where possible:

a $3pq + 4pq - pq$ **b** $m^2 - 4 + 2m^2 + 3$ **c** $g + 2 + 3g - 10$

d $-2a + b - 4a - 3b$ **e** $2ab + b + 3b - 2a$ **f** $x^2 + x^3 - x$

g $2x^2 + 4x^3 + 3x^2 - x^3$ **h** $ab^2 + 2ab + a^2b$ **i** $3x^3 - x^2 + 2x^2 - 3x + x^3$

F ALGEBRAIC PRODUCTS

When two numbers are multiplied together, the result is called a **product**. If we are dealing with integers, we talk about writing a number as a product of **factors**.

We also use these words in algebra.

For example: $-3x \times 5x^2$

$= -3 \times x \times 5 \times x \times x$ {expanding each product}

$= -3 \times 5 \times x \times x \times x$ {changing the order}

$= -15x^3$ {multiplying the numbers and using exponent notation}

We can therefore say that:

- the *product* of $-3x$ and $5x^2$ is $-15x^3$
- $-3x$ and $5x^2$ are *factors* of $-15x^3$.

To find an algebraic product, we:
- multiply the numbers, including any signs
- write the product of the variables using exponent notation where appropriate.

Example 10 Self Tutor

Simplify:

a $2x \times -x^2$ **b** $-4ab \times -2b^2$

a $2x \times -x^2$
$= 2 \times x \times -1 \times x \times x$
$= 2 \times -1 \times x \times x \times x$
$= -2x^3$

b $-4ab \times -2b^2$
$= -4 \times a \times b \times -2 \times b \times b$
$= -4 \times -2 \times a \times b \times b \times b$
$= 8ab^3$

EXERCISE 2F

1 Simplify:

a $2 \times 5x$ **b** $4x \times 3$ **c** $-2 \times 6x$ **d** $-3x \times 7$

e $3x \times x^2$ **f** $6x \times 3x$ **g** $-4d \times d$ **h** $-9x \times -2x^2$

i $-2x \times x^3$ **j** $-5x \times 3x^2$ **k** $(-a)^2$ **l** $3k^2 \times k^2$

m $x \times 3x \times 2x^2$ **n** $5t \times -2t^2 \times t$ **o** $-x \times 4x^3 \times -2x$ **p** $(3x)^3$

Example 11 Self Tutor

Simplify: $5s \times st - 2t \times 3s^2$

$5s \times st - 2t \times 3s^2$
$= 5 \times s \times s \times t - 2 \times t \times 3 \times s \times s$
$= 5s^2t - 6s^2t$
$= -s^2t$ {collecting like terms}

2 Simplify:

a $3 \times 2x + 4x \times 2$ **b** $5 \times 6a - 7b \times 2b$ **c** $6d^2 + 2d \times 5d$

d $x \times 4y + 3y \times 2x$ **e** $9 \times x^2 - 2x \times 4x$ **f** $5x \times y - 4x \times 3y$

g $6a \times 2b + a \times 3b$ **h** $6x \times y - 5x \times 2y$ **i** $5c \times d - 3c \times 4d$

3 Simplify:

a $7x \times 2y - x^2 \times 3y - 3x \times 3y$ **b** $5a^2b \times 3ab + 6ab^2 \times 2ab - 4a \times 4a^2b^2$

c $-3m^2 \times 2n^2 + 5mn^2 \times 2m - n \times 2m^2n$

G ALGEBRAIC QUOTIENTS

$\frac{3x}{7x^3}$ and $\frac{x^2}{5x}$ are examples of **algebraic quotients**, also known as **algebraic fractions**. They can be simplified by cancelling common factors in exactly the same way as for numerical fractions.

For example, just as $\frac{10}{15} = \frac{2 \times \cancel{5}}{3 \times \cancel{5}} = \frac{2}{3}$,

$$\frac{3x^3}{x^4} = \frac{3 \times \cancel{x} \times \cancel{x} \times \cancel{x}}{x \times \cancel{x} \times \cancel{x} \times \cancel{x}} = \frac{3}{x}.$$

Example 12 **Self Tutor**

Simplify: **a** $\frac{10x^4}{5x}$ **b** $\frac{2a^3b}{8ab^2}$

a $$\frac{10x^4}{5x} = \frac{{}^2\cancel{10} \times x \times x \times x \times \cancel{x}^1}{{}_1\cancel{5} \times \cancel{x}_1} = \frac{2x^3}{1} = 2x^3$$

b $$\frac{2a^3b}{8ab^2} = \frac{{}^1\cancel{2} \times a \times a \times \cancel{a}^1 \times \cancel{b}^1}{{}_4\cancel{8} \times \cancel{a}_1 \times b \times \cancel{b}_1} = \frac{a^2}{4b}$$

EXERCISE 2G

1 Simplify:

a $\frac{12x}{4}$ **b** $\frac{6}{3y}$ **c** $\frac{x^3}{x}$ **d** $\frac{10z}{18z}$ **e** $\frac{5x^2}{x}$ **f** $\frac{a}{a^3}$

g $\frac{16y^3}{8y}$ **h** $\frac{6t^2}{15t^3}$ **i** $\frac{x^4}{x^4}$ **j** $\frac{c^3}{7c^5}$ **k** $\frac{3x^5}{12x^2}$ **l** $\frac{8z^2}{6z^4}$

2 Simplify:

a $\frac{x}{xy}$ **b** $\frac{mn}{m}$ **c** $\frac{3a^2}{ab}$ **d** $\frac{x}{2xy^2}$

e $\frac{a^2b}{4a^2b}$ **f** $\frac{9pq}{3q^2}$ **g** $\frac{2x^2y^3}{10x^2y}$ **h** $\frac{6m}{4m^2n}$

i $\frac{7x^3y}{x^2y^4}$ **j** $\frac{ab^2c}{a^2bc}$ **k** $\frac{2x^2z}{y^3z^2}$ **l** $\frac{9pq^3}{15p^2qr^2}$

H ALGEBRAIC COMMON FACTORS

In previous years you should have seen how numbers can be expressed as the product of factors.

A number which is a factor of two or more numbers is called a **common factor** of these numbers.

The **highest common factor** or **HCF** of two or more numbers is the largest factor that is common to all of them.

For example: $63 = 3 \times 3 \times 7$ and $84 = 2 \times 2 \times 3 \times 7$.

$\therefore$ 3 and 7 are the common factors of 63 and 84.

$\therefore$ the highest common factor (HCF) of 63 and 84 is $3 \times 7 = 21$.

The same method can be used to find the HCF of **algebraic products** which contain variables.

Example 13 Self Tutor

Find the highest common factor of:

a $12x$ and $16y$ **b** $8xy^2$ and $6x^2$

a
$$12x = 2 \times 2 \times 3 \times x$$
$$16y = 2 \times 2 \times 2 \times 2 \times y$$
$$\therefore \text{HCF} = 2 \times 2 = 4$$

b
$$8xy^2 = 2 \times 2 \times 2 \times x \times y \times y$$
$$6x^2 = 2 \times 3 \times x \times x$$
$$\therefore \text{HCF} = 2 \times x = 2x$$

EXERCISE 2H

1 Find the highest common factor of:

a 6 and $3x$ **b** $5a$ and $2a$
c $6m$ and $9n$ **d** $5a$ and 32
e $4x$ and $7x$ **f** $12y$ and $6y$
g $15x$ and $20x^2$ **h** $12x^3$ and $8x^2$
i $12m$, $16m^2$, and $20m$

2 Find the HCF of:

a $10a$ and $16b$ **b** $9x$ and $15xy$ **c** x^2y and $3x$
d $2ab$ and $3b^2$ **e** $3xy$ and $8y^2$ **f** $21x^3$ and $14xy^2$
g $6x^2y$ and $18xy^2$ **h** $15xyz$ and $9x^2z$ **i** $6pr^3$ and p^2qr
j $24a^3b$ and $21ac^2$ **k** $26xy^2z$ and $18xz^2$ **l** $2p^2qr^2$ and $14pq^3r^2$

3 Find the HCF of:

a $x(x+1)$ and $2x^2$ **b** $x(x+1)$ and $2(x+1)$
c $2x^2(x+1)$ and $x(x+1)$ **d** $(x-1)x$ and $3(x-1)$
e $2(x-1)(x+1)$ and $6(x-1)$ **f** $4x(x+1)$ and $(x+1)^2(x-1)$
g $3(x+2)(x-1)$ and $9x(x+2)$ **h** $5(x-1)(x+3)$ and $3(x+3)^2$
i $4x(x+1)$ and $6x(x-2)(x+1)$

MULTIPLE CHOICE QUIZ

QUICK QUIZ

REVIEW SET 2A

1 Write using product notation:

a $8 \times t$ **b** $b \times 7 \times a$ **c** $y+y+y+z+z$

2 Write in words:

a $\frac{a+b}{2}$ **b** $a+\frac{b}{2}$ **c** $\sqrt{a}-3$ **d** $\sqrt{a-3}$

3 Write as an algebraic expression:

a the sum of a and the square root of b **b** the square root of the sum of a and b

4 If $a=-2$, $b=3$, and $c=-5$, find the value of:

a $b-2a$ **b** $\sqrt{c^2-b^2}$ **c** $\frac{2(b-c)}{4a}$

5 State the coefficient of a^2 in:

a $3+2a^2$ **b** $a-1-3a^2$ **c** $\frac{3}{4}a^2+a-1$

6 Brothers Lachlan and Simon had \$100 between them to go shopping. Lachlan spent some money, and then Simon spent a quarter of the money that remained.

a If Lachlan spent \$$x$, find an expression for the amount of money that Simon spent.

b Find the amount of money spent by Simon if Lachlan spent:

i \$20 **ii** \$36 **iii** \$60.

7 Consider $3x-2+7x^2-4x$.

a Is this an equation or an expression? **b** How many terms does it contain?

c State the term which is a constant. **d** Identify the like terms.

8 Simplify, where possible:

a $3+x+5$ **b** $2x-7y+3x+y$ **c** $3a^2+4a-a^2$

d t^2-2t+3 **e** $x^3y+2xy^3+4x^3y$ **f** $x^3+5x+2x^2-5x$

9 Simplify:

a $-4x \times 7$ **b** $5x \times 9x$ **c** $(-3d)^2$

10 Simplify:

a $\frac{5d}{d^4}$ **b** $\frac{x^3y}{3x^2y}$ **c** $\frac{16p^2q^2}{4pq^3}$

11 Find the highest common factor of:

a $10m$ and $6m$ **b** $12ab$ and $3ab^2$ **c** $x(x-2)^2$ and $3x^2(x-2)$

12 I think of a number, subtract 5 from it, then square the result. Suppose the number is x.

a Write an expression for this statement.

b Find the value of your expression when:

i $x = 0$ **ii** $x = 4$ **iii** $x = -4$

c Is there an integer value of x that will make this expression negative? Explain your answer.

REVIEW SET 2B

1 Write using product and exponent notation:

a $5 \times k \times k \times k$ **b** $y + y - y \times y \times y \times y \times 3$

c $a \times a \times b \times a - (b + b)$ **d** $x \times x + x \times y \times z \times z \times 4$

2 Write as an algebraic expression:

a 3 more than the square of x **b** the square of the sum of 3 and x

3 If $p = 1$, $q = -2$, and $r = -3$, find the value of $\dfrac{4q - p}{r}$.

4 If $m = 5$ and $n = 3$, find the value of:

a $(m - n)^2$ **b** $\dfrac{2m - 3n}{m - n}$ **c** $\sqrt{\dfrac{m + n}{2}}$

5 Given the expression $3 - 2x^2 - y + x$, state the:

a total number of terms **b** term which is a constant **c** coefficient of y.

6 State the coefficient of x in:

a $6 + 2x - x^2$ **b** $-2x + \dfrac{3}{x}$

7 Where possible, simplify by collecting like terms:

a $10 - x^2 + 6x - 3$ **b** $3t^3 - t^2 + t + 2t^2$

8 Simplify:

a $-3x \times 2x^2$ **b** $2x \times 3y - y \times 4x$ **c** $x \times 3x^2 + 2x \times 4x - 2x^2 \times x$

9 Simplify: **a** $\dfrac{6ab}{15b^4}$ **b** $\dfrac{x^3yz}{5xz^2}$

10 Find the HCF of:

a $8pq^2$ and $14qr$ **b** $2(x + 1)(x - 2)$ and $4(x + 1)^2$

11 During Eid, the students in two Year 3 classes give each other chocolates. There are a students in class 3A, and b students in class 3B.

a Each student in 3A gives one chocolate to every student in 3B. Write an expression for the total number of chocolates given to 3B.

b Each student in 3B gives *two* chocolates to every student in 3A. Write an expression for the total number of chocolates given to 3A.

c Write an expression for the total number of chocolates given out by *both* classes.

d If there are 15 students in 3A and 16 students in 3B, how many chocolates were given out in total?

Chapter 3

Exponents

Contents:

OPENING PROBLEM

Amedeo Avogadro (1776 - 1856) first proposed that the volume of a gas, at a given pressure and temperature, is proportional to the number of atoms or molecules present, regardless of the nature of the gas.

In 1865, **Johann Josef Loschmidt** estimated that one gram of hydrogen contains 6.02×10^{23} atoms. This number of atoms is now called one **mole** of a substance.

French physicist **Jean Perrin**, who earned a Nobel Prize in Physics, named the number in honour of Avogadro in 1909, for his work in the field of molecular theory.

Amedeo Avogadro

Things to think about:

- **a** How can we write 6.02×10^{23} as an ordinary number?
- **b** How many atoms would be in one tonne of hydrogen gas?
- **c** Can you find the mass of 10^{30} atoms of hydrogen?

We have seen how exponent notation can be used to represent repeated multiplication of the same number:

If n is a positive integer, then a^n is the product of n factors of a.

$$a^n = \underbrace{a \times a \times a \times .... \times a}_{n \text{ factors}}$$

In this Chapter we will explore some laws which will help us perform operations with exponents, and which give meaning to zero and negative exponents.

We will also look at some applications of exponents, including standard form and SI units of measurement.

A EXPONENT LAWS

The definition of a^n where n is a positive integer can be used to help simplify more complicated expressions.

For example:

- $a^3 \times a^4 = (a \times a \times a) \times (a \times a \times a \times a) = a^7$
- $\dfrac{a^5}{a^2} = \dfrac{a \times a \times a \times \cancel{a}^1 \times \cancel{a}^1}{{}_1\cancel{a} \times \cancel{a}_1} = a^3$
- $(a^3)^2 = (a \times a \times a) \times (a \times a \times a) = a^6$
- $(a \times b)^2 = (a \times b) \times (a \times b) = a \times a \times b \times b = a^2b^2$
- $\left(\dfrac{a}{b}\right)^3 = \dfrac{a}{b} \times \dfrac{a}{b} \times \dfrac{a}{b} = \dfrac{a \times a \times a}{b \times b \times b} = \dfrac{a^3}{b^3}$

These examples can be generalised to the following **exponent laws** for **positive integer exponents**:

If m and n are positive integers, then:

- $a^m \times a^n = a^{m+n}$ — To **multiply** numbers with the **same base**, keep the base and **add** the exponents.
- $\dfrac{a^m}{a^n} = a^{m-n}$ provided $a \neq 0$ — To **divide** numbers with the **same base**, keep the base and **subtract** the exponents.
- $(a^m)^n = a^{m \times n}$ — When **raising** a **power** to a **power**, keep the base and **multiply** the exponents.
- $(ab)^n = a^n b^n$ — To **expand** a **product** to a power, raise each factor to the power.
- $\left(\dfrac{a}{b}\right)^n = \dfrac{a^n}{b^n}$ provided $b \neq 0$. — To **expand** a **quotient** to a power, raise both the numerator and denominator to the power.

Example 1 — Self Tutor

Simplify using $a^m \times a^n = a^{m+n}$:

a $11^5 \times 11^3$ **b** $a^4 \times a^5$ **c** $x^4 \times x^a$

a $11^5 \times 11^3 = 11^{5+3} = 11^8$

b $a^4 \times a^5 = a^{4+5} = a^9$

c $x^4 \times x^a = x^{4+a} = x^{a+4}$

To multiply numbers with the same base, keep the base and add the exponents.

EXERCISE 3A

1 Simplify using $a^m \times a^n = a^{m+n}$:

a $5^4 \times 5^2$ **b** $6^5 \times 6^6$ **c** $3^6 \times 3$ **d** $(-2)^3 \times (-2)^4$

e $x^2 \times x^3$ **f** $a \times a^5$ **g** $a^3 \times a^7$ **h** $b^6 \times b^2$

i $a^2 \times a^n$ **j** $b^m \times b^9$ **k** $x^a \times x^b$ **l** $p^2 \times p \times p^4$

Example 2 — Self Tutor

Simplify using $\dfrac{a^m}{a^n} = a^{m-n}$:

a $\dfrac{7^8}{7^5}$ **b** $\dfrac{b^6}{b^m}$

a $\dfrac{7^8}{7^5} = 7^{8-5} = 7^3$

b $\dfrac{b^6}{b^m} = b^{6-m}$

To divide numbers with the same base, keep the base and subtract the exponents.

2 Simplify using $\frac{a^m}{a^n} = a^{m-n}$:

a $\frac{3^6}{3^2}$	**b** $\frac{7^9}{7^3}$	**c** $5^7 \div 5^4$	**d** $\frac{4^5}{4}$
e $\frac{a^8}{a^3}$	**f** $\frac{x^7}{x^5}$	**g** $\frac{b^9}{b^2}$	**h** $\frac{p^n}{p^2}$
i $\frac{y^5}{y^b}$	**j** $\frac{x^a}{x^b}$	**k** $x^6 \div x^3$	**l** $b^{2n} \div b$

Example 3 **Self Tutor**

Simplify using $(a^m)^n = a^{mn}$:

a $(2^4)^3$ **b** $(x^3)^5$ **c** $(b^7)^m$

a $(2^4)^3 = 2^{4\times3} = 2^{12}$

b $(x^3)^5 = x^{3\times5} = x^{15}$

c $(b^7)^m = b^{7\times m} = b^{7m}$

For a power to a power, keep the base and multiply the exponents.

3 Simplify using $(a^m)^n = a^{mn}$:

a $(2^3)^2$	**b** $(3^7)^3$	**c** $(5^4)^7$	**d** $(2^a)^b$
e $(a^2)^6$	**f** $(m^4)^2$	**g** $(3^m)^2$	**h** $(q^3)^3$
i $(d^6)^n$	**j** $(x^y)^8$	**k** $(x^q)^p$	**l** $(g^{2a})^5$

4 Simplify using the exponent laws:

a $\frac{b^4}{b}$	**b** $x^6 \times x^3$	**c** $(a^5)^8$	**d** $x^y \div x^3$
e $(z^3)^n$	**f** $t^7 \times t^k \times t$	**g** $(x^2)^m \times x^n$	**h** $c^{3x} \div (c^y)^4$

5 Copy and complete, replacing each $\square$ with a number or operation:

a $(x^{\square})^4 = x^{12}$	**b** $n^3 \square n^2 = n^5$	**c** $(2^5)^{\square} = 2^{15}$
d $k^8 \square k^5 = k^3$	**e** $p^{\square} \times p^4 = p^7$	**f** $\frac{r^9}{r^{\square}} = r^8$

Example 4 **Self Tutor**

Simplify using the exponent laws:

a $3x^2 \times 5x^5y^3$ **b** $\frac{20a^9b^3}{4a^6b}$

a $3x^2 \times 5x^5y^3$
$= 3 \times 5 \times x^2 \times x^5 \times y^3$
$= 15 \times x^{2+5} \times y^3$
$= 15x^7y^3$

b $\frac{20a^9b^3}{4a^6b}$
$= \frac{20}{4} \times a^{9-6} \times b^{3-1}$
$= 5a^3b^2$

6 Simplify using the exponent laws:

a $3c^2 \times 5c^3$ **b** $4x^5y^2 \times 7xy^4$ **c** $\frac{a^3b^5}{ab^2}$ **d** $\frac{18d^8}{3d^3}$

e $\frac{14p^3q^4}{2pq}$ **f** $12st^3 \times 3s^3$ **g** $5a^2b \times (b^3)^5$ **h** $\frac{6x^8y^2}{3x^3y}$

Example 5 **Self Tutor**

Write as a power with a prime number base:

a 4^4 **b** 9×3^a **c** 49^{x+2}

a 4^4
$= (2^2)^4$
$= 2^{2\times 4}$
$= 2^8$

b 9×3^a
$= 3^2 \times 3^a$
$= 3^{2+a}$
$= 3^{a+2}$

c 49^{x+2}
$= (7^2)^{x+2}$
$= 7^{2(x+2)}$

7 Write as a power with a prime number base:

a 16 **b** 27 **c** 125 **d** 8^2

e 9^3 **f** $2^a \times 4$ **g** $7^t \div 7$ **h** $8^k \times 2^k$

i $\frac{32}{2^x}$ **j** $\frac{5^{x+2}}{5^{x-1}}$ **k** $(7^4)^{a-1}$ **l** $3^b \times 3^{2-b}$

m $\frac{3^y}{9^x}$ **n** $\frac{4^y}{8^x}$ **o** $\frac{11^{x+2}}{11^{2-x}}$ **p** $\frac{5^t \times 25^t}{125^{t-1}}$

Example 6 **Self Tutor**

Expand the brackets then simplify:

a $(3a)^2$ **b** $\left(\frac{2x}{y}\right)^3$

a $(3a)^2 = 3^2 \times a^2$
$= 9a^2$

b $\left(\frac{2x}{y}\right)^3 = \frac{2^3 \times x^3}{y^3}$
$= \frac{8x^3}{y^3}$

Each factor within the brackets is raised to the power outside them.

8 Expand the brackets then simplify:

a $(pq)^3$ **b** $(ab)^5$ **c** $(xy)^4$ **d** $(pqr)^5$

e $(3a)^3$ **f** $(5b)^2$ **g** $(2n)^5$ **h** $(4ab)^3$

9 Expand the brackets then simplify:

a $\left(\frac{x}{y}\right)^4$ **b** $\left(\frac{a}{b}\right)^7$ **c** $\left(\frac{m}{2}\right)^3$ **d** $\left(\frac{5}{p}\right)^2$

e $\left(\frac{4x}{y}\right)^2$ **f** $\left(\frac{a}{3b}\right)^3$ **g** $\left(\frac{c}{2d}\right)^6$ **h** $\left(\frac{10s}{t}\right)^4$

10 Find, using the expansion laws:

a $\left(\frac{6}{7}\right)^2$ **b** $\left(\frac{5}{3}\right)^3$ **c** $\left(\frac{3}{4}\right)^4$ **d** $\left(\frac{1}{2}\right)^7$

Example 7 Self Tutor

Expand the brackets then simplify:

a $(5x^4y)^3$ **b** $\left(\frac{a^3}{4b}\right)^2$

a $(5x^4y)^3 = 5^3 \times (x^4)^3 \times y^3$
$= 125 \times x^{4\times 3} \times y^3$
$= 125x^{12}y^3$

b $\left(\frac{a^3}{4b}\right)^2 = \frac{(a^3)^2}{4^2 \times b^2}$
$= \frac{a^{3\times 2}}{16 \times b^2}$
$= \frac{a^6}{16b^2}$

11 Expand the brackets then simplify:

a $(3a^2)^2$ **b** $(xy^3)^3$ **c** $(p^4q)^5$ **d** $(2x^4)^5$
e $(2a^2b)^4$ **f** $(3cd^4)^3$ **g** $(m^3n^5)^6$ **h** $(5x^2y^5)^4$

12 Expand the brackets then simplify:

a $\left(\frac{x^2}{7}\right)^2$ **b** $\left(\frac{x}{y^2}\right)^4$ **c** $\left(\frac{w^2}{2v}\right)^2$ **d** $\left(\frac{2r}{3s^2}\right)^3$
e $\left(\frac{2}{p^2q}\right)^4$ **f** $\left(\frac{a^4}{3n^7}\right)^2$ **g** $\left(\frac{2x^2}{y^6}\right)^3$ **h** $\left(\frac{3x^2y^3}{4z}\right)^4$

B ZERO AND NEGATIVE EXPONENTS

We will now use the exponent laws to give meaning to zero and negative exponents.

THE ZERO EXPONENT LAW

By cancelling common factors, we know that $\frac{a^2}{a^2} = \frac{{}^1\cancel{a} \times \cancel{a}^1}{{}_1\cancel{a} \times \cancel{a}_1} = 1$ provided $a \neq 0$.

Alternatively, using the exponent law $\frac{a^m}{a^n} = a^{m-n}$, we find that $\frac{a^2}{a^2} = a^{2-2} = a^0$ provided $a \neq 0$.

Comparing these statements, we conclude that $\mathbf{a^0 = 1}$ **for all** $a \neq 0$.

THE NEGATIVE EXPONENT LAW

By cancelling common factors, we know that $\frac{a^3}{a^5} = \frac{{}^1\cancel{a} \times {}^1\cancel{a} \times {}^1\cancel{a}}{a \times a \times \cancel{a}_1 \times \cancel{a}_1 \times \cancel{a}_1} = \frac{1}{a^2}$ provided $a \neq 0$.

Using the exponent law $\frac{a^m}{a^n} = a^{m-n}$, we find that $\frac{a^3}{a^5} = a^{3-5} = a^{-2}$ provided $a \neq 0$.

Comparing these statements, we conclude that $a^{-2} = \frac{1}{a^2}$.

More generally:

> If a is any non-zero number and n is an integer, then $\boldsymbol{a^{-n} = \frac{1}{a^n}}$.
>
> We say that a^n and a^{-n} are **reciprocals** of one another.
>
> In particular, notice that $\boldsymbol{a^{-1} = \frac{1}{a}}$.

Using the negative exponent law,

$$\begin{aligned}\left(\tfrac{2}{3}\right)^{-4} = \frac{1}{\left(\frac{2}{3}\right)^4} &= 1 \div \left(\tfrac{2}{3}\right)^4 \\ &= 1 \div \frac{2^4}{3^4} && \{\text{expansion law}\} \\ &= 1 \times \frac{3^4}{2^4} && \{\text{division of fractions}\} \\ &= \left(\tfrac{3}{2}\right)^4 && \{\text{expansion law}\}\end{aligned}$$

So, in general: $\left(\frac{a}{b}\right)^{-n} = \left(\frac{b}{a}\right)^n$ provided $a \neq 0$, $b \neq 0$.

Example 8 **Self Tutor**

Simplify, giving answers in simplest rational form:

a $2^0 - 2^{-2}$ **b** $\left(\frac{3}{5}\right)^{-2}$

a $2^0 - 2^{-2}$
$= 1 - \frac{1}{2^2}$
$= \frac{3}{4}$

b $\left(\frac{3}{5}\right)^{-2} = \left(\frac{5}{3}\right)^2$
$= \frac{5^2}{3^2}$
$= \frac{25}{9}$

EXERCISE 3B

1 Simplify:

a 4^0 **b** 12^0 **c** $\left(\frac{2}{3}\right)^0$ **d** $2^0 + 2^2$
e a^0 **f** $(13^3)^0$ **g** $3y^0$ **h** $(3y)^0$
i $12^0 \times 12$ **j** $d^2 \times d^0$ **k** $2 + 3^0$ **l** $8 - 9^0$

2 Simplify:

a $\frac{c^7}{c^7}$ **b** $\frac{a^2b^3}{b^3}$ **c** $\frac{x^2y}{xy}$ **d** $\frac{3a^3b}{6a^2b}$

3 Evaluate, giving your answer as a fraction or mixed number:

a 2^{-1} **b** 3^{-1} **c** 6^{-1} **d** 8^{-1}
e 3^{-2} **f** 2^{-3} **g** 5^{-2} **h** 10^{-2}
i $3^0 - 3^{-1}$ **j** $7^{-1} + 7^0$ **k** $2^0 + 2^1 + 2^{-1}$ **l** $5 + 5^{-2}$

4 Write as a fraction:

a t^{-1} **b** x^{-2} **c** y^{-4} **d** z^{-5}

5 Write as a power with a negative exponent:

a $\frac{5^2}{5^4}$ **b** $\frac{3}{3^5}$ **c** $\frac{2^{10}}{2^{15}}$ **d** $\frac{7^3}{7^{11}}$

6 Simplify, giving answers in simplest rational form:

a $\left(\frac{1}{3}\right)^{-1}$ **b** $\left(\frac{1}{5}\right)^{-1}$ **c** $\left(\frac{3}{7}\right)^{-1}$ **d** $\left(\frac{5}{2}\right)^{-1}$

e $\left(\frac{1}{10}\right)^{-1}$ **f** $\left(\frac{5}{6}\right)^{-2}$ **g** $\left(2\frac{1}{4}\right)^{-2}$ **h** $\left(\frac{2}{3}\right)^{-3}$

Example 9 **Self Tutor**

Write without brackets or negative exponents:

a $(5x)^{-1}$ **b** $5x^{-1}$ **c** $(3b^2)^{-2}$

a $(5x)^{-1}$
$= \frac{1}{5x}$

b $5x^{-1}$
$= \frac{5}{x}$

c $(3b^2)^{-2}$
$= \frac{1}{(3b^2)^2}$
$= \frac{1}{3^2 \times b^4}$
$= \frac{1}{9b^4}$

In $5x^{-1}$, the exponent -1 refers to the x only.

7 Write without brackets or negative exponents:

a $(2x)^{-1}$ **b** $2x^{-1}$ **c** $3b^{-1}$ **d** $(3b)^{-1}$

e $(2b)^{-2}$ **f** $\left(\frac{2}{b}\right)^{-2}$ **g** $\left(\frac{1}{n}\right)^{-2}$ **h** $(3n^{-2})^{-1}$

i $(ab)^{-1}$ **j** ab^{-1} **k** ab^{-2} **l** $(ab)^{-2}$

m $(2ab)^{-1}$ **n** $2(ab)^{-1}$ **o** $2ab^{-1}$ **p** $\frac{(ab)^2}{b^{-1}}$

8 Write using powers of 2, 3, and 5 only:

a 16 **b** $\frac{1}{16}$ **c** 25 **d** $\frac{1}{25}$ **e** 81

f $\frac{1}{81}$ **g** $\frac{1}{1024}$ **h** $\frac{1}{2^a}$ **i** $\frac{9}{25}$ **j** $\frac{32}{81}$

k $2\frac{2}{5}$ **l** $9\frac{3}{8}$ **m** $\frac{1}{8^x}$ **n** $\frac{1}{81^k}$ **o** $\frac{1}{12^m}$

DISCUSSION

- Can we *always* cancel the common factor of x in $\frac{2x}{x}$?
- Discuss how to simplify $\frac{x}{x}$ using the exponent laws. By considering the case where $x = 0$, explain why 0^0 is undefined.

GLOBAL CONTEXT — LOCATION ARITHMETIC

Global context: Scientific and technical innovation
Statement of inquiry: Changing the way we represent numbers can help us perform calculations.
Criterion: Investigating patterns

GLOBAL CONTEXT

C STANDARD FORM (SCIENTIFIC NOTATION)

There are many situations where we need to describe very large and very small numbers. For example:

- There are about 75 000 000 000 000 cells in the human body.
- A glass of tap water contains about 0.000 04 grams of chlorine.

Numbers with so many digits can be hard to comprehend and operate with. We can use **standard form** or **scientific notation** to write these numbers in a way that is easier to understand.

Standard form involves writing any positive number as a number between 1 inclusive and 10, multiplied by an integer power of 10.

The result has the form $a \times 10^k$ where $1 \leqslant a < 10$ and k is an integer.

INVESTIGATION

In this Investigation we seek to understand the properties of numbers written in standard form.

What to do:

1 Write as a power of 10:

a 100 **b** 1000 **c** 10 **d** 100 000
e 0.1 **f** 0.01 **g** 0.0001 **h** 0.001
i 10 000 **j** 1 **k** 0.000 001 **l** 100 000 000

2 What can we say about the size of 10^k if k is:

a a positive integer **b** 0 **c** a negative integer?

3 These numbers are all written correctly in standard form:

$580 = 5.8 \times 10^2$ $3040 = 3.04 \times 10^3$ $9.2 = 9.2 \times 10^0$
$0.75 = 7.5 \times 10^{-1}$ $0.0006 = 6 \times 10^{-4}$

For each number in standard form:

a Are the significant figures the same and in the same order as the original number?

b Is the decimal point always immediately after the first significant figure?

c Does 10^k tell us the place value of the first significant figure in the original number?

d What can you say about the size of k if the original number is:

i greater than or equal to 10 **ii** between 1 inclusive and 10 **iii** less than 1?

From the **Investigation** you should have concluded that:

When a number is written in standard form:

- its significant figures are the same and in the same order
- the decimal point is immediately after the first significant figure
- 10^k is the *place value* of the first significant figure in the original number.
  - If the original number is greater than or equal to 10, then k is positive.
  - If the original number is between 1 inclusive and 10, then $k = 0$.
  - If the original number is less than 1, then k is negative.

Example 10 — Self Tutor

Write in scientific notation:

a 4 500 000 **b** 0.000 592

a $4\,500\,000 = 4.5 \times 10^6$

b $0.000\,592$
$= 5.92 \div 10^4$
$= 5.92 \times \dfrac{1}{10^4}$
$= 5.92 \times 10^{-4}$

EXERCISE 3C.1

1 The following values are all equal to 53 000. Which of them is written in standard form?

A 53×10^3 **B** 0.53×10^5 **C** 5.3×10^4 **D** $53\,000 \times 10^0$

2 Copy and complete to write the following numbers in scientific notation:

a $376 = 3.76 \times 10^{......}$
b $8000 = ... \times 10^3$
c $0.04 = ... \times 10^{-2}$
d $0.005\,07 = 5.07 \times 10^{......}$
e $9\,040\,000 = ... \times 10^6$
f $0.000\,000\,23 = 2.3 \times 10^{......}$

3 Write in standard form:

a 425 **b** 425 000 **c** 4.25 **d** 0.425
e 20.1 **f** 20 100 **g** 0.002 01 **h** 2 010 000
i 3870 **j** 0.0387 **k** 38 700 000 **l** 0.000 387

4 Write each quantity in scientific notation:

a The circumference of the Earth is approximately 40 075 kilometres.
b Bacteria are single cell organisms, some of which have diameter 0.0004 mm.
c There are typically 40 million bacteria in a gram of soil.
d The probability that your six numbers will be selected for Lotto on Saturday night is 0.000 000 141 62.
e Superfine sheep have wool fibres as low as 0.01 mm in diameter.
f The temperature of the Sun's core is 15 million degrees Celsius.

Example 11 Self Tutor

Write as an ordinary number:

a 8.6×10^2 **b** 2.15×10^{-4}

a 8.6×10^2
$= 8.60 \times 100$
$= 860$

b 2.15×10^{-4}
$= 00002.15 \div 10^4$
$= 0.000\,215$

5 Write as an ordinary number:

a 5×10^4 **b** 3×10^3 **c** 1.8×10^7 **d** 8.1×10^2
e 6.5×10^5 **f** 1.1×10^1 **g** 2.75×10^8 **h** 8×10^6

6 Write as an ordinary number:

a 3×10^{-2} **b** 9×10^{-5} **c** 7×10^{-3} **d** 4.1×10^{-4}
e 8.2×10^{-6} **f** 7.61×10^{-1} **g** 3.25×10^{-7} **h** 2×10^{-8}

7 Write as an ordinary number:

a The wavelength of blue light is about 4.75×10^{-7} m.
b The estimated world population for 2017 was 7.53×10^9 people.
c Physicists in Japan created a model bull which is only 1.2×10^{-5} m long.
d The diameter of the Earth is approximately 1.27×10^4 km.

8 Write in standard form:

a 18.17×10^6 **b** 0.934×10^{11} **c** 0.041×10^{-2}

9 Which number is larger?

a 2.2×10^8 or 5.8×10^6 **b** 7.71×10^{-7} or 3.5×10^{-5}
c 4.9×10^{-3} or 4.9×10^3 **d** 6.2×10^8 or 6.41×10^8
e 1.006×10^{-6} or 1.01×10^{-6} **f** 9×10^{-8} or 0

Example 12 Self Tutor

Evaluate, giving your answer in scientific notation:

a $(5 \times 10^4) \times (4 \times 10^5)$ **b** $(8 \times 10^5) \div (2 \times 10^3)$

a $(5 \times 10^4) \times (4 \times 10^5)$
$= 5 \times 4 \times 10^4 \times 10^5$
$= 20 \times 10^{4+5}$
$= 2 \times 10^1 \times 10^9$
$= 2 \times 10^{10}$

b $(8 \times 10^5) \div (2 \times 10^3)$
$= \dfrac{8 \times 10^5}{2 \times 10^3}$
$= \dfrac{8}{2} \times 10^{5-3}$
$= 4 \times 10^2$

10 Evaluate, giving your answer in scientific notation:

a $(2 \times 10^5) \times (3 \times 10^2)$ **b** $(4 \times 10^3) \times (6 \times 10^3)$ **c** $(8 \times 10^5) \times (5 \times 10^6)$
d $(9 \times 10^9)^2$ **e** $(3 \times 10^4) \times (3 \times 10^{-9})$ **f** $(7 \times 10^{-3})^2$
g $(8 \times 10^2) \div (2 \times 10^3)$ **h** $(6 \times 10^7) \div (3 \times 10^{-4})$ **i** $(2 \times 10^9) \div (8 \times 10^5)$

11 This table shows the coal production of five countries in 2020.

Country	*Coal production* (tonnes)
United States	4.8×10^8
Pakistan	7.7×10^6
China	3.9×10^9
Serbia	4.0×10^7
Romania	1.5×10^7

- **a** Which country produced the most coal?
- **b** Write, in standard form, the coal production of Romania in kilograms.
- **c** How many times more coal did the United States produce than Serbia?

STANDARD FORM ON A CALCULATOR

Calculators automatically use standard form to display very large and very small numbers which could not otherwise be displayed on the screen. However, if we wish, we can tell the calculator to give all of its answers in standard form.

GRAPHICS CALCULATOR INSTRUCTIONS

Example 13 — Self Tutor

Use your calculator to find:

a $(1.67 \times 10^4) \times (2.3 \times 10^7)$ **b** $(9.49 \times 10^{-3}) \div (2.6 \times 10^6)$

a $(1.67 \times 10^4) \times (2.3 \times 10^7) = 3.841 \times 10^{11}$

b $(9.49 \times 10^{-3}) \div (2.6 \times 10^6) = 3.65 \times 10^{-9}$

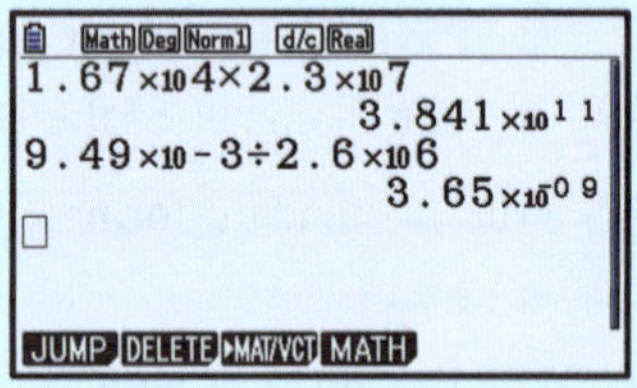

EXERCISE 3C.2

1 Find, in standard form with the decimal part correct to 2 decimal places:

- **a** $0.0003 \times 0.01 \div 5000$
- **b** $375 \times 220 \times 290\,000$
- **c** $8\,760\,000 \times 25\,000$
- **d** $800 \times 740 \times 67\,800$
- **e** $0.002\,12 \div 3\,400\,000$
- **f** $0.019 \times 0.000\,27 \times 0.02$

2 Find, in standard form with the decimal part correct to 2 decimal places:

- **a** $(2.81 \times 10^5) \times (3.4 \times 10^4)$
- **b** $(9.81 \times 10^{-4})^2$
- **c** $\dfrac{3.43 \times 10^{-6}}{7 \times 10^7}$
- **d** $(8.66 \times 10^{-3}) \times (7.5 \times 10^{-5})$
- **e** $\dfrac{1}{2.5 \times 10^6}$
- **f** $(7.59 \times 10^4)^3$

3 Electrons and protons are sub-atomic particles present in every atom.

- **a** Write in standard form:
  - **i** The mass of an electron is about $0.000\,000\,000\,000\,000\,000\,000\,000\,000\,911$ g.
  - **ii** The mass of a proton is about $0.000\,000\,000\,000\,000\,000\,000\,001\,67$ g.
- **b** How many times heavier is a proton than an electron?

4 Answer the following in scientific notation correct to 3 significant figures:

a How many millimetres are there in 479.8 kilometres?

b How many seconds are there in one year? Assume that 1 year = 365 days.

c How many minutes are there in one millennium?

d How many kilograms are there in 0.5 milligrams?

5 If a rocket travels at 3600 km/h, how far will it travel in:

a 1 day **b** 1 week **c** 2 years?

Give your answers in standard form with the decimal part correct to 2 decimal places.
Assume that 1 year = 365 days.

6 Answer the **Opening Problem** on page **36**.

7 The charge of one electron is about 1.602×10^{-19} coulombs.
A bolt of lightning transfers 16.3 coulombs of charge.
How many electrons does it transfer?

D INTERNATIONAL SYSTEM (SI) UNITS

The **International System of Units**, abbreviated SI from the French *le Système International d'Unités*, is the world's most widely used system of measurement.

It is founded on seven base units:

Quantity	*Name*	*Symbol*
Distance	metre	m
Mass	kilogram	kg
Time	second	s
Electric current	ampere	A
Temperature	kelvin	K
Intensity of light	candela	cd
Amount of substance	mole	mol

Other SI units, called **derived units**, are defined in terms of the base units by multiplying or dividing them. The result is often given its own special name.

Some of the common SI derived units are:

Quantity	*Name*	*Symbol*
Area	square metre	m^2
Volume	cubic metre	m^3
Velocity	metres per second	m/s
Angle	radian	rad
Force	newton	N

Quantity	*Name*	*Symbol*
Pressure	pascal	Pa
Energy	joule	J
Power	watt	W
Frequency	hertz	Hz

In addition to the base and derived units, the SI allows the use of other units, such as:

Quantity	*Name*	*Symbol*	*SI equivalent*
Time	minute	min	60 s
	hour	h	3600 s
Mass	gram	g	0.001 kg
	tonne	t	1000 kg
Capacity	litre	L	0.001 m^3
Area	hectare	ha	10 000 m^2
Angle	degree	°	$\frac{\pi}{180}$ rad
Temperature	degree Celsius	°C	$K - 273.15$
Pressure	millibar	mb	100 Pa
Distance at sea	nautical mile	Nm	1.852 km
Speed at sea	knot	kn	1.852 km/h
Energy	kilowatt hour	kWh	3.6 MJ

Smaller or larger multiples of SI units are obtained by combining the base unit with a prefix chosen according to powers of 10. The most commonly used are:

nano	n	$10^{-9} = \frac{1}{1\,000\,000\,000}$
micro	μ	$10^{-6} = \frac{1}{1\,000\,000}$
milli	m	$10^{-3} = \frac{1}{1000}$

kilo	k	$10^3 = 1000$
mega	M	$10^6 = 1\,000\,000$
giga	G	$10^9 = 1\,000\,000\,000$

The SI also accepts the prefix "centi" (10^{-2}) which can be used in conjunction with metre, litre, or gram.

Example 14 **Self Tutor**

Write:

a 642 μm in m

b 2.5×10^{-8} s in ns

a 642 μm
$= 642 \times 10^{-6}$ m
$= 6.42 \times 10^2 \times 10^{-6}$ m
$= 6.42 \times 10^{-4}$ m

b 2.5×10^{-8} s
$= 2.5 \times 10^1 \times 10^{-9}$ s
$= 2.5 \times 10$ ns
$= 25$ ns

EXERCISE 3D

1 Write each length in metres, giving your answer in standard form:

a 6 micrometres
b 4.3 nanometres
c 1.2 millimetres
d 27 nm
e 168 μm
f 0.76 mm

2 Write:

a 3.4×10^{-6} m in micrometres
b 2.7×10^{-9} m in nanometres
c 6.9×10^{-5} m in μm
d 8.5×10^{-7} m in nm

3 Write each time interval in seconds, giving your answer in scientific notation:

a 9 milliseconds
b 4 microseconds
c 2 nanoseconds
d 38 μs
e 410 ms
f 28.5 ns

4 Write:

a 8×10^{-3} s in milliseconds
b 7.8×10^{-5} s in μs
c 1.26×10^{-8} s in ns
d 4.5×10^{-4} s in μs

5 Write each unit of capacity in litres, giving your answer in scientific notation:

a 8 ML
b 5.2 GL
c 31.4 kL
d 172 μL
e 0.4 mL
f 149 kL

6 Write:

a 4.8×10^{3} L in kL
b 1.64×10^{-3} L in mL
c 3.29×10^{8} L in ML
d 6.85×10^{-5} L in μL

7 How many times larger is:

a 1 mm than 1 μm
b 1 ms than 1 ns
c 1 km than 1 mm
d 1 ML than 1 mL?

8 A strand of DNA is 2.5 nm thick, and a piece of paper is 0.1 mm thick.

a Write each thickness in metres.
b How many times thicker is a piece of paper than a strand of DNA?

9 In 2019, the world generated 7.1×10^{9} kWh of nuclear energy each day. Find the total amount of nuclear energy generated in 2019, giving your answer in:

a MWh
b GWh.

10 Lucy's computer can perform 200 million operations per second. Find how long it takes the computer to perform one operation, giving your answer in:

a seconds
b microseconds
c nanoseconds.

11 This table shows the daily water consumption in five countries.

a How much more water is consumed daily in Guinea than in Namibia?
b How many times more water is consumed in Indonesia than in Djibouti?

Country	*Daily water consumption*
Guinea	1.5 GL
Finland	18 GL
Namibia	790 ML
Indonesia	600 GL
Djibouti	50 ML

12 The Environmental Protection Agency has set the acceptable level of lead in public drinking water at 1.5×10^{-5} g/L.

a Write this rate in μg/L.
b Find, in nanograms, the maximum acceptable amount of lead in a 200 mL glass of water.
c A 200 ML source of drinking water contains 2.5 kg of lead. Is this amount of lead within the acceptable limit?

QUICK QUIZ

MULTIPLE CHOICE QUIZ

REVIEW SET 3A

1 Simplify using the exponent laws:

a $q^2 \times q^5$ **b** $\dfrac{a^6}{a^4}$ **c** $(t^{2x})^2$ **d** $(x^{2k})^4 \div x^3$

2 Simplify using the exponent laws:

a $2d^3 \times 5d^6$ **b** $\dfrac{28x^8}{7x^2}$ **c** $3ab^2 \times (a^4)^5$

3 Expand the brackets then simplify:

a $(ab)^3$ **b** $\left(\dfrac{2}{z}\right)^5$ **c** $(5p^3q^2)^3$ **d** $\left(\dfrac{2a^2}{d^3}\right)^4$

4 Simplify:

a 7^0 **b** $6 - \left(\frac{1}{4}\right)^0$ **c** $\dfrac{5a^2b}{a^2}$

5 Evaluate, giving your answer as a fraction or mixed number:

a 4^{-2} **b** 3^{-3} **c** $8^0 - 8^{-1}$ **d** $2^3 + 2^{-4}$

6 Write without brackets or negative exponents:

a $(5c)^{-1}$ **b** $7k^{-2}$ **c** $(4d^2)^{-3}$ **d** $(pq)^{-2}$

7 Write using powers of 2, 3, and 5 only:

a $\frac{1}{8}$ **b** $\dfrac{1}{9^t}$ **c** $1\frac{2}{25}$

8 Write in standard form:

a 1700 **b** 468 200 **c** 0.058 **d** 0.000 606

9 Write as an ordinary number:

a 7.71×10^{-3} **b** 4×10^4

c 1.702×10^{-5} **d** 5.8×10^7

10 The diameter of Jupiter is about 139 800 km. Write this in scientific notation.

11 Evaluate, giving your answer in standard form:

a $(4 \times 10^5) \times (7 \times 10^4)$ **b** $(3 \times 10^{-4})^2$

c $(3 \times 10^3) \div (6 \times 10^8)$

12 The Earth orbits the Sun at a speed of approximately 1.07×10^5 km/h. How far does the Earth move, relative to the Sun, in:

a 1 day **b** 1 week **c** 1 year?

Give your answers in scientific notation with the decimal part correct to 2 decimal places. Assume that $1 \text{ year} = 365$ days.

13 Write, giving your answer in standard form:

a 7 microseconds in seconds **b** 8.52 GL in litres

c 23 nm in metres **d** 504 μg in grams

14 Write:

a 6.7×10^{-6} m in μm **b** 9.42×10^3 L in kL

c 3.8×10^{10} W in GW **d** 1.5×10^{-7} s in ns

15 A grain of sand weighs 350 nanograms. Find the number of grains in 2 kg of sand. Give your answer in scientific notation.

16 A spaceship is travelling to the Alpha Centauri system, which is 4.37 light years from our Sun.

a If one light year is approximately 9.46×10^{12} km, calculate the distance in km between the Sun and Alpha Centauri.

b The Earth is about 1.496×10^{11} m from the Sun.

i Write this distance in kilometres.

ii How many times farther is it from the Sun to Alpha Centauri, than from the Sun to Earth?

REVIEW SET 3B

1 Simplify using the exponent laws:

a $(-5)^4 \times (-5)^6$ **b** $x^{4n} \div x^3$ **c** $(c^5)^k$

2 Copy and complete, replacing each $\square$ with a number:

a $t^{\square} \times t^3 = t^{12}$ **b** $\dfrac{k^8}{k^{\square}} = k^4$

3 Write as a power with a prime number base:

a 25^3 **b** $16^k \times 4^k$ **c** $\dfrac{27^t}{9 \times 3^t}$

4 Write in simplest form, without brackets:

a $(3p)^4$ **b** $\left(\frac{2}{3}\right)^3$ **c** $\left(\dfrac{a^2}{b}\right)^5$

5 Evaluate, giving your answer as a fraction or mixed number:

a 4^{-1} **b** $3^0 + 3^2 + 3^{-1}$ **c** $10^0 - 10^{-3}$

6 Write:

a k^{-3} as a fraction **b** $\dfrac{7^2}{7^7}$ as a power with a negative exponent.

7 Simplify, giving answers in simplest rational form:

a $\left(\frac{4}{5}\right)^{-1}$ **b** $\left(\frac{2}{7}\right)^{-2}$ **c** $\left(3\frac{1}{3}\right)^{-2}$

8 Write without brackets or negative exponents:

a $a^{-2}b$ **b** $(5a^{-2})^2$ **c** $(5p^{-3})^{-1}$

9 Write in scientific notation:

a 263.57 **b** 0.000 511 **c** 863 400 000

10 Write as an ordinary number:

a 2.78×10^0 **b** 3.99×10^7 **c** 2.081×10^{-3}

11 The area of Portugal is approximately 9×10^4 km^2, and the area of Kazakhstan is approximately 2.7×10^6 km^2. How many times larger is Kazakhstan than Portugal?

12 Calculate the following, giving your answer in scientific notation with the decimal part correct to 2 decimal places:

a $(1.91 \times 10^{12}) \times (1.6 \times 10^{-16})$ **b** $\dfrac{8.3 \times 10^{18}}{4.9 \times 10^9}$ **c** $(2.18 \times 10^4)^2$

13 Write 74 ML in litres, giving your answer in standard form.

14 Write:

a 5.3×10^{-6} s in μs **b** 4.57×10^{-7} m in nm

15 A household uses an average of 2.4×10^7 J of gas each day. Find the amount of gas used by the household in a year, giving your answer in GJ. Assume 1 year = 365 days.

16 Thorium is a natural radioactive chemical element. It is named after the Norse god of thunder, Thor.

a The diameter of a Thorium atom is 3.58×10^{-4} μm. Write this in metres.

b Thorium has a half-life of 14.05 billion years. Write this number in scientific notation.

c The *molar mass* of Thorium is 232.04. The mass, in grams, of a single atom of Thorium is found by dividing this molar mass by 6.02×10^{23}.

i Find the mass, in grams, of a single Thorium atom.

ii How many atoms are there in a 10 g block of Thorium?

Chapter 4

Algebra: Expansion

Contents:

OPENING PROBLEM

Each month, Jasmin earns \$10 from her father for washing the family car, and \$18 from her grandfather for mowing his lawns.

Things to think about:

a Can you explain why the total amount Jasmin earns each year is:
  i $(12 \times 10 + 12 \times 18)$ dollars
  ii $12(10 + 18)$ dollars?

b Can you explain *why* $12(10 + 18) = 12 \times 10 + 12 \times 18$?

c Now suppose Jasmin earns $\$x$ each month for washing the family car.
Can you explain why $12(x + 18) = 12x + 216$?

d What mathematical rules can we use to remove brackets from expressions?

When we use algebra to describe real life problems, we often write expressions containing brackets.

In order to work with these expressions, we often want to rewrite them without brackets. The process of writing expressions without brackets is called **expansion**.

A THE DISTRIBUTIVE LAW

In the **Opening Problem**, suppose Jasmin earns $\$x$ each month for washing the family car, and \$18 per month for mowing her grandfather's lawns.

- Each month she earns $(x + 18)$ dollars, so in the whole year she earns $12(x + 18)$ dollars.
- Over a year, she earns $(12 \times x)$ dollars for washing the car,
  and (12×18) dollars for mowing the lawns.

We can hence conclude that $12(x + 18) = 12 \times x + 12 \times 18$.

In the expression $12(x + 18)$, we say that 12 is the **coefficient** of the expression in brackets. We can **expand** the brackets in an expression like this using the **distributive law**:

> When a coefficient is multiplied by an expression in brackets, we remove the brackets by multiplying each term inside them by the coefficient, then adding the results.
>
> $$\boldsymbol{a(b + c) = ab + ac}$$

We can explain why the distributive law works using geometry:

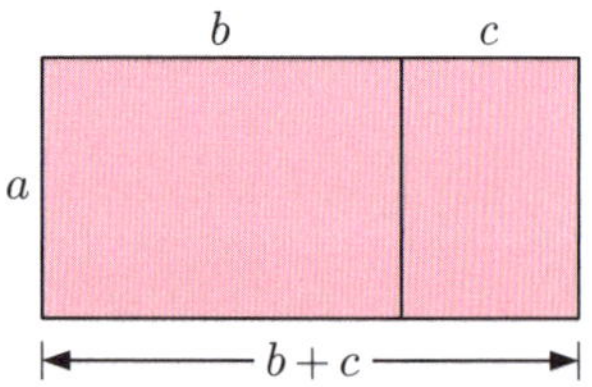

The overall area is $a(b + c)$.

However, the overall area could also be found by adding the areas of the two smaller rectangles. This gives us $ab + ac$.

So, $a(b + c) = ab + ac$. {equating areas}

Example 1 — Self Tutor

Expand using the distributive law:

a $3(4x+1)$ **b** $2x(5-2x)$ **c** $-2x(x-3)$

a
$$\begin{aligned}&3(4x+1)\\&=3\times 4x+3\times 1\\&=12x+3\end{aligned}$$

b
$$\begin{aligned}&2x(5-2x)\\&=2x(5+-2x)\\&=2x\times 5+2x\times -2x\\&=10x-4x^2\end{aligned}$$

c
$$\begin{aligned}&-2x(x-3)\\&=-2x(x+-3)\\&=-2x\times x+-2x\times -3\\&=-2x^2+6x\end{aligned}$$

With practice, you should not need to write all of these steps.

EXERCISE 4A

1 Expand and then simplify:

a $3(x+1)$ **b** $2(5-x)$ **c** $5(x+2)$
d $4(3-x)$ **e** $5(x-y)$ **f** $9(7+x)$
g $x(x+3)$ **h** $m(7-m)$ **i** $a(a+b)$

2 Expand and then simplify:

a $2(5x+3)$ **b** $4(a+2b)$ **c** $3(2x-y)$
d $6(m-4n)$ **e** $5(x^2+4x)$ **f** $6(-x^2+y^2)$
g $a(4a+7)$ **h** $2x(x-5)$ **i** $3p(5p-q)$

3 Expand and then simplify:

a $-2(x+4)$ **b** $-3(x+2)$ **c** $-4(x-3)$
d $-3(2x-1)$ **e** $-(7-2x)$ **f** $-2(x-y)$
g $-a(a-b)$ **h** $-x(3x+8)$ **i** $-2x(2-x)$

4 Expand and then simplify:

a $4(x+5)$ **b** $5(x-7)$ **c** $2(3x-1)$ **d** $x(x-6)$
e $3(4-3x)$ **f** $-4(x+2)$ **g** $x(2x-1)$ **h** $-a(5a+6)$
i $4x(x+1)$ **j** $-c(2c-d)$ **k** $3m(4m+n)$ **l** $-6x(1-7x)$

5 Use areas on this diagram to show that $a(b+c+d)=ab+ac+ad$.

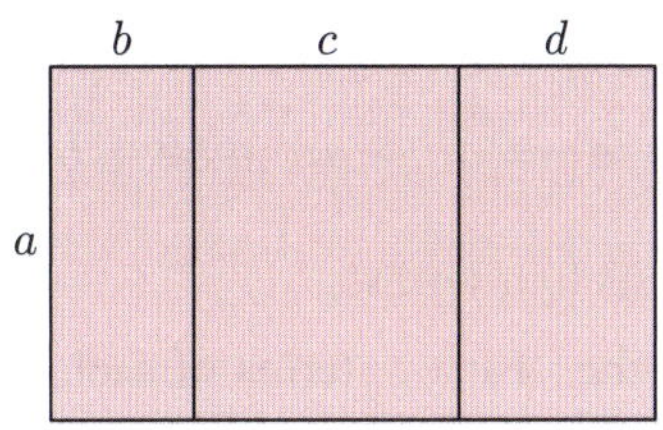

6 Expand using the distributive law:

a $3(x-y+1)$ **b** $-(x^2+x-4)$ **c** $2(a-3b-ab)$
d $3x(2x+y-1)$ **e** $2x(x^2-x-2)$ **f** $-3x(x+1-x^2)$

Example 2 Self Tutor

Expand then simplify by collecting like terms:

a $2(3x-1)+3(5-x)$ **b** $x(2x-1)-2x(5-x)$

a
$$\begin{aligned}&2(3x-1)+3(5-x)\\&=2\times 3x+2\times -1+3\times 5+3\times -x\\&=6x-2+15-3x\\&=3x+13\end{aligned}$$

b
$$\begin{aligned}&x(2x-1)-2x(5-x)\\&=x\times 2x+x\times -1+-2x\times 5+-2x\times -x\\&=2x^2-x-10x+2x^2\\&=4x^2-11x\end{aligned}$$

7 Expand then simplify by collecting like terms:

a $1+2(x+2)$ **b** $13-4(x+3)$ **c** $3(x-2)+5$
d $4(3-x)-10$ **e** $x(x-1)+x$ **f** $2+3(2x-1)$
g $5x-(2+3x)$ **h** $-3x+2(2x+5)$ **i** $2x(3-x)+x^2$
j $3x-4(2-3x)$ **k** $4x-3x(x-1)$ **l** $x^2+x(x-1)$
m $-x^2-x(x-2)$ **n** $2a+(a-2b)$ **o** $2a-(a-2b)$
p $2a(b-a)+3a^2$ **q** $7x^2-5x(x+2)$ **r** $3x^2-2x(1+x)+x$

8 Expand then simplify by collecting like terms:

a $3(x-4)+2(5+x)$ **b** $3(y+1)+6(2-y)$ **c** $2(y-3)-4(2y+1)$
d $-4(x-2)-(3-x)$ **e** $5(2x-1)-(2x+3)$ **f** $2(b-a)+3(a+b)$
g $x(x+4)+2(x-3)$ **h** $x(x+4)-2(x-3)$ **i** $x(x+y)-y(x+y)$
j $3(a-b)-2(2a-b)$ **k** $a(a+b)+2a(b-a)$ **l** $4x(x-3)-2x(5-x)$

B THE PRODUCT $(a+b)(c+d)$

The expression $(a+b)(c+d)$ is the product of the factors $(a+b)$ and $(c+d)$.

We can evaluate this product by using the distributive law several times.

$$\begin{aligned}(a+b)(c+d)&=a(c+d)+b(c+d)\\&=ac+ad+bc+bd\end{aligned}$$

So, $\boldsymbol{(a+b)(c+d)=ac+ad+bc+bd}$

This is sometimes called the **FOIL** rule.

The final result contains four terms:

ac is the product of the **F**irst terms of each bracket.
ad is the product of the **O**uter terms of each bracket.
bc is the product of the **I**nner terms of each bracket.
bd is the product of the **L**ast terms of each bracket.

Example 3 — Self Tutor

Expand then simplify by collecting like terms:

a $(x+3)(x+2)$ **b** $(2x+1)(3x-2)$

a $(x+3)(x+2)$
$= x \times x + x \times 2 + 3 \times x + 3 \times 2$
$= x^2 + 2x + 3x + 6$
$= x^2 + 5x + 6$

b $(2x+1)(3x-2)$
$= 2x \times 3x + 2x \times -2 + 1 \times 3x + 1 \times -2$
$= 6x^2 - 4x + 3x - 2$
$= 6x^2 - x - 2$

With practice you should not need the second line of working.

EXERCISE 4B

1 Consider the figure alongside.
Write an expression for the area of:

a rectangle **1** **b** rectangle **2**
c rectangle **3** **d** rectangle **4**
e the overall rectangle.

What can you conclude?

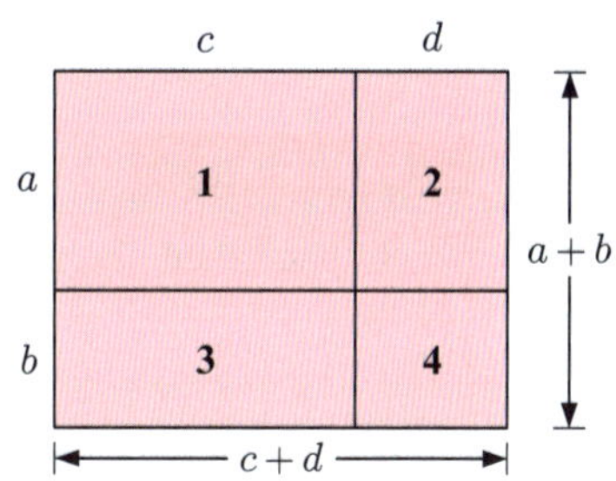

2 Expand then simplify by collecting like terms:

a $(x+1)(x+2)$ **b** $(x+4)(x+3)$ **c** $(x+5)(x-1)$
d $(x-4)(x+2)$ **e** $(x+7)(x-3)$ **f** $(x-4)(x-3)$
g $(x+3)(x+7)$ **h** $(x+5)(x-4)$ **i** $(x-3)(x+6)$
j $(x-5)(x-2)$ **k** $(x-8)(x+3)$ **l** $(2x+1)(3x+4)$
m $(1-2x)(4x+1)$ **n** $(4-x)(2x+3)$ **o** $(3x-2)(1+2x)$
p $(5-3x)(5+x)$ **q** $(7-x)(4x-1)$ **r** $(1-3x)(2x+3)$

Example 4 — Self Tutor

Expand then simplify:

a $(x+3)(x-3)$ **b** $(3x-5)(3x+5)$

a $(x+3)(x-3)$
$= x^2 - 3x + 3x - 9$
$= x^2 - 9$

b $(3x-5)(3x+5)$
$= 9x^2 + 15x - 15x - 25$
$= 9x^2 - 25$

3 Expand then simplify:

a $(x+2)(x-2)$ **b** $(a-5)(a+5)$ **c** $(4+x)(4-x)$
d $(b-4)(b+4)$ **e** $(x+9)(x-9)$ **f** $(6-y)(6+y)$
g $(2x+1)(2x-1)$ **h** $(5a+3)(5a-3)$ **i** $(4+3a)(4-3a)$

Example 5 Self Tutor

Expand then simplify:

a $(3x+1)^2$ **b** $(2x-3)^2$

a $(3x+1)^2$
$= (3x+1)(3x+1)$
$= 9x^2+3x+3x+1$
$= 9x^2+6x+1$

b $(2x-3)^2$
$= (2x-3)(2x-3)$
$= 4x^2-6x-6x+9$
$= 4x^2-12x+9$

4 Expand then simplify:

a $(x+3)^2$ **b** $(x-2)^2$ **c** $(3x-2)^2$ **d** $(5+2x)^2$
e $(1-3x)^2$ **f** $(x+y)^2$ **g** $(3-4x)^2$ **h** $(5x-y)^2$
i $(a-b)^2$ **j** $(4x+2)^2$ **k** $(2x+y)^2$ **l** $(x^2+1)^2$

Example 6 Self Tutor

Expand then simplify: $(x+2)(x-3)+5(x+4)$

$(x+2)(x-3)+5(x+4)$
$= x^2-3x+2x-6+5x+20$
$= x^2+4x+14$

5 Expand then simplify:

a $(x+5)(x+1)+2(x-2)$ **b** $4(x+2)+(x-3)(x+6)$
c $(2a+5)(a-6)+a(a+7)$ **d** $(3x+5)(2x-3)-x(x+3)$
e $(2x+3)(x+2)+(x+6)(x-5)$ **f** $(y+7)(y-4)-(y+1)(y+3)$
g $(k+2)(k-5)-(2k+1)(k-3)$ **h** $(x+2)(x-2)+(x+6)(x-6)$

C THE DIFFERENCE BETWEEN TWO SQUARES

Consider the product $(a+b)(a-b)$.

Using the FOIL rule to expand this product, $(a+b)(a-b)$
$= a^2-ab+ab-b^2$
$= a^2-b^2$

Thus, $\boldsymbol{(a+b)(a-b)=a^2-b^2}$

This is called the **difference between two squares** expansion, because the expression on the right hand side is the difference between the two squares a^2 and b^2.

GEOMETRIC DEMONSTRATION

In the figure alongside,

shaded area = area of large square − area of small square
$= a^2 - b^2$

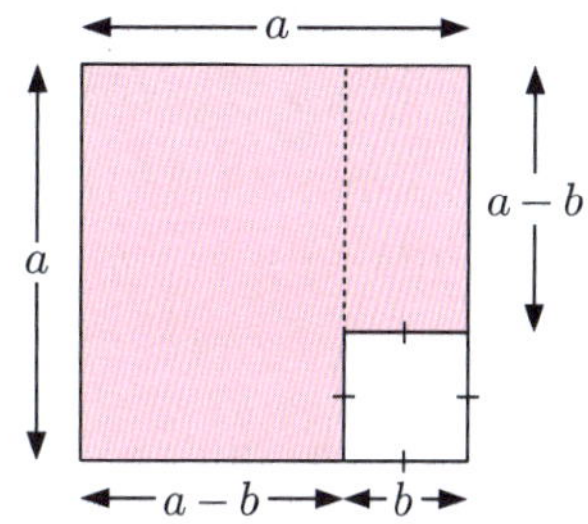

If the rectangle on the right hand side is rotated and placed on top of the remaining shaded area, we form a new rectangle.

$\therefore$ shaded area $= (a+b)(a-b)$

$\therefore$ $(a+b)(a-b) = a^2 - b^2$ {equating areas}

DEMO

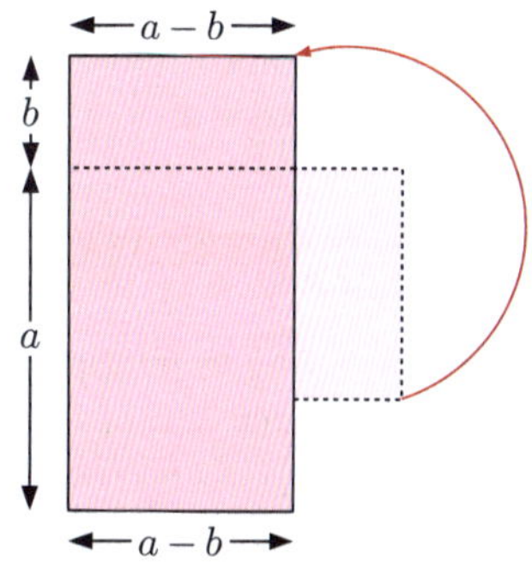

Example 7 **Self Tutor**

Expand then simplify:

a $(x+5)(x-5)$ **b** $(3-y)(3+y)$

a $(x+5)(x-5)$
$= x^2 - 5^2$
$= x^2 - 25$

b $(3-y)(3+y)$
$= 3^2 - y^2$
$= 9 - y^2$

EXERCISE 4C

1 Expand using the rule $(a+b)(a-b) = a^2 - b^2$, then simplify:

a $(x+2)(x-2)$ **b** $(x-2)(x+2)$ **c** $(2+x)(2-x)$
d $(2-x)(2+x)$ **e** $(x+1)(x-1)$ **f** $(1-x)(1+x)$
g $(x+7)(x-7)$ **h** $(c+8)(c-8)$ **i** $(d-5)(d+5)$
j $(x+y)(x-y)$ **k** $(4+d)(4-d)$ **l** $(5-e)(5+e)$

Example 8 **Self Tutor**

Expand then simplify:

a $(2x-3)(2x+3)$ **b** $(5-3y)(5+3y)$ **c** $(3x+4y)(3x-4y)$

a $(2x-3)(2x+3)$
$= (2x)^2 - 3^2$
$= 4x^2 - 9$

b $(5-3y)(5+3y)$
$= 5^2 - (3y)^2$
$= 25 - 9y^2$

c $(3x+4y)(3x-4y)$
$= (3x)^2 - (4y)^2$
$= 9x^2 - 16y^2$

2 Expand then simplify:

- **a** $(2x-1)(2x+1)$
- **b** $(3x+2)(3x-2)$
- **c** $(4y-5)(4y+5)$
- **d** $(2y+5)(2y-5)$
- **e** $(3x+1)(3x-1)$
- **f** $(1-3x)(1+3x)$
- **g** $(2-5y)(2+5y)$
- **h** $(3+4a)(3-4a)$
- **i** $(4+3a)(4-3a)$
- **j** $(2a+b)(2a-b)$
- **k** $(a-2b)(a+2b)$
- **l** $(4x+y)(4x-y)$
- **m** $(4x+5y)(4x-5y)$
- **n** $(2x+3y)(2x-3y)$
- **o** $(7x-2y)(7x+2y)$

3 **a** Use the difference between two squares expansion to show that:

- **i** $43\times 37 = 40^2-3^2$
- **ii** $24\times 26 = 25^2-1^2$

b Evaluate without using a calculator:

- **i** 18×22
- **ii** 49×51
- **iii** 103×97

ACTIVITY 1 THE PRODUCT OF THREE CONSECUTIVE INTEGERS

Con was trying to multiply $19\times 20\times 21$ without a calculator. Aimee told him to "cube the middle integer and then subtract the middle integer" to get the answer.

What to do:

1 Use a calculator to find:

- **a** $19\times 20\times 21$
- **b** 20^3-20

Comment on your answers.

2 Check that Aimee's rule also works for:

- **a** $4\times 5\times 6$
- **b** $9\times 10\times 11$
- **c** $49\times 50\times 51$

3 Let the middle of three consecutive integers be x.
Use the difference between two squares expansion to help prove Aimee's rule.

D THE PERFECT SQUARES EXPANSION

$(a+b)^2$ and $(a-b)^2$ are called **perfect squares**.

Notice that
$$\begin{aligned}(a+b)^2 &= (a+b)(a+b)\\ &= a^2+ab+ab+b^2 \quad \{\text{using "FOIL"}\}\\ &= a^2+2ab+b^2\end{aligned}$$

Thus, we can state the perfect square expansion rule:

$$(a+b)^2 = a^2+2ab+b^2$$

We can remember this rule as follows:

Step 1: Square the *first term*.
Step 2: Add twice the product of the *first* and *last terms*.
Step 3: Add on the square of the *last term*.

Notice that $(a-b)^2 = (a+(-b))^2$
$= a^2 + 2a(-b) + (-b)^2$
$= a^2 - 2ab + b^2$

Once again, we have the sum of the square of the first term, twice the product of the first and last terms, and the square of the last term.

Example 9 **Self Tutor**

Expand then simplify:

a $(x+3)^2$ **b** $(x-5)^2$

a $(x+3)^2$
$= x^2 + 2 \times x \times 3 + 3^2$
$= x^2 + 6x + 9$

b $(x-5)^2$
$= x^2 + 2 \times x \times (-5) + (-5)^2$
$= x^2 - 10x + 25$

EXERCISE 4D

1 Consider the figure alongside.
Write an expression for the area of:

a square **1** **b** rectangle **2** **c** rectangle **3**
d square **4** **e** the overall square.

What can you conclude?

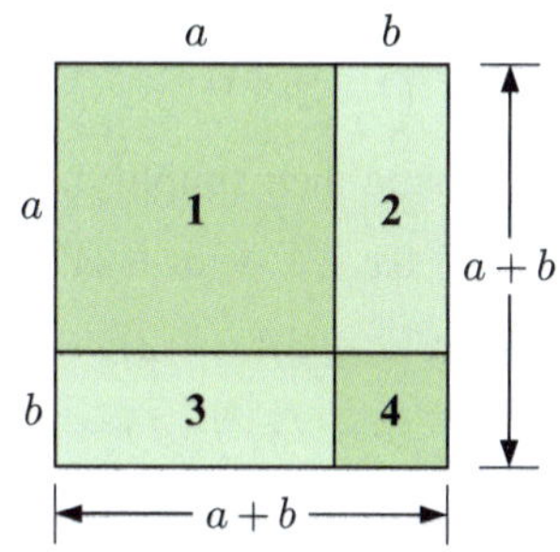

2 Use the rule $(a+b)^2 = a^2 + 2ab + b^2$ to expand $(5+3)^2$.
Hence explain why $(5+3)^2 \neq 5^2 + 3^2$.

3 Expand using the rule $(a+b)^2 = a^2 + 2ab + b^2$, then simplify:

a $(x+5)^2$ **b** $(x+4)^2$ **c** $(x+7)^2$
d $(a+2)^2$ **e** $(3+c)^2$ **f** $(2+x)^2$

4 Expand then simplify:

a $(x-3)^2$ **b** $(x-2)^2$ **c** $(y-8)^2$
d $(a-7)^2$ **e** $(5-x)^2$ **f** $(4-y)^2$

Example 10 **Self Tutor**

Expand then simplify:

a $(5x+1)^2$ **b** $(4-3x)^2$

a $(5x+1)^2$
$= (5x)^2 + 2 \times 5x \times 1 + 1^2$
$= 25x^2 + 10x + 1$

b $(4-3x)^2$
$= 4^2 + 2 \times 4 \times (-3x) + (-3x)^2$
$= 16 - 24x + 9x^2$

5 Expand then simplify:

a $(3x+4)^2$ **b** $(2a-3)^2$ **c** $(3y+1)^2$
d $(2x-5)^2$ **e** $(3y-5)^2$ **f** $(7+2a)^2$
g $(1+5x)^2$ **h** $(7-3y)^2$ **i** $(3+4a)^2$

Example 11 **Self Tutor**

Expand then simplify:

a $(2x^2+3)^2$ **b** $5-(x+2)^2$

a $(2x^2+3)^2$
$= (2x^2)^2 + 2 \times 2x^2 \times 3 + 3^2$
$= 4x^4 + 12x^2 + 9$

b $5-(x+2)^2$
$= 5 - [x^2+4x+4]$
$= 5 - x^2 - 4x - 4$
$= 1 - x^2 - 4x$

The square brackets in the second line remind us that the minus in front of the brackets affects *all* terms within them.

6 Expand then simplify:

a $(x^2+2)^2$ **b** $(y^2-3)^2$ **c** $(3a^2+4)^2$
d $(1-2x^2)^2$ **e** $(x^2+y^2)^2$ **f** $(x^2-a^2)^2$

7 Expand then simplify:

a $3x+1-(x+3)^2$ **b** $5x-2+(x-2)^2$
c $(x+2)(x-2)+(x+3)^2$ **d** $(x+4)(x-4)-(x+3)^2$
e $(3-2x)^2-(x-1)(x+2)$ **f** $(1-3x)^2+(x+2)(x-3)$
g $(2x+3)(2x-3)-(x+1)^2$ **h** $(4x+3)(x-2)-(2-x)^2$
i $(1-x)^2+(x+2)^2$ **j** $(2x-1)^2-(3+x)^2$

8 Jeremy studies the squares of the consecutive integers 6, 7, and 8.

He notices that $\frac{6^2+8^2}{2} = 7^2+1$, or in other words, that the average of the squares of the highest and lowest integers is one more than the square of the middle integer.
Prove that this property holds for the squares of any three consecutive integers.

ACTIVITY 2

What to do:

1 Explain why:
   a every even number can be written in the form $2a$ where a is an integer
   b every odd number can be written in the form $2a+1$ where a is an integer.

2 Prove that the sum of the squares of any two positive even integers is divisible by 4.
Hint: Let one integer be $2a$ and the other be $2b$.

3 What can you say about the sum of the squares of any two positive odd integers?

4 Suppose the sum of the squares of two integers is odd. What can you say about the integers? Prove your result.

E FURTHER EXPANSION

In this Section we expand more complicated expressions by repeated use of the expansion laws.

Consider the expansion of $(a+b)(c+d+e)$.

Now $\quad (a+b)(c+d+e)$ $\qquad$ {Compare with $\square(c+d+e)$

$= (a+b)c + (a+b)d + (a+b)e$ $\qquad$ $= \square c + \square d + \square e$}

$= ac + bc + ad + bd + ae + be$

Notice that there are 6 terms in this expansion. Each term in the first bracket is multiplied by each term in the second bracket.

2 terms in the first bracket $\times$ 3 terms in the second bracket $\longrightarrow$ 6 terms in the expansion.

Example 12 **Self Tutor**

Expand then simplify: $\quad (2x+3)(x^2+4x+5)$

$(2x+3)(x^2+4x+5)$

$= 2x^3 + 8x^2 + 10x$ $\qquad$ {$2x \times$ all terms in 2nd bracket}

$\quad + 3x^2 + 12x + 15$ $\qquad$ {$3 \times$ all terms in 2nd bracket}

$= 2x^3 + 11x^2 + 22x + 15$ $\qquad$ {collecting like terms}

EXERCISE 4E

1 Expand then simplify:

a $(x+2)(x^2+x+1)$ **b** $(x+3)(x^2+x-2)$ **c** $(x+4)(x^2-x+2)$

d $(x+6)(x^2-x-1)$ **e** $(3x+1)(x^2+x+5)$ **f** $(5x-2)(x^2-x+4)$

g $(x+3)(2x^2+x-2)$ **h** $(2x-1)(3x^2-x+4)$ **i** $(2x^2+x-3)(3x+2)$

Example 13 **Self Tutor**

Expand then simplify: $\quad (x+2)^3$

$(x+2)^3 = (x+2) \times (x+2)^2$

$= (x+2)(x^2+4x+4)$ $\qquad$ {perfect square expansion}

$= x^3 + 4x^2 + 4x$ $\qquad$ {$x \times$ all terms in 2nd bracket}

$\quad + 2x^2 + 8x + 8$ $\qquad$ {$2 \times$ all terms in 2nd bracket}

$= x^3 + 6x^2 + 12x + 8$ $\qquad$ {collecting like terms}

2 Expand then simplify:

a $(x+1)^3$ **b** $(x+3)^3$ **c** $(x-1)^3$

d $(x-2)^3$ **e** $(2x+1)^3$ **f** $(2x-3)^3$

Example 14 **Self Tutor**

Expand then simplify: $x(x+1)(x+2)$

$$\begin{aligned} & x(x+1)(x+2) \\ &= (x^2+x)(x+2) && \{x \times \text{all terms in 1st bracket}\} \\ &= x^3+2x^2+x^2+2x && \{\text{expanding remaining factors}\} \\ &= x^3+3x^2+2x && \{\text{collecting like terms}\} \end{aligned}$$

3 Expand then simplify:

a $x(x+2)(x+4)$ **b** $x(x-5)(x+3)$ **c** $x(x-4)(x-1)$

d $2x(x+3)(x-2)$ **e** $2x(x+1)(7-x)$ **f** $-x(x+3)(x-6)$

g $-5x(2x-1)(x+3)$ **h** $-x(1-4x)(x-3)$ **i** $3x^2(x+2)^2$

Example 15 **Self Tutor**

Expand then simplify: $(x+1)(x-2)(x+2)$

$$\begin{aligned} & (x+1)(x-2)(x+2) \\ &= (x+1)(x^2-4) && \{\text{difference between two squares}\} \\ &= x^3-4x+x^2-4 && \{\text{expanding factors}\} \\ &= x^3+x^2-4x-4 \end{aligned}$$

4 Expand then simplify:

a $(x+1)(x+2)(x+4)$ **b** $(x-2)(x+1)(x-3)$ **c** $(x-5)(x+2)(x-1)$

d $(2x-1)(x-4)(x-1)$ **e** $(3x+1)(x+2)(x+3)$ **f** $(4x-1)(4x+1)(x+4)$

g $(1-x)(2-3x)(x+2)$ **h** $(x-5)(1-x)(5x+2)$ **i** $(4-3x)(x+3)^2$

5 **a** Expand then simplify:

i $(x-1)(x^2+x+1)$ **ii** $(x-1)(x^3+x^2+x+1)$

iii $(x-1)(x^6+x^5+x^4+x^3+x^2+x+1)$

b Predict the simplification of $(x-1)(x^n+x^{n-1}+x^{n-2}+....+x^2+x+1)$, where n is a positive integer.

MULTIPLE CHOICE QUIZ

QUICK QUIZ

REVIEW SET 4A

1 Expand using the distributive law:

a $3(2+x)$ **b** $-(4x-2)$ **c** $3x(x+3)$

2 Expand then simplify:

a $7+4(x-2)$ **b** $9x-2(x+4)$ **c** $-4(2x-3)-x(x+1)$

3 Expand then simplify:

a $(x+4)(x+7)$ **b** $(1-x)(5x+6)$ **c** $(4x+1)(5x-4)$

4 Expand then simplify:

a $(x-10)^2$ **b** $(3-4x)^2$ **c** $(x+2y)^2$

5 Expand then simplify:

a $(x+9)(x-9)$ **b** $(3k+2)(3k-2)$ **c** $(1-7x)(1+7x)$

6 Expand then simplify:

a $(x+3)(x+2)+3(x-4)$ **b** $x(x+4)-(x+7)^2$ **c** $(3x-1)^2-(1+x)^2$

7 Expand then simplify:

a $(x+4)(x^2+x-7)$ **b** $(4x-3)(x^2-x-5)$ **c** $(2x^2-x+4)(2x+1)$

d $(2x-1)^3$ **e** $x(x+5)(x-6)$ **f** $(4x-1)(x+1)(x-2)$

8 **a** Explain why the shaded area $= a^2 - b^2$.

b The dotted line divides the shaded area into two identical trapezia.

Find, in terms of a and b:

i the perpendicular height of each trapezium

ii the area of each trapezium.

c Hence show that $(a+b)(a-b) = a^2 - b^2$.

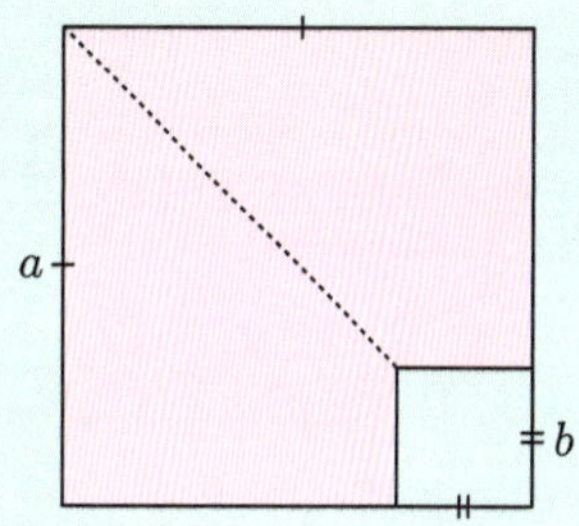

REVIEW SET 4B

1 Expand using the distributive law:

a $-3(x+4)$ **b** $x(y+3)$ **c** $2a(a-5)$

2 Expand then simplify:

a $4(x-2)+3(2x-1)$ **b** $x(x+1)-4(x-2)$ **c** $-(3-x)+x(x+3)$

3 Expand then simplify:

a $(x+9)(x-4)$ **b** $(2x-7)(x+6)$ **c** $(2a-3)(a-4)$

4 Expand then simplify:

a $(y+3)^2$ **b** $(3x+2)^2$ **c** $(4a-b)^2$

5 Expand then simplify:

a $(5-x)(5+x)$ **b** $(3y+4)(3y-4)$ **c** $(4x+1)(1-4x)$

6 Expand then simplify:

a $-4(x+2)-(x+1)^2$
b $(x^2-5)^2$
c $(n-3)(n+5)-(n-2)(n+3)$
d $5(y-3)+(y+5)(y-5)$

7 Expand then simplify:

a $(x^2-3x+4)(2-x)$
b $(x+4)^2(x-2)$
c $(2x+1)(2x-1)(x-3)$
d $(x+4)^3$
e $(x-5)^3$

8 In the expression $2x+3$, the sum of the coefficients is $2+3=5$.

a Expand then simplify $(2x+3)^2$, and find the sum of its coefficients.
b Expand then simplify $(2x+3)^3$, and find the sum of its coefficients.
c Predict the sum of the coefficients in the expansion of:

i $(2x+3)^4$
ii $(2x+3)^n$, where n is a positive integer.

Chapter 5

Sets

Contents:

OPENING PROBLEM

A pharmaceutical company is conducting two clinical trials. Trial A involves people aged from 25 to 40. Trial B involves people aged 30 and above.

Things to think about:

- **a** Can you list the set of valid ages for trial A?
- **b** Martin describes the set of valid ages for trial A as "any number from 25 to 40". Is he correct?
- **c** Can you use mathematical notation to efficiently describe the set of valid ages for trial A?
- **d** Can you describe mathematically the set of ages of people eligible for:
  - **i** trial B
  - **ii** both trial A and trial B
  - **iii** neither trial?

A SETS

A **set** is a collection of numbers or objects.

Each object is called an **element** or **member** of the set.

When we record a set, we write its members within curly brackets, separated by commas.

We often use a capital letter to represent a set so that we can refer to it easily.

For example, if F is the set of all factors of 12, we write $F = \{1, 2, 3, 4, 6, 12\}$.

We use the following **set notation**:

$\in$ means "is an element of" or "is in".
$\notin$ means "is not an element of" or "is not in".
$n(A)$ means "the number of elements in set A".

For example, if $F = \{1, 2, 3, 4, 6, 12\}$ then $4 \in F$, $7 \notin F$, and $n(F) = 6$.

We also define:

- Two sets are **equal** if they contain exactly the same elements.

  For example, $\{4, 7, 10, 13, 14\} = \{10, 14, 4, 13, 7\}$.

- Set A is a **subset** of set B if every element of A is also an element of B. We write $A \subseteq B$.

  For example, if $A = \{3, 8\}$ and $B = \{2, 3, 6, 8, 9\}$, then every element of A is also an element of B. A is a subset of B, and we write $A \subseteq B$.

- The **empty set** $\varnothing$ or $\{ \}$ is a set which contains no elements.

  For example, the set of prime numbers between 24 and 28 is the empty set.
  The empty set is a subset of all other sets.

Example 1 **Self Tutor**

Let A be the set of factors of 27, and B be the set of multiples of 3 less than 30.

a List the elements of A and B.

b True or false? **i** $6 \in A$ **ii** $21 \in B$

c Find: **i** $n(A)$ **ii** $n(B)$

d Is $A \subseteq B$?

a $A = \{1, 3, 9, 27\}$, $B = \{3, 6, 9, 12, 15, 18, 21, 24, 27\}$

b **i** false **ii** true

c **i** $n(A) = 4$ $\{A$ has 4 elements$\}$ **ii** $n(B) = 9$ $\{B$ has 9 elements$\}$

d 1 is an element of A, but is not an element of B.

$\therefore$ A is *not* a subset of B.

EXERCISE 5A

1 Look at the number plate.

a List the elements of the set of:

i colours, C **ii** digits, D **iii** letters, L.

b Find:

i $n(C)$ **ii** $n(D)$ **iii** $n(L)$

2 **a** List the elements of P, the set of factors of 45. Hence state $n(P)$.

b List the elements of Q, the set of letters in the word ANTARCTICA. Hence state $n(Q)$.

3 Let $M = \{$multiples of 4 less than 30$\}$ and $C = \{$composite numbers less than 30$\}$.

a List the elements of M and C.

b True or false? **i** $7 \in M$ **ii** $15 \in C$

c Find: **i** $n(M)$ **ii** $n(C)$

d Is $M \subseteq C$?

4 Decide whether $R \subseteq S$ for:

a $R = \varnothing$ and $S = \{1, 2, 3, 4, 6, 12\}$ **b** $R = \{1, 2, 3, 4, 5\}$ and $S = \{1, 3, 5\}$

c $R = \{b, c, f, n, p\}$ and $S = \{a, b, c, f, n, q\}$

d $R = \{$blue, white$\}$ and $S = \{$red, blue, green, white$\}$.

5 List all the subsets of:

a $\{\square\}$ **b** $\{a, b\}$ **c** $\{-1, 0, 2\}$

6 Consider two sets A and B such that $B \subseteq A$. Find the possible values of x if:

a $A = \{2, 3, 4, 5, 6, 7, 8\}$ and $B = \{3, 5, 7, x\}$

b $A = \{1, x-2, x\}$ and $B = \{1, 2, x\}$

c $A = \{0, 2, -2, 16, -4, 8, 4\}$ and $B = \{x+4, x^2\}$.

7 Suppose two dice are rolled simultaneously.

Let $S = \{$sum of the dice$\}$ and $P = \{$product of the dice$\}$.

List the elements of B, the largest set which is a subset of both S and P.

B COMPLEMENT OF A SET

The **universal set** U is the set of all elements we are considering.

For example, if we are considering the single digit positive whole numbers, the universal set is $U = \{1, 2, 3, 4, 5, 6, 7, 8, 9\}$.

The **complement** of a set A is the set of all elements of U that are *not* elements of A.

The complement of A is written A'. We say that A and A' are **complementary**.

For example, if $U = \{1, 2, 3, 4, 5, 6, 7, 8, 9\}$ and $A = \{1, 3, 4, 6\}$, then $A' = \{2, 5, 7, 8, 9\}$.

EXERCISE 5B

1. Let $U = \{1, 2, 3, 4, 5, 6, 7, 8, 9, 10\}$. Find the complement of:
   - **a** $A = \{1, 4, 5, 7\}$
   - **b** $B = \{8, 5, 2\}$
   - **c** $C = \{1, 2, 3, 4\}$
   - **d** $D = \{2, 4, 6, 8, 10\}$
   - **e** $E = \{1, 2, 3, 4, 5, 7, 8, 9, 10\}$

2. Let $U = \{\text{planets in our solar system}\}$.
   $M = \{\text{Earth, Mars, Jupiter, Saturn, Uranus, Neptune}\}$ is the set of planets with moons.
   $R = \{\text{Jupiter, Saturn, Uranus, Neptune}\}$ is the set of planets with rings.
   - **a** List the elements of M', and explain what the set represents.
   - **b** List the elements of R', and explain what the set represents.

3. Suppose $U = \{1, 2, 3, 4, 5, ..., 24\}$, $F = \{\text{factors of } 24\}$, $M = \{\text{multiples of } 4\}$, and $P = \{\text{prime numbers}\}$.
   - **a** List the elements of:
     i F **ii** F' **iii** M **iv** M' **v** P **vi** P'
   - **b** Find:
     i $n(F)$ **ii** $n(F')$ **iii** $n(M)$ **iv** $n(M')$ **v** $n(P)$ **vi** $n(P')$

4.
   - **a** For any set A within a universal set U, explain why $n(A) + n(A') = n(U)$.
   - **b** Suppose $n(U) = 20$, $n(A) = 7$, and $n(B) = 15$. Find:
     i $n(A')$ **ii** $n(B')$

5. Let $U = \{\text{triangles}\}$, $A = \{\text{acute angled triangles}\}$, $B = \{\text{obtuse angled triangles}\}$, $C = \{\text{right angled triangles}\}$, $D = \{\text{scalene triangles}\}$, $E = \{\text{isosceles triangles}\}$, $F = \{\text{equilateral triangles}\}$.
   - **a** True or false?
     i $F \subseteq U$ **ii** $F \subseteq A$ **iii** $F \subseteq D$ **iv** $F \subseteq E$
   - **b** Which pair of sets is complementary?

6. Suppose A is a set within a universal set U. State the complement of:
   - **a** U
   - **b** $\varnothing$
   - **c** A'

7. Suppose A and B are sets such that $A \subseteq B$. Explain why $B' \subseteq A'$.

C INTERSECTION AND UNION

INTERSECTION

The **intersection** of two sets A and B is the set of elements that are in **both** set A **and** set B.

The intersection of sets A and B is written $\boldsymbol{A \cap B}$.

For example, if $A = \{1, 2, 4, 5, 6, 8\}$ and $B = \{0, 2, 3, 5, 6, 7\}$, then $A \cap B = \{2, 5, 6\}$.

Two sets A and B are **disjoint** if they have no elements in common. In this case $A \cap B = \varnothing$.

UNION

The **union** of two sets A and B is the set of elements that are in **either** set A **or** set B.

The union of sets A and B is written $\boldsymbol{A \cup B}$.

For example, if $A = \{1, 2, 4, 5, 6, 8\}$ and $B = \{0, 2, 3, 5, 6, 7\}$, then $A \cup B = \{0, 1, 2, 3, 4, 5, 6, 7, 8\}$.

Example 2 — Self Tutor

Let $A = \{2, 3, 5, 7, 11\}$ and $B = \{1, 3, 4, 6, 7, 8, 9\}$. Find:

a $A \cap B$ **b** $A \cup B$

a $A \cap B = \{3, 7\}$ {3 and 7 are elements of both sets}

b $A \cup B = \{1, 2, 3, 4, 5, 6, 7, 8, 9, 11\}$ {elements of either A or B}

EXERCISE 5C

1 Find $A \cap B$ and $A \cup B$ for:

a $A = \{a, b, c, d\}$ and $B = \{b, d, f, h\}$

b $A = \{1, 2, 3, 4, 5\}$ and $B = \{4, 5, 6, 7, 8\}$

c $A = \{\text{vanilla, chocolate, strawberry}\}$ and $B = \{\text{vanilla, lemon, chocolate, strawberry, coffee}\}$.

2 Suppose $P = \{1, 3, 5, 7\}$ and $Q = \{2, 4, 6, 8\}$.

a Find $P \cap Q$. **b** Are P and Q disjoint?

3 Suppose $A = \{\text{multiples of 3 less than 30}\}$ and $B = \{\text{factors of 30}\}$.

a List the elements of A and B.

b Find:

i $A \cap B$ **ii** $n(A \cap B)$ **iii** $A \cup B$ **iv** $n(A \cup B)$

4 Let $A = \{\text{prime numbers}\}$, $B = \{\text{square numbers}\}$, $C = \{\text{cubic numbers}\}$, and $D = \{\text{numbers between 80 and 100}\}$.
Which pairs of sets are disjoint?

5 Suppose U is the set of the months of the year.
Let $J = \{\text{months that begin with a "J"}\}$ and $K = \{\text{months containing 31 days}\}$.

a List the elements of J and K.

b Find each set and explain what it represents:

i $J \cap K$ **ii** $J \cup K$ **iii** $(J \cup K)'$

AUGUST

S	M	T	W	T	F	S
		1	2	3	4	5
6	7	8	9	10	11	12
13	14	15	16	17	18	19
20	21	22	23	24	25	26
27	28	29	30	31		

6 Suppose A and B are sets in a universal set U. Explain why:

a $A \cup A' = U$ **b** $A \cap A' = \varnothing$ **c** $A \cap B \subseteq A$ **d** $A \cap B \subseteq A \cup B$

7 Discuss whether the following statements are true or false:

a If two sets are complementary then they must be disjoint.

b If two sets are disjoint then they must be complementary.

c If A and B are disjoint sets, and B and C are disjoint sets, then A and C must be disjoint sets.

8 Suppose $U = \{12, 13, 14, 15, 16, 17, 18\}$, $A = \{\text{even numbers}\}$, and $B = \{\text{multiples of } 3\}$.

a List the elements of:

i A **ii** B **iii** A' **iv** B' **v** $A \cup B$ **vi** $A' \cap B'$

b Show that $(A \cup B)' = A' \cap B'$.

9 Suppose $U = \{\text{whole numbers from 1 to 10}\}$, $P = \{\text{prime numbers}\}$, and $Q = \{\text{odd numbers}\}$.

a List the elements of:

i P **ii** Q **iii** P' **iv** Q' **v** $P \cap Q$ **vi** $P \cup Q$

b Show that $(P \cap Q)' = P' \cup Q'$.

c For any sets P and Q in a universal set U, explain why $(P \cap Q)' = P' \cup Q'$.

DISCUSSION

Suppose X is the set of factors of one unknown number, and Y is the set of factors of another. You are told the elements of $X \cup Y$ and $X \cap Y$.

Will this always be sufficient information to determine the values of the two numbers?

D SPECIAL NUMBER SETS

To help us refer to some special number sets easily, we give them symbols:

- $\mathbb{N} = \{0, 1, 2, 3, 4, 5, 6, 7, ...\}$ is the set of **natural numbers**.

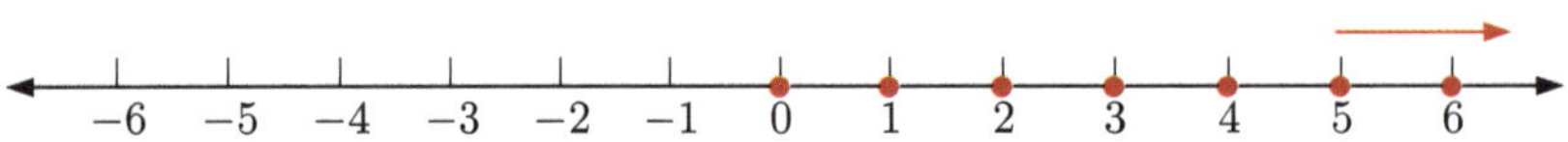

- $\mathbb{Z} = \{0, \pm1, \pm2, \pm3, \pm4, ...\}$ is the set of **integers**.

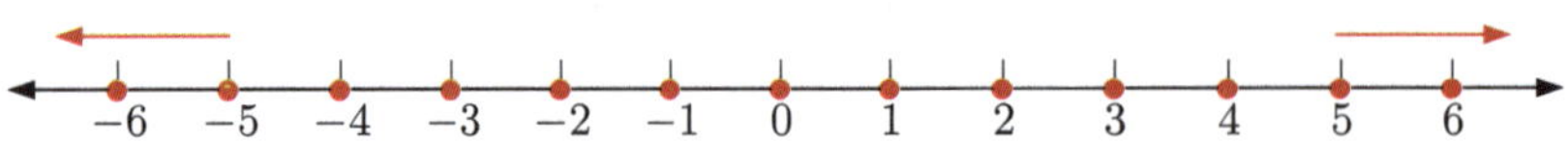

- $\mathbb{Z}^+ = \{1, 2, 3, 4, 5, 6, 7, ...\}$ is the set of **positive integers**.

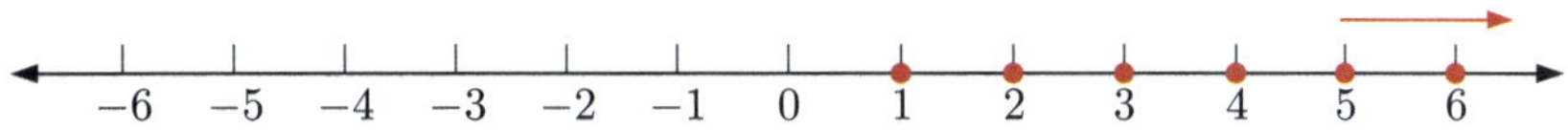

- $\mathbb{Z}^- = \{-1, -2, -3, -4, -5, -6, -7, ...\}$ is the set of **negative integers**.

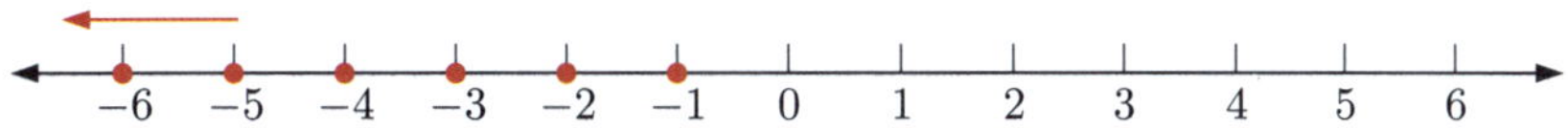

- $\mathbb{Q}$ is the set of **rational numbers**, or numbers which can be written in the form $\frac{p}{q}$ where p and q are integers, $q \neq 0$.

For example:
- $\frac{1}{3}, \frac{15}{32}, \frac{-5}{4}$ are all rational
- $-2\frac{1}{4}$ is rational as $-2\frac{1}{4} = \frac{-9}{4}$
- 5 is rational as $5 = \frac{5}{1}$ or $\frac{20}{4}$
- 0.431 is rational as $0.431 = \frac{431}{1000}$
- $0.\overline{8}$ is rational as $0.\overline{8} = \frac{8}{9}$
- $\sqrt{9}$ is rational as $\sqrt{9} = \frac{3}{1}$
- All decimal numbers that terminate or recur are rational numbers.

- $\mathbb{Q}'$ is the set of **irrational numbers**, or numbers which cannot be written in rational form.

For example: $\sqrt{2}$, $\sqrt[3]{7}$, and π are all irrational.

- $\mathbb{R}$ is the set of **real numbers**, which are all numbers that can be placed on the number line.

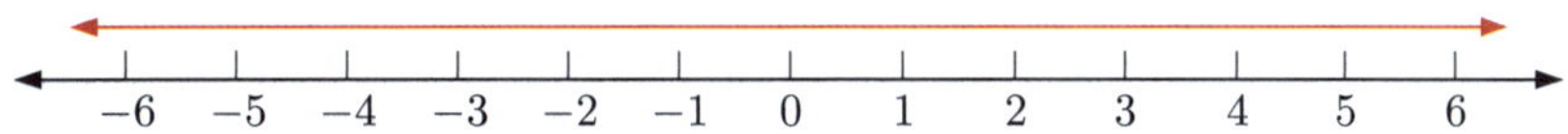

$\mathbb{R}$ includes all rational and irrational numbers.

$\frac{2}{0}$ and $\sqrt{-5}$ cannot be placed on a number line, and so are not real numbers.

All of these special number sets have infinitely many elements. We call them **infinite sets**.

EXERCISE 5D

1 Copy and complete:

Number	$\mathbb{N}$	$\mathbb{Z}$	$\mathbb{Q}$	$\mathbb{R}$
19	✓	✓	✓	✓
$-\frac{2}{3}$				
$\sqrt{7}$				
5.6389				
$\sqrt{16}$				
2π				
-11				
$\frac{6}{0}$				
$\sqrt{-2}$				

2 True or false?

a $5 \in \mathbb{Z}$ **b** $-3 \in \mathbb{N}$ **c** $\sqrt{2} \in \mathbb{R}$ **d** $\frac{3}{7} \in \mathbb{Q}$

e $-8.5 \in \mathbb{Z}^-$ **f** $3\pi \in \mathbb{Q}'$ **g** $6.23 \in \mathbb{R}$ **h** $5\frac{1}{4} \in \mathbb{Q}'$

3 True or false?

a $\mathbb{N} \subseteq \mathbb{Z}$ **b** $\mathbb{Z}^+ \subseteq \mathbb{Q}$ **c** $\mathbb{Z}^- \subseteq \mathbb{Z}$ **d** $\mathbb{R} \subseteq \mathbb{Q}$

4 Which two special number sets are complementary within the universal set:

a $U = \mathbb{Z}$ **b** $U = \mathbb{R}$?

5 Find:

a $\mathbb{Z}^+ \cap \mathbb{Z}^-$ **b** $\mathbb{N} \cup \mathbb{Z}$ **c** $\mathbb{Q}' \cap \mathbb{R}$ **d** $\mathbb{Q}' \cup \mathbb{R}$

6 Let $U = \mathbb{Z}$. Find $(\mathbb{Z}^- \cup \mathbb{Z}^+)'$.

DISCUSSION

Which of the following *may* be rational? If possible, provide examples.

- The sum of two irrational numbers.
- The difference between two irrational numbers.
- The product of two irrational numbers.

E INTERVAL NOTATION

To describe the set of integers between 3 and 8, we can list the set as $\{4, 5, 6, 7\}$ or illustrate the set as points on a number line.

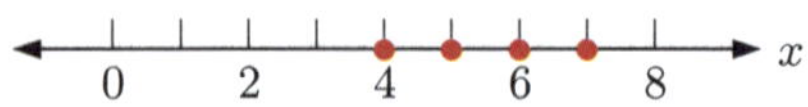

Alternatively, we can write the set using an *interval* on the number line.

Using **interval notation**, we write the set as $\{x \in \mathbb{Z} \mid 3 < x < 8\}$.

We read this as "the set of integers x such that x lies between 3 and 8".

Interval notation is very useful if the set contains a large or infinite number of elements and listing them would be time consuming or impossible.

For example, the number line alongside shows the set of *real numbers* between 3 and 8. Using interval notation, we write $\{x \in \mathbb{R} \mid 3 < x < 8\}$.

Unless stated otherwise, we *assume* we are dealing with *real* numbers. So, the set can also be written as $\{x \mid 3 < x < 8\}$.

Open circles indicate that 3 and 8 are not included.

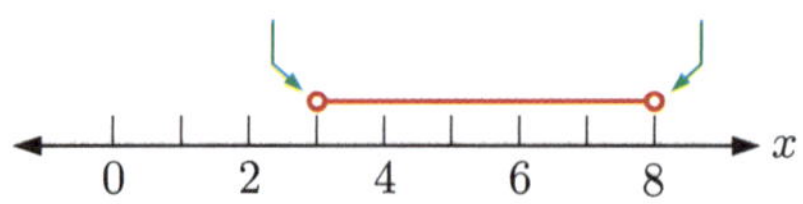

The set of real numbers between 3 and 8 *inclusive* would be written as $\{x \in \mathbb{R} \mid 3 \leqslant x \leqslant 8\}$ or just $\{x \mid 3 \leqslant x \leqslant 8\}$.

Filled in circles indicate that 3 and 8 are included.

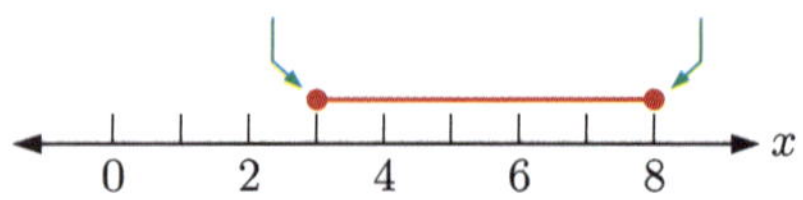

Example 3 — Self Tutor

Represent on a number line:

a $1 \leqslant x < 5$ **b** $x < 0$ or $x \geqslant 4$

a 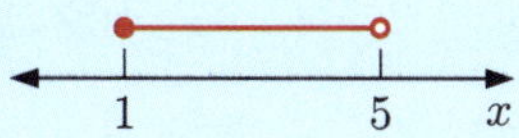

b 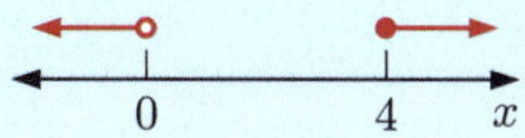

EXERCISE 5E

1 Represent on a number line:

a $x > 5$ **b** $x \geqslant 1$ **c** $x \leqslant 2$

d $x < -1$ **e** $-2 \leqslant x \leqslant 2$ **f** $-3 < x \leqslant 4$

g $1 \leqslant x < 6$ **h** $-1 < x < 0$ **i** $x < 0$ or $x \geqslant 3$

j $x \leqslant -1$ or $x \geqslant 2$ **k** $x < 2$ or $x > 5$ **l** $x \leqslant -2$ or $x > 0$

2 Write in interval notation:

a 

b 

c 

d 

e 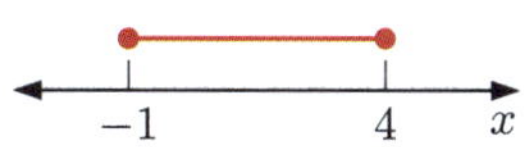

f 

g 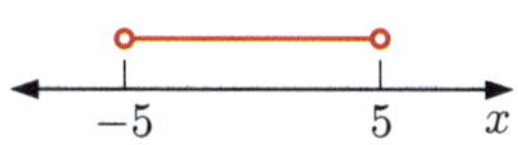

h 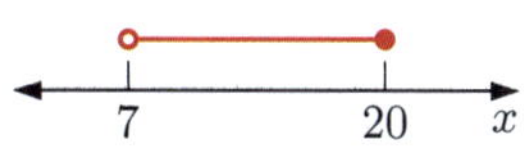

i 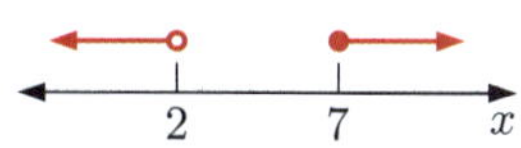

j 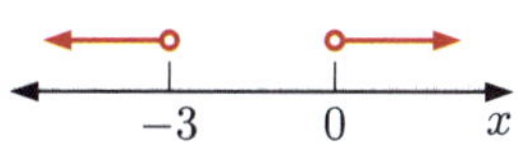

k 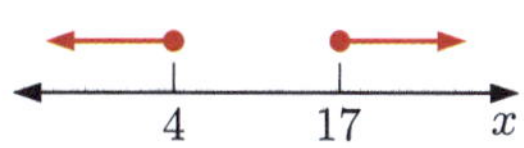

l 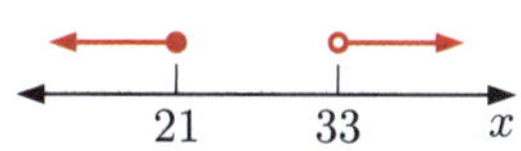

Example 4 — Self Tutor

Write down the meaning of the interval notation, and list the elements of the set:

a $\{x \in \mathbb{Z} \mid -2 \leqslant x \leqslant 1\}$ **b** $\{x \in \mathbb{N} \mid x > 2\}$

a $\{x \in \mathbb{Z} \mid -2 \leqslant x \leqslant 1\}$ is the set of integers greater than or equal to -2 and less than or equal to 1.

$\therefore\ x \in \{-2, -1, 0, 1\}$

b $\{x \in \mathbb{N} \mid x > 2\}$ is the set of natural numbers greater than 2.

$\therefore\ x \in \{3, 4, 5, 6, ...\}$

3 Write down the meaning of the interval notation:

a $\{x \mid x > 4\}$ **b** $\{x \in \mathbb{N} \mid x \leqslant 5\}$ **c** $\{y \in \mathbb{Z} \mid 0 < y < 8\}$

d $\{x \in \mathbb{R} \mid 1 \leqslant x \leqslant 4\}$ **e** $\{t \in \mathbb{Z}^+ \mid 2 < t < 7\}$ **f** $\{n \mid n \leqslant 3$ or $n > 6\}$

4 List the elements of the set:

a $\{x \in \mathbb{Z} \mid 1 < x \leqslant 6\}$ **b** $\{x \in \mathbb{Z} \mid x \leqslant 5\}$ **c** $\{x \in \mathbb{Z} \mid x \geqslant 3\}$

d $\{x \in \mathbb{N} \mid -3 \leqslant x \leqslant 5\}$ **e** $\{x \in \mathbb{Z}^- \mid x \geqslant -4\}$ **f** $\{x \in \mathbb{Z}^+ \mid x > 6\}$

Example 5 Self Tutor

Write in interval notation:

a $\{-3, -2, -1, 0, 1, 2, ...\}$ **b**

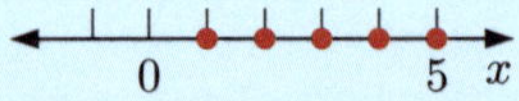

c

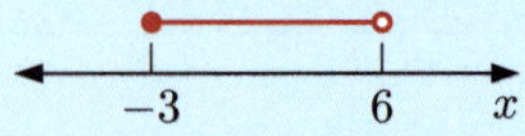

a The elements of the set are integers, so we know $x \in \mathbb{Z}$.

$\{-3, -2, -1, 0, 1, 2, ...\} = \{x \in \mathbb{Z} \mid x \geqslant -3\}$

b $\{x \in \mathbb{N} \mid 1 \leqslant x \leqslant 5\}$ or $\{x \in \mathbb{Z} \mid 1 \leqslant x \leqslant 5\}$ or $\{x \in \mathbb{Z}^+ \mid x \leqslant 5\}$

c $\{x \mid -3 \leqslant x < 6\}$

5 Write in interval notation:

a $\{-5, -4, -3, -2, -1\}$ **b** $\{0, 1, 2, 3, 4, 5\}$ **c** $\{4, 5, 6, 7, 8, ...\}$

d $\{...., -3, -2, -1, 0, 1\}$ **e** $\{-5, -4, -3, -2, -1, 0, 1\}$ **f** $\{...., 41, 42, 43, 44\}$

6 Write in interval notation:

a

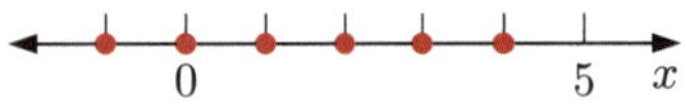

b

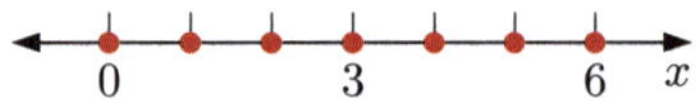

7 Sketch on a number line:

a $\{x \in \mathbb{N} \mid 4 \leqslant x < 8\}$ **b** $\{x \in \mathbb{Z} \mid -5 < x \leqslant 4\}$ **c** $\{x \in \mathbb{Z} \mid x > -5\}$

8 Determine whether each set is finite or infinite:

a $\{x \in \mathbb{Z} \mid 1 \leqslant x \leqslant 10\}$ **b** $\{x \in \mathbb{Z} \mid x < 4\}$ **c** $\{x \in \mathbb{Z}^+ \mid x > 5\}$

d $\{x \in \mathbb{Z}^+ \mid x < 7\}$ **e** $\{x \in \mathbb{N} \mid x \leqslant 9\}$ **f** $\{x \in \mathbb{R} \mid 2 \leqslant x \leqslant 6\}$

9 Let $U = \{x \in \mathbb{Z} \mid 1 \leqslant x \leqslant 30\}$, $P = \{x \mid 1 \leqslant x \leqslant 18\}$, and $Q = \{x \mid 12 \leqslant x \leqslant 25\}$.

a Write in interval notation:

i $P \cap Q$ **ii** $P \cup Q$ **iii** P' **iv** Q'

b Find:

i $n(P)$ **ii** $n(Q)$ **iii** $n(P \cap Q)$ **iv** $n(P \cup Q)$

10 Let $U = \{x \mid 0 \leqslant x \leqslant 10\}$, $A = \{x \in \mathbb{Z} \mid x > 4\}$, $B = \{x \in \mathbb{Q} \mid x \geqslant 5\}$, and $C = \{x \mid x > 4\}$.
Determine whether each statement is true or false:

a $A \subseteq C$ **b** $A \subseteq B$ **c** $C' \subseteq A'$ **d** $B' \subseteq C'$

11 Explain why we cannot represent the set of rational numbers $\mathbb{Q}$ on a number line.

MULTIPLE CHOICE QUIZ

QUICK QUIZ

REVIEW SET 5A

1. **a** List the elements of set C, the set of all colours on the flag of South Africa.

   b Find $n(C)$.

   c True or false?

   i black $\in C$ **ii** green $\notin C$

2. Decide whether $A \subseteq B$ for:

   a $A = \{1, 3, 6\}$, $B = \{1, 2, 3, 5, 6, 8\}$

   b $A = \{b, e, f, j\}$, $B = \{a, b, d, e, g, j, l, m\}$

   c $A = \varnothing$, $B = \{\text{Japan, Singapore, India}\}$

3. Let $U = \{\text{positive whole numbers less than } 10\}$. Find the complement of:

   a $P = \{3, 5, 7, 8\}$ **b** $Q = \{\text{square numbers}\}$ **c** $R = \{\text{composite numbers}\}$

4. Suppose $A = \{a, e, i, o, u\}$ and $B = \{a, b, c, d, e\}$. Find:

   a $A \cup B$ **b** $A \cap B$

5. Suppose $U = \{\text{positive integers} \leqslant 15\}$, $M = \{\text{multiples of } 5\}$, and $F = \{\text{factors of } 10\}$. List the elements of:

   a M **b** F **c** $M \cup F$

   d $M \cap F$ **e** $(M \cap F)'$ **f** $M \cap F'$

6. True or false?

   a $\sqrt{200} \in \mathbb{Q}$ **b** $\mathbb{N} \subseteq \mathbb{Z}^+$ **c** $\mathbb{Z} \subseteq \mathbb{Q}$

7. Find:

   a $\mathbb{N} \cap \mathbb{Z}$ **b** $\mathbb{Q} \cup \mathbb{R}$

8. Consider the set $\{x \in \mathbb{Z}^- \mid x \geqslant -8\}$.

   a Write down the meaning of the interval notation.

   b List the elements of the set. **c** Represent the set on a number line.

9. Write in interval notation:

   a

   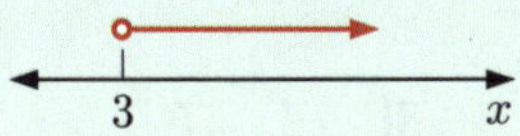

   b

   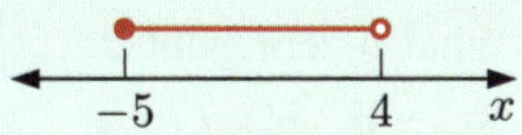

   c

   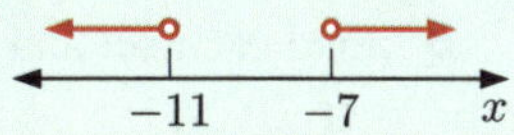

10. Let $U = \{\text{natural numbers} \leqslant 30\}$, $A = \{\text{multiples of } 5\}$, and $B = \{\text{multiples of } 10\}$.

    a Describe the set U using interval notation.

    b Is $A \subseteq B$? Explain your answer.

    c Find $A \cap B'$ and describe what this set means.

11. Let $U = \{x \in \mathbb{Z}^+ \mid x \leqslant 20\}$, $A = \{x \in \mathbb{Z}^+ \mid x > 7\}$, and $B = \{x \in \mathbb{Z}^+ \mid 4 \leqslant x < 16\}$. Write in interval notation:

    a $A \cap B$ **b** $A \cup B$ **c** A' **d** B'

12. Answer the **Opening Problem** on page **68**.

REVIEW SET 5B

1 List the elements of P, the set of prime numbers between 20 and 40.

2 List all the subsets of $\{2, 5, 7\}$.

3 Suppose $n(U) = 18$ and $n(A) = 13$. Find $n(A')$.

4 Suppose $F = \{\text{factors of } 32\}$, and $M = \{\text{multiples of 4 less than } 32\}$.
List the elements of:

a $F \cap M$ **b** $F \cup M$

5 Let $U = \{\text{positive whole numbers less than } 50\}$,
$A = \{\text{positive whole numbers less than } 20\}$,
$B = \{\text{positive whole numbers greater than } 20\}$, and $C = \{\text{odd numbers}\}$.

a List the elements of $A \cap C$. **b** Are A and B complementary?

6 Explain why $\mathbb{N} \subseteq \mathbb{Q}$.

7 True or false?

a $-4.5 \in \mathbb{Z}^-$ **b** $\mathbb{Z}$ and $\mathbb{Q}'$ are disjoint.

8 List the elements of the set:

a $\{x \in \mathbb{N} \mid 3 \leqslant x < 14\}$ **b** $\{x \in \mathbb{Z}^+ \mid x \leqslant 10\}$

9 Write in interval notation:

a $\{-4, -3, -2, -1, 0, 1\}$ **b**

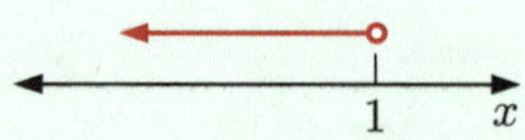

10 Sketch the set $\{x \in \mathbb{R} \mid x \leqslant 3 \text{ or } x > 7\}$ on a number line.

11 Two rival garden stores have various herbs for sale.
"Pete's Plants" sells lavender, rosemary, parsley, thyme, mint, and basil.
"Garden World" sells basil, thyme, sage, lemongrass, and chives.
Let P be the herbs sold at "Pete's Plants", and G be the herbs sold at "Garden World".

a Find $n(P)$ and $n(G)$.

b Find each set and explain what it represents:

i $P \cap G$ **ii** $P \cup G$ **iii** $P' \cap G$

c Are P and G disjoint? Explain your answer.

12 Consider the set A represented by this number line.

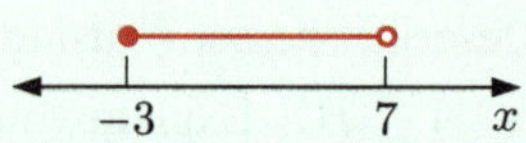

a Is A finite or infinite?

b Describe A:

i in words **ii** in interval notation.

c Write $A \cap \mathbb{N}$ in interval notation.

d Given the universal set $U = \mathbb{R}$, describe A':

i in words **ii** in interval notation **iii** on a number line.

Chapter 6

Linear equations and inequalities

Contents:

OPENING PROBLEM

Mrs May set her class the following challenge:

Find a fraction whose numerator is 4 *more than its denominator, and the value of the fraction is equal to* $\frac{1}{3}$.

"That's impossible!" Stan said, "If the numerator is more than the denominator, the value of the fraction must be greater than 1!"

Things to think about:

- **a** Can you explain why Stan is wrong?
- **b** Can you use algebra to solve the problem?

Equations are a fundamental part of mathematics, and an important tool for problem solving. The use of equations dates back to Ancient Egypt, where the scribe **Ahmes** recorded a series of problems which were solved using equations.

An **equation** is a mathematical statement which indicates that two expressions have the same value.

The expressions are connected by an **equal sign** $=$.

The **left hand side** (LHS) of an equation is on the left of the $=$ sign.

The **right hand side** (RHS) of an equation is on the right of the $=$ sign.

We can often convert a worded problem into an equation, then follow a formal procedure to **solve** the equation and hence the problem.

In other problems, one quantity we are considering may be greater than or less than another quantity. The problem may then be written as a mathematical **inequality**.

An **inequality** is a mathematical statement which *compares* the values of two expressions. It will contain one of the symbols $>$, $<$, $\geqslant$, or $\leqslant$.

A LINEAR EQUATIONS

Linear equations are equations in which the variable is raised only to the power 1.

All linear equations can be written in the form $ax + b = 0$ where a and b are constants, $a \neq 0$, and x is the variable.

For example, $4x - 1 = 7$ is a linear equation.

An equation must always be kept in balance. Whatever we do to one side of the equation, we must do the same to the other side to **maintain the balance**.

To solve simple linear equations, we first think about how the expression involving the unknown has been **built up**.

For the equation $4x - 1 = 7$, the LHS is built up by starting with x, multiplying by 4, then subtracting 1.

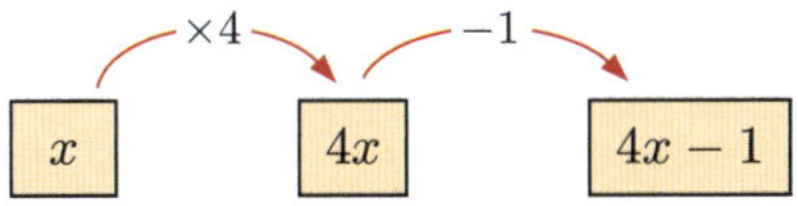

We then **isolate the unknown** by performing **inverse operations** in the **reverse order**.

For the equation $4x - 1 = 7$, we first add 1 to both sides, then divide both sides by 4.

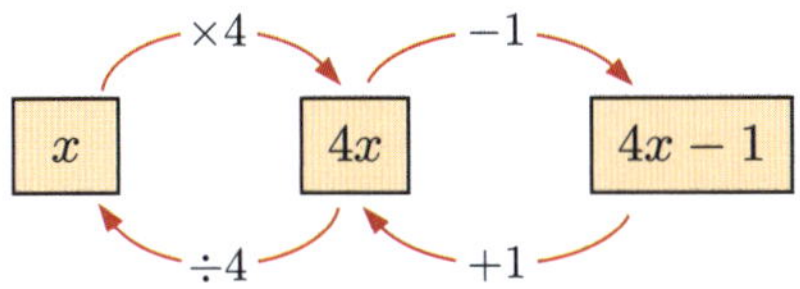

Remember that $+$ and $-$ are inverse operations, and that $\times$ and $\div$ are inverse operations.

You can always check your answer by substituting back into the original equation.

Example 1 **Self Tutor**

Solve for x:

a $4x - 1 = 7$

b $5 - 3x = 6$

a
$$4x - 1 = 7$$
$$\therefore\ 4x - 1 + 1 = 7 + 1 \quad \{+1 \text{ both sides}\}$$
$$\therefore\ 4x = 8$$
$$\therefore\ \frac{4x}{4} = \frac{8}{4} \quad \{\div 4 \text{ both sides}\}$$
$$\therefore\ x = 2$$

Check: $4 \times 2 - 1 = 8 - 1 = 7$ ✓

b
$$5 - 3x = 6$$
$$\therefore\ 5 - 3x - 5 = 6 - 5 \quad \{-5 \text{ both sides}\}$$
$$\therefore\ -3x = 1$$
$$\therefore\ \frac{-3x}{-3} = \frac{1}{-3} \quad \{\div -3 \text{ both sides}\}$$
$$\therefore\ x = -\tfrac{1}{3}$$

Check: $5 - 3 \times (-\frac{1}{3}) = 5 + 1 = 6$ ✓

Example 2 **Self Tutor**

Solve for x:

a $\frac{x}{5} - 3 = -1$

b $\frac{1}{5}(x - 3) = -1$

a
$$\frac{x}{5} - 3 = -1$$
$$\therefore\ \frac{x}{5} - 3 + 3 = -1 + 3 \quad \{+3 \text{ both sides}\}$$
$$\therefore\ \frac{x}{5} = 2$$
$$\therefore\ \frac{x}{5} \times 5 = 2 \times 5 \quad \{\times 5 \text{ both sides}\}$$
$$\therefore\ x = 10$$

Check: $\frac{10}{5} - 3 = 2 - 3 = -1$ ✓

b
$$\tfrac{1}{5}(x - 3) = -1$$
$$\therefore\ \tfrac{1}{5}(x - 3) \times 5 = -1 \times 5 \quad \{\times 5 \text{ both sides}\}$$
$$\therefore\ x - 3 = -5$$
$$\therefore\ x - 3 + 3 = -5 + 3 \quad \{+3 \text{ both sides}\}$$
$$\therefore\ x = -2$$

Check: $\frac{1}{5}((-2) - 3) = \frac{1}{5} \times -5 = -1$ ✓

EXERCISE 6A

1 Solve using an inverse operation:

a $x+7=3$ **b** $6x=42$ **c** $x-8=-3$ **d** $33=-3x$

e $\frac{x}{5}=20$ **f** $5+x=-7$ **g** $3=x-7$ **h** $4=\frac{x}{-8}$

2 Solve for x:

a $4x+9=21$ **b** $7-x=4$ **c** $2x-7=-3$

d $2-3x=23$ **e** $3+2x=27$ **f** $-2x-5=17$

g $5-3x=-13$ **h** $-6=4-5x$ **i** $-5x+3=18$

j $4+7x=-10$ **k** $12=8x-5$ **l** $-5+6x=-3$

3 Solve for x:

a $\frac{x}{2}+7=-3$ **b** $\frac{x+2}{5}=-3$ **c** $\frac{1}{4}(x+7)=5$

d $\frac{3x-4}{6}=-1$ **e** $4(2x-3)=28$ **f** $\frac{2x-7}{3}=4$

g $-5(3x+1)=35$ **h** $\frac{1}{7}(4-x)=-2$ **i** $\frac{1-5x}{3}=-6$

4 Solve for x:

a $-3+2(1-3x)=-19$ **b** $-5(2x+4)+3=-67$

c $\frac{5-3x}{2}-5=-1$ **d** $7-\frac{2x-11}{3}=2$

Example 3 Self Tutor

Solve for x: $3(2x-5)-2(x-1)=3$

$$3(2x-5)-2(x-1)=3$$
$$\therefore\ 3\times 2x+3\times(-5)-2\times x-2\times(-1)=3 \quad \{\text{distributive law}\}$$
$$\therefore\ 6x-15-2x+2=3$$
$$\therefore\ 4x-13=3 \quad \{\text{collecting like terms}\}$$
$$\therefore\ 4x-13+13=3+13 \quad \{+\,13 \text{ both sides}\}$$
$$\therefore\ 4x=16$$
$$\therefore\ \frac{4x}{4}=\frac{16}{4} \quad \{\div\,4 \text{ both sides}\}$$
$$\therefore\ x=4$$

Check: $3(2\times 4-5)-2(4-1)=3\times 3-2\times 3=3$ ✓

5 Solve for x:

a $2x+5+x=14$ **b** $3x-1+4x=27$

c $4(x-2)-2x=-18$ **d** $x-5(x-3)=21$

e $2(x+4)+4(x+3)=14$ **f** $4(x-2)+5(x+1)=-30$

g $3(2x+1)-2(x-2)=23$ **h** $6(x-3)+3(4x-1)=-3$

i $2(3x-4)-3(x-5)=-11$ **j** $3(2x-3)-4(7-2x)=-30$

Example 4 — Self Tutor

Solve for x: $\quad 3x - 4 = 2x + 6$

$$\begin{aligned} 3x - 4 &= 2x + 6 \\ \therefore\ 3x - 4 - 2x &= 2x + 6 - 2x \qquad \{-\,2x \text{ both sides}\} \\ \therefore\ x - 4 &= 6 \\ \therefore\ x - 4 + 4 &= 6 + 4 \qquad \{+\,4 \text{ both sides}\} \\ \therefore\ x &= 10 \end{aligned}$$

Check: LHS $= 3 \times 10 - 4 = 26$, RHS $= 2 \times 10 + 6 = 26$. ✓

6 Solve for x:

- **a** $3x - 4 = 2x + 1$
- **b** $2x + 5 = 11 - x$
- **c** $7 - 2x = 3x + 4$
- **d** $-x = 4x + 9$
- **e** $8 - 3x = 4x + 5$
- **f** $3x - 7 = 2 - 3x$

7 Solve for x:

- **a** $5 - x - 2(x + 3) = x + 4$
- **b** $4 - 3(x + 2) = 5x - 1$
- **c** $3 - 4x - (x + 5) = -7$
- **d** $5(2x - 3) - x = x + 1$
- **e** $4(x + 3) = 7 - (2 - 3x)$
- **f** $3(x - 2) + 5 = 7(x - 3)$
- **g** $5x - 2(3x - 1) = 4(x - 6)$
- **h** $6(x - 1) - 5(3x + 4) = 8(x - 3)$

8 Solve for x, if possible. Explain your results.

- **a** $3(2x - 5) + 10 = 6x - 5$
- **b** $6(x - 2) = 3(2x + 1)$

LEARNING ALGEBRA

B EQUATIONS WITH FRACTIONS

Some equations involving fractions can be rewritten in linear form, even if the variable started in the denominator. In these cases, we:

- write all fractions with the same **lowest common denominator (LCD)**, and then
- **equate numerators**.

For fractions whose denominators involve the variable, the lowest common denominator is found in the same way as for numerical fractions.

For example:

- in $\dfrac{x}{4} = \dfrac{x+1}{6}$ the LCD is 12
- in $\dfrac{5}{3x} = \dfrac{7}{9}$ the LCD is $9x$
- in $\dfrac{x-7}{3} = \dfrac{x}{2x-1}$ the LCD is $3(2x - 1)$
- in $\dfrac{3}{x+1} = \dfrac{2}{x-2}$ the LCD is $(x + 1)(x - 2)$.

Example 5 Self Tutor

Solve for x: $\frac{x}{2} = \frac{3+x}{5}$

$$\frac{x}{2} = \frac{3+x}{5}$$

$\therefore \frac{x}{2} \times \frac{5}{5} = \frac{2}{2} \times \left(\frac{3+x}{5}\right)$ {LCD = 10}

$\therefore 5x = 2(3+x)$ {equating numerators}

$\therefore 5x = 6 + 2x$ {distributive law}

$\therefore 3x = 6$ $\{-2x$ both sides$\}$

$\therefore x = 2$ $\{\div 3$ both sides$\}$

EXERCISE 6B

1 Solve for x:

a $\frac{x}{2} = \frac{5}{9}$
b $\frac{x-1}{5} = \frac{x}{10}$
c $\frac{x}{3} = \frac{x+1}{2}$
d $\frac{x+2}{5} = \frac{3x-1}{4}$
e $\frac{3x}{4} = \frac{6-x}{5}$
f $\frac{2x-5}{4} = \frac{3x-1}{-2}$
g $\frac{2x-1}{4} = \frac{5x+2}{8}$
h $\frac{2x-9}{3} = 2 - x$
i $7 + 2x = \frac{4x-1}{-3}$

Example 6 Self Tutor

Solve for x: $\frac{x}{3} - \frac{1-2x}{6} = -4$

$$\frac{x}{3} - \frac{1-2x}{6} = -4$$

$\therefore \frac{x}{3} \times \frac{2}{2} - \left(\frac{1-2x}{6}\right) = -4 \times \frac{6}{6}$ {LCD = 6}

$\therefore 2x - (1-2x) = -24$ {equating numerators}

$\therefore 2x - 1 + 2x = -24$ {distributive law}

$\therefore 4x - 1 = -24$ {collecting like terms}

$\therefore 4x = -23$ $\{+1$ both sides$\}$

$\therefore x = -\frac{23}{4}$ $\{\div 4$ both sides$\}$

2 Solve for x:

a $\frac{x}{2} - \frac{x}{4} = 3$
b $\frac{x}{3} - 2 = \frac{3x}{4}$
c $\frac{x}{6} + \frac{x+2}{3} = -3$
d $\frac{x+1}{2} + \frac{x-3}{5} = -1$
e $\frac{3x-1}{4} - \frac{x+2}{8} = 2$
f $\frac{x}{6} = 2 - \frac{x+3}{2}$

3 Solve for x:

a $\frac{3x-8}{2} - 1 = \frac{x-4}{8}$
b $\frac{x+4}{6} - \frac{x}{2} = \frac{3x-2}{3}$
c $\frac{x}{6} - \frac{x+4}{5} = \frac{2}{3}$
d $\frac{x+3}{2} + \frac{x-2}{4} = \frac{x+1}{8}$
e $\frac{x-4}{3} - \frac{3x+7}{12} = \frac{x+1}{6}$
f $\frac{3x+1}{4} - \frac{1-2x}{6} = \frac{3x+2}{8}$

Example 7 — Self Tutor

Solve for x: $\frac{5}{2x} = -\frac{3}{4}$

$$\frac{5}{2x} = -\frac{3}{4}$$

$$\therefore \frac{5}{2x} \times \frac{2}{2} = -\frac{3}{4} \times \frac{x}{x} \quad \{\text{LCD} = 4x\}$$

$$\therefore 10 = -3x \quad \{\text{equating numerators}\}$$

$$\therefore x = -\tfrac{10}{3} \quad \{\div -3 \text{ both sides}\}$$

4 Solve for x:

a $\frac{1}{x} = 3$ **b** $\frac{1}{x} = -4$ **c** $\frac{3}{2x} = 1$ **d** $-\frac{1}{3x} = 2$

e $\frac{3}{x} = \frac{2}{7}$ **f** $\frac{6}{5} = \frac{4}{x}$ **g** $\frac{4}{3x} = \frac{8}{7}$ **h** $\frac{7}{2x} = -\frac{1}{8}$

5 Consider the equation $\frac{3}{2x} = \frac{1}{3x}$.

Write the fractions with the same LCD then equate the numerators.

What does this tell you about the equation?

Example 8 — Self Tutor

Solve for x: $\frac{2x+1}{3-x} = \frac{3}{4}$

$$\frac{2x+1}{3-x} = \frac{3}{4}$$

$$\therefore \frac{4}{4} \times \left(\frac{2x+1}{3-x}\right) = \frac{3}{4} \times \left(\frac{3-x}{3-x}\right) \quad \{\text{LCD} = 4(3-x)\}$$

$$\therefore 4(2x+1) = 3(3-x) \quad \{\text{equating numerators}\}$$

$$\therefore 8x + 4 = 9 - 3x \quad \{\text{distributive law}\}$$

$$\therefore 11x + 4 = 9 \quad \{+3x \text{ both sides}\}$$

$$\therefore 11x = 5 \quad \{-4 \text{ both sides}\}$$

$$\therefore x = \tfrac{5}{11} \quad \{\div 11 \text{ both sides}\}$$

6 Solve for x:

a $\frac{1}{x+3} = 1$ **b** $\frac{3}{1-x} = 2$ **c** $\frac{2}{3-x} = \frac{1}{2}$

d $\frac{2x+1}{x-4} = 4$ **e** $\frac{2x}{x+4} = 3$ **f** $\frac{4x+1}{x+2} = -3$

g $\frac{x+4}{1-2x} = \frac{3}{4}$ **h** $\frac{2x-3}{4-5x} = -\frac{2}{3}$ **i** $\frac{3x+2}{4-3x} = 2$

7 Solve for x:

a $\frac{2}{x} = \frac{5}{x+3}$ **b** $\frac{5}{x-8} = \frac{1}{x-2}$ **c** $\frac{3}{x-1} = \frac{7}{x+2}$

PUZZLE — SOLVING EQUATIONS

Match each LHS with a different RHS and solution to form 10 correct equations.

LHS	
$x-15$	$6(x-3)$
$\dfrac{x+6}{3}$	$\dfrac{x}{7}+3$
$10(x+3)$	$\dfrac{27}{2x-7}$
$\dfrac{14}{x+1}$	$8(x-2)$
$\dfrac{x}{5}+4$	$8-3x$

RHS	
-13	7
3	12
70	32
-1	4
6	5

Solution	
$x=1$	$x=2$
$x=3$	$x=4$
$x=5$	$x=6$
$x=7$	$x=8$
$x=9$	$x=10$

PRINTABLE WORKSHEET

C PROBLEM SOLVING

Many problems can be translated into **algebraic equations**.

To solve problems using algebra, we follow these steps:

Step 1: Decide on the unknown quantity and allocate it a variable such as x.

Step 2: Translate the problem into an equation.

Step 3: Solve the equation by isolating the variable.

Step 4: Check that the solution satisfies the original problem.

Step 5: Write the answer in sentence form, describing how the solution relates to the original problem.

Example 9 — Self Tutor

If twice a number is subtracted from 11, the result is 4 more than the number. What is the number?

Let x be the number.

$\therefore\ 11-2x=x+4$

$\therefore\ 11=3x+4$ $\{+\,2x$ both sides$\}$

$\therefore\ 7=3x$ $\{-\,4$ both sides$\}$

$\therefore\ x=2\frac{1}{3}$ $\{\div\,3$ both sides$\}$

So, the number is $2\frac{1}{3}$.

EXERCISE 6C.1

1 Find the number:
   a When a number is doubled, the result is 18.
   b When 6 is added to a number, the result is 11.
   c When four times a number is subtracted from 23, the result is -5.
   d When a number is decreased by 1 and the result is halved, the answer is 45.
   e Two less than twice a number is equal to 6 more than four times the number.
   f If I divide the sum of 8 and a number by 4, the result is one more than a third of the number.

2 Cat food tins are sold in packs of 6. Rachel bought 42 tins in total. How many packs did she buy?

3 Julian's sister owns 3 more shirts than he does. Between them they own 15 shirts. How many shirts does Julian own?

4 A plane flying from Liverpool to Madrid is carrying 30 rows of passengers, as well as 12 crew members. There are 222 people on board the plane, and every seat is taken. How many passengers are in each row?

5 Drew, Paige, and Henry went to a birthday party. Drew ate 7 more sweets than Paige, and Henry ate 2 fewer sweets than Paige. Between them they ate 20 sweets. How many sweets did Paige eat?

6 During a volleyball training session, Keela drank twice as much water as Carol, and Xavier drank 100 mL more water than Keela. Between them they drank 3 litres of water. How much water did Carol drink?

Example 10 **Self Tutor**

What number must be added to both the numerator and the denominator of $\frac{1}{3}$ to obtain a fraction equal to $\frac{7}{8}$?

Let x be the number.

$$\therefore \quad \frac{1+x}{3+x} = \frac{7}{8}$$

$$\therefore \quad \frac{8}{8} \times \left(\frac{1+x}{3+x}\right) = \frac{7}{8} \times \left(\frac{3+x}{3+x}\right) \qquad \{\text{LCD} = 8(3+x)\}$$

$$\therefore \quad 8(1+x) = 7(3+x) \qquad \{\text{equating numerators}\}$$

$$\therefore \quad 8 + 8x = 21 + 7x \qquad \{\text{distributive law}\}$$

$$\therefore \quad 8 + x = 21 \qquad \{-\ 7x \text{ both sides}\}$$

$$\therefore \quad x = 13 \qquad \{-\ 8 \text{ both sides}\}$$

So, the number is 13.

7 What number must be added to both the numerator and the denominator of $\frac{3}{7}$ to obtain a fraction equal to $\frac{7}{9}$?

8 Write a fraction for which the sum of the numerator and denominator is 20, and the value of the fraction is $\frac{2}{3}$.

9 Consider the **Opening Problem** on page **80**. Solve the challenge set by Mrs May.

10 At a restaurant, Table A's bill of \$72 is split equally between the people at the table. Table B has two more people than Table A, and its bill of \$108 is also split equally between the people at the table. Each person at Table A pays the same amount as each person at Table B.
How many people were at Table A?

USING TABLES TO HELP SOLVE PROBLEMS

For more complicated problems, it can be useful to sort the information in a table. This helps us construct the correct equation.

Example 11 — Self Tutor

Cans of sardines are sold in two sizes. Small cans cost \$2 each, and large cans cost \$3 each. If 15 cans of sardines were bought for a total of \$38, how many small cans were bought?

Size	*Cost per can*	*Number bought*	*Value*
small	\$2	x	$\$2x$
large	\$3	$15 - x$	$\$3(15 - x)$
		15	\$38

$2x + 3(15 - x) = 38$
$\therefore\ 2x + 45 - 3x = 38$ {distributive law}
$\therefore\ 45 - x = 38$ {collecting like terms}
$\therefore\ -x = -7$ {-45 both sides}
$\therefore\ x = 7$ {$\div -1$ both sides}

So, 7 small cans were bought.

EXERCISE 6C.2

1 Isaac is going to boarding school. He buys school shirts at \$35 each and trousers at \$49 each. Altogether he buys 9 items, and their total cost is \$357. How many shirts does he buy?

2 I have 36 coins in my pocket, all of which are 50 cent or 1 dollar coins. If their total value is \$32, how many 50 cent coins do I have?

3 Oranges cost 25 cents each and apples cost 30 cents each. I bought 5 more oranges than apples, and the total cost was \$4.55. How many apples did I buy?

4 Ellie is now four times as old as her son. In 5 years' time she will be three times as old as her son. How old is Ellie's son now?

5 Four years ago, Adrian was one quarter of his brother's age. In two years' time, he will be half his brother's age. How old is Adrian now?

6 Lana collects 20 cent and 50 cent coins. She has four more 50 cent coins than 20 cent coins, and the total value of her coins is \$9.70. How many of each type does she have?

7 Tickets to a concert cost \$20, \$35, or \$50 each. The number of \$20 tickets sold was triple the number of \$35 tickets sold. 400 more \$50 tickets were sold than \$35 tickets. If the tickets sold had a total value of \$150 500, how many of each type of ticket was sold?

8 Olivia made her own fruit and nut mix using dried fruit costing \$8 per kilogram, and nuts costing \$12 per kilogram. She made a total of 30 kg of fruit and nut mix costing \$10.80 per kilogram. How many kilograms of each ingredient did she use?

9 Canadian Airways shares pay a yearly return of 8%, while BankCorp shares pay 11%. Tom invests \$1500 more in BankCorp shares than in Canadian Airways shares, and his total yearly earnings from both investments is \$1020. How much did Tom invest in Canadian Airways shares?

MIXTURE PROBLEMS

Example 12 — Self Tutor

How much water should be added to 2 L of 5% cordial mixture to produce a 3% cordial mixture?

Suppose we add x L of water to the mixture.

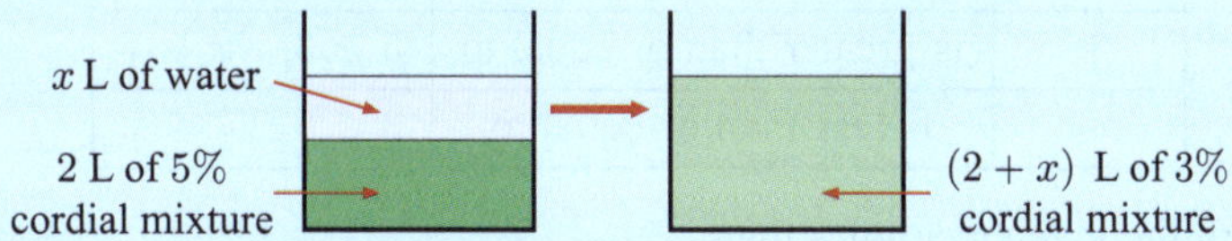

If we consider the total amount of cordial in the mixture:

5% of 2 L $= 3\%$ of $(x+2)$ L

$\therefore\ \frac{5}{100} \times 2 = \frac{3}{100} \times (x+2)$

$\therefore\ \frac{10}{100} = \frac{3(x+2)}{100}$

$\therefore\ 10 = 3(x+2)$ {equating numerators}

$\therefore\ 10 = 3x + 6$ {distributive law}

$\therefore\ 4 = 3x$ {$-$ 6 both sides}

$\therefore\ x = 1\frac{1}{3}$ {$\div$ 3 both sides}

$\therefore\ 1\frac{1}{3}$ L of water must be added to the mixture.

EXERCISE 6C.3

1. How much water must be added to 3 L of 6% cordial mixture to produce a 4% cordial mixture?
2. How much pure cordial must be added to 30 L of 4% cordial mixture to make a 10% cordial mixture?
3. How many litres of 12% weedkiller mixture must be added to 2 L of 7% weedkiller mixture to make a 9% weedkiller mixture?
4. Dominic needs 15 L of 5% saline solution. He has already made up 4% saline and 7% saline mixtures. How much of each should he mix together?

D LINEAR INEQUALITIES

For safety reasons, many rides at theme parks have height restrictions. It is common to see signs which say:

"You must be at least 120 cm tall to go on this ride."

We could write this requirement as an *inequality*:

$$\text{height} \geqslant 120 \text{ cm} \quad \text{or} \quad H \geqslant 120 \text{ cm}.$$

If a person has height $H = 120$, 120.5, 125, or 150 cm, then they can go on the ride. All of these values satisfy $H \geqslant 120$ cm.

An **inequality** is a mathematical statement which *compares* the values of two expressions. It will contain one of the symbols $>$, $<$, $\geqslant$, or $\leqslant$.

Symbol	*Meaning*	*Examples*
$>$	is greater than	$5 > 3$, $x > 7$
$<$	is less than	$3 < 5$, $7 < x$
$\geqslant$	is greater than or equal to	$5 \geqslant 3$, $x \geqslant 7$
$\leqslant$	is less than or equal to	$3 \leqslant 5$, $7 \leqslant x$

You should notice from the given examples that:

If we swap the LHS and RHS of an inequality, the **inequality sign** needs to be **reversed**.

Linear inequalities take the same form as linear equations, except they contain an inequality sign instead of an "equals" sign.

For example, $1 - 3x > 0$ and $2x + 3 \leqslant 9$ are linear inequalities.

We have already used linear inequalities in **interval notation** to describe intervals on the number line.

EXERCISE 6D

1 Write as an inequality:

- **a** the speed S must not exceed 40 km/h
- **b** the age A must be at least 18 years
- **c** a is greater than 3
- **d** b is less than or equal to -3
- **e** d is less than 5
- **f** -20 is greater than or equal to x
- **g** 4 is less than y
- **h** z is greater than or equal to 0.

2 Rewrite each inequality with the variable on the LHS:

a $2 < x$ **b** $5 > b$ **c** $2\frac{1}{2} \leqslant c$ **d** $-7 \geqslant d$ **e** $-19 > a$ **f** $-3 < p$

3 Illustrate on separate number lines:

a $x > 2$ **b** $a > -3$ **c** $b \leqslant 2$ **d** $m \geqslant -1$

e $3 < x$ **f** $2 > a$ **g** $x > 3\frac{1}{2}$ **h** $x \leqslant -1\frac{1}{4}$

4 Given that $a > 5$, decide whether each of the following is a possible value for a:

a 7 **b** -5 **c** 5 **d** 12

5 **a** On the same number line, illustrate the solutions to the inequalities $x > 3$ and $x \leqslant \frac{5}{2}$.

b State the only integer value of x which does not satisfy either of these inequalities.

6 Suppose $p > 25$ and $q \leqslant 10$ where $p, q \in \mathbb{Z}$. Find the *smallest* possible value of $p - q$.

7 Suppose $w \leqslant 36$ and $x < 20$ where $w, x \in \mathbb{Z}$. Find the *largest* possible value of $w + x$.

8 Suppose $a < 19$ and $b \geqslant 6$ where $a, b \in \mathbb{Z}$. Find the *largest* possible value of $\frac{a}{b}$.

9 Suppose $-5 < x \leqslant 10$ and $-3 \leqslant y < 7$ where $x, y \in \mathbb{Z}$. Find the *smallest* possible value of xy.

E SOLVING LINEAR INEQUALITIES

The **solution** of an inequality is the set of values of the unknown which make the inequality true.

In previous years we have compared a mathematical equation to a balanced set of scales. An inequality is like an **unbalanced** set of scales.

For example, consider the inequality $3x + 2 > 7$. We let a square represent x and a circle represent 1.

To solve a linear inequality, we need to **maintain the imbalance**.

We perform inverse operations on both sides of the inequality, in order to isolate the unknown. However we need to make sure this will not change its solution.

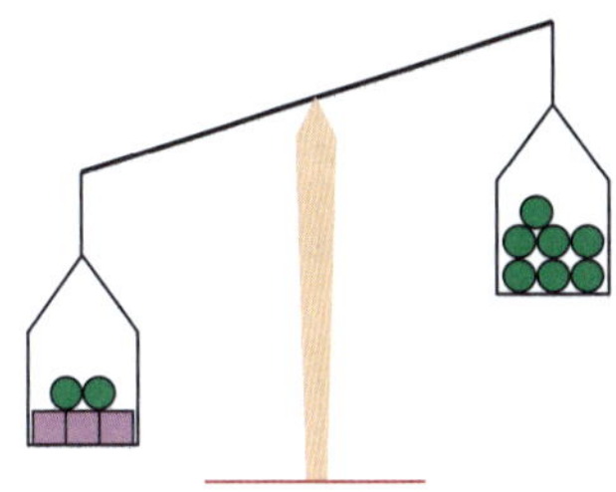

INVESTIGATION — OPERATIONS WITH INEQUALITIES

In this Investigation we will begin with inequalities which we know are true, and then perform various operations to both sides of the inequalities. By checking whether the resulting inequalities are true or false, we will establish some rules for operating with inequalities.

PRINTABLE WORKSHEET

Remember: $a > b$ if a is to the right of b on the number line.

Addition and subtraction

What to do:

1 Copy and complete the following table where T means *true* and F means *false*:

True inequality	*Operation*	*Working*	*Simplify*	*T or F*
$5 > 1$	$+3$	$5+3 > 1+3$	$8 > 4$	T
$11 > 7$	$+6$			
$8 > -3$	$+4$			
$-7 < -4$	$+5$			
$-6 < 5$	$+11$			
$8 > 6$	-2			
$9 > 7$	-4			
$2 > -1$	-2			
$-7 < -5$	-5			

2 If I *add* a number to both sides of an inequality, will it still be true?

3 If I *subtract* a number from both sides of an inequality, will it still be true?

Multiplication and division

What to do:

1 Copy and complete the following table where T means *true* and F means *false*:

True inequality	*Operation*	*Working*	*Simplify*	*T or F*
$5 > 2$	$\times 3$	$5 \times 3 > 2 \times 3$	$15 > 6$	T
$-3 < -2$	$\times 4$			
$4 > 3$	$\times -3$			
$-2 > -5$	$\times -4$			
$7 > 3$	$\times 2$			
$-4 < -2$	$\times 3$			
$2 > -6$	$\times -2$			
$-5 < -3$	$\times -4$			
$6 > 3$	$\div 3$			
$-6 > -8$	$\div 2$			
$6 > 3$	$\div -3$	$\frac{6}{-3} > \frac{3}{-3}$	$-2 > -1$	F
$-6 > -8$	$\div -2$			
$7 < 14$	$\div 7$			
$-10 < -5$	$\div 5$			
$7 < 14$	$\div -7$			

2 If I multiply or divide both sides of an inequality by a *positive* number, will it still be true?

3 If I multiply or divide both sides of an inequality by a *negative* number, will it still be true?

From the **Investigation**, you should have found that when we multiply or divide both sides of an inequality by a *negative* number, the inequality is no longer true.

To illustrate why this occurs, consider the values 5 and 2 on this number line. 5 is to the *right* of 2 on the number line, so $5 > 2$.

Suppose we multiply each value by -1, giving -5 and -2. -5 is to the *left* of -2, so we must **reverse** the inequality signs, giving $-5 < -2$.

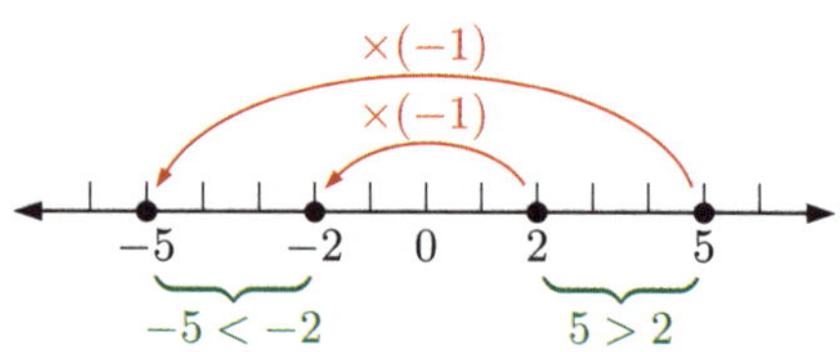

RULES FOR SOLVING INEQUALITIES

- If we **swap** the LHS and RHS, we **reverse** the inequality sign.
- If we **add** or **subtract** both sides by the same number, we **keep** the inequality sign.
- If we **multiply** or **divide** both sides by a **positive** number, we **keep** the inequality sign.
- If we **multiply** or **divide** both sides by a **negative** number, we **reverse** the inequality sign.

Example 13 **Self Tutor**

Solve for x and graph the solution:

a $5 - 2x > 9$ **b** $\frac{x}{4} - 2 \geqslant 3$

a $5 - 2x > 9$

$\therefore\ 5 - 2x - 5 > 9 - 5$ $\{-\ 5$ both sides$\}$

$\therefore\ -2x > 4$

$\therefore\ \frac{-2x}{-2} < \frac{4}{-2}$ $\{\div -2$ both sides$\}$

$\therefore\ x < -2$

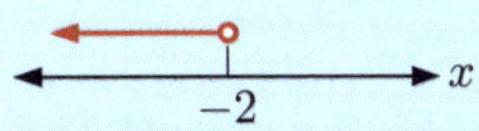

Check: If $x = -5$ then $5 - 2x = 5 - 2(-5) = 15$ and $15 > 9$. ✓

b $\frac{x}{4} - 2 \geqslant 3$

$\therefore\ \frac{x}{4} - 2 + 2 \geqslant 3 + 2$ $\{+\ 2$ both sides$\}$

$\therefore\ \frac{x}{4} \geqslant 5$

$\therefore\ \frac{x}{4} \times 4 \geqslant 5 \times 4$ $\{\times\ 4$ both sides$\}$

$\therefore\ x \geqslant 20$

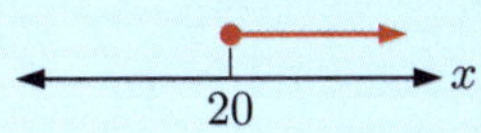

Check: If $x = 32$, then $\frac{x}{4} - 2 = \frac{32}{4} - 2 = 6$ and $6 \geqslant 3$. ✓

EXERCISE 6E

1 Solve the inequality and graph its solution:

a $a + 4 > 6$ **b** $3b \leqslant -9$ **c** $s - 4 < 2$ **d** $\frac{c}{5} < 2$

e $x - 8 \geqslant -3$ **f** $-4b > 16$ **g** $5 + t > 0$ **h** $-\frac{m}{5} \geqslant 12$

2 Solve the inequality and graph its solution:

a $2x + 5 > 11$ **b** $3a - 2 \geqslant 8$ **c** $4a + 9 < 1$

d $7 - 2b < -3$ **e** $16 + 7s > 2$ **f** $3b - 1 \geqslant 0$

g $5n + 7 < -3$ **h** $18 - x > 5$ **i** $11 - 4b \leqslant 4$

3 Solve for x and graph the solution:

a $\frac{x}{3} + 1 > 4$ **b** $\frac{x}{5} - 2 \leqslant -3$ **c** $\frac{3x}{2} + 5 \geqslant -2$

d $1 - \frac{x}{2} < 4$ **e** $3 - \frac{2x}{5} < 2$ **f** $5 - \frac{3x}{4} > -2$

g $\frac{x-3}{2} < 6$ **h** $\frac{4x+3}{5} \leqslant -1$ **i** $\frac{5-3x}{2} \leqslant -6$

4 Solve by first swapping the LHS and RHS:

a $5 > 4 + x$ **b** $2 \leqslant 6c + 14$ **c** $-4 > 2b$

d $7 \leqslant \frac{a}{3}$ **e** $3 < \frac{d-2}{4}$ **f** $6 \geqslant 3(p+2)$

Example 14 Self Tutor

Solve for x and graph the solution: $9 - 2x \geqslant 4x - 3$

$$9 - 2x \geqslant 4x - 3$$
$$\therefore\ 9 - 2x - 4x \geqslant 4x - 3 - 4x \qquad \{-4x \text{ both sides}\}$$
$$\therefore\ 9 - 6x \geqslant -3$$
$$\therefore\ 9 - 6x - 9 \geqslant -3 - 9 \qquad \{-9 \text{ both sides}\}$$
$$\therefore\ -6x \geqslant -12$$
$$\therefore\ \frac{-6x}{-6} \leqslant \frac{-12}{-6} \qquad \{\div -6 \text{ both sides}\}$$
$$\therefore\ x \leqslant 2$$

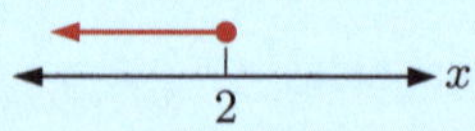

Check: If $x = 0$ then LHS $= 9 - 2(0) = 9$ and RHS $= 4(0) - 3 = -3$ so LHS $\geqslant$ RHS. ✓

5 Solve for x and graph the solution:

a $5x - 3 > 3x + 1$ **b** $2x + 1 \geqslant 4x + 7$ **c** $8x + 6 < 3x + 1$

d $2x + 7 > 7x + 3$ **e** $6x + 2 \leqslant 3x - 7$ **f** $x - 11 \leqslant 6x - 1$

g $3x + 1 \leqslant 7 - x$ **h** $1 - 4x > x + 5$ **i** $2 - 5x > \frac{1-3x}{4}$

6 Solve for x:

a $3(x+5) < -2$ **b** $2(x-4) \geqslant x + 5$ **c** $5(2x-3) \leqslant 4x - 8$

d $4(x-1) > \frac{5x+1}{2}$ **e** $2(3-x) \leqslant \frac{x+4}{3}$ **f** $3(x+2) > \frac{1-2x}{4}$

ACTIVITY — SIR COOL AND THE INEQUALITY CRUSADE

Click on the icon to help Sir Cool rid the world of inequality!

THE ADVENTURE OF SIR COOL

MULTIPLE CHOICE QUIZ

QUICK QUIZ

REVIEW SET 6A

1 Solve for x:

a $56 = -8x$ **b** $4x + 7 = -1$ **c** $\dfrac{5x+4}{3} = -2$

2 Solve for x:

a $2(x-7) - 3(8-x) = 9$ **b** $11 - x + 3(x-2) = x - 4$

3 I think of a number. If I subtract 1 from the number then multiply by 4, the result is two more than three times the number. Find the number.

4 Solve for x:

a $\dfrac{x+3}{5} = \dfrac{x-4}{2}$ **b** $\dfrac{x}{2} - \dfrac{x+3}{8} = \dfrac{5x}{6}$ **c** $\dfrac{4}{3x} = \dfrac{10}{7}$

5 Solve for x:

a $\dfrac{3x+5}{x-2} = -9$ **b** $\dfrac{4}{x+1} = \dfrac{9}{x-3}$

6 What number must be subtracted from both the numerator and denominator of $\frac{5}{9}$ to obtain a fraction equal to $\frac{3}{8}$?

7 Three years ago, Evan was one third of his sister's age. In a year's time, Evan will be half his sister's age. How old is Evan now?

8 Elaine has 88 coins in a jar. They are all 5 cent and 10 cent coins. The total value of the coins is \$7. How many of each type of coin does Elaine have?

9 Simon invests twice as much money in petroleum shares as he invests in technology shares. Petroleum shares pay a yearly return of 12%, while technology shares pay 7% per year. Simon's yearly income from these shares is \$744. How much money has Simon invested in each type of share?

10 Illustrate on separate number lines:

a $x \geqslant 5$ **b** $3 > x$ **c** $c > 4\frac{1}{2}$

11 Solve the inequality and graph its solution:

a $\dfrac{x}{5} - 3 < -1$ **b** $3 - 3x \geqslant -4$ **c** $\dfrac{2x+3}{2} < -3$

12 Solve for x:

a $2 - 5x > x + 4$ **b** $3(x+2) - 1 \leqslant 1 - 4x$

13 A scientist has a 10% ethanol mixture. She needs to make a 4% ethanol mixture for an experiment.

a How much water must be added to 5 L of 10% mixture to produce a 4% ethanol mixture?

b After the experiment there are 3 L of 4% ethanol mixture left over.

i How much pure ethanol should she add to increase the concentration to 10% ethanol?

ii How much ethanol is now in the mixture?

REVIEW SET 6B

1 Solve for x:

a $9-2x=-5$ **b** $\frac{x}{4}-3=-10$

2 Solve for x:

a $7-4x=5x+6$ **b** $6(3x-2)-x=2x+7$

3 At a theme park, Setha went on twice as many rides as Chenda, who went on three more rides than Dara. Together, they went on 25 rides. How many rides did Dara go on?

4 Solve for x:

a $\frac{x-3}{4}=\frac{2x}{5}$ **b** $\frac{x+9}{2x-1}=\frac{3}{4}$ **c** $\frac{5}{x}=\frac{7}{x-3}$

5 Write a fraction for which the sum of the numerator and denominator is 44, and the value of the fraction is $\frac{5}{6}$.

6 When Amanda was born, her mother was 27 years old. Amanda is now one quarter of her mother's age. How old is Amanda now?

7 A supermarket sells tuna in 90 g and 200 g tins. In one day, 112 tins of tuna were sold, with a total mass of 18 kg. How many tins of each type were sold?

8 How many litres of 8% cordial mixture must be added to 5 L of 5% cordial mixture to make a 6% cordial mixture?

9 Write as an inequality:

a k is less than -2 **b** 7 is greater than or equal to y.

10 Solve the inequality and graph its solution:

a $3-2x\geqslant 11$ **b** $-5<\frac{x-4}{3}$ **c** $6(2x+1)\geqslant 6-13x$

11 Miriam has a stamp collection made up of 10 cent and 20 cent stamps. She has five more 20 cent stamps than 10 cent stamps, and the stamps have a total value of \$4.60.

a Find the number of:

i 10 cent stamps **ii** 20 cent stamps.

b Miriam adds 16 more stamps to her collection. Some are 20 cent stamps and others are 50 cent stamps. The new total value for the stamp collection is \$9.90. Find the total number of:

i 50 cent stamps **ii** 20 cent stamps.

12 **a** Solve the inequality $\frac{x}{5}-3>-5$ and graph its solution.

b Solve the inequality $8-3(x+2)\geqslant 1$ and graph its solution.

c Find the values of x which satisfy *both* inequalities.

Chapter 7

Venn diagrams

Contents:

OPENING PROBLEM

50 tourists in Budapest were asked which tourist attractions they had visited. 31 had visited Buda Castle, and 23 had visited Heroes' Square. 10 tourists had not visited either of these places.

Things to think about:

a How can we represent this information using a diagram?

b How many of the tourists had visited:

  i Buda Castle, but not Heroes' Square **ii** both of these places?

In **Chapter 5** we used set notation and interval notation to *describe* the elements of a set. In this Chapter we will use **Venn diagrams** to *illustrate* sets. Venn diagrams allow us to quickly observe the relationship between different sets.

A VENN DIAGRAMS

A **Venn diagram** consists of a universal set U represented by a rectangle, and sets within it that are usually represented by circles.

The Venn diagram alongside shows the set A within the universal set U.

The **complement** of A is the region outside the circle.

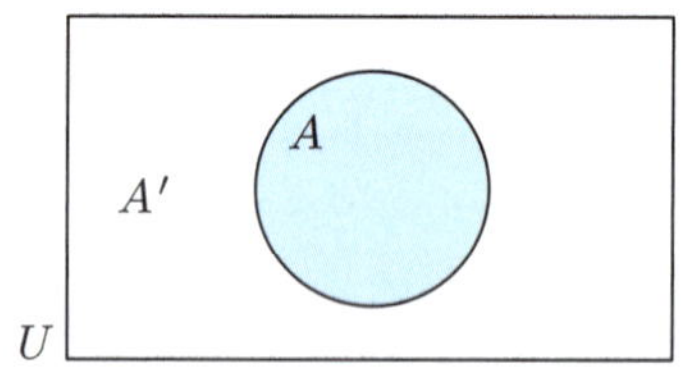

VENN DIAGRAMS FOR TWO SETS

If we are representing two sets A and B on the same Venn diagram, there are several layouts we can choose from.

- If A and B have elements in common, but neither is a subset of the other, the circles for the sets overlap.

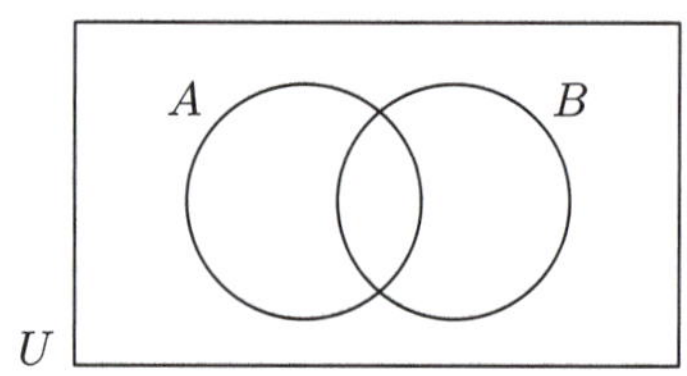

Notice that:

- The **intersection** of A and B is the region which is in *both* circles.
  $A \cap B = \{x \mid x \in A \textbf{ and } x \in B\}$

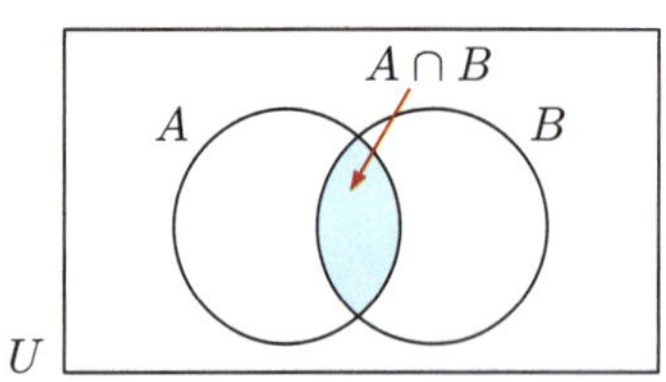

- The **union** of A and B is the region which is in *either* circle.
  $A \cup B = \{x \mid x \in A \textbf{ or } x \in B\}$

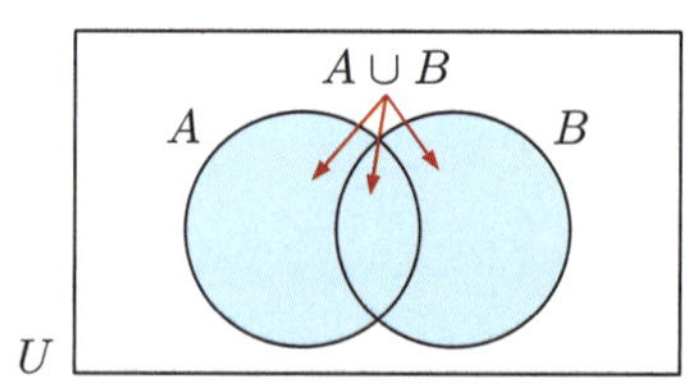

- If A is a **subset** of B, then every element of A is also an element of B.
  We place a circle representing A completely within the circle representing B.

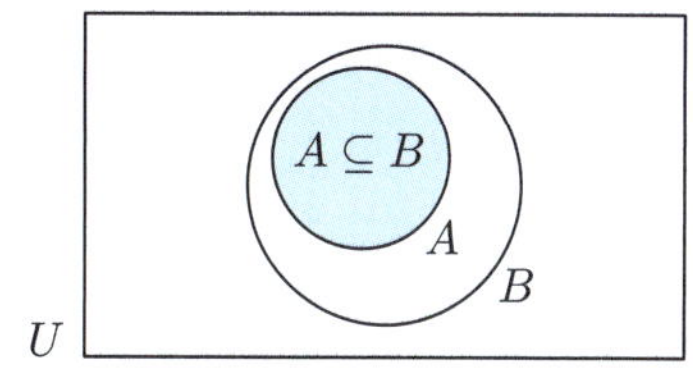

- If A and B are **disjoint** then there are no elements in both A and B.
  We represent A and B using circles which do not overlap.

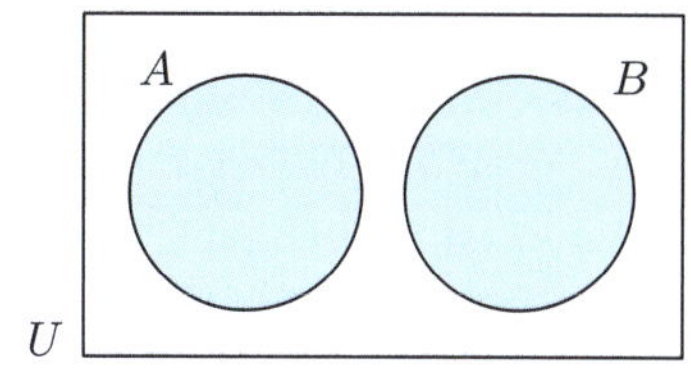

Example 1 — Self Tutor

If we roll an ordinary die, the universal set of the possible outcomes is $U = \{1, 2, 3, 4, 5, 6\}$.
Illustrate on a Venn diagram:

a $A = \{1, 5, 6\}$
b $A = \{1, 2\}$ and $B = \{1, 3, 4\}$
c $A = \{1, 3, 5\}$ and $B = \{3, 5\}$
d $A = \{2, 4, 6\}$ and $B = \{3, 5\}$

a $A = \{1, 5, 6\}$
$A' = \{2, 3, 4\}$

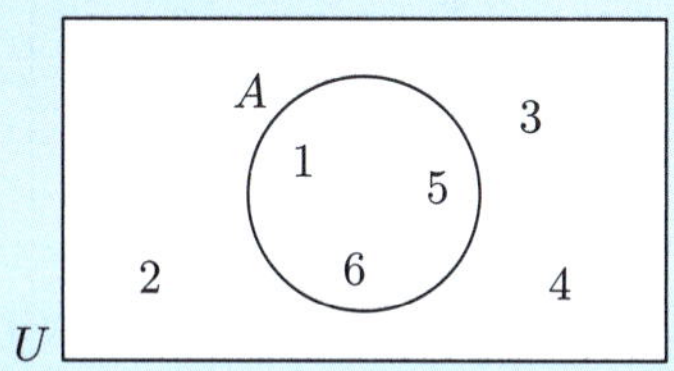

b $A \cap B = \{1\}$

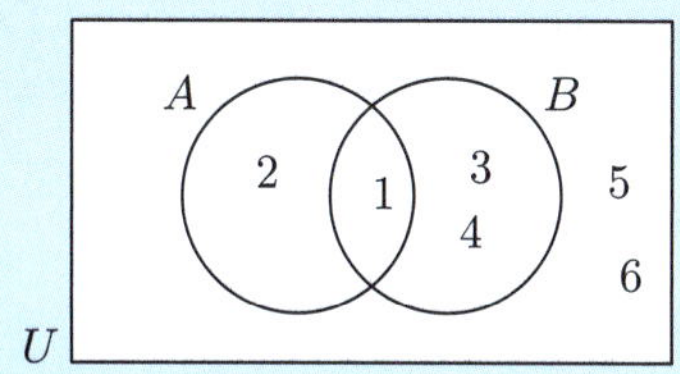

c $A \cap B = \{3, 5\}$, $B \subseteq A$

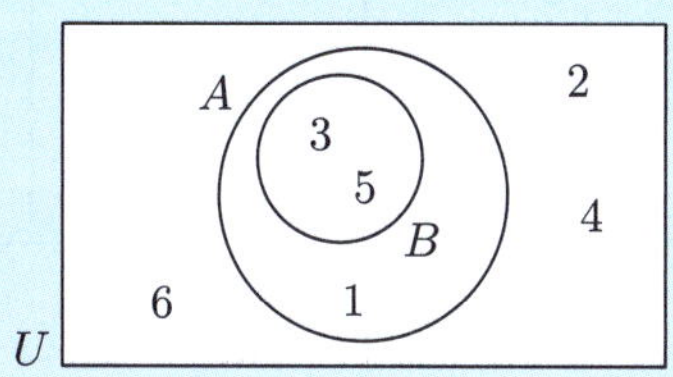

d $A \cap B = \varnothing$, so A and B are disjoint.

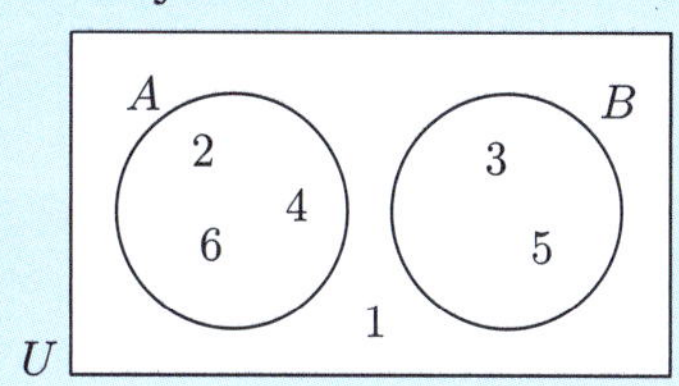

EXERCISE 7A

1 Let $U = \{1, 2, 3, 4, 5, 6, 7, 8, 9, 10\}$. Draw a Venn diagram to represent:

a $A = \{2, 3, 4\}$ **b** $A = \{1, 4, 5, 8\}$ **c** $A = \{2, 4, 5, 6, 7, 9, 10\}$

2

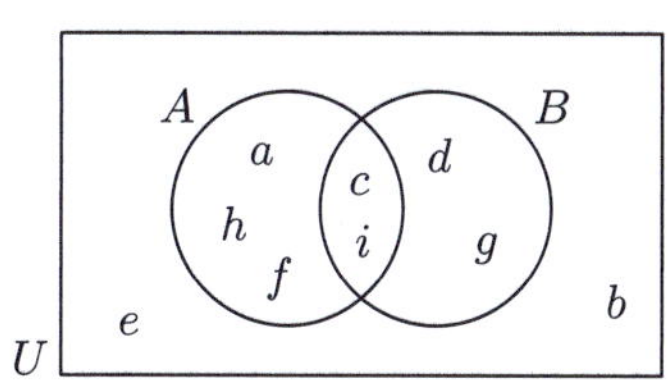

Consider the Venn diagram alongside.
List the elements of:

a A **b** B **c** A' **d** B'
e $A \cap B$ **f** $A \cup B$ **g** U

3 Let $U = \{1, 2, 3, 4, 5, 6, 7, 8, 9\}$. Draw a Venn diagram to represent:

- **a** $A = \{1, 2, 4, 8\}$ and $B = \{2, 3, 4, 5\}$
- **b** $A = \{1, 2, 3, 5, 8\}$ and $B = \{4, 6, 7\}$
- **c** $A = \{1, 2, 3\}$ and $B = \{1, 2, 3, 6\}$
- **d** $A = \{1, 3, 7, 9\}$ and $B = \{1, 2, 4, 5, 7, 8\}$.

4 Let $U = \{1, 2, 3, 4, 5, 6, 7, 8, 9, 10, 11, 12\}$, $A = \{\text{factors of } 12\}$, and $B = \{\text{prime numbers} \leqslant 12\}$.

- **a** List the elements of:
  - **i** A **ii** B **iii** $A \cap B$ **iv** $A \cup B$
- **b** Illustrate A and B on a Venn diagram.

5 Let $U = \{x \in \mathbb{Z}^+ \mid x \leqslant 20\}$, $R = \{\text{prime numbers less than } 20\}$, and $S = \{\text{composite numbers less than } 20\}$.

- **a** List the elements of:
  - **i** R **ii** S **iii** $R \cap S$ **iv** $R \cup S$
- **b** Illustrate R and S on a Venn diagram.

6 A group of students sat a Mathematics test and a History test, both out of 20 marks. The results are shown in the table.

Let M be the students who scored 15 or more for the Mathematics test, and H be the students who scored 15 or more for the History test.

- **a** List the elements of M and H.
- **b** Display the sets on a Venn diagram.
- **c** List the elements of $M \cup H$ and describe what this set represents.
- **d** List the elements of $M \cap H'$ and describe what this set represents.

Student	*Mathematics*	*History*
Paul	12	15
Tess	11	14
Shaun	16	14
Ken	19	15
Goran	13	12
Pam	13	17
Ling	16	18
Stefan	12	17
Reg	14	16
Claude	10	14
Don	15	17
Vivian	20	13

7 Let $U = \{\text{polygons}\}$, $Q = \{\text{quadrilaterals}\}$, and $Z = \{\text{quadrilaterals with all sides equal}\}$.

- **a** Describe the relationship between Q and Z.
- **b** Illustrate Q and Z on a Venn diagram. Indicate the regions in which you would find a rhombus, a kite, and a triangle.

8 Suppose $U = \{x \in \mathbb{Z}^+ \mid x \leqslant 25\}$, $A = \{\text{odd numbers}\}$, $B = \{\text{multiples of } 5\}$, and $C = \{\text{factors of } 50\}$.

- **a** Represent A, B, and C on a Venn diagram like the one shown.
- **b** List the elements which are in:
  - **i** both A and B
  - **ii** A but not C
  - **iii** B and C but not A.

PRINTABLE VENN DIAGRAM

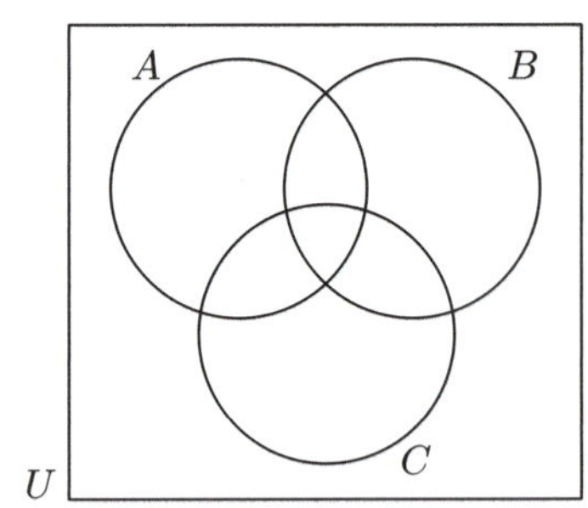

ACTIVITY

Click on the icon to practise constructing Venn diagrams.

You will need to choose the most appropriate layout for the Venn diagram.

VENN DIAGRAM CONSTRUCTION

B VENN DIAGRAM REGIONS

We can use Venn diagrams to help illustrate regions such as the union or intersection of sets.

Example 2 **Self Tutor**

On separate Venn diagrams, shade these regions for two intersecting sets A and B:

a $A \cup B$ **b** $A' \cap B$ **c** $(A \cap B)'$

a

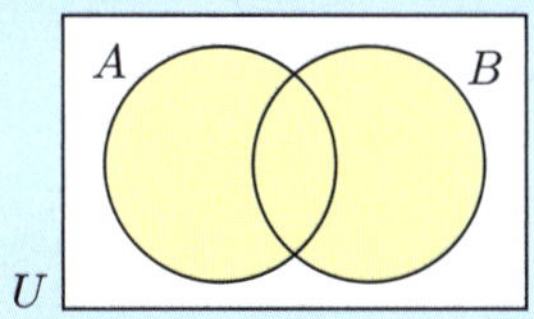

$A \cup B$ means in A, B, or both.

b

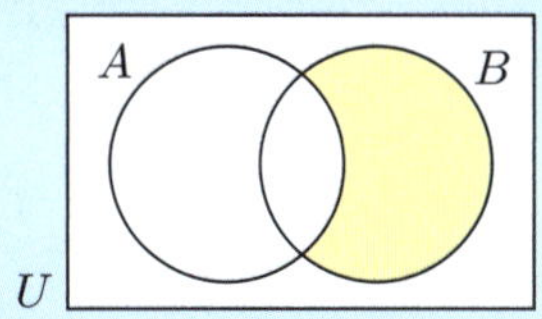

$A' \cap B$ means outside A, intersected with B.

c

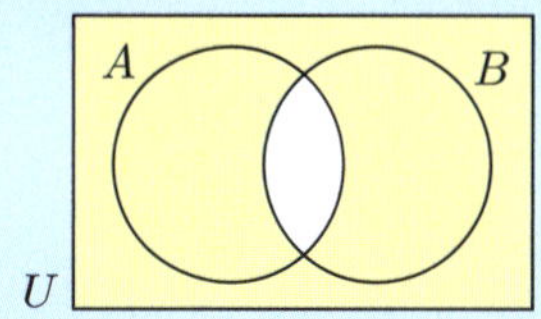

$(A \cap B)'$ means outside the intersection of A and B.

EXERCISE 7B

1

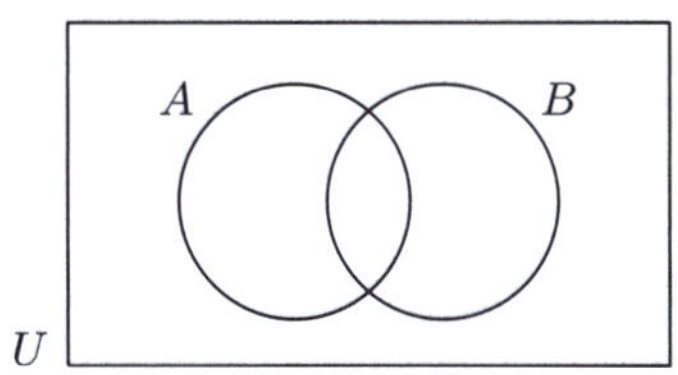

On separate Venn diagrams, shade these regions:

a in A **b** not in A
c in neither A nor B **d** in A and B
e in either A or B, but not both.

2

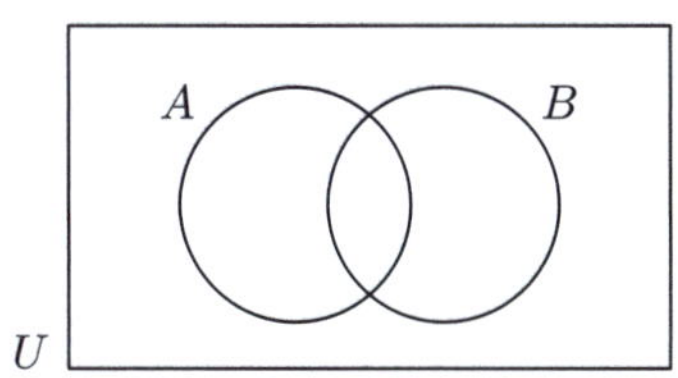

On separate Venn diagrams, shade:

a B **b** B'
c $A \cap B'$ **d** $A \cup B'$
e $(A \cap B)'$ **f** $(A \cup B)'$

PRINTABLE VENN DIAGRAMS

3 Describe in words, the shaded region:

a

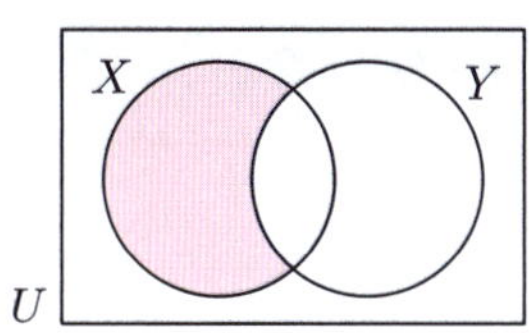

b

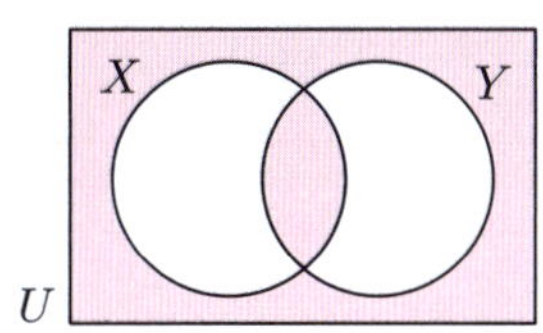

c

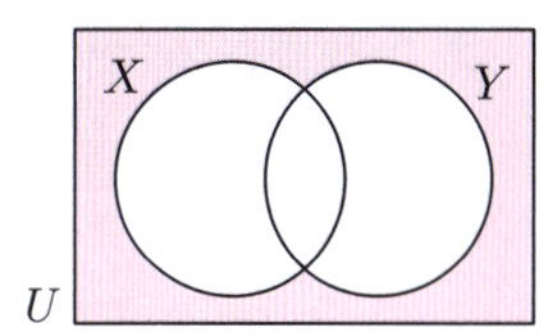

4 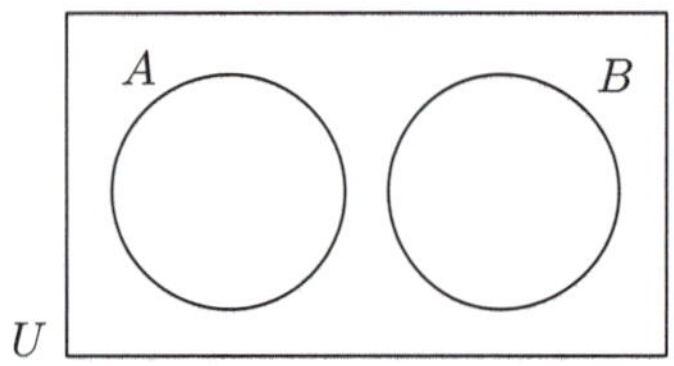

Suppose A and B are disjoint sets.
On separate Venn diagrams, shade:

a A **b** B' **c** $A \cap B$
d $A \cup B$ **e** $A \cap B'$ **f** $(A \cup B)'$

5 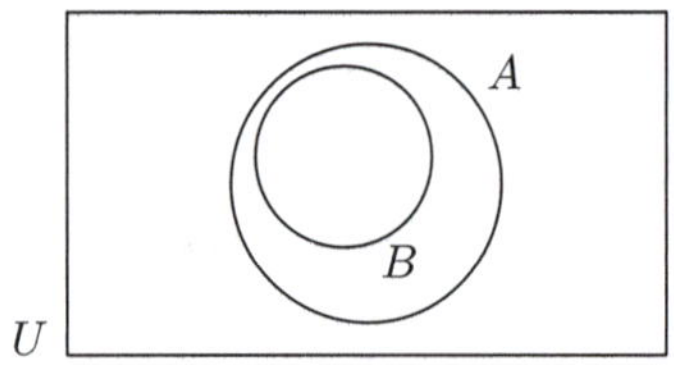

Suppose $B \subseteq A$.
On separate Venn diagrams, shade:

a A **b** B **c** A'
d $A \cap B$ **e** $A \cup B$ **f** $A \cap B'$

6 On separate Venn diagrams, shade:

a C **b** B'
c $A \cap B$ **d** $B \cup C$
e $A \cap B'$ **f** $A \cap B \cap C$

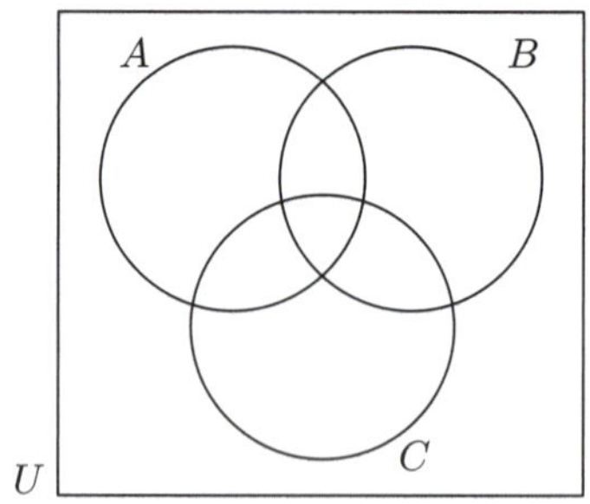

C NUMBERS IN REGIONS

For many problems involving Venn diagrams, we are more interested in *how many* elements are in each region, rather than which elements these are.

To indicate the **number of elements** in a region of a Venn diagram, we write the number in brackets.

Example 3 Self Tutor

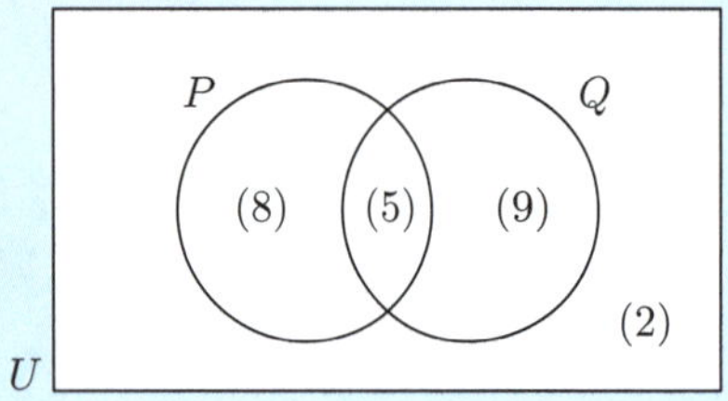

State the number of elements in:

a P **b** Q'
c $P \cup Q$ **d** P but not Q
e Q but not P **f** neither P nor Q.

a $n(P) = 8 + 5 = 13$
b $n(Q') = 8 + 2 = 10$
c $n(P \cup Q) = 8 + 5 + 9 = 22$
d $n(P \text{ but not } Q) = 8$
e $n(Q \text{ but not } P) = 9$
f $n(\text{neither } P \text{ nor } Q) = 2$

EXERCISE 7C

1 State the number of elements in:

a B **b** A'
c $A \cup B$ **d** A but not B
e B but not A **f** U

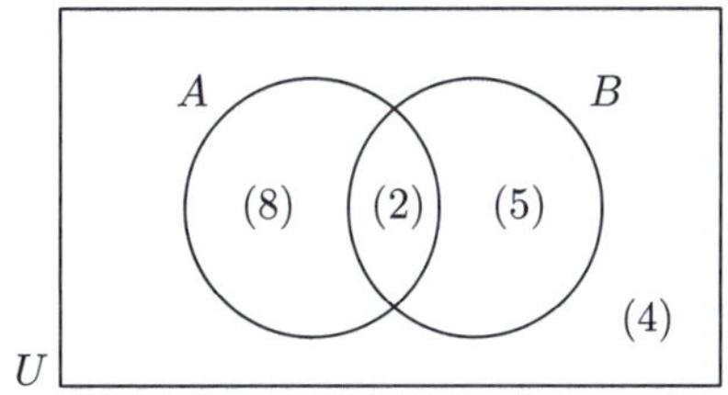

2

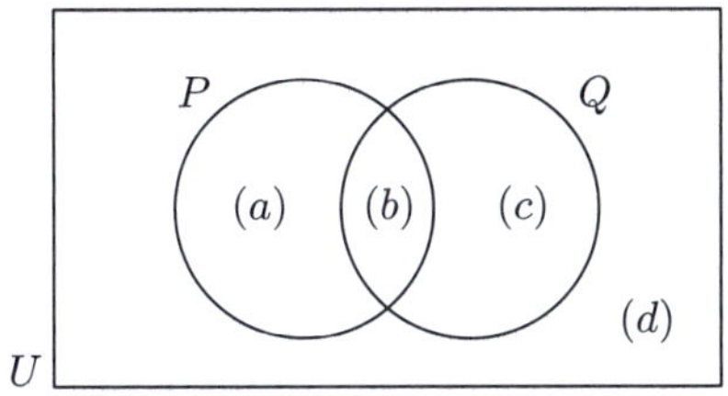

Write, in terms of a, b, c, and d, the number of elements in:

a Q **b** P'
c P or Q but not both **d** P and Q
e $(P \cap Q)'$ **f** $(P \cup Q)'$

3 State the number of elements in:

a X' **b** X and Y
c $X \cup Y$ **d** $X \cap Y'$
e $Y \cap X'$ **f** $X' \cap Y'$

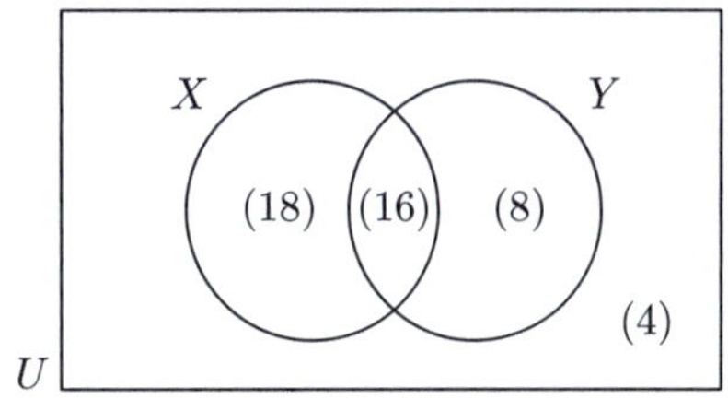

4 **a** Find, in terms of a:

i $n(Q)$ **ii** $n(P \cup Q)$
iii $n(Q')$ **iv** $n(U)$

b Given that $n(U) = 43$, find a.

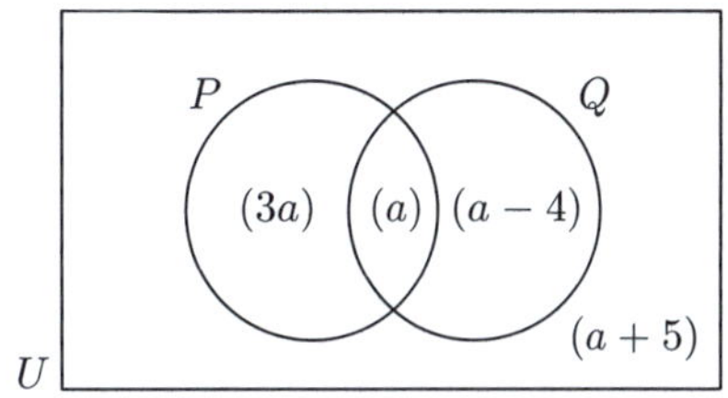

5

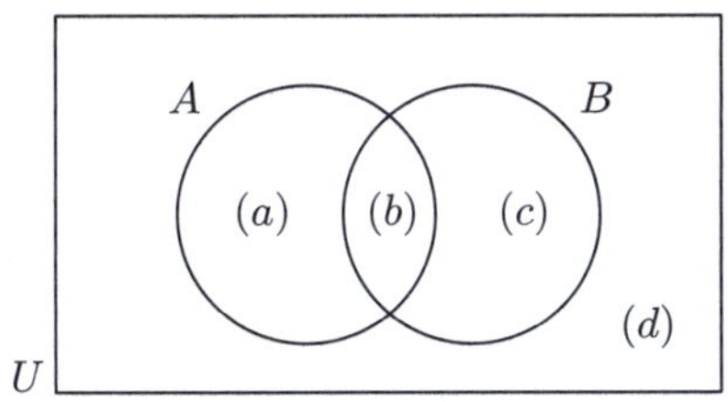

a Use the Venn diagram to show that $n(A \cup B) = n(A) + n(B) - n(A \cap B)$.

b Suppose A and B are disjoint sets. Explain why $n(A \cup B) = n(A) + n(B)$.

Example 4 **Self Tutor**

Suppose $n(A) = 13$, $n(B) = 10$, $n(A \cap B) = 7$, and $n(U) = 29$.
Use this information to construct a Venn diagram for sets A and B.

We are given $n(A \cap B) = 7$
$\therefore\ n(A \cap B') = 13 - 7 = 6$
and $n(A' \cap B) = 10 - 7 = 3$
$\therefore\ n(A' \cap B') = 29 - 7 - 6 - 3 = 13$

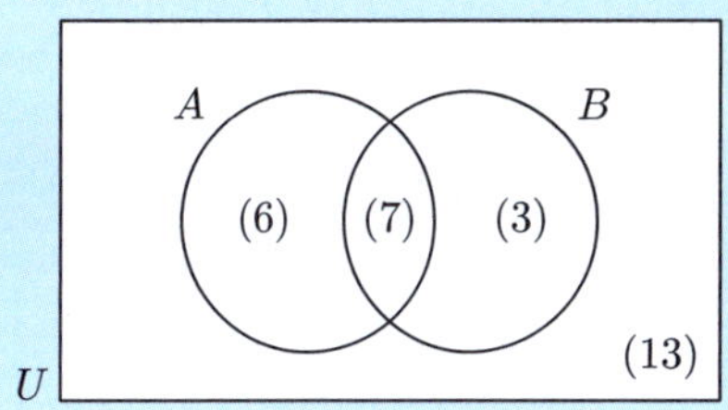

6 Suppose $n(A) = 8$, $n(B) = 9$, $n(A \cap B) = 2$, and $n(U) = 20$.

a Copy and complete the Venn diagram.

b Find:

i $n(A \cup B)$ **ii** $n(B \cap A')$

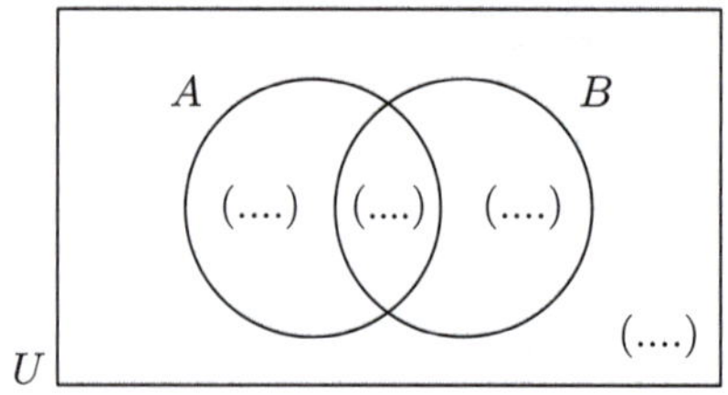

7 Suppose $n(P) = 14$, $n(Q) = 20$, $n(P \cap Q) = 11$, and $n(U) = 30$.

a Use this information to construct a Venn diagram for sets P and Q.

b Hence find the number of elements in:

i P but not Q **ii** P or Q but not both.

8 Suppose $n(U) = 35$, $n(N) = 16$, $n(N \cap R) = 4$, and $n(N \cup R) = 31$. Find:

a $n(R)$ **b** $n((N \cup R)')$ **c** $n(N' \cap R)$

Example 5 **Self Tutor**

Suppose $n(U) = 18$, $n(A) = 10$, $n(B) = 7$, and $n(A' \cap B') = 5$.
Find the number of elements in:

a $A \cap B$ **b** A or B but not both.

Let $n(A \cap B) = x$

$\therefore n(A \cap B') = 10 - x$ and $n(A' \cap B) = 7 - x$.

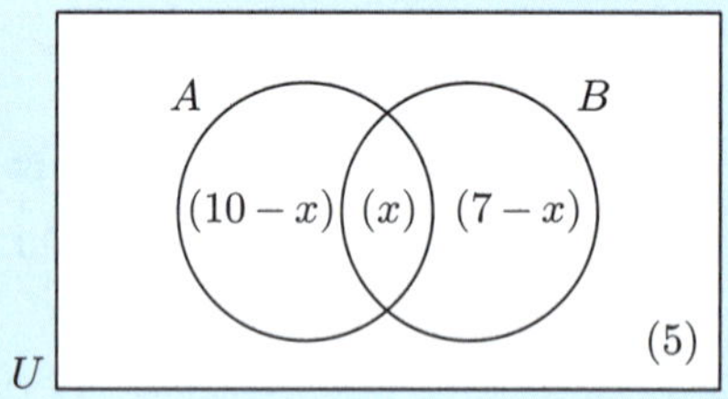

Since $n(U) = 18$,

$$(10 - x) + x + (7 - x) + 5 = 18$$
$$\therefore 22 - x = 18$$
$$\therefore x = 4$$

a $n(A \cap B) = 4$, so there are 4 elements in $A \cap B$.

b $n(A \cap B') = 10 - 4 = 6$ and $n(A' \cap B) = 7 - 4 = 3$
So, there are $6 + 3 = 9$ elements in A or B but not both.

9 Suppose $n(U) = 28$, $n(A) = 13$, $n(B) = 17$, and $n(A' \cap B') = 6$.

a Represent this information on a Venn diagram.

b Hence find the number of elements in:

i $A \cap B$ **ii** A but not B.

10 Suppose $n(U) = 55$, $n(P) = 26$, $n(Q) = 35$, and $n(P' \cap Q') = 9$.
Find the number of elements in:

a $P \cap Q$ **b** P or Q but not both.

11 Suppose $n(A) = 15$, $n(A \cup B) = 25$, $n(A' \cup B') = 23$, and $n(U) = 30$.
Find the number of elements in:

a B **b** A or B but not both.

D PROBLEM SOLVING WITH VENN DIAGRAMS

By considering the number of elements in different regions, we can use Venn diagrams to solve problems.

Example 6 — Self Tutor

A sports club has 75 members.
42 members play tennis.
53 members play hockey.
27 members play both of these sports.

a Place this information on a Venn diagram.
b Hence determine the number of members who play:
  i neither sport **ii** at least one sport.

a Let T be the set of members who play tennis and H be the set of members who play hockey.

$n(T \cap H) = 27$

$\therefore\ n(T \cap H') = 42 - 27 = 15$

and $n(T' \cap H) = 53 - 27 = 26$

$\therefore\ n(T' \cap H') = 75 - 27 - 15 - 26 = 7$

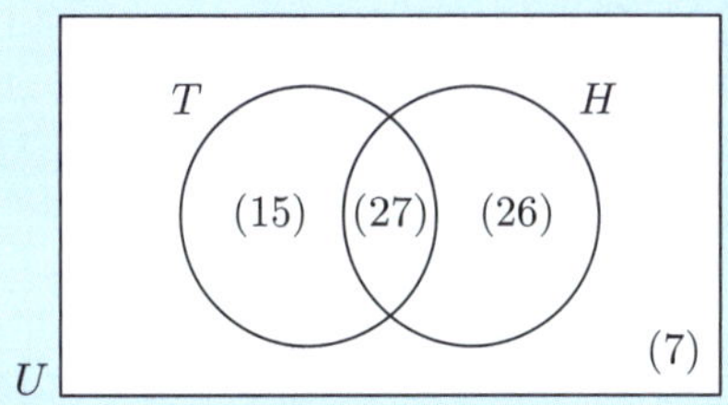

b **i** 7 members play neither sport. **ii** $75 - 7 = 68$ members play at least one sport.

EXERCISE 7D

1 This Venn diagram describes the member participation of an outdoor adventure club in rock climbing (R) and kayaking (K). Determine the number of members:

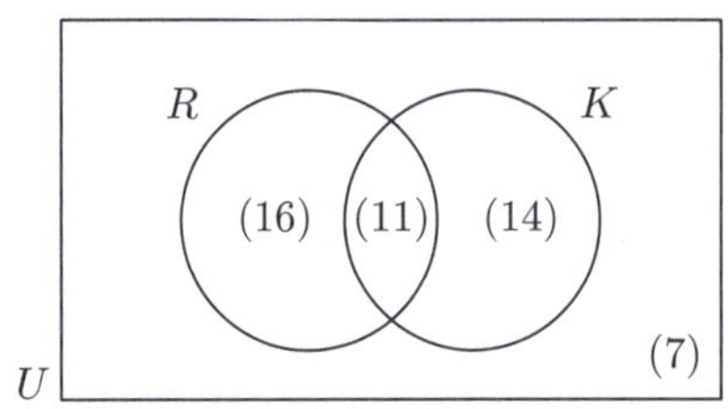

  a in the club **b** who go rock climbing
  c who do not go kayaking
  d who rock-climb but do not go kayaking
  e who do exactly one of these activities.

2 64 people participated in a baking competition. 29 entrants baked a pie, 38 entrants baked a cake, and 12 entrants baked both.
  a Place this information on a Venn diagram.
  b How many entrants:
    i baked a pie but not a cake **ii** baked a cake but not a pie
    iii baked neither a cake nor a pie?

3 Lightpark High has 65 Year 9 students. 24 students play a musical instrument, 40 students play a sport, and 13 students play both.
  a Place this information on a Venn diagram.
  b How many students:
    i play either a musical instrument or a sport but not both
    ii play neither a musical instrument nor a sport?

4 A team of 24 swimmers took part in a competition. 15 competed in freestyle, 11 competed in backstroke, and 6 competed in both of these strokes.

a Place this information on a Venn diagram.

b Hence determine the number of swimmers who competed in:

i backstroke but not freestyle
ii at least one of these strokes
iii freestyle but not backstroke
iv neither stroke
v exactly one of these strokes.

Example 7 **Self Tutor**

In a class of 24 boys, 16 play football (F) and 11 play baseball (B). If 2 play neither game, how many play both games?

Let $n(F \cap B) = x$

$\therefore \; n(F \cap B') = 16 - x$ and $n(F' \cap B) = 11 - x$.

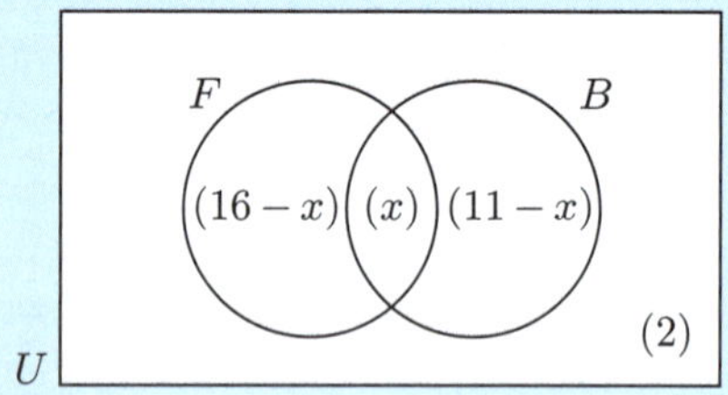

We know 2 play neither game.

Since there are 24 boys in total,

$$(16 - x) + x + (11 - x) + 2 = 24$$
$$\therefore \; 29 - x = 24$$
$$\therefore \; x = 5$$

So, 5 play both games.

5 In a building with 58 apartments, 45 households have children, 19 have pets, and 5 have neither children nor pets.

a Draw a Venn diagram to display this information.

b Hence find the number of households which:

i do not have children
ii have children or pets or both
iii have children or pets but not both
iv have pets but not children
v have children but not pets.

6 At their beachhouse, Anna and Ben have 37 books. Anna has read 21 of them, and Ben has read 26 of them. 2 of the books have not been read at all. Find the number of books which have been read by:

a both Anna and Ben
b Ben but not Anna.

7 41% of students in a Genevan school have been on a holiday to Paris, 46% have been to Rome, and 37% have been to neither Paris nor Rome.

a Place this information on a Venn diagram.

b What percentage of students have been on a holiday to:

i both Paris and Rome
ii either Paris or Rome
iii Paris only
iv Paris or Rome but not both?

8 On a particular day, 500 people visited a carnival. 300 people rode the Ferris wheel, and 350 people rode the roller coaster. Each person rode at least one of these attractions. Find the number of people who rode:

a both attractions **b** the Ferris wheel but not the roller coaster.

9 There are 46 shops in the local mall. 25 shops sell clothes, 16 sell shoes, and 34 sell at least one of these items. Find how many shops sell:

a both clothes and shoes **b** neither clothes nor shoes **c** clothes but not shoes.

10 All of the guests at a barbecue ate either sausages or chicken shashliks. 18 guests ate sausages, 15 ate both sausages and chicken shashliks, and 24 ate either sausages or chicken shashliks but not both. How many guests attended the barbecue?

11 A group of 28 workers are repairing a road. 9 use machinery, 15 do not use shovels, and 7 use neither machinery nor shovels. How many workers use both machinery and shovels?

12 A class of 28 students were asked to draw a quadrilateral. Let P be the set of quadrilaterals drawn with at least one pair of parallel sides, E be the set with at least two equal sides, and D be the set whose diagonals intersect at right angles.
$n(P) = 18$, $n(E) = 18$, $n(D) = 12$, $n(P \cap E) = 14$, $n(P \cap D) = 9$, $n(D \cap E) = 10$, and $n(P' \cap E' \cap D') = 5$.

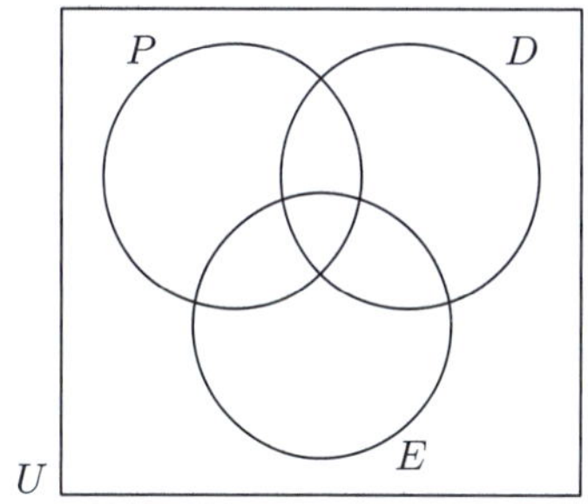

a Find $n(P \cap E \cap D)$ and describe the shapes in this set.

b Draw an example of a quadrilateral for each region of the Venn diagram.

MULTIPLE CHOICE QUIZ

QUICK QUIZ

REVIEW SET 7A

1 Let $U = \{1, 2, 3, 4, 5, 6, 7, 8, 9, 10, 11, 12\}$. Draw a Venn diagram to represent:

a $A = \{1, 4, 5, 8, 10, 12\}$ **b** $A = \{2, 4, 7, 9\}$

2

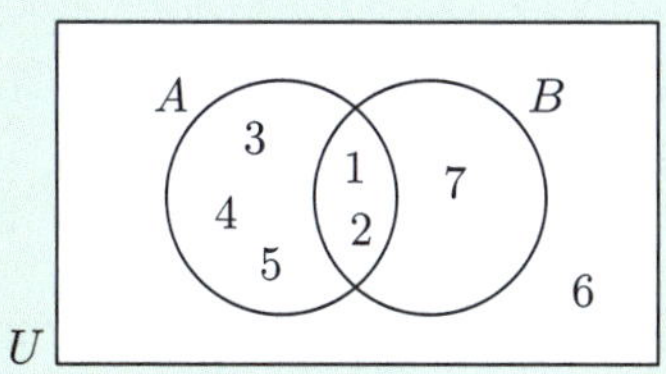

Consider the Venn diagram alongside.

a List the elements of:

i A **ii** B **iii** U
iv $A \cup B$ **v** $A \cap B$

b Find:

i $n(A)$ **ii** $n(B)$ **iii** $n(A \cup B)$

3 Suppose $U = \{a, b, c, d, e, f, g, h\}$, $A = \{a, b, d, e\}$, and $B = \{a, b, g\}$.

a List the elements of:

i A' **ii** $A \cap B$ **iii** $A \cup B$

b Illustrate A and B on a Venn diagram.

4 Describe in words, the shaded region:

a 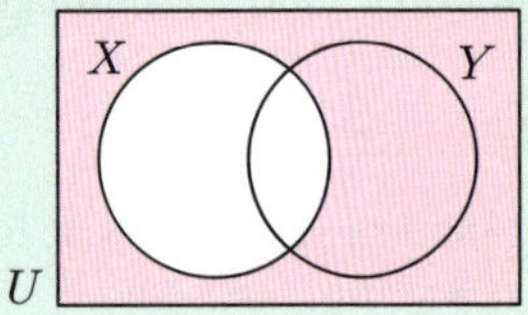

b 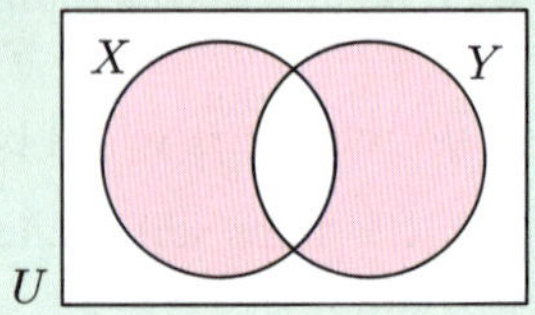

c 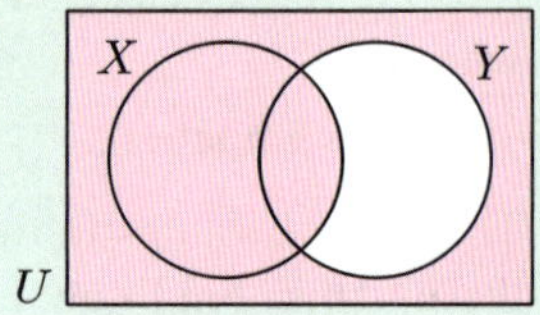

5 On separate Venn diagrams, shade:

a C' **b** $B \cap C$ **c** $A \cup B \cup C$

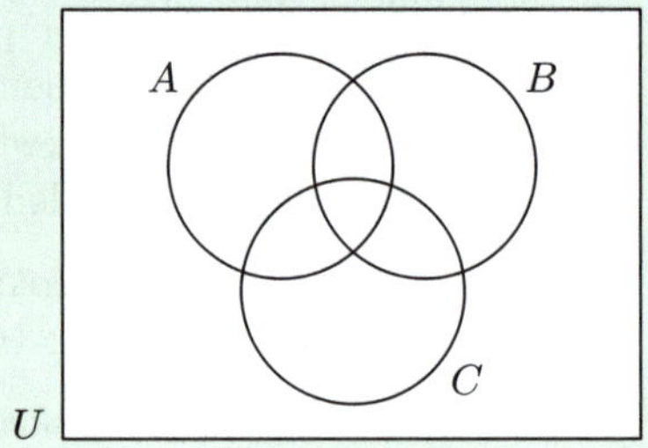

6 State the number of elements in:

a M' **b** $M \cap N$ **c** M but not N.

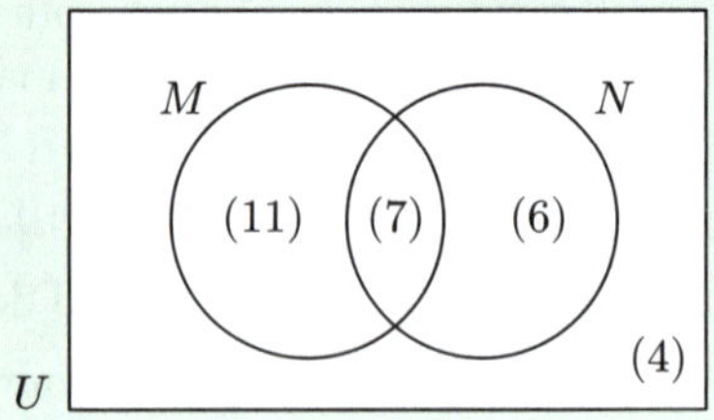

7 Suppose $n(A) = 16$, $n(B) = 13$, $n(A \cap B) = 9$, and $n(U) = 30$.

a Represent this information on a Venn diagram.

b Hence find the number of elements in:

i B but not A **ii** neither A nor B.

8 In a class of 27 students, 13 students have brothers, 16 have sisters, and 7 have both.

a Place this information on a Venn diagram.

b How many students have:

i neither brothers nor sisters **ii** sisters only?

9 Zhe and Michelle are competitive chefs. In the 34 competitions they have both entered, Zhe has won 9 awards, and Michelle has won 7 awards. They have both missed out on 22 occasions.

a Place this information on a Venn diagram.

b Hence determine the number of times that:

i Zhe has won an award but Michelle has not

ii Michelle has won an award but Zhe has not.

10 In the last 40 days, Jackson practised the piano on 20 days, and practised the violin on 23 days. On 19 days he practised the piano or the violin but not both.

a Place this information on a Venn diagram.

b On how many days did Jackson practise both the piano and the violin?

11 Answer the **Opening Problem** on page **98**.

12 A vegan restaurant has 17 dishes on its dinner menu. 15 of the dishes are gluten free, 13 dishes are onion free, and 1 dish is neither gluten free nor onion free.

a Place this information on a Venn diagram.

b How many dishes are:

i gluten free but not onion free **ii** onion free but not gluten free

iii both gluten free and onion free?

REVIEW SET 7B

1 Suppose $U = \{x \in \mathbb{Z}^+ \mid x \leqslant 12\}$ and $A = \{\text{multiples of 3 which are} \leqslant 12\}$. Show A on a Venn diagram.

2 Consider the Venn diagram alongside. List the elements of:

a P **b** Q'

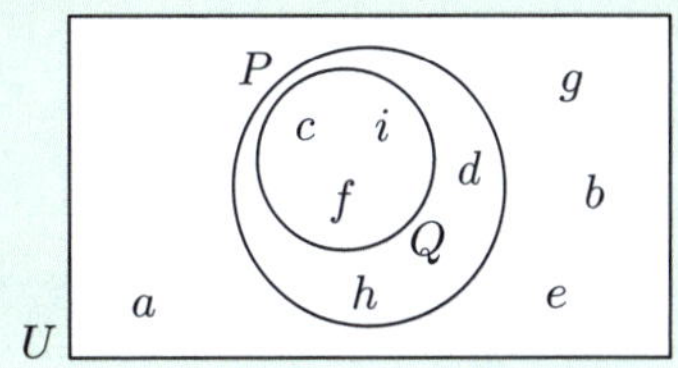

3 Show these sets on a Venn diagram:

a $U = \{10, 11, 12, 13, 14, 15\}$, $A = \{10, 12, 14\}$, $B = \{11, 12, 13\}$

b $U = \{\text{quadrilaterals}\}$, $S = \{\text{squares}\}$, $R = \{\text{rectangles}\}$

4 Suppose $U = \{x \in \mathbb{Z}^+ \mid x \leqslant 25\}$, $A = \{\text{square numbers}\}$, and $B = \{\text{even numbers}\}$. Represent this information on a Venn diagram.

5 On separate Venn diagrams like the one shown, shade the region representing:

a B' **b** $(A \cup B)'$

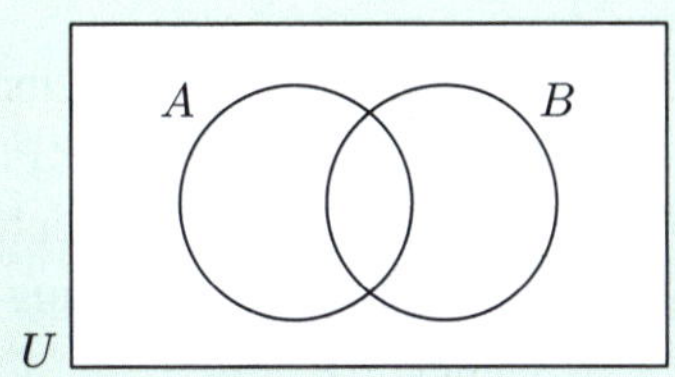

6 For the Venn diagram shown, find the number of elements in:

a $A \cap B$ **b** $A \cup C$ **c** U

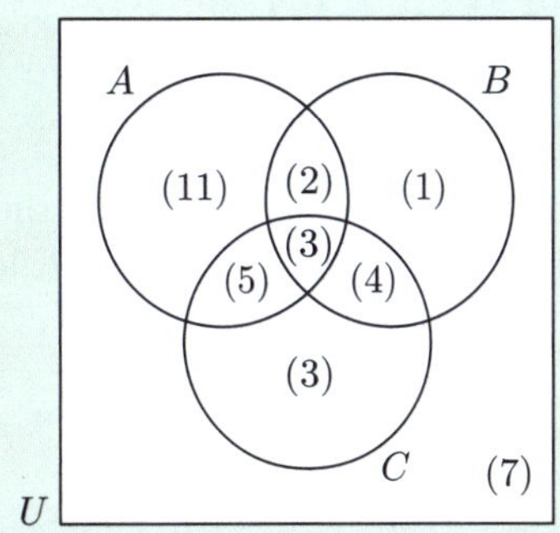

7 Suppose $n(U) = 30$, $n(A) = 16$, $n(B) = 19$, and $n(A' \cap B') = 6$. Find the number of elements in:

a $A \cap B$ **b** B but not A.

8 **a** Find $n(Y \cup Z)$ in terms of x.

b Find x if $n(U) = 21$.

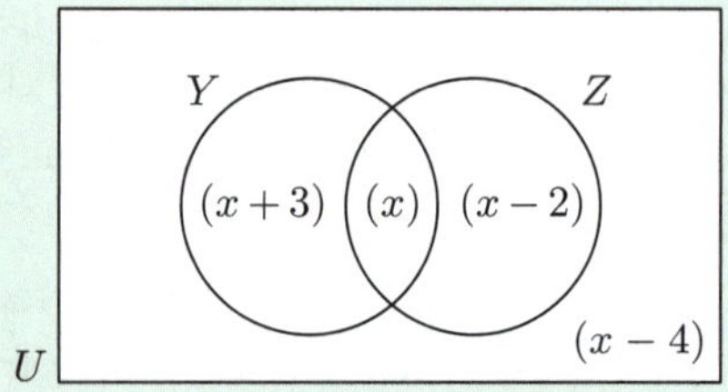

9 A Year 9 class has 25 students. On one day, 7 students forgot their diary, 5 were late, and 2 were late *and* forgot their diary.

a Draw a Venn diagram to represent this information.

b Find the number of students who were on time *and* remembered their diary.

10 Of the 70 apartments in a beachside building, 44 have balconies, 32 have beach views, and 57 have at least one of these features. How many apartments have:

a a balcony and a beach view **b** a balcony but not a beach view?

11 6000 people in Los Angeles auditioned for a reality TV show. 180 were selected for the second round of auditions, and 16 made it to the final.

a Place this information on a Venn diagram.

b How many people:

i made it to the second round, but did not make the final

ii did not make the second round?

c What percentage of the people who auditioned made it to the final?

12 Let U = {letters of the English alphabet},
A = {letters in the word PERPENDICULAR},
B = {first 12 letters of the alphabet}, and
C = {letters used when writing the first six counting numbers}.

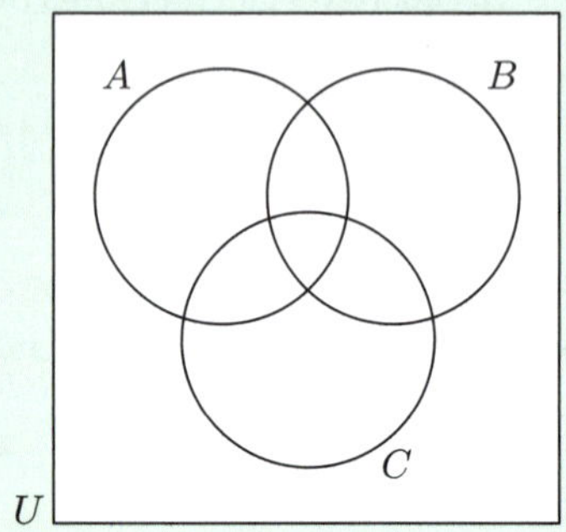

a Represent A, B, and C on a Venn diagram like the one shown.

b Find $n(A \cap B)$.

c On your Venn diagram, shade the region $A \cap C'$, and describe what this region represents.

d List the elements of $(A \cup B \cup C)'$.

Chapter 8

Surds and other radicals

Contents:

OPENING PROBLEM

A class of students was asked to each draw a rectangle with area 12 cm^2.

The students drew several different rectangles including 1 cm $\times$ 12 cm, 2 cm $\times$ 6 cm, and 3 cm $\times$ 4 cm.

Xiang decided to draw a square. He reasoned that $\sqrt{12} \times \sqrt{12} = 12$, so the square should have sides of length $\sqrt{12}$ cm.

Things to think about:

- **a** Between which two consecutive integers does $\sqrt{12}$ lie?
- **b** Is $\sqrt{12}$ a rational number?
- **c** What is the perimeter of the square? Can it be written in the form $\sqrt{k}$ where $k \in \mathbb{Z}$?

A SQUARE ROOTS

In previous years we have seen that:

The **square root** of the number a is the non-negative number which, when squared, gives a.

We write the square root of a as $\sqrt{a}$.

$$\sqrt{a} \times \sqrt{a} = a$$

The square root of a is a real number only if $a \geqslant 0$.

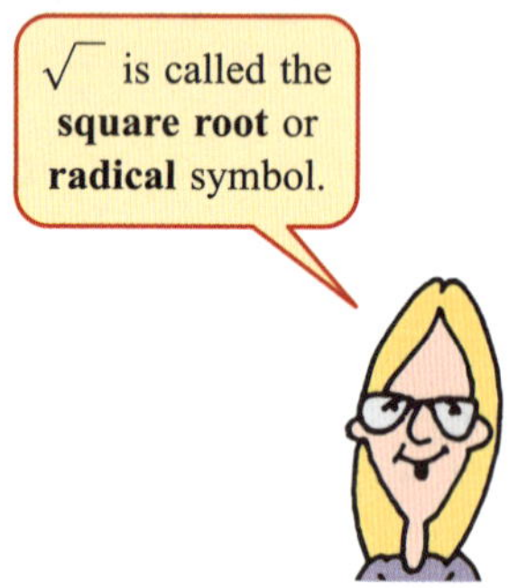

For example:

- Since $3^2 = 9$, $\sqrt{9} = 3$.
- Since $\left(\frac{1}{4}\right)^2 = \frac{1}{16}$, $\sqrt{\frac{1}{16}} = \frac{1}{4}$.
- There is no real number whose square is -4, so $\sqrt{-4}$ is not a real number.

RADICALS AND SURDS

Any number written with the radical symbol $\sqrt{}$ is called a **radical**.

Some radicals are rational numbers, since they can be written in the form $\frac{p}{q}$ where $p, q \in \mathbb{Z}$, $q \neq 0$.

For example, $\sqrt{49} = \frac{7}{1}$ and $\sqrt{\frac{1}{16}} = \frac{1}{4}$ are both rational. We write $\sqrt{49} \in \mathbb{Q}$ and $\sqrt{\frac{1}{16}} \in \mathbb{Q}$.

Most radicals are irrational numbers. They cannot be written in the form $\frac{p}{q}$ where $p, q \in \mathbb{Z}$, $q \neq 0$, and we cannot write their value exactly as a decimal.

For example, $\sqrt{3} \approx 1.732\,050\,8....$ is irrational. We write $\sqrt{3} \in \mathbb{Q}'$.

A radical which is irrational is called a **surd**.

EXERCISE 8A

1 Evaluate:

a $\sqrt{4}$ **b** $\sqrt{9}$ **c** $\sqrt{1}$ **d** $\sqrt{36}$
e $\sqrt{0}$ **f** $\sqrt{121}$ **g** $\sqrt{81}$ **h** $\sqrt{144}$

2 Evaluate:

a $\sqrt{\frac{1}{9}}$ **b** $\sqrt{\frac{1}{25}}$ **c** $\sqrt{\frac{1}{64}}$ **d** $\sqrt{\frac{4}{9}}$

3 Use your calculator to find, correct to 3 decimal places:

a $\sqrt{3}$ **b** $\sqrt{14}$ **c** $\sqrt{22}$ **d** $\sqrt{31}$
e $\sqrt{72}$ **f** $\sqrt{130}$ **g** $\sqrt{146}$ **h** $\sqrt{200}$
i $\sqrt{2.18}$ **j** $\sqrt{0.09}$ **k** $\sqrt{\frac{5}{8}}$ **l** $\sqrt{5.49}$

4 State whether the following radicals are surds. If the radical is a surd, find its value correct to 3 decimal places. If the radical is not a surd, find its exact value.

a $\sqrt{10}$ **b** $\sqrt{100}$ **c** $\sqrt{65}$ **d** $\sqrt{7}$
e $\sqrt{27}$ **f** $\sqrt{256}$ **g** $\sqrt{\frac{1}{4}}$ **h** $\sqrt{\frac{9}{16}}$
i $\sqrt{0.23}$ **j** $\sqrt{0.04}$ **k** $\sqrt{1.08}$ **l** $\sqrt{2.25}$

Example 1 **Self Tutor**

Between which two consecutive integers does $\sqrt{40}$ lie?

The square numbers either side of 40 are 36 and 49.

$\therefore$ $\sqrt{40}$ is between $\sqrt{36}$ and $\sqrt{49}$.

$\therefore$ $\sqrt{40}$ is between 6 and 7.

5 Between which two consecutive integers do the following values lie?

a $\sqrt{2}$ **b** $\sqrt{19}$ **c** $\sqrt{7}$ **d** $\sqrt{24}$
e $\sqrt{50}$ **f** $\sqrt{107}$ **g** $\sqrt{99}$ **h** $\sqrt{68}$

Use your calculator to check your answers.

Example 2 **Self Tutor**

Evaluate:

a $(\sqrt{11})^2$ **b** $(2\sqrt{3})^2$ **c** $\sqrt{5}\times 3\sqrt{5}$

a $(\sqrt{11})^2$
$= \sqrt{11}\times\sqrt{11}$
$= 11$

b $(2\sqrt{3})^2$
$= 2\sqrt{3}\times 2\sqrt{3}$
$= 2\times 2\times\sqrt{3}\times\sqrt{3}$
$= 4\times 3$
$= 12$

c $\sqrt{5}\times 3\sqrt{5}$
$= \sqrt{5}\times 3\times\sqrt{5}$
$= 3\times 5$
$= 15$

6 Evaluate:

a $(\sqrt{5})^2$ **b** $(\sqrt{13})^2$ **c** $(\sqrt{19})^2$ **d** $(\sqrt{20})^2$

e $(3\sqrt{3})^2$ **f** $(4\sqrt{5})^2$ **g** $(6\sqrt{2})^2$ **h** $(2\sqrt{6})^2$

i $(5\sqrt{2})^2$ **j** $(4\sqrt{3})^2$ **k** $(2\sqrt{11})^2$ **l** $(3\sqrt{7})^2$

m $(-\sqrt{5})^2$ **n** $(-\sqrt{3})^2$ **o** $(-2\sqrt{2})^2$ **p** $(-3\sqrt{5})^2$

7 Evaluate:

a $2\sqrt{3}\times\sqrt{3}$ **b** $\sqrt{7}\times 2\sqrt{7}$ **c** $3\sqrt{2}\times 2\sqrt{2}$ **d** $2\sqrt{11}\times 3\sqrt{11}$

e $-\sqrt{5}\times 2\sqrt{5}$ **f** $\sqrt{2}\times(-3\sqrt{2})$ **g** $(-4\sqrt{2})\times(-2\sqrt{2})$ **h** $(2\sqrt{3})^3\times\sqrt{3}$

8 Evaluate:

a $(\sqrt{7})^2$ **b** $(\sqrt{7})^4$ **c** $\left(\frac{1}{\sqrt{7}}\right)^2$ **d** $\left(\frac{3}{\sqrt{7}}\right)^2$

e $\left(\frac{7}{\sqrt{7}}\right)^2$ **f** $\left(\frac{\sqrt{7}}{7}\right)^2$ **g** $\left(\frac{7}{2\sqrt{7}}\right)^2$ **h** $\left(\frac{1}{\sqrt{7}}\right)^4$

B PROPERTIES OF RADICALS

INVESTIGATION PROPERTIES OF RADICALS

Notice that
$$\begin{aligned}(\sqrt{2}\times\sqrt{3})^2 &= (\sqrt{2})^2\times(\sqrt{3})^2 \quad \{\text{using } (ab)^n = a^n b^n\}\\ &= 2\times 3\\ &= 6\\ \therefore\ \sqrt{2}\times\sqrt{3} &= \sqrt{6} \quad \{\text{definition of a square root}\}\end{aligned}$$

What to do:

1 Use the procedure above to show that:

a $\sqrt{2}\times\sqrt{7}=\sqrt{14}$ **b** $\sqrt{3}\times\sqrt{7}=\sqrt{21}$ **c** $\sqrt{5}\times\sqrt{6}=\sqrt{30}$

2 Assuming a and b are positive, can you write $\sqrt{a}\times\sqrt{b}$ as a single radical?

3 Since $\sqrt{2}\times\sqrt{3}=\sqrt{6}$, we can deduce that $\frac{\sqrt{6}}{\sqrt{2}}=\sqrt{3}$ and $\frac{\sqrt{6}}{\sqrt{3}}=\sqrt{2}$.

Use the results from **1** to find:

a $\frac{\sqrt{14}}{\sqrt{2}}$ **b** $\frac{\sqrt{14}}{\sqrt{7}}$ **c** $\frac{\sqrt{21}}{\sqrt{3}}$ **d** $\frac{\sqrt{21}}{\sqrt{7}}$ **e** $\frac{\sqrt{30}}{\sqrt{5}}$ **f** $\frac{\sqrt{30}}{\sqrt{6}}$

4 Assuming a and b are positive, can you write $\frac{\sqrt{a}}{\sqrt{b}}$ as a single radical?

From the **Investigation** you should have discovered that:

For positive numbers a and b,
$\sqrt{a}\times\sqrt{b}=\sqrt{ab}$ and $\frac{\sqrt{a}}{\sqrt{b}}=\sqrt{\frac{a}{b}}$.

Example 3 — Self Tutor

Simplify using the properties of radicals:

a $\sqrt{3}\times\sqrt{5}$ **b** $\dfrac{\sqrt{18}}{\sqrt{3}}$ **c** $\sqrt{\frac{4}{9}}$

a $\sqrt{3}\times\sqrt{5}$
$=\sqrt{3\times 5}$
$=\sqrt{15}$

b $\dfrac{\sqrt{18}}{\sqrt{3}}=\sqrt{\frac{18}{3}}$
$=\sqrt{6}$

c $\sqrt{\frac{4}{9}}=\dfrac{\sqrt{4}}{\sqrt{9}}$
$=\frac{2}{3}$

EXERCISE 8B

1 Write as a single radical:

a $\sqrt{5}\times\sqrt{2}$ **b** $\sqrt{3}\times\sqrt{2}$ **c** $\sqrt{5}\times\sqrt{7}$
d $\sqrt{6}\times\sqrt{11}$ **e** $\sqrt{10}\times\sqrt{23}$ **f** $\sqrt{17}\times\sqrt{3}$
g $\sqrt{5}\times\sqrt{13}$ **h** $\sqrt{2}\times\sqrt{5}\times\sqrt{7}$ **i** $\sqrt{3}\times\sqrt{5}\times\sqrt{2}$

2 Evaluate $\sqrt{4}\times\sqrt{9}$ by:

a evaluating the square roots first **b** applying the rule $\sqrt{a}\times\sqrt{b}=\sqrt{ab}$.

3 Simplify:

a $2\sqrt{2}\times\sqrt{3}$ **b** $\sqrt{3}\times 5\sqrt{2}$ **c** $2\sqrt{5}\times 3\sqrt{2}$
d $-\sqrt{3}\times 2\sqrt{5}$ **e** $3\sqrt{7}\times 2\sqrt{3}$ **f** $4\sqrt{2}\times(-3\sqrt{5})$

4 Write as a single radical:

a $\dfrac{\sqrt{12}}{\sqrt{6}}$ **b** $\dfrac{\sqrt{20}}{\sqrt{10}}$ **c** $\dfrac{\sqrt{39}}{\sqrt{3}}$ **d** $\dfrac{\sqrt{91}}{\sqrt{13}}$ **e** $\dfrac{\sqrt{68}}{\sqrt{2}}$ **f** $\dfrac{\sqrt{7}\times\sqrt{6}}{\sqrt{2}}$

5 Evaluate using the properties of radicals:

a $\sqrt{\frac{9}{16}}$ **b** $\sqrt{\frac{36}{25}}$ **c** $\sqrt{\frac{16}{81}}$ **d** $\sqrt{\frac{9}{64}}$ **e** $\sqrt{\frac{100}{49}}$ **f** $\sqrt{\frac{121}{81}}$

6 Evaluate by first writing the fraction in lowest terms:

a $\sqrt{\frac{18}{8}}$ **b** $\sqrt{\frac{12}{27}}$ **c** $\sqrt{\frac{8}{50}}$ **d** $\sqrt{\frac{45}{125}}$ **e** $\sqrt{\frac{125}{80}}$ **f** $\sqrt{\frac{200}{18}}$

Example 4 — Self Tutor

Write $\sqrt{32}$ in the form $k\sqrt{2}$.

$\sqrt{32}=\sqrt{16\times 2}$
$=\sqrt{16}\times\sqrt{2}$ {using $\sqrt{ab}=\sqrt{a}\times\sqrt{b}$}
$=4\sqrt{2}$

7 Write in the form $k\sqrt{2}$:

a $\sqrt{8}$ **b** $\sqrt{18}$ **c** $\sqrt{50}$ **d** $\sqrt{98}$
e $\sqrt{200}$ **f** $\sqrt{288}$ **g** $\sqrt{20\,000}$ **h** $\sqrt{\frac{1}{2}}$

8 Write in the form $k\sqrt{3}$:

a $\sqrt{12}$ **b** $\sqrt{27}$ **c** $\sqrt{75}$ **d** $\sqrt{\frac{1}{3}}$

9 Write in the form $k\sqrt{5}$:

a $\sqrt{20}$ **b** $\sqrt{45}$ **c** $\sqrt{125}$ **d** $\sqrt{\frac{1}{5}}$

DISCUSSION

As a class, give examples which show that, in general, for positive numbers a, b:

- $\sqrt{a+b} \neq \sqrt{a} + \sqrt{b}$
- $\sqrt{a-b} \neq \sqrt{a} - \sqrt{b}$

C SIMPLEST SURD FORM

We have seen that if a radical is *not* a surd, we can write it in an exact form without a radical symbol.

We cannot do this for a surd. However, we can use the properties of radicals to write the surd so the number under the radical symbol is as small as possible.

If a surd is written so the number under the radical symbol is the smallest possible integer, we say it is in **simplest surd form**.

To write a surd in simplest surd form, we look for the highest perfect square factor under the radical symbol.

Example 5 Self Tutor

Write $\sqrt{98}$ in simplest surd form.

$\sqrt{98}$

$= \sqrt{49 \times 2}$ {49 is the highest perfect square factor of 98}

$= \sqrt{49} \times \sqrt{2}$

$= 7\sqrt{2}$

EXERCISE 8C

1 Write in simplest surd form:

a $\sqrt{8}$ **b** $\sqrt{20}$ **c** $\sqrt{18}$ **d** $\sqrt{32}$

e $\sqrt{27}$ **f** $\sqrt{52}$ **g** $\sqrt{48}$ **h** $\sqrt{50}$

i $\sqrt{108}$ **j** $\sqrt{75}$ **k** $\sqrt{45}$ **l** $\sqrt{99}$

m $\sqrt{60}$ **n** $\sqrt{88}$ **o** $\sqrt{140}$ **p** $\sqrt{275}$

2 Show that $\dfrac{\sqrt{96}}{\sqrt{6}} = 4$ by:

a using the rule $\dfrac{\sqrt{a}}{\sqrt{b}} = \sqrt{\dfrac{a}{b}}$

b first writing $\sqrt{96}$ in simplest surd form.

3 **a** Write 2028 in prime factored form.

b Hence write $\sqrt{2028}$ in simplest surd form.

D CUBE AND HIGHER ROOTS

In previous years, we have seen that:

The **cube root** of the number a is the number which, when cubed, gives a.

We write the cube root of a as $\sqrt[3]{a}$.

$$\sqrt[3]{a} \times \sqrt[3]{a} \times \sqrt[3]{a} = a$$

For example:

- Since $2^3 = 8$, $\sqrt[3]{8} = 2$.
- Since $(-3)^3 = -27$, $\sqrt[3]{-27} = -3$.

In general we define:

The ***n*th root of a** is the number $\sqrt[n]{a}$ such that $(\sqrt[n]{a})^n = a$.

If n is even then we define $\sqrt[n]{a}$ as the *non-negative* number with this property, and $\sqrt[n]{a}$ is only real for $a \geqslant 0$.

For example:

- Since $2^4 = 16$, $\sqrt[4]{16} = 2$.
- $\sqrt[4]{-8}$ is not a real number.

Example 6 **Self Tutor**

Evaluate: **a** $\sqrt[3]{\frac{1}{64}}$ **b** $\sqrt[3]{-\frac{27}{8}}$

a $\sqrt[3]{\frac{1}{64}}$
$= \frac{\sqrt[3]{1}}{\sqrt[3]{64}}$
$= \frac{1}{4}$

b $\sqrt[3]{-\frac{27}{8}}$
$= \frac{\sqrt[3]{-27}}{\sqrt[3]{8}}$
$= \frac{-3}{2}$
$= -\frac{3}{2}$

EXERCISE 8D

1 Evaluate:

a $\sqrt[3]{1}$ **b** $\sqrt[3]{-1}$ **c** $\sqrt[3]{8}$ **d** $\sqrt[3]{-8}$ **e** $\sqrt[3]{0}$

f $\sqrt[3]{-64}$ **g** $\sqrt[3]{64}$ **h** $\sqrt[3]{-216}$ **i** $\sqrt[3]{-1000}$ **j** $\sqrt[3]{343}$

2 Evaluate:

a $\sqrt[4]{81}$ **b** $\sqrt[5]{32}$ **c** $\sqrt[4]{625}$ **d** $\sqrt[5]{243}$ **e** $\sqrt[7]{128}$

3 Use your calculator to find, correct to 3 decimal places:

a $\sqrt[3]{4}$ **b** $\sqrt[3]{-16}$ **c** $\sqrt[3]{100}$

d $\sqrt[4]{450}$ **e** $\sqrt[3]{-3000}$

GRAPHICS CALCULATOR INSTRUCTIONS

4 Decide whether each of the following is a real number:

a $\sqrt[3]{-15}$ **b** $\sqrt[4]{-15}$ **c** $\sqrt[4]{5}$ **d** $\sqrt[5]{-12}$ **e** $\sqrt[6]{-12}$

5 Evaluate:

a $\sqrt[3]{\frac{1}{125}}$ **b** $\sqrt[3]{\frac{-8}{27}}$ **c** $\sqrt[3]{\frac{343}{64}}$ **d** $\sqrt[3]{-\frac{125}{27}}$

6 **a** Write 216 in prime factored form.

b Hence evaluate $\sqrt[3]{216}$.

E POWER EQUATIONS

A **power equation** is an equation of the form $x^n = k$ where $n \neq 0$.

In this Section we consider the cases $n = 2$ and $n = 3$.

EQUATIONS OF THE FORM $x^2 = k$

Consider the equation $x^2 = 7$.

We know that $\sqrt{7} \times \sqrt{7} = 7$

and $-\sqrt{7} \times -\sqrt{7} = 7$.

So, $x = \sqrt{7}$ and $x = -\sqrt{7}$ are both solutions of the equation.

We write the solutions as $x = \pm\sqrt{7}$, which reads "plus or minus the square root of 7".

Now consider the equation $x^2 = -7$.

We know that when any non-zero real number is multiplied by itself, the result is positive.

So, there are no real numbers which satisfy $x^2 = -7$.

In general, we conclude that:

$$\text{If } x^2 = k \text{ then } \begin{cases} x = \pm\sqrt{k} & \text{if } k > 0 \\ x = 0 & \text{if } k = 0 \\ x \text{ has no real solutions if } k < 0. \end{cases}$$

EQUATIONS OF THE FORM $x^3 = k$

In contrast to $x^2 = k$, there is always exactly one real solution of $x^3 = k$.

We know that $\sqrt[3]{7} \times \sqrt[3]{7} \times \sqrt[3]{7} = 7$, so $x = \sqrt[3]{7}$ is the solution of $x^3 = 7$.

We know that $\sqrt[3]{-7} \times \sqrt[3]{-7} \times \sqrt[3]{-7} = -7$, so $x = \sqrt[3]{-7}$ is the solution of $x^3 = -7$.

In general,

$$\text{If } x^3 = k \text{ then } x = \sqrt[3]{k}.$$

Example 7 Self Tutor

Solve for x:

a $x^2 = 4$ **b** $x^2 = -5$ **c** $x^3 = 45$

a $x^2 = 4$
$\therefore\ x = \pm\sqrt{4}$
$\therefore\ x = \pm 2$

b $x^2 = -5$ has no real solutions.

c $x^3 = 45$
$\therefore\ x = \sqrt[3]{45}$
$\therefore\ x \approx 3.56$

EXERCISE 8E

1 Solve for x:

a $x^2 = 9$ **b** $x^2 = 49$ **c** $x^2 = 36$ **d** $x^2 = 0$
e $x^2 = 1$ **f** $x^2 = 17$ **g** $x^2 = 23$ **h** $x^2 = 100$
i $x^2 = -4$ **j** $x^2 = -7$ **k** $x^2 = 27$ **l** $x^2 = -27$

2 Solve for x:

a $x^3 = 8$ **b** $x^3 = 27$ **c** $x^3 = 0$ **d** $x^3 = 64$
e $x^3 = -1$ **f** $x^3 = -125$ **g** $x^3 = \frac{1}{8}$ **h** $x^3 = 36$

Example 8 Self Tutor

Solve for x:

a $x^2 + 4 = 9$ **b** $x^3 + 27 = 0$

a $x^2 + 4 = 9$
$\therefore\ x^2 = 5$
$\therefore\ x = \pm\sqrt{5}$

b $x^3 + 27 = 0$
$\therefore\ x^3 = -27$
$\therefore\ x = \sqrt[3]{-27}$
$\therefore\ x = -3$

3 Solve for x:

a $x^2 + 5 = 9$ **b** $x^2 + 16 = 25$ **c** $x^2 + 2 = 27$
d $x^2 + 7 = 23$ **e** $8 + x^2 = 44$ **f** $x^2 + 14 = 39$
g $10 + x^2 = 60$ **h** $x^2 + 5 = 49$ **i** $6 + x^2 = 18$

4 Solve for x:

a $x^3 - 1 = 0$ **b** $x^3 + 8 = 0$ **c** $x^3 - 216 = 0$
d $x^3 - 1\,000\,000 = 0$ **e** $x^3 - 16 = 0$ **f** $x^3 + 800 = 0$

5 Solve for x:

a $2x^2 = 18$ **b** $3x^2 = 48$ **c** $5x^2 = 20$
d $x^2 + x^2 = 32$ **e** $x^2 + 2x^2 = 90$ **f** $x^2 + 5x^2 = 120$
g $x^2 + (2x)^2 = 45$ **h** $x^2 + (3x)^2 = 70$ **i** $x^2 + (2x)^2 = 1$

F OPERATIONS WITH RADICALS

We can perform operations with radicals just as we do with ordinary numbers. In particular:

- We can add and subtract "like" radicals in the same way that we add and subtract "like" algebraic terms.
- We can use the usual rules for expanding brackets.

Example 9 Self Tutor

Simplify:

a $3\sqrt{2}+4\sqrt{2}$ **b** $5\sqrt{3}-6\sqrt{3}$

a $3\sqrt{2}+4\sqrt{2}$
$=7\sqrt{2}$
{Compare with $3x+4x=7x$}

b $5\sqrt{3}-6\sqrt{3}$
$=-1\sqrt{3}$
$=-\sqrt{3}$
{Compare with $5x-6x=-x$}

We write the coefficient in front of the radical. So, we write $5\sqrt{2}$ not $\sqrt{2}\,5$.

EXERCISE 8F

1 Simplify:

a $\sqrt{2}+\sqrt{2}$ **b** $\sqrt{2}-\sqrt{2}$ **c** $3\sqrt{2}-2\sqrt{2}$
d $2\sqrt{3}-\sqrt{3}$ **e** $5\sqrt{7}+2\sqrt{7}$ **f** $3\sqrt{5}-6\sqrt{5}$
g $6\sqrt{2}-9\sqrt{2}$ **h** $-\sqrt{5}+7\sqrt{5}$ **i** $3\sqrt{2}+4\sqrt{2}-\sqrt{2}$
j $3\sqrt{2}-5\sqrt{2}-\sqrt{2}$ **k** $3\sqrt{3}-\sqrt{3}+2\sqrt{3}$ **l** $-3\sqrt{5}+7\sqrt{5}-10$

Example 10 Self Tutor

Simplify:

a $4\sqrt{5}+3\sqrt{7}-\sqrt{5}+5\sqrt{7}$ **b** $5\sqrt{2}-3\sqrt{8}$

a "like" radicals
$4\sqrt{5}+3\sqrt{7}-\sqrt{5}+5\sqrt{7}$
"like" radicals
$=3\sqrt{5}+8\sqrt{7}$

b $5\sqrt{2}-3\sqrt{8}$
$=5\sqrt{2}-3\sqrt{4\times 2}$
$=5\sqrt{2}-6\sqrt{2}$
$=-\sqrt{2}$

2 Simplify:

a $3\sqrt{2}+2\sqrt{3}-\sqrt{2}+5\sqrt{3}$ **b** $7\sqrt{2}-4\sqrt{3}-2\sqrt{2}+3\sqrt{3}$
c $-6\sqrt{2}-2\sqrt{3}-\sqrt{2}+6\sqrt{3}$ **d** $2\sqrt{5}+4\sqrt{2}+9\sqrt{5}-9\sqrt{2}$
e $3\sqrt{2}-5\sqrt{7}-\sqrt{2}-5\sqrt{7}$ **f** $3\sqrt{2}+4\sqrt{11}-\sqrt{2}-\sqrt{11}$
g $6-2\sqrt{2}-\sqrt{2}-5$ **h** $5\sqrt{3}-5+4\sqrt{3}-8$

3 Simplify:

a $8\sqrt{2}-\sqrt{8}$ **b** $2\sqrt{3}+3\sqrt{12}$ **c** $14\sqrt{5}-2\sqrt{125}$

4 Show that:

a $\sqrt{12}+\sqrt{27}=\sqrt{75}$ **b** $\sqrt{80}-\sqrt{5}=\sqrt{45}$

Example 11 Self Tutor

Expand and simplify:

a $2(2+\sqrt{3})$ **b** $\sqrt{2}(5-2\sqrt{2})$

a $2(2+\sqrt{3})$
$=2\times 2+2\times\sqrt{3}$
$=4+2\sqrt{3}$

b $\sqrt{2}(5-2\sqrt{2})$
$=\sqrt{2}\times 5+\sqrt{2}\times(-2\sqrt{2})$
$=5\sqrt{2}-4$

5 Expand and simplify:

a $4(3+\sqrt{2})$ **b** $3(\sqrt{2}+\sqrt{3})$ **c** $5(4-\sqrt{7})$

d $6(\sqrt{11}-4)$ **e** $\sqrt{2}(1+\sqrt{2})$ **f** $\sqrt{2}(\sqrt{2}-5)$

g $\sqrt{3}(2+2\sqrt{3})$ **h** $\sqrt{3}(\sqrt{3}-\sqrt{2})$ **i** $\sqrt{5}(6-\sqrt{5})$

j $\sqrt{5}(2\sqrt{5}-1)$ **k** $\sqrt{5}(2\sqrt{5}+\sqrt{3})$ **l** $\sqrt{7}(2+\sqrt{7}+\sqrt{2})$

Example 12 Self Tutor

Expand and simplify:

a $-\sqrt{3}(2+\sqrt{3})$ **b** $-\sqrt{2}(\sqrt{2}-\sqrt{3})$

a $-\sqrt{3}(2+\sqrt{3})$
$=(-\sqrt{3})\times 2+(-\sqrt{3})\times\sqrt{3}$
$=-2\sqrt{3}-3$

b $-\sqrt{2}(\sqrt{2}-\sqrt{3})$
$=(-\sqrt{2})\times\sqrt{2}+(-\sqrt{2})\times(-\sqrt{3})$
$=-2+\sqrt{6}$

6 Expand and simplify:

a $-\sqrt{2}(4+\sqrt{2})$ **b** $-\sqrt{2}(3-\sqrt{2})$ **c** $-\sqrt{2}(\sqrt{2}-\sqrt{7})$

d $-\sqrt{3}(3+\sqrt{3})$ **e** $-\sqrt{3}(5-\sqrt{3})$ **f** $-\sqrt{3}(2\sqrt{3}+\sqrt{5})$

g $-\sqrt{5}(2\sqrt{2}-\sqrt{3})$ **h** $-2\sqrt{2}(\sqrt{2}+\sqrt{3})$ **i** $-5\sqrt{2}(4-2\sqrt{2})$

j $-\sqrt{7}(2\sqrt{7}+4)$ **k** $-\sqrt{11}(2-\sqrt{11})$ **l** $-2\sqrt{3}(1-2\sqrt{2})$

Example 13 Self Tutor

Expand and simplify: $(3+\sqrt{5})(1-\sqrt{5})$

$(3+\sqrt{5})(1-\sqrt{5})$
$=3\times 1+3\times(-\sqrt{5})+\sqrt{5}\times 1+\sqrt{5}\times(-\sqrt{5})$
$=3-3\sqrt{5}+\sqrt{5}-5$
$=-2-2\sqrt{5}$

7 Expand and simplify:

a $(1+\sqrt{2})(5+\sqrt{2})$ **b** $(3+\sqrt{2})(3+\sqrt{2})$ **c** $(\sqrt{2}+2)(\sqrt{2}-1)$

d $(4-\sqrt{3})(3+\sqrt{3})$ **e** $(2+\sqrt{3})(2-\sqrt{3})$ **f** $(\sqrt{7}+2)(\sqrt{7}-3)$

g $(\sqrt{11}+\sqrt{2})(\sqrt{11}-\sqrt{2})$ **h** $(\sqrt{2}+1)(3-\sqrt{2})$ **i** $(6-2\sqrt{2})(2+\sqrt{2})$

Example 14 Self Tutor

Expand and simplify:

a $(\sqrt{2}+3)^2$ **b** $(\sqrt{5}-\sqrt{3})^2$

a
$$\begin{aligned} &(\sqrt{2}+3)^2 \\ &= (\sqrt{2})^2 + 2\sqrt{2}(3) + 3^2 \\ &= 2 + 6\sqrt{2} + 9 \\ &= 11 + 6\sqrt{2} \end{aligned}$$

b
$$\begin{aligned} &(\sqrt{5}-\sqrt{3})^2 \\ &= (\sqrt{5}+-\sqrt{3})^2 \\ &= (\sqrt{5})^2 + 2\sqrt{5}(-\sqrt{3}) + (-\sqrt{3})^2 \\ &= 5 - 2\sqrt{15} + 3 \\ &= 8 - 2\sqrt{15} \end{aligned}$$

8 Expand and simplify:

a $(1+\sqrt{3})^2$ **b** $(\sqrt{2}+5)^2$ **c** $(3-\sqrt{2})^2$

d $(1+\sqrt{7})^2$ **e** $(\sqrt{3}-\sqrt{2})^2$ **f** $(4-\sqrt{5})^2$

g $(\sqrt{3}+\sqrt{5})^2$ **h** $(3-\sqrt{6})^2$ **i** $(\sqrt{6}-\sqrt{3})^2$

j $(2\sqrt{2}+3)^2$ **k** $(3-2\sqrt{2})^2$ **l** $(3-5\sqrt{2})^2$

Example 15 Self Tutor

Expand and simplify:

a $(4+\sqrt{2})(4-\sqrt{2})$ **b** $(2\sqrt{2}+3)(2\sqrt{2}-3)$

a
$$\begin{aligned} &(4+\sqrt{2})(4-\sqrt{2}) \\ &= 4^2 - (\sqrt{2})^2 \\ &= 16 - 2 \\ &= 14 \end{aligned}$$

b
$$\begin{aligned} &(2\sqrt{2}+3)(2\sqrt{2}-3) \\ &= (2\sqrt{2})^2 - 3^2 \\ &= 4 \times 2 - 9 \\ &= 8 - 9 \\ &= -1 \end{aligned}$$

9 Expand and simplify:

a $(3+\sqrt{2})(3-\sqrt{2})$ **b** $(\sqrt{3}-1)(\sqrt{3}+1)$ **c** $(5+\sqrt{3})(5-\sqrt{3})$

d $(\sqrt{3}-4)(\sqrt{3}+4)$ **e** $(\sqrt{7}-3)(\sqrt{7}+3)$ **f** $(2+\sqrt{2})(2-\sqrt{2})$

g $(\sqrt{7}-5)(\sqrt{7}+5)$ **h** $(2\sqrt{5}+6)(2\sqrt{5}-6)$ **i** $(3\sqrt{2}+2)(3\sqrt{2}-2)$

j $(\sqrt{5}-\sqrt{2})(\sqrt{5}+\sqrt{2})$ **k** $(\sqrt{3}-\sqrt{7})(\sqrt{3}+\sqrt{7})$ **l** $(2\sqrt{2}+1)(2\sqrt{2}-1)$

G DIVISION WITH SURDS

When an expression involves division by a surd, we can write the expression with an **integer denominator** which does *not* contain surds. This process is called **rationalising the denominator**.

If the denominator has only one term, we can remove the surd by using the definition $\sqrt{a} \times \sqrt{a} = a$.

For any fraction of the form $\dfrac{b}{c\sqrt{a}}$, $a > 0$, we can remove the radical from the denominator by multiplying by $\dfrac{\sqrt{a}}{\sqrt{a}}$.

Example 16 **Self Tutor**

Write with an integer denominator:

a $\dfrac{9}{\sqrt{2}}$ **b** $\dfrac{6}{\sqrt{3}}$ **c** $\dfrac{4}{3\sqrt{2}}$

a $\dfrac{9}{\sqrt{2}} = \dfrac{9}{\sqrt{2}} \times \dfrac{\sqrt{2}}{\sqrt{2}}$
$= \dfrac{9\sqrt{2}}{2}$

b $\dfrac{6}{\sqrt{3}} = \dfrac{6}{\sqrt{3}} \times \dfrac{\sqrt{3}}{\sqrt{3}}$
$= \dfrac{6\sqrt{3}}{3}$
$= 2\sqrt{3}$

c $\dfrac{4}{3\sqrt{2}} = \dfrac{4}{3\sqrt{2}} \times \dfrac{\sqrt{2}}{\sqrt{2}}$
$= \dfrac{4\sqrt{2}}{6}$
$= \dfrac{2\sqrt{2}}{3}$

EXERCISE 8G

1 Write with an integer denominator:

a $\dfrac{1}{\sqrt{3}}$ **b** $\dfrac{3}{\sqrt{3}}$ **c** $\dfrac{9}{\sqrt{3}}$ **d** $\dfrac{11}{\sqrt{3}}$

e $\dfrac{2}{\sqrt{2}}$ **f** $\dfrac{6}{\sqrt{2}}$ **g** $\dfrac{12}{\sqrt{2}}$ **h** $\dfrac{\sqrt{3}}{\sqrt{2}}$

i $\dfrac{5}{\sqrt{5}}$ **j** $\dfrac{15}{\sqrt{5}}$ **k** $\dfrac{-3}{\sqrt{5}}$ **l** $\dfrac{200}{\sqrt{5}}$

m $\dfrac{7}{\sqrt{7}}$ **n** $\dfrac{21}{\sqrt{7}}$ **o** $\dfrac{2}{\sqrt{11}}$ **p** $\dfrac{26}{\sqrt{13}}$

2 Rationalise the denominator:

a $\sqrt{\frac{5}{3}}$ **b** $\sqrt{\frac{7}{2}}$ **c** $\sqrt{\frac{3}{10}}$ **d** $\sqrt{\frac{7}{8}}$

3 Rationalise the denominator:

a $\dfrac{1}{4\sqrt{2}}$ **b** $\dfrac{1}{3\sqrt{5}}$ **c** $\dfrac{2}{3\sqrt{2}}$ **d** $\dfrac{10}{7\sqrt{2}}$

e $\dfrac{\sqrt{2}}{3\sqrt{3}}$ **f** $\dfrac{\sqrt{3}}{2\sqrt{5}}$ **g** $\dfrac{\sqrt{2}}{4\sqrt{6}}$ **h** $\dfrac{5\sqrt{3}}{3\sqrt{2}}$

ACTIVITY — APPROXIMATING SURDS

In this Activity we will explore a method for approximating surds. Many calculators use this method for calculating values such as $\sqrt{2}$ and $\sqrt[3]{7}$.

To approximate the value of $\sqrt{2}$, we follow these steps:

Step 1: Make an initial guess of the value of $\sqrt{2}$, and call this guess x.

Step 2: Calculate $\frac{2}{x}$.

Step 3: Find the average of x and $\frac{2}{x}$. This is your new guess for the value of $\sqrt{2}$.

Step 4: Repeat *steps 2* and *3* with the new guess x.

With an initial guess of $x = 1$, the results of this process are shown alongside.

As the process is repeated several times, the values of x get closer to $\sqrt{2} \approx 1.414\,213\,562$.

x	$\frac{2}{x}$	*Average of x and $\frac{2}{x}$*
1	2	1.5
1.5	1.333 333	1.416 667
1.416 667	1.411 765	1.414 216
1.414 216	1.414 211	1.414 214
⋮	⋮	⋮

What to do:

1. Explain why this method works.
2. Use a similar method to approximate $\sqrt{3}$. Use an initial guess of $x = 2$.
3. Click on the spreadsheet icon. Use the software to approximate other square roots. Experiment with different initial guesses, and investigate the effect this has on the accuracy of the approximations.
4. Adapt this method to approximate cube roots instead of square roots.

SPREADSHEET

MULTIPLE CHOICE QUIZ

QUICK QUIZ

REVIEW SET 8A

1. State whether the following radicals are surds. If the radical is a surd, find its value correct to 3 decimal places. If the radical is not a surd, find its exact value.

   a $\sqrt{24}$ **b** $\sqrt{16}$ **c** $\sqrt{1.5}$

2. Between which two consecutive integers do the following values lie?

   a $\sqrt{21}$ **b** $\sqrt{110}$

3 Evaluate:

a $(\sqrt{7})^2$ **b** $(3\sqrt{5})^2$ **c** $4\sqrt{2}\times 5\sqrt{2}$

4 Write as a single radical:

a $\sqrt{3}\times\sqrt{5}$ **b** $\dfrac{\sqrt{42}}{\sqrt{6}}$ **c** $\dfrac{\sqrt{11}\times\sqrt{6}}{\sqrt{3}}$

5 Evaluate $\sqrt{\frac{1000}{490}}$ by first writing the fraction in lowest terms.

6 Write in simplest surd form:

a $\sqrt{40}$ **b** $\sqrt{180}$

7 Evaluate:

a $\sqrt[3]{27}$ **b** $\sqrt[4]{10\,000}$ **c** $\sqrt[3]{\frac{8}{125}}$

8 Solve for x:

a $x^2 = 64$ **b** $x^3 = 81$ **c** $x^3 = -10$

9 Solve for x:

a $15 + x^2 = 50$ **b** $8x^2 = 72$

10 **a** Find the value of 128×99.

b Write in simplest surd form: **i** $\sqrt{128}$ **ii** $\sqrt{99}$

c Hence write $\sqrt{12\,672}$ in simplest surd form.

11 Simplify:

a $8\sqrt{11} - 10\sqrt{11} + 5\sqrt{11}$ **b** $\sqrt{7} - 2\sqrt{3} + 5\sqrt{7} - \sqrt{3}$ **c** $4\sqrt{2} + 3\sqrt{8}$

12 Expand and simplify:

a $2(\sqrt{3}+1)$ **b** $\sqrt{2}(3-\sqrt{2})$ **c** $(1+\sqrt{7})^2$

d $(2-3\sqrt{5})^2$ **e** $(5+\sqrt{6})(5-\sqrt{6})$ **f** $(3+\sqrt{2})(1-\sqrt{2})$

13 Write with an integer denominator:

a $\dfrac{10}{\sqrt{5}}$ **b** $\sqrt{\frac{2}{3}}$ **c** $\dfrac{8}{3\sqrt{2}}$

REVIEW SET 8B

1 Use your calculator to find $\sqrt{73}$ correct to 3 decimal places.

2 Between which two consecutive integers do the following values lie?

a $\sqrt{70}$ **b** $\sqrt{150}$

3 Evaluate:

a $(3\sqrt{7})^2$ **b** $2\sqrt{5}\times 3\sqrt{5}$ **c** $3\sqrt{7}\times(-\sqrt{7})$

4 Write as a single radical:

a $\sqrt{5}\times\sqrt{11}$ **b** $\dfrac{\sqrt{8}\times\sqrt{3}}{\sqrt{6}}$

5 Evaluate using the properties of radicals:

a $\sqrt{\frac{4}{25}}$ **b** $\sqrt{\frac{81}{16}}$ **c** $\sqrt{\frac{64}{121}}$

6 Write in simplest surd form:

a $\sqrt{125}$ **b** $\sqrt{90}$ **c** $\sqrt{300}$

7 Evaluate $\sqrt[3]{-\frac{8}{343}}$.

8 Use your calculator to find, correct to 3 decimal places:

a $\sqrt[3]{11}$ **b** $\sqrt[3]{-50}$ **c** $\sqrt[4]{300}$

9 Solve for x:

a $x^2 = -2$ **b** $x^3 = 50$ **c** $x^3 + 70 = 0$ **d** $x^2 + (4x)^2 = 85$

10 Answer the **Opening Problem** on page **112**.

11 **a** Is $5\sqrt{3}$ rational or irrational?

b Evaluate:

i $(5\sqrt{3})^2$ **ii** $(5\sqrt{3})^3$ **iii** $(5\sqrt{3})^4$

c For what values of $n \in \mathbb{N}$ is $(5\sqrt{3})^n$ rational?

d Solve the equation $\frac{x^3}{375} + \sqrt{3} = 0$.

12 Simplify:

a $5\sqrt{2} - \sqrt{5} + \sqrt{2} + 3\sqrt{5}$ **b** $\sqrt{75} - 2\sqrt{3}$

13 Expand and simplify:

a $\sqrt{7}(\sqrt{2} - 1)$ **b** $-\sqrt{3}(2\sqrt{3} + \sqrt{5})$ **c** $(2 - \sqrt{3})(5 + 2\sqrt{3})$

d $(\sqrt{3} + \sqrt{2})^2$ **e** $(2 - \sqrt{5})(2 + \sqrt{5})$ **f** $(1 + 2\sqrt{2})(\sqrt{2} - 2)$

14 Write with an integer denominator:

a $\frac{8}{\sqrt{3}}$ **b** $\sqrt{\frac{7}{13}}$ **c** $\frac{2\sqrt{5}}{7\sqrt{2}}$

Chapter 9

Pythagoras' theorem

Contents:

OPENING PROBLEM

Suppose four identical right angled triangles with side lengths a, b, and c are drawn within a square as shown.

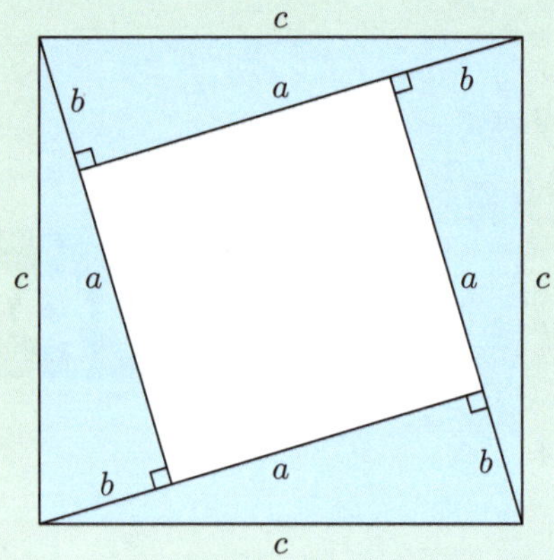

Things to think about:

a Can you explain why the white quadrilateral is a square with side length $(a - b)$?

b Can you write an expression for the area of the large square in terms of:

  i c **ii** a and b?

c What relationship exists between a, b, and c?

A **right angled triangle** is a triangle which has a right angle as one of its angles.

The side **opposite** the right angle is called the **hypotenuse**. It is the **longest** side of the triangle.

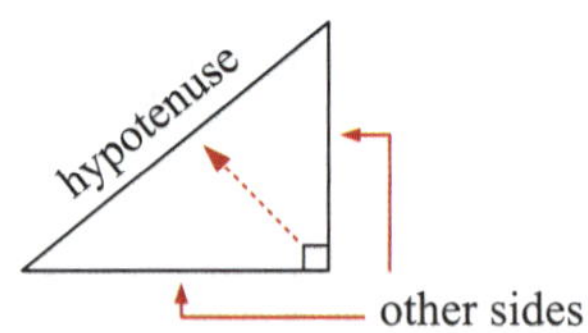

A PYTHAGORAS' THEOREM

In the **Opening Problem**, you should have proven **Pythagoras' theorem**:

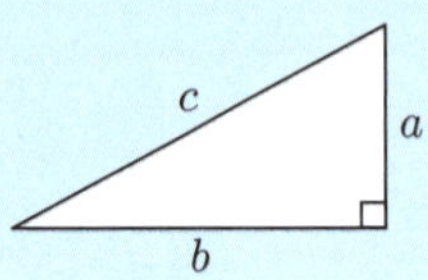

In a right angled triangle with hypotenuse of length c, and other sides of length a and b,

$$c^2 = a^2 + b^2.$$

In geometric form, Pythagoras' theorem states:

In any right angled triangle, the area of the square on the hypotenuse is equal to the sum of the areas of the squares on the other two sides.

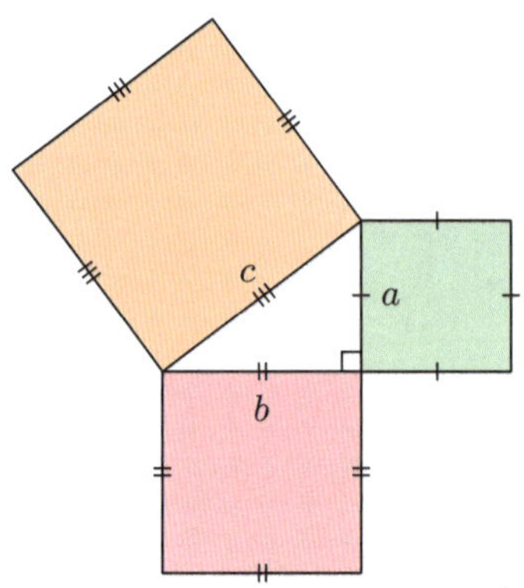

To find side lengths in triangles using Pythagoras' theorem, we need to solve a power equation. Since side lengths must be positive, we can always reject the negative solution.

Example 1 **Self Tutor**

Find the length of the hypotenuse in this right angled triangle, rounding your answer to 2 decimal places.

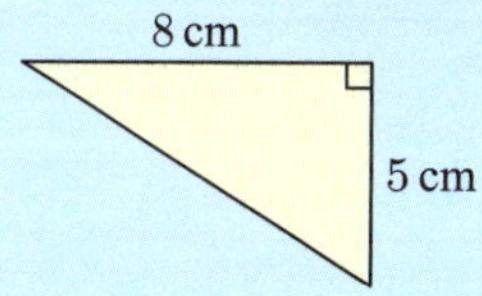

Let the hypotenuse have length x cm.

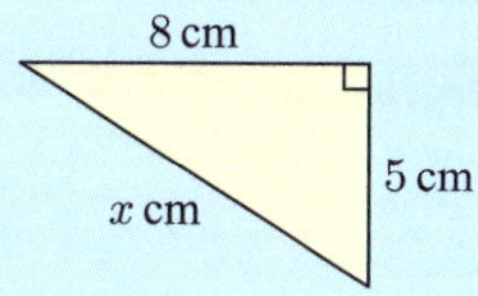

$\therefore\ x^2 = 8^2 + 5^2$ {Pythagoras}

$\therefore\ x^2 = 64 + 25$

$\therefore\ x^2 = 89$

$\therefore\ x = \sqrt{89}$ {since $x > 0$}

$\therefore\ x \approx 9.43$

The hypotenuse is about 9.43 cm long.

EXERCISE 9A

1 Find the length of the hypotenuse in each right angled triangle. Round your answer to 2 decimal places if necessary.

a
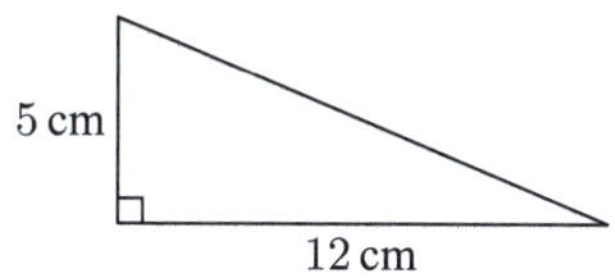

b
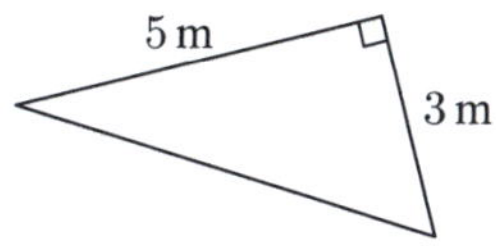

c
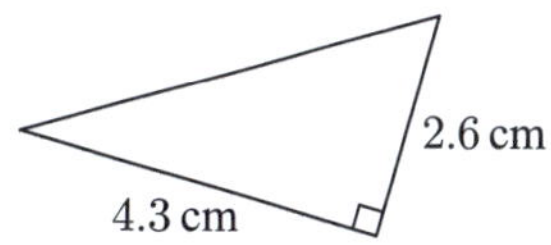

2 Find the length of the hypotenuse. Leave your answer in simplest surd form.

a
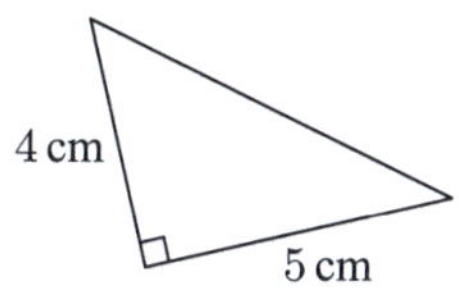

b
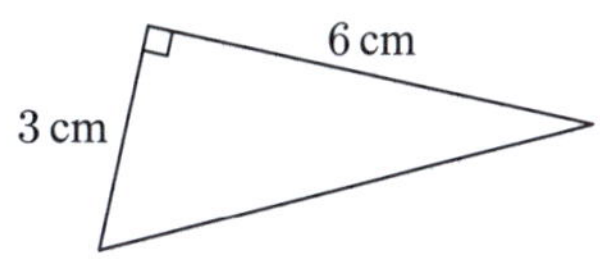

c
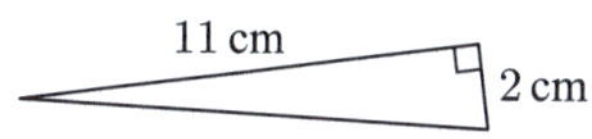

Example 2 **Self Tutor**

Find the length of the third side of this right angled triangle, giving your answer in simplest surd form.

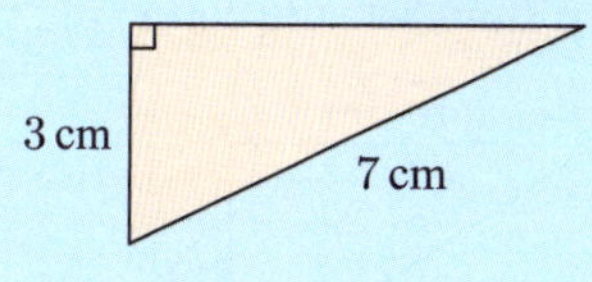

Let the unknown side have length x cm.

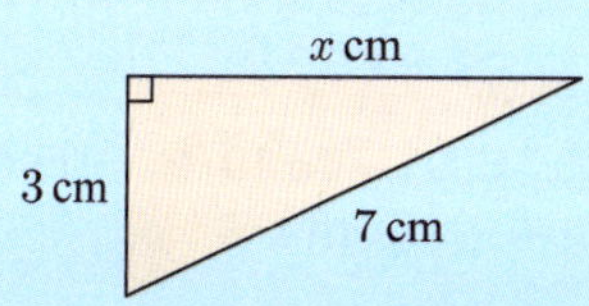

$\therefore\ x^2 + 3^2 = 7^2$ {Pythagoras}

$\therefore\ x^2 + 9 = 49$

$\therefore\ x^2 = 40$

$\therefore\ x = \sqrt{40}$ {since $x > 0$}

$\therefore\ x = 2\sqrt{10}$

The third side is $2\sqrt{10}$ cm long.

3 Find the length of the third side, giving your answer in simplest surd form where necessary:

a
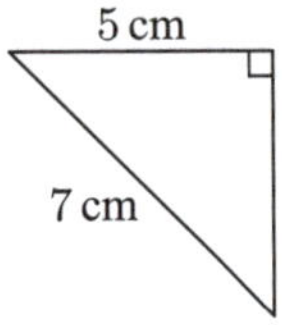

b
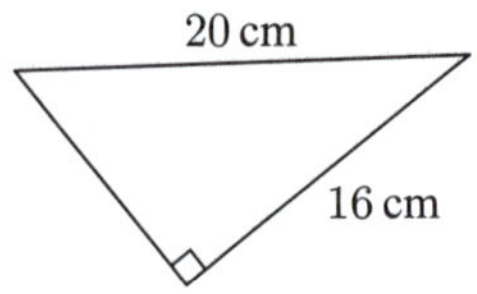

c
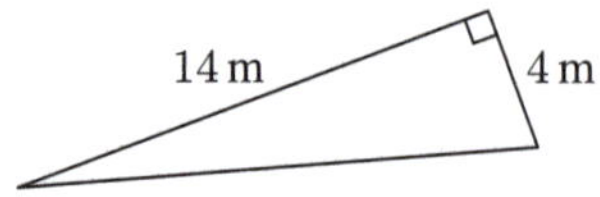

4 Find the length of the third side, rounding your answer to 2 decimal places:

a
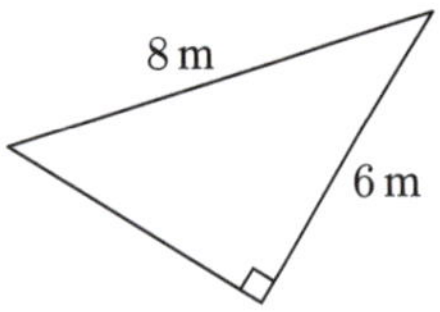

b
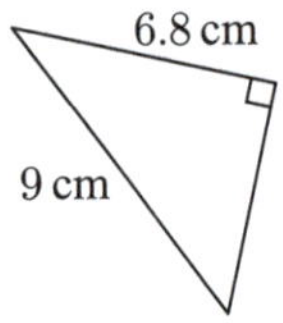

c
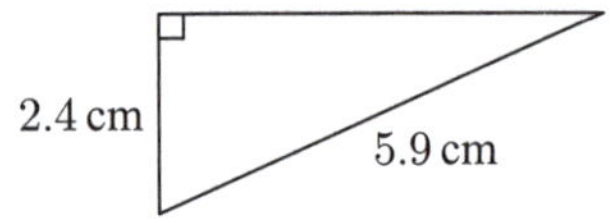

5 Find the length of the third side in each right angled triangle. Leave your answer in simplest surd form.

a
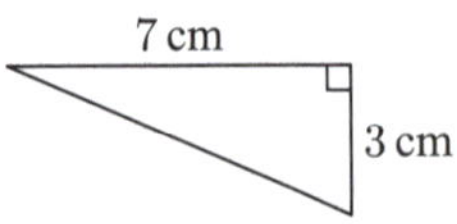

b
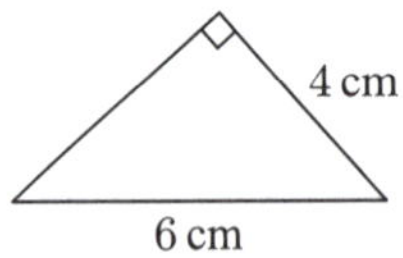

c
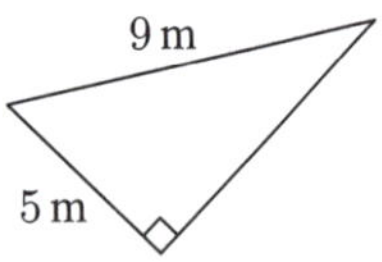

d
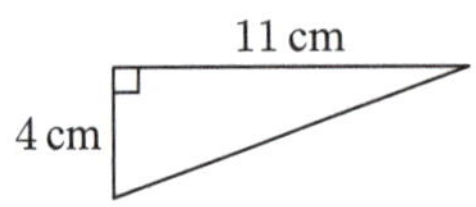

e
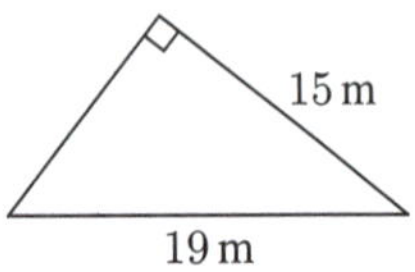

f
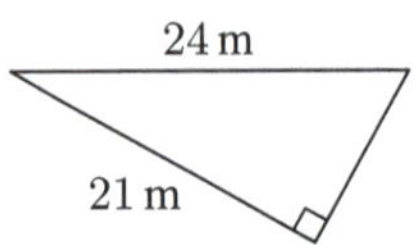

Example 3 Self Tutor

Find x in simplest surd form:

a
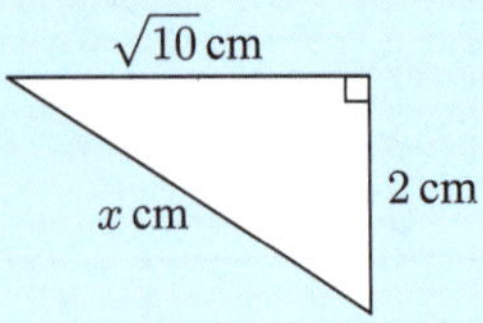

b
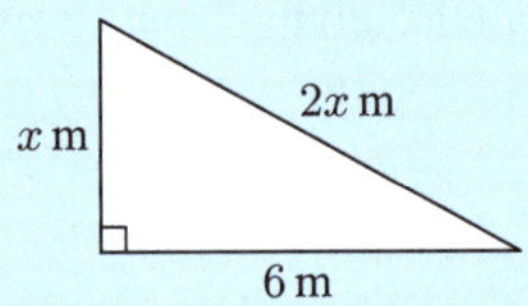

a

$x^2 = 2^2 + (\sqrt{10})^2$ {Pythagoras}

$\therefore\ x^2 = 4 + 10$

$\therefore\ x^2 = 14$

$\therefore\ x = \sqrt{14}$ {since $x > 0$}

b

$(2x)^2 = x^2 + 6^2$ {Pythagoras}

$\therefore\ 3x^2 = 36$

$\therefore\ x^2 = 12$

$\therefore\ x = \sqrt{12}$ {since $x > 0$}

$\therefore\ x = 2\sqrt{3}$

6 Find x in simplest surd form:

a
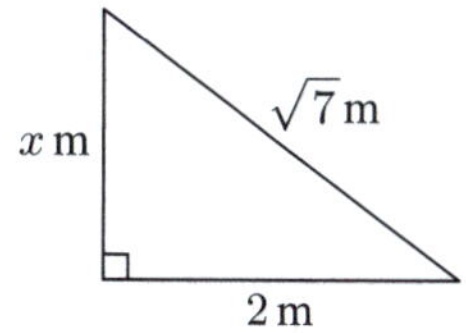

b
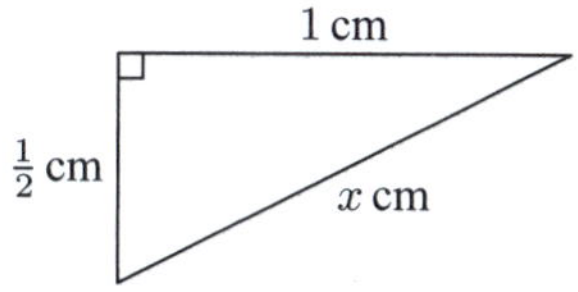

c
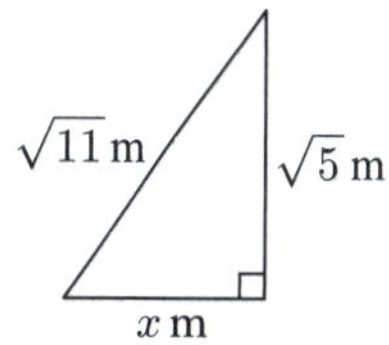

d
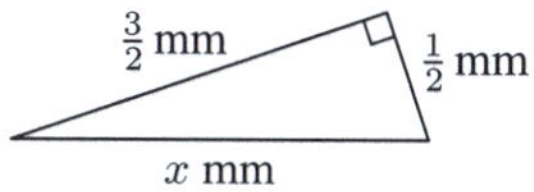

e
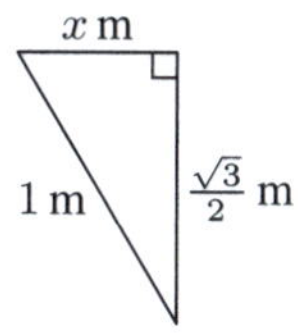

f
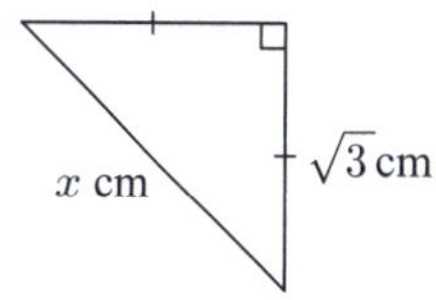

7 Find the value of x:

a
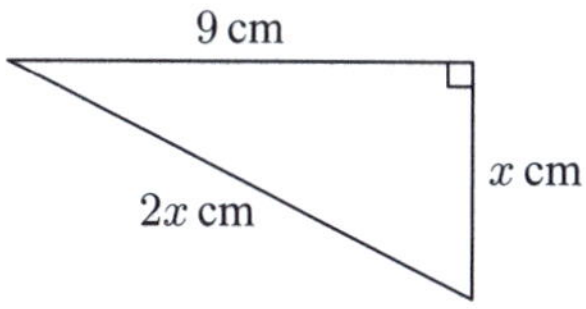

b
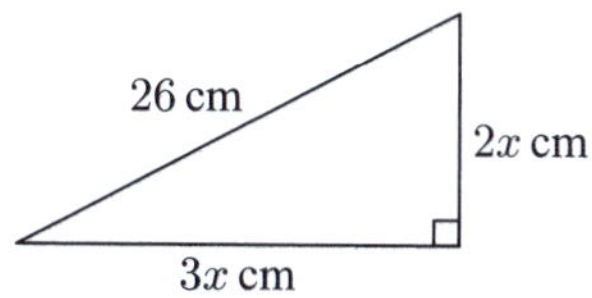

c
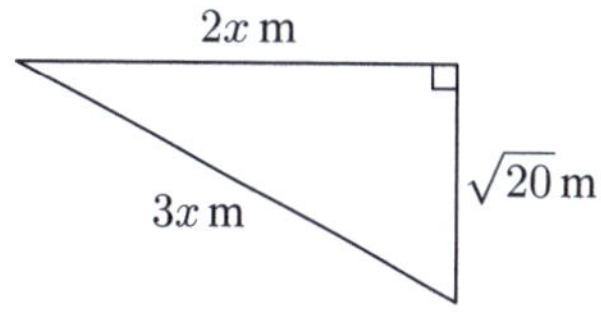

Example 4 **Self Tutor**

Find the unknowns:

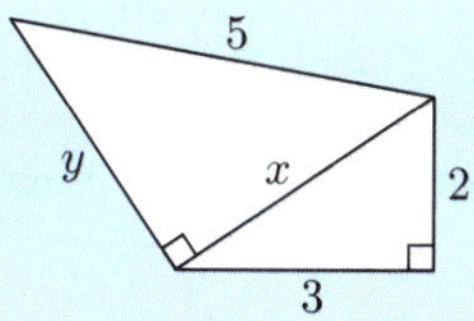

$x^2 = 3^2 + 2^2$ {Pythagoras}
$\therefore\ x^2 = 9 + 4$
$\therefore\ x^2 = 13$
$\therefore\ x = \sqrt{13}$ {since $x > 0$}

$y^2 + (\sqrt{13})^2 = 5^2$ {Pythagoras}
$\therefore\ y^2 + 13 = 25$
$\therefore\ y^2 = 12$
$\therefore\ y = \sqrt{12}$ {since $y > 0$}
$\therefore\ y = 2\sqrt{3}$

8 Find the unknown lengths:

a
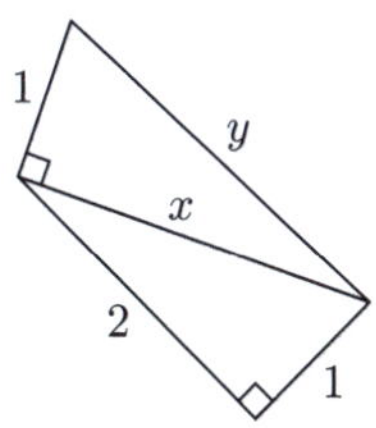

b
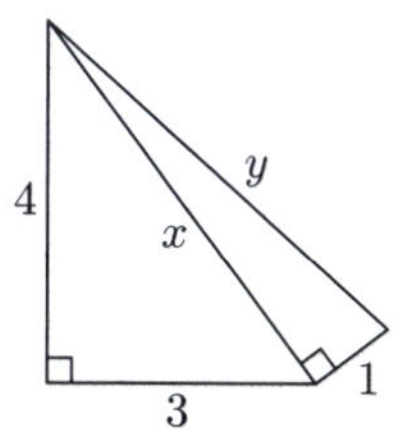

c
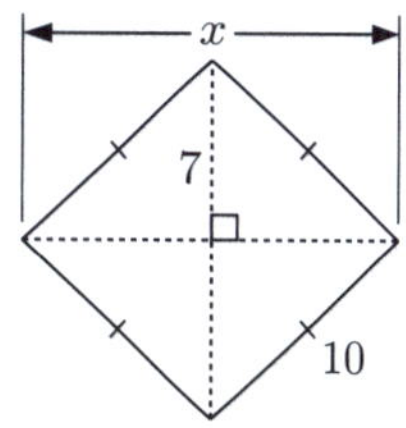

d
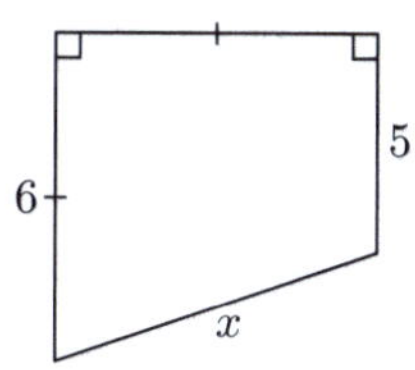

e
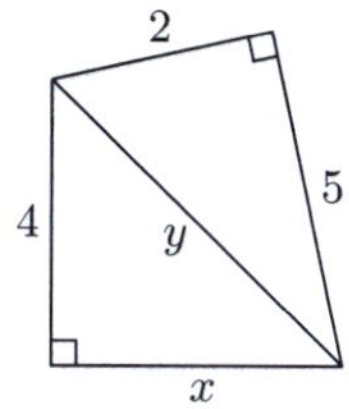

f
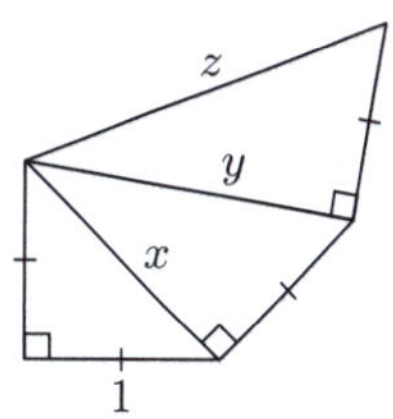

ACTIVITY — IRRATIONALS ON THE NUMBER LINE

Look at the number line and grid shown.

The right angled triangle is drawn with one vertex at 0 and one side along the number line.

Since $4^2 + 7^2 = 16 + 49 = 65$, the hypotenuse has length $\sqrt{65}$ units.

Placing a compass needle at 0 on the number line and setting the radius equal to the hypotenuse, we draw an arc around to place $\sqrt{65}$ on the number line.

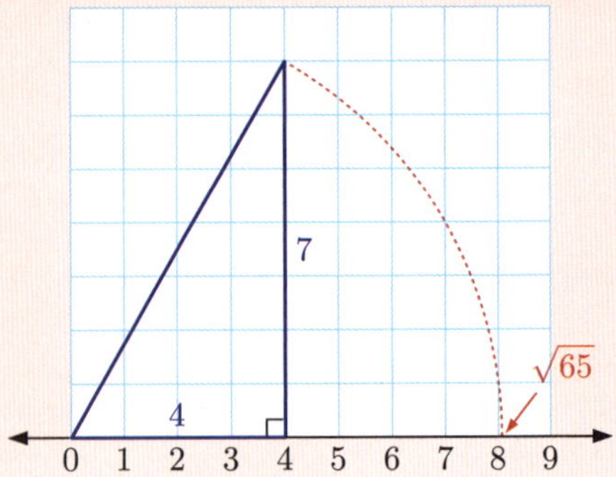

What to do:

1. Find two integers a and b such that $a^2 + b^2$ equals:

   a 5 **b** 13 **c** 20 **d** 29
   e 34 **f** 52 **g** 73 **h** 85

TEMPLATE

2. Draw a number line on grid paper or print the template.
   Use the method above and your results from **1** to plot, on the same number line:

   a $\sqrt{5}$ **b** $\sqrt{13}$ **c** $\sqrt{20}$ **d** $\sqrt{29}$
   e $\sqrt{34}$ **f** $\sqrt{52}$ **g** $\sqrt{73}$ **h** $\sqrt{85}$

3. By drawing one or more triangles and using your compass, plot on the same number line:

   a $2\sqrt{8}$ **b** $\sqrt{17} + 2$ **c** $\sqrt{17} - 2$ **d** $\sqrt{45} + \sqrt{20}$

B PYTHAGOREAN TRIPLES

The simplest right angled triangle with sides of integer length is the **3-4-5 triangle**.

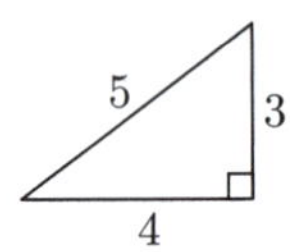

The numbers 3, 4, and 5 satisfy the rule $3^2 + 4^2 = 5^2$.

> The set of positive integers $\{a, b, c\}$ is a **Pythagorean triple** if it obeys the rule $a^2 + b^2 = c^2$.

When we write down a Pythagorean triple, we list the numbers in ascending order.

Example 5 — Self Tutor

Find k given that $\{10, k, 26\}$ is a Pythagorean triple.

Now, $10^2 + k^2 = 26^2$
$\therefore\ 100 + k^2 = 676$
$\therefore\ k^2 = 576$
$\therefore\ k = 24$ {since $k > 0$}

EXERCISE 9B

1 Show that these sets of integers are Pythagorean triples:

a $\{5, 12, 13\}$ **b** $\{7, 24, 25\}$ **c** $\{8, 15, 17\}$

2 Determine whether the following are Pythagorean triples:

a $\{6, 8, 10\}$ **b** $\{16, 30, 34\}$ **c** $\{5, 6, 7\}$

d $\{14, 48, 50\}$ **e** $\{1, 2, 3\}$ **f** $\{20, 48, 52\}$

3 Find k given that the following are Pythagorean triples:

a $\{9, 12, k\}$ **b** $\{k, 24, 26\}$ **c** $\{20, k, 29\}$

d $\{15, 20, k\}$ **e** $\{k, 45, 51\}$ **f** $\{11, k, 61\}$

4 **a** Explain why there are infinitely many Pythagorean triples of the form $\{3k, 4k, 5k\}$ where $k \in \mathbb{Z}^+$.

b Write down the Pythagorean triple generated by letting:

i $k = 2$ **ii** $k = 3$ **iii** $k = 4$ **iv** $k = 5$

5 Suppose $\{a, b, c\}$ is a Pythagorean triple.

a What can you say about a triangle with side lengths ka, kb, and kc, where $k > 0$? Justify your answer.

b Explain why $\{ka, kb, kc\}$ will always be a Pythagorean triple for any $k \in \mathbb{Z}^+$.

c Does the set of Pythagorean triples described in **b** necessarily contain the *only* Pythagorean triples with numbers in the ratio $a : b : c$? Explain your answer.

6 **a** Show that $\{2m, m^2 - 1, m^2 + 1\}$ is a Pythagorean triple for any $m \in \mathbb{Z}^+$, $m > 1$.

b Write down the Pythagorean triple generated by letting:

i $m = 3$ **ii** $m = 6$ **iii** $m = 7$

c Do the triangles generated by $\{2m, m^2 - 1, m^2 + 1\}$ all have side lengths in the same ratio?

C PROBLEM SOLVING

Right angles can be found in many common geometric figures.

For example:

- Each corner of a **rectangle** is a right angle. We can construct a diagonal to form a right angled triangle.

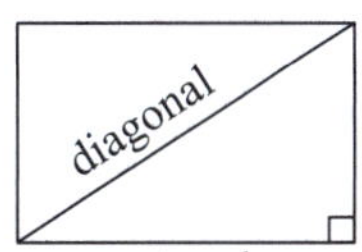

rectangle

- In a **square** and a **rhombus**, the diagonals bisect each other at right angles.

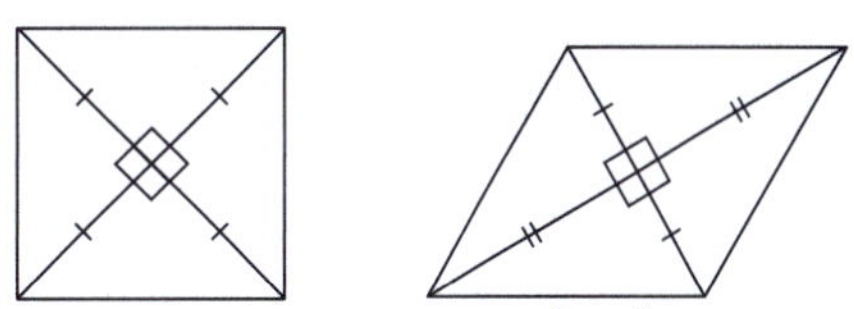

square rhombus

- In an **isosceles triangle** and an **equilateral triangle**, the altitude bisects the base at right angles.

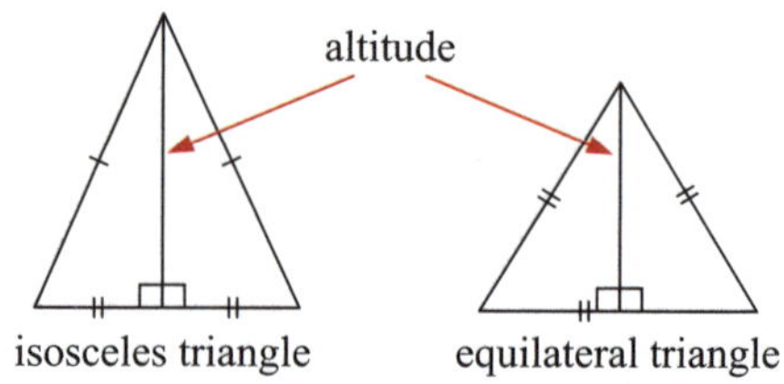

Whenever we find a right angled triangle, we can apply Pythagoras' theorem to help find unknown side lengths.

The problem solving approach involves the following steps:

Step 1: Draw a neat, clear diagram of the situation.
Step 2: Mark known lengths and right angles on the diagram.
Step 3: Use a variable such as x to represent the unknown length.
Step 4: Write down Pythagoras' theorem for the given information.
Step 5: Solve the equation.
Step 6: Where necessary, write your answer in sentence form.

Example 6 **Self Tutor**

An A-frame ladder is 2.1 m long. When it is set up, its feet are 1.4 m apart. Find the height of the ladder once it is set up.

Let the height of the ladder be h m.

$$h^2 + 0.7^2 = 2.1^2 \quad \{\text{Pythagoras}\}$$
$$\therefore\ h^2 + 0.49 = 4.41$$
$$\therefore\ h^2 = 3.92$$
$$\therefore\ h = \sqrt{3.92} \quad \{\text{since } h > 0\}$$
$$\therefore\ h \approx 1.98$$

The ladder is about 1.98 m high.

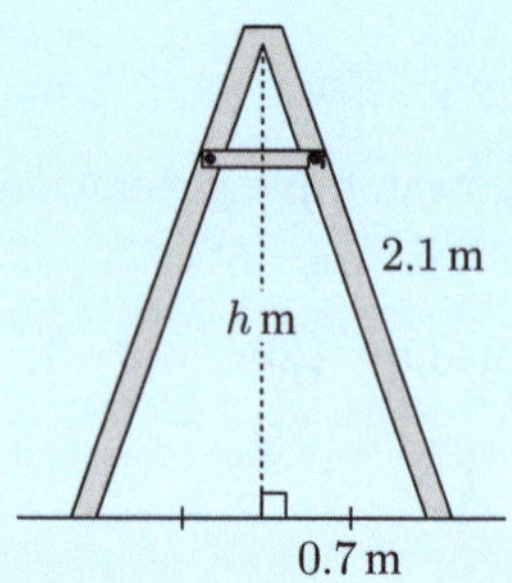

EXERCISE 9C

1 The rectangular frame of a gate is 1.2 m by 2 m. Find, to the nearest centimetre, the length of the diagonal support across the gate.

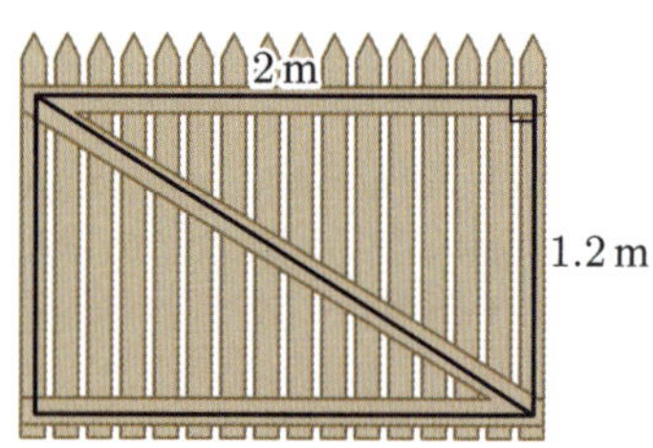

2 A garden shed has a rectangular base 4 m by 2.5 m. A steel rod runs across its diagonal just under the ceiling to keep the frame rectangular. Find the length of the rod.

3 Find, in simplest surd form, the length of a diagonal of a square courtyard with side length 4 m.

4 On the grid there are four towns A, B, C, and D. The grid lines are 5 km apart. Find, in simplest surd form, the distance from:

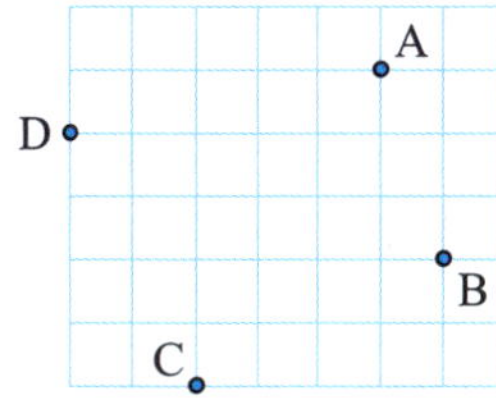

a A to B **b** B to C **c** C to D
d D to A **e** A to C **f** B to D.

5 A 3.6 m extension ladder is leant against a building with its feet 1.6 m away. Will the ladder reach a window sill 3.2 m above the ground? Explain your answer.

6 Find the height of the illustrated tent.

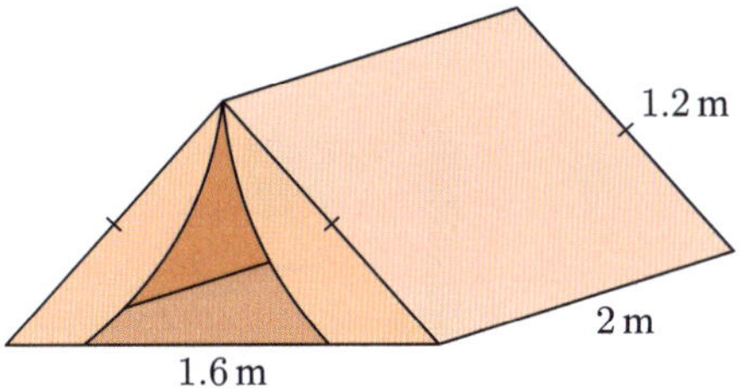

7 An equilateral triangle has sides of length 5 cm. Find the altitude of the triangle, giving your answer in simplest surd form.

8 Each corner of a square marquee is anchored to the ground in two places by a single rope as shown.
Find the total length of the ropes which anchor the marquee.

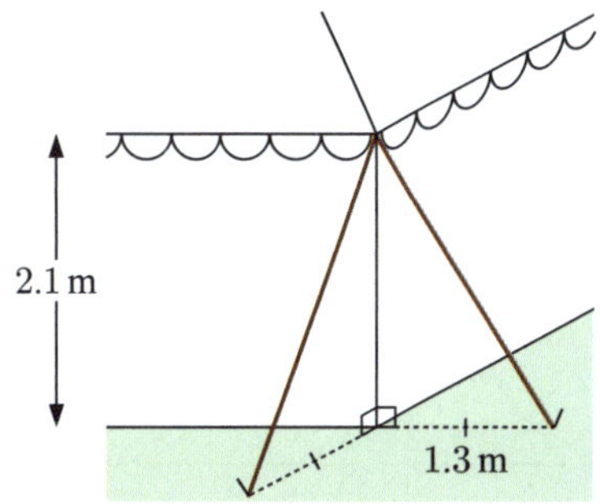

9 Using the map of a ski resort, Diane can see that the cable car travels a horizontal distance of 1.3 km. The cable moves at 6 m/s, and the journey takes 4 minutes. Find the altitude gained by the cable car.

Example 7 **Self Tutor**

A beach volleyball court is twice as long as it is wide.
A diagonal of the court is 19 m long. Find the width of the court.

Let the width of the court be x m.
$\therefore$ the length of the court is $2x$ m.

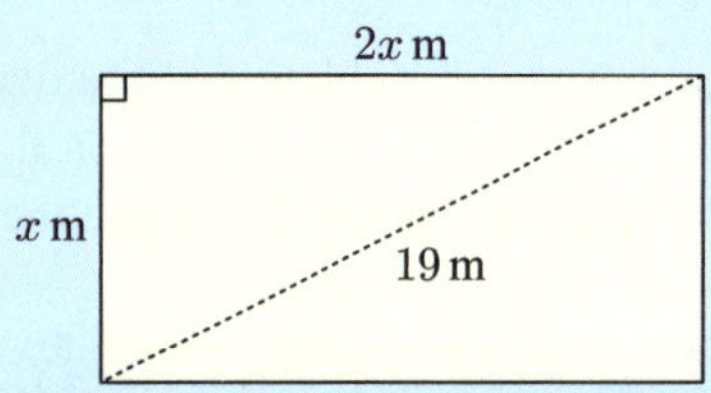

$x^2 + (2x)^2 = 19^2$ {Pythagoras}
$\therefore\ x^2 + 4x^2 = 361$
$\therefore\ 5x^2 = 361$
$\therefore\ x^2 = 72.2$
$\therefore\ x = \sqrt{72.2}$ {since $x > 0$}
$\therefore\ x \approx 8.50$

The court is about 8.50 m wide.

10 A fielder throws the baseball 38.2 m from 3rd base to 1st base.
How far must the 1st base fielder throw the baseball to home base?

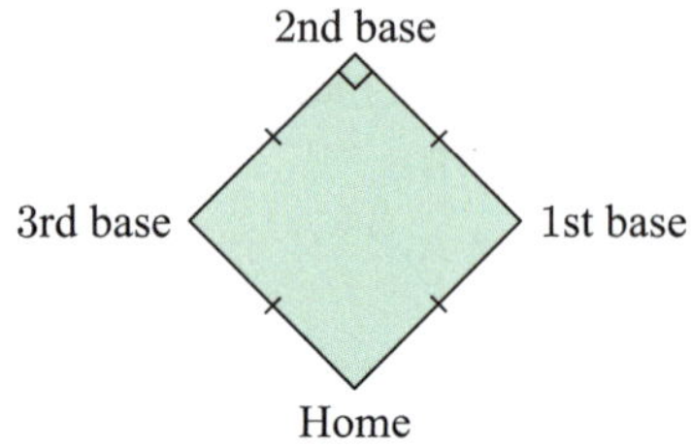

11 The shorter side of a hockey field is 40% shorter than the longer side. The diagonal of the field is 107 m long. Find the length of the longer side.

12 During a storm, a tree snaps one third of the way up from its base. The top of the tree touches the ground 6 m from the base. Find the height of the tree before the storm.

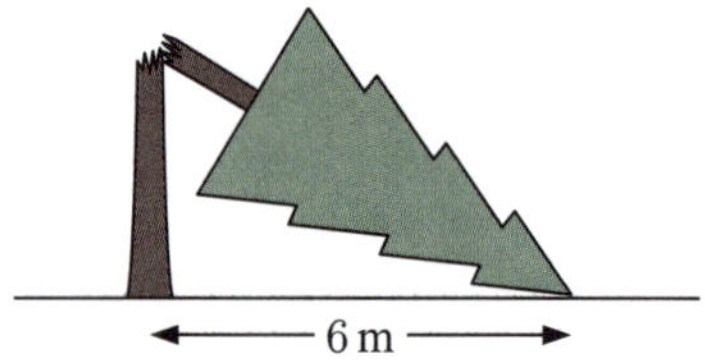

13 A yacht sails 2.5 km northeast then 1.4 km due east. How far is it from its starting point?

Example 8 **Self Tutor**

A cone is 7 cm high and has base radius 3 cm.
Find the slant height of the cone.

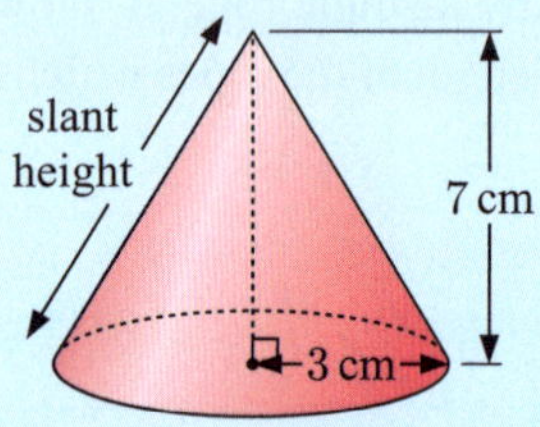

Let the slant height be s cm.

$\therefore\ s^2 = 7^2 + 3^2$ {Pythagoras}

$\therefore\ s^2 = 49 + 9$

$\therefore\ s^2 = 58$

$\therefore\ s = \sqrt{58}$ {since $s > 0$}

$\therefore$ the slant height is $\sqrt{58} \approx 7.62$ cm.

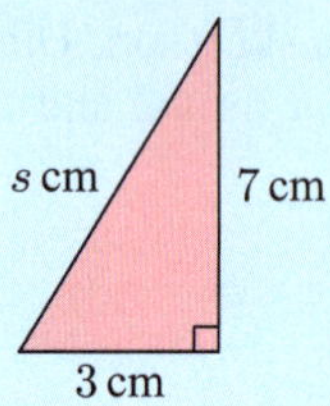

14 A cone has slant height 9 cm, and base radius 5 cm. Find the height of the cone.

15 A cylindrical barrel is 1 m high, and has diameter 75 cm. A 1.65 m long star picket is placed in the barrel on the diagonal as shown.
What length of the star picket is outside the barrel?

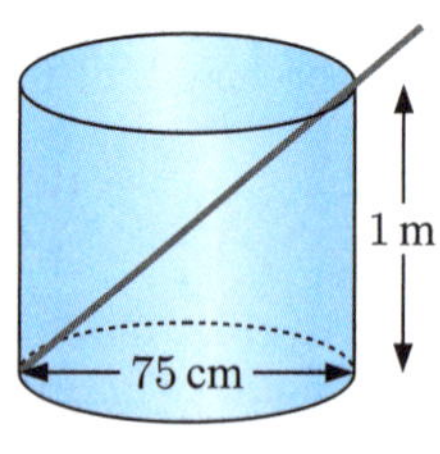

Example 9 **Self Tutor**

A room has floor dimensions 5 m by 4 m. The height of the room is 3 m. Find the distance from a corner point on the floor to the opposite corner point on the ceiling.

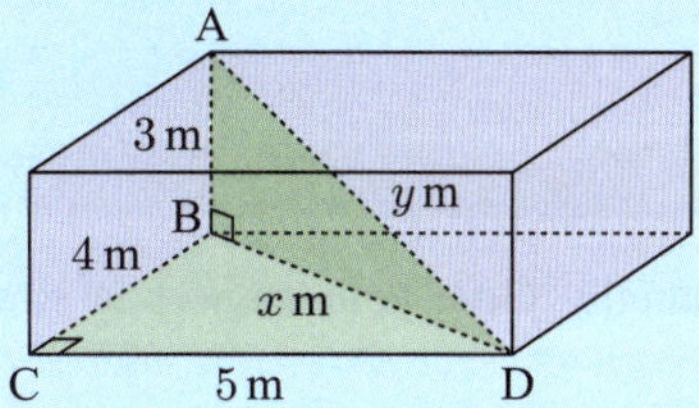

Let BD $= x$ m and AD $= y$ m.

In $\triangle$BCD, $x^2 = 4^2 + 5^2$ {Pythagoras}

In $\triangle$ABD, $y^2 = x^2 + 3^2$ {Pythagoras}

$\therefore\ y^2 = (4^2 + 5^2) + 3^2$

$\therefore\ y^2 = 16 + 25 + 9$

$\therefore\ y^2 = 50$

$\therefore\ y = \sqrt{50}$ {since $y > 0$}

$\therefore$ the required distance is $\sqrt{50} \approx 7.07$ m.

16 A cube has sides of length 1 cm.
Find the length of a diagonal of the cube, giving your answer in simplest surd form.

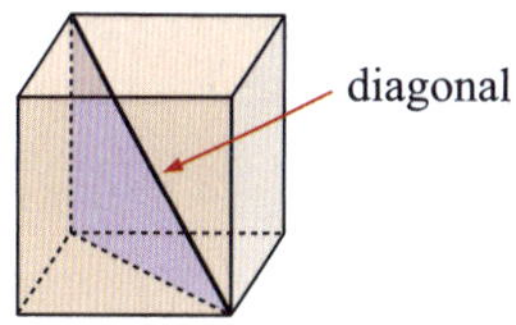

17 A room has floor dimensions 4 m by 3 m. The height of the room is 3 m. Find the distance from a corner point on the floor to the opposite corner point on the ceiling.

18 An aquarium has the dimensions as shown.
Find the length of the diagonal [XY].

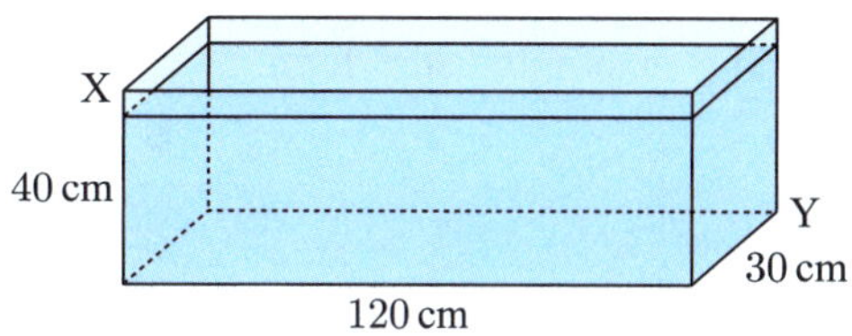

19 The edges of this square-based pyramid are all 10 m long.
The apex E is directly above the centre of the base M.
Find the height of the pyramid.

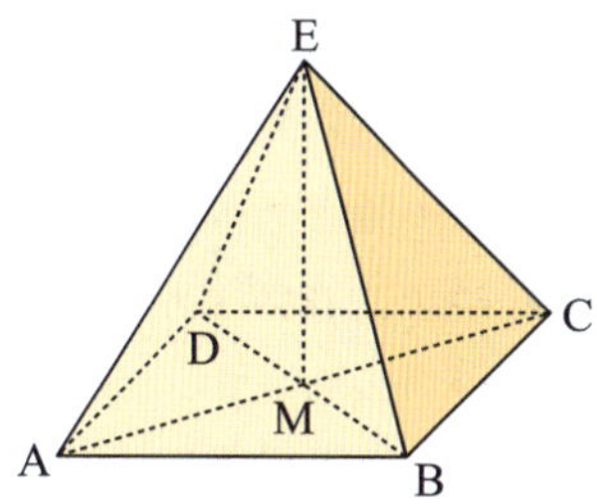

20

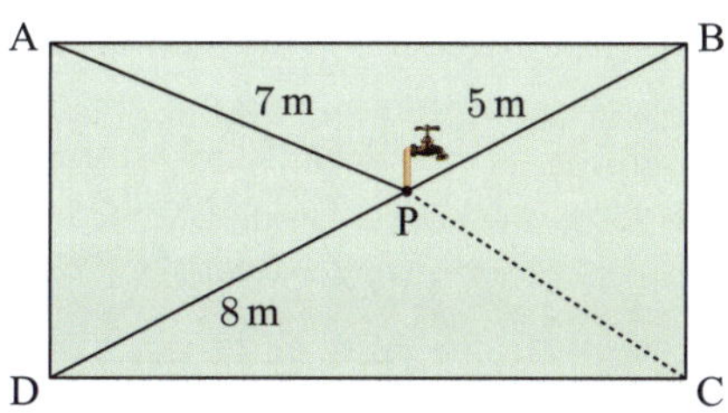

ABCD is a rectangular garden plot.
P is a water tap.
Find the distance from the tap to corner C.

MULTIPLE CHOICE QUIZ

QUICK QUIZ

REVIEW SET 9A

1 Find the length of the hypotenuse in each right angled triangle. Give your answer in simplest surd form where necessary.

a

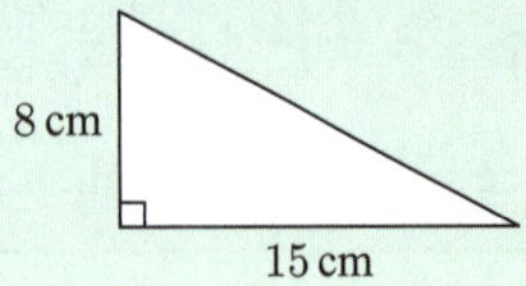

b

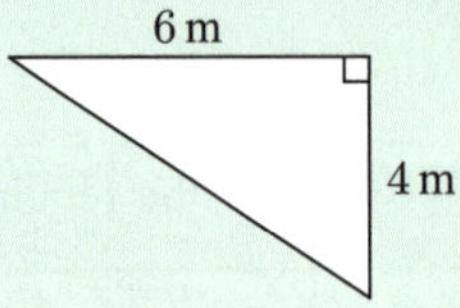

c

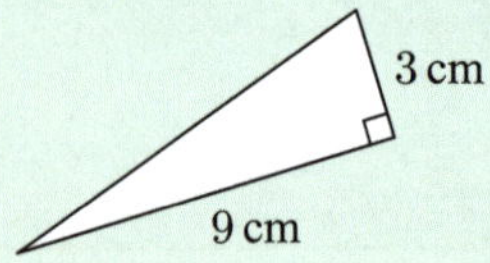

2 Find x and y, rounding your answers to 2 decimal places.

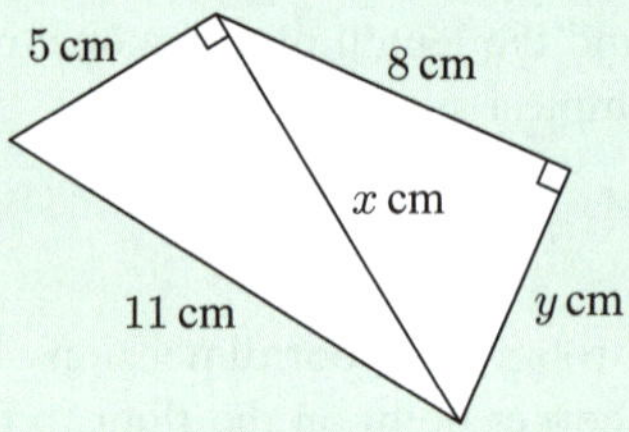

3 Find x in simplest surd form:

a

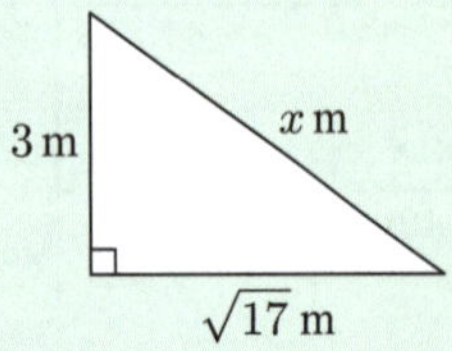

b

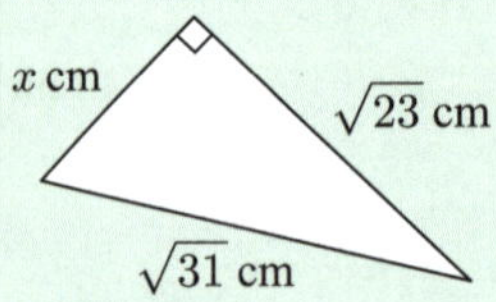

c

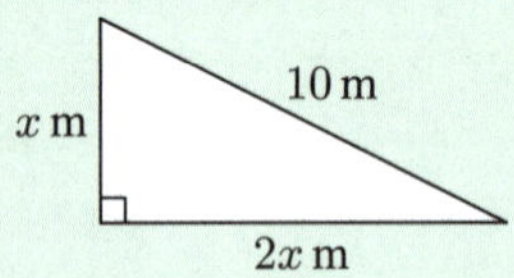

4 Determine whether the following are Pythagorean triples:

a $\{8, 10, 12\}$ **b** $\{9, 40, 41\}$ **c** $\{16, 20, 24\}$

5 Jordan jogged 3 km north, 6 km east, then 8 km south. How far is he from his starting point?

6 How high is the roof above the walls in the roof structure shown?

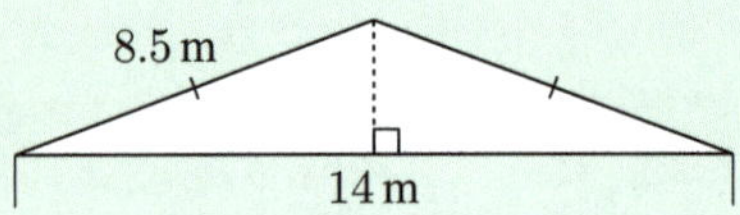

7 An equilateral triangle has height 5 cm. Find the length of the equal sides of the triangle.

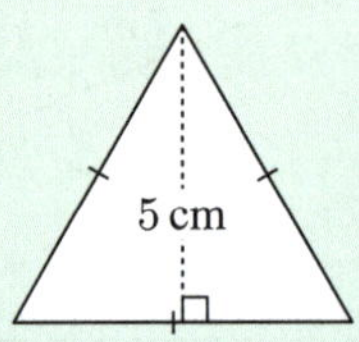

8 The slant height of a cone is three times as long as its base radius. The cone is 12 cm high. Find the slant height of the cone.

9 When viewed from above, a clothesline has the design shown.
Each of the arms is 1.6 m long.

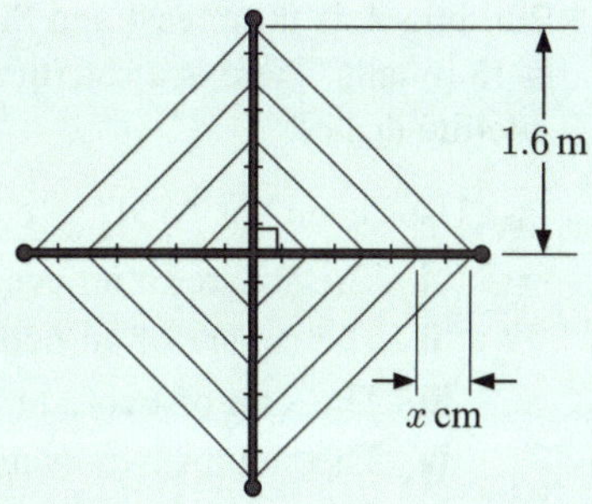

a Find the value of x.
b Calculate the *total* length of:
  i the innermost line
  ii the outermost line.
c Find the total length of a cord needed for the clothesline.

REVIEW SET 9B

1 Find the length of the third side of the triangle, rounding your answer to 3 significant figures:

a
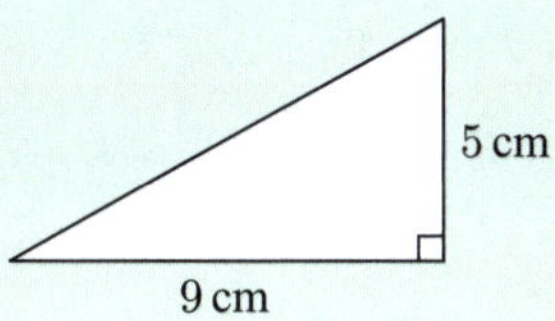

b
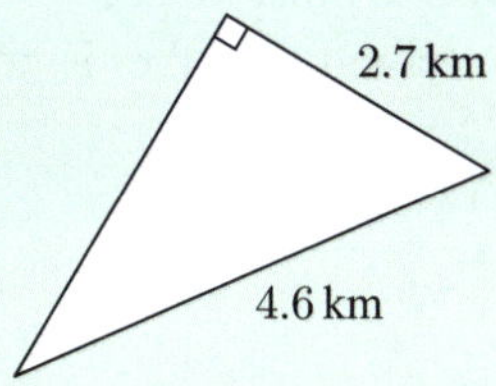

c
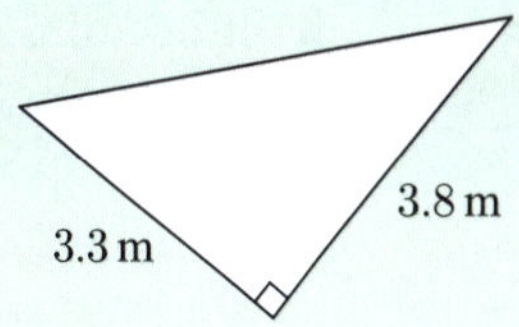

2 Find the length of the third side in simplest surd form.

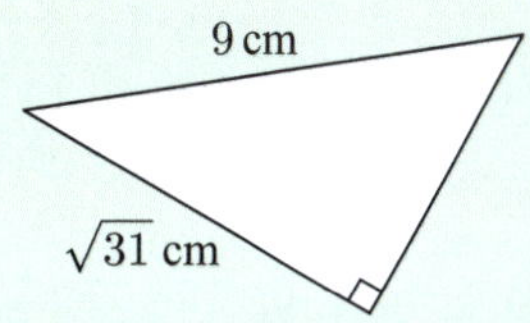

3 Find the unknown lengths in the given figure:

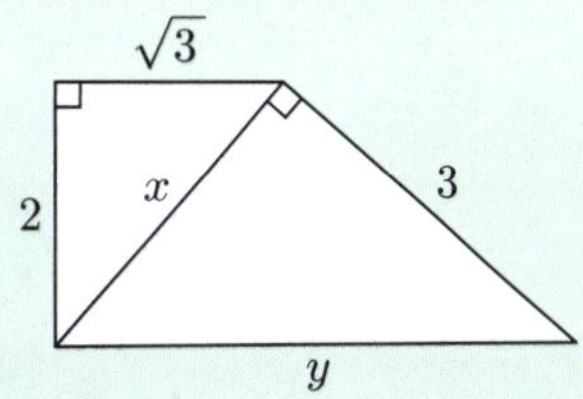

4 Find k given that $\{11, 60, k\}$ is a Pythagorean triple.

5 The ratio of a screen's width to its height is $4 : 3$. The screen's diagonal is 90 cm long. Find the dimensions of the screen.

6 A rhombus has sides of length 20 cm. One diagonal is 24 cm long. Find the length of the other diagonal.

7 A flagpole is held vertical by three cables.
Each cable is fixed to the ground 2.6 m from the base of the flagpole, and attached 3.5 m up the flagpole.
Find the total length of the three cables.

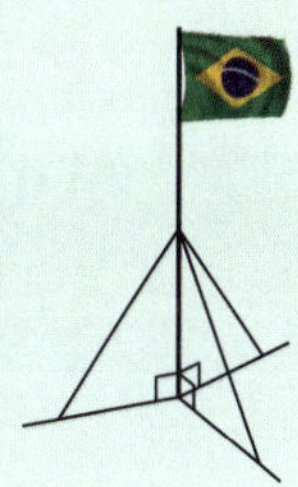

8 Building A is 60 m east and 30 m north of building B. Building A is 70 m tall and building B is 45 m tall. There is a satellite dish on the top of each building. Find the distance between the satellite dishes.

9 **a** Use "even" or "odd" to copy and complete each statement:

- **i** The square of an even number is ......
- **ii** The square of an odd number is ......
- **iii** The sum of two odd numbers is ......
- **iv** The sum of two even numbers is ......
- **v** The sum of an odd number and an even number is ......

b Explain why it is impossible for all three numbers in a Pythagorean triple to be odd.

c Give an example of a Pythagorean triple which contains all even numbers.

d Suppose $\{a, b, c\}$ is a Pythagorean triple.
Use "even" or "odd" to copy and complete each statement:

- **i** If a and b are both even, c must be ......
- **ii** If one of a and b is even and the other is odd, then c must be ......

Chapter 10

Formulae

Contents:

OPENING PROBLEM

A **centripetal force** is a force acting on an object to keep it moving in a circular path.

The size of a centripetal force F can be found using the **formula** $F = \frac{mv^2}{r}$ newtons, where:

- m is the mass of the object in kilograms
- v is the speed of the object in m/s
- r is the radius of the circular path in metres.

Things to think about:

a What does it mean to say that F is the *subject* of the formula?

b Anita is a hammer thrower. She swings the 4 kg hammer at a radius of 1.2 m with speed 18 m/s.

i Can you find the size of the centripetal force which Anita applies on the hammer?

ii What speed must Anita reach with the hammer to increase the centripetal force to 1500 newtons?

c Can you *rearrange* the formula so that v is the subject? Can you explain why this may be useful?

A **formula** is an equation which connects two or more variables.

The plural of formula is **formulae**.

The **subject** of a formula is the variable which appears by itself on one side of the equation.

In the **Opening Problem**, the centripetal force F is by itself on the LHS of the equation. We say that F is the *subject* of the formula, and that F is written *in terms of* m, v, and r.

A FORMULA CONSTRUCTION

To **construct** or **derive** a formula to connect related variables, we often start with numerical examples. Once we understand the specific situation, we can **generalise** the result.

Example 1 Self Tutor

Write a formula for the total cost $\$C$ of a taxi trip which includes:

a a fixed charge of \$3, plus \$0.55 per km for 12 km

b a fixed charge of \$3, plus \$0.55 per km for k km

c a fixed charge of \$3, plus $\$d$ per km for k km

d a fixed charge of $\$F$, plus $\$d$ per km for k km.

a $C = 3 + 0.55 \times 12$

b $C = 3 + 0.55 \times k$
$\therefore C = 3 + 0.55k$

c $C = 3 + d \times k$
$\therefore C = 3 + dk$

d $C = F + d \times k$
$\therefore C = F + dk$

EXERCISE 10A

1 A mobile gardener charges a \$30 call-out fee, plus \$60 per hour thereafter. Write a formula for the gardener's total fee $\$F$ for a service lasting:

a 2 hours **b** h hours.

2 Write a formula for the amount $£A$ in a savings account, given monthly deposits of:

a £250 over 6 months **b** £250 over m months **c** $£D$ over m months.

3 Write a formula for the number of students F who failed a test if:

a 50 students took the test, and 40 students passed

b 50 students took the test, and p students passed

c n students took the test, and p students passed.

4 Write a formula for the total mass M kg of:

a a 20 kg wheelbarrow containing 12 bricks weighing 5 kg each

b a 20 kg wheelbarrow containing b bricks weighing 5 kg each

c a 20 kg wheelbarrow containing b bricks weighing n kg each

d a w kg wheelbarrow containing b bricks weighing n kg each.

5 Write a formula for the amount $\$A$ in Chris' wallet if initially he had:

a \$100, and he bought 6 books costing \$12 each

b \$100, and he bought b books costing \$12 each

c \$100, and he bought b books costing $\$p$ each

d $\$d$, and he bought b books costing $\$p$ each.

6 Helen works at a garden centre. When it is rainy, she works a 5 hour day. When it is fine, the centre is busier, so she works an 8 hour day. Write a formula for the total number of hours H that Helen works if:

a she works for 6 days, 2 of which are rainy

b she works for 7 days, r of which are rainy

c she works for d days, r of which are rainy.

7 An art gallery has a series of paintings to be hung along a wall. The gaps between the paintings and at either end of the wall are equal.

Write a formula for the length of the gap G metres between each painting, if:

a 3 paintings, each 1.2 metres wide, are to be hung along a 6 metre wall

b 3 paintings, each w metres wide, are to be hung along a 6 metre wall

c 3 paintings, each w metres wide, are to be hung along a d metre wall

d n paintings, each w metres wide, are to be hung along a d metre wall.

B SUBSTITUTING INTO FORMULAE

Suppose a formula contains two or more variables and we know the value of all but one of them. We can **substitute** the known values into the formula, leaving an equation which can be solved to find the value of the remaining variable.

Example 2 Self Tutor

When an object is falling under gravity, its downward speed after time t is given by $v = u + gt$, where u is its initial speed downwards and $g = 9.8$ is the acceleration due to gravity.

Find the value of:

a v if $u = 15$ and $t = 10$

b t if $v = 45$ and $u = 20$.

a When $u = 15$, $t = 10$, and $g = 9.8$,

$v = 15 + 9.8 \times 10$

$\therefore\ v = 15 + 98$

$\therefore\ v = 113$

b When $v = 45$, $u = 20$, and $g = 9.8$,

$45 = 20 + 9.8t$

$\therefore\ 25 = 9.8t$

$\therefore\ \dfrac{25}{9.8} = t$

$\therefore\ t \approx 2.55$

EXERCISE 10B

1 Let $D = 4n - 5$. Find the value of:

a D when $n = 3$

b n when $D = 35$.

2 Let $F = g + 5h$. Find the value of:

a F when $g = 2$ and $h = 3$

b g when $F = 32$ and $h = 5$

c h when $F = 21$ and $g = 9$.

3 The cost of hiring a scooter for t minutes is given by $C = 0.8 + 0.3t$ dollars.

a Find the cost of hiring a scooter for:

i 10 minutes

ii 24 minutes.

b Elizabeth spent \$20 on scooter hire. For how long did she hire the scooter?

4 A motorcade with c cars and m motorcycles contains a total of $W = 4c + 2m$ wheels.

a Find the total number of wheels in a motorcade with 8 cars and 3 motorcycles.

b Find m if $W = 50$ and $c = 9$.

c Find c if $W = 80$ and $m = 12$.

5 The momentum P of an object is given by $P = mv$ kg m/s, where m is the mass of the object in kg, and v is the speed of the object in m/s.

a Find the momentum of:

i a 1500 kg car travelling at 15 m/s

ii a 3 g bullet travelling at 500 m/s.

b Sidney hit a 170 g hockey puck with his hockey stick. The puck travelled with momentum 6.8 kg m/s. At what speed was it travelling?

6 Consider the formula $T = 3a^2 + 2$.

a Find the value of T when:

i $a = 2$ **ii** $a = 5$ **iii** $a = -4$

b Find the possible values of a when:

i $T = 5$ **ii** $T = 29$ **iii** $T = 194$

7 If n dots are placed on each side of a square, there are $S = 4n^3$ triangles which can be drawn within the square by connecting the dots.

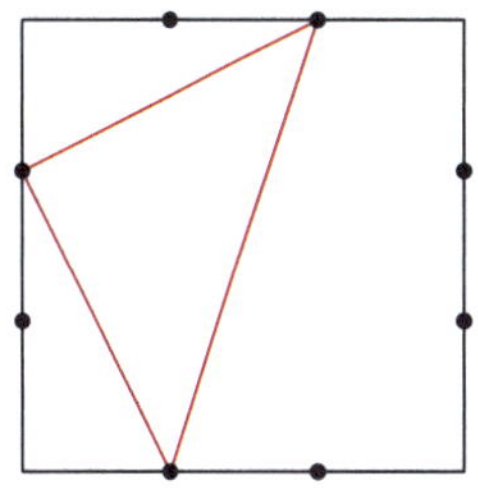

a Find the value of S when:

i $n = 2$ **ii** $n = 6$

b Find the value of n such that:

i $S = 108$ **ii** $S = 500$

8 Fulvio's formula for the price of his pizzas is $P = F + \frac{r^2}{32}$ euros, where r is the radius of the pizza in cm, and F is the profit in euros. Find:

a the price of a 12 cm radius pizza which makes €3 profit

b the profit if a 15 cm radius pizza sells for €9.90

c the radius of a €19.90 pizza which makes €7.50 profit.

9 Heron's formula for the area of a triangle with side lengths a, b, and c is $A = \sqrt{s(s-a)(s-b)(s-c)}$ where $s = \frac{a+b+c}{2}$.

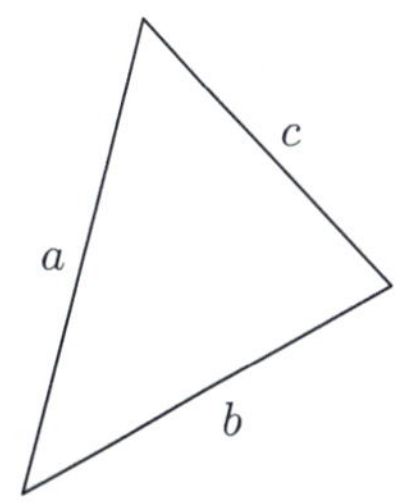

Use Heron's formula to find the area of:

a a triangular garden plot with side lengths 6 m, 7 m, and 8 m

b an isosceles triangle with side lengths 3 cm, 3 cm, and 2 cm.

10 A block is n cubes long, n cubes deep, and n cubes high. The case where $n = 6$ is illustrated.

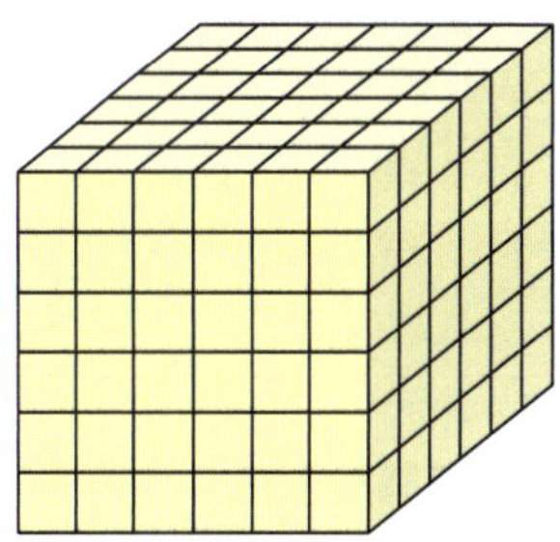

a Write a formula for the total number of cubes C in terms of n.

b Find:

i C if $n = 3$ **ii** n if $C = 216$.

c Let F be the number of cubes with only one exposed face.

i Explain why $F = 6(n-2)^2$.

ii Find F when $n = 5$.

d Let G be the number of cubes with exactly two exposed faces. Explain why $G = 12(n-2)$.

e If $G = 60$, find the value of:

i n **ii** F **iii** C

f Let H be the number of cubes with no exposed faces.

i Write a formula for H in terms of n.

ii Find H when $n = 10$.

ACTIVITY 1 FALLING OBJECTS

When an object is dropped, the distance it falls in the first t seconds is approximately $D = 4.9t^2$ metres.

In this Activity you will conduct an experiment to check this formula.

Click on the icon to obtain a printable worksheet.

WORKSHEET

GAME

Click on the icon to obtain the instructions and scoresheet for a dice game involving formula substitution.

GAME

C REARRANGING FORMULAE

We can **rearrange** a formula to make one of the other variables the subject using **inverse operations** on both sides. This ensures that the formula is still true.

You can always swap the sides of an equation if you wish.

Example 3 Self Tutor

Make x the subject of:

a $M = 3x + k$ **b** $n = \frac{a}{x}$

a
$M = 3x + k$
$\therefore\ 3x + k = M$ {swapping sides}
$\therefore\ 3x = M - k$ {$-\,k$ both sides}
$\therefore\ x = \frac{M-k}{3}$ {$\div\, 3$ both sides}

b
$n = \frac{a}{x}$
$\therefore\ nx = a$ {$\times\, x$ both sides}
$\therefore\ x = \frac{a}{n}$ {$\div\, n$ both sides}

EXERCISE 10C

1 Make x the subject of:

a $y = 5x$ **b** $y = x - 3$ **c** $y = 3x + 2$

d $y = \frac{1}{2}x - 1$ **e** $y = 3 - x$ **f** $y = \frac{2-x}{3}$

2 Make x the subject of:

a $B = 7x + n$ **b** $T = 3xy$ **c** $G = x - 5p$

d $K = l - 9x$ **e** $P = wx + yz$ **f** $C = 4xy^2 - z$

3 Make y the subject of:

a $2x + 3y = 9$ **b** $3x - y = 4$ **c** $4x + 2y = 3$

4 The ideal gas law $PV = nRT$ connects the amount n of gas with its pressure P, volume V, and temperature T. R is the universal gas constant.
Rearrange this formula to make T the subject.

5 Make a the subject of:

a $x = \frac{a}{y}$ **b** $K = \frac{c}{a}$ **c** $M = \frac{abc}{n}$

d $T = \frac{a+s}{p}$ **e** $N = \frac{b+d}{2a}$ **f** $F = \frac{3a-d}{2g}$

6 The density D of an object with mass M and volume V is given by $D = \frac{M}{V}$.
Rearrange this formula to make:

a M the subject **b** V the subject.

7 Make p the subject of:

a $\frac{2}{r} = \frac{p}{5}$ **b** $\frac{q}{r} = \frac{p}{s}$ **c** $\frac{q}{s} = \frac{r}{p}$

d $\frac{q}{p} = \frac{r}{s}$ **e** $\frac{s}{r} = \frac{r}{p}$ **f** $\frac{r}{2p} = \frac{s}{r^2}$

Example 4 Self Tutor

Make x the subject of $P = \frac{t}{x^2}$, given $x > 0$.

$$P = \frac{t}{x^2}$$
$$\therefore\ x^2P = t \quad \{\times x^2 \text{ both sides}\}$$
$$\therefore\ x^2 = \frac{t}{P} \quad \{\div P \text{ both sides}\}$$
$$\therefore\ x = \sqrt{\frac{t}{P}} \quad \{\text{since } x > 0\}$$

8 Make x the subject of each formula, given $x > 0$:

a $D = x^2 + 1$ **b** $K = 5x^2$ **c** $F = \frac{x^2}{y}$

d $N = \frac{7z}{x^2}$ **e** $C = 2x^2 + y^2$ **f** $T = 3x^2z - w$

9 When an object is dropped, the distance s m it falls in the first t seconds is given by $s = \frac{1}{2}gt^2$ where g is the acceleration due to gravity.
Rearrange this formula to make t the subject.

10 The body-mass index B of a person with mass M kg and height H metres, is given by $B = \frac{M}{H^2}$.
Rearrange this formula to make:

a M the subject **b** H the subject.

11 Make x the subject of:

a $xy = 2x - 3$ **b** $\frac{3x}{x-2} = y$ **c** $y = \frac{3}{\sqrt{5-x}}$

d $\frac{x+2y}{x-y} = \frac{5}{7}$ **e** $x^2y = \frac{x^2+3}{y}$, $x > 0$ **f** $\frac{4x}{2x+y^3} = \frac{2}{xy+1}$

12 Make u the subject of $\frac{1}{f} = \frac{1}{u} + \frac{1}{v}$.

D REARRANGEMENT AND SUBSTITUTION

In the Section on formula substitution, the known variables were replaced by numbers, and we then solved an equation to find the unknown.

In situations when we need to perform this process several times, it is quicker to **rearrange** the formula first, and then **substitute**.

Example 5 — Self Tutor

The distance travelled in t seconds by an object with initial speed u m/s and acceleration a m/s^2 is given by $x = ut + \frac{1}{2}at^2$ metres.

a Rearrange this formula to make a the subject.

b Find the acceleration required such that:

  i an object with initial speed 2 m/s travels 40 m in 4 seconds

  ii an object with initial speed 5 m/s travels 180 m in 10 seconds.

a
$$x = ut + \tfrac{1}{2}at^2$$
$$\therefore\ x - ut = \tfrac{1}{2}at^2 \qquad \{-\,ut \text{ both sides}\}$$
$$\therefore\ 2(x - ut) = at^2 \qquad \{\times\, 2 \text{ both sides}\}$$
$$\therefore\ a = \frac{2(x-ut)}{t^2} \qquad \{\div\, t^2 \text{ both sides}\}$$

b **i** When $u = 2$, $x = 40$, and $t = 4$, $a = \dfrac{2(40 - 2\times 4)}{4^2} = \dfrac{2\times 32}{16} = 4$

So, the acceleration required is 4 m/s^2.

ii When $u = 5$, $x = 180$, and $t = 10$, $a = \dfrac{2(180 - 5\times 10)}{10^2} = \dfrac{2\times 130}{100} = 2.6$

So, the acceleration required is 2.6 m/s^2.

EXERCISE 10D

1 **a** Make t the subject of the formula $M = \dfrac{4t}{s^3}$.

b Hence find the value of t when:

  i $M = 12$ and $s = 3$ **ii** $M = 15$ and $s = 2$ **iii** $M = 2$ and $s = 5$.

2 The average speed of a car which travels d kilometres in t hours is given by $s = \dfrac{d}{t}$ km/h.

a Make d the subject of the formula. Hence find the distance travelled by a car travelling at an average speed of:

  i 60 km/h for 3 hours **ii** 80 km/h for $1\frac{1}{2}$ hours **iii** 95 km/h for 1 h 20 min.

b Make t the subject of the formula. Hence find the time required for a car to travel:

  i 180 km at an average speed of 60 km/h

  ii 140 km at an average speed of 84 km/h

  iii 220 km at an average speed of 100 km/h.

3 The energy of a photon of light is given by $E = hf$ joules where f is the frequency of the photon in hertz, and $h \approx 6.63 \times 10^{-34}$ is Planck's constant.

- **a** Rearrange the formula to make f the subject.
- **b** Hence find the frequency of:
  - **i** a photon of green light with energy 3.98×10^{-19} joules
  - **ii** a photon of infrared light with energy 4.97×10^{-22} joules
  - **iii** a photon of ultraviolet light with energy 9.92×10^{-18} joules.

  Write your answers using scientific notation.

4 Pythagoras' theorem states that if a right angled triangle has hypotenuse of length c, and its other sides have lengths a and b, then the side lengths are connected by the formula $c^2 = a^2 + b^2$.

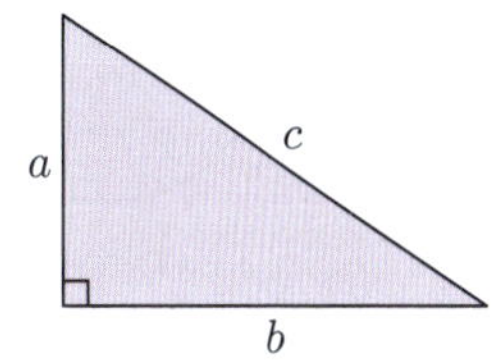

- **a** Make a the subject of the formula.
- **b** *Hence* find a in each triangle:

i

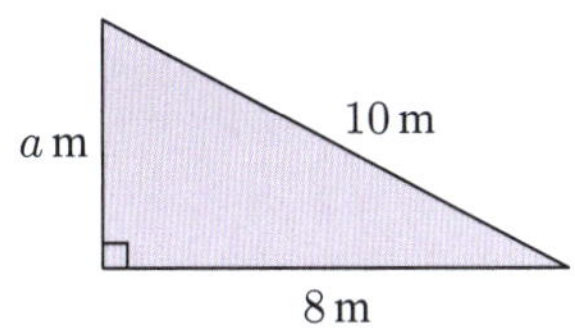

ii

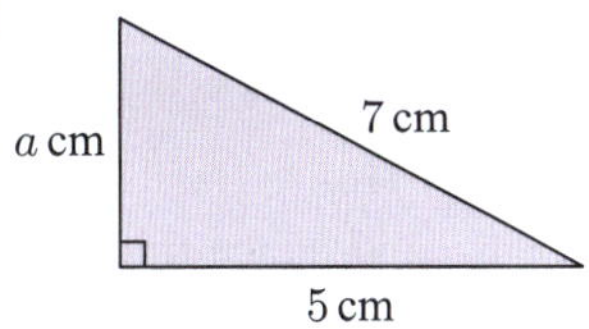

iii

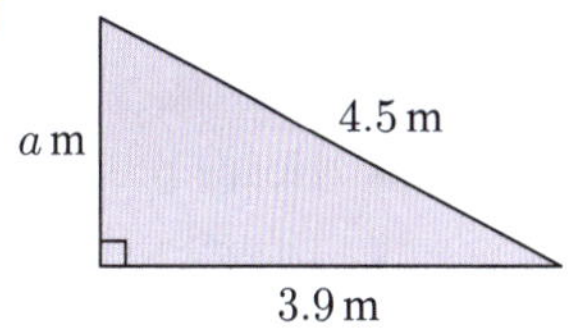

5 In the **Opening Problem**, the centripetal force acting on an object moving in a circular path is given by $F = \dfrac{mv^2}{r}$ newtons, where m is the mass of the object in kilograms, v is the speed of the object in m/s, and r is the radius of the circular path in metres.

- **a** Anita swings the 4 kg hammer at a radius of 1.2 m with speed 18 m/s. Find the size of the centripetal force acting on the hammer.
- **b** Rearrange the formula to make v the subject.
- **c** What speed must Anita reach with the hammer to increase the centripetal force to:
  - **i** 1500 newtons
  - **ii** 2000 newtons
  - **iii** 3000 newtons?

ACTIVITY 2 — PREDICTING FORMULAE

When we observe a pattern in a few simple cases, it is tempting to predict that the pattern is true in general. In some cases our prediction may be correct. However, in other cases the pattern may not continue as we expect. We therefore need to be careful when making predictions, and we seek to *prove* that our prediction is always true.

What to do:

1 Consider this pattern:

$$1 = 1^2$$
$$1 + 3 = 4 = 2^2$$
$$1 + 3 + 5 = 9 = 3^2$$
$$1 + 3 + 5 + 7 = 16 = 4^2$$

Assuming this pattern continues, we predict that "the sum of the first n odd integers is n^2".

- **a** Is it possible to check *all* of the possible cases to be certain that the prediction is true?

b The prediction *is* in fact true for all $n \in \mathbb{Z}^+$.
One way to *prove* the result is to notice that the nth odd integer is $2n-1$.
We let S be the sum of the first n odd integers.

$$\therefore\ S = 1 + 3 + 5 + 7 + \dots\dots + 2n-3 + 2n-1$$

$$\text{and}\ \ S = 2n-1 + 2n-3 + \dots\dots\ \dots\dots + 3 + 1$$

Add these equations vertically to complete the proof.

2

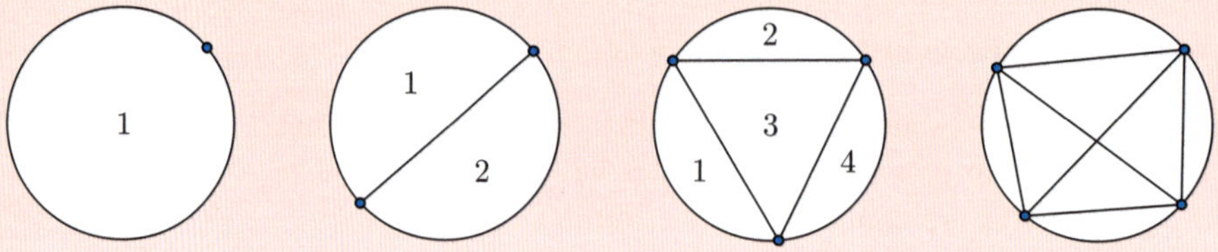

$n=1$ $\qquad$ $n=2$ $\qquad$ $n=3$ $\qquad$ $n=4$

The above figures show the first four cases of n points being placed on a circle's circumference, and each point joined with a line to every other point. Let R_n be the number of regions formed. So, $R_1 = 1$, $R_2 = 2$, and $R_3 = 4$.

a Find R_4.

b Hence predict a formula for R_n.

c Draw the case for $n = 5$, and hence find R_5.

d Do you believe your formula for R_n will be true for all $n \in \mathbb{Z}^+$?

3 Consider the sum of the first n prime numbers, then add 2.
The table below shows the result for the first 6 values of n, as well as whether the result is prime or composite:

n	*Sum*	*Value*	*Prime/Composite*
1	$2+2$	4	composite
2	$2+3+2$	7	prime
3	$2+3+5+2$	12	composite
4	$2+3+5+7+2$	19	prime
5	$2+3+5+7+11+2$	30	composite
6	$2+3+5+7+11+13+2$	43	prime

a Continue this table up to $n = 10$.

b Use the word "prime" or "composite" to complete these predictions:

i If n is odd, the value is ...... $\qquad$ **ii** If n is even, the value is ......

c For each prediction in **b**, either prove that it is always true, or find a counter-example to prove it false.

QUICK QUIZ

MULTIPLE CHOICE QUIZ

REVIEW SET 10A

1 A gardener charges a fixed amount for travel and then an amount per hour for the time he works. Write a formula for the total charge C euros if the fixed amount is:

a €40, and he works for n hours at €20 per hour

b €F, and he works for n hours at €a per hour.

2 Consider an object moving with initial speed u m/s and acceleration a m/s^2. Its speed after t seconds is given by the formula $v = u + at$ m/s.

a A ball is dropped with initial speed 1 m/s and acceleration 9.8 m/s^2. Find its speed after 5 seconds.

b A motorcycle initially travelling at 5 m/s, accelerates at 3.125 m/s^2. How long will it take for the motorcycle to reach a speed of 20 m/s?

3 Consider the formula $P = k^3 - 4$.

a Find the value of P when:

i $k = 3$ **ii** $k = 5$

b Find the value of k when:

i $P = 60$ **ii** $P = 996$

4 Make x the subject of the formula:

a $mx + c = y$ **b** $a = \frac{2}{x}$ **c** $N = \frac{5-2x}{p}$

5 Make y the subject of:

a $5x - 3y = 8$ **b** $\frac{3}{y} = \frac{a}{b}$ **c** $K = x^2 + 3y^2$, given $y > 0$.

6 The intensity of a sound at a distance d m from the source is given by $I \approx 0.08\frac{P}{d^2}$ W/m^2, where P is the power of the sound in watts (W).

a Rearrange the formula to make d the subject.

b The sound of a jet engine has power 100 000 W. How far away from the jet engine should you be so that the intensity of the sound is:

i 1 W/m^2 **ii** 0.1 W/m^2?

7 The monthly repayment for a simple interest loan is given by $R = \frac{P(in+1)}{12n}$, where P is the principal, i is the interest rate per annum expressed as a decimal, and n is the duration of the loan in years.

a Find the monthly repayment for a simple interest loan of \$20 000 at 4% p.a. interest to be repaid over 5 years.

b **i** Rearrange the formula to make P the subject.

ii Kyle takes out a loan at 3% p.a. simple interest over 4 years. His monthly repayments are \$250. How much money did he borrow?

c **i** Rearrange the formula to make i the subject.

ii Dev needs to take out a loan of \$8000. She would like to repay the loan in 2.5 years, and she can afford to repay \$290 per month. What annual rate of simple interest will she need?

REVIEW SET 10B

1 It costs 60 cents to photocopy a page in black and white, and \$1.40 to photocopy a page in colour. Write a formula for the balance on Penny's photocopying card B dollars if she:

a loads \$20 onto the card, then photocopies 8 pages in black and white, and 5 pages in colour

b loads \$$l$ onto the card, then photocopies b pages in black and white, and c pages in colour.

2 The period of a pendulum is the time for one complete swing. It can be estimated using the formula $T = \frac{1}{5}\sqrt{l}$ seconds, where l is the length of the pendulum in cm.

Estimate the period of a pendulum with length 74 cm.

3 Suppose $A = \dfrac{b^3 + c}{2}$. Find:

a A when $b = 2$ and $c = 10$

b b when $A = 11$ and $c = -5$.

4 Make x the subject of the formula:

a $y = 3q - 2x$

b $3y = \dfrac{4}{x}$

5 Make b the subject of:

a $2a + 3b = c$

b $a - \dfrac{4}{b} = c$

c $\dfrac{a}{c} = \dfrac{4}{b}$

6 The *inradius* of a triangle is the radius r of the largest circle which can fit inside the triangle.

The inradius of a triangle with area a and perimeter p is given by $r = \dfrac{2a}{p}$.

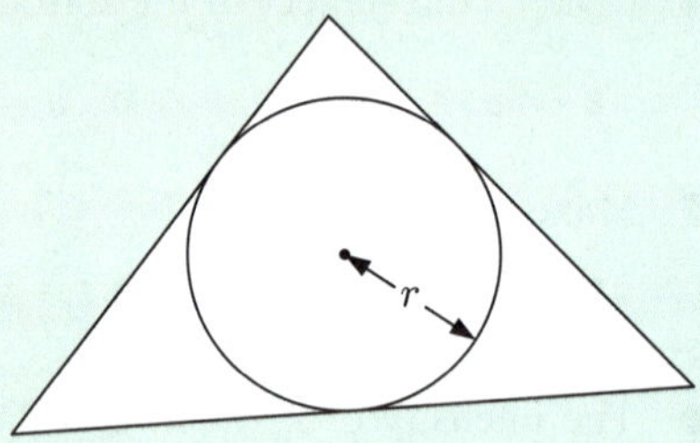

a Rearrange this formula to make p the subject.

b Find the perimeter of a triangle with area 54 cm^2 and inradius 3 cm.

7 The energy output from a square solar panel each day is given by $E = s^2 \times H \times c$ where:

- E is the energy in kilowatt hours (kWh)
- s is the side length of the panel in m
- H is the daily solar radiation received by the panel in kWh/m^2
- c is an efficiency constant.

a Lucy's house has a square solar panel with side length 2.5 m and $c = 0.12$. Her house receives 5.85 kWh/m^2 of solar radiation per day during summer, and 4.35 kWh/m^2 per day during winter.

Find the amount of energy produced by the panel per day in:

i summer

ii winter.

b **i** Rearrange the formula to make H the subject.

ii Bill's square solar panel has side length 2 m and $c = 0.14$. How much solar radiation must the panel receive each day for it to produce 4 kWh of energy daily?

c **i** Rearrange the formula to make s the subject.

ii Tanya's square solar panel has $c = 0.1$. Yesterday the panel received 6.25 kWh/m^2 of solar radiation, and the panel produced 6.4 kWh of energy. Find the side length of the panel.

Chapter 11

Financial mathematics

Contents:

OPENING PROBLEM

Michelle bought an antique clock for $1500.

Its value increased by 4.8% each year until she sold it 5 years later.

Things to think about:

- **a** What do we call it when the value of an item increases over time?
- **b** How can we most efficiently calculate the price Michelle sold the clock for?
- **c** What profit did Michelle make on her investment?

Money plays a large role in our everyday lives. Performing calculations involving money allows us to make sensible decisions when we run a business, buy or sell items, take out a loan, or invest money.

A PERCENTAGE INCREASE OR DECREASE

In problems involving money, prices are often increased or decreased by a **percentage**. We can apply percentage change using a **multiplier**.

Before we change a quantity, its original amount is regarded as 100%.

- If we **increase** a quantity by 15%, then this is **added**, and in total we have $100\% + 15\% = 115\%$ of the original amount.
  So, to increase a quantity by 15%, we multiply the original amount by 115% or 1.15.
  The value 1.15 is called the **multiplier**.
- If we **decrease** a quantity by 15%, then this is **subtracted**, and we are left with $100\% - 15\% = 85\%$ of the original amount.
  So, to decrease a quantity by 15%, we multiply the original amount by 85% or 0.85.
  The value 0.85 is called the **multiplier**.

Example 1 **Self Tutor**

Use a multiplier to:

a increase $4200 by 22%

b decrease $183 by 7%.

a To increase by 22%, we multiply by $100\% + 22\% = 122\%$.

$\therefore$ the new amount $= 122\%$ of $4200
$= 1.22 \times \$4200$
$= \$5124$

b To decrease by 7%, we multiply by $100\% - 7\% = 93\%$.

$\therefore$ the new amount $= 93\%$ of $183
$= 0.93 \times \$183$
$= \$170.19$

In summary, when calculating the change in a quantity, the

original amount × multiplier = new amount

EXERCISE 11A

1 Find the multiplier corresponding to:

- **a** an increase of 25%
- **b** a decrease of 10%
- **c** a decrease of 19%
- **d** an increase of 8.2%
- **e** a decrease of $7\frac{1}{2}\%$
- **f** an increase of 150%.

2 Write down the percentage change corresponding to the multiplier:

- **a** 1.2
- **b** 0.7
- **c** 1.45
- **d** 0.45
- **e** 1.13
- **f** 0.96
- **g** 2.11
- **h** 0.935

3
- **a** Increase \$50 by 5%.
- **b** Decrease \$40 by 30%.
- **c** Decrease £780 by 20%.
- **d** Increase \$68 by 15%.
- **e** Increase \$1600 by 18%.
- **f** Decrease €4500 by 7.5%.

4 Find the percentage change that occurs when:

- **a** \$80 increases to \$100
- **b** €9000 decreases to €7200
- **c** \$85 reduces to \$82
- **d** £96 increases to £108
- **e** €75 000 increases to €86 000
- **f** RM3500 reduces to RM2400.

5 A restaurant applies a 10% surcharge on Sundays, so the price of each item is 10% higher than normal. Lynette's favourite dish costs \$21.90 during the week. How much will it cost on Sundays?

6 At a hardware store, 5 litre paint cans cost \$40. Given that 3 litre paint cans are 35% cheaper, find the cost of the 3 litre cans.

7 Last year a charity received \$562 000 in donations. This year they only received \$497 000. By what percentage have the donations decreased?

8 Jessi's rent increases by 5% at the end of each year. If she currently pays \$220 per week, how much will Jessi be paying two years from now?

B BUSINESS CALCULATIONS

In previous years we have used percentage change to solve problems involving discount, profit and loss, and tax. These calculations are all needed in daily business.

In general:

- A business buys, or produces an item for a particular **cost price**.
- The business applies a **mark-up** to calculate the **marked price** at which the item is offered for sale.
- The business may offer a **discount** from the marked price, to entice customers to buy the item.
- The business sells the item for a particular **selling price**.
- Depending on the item, all of these prices may include a **value-added tax (VAT)** or **goods and services tax (GST)**.
- The value **selling price − cost price** is:
  - **profit** if it is positive
  - **loss** if it is negative
  - zero if the business **breaks even**.

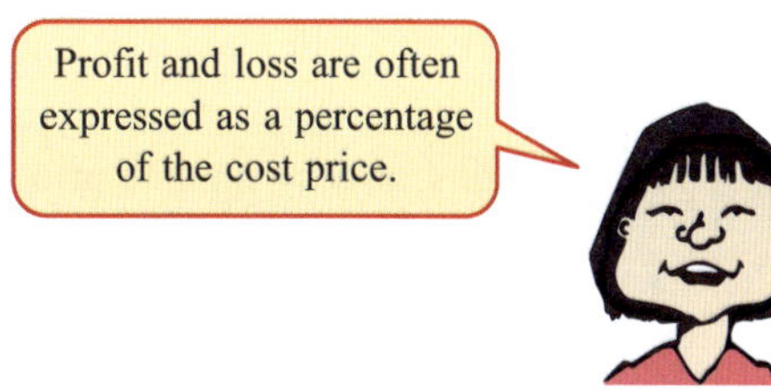

ACTIVITY 1 USING THE PERCENTAGE CHANGE FORMULA

For each of the business calculations we have described, we need to be able to recognise:

- whether it is a percentage increase or decrease
- which variables are the "new amount" and "old amount" in the percentage change formula.

What to do:

1 Copy and complete:

	increase or decrease?	*old amount*	*new amount*
mark-up			
discount			
profit			
loss			

2 GST is added to an *amount excluding GST* to produce an *amount including GST*.
Write a formula connecting the variables, assuming GST is 10%.

Example 2 Self Tutor

Find the selling price of:

a a mirror bought for \$28, and sold at a 60% profit

b a lawnmower bought for \$450, and sold at a 15% loss.

a selling price = cost price × multiplier
= \$28 × 1.6
= \$44.80

The mirror is sold for \$44.80.

b selling price = cost price × multiplier
= \$450 × 0.85
= \$382.50

The lawnmower is sold for \$382.50.

EXERCISE 11B

1 A store bought a shirt for \$15, and sold it at a 60% profit. Find the selling price of the shirt.

2 A bicycle was bought for \$300 and sold at a 25% loss. Find the selling price of the bicycle.

3 A salesman buys a car for \$4800 and marks it up by 35%. Find the marked price of the car.

4 A surfboard manufactured for \$300 is sold to a retail shop for a 40% profit. The shop then applies a 20% mark-up. What price does the customer pay?

Example 3 Self Tutor

A dress marked at \$145 is discounted by 60% for the end of season sale.
Find the selling price of the dress.

selling price = marked price × multiplier
= \$145 × 0.4
= \$58

The dress is sold for \$58.

5 Find the selling price of:

- **a** a business shirt marked at \$58 which is discounted 15%
- **b** a handbag marked at £155 which is discounted 22%
- **c** a refrigerator marked at €850 which is discounted by 18%.

6 An art supplies store buys canvases for \$40, and applies a 40% mark-up. During a sale, a discount is offered of 12.5% for non-members, and 15% for members. Find the difference in the selling price for non-members and members.

Example 4 Self Tutor

A carton of brussels sprouts costs a farmer \$3.35 to produce. He sells the carton to a supermarket for \$4.35. Find the percentage profit.

selling price = cost price × multiplier

$\therefore$ multiplier $= \dfrac{\text{selling price}}{\text{cost price}} = \dfrac{\$4.35}{\$3.35} \approx 1.30$

So, the farmer makes a profit of about 30%.

7 A jewellery store buys a bracelet for \$70 and sells it for \$126. Find the percentage profit.

8 Felipe paid €18 200 for a boat, but because of financial difficulties he was soon forced to sell it for €13 600. Find the percentage loss.

9 Find the percentage discount when:

- **a** a wok is discounted from \$75 to \$69
- **b** a vacuum cleaner with recommended retail price \$160 is discounted to \$124.

10 A department store buys a fan for \$18, marks it up by 30%, then sells it with a 25% discount.

- **a** Find the selling price of the fan.
- **b** Did the department store make a profit or a loss? Explain your answer.
- **c** Write the profit or loss as a percentage of the cost price.

11 A rare book is bought by an antiques dealer for \$200, and he marks its price up by 50% for sale. One of his regular customers is interested in buying the book, so the dealer offers the customer a discounted price of \$252. Find the percentage discount given to the customer.

Example 5 Self Tutor

An electrical goods store buys a TV set at wholesale price. The price of the TV is increased by 25% for sale by the store. Its selling price is now \$550.

For what price did the store buy the TV set?

cost price × multiplier = selling price

$\therefore$ cost price × 1.25 = \$550

$\therefore$ cost price $= \dfrac{\$550}{1.25} = \$440$

So, the store bought the TV set for \$440.

12 A shop applies a 45% mark-up on fresh fish, and sells it for \$45.50 per kg. Find the cost price of the fish.

13 A litre of milk sells for \$1.47 after being marked up by 40%. What was the cost price?

14 A swimsuit is sold for \$44 after a 45% discount is applied. Find the marked price of the swimsuit.

15 A fish shop is forced to sell prawns for \$22.08 per kg, which is a 36% loss. Find the cost price of the prawns.

Example 6 — Self Tutor

A washing machine is sold for \$540 plus 10% GST.

a Find the selling price of the washing machine.

b How much is the GST?

---

a new amount = original amount × multiplier
$= \$540 \times 1.1$
$= \$594$

The washing machine sells for \$594.

b GST = \$594 − \$540
$= \$54$

16 A double bed is sold for €900 plus 20% VAT.

a Find the selling price of the bed.

b How much is the VAT?

17 A marine shop owner in Sydney, Australia has bought a boat and wants to receive \$4500 from selling it. The GST in Australia is 10%.

a For what price must the boat be sold so that the owner will keep \$4500?

b How much GST will be paid on the sale?

18 The price tag on a coat says €226.25. This includes 25% VAT.

a How much will the store keep from the sale of the coat?

b How much of the sale price is paid to the government as VAT?

DISCUSSION

If a price is increased by 10% and the result is increased by 15%, we multiply by 1.1 and then by 1.15.

- By what overall percentage is the original amount increased?
- How can we use multipliers to:
  - apply more than one percentage change in a single step
  - find the overall percentage change?

ACTIVITY 2

Business calculations in the real world are quite complicated because the business owner needs to consider all of the costs associated with their business.

For example, if a clothing store imports a skirt from China for \$6.40 and sells it for \$69.95, the store is not actually making \$63.55 profit.

In this Activity we consider a more realistic model for the sale of this skirt.

What to do:

Consider a store in the USA, where VAT is 10%.

1 The store buys the skirt for \$6.40. It needs to pay VAT on this amount when the skirt arrives in the USA. The freight costs \$1.32 including VAT.
Find the total cost of the imported item, including VAT.

2 The store rents its building. It pays for electricity and water. It pays staff wages. It pays for fitting out the store.
To estimate the component of these costs which can be associated with the skirt, the business owner considers the following:

- The income from the skirt is \$69.95 out of the store's total annual income of \$883 200.
- The store's annual expenses *excluding* the direct cost of importing items, are \$439 700.

a What percentage of the store's income comes from the skirt?

b Find the same percentage of the store's annual expenses excluding the direct cost of importing items.

3 Combine your results from **1** and **2** to estimate the total cost price of the skirt.

4 Find the difference between the selling price \$69.95 and the total cost price you calculated in **3**.

5 The value you calculated in **4** is *not* the profit to the store, because the amount includes 10% VAT which needs to be paid to the government.
Find the actual profit made by the store on the skirt.

6 **a** For what price does the store need to sell the skirt, in order to break even?

b What percentage discount could the store offer on the skirt, without making a loss?

C APPRECIATION AND DEPRECIATION

When the value of an investment or an item such as a house increases, we say it **appreciates** in value.

You may have noticed that the prices of everyday goods and services also increase over time. This is known as **inflation**, and is a form of appreciation.

Example 7 Self Tutor

The inflation rate over the next three years is predicted to be 2%, then 3%, then 3.5%. If an item currently costs \$140, and its cost rises in line with the predicted inflation, find the item's cost in 3 years' time.

$$\text{Cost in 3 years} = \$140 \times 1.02 \times 1.03 \times 1.035$$
$$= \$152.23$$

Electrical equipment, furniture, vehicles, and machinery all lose value over time. This may be because they become damaged or worn, or because their technology is no longer the latest and best available. We say that an item **depreciates** in value.

Example 8 Self Tutor

A computer is bought for \$1500. It depreciates by 25% in the first year, then 30% in the second year. Find the value of the computer after 2 years.

Value after 2 years $= \$1500 \times 0.75 \times 0.7$
$= \$787.50$

EXERCISE 11C

1. The inflation rate is predicted to be 3% and 3.5% over the next two years. What would you expect a bicycle costing \$500 today to cost in two years' time?

2. Three years ago, an ornate vase was valued at \$2000. Over the next three years, it appreciated by 7%, 5%, and 8%. Find the present value of the vase.

3. Sophie bought a car for \$5000. Over the next two years it depreciated by 20% and 15%. Find the value of the car after two years.

4. A photocopier is bought for £3000, and depreciates by 15% each year.
   a. Find the value of the photocopier after 3 years.
   b. How many years will it take for the value of the photocopier to fall below £1500?

5. Kevin and Don each bought a motorbike for €8000. Kevin's motorbike depreciated at 22% per year, and Don's motorbike depreciated at 12% per year. Find the difference between the values of the motorbikes after 3 years.

Example 9 Self Tutor

Over the last two years, an antique dresser has increased in value by 2.6% and then 4.5%. The dresser is now worth \$6718. Find:

a. the value of the dresser two years ago
b. the overall percentage appreciation of the dresser.

a. Suppose the value of the dresser two years ago was $\$x$.

$\therefore$ the current value $= \$x \times 1.026 \times 1.045$

$\therefore \$6718 = \$x \times 1.072\,17$

$\therefore x = \dfrac{6718}{1.072\,17} \approx 6265.80$

The value of the dresser two years ago was about \$6265.80.

b. Using the overall multiplier 1.072 17, the overall appreciation is about 7.22%.

6. Two years ago, Mariah bought a designer handbag. Over the next two years, it depreciated by 10%, then appreciated by 5%.
   If the handbag is valued at \$1512 now, find:
   a. the value of the handbag last year
   b. the price Mariah paid for the handbag.

7 Over the last two years, a collectable figurine has increased in value by 15% and then 11%. The figurine is now worth \$346. Find:

- **a** the value of the figurine two years ago
- **b** the overall percentage appreciation of the figurine.

8 Jamie bought a piece of cricket memorabilia. Over the next 4 years the item appreciated by 5%, depreciated by 4%, depreciated by 8%, and appreciated by 6%. Find the overall percentage appreciation or depreciation of the item over the 4 years.

9 An apartment appreciated by 5% each year for 2 years, then depreciated by 5% each year for 2 years. Find the overall percentage appreciation or depreciation of the apartment.

10 The table below shows the inflation rates in Belgium each year from 2017 to 2020.

Year	2017	2018	2019	2020
Inflation rate	2.22%	2.31%	1.25%	0.43%

Estimate the cost of:

- **a** a dining table in 2019 if it cost €850 in 2016
- **b** an electric scooter in 2018 if it cost €386 in 2020.

11 Phillipe bought some shares for \$150. After 3 years, the shares were valued at \$195. Find:

- **a** the overall percentage appreciation of the shares
- **b** the *average* percentage appreciation each year.

D SIMPLE INTEREST

If you **borrow** money from a bank, you must repay the loan in full, and also pay an additional charge called **interest**.

Likewise, if you **invest** money in a bank account, the bank pays **interest** on the amount you have deposited.

One method for calculating interest is called **simple interest**.

> **Simple interest** is interest that is calculated each year as a fixed percentage of the original amount borrowed or invested.
>
> The fixed percentage is called the **interest rate**, and is usually written as a percentage **per annum** (p.a.), which means "per year".

Suppose \$5000 is invested for 6 years at 8% per annum simple interest.

$$\begin{aligned}\text{The simple interest paid each year} &= 8\% \text{ of } \$5000\\ &= 0.08 \times \$5000 = \$400\\ \therefore\ \text{the simple interest paid for 6 years} &= 0.08 \times \$5000 \times 6 = \$2400\end{aligned}$$

These observations lead to the **simple interest formula**:

> The **simple interest** I can be calculated using the formula $\boldsymbol{I = Pin}$
>
> where P is the **principal** or **initial amount** borrowed or invested,
>
> i is the **flat rate of interest per annum (p.a.)**, and
>
> n is the **time** or **duration** of the loan or investment in **years**.

Example 10 — Self Tutor

Find the simple interest payable on a loan of $60 000 at 9% p.a. over:

a 4 years **b** 18 months.

a $P = 60\,000$
$i = 9\% = 0.09$
$n = 4$
Now $I = Pin$
$\therefore\ I = 60\,000 \times 0.09 \times 4$
$\therefore\ I = 21\,600$
$\therefore$ the simple interest is $21 600.

b $P = 60\,000$
$i = 9\% = 0.09$
$n = \frac{18}{12} = 1.5$
Now $I = Pin$
$\therefore\ I = 60\,000 \times 0.09 \times 1.5$
$\therefore\ I = 8100$
$\therefore$ the simple interest is $8100.

EXERCISE 11D

1 Find the simple interest payable on an investment of:

a $3000 at 11% p.a. for 2 years
b $5500 at $7\frac{1}{2}\%$ p.a. for 5 years
c $25 000 at 6.2% p.a. for 10 years
d $15 000 at 5.75% p.a. for $4\frac{1}{2}$ years.

2 Calculate the simple interest payable on a loan of $48 000 at 4.5% p.a. for 6 years.

3 Find the simple interest payable on a loan of:

a $2000 at 4.2% p.a. over 9 months
b $8500 at 8% p.a. over 3 months
c $2 750 000 at 7% p.a. over 18 months
d $60 000 at 3.25% p.a. over 10 months.

Example 11 — Self Tutor

Alicia borrows $9000 for 6 years at 3.8% p.a. simple interest.

a Calculate the total amount she must repay.

b If Alicia repays the loan in equal monthly repayments over the 6 years, find the amount she must repay each month.

a $P = 9000$
$i = 3.8\% = 0.038$
$n = 6$

Now $I = Pin$
$= 9000 \times 0.038 \times 6$
$= 2052$

The total amount to be repaid $=$ initial amount $+$ interest
$= \$9000 + \$2052$
$= \$11\,052$

b Alicia repays the loan over 6 years $= 72$ months.

$\therefore$ each monthly repayment $= \dfrac{\$11\,052}{72}$
$= \$153.50$

4 Calculate the total amount to be repaid if:

- **a** \$14 000 is borrowed for 2 years at 6.5% p.a. simple interest
- **b** \$8200 is borrowed for 4 months at 3.8% p.a. simple interest.

5 Calculate the balance in an account if:

- **a** \$7500 is deposited for $2\frac{1}{2}$ years at 2.2% p.a. simple interest
- **b** \$23 000 is deposited for 4 months at 1.5% p.a. simple interest.

6 Kyle borrows \$25 000 at 6% p.a. simple interest over 4 years. Find:

- **a** the total amount needed to repay the loan
- **b** his equal monthly repayments.

7 Cameron borrows \$15 000 from a bank to renovate his house. He borrows the money at 9% p.a. simple interest over 3 years. What are his equal monthly repayments?

8 An electric guitar with all attachments is advertised at \$2400. Ben can afford to pay a deposit of \$600, but he then has to borrow the remainder at 12% p.a. simple interest over 18 months. What are his equal monthly repayments?

9 Rearrange the simple interest formula $I = Pin$ to make the subject:

- **a** P
- **b** i
- **c** n

10 \$1260 interest is charged on a loan at 4% p.a. simple interest over 7 years. Find the amount originally borrowed.

11 A simple interest investment at 6.5% p.a. earns \$975 interest over 4 years. Find the amount originally invested.

12 An investor wants to earn \$3500 interest in 8 months. How much would he need to invest given an interest rate of $7\frac{3}{4}$% p.a. simple interest?

13 Jenny took out a loan at 8% p.a. simple interest for 6 years. She paid a total of \$1680 interest on the loan.

- **a** How much money did she borrow?
- **b** Find her equal monthly repayments.

14 Find the simple interest rate if:

- **a** \$1960 interest is earned on an investment of \$3500 over 8 years
- **b** \$1000 interest is charged on a loan of \$8800 over 20 months.

15 Edward deposits \$3000 in an account paying simple interest. After 18 months, he closes the account and withdraws the total balance of \$3500. Find the simple interest rate for Edward's investment.

16 Find the time taken for an investment of \$6500 at 4.8% p.a. simple interest to earn \$2808 interest.

17 Brian borrows \$75 000 for home renovations. He agrees to pay 3.2% p.a. simple interest. The total repayments on Brian's loan are \$90 000. What is the duration of Brian's loan?

18 Helen borrowed \$80 000 at 7.3% p.a. simple interest to start her crafts business. In total, her repayments will be \$117 960.

- **a** How long will it take Helen to repay the loan?
- **b** Find Helen's equal monthly repayments.

19 Jacinta borrows \$35 000 at 2.6% p.a. simple interest. Her equal monthly repayments are \$909.17. Find the duration of Jacinta's loan.

E COMPOUND INTEREST

A more common method for calculating interest is **compound interest**.

If you leave your money in the bank for a period of time, the interest is automatically added to your account.

With compound interest, any interest that is added to your account will also earn interest in the next time period.

Compound interest is calculated as a percentage of the total amount at the end of the previous compounding period.

Suppose $8000 is placed in an account earning 5% p.a. compound interest.

We can show this in a table:

Year	*Amount at beginning of year*	*Compound interest*	*Amount at end of year*
1	$8000	5% of $8000 = $400	$8000 + $400 = $8400
2	$8400	5% of $8400 = $420	$8400 + $420 = $8820
3	$8820	5% of $8820 = $441	$8820 + $441 = $9261

After 3 years there is a total of $9261 in the account. We have earned $1261 in compound interest.

Notice that the amount of compound interest increases each compounding period. This is because compound interest allows you to earn *interest on interest*.

EXERCISE 11E.1

1 Suppose $16 000 is invested at 6% p.a. compound interest for 4 years.

a Copy and complete:

Year	*Initial amount*	*Interest*	*Final amount*
1	$16 000	6% of $16 000 = $960	
2			
3			
4			

Round your calculations to the nearest cent.

b Hence find:

i the final value of the investment after 4 years

ii the total interest earned in the 4 year period.

2 Consider an investment of $10 000 for 3 years.

a Calculate the interest payable on the investment at:

i 4.5% p.a. simple interest   ii 4.5% p.a. compound interest.

b Which method of calculating interest pays more? Explain why this occurs.

3 Use a table to determine the final value of:

a a $4000 investment at 10% p.a. compound interest for 3 years

b a $20 000 investment at 3.8% p.a. compound interest for 4 years.

4 Use a table to help determine the interest earned by an investment of $5000 at 7.5% p.a. compound interest for 5 years.

ACTIVITY 3 COMPOUND INTEREST SPREADSHEET

A **spreadsheet** is a convenient tool which allows us to construct a compound interest table very rapidly.

Click on the icon to obtain instructions for this Activity.

ACTIVITY

THE COMPOUND INTEREST FORMULA

Consider again an investment of \$8000 placed in an account earning 5% p.a. compound interest for 3 years. Each year, the amount in the account increases by 5%, so the value after 3 years is

$$\begin{aligned} &\$8000 \times 1.05 \times 1.05 \times 1.05 \\ &= \$8000 \times (1.05)^3 \\ &= \$9261 \end{aligned}$$

Notice that $1.05 = 1 + 0.05 = 1 + i$ where i is the annual interest rate written as a decimal.

These observations lead to the **compound interest formula**:

> The **future value** F of a compound interest investment is given by
>
> $$\boldsymbol{F = P(1+i)^n}$$
>
> where P is the **present value** or amount initially invested
> i is the **rate of interest per annum (p.a.)**, and
> n is the **duration** of the investment in *years*.
>
> The **interest** earned over the duration of the investment is given by $\boldsymbol{F - P}$.

Example 12 Self Tutor

\$6000 is invested for 4 years at 8% p.a. compound interest. Find:

a the value of the investment after 4 years **b** the interest earned.

a $P = 6000, \quad i = 8\% = 0.08, \quad n = 4$

Now $F = P(1+i)^n$

$\therefore \; F = 6000 \times (1.08)^4$

$\therefore \; F = 8162.93$

The value after 4 years is \$8162.93.

b Interest earned $= F - P$

$= 8162.93 - 6000$

$= 2162.93$

So, \$2162.93 interest is earned.

EXERCISE 11E.2

1 Sunil invested \$4000 at 5% p.a. compound interest. Find the value of the investment after 2 years.

2 Cassandra invested \$7000 at 3.5% p.a. compound interest for 2 years. Find:

a the value of the investment after 2 years **b** the interest earned.

3 Donna borrows \$6500 for 3 years at 4% p.a. compound interest.

a Find the total amount Donna must repay after 3 years.

b In total, how much interest will Donna be charged?

4 How much compound interest is earned by investing \$9000 for 3 years at 4.5% p.a.?

5 Suppose you have \$8000 to invest for 3 years. You have been offered two investment options:

Option 1: Invest at 9% p.a. simple interest.

Option 2: Invest at 8% p.a. compound interest.

- **a** Calculate the final value of each investment, and decide which option you would choose.
- **b** Would you change your decision if you were investing for 5 years?

6 Helen invests \$2000 in an account which offers 6% p.a. compound interest in the first 2 years, and then 3.5% p.a. compound interest for each subsequent year.

- **a** Find the value of the investment after 4 years.
- **b** Would Helen have been better off investing in an account which paid a constant compound interest rate of 5% p.a.?

7
- **a** Rearrange the compound interest formula to make P the subject.
- **b** Hence find the initial investment needed so that the final value will be:
  - **i** \$8427 after 2 years at 6% p.a. compound interest
  - **ii** \$132 487.08 after 6 years at 5.7% p.a. compound interest.

8 Abdul requires \$20 000 in 4 years' time. His bank offers an interest rate of 6.8% p.a. compound interest. How much does Abdul need to invest now?

9 Terry invests a sum of money at 3.7% p.a. compound interest. After 4 years he withdraws \$2000, and he deposits the remaining balance into an account earning 4.5% p.a. compound interest. After 3 more years, there is \$5635.66 in the account. How much did Terry initially invest?

10 A compound interest investment of \$7000 amounted to \$8300 after 3 years. Find the rate of interest per annum.

GLOBAL CONTEXT — PAYING OFF A HOME LOAN

Global context: Globalisation and sustainability

Statement of inquiry: Decision-making can be improved by using a model to investigate possible options.

Criterion: Applying mathematics in real-life contexts

GLOBAL CONTEXT

MULTIPLE CHOICE QUIZ

QUICK QUIZ

REVIEW SET 11A

1 Find the multiplier corresponding to:

- **a** an increase of 37%
- **b** a decrease of 9.5%.

2 Find the percentage change that occurs when:

a \$40 decreases to \$34 **b** €160 increases to €190.

3 Eito purchased a pair of shoes for \$200 and sold them at a 45% profit. Find the selling price of the shoes.

4 Moira bought a car for \$4500 but had to sell it for \$4000 soon after. Find the percentage loss.

5 A heater is sold for \$68 after a 15% discount is applied. Find the marked price of the heater.

6 A publisher sells a book for \$20 to a retailer. The retailer marks up the price by 75%, then adds a further 17.5% VAT. What price does the customer pay?

7 I bought a house for \$420 000. Over the next three years it increased in value by 9%, decreased by 3%, and increased by 12%. Find the value of the house after these three years.

8 Over the last two years, an antique vase has increased in value by 6% and then 4%. The vase is now worth \$620. Find:

a the value of the vase two years ago

b the overall percentage appreciation of the vase.

9 When Luke completed his university degree three years ago, he had a student loan of €22 000 to repay. Over the next three years, the inflation rate was 2.2%, 1.9%, and 2.6%.

a If Luke's loan rose in line with inflation, find the amount he has to repay now.

b Luke will receive a 10% discount on his loan if he repays the loan in a single payment. How much will he need to pay to repay the loan in a single payment?

10 Raj borrows \$5000 from a bank at 8.5% p.a. simple interest. He will repay the money over a 4 year period, with equal monthly repayments.

a Find the total amount that Raj must repay the bank.

b Find Raj's equal monthly repayments.

11 Sally has \$12 000 to invest for 10 years. She is offered an investment which pays 5% p.a. compound interest.

a Find the value of the investment after 10 years.

b How much interest will she earn?

12 Kelly would like to earn \$3000 interest on an investment of \$17 000 over 3 years. Find the minimum rate of simple interest that will achieve this target.

13 Four years ago, Alexander invested a sum of money in an account paying 5.2% p.a. compound interest. There is currently \$7593.72 in the account. How much did Alexander invest initially?

14 Carlos has €4000 to invest over 5 years. Banks A and B offer the following rates:

Bank A: 8% p.a. simple interest

Bank B: 6.7% p.a. compound interest

a Find the final value of Carlos' investment if he uses:

i bank A **ii** bank B.

b Which bank should Carlos invest his money with? Explain your answer.

c Would Carlos' decision change if he were investing for 7 years?

REVIEW SET 11B

1 Write down the percentage change corresponding to the multiplier:

a 1.7 **b** 0.71 **c** 2.45

2 **a** Increase \$70 by 12%. **b** Decrease £2500 by 36%.

3 A laptop was bought for \$550 and sold at a 35% loss. Find the selling price of the laptop.

4 Find the percentage discount when a speaker is discounted from \$175 to \$147.

5 A furniture store bought a chair for \$380, marked it up by 35%, and then discounted it by 25%. Find the selling price of the chair.

6 The price tag on a cupboard says \$253, including 10% GST.

a How much will the store keep from the sale of the cupboard?

b How much of the sale price is paid to the government as GST?

7 The annual rate of inflation is predicted to be 3% next year, then 3.5% in the year after that. If an item currently costs \$50, estimate its cost in two years' time.

8 A truck depreciates by 15% each year for three years. Find the overall percentage depreciation of the truck.

9 Find the simple interest earned by an investment of \$4000 at 7% p.a. for 18 months.

10 How long would it take to earn \$5000 interest on an investment of \$22 500 at a simple interest rate of 9.5% p.a.?

11 You have \$5000 to invest for 4 years. Which of these options should you take? Explain your answer.

A 7% p.a. simple interest **B** 6% p.a. compound interest

12 Denise invests £30 000 in an account paying 4.6% p.a. compound interest.

a Find the value of the investment after 3 years.

b How many years will it take for the investment to be worth at least £40 000?

13 Jenna bought a painting for €4000. The painting appreciated by 5% each year for 2 years.

a Find the overall percentage appreciation of the painting.

b Find the value of the painting after 2 years.

c Jenna sold the painting after 2 years. She used the money to purchase solar panels for her home, with €200 left over.

i Find the price Jenna paid for the solar panels.

ii Jenna bought the solar panels during a 25% discount sale. Find the marked price of the solar panels.

14 Paula has \$8000 to invest over 4 years. She places it in an account which pays 6% p.a. compound interest.

a Find the value of the investment after 4 years.

b How much interest will she earn?

c What rate of simple interest would be required to earn the same amount of interest in this time?

Chapter 12

Measurement: Length and area

Contents:

OPENING PROBLEM

This diagram shows the required dimensions for a helicopter landing site.

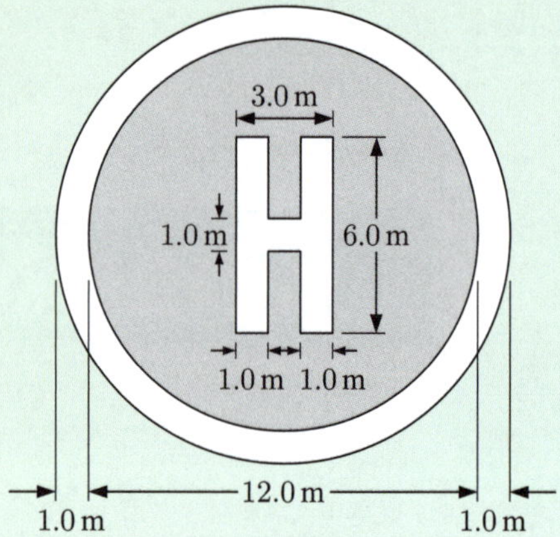

Things to think about:

- **a** What is the distance around the outer boundary of the landing site?
- **b** What is the area of the "H" symbol?
- **c** What percentage of the landing site is painted white?

A UNITS OF LENGTH

The base unit of length in the metric system is the **metre** (m). Other units of length based on the metre are connected as follows:

1 kilometre (km) = 1000 metres (m)
1 metre (m) = 100 centimetres (cm)
1 centimetre (cm) = 10 millimetres (mm)

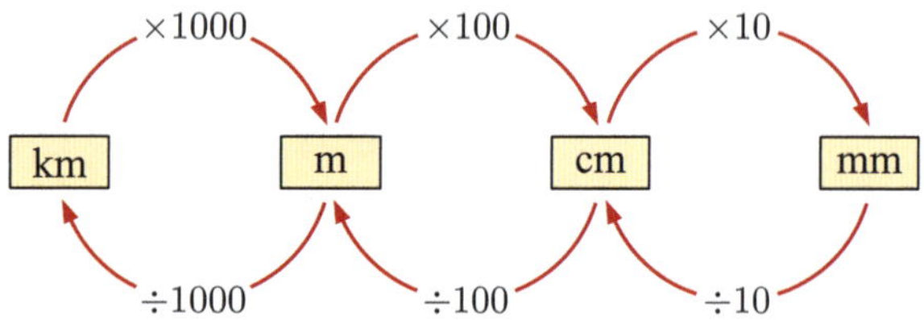

Notice that when we convert:

- larger units into smaller units we **multiply**
- smaller units into larger units we **divide**.

Example 1 — Self Tutor

Convert: **a** 39.4 m into cm **b** 567 mm into cm.

a 39.4 m
$= (39.4 \times 100)$ cm
$= 3940$ cm

b 567 mm
$= (567 \div 10)$ cm
$= 56.7$ cm

EXERCISE 12A

1 Convert:

a 52 km into m	**b** 480 cm into m	**c** 115 cm into mm
d 54 mm into cm	**e** 1.65 m into cm	**f** 5280 m into km
g 4.21 km into m	**h** 42.5 cm into m	**i** 2000 mm into m
j 6.3 m into mm	**k** 7 000 000 mm into km	**l** 0.25 km into mm.

2 Mike ran 8.5 km while Mal walked 3200 m. How much farther did Mike run than Mal walk?

3 I have 45 coils of fencing wire. Each coil is 275 m long. How many kilometres of wire do I have?

4 John has 3.85 km of copper wire. His machine will cut it into 1.5 cm lengths for use in electric toasters. How many lengths will be cut?

5 When Lucy walks, the length of her stride is 80 cm. Every morning she walks 1.7 km to school. How many steps does she take?

6 The height of a pack of 52 cards is 1.95 cm. Find:

- a the thickness, in millimetres, of a single card
- b the height, in metres, of 60 packs stacked on top of one another
- c the number of cards required to form a stack 12 cm high.

B PERIMETER

The **perimeter** of a closed figure is the total distance around its **boundary**.

POLYGONS

The perimeter of a polygon is the sum of the lengths of its sides.

Example 2 Self Tutor

Find the perimeter of the figure:

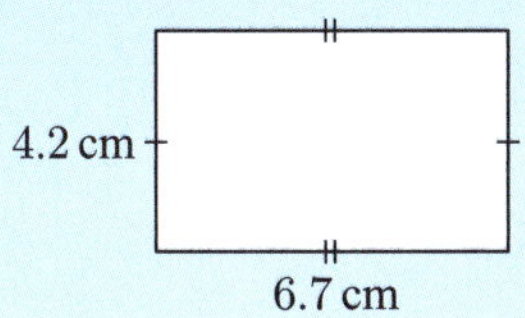

$$\begin{aligned}\text{Perimeter} &= (2 \times 4.2 + 2 \times 6.7) \text{ cm} \\ &= (8.4 + 13.4) \text{ cm} \\ &= 21.8 \text{ cm}\end{aligned}$$

CIRCLES

In the case of the circle, the perimeter is given the special name **circumference**.

Circumference $\boldsymbol{C = 2\pi r}$
or $\boldsymbol{C = \pi d}$

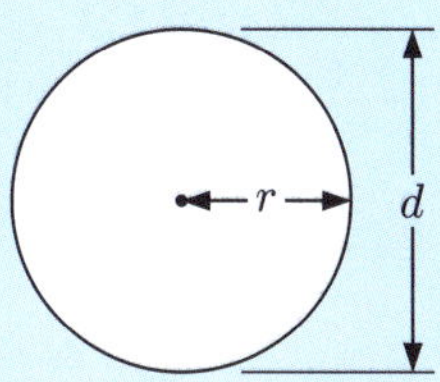

ARC LENGTH

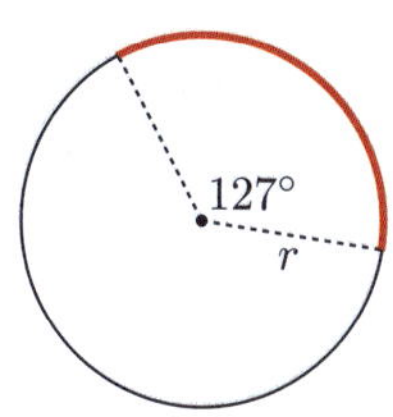

Consider the red arc shown. The angle at the centre of the circle is 127°.

Since there are 360° in the whole circle, the arc makes up $\frac{127}{360}$ of the circle's circumference.

So, the length of the arc is $\frac{127}{360} \times 2\pi r$.

Arc length $l = \dfrac{\theta}{360} \times 2\pi r$

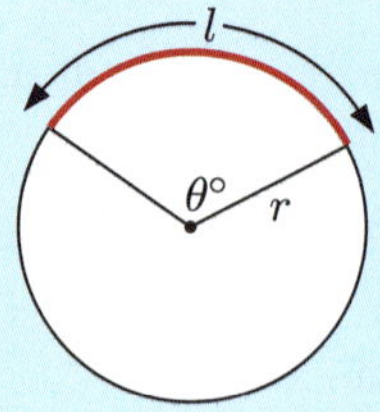

Example 3 — Self Tutor

Find the perimeter of the following figures:

a

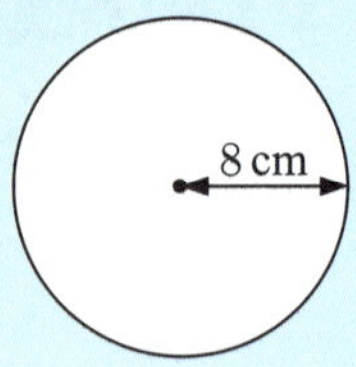

b

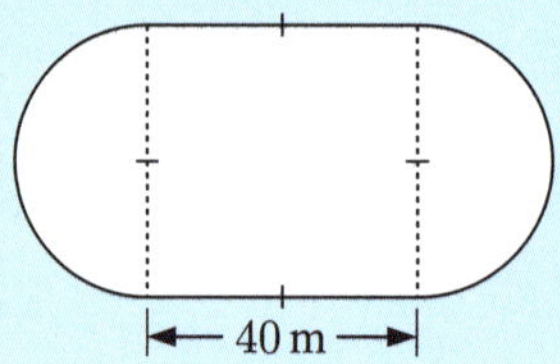

c

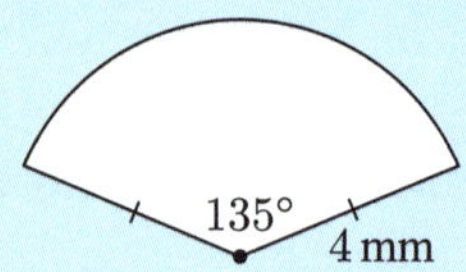

a Circumference
$= 2\pi r$
$= 2\pi \times 8$ cm
$= 16\pi$ cm
≈ 50.3 cm

b Perimeter
$=$ circumference of circle + straight sides
$= (\pi \times 40 + 2 \times 40)$ m
$= (40\pi + 80)$ m
≈ 206 m

c Perimeter
$=$ length of arc + straight sides
$= \left(\frac{135}{360} \times 2 \times \pi \times 4 + 2 \times 4\right)$ mm
≈ 17.4 mm

EXERCISE 12B

1 Find the perimeter of each figure:

a

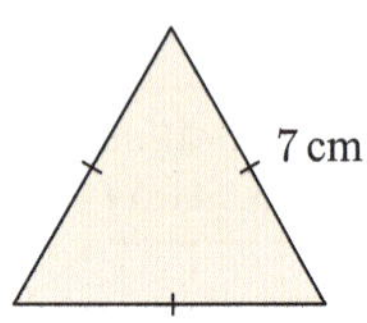

b

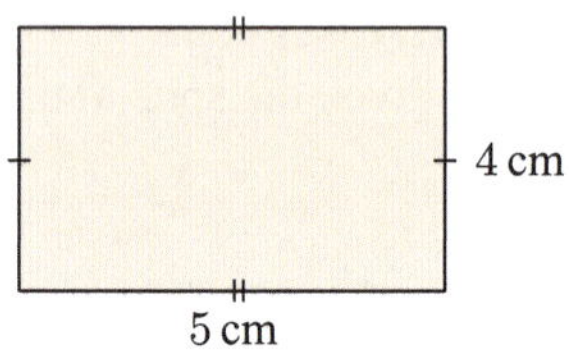

c

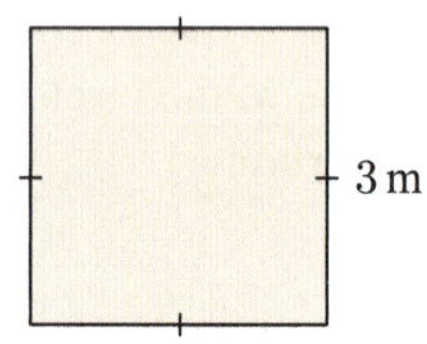

d

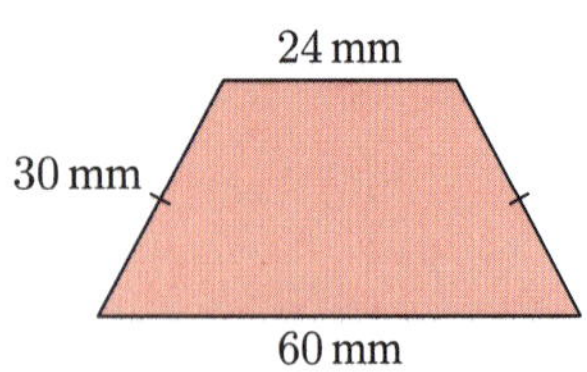

e

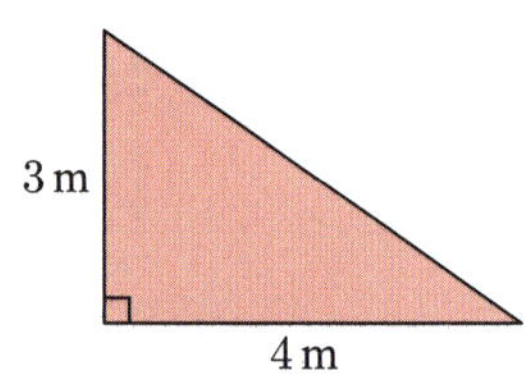

f

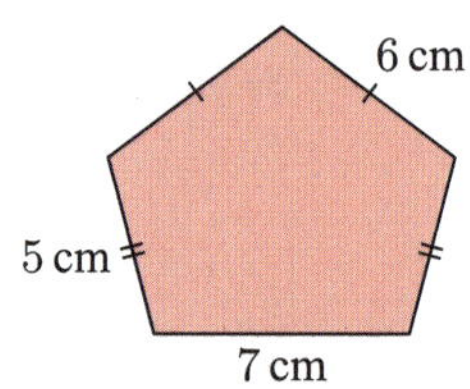

g

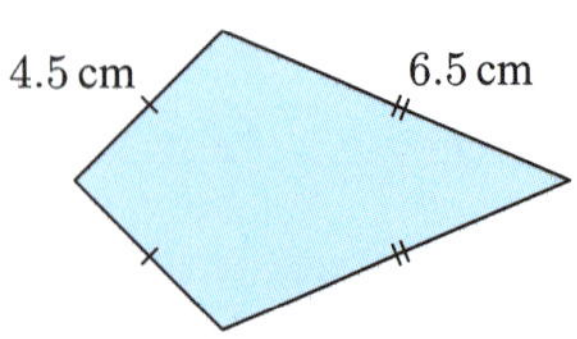

h

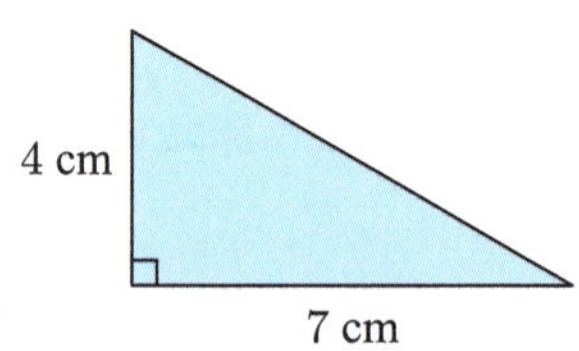

i

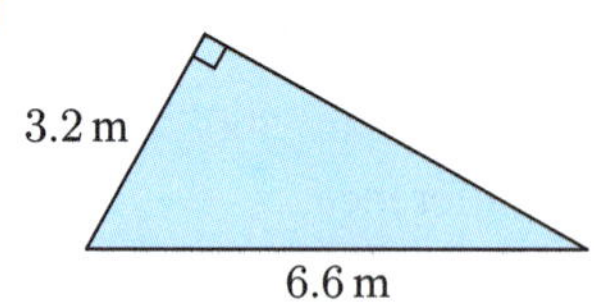

2 Find the perimeter of each figure:

a

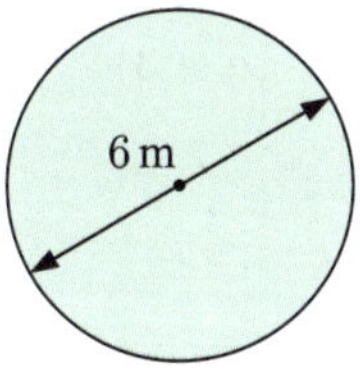

b

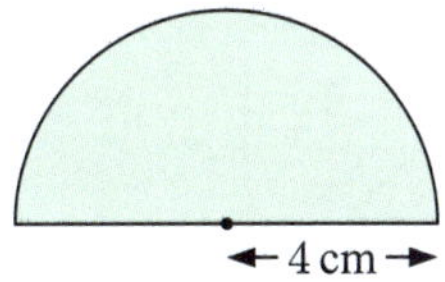

c

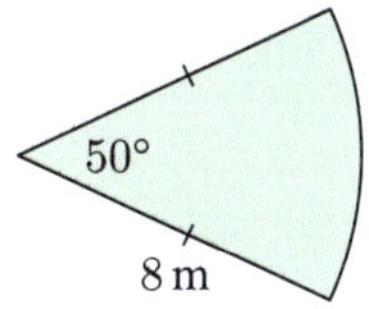

d

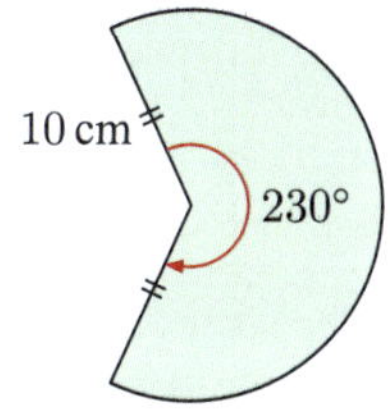

3 Find:

- **a** the radius of a circle with circumference 18π m
- **b** the side length of a square with perimeter 76 mm
- **c** the circumference of a circle with radius 4.2 cm
- **d** the diameter of a circle with circumference 18.6 cm
- **e** the perimeter of a semi-circle with diameter 1.2 m.

4 Find the perimeter of each figure:

a

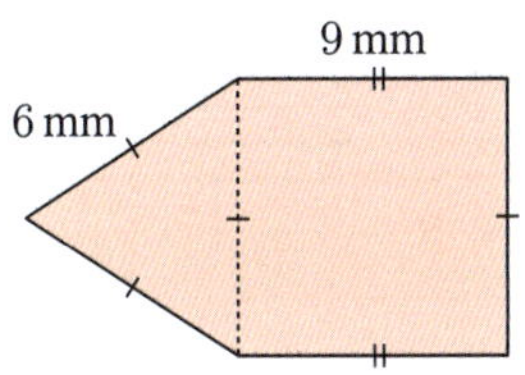

b

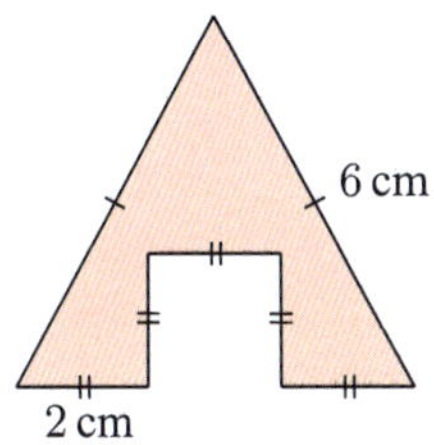

c

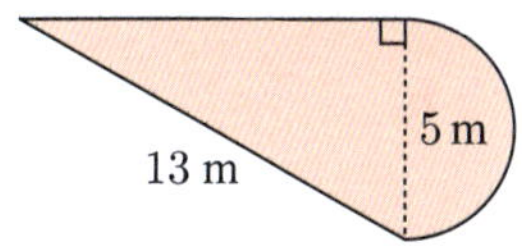

d

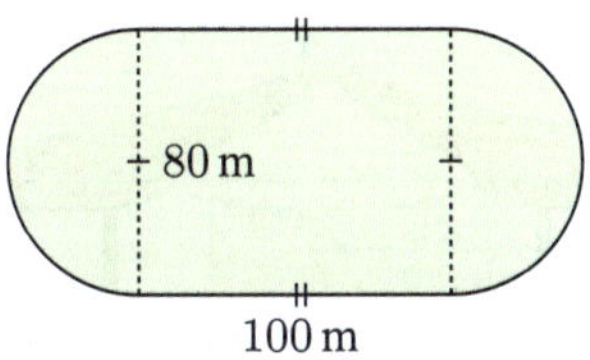

e

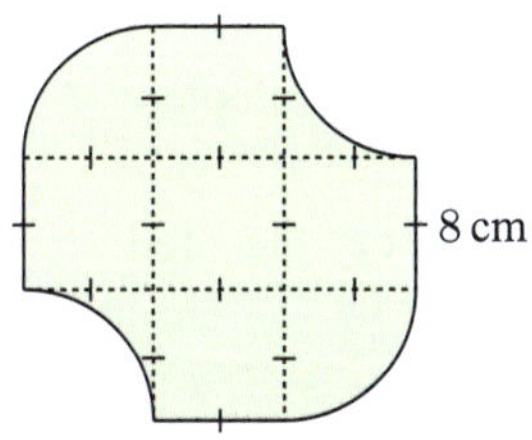

f

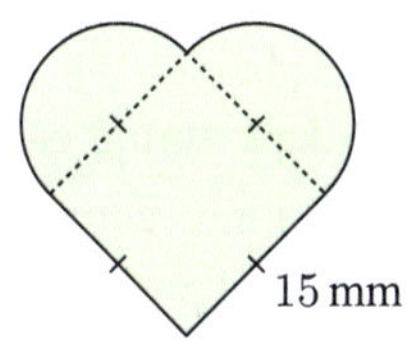

g

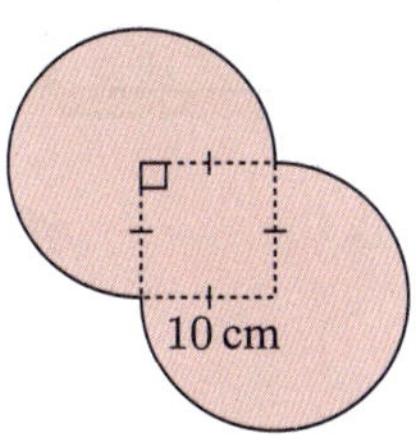

h

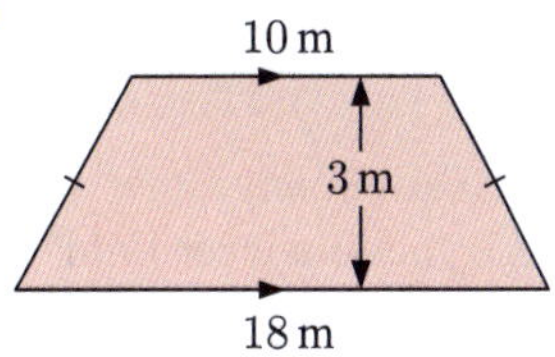

i

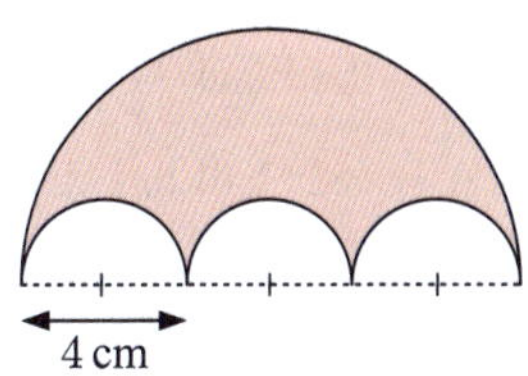

5 Find the external perimeter of the house in this plan.

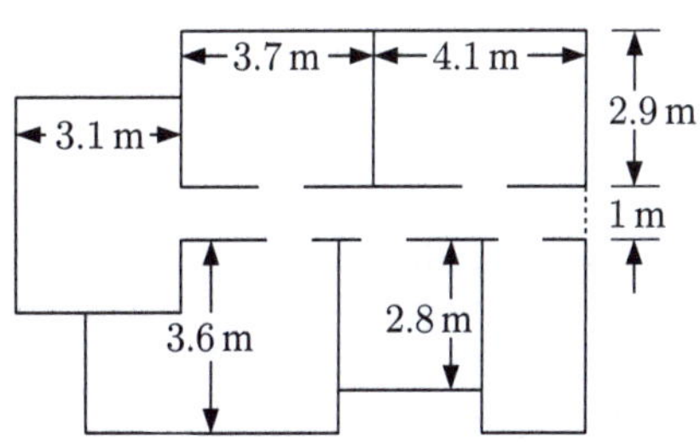

Example 4 **Self Tutor**

Write a formula for the perimeter P of:

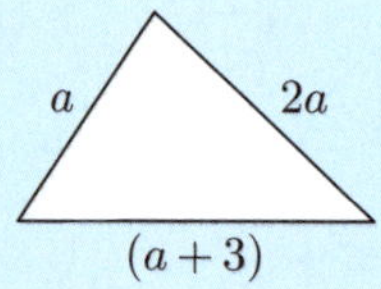

$P = a + 2a + (a+3)$

$\therefore\ P = 4a + 3$

6 Write a formula for the perimeter P of each figure:

a

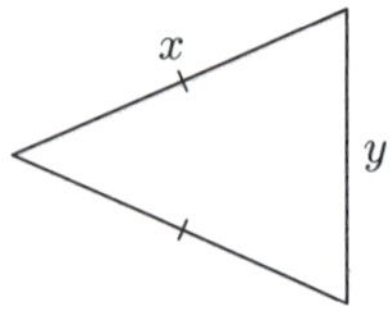

b

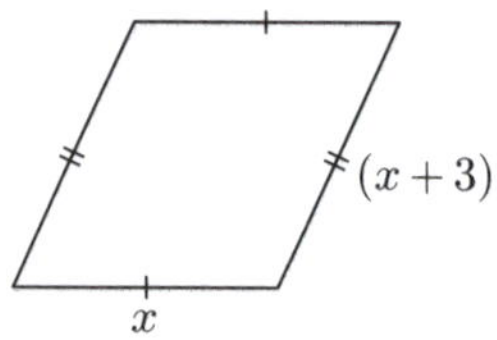

c

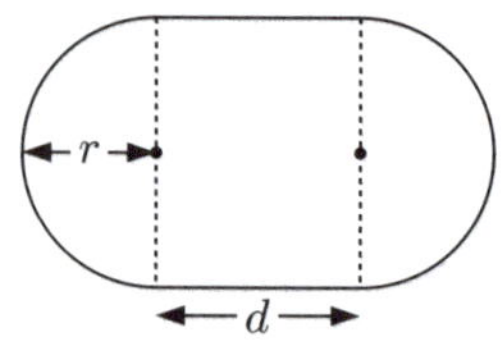

7 An Olympic sized boxing ring has dimensions 6.1 m by 6.1 m, and requires four ropes around each side as shown. The rope costs \$39.50 per metre. Find:

- **a** the perimeter of the boxing ring
- **b** the total length of rope around the ring
- **c** the total cost of the rope.

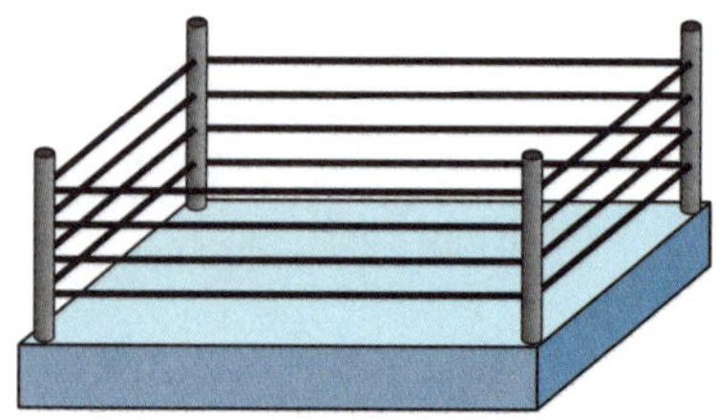

8 A rectangular room is 3.5 m by 4.8 m. It has a door 80 cm wide, and a window 1.3 m wide. A wallpaper frieze runs around the room at eye level, so it is broken for the door and window. Find the total length of frieze used in the room.

9 The horses on this merry-go-round complete 20 laps during each ride.

- **a** Find, correct to 1 decimal place, the total distance travelled during a ride by:
  - **i** the black horse
  - **ii** the white horse.
- **b** How much farther than the white horse does the black horse travel?

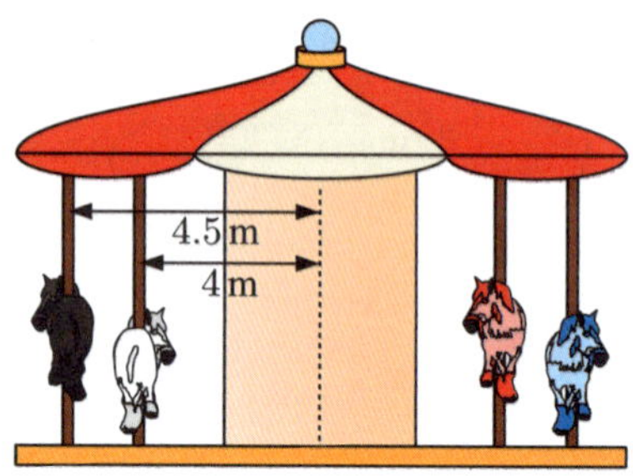

10 A soccer ball has diameter 24 cm. How many times must it roll along the ground to travel the length of a 100 metre soccer field?

11 A cone is made from this net.
Find:

- **a** the length of the arc
- **b** the perimeter of the net
- **c** the radius of the cone.

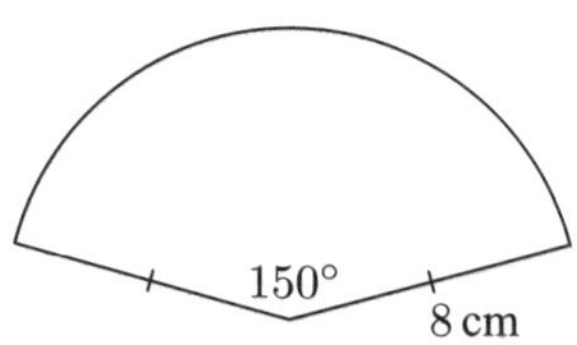

12 Three plastic pipes, each with diameter 10 cm, are held together by straps. Find the length of each strap to the nearest centimetre.

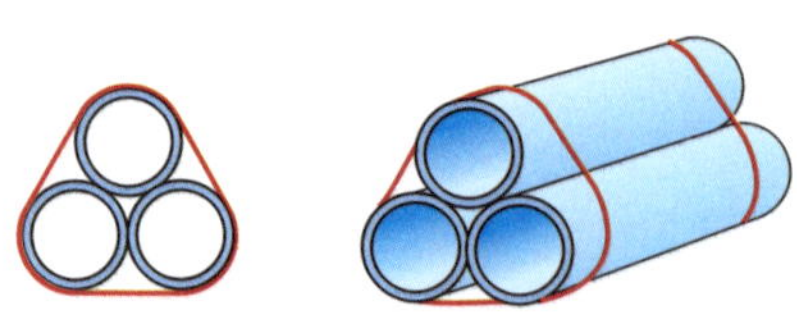

13 A cyclist rides a bicycle with 88 pedal revolutions per minute. Each pedal revolution results in 3.2 revolutions of the wheels. The wheels have diameter 70 cm. How long will it take the cyclist to travel 40 km?

C UNITS OF AREA

Area is the measurement of the size of a surface.

In the metric system, area is measured using the *squares* of the length units. We also use a unit called a **hectare** (ha).

$$1 \text{ mm}^2 = 1 \text{ mm} \times 1 \text{ mm}$$
$$1 \text{ cm}^2 = 10 \text{ mm} \times 10 \text{ mm} = 100 \text{ mm}^2$$
$$1 \text{ m}^2 = 100 \text{ cm} \times 100 \text{ cm} = 10\,000 \text{ cm}^2$$
$$1 \text{ ha} = 100 \text{ m} \times 100 \text{ m} = 10\,000 \text{ m}^2$$
$$1 \text{ km}^2 = 1000 \text{ m} \times 1000 \text{ m} = 1\,000\,000 \text{ m}^2 \text{ or } 100 \text{ ha}$$

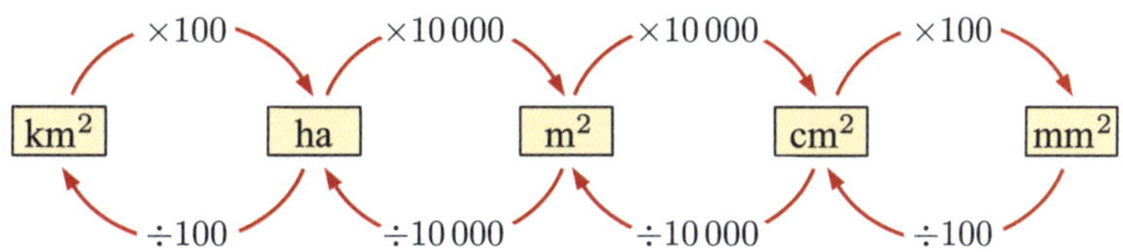

Example 5 Self Tutor

Convert: **a** 6 m^2 into cm^2 **b** $18\,500 \text{ m}^2$ into ha.

a 6 m^2
$= (6 \times 10\,000) \text{ cm}^2$
$= 60\,000 \text{ cm}^2$

b $18\,500 \text{ m}^2$
$= (18\,500. \div 10\,000) \text{ ha}$
$= 1.85 \text{ ha}$

EXERCISE 12C

1 Convert:

a 23 mm^2 into cm^2 **b** 3.6 ha into m^2 **c** 726 cm^2 into m^2
d 7.6 m^2 into mm^2 **e** 8530 m^2 into ha **f** 0.354 ha into cm^2
g 13.54 cm^2 into mm^2 **h** 432 m^2 into cm^2 **i** $0.004\,82 \text{ m}^2$ into mm^2.

2 Convert:

a 3 km^2 into m^2 **b** 0.7 km^2 into ha **c** 660 ha into m^2
d 660 ha into km^2 **e** 0.05 m^2 into cm^2 **f** 25 cm^2 into m^2
g 5.2 mm^2 into cm^2 **h** 0.72 km^2 into mm^2.

3 Sam purchased 2 m^2 of material. She needs to cut it into rectangles of area 200 cm^2 for a patchwork quilt. This can be done with no waste. How many rectangles can Sam cut out?

4 I have purchased a 4.2 ha property. Council regulations allow me to have 5 free range chickens for every 100 m^2. How many free range chickens am I allowed to have?

D AREA OF POLYGONS

In previous years you should have seen the following area formulae:

Shape	*Figure*	*Formula*
Rectangle	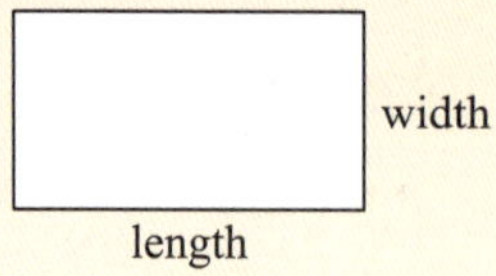	**Area = length × width**
Triangle	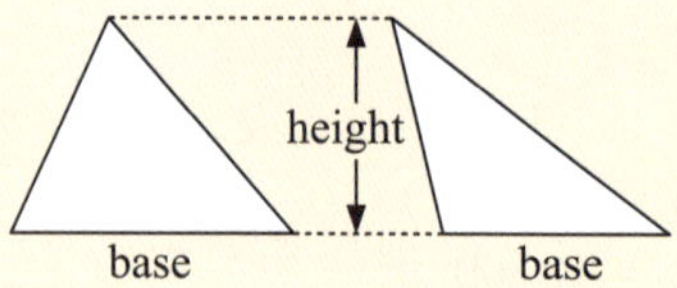	**Area** $= \frac{1}{2} \times$ **base** $\times$ **height**
Parallelogram	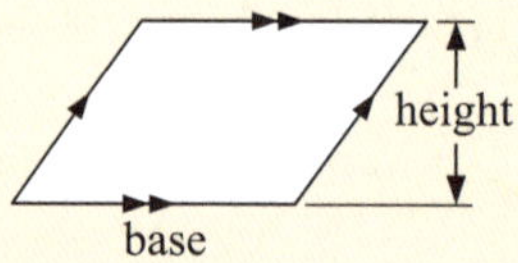	**Area = base × height**
Trapezium	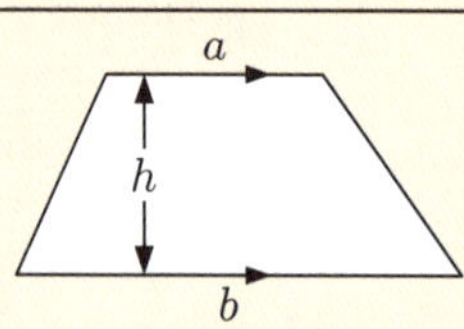	**Area** $= \left(\frac{a+b}{2}\right) \times h$
Kite	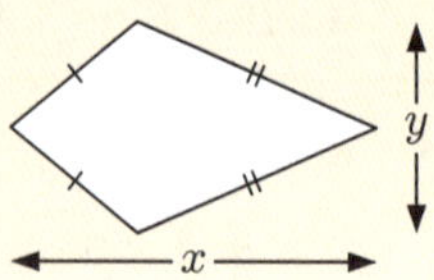	**Area** $= \frac{1}{2} \times x \times y$

DEMO

DEMO

DEMO

DEMO

Example 6 Self Tutor

Find the area of each figure:

a
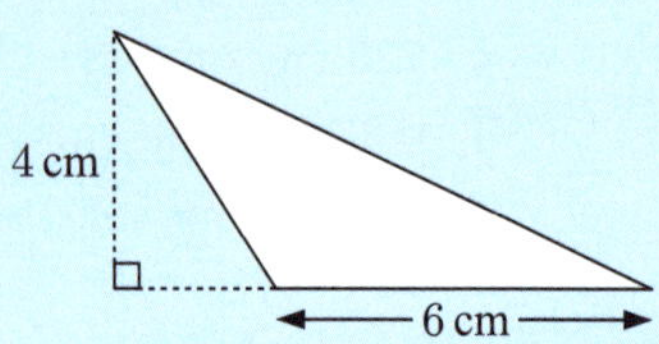

b
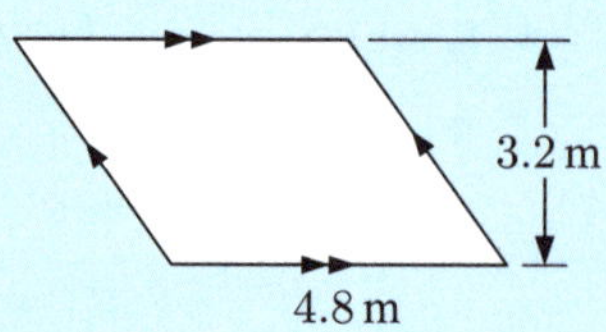

c
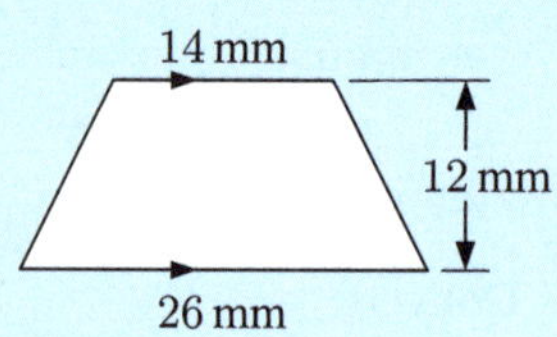

a Area

$= \frac{1}{2} \times \text{base} \times \text{height}$

$= \frac{1}{2} \times 6 \times 4 \text{ cm}^2$

$= 12 \text{ cm}^2$

b Area

$= \text{base} \times \text{height}$

$= 4.8 \times 3.2 \text{ m}^2$

$= 15.36 \text{ m}^2$

c Area

$= \left(\frac{a+b}{2}\right) \times h$

$= \left(\frac{14+26}{2}\right) \times 12 \text{ mm}^2$

$= 240 \text{ mm}^2$

EXERCISE 12D

1 Find the area of each figure:

a
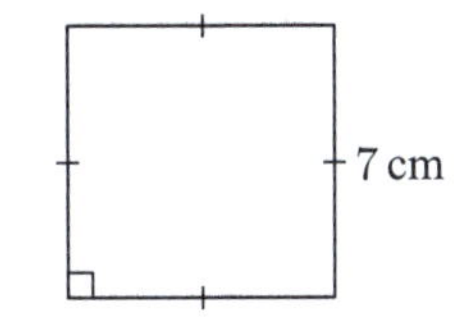

b
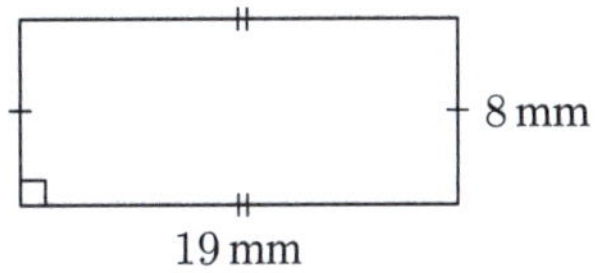

c
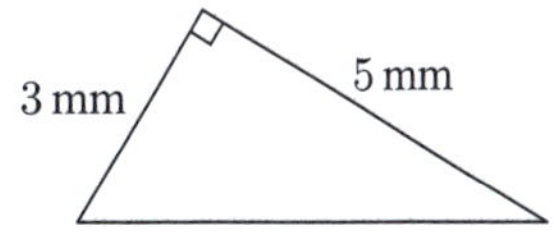

d
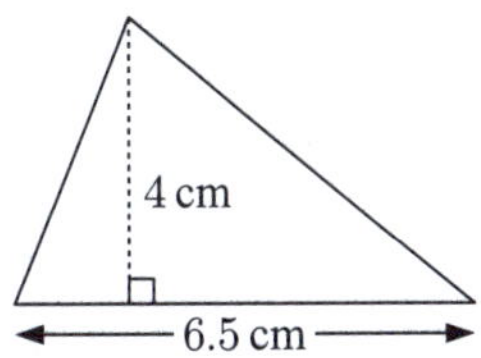

e
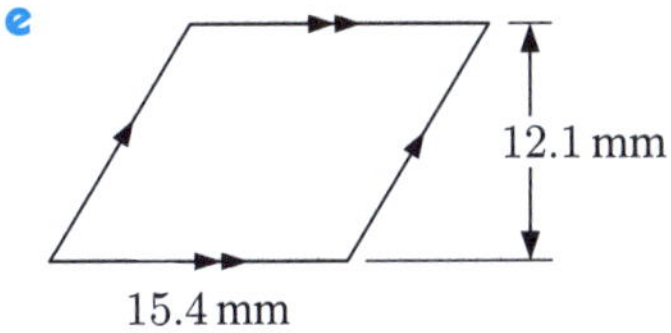

f
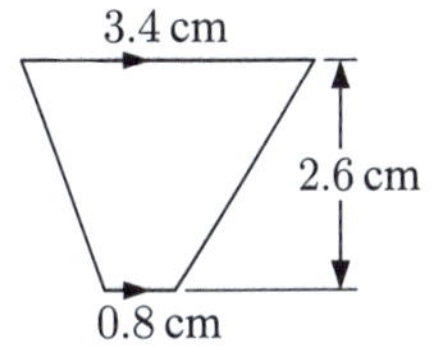

g
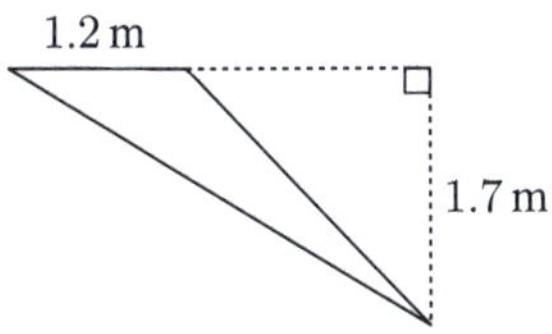

h
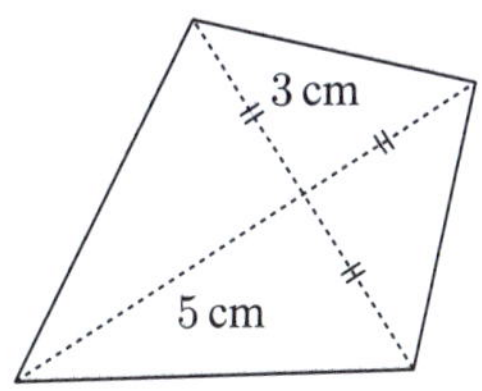

i
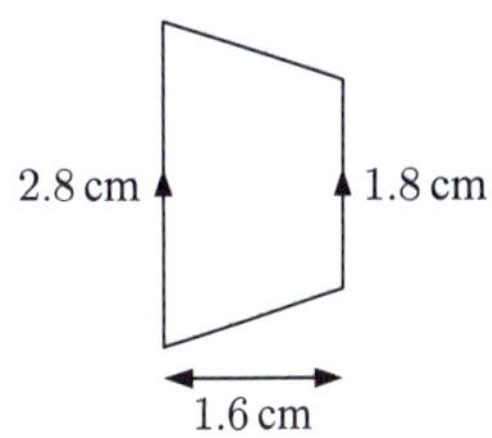

2 Find the area of each figure:

a
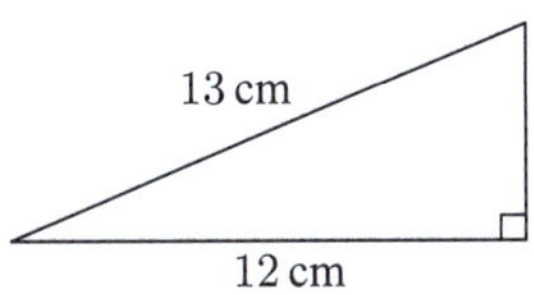

b
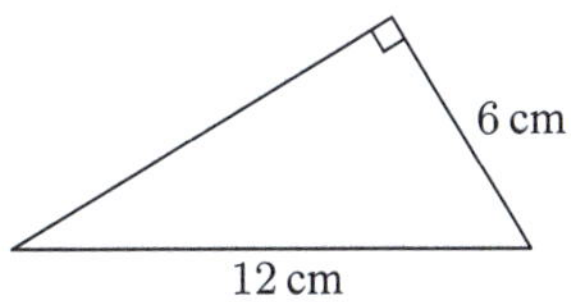

c
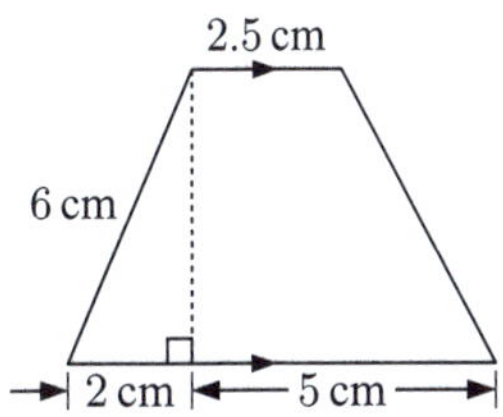

3 Calculate the height h of each figure:

a
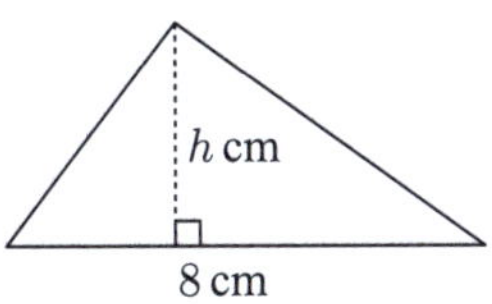

Area = 24 cm^2

b
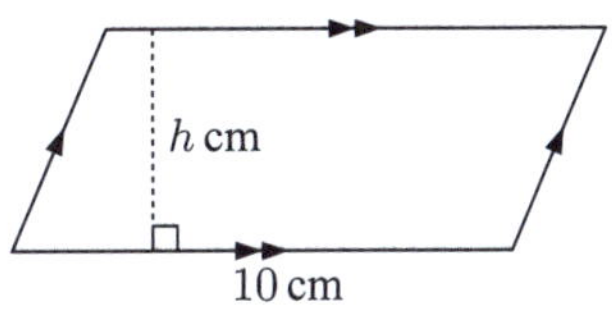

Area = 36 cm^2

c
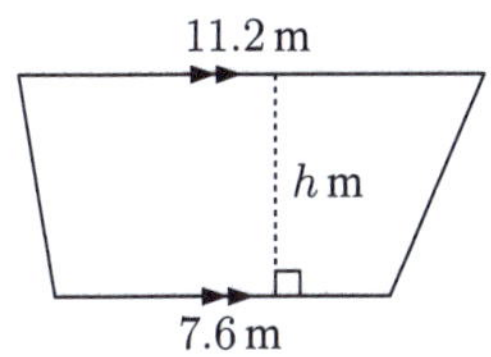

Area = 47 m^2

4 A council orders 25 road signs to warn motorists to watch out for moose.

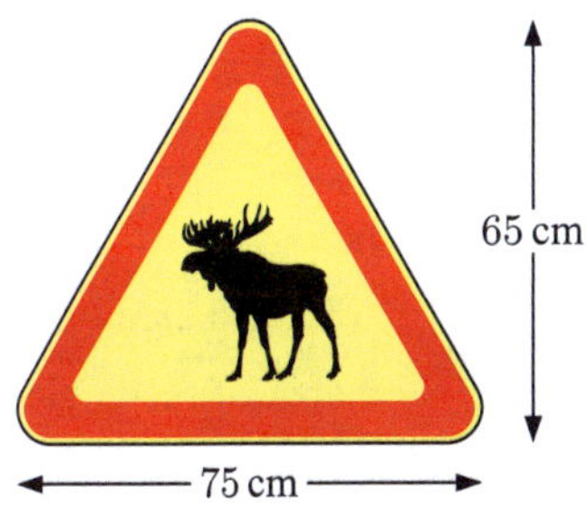

a By assuming the signs are triangular, find the area of metal sheeting (in m^2) required for the 25 signs.

b To allow for wastage when the signs are cut, an extra 20% of the metal is needed. How much metal needs to be purchased?

c The metal costs $28.40 per m^2. Find its total cost.

5 A 15 cm by 20 cm rectangle has the same perimeter as a square. Which figure has the greater area? Explain your answer.

6 A 4.2 m by 3.5 m tablecloth is used to cover a square table with sides of length 3.1 m. Find the area of the tablecloth which overhangs the edges.

7 A square tile has area 256 cm^2. How many tiles are needed to cover a floor $4 \text{ m} \times 2.4 \text{ m}$?

8 Find the area of a rhombus which has diagonals of length 12 cm and 8 cm.

9 One diagonal of a rhombus is twice the length of the other. Given that the rhombus has area 32 cm^2, find the length of the shorter diagonal.

10 A chessboard consists of 5 cm squares of blackwood for the black squares, and maple for the white squares. The squares are surrounded by an 8 cm wide blackwood border. Determine the percentage of the board which is made of maple.

11 The area of trapezium ABCD is 204 cm^2. Find the area of triangle DBC.

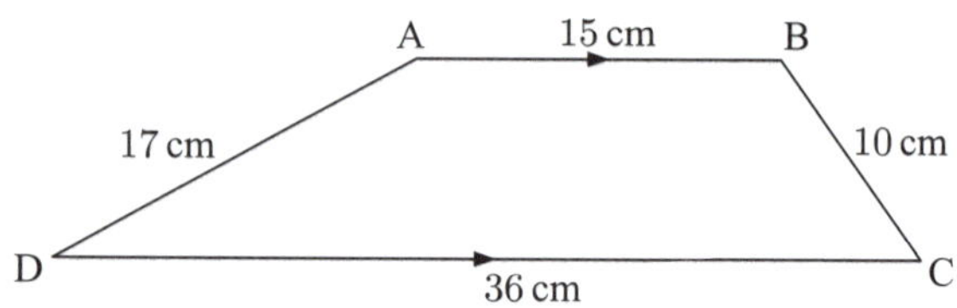

12 Find the area of a kite with diagonals of length 16 cm and 10 cm.

13 Find the area of an equilateral triangle with side length 4.8 cm.

14 Find the side length of an equilateral triangle with area 325 mm^2.

15 Write a formula for the area A of each shape:

a

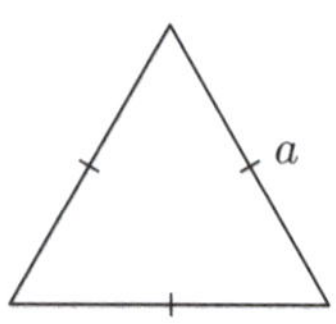

b

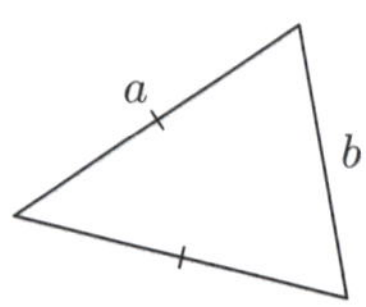

c

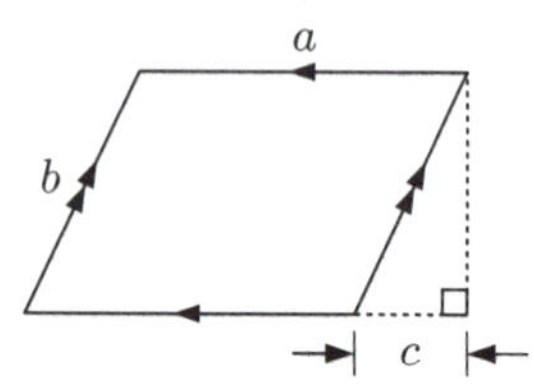

Example 7 **Self Tutor**

Find the shaded area:

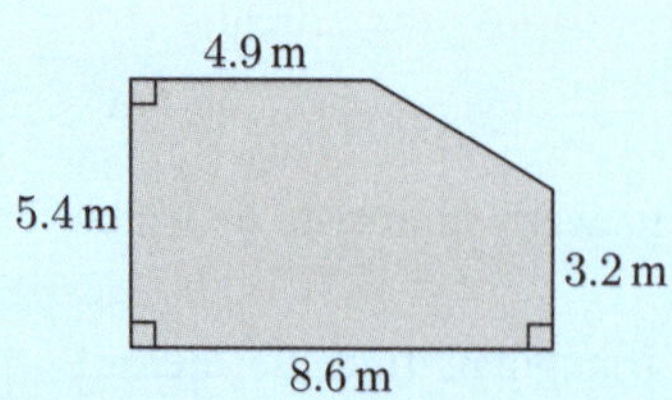

Shaded area

$= \text{area of rectangle} - \text{area of triangle}$

$= (8.6 \times 5.4) \text{ m}^2 - \left(\frac{1}{2} \times 3.7 \times 2.2\right) \text{ m}^2$

$= 42.37 \text{ m}^2$

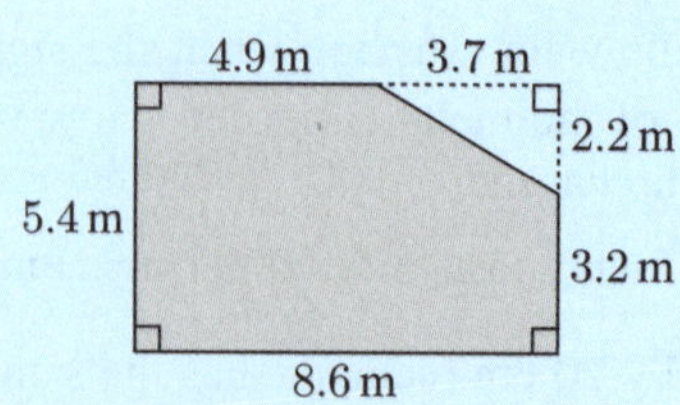

16 Find the shaded area:

a

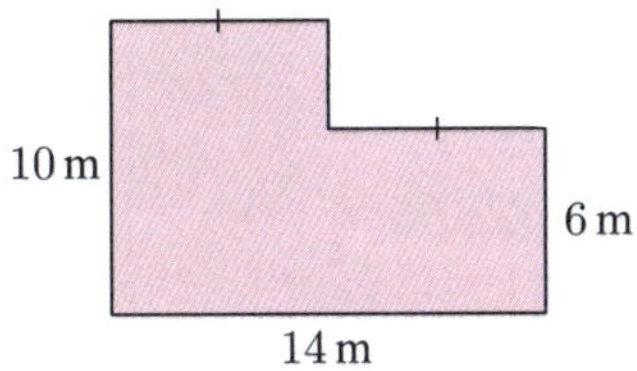

b

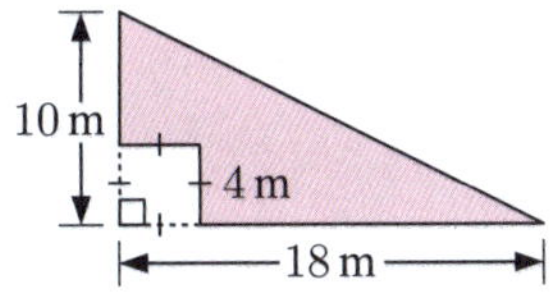

c

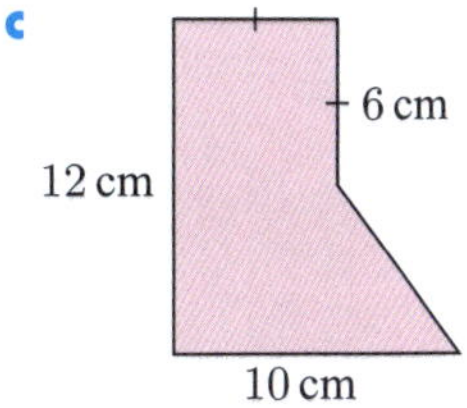

d

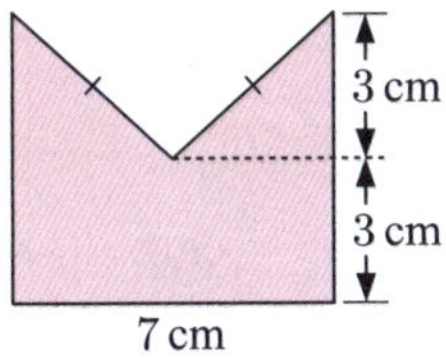

e

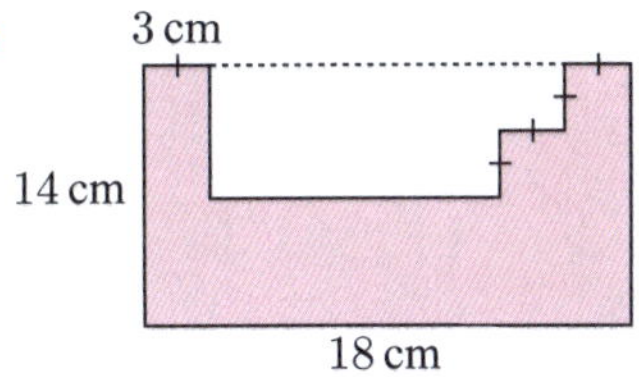

f

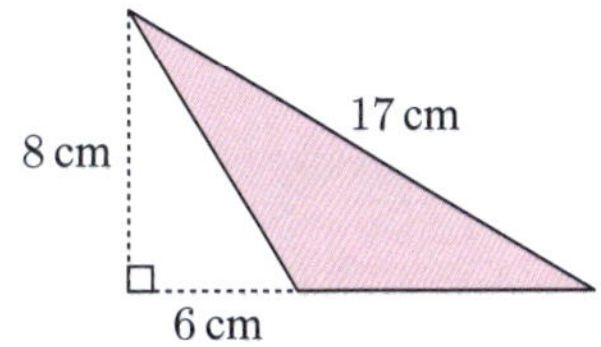

17 A regular hexagon and an equilateral triangle have the same perimeter. Find the ratio of their areas.

E AREA OF CIRCLES AND SECTORS

Consider drawing concentric circles within a circle with radius r.

We cut along a radius, then "unwrap" the rings to form a triangle.

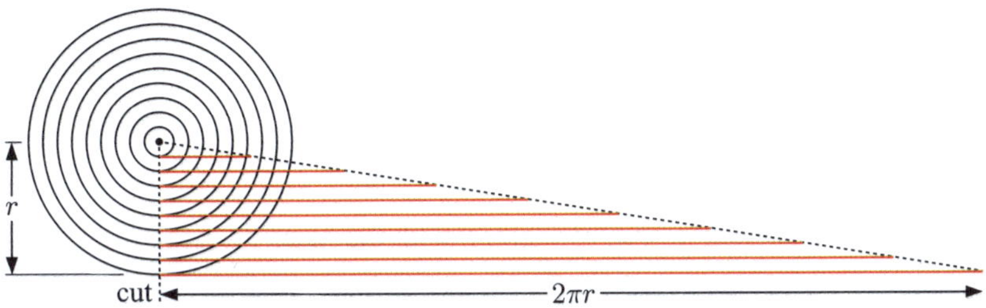

The height of the triangle is the radius of the circle.

The length of the triangle is the circumference of the circle.

DEMO

So, the area of the circle $= \frac{1}{2} \times \text{length} \times \text{height of the triangle}$

$= \frac{1}{2} \times 2\pi r \times r$

$= \pi r^2$

Area of circle $= \pi r^2$

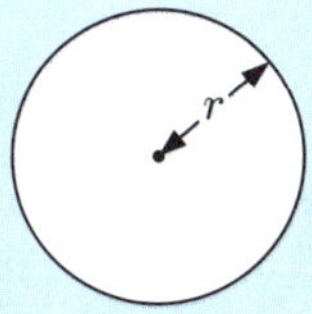

The area of a **sector** is a fraction of the area of the circle it is taken from.

Area of sector $= \dfrac{\theta}{360} \times \pi r^2$

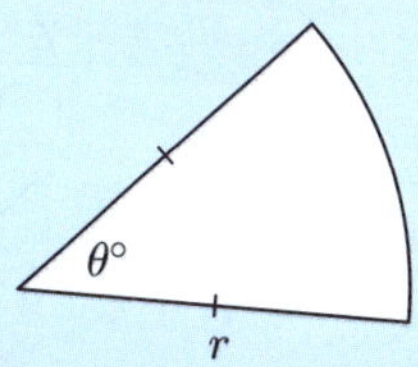

Example 8 **Self Tutor**

Find, to 1 decimal place, the shaded area:

a

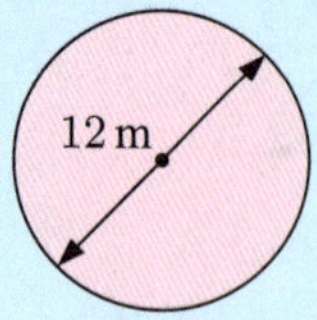

b

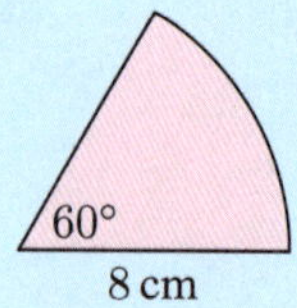

a Area $= \pi r^2$
$= \pi \times 6^2$
$= 36\pi$
$\approx 113.1 \text{ m}^2$

b Area $= \frac{\theta}{360} \times \pi r^2 \text{ cm}^2$
$= \frac{60}{360} \times \pi \times 8^2 \text{ cm}^2$
$\approx 33.5 \text{ cm}^2$

EXERCISE 12E

1 Find the area of each shape:

a

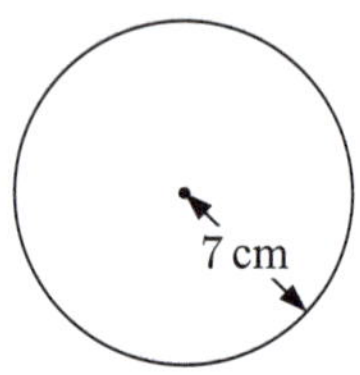

b

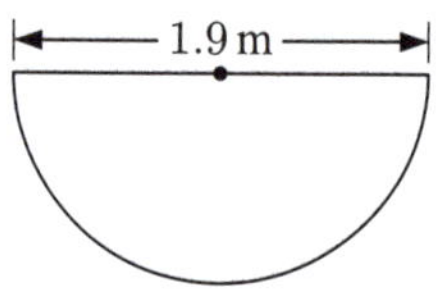

c

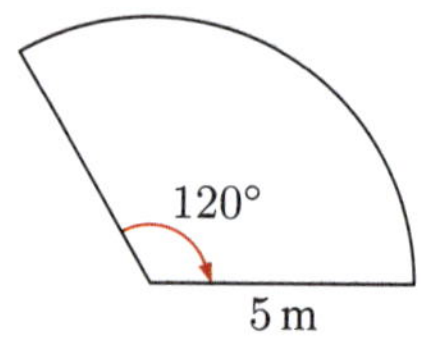

d

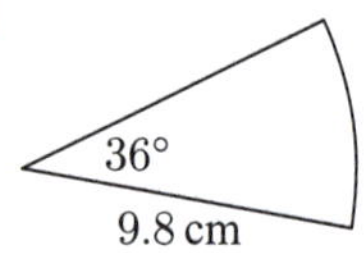

2 Find, in terms of π, the area of a circle with radius 3 cm.

3 Calculate, to 2 decimal places, the area of a circle with diameter 9.7 cm.

4 Calculate the area of a sector with radius 8.7 cm and angle 210°.

5 Find, in terms of π, the:

a length of arc AB

b perimeter of sector OAB

c area of sector OAB.

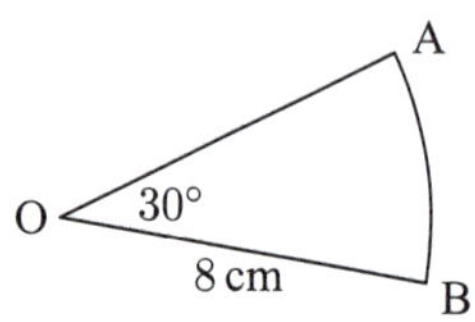

6 Calculate the radius of a circle with:

a area $9\pi \text{ m}^2$

b area 20 cm^2.

7 Find the angle of a sector with area 30 cm^2 and radius 12 cm.

8 A circular golfing green has diameter 20 m. The hole must be positioned on the green at least 3 metres from its edge. Find the area of the green on which the hole can be positioned.

9 Find the shaded area:

a

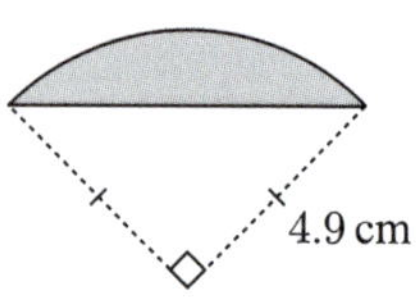

b

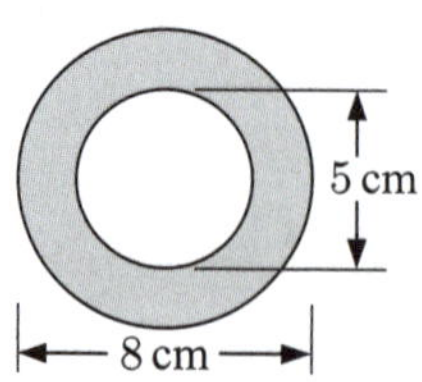

c

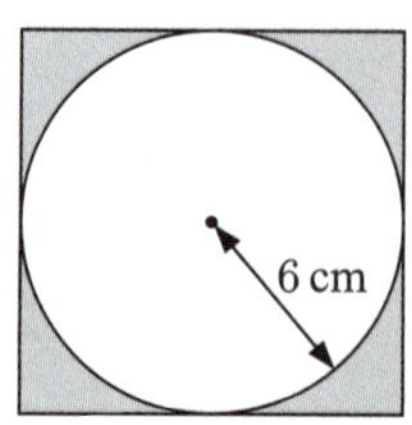

d

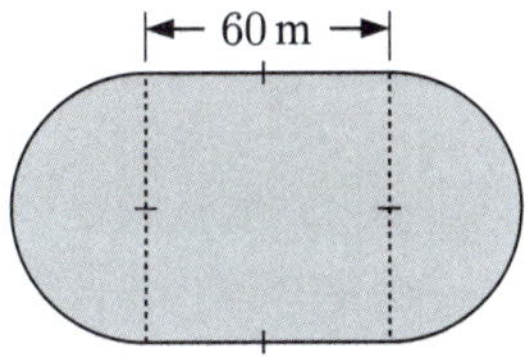

e

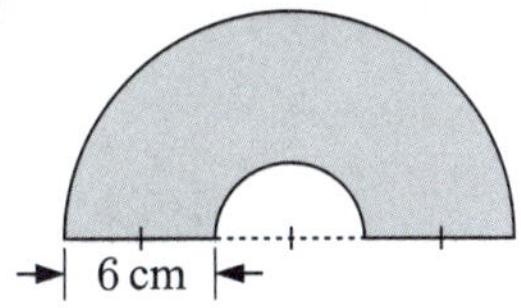

f

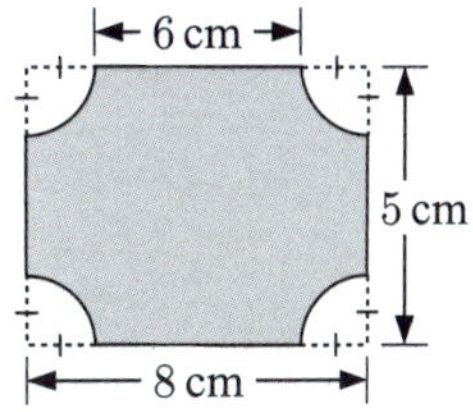

10 A door is in the shape of a rectangle surmounted by a semi-circle. The width of the door is 1.2 m, and the height of the door is 2.5 m. Find the total area of the door.

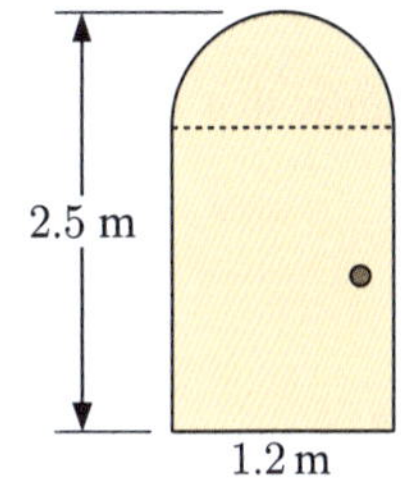

11 A restaurant uses square tables with sides of length 1.3 m, and round tablecloths with diameter 2 m. Determine the percentage of each tablecloth which overhangs its table.

12 6 identical metal discs are stamped out of an 18 cm by 12 cm sheet of tin as illustrated. What percentage of the tin is wasted?

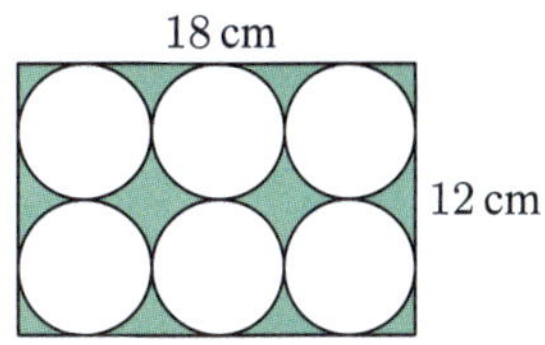

13 A gravel path 1 m wide is placed around a circular garden bed of diameter 2 m. The gravel costs \$7.90 per m^2.

a Find the area of the path.

b Find the cost of the gravel.

14 Answer the **Opening Problem** on page **170**.

Example 9 Self Tutor

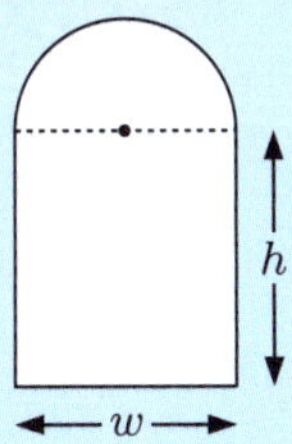

The illustrated door consists of a semi-circle and a rectangle. Find a formula for the area of the door in terms of the width w and height h of the rectangular part.

The area of a rectangle = height × width

$= hw$

The radius of the semi-circle is $\frac{w}{2}$.

$\therefore$ the area of the semi-circle $= \frac{1}{2} \times \pi r^2$

$= \frac{1}{2} \times \pi \times \left(\frac{w}{2}\right)^2$

$= \frac{1}{2} \times \pi \times \frac{w^2}{4}$

$= \frac{1}{8}\pi w^2$

$\therefore$ the total area is $A = hw + \frac{1}{8}\pi w^2$

15 Find a formula for the area A of the following regions:

a

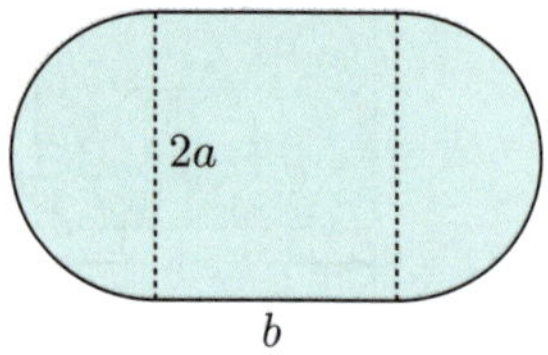

b

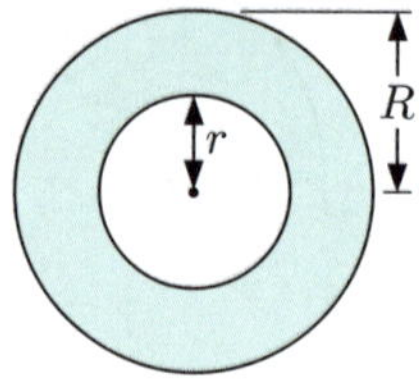

c

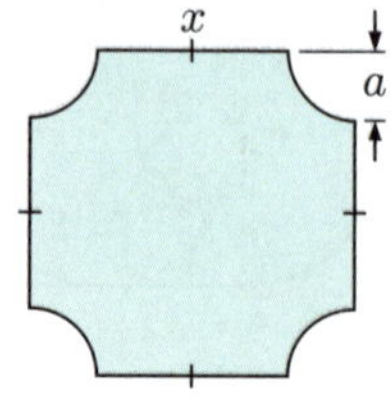

16 Show that the two shaded regions have equal area.

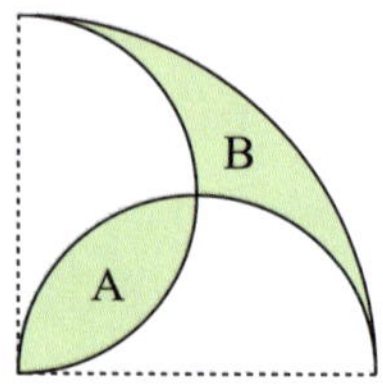

MULTIPLE CHOICE QUIZ

QUICK QUIZ

REVIEW SET 12A

1 Convert:

a 6.7 km into m **b** 9.3 mm into cm **c** 2.59 ha into m^2.

2 Find the perimeter of each figure:

a

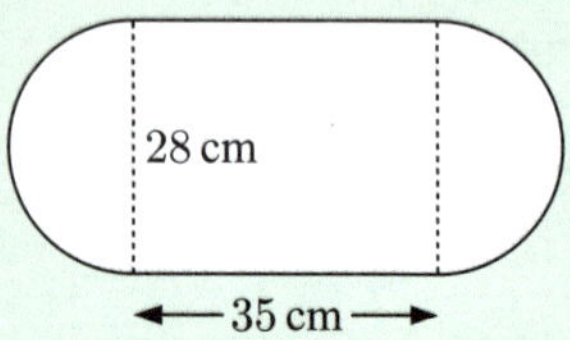

b

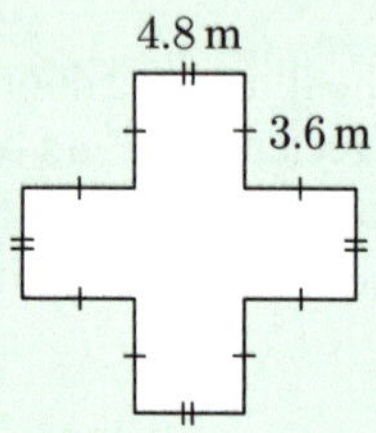

c

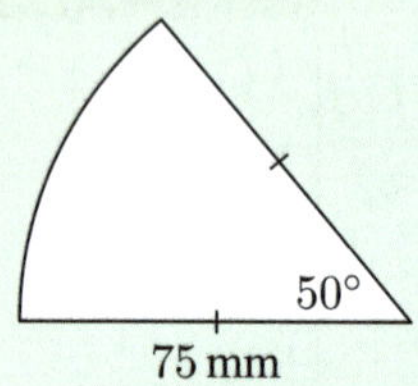

3 Find the perimeter of this racing track.

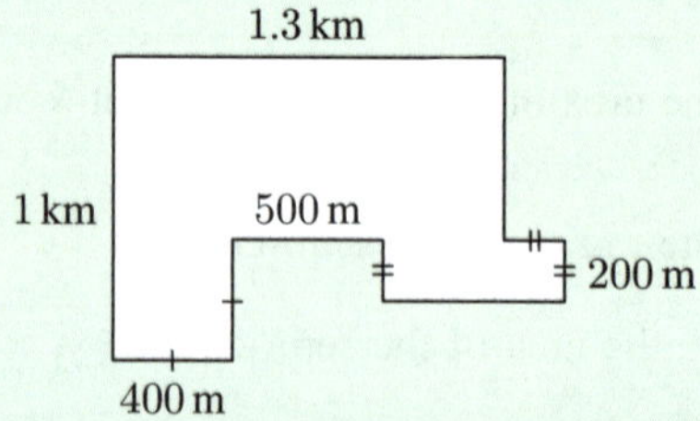

4 A circular playing field has radius 80 metres. The field is surrounded by a fence 10 metres from the edge of the field. Find the length of the fence.

5 In Germany, Frankfurt Airport occupies 23 km^2 of land, whereas Stuttgart Airport occupies 400 ha. How many times larger is Frankfurt Airport than Stuttgart Airport?

6 Find the area of each figure:

a

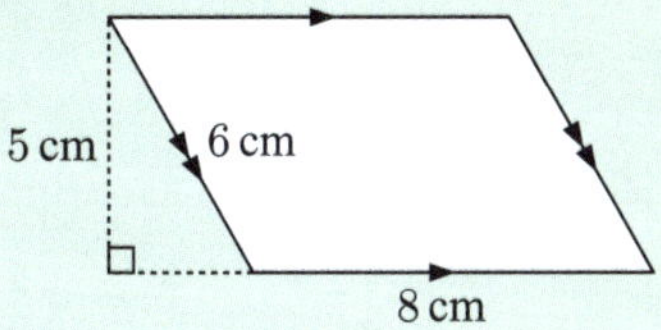

b

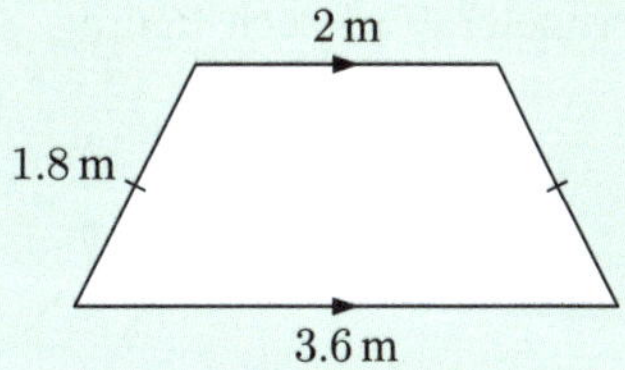

c

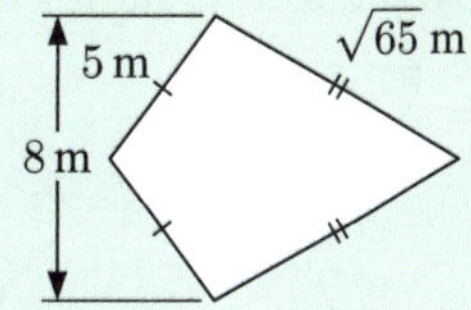

7 Find the area of:

- **a** a sector with radius 10 cm and angle 120°
- **b** a right angled triangle with base 5 cm and hypotenuse 13 cm
- **c** a circle with diameter 15 cm.

8 A circle has area 8 m^2. Find its circumference.

9 Write a formula for the area A of each figure:

a

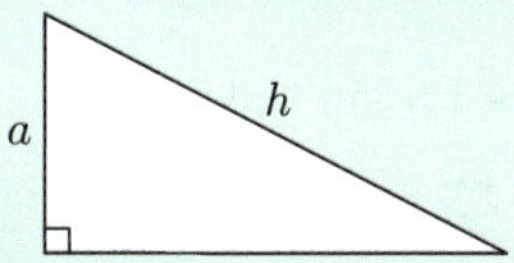

b

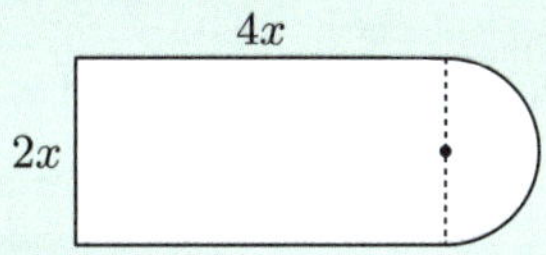

c

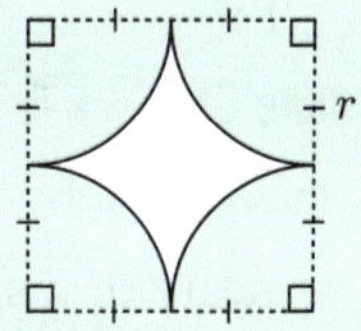

10 In the semi-circle alongside, ACD is a sector of a circle with centre A.
Show that the two shaded regions have equal area.

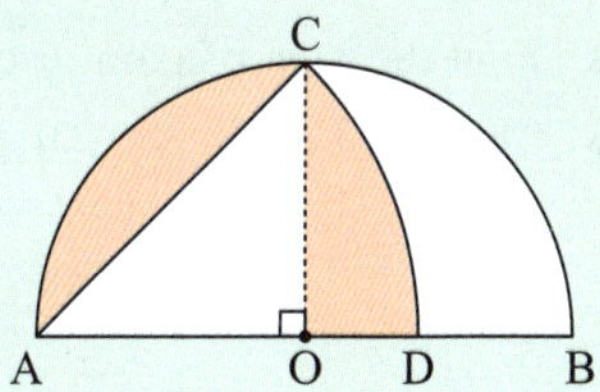

11 A new food warehouse has floor area 1.4 ha.

- **a** Write this area in m^2.
- **b** The concrete floor costs \$90 per square metre. Find its total cost.
- **c** The warehouse is rectangular with length 125 m.
  - **i** Find the width of the warehouse.
  - **ii** Hence find the perimeter of the warehouse.

REVIEW SET 12B

1 Convert:

a 7260 m into km **b** 340 cm^2 into mm^2 **c** 38 200 mm^2 into m^2.

2 12 boxes with height 28 cm are stacked on top of each other.
Find the height of the stack in metres.

3 Find the perimeter of this figure.

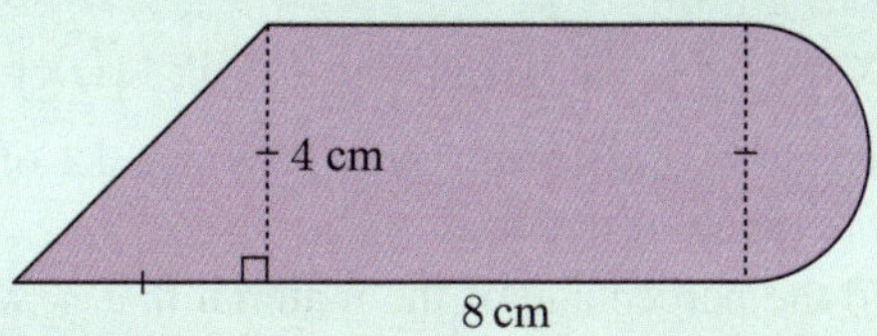

4 Find the angle of a sector with arc length 32 cm and radius 7 cm.

5 Write a formula for the perimeter P of each figure:

a

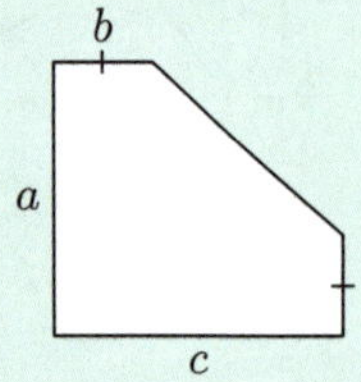

b

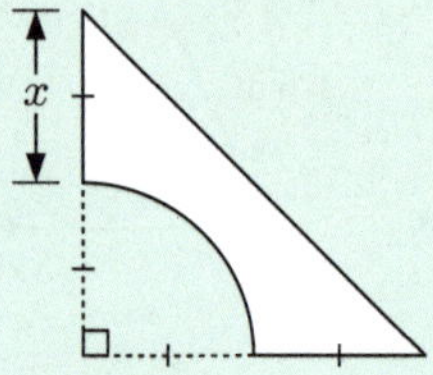

6 Find the shaded area:

a

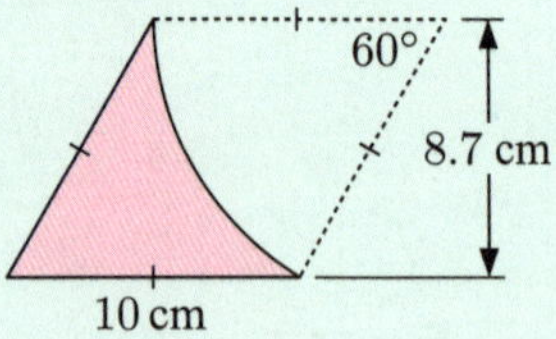

b

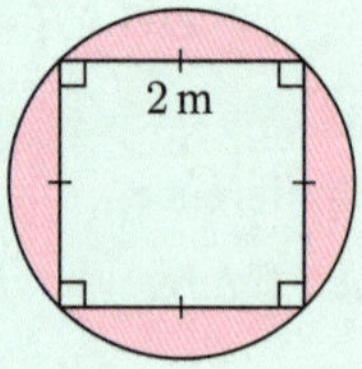

c

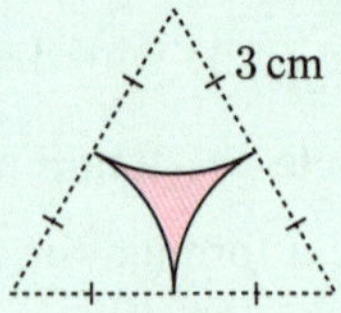

7 Calculate the area in hectares of a rectangular field with sides 1.6 km and 700 m.

8 How many 8 cm by 10 cm tiles are needed to cover a 4.4 m by 6 m floor?

9 The boundary of a sports field has the dimensions shown.

a Find the area of paint used for the boundary.

b The paint costs \$31.50 per m^2. How much does the boundary cost to paint?

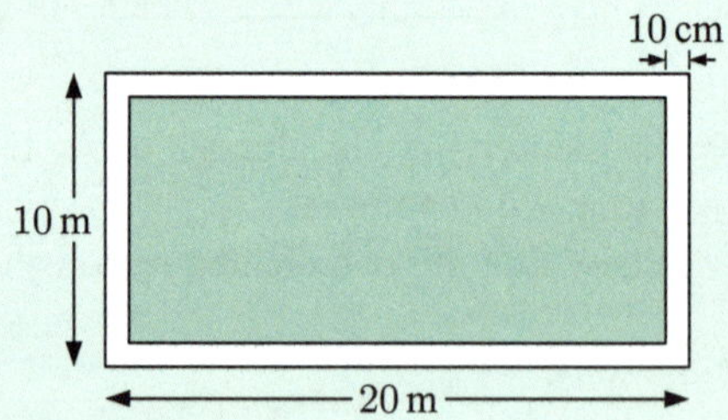

10 Find the side length of an equilateral triangle with area 80 cm^2.

11 The shaded regions have equal area. Find the value of θ.

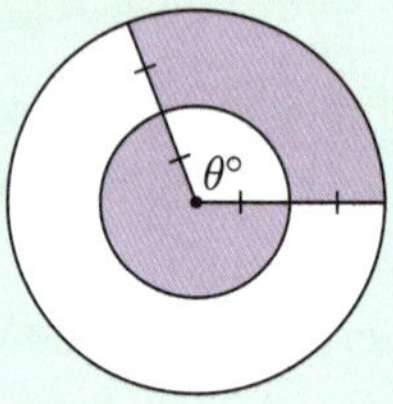

12 A kite manufacturer produces kites with the dimensions shown alongside.

a Find the perimeter of the kite.

b Find the area of the kite.

c **i** Find the length of the long diagonal.

ii Hence find the length of the short diagonal.

d Find the length of:

i [AX] **ii** [BX] **iii** [DX]

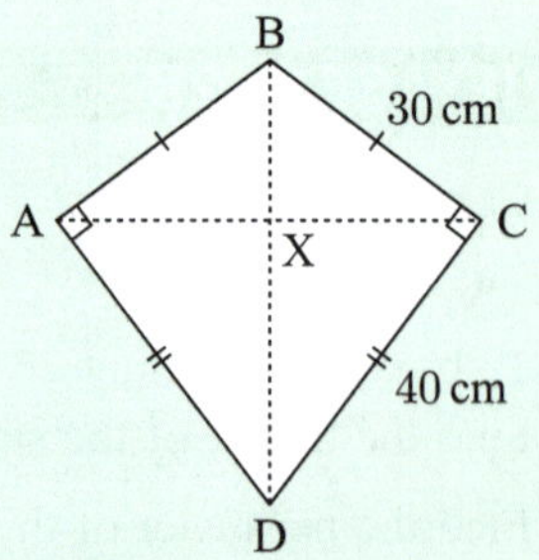

e **i** Illustrate *two* different ways in which 12 of these kites can be cut from a 1.7 m by 1.2 m piece of material.

ii Find the percentage of the material that is wasted.

Chapter 13

Measurement: Surface area, volume, and capacity

Contents:

OPENING PROBLEM

A hollow spherical glass bauble contains a winter snowman scene. The bauble has internal diameter 6.8 cm and external diameter 7.0 cm.

Things to think about:

- **a** What is the outer *surface area* of the bauble?
- **b** What *volume* of glass was used to make the bauble?
- **c** What is the *capacity* of the interior of the bauble?

In this Chapter we will extend our study of measurement to consider 3-dimensional objects.

In particular, we consider:

- **surface area**, which is the total area of an object's surfaces
- **volume**, which is the amount of space a solid occupies
- **capacity**, which is the amount of fluid which can fit in a container.

A SURFACE AREA OF A SOLID WITH PLANAR FACES

For any solid with flat or **planar** faces, we can draw a **net** which can be folded to form the solid.

The surface area of a solid with planar faces is the area of its net.

Example 1

Self Tutor

Find the surface area of:

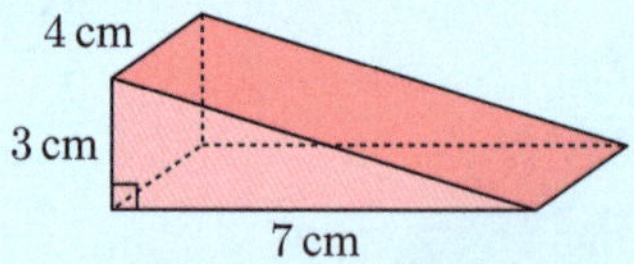

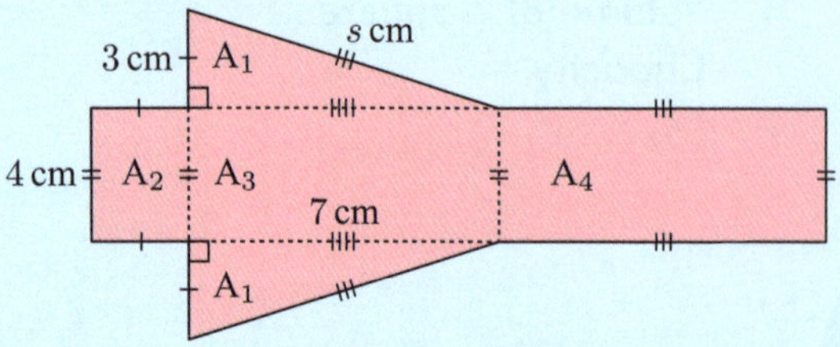

Let the slant edge have length s cm.

$s^2 = 7^2 + 3^2$ {Pythagoras}

$\therefore\ s^2 = 58$

$\therefore\ s = \sqrt{58}$ {since $s > 0$}

Surface area $= 2 \times A_1 + A_2 + A_3 + A_4$

$= (2 \times \frac{1}{2} \times 3 \times 7 + 3 \times 4 + 7 \times 4 + \sqrt{58} \times 4)\text{ cm}^2$

$\approx 91.5\text{ cm}^2$

EXERCISE 13A

1 Find the surface area of a cube with side length:

a 5 cm **b** 9.5 mm

2 Find the surface area of:

a

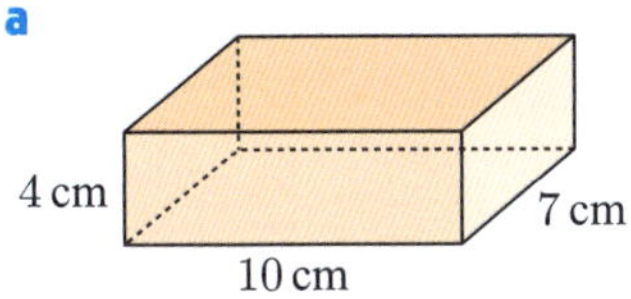

b

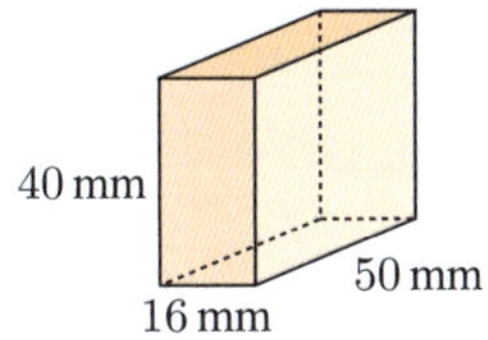

c

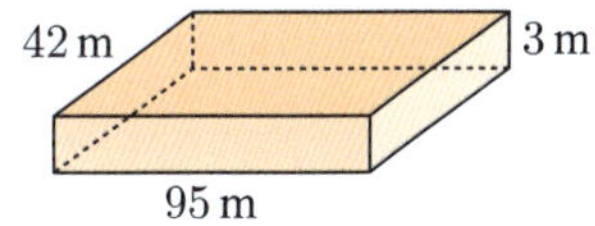

3 Find the surface area of:

a

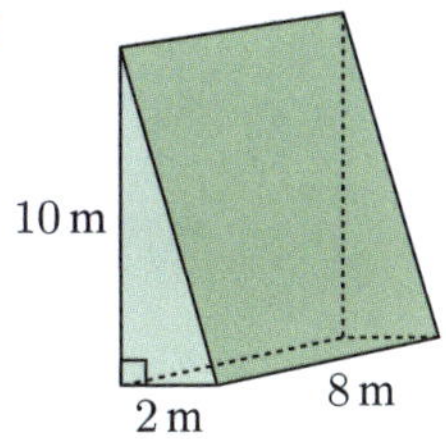

b

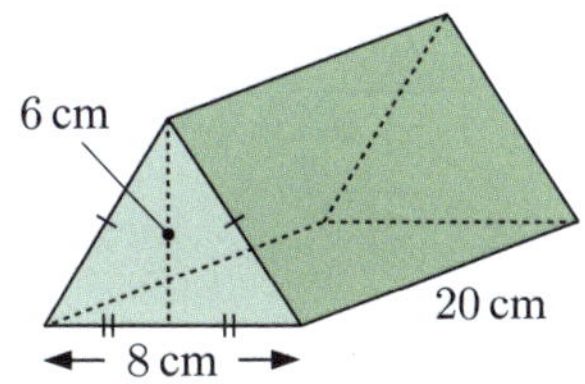

c

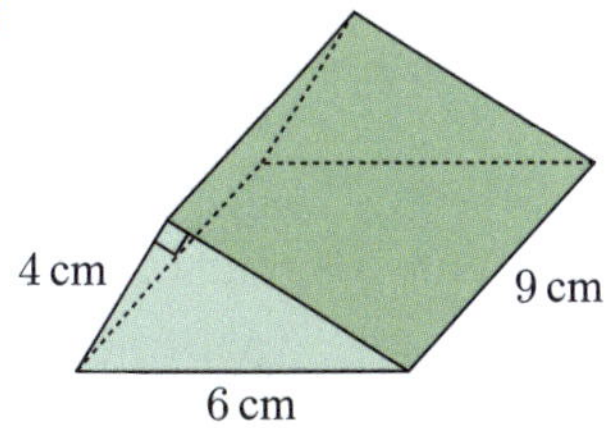

4 A metal pencil box is 20 cm by 15 cm by 8 cm high. Find the total area of metal used to make the pencil box.

Example 2

Self Tutor

Find the surface area of:

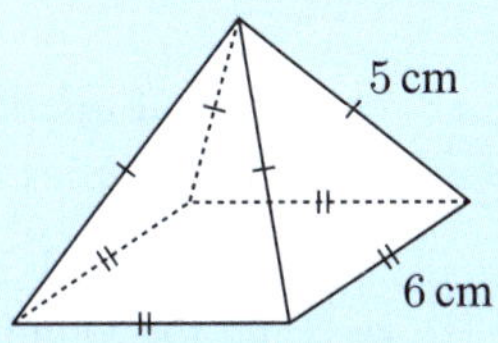

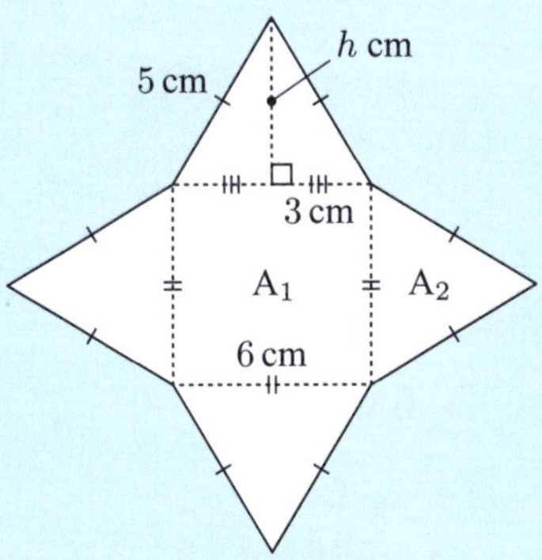

Let the height of the triangular faces be h cm.

$$h^2 + 3^2 = 5^2 \quad \{\text{Pythagoras}\}$$
$$\therefore \; h^2 + 9 = 25$$
$$\therefore \; h^2 = 16$$
$$\therefore \; h = 4 \quad \{\text{since } h > 0\}$$

$$\begin{aligned}\text{Surface area} &= A_1 + 4 \times A_2 \\ &= (6 \times 6 \; + \; 4 \times \tfrac{1}{2} \times 6 \times 4) \text{ cm}^2 \\ &= (36 + 48) \text{ cm}^2 \\ &= 84 \text{ cm}^2\end{aligned}$$

5 Find the surface area of:

a

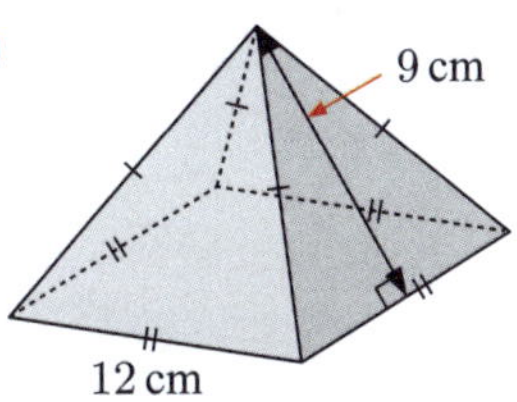

b

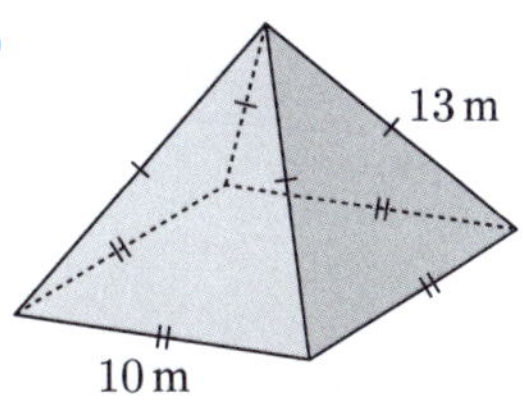

c

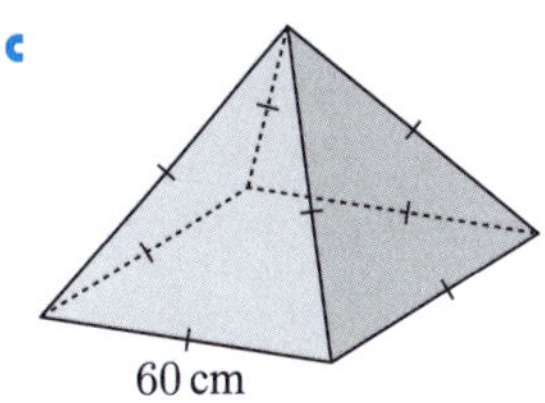

6 Find the surface area of:

a

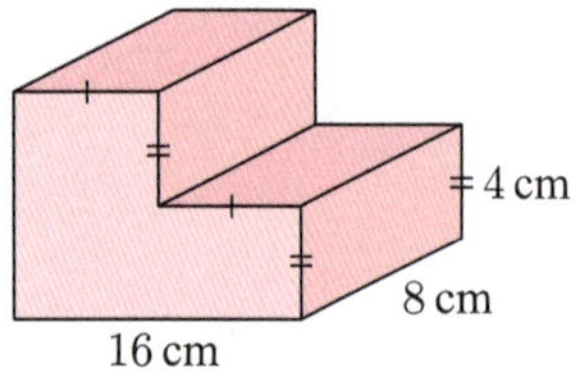

b

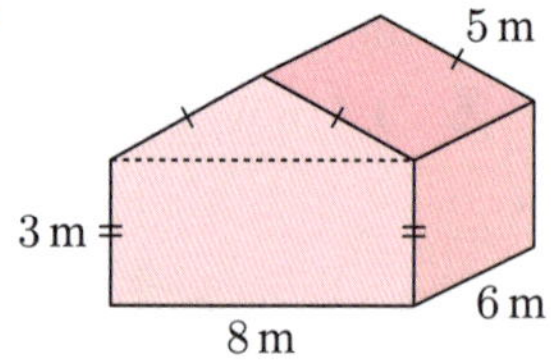

c

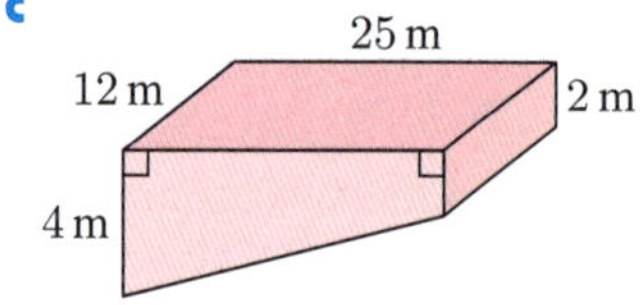

d

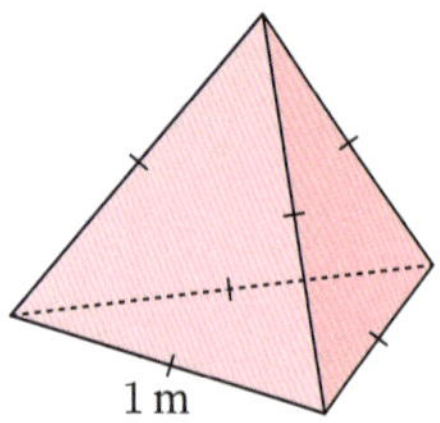

e

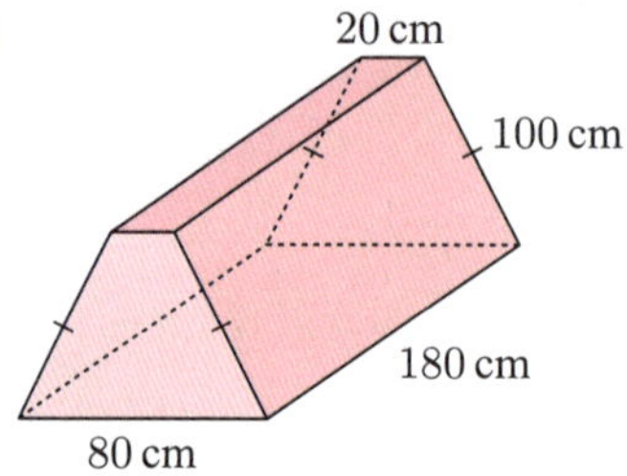

7 Write a formula for the surface area A of:

a

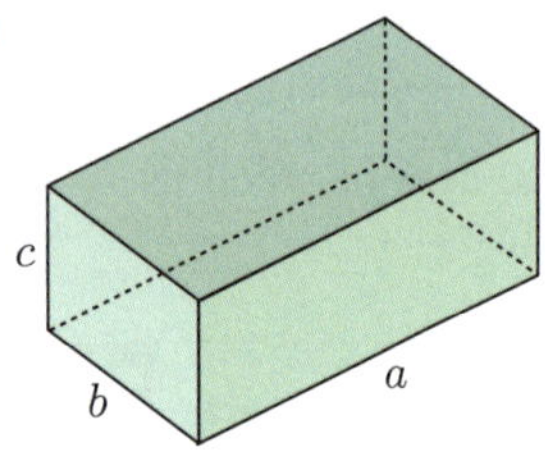

b

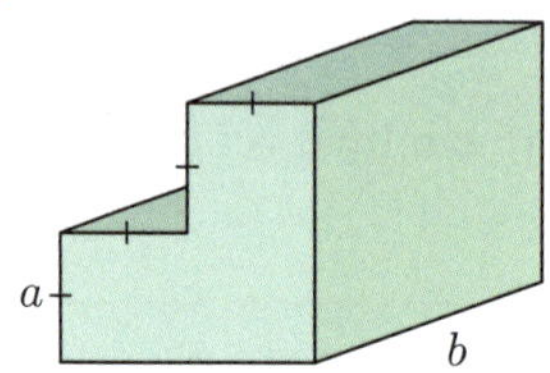

c

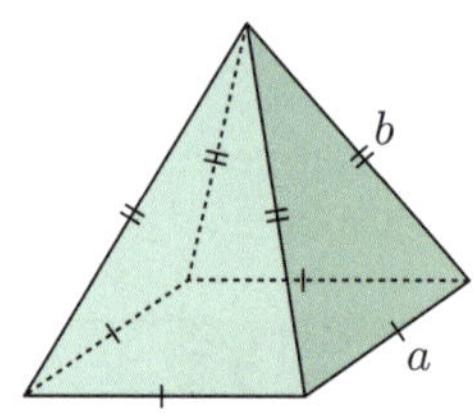

8

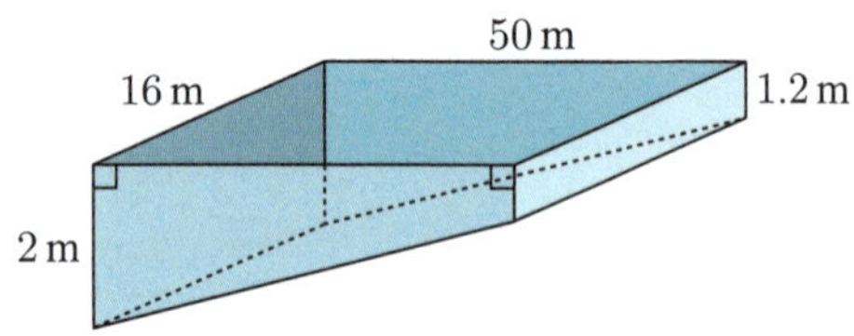

The base and internal walls of this swimming pool are tiled. The tiles cost \$25 per m^2.

a Find the total area of tiles.

b Find the value of the tiles.

9 A marquee with the dimensions shown is made from canvas costing \$31.50 per square metre. The marquee has no floor. Find the total cost of the canvas.

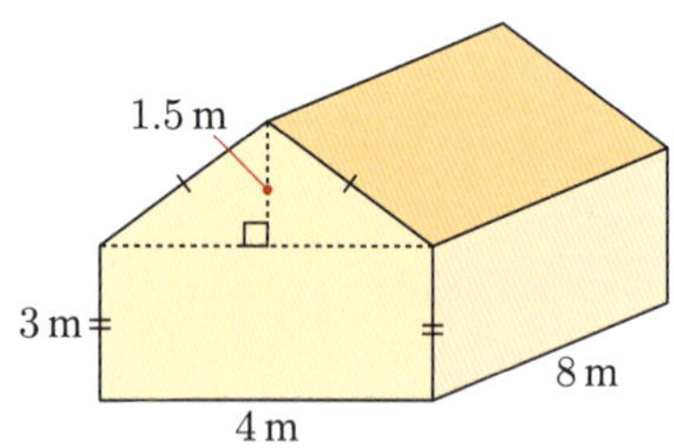

10 An aviary has the shape of a hexagonal prism. Find the area of netting required to cover the aviary, including the floor.

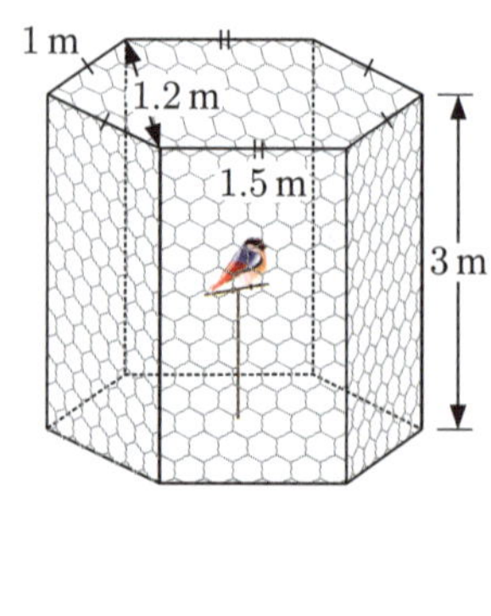

B SURFACE AREA OF A CYLINDER

This cylinder with radius r and height h has no top or bottom.

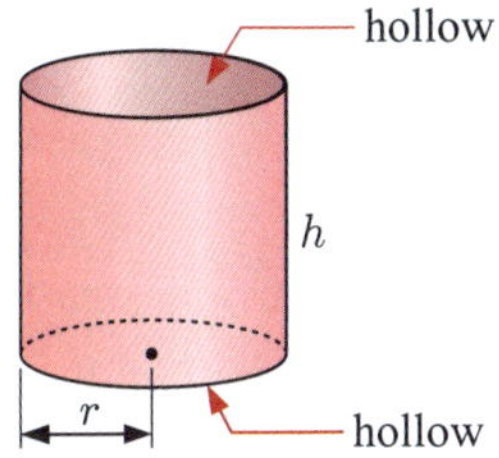

If the cylinder is cut, opened out, and flattened, it takes the shape of a rectangle.

You can check this by peeling the label off a cylindrical can.

The length of the rectangle is the circumference of the cylinder, which is $2\pi r$.

So, the area of the *curved surface* of a cylinder is

$$\begin{aligned} A &= \text{area of rectangle} \\ &= \text{length} \times \text{width} \\ &= 2\pi r \times h \\ &= 2\pi r h \end{aligned}$$

DEMO

Hollow cylinder (no ends)	**Open cylinder** (one end)	**Solid cylinder** (two ends)
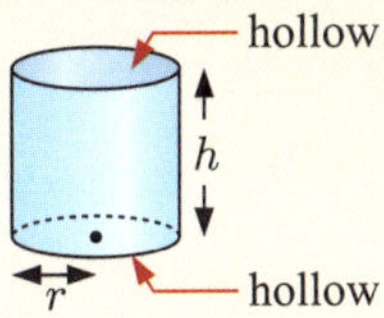	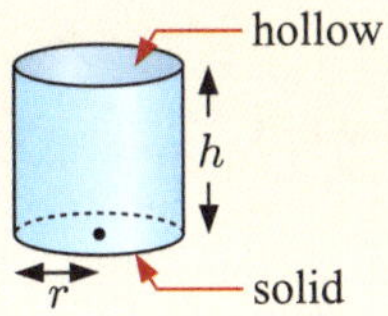	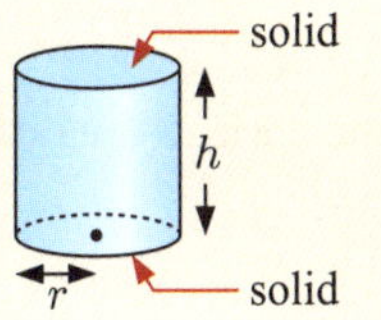
$A = 2\pi rh$	$A = 2\pi rh + \pi r^2$	$A = 2\pi rh + 2\pi r^2$

Notice how these formulae include the *exterior* surfaces only. We ignore any exposed *interior* surfaces unless they are specifically required by the context of a problem.

EXERCISE 13B

1 Find the outer surface area of each cylinder:

a solid

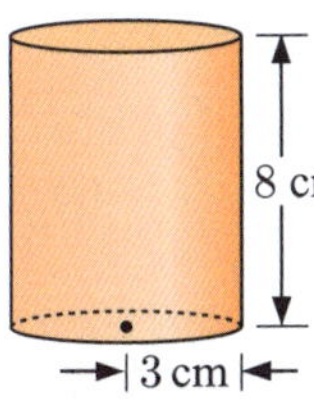

b can (no top)

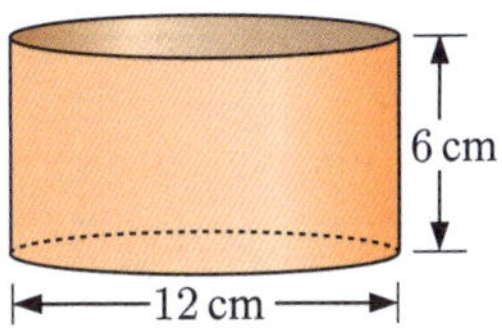

c solid

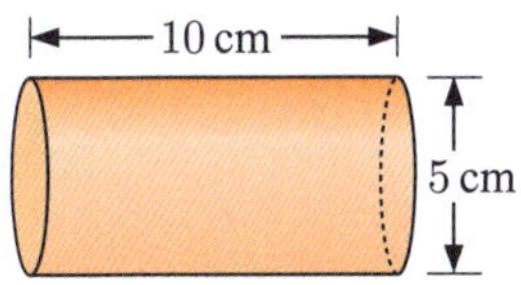

d well (no top)

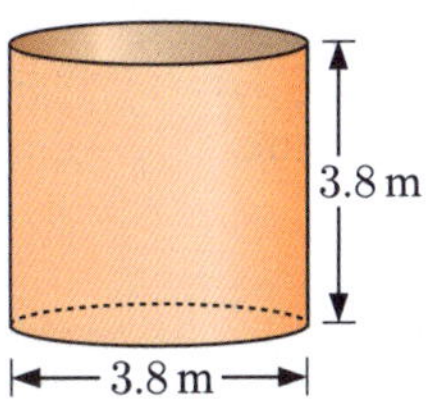

e solid

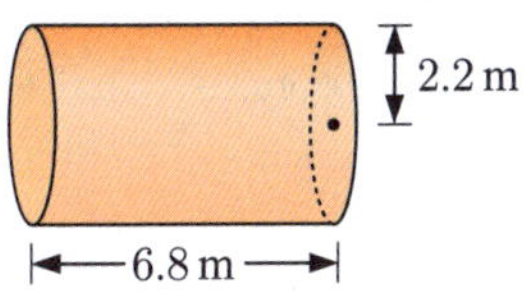

f hollow throughout

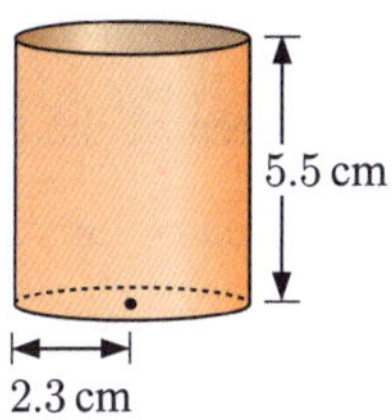

2 A cylindrical drinking glass is 12 cm high and has radius 2.4 cm. Find the exterior surface area of the glass.

3 A trough 1.2 m long has a semi-circular cross-section with radius 28 cm. Find the exterior surface area of the trough.

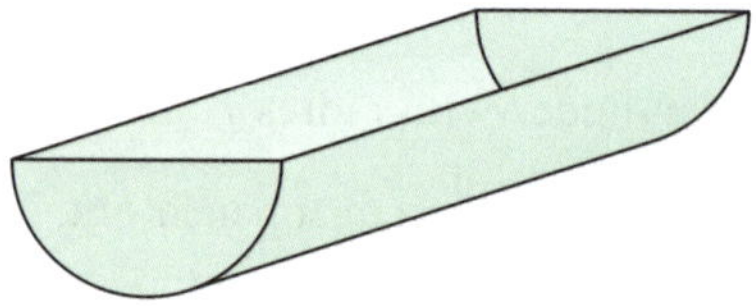

4 **a** Rearrange the formula $A = 2\pi rh + 2\pi r^2$ to make h the subject.

b Hence find the length of:

i an aluminium can with radius 3.2 cm and surface area 223 cm^2

ii a steel rod with radius 15 mm and surface area 0.12 m^2.

5 The curved surface of a bollard is painted with red and white candy stripes with equal width. The bollard is 1.4 m tall and has radius 16 cm. Its top is painted white.
Find the area of:

a red paint **b** white paint.

6 Find the surface area of this jewellery box:

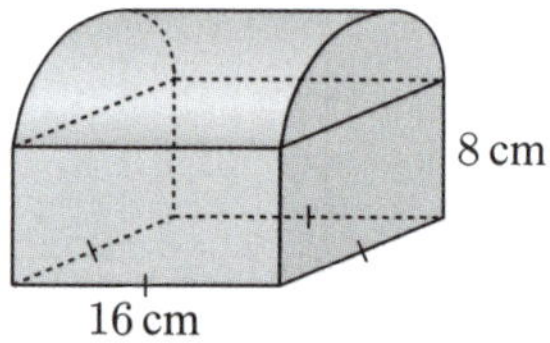

C SURFACE AREA OF A CONE

INVESTIGATION 1 THE CURVED SURFACE OF A CONE

In previous years, you should have seen that the net for a cone is made up of a circle and a sector.

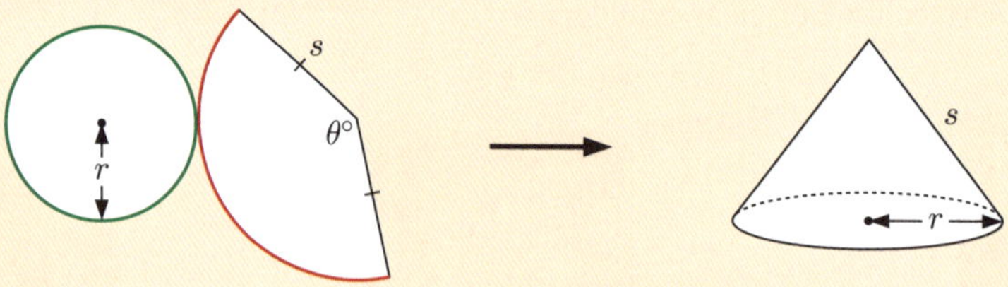

What to do:

1 Explain why the circumference of the green circle is equal to the length of the red arc.

2 Hence show that $\dfrac{\theta}{360} = \dfrac{r}{s}$.

3 The area of the curved surface of the cone is equal to the area of the sector.

So, area of curved surface $= \dfrac{\theta}{360} \times \pi s^2$.

Write a formula for the area of the curved surface in terms of r and s only.

From the **Investigation**, you should have found that the area of the curved surface of a cone with radius r and slant height s is $\pi r s$.

This leads to the formulae:

Hollow cone (no end)	**Solid cone** (closed end)
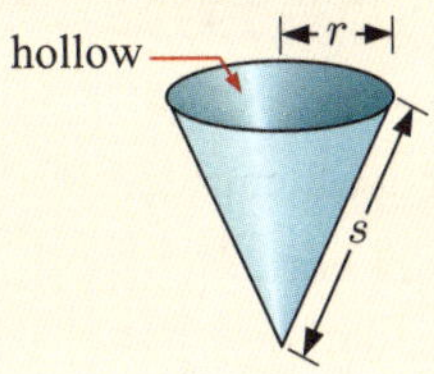	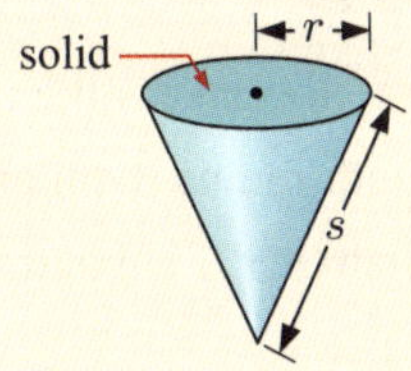
$A = \pi r s$	$A = \pi r s + \pi r^2$

The formulae include the *exterior* surfaces only.

Example 3 — Self Tutor

At the local show, hot chips are sold in a paper cone with radius 6 cm and depth 15 cm. Find the surface area of the paper cone.

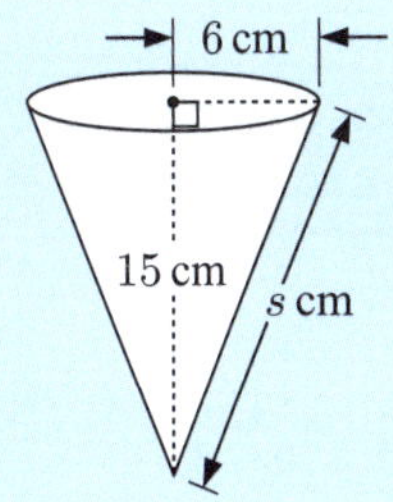

Let the slant height of the paper cone be s cm.

$s^2 = 6^2 + 15^2$ {Pythagoras}

$\therefore\ s^2 = 261$

$\therefore\ s = \sqrt{261}$ {since $s > 0$}

Now $A = \pi r s$

$\therefore\ A = \pi \times 6 \times \sqrt{261}\ \text{cm}^2$

$\therefore\ A \approx 304.5\ \text{cm}^2$

The surface area of the paper cone is about 305 cm^2.

EXERCISE 13C

1 Find the surface area of each cone:

a solid

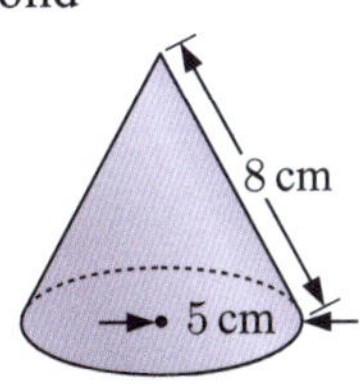

b hollow

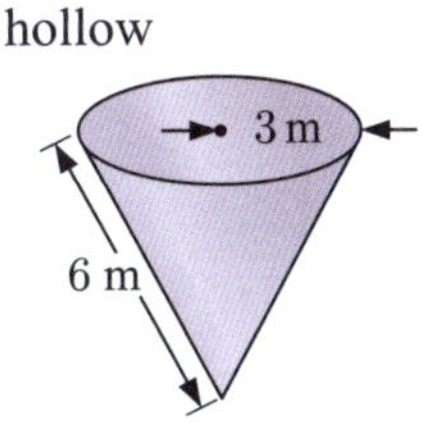

c solid

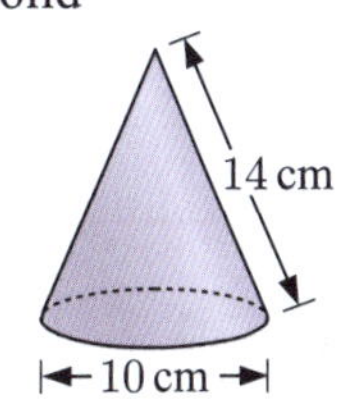

d hollow

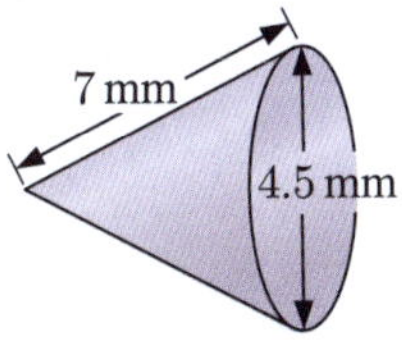

e solid

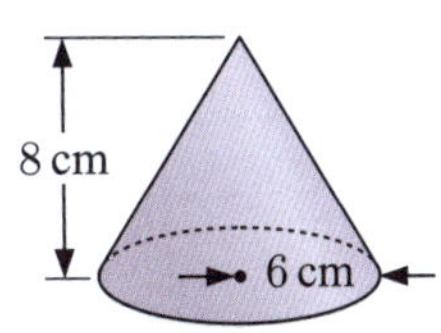

f hollow

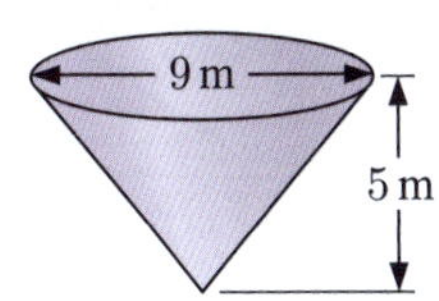

2 In a market stall, a spice is piled in a cone with radius 15 cm and slant height 26 cm. Find the surface area of the spice exposed to the air.

3 An ice cream cone has radius 2.1 cm and height 9.4 cm. Find the exterior surface area of the cone.

4 **a** Rearrange the formula $A = \pi rs + \pi r^2$ to make s the subject.

b Find the *slant height* of a solid cone with radius 3 cm and surface area 82 cm^2.

c Find the *height* of a solid cone with diameter 5 cm and surface area 54 cm^2.

D SURFACE AREA OF A SPHERE

Archimedes of Syracuse (287 BC - 212 BC) was first to prove a formula for the surface area of a sphere. He proved that the surface area of a sphere is equal to the area of the curved surface of the cylinder that exactly encloses it.

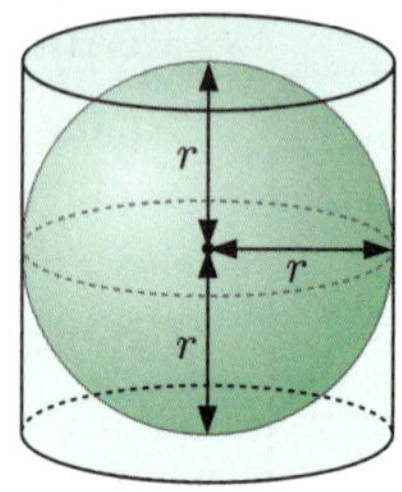

The cylinder exactly enclosing a sphere with radius r, has radius r and height $2r$.

$\therefore$ its surface area $= 2\pi r \times 2r$
$= 4\pi r^2$

$$A = 4\pi r^2$$

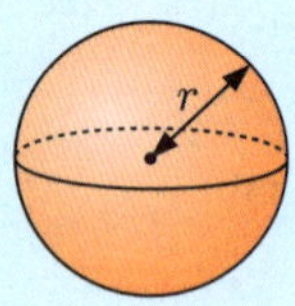

Example 4 **Self Tutor**

Find the surface area of the sphere:

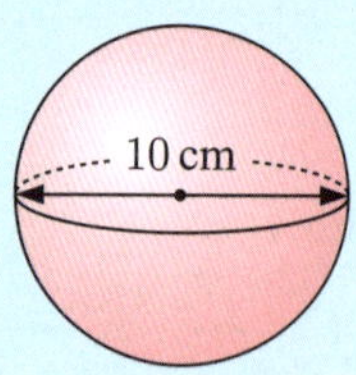

Surface area
$= 4\pi r^2$
$= 4 \times \pi \times 5^2$ cm^2
≈ 314 cm^2

EXERCISE 13D

1 Find the surface area of each solid. Round your answer to 3 significant figures.

a

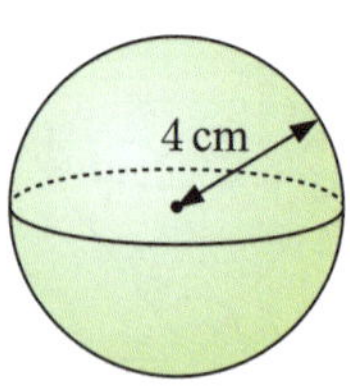

b

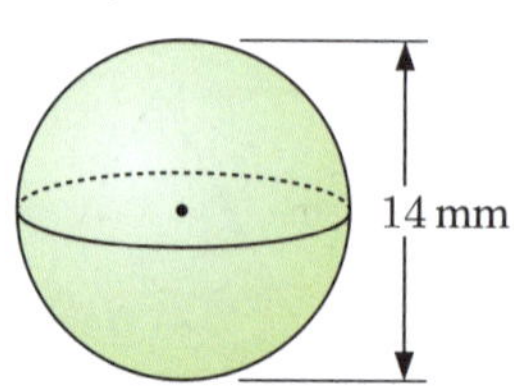

c

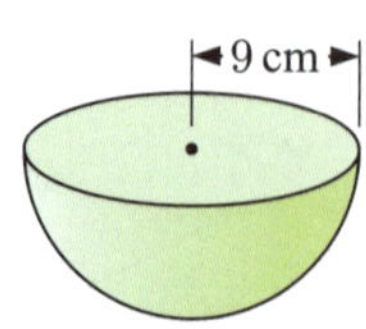

2 Find the surface area of a billiard ball with radius 2.6 cm.

3 Find, in terms of π, the surface area of:

a a sphere with diameter 3 m **b** a hemisphere with radius 8 cm.

4 Foam balls with diameter 15 cm are to be brushed with glue so they are ready for decorating. The pot of glue will cover 2 m^2. How many foam balls can be brushed with glue?

5 A spherical snowball has surface area 270 cm^2. Find its radius.

6 A sphere has the same surface area as a 10 cm by 8 cm by 5 cm box. Find the diameter of the sphere.

7 Find, correct to 1 decimal place, the surface area of:

a

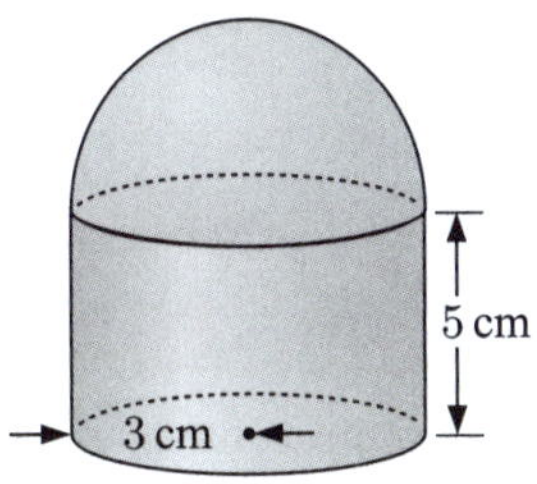

b

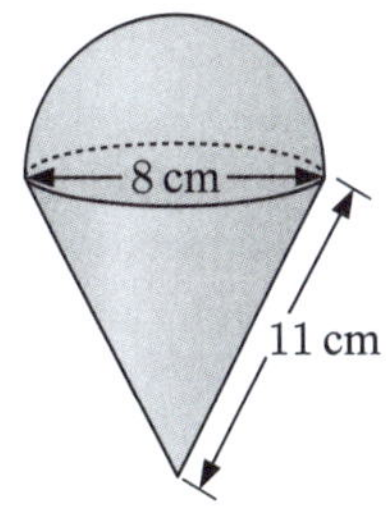

8 A grapefruit was cut into equal wedges.
The total surface area of the wedges is three times the surface area of the original grapefruit.
Into how many wedges was the grapefruit cut?

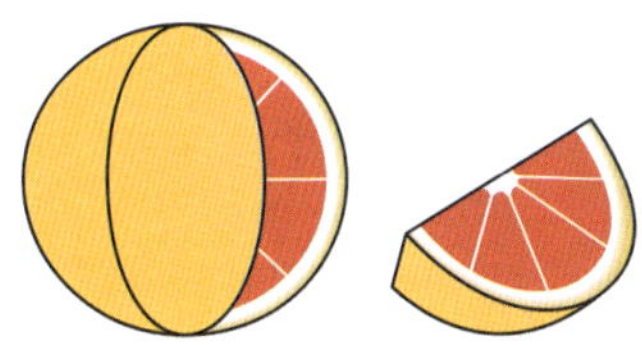

E UNITS OF VOLUME

In the metric system, volume is measured using the *cubes* of the length units.

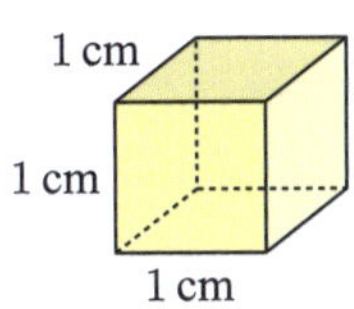

1 cm^3
$= 10 \text{ mm} \times 10 \text{ mm} \times 10 \text{ mm}$
$= 1000 \text{ mm}^3$

Likewise, 1 m^3
$= 100 \text{ cm} \times 100 \text{ cm} \times 100 \text{ cm}$
$= 1\,000\,000 \text{ cm}^3$

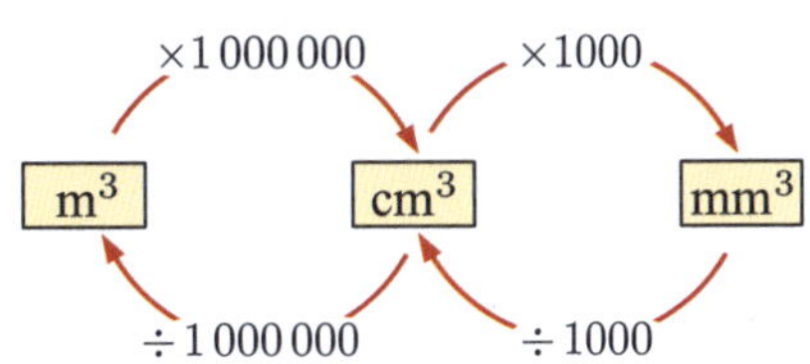

Example 5 Self Tutor

Convert: **a** 0.481 m^3 into cm^3 **b** $19\,720 \text{ mm}^3$ into cm^3.

a 0.481 m^3
$= 0.481 \times 1\,000\,000 \text{ cm}^3$
$= 481\,000 \text{ cm}^3$

b $19\,720 \text{ mm}^3$
$= 19\,720 \div 1000 \text{ cm}^3$
$= 19.72 \text{ cm}^3$

EXERCISE 13E

1 Convert:

a 0.25 cm^3 into mm^3
b 0.083 m^3 into cm^3
c 598 mm^3 into cm^3
d 9810 cm^3 into m^3
e $635\,100\text{ cm}^3$ into m^3
f 81.5 cm^3 into mm^3.

2 Write as a percentage:

a 850 mm^3 out of 4 cm^3
b $30\,000\text{ cm}^3$ out of 0.7 m^3

3 A pallet of 460 books has total volume 1.2 m^3. Find the volume of each book in cm^3.

4 A set of 200 dice, each with volume 600 mm^3, fits perfectly inside a box. Find the volume of the box in cm^3.

5 A toy store has 400 packets of marbles on display. Each packet contains 80 marbles, and each marble contains 850 mm^3 of glass. Find the total volume of glass in the marbles on display, giving your answer in cm^3.

6 A lolly manufacturer ships chocolate bars in $25\,000\text{ cm}^3$ boxes. The average volume of a chocolate bar is 50 cm^3.
Last year, the manufacturer made 554 million chocolate bars.

a Estimate the total volume of chocolate bars made last year. Write your answer in m^3.
b How many boxes of chocolate bars were shipped last year? Write your answer using scientific notation.

F VOLUME OF A SOLID OF UNIFORM CROSS-SECTION

In the solid alongside, any vertical slice parallel to the front face will be the same size and shape as that face. Solids like this are called **solids of uniform cross-section**. In this case, the cross-section is a hexagon.

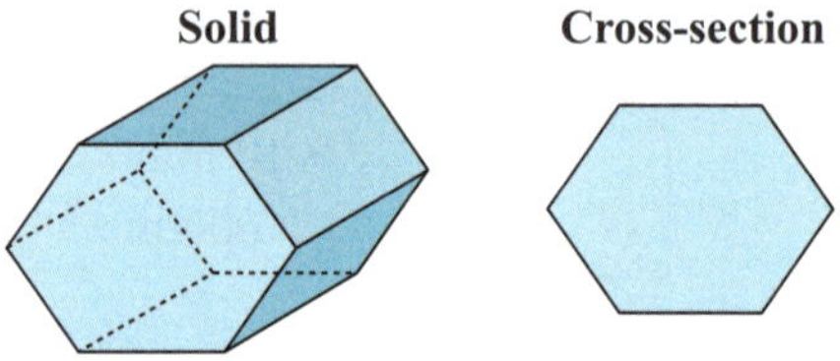

All solids of uniform cross-section contain two identical end faces.

Volume of solid of uniform cross-section = area of end × length

We can use more specific formulae for some special solids of uniform cross-section.

- For a rectangular prism, the cross-section is a rectangle.

Volume of rectangular prism
= length × width × height

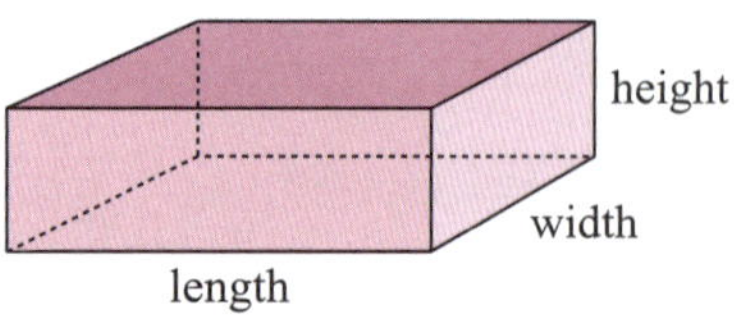

- For a cylinder, the cross-section is a circle.

  Volume = area of end × length
  = area of circle × length
  = $\pi r^2 \times l$

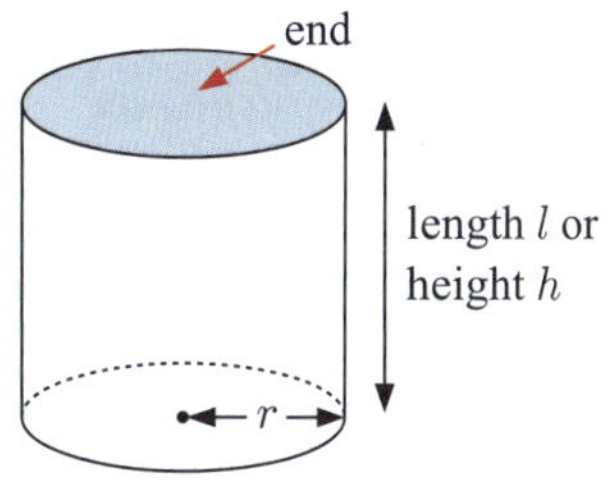

Volume of cylinder $V = \pi r^2 l$
or $V = \pi r^2 h$

Example 6 — Self Tutor

Find the volume of each solid:

a

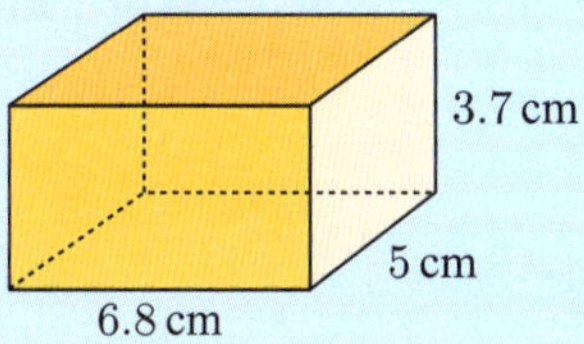

b

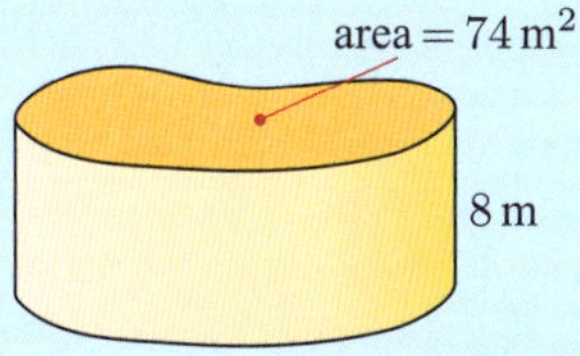

c

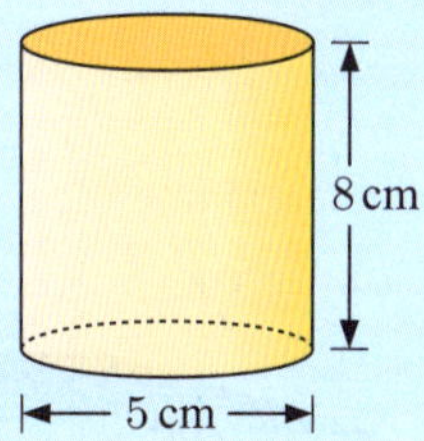

a Volume
= length × width × height
= 6.8 cm × 5 cm × 3.7 cm
= 125.8 cm^3

b Volume
= area of cross-section × height
= 74 m^2 × 8 m
= 592 m^3

c The base has diameter 5 cm, so the radius is 2.5 cm.

$V = \pi r^2 h$
$= \pi \times 2.5^2 \times 8$ cm^3
≈ 157 cm^3

EXERCISE 13F

1 Find the volume of each rectangular prism:

a

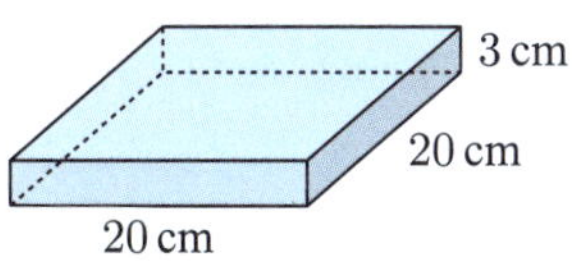

b

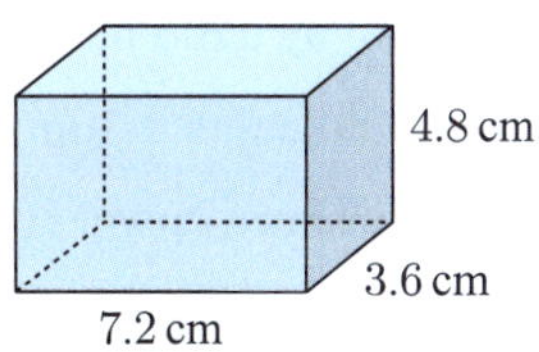

c

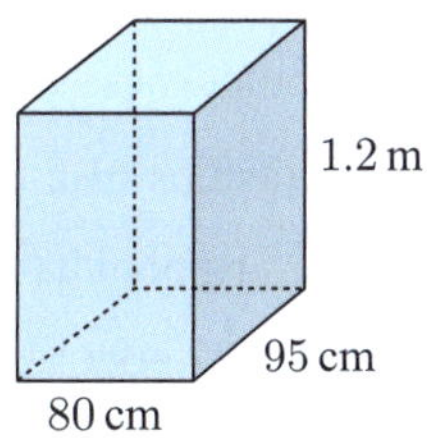

2 Find, rounded to 3 significant figures, the volume of each cylinder:

a

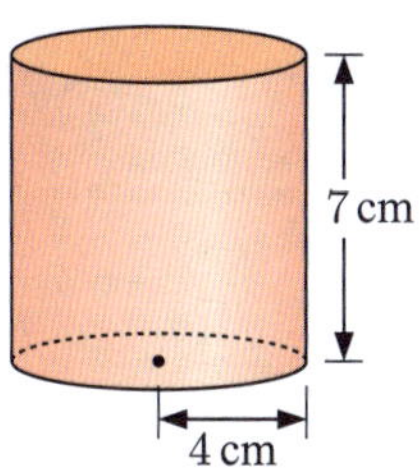

b

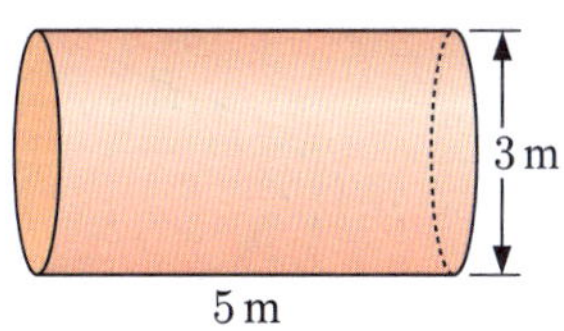

c

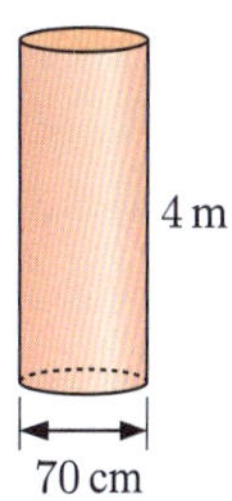

3 Find the volume of each solid:

a

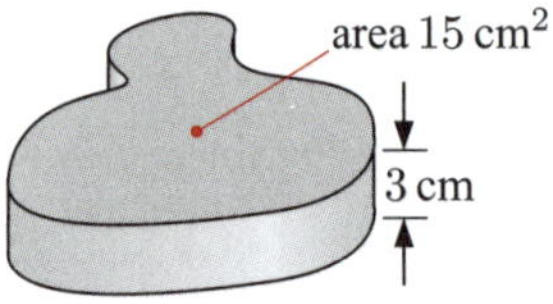

b

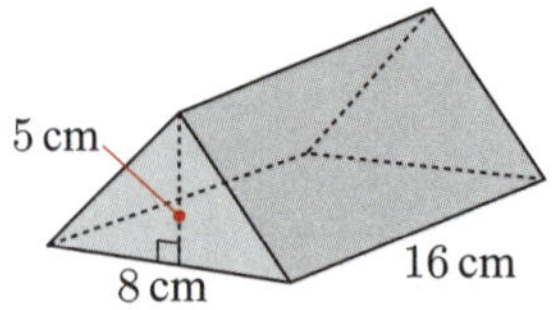

c

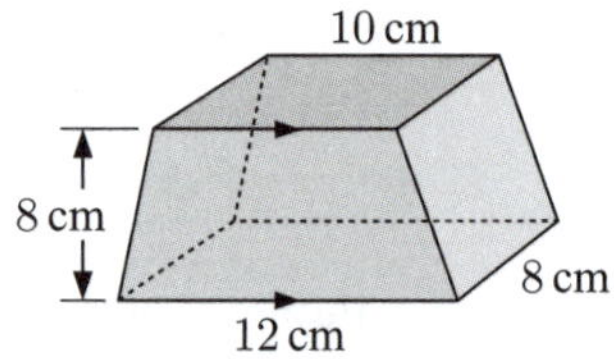

d

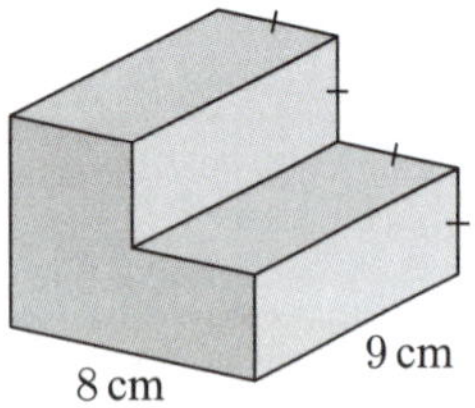

e

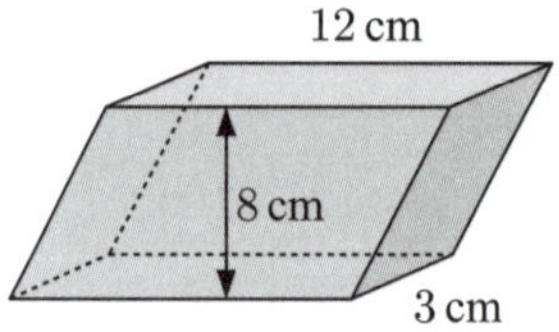

f

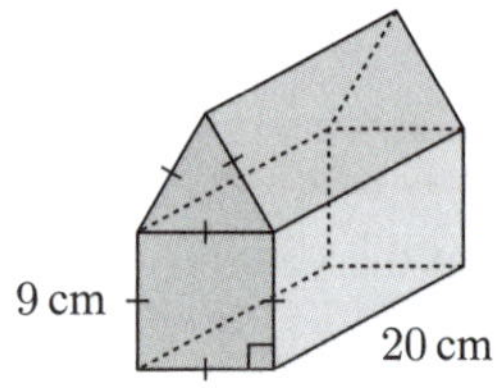

4 Find the volume of:

a this brick

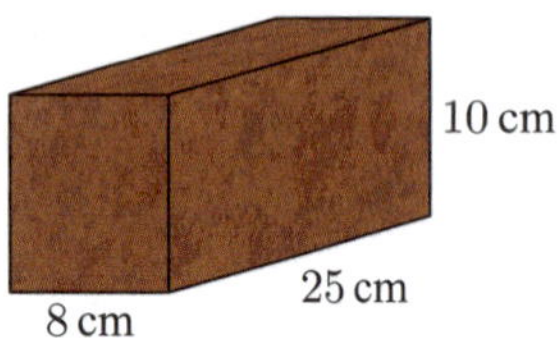

b this coin

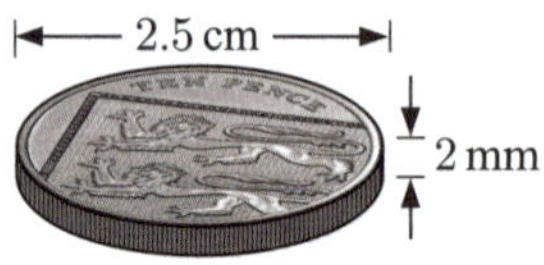

c this wedge of cheese

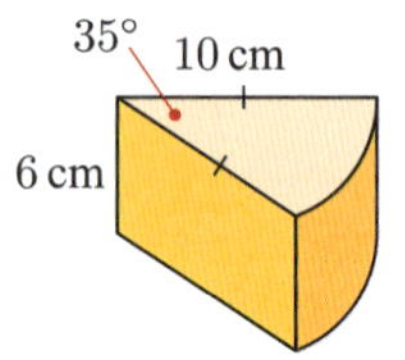

d this picket

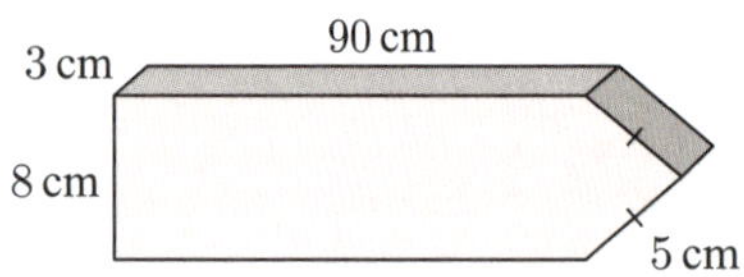

e this dumbbell

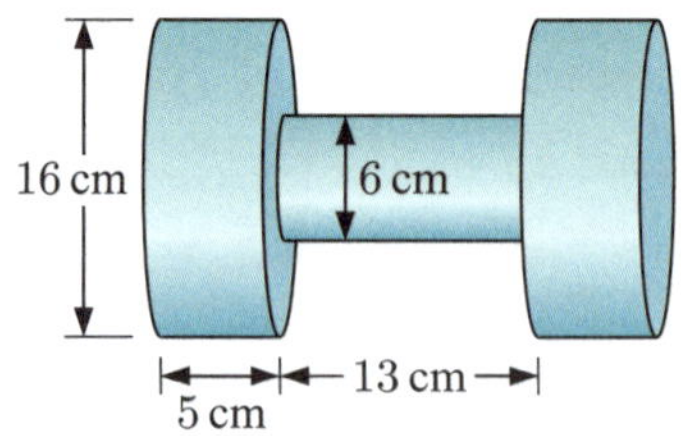

5 Find the side length of a cube with volume 2.744 cm^3.

6 Find the height of a cylinder with diameter 2.4 cm and volume 3.68 cm^3.

7 Find the thickness of a board 2400 mm by 1200 mm, with volume 4.6×10^7 mm^3.

8 Find the radius of a cylindrical rod with length 1.3 m and volume 915 cm^3.

9 A swimming pool has the internal dimensions shown.

a Find the area of a trapezium-shaped side.

b Determine the volume of water required to fill the pool.

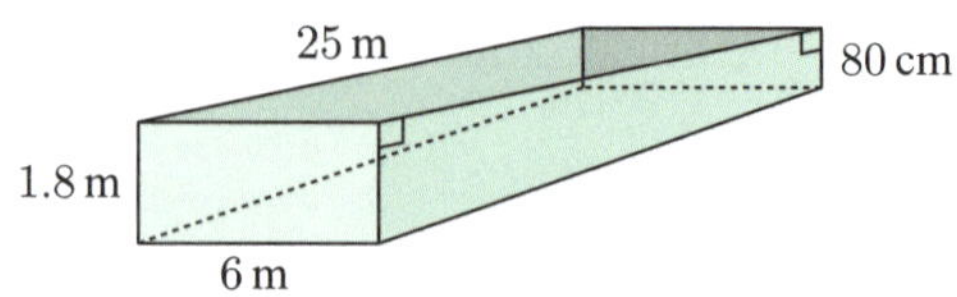

10 Write a formula for the volume V of each object:

a

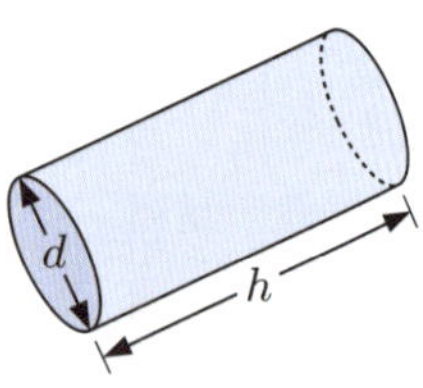

b

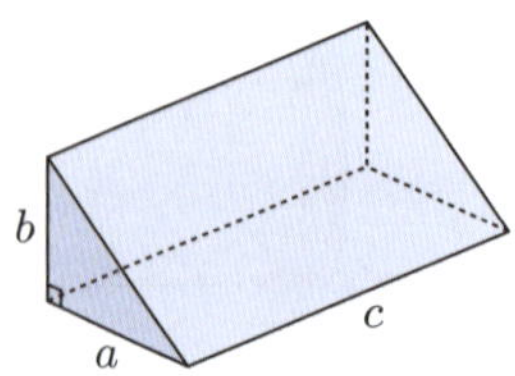

c

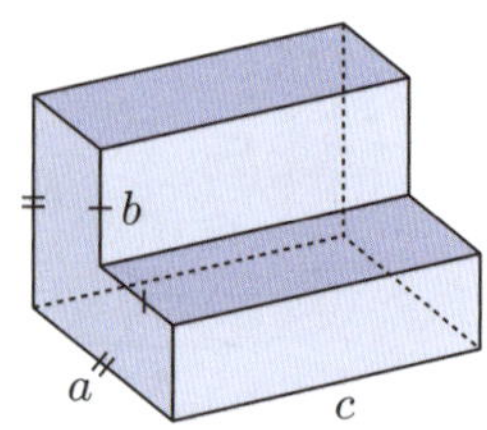

11 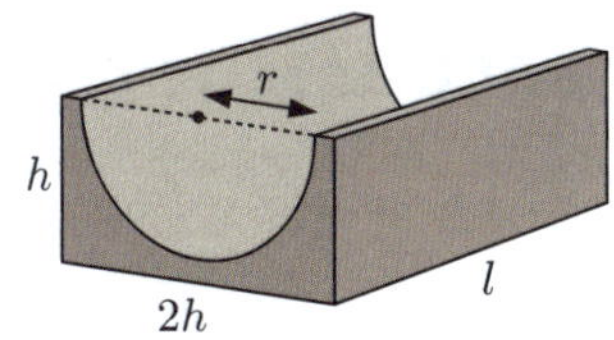

A half-pipe for skateboarding is made with the dimensions shown. Show that the volume of concrete used is given by the formula $V = l(2h^2 - \frac{1}{2}\pi r^2)$.

12 A concrete tank has an external diameter of 5 m and an internal height of 3 m. The walls and base of the tank are 20 cm thick. The concrete costs \$142 per m^3.

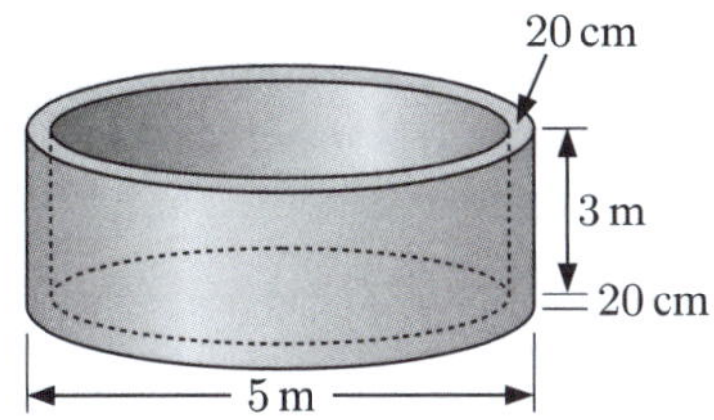

a Find the volume of concrete required to make:
  i the base **ii** the walls.

b Find the total volume of concrete required.

c Find the cost of the concrete required to make the tank.

13 A farrier makes four steel horse shoes 8 mm thick and 14 mm wide. The other dimensions are shown in the diagram.
Find the total volume of the shoes.

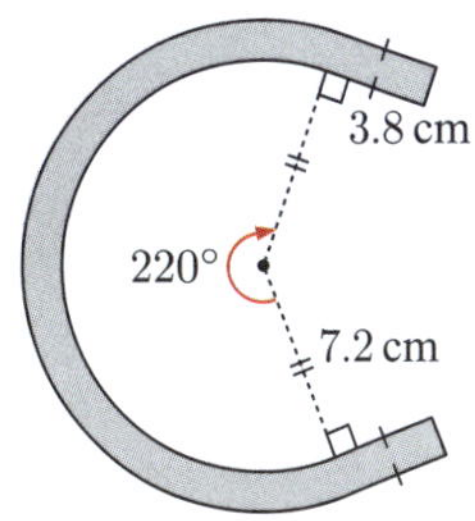

ACTIVITY **COMPARING PRISMS**

In this Activity we consider how the shape of a prism affects its surface area to volume ratio.

ACTIVITY

G VOLUME OF A TAPERED SOLID

Pyramids and cones are known as **tapered solids**. They have a flat base, and come to a point called the **apex**.

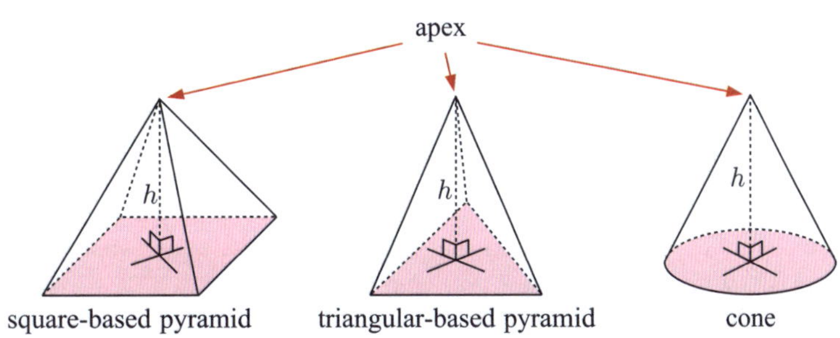

DEMO

Tapered solids do *not* have uniform cross-sections.

Their cross-sections always have the same *shape*, but not the same *size*.

INVESTIGATION 2 THE VOLUME OF A SQUARE-BASED PYRAMID

In this Investigation we will discover a formula for the volume of a square-based pyramid by comparing it with the square-based prism with the same base area and height.

What to do:

1 We first consider square-based pyramids in which the base length is twice its height, as shown in figure **A**.

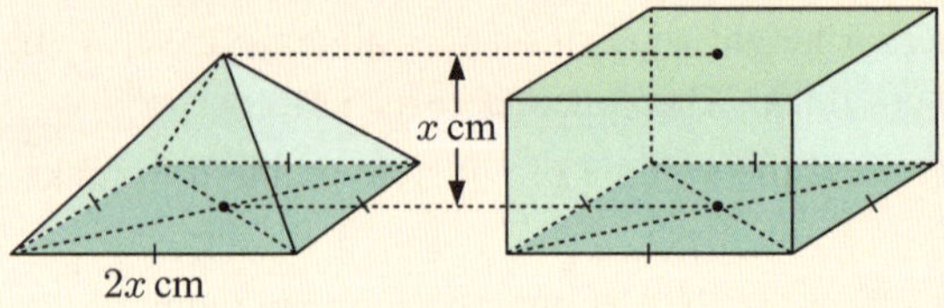

A **B**

NETS

a Click on the icon to obtain a printable sheet containing nets for six identical square-based pyramids.

b Print the nets, cut them out, and construct the pyramids using sticky tape.

c Arrange the pyramids to form a cube.

d Explain why the volume of the cube is:

  i six times the volume of figure **A**

  ii twice the volume of figure **B**.

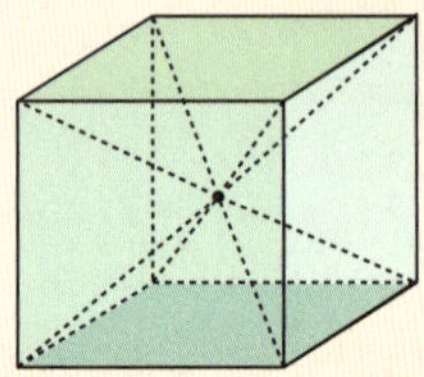

e Hence determine the relationship between the volume of the square-based pyramid and the volume of the rectangular prism in figure **B**.

2 Now consider another arrangement of square-based pyramids.

NETS

a Click on the icon to obtain a printable sheet containing nets for three identical square-based pyramids.

b Print the nets, cut them out, and construct the pyramids using sticky tape. Notice that the apex of each pyramid is not above the centre of the base, but instead is above one of the vertices.

c Arrange the pyramids to form a square-based prism. Notice that the height of the prism is the same as the height of each pyramid.

d What is the relationship between the volume of the pyramid and the volume of the prism?

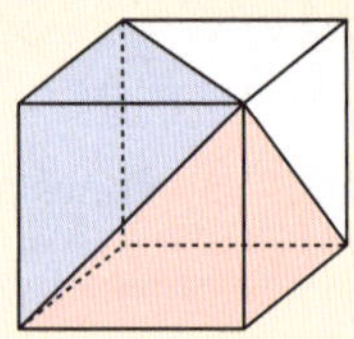

3 Given that the volume of a prism is $V =$ area of base $\times$ height, predict a formula for the volume of a pyramid.

From the **Investigation**, you should have found that the volume of a square-based pyramid is one third of the volume of the prism with the same base area and height.

More generally, for any tapered solid:

$$\textbf{Volume of tapered solid} = \frac{1}{3} \times \textbf{area of base} \times \textbf{height}$$

Example 7 — Self Tutor

Find the volume of each tapered solid:

a

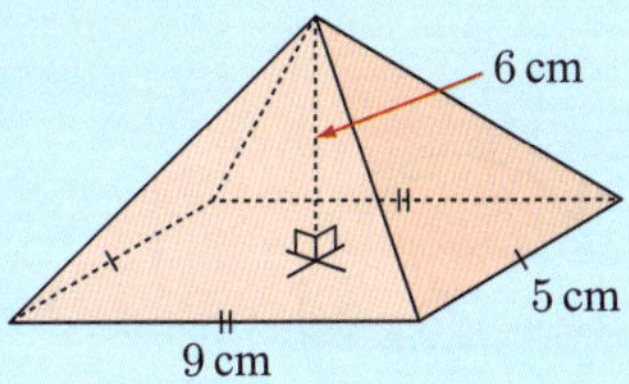

b

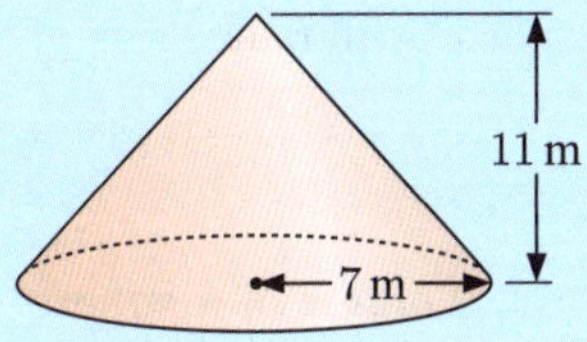

a Volume
$= \frac{1}{3} \times \text{area of base} \times \text{height}$
$= \frac{1}{3} \times 9 \times 5 \times 6 \text{ cm}^3$
$= 90 \text{ cm}^3$

b Volume
$= \frac{1}{3} \times \text{area of base} \times \text{height}$
$= \frac{1}{3} \times \pi \times 7^2 \times 11 \text{ m}^3$
$\approx 564 \text{ m}^3$

EXERCISE 13G

1 Find the volume of each tapered solid:

a

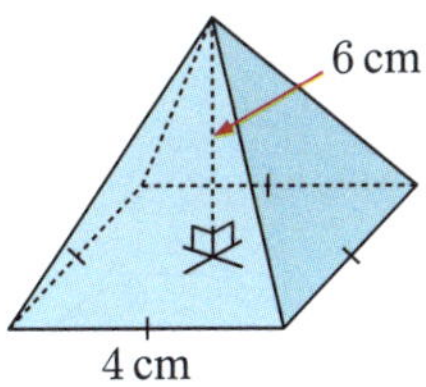

b

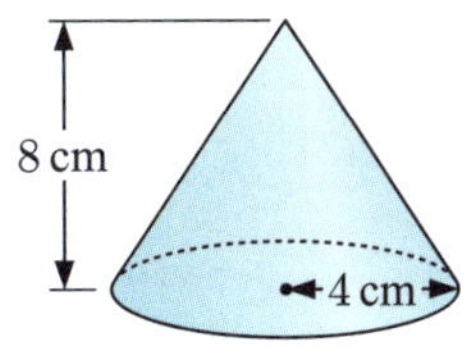

c

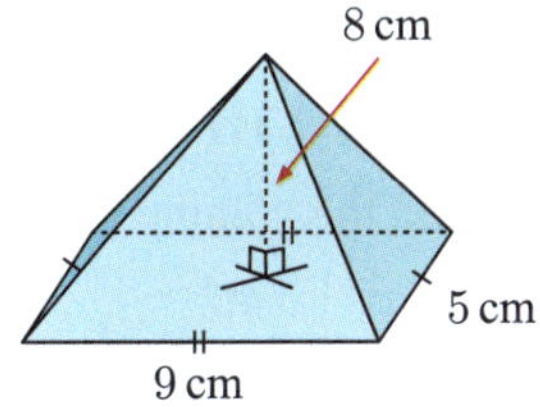

2 Find the volume of:

a

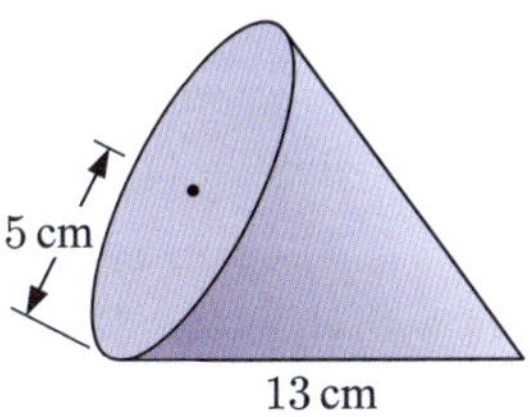

b

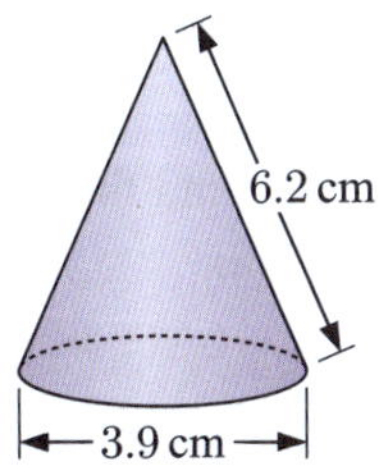

c

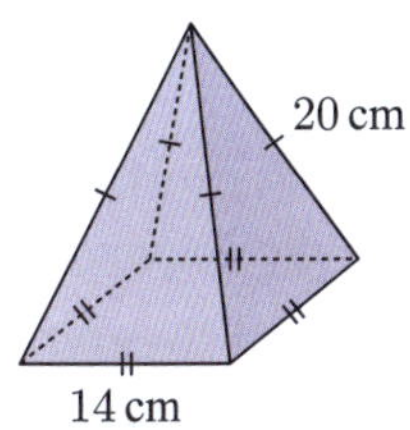

3 The volume of a cone with radius r and height h is given by $V = \frac{1}{3}\pi r^2 h$.

  a Rearrange the formula to make r the subject.

  b Hence find the base radius of a cone with:

    i height 6 cm and volume 100 cm^3

    ii height 10 cm and volume 200 cm^3.

4 A wax crayon has the dimensions shown. Find the total volume of wax required to make a set of 24 crayons.

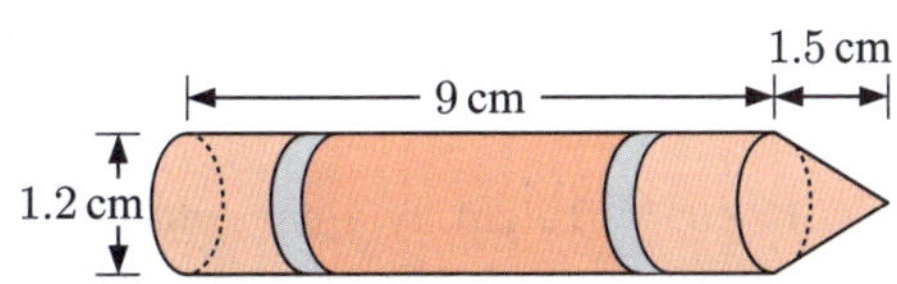

5 A square-based pyramid has volume 864 cm^3. Its base has side length 12 cm. Find the surface area of the pyramid.

H VOLUME OF A SPHERE

The volume of a sphere with radius r is given by:

Volume of a sphere $= \frac{4}{3}\pi r^3$

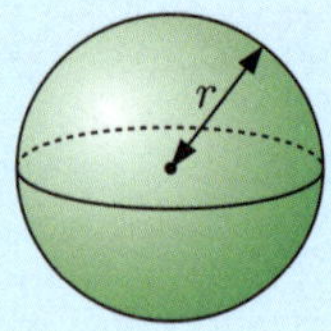

Example 8 | Self Tutor

Find the volume of this sphere:

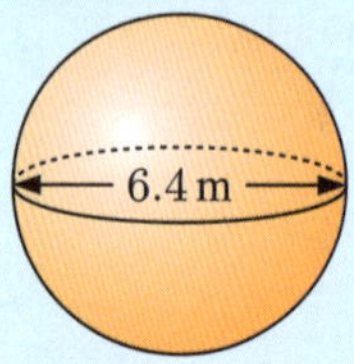

Volume $= \frac{4}{3}\pi r^3$

$= \frac{4}{3} \times \pi \times 3.2^3 \text{ m}^3$

$\approx 137 \text{ m}^3$

EXERCISE 13H

1 Find the volume of each solid:

a

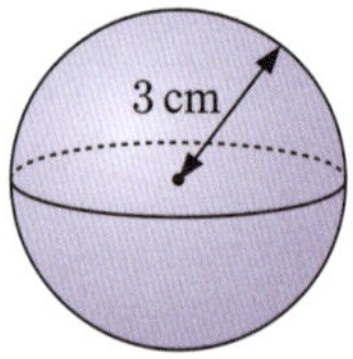

b

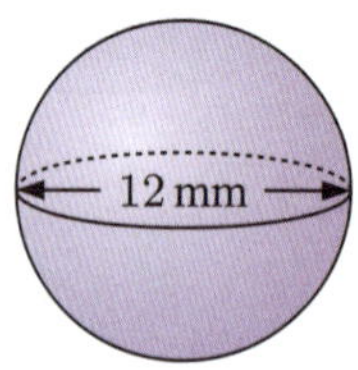

c

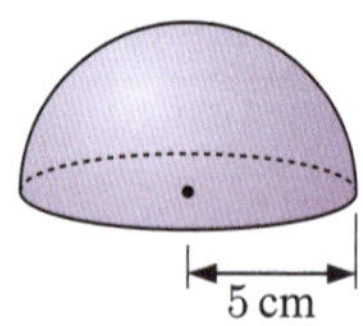

2 A spherical balloon has radius 15 cm. Find the volume of air inside the balloon.

3 Find the volume of:

a a sphere with diameter 9.2 m

b a hemisphere with diameter 13.8 cm.

4 How many chocolate spheres with radius 2.5 cm can be made from 1.3 m^3 of chocolate?

5 Spherical hollow glass baubles have internal diameter 4.9 cm and external diameter 5.0 cm. Find the volume of glass in each bauble.

6 The volume of a sphere with radius r is given by $V = \frac{4}{3}\pi r^3$.

a Make r the subject of the formula.

b Hence find the radius of a sphere which has volume:

i 40 cm^3

ii 800 cm^3

iii 1 000 000 cm^3.

7 A 15 cm by 10 cm by 4 cm gold ingot is melted down and cast into spheres of radius 1 cm.

a How many spheres can be cast?

b What volume of gold is left over?

c The remaining gold is cast into a smaller sphere. Find the radius of this sphere in millimetres.

I CAPACITY

The **capacity** of a container is the amount of fluid it can contain.

The units for capacity are:

- millilitres (mL)
- litres (L)
- kilolitres (kL)
- megalitres (ML).

1 ML = 1000 kL
1 kL = 1000 L
1 L = 1000 mL

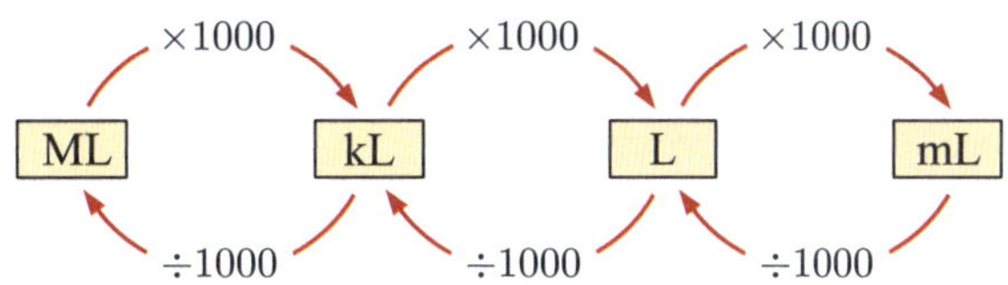

CONNECTING CAPACITY AND VOLUME

The units of capacity and volume are related:

$1 \text{ mL} \equiv 1 \text{ cm}^3$
$1 \text{ L} \equiv 1000 \text{ cm}^3$
$1 \text{ kL} \equiv 1 \text{ m}^3$
$1 \text{ ML} \equiv 1000 \text{ m}^3$

Example 9 Self Tutor

A trumpet case is a solid of uniform cross-section. Its cross-section is a rectangle whose corners have been rounded using quarter circles with radius 3 cm.

Find the capacity of the trumpet case in litres.

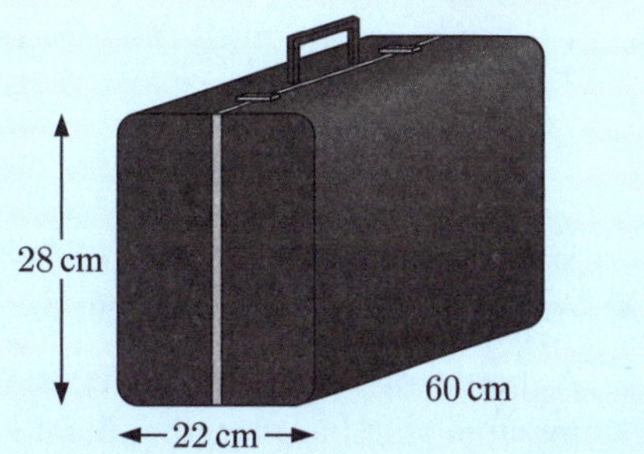

Area of cross-section
= area of 4 quarter circles + area of 2 small rectangles + area of large rectangle
$= (\pi \times 3^2 + 2 \times 3 \times 16 + 22 \times 22) \text{ cm}^2$
$\approx 608.3 \text{ cm}^2$

$\therefore$ volume $\approx 608.3 \times 60 \text{ cm}^3$
$\approx 36\,496 \text{ cm}^3$

$\therefore$ capacity $\approx 36\,496$ mL
$\approx 36\,496 \div 1000$ L
≈ 36.5 L

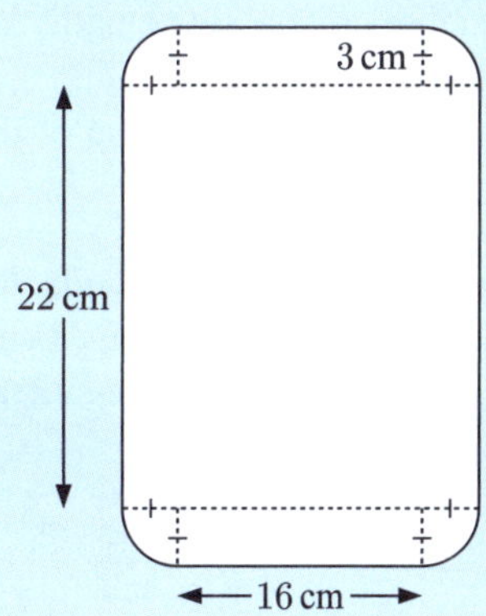

EXERCISE 13I

1. Convert:
   - a 5.6 L into mL
   - b 524 kL into L
   - c 280 mL into L
   - d 6300 kL into ML
   - e 3.5 ML into kL
   - f 83 L into kL
   - g 0.31 L into mL
   - h 743 000 L into ML
   - i 0.84 kL into mL.

2. A villager fills a 15 L clay pot of water from a well each day. How much water, in kilolitres, does she draw from the well each year? Assume that 1 year = 365 days.

3. The average adult has 5.2 litres of blood in their body. During a blood donation, approximately 470 mL of blood is taken. What percentage of an adult's blood is taken during a donation?

4. Find the capacity of each container:

a
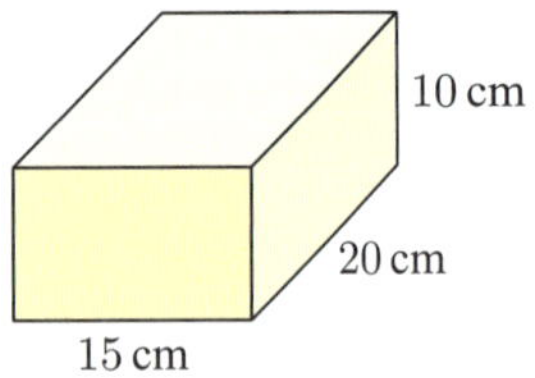

b
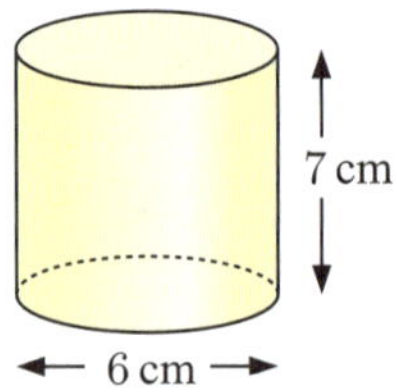

c
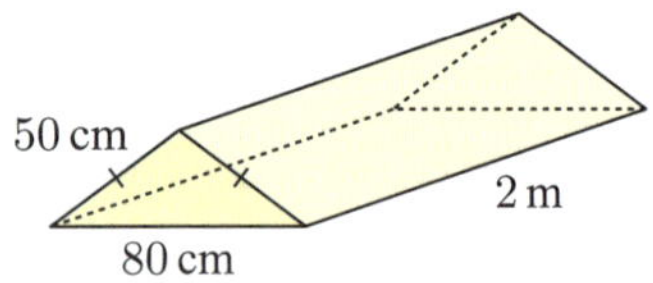

5. A car has a rectangular prism petrol tank with dimensions 48 cm by 56 cm by 20 cm.
   - a Find the capacity of the petrol tank.
   - b The car consumes petrol at an average rate of 8.7 litres per 100 km. How far can the car travel on a full tank of petrol?

6. Find the radius of a spherical balloon with capacity 1.8 L.

7. This pot is filled 6 cm from the top with soup. How many 1.8 L tubs can be filled from the pot?

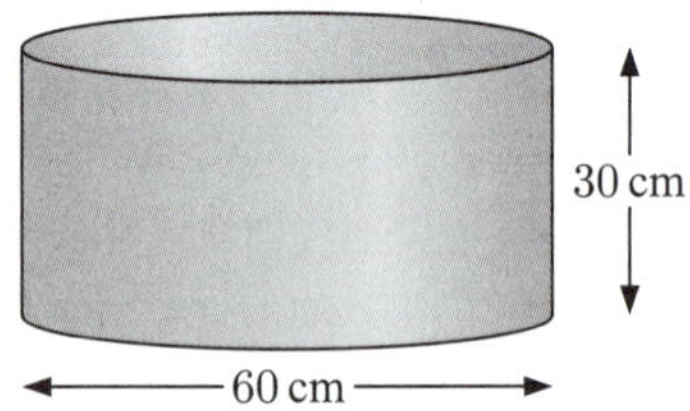

8. Answer the **Opening Problem** on page **186**.

9. A bottle of Sphynx cologne is a *frustum* in which a pyramid with base length 1.5 cm has been removed from the top of a larger pyramid. The height of each pyramid is equal to its base side length. When full, the bottle contains 200 mL of cologne.
   Find the area of the base of the bottle.

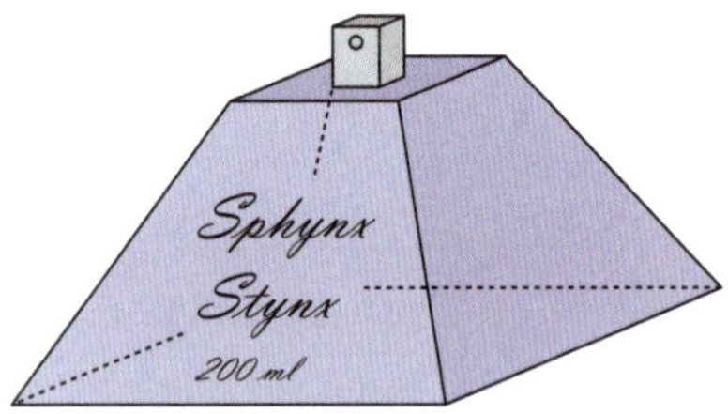

10. This suitcase is 50 cm wide, 60 cm high, and 30 cm deep. Its corners and edges have been rounded with radius 5 cm. Find the capacity of the suitcase in litres.

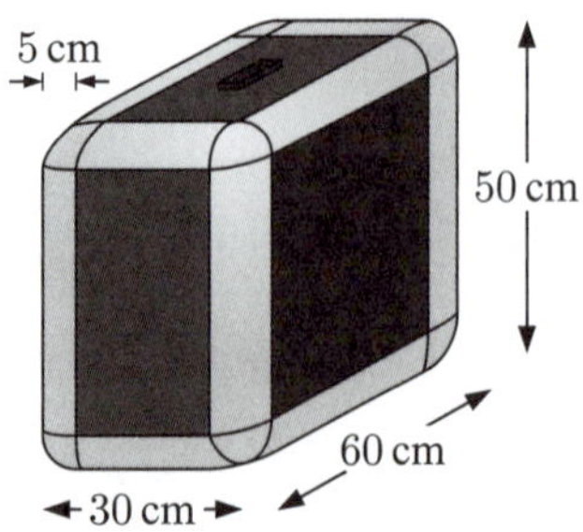

GLOBAL CONTEXT — FAST FASHION

Global context: Fairness and development

Statement of inquiry: Performing measurements and calculations can help us to responsibly manage natural resources.

Criterion: Applying mathematics in real-life contexts

GLOBAL CONTEXT

MULTIPLE CHOICE QUIZ

QUICK QUIZ

REVIEW SET 13A

1 Find the surface area of:

a

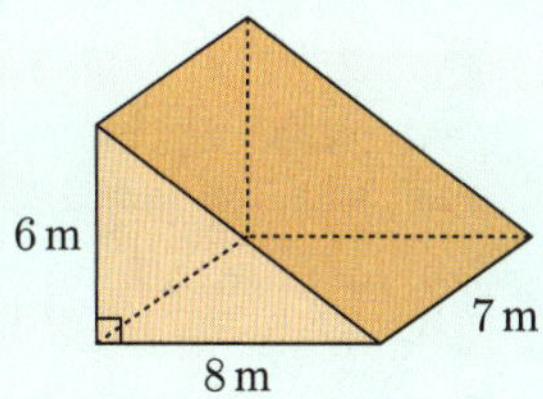

b

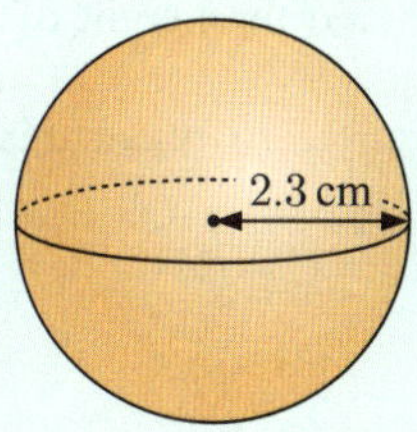

c

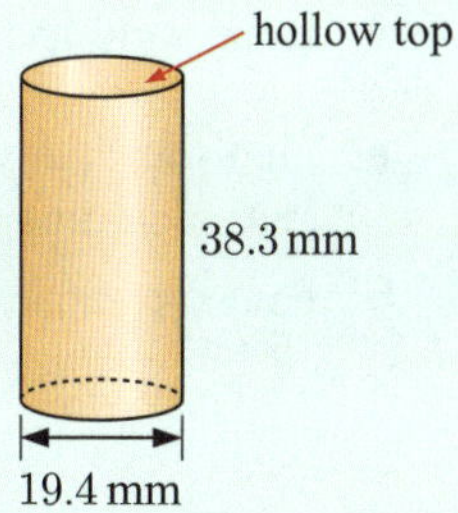

2 Convert:

a 0.2 cm^3 into mm^3 **b** 57 mm^3 into cm^3 **c** 9.04 m^3 into cm^3.

3 Find the surface area and volume of:

a

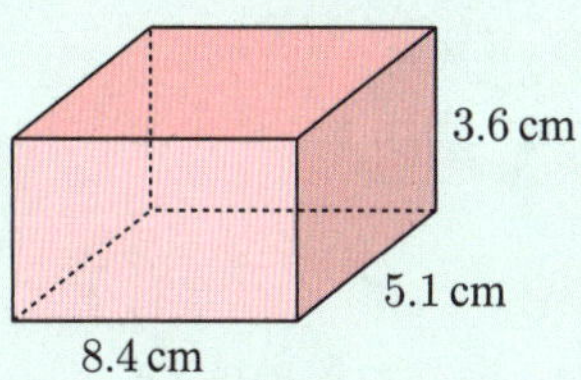

b

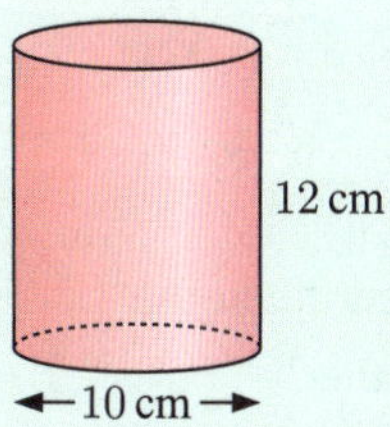

c

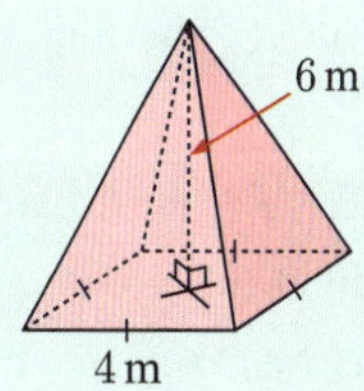

4 Find the volume of:

a a cone with diameter 9 cm and height 8 cm

b a hemisphere with radius 2.8 m.

5 A solid cone has base radius 2 m and surface area $10\pi\text{ m}^2$. Find the slant height of the cone.

6 The surface area of a sphere is 86.26 cm^2. Find the diameter of the sphere correct to 3 significant figures.

7 A dam contained 6.4 ML of water. A storm added 890 kL of water to the dam. How much water does the dam now contain?

8 This diagram shows the cross-section of a railway cutting that needs to be excavated. The cutting will be 56 m long. Find the volume of soil that needs to be excavated.

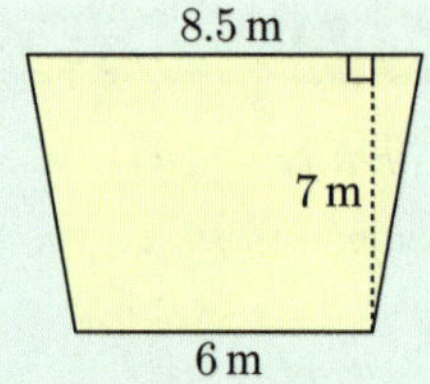

9 Write a formula for the surface area A of each solid:

a

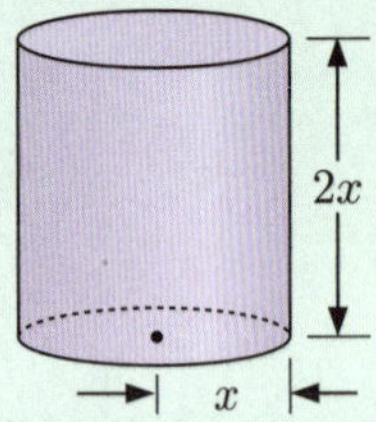

b

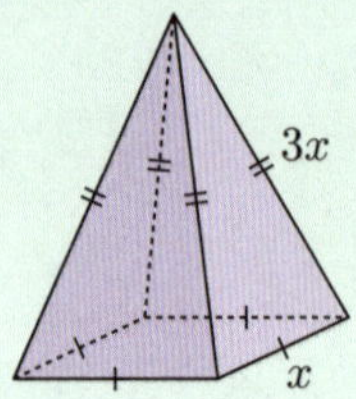

c

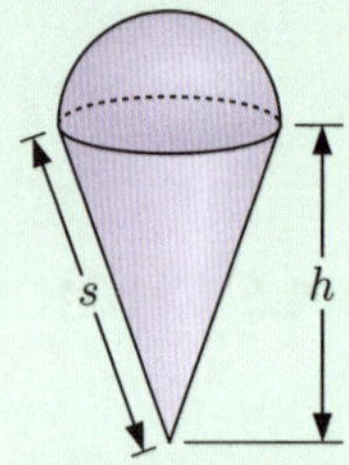

10 A conical heap of sand is twice as wide as it is high.

- **a** Suppose the heap of sand is h m high. Write a formula for the volume V of the sand in terms of h.
- **b** Rearrange the formula to make h the subject.
- **c** If the sand has volume 5 m^3, find the height of the heap.

11
- **a** Find the area of material required to make this tent, including the floor.
- **b** Find the capacity of the tent.

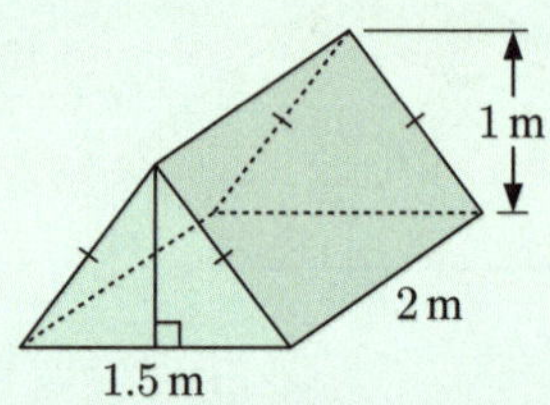

12 A sphere with radius r is cut into n identical wedges. Write a formula for their *total* surface area S.

13 A school hall contains 40 lockers, each with interior 40 cm × 60 cm × 60 cm. Find, in kL, the total capacity of the lockers.

14 Ben makes figurines, each of which requires 6.75 cm^3 of resin. Ben has melted 1.5 L of resin to make his latest batch.

- **a** Find the volume of one figurine in mm^3.
- **b** How many mL of resin is this?
- **c** How many *complete* figurines can Ben make?
- **d** Ben melted the 1.5 L of resin in a cylindrical container with diameter 8 cm. Find the depth of resin in the container.

15 A petrol tank is constructed as shown, as a cylinder with hemispherical ends.

- **a** Find the capacity of the tank, in kL.
- **b** Suppose petrol is pumped out of the full tank at the rate of 300 L per minute. How long would it take to empty the tank?

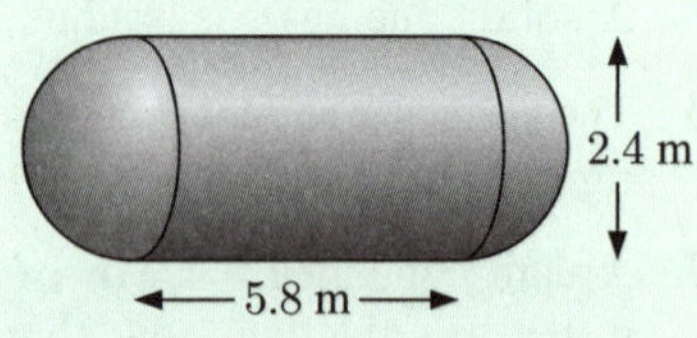

REVIEW SET 13B

1 Write 750 mm^3 as a percentage of 5 cm^3.

2 Convert:

a 820 mL into L **b** 3.07 ML into kL **c** 0.72 kL into mL.

3 Find the surface area of each cone:

a solid

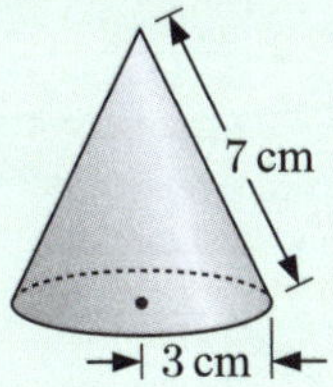

b hollow

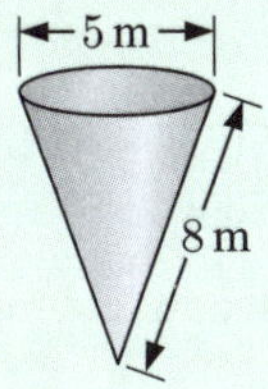

c solid

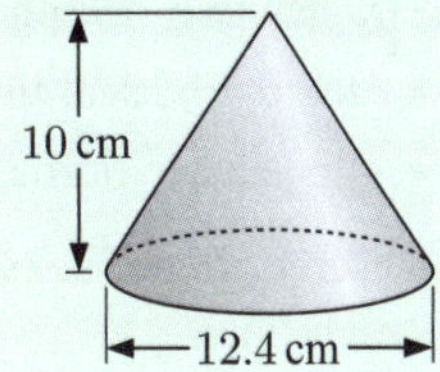

4 Find the surface area and volume of:

a

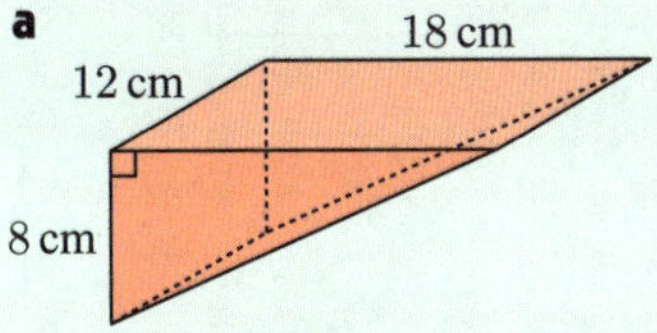

b

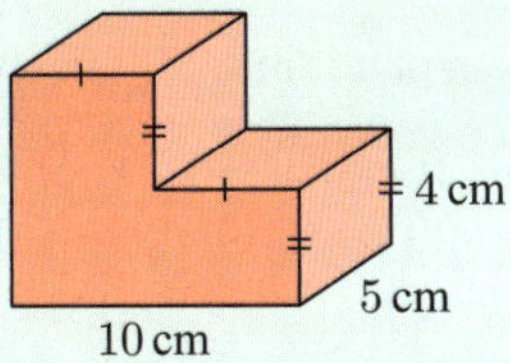

c

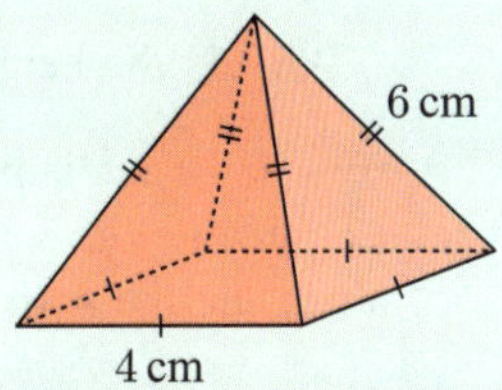

5 How many cubes of side length 5 cm can be painted with 2 litres of paint, given that each litre of paint covers 12 m^2?

6 Find the diameter of a cylinder with height 6 cm and volume 300 cm^3.

7 A cylindrical tree trunk with length 2.5 m and diameter 60 cm is sliced in half.

a Find the surface area of this slice.

b Find the volume of wood in this slice.

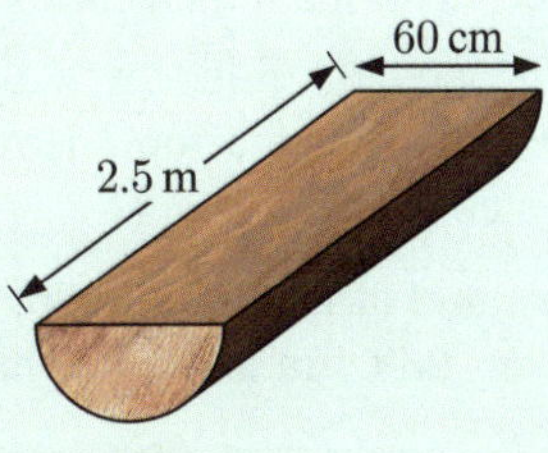

8 Soup cans have base diameter 7 cm and height 10 cm. How many soup can labels can be made from 5 m^2 of paper?

9 A wedge with angle $\theta°$ is cut from the centre of a cylindrical cake with radius r and height h. Write a formula for its:

a surface area A **b** volume V.

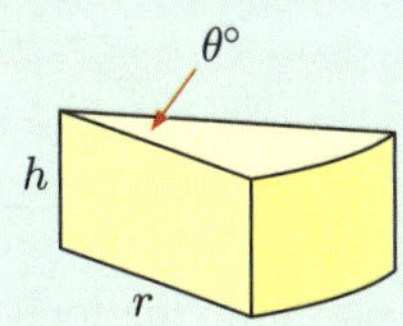

10 The moon is approximately spherical with radius 1737 km. Estimate:

a the distance around the equator of the moon

b the surface area of the moon

c the volume of the moon.

11 Write a formula for the volume V of:

a 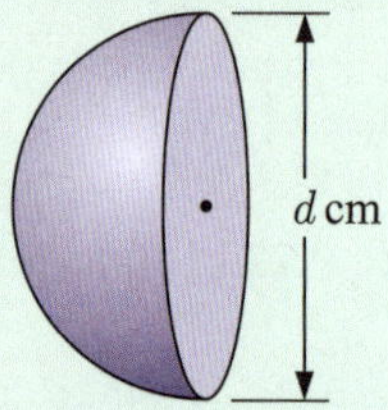

b 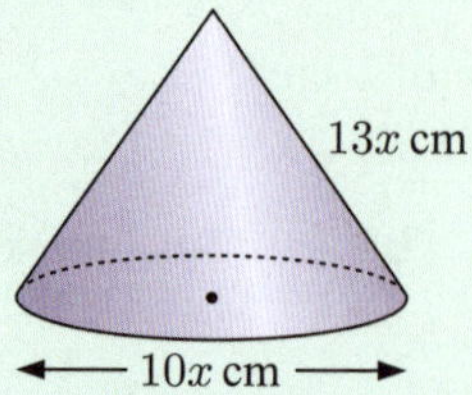

c 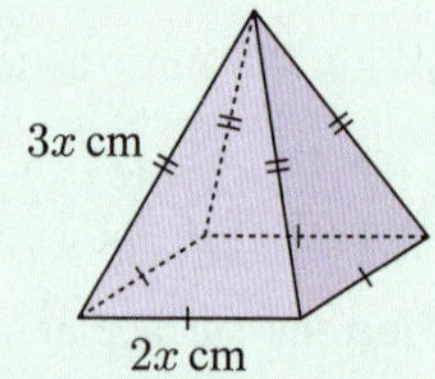

12 Keira has two baking tins:

- the cylindrical tin has radius 12 cm and height 6 cm
- the square tin has side length 15 cm.

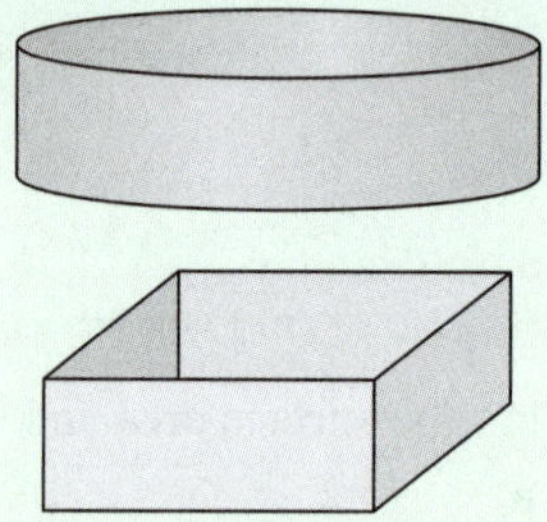

The same area of baking paper is needed to line the inside of each tin.

a What area of baking paper is needed to line the inside of the cylindrical tin?

b What is the height of the square tin?

13 A zorb ball has diameter 3 m. It contains 30 L of water and a person with volume 75 000 cm^3. Find the volume of air inside the zorb ball.

14 **a** Find the capacity of a swimming pool with the dimensions shown.

b The pool is filled to a depth of 10 cm from the top. How much water does it contain?

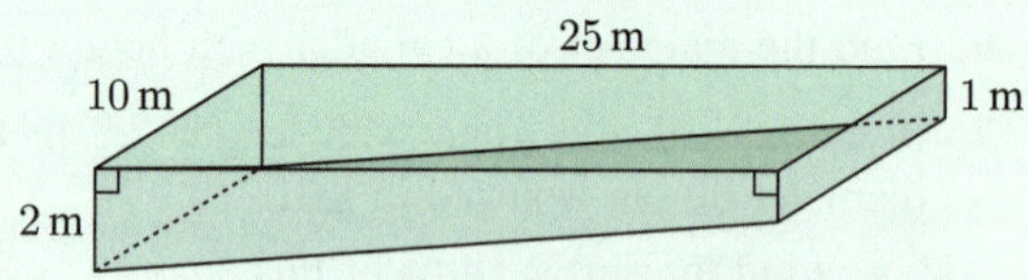

15 The Pyramid of Khufu in Egypt has a square base with sides 230.4 m and a height of 138.8 m.

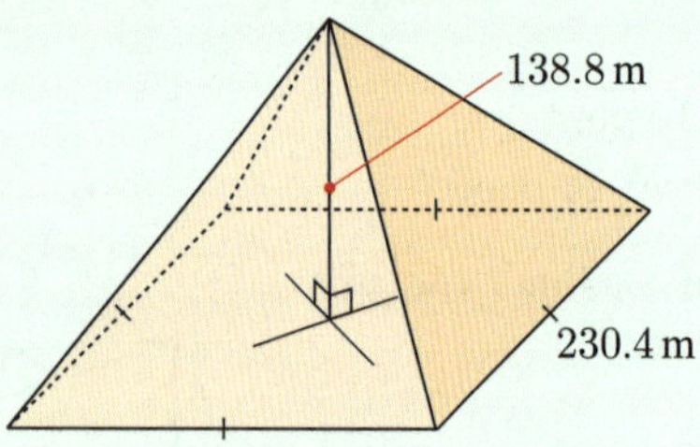

a Find the total volume of the pyramid.

b Suppose the pyramid is solid stone. If the stone has density 2.67 tonnes per m^3, find the total mass of stone used. Give your answer in scientific notation.

c The King's Chamber in the pyramid is rectangular and measures 10.47 m by 5.23 m by 5.97 m. Find the capacity of the chamber.

Chapter 14

Algebra: Factorisation

Contents:

OPENING PROBLEM

A square garden plot is surrounded by a path 1 m wide. The sides of the garden plot have length x m, as shown.

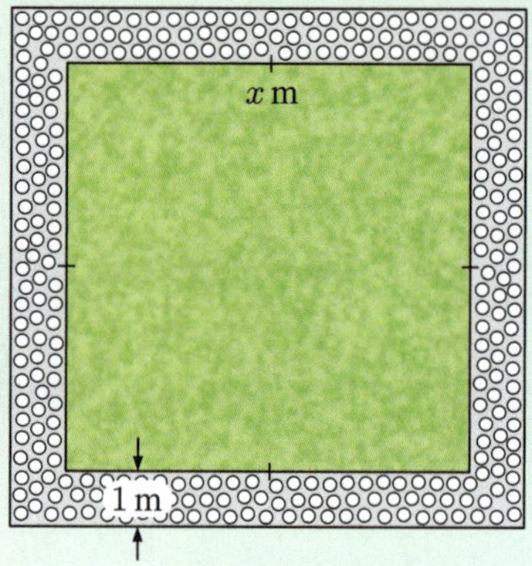

The landscaper needs to know the area of the path so that he can order enough pebbles to cover it.

Things to think about:

a Can you explain why the area of the path can be expressed as:
  i $[(x+2)^2 - x^2]$ m^2 **ii** $(4x+4)$ m^2?

b Can you use an expansion law in *reverse* to explain why $(x+2)^2 - x^2 = 2(2x+2)$?

In **Chapter 4** we studied algebraic **expansion** which allows us to write expressions without brackets. The result is a sum of algebraic terms.

For example, using the distributive law, $5(x-1) = 5x - 5$.

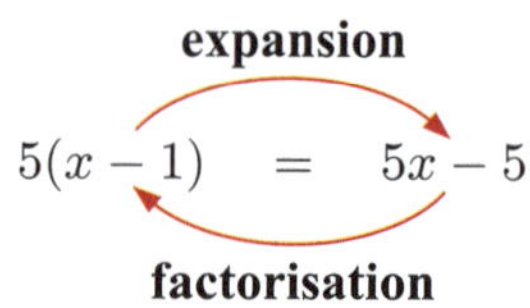

The reverse process of expansion is called *factorisation.*

When we factorise an expression, we *insert* brackets.

$5(x-1)$ is the product of two factors: 5 and $(x-1)$.

Factorisation is the process of writing an expression as a **product** of its **factors**.

Since factorisation is the reverse process of expansion, we can perform it by applying the expansion laws in reverse.

A COMMON FACTORS

When we apply the distributive law $a(b+c) = ab + ac$, the factor a is a common factor of each term in the expansion.

If the HCF of the terms in an expression is not 1, we write it in front of a set of brackets and use the reverse of the distributive law.

$$\boldsymbol{ab + ac = a(b+c)}$$

For example: $6x^2$ and $2xy$ have HCF $= 2x$.

$$\therefore \; 6x^2 + 2xy = 2x \times 3x + 2x \times y$$
$$= 2x(3x+y)$$

An expression is **fully factorised** if none of its algebraic factors can be factorised further.

For example: $18x + 12 = 3(6x+4)$ is not fully factorised since $(6x+4)$ still has the common factor 2.

$18x + 12 = 6(3x+2)$ is fully factorised.

This is why we must write the *highest* common factor in front of the brackets.

EXERCISE 14A

1 Copy and complete:

a $3x+6=3(x+......)$ **b** $5x-15=5(x-......)$

c $6a+12=6(......+2)$ **d** $20-5m=5(......-m)$

e $4x+8x^2=4x(......+2x)$ **f** $3x^2-6x=3x(x-......)$

g $3p+9=3(......+......)$ **h** $6x-24=6(......-......)$

i $4m+8mn=......(1+2n)$ **j** $15a-5a^2=......(3-a)$

k $cd+de=d(......+......)$ **l** $4ab-6bc=......(2a-3c)$

Example 1 Self Tutor

Fully factorise:

a $4-2x$ **b** $3ab+ac$

a $4-2x$
$=2\times 2-2\times x$
$=2(2-x)$ {HCF $=2$}

b $3ab+ac$
$=3\times a\times b+a\times c$
$=a(3b+c)$ {HCF $=a$}

2 Fully factorise:

a $2x+6$ **b** $2x-6$ **c** $5a+10$ **d** $5a-10$

e $4+6x$ **f** $4-6x$ **g** $a+ab$ **h** $a-ab$

3 Fully factorise:

a $3x-6$ **b** $8d+40$ **c** $10x-30$ **d** $4+2x$

e $5p+5q$ **f** $ab+ad$ **g** $42-35x$ **h** $gh+h$

i $y+yz$ **j** $p-pq$ **k** $3e+ef$ **l** $9x-xy$

m $jk-3hk$ **n** $rs-st$ **o** $3xy-2yz$ **p** $4mn+lm$

Example 2 Self Tutor

Fully factorise:

a $8x^2+12x$ **b** $3y^2-6xy$

a $8x^2+12x$
$=2\times 4\times x\times x+3\times 4\times x$
$=4x(2x+3)$ {HCF $=4x$}

b $3y^2-6xy$
$=3\times y\times y-2\times 3\times x\times y$
$=3y(y-2x)$ {HCF $=3y$}

4 Fully factorise:

a x^2+5x **b** $5x-2x^2$

c $4x^2+8x$ **d** $14x-7x^2$

e $6x^2+12x$ **f** x^3+9x^2

g x^2y+xy^2 **h** $4x^3-6x^2$

i $9x^3-18xy$ **j** $6xy^2+9xy$

5 Fully factorise:

a $a^3 + a^2 + a$ **b** $2a^2 + 4a + 8$ **c** $3a^3 - 6a^2 + 9a$

6 Explain why each expression is not *fully* factorised:

a $x(6 - 2x)$ **b** $6(2x^2 - 3xy)$

Example 3 Self Tutor

Fully factorise:

a $-2a + 6ab$ **b** $-2x^2 - 4x$

a $-2a + 6ab$
$= 6ab - 2a$ {Write with $6ab$ first.}
$= 2 \times 3 \times a \times b - 2 \times a$
$= 2a(3b - 1)$ {HCF $= 2a$}

b $-2x^2 - 4x$
$= -2 \times x \times x + -2 \times 2 \times x$
$= -2x(x + 2)$ {HCF $= -2x$}

7 Fully factorise:

a $-9a + 9b$ **b** $-3 + 6b$ **c** $-8a + 4b$ **d** $-7c + cd$
e $-a + ab$ **f** $-6x^2 + 12x$ **g** $-5x + 15x^2$ **h** $-2b^2 + 4ab$

8 Fully factorise:

a $-6a - 6b$ **b** $-4 - 8x$ **c** $-3y - 6z$ **d** $-9c - cd$
e $-x - xy$ **f** $-5x^2 - 20x$ **g** $-12y - 3y^2$ **h** $-18a^2 - 9ab$

Example 4 Self Tutor

Fully factorise:

a $2(x + 3) + x(x + 3)$ **b** $x(x + 4) - (x + 4)$

a $2(x + 3) + x(x + 3)$ {HCF $= (x + 3)$}
$= (x + 3)(2 + x)$

b $x(x + 4) - (x + 4)$ {HCF $= (x + 4)$}
$= x(x + 4) - 1(x + 4)$
$= (x + 4)(x - 1)$

9 Fully factorise:

a $2(x - 7) + x(x - 7)$ **b** $a(x + 3) + b(x + 3)$ **c** $4(x + 2) - x(x + 2)$
d $x(x + 9) + (x + 9)$ **e** $a(b + 4) - (b + 4)$ **f** $a(b + c) + d(b + c)$
g $a(m + n) - b(m + n)$ **h** $x(x + 3) - x - 3$ **i** $2x(x - 2) + 2 - x$

Example 5 Self Tutor

Fully factorise $(x - 1)(x + 2) + 3(x - 1)$

$(x - 1)(x + 2) + 3(x - 1)$ {HCF $= (x - 1)$}
$= (x - 1)[(x + 2) + 3]$
$= (x - 1)(x + 5)$

We use square brackets to help distinguish them from the round brackets.

10 Fully factorise:

- **a** $(x+3)(x-5)+4(x+3)$
- **b** $5(x-7)+(x-7)(x+2)$
- **c** $(x+6)(x+4)-8(x+6)$
- **d** $(x-2)^2-6(x-2)$
- **e** $(x+2)^2-(x+2)(x+1)$
- **f** $5(a+b)-(a+b)(a+1)$
- **g** $3(a-2)^2-6(a-2)$
- **h** $(x+4)^2+3(x+4)(x-1)$
- **i** $x(x-1)-6(x-1)(x-5)$
- **j** $3(x+5)-4(x+5)^2$

11 Write a fully factorised expression for the total area of:

- **a** the blue rectangles
- **b** the red rectangles
- **c** the figure.

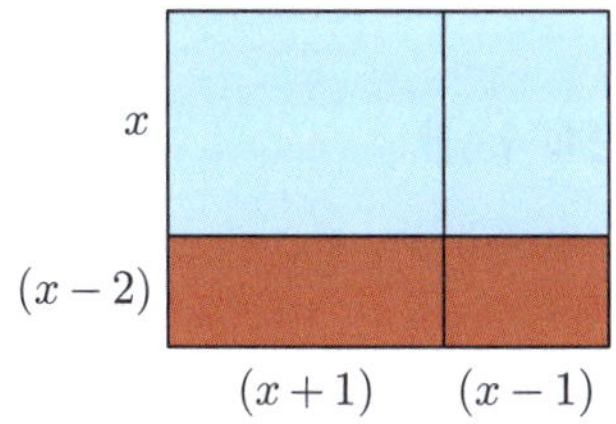

B DIFFERENCE BETWEEN TWO SQUARES FACTORISATION

INVESTIGATION THE DIFFERENCE BETWEEN TWO SQUARES

In the diagram alongside, a square with side length b has been cut from a square with side length a.

WORKSHEET

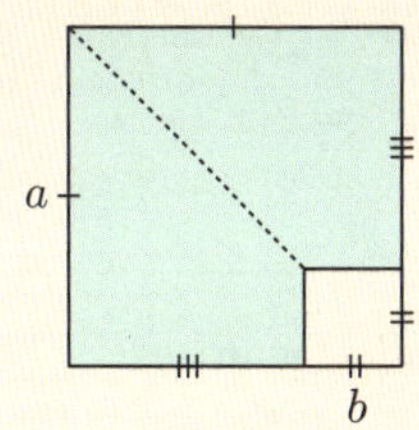

What to do:

1. Explain why the green shaded area is given by a^2-b^2.
2. Copy the above diagram, or print the **worksheet** by clicking on the icon. Cut along the dotted line.
3. Rearrange the two trapezia to form a rectangle as shown.
4. Write, in terms of a and b, the length of:
   - **a** [AB]
   - **b** [BC]
5. Hence write the area of the rectangle in the form (......)(......).
6. What can be deduced by equating the expressions in **1** and **5**?

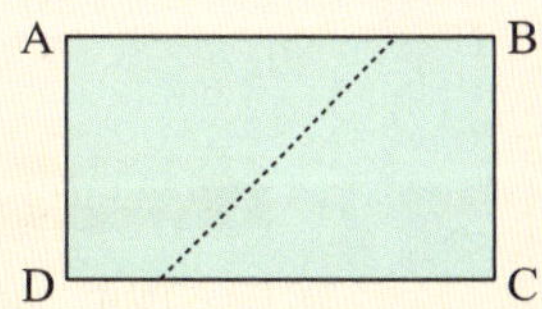

We know the expansion of $(a+b)(a-b)$ is a^2-b^2.

$\therefore$ the factorisation of a^2-b^2 is $(a+b)(a-b)$.

The **difference** between a^2 and b^2 is a^2-b^2.

$$\boldsymbol{a^2-b^2=(a+b)(a-b)}$$

By contrast, the **sum** of two squares does not factorise into real linear factors.

For example:

- x^2-4 is the difference between two squares, so $x^2-4=(x+2)(x-2)$.
- x^2+4 is the sum of two squares, which does not factorise into real linear factors.

Example 6 — Self Tutor

Fully factorise:

a $36 - x^2$ **b** $1 - 25y^2$

a $36 - x^2$
$= 6^2 - x^2$
$= (6 + x)(6 - x)$

b $1 - 25y^2$
$= 1^2 - (5y)^2$
$= (1 + 5y)(1 - 5y)$

EXERCISE 14B

1 Fully factorise:

a $d^2 - e^2$ **b** $p^2 - q^2$ **c** $q^2 - p^2$ **d** $x^2 - y^2$
e $x^2 - 9$ **f** $x^2 - 100$ **g** $25 - x^2$ **h** $y^2 - 49$
i $81 - a^2$ **j** $4b^2 - 1$ **k** $1 - 4x^2$ **l** $16x^2 - 9$
m $36y^2 - 49$ **n** $81 - d^2$ **o** $121 - 9y^2$

Example 7 — Self Tutor

Fully factorise:

a $2x^2 - 18$ **b** $x^3 - xy^2$

a $2x^2 - 18$
$= 2(x^2 - 9)$
$= 2(x + 3)(x - 3)$

b $x^3 - xy^2$
$= x(x^2 - y^2)$
$= x(x + y)(x - y)$

2 Fully factorise:

a $3x^2 - 75$ **b** $8x^2 - 32$ **c** $5a^2 - 20$ **d** $5x^2 - 5$
e $9b^2 - 900$ **f** $80 - 5t^2$ **g** $6k^2 - 24$ **h** $15 - 15y^2$
i $r^3 - 49r$ **j** $z^3 - z$ **k** $x^5 - x^3$ **l** $x^3y - xy^3$

Example 8 — Self Tutor

Fully factorise:

a $4a^2 - 9b^2$ **b** $x^2y^2 - 16$

a $4a^2 - 9b^2$
$= (2a)^2 - (3b)^2$
$= (2a + 3b)(2a - 3b)$

b $x^2y^2 - 16$
$= (xy)^2 - 4^2$
$= (xy + 4)(xy - 4)$

3 Fully factorise:

a $25a^2 - b^2$ **b** $x^2 - 81y^2$ **c** $16x^2 - 49y^2$ **d** $25p^2 - 36q^2$
e $81j^2 - k^2$ **f** $s^2t^2 - 9$ **g** $49x^2 - y^2z^2$ **h** $4d^2 - 25e^2$

4 Fully factorise:

a $8x^2 - 2$ **b** $12 - 27x^2$ **c** $2y^2 - 8x^2$ **d** $9a - 16a^3$

Example 9 Self Tutor

Factorise using the difference between two squares:

a $(x+2)^2-9$ **b** $25-(x-2)^2$

a
$$\begin{aligned}&(x+2)^2-9\\&=(x+2)^2-3^2\\&=[(x+2)+3][(x+2)-3]\\&=[x+2+3][x+2-3]\\&=(x+5)(x-1)\end{aligned}$$

b
$$\begin{aligned}&25-(x-2)^2\\&=5^2-(x-2)^2\\&=[5+(x-2)][5-(x-2)]\\&=[5+x-2][5-x+2]\\&=(x+3)(7-x)\end{aligned}$$

5 Factorise using the difference between two squares:

a $(x+2)^2-4$ **b** $(x-3)^2-36$ **c** $100-(x-11)^2$

d $81-(x-8)^2$ **e** $(x+7)^2-1$ **f** $9-(x+4)^2$

6 Answer the **Opening Problem** on page **208**.

7 Factorise using the difference between two squares:

a $(x+3)^2-(x-1)^2$ **b** $(2x-1)^2-(x+1)^2$ **c** $(1-3x)^2-(2x+1)^2$

C PERFECT SQUARES FACTORISATION

The **expansions** of the perfect squares $(a+b)^2$ and $(a-b)^2$ are:

$$\begin{aligned}(a+b)^2&=(a+b)(a+b)\\&=a^2+ab+ab+b^2\\&=a^2+2ab+b^2\end{aligned}\qquad\text{and}\qquad\begin{aligned}(a-b)^2&=(a-b)(a-b)\\&=a^2-ab-ab+b^2\\&=a^2-2ab+b^2\end{aligned}$$

$\therefore$ the **factorisations** for these perfect squares are:

$$\boldsymbol{a^2+2ab+b^2=(a+b)^2}\qquad\text{and}\qquad\boldsymbol{a^2-2ab+b^2=(a-b)^2}$$

So, if an expression contains two squares a^2 and b^2, and a middle term $\pm 2ab$, the *sign* of the middle term indicates the *sign* within the perfect square.

Example 10 Self Tutor

Determine whether each expression is a perfect square:

a $x^2+12x+36$ **b** x^2-4x+9

a $x^2+12x+36=x^2+12x+6^2$

The expression contains two squares x^2 and 6^2, and the middle term $12x=2\times x\times 6$.

$\therefore$ $x^2+12x+36$ is a perfect square.

b $x^2-4x+9=x^2-4x+3^2$

The expression contains two squares x^2 and 3^2, but the middle term $-4x\neq -2\times x\times 3$.

$\therefore$ x^2-4x+9 is *not* a perfect square.

EXERCISE 14C

1 Determine whether each expression is a perfect square:

a $x^2+8x+16$ **b** $x^2+14x+36$ **c** $9+6a+a^2$
d $x^2-12x+49$ **e** t^2-2t+1 **f** $x^2-10x+100$

Example 11 Self Tutor

Use perfect squares rules to fully factorise:

a $x^2+20x+100$ **b** $x^2-8x+16$

a $x^2+20x+100$
$=x^2+2\times x\times 10+10^2$
$=(x+10)^2$

b $x^2-8x+16$
$=x^2-2\times x\times 4+4^2$
$=(x-4)^2$

2 Use perfect squares rules to fully factorise:

a x^2-2x+1 **b** $x^2+10x+25$ **c** x^2+4x+4
d $x^2-14x+49$ **e** $x^2+22x+121$ **f** $x^2-20x+100$
g $x^2+12x+36$ **h** x^2-6x+9 **i** $x^2+24x+144$
j $49+14x+x^2$ **k** $x^2-18x+81$ **l** $x^2-2xy+y^2$

Example 12 Self Tutor

Fully factorise:

a $16x^2+24x+9$ **b** $4x^2-20x+25$

a $16x^2+24x+9$
$=(4x)^2+2\times 4x\times 3+3^2$
$=(4x+3)^2$

b $4x^2-20x+25$
$=(2x)^2-2\times 2x\times 5+5^2$
$=(2x-5)^2$

3 Determine whether each expression is a perfect square:

a $9x^2+6x+1$ **b** $16x^2+28x+49$ **c** $x^2-2xy+4y^2$

4 Fully factorise:

a $4x^2-4x+1$ **b** $25x^2+20x+4$ **c** $64x^2-48x+9$
d $9x^2+42x+49$ **e** $36x^2-60x+25$ **f** $16x^2-40x+25$
g $9x^2-24x+16$ **h** $49x^2+14x+1$ **i** $4x^2+9y^2-12xy$

Example 13 Self Tutor

Fully factorise:

a $3x^2-18x+27$ **b** $-2x^2+8x-8$

a $3x^2-18x+27$
$=3(x^2-6x+9)$
$=3(x^2-2\times x\times 3+3^2)$
$=3(x-3)^2$

b $-2x^2+8x-8$
$=-2(x^2-4x+4)$
$=-2(x^2-2\times x\times 2+2^2)$
$=-2(x-2)^2$

5 Fully factorise:

- **a** $2x^2 + 12x + 18$
- **b** $2x^2 - 4x + 2$
- **c** $3x^2 + 12x + 12$
- **d** $4x^2 - 32x + 64$
- **e** $5x^2 + 50x + 125$
- **f** $-x^2 + 12x - 36$
- **g** $-x^2 - 16x - 64$
- **h** $-2x^2 + 40x - 200$
- **i** $-3x^2 + 30x - 75$

6 Evaluate $30^2 + 2 \times 30 \times 1 + 1^2$, and hence find the square root of 961.

7 Fully factorise:

- **a** $x^3 + 2x^2 + x$
- **b** $x^4 + 2x^2 + 1$
- **c** $x^4 - 2x^2 + 1$

D FACTORISING $x^2 + bx + c$

A **quadratic trinomial** is an algebraic expression of the form $ax^2 + bx + c$ where x is a variable and a, b, c are constants, $a \neq 0$.

In this Section we learn to factorise quadratic trinomials for which $a = 1$. They have the form $x^2 + bx + c$.

Consider the expansion

$$\begin{aligned}(x+5)(x+3) &= x^2 + 3x + 5x + 5 \times 3 \quad \{\text{using FOIL}\}\\ &= x^2 + (3+5)x + 5 \times 3\\ &= x^2 + (\textbf{sum} \text{ of 5 and 3})x + (\textbf{product} \text{ of 5 and 3})\\ &= x^2 + 8x + 15\end{aligned}$$

So, to factorise $x^2 + 8x + 15$, we need two numbers whose sum is 8 and whose product is 15. The numbers are 5 and 3, so $x^2 + 8x + 15 = (x+5)(x+3)$.

In general: $\boldsymbol{x^2 + bx + c = (x+p)(x+q)}$ where $p + q = b$ and $pq = c$.

This process is sometimes called **sum and product** factorisation.

Example 14 Self Tutor

Fully factorise: $x^2 + 11x + 24$

sum $= 11$ and product $= 24$

We consider the factor pairs of 24:

Factor pair	1×24	2×12	3×8	4×6
Sum of factors	25	14	11	10

$\therefore$ the numbers are 3 and 8.

$\therefore$ $x^2 + 11x + 24 = (x+3)(x+8)$

If we try all of the factor pairs with a particular product c but find none with the correct sum b, we cannot use this method of factorisation.

EXERCISE 14D

1 Find two numbers which have:

- **a** sum 7 and product 12
- **b** sum 7 and product 10
- **c** sum 10 and product 16
- **d** sum 11 and product 18
- **e** sum 4 and product -21
- **f** sum -4 and product -21
- **g** sum -4 and product -12
- **h** sum 13 and product -30.

2 Fully factorise:

- **a** $x^2 + 4x + 3$
- **b** $x^2 + 14x + 24$
- **c** $x^2 + 10x + 21$
- **d** $x^2 + 15x + 54$
- **e** $x^2 + 9x + 20$
- **f** $x^2 + 8x + 12$
- **g** $x^2 + 10x + 24$
- **h** $x^2 + 9x + 14$
- **i** $x^2 + 6x + 8$

Example 15 **Self Tutor**

Fully factorise: $x^2 - 7x + 12$

sum $= -7$ and product $= 12$

$\therefore$ the numbers are -3 and -4

$\therefore$ $x^2 - 7x + 12 = (x-3)(x-4)$

3 Fully factorise:

- **a** $x^2 - 3x + 2$
- **b** $x^2 - 4x + 3$
- **c** $x^2 - 5x + 6$
- **d** $x^2 - 14x + 33$
- **e** $x^2 - 16x + 39$
- **f** $x^2 - 19x + 48$
- **g** $x^2 - 11x + 28$
- **h** $x^2 - 14x + 24$
- **i** $x^2 - 20x + 36$

Example 16 **Self Tutor**

Fully factorise:

a $x^2 - 2x - 15$ **b** $x^2 + x - 6$

a sum $= -2$ and product $= -15$

$\therefore$ the numbers are -5 and $+3$

$\therefore$ $x^2 - 2x - 15 = (x-5)(x+3)$

b sum $= 1$ and product $= -6$

$\therefore$ the numbers are $+3$ and -2

$\therefore$ $x^2 + x - 6 = (x+3)(x-2)$

4 Fully factorise:

- **a** $x^2 - 7x - 8$
- **b** $x^2 + 4x - 21$
- **c** $x^2 - x - 2$
- **d** $x^2 - 2x - 8$
- **e** $x^2 + 5x - 24$
- **f** $x^2 - 3x - 10$
- **g** $x^2 + 3x - 54$
- **h** $x^2 + x - 72$
- **i** $x^2 - 4x - 21$
- **j** $x^2 - x - 6$
- **k** $x^2 - 7x - 60$
- **l** $x^2 + 7x - 60$
- **m** $x^2 + 3x - 18$
- **n** $x^2 - 7x - 18$
- **o** $x^2 - 12x - 45$

5 Explain why these quadratic trinomials cannot be factorised using the "sum and product" method:

- **a** $x^2 + 3x - 2$
- **b** $x^2 + 2$
- **c** $x^2 - 8x + 4$

6 Fully factorise:

a x^2+7x+6	**b** $x^2-2x-63$	**c** $x^2-11x+18$
d $x^2+6x-16$	**e** x^2-5x+4	**f** $x^2+12x+35$
g x^2-x-20	**h** $x^2-9x-22$	**i** $x^2+8x-48$
j $x^2-3x-28$	**k** x^2+13x	**l** $x^2-14x+49$

Example 17 Self Tutor

Fully factorise by first removing a common factor:

a $3x^2-9x+6$ **b** $-2x^2+2x+12$

a $3x^2-9x+6$

$= 3(x^2-3x+2)$ {HCF = 3}

$= 3(x-2)(x-1)$ {sum $= -3$ and product $= 2$

$\therefore$ the numbers are -2 and -1}

b $-2x^2+2x+12$

$= -2(x^2-x-6)$ {HCF = -2}

$= -2(x-3)(x+2)$ {sum $= -1$ and product $= -6$

$\therefore$ the numbers are -3 and 2}

7 Fully factorise by first removing a common factor:

a $2x^2-6x-8$	**b** $3x^2+9x-12$	**c** $5x^2+10x-15$
d $4x^2+4x-80$	**e** $2x^2-4x-30$	**f** $3x^2+12x-63$
g $-2x^2+2x+40$	**h** $-3x^2+12x-12$	**i** $-7x^2-21x+28$
j $-x^2-3x-2$	**k** $-x^2+5x-6$	**l** $-x^2+9x-18$
m $5x^2+15x-50$	**n** $-2x^2-8x+42$	**o** $-x^2+4x+32$

E MISCELLANEOUS FACTORISATION

In this Section you will need to decide which methods of factorisation are most appropriate.

Things to look for when factorising:

- common factors
- difference between two squares $a^2-b^2=(a+b)(a-b)$
- perfect squares $a^2+2ab+b^2=(a+b)^2$ or $a^2-2ab+b^2=(a-b)^2$
- sum and product type $x^2+bx+c=(x+p)(x+q)$ where $p+q=b$ and $pq=c$.

EXERCISE 14E

1 Fully factorise, if possible:

a x^2+49x	**b** x^2+49	**c** $-x^2-49x$	**d** $-x^2+49$
e $4a^2+16a$	**f** $4a^2-16a$	**g** $4a^2+16$	**h** $4a^2-16$
i st^2-2st	**j** s^2t-st	**k** s^2t^2-1	**l** $st+2s$

2 Fully factorise:

a $3b^2 + 15$ **b** $h^2 + 2h^3$ **c** $-2x^2 - 18$

d $r^2 + r - 42$ **e** $(x+1)^2 - 2(x+1)$ **f** $81y^2 - 49x^2$

g $2(x-5) - x(x-5)$ **h** $y^2 - 7y + 12$ **i** $s(s+t) - t(s+t)$

3 Fully factorise:

a $x^2 - 12x + 32$ **b** $x^2 - 3x - 54$ **c** $x^2 - 3x - 18$

d $x^2 - 169$ **e** $x^2 - 10x + 25$ **f** $x^2 + 11x + 30$

g $x^2 + 7x - 18$ **h** $x^2 + 12x + 36$ **i** $x^2 - 5x - 24$

j $5x^2 - 125$ **k** $3x^2 + 6x - 72$ **l** $4x^2 - 8x - 60$

m $3x^2 - 42x + 99$ **n** $9x^2 - 9x - 18$ **o** $-x^2 + 9x - 14$

p $6x^2 - 36x + 54$ **q** $-x^2 - 13x - 36$ **r** $2x^2 - 32x + 128$

s $-2x^2 - 14x + 36$ **t** $20x - 5x^3$ **u** $x^3 + 4x^2 + 4x$

GAME

Shoot the correct expression to win your prize!

GAME

F EXPRESSIONS WITH FOUR TERMS

Some expressions with four terms do not have an overall common factor, but can be factorised by pairing the four terms.

For example, $\underbrace{ab + ac} + \underbrace{bd + cd}$

$= a(b+c) + d(b+c)$ {factorising each pair separately}

$= (b+c)(a+d)$ {removing the common factor $(b+c)$}

Sometimes we need to reorder the terms first.

Example 18 **Self Tutor**

Fully factorise: $ax + by + bx + ay$

$ax + by + bx + ay$

$= \underbrace{ax + ay} + \underbrace{bx + by}$ {writing terms with common factor a together}

$= a(x+y) + b(x+y)$ {factorising each pair}

$= (x+y)(a+b)$ {$(x+y)$ is a common factor}

EXERCISE 14F

1 Fully factorise:

a $bx + cx + by + cy$ **b** $px + 3q + qx + 3p$ **c** $6ax + 3bx + b + 2a$

d $am - bn - an + bm$ **e** $3dr + r - 3ds - s$ **f** $2ac - 5a + 2bc - 5b$

Example 19 **Self Tutor**

Fully factorise: $2x^2 - 15 + 3x - 10x$

$$\begin{aligned} & 2x^2 - 15 + 3x - 10x \\ &= \underbrace{2x^2 - 10x} + \underbrace{3x - 15} && \{\text{writing terms with common factor } 2x \text{ together}\} \\ &= 2x(x-5) + 3(x-5) && \{\text{factorising each pair}\} \\ &= (x-5)(2x+3) && \{(x-5) \text{ is a common factor}\} \end{aligned}$$

2 Fully factorise:

a $x^2 + 5x + 7x + 35$	**b** $x^2 - 2x - 6x + 12$	**c** $x^2 + 3x + 9 + 3x$
d $x^2 + 8 + 8x + x$	**e** $2x^2 + 3x + 3 + 2x$	**f** $3x^2 + x - 3x - 1$
g $2x^2 + 3x + 10x + 15$	**h** $6x^2 - 3x - 2 + 4x$	**i** $4x^2 + x + 8x + 2$
j $6x^2 + 3x + 10x + 5$	**k** $6x^2 - 4x + 9x - 6$	**l** $4x^2 + x - 8x - 2$
m $3x^2 + 4x + 33x + 44$	**n** $18x^2 + 3x - 12x - 2$	**o** $10x^2 + 4x - 35x - 14$

G FACTORISING $ax^2 + bx + c$, $a \neq 1$

So far we have considered the factorisation of quadratic trinomials of the form $ax^2 + bx + c$ where:

- $a = 1$, for example $x^2 + 5x + 6 = (x+3)(x+2)$
- a is a common factor, for example $2x^2 + 10x + 12 = 2(x^2 + 5x + 6)$
  $= 2(x+3)(x+2)$
- we have a perfect square or difference between two squares,
  for example $4x^2 - 9 = (2x)^2 - 3^2$
  $= (2x+3)(2x-3)$

Quadratic trinomials such as $8x^2 + 22x + 15$ do not fall into any of these categories. We therefore need a method for factorising this type of expression.

FACTORISATION BY "SPLITTING" THE MIDDLE TERM

Using the FOIL rule, we see that

$$\begin{aligned} & (2x+3)(4x+5) \\ &= 8x^2 + 10x + 12x + 15 \\ &= 8x^2 + 22x + 15 \end{aligned}$$

We can *reverse* the process to factorise the quadratic trinomial $8x^2 + 22x + 15$.

$$\begin{aligned} & 8x^2 + 22x + 15 \\ &= 8x^2 + 10x + 12x + 15 && \{\text{"splitting" the middle term}\} \\ &= (8x^2 + 10x) + (12x + 15) && \{\text{grouping in pairs}\} \\ &= 2x(4x+5) + 3(4x+5) && \{\text{factorising each pair separately}\} \\ &= (4x+5)(2x+3) && \{\text{completing the factorisation}\} \end{aligned}$$

But how do we know how to correctly "split" the middle term? How do we know that $22x$ should be written as $+10x + 12x$ rather than $+15x + 7x$ or $+20x + 2x$?

When looking at $8x^2 + 10x + 12x + 15$, we notice that $8 \times 15 = 120$ and $10 \times 12 = 120$.

So, for $8x^2 + 22x + 15$, we need two numbers whose *sum* is 22 and whose *product* is $8 \times 15 = 120$. These numbers are 10 and 12.

If we chose a different pair of numbers with sum 22, we would not be able to factorise the expression in pairs.

To factorise $ax^2 + bx + c$ by "splitting" the middle term:

Step 1: Find two numbers p and q such that $p + q = b$ and $pq = ac$.

Step 2: Replace bx with $px + qx$.

Step 3: Complete the factorisation.

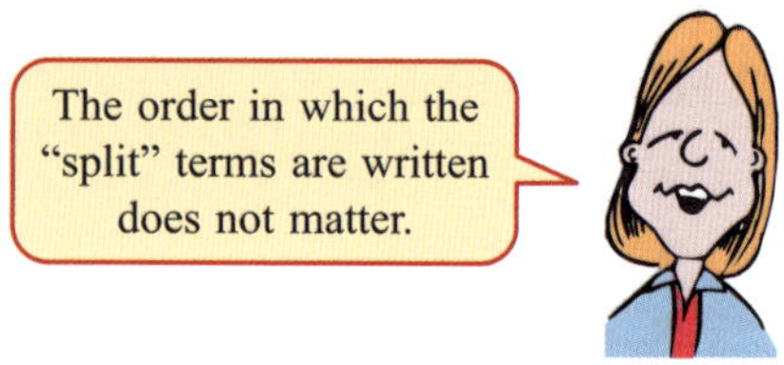

Example 20 **Self Tutor**

Factorise by "splitting" the middle term:

a $6x^2 + 19x + 10$ **b** $3x^2 - x - 10$

a In $6x^2 + 19x + 10$, $ac = 60$ and $b = 19$.
We need two numbers with product 60 and sum 19. These are 4 and 15.

$\therefore$ $6x^2 + 19x + 10$
$= 6x^2 + 4x + 15x + 10$ {"splitting" the middle term}
$= 2x(3x + 2) + 5(3x + 2)$ {factorising in pairs}
$= (3x + 2)(2x + 5)$ {taking out the common factor}

b In $3x^2 - x - 10$, $ac = -30$ and $b = -1$.
We need two numbers with product -30 and sum -1. These are 5 and -6.

$\therefore$ $3x^2 - x - 10$
$= 3x^2 + 5x - 6x - 10$ {"splitting" the middle term}
$= x(3x + 5) - 2(3x + 5)$ {factorising in pairs}
$= (3x + 5)(x - 2)$ {taking out the common factor}

EXERCISE 14G

1 Factorise $2x^2 + 9x + 4$ by "splitting" the middle term into:

a $+ 8x + x$ **b** $+ x + 8x$

2 Fully factorise:

a $2x^2 + 7x + 6$ **b** $2x^2 + 15x + 25$ **c** $2x^2 + 11x + 15$
d $3x^2 + 7x + 2$ **e** $3x^2 + 14x + 15$ **f** $6x^2 + 19x + 10$

3 Factorise $4x^2 + 4x - 15$ by "splitting" the middle term into:

a $+ 10x - 6x$ **b** $- 6x + 10x$

4 Fully factorise:

- **a** $2x^2 + 5x - 3$ **b** $2x^2 + x - 3$ **c** $3x^2 + 4x - 4$
- **d** $3x^2 - 8x - 3$ **e** $5x^2 - 8x - 4$ **f** $6x^2 + 7x - 3$

5 Fully factorise:

- **a** $2x^2 - 11x + 5$ **b** $2x^2 - 11x + 12$ **c** $2x^2 - 9x + 10$
- **d** $3x^2 - 14x + 8$ **e** $10x^2 - 9x + 2$ **f** $9x^2 - 15x + 4$

6 Suppose we "split" the middle term of $ax^2 + bx + c$ where $a > 0$, $a \neq 1$ into $px + qx$. What can we say about p and q if:

- **a** b and c are positive
- **b** c is positive but b is negative
- **c** c is negative but b is positive
- **d** b and c are negative?

7 Fully factorise:

- **a** $12x^2 + 13x - 4$ **b** $9x^2 + 9x - 4$ **c** $8x^2 + 2x - 3$
- **d** $6x^2 - 19x + 10$ **e** $10x^2 - 13x - 3$ **f** $20x^2 + 31x + 12$

8 Fully factorise:

- **a** $4x^2 + 2x - 12$ **b** $18x^2 + 21x + 6$ **c** $8x^2 - 16x + 6$

MULTIPLE CHOICE QUIZ

QUICK QUIZ

REVIEW SET 14A

1 Fully factorise:

- **a** $3x - 18$ **b** $2ab + 10ac$ **c** $xy - 6yz$ **d** $-2x^2 - 32x$

2 Fully factorise:

- **a** $d(t+2) - 4(t+2)$ **b** $(x-1)^2 - (x-1)$ **c** $2x(x+3) - 5(x+3)$

3 Fully factorise:

- **a** $x^2 - 25$ **b** $100 - k^2$ **c** $9 - 16x^2$
- **d** $9a^2 - 4b^2$ **e** $6x^2 - 24$ **f** $16m^2 - 100n^2$

4
- **a** Write expressions for the areas A_1, A_2, A_3, and A_4.
- **b** Explain why $a^2 = (a-b)^2 + 2ab - b^2$.
- **c** Hence show that $a^2 - 2ab + b^2 = (a-b)^2$.

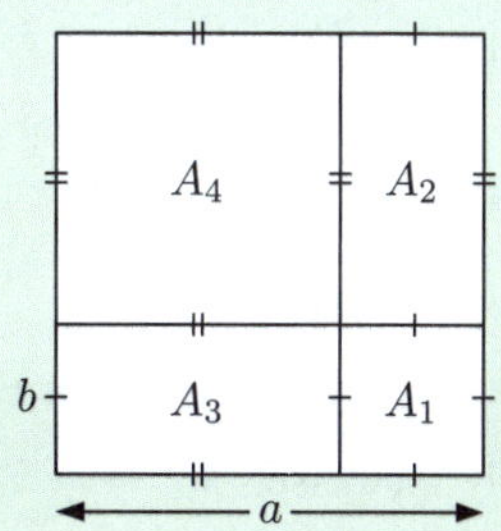

5 Determine whether each expression is a perfect square:

- **a** $x^2 - 6x + 36$ **b** $16x^2 + 24x + 9$

6 Fully factorise:

a $x^2 - 8x + 16$ **b** $x^2 + 14x + 49$ **c** $4x^2 + 20x + 25$

7 Explain why $x^2 - 2x + 4$ cannot be factorised using the "sum and product" method.

8 Fully factorise:

a $x^2 + 12x + 27$ **b** $x^2 - 16x + 48$ **c** $x^2 + 9x - 22$

d $x^2 - 10x + 24$ **e** $x^2 - 13x + 36$ **f** $-3x^2 - 6x + 24$

9 Fully factorise:

a $ax + 2a + 2b + bx$ **b** $x^2 - 8x + 24 - 3x$ **c** $6x^2 + 9x - 2x - 3$

10 Fully factorise:

a $2x^2 + 17x + 8$ **b** $2x^2 + 15x - 8$ **c** $3x^2 - 17x + 10$

d $6x^2 - 11x - 10$ **e** $12x^2 + 5x - 2$ **f** $12x^2 - 2x - 4$

REVIEW SET 14B

1 Fully factorise:

a $x^2 - 3x$ **b** $3mn + 6n^2$ **c** $ax^3 + 2ax^2$ **d** $-3x^2 + 18x$

2 Fully factorise:

a $8(x+1) + x(x+1)$ **b** $(a-2)(a+5) - 6(a+5)$ **c** $3(g+1)^2 - 9(g+1)$

3 Fully factorise:

a $9 - x^2$ **b** $3x^2 - 48$ **c** $25y^3 - 36y$

4 Fully factorise:

a $(x+3)^2 - 49$ **b** $4 - (x-5)^2$ **c** $(2x+3)^2 - (x-2)^2$

5 Fully factorise:

a $x^2 + 6x + 9$ **b** $x^2 - 28x + 196$ **c** $9x^2 - 6x + 1$

6 **a** Show that $4x^2 - 20x + 9$ is not a perfect square.

b Find the value of k such that the expression $4x^2 - 20x + k$ is a perfect square.

c Fully factorise the expression in this case.

7 Fully factorise:

a $5x^2 - 20x + 20$ **b** $-3x^2 + 18x - 27$ **c** $6x^2 - 120x + 600$

8 Fully factorise:

a $x^2 + 8x + 15$ **b** $x^2 + 2x - 15$ **c** $x^2 - 2x - 15$

d $4x^2 + 8x - 12$ **e** $2x^2 - 2x - 60$ **f** $3x^2 - 30x + 48$

9 Fully factorise:

a $mx + nx - my - ny$ **b** $2x^2 + 3x + 8x + 12$ **c** $5x^2 - 2 + x - 10x$

10 Fully factorise:

a $3x^2 - 17x - 6$ **b** $3x^2 - 19x + 6$ **c** $8x^2 + 18x - 5$

d $12x^2 + 7x + 1$ **e** $12x^2 - 23x - 2$ **f** $6x^2 + 9x - 42$

Chapter 15

Algebraic fractions

Contents:

OPENING PROBLEM

Robert and Trevor are attending cooking classes. They have each baked a cake to take home. The cakes are the same size.

Robert shares his cake equally between the r members of his family, including his son Ronnie.

Trevor shares his cake equally between the t members of his family, including his daughter Tania.

Things to think about:

a What fraction of a cake does Ronnie receive?

b Given that there are more people in Trevor's family than in Robert's family, can you write an expression for:

i the *difference* between the size of the pieces Ronnie and Tania receive

ii the amount that Tania receives as a *fraction* of the amount that Ronnie receives?

Algebraic fractions are fractions which contain at least one variable or unknown.

The variable may be in the numerator, the denominator, or both the numerator and denominator.

For example, $\frac{x}{2}$, $\frac{4}{1-a}$, and $\frac{2p-1}{3q}$ are all algebraic fractions.

Algebraic fractions are sometimes called **rational expressions**.

In this Chapter we will see how we can simplify, add, subtract, multiply, and divide algebraic fractions in the same way as we do ordinary fractions.

A EVALUATING ALGEBRAIC FRACTIONS

To **evaluate** an algebraic fraction, we replace the variables with their known values. We then give our answer in lowest terms.

Example 1 — Self Tutor

If $a = 2$, $b = -3$, and $c = -5$, evaluate:

a $\frac{a-b}{c^2}$ **b** $\frac{a-c-b}{b-a}$

a $\frac{a-b}{c^2}$

$= \frac{2-(-3)}{(-5)^2}$

$= \frac{2+3}{25}$

$= \frac{5}{25}$

$= \frac{1}{5}$

b $\frac{a-c-b}{b-a}$

$= \frac{2-(-5)-(-3)}{(-3)-2}$

$= \frac{2+5+3}{-3-2}$

$= \frac{10}{-5}$

$= -2$

EXERCISE 15A

1 If $a = 3$, $b = 2$, and $c = 6$, evaluate:

a $\frac{c}{2}$ **b** $\frac{c}{a}$ **c** $\frac{a}{c}$ **d** $\frac{c}{b-a}$ **e** $\frac{a+c}{b}$

f $\frac{ab}{c}$ **g** $\frac{a^2}{b}$ **h** $\frac{c^2}{a}$ **i** $\frac{ab^2}{c}$ **j** $\frac{(ab)^2}{c}$

2 If $a = 2$, $b = -3$, and $c = -4$, evaluate:

a $\frac{c}{a}$ **b** $\frac{a}{c}$ **c** $\frac{-1}{b}$ **d** $\frac{c^2}{a}$ **e** $\frac{c}{a+b}$

f $\frac{a-c}{2b}$ **g** $\frac{b}{c-a}$ **h** $\frac{a-c}{a+c}$ **i** $\frac{c-a}{b^2}$ **j** $\frac{a^2}{c-b}$

3 If $x = 4$, $y = -2$, and $z = 3$, evaluate:

a $\frac{x+2z}{y}$ **b** $\frac{x^2}{z-2y}$ **c** $\frac{x-y}{x+y}$ **d** $\frac{2z}{y-x}$

e $\frac{\sqrt{x-z}}{y^2}$ **f** $\frac{x(x-y)}{\sqrt{3z}}$ **g** $\frac{-xy}{z^3-3}$ **h** $\frac{x^2-y^2}{xz}$

B SIMPLIFYING ALGEBRAIC FRACTIONS

To write a fraction with integer numerator and denominator in **lowest terms**, we divide the numerator and denominator by their **highest common factor**.

In practice, we often achieve this by **cancelling common factors**.

For example: $\frac{6}{15} = \frac{2 \times \cancel{3}^1}{5 \times \cancel{3}_1}$ {cancelling the HCF 3}

$= \frac{2}{5}$

The same principle can be applied to simplify algebraic fractions.

For example: $\frac{2xy}{4x} = \frac{{}^1\cancel{2} \times {}^1\cancel{x} \times y}{{}_1\cancel{2} \times 2 \times \cancel{x}_1}$ {cancelling the HCF $2x$}

$= \frac{y}{2}$

Fractions such as $\frac{2a}{5bc}$ cannot be simplified since the numerator and denominator do not have any common factors.

ERRORS IN CANCELLING

A common **error** in cancelling is to cancel terms, rather than common factors.

For example, this cancelling is **incorrect**:

$$\frac{x - \cancel{4}^1}{\cancel{8}_2} = \frac{x-1}{2}$$

The numerator $x - 4$ cannot be written as the product of factors other than $1 \times (x - 4)$. x and -4 are *terms* of the expression, not factors.

Example 2 — Self Tutor

Simplify, if possible:

a $\dfrac{3x}{x^2}$ **b** $\dfrac{2xy^2}{8y}$ **c** $\dfrac{4x^2}{-2x}$ **d** $\dfrac{x+6}{x}$

a $\dfrac{3x}{x^2} = \dfrac{3 \times \cancel{x}^1}{x \times \cancel{x}_1} = \dfrac{3}{x}$

b $\dfrac{2xy^2}{8y} = \dfrac{{}^1\cancel{2} \times x \times \cancel{y}^1 \times y}{{}_4\cancel{8} \times \cancel{y}_1} = \dfrac{xy}{4}$

c $\dfrac{4x^2}{-2x} = \dfrac{{}^2\cancel{4} \times \cancel{x}^1 \times x}{-1 \times \cancel{2}_1 \times \cancel{x}_1} = \dfrac{2x}{-1} = -2x$

d $\dfrac{x+6}{x}$ cannot be simplified as there are no common factors.

EXERCISE 15B.1

1 Simplify:

a $\dfrac{4}{2x}$ **b** $\dfrac{6x}{2}$ **c** $\dfrac{3x}{x}$ **d** $\dfrac{4a}{8a}$ **e** $\dfrac{bc}{b}$

f $\dfrac{a}{ab}$ **g** $\dfrac{xy}{xz}$ **h** $\dfrac{a^2}{a}$ **i** $\dfrac{a}{a^2}$ **j** $\dfrac{p^2}{p^2}$

k $\dfrac{p}{2p^2}$ **l** $\dfrac{2x^2}{4x}$ **m** $\dfrac{2x}{3xy^2}$ **n** $\dfrac{xy}{2x^2y}$ **o** $\dfrac{12xy^2}{3x^2}$

2 Simplify:

a $\dfrac{a^2}{-2a}$ **b** $\dfrac{-a}{2a^2}$ **c** $\dfrac{-a^3}{-4a}$ **d** $\dfrac{4a^2}{-a^2}$ **e** $\dfrac{-x}{3x^3}$

f $\dfrac{-6t}{-3t^2}$ **g** $\dfrac{4d^2}{-2d}$ **h** $\dfrac{-ab^2}{2ab}$ **i** $\dfrac{-4ab^2}{6a^2b}$ **j** $\dfrac{-4x^2y}{-2xy^3}$

3 Simplify, if possible:

a $\dfrac{2t}{2}$ **b** $\dfrac{2+t}{2}$ **c** $\dfrac{xy}{x}$ **d** $\dfrac{x+y}{x}$ **e** $\dfrac{ac}{bc}$

f $\dfrac{a+c}{b+c}$ **g** $\dfrac{x}{x-2}$ **h** $\dfrac{5a}{9b}$ **i** $\dfrac{14c}{8d}$ **j** $\dfrac{4x+1}{2}$

Example 3 — Self Tutor

Simplify:

a $\dfrac{(3x)^2}{x}$ **b** $\dfrac{4(x+8)}{2}$ **c** $\dfrac{3(x-1)^2}{x-1}$

a $\dfrac{(3x)^2}{x} = \dfrac{3x \times 3x}{x} = \dfrac{3 \times x \times 3 \times \cancel{x}^1}{\cancel{x}_1} = 9x$

b $\dfrac{4(x+8)}{2} = \dfrac{{}^2\cancel{4} \times (x+8)}{\cancel{2}_1} = 2(x+8)$

c $\dfrac{3(x-1)^2}{x-1} = \dfrac{3(x-1)\cancel{(x-1)}^1}{\cancel{x-1}_1} = 3(x-1)$

4 Simplify:

a $\dfrac{(5a)^2}{a}$ b $\dfrac{(5a)^2}{5}$ c $\dfrac{5a}{(-a)^2}$ d $\dfrac{(2a)^2}{-a}$

e $\dfrac{(-3b)^2}{6}$ f $\dfrac{(-4c)^2}{8}$ g $\dfrac{(3d)^2}{d^2}$ h $\dfrac{(e^3)^2}{e^3}$

5 Simplify:

a $\dfrac{2(a+3)}{2}$ b $\dfrac{4(x+2)}{2}$ c $\dfrac{8(b+5)}{2}$ d $\dfrac{2(d-3)}{6}$

e $\dfrac{3}{6(z+3)}$ f $\dfrac{4}{2(x+1)}$ g $\dfrac{12}{4(2-x)}$ h $\dfrac{9}{6(1+x)}$

6 Simplify, if possible:

a $\dfrac{x+y}{x+y}$ b $\dfrac{x-y}{x+y}$ c $\dfrac{x-y}{x-y}$

7 Simplify:

a $\dfrac{10(a+2)}{2(a+2)}$ b $\dfrac{3(x+4)}{9(x+4)}$ c $\dfrac{12(a-3)}{6(a-3)}$

d $\dfrac{6(x-2)}{8(x-2)}$ e $\dfrac{(x+y)(x-y)}{3(x-y)}$ f $\dfrac{2xy(x-y)}{6x(x-y)}$

g $\dfrac{5(y+2)^2}{15(y+2)}$ h $\dfrac{12(p+q)^2}{8(p+q)}$ i $\dfrac{x(x+1)(x+2)}{3x(x+2)}$

DISCUSSION

The algebraic fraction $\dfrac{2x}{x}$ has the common factor x in its numerator and denominator.

Can we *always* cancel the common factor x to say $\dfrac{2x}{x} = 2$, or is there a particular value of x which would make this cancellation incorrect?

FACTORISATION AND SIMPLIFICATION

It is often necessary to **factorise** either the numerator or denominator before we can simplify.

Example 4 Self Tutor

Factorise, then simplify if possible:

a $\dfrac{3a-9}{3}$ b $\dfrac{4x+6}{3}$

a $\dfrac{3a-9}{3}$

$= \dfrac{{}^1\cancel{3}(a-3)}{\cancel{3}_1}$

$= a-3$

b $\dfrac{4x+6}{3}$

$= \dfrac{2(2x+3)}{3}$

which cannot be simplified as there are no common factors.

EXERCISE 15B.2

1 Factorise, then simplify if possible:

a $\dfrac{2x+10}{2}$ **b** $\dfrac{6x+12}{5}$ **c** $\dfrac{2x-6}{3}$ **d** $\dfrac{2x-6}{2}$

e $\dfrac{12a+4}{2}$ **f** $\dfrac{15a+5}{3}$ **g** $\dfrac{5b-10}{2}$ **h** $\dfrac{5b-10}{5}$

i $\dfrac{3x-15}{3}$ **j** $\dfrac{3x-15}{5}$ **k** $\dfrac{8y+4}{3}$ **l** $\dfrac{8y+4}{4}$

m $\dfrac{4x+16}{2}$ **n** $\dfrac{4x+16}{4}$ **o** $\dfrac{4x+16}{8}$ **p** $\dfrac{x^2+2x}{x}$

Example 5 Self Tutor

Factorise, then simplify: **a** $\dfrac{4x+8}{5x+10}$ **b** $\dfrac{ab-ac}{b-c}$

a $\dfrac{4x+8}{5x+10}$ ← HCF is 4 (numerator), HCF is 5 (denominator)

$= \dfrac{4(x+2)}{5(x+2)}$

$= \dfrac{4\cancel{(x+2)}^{1}}{5\cancel{(x+2)}_{1}}$

$= \frac{4}{5}$

b $\dfrac{ab-ac}{b-c}$ ← HCF is a

$= \dfrac{a(b-c)}{1(b-c)}$

$= \dfrac{a\cancel{(b-c)}^{1}}{1\cancel{(b-c)}_{1}}$

$= a$

2 Factorise, then simplify:

a $\dfrac{2x+4}{3x+6}$ **b** $\dfrac{2x-10}{3x-15}$ **c** $\dfrac{4x+12}{2x+6}$ **d** $\dfrac{5x-20}{3x-12}$

e $\dfrac{7x-14}{x-2}$ **f** $\dfrac{x+4}{2x+8}$ **g** $\dfrac{xy+3x}{y+3}$ **h** $\dfrac{12-3x}{4-x}$

3 Factorise, then simplify:

a $\dfrac{ab+bc}{b}$ **b** $\dfrac{ab+bc}{a+c}$ **c** $\dfrac{x-y}{ax-ay}$ **d** $\dfrac{ab+bc+b}{b}$

Example 6 Self Tutor

Factorise, then simplify: **a** $\dfrac{x^2-3x}{x}$ **b** $\dfrac{2x^2+4x}{x^2+2x}$

a $\dfrac{x^2-3x}{x}$ ← HCF is x

$= \dfrac{x(x-3)}{x}$

$= \dfrac{{}^{1}\cancel{x}(x-3)}{\cancel{x}_{1}}$

$= x-3$

b $\dfrac{2x^2+4x}{x^2+2x}$ ← HCF is $2x$ (numerator), HCF is x (denominator)

$= \dfrac{2x(x+2)}{x(x+2)}$

$= \dfrac{2\cancel{x}^{1}\cancel{(x+2)}^{1}}{{}_{1}\cancel{x}\cancel{(x+2)}_{1}}$

$= 2$

4 Factorise, then simplify:

a $\dfrac{x^2+3x}{x}$ b $\dfrac{4x^2-8x}{4x}$ c $\dfrac{3x^2+9x}{x+3}$

d $\dfrac{5x^2-10x}{x-2}$ e $\dfrac{4x^2-12x}{x^2-3x}$ f $\dfrac{7x^2+14x}{2x^2+4x}$

g $\dfrac{4x^2-20x}{2x-10}$ h $\dfrac{x^2+2x}{x+2}$ i $\dfrac{x^3-x}{x}$

j $\dfrac{x^3-x}{x^2-1}$ k $\dfrac{ab^2-cb^2}{a-c}$ l $\dfrac{3xy+y^2}{6x+2y}$

Example 7 Self Tutor

Factorise, then simplify:

a $\dfrac{y-x}{2x-2y}$ b $\dfrac{x^2y-2xy}{2-x}$

a $\dfrac{y-x}{2x-2y}$

$= \dfrac{-\cancel{(x-y)}^{1}}{2\cancel{(x-y)}_{1}}$

$= -\frac{1}{2}$

b $\dfrac{x^2y-2xy}{2-x}$

$= \dfrac{xy\cancel{(x-2)}^{1}}{-\cancel{(x-2)}_{1}}$

$= -xy$

5 Factorise, then simplify if possible:

a $\dfrac{m-n}{n-m}$ b $\dfrac{a+b}{b-a}$ c $\dfrac{2x-2y}{y-x}$ d $\dfrac{4a-4c}{3c-3a}$

e $\dfrac{p-2q}{6q-3p}$ f $\dfrac{2a-3b}{12b-8a}$ g $\dfrac{1-r}{r^2-r}$ h $\dfrac{xy^2-y}{2-2xy}$

Example 8 Self Tutor

Factorise, then simplify:

a $\dfrac{x^2-4}{x-2}$ b $\dfrac{2a+2b}{a^2-b^2}$

a $\dfrac{x^2-4}{x-2}$

$= \dfrac{(x+2)\cancel{(x-2)}^{1}}{\cancel{(x-2)}_{1}}$

$= x+2$

b $\dfrac{2a+2b}{a^2-b^2}$

$= \dfrac{2\cancel{(a+b)}^{1}}{{}_{1}\cancel{(a+b)}(a-b)}$

$= \dfrac{2}{a-b}$

6 Factorise, then simplify:

a $\dfrac{x-1}{x^2-1}$ b $\dfrac{a^2-1}{a+1}$ c $\dfrac{2+m}{4-m^2}$ d $\dfrac{x^2-y^2}{x+y}$

e $\dfrac{x^2-1}{3x+3}$ f $\dfrac{x^2-16}{4-x}$ g $\dfrac{a^2-b^2}{ab-b^2}$ h $\dfrac{x^2-2xy}{2y^2-xy}$

Example 9 **Self Tutor**

Factorise, then simplify: $\dfrac{x^2+x-6}{x+3}$

$\dfrac{x^2+x-6}{x+3}$ ← Two numbers with sum 1 and product -6 are $+3$ and -2.

$= \dfrac{{}^{1}\cancel{(x+3)}(x-2)}{\cancel{(x+3)}_{1}}$

$= x-2$

7 Factorise, then simplify:

a $\dfrac{x^2-5x+6}{x-3}$ **b** $\dfrac{x^2+6x-7}{x-1}$ **c** $\dfrac{x+2}{x^2-3x-10}$ **d** $\dfrac{1-x}{x^2-3x+2}$

e $\dfrac{x^2+x}{x^2-3x-4}$ **f** $\dfrac{x^2+8x+12}{x^2+6x+8}$ **g** $\dfrac{x^2-x-6}{x^2-7x+12}$ **h** $\dfrac{x^2-6x+9}{x^2-3x}$

i $\dfrac{1-x^2}{x^2+4x+3}$ **j** $\dfrac{2x^2-18}{x^2+5x+6}$ **k** $\dfrac{x^2+7x+12}{x^2-x-12}$ **l** $\dfrac{x^2+10x+25}{25-x^2}$

PUZZLE

Click on the icon to load a dynamic puzzle for algebraic fractions.

RATIONAL MAZE

C MULTIPLYING ALGEBRAIC FRACTIONS

Since variables are used in algebraic fractions to represent unknown numbers, we can perform operations with algebraic fractions in the same way as numerical fractions.

To **multiply** two or more fractions, we multiply the numerators together and multiply the denominators together.

$$\frac{a}{b}\times\frac{c}{d}=\frac{a\times c}{b\times d}$$

To write the result in lowest terms, we cancel any common factors in the numerator and denominator.

Example 10 **Self Tutor**

Simplify: **a** $\dfrac{a}{b}\times\dfrac{b}{2}$ **b** $\dfrac{n}{3}\times 6$

a $\dfrac{a}{b}\times\dfrac{b}{2}$

$= \dfrac{a\times \cancel{b}^{1}}{{}_{1}\cancel{b}\times 2}$

$= \dfrac{a}{2}$

b $\dfrac{n}{{}_{1}\cancel{3}}\times \cancel{6}^{2}$

$= \dfrac{2n}{1}$

$= 2n$

EXERCISE 15C

1 Simplify:

a $\frac{1}{2}\times\frac{x}{2}$ **b** $\frac{b}{3}\times\frac{3}{2}$ **c** $\frac{3}{a}\times\frac{2}{3}$ **d** $\frac{x}{2}\times\frac{x}{3}$

e $\frac{a}{4}\times\frac{2}{a}$ **f** $\frac{x}{2}\times 6$ **g** $\frac{2}{x}\times 6$ **h** $8\times\frac{n}{2}$

i $\frac{m}{14}\times 2$ **j** $\frac{a}{b}\times\frac{3}{4}$ **k** $\frac{d}{5}\times\frac{5}{d}$ **l** $\frac{c}{d}\times\frac{2d}{c}$

m $\frac{4a}{5}\times\frac{b}{4}$ **n** $\left(\frac{x}{3}\right)^2$ **o** $p\times\frac{1}{p}$ **p** $\frac{x^3}{3}\times\frac{6}{x^2}$

q $\left(\frac{m}{2}\right)^2$ **r** $4\times\frac{1}{c}$ **s** $\frac{1}{x}\times 8$ **t** $\frac{2}{b}\times\frac{1}{b}$

u $\frac{a}{2}\times\frac{a}{4}$ **v** $\left(\frac{1}{m}\right)^3$ **w** $\frac{2}{a}\times\frac{a}{3}$ **x** $\frac{a}{b}\times\frac{b}{c}\times\frac{c}{a}$

2 Simplify:

a $\frac{x}{2y}\times\frac{3}{x}$ **b** $\frac{3}{x^2}\times\frac{x}{9y}$ **c** $\frac{2x}{y}\times\frac{xy}{4}$

d $\frac{x^2}{8y}\times\frac{2y^2}{x}$ **e** $\frac{3y^2}{x}\times\frac{2x}{3y^2}$ **f** $\frac{4x^2}{5y}\times\frac{10}{3xy}$

3 Simplify:

a $\frac{x}{x+1}\times\frac{x-1}{x}$ **b** $\frac{x+2}{x-3}\times\frac{x-3}{x+2}$ **c** $\frac{x^2-1}{2x^2}\times\frac{x}{x+1}$

D DIVIDING ALGEBRAIC FRACTIONS

To **divide** by a fraction, we multiply by the **reciprocal** of that fraction.

$$\frac{a}{b}\div\frac{c}{d}=\frac{a}{b}\times\frac{d}{c}$$

Example 11 **Self Tutor**

Simplify: **a** $\frac{2}{a}\div\frac{a}{b}$ **b** $4\div\frac{2}{x}$

a $\frac{2}{a}\div\frac{a}{b}$
$=\frac{2}{a}\times\frac{b}{a}$
$=\frac{2\times b}{a\times a}$
$=\frac{2b}{a^2}$

b $4\div\frac{2}{x}$
$=\frac{4}{1}\times\frac{x}{2}$
$=\frac{{}^{2}\cancel{4}\times x}{1\times\cancel{2}_{1}}$
$=\frac{2x}{1}$
$=2x$

EXERCISE 15D

1 Simplify:

a $\frac{a}{b} \div \frac{1}{b}$ **b** $\frac{a}{b} \div b$ **c** $b \div \frac{a}{b}$ **d** $b \div \frac{b}{a}$

e $\frac{c}{3} \div \frac{c}{4}$ **f** $\frac{c}{3} \div \frac{4}{c}$ **g** $\frac{3}{c} \div \frac{4}{c}$ **h** $\frac{3}{c} \div \frac{c}{4}$

i $a \div \frac{1}{a}$ **j** $\frac{1}{x} \div 3$ **k** $5 \div \frac{1}{x}$ **l** $\frac{1}{b} \div b$

2 Simplify:

a $\frac{4a}{3} \div \frac{a}{6}$ **b** $\frac{21}{b} \div \frac{7}{2b}$ **c** $\frac{9}{2x} \div \frac{18}{x}$ **d** $\frac{a^2}{2} \div \frac{a}{6}$

e $\frac{3a^2}{5} \div \frac{a}{10}$ **f** $\frac{x}{6} \div \frac{x^2}{15}$ **g** $\frac{a}{b} \div \frac{ab}{c}$ **h** $\frac{2a}{b} \div \frac{a^2}{8}$

ACTIVITY

Click on the icon for more practice.

LEARNING ALGEBRA

E ADDING AND SUBTRACTING ALGEBRAIC FRACTIONS

The rules for adding and subtracting algebraic fractions are identical to those for numerical fractions.

To **add** two or more fractions, we obtain the *lowest common denominator* and then add the resulting numerators. $\frac{a}{c} + \frac{b}{c} = \frac{a+b}{c}$

To **subtract** two or more fractions, we obtain the *lowest common denominator* and then subtract the resulting numerators. $\frac{a}{c} - \frac{d}{c} = \frac{a-d}{c}$

To find the lowest common denominator of numerical fractions, we look for the **lowest common multiple of the denominators**.

For example, when finding $\frac{3}{4} + \frac{1}{6}$, the lowest common denominator is 12.

The same method is used when there are variables in the denominator.

For example:

- when finding $\frac{2}{3} + \frac{1}{x}$, the lowest common denominator is $3x$
- when finding $\frac{2}{a} + \frac{3}{2a}$, the lowest common denominator is $2a$
- when finding $\frac{3}{4a} + \frac{5}{6b}$, the lowest common denominator is $12ab$.

Example 12 — Self Tutor

Simplify by writing as a single fraction:

a $\frac{a}{6}+\frac{2a}{3}$ **b** $\frac{2x}{3}-\frac{x}{2}$

a $\frac{a}{6}+\frac{2a}{3}$

$=\frac{a}{6}+\frac{2a\times 2}{3\times 2}$ {LCD = 6}

$=\frac{a}{6}+\frac{4a}{6}$

$=\frac{a+4a}{6}$

$=\frac{5a}{6}$

b $\frac{2x}{3}-\frac{x}{2}$

$=\frac{2x\times 2}{3\times 2}-\frac{x\times 3}{2\times 3}$ {LCD = 6}

$=\frac{4x}{6}-\frac{3x}{6}$

$=\frac{4x-3x}{6}$

$=\frac{x}{6}$

EXERCISE 15E

1 Simplify by writing as a single fraction:

a $\frac{b}{3}+\frac{b}{3}$ **b** $\frac{a}{5}-\frac{4}{5}$ **c** $\frac{2x}{3}+\frac{1}{3}$ **d** $\frac{3}{4}-\frac{x}{4}$

2 Simplify by writing as a single fraction:

a $\frac{b}{2}+\frac{b}{3}$ **b** $\frac{a}{3}-\frac{a}{6}$ **c** $\frac{5a}{6}+\frac{a}{8}$ **d** $\frac{a}{4}-\frac{b}{3}$

e $\frac{n}{5}+\frac{3n}{2}$ **f** $\frac{p}{7}-\frac{2p}{3}$ **g** $\frac{2k}{3}+\frac{3k}{4}$ **h** $\frac{3s}{8}-\frac{s}{2}$

Example 13 — Self Tutor

Simplify by writing as a single fraction:

a $3+\frac{x}{7}$ **b** $\frac{c}{3}-2$

a $3+\frac{x}{7}=\frac{3}{1}+\frac{x}{7}$

$=\frac{3\times 7}{1\times 7}+\frac{x}{7}$ {LCD = 7}

$=\frac{21+x}{7}$

b $\frac{c}{3}-2=\frac{c}{3}-\frac{2}{1}$

$=\frac{c}{3}-\frac{2\times 3}{1\times 3}$ {LCD = 3}

$=\frac{c-6}{3}$

3 Simplify by writing as a single fraction:

a $\frac{n}{3}+1$ **b** $5+\frac{y}{4}$ **c** $7-\frac{d}{2}$ **d** $\frac{n}{3}+n$

e $x+\frac{x}{2}$ **f** $\frac{m}{3}-2$ **g** $1-\frac{2x}{3}$ **h** $\frac{x}{4}+2x$

Example 14 **Self Tutor**

Simplify by writing as a single fraction:

a $\dfrac{3}{x}+\dfrac{2}{y}$ **b** $\dfrac{2}{3a}-\dfrac{1}{2a}$

a $\dfrac{3}{x}+\dfrac{2}{y}$

$=\dfrac{3\times y}{x\times y}+\dfrac{2\times x}{y\times x}$ {LCD $= xy$}

$=\dfrac{3y}{xy}+\dfrac{2x}{xy}$

$=\dfrac{3y+2x}{xy}$

b $\dfrac{2}{3a}-\dfrac{1}{2a}$

$=\dfrac{2\times 2}{3a\times 2}-\dfrac{1\times 3}{2a\times 3}$ {LCD $= 6a$}

$=\dfrac{4}{6a}-\dfrac{3}{6a}$

$=\dfrac{1}{6a}$

4 Simplify by writing as a single fraction:

a $\dfrac{2}{x}+\dfrac{3}{x}$ **b** $\dfrac{1}{x}+\dfrac{3}{2x}$ **c** $\dfrac{1}{2}+\dfrac{1}{x}$ **d** $\dfrac{1}{x}-\dfrac{1}{2}$

e $\dfrac{5}{p}+\dfrac{3}{q}$ **f** $\dfrac{7}{r}+\dfrac{2}{s}$ **g** $\dfrac{4}{p}+\dfrac{3}{q}$ **h** $\dfrac{2t}{s}-\dfrac{t}{2s}$

i $\dfrac{a}{b}+\dfrac{2}{c}$ **j** $\dfrac{5}{b}-\dfrac{1}{2b}$ **k** $\dfrac{a}{b}+\dfrac{3}{5}$ **l** $\dfrac{a}{b}+\dfrac{c}{3b}$

5 Simplify by writing as a single fraction:

a $\dfrac{3}{a}+1$ **b** $1-\dfrac{2}{b}$ **c** $c+\dfrac{1}{c}$ **d** $\dfrac{2}{d}-3$

e $4+\dfrac{3}{p}$ **f** $7-\dfrac{2}{q}$ **g** $8+\dfrac{1}{3r}$ **h** $4-\dfrac{3}{2s}$

6 Simplify by writing as a single fraction:

a $\dfrac{5}{2p}+\dfrac{3}{2p}$ **b** $\dfrac{7}{2q}-\dfrac{4}{3q}$ **c** $\dfrac{2}{x}+\dfrac{5}{y}$ **d** $\dfrac{4}{3x}-\dfrac{2}{x}$

e $\dfrac{a}{6}+3a$ **f** $3x-\dfrac{x}{2}$ **g** $\dfrac{3}{a}+\dfrac{2}{a^2}$ **h** $\dfrac{5}{b^2}+\dfrac{1}{b}$

i $\dfrac{6}{d}-\dfrac{1}{d^2}$ **j** $\dfrac{5}{x^2}-\dfrac{4}{x}$ **k** $1+\dfrac{1}{x^2}$ **l** $x+\dfrac{4}{x^2}$

m $5-\dfrac{a^2}{4}$ **n** $6-\dfrac{2}{b^2}$ **o** $x+\dfrac{x^2}{2}$ **p** $x-\dfrac{1}{x^2}$

q $4x+\dfrac{2}{x}$ **r** $6x-\dfrac{1}{x}$ **s** $\dfrac{3x}{4}-\dfrac{1}{x}$ **t** $\dfrac{1}{2x}+x$

Example 15 **Self Tutor**

Write $\dfrac{c+8}{4}$ as a sum of two parts in lowest terms.

$\dfrac{c+8}{4}=\dfrac{c}{4}+\dfrac{8}{4}$

$=\dfrac{c}{4}+2$

7 Write as a sum of two parts in lowest terms:

a $\dfrac{a+4}{2}$ **b** $\dfrac{2x+3}{2}$ **c** $\dfrac{4x+3}{2}$ **d** $\dfrac{6-3x}{2}$

e $\dfrac{2x^2+4x}{4}$ **f** $\dfrac{x^2-6x}{x^2}$ **g** $\dfrac{x^2-6x}{6}$ **h** $\dfrac{5-3x^2}{x}$

8 Answer the **Opening Problem** on page **224**.

9 Corey has two sisters, Payton and Jade. Payton is the eldest child, so she receives the most pocket money. Payton receives twice as much as Corey. Jade receives two thirds as much as Corey.

Suppose Corey receives x dollars each week.

a Write an expression for the weekly pocket money received by:

i Payton **ii** Jade.

b Write a single fraction for the *total* amount of pocket money given to the children each week.

c If Corey receives $15 each week, find the total amount of pocket money given to the children.

QUICK QUIZ

MULTIPLE CHOICE QUIZ

REVIEW SET 15A

1 If $a = 3$, $b = -2$, and $c = 4$, evaluate:

a $\dfrac{a}{2b}$ **b** $\dfrac{3b}{a-c}$ **c** $\dfrac{a-b}{c^2-2b}$ **d** $\dfrac{\sqrt{a-3b}}{bc}$

2 Simplify:

a $\dfrac{3a^3}{6a}$ **b** $\dfrac{4x^3y}{8xy^2}$ **c** $\dfrac{(-2x)^2}{x^2}$

3 Explain why $\dfrac{3+x}{x}$ cannot be simplified.

4 Simplify:

a $\dfrac{12}{3(x-3)}$ **b** $\dfrac{(a+b)(a-b)}{5a(a+b)}$

5 Factorise, then simplify:

a $\dfrac{xy-6y}{5x-30}$ **b** $\dfrac{x^2+4x+4}{x^2-4}$ **c** $\dfrac{x^2-14x+45}{x^2-12x+27}$

6 Simplify:

a $\dfrac{5}{x} \times 3$ **b** $\dfrac{6a}{5} \times \dfrac{10}{a}$ **c** $\dfrac{12}{x^2} \times \dfrac{xy}{9}$

7 Simplify:

a $\frac{a}{b} \div \frac{b}{3}$ b $12 \div \frac{2}{x}$ c $\frac{4c^2}{d} \div \frac{2d}{c}$

8 Simplify by writing as a single fraction:

a $\frac{a}{4} + \frac{3a}{7}$ b $3 - \frac{2n}{3}$ c $\frac{3x}{4} - \frac{5x}{6}$

9 Simplify by writing as a single fraction:

a $\frac{9}{x} - \frac{3}{2x}$ b $\frac{x}{4} + 5x$ c $\frac{2x}{3} - \frac{1}{x}$

10 Write as a sum of two parts in lowest terms:

a $\frac{c+6}{2}$ b $\frac{9x-1}{3}$ c $\frac{x^2-2x}{2}$

11 I have two pizzas A and B which are the same size, but have different toppings.

a Suppose I cut pizza A into 4 equal pieces, and pizza B into 6 equal pieces. I take a pieces from pizza A and b pieces from pizza B. Write, in lowest terms, the fraction of a whole pizza that I have taken.

b Suppose I cut pizza A into m equal pieces and pizza B into n equal pieces. I take 2 pieces from pizza A and 3 pieces from pizza B.

i Write an expression for the fraction of a whole pizza that I have taken.

ii Evaluate this fraction given $m = 8$ and $n = 10$.

REVIEW SET 15B

1 If $x = -3$, $y = 5$, and $z = -1$, evaluate:

a $\frac{y-2z}{x}$ b $\frac{y^2}{x+2y}$ c $\frac{y(x-z^2)}{\sqrt{y+z}}$

2 Simplify:

a $\frac{ab}{ac}$ b $\frac{-x^2}{5xy^2}$ c $\frac{(4t)^2}{-t}$

3 Simplify:

a $\frac{4(a-4)}{12}$ b $\frac{10(x-1)}{6(x-1)}$ c $\frac{8(k+3)}{20(k+3)^2}$

4 Factorise, then simplify:

a $\frac{4x+20}{4}$ b $\frac{3y^2+12y}{y+4}$ c $\frac{x-2}{2-x}$

5 Factorise, then simplify:

a $\frac{5x^2-25x}{2x-10}$ b $\frac{x^2-2x-15}{x+3}$ c $\frac{x^2+3x-10}{2x^2-4x}$

6 Simplify:

a $\frac{m}{6} \times \frac{3}{m}$ b $\frac{k}{4} \times 10$ c $\left(\frac{5}{p}\right)^2$ d $\frac{2x^2}{y} \times \frac{y^2}{8x}$

7 Simplify:

a $\frac{x}{8} \div \frac{x}{2}$ **b** $\frac{3}{t} \div t$ **c** $\frac{2m}{5} \div \frac{m}{3}$ **d** $\frac{a^2}{2b} \div \frac{3}{ab}$

8 Simplify by writing as a single fraction:

a $\frac{x}{5} + \frac{3x}{5}$ **b** $\frac{4x}{5} - \frac{x}{10}$ **c** $6 + \frac{n}{2}$

9 Simplify by writing as a single fraction:

a $\frac{3}{x} + \frac{1}{3x}$ **b** $\frac{5}{a} - \frac{4}{b}$ **c** $x^2 + \frac{4}{x}$

10 Write $\frac{4x^2 - 12x}{x^2}$ as a sum of two parts in lowest terms.

11 Simplify:

a $\frac{x}{3} + 2$ **b** $2 - \frac{x}{3}$ **c** $\frac{3a}{5} \times \frac{10}{a^2}$

d $\frac{3a}{5} \div \frac{10}{a^2}$ **e** $\frac{5}{2x} - \frac{2}{x}$ **f** $\frac{5}{2x} + \frac{2}{x^2}$

12 **a** Write as a single fraction:

i $a - \frac{9}{a}$ **ii** $1 - \frac{a}{3}$

b Hence simplify $\left(a - \frac{9}{a}\right) \div \left(1 - \frac{a}{3}\right)$.

c Evaluate $\left(a - \frac{9}{a}\right) \div \left(1 - \frac{a}{3}\right)$ if:

i $a = 1$ **ii** $a = 3$ **iii** $a = 5$

HISTORICAL NOTE — SRINIVASA RAMANUJAN 1887 - 1920

Ramanujan was born in Erode, India in 1887. His parents were poor, but were able to send him to school. He was fascinated by mathematics.

Early attempts to study at University failed because he was still extremely poor, and because mathematics was the only subject at which he excelled.

Ramanujan taught himself from books and worked at home on mathematical research. He carefully recorded his work and results, but only rarely did he provide proofs for his theories.

Fortunately, he was able to obtain a position at the Madras Port Trust Office, a job that paid a small wage and left him with enough time to continue his research. He was able to take away used wrapping paper on which to write his mathematics.

Eventually, Ramanujan obtained a grant from Madras University which gave him access to the library and research facilities, and he was able to undertake his studies and research in a logical way. As a result, Ramanujan was awarded a scholarship to Cambridge University in England. Some of his work revealed amazing discoveries, but it also revealed a lack of background knowledge, and Ramanujan spent most of his time improving his basic knowledge and establishing proofs for some of his discoveries.

The English climate and food were difficult for Ramanujan, but he continued working. 32 important papers of his work were published between 1914 and 1921 even though he was ill with tuberculosis.

In 1918 Ramanujan was made a Fellow of the Royal Society and was awarded Fellowship of Trinity College. Sadly, he was too ill to accept the position of professor of mathematics at Madras University. He died in 1920 in Kumbakonam, India.

The famous English mathematician **G. H. Hardy** wrote of Ramanujan:

"One gift he has which no-one can deny: profound and invincible originality. He would probably have been a greater mathematician if he had been caught and tamed a little in his youth; he would have discovered more that was new, and that, no doubt, of greater importance. On the other hand he would have been less of a Ramanujan, and more of a European professor, and the loss might have been greater than the gain."

Hardy's work with Ramanujan was depicted in the 2015 film *The Man Who Knew Infinity*.

Chapter 16

Coordinate geometry

Contents:

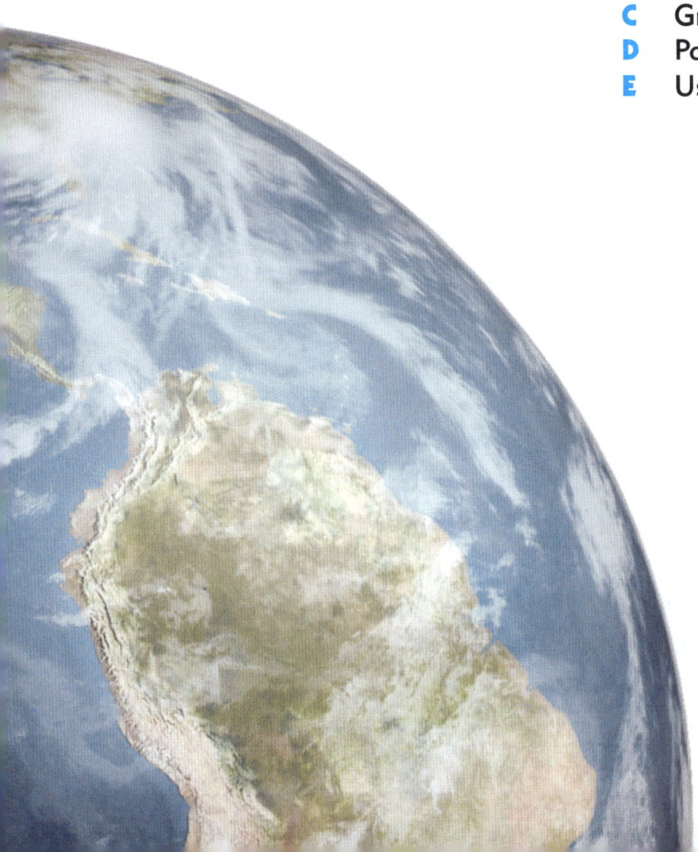

OPENING PROBLEM

On the given map, Peta lives at point P and Russell lives at point R. Each grid unit represents 100 m.

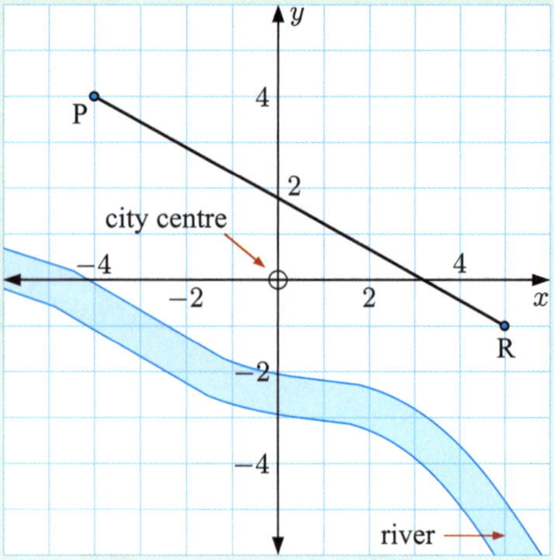

Things to think about:

a How can we describe the *positions* of P and R using ordered pairs of numbers?

b How far is Peta's house from Russell's house?

c Which point is midway between Peta's house and Russell's house?

The number grid alongside is a **Cartesian plane**, named after **René Descartes**.

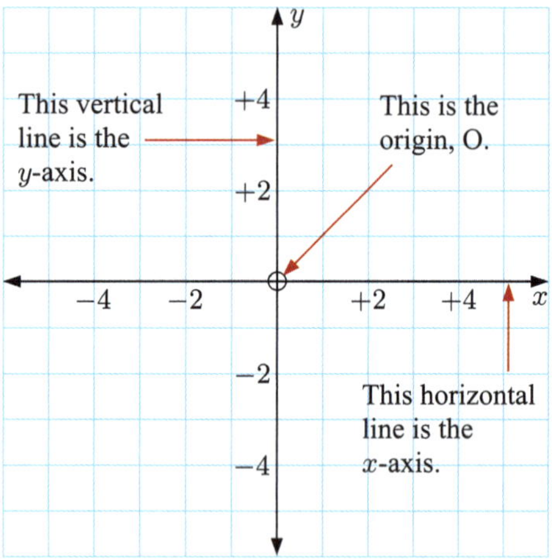

We start with a point of reference O called the **origin**.

Through the origin we draw two fixed number lines or **axes** which are perpendicular to each other.

- The horizontal axis is called the ***x*-axis**. It is positive to the right of O and negative to the left of O.
- The vertical axis is called the ***y*-axis**. It is positive above O and negative below O.

QUADRANTS

The x and y-axes divide the Cartesian plane into four regions called **quadrants**. These quadrants are numbered in an **anticlockwise direction** as shown:

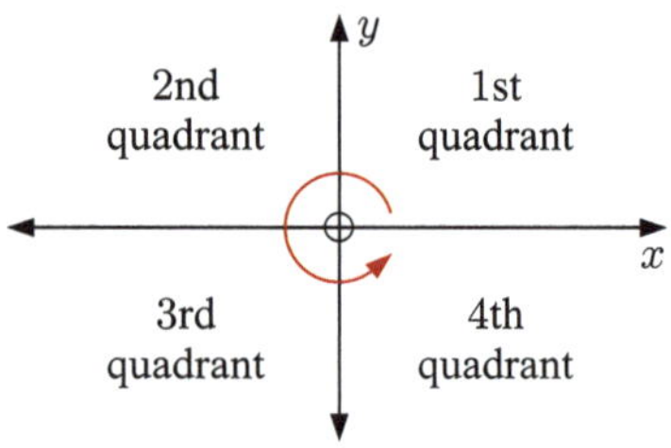

COORDINATES

We can describe any point on the Cartesian plane using an **ordered pair** of numbers called **coordinates**.

- The coordinates of a point are written (x-coordinate, y-coordinate).
- The x-coordinate gives the horizontal position along the x-axis.
- The y-coordinate gives the vertical position along the y-axis.

A THE DISTANCE BETWEEN TWO POINTS

On the Cartesian plane alongside we see the points A(2, 4) and B(7, 1).

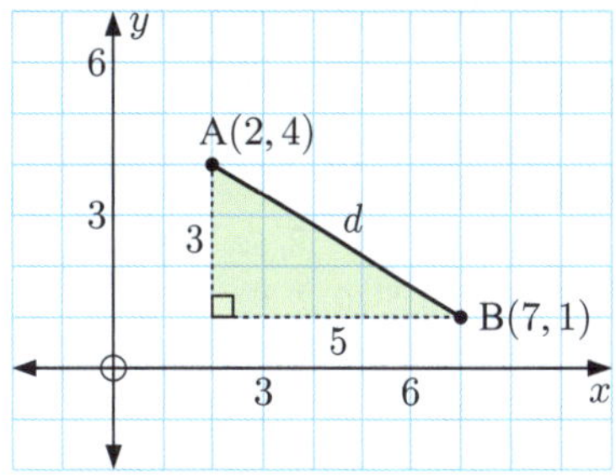

Suppose the distance between these points is d units.

We can use the horizontal and vertical grid lines to help construct a right angled triangle with hypotenuse [AB].

The horizontal distance or **x-step** between the points is 5 units. The vertical distance or **y-step** between the points is 3 units.

$\therefore \ d^2 = 5^2 + 3^2$ {Pythagoras}

$\therefore \ d^2 = 34$

$\therefore \ d = \sqrt{34}$ {since $d > 0$}

So, the distance between A and B is $\sqrt{34}$ units.

EXERCISE 16A.1

1 Find the distance between each pair of points, using Pythagoras' theorem where necessary:

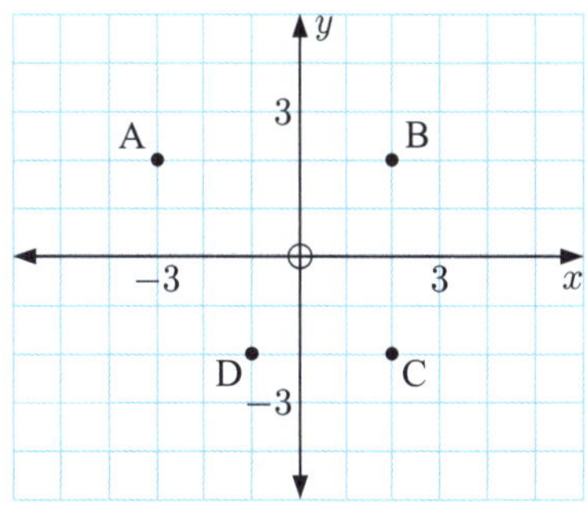

- **a** A and B
- **b** B and C
- **c** C and D
- **d** A and C
- **e** B and D
- **f** O and A.

2 By plotting points and using Pythagoras' theorem, find the distance between:

- **a** O(0, 0) and M(3, 5)
- **b** C(1, 2) and D(3, 7)
- **c** A(−1, −2) and B(2, 3)
- **d** A(1, 4) and B(2, −1)
- **e** P(3, 5) and Q(−1, 4)
- **f** R(−2, 0) and S(0, 3).

3 Two planes are travelling from A(−5, −4) to B(7, 5). The first plane flies via C(0, 2), while the second plane flies via D(4, 0). Each grid unit represents 10 km.

- **a** Plot the points A, B, C, and D on a number plane.
- **b** Find the distance between A and B.
- **c** Find, correct to 1 decimal place, the total distance travelled by:
  - **i** the first plane
  - **ii** the second plane.
- **d** Which plane travels the shorter distance?

4
- **a** Plot A(−3, −2), B(5, 2), C(−1, 3), and D(7, −1) on the same set of axes. Draw [AB] and [CD].
- **b** State the coordinates of the point X where the line segments intersect.
- **c** Find the distance from X to the closest of the points A, B, C, or D.

THE DISTANCE FORMULA

Instead of graphing points and then applying Pythagoras' theorem, we establish the **distance formula** for the distance between two general points.

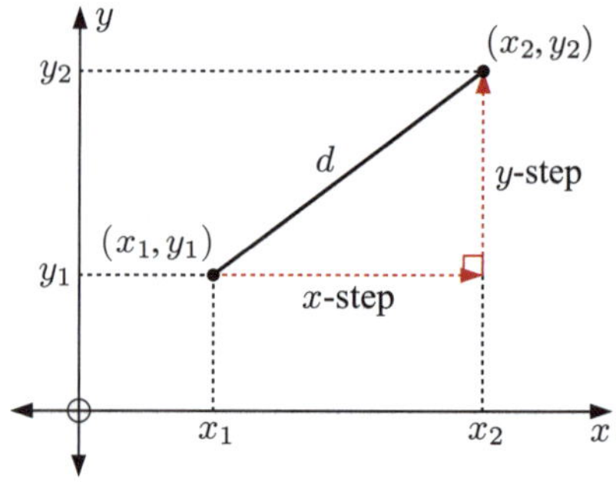

To go from (x_1, y_1) to (x_2, y_2), we find the

$$x\text{-step} = x_2 - x_1$$
$$\text{and } y\text{-step} = y_2 - y_1.$$

$\therefore\ d^2 = (x\text{-step})^2 + (y\text{-step})^2$ {Pythagoras}

$\therefore\ d = \sqrt{(x\text{-step})^2 + (y\text{-step})^2}$

$\therefore\ d = \sqrt{(x_2 - x_1)^2 + (y_2 - y_1)^2}$

The distance d between two points (x_1, y_1) and (x_2, y_2) is given by

$$\boldsymbol{d = \sqrt{(x_2 - x_1)^2 + (y_2 - y_1)^2}}.$$

Example 1 — Self Tutor

Find the distance from A(−3, 2) to B(2, 4).

A(−3, 2): $x_1 = -3$, $y_1 = 2$ B(2, 4): $x_2 = 2$, $y_2 = 4$

$$\begin{aligned} \text{AB} &= \sqrt{(x_2 - x_1)^2 + (y_2 - y_1)^2} \\ &= \sqrt{(2 - -3)^2 + (4 - 2)^2} \\ &= \sqrt{5^2 + 2^2} \\ &= \sqrt{25 + 4} \\ &= \sqrt{29} \text{ units} \end{aligned}$$

We can use the distance formula to classify triangles as equilateral, isosceles, or scalene.

We can also test whether a triangle is right angled by applying the converse of Pythagoras' theorem:

If a triangle has sides of length a, b, and c units where $a^2 + b^2 = c^2$, then the triangle is right angled.

EXERCISE 16A.2

1 Use the distance formula to find the distance between:

- **a** O(0, 0) and P(3, −1)
- **b** A(2, 1) and B(4, 4)
- **c** C(−2, 1) and D(2, 5)
- **d** E(4, 3) and F(−1, −1)
- **e** G(0, −3) and H(1, 4)
- **f** I(0, 2) and J(−3, 0)
- **g** P(3, 9) and Q(11, −1)
- **h** R($-\sqrt{2}$, 3) and S($\sqrt{2}$, −1).

2 Write a formula for the distance d between O(0, 0) and P(a, b).

Example 2 Self Tutor

The points P(1, 0), Q(3, 1), and R(−3, 8) form a triangle PQR.

a Classify the triangle as equilateral, isosceles, or scalene.

b Determine whether the triangle is right angled.

a 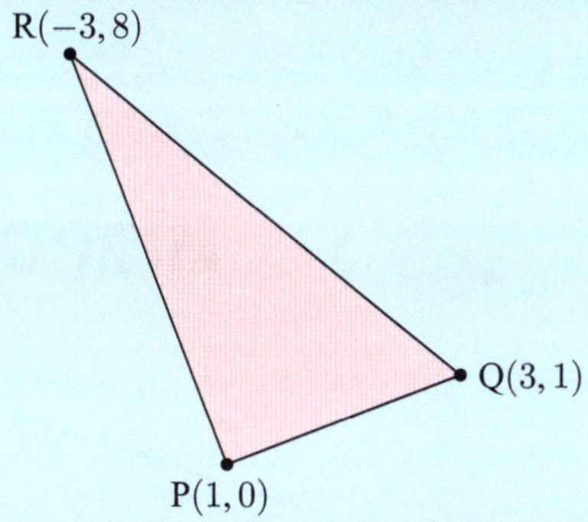

$$\begin{aligned} PQ &= \sqrt{(3-1)^2+(1-0)^2} \\ &= \sqrt{2^2+1^2} \\ &= \sqrt{5} \text{ units} \end{aligned}$$

$$\begin{aligned} PR &= \sqrt{(-3-1)^2+(8-0)^2} \\ &= \sqrt{(-4)^2+8^2} \\ &= \sqrt{80} \text{ units} \end{aligned}$$

$$\begin{aligned} QR &= \sqrt{(-3-3)^2+(8-1)^2} \\ &= \sqrt{(-6)^2+7^2} \\ &= \sqrt{85} \text{ units} \end{aligned}$$

Since none of the side lengths are equal, triangle PQR is scalene.

b The shortest sides are [PQ] and [PR].

Now $\begin{aligned} PQ^2 + PR^2 &= 5 + 80 \\ &= 85 \\ &= QR^2 \end{aligned}$

Using the converse of Pythagoras' theorem, triangle PQR is right angled. The right angle is at P, opposite the longest side.

3 Classify the triangle formed by each set of points as equilateral, isosceles, or scalene:

a A(0, −2), B(1, 2), C(−3, 1)

b X(1, 3), Y(−1, 0), Z(3, −4)

c H(3, −2), I(1, 4), J(−3, 0)

d W($2\sqrt{3}$, 0), X(0, 6), Y($-2\sqrt{3}$, 0)

4 Determine whether triangle ABC is right angled. If it is, state where the right angle occurs.

a A(3, 2), B(1, 5), C(9, 6)

b A(−3, −5), B(0, 4), C(1, −4)

c A(−5, 4), B(1, 6), C(−1, 3)

d A(−7, 5), B(−6, 1), C(2, 3)

5 Find, correct to 1 decimal place, the perimeter of triangle PQR.

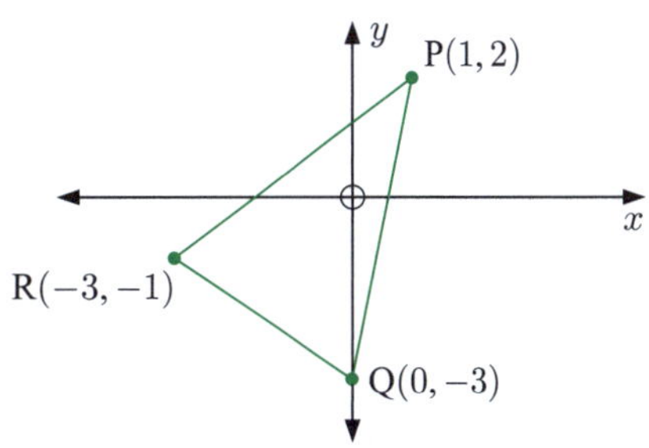

6 Consider triangle LMN with vertices L(−1, 2), M(3, 4), and N(5, 0).

a Show that the triangle is isosceles.

b Show that the triangle is right angled at M.

c Find the size of $M\widehat{L}N$.

d Find the area of the triangle.

7 Quadrilateral WXYZ has vertices W(−3, 2), X(−2, 4), Y(4, 1), and Z(3, −1).

- **a** Show that WX = YZ and XY = WZ.
- **b** Find the length of the diagonal [WY].
- **c** Hence show that $W\widehat{X}Y$ is a right angle.
- **d** Classify quadrilateral WXYZ.
- **e** Illustrate the quadrilateral on a number plane.
- **f** Find the area of the quadrilateral.

B MIDPOINTS

The **midpoint** of line segment [AB] is the point on [AB] which is equidistant from A and B.

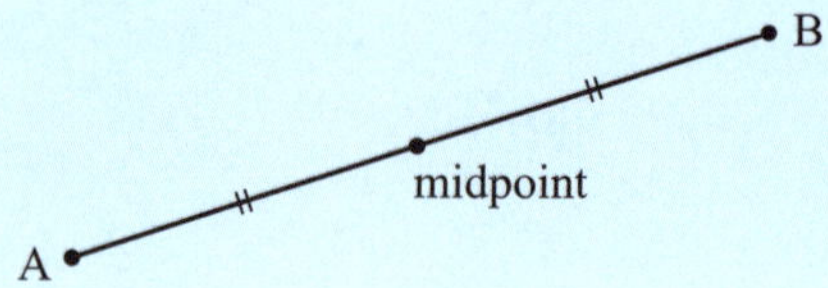

Consider the points A(1, −2) and B(5, 3). From the diagram we see that the midpoint of [AB] is M(3, $\frac{1}{2}$).

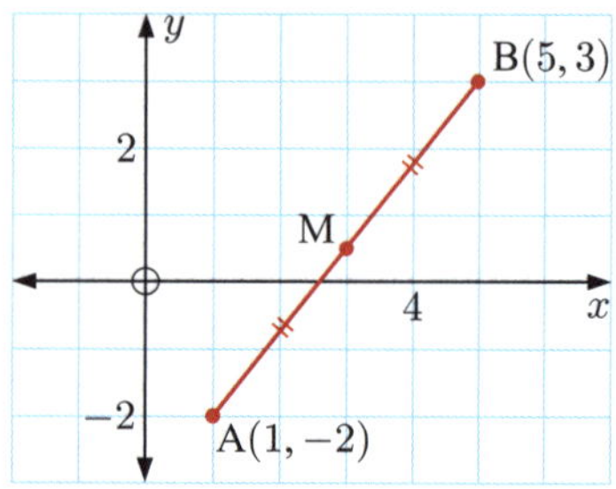

The x-coordinate of M is the *average* of the x-coordinates of A and B.

$\therefore$ the x-coordinate of M $= \dfrac{1+5}{2} = 3$

The y-coordinate of M is the *average* of the y-coordinates of A and B.

$\therefore$ the y-coordinate of M $= \dfrac{-2+3}{2} = \dfrac{1}{2}$

If A(x_1, y_1) and B(x_2, y_2) are two points, then the **midpoint** of [AB] has coordinates $\left(\dfrac{\boldsymbol{x_1+x_2}}{\mathbf{2}}, \dfrac{\boldsymbol{y_1+y_2}}{\mathbf{2}}\right)$.

DEMO

Example 3 Self Tutor

Find the midpoint M of [AB] given A(−2, 3) and B(4, 8).

A(−2, 3) with $x_1 = -2$, $y_1 = 3$ and B(4, 8) with $x_2 = 4$, $y_2 = 8$.

The x-coordinate of M

$= \dfrac{x_1+x_2}{2}$

$= \dfrac{-2+4}{2}$

$= 1$

The y-coordinate of M

$= \dfrac{y_1+y_2}{2}$

$= \dfrac{3+8}{2}$

$= \frac{11}{2} = 5\frac{1}{2}$

So, the midpoint is M(1, $5\frac{1}{2}$).

EXERCISE 16B

1 **a** Use the midpoint formula to find the midpoint M of [AB].

b Use the distance formula to verify that AM = BM.

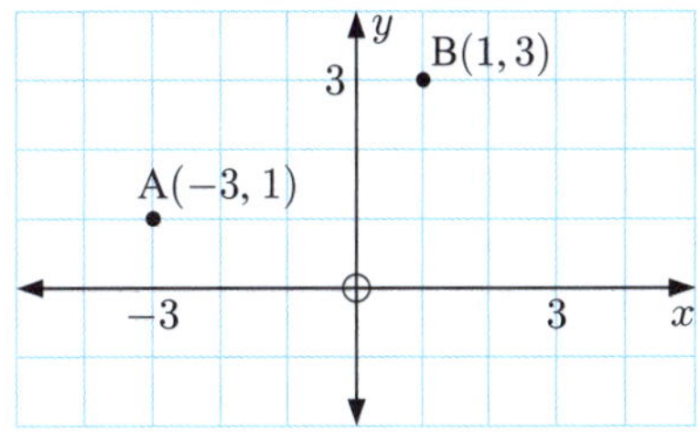

2 Find the midpoint M of [AB] given:

a A(5, 2) and B(1, 8)
b A(4, 1) and B(4, −1)
c A(1, 0) and B(−3, 2)
d A(6, 0) and B(0, 3)
e A(0, −1) and B(−3, 5)
f A(−2, 4) and B(4, −2)
g A(−4, 3) and B(9, 5)
h A(−5, 1) and B(−2, 6)
i A(a, 5) and B(2, −1)
j A(a, b) and B($-a$, $3b$).

3 Find the coordinates of:

a C **b** B **c** D

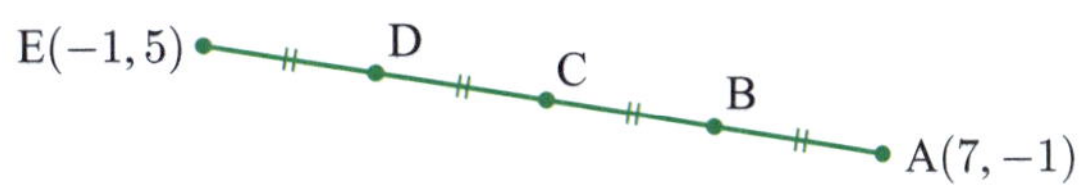

4 In triangle ABC, M is the midpoint of [AB].
Find the coordinates of N, the midpoint of [CM].

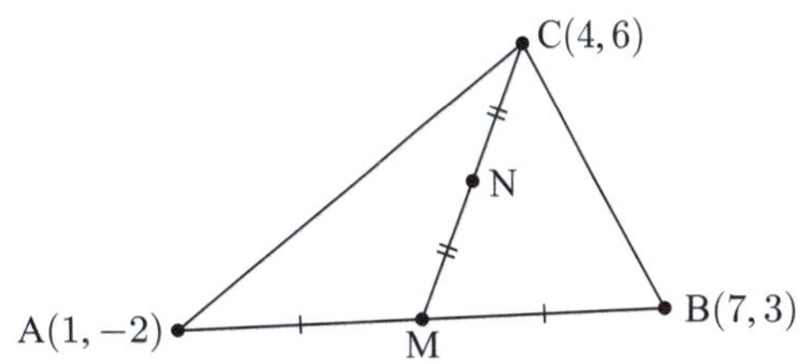

Example 4 **Self Tutor**

Suppose M is the midpoint of [PQ] where P is (3, 1) and M is (1, −3).
Find the coordinates of Q using:

a the midpoint formula **b** equal steps.

a Suppose Q has coordinates (a, b).

$\therefore \dfrac{a+3}{2} = 1$ and $\dfrac{b+1}{2} = -3$

$\therefore a+3 = 2$ and $b+1 = -6$

$\therefore a = -1$ and $b = -7$

So, Q is $(-1, -7)$.

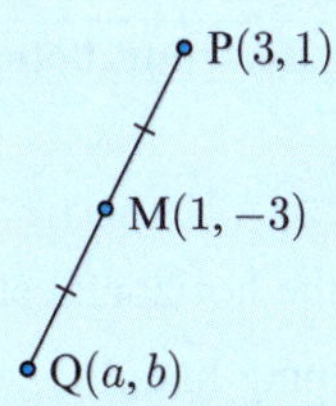

b x-step: $3 \xrightarrow{-2} 1 \xrightarrow{-2} -1$

y-step: $1 \xrightarrow{-4} -3 \xrightarrow{-4} -7$

So, Q is $(-1, -7)$.

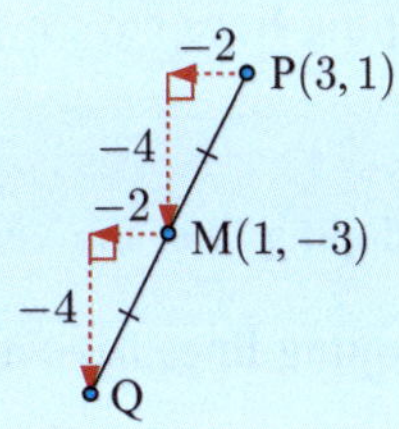

5 Suppose M is the midpoint of [PQ]. Find the coordinates of Q given:

- **a** P is $(-1, 3)$ and M is $(-1, 7)$
- **b** P is $(-1, 0)$ and M is $(0, -5)$
- **c** M is $(2, 1\frac{1}{2})$ and P is $(-2, 3)$
- **d** M is $(-1, 3)$ and P is $(-\frac{1}{2}, 0)$.

6 Suppose T is the midpoint of [AB]. Find the coordinates of A given:

- **a** T is $(4, -3)$ and B is $(-2, 3)$
- **b** B is $(0, 4)$ and T is $(-3, -2)$.

7 Find the coordinates of:

- **a** B
- **b** D

8 A circle has diameter [AB]. The centre of the circle is $(1, 3)$ and B is $(4, -1)$.

- **a** Find the coordinates of A.
- **b** Find the length of the diameter.

9 ABCD is a parallelogram. Diagonals [AC] and [BD] bisect each other at X. Find the coordinates of:

- **a** X
- **b** C

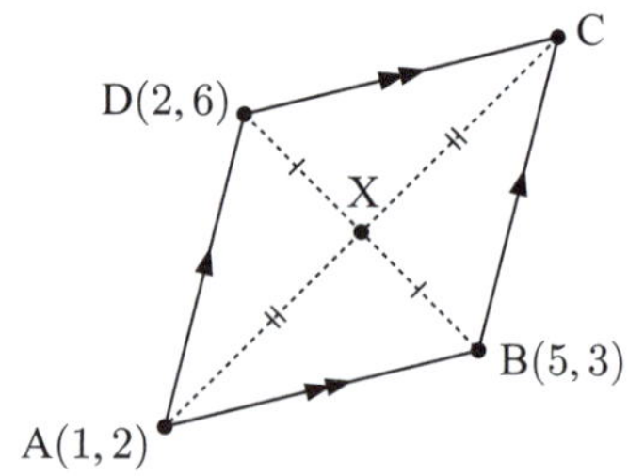

C GRADIENT

When looking at line segments drawn on a set of axes, it is clear that different line segments are inclined to the horizontal at different angles. Some appear to be *steeper* than others.

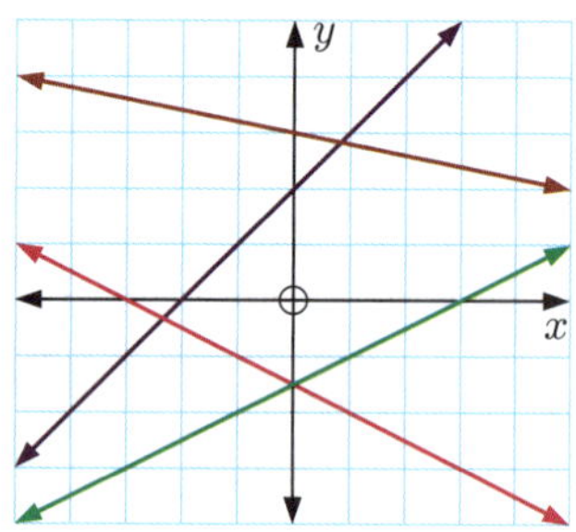

The **gradient** of a line is a measure of its steepness.

We can calculate the rate at which a line rises or falls by choosing two points on the line, then dividing the vertical step between them by the horizontal step between them.

$$\textbf{gradient} = \frac{\textbf{vertical step}}{\textbf{horizontal step}} \quad \text{or} \quad \frac{y\textbf{-step}}{x\textbf{-step}}$$

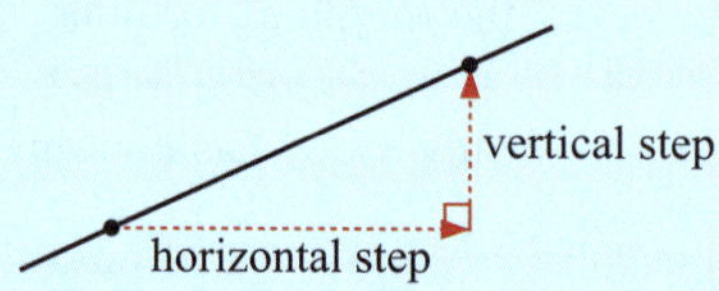

In general, we choose the horizontal step to be positive.

- For an upward sloping line, the vertical step is also positive.

Upward sloping lines have a **positive gradient**.

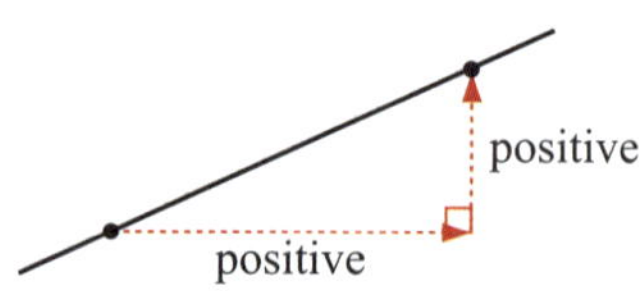

- For a downward sloping line, the vertical step is negative.

Downward sloping lines have a **negative gradient**.

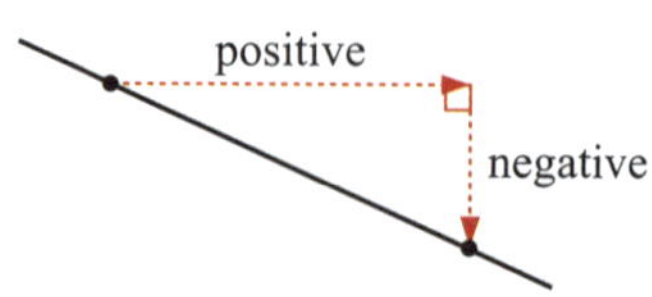

DISCUSSION

Why do you think we define gradient as $\dfrac{y\text{-step}}{x\text{-step}}$ rather than $\dfrac{x\text{-step}}{y\text{-step}}$?

Example 5

Self Tutor

Find the gradient of each line segment:

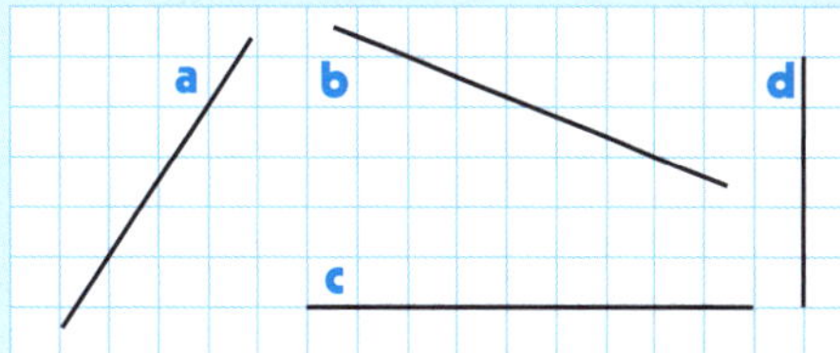

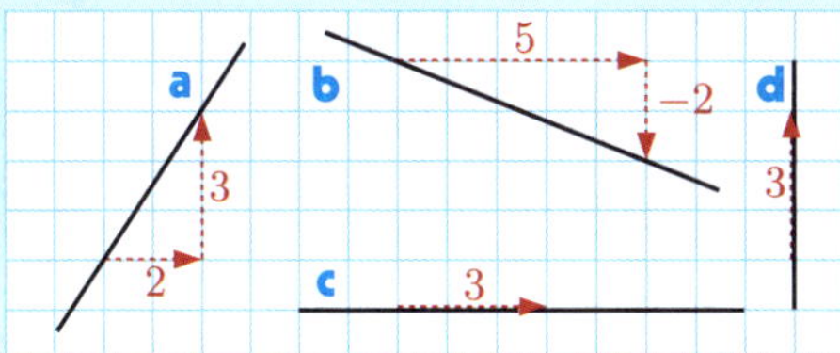

a gradient $= \frac{3}{2}$

b gradient $= \frac{-2}{5} = -\frac{2}{5}$

c gradient $= \frac{0}{3} = 0$

d gradient $= \frac{3}{0}$ which is undefined

From the Example, we can see that:

- For a **horizontal line**, the vertical step is 0, so the **gradient of a horizontal line is 0**.
- For a **vertical line**, the horizontal step is 0, so the **gradient of a vertical line is undefined**.

EXERCISE 16C.1

1 Find the gradient of each line:

a

b

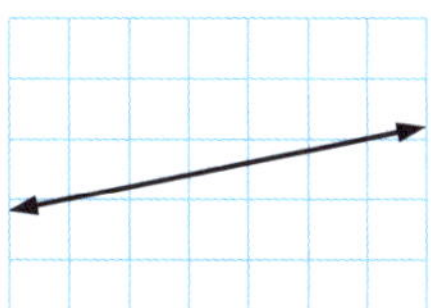

c

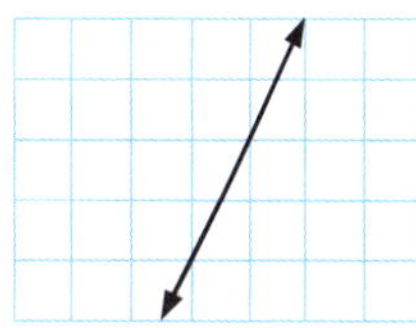

d

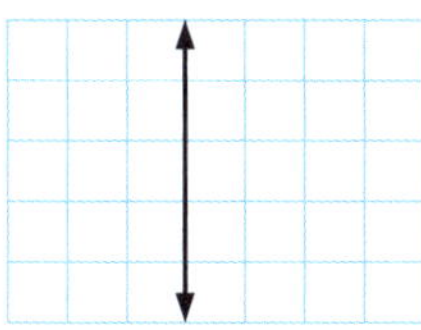

e

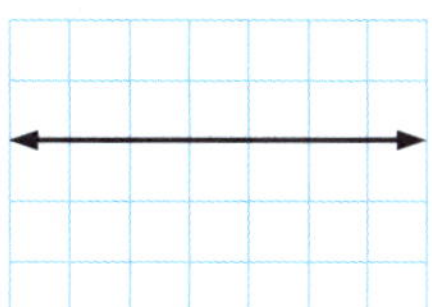

f

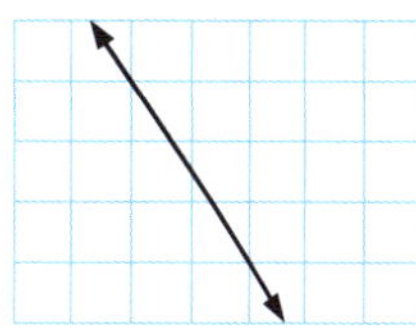

g

h

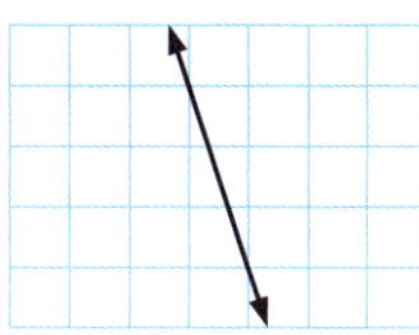

i

2

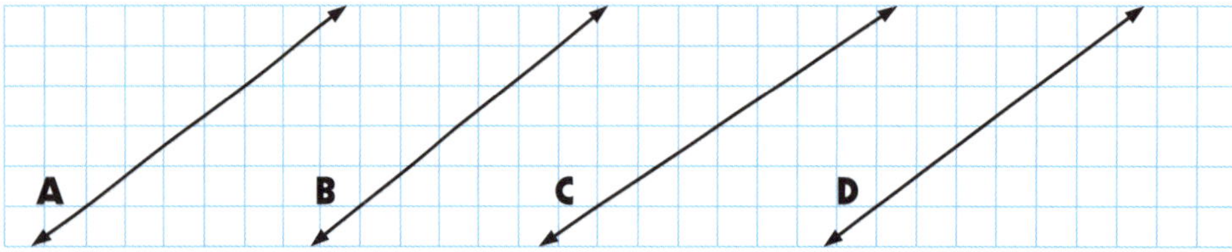

a Which two lines have the same gradient? **b** Which line is steepest?

Example 6 — Self Tutor

Draw a line through the point $(2, 4)$ with gradient $-\frac{2}{3}$.

$\text{gradient} = -\frac{2}{3} = \frac{-2}{3}$ ← y-step / ← x-step

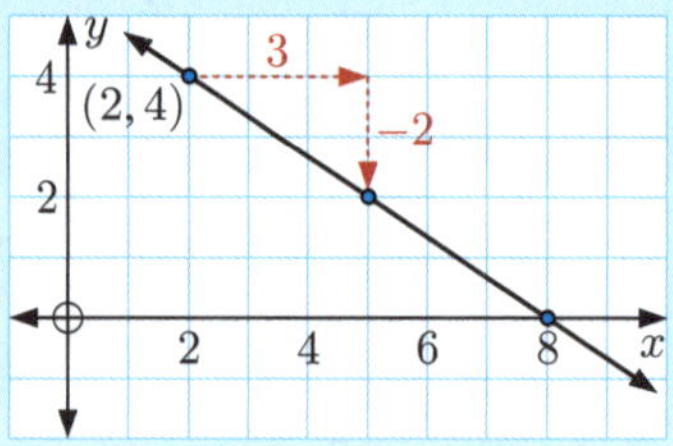

3 On grid paper, draw a line segment with gradient:

a $\frac{2}{3}$ **b** -1 **c** $\frac{4}{3}$ **d** $-\frac{2}{3}$ **e** 4 **f** $-\frac{6}{5}$

4 Draw a line through the point:

a $(3, 2)$ with gradient $\frac{1}{3}$ **b** $(-1, 4)$ with gradient -2 **c** $(-2, -4)$ with gradient 3

d $(1, 3)$ with gradient $-\frac{2}{3}$ **e** $(1, -2)$ with gradient $\frac{5}{3}$ **f** $(0, 4)$ with gradient $-\frac{4}{3}$.

5 On the same set of axes, draw lines through $(0, 0)$ with gradients 0, $\frac{1}{3}$, 1, $\frac{3}{2}$, and 5.

6 On the same set of axes, draw lines through $(3, 4)$ with gradients 0, $-\frac{1}{2}$, -1, $-\frac{3}{4}$, and -4.

7 On the same set of axes, draw a line through $(-2, 1)$ with gradient $\frac{1}{4}$, and a line through $(3, -1)$ with gradient -3. Label the point where the lines intersect.

THE GRADIENT FORMULA

INVESTIGATION 1 — THE GRADIENT FORMULA

We have seen how to calculate the gradient of the line segment between two points using horizontal and vertical steps. In this Investigation we develop this method into a **formula**.

What to do:

1

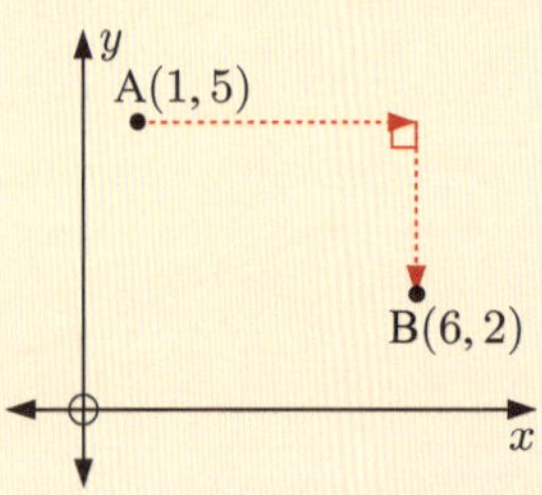

a Write an *expression* for the:

i horizontal step between A and B

ii vertical step between A and B.

b Write an *expression* for the gradient of the line using $\text{gradient} = \frac{\text{vertical step}}{\text{horizontal step}}$.

2

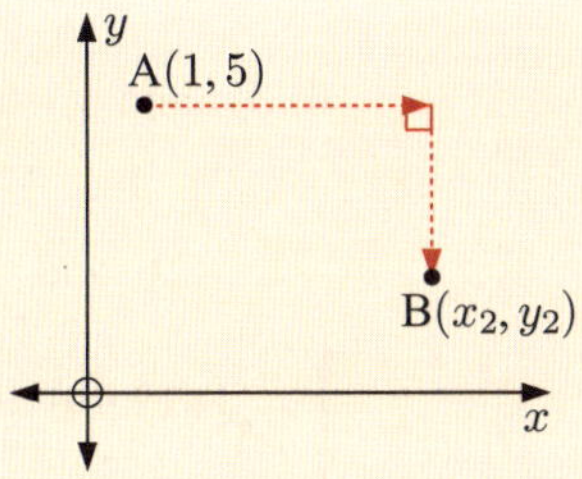

a Write an *expression* for the:

i horizontal step between A and B

ii vertical step between A and B.

b Write an *expression* for the gradient of the line using gradient $= \dfrac{\text{vertical step}}{\text{horizontal step}}$.

3

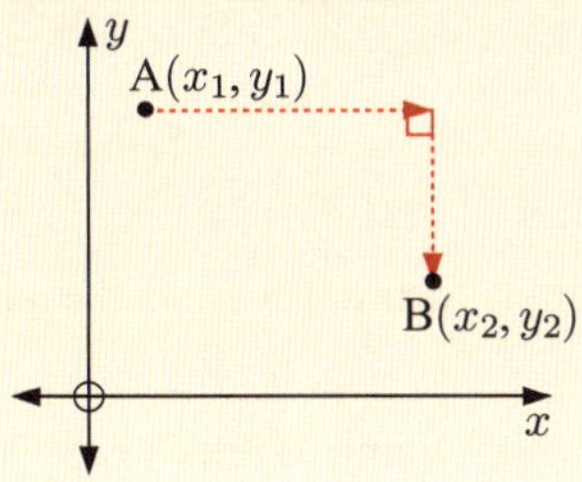

a Write an *expression* for the:

i horizontal step between A and B

ii vertical step between A and B.

b Write an *expression* for the gradient of the line using gradient $= \dfrac{\text{vertical step}}{\text{horizontal step}}$.

From the **Investigation**, you should have found that:

The **gradient** of the line through (x_1, y_1) and (x_2, y_2) is $\dfrac{y_2 - y_1}{x_2 - x_1}$.

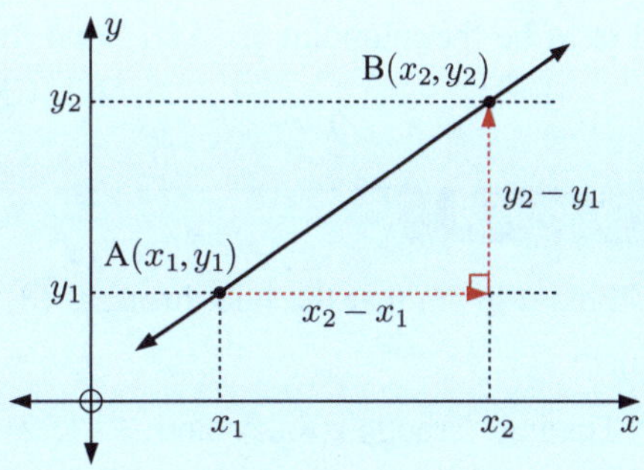

Example 7 — Self Tutor

Find the gradient of [PQ] given P(1, 3) and Q(4, −2).

P(1, 3): x_1 y_1 Q(4, −2): x_2 y_2

$$\text{gradient of [PQ]} = \frac{y_2 - y_1}{x_2 - x_1} = \frac{-2 - 3}{4 - 1} = -\frac{5}{3}$$

EXERCISE 16C.2

1 Find the gradient of the line segment joining:

a O(0, 0) and P(2, 5) **b** O(0, 0) and P(1, −4) **c** O(0, 0) and P(a, b).

2 Find the gradient of [PQ] given:

a P(1, 4) and Q(8, 6) **b** P(2, 7) and Q(6, −9) **c** P(−4, 1) and Q(8, −7).

3 Find the gradient of the line segment connecting:

a $(4, 7)$ and $(3, 2)$ **b** $(6, 1)$ and $(7, 5)$ **c** $(6, 3)$ and $(-2, 1)$

d $(1, -6)$ and $(4, -3)$ **e** $(5, -1)$ and $(5, 5)$ **f** $(-4, 3)$ and $(-1, 3)$

g $(-3, -2)$ and $(-1, 5)$ **h** $(0, 2)$ and $(2, -5)$ **i** $(-2, -2)$ and $(-1, 0)$.

4 For the triangle ABC shown, find the gradient of:

a [AB] **b** [BC] **c** [CA]

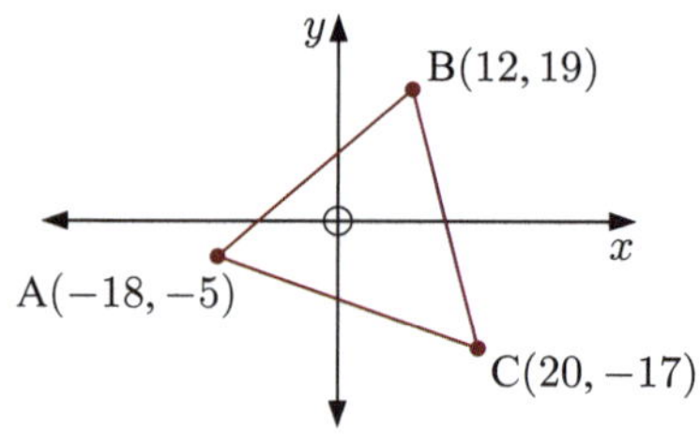

5 **a** Consider the points A$(1, 3)$ and B$(4, 5)$. Use the gradient formula to find the gradient of:

i [AB] **ii** [BA]

b Explain why the gradient formula ensures that for *any* points A and B, the gradients of [AB] and [BA] are equal.

6 A quadrilateral ABCD has vertices A$(-4, 2)$, B$(2, 5)$, C$(5, 4)$, and D$(2, -4)$.

a Find the gradient of each side of the quadrilateral.

b Let M be the midpoint of [AD]. Find the gradient of:

i [BM] **ii** [CM]

Example 8 **Self Tutor**

Find a given that the line through $(3, a)$ and $(-1, 2)$ has gradient $-\frac{3}{2}$.

The line through $(3, a)$ and $(-1, 2)$ has gradient $= \dfrac{2-a}{-1-3} = \dfrac{2-a}{-4} = \dfrac{a-2}{4}$

$$\therefore \frac{a-2}{4} = -\frac{3}{2}$$
$$\therefore a - 2 = -6$$
$$\therefore a = -4$$

7 Find a given that the line segment joining:

a $(1, 2)$ and $(5, a)$ has gradient $\frac{1}{2}$ **b** $(-1, a)$ and $(2, -3)$ has gradient -2

c $(-2, 3)$ and $(a, -1)$ has gradient $-\frac{3}{4}$ **d** $(a, 2)$ and $(-4, -1)$ has gradient $\frac{4}{3}$

e $(-1, 3)$ and (a, a) has gradient -1.

ACTIVITY 1 MOVING STAIRCASES

In a magical school for mathematics, the staircases are always moving.

Click on the icon to help our young student find his way around the school.

GAME

ACTIVITY 2 RATES

In real life we use the word *gradient* to measure a slope.

For example, the sign alongside indicates to drivers that there is an uphill climb ahead. The gradient of the road indicates how fast it climbs compared with horizontal distance travelled.

In mathematics, when we graph one variable against another, the gradient of the graph tells us the **rate** at which the variable on the vertical axis is changing compared with the variable on the horizontal axis.

What to do:

1 The graph alongside shows the distance run by a sprinter over time.

a Find and interpret the gradient of the line.

b Is the speed of the runner constant? Explain your answer.

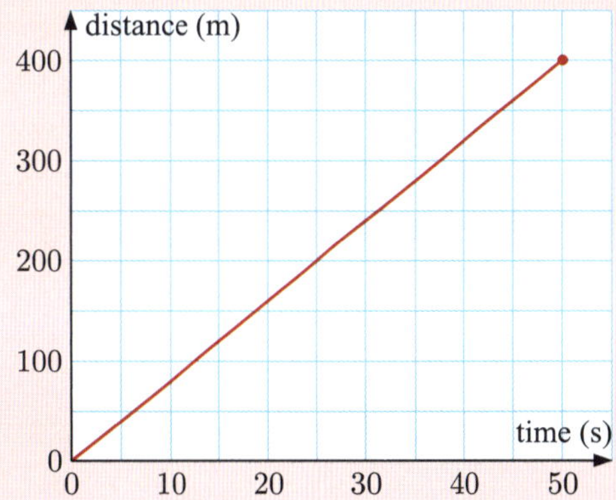

2 This graph shows the distance travelled by a truck driver over time during a journey.

a Determine the average speed from towns:

i A to B **ii** C to D.

b Find the average speed for the whole trip.

c Determine the time interval(s) over which the average speed was greatest.

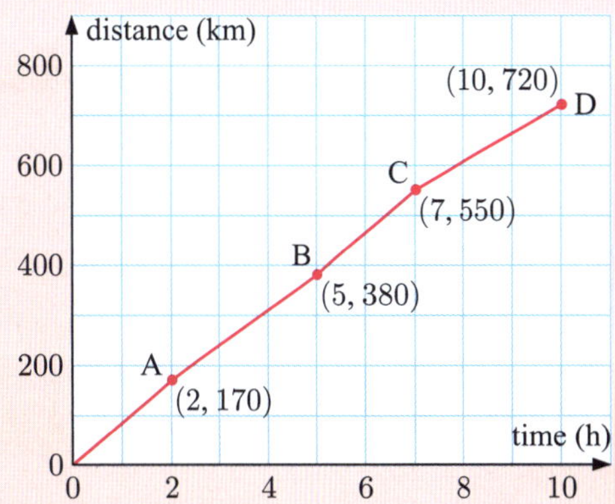

3 The graph alongside indicates the wage paid to a security guard.

a What does the intercept on the vertical axis mean?

b Find and interpret the gradient of the line.

c Determine the wage for a security guard working:

i 6 hours **ii** 15 hours.

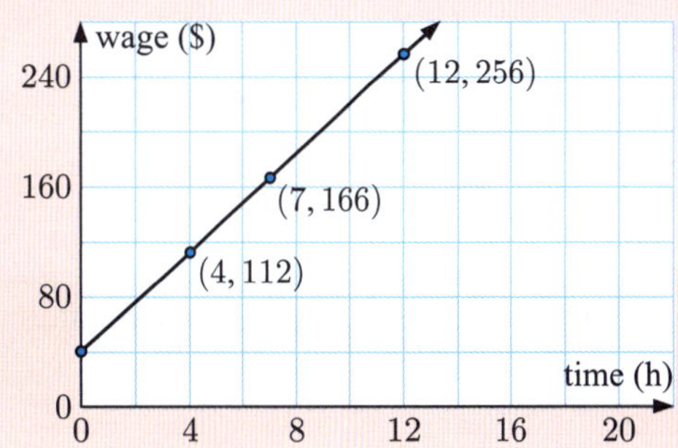

4 This graph shows the fuel consumption and distance travelled for a vehicle at speeds 60 km/h (graph A) and 90 km/h (graph B).

a Find and interpret the gradient of each line.

b At which speed would you say the vehicle is more fuel efficient?

c Suppose fuel costs $1.40 per litre. Compare the cost of travelling 1000 km at 90 km/h with the cost at 60 km/h.

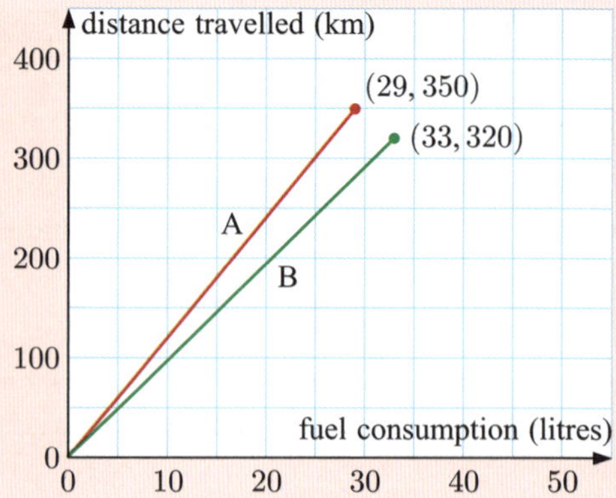

5 This graph shows the amount charged by a plumber according to the time taken to do a job.

a What does the value at A indicate?

b Find and interpret the gradient of:

i [AB]	**ii** [BC]
iii [AC]	**iv** [OC]

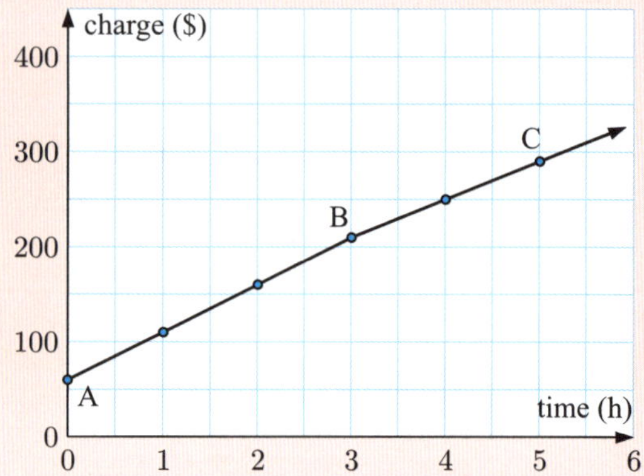

D PARALLEL AND PERPENDICULAR LINES

Two lines are **parallel** if there is always the same distance between them.

Parallel lines never intersect.

Two lines are **perpendicular** if they are at right angles to one another.

INVESTIGATION 2 PARALLEL LINES

In this Investigation we will discover the relationship between the gradients of parallel lines.

What to do:

1 Look at the lines alongside.

a Use a ruler to measure the vertical distance between the lines at several locations. Are the lines parallel?

b Find the gradient of each line. What do you notice?

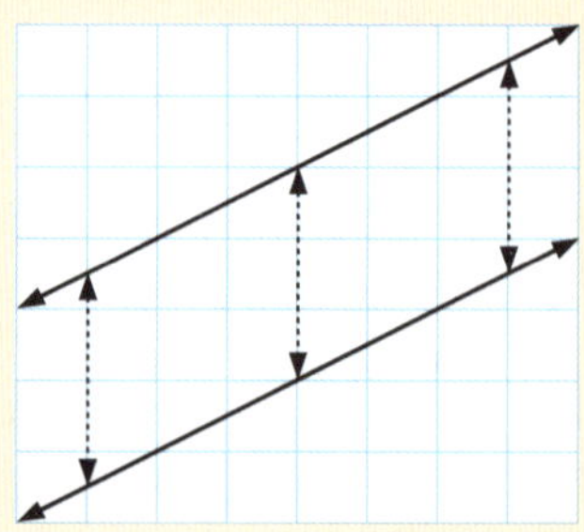

2 On graph paper, draw several pairs of lines, each pair having the same gradient. Determine whether each pair of lines is parallel.

3 Is the gradient of every horizontal line the same?

4 What can you say about the gradient of every vertical line?

GRAPH PAPER

INVESTIGATION 3 PERPENDICULAR LINES

In this Investigation we will discover the relationship between the gradients of perpendicular lines.

What to do:

1 Look at the lines alongside.

  a Use a protractor to measure the angle between the two lines.

  b Find the gradient of each line. What do you notice?

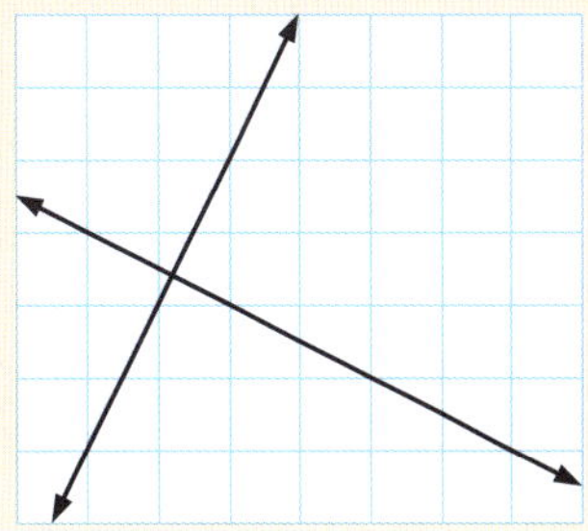

2 The **negative reciprocal** of $\frac{a}{b}$ is $-\frac{b}{a}$.

On graph paper, draw several pairs of lines, each pair having gradients which are negative reciprocals.

Measure the angle between each pair of lines.

3 Is a horizontal line always perpendicular to a vertical line?

From the **Investigations**, you should have found that:

For two lines which are neither horizontal nor vertical:

- the lines are **parallel** if and only if they have the **same gradient**
- the lines are **perpendicular** if and only if their gradients are **negative reciprocals**.

DISCUSSION

Discuss whether the rules above are valid for horizontal and vertical lines.

Example 9 Self Tutor

Line L has gradient $\frac{2}{5}$. Find the gradient of any line which is:

a parallel to L **b** perpendicular to L.

a Parallel lines have equal gradients, so a line parallel to L will also have gradient $\frac{2}{5}$.

b The negative reciprocal of $\frac{2}{5}$ is $-\frac{5}{2}$.

$\therefore$ a line perpendicular to L will have gradient $-\frac{5}{2}$.

EXERCISE 16D

1 Line M has gradient $\frac{3}{5}$ and line N has gradient 6. Find the gradient of any line which is:

a parallel to M
b perpendicular to M
c parallel to N
d perpendicular to N.

2 Find the gradient of a line which is perpendicular to a line with gradient:

a $\frac{4}{5}$
b $\frac{1}{3}$
c $-\frac{3}{7}$
d 2
e -11
f 0
g undefined
h $-1\frac{1}{2}$

3 Consider the points A(−2, 4), B(4, 8), C(7, 5), and D(−11, −7).

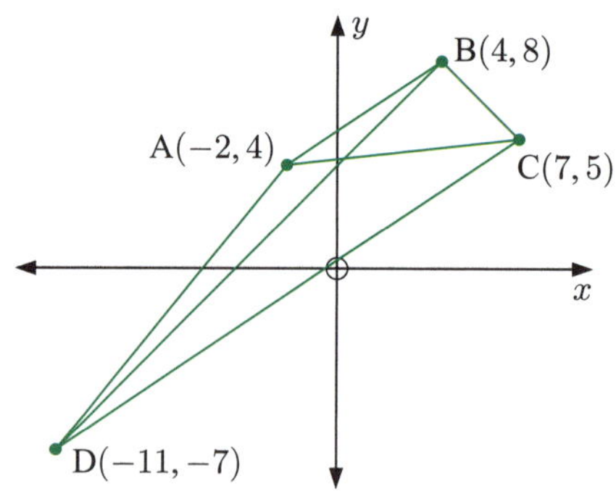

a Find the gradient of:
i [AB] **ii** [BC] **iii** [CD]
iv [DA] **v** [AC] **vi** [BD]

b Which of the line segments are:
i parallel **ii** perpendicular?

4 A(3, 1), B(2, −4), and C(7, −5) are three points in the Cartesian plane.

a Find the gradient of: **i** [AB] **ii** [BC]
b Classify triangle ABC.

Example 10 — Self Tutor

Find a given that the line joining A(2, a) and B(5, −1) is perpendicular to a line with gradient $\frac{1}{2}$.

Gradient of [AB] $= -2$ {perpendicular to line with gradient $\frac{1}{2}$}

$$\therefore \frac{-1-a}{5-2} = -2$$
$$\therefore \frac{-(1+a)}{3} = -2$$
$$\therefore \frac{1+a}{3} = 2$$
$$\therefore 1+a = 6$$
$$\therefore a = 5$$

5 Find a given that the line joining:

a A(3, 2) and B(−1, a) is parallel to a line with gradient −1
b C(−1, 1) and D(a, 3) is parallel to a line with gradient 2.

6 Find t given that the line joining:

a M(2, −1) and N(−1, t) is perpendicular to a line with gradient $-\frac{3}{2}$
b P(−1, t) and Q(2, 5) is perpendicular to a line with gradient $\frac{3}{4}$.

E USING COORDINATE GEOMETRY

Coordinate geometry is a powerful tool which can be used to:

- **check** the truth of a geometrical fact
- **prove** geometrical facts using general cases.

Example 11 Self Tutor

Consider the points A(1, 0), B(2, 4), and C(5, 1).

a Show that triangle ABC is isosceles.
b Find the midpoint M of [BC].
c Use gradients to show that [AM] and [BC] are perpendicular.
d Illustrate **a**, **b**, and **c** on a set of axes.

a AB
$= \sqrt{(2-1)^2+(4-0)^2}$
$= \sqrt{1+16}$
$= \sqrt{17}$ units

BC
$= \sqrt{(5-2)^2+(1-4)^2}$
$= \sqrt{9+9}$
$= \sqrt{18}$ units

AC
$= \sqrt{(5-1)^2+(1-0)^2}$
$= \sqrt{16+1}$
$= \sqrt{17}$ units

Since AB = AC, $\triangle$ABC is isosceles.

b M is $\left(\frac{2+5}{2}, \frac{4+1}{2}\right)$ which is $\left(\frac{7}{2}, \frac{5}{2}\right)$.

c gradient of [AM] $= \dfrac{\frac{5}{2}-0}{\frac{7}{2}-1} = \dfrac{\frac{5}{2}}{\frac{5}{2}} = 1$

gradient of [BC] $= \dfrac{1-4}{5-2} = -\dfrac{3}{3} = -1$

Since the gradients are negative reciprocals, [AM] and [BC] are perpendicular.

d 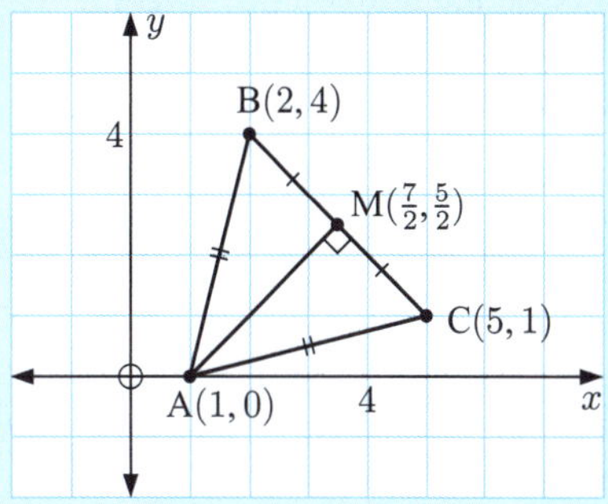

EXERCISE 16E

1 Consider the points A(−1, 0), B(1, 4), and C(3, 2).
   a Show that triangle ABC is isosceles.
   b Find the midpoint M of [BC].
   c Use gradients to show that [AM] and [BC] are perpendicular.
   d Illustrate **a**, **b**, and **c** on a set of axes.

2 Triangle ABC has vertices A(1, 3), B(5, 1), and C(3, −3).
M is the midpoint of [AB] and N is the midpoint of [BC]. Show that [MN] is:
   a parallel to [AC]
   b half the length of [AC].

3 Consider the points A(7, 7), B(16, 7), C(1, −2), and D(−8, −2).

a Show that:

i [AB] is parallel to [CD] **ii** [AD] is parallel to [BC].

b What kind of quadrilateral is ABCD?

4

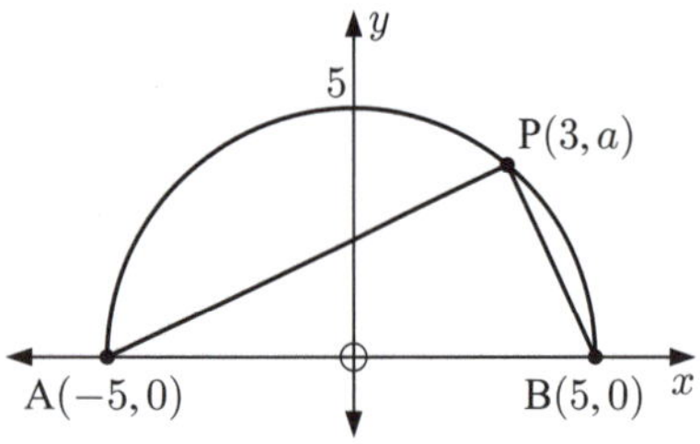

[AB] is the diameter of a semi-circle.
P(3, a) lies on the semi-circle.

a Find the radius of the semi-circle.

b Find a.

c Hence find the gradient of:

i [AP] **ii** [BP]

d Show that $\widehat{APB}$ is a right angle.

INVESTIGATION 4 THE MIDPOINTS OF A TRIANGLE THEOREM

Suppose we place the origin at one vertex O of a triangle, and orient the axes so a second vertex A($2a$, 0) lies on the positive x-axis.

The third vertex is B($2b$, $2c$).

Let M be the midpoint of [OA] and N be the midpoint of [OB].

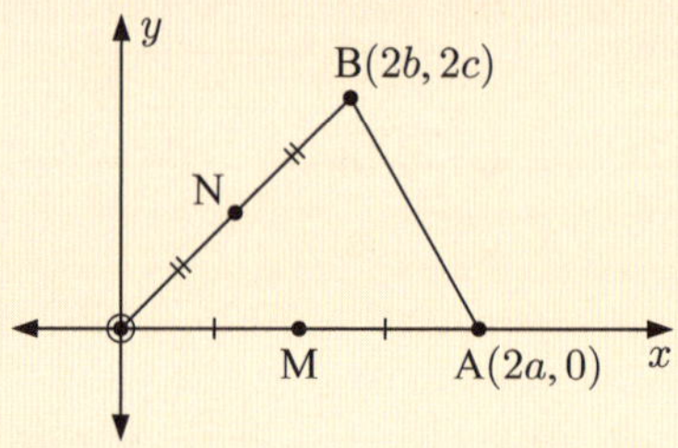

What to do:

1. Find the coordinates of: **a** M **b** N
2. Write an expression for the gradient of: **a** [MN] **b** [AB]
3. Write an expression for the length of: **a** [MN] **b** [AB]
4. Hence find $\dfrac{MN}{AB}$ in simplest form.
5. Copy and complete:

   The line segment between the midpoints of two sides of a triangle is ...... to the third side and ...... its length.

MULTIPLE CHOICE QUIZ

QUICK QUIZ

REVIEW SET 16A

1 Given the points A(7, −3) and B(−5, 3), find:

a the distance between A and B **b** the midpoint of [AB].

2 The points $P(2, 1)$, $Q(-4, -3)$, and $R(-2, 7)$ form a triangle PQR.

a Classify the triangle as equilateral, isosceles, or scalene.

b Determine whether triangle PQR is right angled. If it is, state where the right angle occurs.

3 Suppose M is the midpoint of [CD] as shown. Find the coordinates of D.

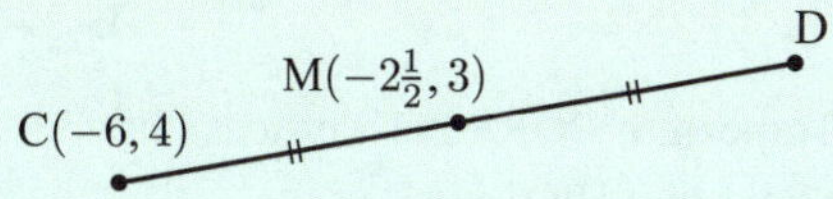

4 Draw a line through the point:

a $(-1, 2)$ with gradient -2 **b** $(2, 3)$ with gradient $\frac{1}{3}$.

5 Find the gradient of the line segment joining the points:

a $(4, -2)$ and $(6, 3)$ **b** $(-2, 7)$ and $(-2, 5)$.

6 Line L has gradient $-\frac{1}{4}$. Find the gradient of any line which is:

a parallel to L **b** perpendicular to L.

7 Find a given that the line joining $(-2, -1)$ and $(1, a)$ is:

a parallel to a line with gradient 2

b perpendicular to a line with gradient $-\frac{4}{3}$.

8

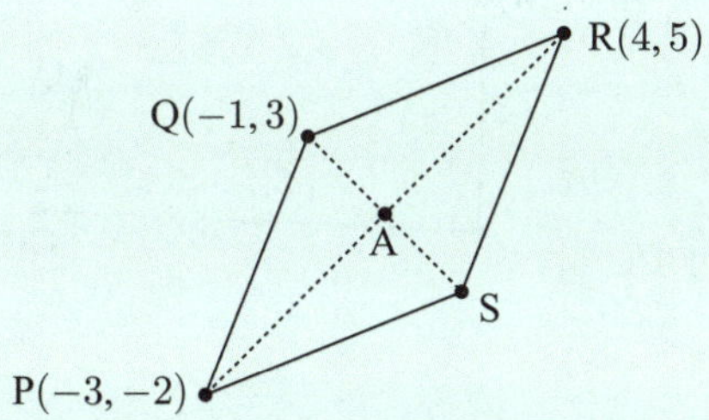

PQRS is a rhombus. Diagonals [QS] and [PR] bisect each other at A.

a Find the coordinates of:

i A **ii** S

b Show that [QS] and [PR] are perpendicular.

9 A quadrilateral has vertices $A(3, 2)$, $B(9, -2)$, $C(5, -8)$, and $D(-1, -4)$.

a Find the length of:

i [AB] **ii** [BC] **iii** [CD] **iv** [DA]

b Find the lengths of the diagonals [AC] and [BD].

c Find the midpoints of the diagonals [AC] and [BD].

d Classify the quadrilateral.

REVIEW SET 16B

1 Given $A(4, 1)$, $B(-1, 3)$, and $C(-2, -2)$, find:

a the shortest side of triangle ABC **b** the midpoint of [AC].

2 $M(7, \frac{1}{2})$ is the midpoint of [PQ], where P is $(2, -1)$. Find the coordinates of Q.

3 On the same set of axes, draw lines with gradients $\frac{1}{2}$, 1, -2, and 3 which pass through the point $(1, 3)$.

4 Find the gradient of the line segment joining:

a $(4, -3)$ and $(2, -6)$ **b** $(-3, 5)$ and $(1, 2)$.

5 Find t given that the line segment joining:

a $(t, 1)$ and $(4, -3)$ has gradient $\frac{2}{3}$ **b** $(1, 6)$ and $(-t, t)$ has gradient 1.

6 Find the gradient of a line which is perpendicular to a line with gradient:

a -2 **b** $\frac{3}{4}$ **c** $-3\frac{1}{3}$

7 Trapezium PQRS has vertices P$(-5, -2)$, Q$(3, 4)$, R$(10, 3)$, and S$(-2, -6)$.
Show that [PQ] is:

a parallel to [SR] **b** perpendicular to [PS].

8 A triangle has vertices O$(0, 0)$, A$(\sqrt{3}, -1)$, and B$(0, -2)$.

a Find the midpoint M of [OB].

b Show that triangle OMA is right angled at M.

c Find the length of:

i [AB] **ii** [OA] **iii** [OB]

d Classify triangle OAB.

e Illustrate the information you have found on a number plane.

9 P, Q, R, and S are the midpoints of [AB], [BC], [CD], and [AD] respectively.

a Find the coordinates of:

i P **ii** Q **iii** R **iv** S

b Show that:

i [PQ] is parallel to [SR]

ii [QR] is parallel to [PS].

c What type of quadrilateral is PQRS?

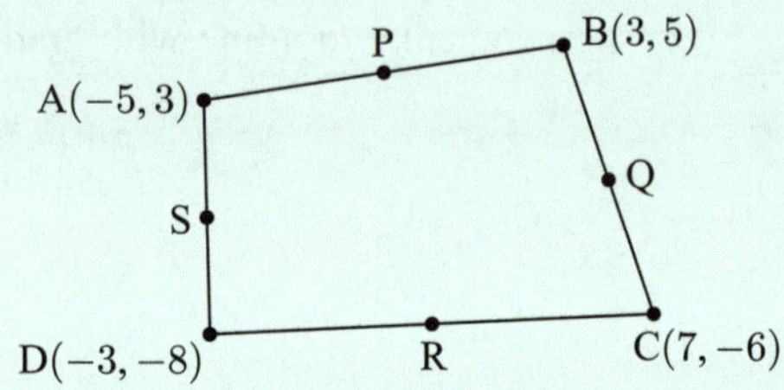

Chapter 17

Straight lines

Contents:

OPENING PROBLEM

Lifeguards from the tower at O have placed flags along the shoreline at (5, 1) and (1, 4). A wave parallel to the shore breaks when it reaches (3, 4).

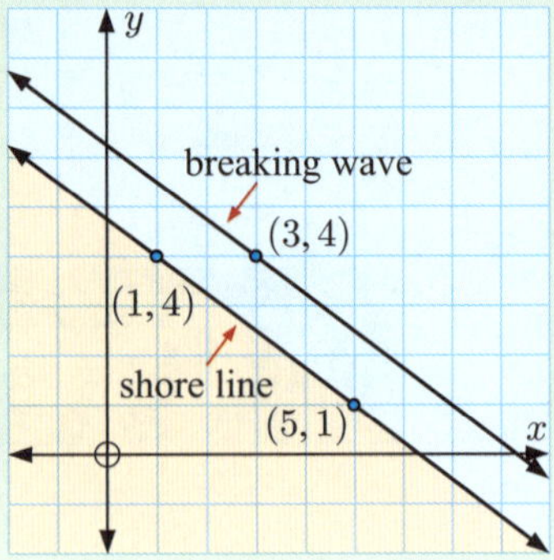

Things to think about:

a What information do we need about a straight line in order to find its equation?

b Can you find the equation of:

i the shoreline

ii the breaking wave?

We have seen previously that a *linear* equation is an equation which can be written in the form $ax + b = 0$ where $a \neq 0$.

The word "linear" comes from the relationship of this form to a straight line.

The **equation of a line** is a rule which connects the x and y coordinates of *all* points on the line.

For any non-vertical line, the equation can be written in the form $y = ax + b$.

A VERTICAL AND HORIZONTAL LINES

Vertical and horizontal lines are defined by specifying a constant value for one coordinate. The other coordinate can take any value.

VERTICAL LINES

For all points on a vertical line, the x-coordinate is constant. The y-coordinate can take any value.

The graph alongside shows the vertical lines $x = -1$ and $x = 3$.

All **vertical** lines have equations of the form $x = a$.

The gradient of a vertical line is **undefined**.

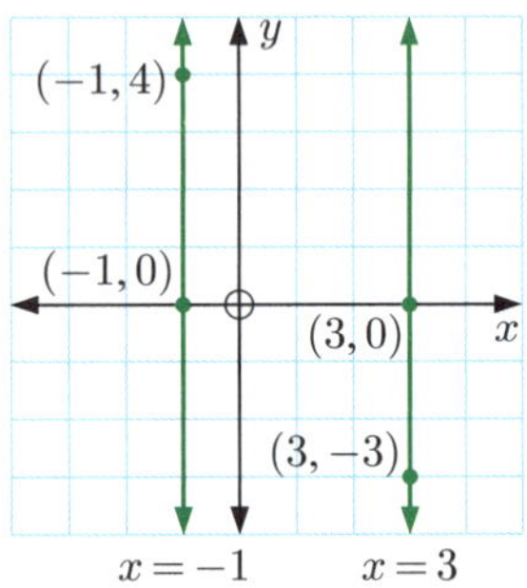

HORIZONTAL LINES

For all points on a horizontal line, the y-coordinate is constant. The x-coordinate can take any value.

The graph alongside shows the horizontal lines $y = 1$ and $y = -2$.

All **horizontal** lines have equations of the form $y = c$.

The gradient of a horizontal line is **zero**.

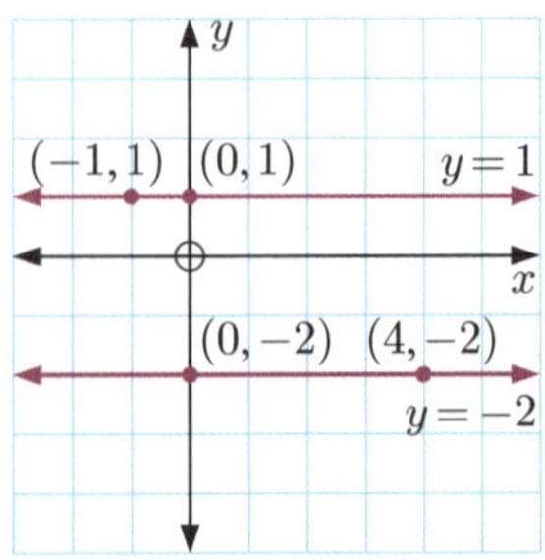

EXERCISE 17A

1 Draw the graph of each line and state its gradient:

a $x = 1$ **b** $y = 3$ **c** $x = -4$ **d** $y = 2.5$ **e** $x = 0$

2 Find the equation of:

a the x-axis **b** the y-axis

c a line parallel to the x-axis and three units below it

d a line parallel to the y-axis and 4 units to the right of it.

3 Find the equation of:

a the line with zero gradient that passes through $(-1, 3)$

b the line with undefined gradient that passes through $(4, -2)$.

4 Suppose l_1 is the line with equation $x = 5$, and l_2 is the line with equation $y = -3$. l_1 cuts the x-axis at A, and l_2 cuts the y-axis at B.

a Graph l_1 and l_2 on the same set of axes. Label points A and B.

b Find the coordinates of A and B.

c Find the gradient of the line segment [AB].

5 **a** On the same set of axes, graph the lines $x = -3$, $x = 2$, $y = 4$, and $y = -1$.

b Consider the quadrilateral formed by the points where these lines intersect.

i What type of quadrilateral is it?

ii Find the area of the quadrilateral.

iii Find the gradients of the diagonals of the quadrilateral.

B POINTS ON A LINE

A point lies on a line if its coordinates satisfy the equation of the line.

Example 1 Self Tutor

Determine whether:

a $(3, 7)$ lies on the line with equation $y = 2x + 1$

b $(4, -2)$ lies on the line with equation $5x + 3y = 16$.

a When $x = 3$, we have

$$\begin{aligned} y &= 2(3) + 1 \\ &= 6 + 1 \\ &= 7 \quad ✓ \end{aligned}$$

So, $(3, 7)$ does lie on the line.

b Substituting $x = 4$ and $y = -2$ into the LHS gives

$$\begin{aligned} &5(4) + 3(-2) \\ &= 20 - 6 \\ &= 14 \quad ✗ \end{aligned}$$

Since LHS $\neq$ RHS, $(4, -2)$ does *not* lie on the line.

EXERCISE 17B

1 Determine whether the point:

a $(2,\ 5)$ lies on the line $y = 3x + 1$

b $(3,\ -1)$ lies on the line $y = -x + 2$

c $(-4,\ 3)$ lies on the line $y = -2x - 5$

d $\left(-2,\ -\frac{3}{2}\right)$ lies on the line $y = \frac{3}{2}x - 4$

e $(2,\ -1)$ lies on the line $3x + y = 5$

f $(-3,\ 4)$ lies on the line $2x - 5y = -14$.

2 Find a given that:

a $(3,\ a)$ lies on the line $y = \frac{1}{2}x + \frac{1}{2}$

b $(-2,\ a)$ lies on the line $y = -3x + 7$

c $(a,\ 4)$ lies on the line $y = 2x - 6$

d $(a,\ -1)$ lies on the line $y = -x + 3$

e $(2,\ a)$ lies on the line $2x + y = 7$

f $(3,\ a)$ lies on the line $x + 2y = -1$

g $(a,\ 1)$ lies on the line $3x + 2y = 8$

h $(a,\ -3)$ lies on the line $2x - 3y = 5$.

3 Determine whether $(5,\ -2)$ lies on, above, or below the line $y = -2x + 3$.

C AXES INTERCEPTS

The **x-intercept** of a line is the x-coordinate of the point where the line cuts the x-axis.

The **y-intercept** of a line is the y-coordinate of the point where the line cuts the y-axis.

These values are together called the **axes intercepts** of a line.

For example, on the graph alongside:

- The line cuts the x-axis at $(-4,\ 0)$, so the x-intercept is -4.
- The line cuts the y-axis at $(0,\ 5)$, so the y-intercept is 5.

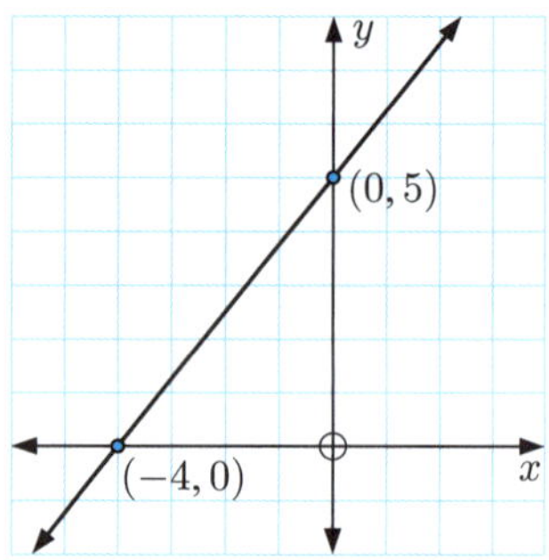

Example 2 — Self Tutor

Draw the line with x-intercept 3 and y-intercept -2.
Find the gradient of the line.

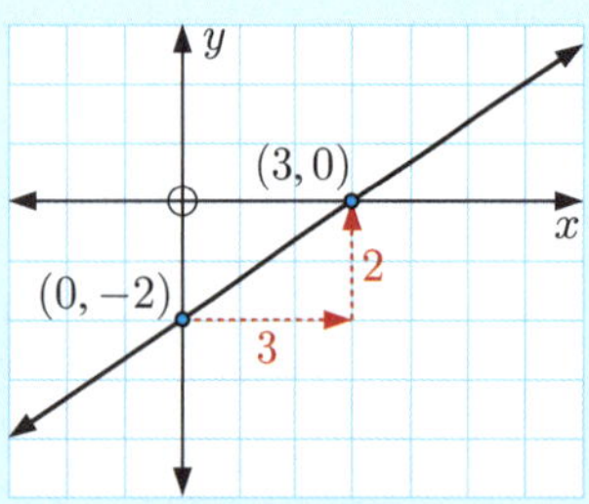

The line passes through $(3,\ 0)$ and $(0,\ -2)$.

Its gradient is $\frac{2}{3}$.

EXERCISE 17C

1 Find the x and y-intercept of each line, and hence find its gradient:

a 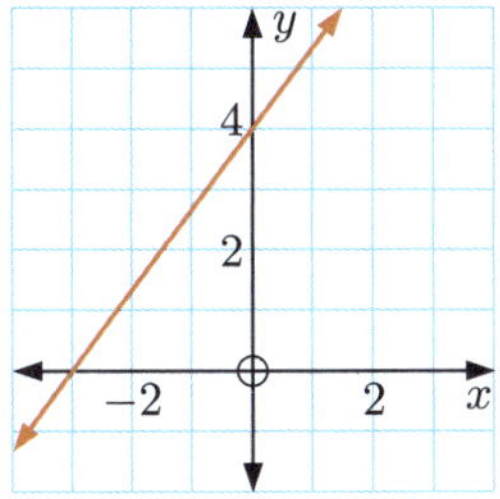

b 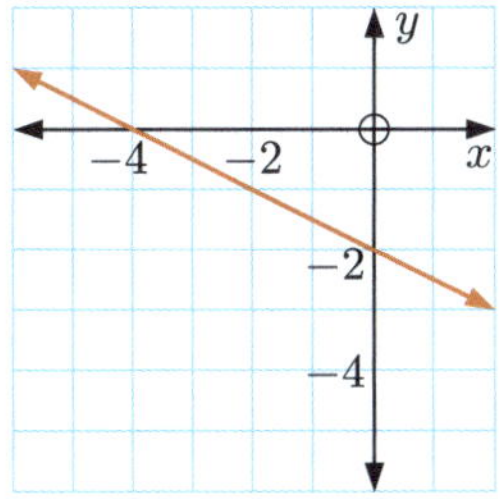

c 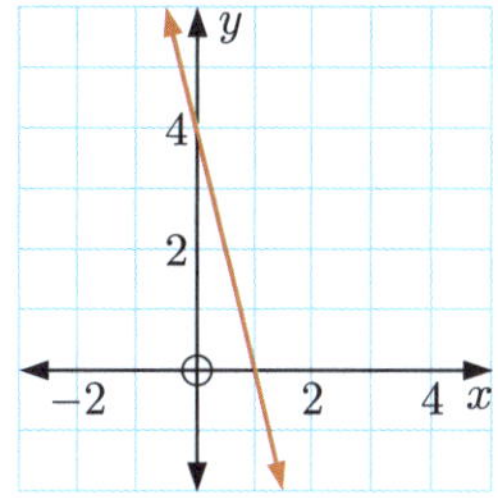

2 Draw the line with these axes intercepts, and hence determine its gradient:

a x-intercept 5 and y-intercept 3

b x-intercept -1 and y-intercept 4

c x-intercept -3 and y-intercept -2

d x-intercept 6 and y-intercept -2.

3 Find the y-intercept of each straight line by letting $x = 0$:

a $y = x + 5$ **b** $y = x - 3$ **c** $y = 2x$ **d** $y = 5 - x$

e $y = -2x + 1$ **f** $x + y = 2$ **g** $2x - y = 4$ **h** $3x + 2y = 12$

4 Find the x-intercept of each straight line by letting $y = 0$:

a $y = x - 4$ **b** $y = x + 2$ **c** $y = -x + 1$ **d** $y = -2x - 6$

e $y = 3x - 12$ **f** $y = -3x + 9$ **g** $x + 2y = 6$ **h** $2x - 3y = -8$

5 Consider the line with equation $y = mx + c$, $m \neq 0$:

a Find the y-intercept of the line by letting $x = 0$.

b Find the x-intercept of the line by letting $y = 0$.

c Use the points where the line cuts the axes to find the gradient of the line.

D GRAPHING FROM A TABLE OF VALUES

Consider the equation $y = 2x + 1$. We can choose any value we like for x and use our equation to find the corresponding value for y.

We can hence construct a **table of values** for points on the line.

x	-3	-2	-1	0	1	2	3
y	-5	-3	-1	1	3	5	7

$$y = 2 \times -3 + 1 = -6 + 1 = -5$$

$$y = 2 \times 2 + 1 = 4 + 1 = 5$$

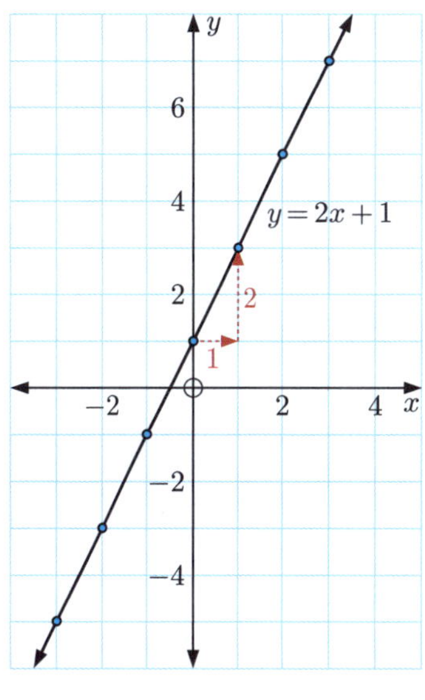

From this table we plot the points $(-3, -5)$, $(-2, -3)$, $(-1, -1)$, $(0, 1)$, and so on.

The tabled points are collinear and we can connect them with a straight line.

The gradient of the line is $\frac{y\text{-step}}{x\text{-step}} = \frac{2}{1} = 2$.

Example 3 — Self Tutor

Consider the line with equation $y = x - 2$.

a Construct a table of values using $x = -3, -2, -1, 0, 1, 2$, and 3.

b Hence draw the graph of $y = x - 2$.

c Find the gradient and axes intercepts of the line.

a

x	-3	-2	-1	0	1	2	3
y	-5	-4	-3	-2	-1	0	1

b

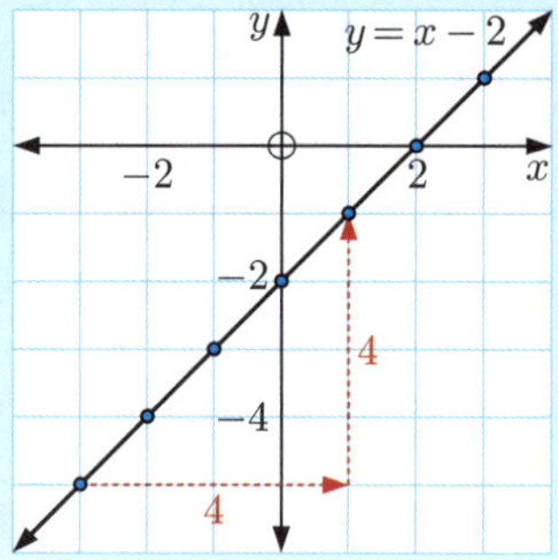

c Using $(-3, -5)$ and $(1, -1)$,

the gradient $= \dfrac{y\text{-step}}{x\text{-step}} = \dfrac{4}{4} = 1$.

The x-intercept is 2.

The y-intercept is -2.

EXERCISE 17D

1 For each of the following equations:

i Construct a table of values using $x = -3, -2, -1, 0, 1, 2$, and 3.

ii Draw the graph of the straight line.

iii Find the gradient and axes intercepts of the line.

a $y = x$ **b** $y = -x$ **c** $y = 2x$

d $y = -2x$ **e** $y = 2x - 1$ **f** $y = \frac{1}{2}x + 2$

g $y = -\frac{1}{2}x + 2$ **h** $y = 2 - x$ **i** $y = -\frac{3}{2}x - 1$

GRAPHING PACKAGE

2 Consider the straight line $y = mx + c$ where $m \neq 0$.

a Copy and complete this table of values:

x	-2	0	2	$-\frac{c}{m}$
y				

b State the axes intercepts of the straight line.

c Use the points with x-coordinates -2 and 2 to find the gradient of the line.

d Copy and complete:

A straight line with equation $y = mx + c$ has gradient ...... and y-intercept ......

E GRADIENT-INTERCEPT FORM

From the previous Sections you should have found that:

> $\boldsymbol{y = mx + c}$ is the equation of a straight line with gradient m and y-intercept c.

We call this the **gradient-intercept form** of a line.

Example 4 Self Tutor

State the gradient and y-intercept of the line with equation:

a $y = 2x - 3$ **b** $y = -3x + 5$

a $y = 2x - 3$ has $m = 2$ and $c = -3$
$\therefore$ the gradient is 2 and the y-intercept is -3.

b $y = -3x + 5$ has $m = -3$ and $c = 5$
$\therefore$ the gradient is -3 and the y-intercept is 5.

To draw the graph of $y = mx + c$:

- Use the y-intercept c to plot the point $(0, c)$.
- Starting from $(0, c)$, use horizontal and vertical steps from the gradient m to locate another point on the line.
- Join the two points and extend the line in both directions.

Example 5 Self Tutor

Draw the graph of $y = -\frac{3}{4}x + 5$.

For $y = -\frac{3}{4}x + 5$:

- the y-intercept is $c = 5$
- the gradient is $m = \frac{-3}{4}$ ← y-step, ← x-step

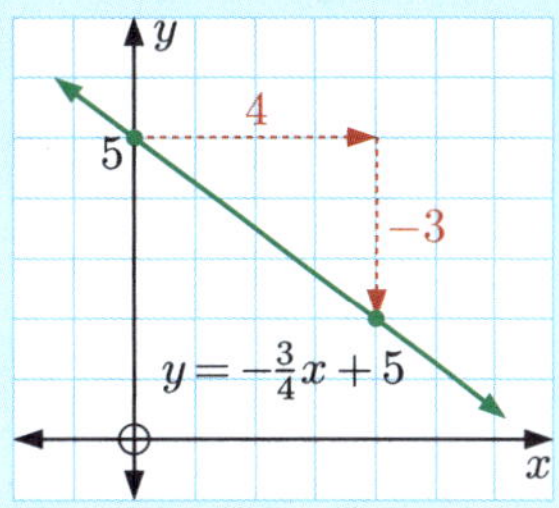

EXERCISE 17E

1 State the gradient and y-intercept of the line with equation:

a $y = 2x - 4$ **b** $y = -4x + 1$ **c** $y = \frac{2}{3}x - 3$

d $y = 7 - \frac{1}{3}x$ **e** $y = \frac{1}{2} + \frac{3}{4}x$ **f** $y = -\frac{3}{2}x - \frac{1}{4}$

g $y = \frac{x - 3}{2}$ **h** $y = \frac{5x + 1}{4}$ **i** $y = \frac{5 - 2x}{3}$

2 Draw the graph of:

a $y = x - 2$ **b** $y = 2x + 1$ **c** $y = 3x - 1$

d $y = \frac{2}{3}x$ **e** $y = \frac{4}{3}x - 2$ **f** $y = \frac{3x + 1}{2}$

3 Draw the graph of:

a $y = -x + 4$ **b** $y = -2x + 2$ **c** $y = -\frac{1}{2}x - 1$

d $y = -\frac{2}{5}x - 3$ **e** $y = 5 - 2x$ **f** $y = \frac{-x + 4}{3}$

4 Consider the line with equation $y = 3x - 5$.

a Find the:

i gradient **ii** y-intercept **iii** x-intercept.

b Draw the graph of the line.

c Find the area of the triangle formed by the line and the two axes.

5 Consider the line with equation $y = -\frac{2}{3}x + 2$.

a Find the:

i gradient **ii** y-intercept **iii** x-intercept.

b Draw the graph of the line.

c Does the point $\left(4, -\frac{2}{3}\right)$ lie on the line?

6 On the same set of axes, draw the graphs of $y = 2x - 5$ and $y = 1 - x$.
Label the axes intercepts and the point where the lines intersect.

7 The cost of hiring a squash court for t hours is given by $C = 28t$ dollars.

a Graph C against t for $t \geqslant 0$.

b State and interpret the gradient of the graph.

8 The cost of hiring a scooter for t minutes is given by $C = 2 + \frac{1}{2}t$ dollars.

a Graph C against t for $t \geqslant 0$.

b State and interpret the gradient of the graph.

c State and interpret the C-intercept of the graph.

F GENERAL FORM

The **general form** of the equation of a straight line is $ax + by = d$ where a, b, and d are constants.

For example, the equations $4x + 5y = 3$ and $x - 3y = -4$ are in general form.

Equations in general form are usually written with a positive coefficient of x.

We can rearrange equations to convert between gradient-intercept form and general form.

Example 6 **Self Tutor**

Write the equation:

a $y = -\frac{1}{4}x + \frac{3}{4}$ in general form **b** $2x + 5y = 7$ in gradient-intercept form.

a
$y = -\frac{1}{4}x + \frac{3}{4}$
$\therefore\ 4y = -x + 3$ {$\times$ 4 both sides}
$\therefore\ x + 4y = 3$ {$+\,x$ both sides}

b
$2x + 5y = 7$
$\therefore\ 5y = -2x + 7$ {$-\,2x$ both sides}
$\therefore\ y = -\frac{2}{5}x + \frac{7}{5}$ {$\div$ 5 both sides}

The general form of an equation is useful because:

- it allows us to write many equations without fractions
- it can be used for *all* straight lines, including vertical lines
- it allows us to quickly calculate both axes intercepts.

To draw the graph of $ax + by = d$:

- Find the y-intercept by letting $x = 0$.
- Find the x-intercept by letting $y = 0$.
- Join the points where the line cuts the axes and extend the line in both directions.

Example 7 **Self Tutor**

Draw the graph of $4x - 3y = 12$.

When $x = 0$, $-3y = 12$
$\therefore y = -4$

So, the y-intercept is -4.

When $y = 0$, $4x = 12$
$\therefore x = 3$

So, the x-intercept is 3.

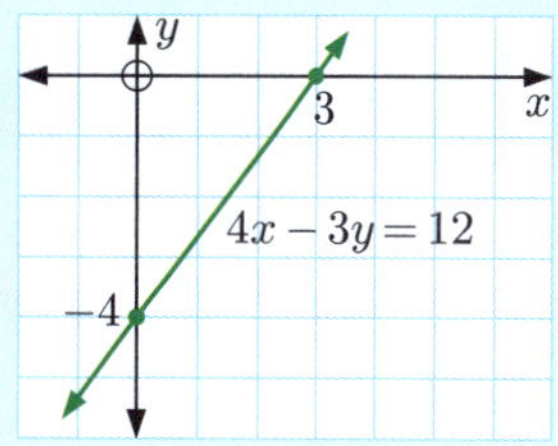

EXERCISE 17F

1 Write in general form:

a $y = -2x + 11$ **b** $y = 3x - 4$ **c** $y = -\frac{1}{5}x + \frac{6}{5}$

d $y = -\frac{6}{7}x + \frac{5}{7}$ **e** $y = \frac{5}{8}x - \frac{1}{8}$ **f** $y = \frac{4}{9}x + 2$

2 For each of the following lines:

i Write the equation in gradient-intercept form.
ii State the gradient of the line.

a $3x + y = 5$ **b** $2x + 5y = 10$ **c** $7x + 4y = -9$

d $6x - y = 1$ **e** $5x - 11y = 2$ **f** $9x - 2y = -5$

3 Draw the graph of:

a $x + 3y = 6$ **b** $3x - 2y = 12$ **c** $2x + 5y = 10$

d $4x + 3y = 6$ **e** $x + y = 5$ **f** $x - y = -3$

g $3x - y = -6$ **h** $7x + 2y = 14$ **i** $4x + 9y = -18$

4 Consider the line with equation $3x - 5y = 15$.

a Find the axes intercepts.

b Determine whether each point lies on the line: **i** $(-5, -6)$ **ii** $(1, -2)$

c Draw the graph of the line, showing the information you have found.

5 **a** Draw the graph of $3x + 4y = 24$.

b Which of these lines is parallel to $3x + 4y = 24$?

A $y = \frac{4}{3}x + 1$ **B** $y = -\frac{4}{3}x - 5$ **C** $y = \frac{3}{4}x + 6$ **D** $y = -\frac{3}{4}x + 7$

c Which of these lines is perpendicular to $3x + 4y = 24$?

A $y = \frac{4}{3}x + 1$ **B** $y = -\frac{4}{3}x - 5$ **C** $y = \frac{3}{4}x + 6$ **D** $y = -\frac{3}{4}x + 7$

6 Suzanne buys x carrots and y potatoes. The carrots cost 20 cents each and the potatoes cost 50 cents each. The total cost is \$5.

a Explain why $2x + 5y = 50$.

b Draw the graph of $2x + 5y = 50$.

c Explain the significance of the: **i** x-intercept **ii** y-intercept.

7 Consider the graph of a line in the general form $ax + by = d$.

a By rearranging the equation into gradient-intercept form, find the gradient of the line.

b For what values of b is the gradient defined?

G FINDING THE EQUATION OF A LINE

To determine the equation of a line, we need to know either:
- its gradient and at least one point which lies on the line, *or*
- two points which lie on the line.

GIVEN THE GRADIENT AND y-INTERCEPT

If we know the gradient m and y-intercept c of a line, we can write down the equation of the line immediately as $y = mx + c$.

Example 8 Self Tutor

Find the equation of the line with gradient 5 and y-intercept -3.

The line has gradient $m = 5$ and y-intercept $c = -3$.

So, the equation of the line is $y = 5x - 3$.

EXERCISE 17G.1

1 Write down the equation of the line with:

a gradient 3 and y-intercept 4

b gradient 7 and y-intercept -1

c gradient -1 and y-intercept $\frac{1}{3}$

d gradient $\frac{1}{2}$ and y-intercept 0

e gradient 0 and y-intercept 5

f gradient $-\frac{5}{2}$ and y-intercept $-\frac{2}{3}$.

2 A straight line has gradient $\frac{1}{4}$ and y-intercept -6.

a Find the equation of the line.

b Find the x-intercept of the line.

3 A broken tap allows water to flow from a tank at a constant rate of 8 litres per minute. The graph alongside shows the amount of water in the tank over time.

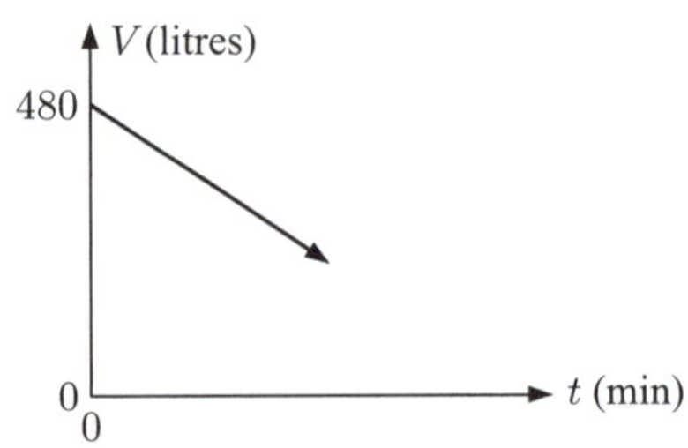

a Explain why the gradient of the graph must be -8.

b State and interpret the V-intercept.

c Write a formula for V in terms of t.

d Find and interpret the t-intercept.

GIVEN THE GRADIENT AND A POINT

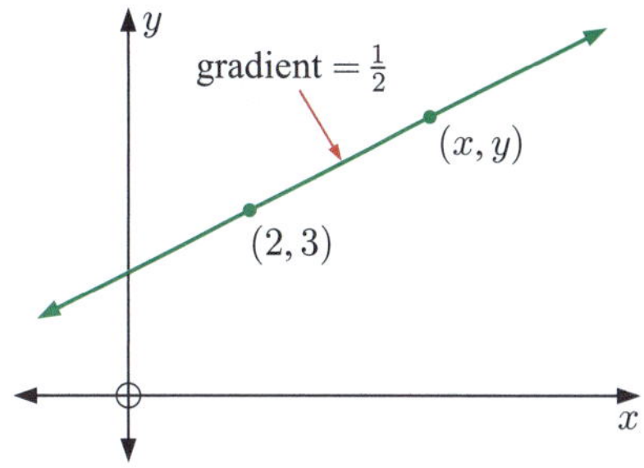

Suppose that a line has gradient $\frac{1}{2}$, and passes through the point (2, 3).

For any point (x, y) which lies on the line, the gradient between (2, 3) and (x, y) is $\frac{y-3}{x-2}$.

$\therefore$ the line has equation $\frac{y-3}{x-2} = \frac{1}{2}$

which can be written as $y - 3 = \frac{1}{2}(x-2)$.

We can rearrange this to find the equation of the line in either:

- gradient-intercept form

$$y - 3 = \tfrac{1}{2}(x-2)$$
$$\therefore\ y - 3 = \tfrac{1}{2}x - 1$$
$$\therefore\ y = \tfrac{1}{2}x + 2$$

- general form

$$y - 3 = \tfrac{1}{2}(x-2)$$
$$\therefore\ 2(y-3) = 1(x-2)$$
$$\therefore\ 2y - 6 = x - 2$$
$$\therefore\ x - 2y = -4$$

If a straight line has gradient m and passes through (a, b), then its equation in **point-gradient** form is

$$\boldsymbol{y - b = m(x - a)}.$$

We can rearrange the equation into either **gradient-intercept form** or **general form**.

Example 9 **Self Tutor**

Find, in gradient-intercept form, the equation of the line with gradient 5 that passes through $(-1, 3)$.

The equation of the line is
$$y - 3 = 5(x - -1)$$
$$\therefore\ y - 3 = 5(x+1)$$
$$\therefore\ y - 3 = 5x + 5$$
$$\therefore\ y = 5x + 8$$

EXERCISE 17G.2

1 Find, in gradient-intercept form, the equation of the line which has gradient:

- **a** 2 and which passes through (1, 8)
- **b** -3 and which passes through $(4, -16)$
- **c** 5 and which passes through $(-3, -1)$
- **d** $\frac{3}{4}$ and which passes through (8, 12)
- **e** 0 and which passes through $(-2, 4)$
- **f** $-\frac{1}{2}$ and which passes through $(5, -9)$
- **g** $-\frac{1}{3}$ and which passes through (1, 4)
- **h** $\frac{2}{3}$ and which passes through $(-2, 5)$.

2 Find, in gradient-intercept form, the equation of the line which is:

- **a** parallel to the line $y = 3x + 4$ and which passes through (2, 4)
- **b** parallel to the line $y = -\frac{1}{4}x - 2$ and which passes through $(8, -3)$
- **c** perpendicular to the line $y = -\frac{1}{2}x + 3$ and which passes through (3, 5).

Example 10 — Self Tutor

Find, in general form, the equation of the line with gradient $\frac{3}{4}$ that passes through $(5, -2)$.

The equation of the line is $y - -2 = \frac{3}{4}(x-5)$

$\therefore\ 4(y+2) = 3(x-5)$

$\therefore\ 4y + 8 = 3x - 15$

$\therefore\ 3x - 4y = 23$

3 Find, in general form, the equation of the line:

a through $(4, 1)$ with gradient $\frac{1}{2}$

b through $(5, 0)$ with gradient $\frac{3}{4}$

c through $(3, -2)$ with gradient 3

d through $(2, -3)$ with gradient $-\frac{3}{4}$

e through $(3, -2)$ with gradient -2

f through $(0, 4)$ with gradient -3.

4 Find, in general form, the equation of the line through $(4, -1)$ which is perpendicular to a line with gradient -3.

GIVEN TWO POINTS ON THE LINE

If we are given two points on a straight line, we use them to find the gradient. We can then use the gradient and *either* point to find the equation of the line.

Example 11 — Self Tutor

Find the equation of the illustrated line:

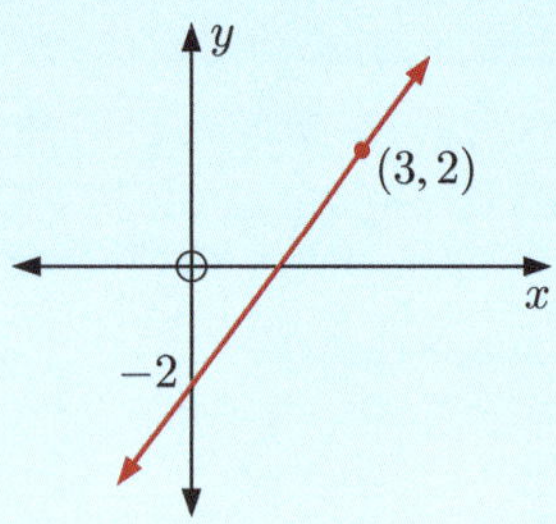

The line passes through $(0, -2)$ and $(3, 2)$.

$\therefore$ the gradient $m = \dfrac{2-(-2)}{3-0} = \dfrac{4}{3}$

and the y-intercept $c = -2$

$\therefore$ the equation is $y = \frac{4}{3}x - 2$.

EXERCISE 17G.3

1 Find the equation of the illustrated line:

a

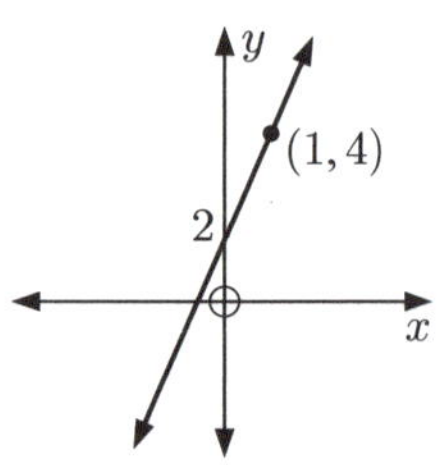

b

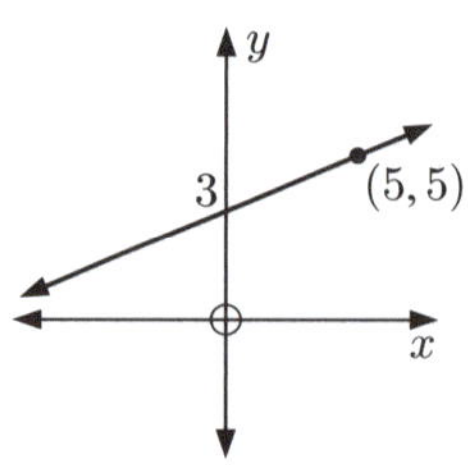

c

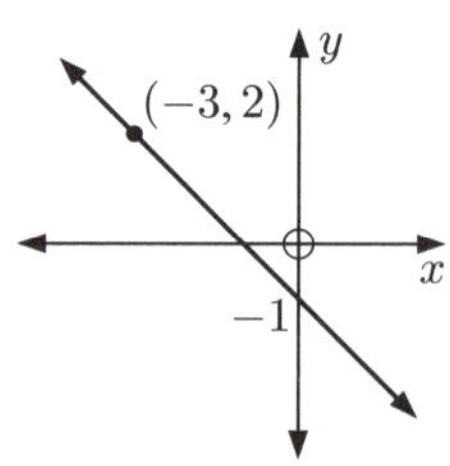

d

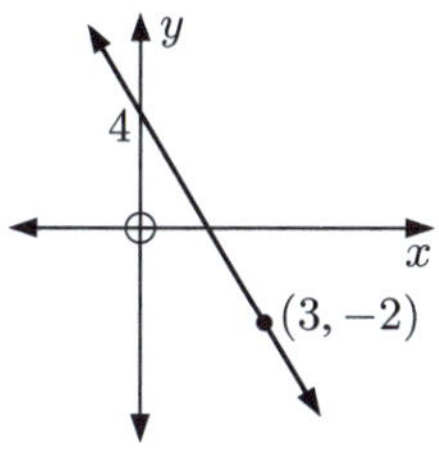

e

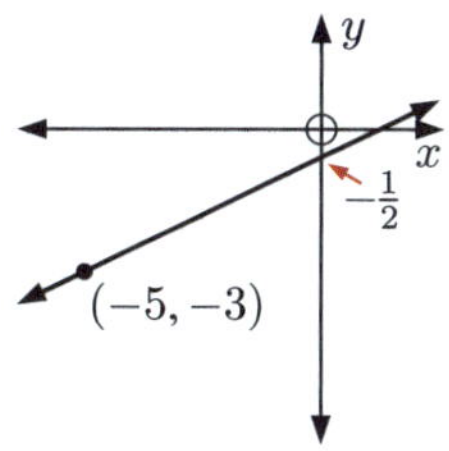

f

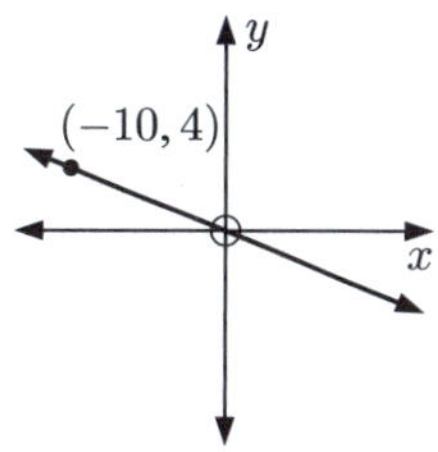

2 Find the equation of the line which has:

- **a** y-intercept 4 and x-intercept -3
- **b** y-intercept -5 and passes through $(3, 7)$
- **c** y-intercept 7 and passes through $(5, -8)$
- **d** y-intercept -1 and passes through $(-8, -4)$.

3 Find the equation of the illustrated line:

a

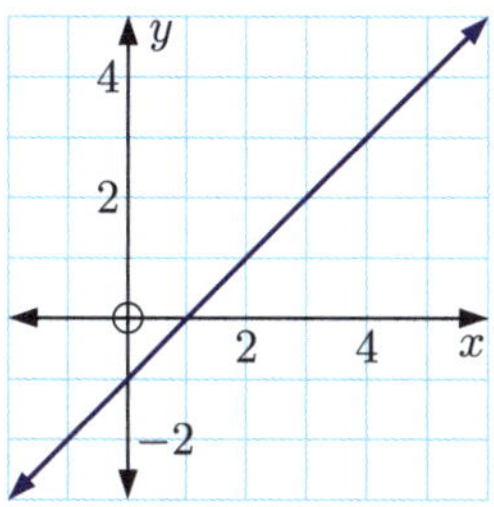

b

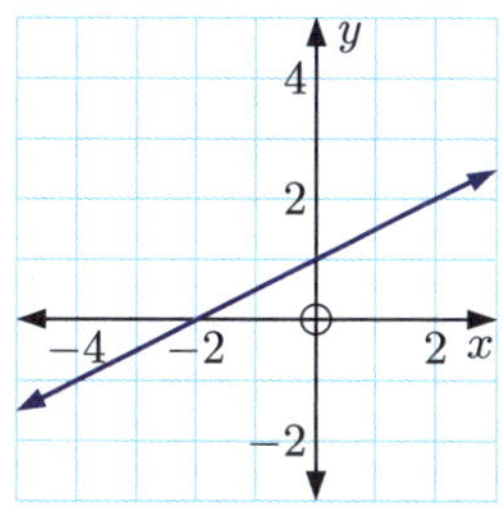

c

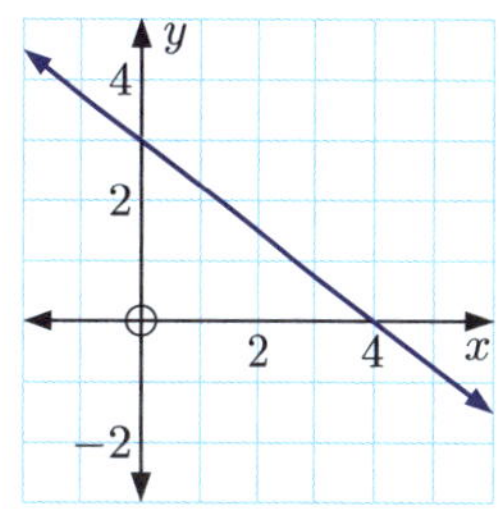

d

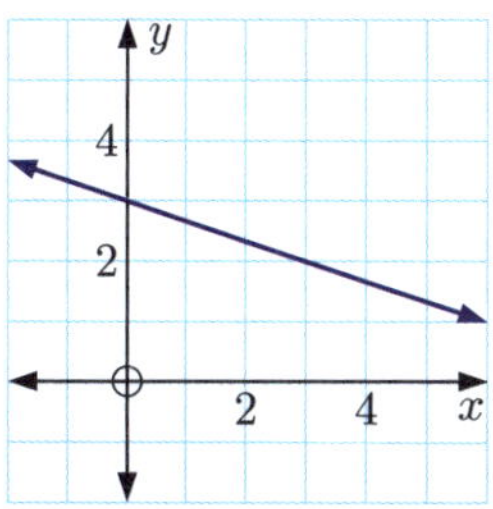

e

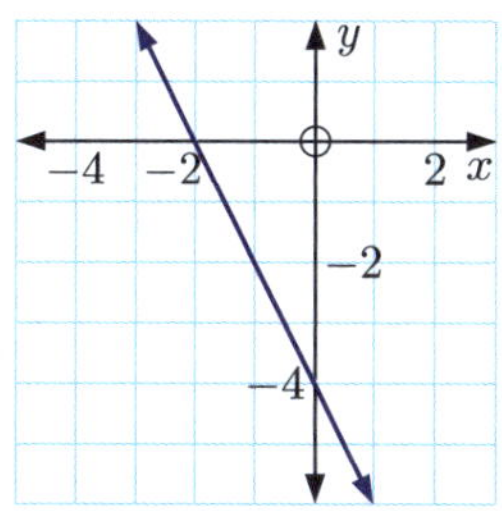

f

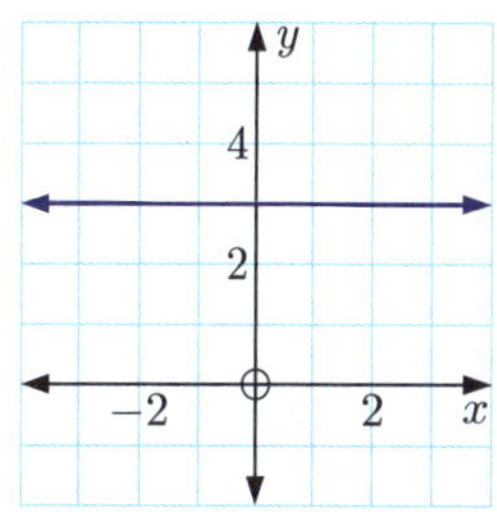

Example 12 **Self Tutor**

Find the equation of the line which passes through $(-2, 1)$ and $(3, 16)$.

The line has gradient $m = \dfrac{16-1}{3--2} = \dfrac{15}{5} = 3$.

Using the point $(-2, 1)$, the equation of the line is

$$y - 1 = 3(x - -2)$$
$$\therefore\ y - 1 = 3(x + 2)$$
$$\therefore\ y - 1 = 3x + 6$$
$$\therefore\ y = 3x + 7$$

You can substitute the coordinates of *either* point into the equation.

4 Find, in gradient-intercept form, the equation of the line which passes through:

- **a** $(2, 7)$ and $(5, 13)$
- **b** $(3, 6)$ and $(6, 0)$
- **c** $(-3, -4)$ and $(6, 2)$
- **d** $(-4, -3)$ and $(3, -3)$
- **e** $(-6, -14)$ and $(2, -2)$
- **f** $(-8, 11)$ and $(-2, 3)$.

Check your answer by making sure *both* points satisfy the equation of the line.

5 Find the equation of the illustrated line:

a
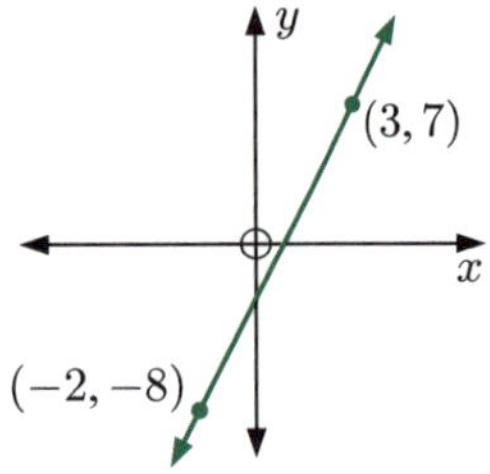

b
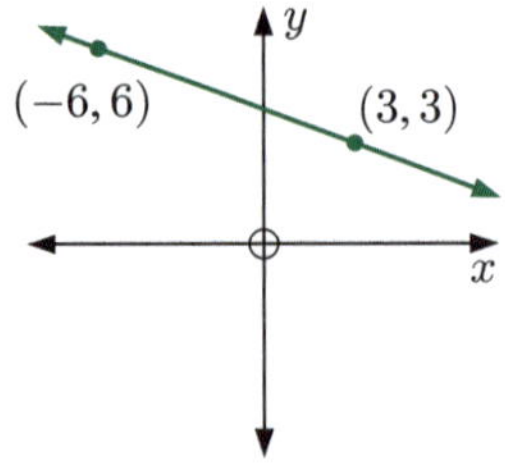

c
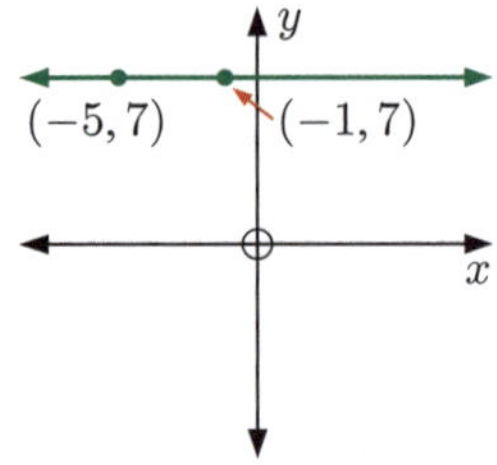

d
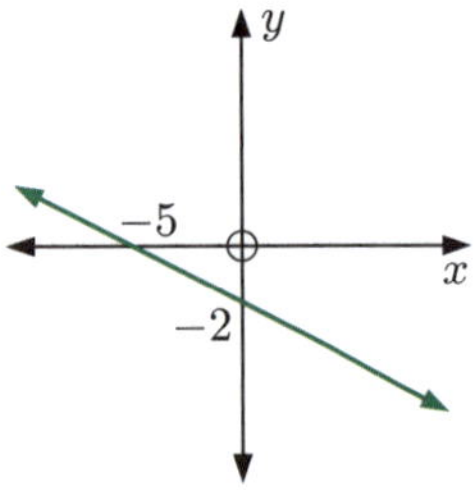

e
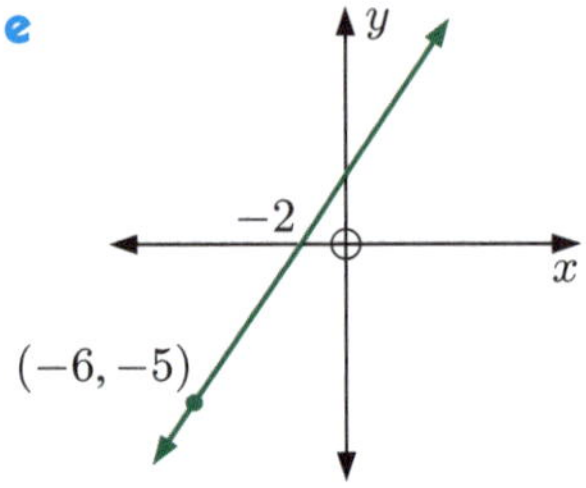

f
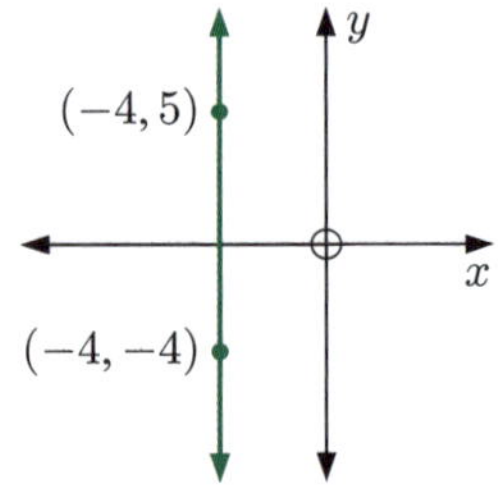

6 Find, in general form, the equation of the line which passes through:

a $(-1, 7)$ and $(2, -2)$ **b** $(5, 2)$ and $(-1, 2)$ **c** $(3, -1)$ and $(3, 4)$

d $(3, -1)$ and $(0, 4)$ **e** $(2, 6)$ and $(-2, 1)$ **f** $(4, -1)$ and $(-1, -2)$.

7 Find the rule connecting the variables:

a
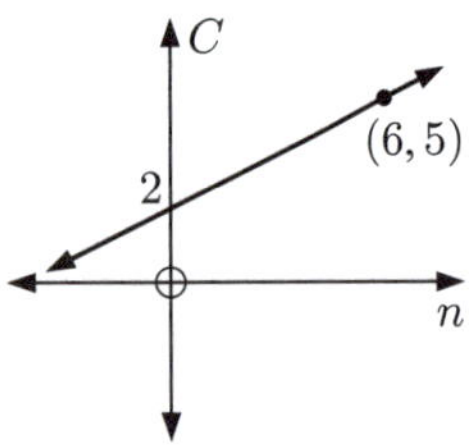

b
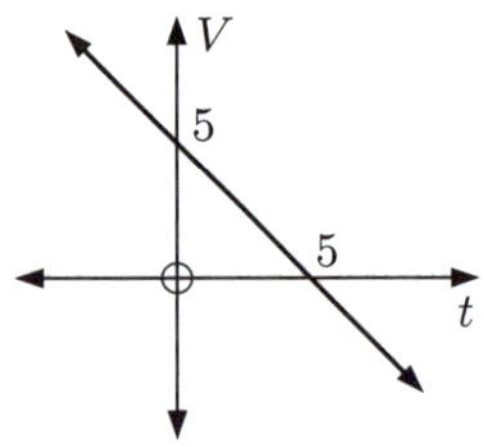

c
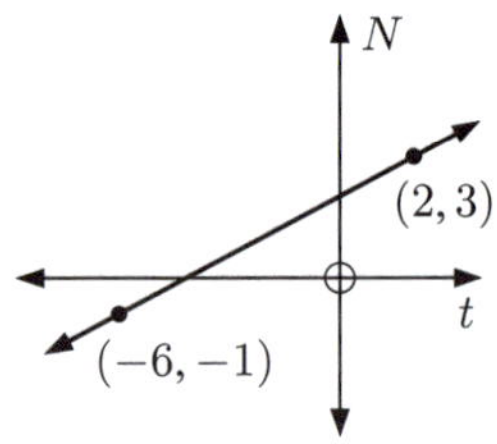

d
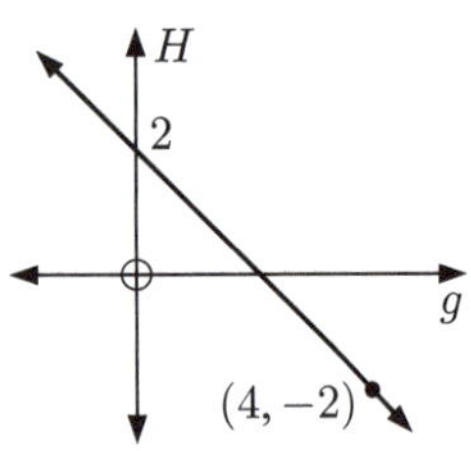

e
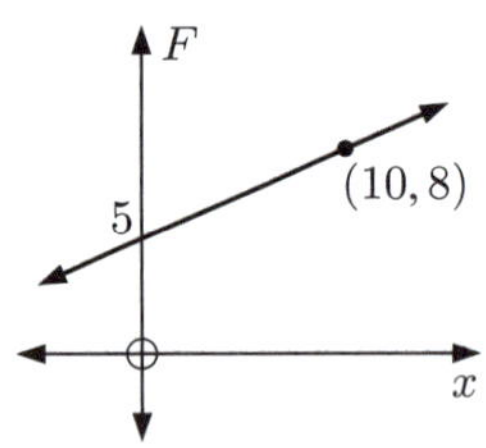

f
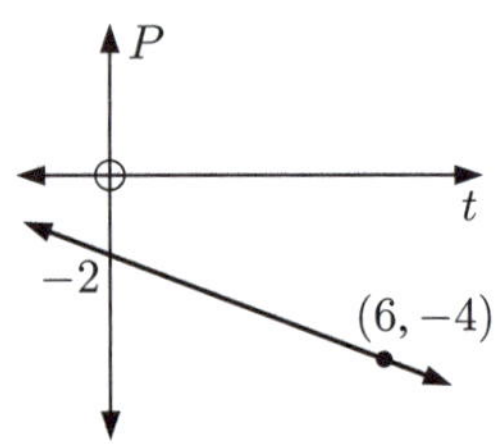

8 ABC is an equilateral triangle.

a Find the coordinates of C.

b Find the coordinates of M and N, the midpoints of line segments [BC] and [AC] respectively.

c Find the equations of all lines of symmetry of triangle ABC.

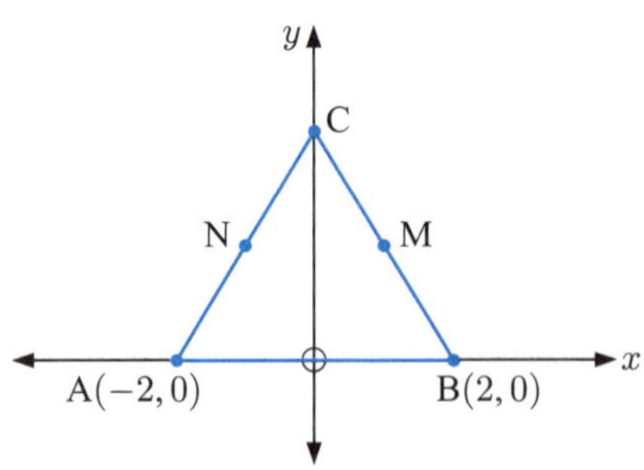

ACTIVITY

Click on the icon to practise your skills matching straight line graphs with their properties.

MATHS CARDS

QUICK QUIZ

MULTIPLE CHOICE QUIZ

REVIEW SET 17A

1 On the same set of axes, graph the lines $x = -2\frac{1}{2}$ and $y = -4$.

2 Determine whether the point:

a $(5, -1)$ lies on the line $y = -2x + 9$

b $(-3, -2)$ lies on the line $2x - 3y = -12$.

3 Draw the line with x-intercept -2 and y-intercept -4, and determine its gradient.

4 For the line with equation $y = -\frac{1}{2}x - 1$:

a Construct a table of values using $x = -3, -2, -1, 0, 1, 2$, and 3.

b Draw the graph of the straight line.

c Find the gradient and axes intercepts of the line.

5 State the gradient and y-intercept of the line with equation:

a $y = 4x - 3$ **b** $y = 2 - x$ **c** $y = \dfrac{3x - 1}{2}$

6 Write:

a $y = \frac{3}{2}x + 4$ in general form

b $5x - 4y = 40$ in gradient-intercept form.

7 Find the axes intercepts and gradient of the line $2x + 3y = 6$.

8 Draw the graph of:

a $y = -3x$ **b** $x + y = 7$ **c** $y = \frac{3}{2}x - 2$

d $2x - 3y = 30$ **e** $y = -\frac{1}{4}x + 5$ **f** $4x + 3y = -36$

9 Find the equation of the line:

a with gradient 2 and y-intercept -3

b which passes through $(-4, 6)$ and $(5, -3)$

c which is parallel to the line $y = \frac{3}{2}x - 1$ and which passes through $(-2, -8)$.

10 Jin buys x bats costing \$10 each and y dozen balls costing \$5 per dozen. He pays \$140 in total.

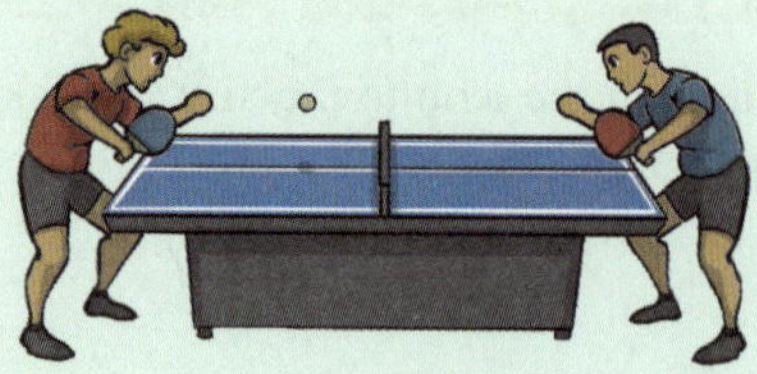

a Explain why $10x + 5y = 140$.

b Draw the graph of $10x + 5y = 140$.

c Explain the significance of the:

i x-intercept **ii** y-intercept.

11 Find, in general form, the equation of the line which:

a has x-intercept -2 and y-intercept -5

b passes through $(-2, -3)$ and $(1, 5)$

c is parallel to the line $2x - 3y = 10$ and which passes through $(3, -4)$.

12 Find the equation of each line:

a

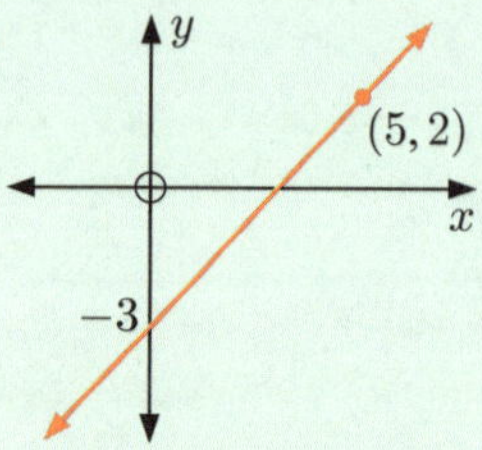

b

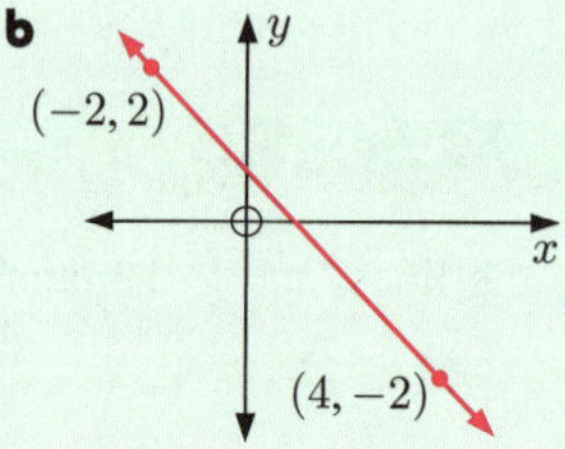

c

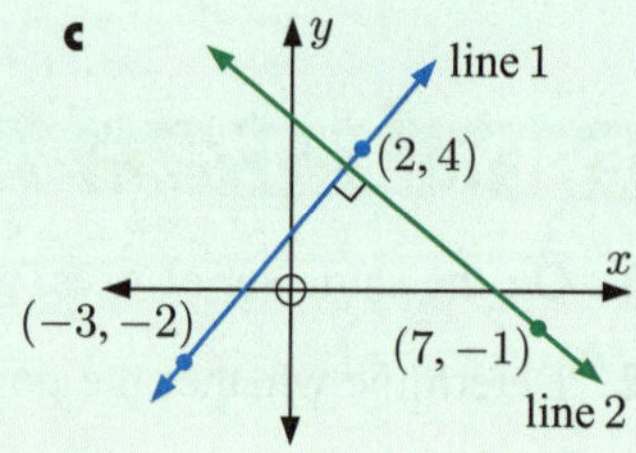

13 The graph alongside shows the cost of hiring a hall for various lengths of time.

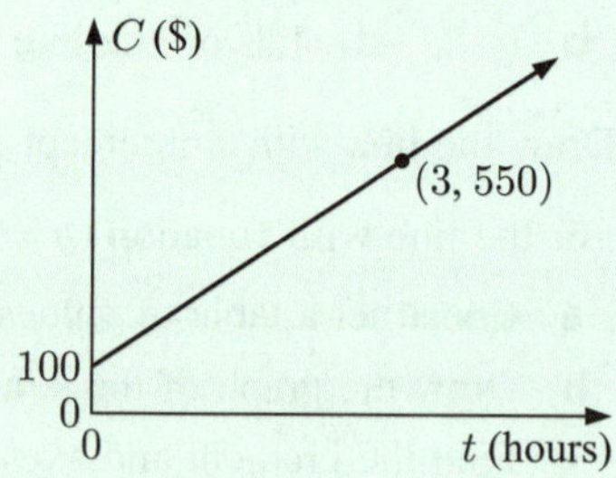

a Find and interpret the gradient of the line.

b State and interpret the C-intercept of the line.

c Write a formula for C in terms of t.

d Find the cost of hiring the hall for 8 hours.

14 **a** Line 1 has gradient $\frac{1}{3}$ and passes through $P(-3, 1)$.

i Find the equation of line 1.

ii Show that line 1 passes through $Q(6, 4)$.

b Line 2 also passes through Q, and is perpendicular to line 1.

i Find the equation of line 2.

ii Find k given that line 2 passes through $R(7, k)$.

c Line 3 passes through P and R. Find the equation of line 3.

d Draw the three lines on the same set of axes.

e Find the area of triangle PQR.

REVIEW SET 17B

1 Find the equation of:

- **a** the line with zero gradient that passes through $(2, -3)$
- **b** the line with undefined gradient that passes through $(5, -1)$.

2 Find k given that:

- **a** $(k, 5)$ lies on the line with equation $3x - y = -8$
- **b** $(-4, k)$ lies on the line with equation $2x + 5y = 12$.

3 Find the:

- **a** y-intercept of the line $3x - 5y = 30$
- **b** x-intercept of the line $2x + 7y = -70$.

4 For the line with equation $y = \frac{3}{2}x + 3$:

- **a** Construct a table of values using $x = -3, -2, -1, 0, 1, 2$, and 3.
- **b** Draw the graph of the straight line.
- **c** Find the gradient and axes intercepts of the line.

5 State the gradient and y-intercept of the line with equation:

- **a** $y = 5 - 2x$
- **b** $y = \frac{7}{2}x - 3$
- **c** $y = \dfrac{1 + 4x}{3}$

6

- **a** Write the equation $4x - 3y = 24$ in gradient-intercept form.
- **b** State the gradient of the line.

7

- **a** Find the axes intercepts of the line with equation $2x - 5y = 10$.
- **b** Draw the graph of the line.

8 Draw the graph of:

- **a** $y = 2x - 4$
- **b** $y = 6 - x$
- **c** $y = -\frac{1}{3}x + 2$
- **d** $y = \frac{3}{4}x - 3$
- **e** $x - 4y = 12$
- **f** $3x + 2y = -18$

9 Tina is cycling home from school. Her distance from home after t minutes is given by $D = 10 - \frac{1}{3}t$ kilometres.

- **a** Graph D against t for $t \geqslant 0$.
- **b** State and interpret the D-intercept of the graph.
- **c** State and interpret the gradient of the graph.
- **d** Find and interpret the t-intercept.

10 Find the equation of the line:

- **a** with gradient -2 and y-intercept 3
- **b** with gradient $\frac{2}{3}$ and which passes through $(-3, 4)$
- **c** which passes through $(1, -2)$ and $(3, 4)$.

11 Find, in general form, the equation of the line which:

- **a** has gradient $-\frac{4}{7}$ and which passes through $(-2, 2)$
- **b** is perpendicular to the line $y = 2x - 3$ and which passes through $(4, 1)$.

12 Find the rule connecting the variables:

a 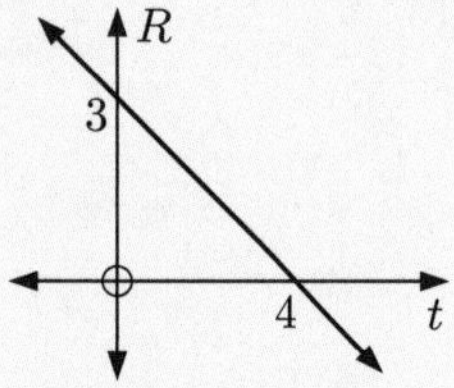

b 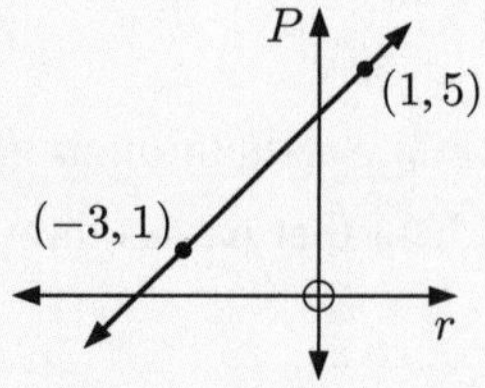

c 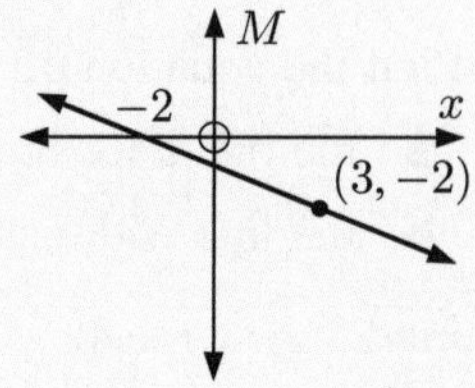

13 Consider the line with equation $5x + 2y = 20$.

a Find the:

i x-intercept **ii** y-intercept.

b Determine whether these points lie on the line:

i $(6, -5)$ **ii** $(2, 4)$

c Find the value of b such that $(1, b)$ lies on the line.

d Draw the graph of the line, showing your results from **a**, **b**, and **c**.

14 OABC is a rhombus.

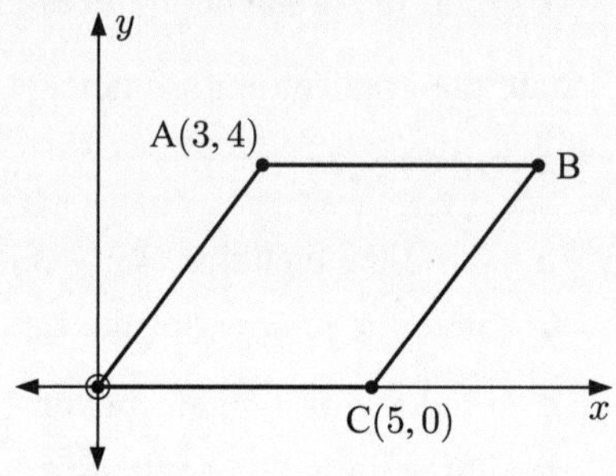

a Find the length of each side of the rhombus.

b Find the coordinates of B.

c Find the equation of:

i (OB) **ii** (AC)

d Verify that the diagonals of the rhombus are perpendicular.

e Find the midpoint M of [OB].

f Show that M lies on the diagonal [AC].

Chapter 18

Simultaneous equations

Contents:

OPENING PROBLEM

At the summer sales, Cassandra buys a dress and a skirt for a total of \$30. The dress cost \$8 more than the skirt.

Suppose the dress cost $\$x$, and the skirt cost $\$y$.

Things to think about:

a Can you explain why:
  i $x+y=30$ **ii** $x-y=8$?

b How many solutions are there to the equation:
  i $x+y=30$ **ii** $x-y=8$?

c Which solution satisfies *both* equations at the same time?

The linear equations in the **Opening Problem** connect *two* variables. You should recognise that each equation will generate a straight line in the Cartesian plane, so each equation has *infinitely many* solutions.

To find a solution which satisfies *both* equations at the same time, we need to solve the equations *simultaneously*. We say they are **simultaneous equations**.

There is in fact only *one* solution to the simultaneous equations
$$\begin{cases} x+y=30 \\ x-y=8 \end{cases}.$$

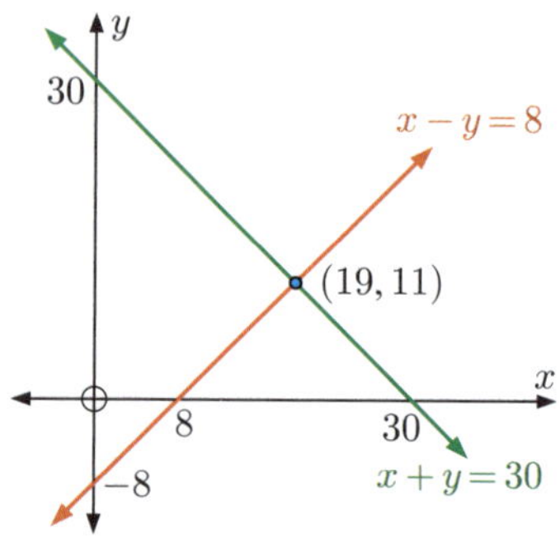

The solution occurs at the point where the two lines intersect.

The solution $x=19$, $y=11$ satisfies both equations at the same time, since $19+11=30$
and $19-11=8$.

So, Cassandra paid \$19 for the dress and \$11 for the skirt.

A SOLUTION BY TRIAL AND ERROR

One way to solve simultaneous equations is by **trial and error**. We list possible solutions to *one* of the equations, and look for the solution which also satisfies the *other* equation.

Example 1

Self Tutor

Solve $\begin{cases} x+y=8 \\ 2x-y=7 \end{cases}$ simultaneously by trial and error.

We consider solutions to $x+y=8$, and find the solution which also satisfies $2x-y=7$.

x	y	$x+y$	$2x-y$
0	8	8	−8
1	7	8	−5
2	6	8	−2
3	5	8	1
4	4	8	4
5	3	8	7 ✓

The solution is $x=5$, $y=3$.

EXERCISE 18A

1 Determine whether the given values satisfy the pair of simultaneous equations:

a $\begin{cases} x - y = 3 \\ 2x + y = 11 \end{cases}$ $(x = 5,\ y = 2)$

b $\begin{cases} x + y = 9 \\ 2x - y = 6 \end{cases}$ $(x = 5,\ y = 4)$

c $\begin{cases} a + b = 2 \\ a - b = 8 \end{cases}$ $(a = 5,\ b = -3)$

d $\begin{cases} 2p + q = 7 \\ 3p + 2q = 10 \end{cases}$ $(p = 4,\ q = -1)$

2 Given that the solutions are integers, solve by trial and error:

a $\begin{cases} x + y = 4 \\ 3x + 5y = 14 \end{cases}$

b $\begin{cases} x + y = 11 \\ 4x + 3y = 40 \end{cases}$

c $\begin{cases} y = x + 2 \\ 9x - 4y = 7 \end{cases}$

d $\begin{cases} y = 6 + x \\ 8x - 3y = -3 \end{cases}$

e $\begin{cases} x + y = 4 \\ 2x - y = 5 \end{cases}$

f $\begin{cases} x + y = 6 \\ 2x + y = 10 \end{cases}$

g $\begin{cases} a - b = 1 \\ 2a + 3b = 2 \end{cases}$

h $\begin{cases} p - q = 3 \\ 5p + 2q = 29 \end{cases}$

i $\begin{cases} 3x + 2y = 17 \\ -y + x = -21 \end{cases}$

DISCUSSION

1 Attempt to solve these simultaneous equations by trial and error:

a $\begin{cases} 3x + 7y = 1 \\ 6x - 14y = -4 \end{cases}$

b $\begin{cases} 2x - y = 8 \\ -4x + 2y = 9 \end{cases}$

c $\begin{cases} x - y = 2 \\ 2y - 2x = -4 \end{cases}$

2 Discuss your answers to **1** with your class.

- For **a** and **b**, did anyone find a solution? *Is* there a solution?
- For **c**, did you all find the same solution? Are all of the solutions you found valid?
- Discuss the problems with using trial and error to solve simultaneous equations.

B GRAPHICAL SOLUTION

Suppose we are given two linear equations involving x and y. The solutions to each equation will form a straight line. If we graph the two equations on the same set of axes, any **point of intersection** corresponds to a **simultaneous solution** of the equations.

- If the lines are not parallel, the lines will meet at exactly one point. The simultaneous equations have *exactly one solution.*

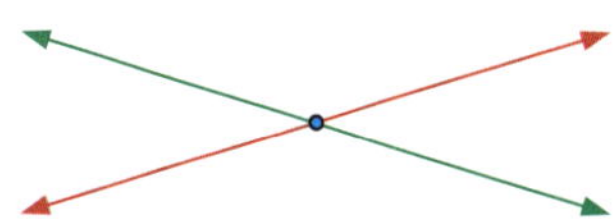

- If the lines are parallel but not identical, the lines will never meet. The simultaneous equations have *no solutions.*

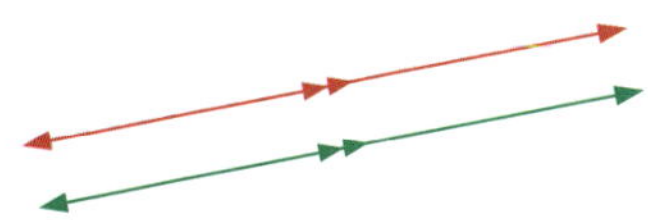

- If the lines are identical, there will be infinitely many points of intersection. The simultaneous equations have *infinitely many solutions.*

Example 2 **Self Tutor**

Solve graphically: $\begin{cases} y = 2x + 5 \\ y = -x - 1 \end{cases}$

We draw the graphs of $y = 2x + 5$ and $y = -x - 1$ on the same set of axes.

The graphs meet at the point $(-2, 1)$.

$\therefore$ the solution is $x = -2$, $y = 1$.

Check: Substituting these values into:

- $y = 2x + 5$ gives $1 = 2(-2) + 5$ ✓
- $y = -x - 1$ gives $1 = -(-2) - 1$ ✓

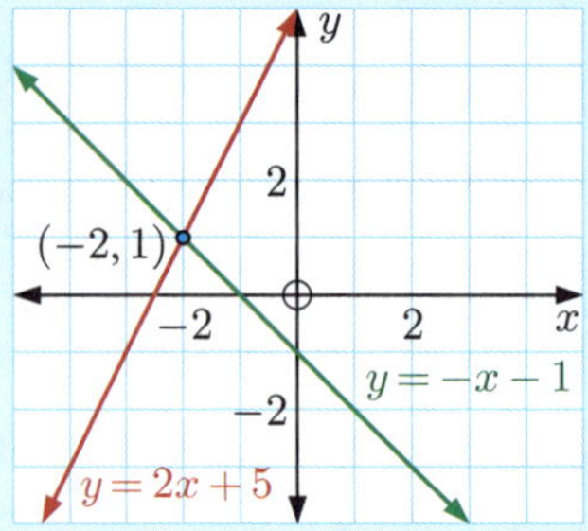

EXERCISE 18B

1 Solve graphically:

a $\begin{cases} y = x \\ y = 2 - x \end{cases}$ **b** $\begin{cases} y = x + 3 \\ y = -2x \end{cases}$ **c** $\begin{cases} y = 3x + 2 \\ y = x - 4 \end{cases}$

d $\begin{cases} y = -2x + 3 \\ y = 5 - x \end{cases}$ **e** $\begin{cases} y = 2 - 3x \\ y = x - 2 \end{cases}$ **f** $\begin{cases} y = 3x + 4 \\ y = 2x + 1 \end{cases}$

Check your answers by substituting into the original equations.

2 Solve graphically:

a $\begin{cases} y = x + 1 \\ 2x - y = 2 \end{cases}$ **b** $\begin{cases} y = x + 4 \\ x + y = 2 \end{cases}$ **c** $\begin{cases} y = -2x + 1 \\ 3x - 2y = 12 \end{cases}$

d $\begin{cases} x - 2y = 6 \\ y = 3x + 2 \end{cases}$ **e** $\begin{cases} x + y = 6 \\ x - 3y = -6 \end{cases}$ **f** $\begin{cases} 3x - 4y = 24 \\ x + 2y = -2 \end{cases}$

3 Graph each pair of lines to find how many solutions the simultaneous equations have. In each case explain your answer.

a $\begin{cases} y = 2x + 3 \\ y = 2x - 2 \end{cases}$ **b** $\begin{cases} x + y = 2 \\ y = 2 - x \end{cases}$ **c** $\begin{cases} y = 3x - 2 \\ 6x - 2y = 12 \end{cases}$

DISCUSSION

Discuss the advantages and disadvantages of solving simultaneous equations by graphical methods.

C SOLUTION BY EQUATING VALUES OF y

We now consider some **algebraic** methods for solving linear simultaneous equations. Algebraic methods are often quicker than trial and error or graphing. They are also more accurate if the solutions are not integers.

If y is the subject of *both* equations, we equate the values of y, then solve to find the value of x which makes the y values equal.

Example 3 — Self Tutor

Solve the simultaneous equations: $\begin{cases} y = 2x - 1 \\ y = x + 3 \end{cases}$

If $y = 2x - 1$ and $y = x + 3$, then

$2x - 1 = x + 3$ {equating ys}
$\therefore\ x - 1 = 3$ {$-\,x$ both sides}
$\therefore\ x = 4$ {$+\,1$ both sides}

Now $y = x + 3$
$\therefore\ y = 4 + 3$
$\therefore\ y = 7$

The solution is $x = 4$, $y = 7$.

Check: In $y = 2x - 1$, $y = 2 \times 4 - 1 = 8 - 1 = 7$ ✓

EXERCISE 18C

1 Solve by equating values of y:

a $\begin{cases} y = x + 1 \\ y = 2x - 3 \end{cases}$ **b** $\begin{cases} y = x + 4 \\ y = -x - 2 \end{cases}$ **c** $\begin{cases} y = x + 2 \\ y = -2x + 5 \end{cases}$

d $\begin{cases} y = -2x - 4 \\ y = x - 4 \end{cases}$ **e** $\begin{cases} y = -x + 4 \\ y = 2x - 8 \end{cases}$ **f** $\begin{cases} y = -2 - x \\ y = 2x + 13 \end{cases}$

2 Find the point of intersection of the two lines:

a $y = 2x - 10$ and $y = 4x - 18$ **b** $y = 5 - 3x$ and $y = 10 - 6x$

c $y = 3x + 7$ and $y = -x - 6$ **d** $y = 6x - 2$ and $y = 3 - 4x$

3 **a** Try to solve by equating values of y: $\begin{cases} y = 3x + 4 \\ y = 3x - 2 \end{cases}$

b Graph $y = 3x + 4$ and $y = 3x - 2$ on the same set of axes.

c Explain what happened in **a**.

D SOLUTION BY SUBSTITUTION

The method of **substitution** is used when a variable is given as the subject of *one* of the equations.

For example, in $\begin{cases} y = 2x + 3 \\ 3x - 4y = 8 \end{cases}$ we see that y is the subject of the first equation.

We substitute the expression for y on the RHS into the second equation, then solve to find the x value of the simultaneous solution.

Example 4 — Self Tutor

Solve simultaneously by substitution: $\begin{cases} y = 2x + 3 \\ 3x - 4y = 8 \end{cases}$

$y = 2x + 3$ .... (1)
$3x - 4y = 8$ .... (2)

Substituting (1) into (2) gives $3x - 4(2x + 3) = 8$

$\therefore\ 3x - 8x - 12 = 8$

$\therefore\ -5x - 12 = 8$

$\therefore\ -5x = 20$

$\therefore\ x = -4$

Substituting $x = -4$ into (1) gives $y = 2(-4) + 3$

$\therefore\ y = -8 + 3$

$\therefore\ y = -5$

The solution is $x = -4,\ y = -5$.

Check: In (2), $3x - 4y = 3(-4) - 4(-5) = -12 + 20 = 8$ ✓

EXERCISE 18D

1 Solve simultaneously by substitution:

a $\begin{cases} y = x + 2 \\ 3x + 2y = 19 \end{cases}$ **b** $\begin{cases} y = 2x + 1 \\ 5x - 4y = 2 \end{cases}$ **c** $\begin{cases} 4x + 3y = -14 \\ y = -3x - 3 \end{cases}$

d $\begin{cases} x = 3y + 2 \\ 2x - 5y = 1 \end{cases}$ **e** $\begin{cases} 7x - 2y = -6 \\ x = y - 3 \end{cases}$ **f** $\begin{cases} y = 3x + 2 \\ 4x + 2y = -1 \end{cases}$

2 Solve simultaneously by substitution:

a $\begin{cases} 3x - 7y = 11 \\ x = 4y + 6 \end{cases}$ **b** $\begin{cases} y = 8 - 4x \\ 12x - y = 12 \end{cases}$ **c** $\begin{cases} 2x - 5y = 0 \\ y = 5 - 3x \end{cases}$

3 Find the point of intersection of the two lines:

a $y = 3 + x$ and $5x - 2y = 0$ **b** $x + 3y = 6$ and $y = x - 2$

c $x = 5 - y$ and $4x + y = 10$ **d** $2x + 3y = 12$ and $x = \frac{1}{3}(y - 4)$

4 **a** Try to solve using substitution: $\begin{cases} y = 3x - 5 \\ 12x - 4y = 7 \end{cases}$

b Graph $y = 3x - 5$ and $12x - 4y = 7$ on the same set of axes.

c Explain what happened in **a**.

5 **a** Try to solve using substitution: $\begin{cases} 6x + 3y = 15 \\ y = 5 - 2x \end{cases}$

b Graph $6x + 3y = 15$ and $y = 5 - 2x$ on the same set of axes.

c Explain what happened in **a**.

E SOLUTION BY ELIMINATION

Consider the simultaneous equations $\begin{cases} 3x + 2y = 5 \\ x - 2y = 3\,. \end{cases}$

In this case, neither variable is given as the subject of an equation. We therefore use a method called **elimination**.

In this method, we make the coefficients of x (or y) the **same size** but **opposite in sign**.

We then **add** the equations to **eliminate** one of the variables.

The method of elimination uses the fact that:

If $a = b$ and $c = d$ then $a + c = b + d$.

This allows us to add the equations together without changing their solutions.

Example 5 **Self Tutor**

Solve simultaneously by elimination: $\begin{cases} 3x + 2y = 5 \\ x - 2y = 3 \end{cases}$

The coefficients of y are the same size but opposite in sign.

$\therefore$ we **add** the LHSs and the RHSs to eliminate y.

$$\begin{aligned} 3x + 2y &= 5 \quad .... (1) \\ x - 2y &= 3 \quad .... (2) \\ \hline \text{Adding,} \quad 4x \qquad &= 8 \\ \therefore \ x &= 2 \end{aligned}$$

Substituting $x = 2$ into (1) gives
$$\begin{aligned} 3(2) + 2y &= 5 \\ \therefore \ 6 + 2y &= 5 \\ \therefore \ 2y &= -1 \\ \therefore \ y &= -\tfrac{1}{2} \end{aligned}$$

The solution is $x = 2$, $y = -\frac{1}{2}$.

Check: In (2), $x - 2y = 2 - 2(-\frac{1}{2}) = 2 + 1 = 3$ ✓

EXERCISE 18E.1

1 What equation results when the following are added vertically?

a $\begin{cases} 5x + 3y = 12 \\ x - 3y = -6 \end{cases}$

b $\begin{cases} 2x + 5y = -4 \\ -2x - 6y = 12 \end{cases}$

c $\begin{cases} 4x - 6y = 9 \\ x + 6y = -2 \end{cases}$

d $\begin{cases} 12x + 15y = 33 \\ -18x - 15y = -63 \end{cases}$

e $\begin{cases} 5x + 6y = 12 \\ -5x + 2y = -8 \end{cases}$

f $\begin{cases} -7x + y = -5 \\ 7x - 3y = -11 \end{cases}$

2 Solve simultaneously by elimination:

a $\begin{cases} 2x + y = 3 \\ 3x - y = 7 \end{cases}$ **b** $\begin{cases} 4x + 3y = 7 \\ 6x - 3y = -27 \end{cases}$ **c** $\begin{cases} 2x + 5y = 16 \\ -2x - 7y = -20 \end{cases}$

d $\begin{cases} 3x + 5y = -11 \\ -3x - 2y = 8 \end{cases}$ **e** $\begin{cases} 4x - 7y = 41 \\ 3x + 7y = -6 \end{cases}$ **f** $\begin{cases} -4x + 3y = -25 \\ 4x - 5y = 31 \end{cases}$

MULTIPLYING EQUATIONS BY A CONSTANT

If the coefficients of x (or y) are *not* the same size and opposite in sign, we first need to **multiply** one or both equations by a constant. This will not change the solutions to the equations.

Example 6 Self Tutor

Solve $\begin{cases} 3x + 2y = -2 \\ 5x - y = 27 \end{cases}$ using the method of elimination.

$3x + 2y = -2$ .... (1)
$5x - y = 27$ .... (2)

We have $+2y$ in (1), so we obtain $-2y$ from (2) by multiplying both sides of (2) by 2.

$$\begin{aligned} 3x + 2y &= -2 \quad \{(1)\} \\ 10x - 2y &= 54 \quad \{(2) \times 2\} \end{aligned}$$

Adding, $13x = 52$

$\therefore x = 4$

Substituting $x = 4$ into (1), $3(4) + 2y = -2$

$\therefore 12 + 2y = -2$

$\therefore 2y = -14$

$\therefore y = -7$

The solution is $x = 4$, $y = -7$.

Check: In (2), $5x - y = 5(4) - (-7) = 20 + 7 = 27$ ✓

EXERCISE 18E.2

1 Give the equation that results when both sides of the equation:

a $3x + 4y = 2$ are multiplied by 3
b $x - 4y = 7$ are multiplied by -2
c $5x - y = -3$ are multiplied by 5
d $7x + 3y = -4$ are multiplied by -3
e $-2x - 5y = 1$ are multiplied by -4
f $3x - y = -1$ are multiplied by -1.

2 Solve simultaneously by elimination:

a $\begin{cases} x - 2y = -5 \\ 2x + y = -5 \end{cases}$ **b** $\begin{cases} 4x - 3y = 6 \\ -2x + 5y = 4 \end{cases}$ **c** $\begin{cases} 2x - y = 9 \\ x + 4y = 36 \end{cases}$

d $\begin{cases} 3x + 4y = 6 \\ x - 3y = -11 \end{cases}$ **e** $\begin{cases} 4x + 3y = 17 \\ 5x - 9y = 34 \end{cases}$ **f** $\begin{cases} 2x - 7y = -5 \\ 6x + 5y = -15 \end{cases}$

g $\begin{cases} 3x + 2y = 11 \\ 9x - 5y = 22 \end{cases}$ **h** $\begin{cases} 5x + 8y = 8 \\ 9x + 2y = 33 \end{cases}$

Example 7 **Self Tutor**

Solve simultaneously by elimination: $\begin{cases} 5x + 3y = 12 \\ 7x + 2y = 19 \end{cases}$

$5x + 3y = 12$ .... (1)
$7x + 2y = 19$ .... (2)

Since the LCM of 3 and 2 is 6, we eliminate y by multiplying (1) by 2 and (2) by -3:

$$\begin{aligned} 10x + 6y &= 24 \quad \{(1) \times 2\} \\ -21x - 6y &= -57 \quad \{(2) \times -3\} \\ \hline \text{Adding,} \quad -11x \quad &= -33 \\ \therefore \quad x &= 3 \end{aligned}$$

Substituting $x = 3$ into (1), $5(3) + 3y = 12$

$$\begin{aligned} \therefore \quad 15 + 3y &= 12 \\ \therefore \quad 3y &= -3 \\ \therefore \quad y &= -1 \end{aligned}$$

The solution is $x = 3$, $y = -1$.

Check: In (2), $7x + 2y = 7(3) + 2(-1) = 21 - 2 = 19$ ✓

3 Solve simultaneously by elimination:

a $\begin{cases} 2x + 3y = 7 \\ 3x - 2y = 4 \end{cases}$ **b** $\begin{cases} 2x - 5y = 4 \\ 3x - 4y = 20 \end{cases}$ **c** $\begin{cases} 4x - 3y = 6 \\ 6x + 7y = 32 \end{cases}$

d $\begin{cases} 2x + 5y = 20 \\ 3x + 2y = 19 \end{cases}$ **e** $\begin{cases} 3x - 2y = 10 \\ 4x + 3y = 19 \end{cases}$ **f** $\begin{cases} -4x + 3y = -5 \\ 3x + 2y = -9 \end{cases}$

g $\begin{cases} 7x - 3y = 29 \\ 3x + 4y = -14 \end{cases}$ **h** $\begin{cases} 3x + 4y + 11 = 0 \\ 5x + 6y + 7 = 0 \end{cases}$

4 Consider the simultaneous equations $\begin{cases} 2x + 2y = 12 \\ 5x - 10y = 15\,. \end{cases}$

a Can you *divide* an equation by a non-zero constant without changing its solution? Explain your answer.

b Divide *one* equation by a non-zero constant so you will be able to eliminate y.

c Hence solve the simultaneous equations.

5 Try to solve these simultaneous equations by elimination. Discuss your results.

a $\begin{cases} 3x + y = 8 \\ 6x + 2y = 16 \end{cases}$ **b** $\begin{cases} 2x + 5y = 8 \\ 4x + 10y = -1 \end{cases}$

F PROBLEM SOLVING

When given a problem in words, we need to interpret the information and use it to write equations involving the unknowns. We can then choose the most appropriate technique to solve the equations simultaneously, and hence answer the original problem.

Example 8 **Self Tutor**

Two numbers have sum 45 and difference 13. Find the numbers.

Let x be the larger number and y be the smaller number.

$$\begin{array}{rrl} \text{The sum of } x \text{ and } y \text{ is} & x + y = 45 & \text{.... (1)} \\ \text{The difference between } x \text{ and } y \text{ is} & x - y = 13 & \text{.... (2)} \\ \hline \text{Adding,} & 2x \quad = 58 & \\ & \therefore\ x = 29 & \end{array}$$

Substituting $x = 29$ into (1) gives $29 + y = 45$

$$\therefore\ y = 16$$

The numbers are 29 and 16.

Check: (1) $29 + 16 = 45$ ✓ (2) $29 - 16 = 13$ ✓

EXERCISE 18F

1. The difference between two numbers is 84 and their sum is 278. What are the numbers?
2. Two numbers have sum 200 and difference 37. Find the numbers.
3. Find two numbers whose difference is 8 and whose average is 13.

Example 9 **Self Tutor**

At a clearance sale, all CDs are sold for one price and all DVDs are sold for another price. Marisa bought 3 CDs and 2 DVDs for a total of \$34.50. Nico bought 2 CDs and 5 DVDs for a total of \$56. Find the cost of each item.

Let x cents be the cost of one CD, and y cents be the cost of one DVD.

3 CDs and 2 DVDs cost \$34.50, so $3x + 2y = 3450$ .... (1)

2 CDs and 5 DVDs cost \$56, so $2x + 5y = 5600$ .... (2)

Since the LCM of 3 and 2 is 6, we eliminate x by multiplying (1) by 2 and (2) by -3.

$$\begin{array}{rrl} & 6x + \ 4y = 6900 & \{(1) \times 2\} \\ & -6x - 15y = -16\,800 & \{(2) \times -3\} \\ \hline \text{Adding,} & -11y = -9900 & \\ & \therefore\ y = 900 & \end{array}$$

Substituting $y = 900$ into (1) gives $3x + 2(900) = 3450$

$$\therefore\ 3x + 1800 = 3450$$
$$\therefore\ 3x = 1650$$
$$\therefore\ x = 550$$

The cost of one CD is \$5.50, and the cost of one DVD is \$9.

4. Two hammers and a screwdriver cost a total of \$34. A hammer and 3 screwdrivers cost a total of \$32. Find the price of each tool.
5. Four adults and three children see a theatre show for \$148. Two adults and five children are charged \$116 for the same performance. Find the price of an adult's ticket and a child's ticket.

6 Four nectarines and three peaches cost \$2.90. Three nectarines and a peach cost \$1.90. Find the cost of each fruit.

7 A purse contains \$3.75 in 5 cent and 20 cent coins. There are 33 coins altogether. How many of each type of coin are in the purse?

8 A yard contains rabbits and pheasants. There are 35 heads and 98 feet in total. How many rabbits and pheasants are in the yard?

9 Milk is sold in one litre and two litre cartons. A shop owner orders 120 litres of milk and receives 97 cartons. How many of each type did she receive?

10 The figure alongside is a rectangle.

- **a** Find the values of x and y.
- **b** Hence find the area of the rectangle.

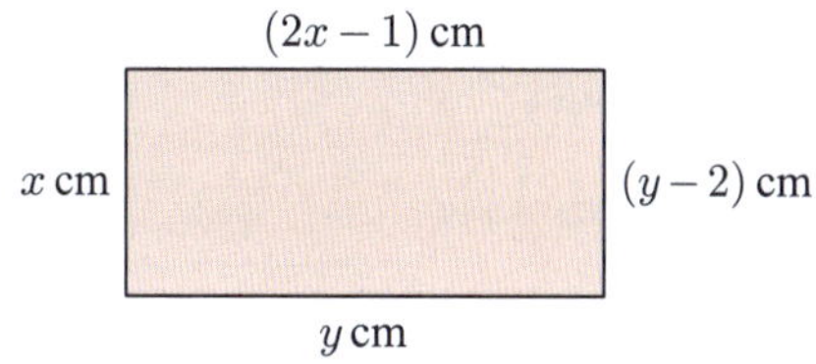

11 KLM is an isosceles triangle.

- **a** Find the values of x and y.
- **b** Hence find the measure of $L\widehat{K}M$.

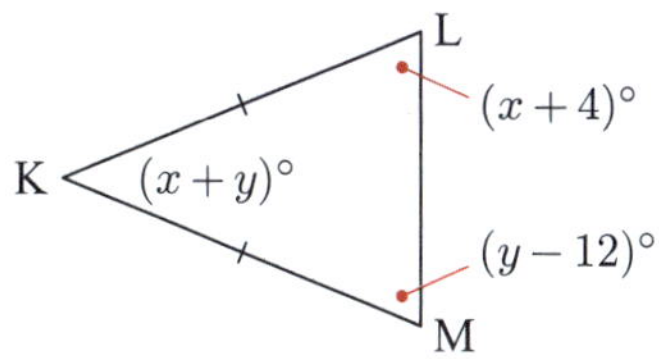

12 The triangle alongside is equilateral.

- **a** Find the values of x and y.
- **b** Hence find the perimeter of the triangle.

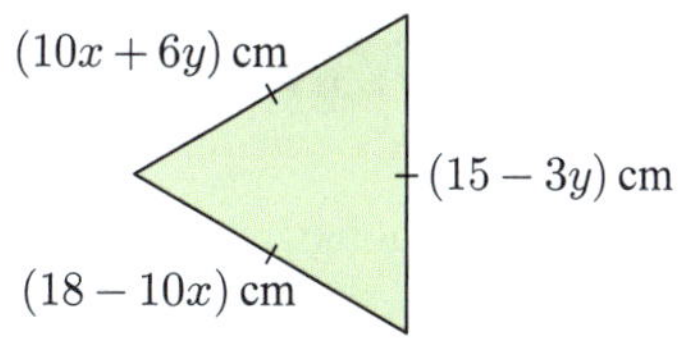

13 On the Celsius temperature scale, ice melts at $0°$C and water boils at $100°$C.
On the Fahrenheit temperature scale, ice melts at $32°$F and water boils at $212°$F.
There is a linear relationship between temperatures in Celsius (C) and Fahrenheit (F). It has the form $C = aF + b$ where a and b are constants.

- **a** Find the values of a and b as fractions in lowest terms.
- **b** Convert $77°$F into degrees Celsius.

14 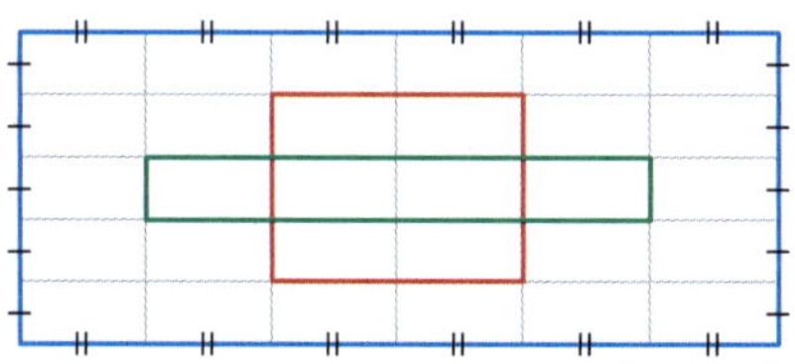

The red rectangle has perimeter 58 m, and the green rectangle has perimeter 66 m.
Find the perimeter of the blue rectangle.

15 [AM] has length $\sqrt{3}$ cm and [CN] has length $\sqrt{7}$ cm.
Find the exact length of [AC].

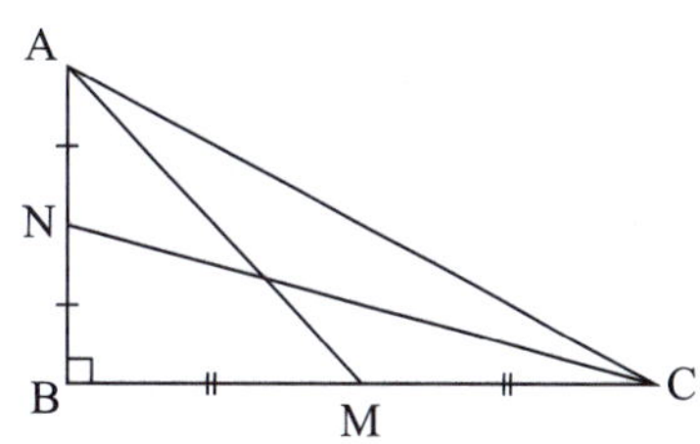

PUZZLE

Each of the 4 different symbols represents a number. The sum for each row and column is given, except for one value.

□	○	▽	17
□	◇	□	9
○	○	▽	22
?	19	18	

1 **a** Determine the correct value for each symbol.

b Hence find the unknown value in the puzzle.

2 Is there a way to find the unknown value *without* finding the value of each symbol first?

MULTIPLE CHOICE QUIZ

QUICK QUIZ

REVIEW SET 18A

1 Given that the solutions are integers, solve by trial and error: $\begin{cases} y = 2x + 1 \\ x + y = 7 \end{cases}$

2 Solve graphically: $\begin{cases} y = 2x - 1 \\ y = x - 2 \end{cases}$

3 Solve by equating values of y:

a $\begin{cases} y = 2x + 3 \\ y = x - 2 \end{cases}$ **b** $\begin{cases} y = 1 - 2x \\ y = 3x + 3 \end{cases}$

4 Solve by substitution:

a $\begin{cases} y = 11 - 3x \\ 4x + 3y = -7 \end{cases}$ **b** $\begin{cases} 2x + 7y = -4 \\ x = 4y + 3 \end{cases}$

5 Solve by elimination:

a $\begin{cases} 3x - 2y = 14 \\ 2x + 2y = 6 \end{cases}$ **b** $\begin{cases} 2x - 3y = 18 \\ 4x + 5y = -8 \end{cases}$ **c** $\begin{cases} 3x + 7y = -1 \\ 5x + 2y = -21 \end{cases}$

6 Solve simultaneously:

a $\begin{cases} 3x - 2y = 16 \\ y = 2x - 10 \end{cases}$ **b** $\begin{cases} 2x + 3y = 9 \\ 4x - y = -17 \end{cases}$

7 Two pencils and a ruler cost 98 cents in total. One pencil and two rulers cost \$1.24 in total. Find the cost of each item.

8 Flour is sold in 5 kg and 2 kg packets. The 5 kg packets cost \$2.75 each, and the 2 kg packets cost \$1.25 each. If the shop sells 67 kg of flour for a total of \$38.50, how many of each size of packet was sold?

9 **a** Rearrange $4x + y = 29$ to make y the subject of the formula.

b Hence solve by substitution: $\begin{cases} 4x + y = 29 \\ 2x - 3y = 25 \end{cases}$

10 The figure alongside is a rectangle.

a Find the values of x and y.

b Hence find the area of the rectangle.

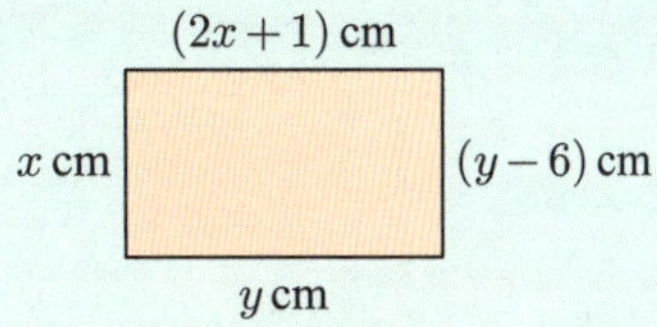

11 Jason paints signs on buildings. He can paint letters which contain all straight edges quite easily, but letters which contain a curved edge take longer.
It takes Jason 25 minutes to paint the word CHEMIST, and 43 minutes to paint the word SUPERMARKET.

a On average, how long does Jason take to paint each:

i straight edged letter **ii** curved edged letter?

b How long would it take Jason to paint the word ACCOUNTANT?

REVIEW SET 18B

1 Given that the solutions are integers, solve by trial and error: $\begin{cases} 5x + y = 4 \\ y = 3x - 12 \end{cases}$

2 Solve graphically: $\begin{cases} y = 3x - 2 \\ y = 2x + 1 \end{cases}$

3 If possible, solve by equating values of y:

a $\begin{cases} y = 16 - 3x \\ y = 2x - 4 \end{cases}$ **b** $\begin{cases} y = -5x + 1 \\ y = -5x - 1 \end{cases}$

4 Solve by substitution:

a $\begin{cases} y = 2x - 3 \\ 3x - 2y = 4 \end{cases}$ **b** $\begin{cases} x = 4 - 3y \\ 5x + 6y = 17 \end{cases}$

5 Solve by elimination:

a $\begin{cases} 5x + 3y = 7 \\ 4x - 3y = 11 \end{cases}$ **b** $\begin{cases} 2x - y = 5 \\ 3x - 2y = 11 \end{cases}$ **c** $\begin{cases} 3x + 5y = 1 \\ 4x - 3y = 11 \end{cases}$

6 Solve simultaneously:

a $\begin{cases} 3x + 4y = -4 \\ y = -2x + 5 \end{cases}$ **b** $\begin{cases} 3x - 5y = 11 \\ 4x + 3y = 5 \end{cases}$

7 The difference between two numbers is 11 and their sum is 85. Find the numbers.

8 Sally has only 10 cent and 50 cent coins in her purse. She has 21 coins altogether with a total value of \$5.30. How many of each coin type does she have?

9 A bus company uses two different sizes of bus. One day it used 7 small buses and 5 large buses to transport 331 students to a swimming carnival. The next day it used 4 small buses and 9 large buses to take 398 people to a concert. On both days, all of the buses were full.
How many people can each bus carry?

10

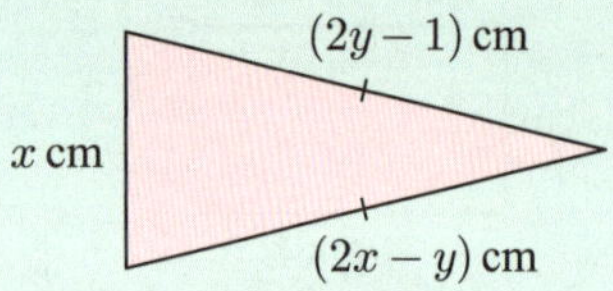

The perimeter of this triangle is 25 cm.

a Find the values of x and y.

b Hence find the length of the equal sides of the triangle.

11 **a** State the gradient of the line $y = 2x + 1$.

b Write, in terms of k, the gradient of the line $kx - 3y = 3$.

c Find the values of k for which the simultaneous equations $\begin{cases} y = 2x + 1 \\ kx - 3y = 3 \end{cases}$ have exactly one solution.
Explain your answer graphically.

d For the value of k for which there is *not* exactly one solution, explain whether there are no solutions or infinitely many solutions.

Chapter 19

Transformation geometry

Contents:

OPENING PROBLEM

Consider the triangles on the number plane.

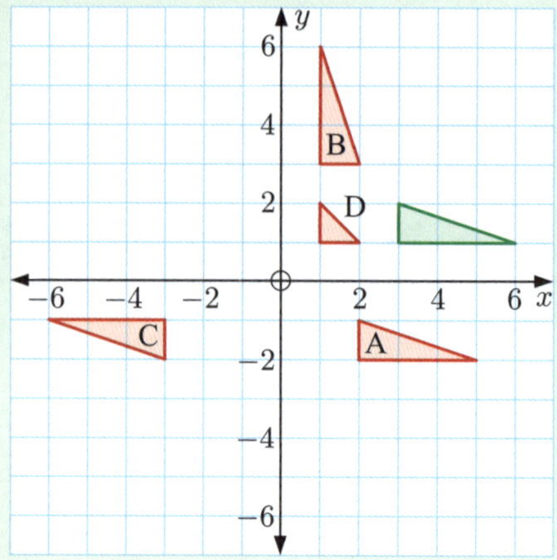

Things to think about:

a What transformation would map the green triangle onto:

- **i** triangle A
- **ii** triangle B
- **iii** triangle C
- **iv** triangle D?

b Can you draw the image if the green triangle is rotated about the origin through $90°$, and then translated 3 units downwards?

A change in the size, shape, orientation, or position of a figure is called a **transformation**.

Many trees, plants, flowers, animals, and insects are **symmetrical** in some way. Such symmetry results from a reflection, so we can describe symmetry using transformations.

Whenever we perform a transformation, the original figure is called the **object** and the new figure is called the **image**.

If A is a point on the object, the corresponding point on the image is labelled A′.

In this Chapter we will consider:

- **translations** where every point moves a fixed distance in a given direction
- **reflections** or mirror images
- **rotations** through a given angle about a point O
- **enlargements** and **reductions** with centre O
- **vertical stretches** with fixed x-axis
- **horizontal stretches** with fixed y-axis.

A TRANSLATIONS

In a **translation**, every point on an object is moved the same distance in the same direction to form the image.

For example, this object has been translated 4 units right and 2 units down.

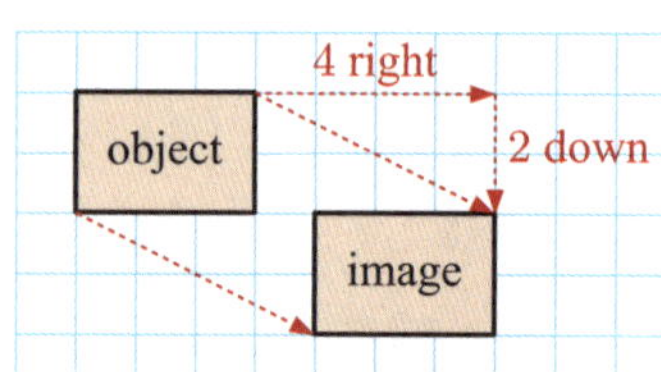

EXERCISE 19A

1 Describe each translation from the object A to the image A′.

a

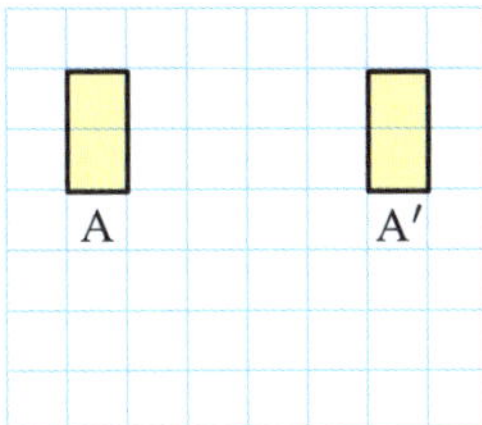

b

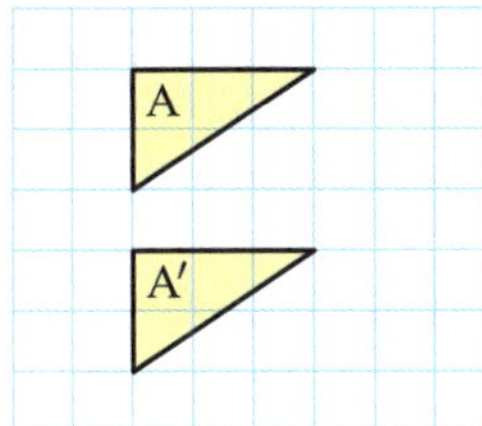

c

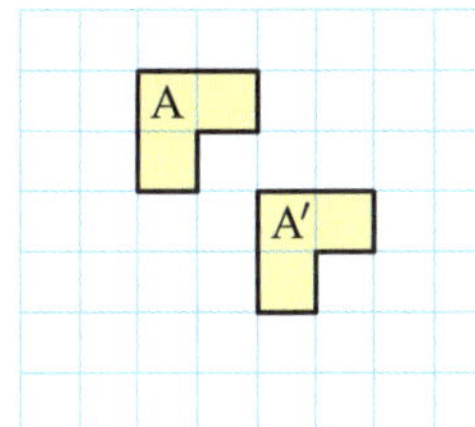

d

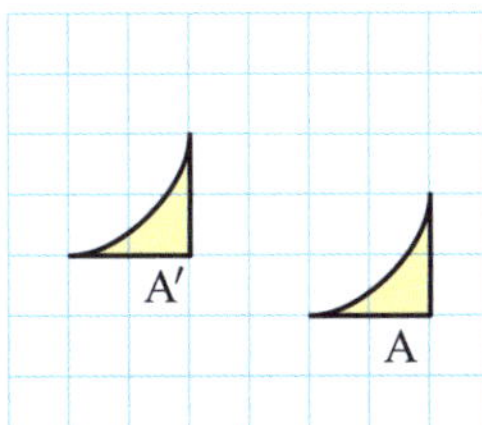

e

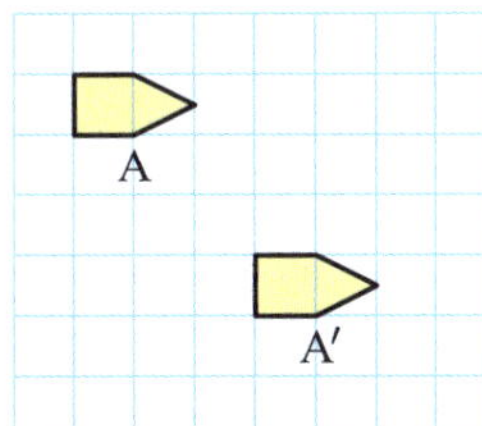

f

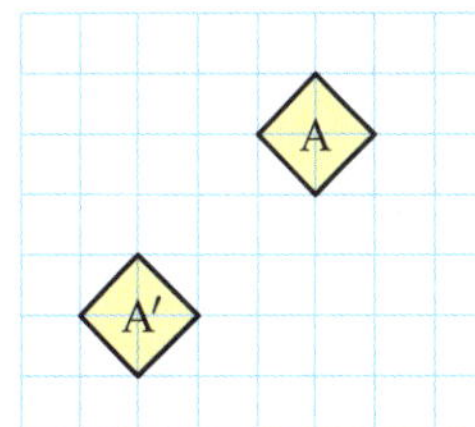

2 Translate each figure as directed:

a

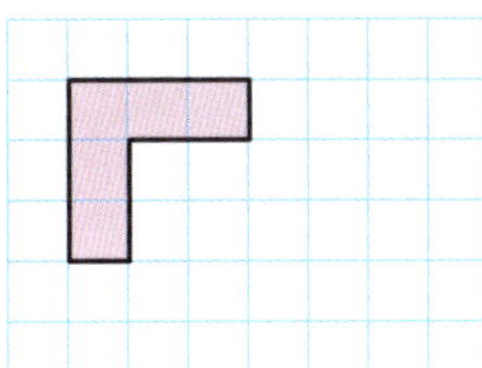

3 units right,
2 units down

b

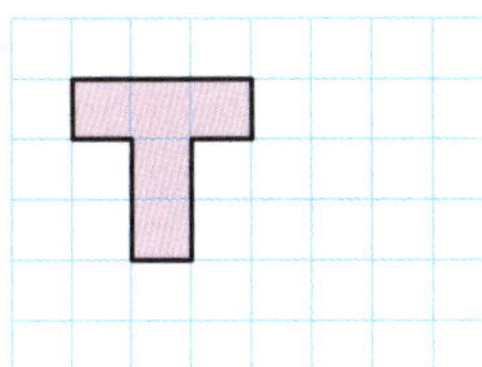

4 units right

c

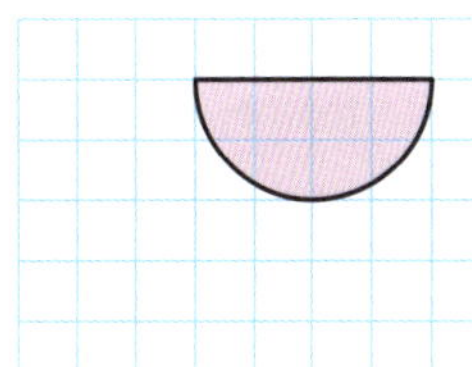

3 units down

PRINTABLE DIAGRAMS

d

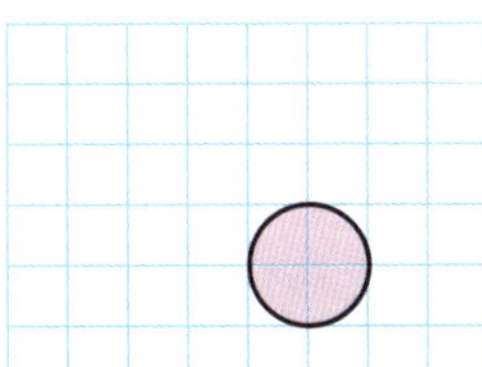

3 units left,
1 unit up

e

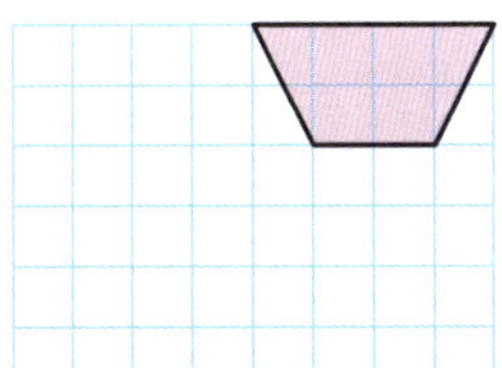

1 unit left,
4 units down

f

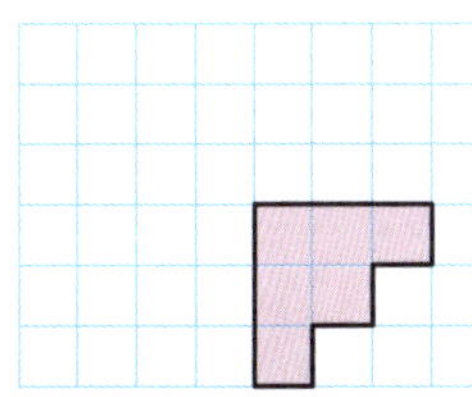

3 units left,
3 units up

3 Decide whether each transformation is a translation. If it is a translation, describe the translation.

a

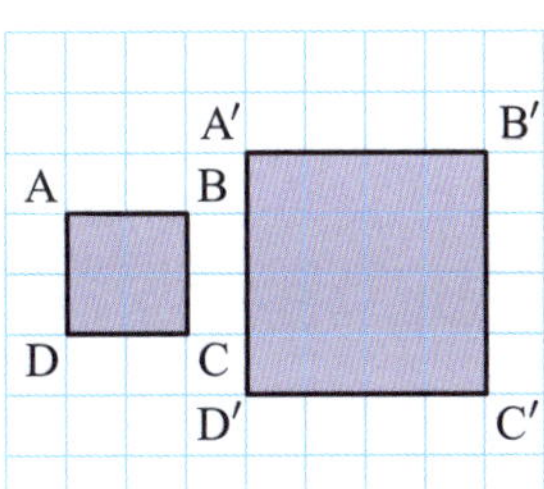

b

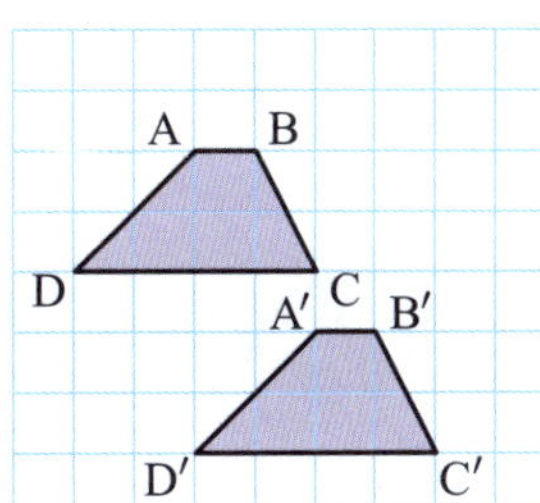

c

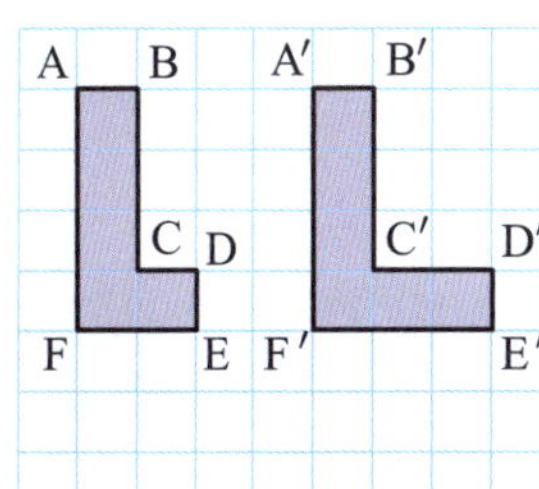

4 Consider the figures alongside.

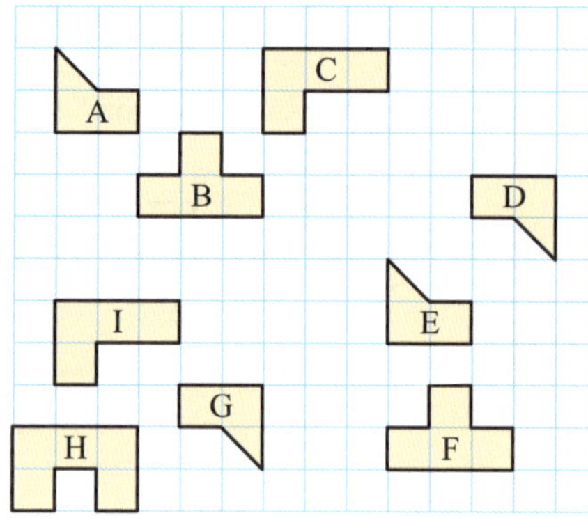

a Which figure is a translation of figure C? Describe the translation from figure C to this figure.

b Which figure is a translation of figure G? Describe the translation from figure G to this figure.

c Which figure cannot be translated to any other figure?

5 Figure A is translated 5 units right and 3 units down to figure B. Figure B is then translated 2 units left and 7 units up to figure C.

Describe the single translation which is needed to move:

a figure B to figure A
b figure C to figure B
c figure A to figure C
d figure C to figure A.

6 For each figure below:

i Translate the figure as directed.
ii State the coordinates of the vertices of the image.

a

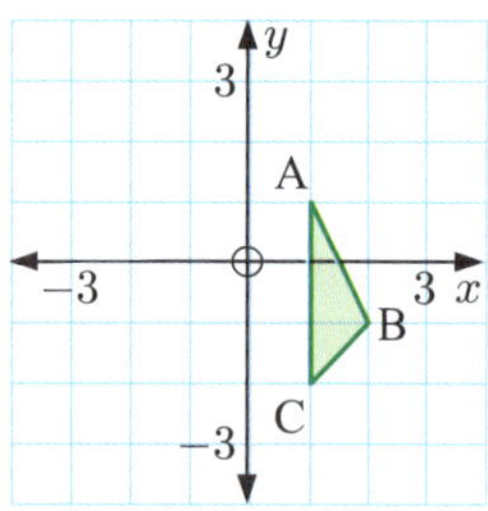

4 units left

b

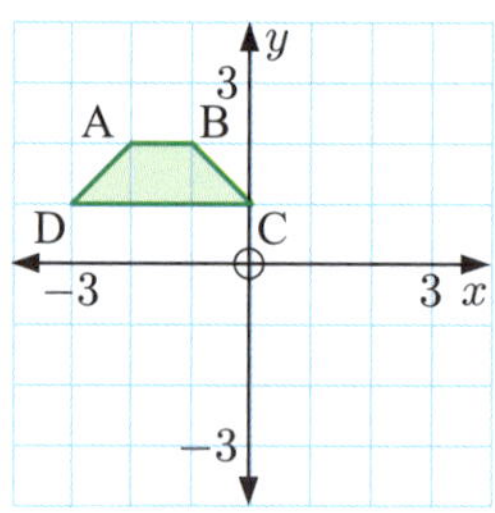

2 units right, 1 unit down

c

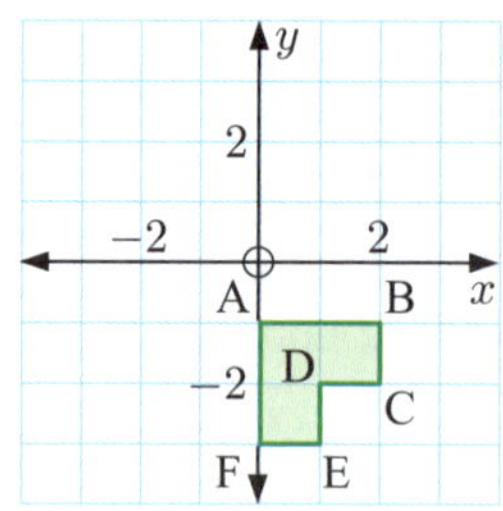

3 units left, 4 units up

7 **a** Write down the coordinates of P, Q, and R.

b Triangle PQR is translated to triangle P′Q′R′ such that P′ and Q′ both lie on an axis.

Find the *two* possible coordinates of R′.

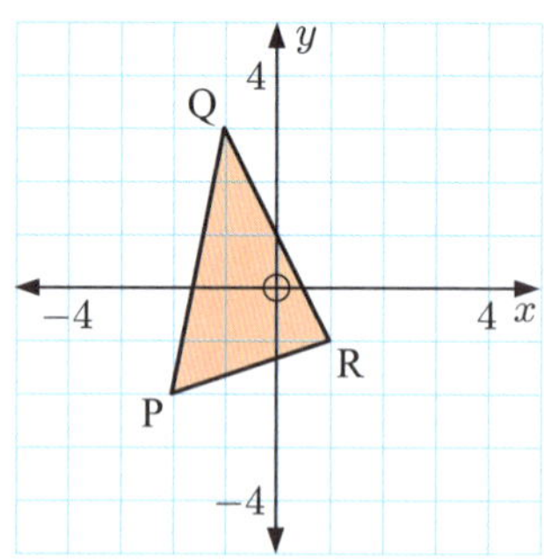

8 **a** Draw triangle ABC with vertices A(−1, 3), B(4, 1), and C(0, −2). Translate the figure 4 units right and 2 units down to form the image A′B′C′.

b State the coordinates of the image vertices A′, B′, and C′.

c Find the gradients of [AA′], [BB′], and [CC′]. Comment on your results.

d Find the distances AA′, BB′, and CC′. Comment on your results.

9 An object is translated h units to the right and k units up.

- **a** Suppose $A(x, y)$ is a point on the object. Find the coordinates of A' on the image.
- **b** Write an expression for the distance between each point on the object and its corresponding point on the image.
- **c** Describe the translation needed to map the image back onto the object.

10 Suppose an object is translated 4 units right and 5 units down. Find the point on the object corresponding to the image point:

- **a** $(7, 1)$
- **b** $(-2, -3)$

B REFLECTIONS

When we **reflect** an object in a **mirror line**, the image is its **reflection**.

To reflect an object, we draw lines at right angles to the mirror line which pass through important points on the object.

Each image point is the same distance from the mirror line as the object point, but is on the other side of the mirror line.

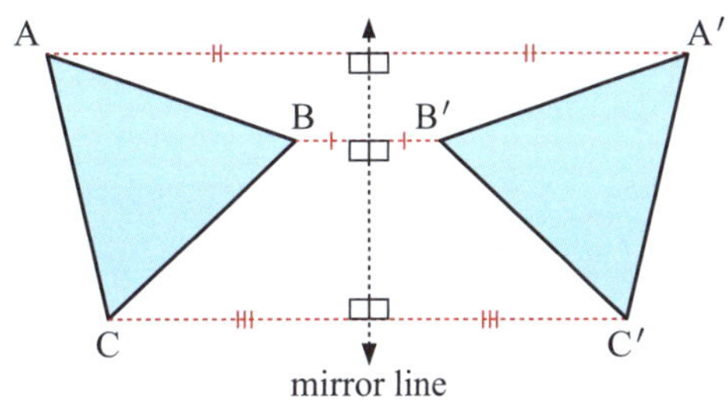

DEMO

DISCUSSION

Suppose the vertices of a polygon are labelled in clockwise order.
In what order will the vertices of its reflection be labelled?

EXERCISE 19B

1 Reflect each object in the given mirror line.

a
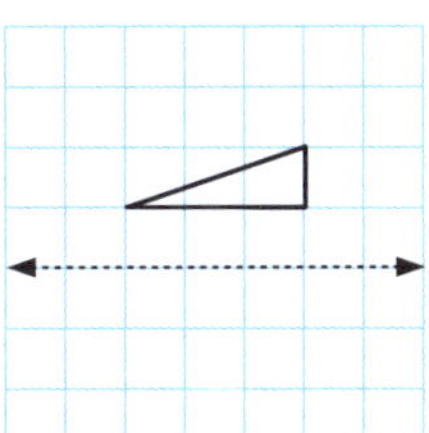

b
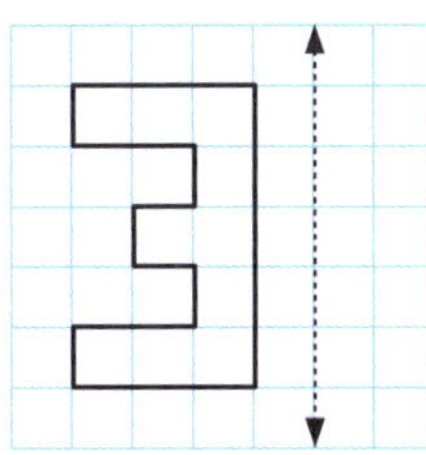

c
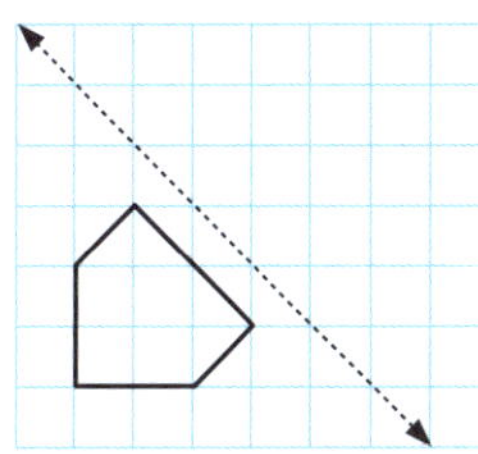

PRINTABLE DIAGRAMS

d
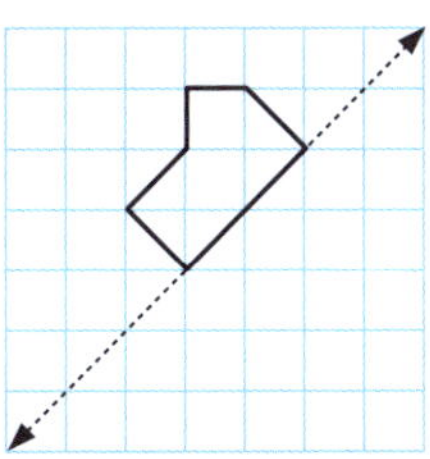

e
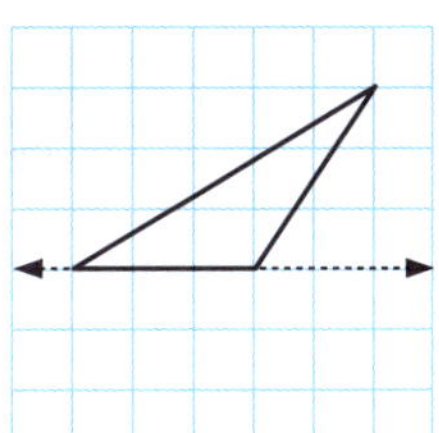

f

2 Is Q a reflection of P in the given mirror line? If not, copy P and the mirror line, then draw the correct reflection.

a

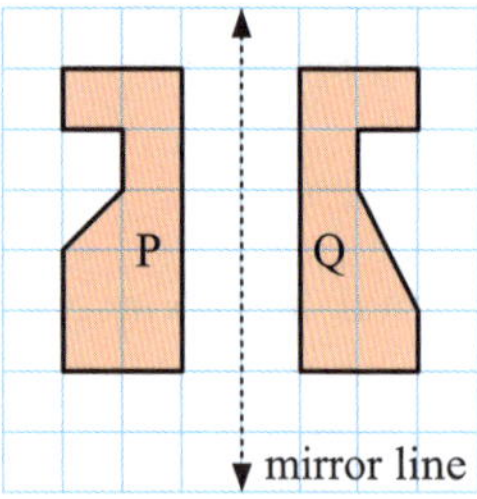

b

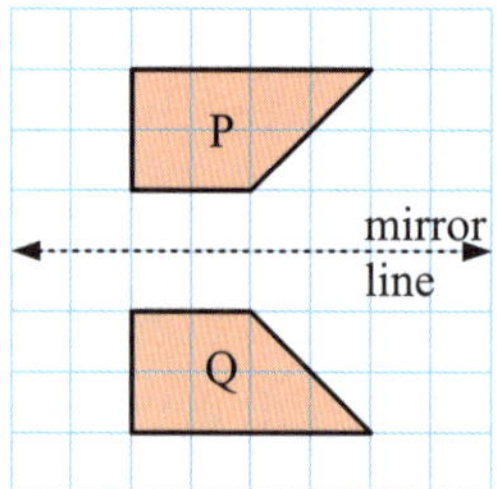

c

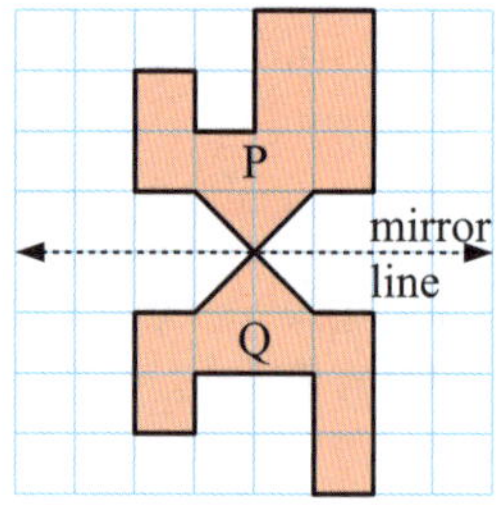

Example 1 **Self Tutor**

Reflect this figure in:

a the y-axis

b the line $y = x$.

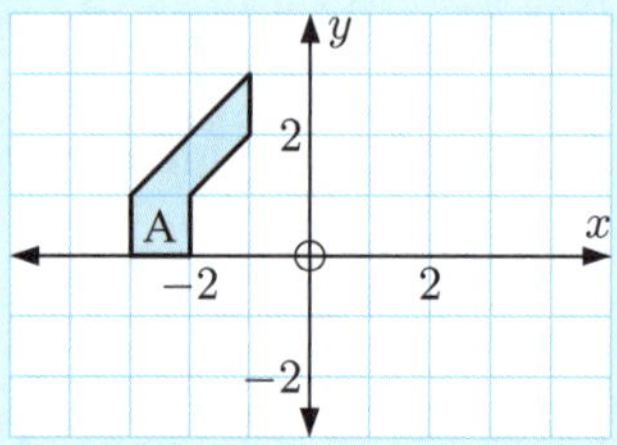

a

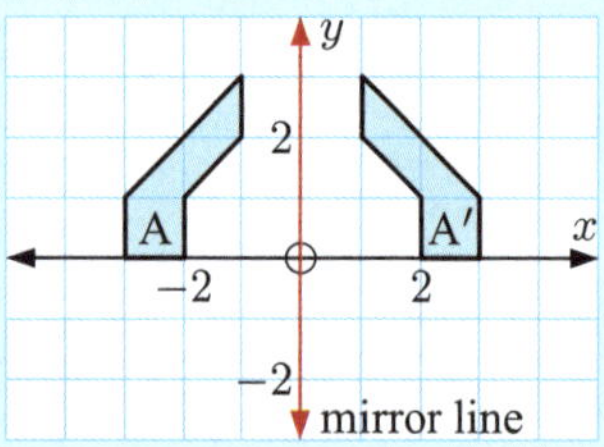

b

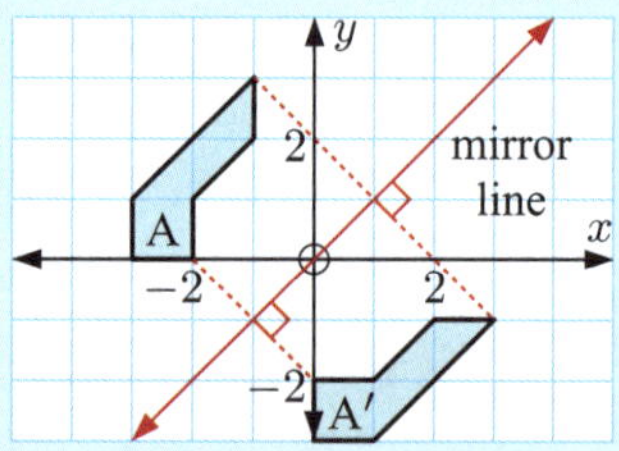

3 Reflect each figure in the line indicated:

a

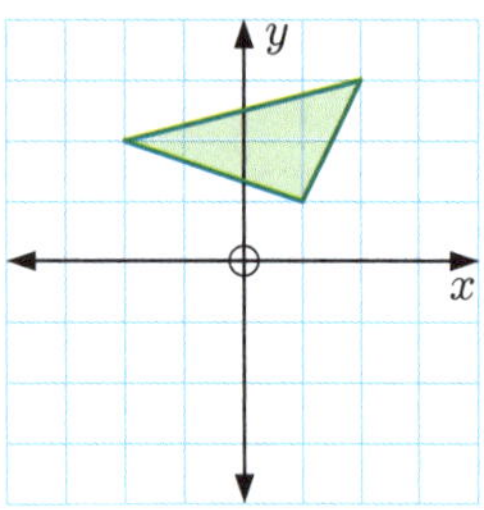

x-axis

b

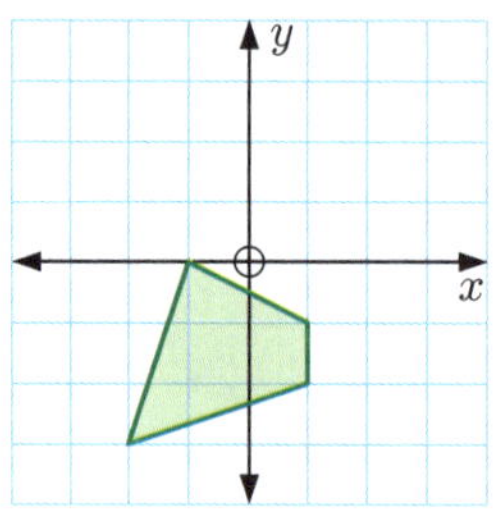

x-axis

c

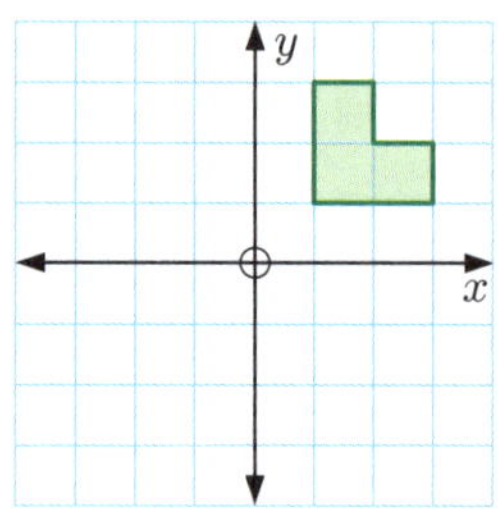

y-axis

d

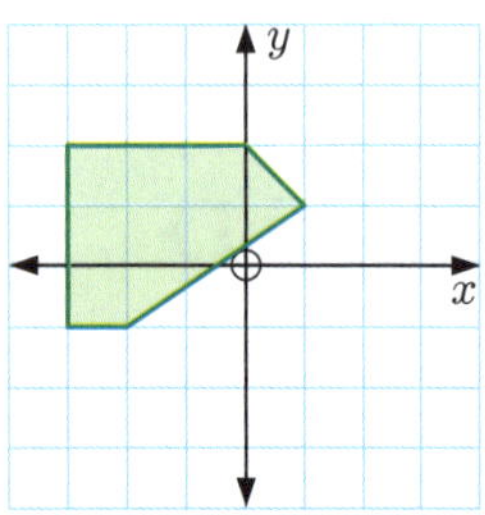

y-axis

e

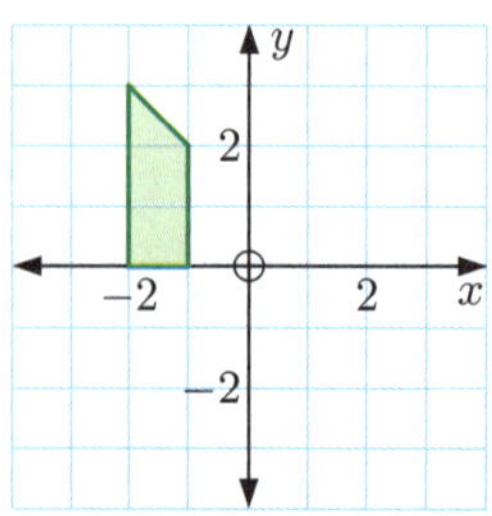

$y = x$

f

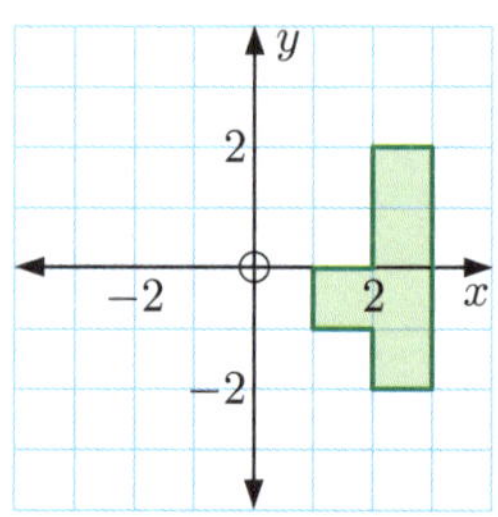

$y = x$

g

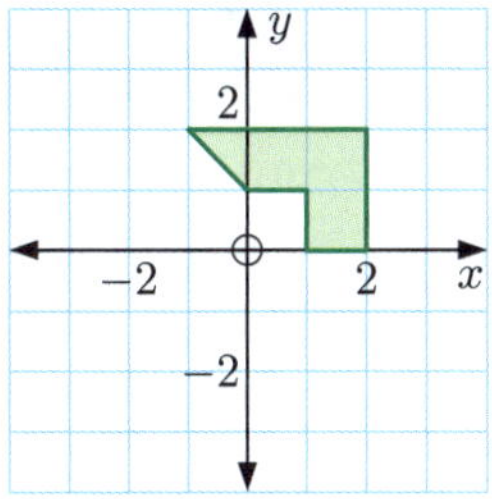

$y = -x$

h

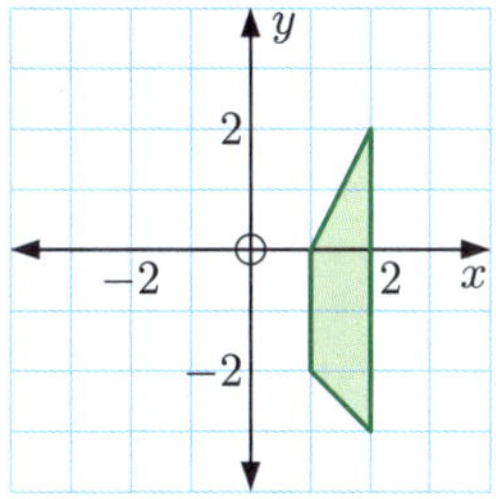

$y = -x$

4 **a** Which of these figures are reflections of each other? State which axis is the mirror line for this reflection.

b Which of the figures are translations of each other? Describe the translation in this case.

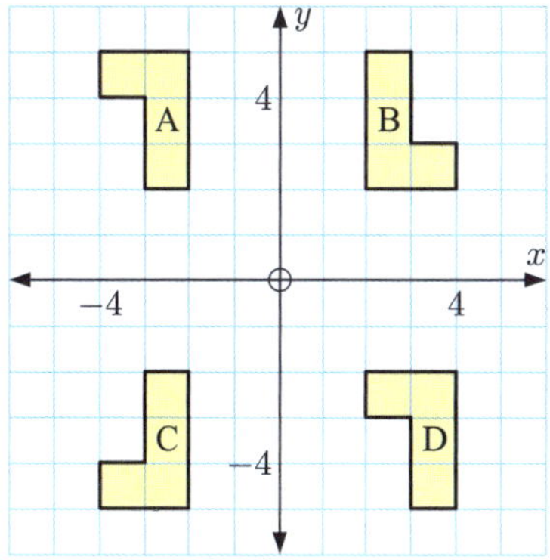

5 **a** A quadrilateral has vertices $A(1, 5)$, $B(3, 6)$, $C(4, 3)$, and $D(1, 2)$.
For each reflection below, plot ABCD and its image on the same set of axes, and state the coordinates of the image vertices.

i a reflection in the x-axis
ii a reflection in the y-axis
iii a reflection in the line $y = x$
iv a reflection in the line $y = -x$

b Write down the coordinates of the image point P′ when an object point $P(x, y)$ is reflected in:

i the x-axis
ii the y-axis
iii the line $y = x$
iv the line $y = -x$.

6 A service track must be built from A to the pipeline and then to B by the shortest route.
Simran's solution is as follows:

Step 1: Use the pipeline as a mirror line to reflect B.
Step 2: Join [AB′], and let [AB′] meet the pipeline at X.
Step 3: Join [BX].
Step 4: AX + XB is the shortest distance.

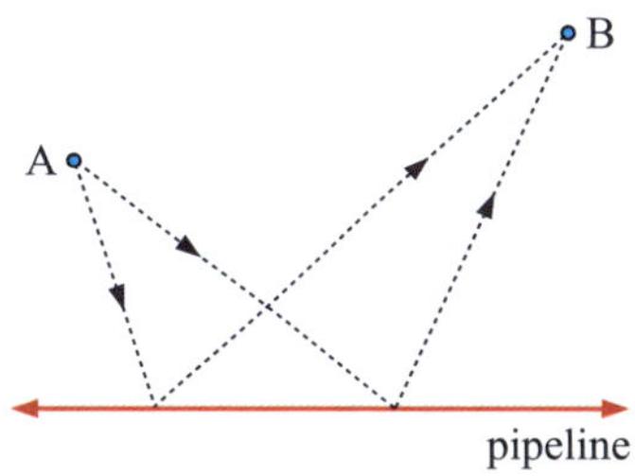

a Draw a diagram to illustrate Simran's solution.

b What can be said about triangle BXB′? Explain your answer.

c Explain why Simran's solution is correct.
Hint: Let Y be any point on the pipeline other than X. Join [AY] and [YB′].

7 When an object is reflected in a particular mirror line, its image is identical to the object. What does this tell us about the object?

C ROTATIONS

A **rotation** turns an object about a point and through a given angle.

The point about which the figure rotates is called the **centre of rotation**, and is often labelled O. In any rotation, each point and its image are the same distance from O.

In mathematics we rotate in an *anticlockwise* direction unless we are told otherwise.

Example 2 — Self Tutor

Rotate each figure about O through the angle indicated:

a

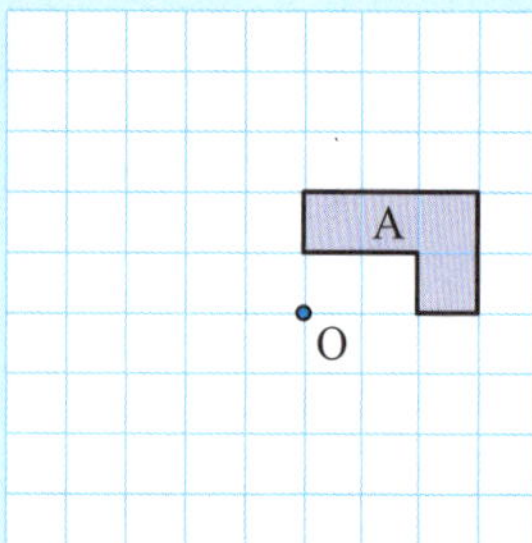

180°

b

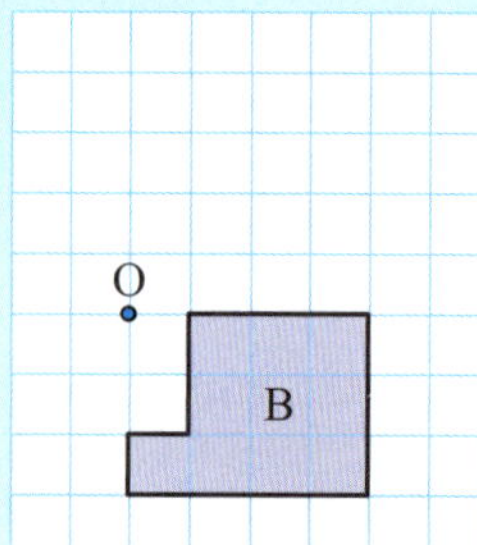

90°

c

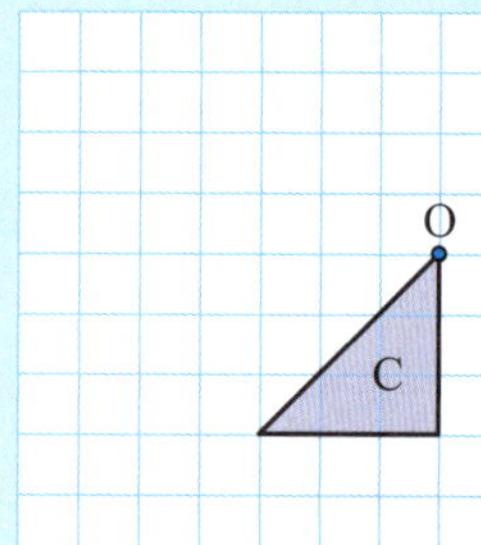

90° clockwise

a

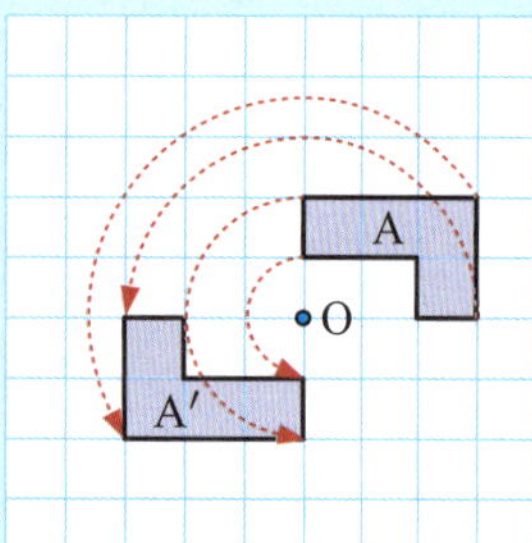

b

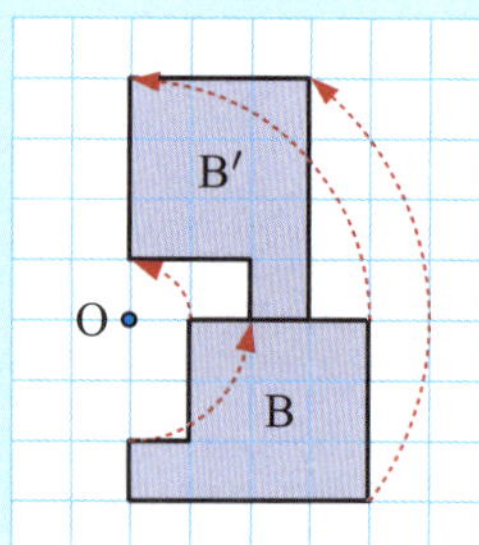

c

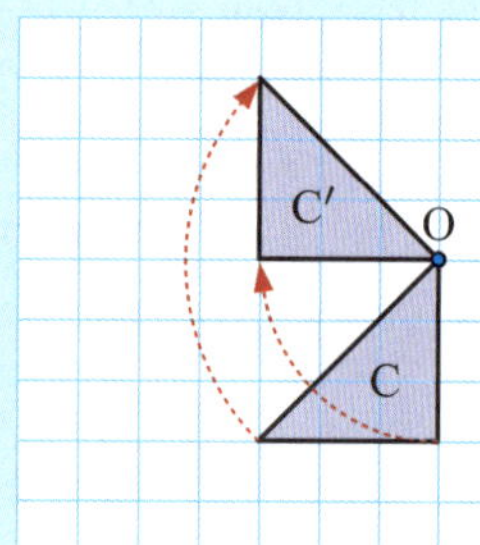

EXERCISE 19C

1 Derek is sitting in chair A.
Which chair will he move to if he rotates anticlockwise about O through:

a 90° **b** 270° **c** 180°?

2 A figure is rotated 75° clockwise about a point O. Describe a rotation which will return the figure to its original position.

3 A figure is rotated 180° about a point O. Describe a rotation which will return the figure to its original position.

4 Rotate each figure about O through the angle indicated:

a

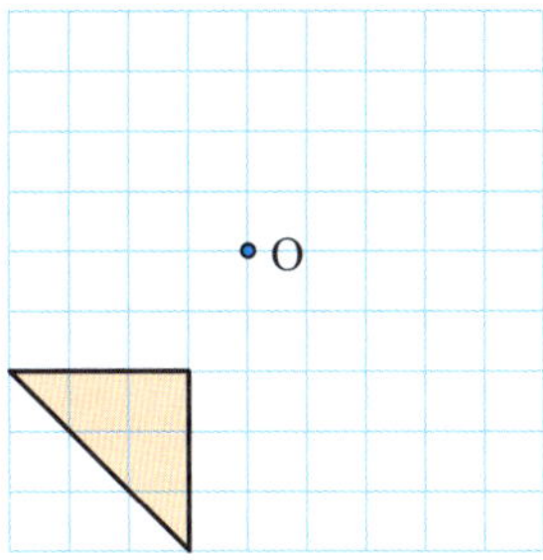

90°

b

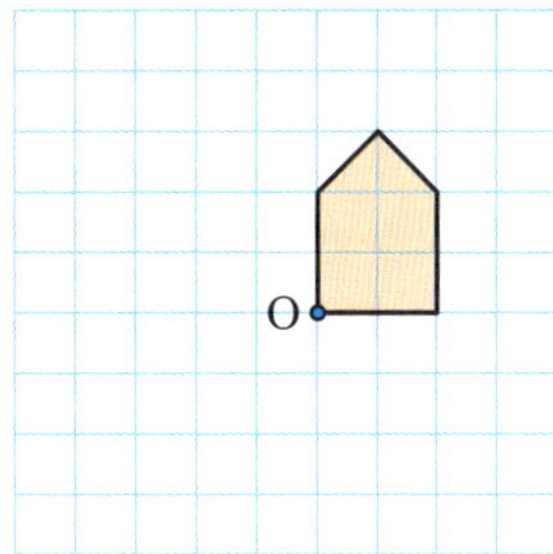

180°

c

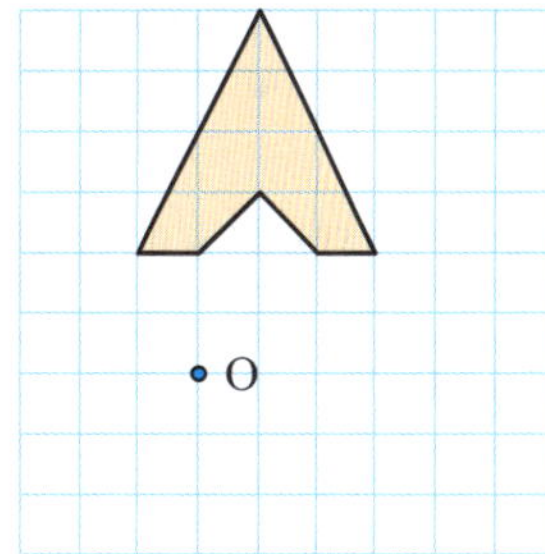

90° clockwise

d

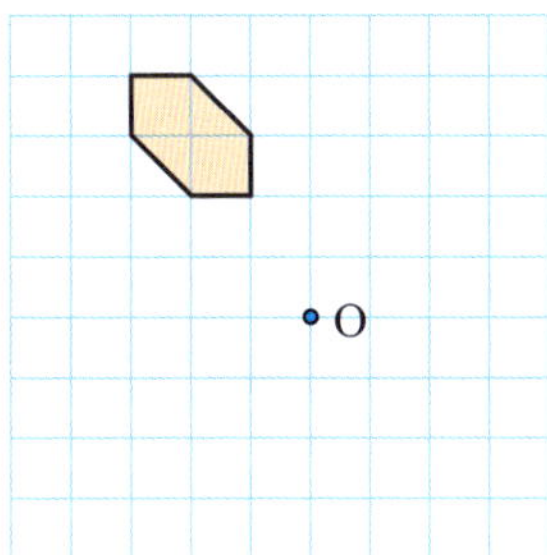

180°

e

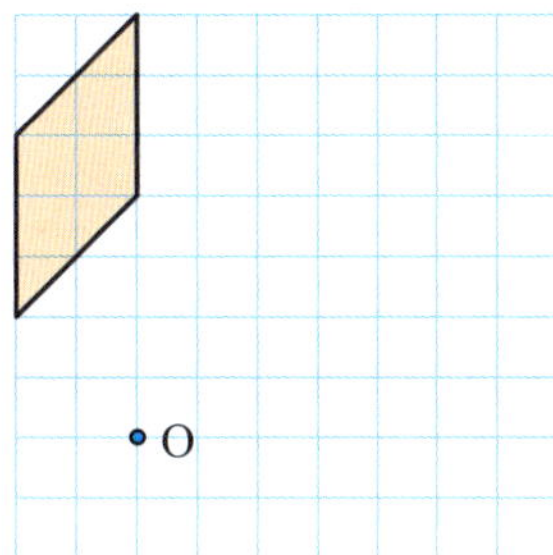

90° clockwise

f

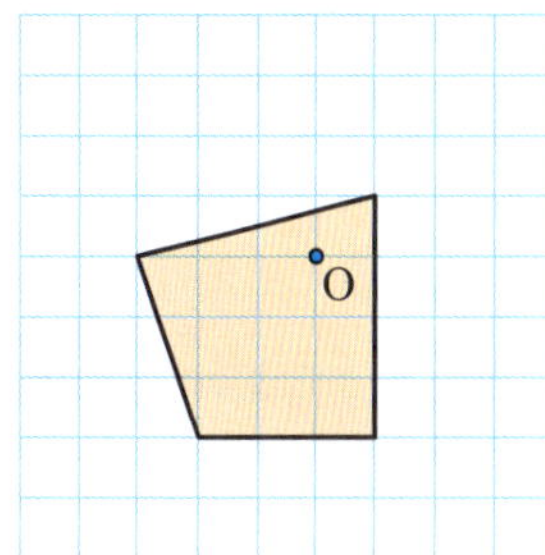

90°

5 **a** Which figure is a rotation of figure A about O?

b Determine the angle of rotation from figure A to this figure.

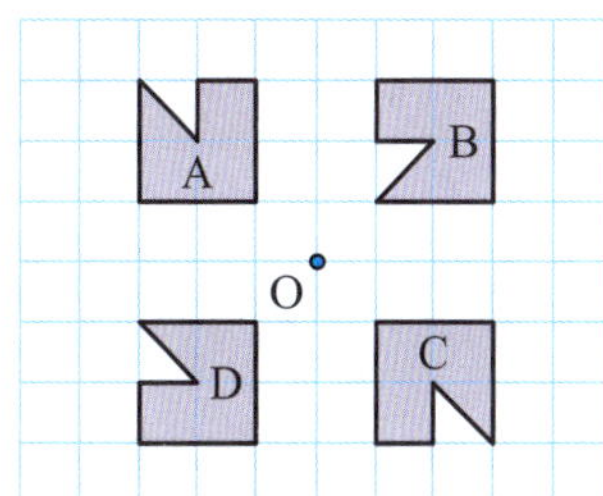

6 **a** State the coordinates of the vertices of triangle ABC.

b Rotate the triangle 90° clockwise about the origin O.

c State the coordinates of the vertices of the image.

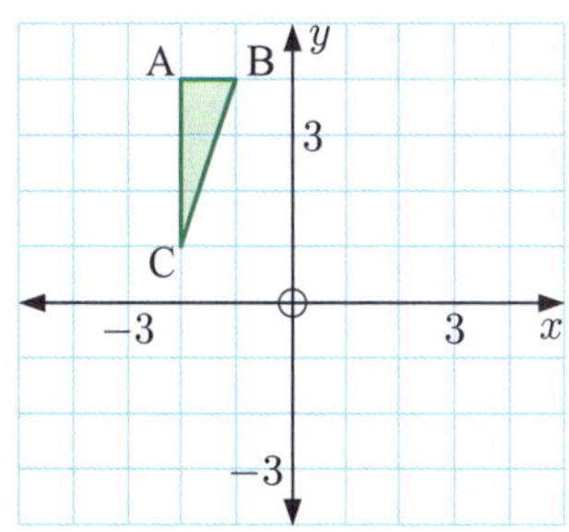

7 **a** State the coordinates of the vertices of triangle PQR.

b Rotate the triangle 90° about the origin O.

c State the coordinates of the vertices of the image.

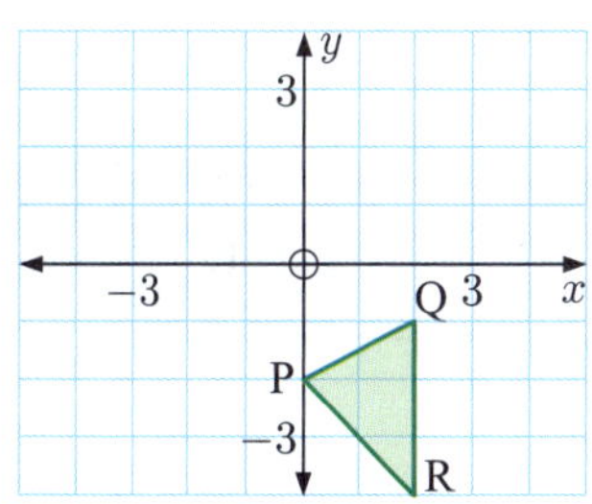

8 **a** A quadrilateral has vertices $A(-4, -3)$, $B(-2, 1)$, $C(-1, -2)$, and $D(-3, -4)$.
For each rotation about O below, plot ABCD and its image on the same set of axes, and state the coordinates of the image vertices.
i $90°$ **ii** $180°$ **iii** $90°$ clockwise

b Write down the coordinates of the image point P' when $P(x, y)$ is rotated about O through:
i $90°$ **ii** $180°$ **iii** $90°$ clockwise.

9 In triangle ABC, O is the midpoint of [AC].

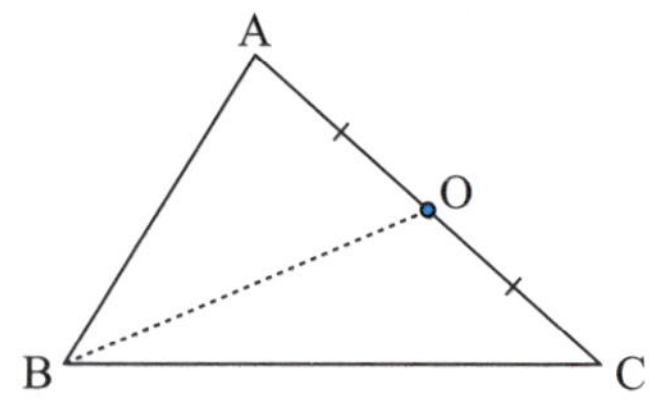

a Redraw the figure and rotate it through $180°$ about O.
b Name the resulting figure which is made up of the original and its image.
c List the geometric properties of the resulting figure.
d What special name is given to the resulting figure if:
i $A\widehat{B}C$ is a right angle **ii** $AB = BC$
iii $A\widehat{B}C$ is a right angle *and* $AB = BC$?

10 When an object is rotated $90°$ about point O, the image is identical to the object.
What does this tell us about the object? Explain your answer.

DISCUSSION

When an object is translated, reflected, or rotated, which of these properties will *not* change?

- size
- shape
- position
- orientation.

D ENLARGEMENTS AND REDUCTIONS

The illustrated shapes A, B, and C are not the same *size*, but they have the same *shape*. Their side lengths are in the same ratio.

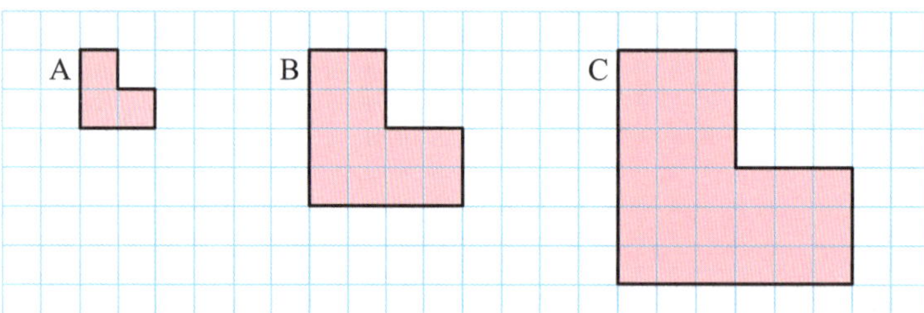

When an object is enlarged or reduced, all of its lengths are enlarged or reduced by the same **scale factor**.
For example:

- The side lengths in figure B are all twice the corresponding side lengths in figure A. We say that B is an **enlargement** of A with scale factor 2.
- The side lengths in figure B are all $\frac{2}{3}$ of the side lengths in figure C. We say that B is a **reduction** of C with scale factor $\frac{2}{3}$.

If the scale factor is greater than 1, we have an **enlargement**.
If the scale factor is less than 1, we have a **reduction**.

EXERCISE 19D.1

1 State whether each transformation from A to B is an enlargement or reduction, and find the scale factor.

a 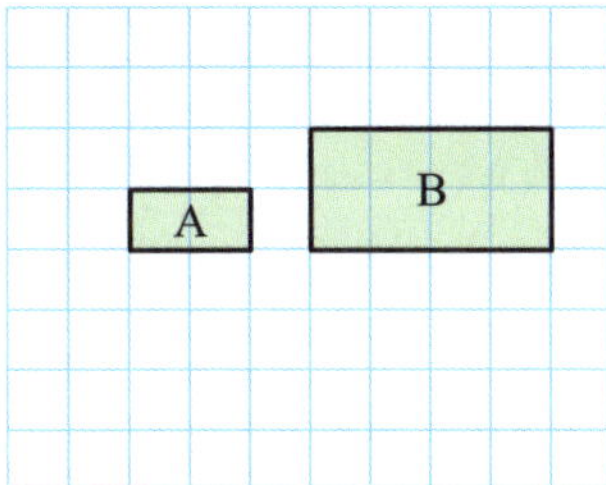

b 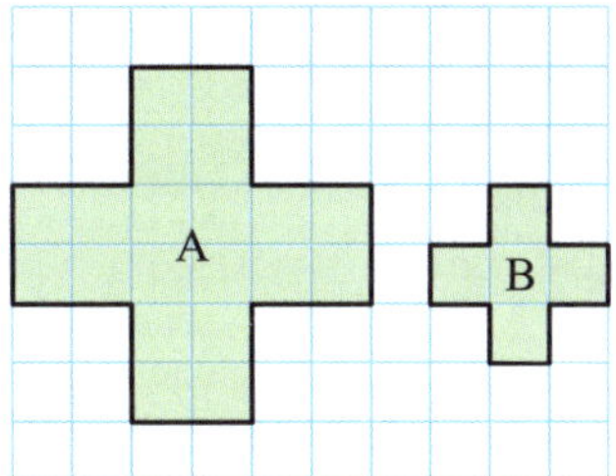

c 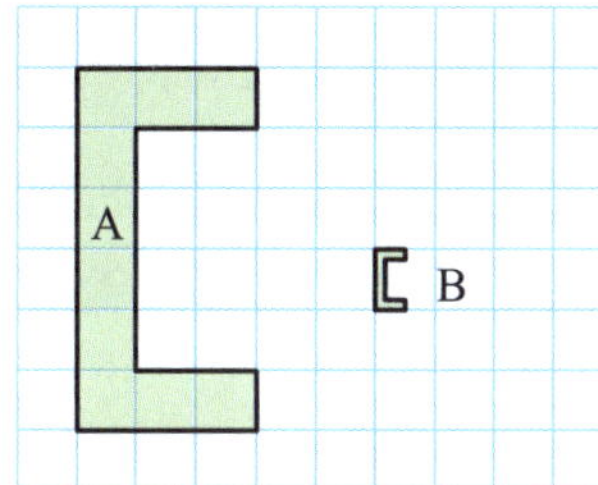

d 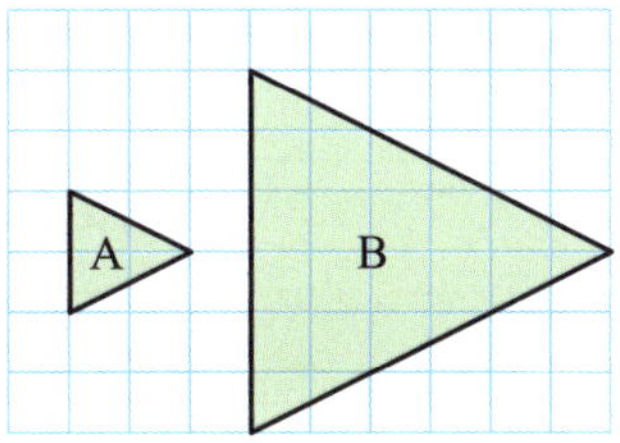

e 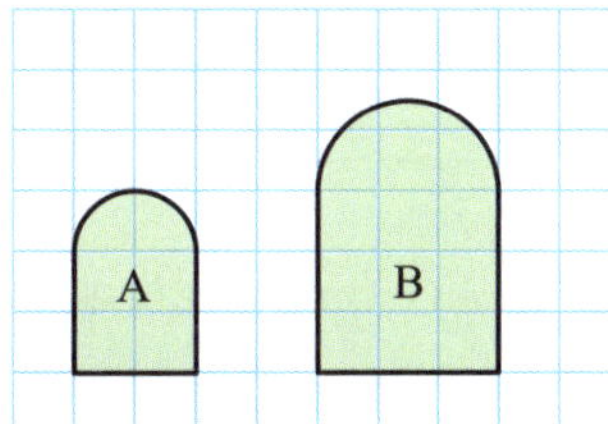

f 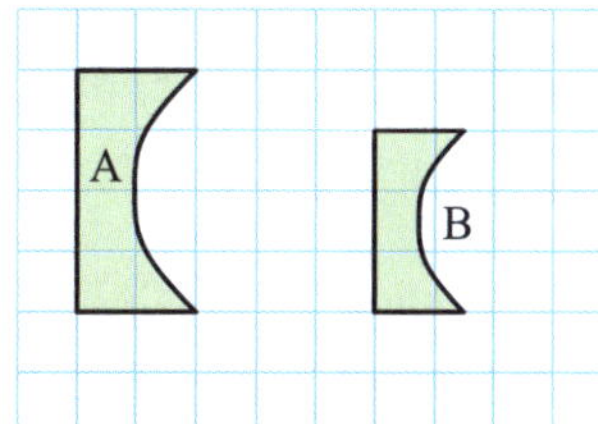

2 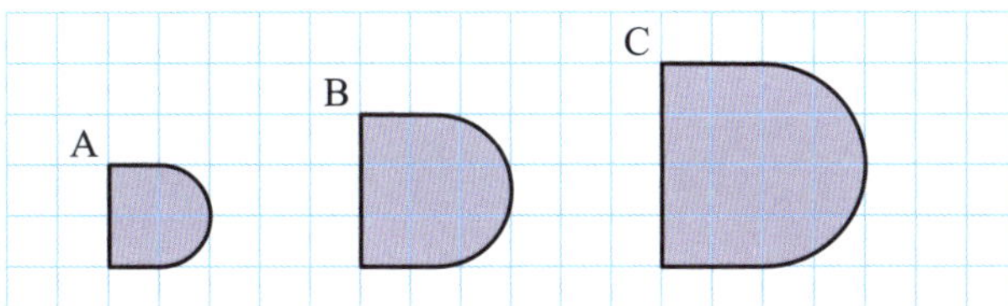

State the scale factor for the enlargement or reduction from:

- a A to B
- b B to A
- c A to C
- d C to A
- e B to C
- f C to B.

INVESTIGATION 1

In the figure alongside, the green triangle has been enlarged to produce images A and B.

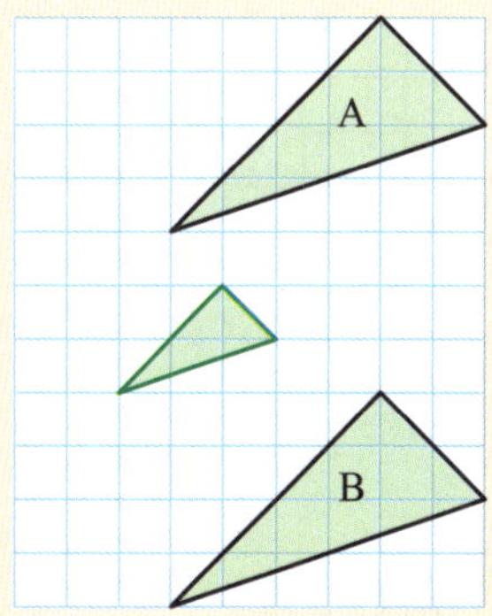

What to do:

1 Was the scale factor for both enlargements the same?

2 If you were asked to enlarge the green triangle with scale factor 2, do you have enough information to know *where* to draw the image?

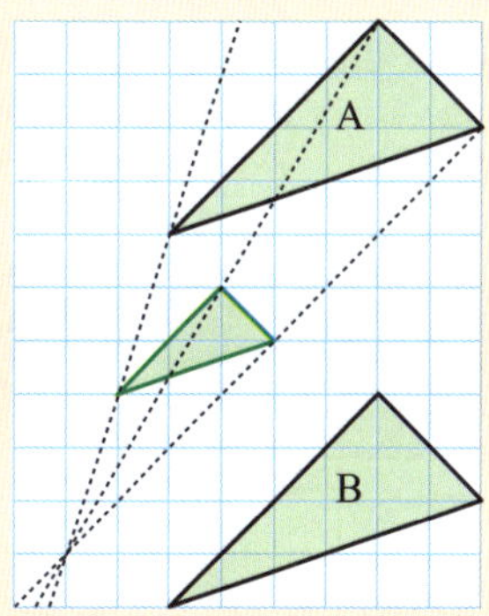

3 Suppose we draw lines through each vertex of image A and its corresponding object point.
What do you notice about these lines?

4 Copy the green triangle and image B onto grid paper. Draw lines through each vertex of B and its corresponding object point. Do you observe the same thing as with image A?

5 Describe how you could use your observations in **3** and **4** to specify exactly where the image of an enlargement or reduction should be drawn.

From the **Investigation**, you should have found that:

In any enlargement, the lines through each object point and its corresponding image meet at a unique point. We call this point the **centre of enlargement** or just **centre**.

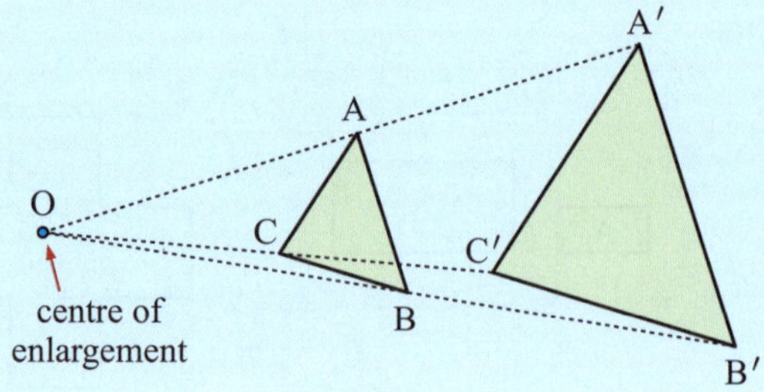

In order to enlarge or reduce an object and specify the position of the image, we need to know both the scale factor *and* the centre.

Example 3 — Self Tutor

For the given scale factor and centre O, draw the image of the figure in red.

a scale factor 2

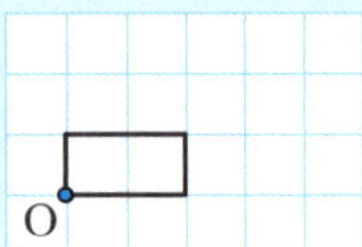

b scale factor 3

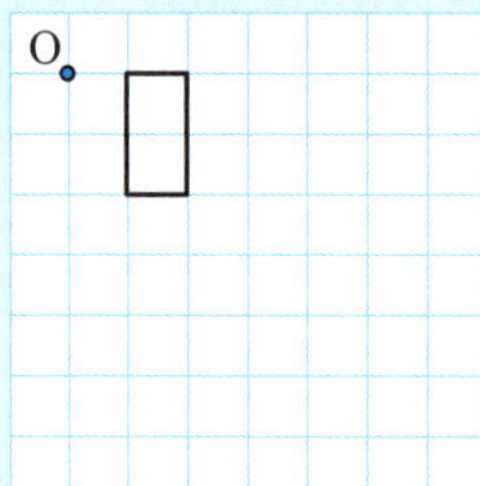

c scale factor $\frac{1}{2}$

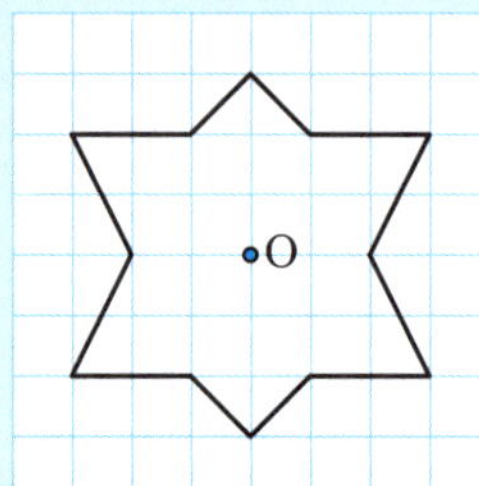

a

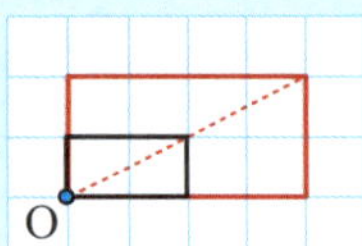

b

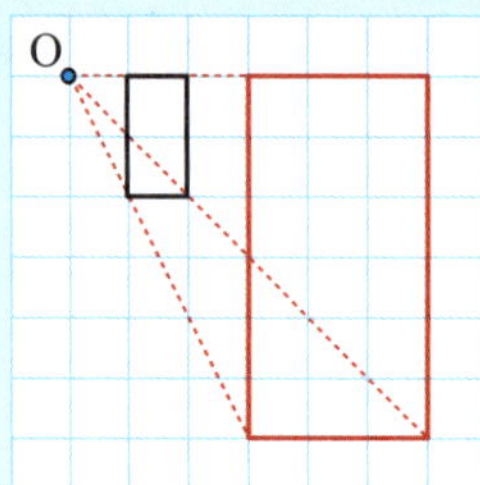

c

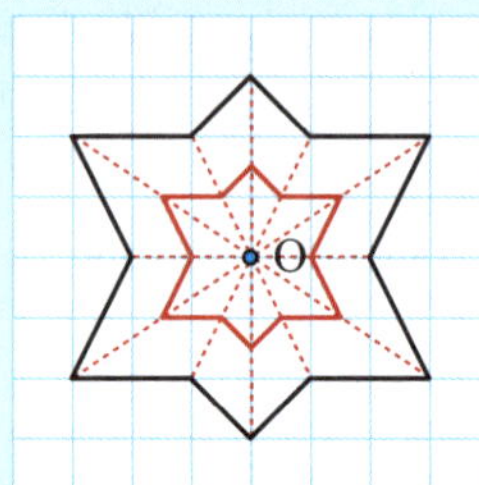

EXERCISE 19D.2

PRINTABLE DIAGRAMS

1 In each diagram, the black figure has been enlarged or reduced to form the red figure. Copy the diagram and locate the centre O.

a

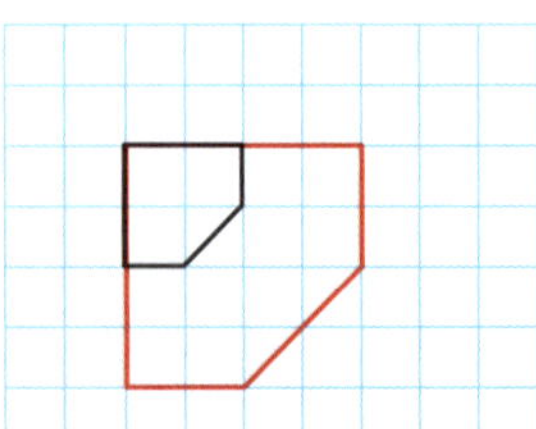

b

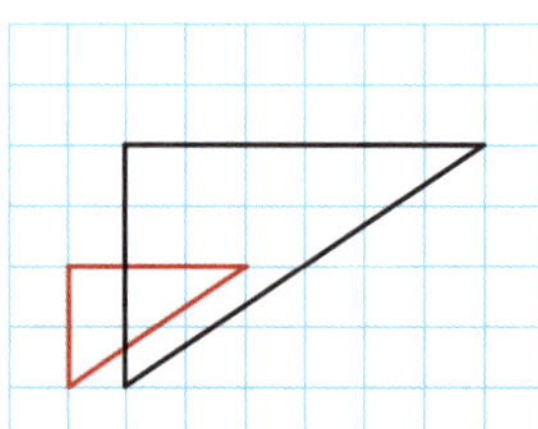

c

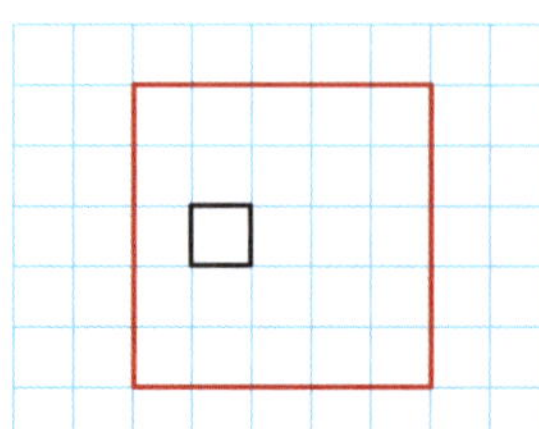

2 Draw the image of each figure for the given scale factor and centre O:

a scale factor 2

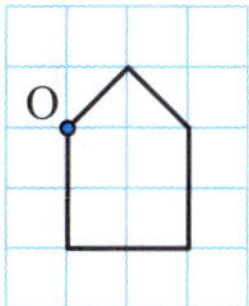

b scale factor 3

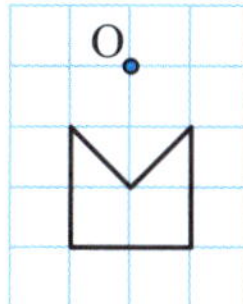

c scale factor $\frac{1}{2}$

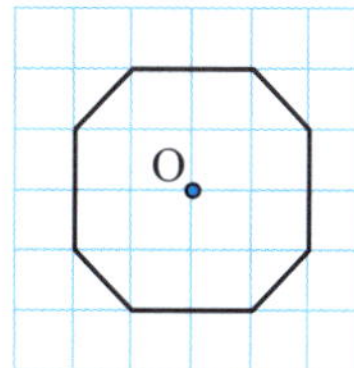

d scale factor 2

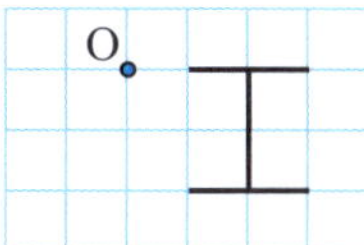

e scale factor $\frac{1}{3}$

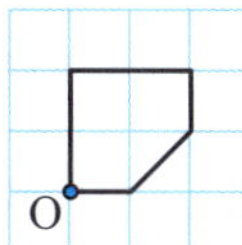

f scale factor 3

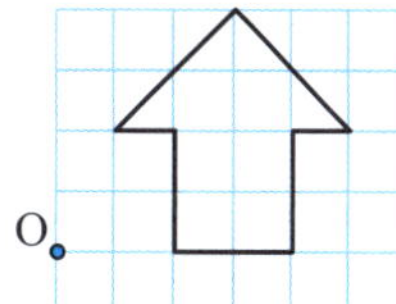

3 **a** Draw the image of each figure for the given scale factor and centre O:

i scale factor 2

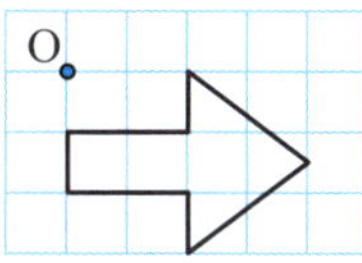

ii scale factor $\frac{1}{2}$

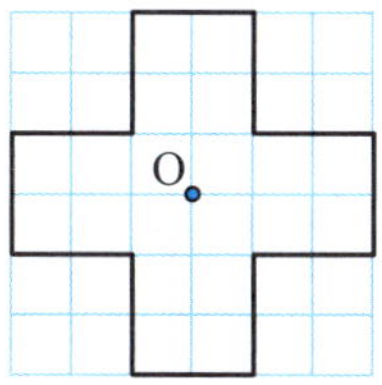

iii scale factor 3

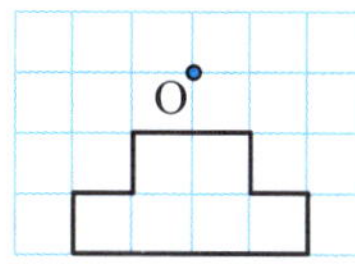

b Copy and complete the table below for each diagram in **a**.

	Area of object	*Area of image*	$\dfrac{\textit{Area of image}}{\textit{Area of object}}$	*Scale factor, k*	k^2
i					
ii					
iii					

c Comment on your results in **b**.

4 **a** A quadrilateral has vertices A(2, 4), B(4, 4), C(4, 0), and D(2, 1). For each transformation below, plot ABCD and its image on the same set of axes, and state the coordinates of the image vertices.

i an enlargement with scale factor 2 and centre O

ii a reduction with scale factor $\frac{1}{2}$ and centre O

b Write down the coordinates of the image point P′ when an object point P(x, y) is enlarged or reduced with scale factor k and centre O.

c Hence write down the coordinates of the image of:

i (2, 3) under an enlargement with scale factor $\frac{5}{2}$ and centre O

ii (−6, 9) under a reduction with scale factor $\frac{2}{3}$ and centre O.

GLOBAL CONTEXT — LINEAR PERSPECTIVE

Global context: Orientation in space and time

Statement of inquiry: Understanding geometric relationships can enhance creativity.

Criterion: Applying mathematics in real-life contexts

E STRETCHES

If we enlarge or reduce an object with centre O, the effect is to stretch the object in both the x and y directions.

In this Section we consider stretches in one direction only:

- In a *vertical* stretch, the distance from each object point to the x-axis is multiplied by the scale factor to give the image point.
- In a *horizontal* stretch, the distance from each object point to the y-axis is multiplied by the scale factor to give the image point.

INVESTIGATION 2 — STRETCHES

In this Investigation we explore the effect of stretches on individual points and shapes.

What to do:

1 In the diagram, the black figure has been stretched vertically with scale factor 2 to give the red figure.

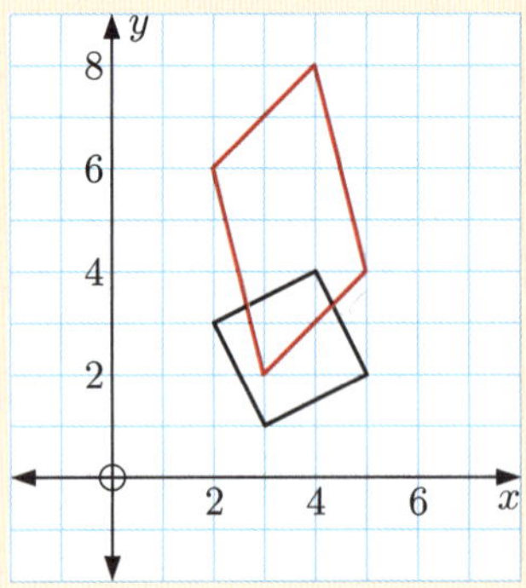

a Has the *shape* of the object been affected by the transformation?

b Has the x-coordinate of any of the vertices been affected by the stretch?

c What has happened to the y-coordinate of each point on the object?

d State the coordinates of the image point P′ when an object point P(x, y) is vertically stretched with scale factor k.

2 In the diagram, the black figure has been stretched horizontally with scale factor $\frac{2}{3}$ to give the red figure.

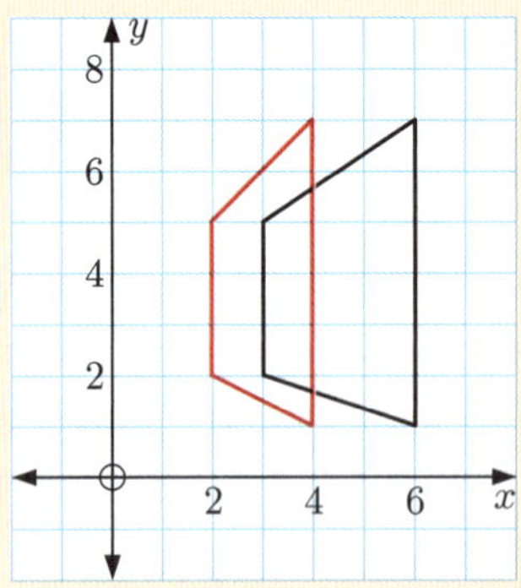

a Has the *shape* of the object been affected by the transformation?

b Has the y-coordinate of any of the vertices been affected by the stretch?

c What has happened to the x-coordinate of each point on the object?

d State the coordinates of the image point P′ when an object point P(x, y) is horizontally stretched with scale factor k.

From the **Investigation**, you should have found that:

- When an object is stretched in one direction only, the shape of the object is changed. Angles in the figure are changed by the stretch.
- The image of $\mathrm{P}(x, y)$ under a vertical stretch with scale factor k is $\mathrm{P}'(x, ky)$.
- The image of $\mathrm{P}(x, y)$ under a horizontal stretch with scale factor k is $\mathrm{P}'(kx, y)$.

EXERCISE 19E

1 Describe the stretch which maps the object onto its image:

a OAB onto OAB′

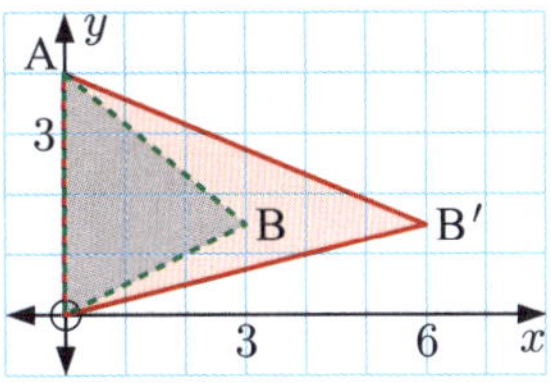

b ABC onto A′B′C′

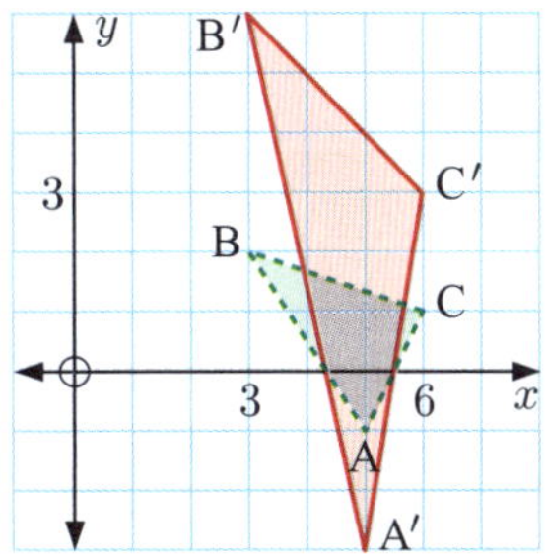

c OABC onto OA′B′C

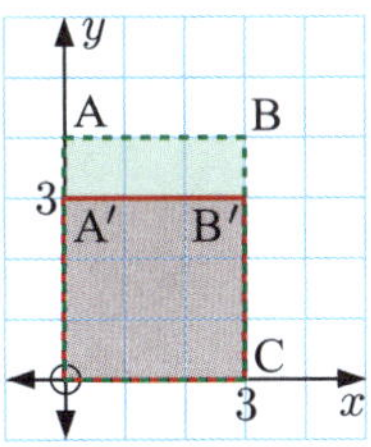

d ABCD onto A′B′C′D′

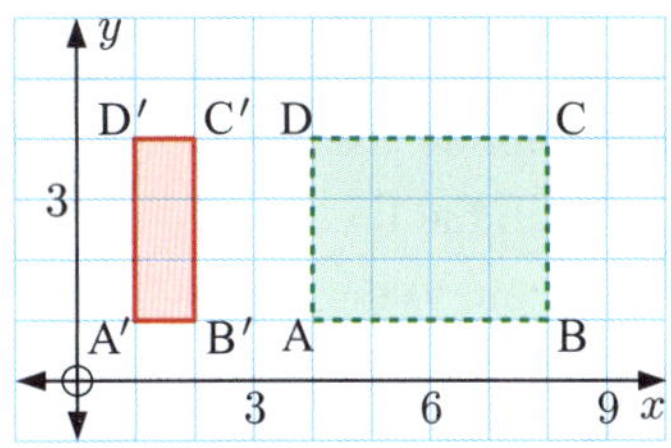

2 Copy the figure and draw its image under the given stretch:

a vertical stretch with scale factor 2

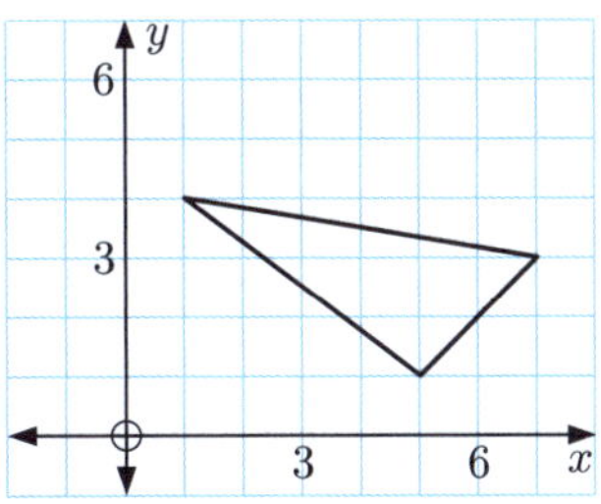

b horizontal stretch with scale factor $\frac{5}{2}$

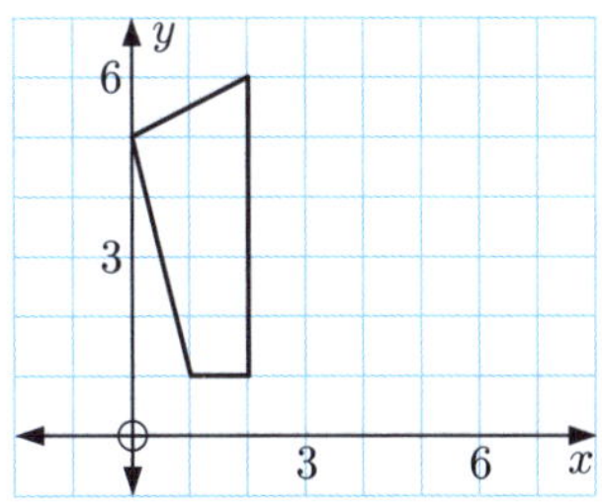

c horizontal stretch with scale factor $\frac{1}{3}$

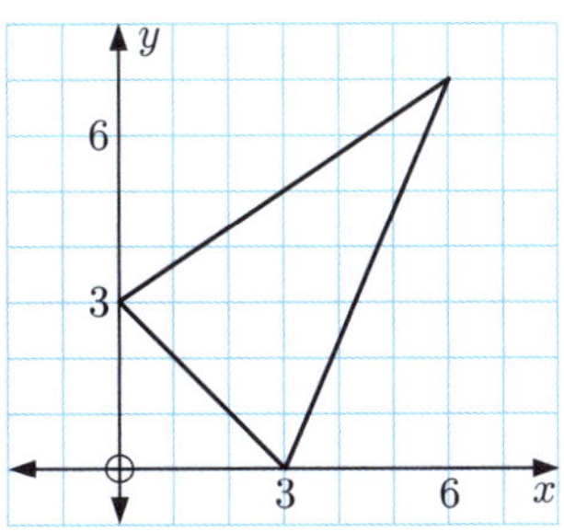

d vertical stretch with scale factor $\frac{3}{4}$

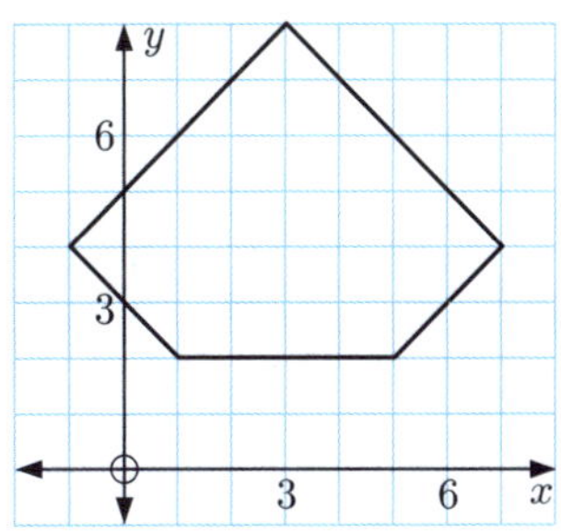

3 Describe the single transformation which maps:

a A onto B
b A onto C
c A onto D
d B onto D
e D onto C
f D onto A.

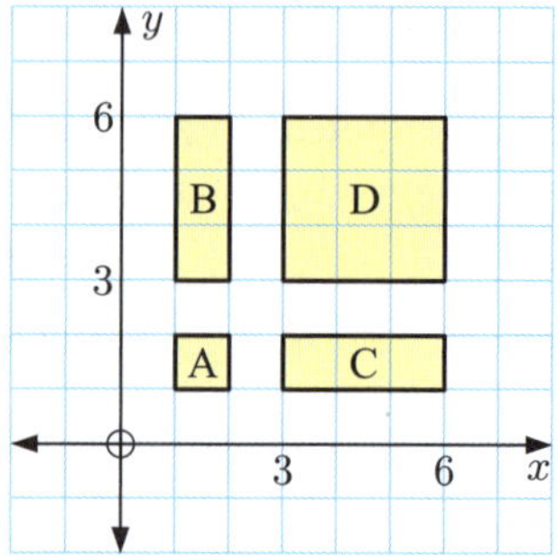

4 Sketch the image of a circle with centre $(4, 0)$ and radius 4 units under:

a a vertical stretch with scale factor $\frac{3}{2}$
b a horizontal stretch with scale factor $\frac{3}{4}$.

5 Triangle PQR has vertices $P(1, 1)$, $Q(1, 4)$, and $R(3, 2)$. The triangle is vertically stretched with scale factor k. Find the value of k such that the image:

a has area 12 units^2
b is a right angled triangle.

DISCUSSION

In any transformation, an **invariant point** is a point which does not move.

Discuss what points, if any, would be invariant under:

- a translation
- a reflection in the x-axis
- a reflection in the line $y = -x$
- a rotation about O through $90°$
- an enlargement with scale factor k and centre O
- a vertical stretch with scale factor k
- a horizontal stretch with scale factor k.

F COMBINATIONS OF TRANSFORMATIONS

In this Section we perform several transformations on a figure, one after another.

Example 4 Self Tutor

Reflect figure A in the mirror line, then rotate the result $90°$ about O.

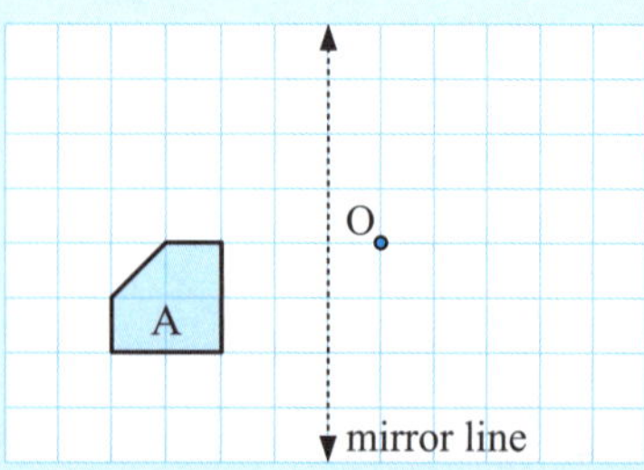

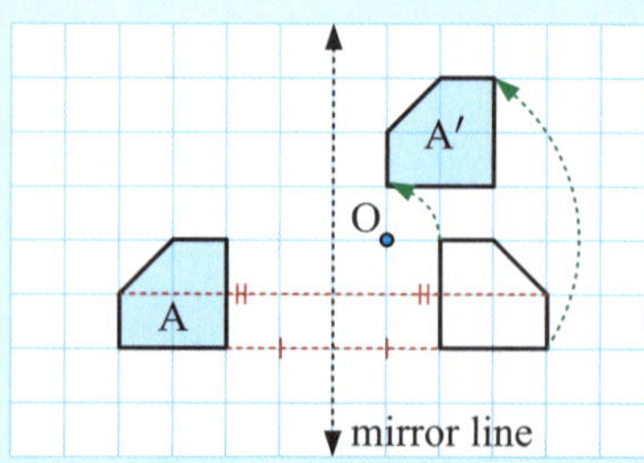

EXERCISE 19F

1 Find the image when the figure alongside is:

a translated 2 units right and 1 unit up, then reflected in the mirror line

b reflected in the mirror line, then rotated 90° about O.

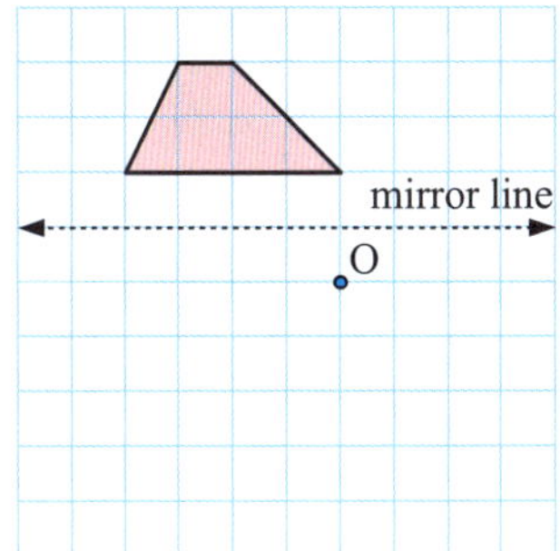

2 **a** Rotate A 90° clockwise about O, then enlarge the result with scale factor 2 and centre O to give A′.

b Describe the combination of transformations which maps A′ back onto A.

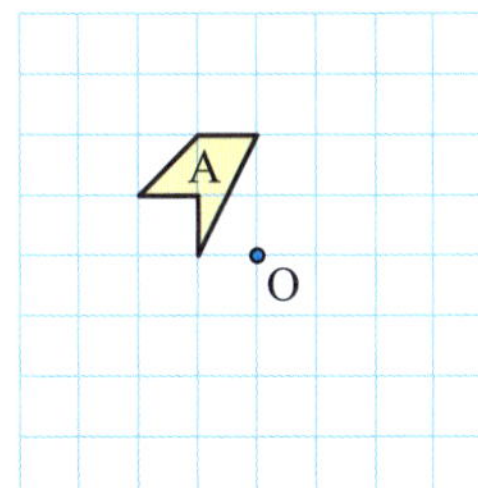

3 Describe how to transform A to B using:

a a reduction with centre O, followed by a translation

b a translation followed by a reduction with centre O.

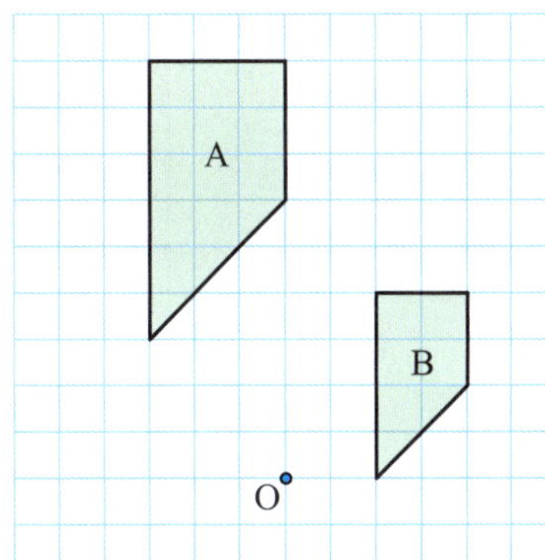

4 Translate quadrilateral PQRS 3 units left and 4 units down, then rotate the result 90° clockwise about O.
State the coordinates of the image vertices.

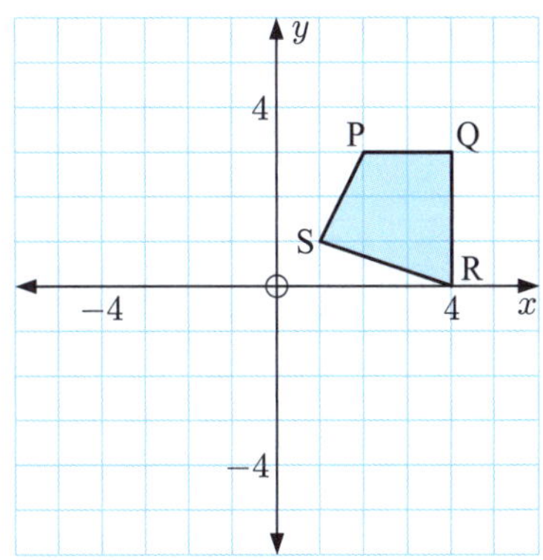

5 **a** A triangle has vertices A(2, 1), B(4, 2), and C(2, 4).
The triangle is reflected in the x-axis, then stretched horizontally with scale factor 2.
Plot ABC and its image on the same set of axes, and state the coordinates of the image vertices.

b Write down the coordinates of the image point P′ when an object point P(x, y) is reflected in the x-axis, then stretched horizontally with scale factor 2.

HISTORICAL NOTE

Sir D'Arcy Wentworth Thompson (1860 - 1948) was a mathematical biologist from Scotland, most noted for his book *On Growth and Form*. He emphasised the similarities between some plants and animals which develop because a particular characteristic is physically beneficial to both organisms.

In the chapter *The Comparison of Related Forms*, Thompson described how differences in the forms of related animals could be described using mathematical transformations. For example:

- a parrotfish can be transformed into an angelfish by using a curved grid

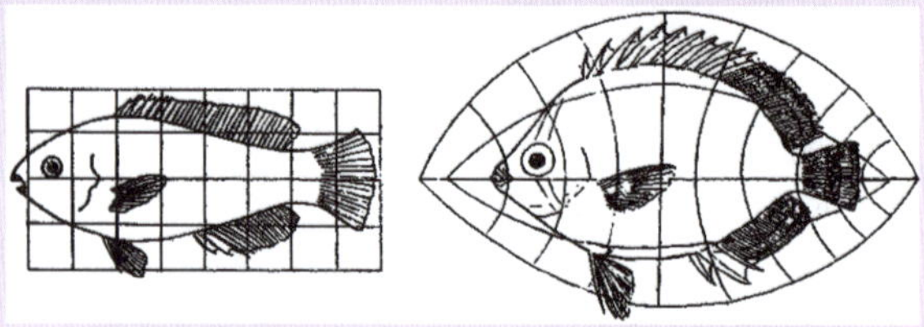

parrotfish angelfish

- a wreckfish can be transformed into a short bigeye by adding perspective to the grid.

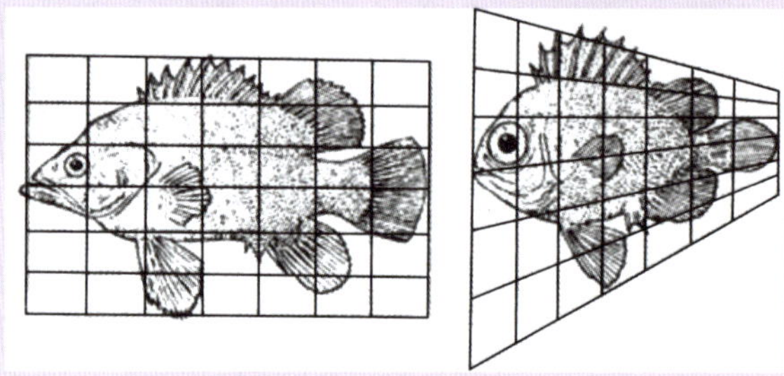

wreckfish short bigeye

MULTIPLE CHOICE QUIZ

QUICK QUIZ

REVIEW SET 19A

1 Translate each figure as directed:

a

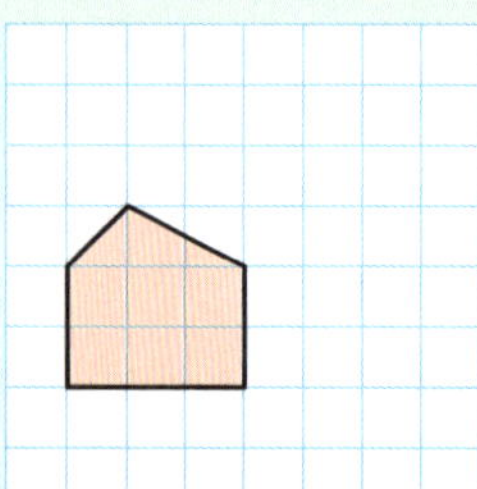

3 units right

b

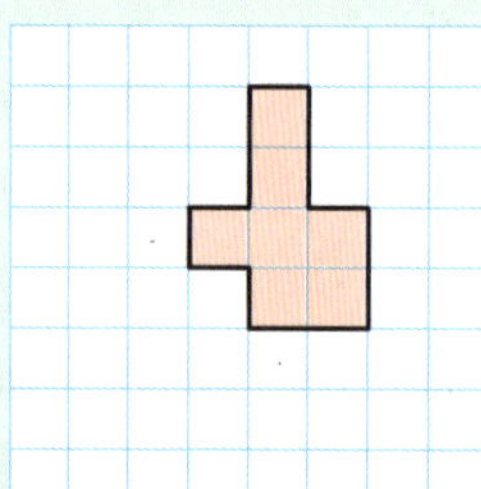

2 units left, 2 units down

c

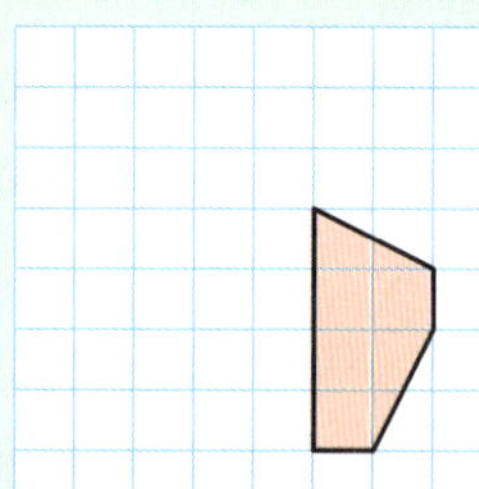

5 units left, 1 unit up

2 Suppose an object is translated 5 units to the right and 3 units down. Find:

a the point on the image corresponding to the object point $(-2, 6)$

b the point on the object corresponding to the image point $(-3, -7)$.

3 Reflect each figure in the mirror line:

a

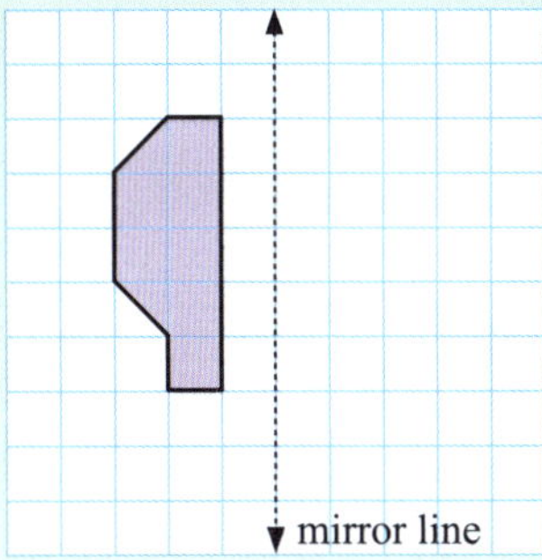

b

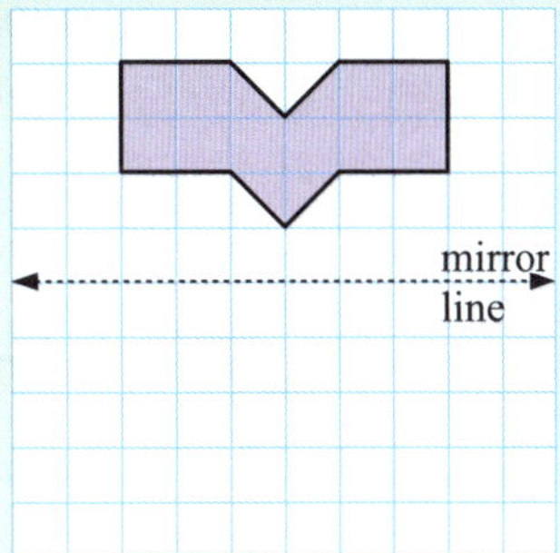

c

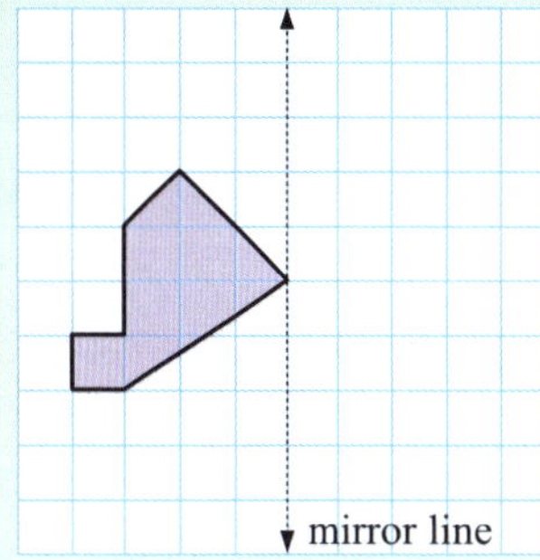

4 Rotate each figure about O through the angle indicated:

a

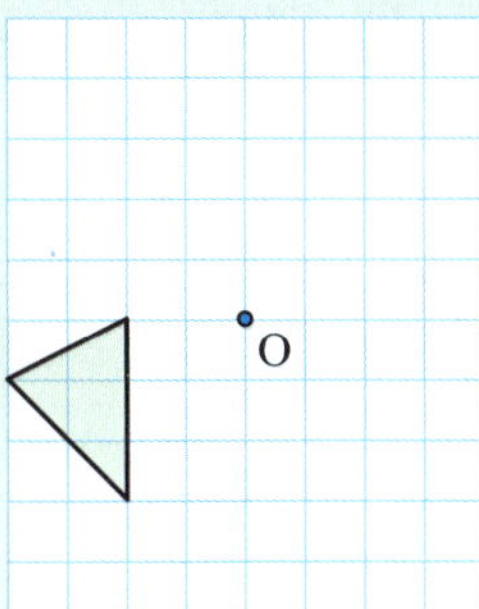

180°

b

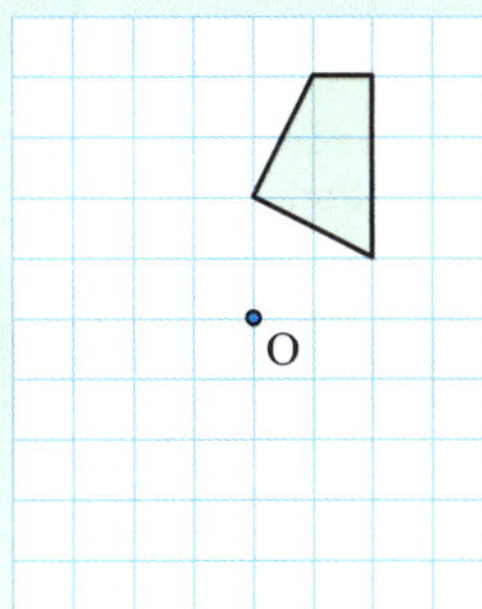

90°

c

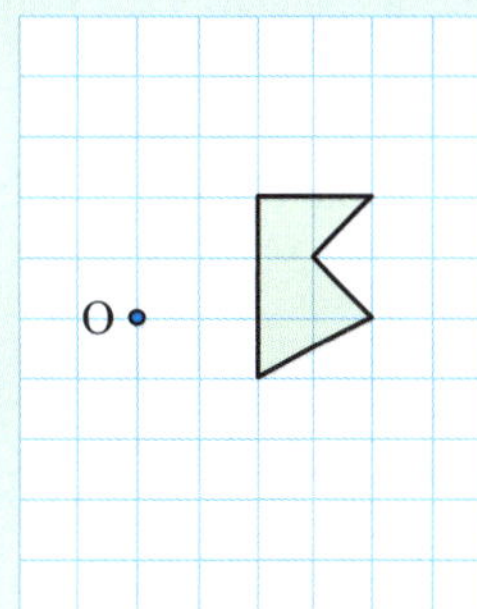

90° clockwise

5 **a** State the coordinates of the vertices of the quadrilateral PQRS.

b Rotate the quadrilateral 90° about the origin O.

c State the coordinates of the vertices of the image.

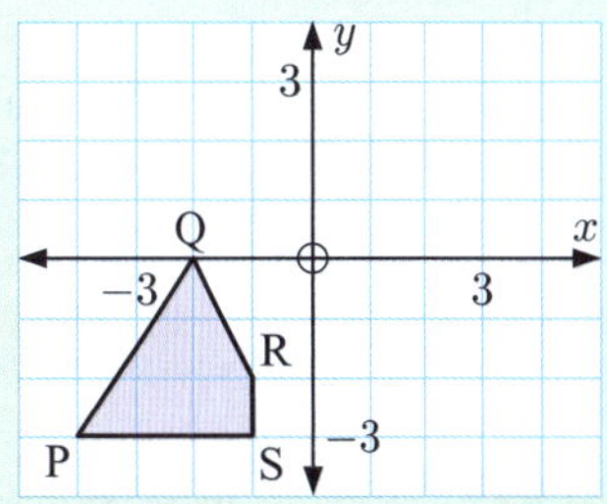

6 State whether each transformation from A to B is an enlargement or reduction, and find the scale factor.

a

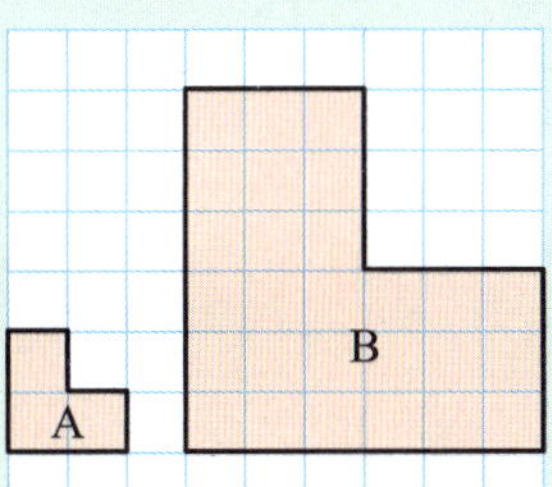

b

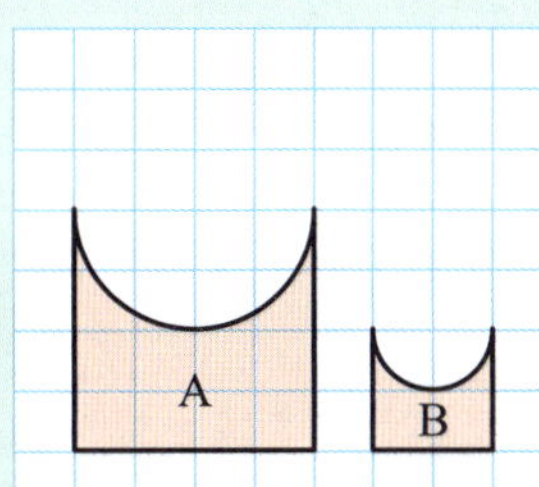

c

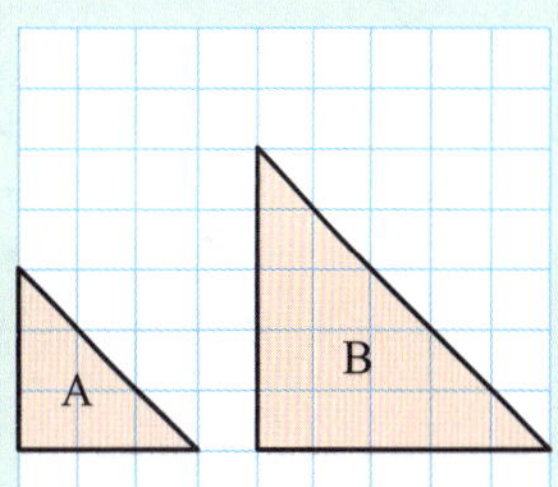

7 Write down the coordinates of the image of:

a $(3, -5)$ under an enlargement with scale factor 4 and centre O

b $(10, -2)$ under a reduction with scale factor $\frac{2}{5}$ and centre O.

8 Describe the stretch which maps the object onto its image:

a OAB onto OAB′

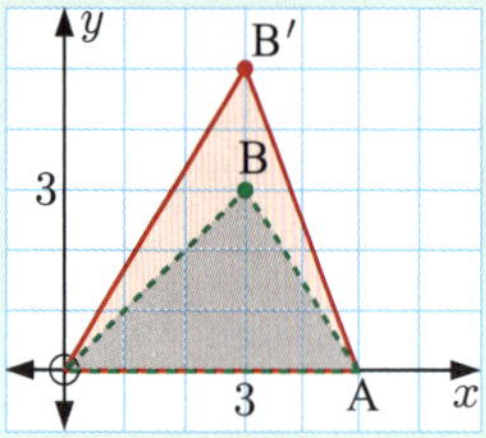

b ABCD onto A′B′C′D′

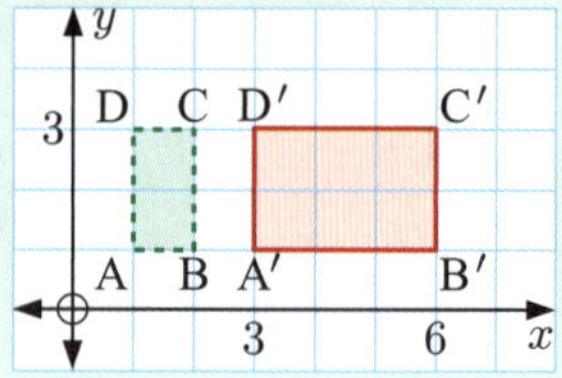

9 Triangle PQR has vertices P(2, 1), Q(2, 5), and R(6, 3). The triangle is horizontally stretched with scale factor k. Find the value of k such that the image:

a has area 40 units^2

b is an equilateral triangle.

10 Find the image when the figure below is rotated 90° clockwise about O, then reflected in the mirror line.

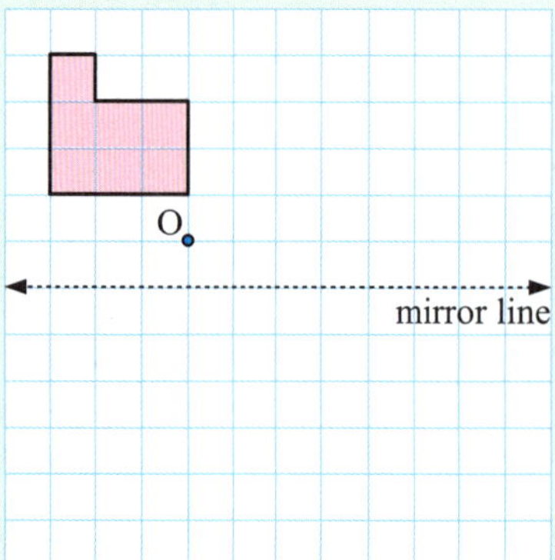

11 Reflect A in the line $y = -x$, then vertically stretch the result with scale factor $\frac{1}{2}$. State the coordinates of the image vertices.

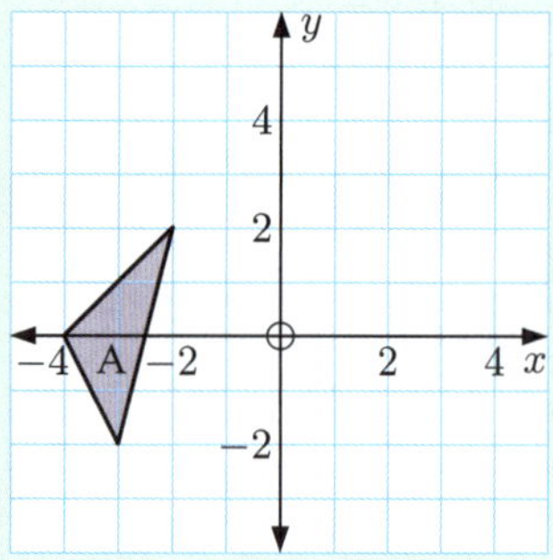

REVIEW SET 19B

1 Describe each translation from A to A′.

a

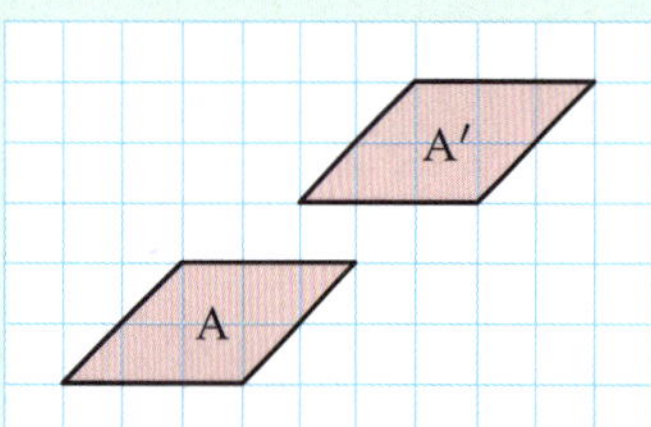

b

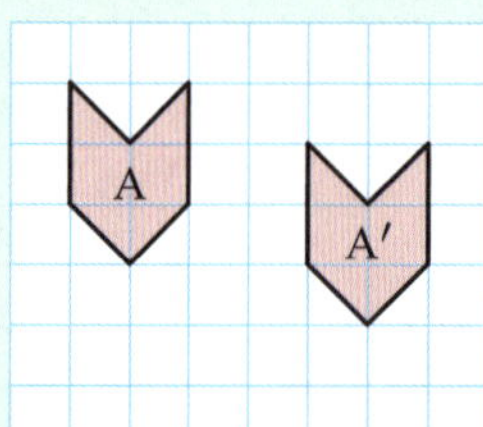

c

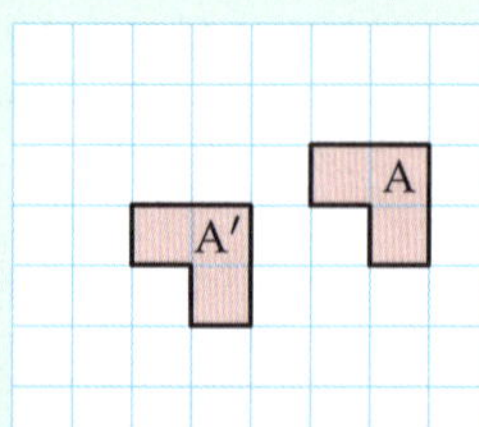

2 **a** Translate the quadrilateral ABCD 2 units right and 3 units down.

b State the coordinates of the vertices of the image.

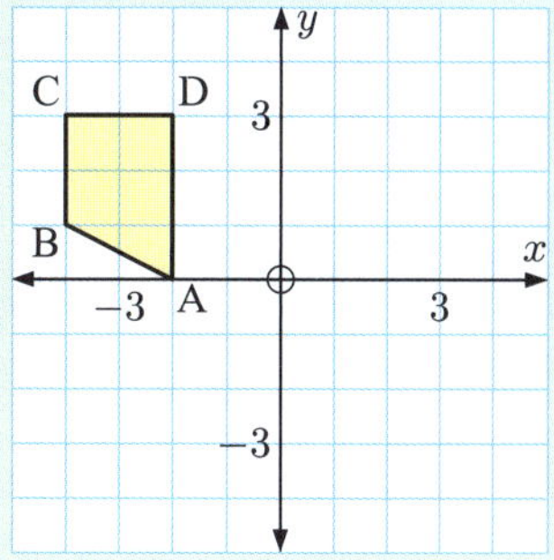

3 Reflect each figure in the line indicated:

a

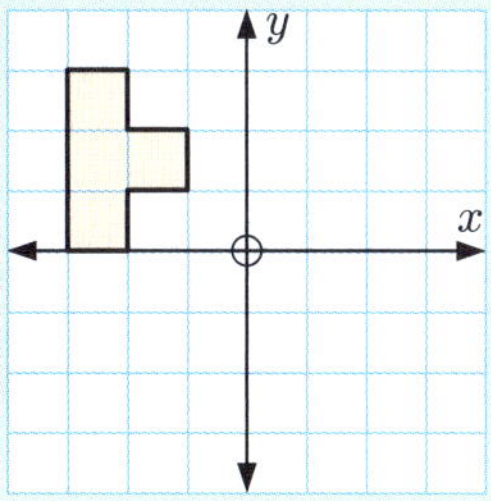

x-axis

b

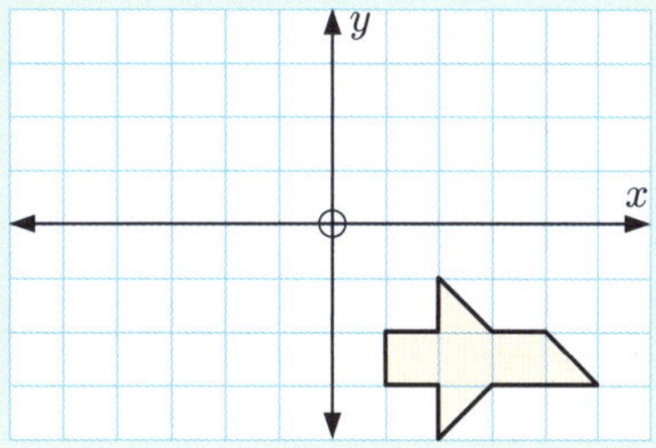

y-axis

c

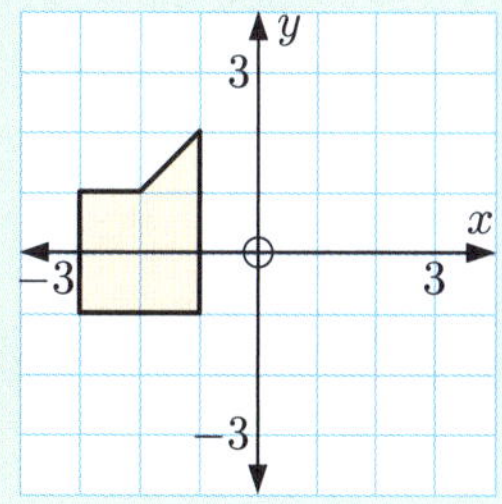

$y = x$

4 A quadrilateral has vertices A(−1, 3), B(1, 4), C(5, 0), and D(3, −2). The quadrilateral is reflected in the line $y = -x$.
Plot ABCD and its image on the same set of axes, and state the coordinates of the image vertices.

5 **a** Which figure is a rotation of figure A about O?

b Determine the angle of rotation from figure A to this figure.

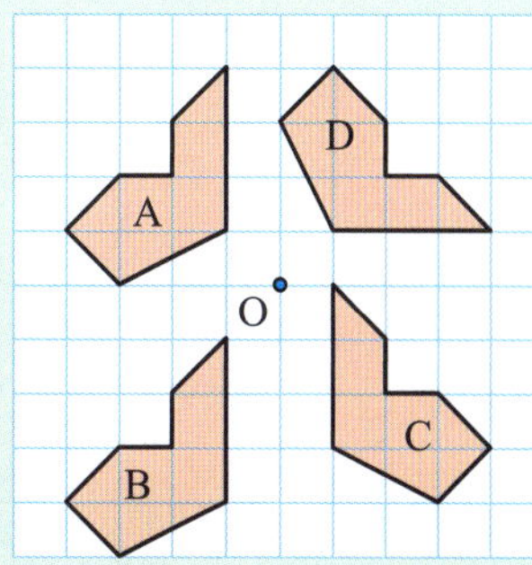

6 A triangle ABC has vertices A(−4, 1), B(−1, 1), and C(−3, 3).

a Plot triangle ABC on a Cartesian plane.

b Rotate the triangle 180° about the origin O.

c State the coordinates of the vertices of the image.

d Describe a rotation that will return the triangle to its original position.

7 In this diagram, the black figure has been reduced to form the red figure.
Copy the diagram and locate the centre O.

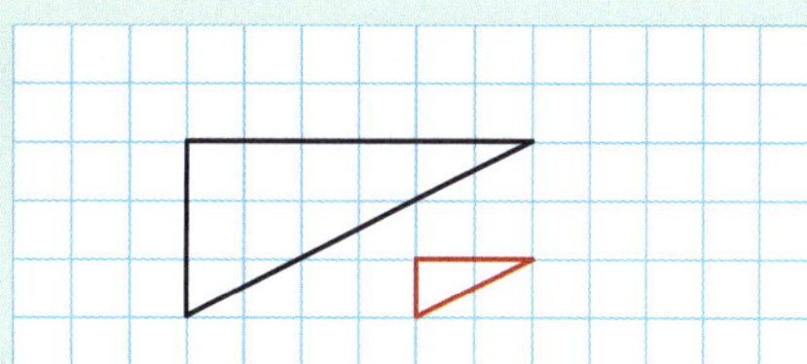

8 Draw the image of each figure for the given scale factor and centre O:

a scale factor 3

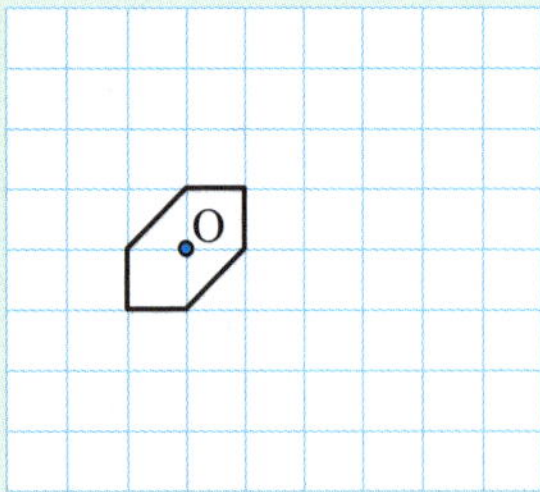

b scale factor $\frac{1}{3}$

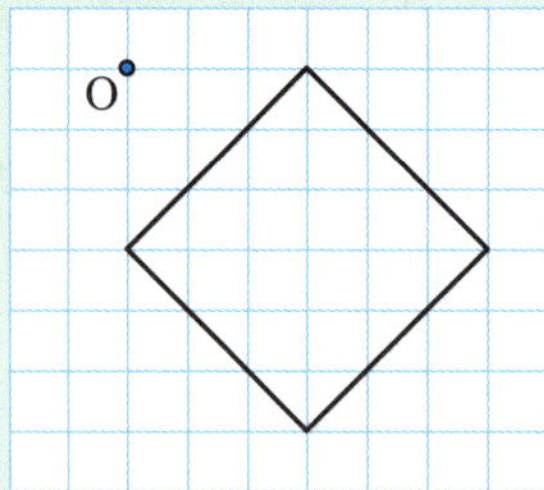

c scale factor $\frac{1}{2}$

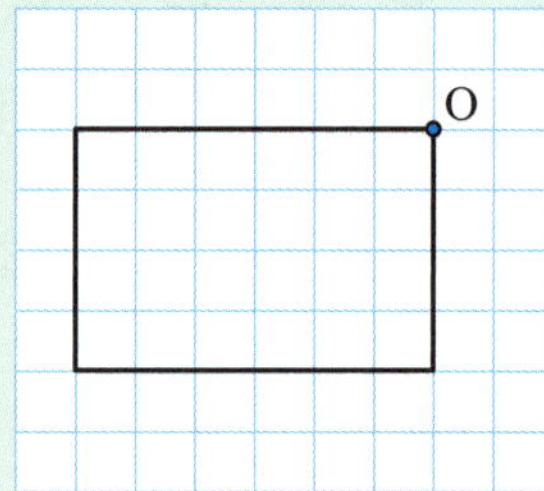

9 Describe the single transformation which maps:

a A onto B
b A onto C
c A onto D
d C onto D
e D onto B
f D onto A.

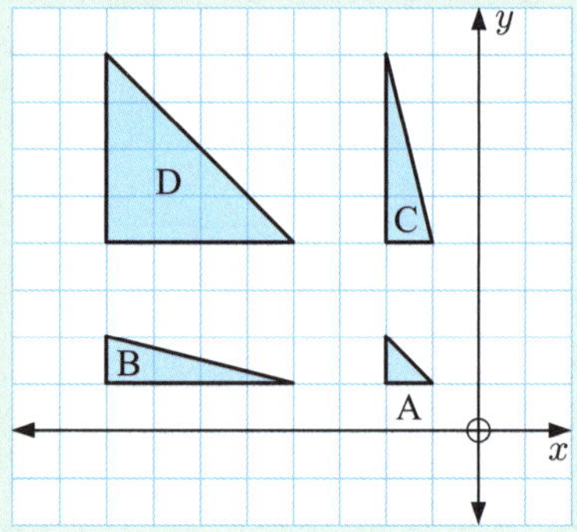

10 Copy the figure and draw its image under the given stretch:

a vertical stretch with scale factor $\frac{1}{2}$

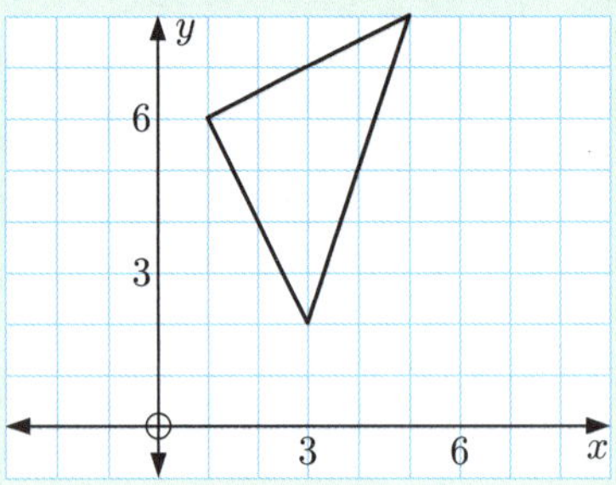

b horizontal stretch with scale factor $\frac{4}{3}$

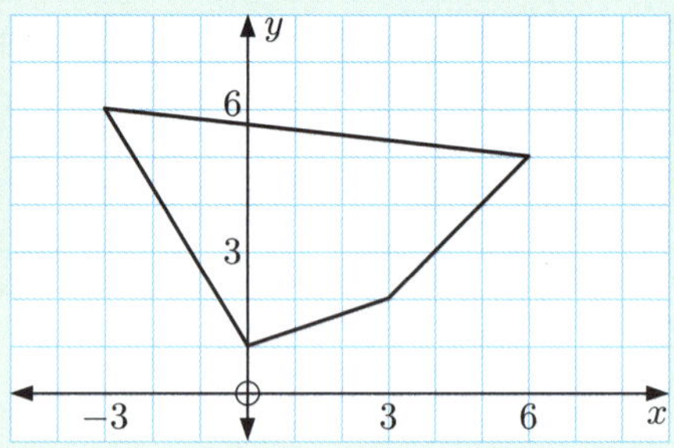

11 **a** Reduce A with scale factor $\frac{2}{3}$ and centre O, then translate the result 3 units right and 1 unit down to give A′.

b Describe the combination of transformations which maps A′ back onto A.

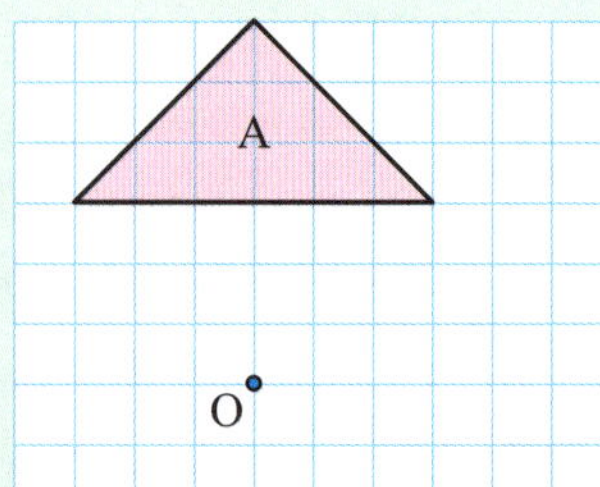

12 Triangle A has vertices P(1, 0), Q(4, 1), and R(4, 2).
The triangle is rotated 90° about O, then reflected in the y-axis to give triangle B.

a Plot triangle A and triangle B on the same set of axes, and state the coordinates of the vertices of triangle B.

b Describe a *single* transformation which maps A onto B.

Chapter 20

Quadratic equations

Contents:

OPENING PROBLEM

Matheus has a piece of paper $15 \text{ cm} \times 20 \text{ cm}$. His friend Gabriela challenges him to divide it into two sections of equal area, one of which is an L-shaped region with constant width x cm, as shown.

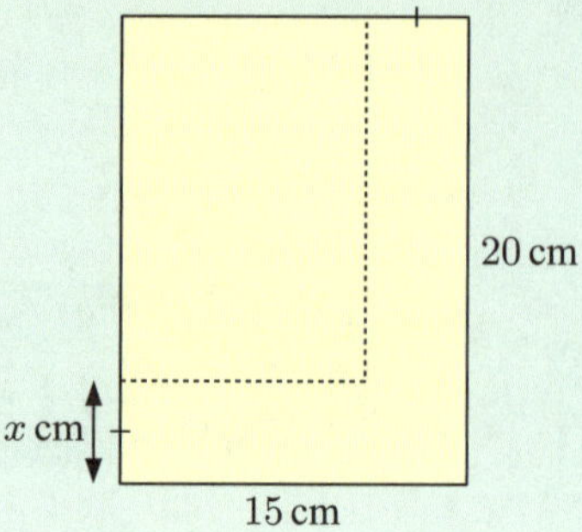

Things to think about:

- **a** What possible values can x take?
- **b** Can you explain why the area of the L-shaped region is given by $[15x + x(20 - x)]$ cm^2?
- **c** Can you hence explain why $35x - x^2 = 150$?
- **d** How many solutions does the equation $35x - x^2 = 150$ have?
- **e** What value of x should Matheus use?

A QUADRATIC EQUATIONS

We have previously solved many *linear* equations which can be written in the form $ax + b = 0$, $a \neq 0$. Linear equations have only *one* solution.

The equation in the **Opening Problem** is *not* a linear equation. We can write it as $x^2 - 35x + 150 = 0$ which has the form $ax^2 + bx + c = 0$, $a \neq 0$.

A **quadratic equation** is an equation which can be written in the form $ax^2 + bx + c = 0$, where a, b, and c are constants, $a \neq 0$.

For example:

- $2x^2 - 3x + 5 = 0$ and $x^2 - 3 = 2x$ are quadratic equations
- $x^2 + 5 + \frac{1}{x} = 0$ is not a quadratic equation.

Unlike linear equations, quadratic equations may have two, one, or zero solutions.

For example, look at the equations and their solutions in the table below:

Equation	$ax^2 + bx + c = 0$ *form*	*Solutions*	
$x^2 - 4 = 0$	$x^2 + 0x - 4 = 0$	$x = 2$ or $x = -2$	**two solutions**
$(x-2)^2 = 0$	$x^2 - 4x + 4 = 0$	$x = 2$	**one solution**
$x^2 + 4 = 0$	$x^2 + 0x + 4 = 0$	no real solutions	**zero solutions**

We can verify that the solutions to a quadratic equation are correct by substituting them into the equation.

For example, the quadratic equation $x^2 + 3x - 10 = 0$ has the solutions $x = 2$ and $x = -5$.

If $x = 2$,
$$\begin{aligned} & x^2 + 3x - 10 \\ &= 2^2 + 3 \times 2 - 10 \\ &= 4 + 6 - 10 \\ &= 0 \ \checkmark \end{aligned}$$

If $x = -5$,
$$\begin{aligned} & x^2 + 3x - 10 \\ &= (-5)^2 + 3 \times (-5) - 10 \\ &= 25 - 15 - 10 \\ &= 0 \ \checkmark \end{aligned}$$

EXERCISE 20A

1 State whether each of the following is a quadratic equation:

a $x^2 - x + 6 = 0$ **b** $2x^3 + 5x - 4 = 0$ **c** $4x - 7 = 8x + 2$

d $-6x^2 - 11 = 0$ **e** $3x^2 + x - \frac{2}{x} = 0$ **f** $\frac{2}{3}x^2 - \frac{1}{2}x = 0$

2 For each quadratic equation, determine which of the numbers in brackets are solutions.

a $x^2 - 7x + 6 = 0$ $\{1, 2, 4, 6, 7\}$ **b** $x^2 + 2x - 8 = 0$ $\{-5, -4, -1, 1, 2\}$

c $x^2 - 5x = 0$ $\{-2, 0, 3, 5, 7\}$ **d** $x^2 - 6x + 9 = 0$ $\{-3, 1, 3, 4, 6\}$

e $2x^2 + 3x - 2 = 0$ $\{-2, -1, -\frac{1}{2}, \frac{1}{2}, 1\}$

3 Explain why the quadratic equation $x^2 + 5 = 0$ has no real solutions.

B EQUATIONS OF THE FORM $x^2 = k$

In **Chapter 8** we studied **power equations** which can be written in the form $x^n = k$ where $n \neq 0$.

You should recognise that when $n = 2$, we also have a *quadratic* equation.

$$\text{If } x^2 = k \text{ then } \begin{cases} x = \pm\sqrt{k} & \text{if } k > 0 \\ x = 0 & \text{if } k = 0 \\ x \text{ has no real solutions if } k < 0. \end{cases}$$

This principle can be extended to other perfect squares, allowing us to solve equations of the form $(ax + b)^2 = k$ without having to expand the LHS.

Example 1 Self Tutor

Solve for x:

a $7 - 2x^2 = 3$ **b** $(x - 2)^2 = 25$ **c** $(x + 1)^2 = 7$

a
$$\begin{aligned} 7 - 2x^2 &= 3 \\ \therefore -2x^2 &= -4 \\ \therefore x^2 &= 2 \\ \therefore x &= \pm\sqrt{2} \end{aligned}$$

b
$$\begin{aligned} (x - 2)^2 &= 25 \\ \therefore x - 2 &= \pm\sqrt{25} \\ \therefore x - 2 &= \pm 5 \\ \therefore x &= 2 \pm 5 \\ \therefore x &= 7 \text{ or } -3 \end{aligned}$$

c
$$\begin{aligned} (x + 1)^2 &= 7 \\ \therefore x + 1 &= \pm\sqrt{7} \\ \therefore x &= -1 \pm \sqrt{7} \end{aligned}$$

In **c**, the two solutions are $x = -1 + \sqrt{7}$ and $x = -1 - \sqrt{7}$.

EXERCISE 20B

1 Solve for x:

a $x^2 = 16$ **b** $x^2 = 19$ **c** $2x^2 = 18$

d $12x^2 = 72$ **e** $3x^2 = -12$ **f** $4x^2 = 0$

g $2x^2 + 1 = 51$ **h** $1 - 3x^2 = 10$ **i** $13 = 7 + 2x^2$

2 Solve for x:

a $(x-2)^2=9$	**b** $(x+4)^2=25$	**c** $(x+3)^2=-1$
d $(x-4)^2=2$	**e** $(x+3)^2=-7$	**f** $(x-2)^2=0$
g $(x+5)^2=3$	**h** $(x-6)^2=2$	**i** $(x+8)^2=13$
j $(3x+5)^2=0$	**k** $(5x+1)^2=4$	**l** $4(2x-1)^2=8$

C THE NULL FACTOR LAW

DISCUSSION

- What result do you get when you multiply any number by zero?
- If you multiply two *non-zero* numbers together, can the result ever be zero?
- What can you say about two numbers if their product is zero?

The **null factor law** states that:

> If the product of two or more numbers is zero, then at least one of them must be zero.
> So, if $ab=0$ then $a=0$ or $b=0$.

For example:
- If $2xy=0$ then $x=0$ or $y=0$.
- If $x(x-2)=0$ then $x=0$ or $x-2=0$.
- If $xyz=0$ then $x=0$, $y=0$, or $z=0$.

Therefore, if we are given an equation where the LHS is fully factorised and the RHS is zero, we can use the null factor law to find its solutions.

Example 2 Self Tutor

Solve for x:

a $5x(x+2)=0$ **b** $(x+4)(x-1)=0$

a $5x(x+2)=0$

$\therefore\ 5x=0$ or $x+2=0$ {null factor law}

$\therefore\ x=0$ or $x=-2$ {solving linear equations}

b $(x+4)(x-1)=0$

$\therefore\ x+4=0$ or $x-1=0$ {null factor law}

$\therefore\ x=-4$ or $x=1$ {solving linear equations}

EXERCISE 20C

1 Explain what can be deduced from:

a $ac=0$	**b** $xy=0$	**c** $abc=0$	**d** $3x=0$
e $x(x-3)=0$	**f** $x^2=0$	**g** $x(x+6)=0$	**h** $x^2y=0$

2 Solve for x:

a $2x(x-1)=0$ **b** $x(x+5)=0$ **c** $3x(x+2)=0$

d $(x-1)^2=0$ **e** $-x(x-4)=0$ **f** $-2x(x+3)=0$

g $x(2x+1)=0$ **h** $3x(4x-3)=0$ **i** $-x(3x+5)=0$

3 Solve for x:

a $(x-1)(x-5)=0$ **b** $(x+2)(x-4)=0$ **c** $(x+3)(x+7)=0$

d $(x+7)(x-11)=0$ **e** $2x(x-8)=0$ **f** $(x+12)(x-5)=0$

g $-3x(x+7)=0$ **h** $(2x+1)(x-3)=0$ **i** $(x+6)(3x-1)=0$

j $(2x+1)(x+6)=0$ **k** $4(x-3)^2=0$ **l** $(x-31)(x+11)=0$

m $(x+4)(4x-1)=0$ **n** $-3x(7x+3)=0$ **o** $(2-x)(3x+4)=0$

D SOLVING BY FACTORISATION

We have seen that if we can write an equation with its LHS fully factorised and RHS zero, then we can solve it using the null factor law.

This gives us a method for solving quadratic equations:

Step 1: If necessary, rearrange the equation so one side is **zero**.
Step 2: **Fully factorise** the other side (usually the LHS).
Step 3: Apply the **null factor law**.
Step 4: **Solve** the resulting linear equations.

COMMON FACTORS

When we factorise an expression, the first thing we look for are common factors.

Example 3 Self Tutor

Solve for x: $x^2=4x$

$$x^2=4x$$
$$\therefore\ x^2-4x=0 \quad \{\text{making the RHS}=0\}$$
$$\therefore\ x(x-4)=0 \quad \{\text{factorising the LHS}\}$$
$$\therefore\ x=0 \text{ or } x-4=0 \quad \{\text{null factor law}\}$$
$$\therefore\ x=0 \text{ or } 4$$

WARNING ON INCORRECT CANCELLING

In the equation $x^2=4x$ from **Example 3**, we may be tempted to divide both sides by the common factor x, giving $x=4$.

While $x=4$ is a solution to the equation, we have lost the solution $x=0$.

From this example we conclude that:

We must never cancel a common factor involving a variable from both sides of an equation unless we are *sure* that this factor cannot be zero.

EXERCISE 20D.1

1 Solve for x:

a $x^2 - x = 0$ **b** $x^2 - 13x = 0$ **c** $x^2 + 8x = 0$
d $x^2 + 3x = 0$ **e** $2x + x^2 = 0$ **f** $5x - x^2 = 0$
g $12x - x^2 = 0$ **h** $x^2 + 7x = 0$ **i** $x^2 - 4x = 0$
j $2x^2 - 7x = 0$ **k** $3x^2 - 15x = 0$ **l** $2x^2 + 8x = 0$

2 Solve for x:

a $x^2 = 3x$ **b** $x^2 = 10x$ **c** $x^2 = -6x$
d $2x^2 = -7x$ **e** $5x = x^2$ **f** $8x = x^2$
g $3x^2 = 18x$ **h** $5x^2 = 6x$ **i** $7x + x^2 = x$
j $x^2 - 4 = 2x - 4$ **k** $2x^2 + x = 3x$ **l** $10x^2 - x = 2x + x^2$

3 Solve for x:

a $x(x-2) = 3(x-2)$ **b** $5(x+4) = x(x+4)$ **c** $x(6-x) = x - 6$

THE DIFFERENCE BETWEEN TWO SQUARES

Equations such as $x^2 - 4 = 0$ and $4x^2 - 25 = 0$ can be rearranged into the form $x^2 = k$ and then solved.

However, we can also solve them by factorising the LHS as the difference between two squares. This is a useful technique which can also be applied to more complicated equations.

Example 4 Self Tutor

Solve for x:

a $x^2 - 4 = 0$ **b** $4x^2 - 25 = 0$

a
$$x^2 - 4 = 0$$
$$\therefore (x+2)(x-2) = 0$$
$$\therefore x + 2 = 0 \text{ or } x - 2 = 0$$
$$\therefore x = -2 \text{ or } x = 2$$
$$\therefore x = \pm 2$$

b
$$4x^2 - 25 = 0$$
$$\therefore (2x)^2 - 5^2 = 0$$
$$\therefore (2x+5)(2x-5) = 0$$
$$\therefore 2x + 5 = 0 \text{ or } 2x - 5 = 0$$
$$\therefore 2x = -5 \text{ or } 2x = 5$$
$$\therefore x = \pm\tfrac{5}{2}$$

EXERCISE 20D.2

1 Solve for x:

a $x^2 - 16 = 0$ **b** $x^2 - 49 = 0$ **c** $x^2 - 144 = 0$
d $2x^2 - 8 = 0$ **e** $3x^2 - 27 = 0$ **f** $5x^2 - 20 = 0$
g $-3x^2 + 12 = 0$ **h** $-2x^2 + 8 = 0$ **i** $x^2 + 4 = 0$

2 Solve for x:

a $9x^2 - 4 = 0$ **b** $1 - 9x^2 = 0$ **c** $4 - 25x^2 = 0$
d $4x^2 - 1 = 0$ **e** $4x^2 - 9 = 0$ **f** $16x^2 - 25 = 0$

3 Solve for x:

a $x^2 - (x-2)^2 = 0$ **b** $(x+5)^2 - (x-1)^2 = 0$

c $(2x+1)^2 = (x-3)^2$ **d** $(3x-4)^2 = (x+2)^2$

OTHER FACTORISATIONS

In addition to common factors and the difference between two squares, we also look for **perfect squares** or else **sum and product** factorisations.

Example 5 Self Tutor

Solve for x:

a $x^2 = x + 12$ **b** $x^2 + 4 = 4x$

a
$$x^2 = x + 12$$
$$\therefore\ x^2 - x - 12 = 0 \quad \{\text{making RHS} = 0\}$$
$$\therefore\ (x-4)(x+3) = 0 \quad \{\text{the numbers } -4 \text{ and } 3 \text{ have sum } -1 \text{ and product } -12\}$$
$$\therefore\ x - 4 = 0 \ \text{ or } \ x + 3 = 0$$
$$\therefore\ x = 4 \text{ or } -3$$

b
$$x^2 + 4 = 4x$$
$$\therefore\ x^2 - 4x + 4 = 0 \quad \{\text{making RHS} = 0\}$$
$$\therefore\ x^2 - 2 \times x \times 2 + 2^2 = 0$$
$$\therefore\ (x-2)^2 = 0 \quad \{\text{perfect square factorisation}\}$$
$$\therefore\ x - 2 = 0$$
$$\therefore\ x = 2$$

EXERCISE 20D.3

1 Solve for x:

a $x^2 - 7x + 10 = 0$ **b** $x^2 + 6x + 8 = 0$

c $x^2 + 11x + 10 = 0$ **d** $x^2 - 8x + 12 = 0$

e $x^2 - 5x + 4 = 0$ **f** $x^2 - 11x + 24 = 0$

g $x^2 + 10x + 25 = 0$ **h** $x^2 - 3x - 18 = 0$

i $x^2 + 7x - 18 = 0$ **j** $x^2 - 22x + 121 = 0$

k $x^2 - 6x + 9 = 0$ **l** $x^2 - 5x - 6 = 0$

m $x^2 + 11x - 60 = 0$ **n** $x^2 + 18x - 63 = 0$

o $x^2 - 12x - 64 = 0$ **p** $x^2 - 19x + 70 = 0$

2 Solve for x:

a $x^2 - 14x = 15$ **b** $x^2 + 2 = 3x$ **c** $x^2 = 3x + 28$

d $x^2 = 20 + x$ **e** $8 = x^2 + 7x$ **f** $x^2 = 5x + 24$

g $x^2 = 4x + 45$ **h** $x^2 + 1 = 2x$ **i** $x^2 = 19x + 20$

3 Solve for x:

a $x(x+2) = 15$ **b** $x(x-2) = 5(x+12)$ **c** $2x - 6 = x(x-5)$

d $x^2 - 4 = x + 2$ **e** $2(x+5) = x^2 + 11$ **f** $5 - x^2 = 2x - 3$

Example 6 — Self Tutor

Solve for x: $3x^2 + 21x + 30 = 0$

$$3x^2 + 21x + 30 = 0 \quad \{\text{HCF} = 3\}$$
$$\therefore\ 3(x^2 + 7x + 10) = 0$$
$$\therefore\ 3(x+2)(x+5) = 0 \quad \{\text{the numbers 2 and 5 have sum 7 and product 10}\}$$
$$\therefore\ x + 2 = 0 \text{ or } x + 5 = 0$$
$$\therefore\ x = -2 \text{ or } -5$$

4 Solve for x:

a $2x^2 + 4x - 30 = 0$ **b** $-x^2 + 12x - 36 = 0$ **c** $3x^2 + 3x - 60 = 0$

d $-3x^2 + 21x - 36 = 0$ **e** $5x^2 - 5x - 210 = 0$ **f** $-4x^2 - 32x - 48 = 0$

5 Solve for x:

a $2x^2 + 2x = 24$ **b** $3x^2 = 30x - 48$ **c** $5x^2 = 20(x + 8)$

ACTIVITY

Click on the icon for more practice solving quadratic equations.

LEARNING ALGEBRA

PUZZLE — QUADRATIC EQUATIONS MAZE

Your task is to find your way from the Start to the Finish of the maze. You may move horizontally or vertically, but not diagonally.

You may only move from one square to another if the quadratic equations in the two squares share a common solution.

PRINTABLE MAZE

Start (top right square)

$x^2 + x - 18 = 4x$	$2x^2 - 12x = 0$	$x^2 - x = 2$	$x^2 - 8x + 15 = 0$
$(x+3)(x-7) = 0$	$x^2 = x$	$x^2 + 4 = 5x$	$(x-3)(x+2) = 0$
$x^2 + 8x + 12 = x$	$x^2 + 12 = 7x$	$-x^2 + 8x - 16 = 0$	$5x^2 - 40 = 10x$
$-2x^2 + 10x + 72 = 0$	$x(x-7) = 18$	$x^2 - 25 = 0$	$x^2 - 9x = 70$
$2x^2 - x - 20 = x^2$	$x^2 = -4(x+1)$	$3x^2 - 30 = 9x$	$x^2 - 8x + 42 = 9x$

Finish (bottom right square)

E PROBLEM SOLVING

When practical problems are converted into algebraic form, a quadratic equation may result. To solve these problems, follow these steps:

Step 1: Carefully **read the question** until you understand the problem. A **sketch** may be useful.
Step 2: Decide on the **unknown** quantity and label it with a variable such as x.
Step 3: Use the information given to write an **equation**.
Step 4: **Solve** the equation.
Step 5: **Check** that any solutions satisfy the equation and are reasonable.
Step 6: Write your answer to the question in **sentence form**.

Example 7 Self Tutor

The difference between a number and its square is 20. Find the number.

Let the number be x, so its square is x^2.

$x^2 - x = 20$
$\therefore\ x^2 - x - 20 = 0$ {making RHS = 0}
$\therefore\ (x-5)(x+4) = 0$ {factorising}
$\therefore\ x = 5$ or $x = -4$

So, the number is -4 or 5.

Check: The square of -4 is 16, and the difference between -4 and 16 is 20. ✓
The square of 5 is 25, and the difference between 5 and 25 is 20. ✓

Example 8 Self Tutor

Two numbers have a sum of 10, and the sum of their squares is 58. Find the numbers.

Let one of the numbers be x, so the other number is $10 - x$.

$x^2 + (10-x)^2 = 58$
$\therefore\ x^2 + 100 - 20x + x^2 = 58$ {expanding}
$\therefore\ 2x^2 - 20x + 42 = 0$ {making RHS = 0}
$\therefore\ 2(x^2 - 10x + 21) = 0$ {HCF = 2}
$\therefore\ 2(x-3)(x-7) = 0$ {factorising}
$\therefore\ x - 3 = 0$ or $x - 7 = 0$
$\therefore\ x = 3$ or 7

If $x = 3$, $10 - x = 7$.
If $x = 7$, $10 - x = 3$.

$\therefore$ the numbers are 3 and 7.

EXERCISE 20E

1 The sum of a number and its square is 42. Find the number.

2 When a number is squared, the result is five times the original number. Find the number.

3 When a number is subtracted from its square, the result is 56. Find the number.

4 Two numbers have a sum of 9, and the sum of their squares is 153. Find the numbers.

5 The product of two consecutive integers is 156. Find the integers.

6 The sum of the squares of two consecutive odd numbers is 290. Find the numbers.

7 Two numbers differ by 4 and their product is 221. Find the numbers.

Example 9 **Self Tutor**

A rectangle has length 3 cm greater than its width, and the area of the rectangle is 28 cm^2.
Find the dimensions of the rectangle.

If the rectangle has width x cm, then its length is $(x+3)$ cm.

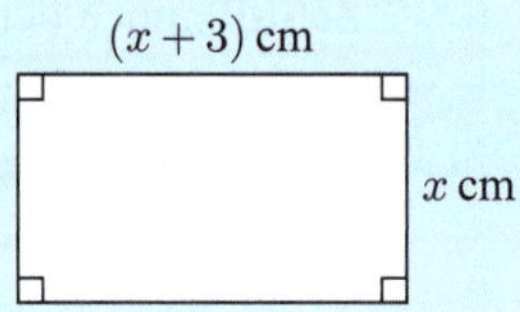

$\therefore\ x(x+3) = 28$

$\therefore\ x^2 + 3x = 28$ {expanding}

$\therefore\ x^2 + 3x - 28 = 0$ {making RHS $= 0$}

$\therefore\ (x+7)(x-4) = 0$ {factorising}

$\therefore\ x + 7 = 0$ or $x - 4 = 0$

$\therefore\ x = -7$ or 4

$\therefore\ x = 4$ {lengths must be positive}

$\therefore$ the rectangle is 4 cm $\times$ 7 cm.

8 The length of a rectangle is 4 cm more than its width, and the area of the rectangle is 96 cm^2. Find the width of the rectangle.

9

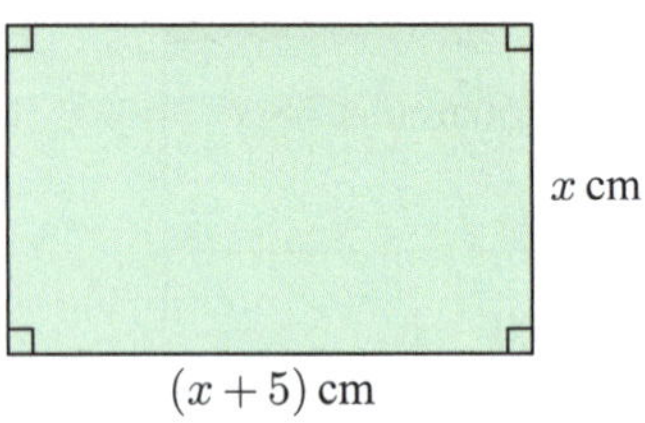

a Write an expression for this rectangle's:
  i area   ii perimeter.

b If the perimeter is 21.6 cm, find the area of the rectangle.

c If the area is 176 cm^2, find the perimeter of the rectangle.

10 A triangle has altitude 4 cm less than its base. The area of the triangle is $38\frac{1}{2}$ cm^2. Find the length of its base.

11 A rectangle has sides which differ in length by 3 cm. Its area is 154 cm^2. Find its perimeter.

12

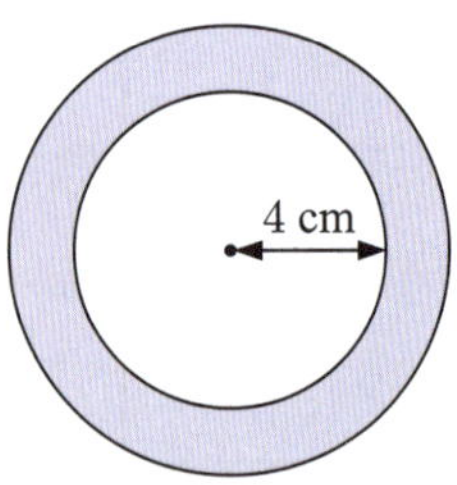

The shaded annulus has area 33π cm^2.
Find the radius of the outer circle.

Example 10 Self Tutor

Find the value of x:

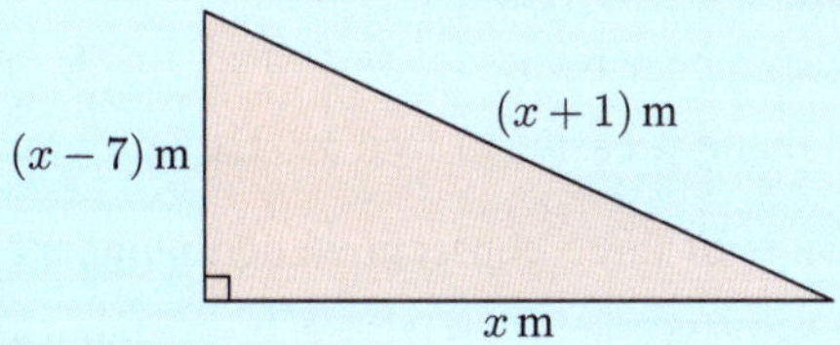

Using Pythagoras' theorem,

$$x^2 + (x-7)^2 = (x+1)^2$$
$$\therefore\ \cancel{x^2} + x^2 - 14x + 49 = \cancel{x^2} + 2x + 1$$
$$\therefore\ x^2 - 16x + 48 = 0 \qquad \{\text{making RHS} = 0\}$$
$$\therefore\ (x-4)(x-12) = 0 \qquad \{\text{factorising LHS}\}$$
$$\therefore\ x - 4 = 0 \ \text{ or } \ x - 12 = 0$$
$$\therefore\ x = 4 \text{ or } 12$$

But $x - 7 > 0$ as lengths must be positive.

$$\therefore\ x = 12$$

13 Find the value of x:

a

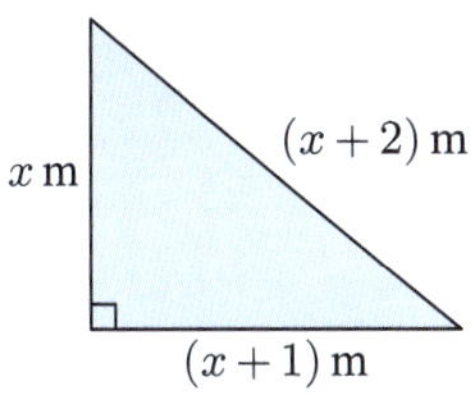

b

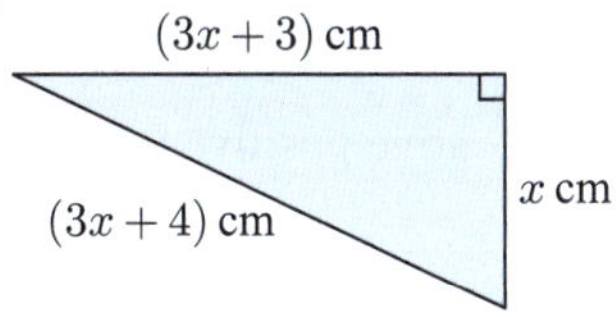

c

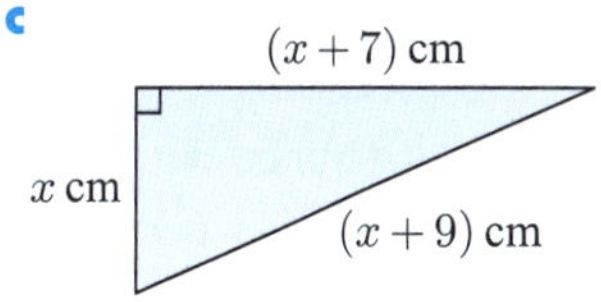

d

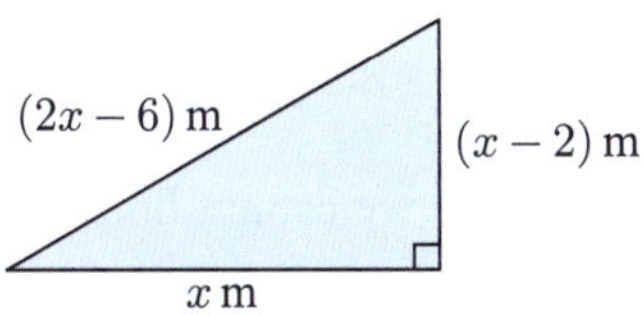

e

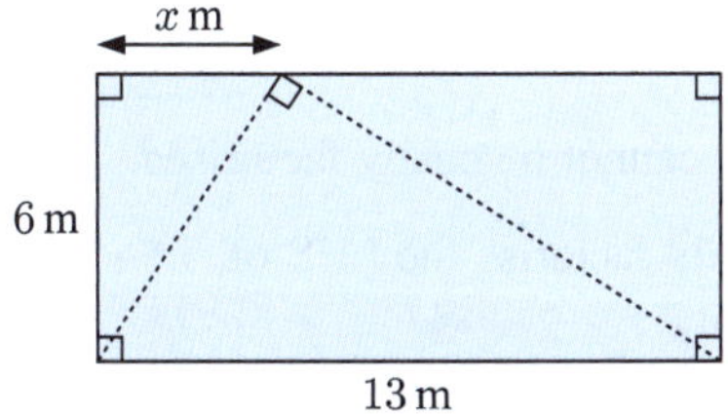

14 In a right angled triangle, the hypotenuse is 6 cm longer than the shortest side. The third side is the average of the other two. Find the:

a perimeter of the triangle **b** area of the triangle.

15 In the figure alongside, the two shaded triangles have equal area. Find the length of [BX].

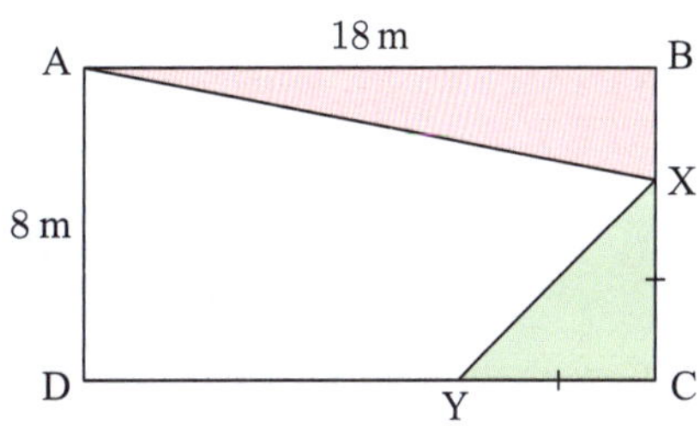

16 Consider the equation $x^2 + bx + c = 0$. Find the values of b and c if the equation has:

a solutions $x = 2$ and $x = -5$ **b** only one solution $x = -\frac{3}{2}$.

17 The sum of the squares of *three* consecutive integers is 110.

a Find the three integers by letting:

i the *smallest* integer be x **ii** the *middle* integer be x.

b Which method made it easier to find x? Explain your answer.

18 The sum of the squares of three consecutive odd numbers is 371. Find the numbers.

19

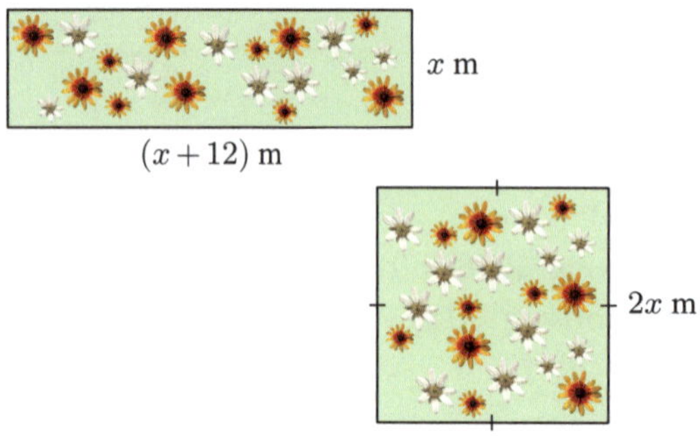

Alongside are the dimensions of two flower beds with equal area.

a Find their dimensions.

b Find the total length of border needed to enclose the two flower beds.

20 A circle is drawn so that it just fits in the corner of a page. Point A is 1 cm from one edge of the page, and 2 cm from the other edge.
Find the radius of the circle.

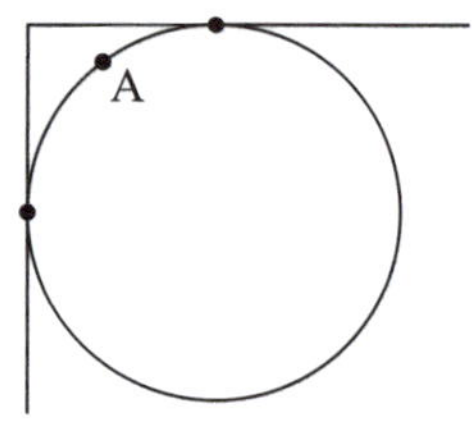

21 The difference between two consecutive positive perfect cubes is 217.
Find the larger perfect cube.

F COMPLETING THE SQUARE

Many quadratic expressions cannot be easily factorised.

For example, we cannot easily factorise the LHS of $x^2 + 10x - 1 = 0$ since we cannot find two integers with sum 10 and product -1.

An alternative method is to rewrite the equation in the form $(x - h)^2 = k$. We call this process **completing the square**.

For the equation $x^2 + 10x - 1 = 0$, we use this procedure:

$$x^2 + 10x - 1 = 0$$
$$\therefore \quad x^2 + 10x = 1 \qquad \{\text{moving constant to RHS}\}$$
$$\therefore \quad x^2 + 10x + 25 = 1 + 25 \qquad \{\text{completing a perfect square on LHS}\}$$
$$\therefore \quad (x + 5)^2 = 26 \qquad \{\text{factorising LHS}\}$$
$$\therefore \quad x + 5 = \pm\sqrt{26}$$
$$\therefore \quad x = -5 \pm \sqrt{26}$$

DISCUSSION

How do we know what number to add to both sides to complete the perfect square?

The perfect square expansion law tells us that $(x+a)^2 = x^2 + 2ax + a^2$.

> To complete a perfect square, the number we must add to both sides is found by **halving the coefficient of x, then squaring this value.**

Example 11 — Self Tutor

For each equation:

- **i** What number must be added to both sides to complete a perfect square on the LHS?
- **ii** Hence write the equation in the form $(x-h)^2 = k$.

a $x^2 + 2x = 6$ **b** $x^2 - 8x = -3$

a **i** In $x^2 + 2x = 6$, half the coefficient of x is $\frac{2}{2} = 1$.

$\therefore$ we add 1^2 to both sides.

ii The equation becomes $x^2 + 2x + 1^2 = 6 + 1^2$

$\therefore (x+1)^2 = 6 + 1$

$\therefore (x+1)^2 = 7$

b **i** In $x^2 - 8x = -3$, half the coefficient of x is $\frac{-8}{2} = -4$.

$\therefore$ we add $(-4)^2 = 4^2$ to both sides.

ii The equation becomes $x^2 - 8x + 4^2 = -3 + 4^2$

$\therefore (x-4)^2 = -3 + 16$

$\therefore (x-4)^2 = 13$

EXERCISE 20F

1 For each equation:

- **i** What number must be added to both sides to complete a perfect square on the LHS?
- **ii** Hence write the equation in the form $(x-h)^2 = k$.

a $x^2 + 2x = 2$ **b** $x^2 - 2x = 4$ **c** $x^2 + 6x = -4$

d $x^2 - 6x = 7$ **e** $x^2 + 10x = -23$ **f** $x^2 + 8x = -13$

g $x^2 + 6x = 2$ **h** $x^2 + 3x = -1$ **i** $x^2 - 7x = 5$

Example 12 — Self Tutor

Solve for x by completing the square, giving your answer in simplest surd form: $x^2 - x - 7 = 0$

$$x^2 - x - 7 = 0$$

$$\therefore x^2 - x = 7$$

$$\therefore x^2 - x + \left(\tfrac{1}{2}\right)^2 = 7 + \left(\tfrac{1}{2}\right)^2 \quad \text{\{completing the square\}}$$

$$\therefore \left(x - \tfrac{1}{2}\right)^2 = 7 + \tfrac{1}{4}$$

$$\therefore \left(x - \tfrac{1}{2}\right)^2 = \tfrac{29}{4}$$

$$\therefore x - \tfrac{1}{2} = \pm\sqrt{\tfrac{29}{4}}$$

$$\therefore x = \tfrac{1}{2} \pm \tfrac{\sqrt{29}}{2}$$

2 Solve for x by completing the square, giving your answer in simplest surd form:

a $x^2 - 4x - 2 = 0$ **b** $x^2 + 2x - 6 = 0$ **c** $x^2 - 2x - 2 = 0$

d $x^2 + 4x - 1 = 0$ **e** $x^2 + 6x - 11 = 0$ **f** $x^2 - 6x + 1 = 0$

g $x^2 + 8x + 5 = 0$ **h** $x^2 + x - 3 = 0$ **i** $x^2 - 3x - 5 = 0$

Example 13 **Self Tutor**

If possible, solve for x by completing the square: $x^2 - 2x + 7 = 0$

$$x^2 - 2x + 7 = 0$$
$$\therefore\ x^2 - 2x = -7$$
$$\therefore\ x^2 - 2x + 1^2 = -7 + 1^2 \quad \{\text{completing the square}\}$$
$$\therefore\ (x-1)^2 = -6$$

$\therefore$ there are no real solutions as $(x-1)^2 \geqslant 0$ for all $x \in \mathbb{R}$.

3 If possible, solve for x by completing the square:

a $x^2 - 2x + 3 = 0$ **b** $x^2 + 4x + 1 = 0$ **c** $x^2 - 4x + 5 = 0$

d $x^2 + 6x - 2 = 0$ **e** $x^2 - x - 4 = 0$ **f** $x^2 + 5x + 8 = 0$

4 Solve $2x^2 + 6x - 1 = 0$ by first dividing both sides by 2, and then completing the square.

5 Solve for x by completing the square:

a $2x^2 + 8x + 1 = 0$ **b** $3x^2 - 6x + 2 = 0$ **c** $5x^2 + 10x + 1 = 0$

MULTIPLE CHOICE QUIZ

QUICK QUIZ

REVIEW SET 20A

1 State whether each of the following is a quadratic equation:

a $4x^2 - 8x = 0$ **b** $x^2 = x + \dfrac{1}{x}$ **c** $2x + 7 = 0$

2 Which of the numbers $\{-3, -2, -1, 0, 1, 2, 3, 4\}$ satisfy the quadratic equation $x^2 = 6 + x$?

3 Solve for x:

a $3x^2 = 48$ **b** $65 - 2x^2 = 15$

4 Solve for x:

a $(x+1)^2 = 36$ **b** $(x-3)^2 = -16$ **c** $(3x-1)^2 = 5$

5 Explain what can be deduced from:

a $mn = 0$ **b** $-7p = 0$ **c** $ab^2 = 0$

6 Solve for x:

a $x(3x+2)=0$ **b** $2(x+2)(x-8)=0$ **c** $-2(2x-1)^2=0$

7 Solve for x:

a $4x^2+8x=0$ **b** $x(x+3)=4(x+3)$ **c** $4x^2-81=0$

8 Solve for x:

a $(2x+3)^2-(x+1)^2=0$ **b** $x^2-5x-24=0$

c $x^2=7x+18$ **d** $x(x-3)=10$

e $2x^2=8(x+3)$ **f** $(x+2)(x-3)=2x(x-3)-2$

g $9x^2-27x-90=0$ **h** $(x+1)(x-1)=2(7x-17)$

9 When the square of a number is subtracted from the original number, the result is -110. Find the number.

10 The difference between two numbers is 3, and the sum of their squares is 5. Find the numbers.

11 A square has sides of length $3x$ cm. A rectangle is $2x$ cm by $(x+7)$ cm. The area of the square is twice the area of the rectangle. Find the dimensions of each figure.

12 Find the value of x, and hence find the perimeter of the triangle.

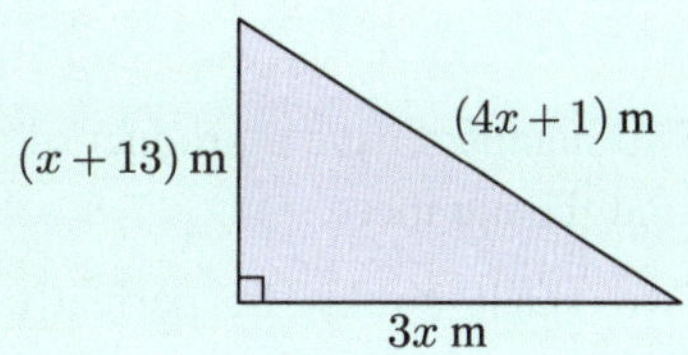

13 Answer the **Opening Problem** on page **314**.

14 Solve for x by completing the square:

a $x^2-2x-10=0$ **b** $x^2+5x-2=0$

15 **a** Solve the equation $x^2+3x=1$ by completing the square.

b Show that the equation $x^2+3x=-4$ has no real solutions.

c For what value(s) of k does the equation $x^2+3x=k$ have:

i two real solutions **ii** one real solution **iii** no real solutions?

REVIEW SET 20B

1 What can be deduced from $2pqr=0$?

2 Determine which of the numbers $\{-\frac{3}{5}, -1, 0, 1, \frac{3}{5}\}$ are solutions to the quadratic equation $5x^2-2x-3=0$.

3 Explain why the quadratic equation $x^2+9=0$ has no real solutions.

4 Solve for x:

a $4x^2-15=85$ **b** $(2x+3)^2=3$ **c** $2x^2+25=16$

5 Solve for x:

a $-(x-5)(x+3)=0$ **b** $5x(4x+3)=0$ **c** $-3(5-x)(2x-1)=0$

6 Solve for x:

a $2x^2 + x = 0$ **b** $7x^2 = 14x$ **c** $5(x-2) = 3x(2-x)$

7 Solve for x:

a $4x^2 - 64 = 0$ **b** $25x^2 - 9 = 0$

8 Solve for x:

a $x^2 = 14x - 33$ **b** $x^2 + 8x + 16 = 0$

9 Solve for x:

a $x^2 - 9 = x + 3$ **b** $x(x+4) = 8x - 3$

10 Solve for x:

a $5x^2 - 20x = 25$ **b** $x(x-2) = 48$ **c** $2x^2 + 4x = x^2 + 45$

11 The sum of the squares of three consecutive even numbers is 308. What are the integers?

12 The rectangle's area is 60 m^2 more than the area of the square. Find the area of the rectangle.

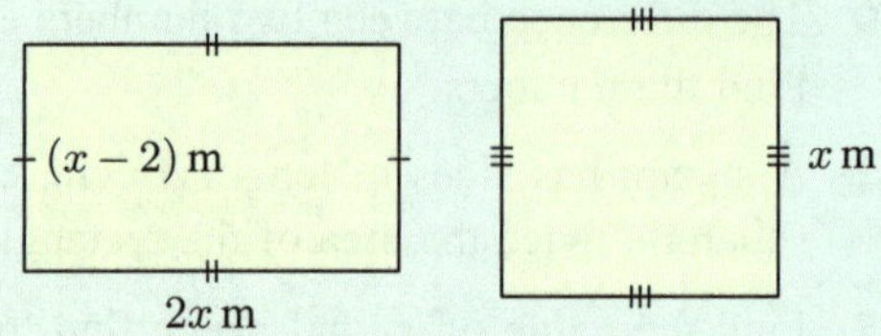

13 Two numbers have a product of -6. One of the numbers is 5 more than the other. Find the numbers.

14 Two small circles fit perfectly inside a large circle of diameter 20 m. The total shaded area is 52π m^2. Find the radius of each of the smaller circles.

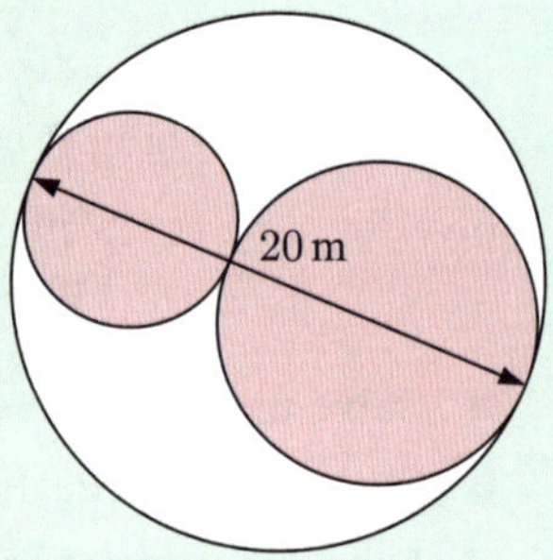

15 If possible, solve for x by completing the square:

a $x^2 - 10x - 2 = 0$ **b** $x^2 - 5x + 7 = 0$

Chapter 21

Quadratic functions

Contents:

OPENING PROBLEM

Stunt motorcyclist Marvin is attempting a world record jump of 110 metres.

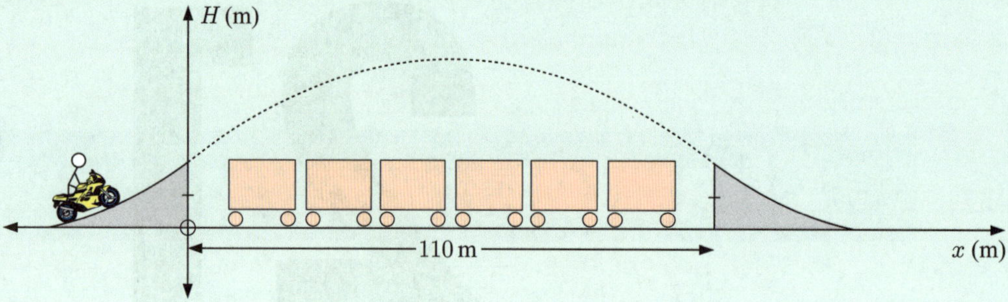

His height above ground at the point where he has travelled x metres horizontally from the take-off ramp, is given by $H = -0.009x^2 + x + 6$ metres, for $0 \leqslant x \leqslant 110$.

Things to think about:

- **a** How can we *describe* the relationship between H and x?
- **b** How high is Marvin above the ground:
  - **i** the instant he leaves the take-off ramp
  - **ii** after travelling 80 metres horizontally?
- **c** Will Marvin safely reach the landing ramp?
- **d** Can you use technology to find the maximum height of Marvin's jump?

In **Chapter 17** we studied straight lines. We described the relationship $y = mx + c$ between the variables as *linear*.

We now consider relationships which can be written as *quadratic* expressions, and the graphs of these relationships which are curves called *parabolas*.

HISTORICAL NOTE — CONIC SECTIONS

Conic sections are curves which can be obtained by cutting a cone with a plane. The Ancient Greek mathematicians were fascinated by conic sections.

The name *parabola* comes from the Greek word for "thrown" because when an object is thrown, its path makes a parabolic arc.

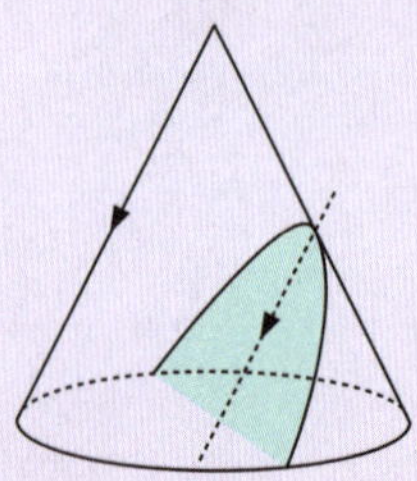

There are many other examples of parabolas in everyday life. For example, parabolic mirrors are used in car headlights, heaters, satellite dishes, and radio telescopes, because of their special geometric properties.

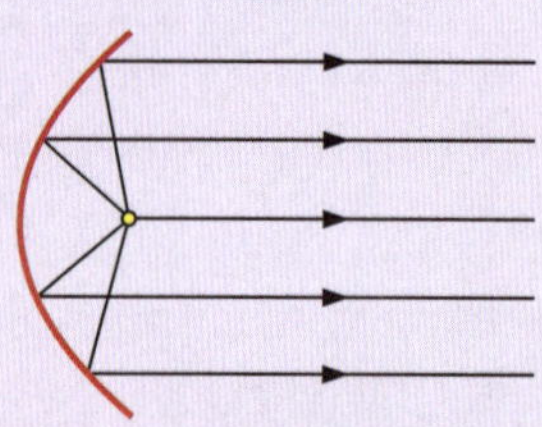

You may like to explore the conic sections for yourself using the software.

CONIC SECTIONS

A QUADRATIC FUNCTIONS

A **quadratic function** is a relationship between two variables which can be written in the form $y = ax^2 + bx + c$ where a, b, c are constants, $a \neq 0$.

For example:

- $y = 3x^2 - 2x + 1$ is a quadratic function with $a = 3$, $b = -2$, $c = 1$
- $y = 2x^2 + 3x$ is a quadratic function with $a = 2$, $b = 3$, $c = 0$
- $y = x^3 - 4x + 7$ is *not* a quadratic function since it has an x^3 term.

FINDING y GIVEN x

For any given value of x, the corresponding value of y can be found by substituting into the function.

Example 1 Self Tutor

Suppose $y = 2x^2 - 3x + 4$. Find the value of y when:

a $x = 2$ **b** $x = -1$

a When $x = 2$,
$$\begin{aligned} y &= 2(2)^2 - 3(2) + 4 \\ &= 8 - 6 + 4 \\ &= 6 \end{aligned}$$

b When $x = -1$,
$$\begin{aligned} y &= 2(-1)^2 - 3(-1) + 4 \\ &= 2 \times 1 + 3 + 4 \\ &= 2 + 3 + 4 \\ &= 9 \end{aligned}$$

FINDING x GIVEN y

When we substitute a value for y, we are left with a quadratic equation which we need to solve for x. Since the equation is quadratic, there may be 0, 1, or 2 possible values for x for any given value of y.

Example 2 Self Tutor

Suppose $y = x^2 + 4x - 5$. Find the value(s) of x for which:

a $y = 0$ **b** $y = -9$

a When $y = 0$,
$$\begin{aligned} x^2 + 4x - 5 &= 0 \\ \therefore\ (x+5)(x-1) &= 0 \\ \therefore\ x = -5 \text{ or } x &= 1 \end{aligned}$$

b When $y = -9$,
$$\begin{aligned} x^2 + 4x - 5 &= -9 \\ \therefore\ x^2 + 4x + 4 &= 0 \\ \therefore\ (x+2)^2 &= 0 \\ \therefore\ x &= -2 \end{aligned}$$

EXERCISE 21A

1 If the function is quadratic, state the values of a, b, and c. If it is not, explain why.

a $y = 2x^2 + x + 4$ **b** $y = 3x + 8$ **c** $y = -x^2 - x - 1$

d $y = x^2$ **e** $y = \frac{1}{2}x^2 - 6$ **f** $y = -2x + 5x^3$

g $y = -3x^2$ **h** $y = -x^{-2} - 4x$ **i** $2y + 4x^2 + 5 = 0$

2 Suppose $y = x^2 + 2x - 5$. Find the value of y when:

a $x = 3$ **b** $x = 0$ **c** $x = -2$

3 Suppose $y = 3x^2 - x + 4$. Find the value of y when:

a $x = 0$ **b** $x = 2$ **c** $x = -3$

4 Suppose $y = -2x^2 - 4x + 7$. Find the value of y when:

a $x = 1$ **b** $x = -3$ **c** $x = \frac{1}{2}$

5 Find the value of k if:

a $(2, k)$ lies on $y = x^2 + 3x - 7$ **b** $(-3, k)$ lies on $y = 4x - 3x^2$

c $(5, k)$ lies on $y = -x^2 + 6x - 12$ **d** $(-\frac{1}{2}, k)$ lies on $y = 6x^2 + x + 3$.

6 Determine whether each point lies on the graph of $y = 2x^2 - 3x - 1$:

a $(0, -1)$ **b** $(3, 10)$ **c** $(-2, 13)$

7 Determine whether each point lies on the graph of $y = -x^2 + 5x + 2$:

a $(0, -2)$ **b** $(2, 7)$ **c** $(-1, -4)$

8 Suppose $y = x^2 + 2x + 3$. Find the value(s) of x for which:

a $y = 3$ **b** $y = 2$ **c** $y = 18$

9 Suppose $y = x^2 - 4x + 1$. Find the value(s) of x for which:

a $y = 6$ **b** $y = 1$ **c** $y = -3$

10 Suppose $y = 3x^2 + 15x - 2$. Find the value(s) of x for which:

a $y = -2$ **b** $y = -20$ **c** $y = 40$

11 Find the value(s) of k for which:

a $(k, 7)$ lies on $y = x^2 + 3x - 3$ **b** $(k, 29)$ lies on $y = 2x^2 + 20x + 7$.

12 A cannonball is fired into the air. Its height after t seconds is given by $H = 60t - 5t^2$ metres.

a Find the height of the cannonball after:

i 1 second **ii** 3 seconds.

b Find the time(s) at which the height of the cannonball is:

i 100 m **ii** 160 m.

c How long does it take before the cannonball hits the ground?

13 When a pebble is tossed from a bridge, its height above the water below after t seconds is given by $H = -5t^2 + 5t + 10$ metres.
How long does it take for the pebble to reach the water?

14 Bjorn makes wooden display cabinets. The profit when he makes x cabinets per month is given by $P = 3200x - 200x^2 - 3200$ dollars.

 a Calculate Bjorn's profit or loss if each month he makes:

  i no cabinets **ii** 4 cabinets **iii** 10 cabinets.

 b How many cabinets must Bjorn make per month to earn \$4600 in profit?

B GRAPHS OF QUADRATIC FUNCTIONS

The graphs of all quadratic functions are **parabolas**.

The simplest quadratic function is $y = x^2$. Its graph can be drawn from a table of values.

x	-3	-2	-1	0	1	2	3
y	9	4	1	0	1	4	9

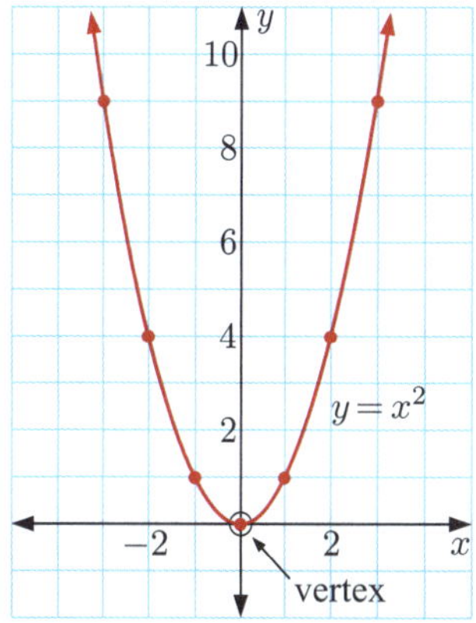

We can see that the graph:

- is **symmetrical** about the y-axis
- has a minimum turning point at $(0, 0)$ which we call the **vertex** of the parabola.

There are no points on the graph below the vertex since $x^2 \geqslant 0$ for all $x \in \mathbb{R}$.

Example 3 **Self Tutor**

Draw the graph of $y = x^2 - 2x - 1$ using a table of values from $x = -3$ to $x = 3$.

When $x = -3$, $y = (-3)^2 - 2(-3) - 1$
$= 9 + 6 - 1$
$= 14$

We complete a table of values by performing similar calculations for the other values of x.

x	-3	-2	-1	0	1	2	3
y	14	7	2	-1	-2	-1	2

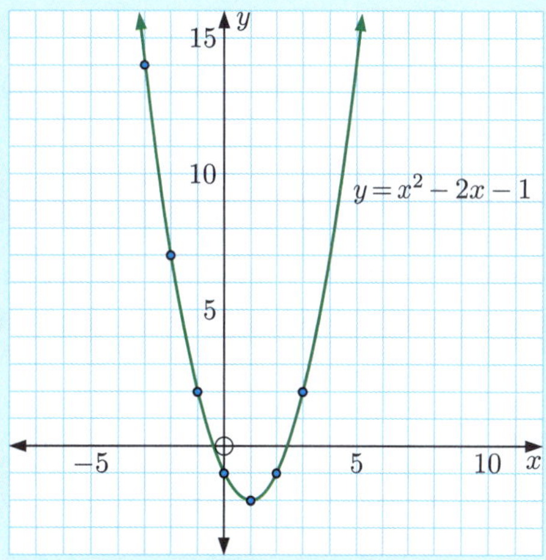

EXERCISE 21B

1 **a** Copy and complete this table for the function $y = x^2 + 3x - 1$.

x	-3	-2	-1	0	1	2	3
y							

 b Hence draw the graph of $y = x^2 + 3x - 1$.

2 Use a table of values from $x = -3$ to $x = 3$ to draw the graph of:

a $y = x^2 - 2x$ **b** $y = x^2 - 2x - 2$ **c** $y = 2x^2$

d $y = 2x^2 + 1$ **e** $y = \frac{1}{2}x^2 - 2$ **f** $y = -x^2$

g $y = -x^2 + x$ **h** $y = -x^2 + 2x + 1$ **i** $y = -2x^2 + 6$

Use technology to check your answers.

GRAPHING PACKAGE

GRAPHICS CALCULATOR INSTRUCTIONS

3 Consider the function $y = x^2 - 2x - 3$.

a Use a table of values from $x = -2$ to $x = 4$ to draw the graph of the function.

b About which vertical line is the parabola symmetric?

c State the coordinates of the vertex of the parabola.

d How many points on the graph are there with y-coordinate:

i 1 **ii** -4 **iii** -6?

e Suppose $x^2 - 2x - 3 = k$. For what values of k does the equation have:

i two solutions **ii** one solution **iii** no solutions?

4 **a** Use tables of values to graph $y = x^2 + x + 2$ and $y = 2x^2$ on the same set of axes.

b *Hence* find the values of x for which $x^2 + x + 2 = 2x^2$.

c Solve algebraically: $x^2 + x + 2 = 2x^2$.

C USING TRANSFORMATIONS TO GRAPH QUADRATICS

In this Section we will learn how to graph quadratic functions without finding a table of values. We do this by transforming the graph of $y = x^2$.

INVESTIGATION 1 GRAPHS OF QUADRATIC FUNCTIONS

In this Investigation you can use the **graphing package** or your **graphics calculator** to draw the graphs.

GRAPHING PACKAGE

GRAPHICS CALCULATOR INSTRUCTIONS

What to do:

1 **a** Graph the two functions on the same set of axes and label the vertex of each graph.

i $y = x^2$ and $y = x^2 + 1$ **ii** $y = x^2$ and $y = x^2 - 1$

iii $y = x^2$ and $y = x^2 + 3$ **iv** $y = x^2$ and $y = x^2 - 3$

b For graphs of the form $y = x^2 + k$, what effect does the value of k have on:

i the position of the graph **ii** the shape of the graph?

c What transformation is needed to graph $y = x^2 + k$ from $y = x^2$?

2 **a** Graph the two functions on the same set of axes and label the vertex of each graph.

i $y = x^2$ and $y = (x-1)^2$ **ii** $y = x^2$ and $y = (x+1)^2$

iii $y = x^2$ and $y = (x-3)^2$ **iv** $y = x^2$ and $y = (x+3)^2$

b For graphs of the form $y = (x-h)^2$, what effect does the value of h have on:

i the position of the graph **ii** the shape of the graph?

c What transformation is needed to graph $y = (x-h)^2$ from $y = x^2$?

3 **a** Graph the two functions on the same set of axes and label the vertex of each graph.

i $y = x^2$ and $y = (x-1)^2 + 2$ **ii** $y = x^2$ and $y = (x+2)^2 - 3$

iii $y = x^2$ and $y = (x-3)^2 - 1$ **iv** $y = x^2$ and $y = (x+4)^2 + 5$

b What transformation is needed to graph $y = (x-h)^2 + k$ from $y = x^2$?

c State the coordinates of the vertex of $y = (x-h)^2 + k$.

4 **a** Graph the two functions on the same set of axes.

i $y = x^2$ and $y = 2x^2$ **ii** $y = x^2$ and $y = 3x^2$

iii $y = x^2$ and $y = -x^2$ **iv** $y = x^2$ and $y = \frac{1}{2}x^2$

v $y = x^2$ and $y = -2x^2$ **vi** $y = x^2$ and $y = -3x^2$

b For graphs of the form $y = ax^2$, what effect does the value of a have on:

i the shape of the graph **ii** the direction in which the graph opens?

c What transformation is needed to graph $y = -ax^2$ from $y = ax^2$?

From the **Investigation**, you should have discovered:

GRAPHS OF THE FORM $y = (x-h)^2 + k$

- Graphs of the form $y = (x-h)^2 + k$ have the same shape as the graph of $y = x^2$.
- The graph of $y = x^2$ is translated h units horizontally and k units vertically to give the graph of $y = (x-h)^2 + k$.
  - If $k > 0$, the graph moves upwards.
  - If $k < 0$, the graph moves downwards.
  - If $h > 0$, the graph moves to the right.
  - If $h < 0$, the graph moves to the left.
- The vertex of $y = (x-h)^2 + k$ is (h, k).

GRAPHS OF THE FORM $y = ax^2$, $a \neq 0$

- The vertex of $y = ax^2$ is $(0, 0)$.
- If $a > 0$, $y = ax^2$ opens upwards. 
- If $a < 0$, $y = ax^2$ opens downwards. 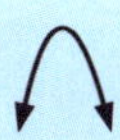
- If $|a| > 1$, $y = ax^2$ is "thinner" than $y = x^2$.
- If $|a| < 1$, $y = ax^2$ is "wider" than $y = x^2$.
- $y = -ax^2$ is a reflection of $y = ax^2$ in the x-axis.

Example 4 **Self Tutor**

Sketch each function on the same set of axes as $y = x^2$, and label its vertex.

a $y = x^2 + 2$

b $y = (x - 3)^2$

a We draw $y = x^2$ then translate it 2 units upwards.

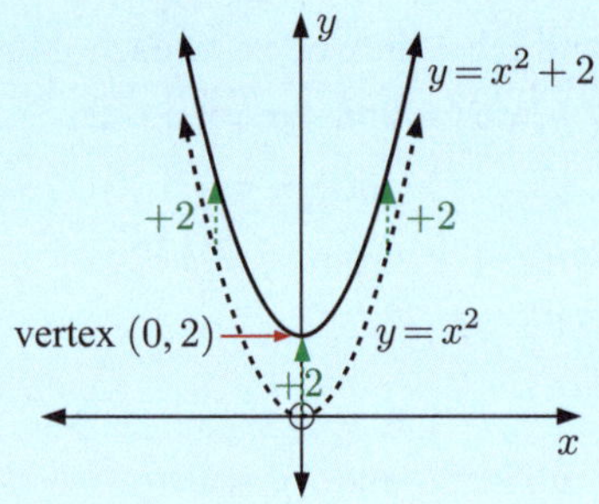

b We draw $y = x^2$ then translate it 3 units to the right.

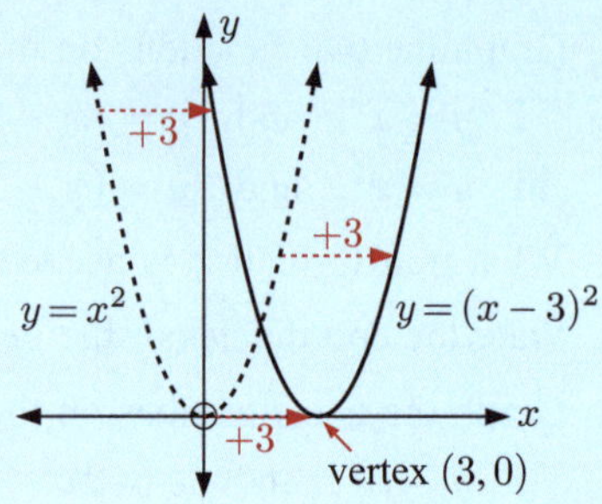

EXERCISE 21C

1 Sketch each function on the same set of axes as $y = x^2$, and label its vertex.

a $y = x^2 + 1$ **b** $y = x^2 + 3$ **c** $y = x^2 - 1$

d $y = x^2 - 4$ **e** $y = x^2 + 6$ **f** $y = x^2 - 5$

2 Sketch each function on the same set of axes as $y = x^2$, and label its vertex.

a $y = (x - 1)^2$ **b** $y = (x + 2)^2$ **c** $y = (x - 4)^2$

d $y = (x + 3)^2$ **e** $y = (x + 4)^2$ **f** $y = (x - 5)^2$

3 State the coordinates of the vertex of each quadratic function:

a $y = x^2 + 2$ **b** $y = (x + 1)^2$ **c** $y = x^2 - 3$ **d** $y = (x - 6)^2$

Example 5 **Self Tutor**

Sketch each function on the same set of axes as $y = x^2$, and label its vertex.

a $y = (x - 3)^2 - 2$

b $y = (x + 1)^2 + 4$

a We draw $y = x^2$ then translate it 3 units to the right and 2 units downwards.

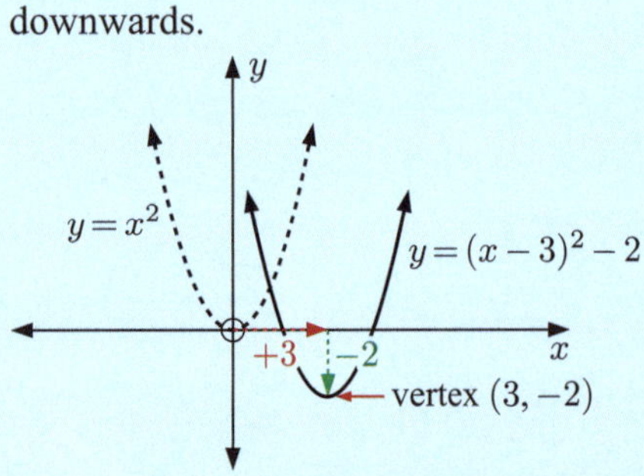

b We draw $y = x^2$ then translate it 1 unit to the left and 4 units upwards.

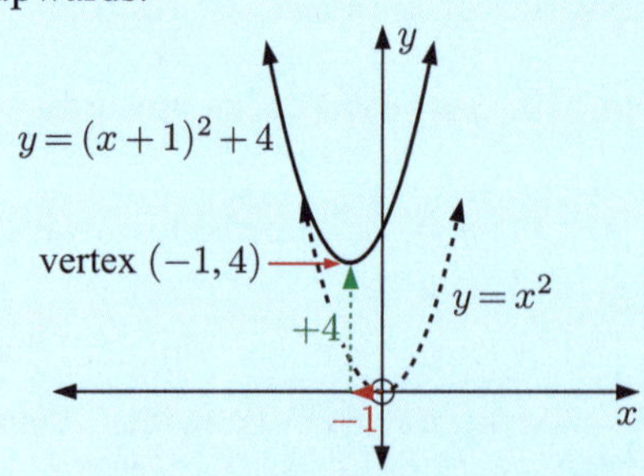

4 Sketch each function on the same set of axes as $y = x^2$, and label its vertex.

a $y = (x-2)^2 + 1$ **b** $y = (x+3)^2 + 4$ **c** $y = (x-1)^2 - 3$

d $y = (x+4)^2 - 2$ **e** $y = (x-4)^2 - 5$ **f** $y = (x+5)^2 + 1$

5 State the coordinates of the vertex of:

a $y = (x-1)^2 + 5$ **b** $y = (x-3)^2 - 1$ **c** $y = (x+3)^2 - 6$

6 **a** Show that $(x-2)^2 - 3 = x^2 - 4x + 1$.

b Graph $y = x^2$ and $y = x^2 - 4x + 1$ on the same set of axes.

Example 6 **Self Tutor**

Sketch $y = 2x^2$ and $y = -2x^2$ on the same set of axes as $y = x^2$.

$y = 2x^2$ is "thinner" than $y = x^2$.

$y = -2x^2$ has the same shape as $y = 2x^2$, but opens downwards.

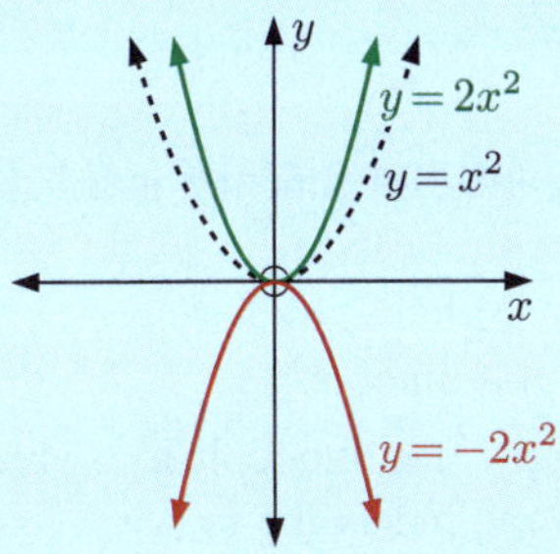

7 Sketch $y = x^2$ and $y = -x^2$ on the same set of axes.

8 Sketch on the same set of axes as $y = x^2$:

a $y = 3x^2$ and $y = -3x^2$ **b** $y = \frac{1}{2}x^2$ and $y = -\frac{1}{2}x^2$

9 Match each quadratic function with its graph:

a $y = -\frac{1}{3}x^2$

b $y = (x-2)^2$

c $y = (x+4)^2$

d $y = \frac{3}{2}x^2$

e $y = x^2 - 2$

f $y = -x^2$

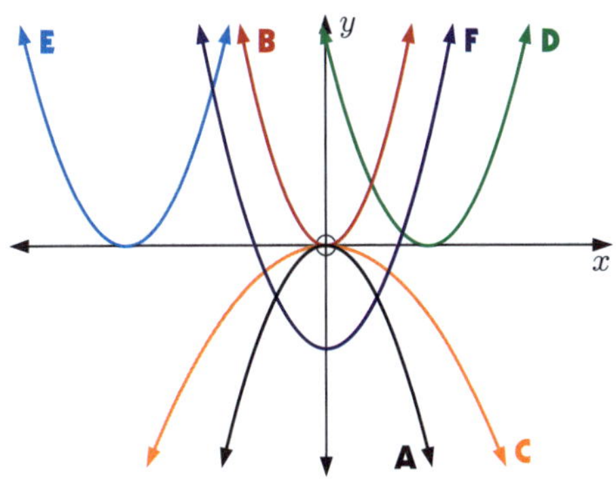

10 For each function:

i By completing the square, write the function in the form $y = (x-h)^2 + k$.

ii Hence graph the function on the same set of axes as $y = x^2$, and label its vertex.

a $y = x^2 + 4x - 2$ **b** $y = x^2 - 8x + 19$ **c** $y = x^2 + 3x + 3$

11 Consider the function $y = -(x-3)^2 + 1$.

a Explain why $y \leqslant 1$ for all x.

b Find the x-intercepts of the function.

c What translation is needed to graph $y = -(x-3)^2 + 1$ from $y = -x^2$?

d Graph $y = -x^2$ and $y = -(x-3)^2 + 1$ on the same set of axes.

12 The equation of a quadratic function in completed square form is $y = a(x-h)^2 + k$.

a Deduce the coordinates of the vertex. Explain your answer.

b Hence find the coordinates of the vertex of:

i $y = 3(x-4)^2 + 2$ **ii** $y = 2x^2 + 12x + 14$ **iii** $y = -x^2 - 8x - 17$

D AXES INTERCEPTS

- An **x-intercept** of a function is a value of x where its graph meets the x-axis.
  x-intercepts are found by letting y be 0 in the equation of the function.
- A **y-intercept** of a function is a value of y where its graph meets the y-axis.
  y-intercepts are found by letting x be 0 in the equation of the function.

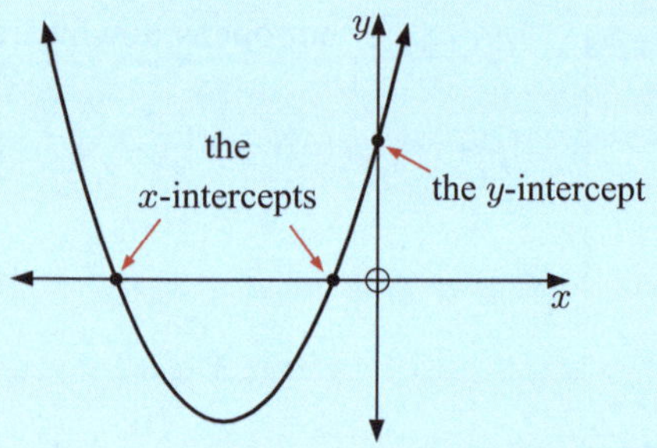

INVESTIGATION 2 AXES INTERCEPTS

What to do:

1 **a** Use technology to draw the graph of each function, including its y-intercept.

i $y = x^2 - 3x + 2$ **ii** $y = -x^2 + x + 5$

iii $y = x^2 - 3x$ **iv** $y = 2x^2 - 4x - 7$

v $y = -3x^2 - x + 4$ **vi** $y = 4x^2 - 6$

b State the y-intercept of a quadratic function of the form $y = ax^2 + bx + c$.

2 **a** Use technology to draw the graph of each function, including its x-intercepts.

i $y = (x-1)(x-3)$ **ii** $y = (x-2)(x-7)$ **iii** $y = -(x+4)(x-5)$

iv $y = 2(x+3)(x-3)$ **v** $y = -3x(x+5)$ **vi** $y = 5(x-2)(x-6)$

b State the x-intercepts of a quadratic function of the form $y = a(x-\alpha)(x-\beta)$.

3 **a** Use technology to draw the graph of each function, including its x-intercepts.

i $y = (x-4)^2$ **ii** $y = -(x+6)^2$ **iii** $y = 3(x-2)^2$

b State the x-intercepts of a quadratic function of the form $y = a(x-\alpha)^2$.

From the **Investigation** you should have found:

- For a quadratic function in the form $y = ax^2 + bx + c$, the y-intercept is the constant term c.

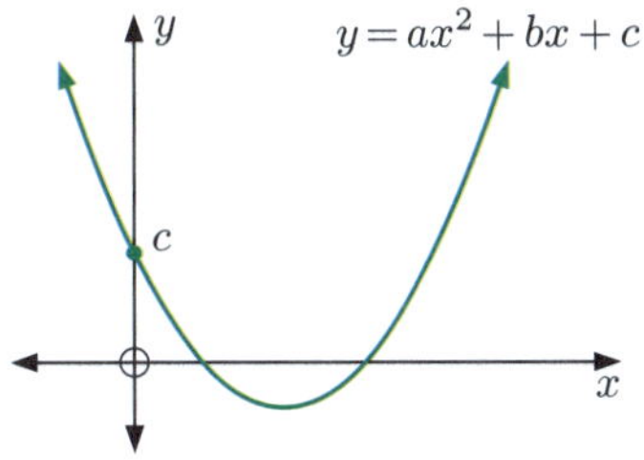

Proof:

Consider $y = ax^2 + bx + c$.
If we let $x = 0$ then $y = a(0)^2 + b(0) + c = c$
$\therefore$ the y-intercept is c.

- A quadratic function with factored form $y = a(x-\alpha)(x-\beta)$ has x-intercepts α and β.

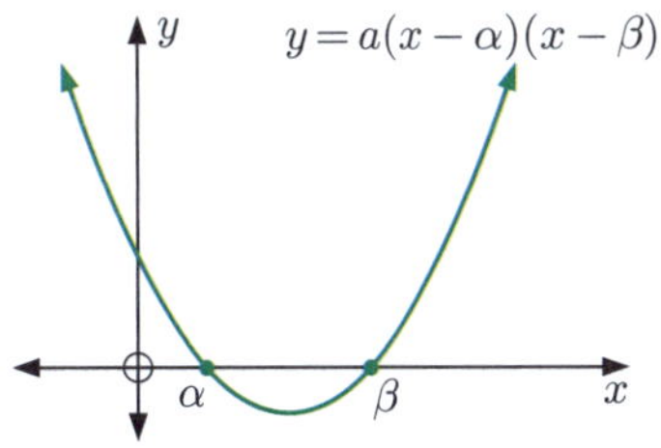

Proof: Consider $y = a(x-\alpha)(x-\beta)$.
If we let $y = 0$ then $a(x-\alpha)(x-\beta) = 0$
$\therefore$ $x - \alpha = 0$ or $x - \beta = 0$
$\therefore$ $x = \alpha$ or $x = \beta$
$\therefore$ the x-intercepts are α and β.

- A quadratic function with factored form $y = a(x-\alpha)^2$ *touches* the x-axis at $x = \alpha$. The only x-intercept is α.

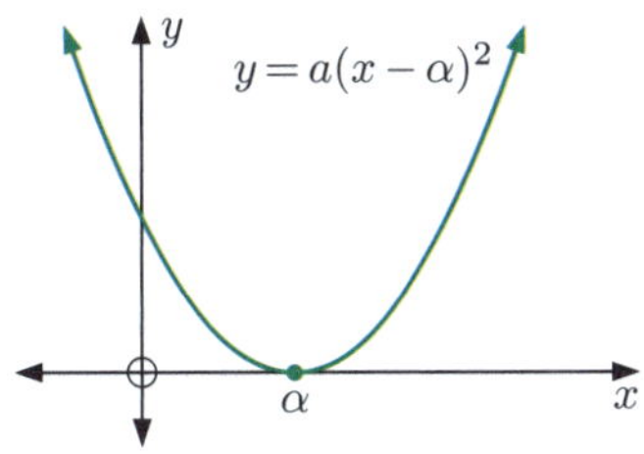

Example 7 — Self Tutor

Find the x-intercept(s) of:

a $y = (x+1)(x-2)$

b $y = -(x-2)^2$

a When $y = 0$, $(x+1)(x-2) = 0$
$\therefore$ $x = -1$ or $x = 2$
$\therefore$ the x-intercepts are -1 and 2.

b When $y = 0$, $-(x-2)^2 = 0$
$\therefore$ $x = 2$
$\therefore$ the x-intercept is 2.

FACTORISING TO FIND x-INTERCEPTS

For any quadratic function of the form $y = ax^2 + bx + c$, the x-intercepts can be found by solving the equation $ax^2 + bx + c = 0$.

The quadratic equation may have two, one, or no real solutions. These solutions correspond to the two, one, or no x-intercepts of the graph of the corresponding quadratic function.

Example 8 **Self Tutor**

Find the x-intercept(s) of the quadratic function:

a $y = x^2 + 4x + 4$

b $y = -x^2 + 2x + 3$

a When $y = 0$,
$x^2 + 4x + 4 = 0$
$\therefore \ (x+2)^2 = 0$
$\therefore \ x = -2$
$\therefore$ the x-intercept is -2.

b When $y = 0$,
$-x^2 + 2x + 3 = 0$
$\therefore \ x^2 - 2x - 3 = 0$
$\therefore \ (x-3)(x+1) = 0$
$\therefore \ x = 3$ or $x = -1$
$\therefore$ the x-intercepts are -1 and 3.

EXERCISE 21D

1 State the y-intercept of:

a $y = x^2 + x + 1$ **b** $y = x^2 - 2x + 3$ **c** $y = 3x^2 + 4x - 2$
d $y = -x^2 - x + 4$ **e** $y = -2x^2 + x + 6$ **f** $y = 4x^2 - 2x + 1$
g $y = 4 + x^2$ **h** $y = 3 + x - 5x^2$ **i** $y = 2x - x^2 + 7$

2 State the x-intercept(s) of:

a $y = (x-1)(x+3)$ **b** $y = -(x+2)(x-3)$ **c** $y = (x-3)(x+7)$
d $y = 3(x-2)(x-5)$ **e** $y = (x+5)^2$ **f** $y = -2(x-7)^2$

3 Prove that the only x-intercept of $y = a(x-\alpha)^2$ is α.

4 By expanding the brackets, find the y-intercept of:

a $y = (x+1)(x-2)$ **b** $y = (x-1)(x+3)$ **c** $y = -(x+2)(x-1)$
d $y = x(x-3)$ **e** $y = -(x-4)(x+1)$ **f** $y = -2(x-5)(x-2)$

5 Find the y-intercept of:

a $y = a(x-\alpha)(x-\beta)$ **b** $y = a(x-\alpha)^2$

6 Find the x-intercept(s) of:

a $y = x^2 - 4$ **b** $y = x^2 - 9$ **c** $y = x^2 - 7x + 12$
d $y = x^2 - x - 6$ **e** $y = 3x - x^2$ **f** $y = -x^2 + 7x - 10$
g $y = x^2 - 6x + 9$ **h** $y = 2x^2 + 4x + 2$ **i** $y = -x^2 - 5x + 6$
j $y = 3x^2 + 12x - 15$ **k** $y = -x^2 + 6x + 16$ **l** $y = -2x^2 - 8x + 42$

7 Find the axes intercepts of:

a $y = \frac{1}{2}(x+2)^2$ **b** $y = 3(x-1)(x+2)$ **c** $y = -x^2 + 5x + 24$

8 Find the x-intercepts of:

a $y = x^2 + 3x - 5$ **b** $y = 2x^2 + 7x - 4$
c $y = -3x^2 + 5x + 2$ **d** $y = x^2 - 4x - 8$
e $y = 4x^2 - 8x - 5$ **f** $y = -2x^2 + 4x + 7$

E USING AXES INTERCEPTS TO GRAPH QUADRATICS

Once we have found the axes intercepts of a quadratic function, we can use them to sketch the graph of the function.

Example 9 — Self Tutor

Graph each quadratic function using its axes intercepts:

a $y = -(x+1)(x-3)$ **b** $y = (x-2)^2$ **c** $y = -x^2 - 3x + 4$

a When $y = 0$, $-(x+1)(x-3) = 0$

$\therefore$ $x = -1$ or $x = 3$

$\therefore$ the x-intercepts are -1 and 3.

When $x = 0$, $y = -(1)(-3) = 3$

$\therefore$ the y-intercept is 3.

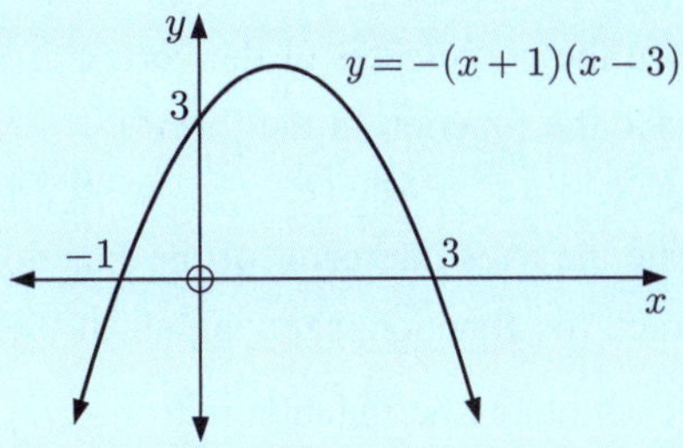

b When $y = 0$, $(x-2)^2 = 0$

$\therefore$ $x = 2$

$\therefore$ the x-intercept is 2.

When $x = 0$, $y = (-2)^2 = 4$

$\therefore$ the y-intercept is 4.

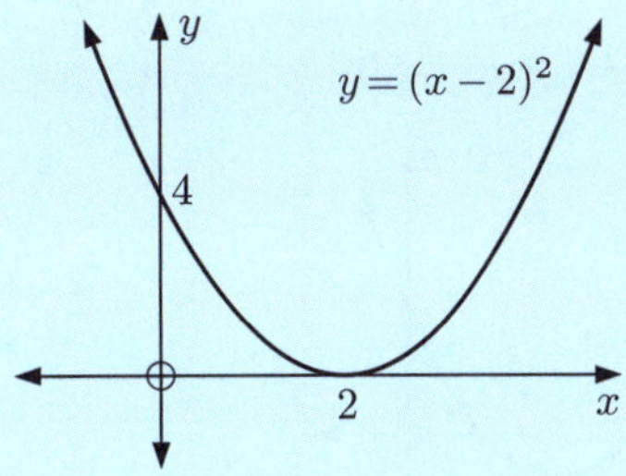

c The y-intercept is 4.

When $y = 0$, $-x^2 - 3x + 4 = 0$

$\therefore$ $x^2 + 3x - 4 = 0$

$\therefore$ $(x+4)(x-1) = 0$

$\therefore$ $x = -4$ or $x = 1$

$\therefore$ the x-intercepts are -4 and 1.

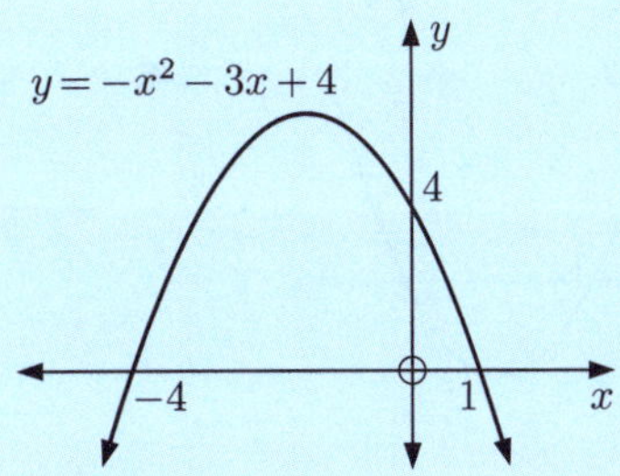

EXERCISE 21E

1 Sketch the graph of the quadratic function which has:

a x-intercepts -2 and 3, and y-intercept -5

b x-intercepts -4 and 1, and y-intercept 6

c x-intercepts 2 and 3, and y-intercept 4

d x-intercept -2, and y-intercept 7

e x-intercepts -8 and -3, and y-intercept -10

f x-intercept 4, and y-intercept -12.

If a quadratic function has only one x-intercept then its graph *touches* the x-axis.

2 Graph each quadratic function using its axes intercepts:

- **a** $y=(x-1)(x-3)$
- **b** $y=-(x-1)(x-5)$
- **c** $y=(x+2)(x-4)$
- **d** $y=(x+1)(x+3)$
- **e** $y=(x-1)^2$
- **f** $y=3(x+2)(x-3)$
- **g** $y=-2(x+4)(x-2)$
- **h** $y=-(x+1)(x+4)$
- **i** $y=-\frac{1}{3}(x+3)^2$

3 Graph each quadratic function using its axes intercepts:

- **a** $y=x^2+3x-10$
- **b** $y=-x^2-4x-4$
- **c** $y=x^2+7x+6$
- **d** $y=-x^2+5x+14$
- **e** $y=2x^2-20x+50$
- **f** $y=3x^2-3x-36$

4 Graph each quadratic function using its axes intercepts and the point with x-coordinate 2:

- **a** $y=x(x-3)$
- **b** $y=-x(x+2)$
- **c** $y=-x^2+x$

5 Consider the function $y=(x-4)^2-4$.

- **a** State the coordinates of the vertex of its graph.
- **b** Write the function in the form:
  - **i** $y=ax^2+bx+c$
  - **ii** $y=a(x-\alpha)(x-\beta)$
- **c** State the axes intercepts of the function.
- **d** Graph the function, showing all of the features you have found.

6 Match each quadratic function with its graph:

- **a** $y=(x+5)(x-3)$
- **b** $y=-(x+3)^2$
- **c** $y=x(x-4)$
- **d** $y=2x^2+14x+20$
- **e** $y=-x^2-2x+15$
- **f** $y=3x^2-18x+15$

A

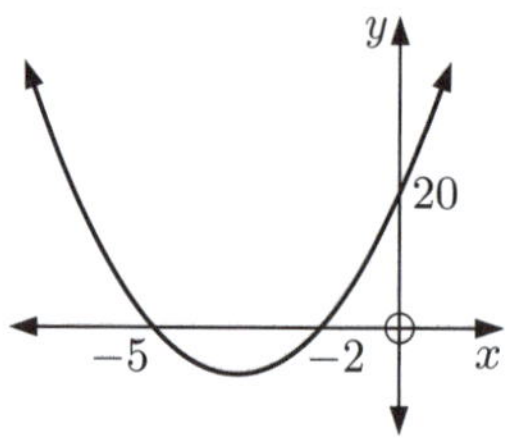

B

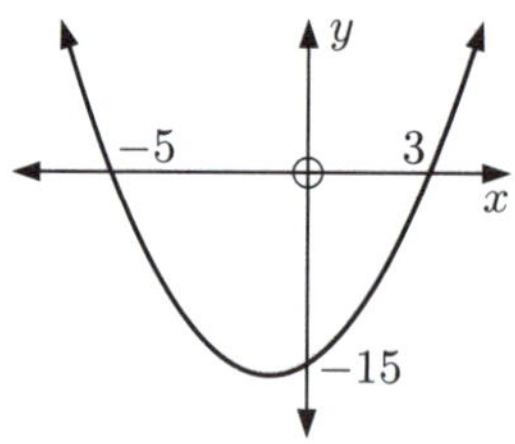

C

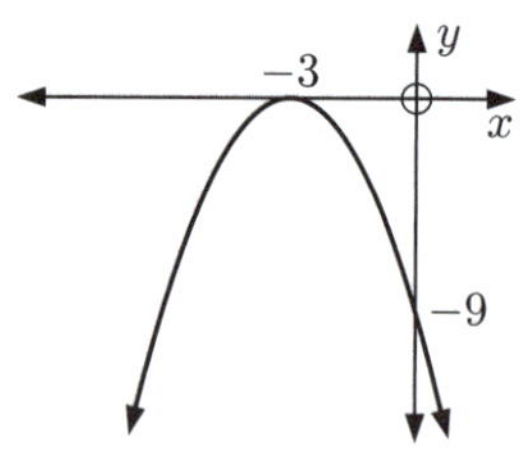

D

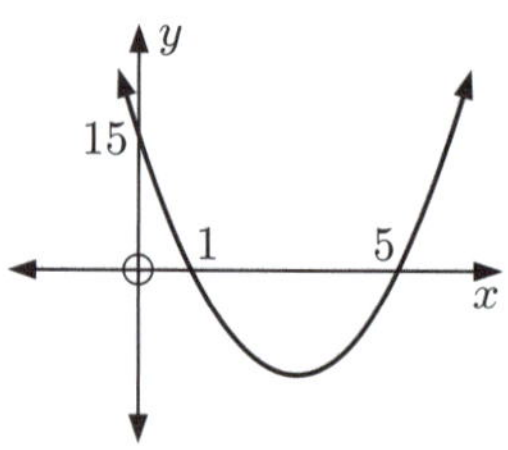

E

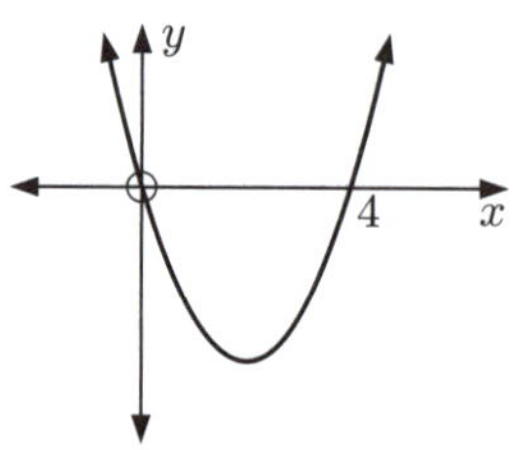

F

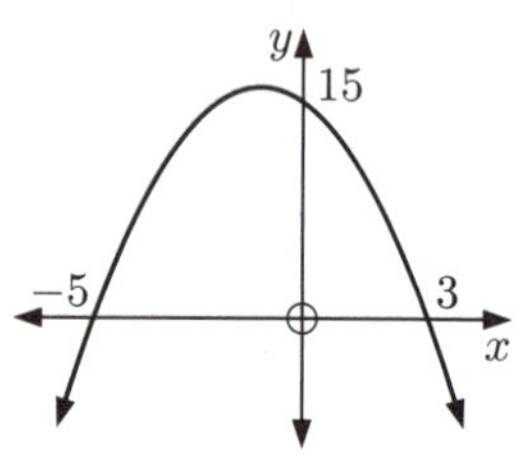

7 Graph each function using its axes intercepts:

- **a** $y=x^2+4x+2$
- **b** $y=3x^2-8x-3$
- **c** $y=-4x^2-12x+7$
- **d** $y=x^2-5x-2$
- **e** $y=6x^2+11x-10$
- **f** $y=-x^2+8x-6$

ACTIVITY — QUADRATIC SHOOTER

Click on the icon to play a game matching properties of quadratics to their graphs.

QUADRATIC SHOOTER

F PROJECTILE MOTION

A **projectile** is an object which is released into the air, and then moves under the force of gravity only.

HISTORICAL NOTE GALILEO'S ANALYSIS OF PROJECTILE MOTION

Galileo Galilei (1564 - 1642) was an Italian astronomer, physicist, and engineer from Pisa.

In his 1638 work *Dialogues of the Two New Sciences*, Galileo recorded his experiments rolling a ball down a ramp at different angles and the flight of the ball from the end of the ramp. His observations led him to the conclusion that the trajectory of a projectile is a *parabola*.

Galileo's work showed how systematic experimentation and observation could enable the discovery of the mathematical structures which govern nature. He is often referred to with titles such as the "father of modern physics" and the "father of the scientific method".

Galileo Galilei

Since the trajectory of a projectile is a parabola, we can use a quadratic function to model the height of a projectile above the ground as a function of time, or as a function of the horizontal distance the projectile has travelled.

In the following Exercise, you should use your **graphics calculator** or the **graphing package** to find the vertex and axes intercepts of quadratic functions.

GRAPHICS CALCULATOR INSTRUCTIONS

GRAPHING PACKAGE

Example 10

Self Tutor

When a softball is struck, its height above the ground after t seconds is given by $H = -4.9t^2 + 24t + 1.2$ m.

a Use technology to help sketch the function from $t = 0$ until the softball lands on the ground. Label the H-intercept, vertex, and t-intercept.

b Find:

- **i** the maximum height reached by the softball
- **ii** the time taken for the softball to land.

a

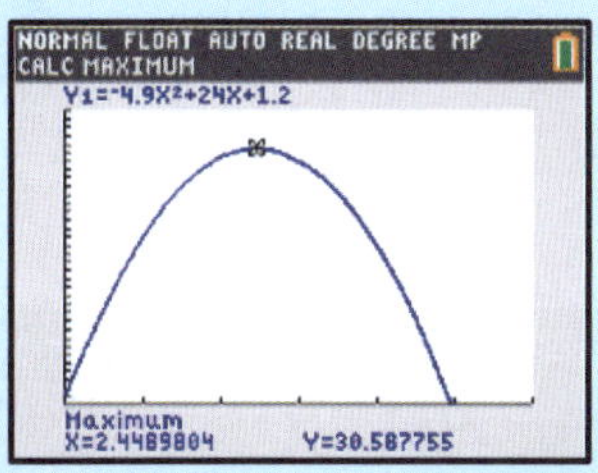

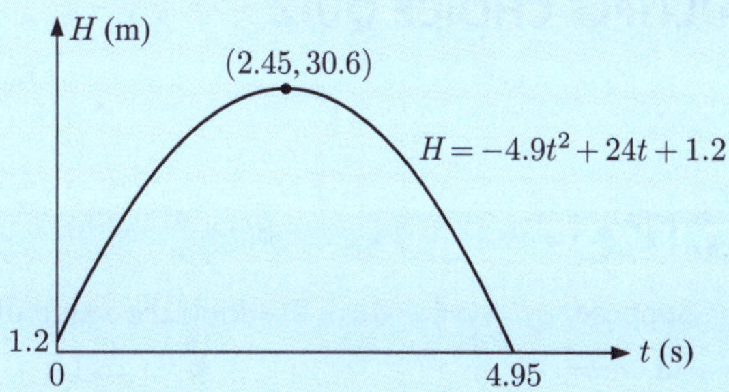

b **i** The maximum height reached was about 30.6 m.

ii It took about 4.95 seconds for the softball to land.

EXERCISE 21F

1. When a cricket ball is thrown, its height above the ground after t seconds is given by $H = -4.9t^2 + 18t + 1.6$ m.
   - **a** Use technology to help sketch the function from $t = 0$ until the cricket ball lands on the ground. Label the H-intercept, vertex, and t-intercept.
   - **b** Find the maximum height reached by the cricket ball.
   - **c** Find the time taken for the ball to reach its maximum height.
   - **d** Find the time taken for the cricket ball to land.

2. When a diver springs from a diving board, the height of her head above the water after t seconds is given by $H = -4.9t^2 + 5.1t + 4.7$ m.
   - **a** Use technology to help sketch the function from $t = 0$ until the diver's head reaches the water. Label the H-intercept, vertex, and t-intercept.
   - **b** Find the maximum height reached by the diver's head.
   - **c** Find the time taken for the diver's head to reach the water.

3. When a baseball is struck, its height above the ground after travelling x m horizontally is given by $H = -0.006x^2 + 0.68x + 1.4$ m.
   - **a** Use technology to help sketch the function from $x = 0$ until the baseball lands on the ground. Label the H-intercept, vertex, and x-intercept.
   - **b** Find the maximum height reached by the baseball.
   - **c** Find the horizontal distance travelled by the baseball until it landed.

4. Answer the **Opening Problem** on page **330**.

GLOBAL CONTEXT — NEWTONIAN MECHANICS

Global context: Scientific and technical innovation

Statement of inquiry: Formulating mathematical laws can help us to understand the world.

Criterion: Applying mathematics in real-life contexts

GLOBAL CONTEXT

QUICK QUIZ

MULTIPLE CHOICE QUIZ

REVIEW SET 21A

1. Suppose $y = 4x^2 - 3x + 2$. Find the value of y when:
   - **a** $x = -2$
   - **b** $x = 1$
   - **c** $x = 3$

2. Find the value(s) of t if:
   - **a** $(-1, t)$ lies on $y = 2x^2 - 5x + 4$
   - **b** $(t, -2)$ lies on $y = x^2 - 3x - 2$.

3 Use a table of values from $x = -3$ to $x = 3$ to draw the graph of $y = x^2 - 2x + 3$.

4 Sketch each function on the same set of axes as $y = x^2$, and label its vertex.

a $y = x^2 - 2$ **b** $y = (x+1)^2$

5 Sketch each function on the same set of axes as $y = x^2$:

a $y = 4x^2$ and $y = -4x^2$ **b** $y = \frac{1}{3}x^2$ and $y = -\frac{1}{3}x^2$

6 Consider the function $y = x^2 - 6x + 2$.

a By completing the square, write the function in the form $y = (x-h)^2 + k$.

b Hence graph the function on the same set of axes as $y = x^2$, and label its vertex.

7 Find the y-intercept of:

a $y = 3x^2 - x - 4$ **b** $y = -2(x-5)(x+1)$ **c** $y = (2x-3)^2$

8 Find the x-intercept(s) of:

a $y = (x+2)(x-3)$ **b** $y = -x^2 + x + 12$ **c** $y = 3x^2 + 24x + 48$

9 Graph each quadratic function using its axes intercepts:

a $y = (x+3)(x-2)$ **b** $y = -\frac{1}{2}(x+6)^2$ **c** $y = x^2 + 2x - 8$

10 Graph the quadratic function $y = -x(x+5)$ using its axes intercepts and the point with x-coordinate 2.

11 Consider the quadratic function $y = (x-3)^2 - 16$.

a State the coordinates of the vertex.

b Write the function in the form $y = ax^2 + bx + c$.

c Hence find the axes intercepts of the function.

d Sketch the graph of the function using the vertex and axes intercepts.

12 When a large rock is ejected from a volcano, its height above ground after travelling x m horizontally is given by $H = -0.012x^2 + 0.95x + 50$ m.

a Use technology to help sketch the function from $x = 0$ until the rock lands on the ground. Label the H-intercept, vertex, and x-intercept.

b Find the horizontal distance travelled by the rock until it landed.

13 When a tennis ball is thrown upwards, its height above the ground after t seconds is given by $H = -4.9t^2 + 20t + 1$ m.

a Use technology to help sketch the function from $t = 0$ until the ball lands on the ground. Label the H-intercept, vertex, and t-intercept.

b Find the maximum height reached by the ball.

c Find the time taken for the ball to land.

REVIEW SET 21B

1 Suppose $y = x^2 - 4x + 14$. Find:

a the value of y when $x = 3$ **b** the value(s) of x for which $y = 19$.

2 Find the value(s) of a for which $(a, 3)$ lies on $y = x^2 + 3x + 5$.

3 Use a table of values from $x = -3$ to $x = 3$ to draw the graph of $y = -x^2 - x + 2$.

4 Sketch each function on the same set of axes as $y = x^2$, and label its vertex.

a $y = (x-3)^2$ **b** $y = x^2 + 5$

5 State the coordinates of the vertex of:

a $y = (x+4)^2$ **b** $y = x^2 - 5$ **c** $y = (x-6)^2 + 3$

6 **a** Show that $(x+2)^2 - 1 = x^2 + 4x + 3$.

b Graph $y = x^2$ and $y = x^2 + 4x + 3$ on the same set of axes.

7 Find the y-intercept of:

a $y = -2x(3-x)$ **b** $y = -3x^2 + 2x + 12$ **c** $y = (x-7)(x+3)$

8 Find the x-intercepts of:

a $y = -(x+3)(x-6)$ **b** $y = 32 - x^2$ **c** $y = x^2 + x - 6$

9 Graph each quadratic function using its axes intercepts:

a $y = -(x+1)(x-4)$ **b** $y = x^2 - 6x - 16$ **c** $y = 2x^2 - 18x + 36$

10 Graph each function using its axes intercepts:

a $y = 4x^2 + 5x - 6$ **b** $y = x^2 + 2x - 2$ **c** $y = 6x^2 + x - 2$

11 **a** Use tables of values to graph $y = x^2 - x - 5$ and $y = 2x^2 - 11$ on the same set of axes.

b *Hence* find the values of x for which $x^2 - x - 5 = 2x^2 - 11$.

c Solve algebraically: $x^2 - x - 5 = 2x^2 - 11$.

12 Consider the function $y = (x+4)^2 - 1$.

a State the coordinates of the vertex of its graph.

b Write the function in the form: **i** $y = ax^2 + bx + c$ **ii** $y = a(x-\alpha)(x-\beta)$

c State the axes intercepts of the function.

d Graph the function, showing the features you have found.

13 A rugby player is taking a field goal kick 40 m from the goal line. To score a field goal, the ball must pass over a 3 m high crossbar on the goal line.
The player kicks the ball directly at the goal, and the height of the ball above the ground after travelling x m horizontally is given by
$H = -0.02x^2 + 0.9x + 0.2$ m.

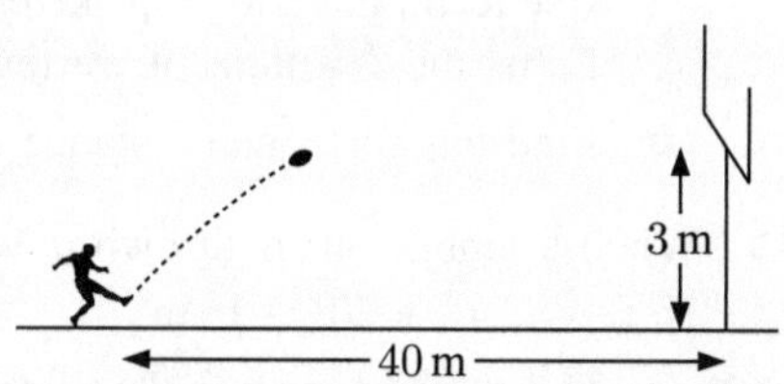

a Use technology to help sketch the function from $x = 0$ until the ball lands on the ground. Label the H-intercept, vertex, and x-intercept.

b Find the maximum height reached by the ball.

c Find the horizontal distance travelled by the ball until it landed.

d Did the player score a field goal? Explain your answer.

Chapter 22

Congruence and similarity

Contents:

OPENING PROBLEM

Jacob makes tanks for storing grain. His tanks have different *sizes*, but all have the same *shape*. His smallest tank has height 2 m and capacity 7 kL.

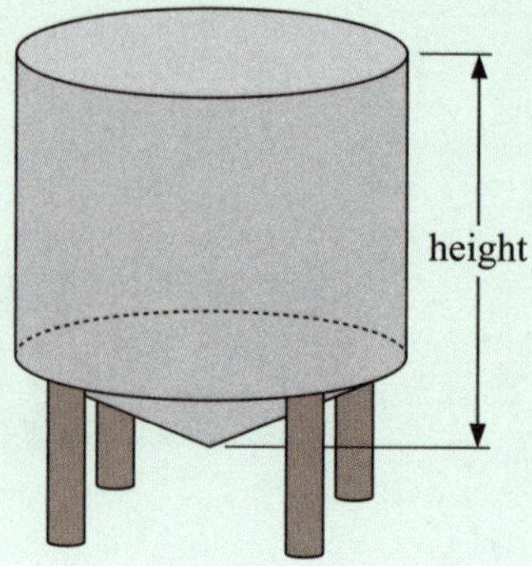

Things to think about:

a What do we mean by saying two objects are different in *size* but have the same *shape*?

b Jacob wants to manufacture a tank which is 4 m high.

  i How many times higher is the new tank than the original tank?

  ii What will the capacity of the new tank be?

A CONGRUENCE

Congruent figures are identical in size and shape. They do not need to have the same position or orientation.

When we perform a **translation**, **rotation**, or **reflection**, the object and image are *congruent*.

When two figures are congruent, we:

- write corresponding vertices in the same order
- use the symbol $\cong$ to mean "is congruent to".

For example, for these congruent figures we write $ABCD \cong PQRS$ rather than $ABCD \cong RSPQ$.

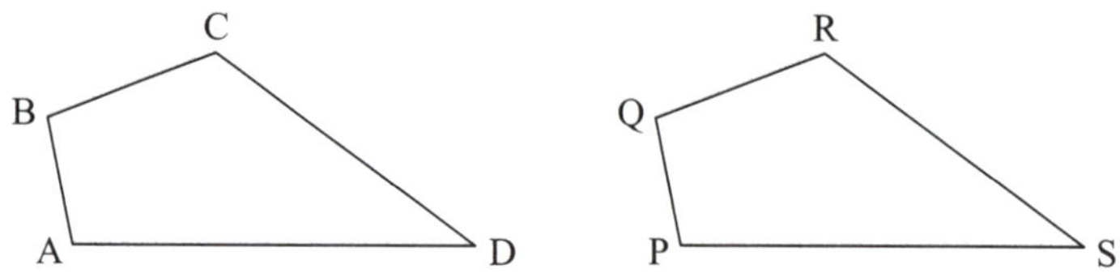

EXERCISE 22A

1 Decide whether each pair of figures is congruent:

a

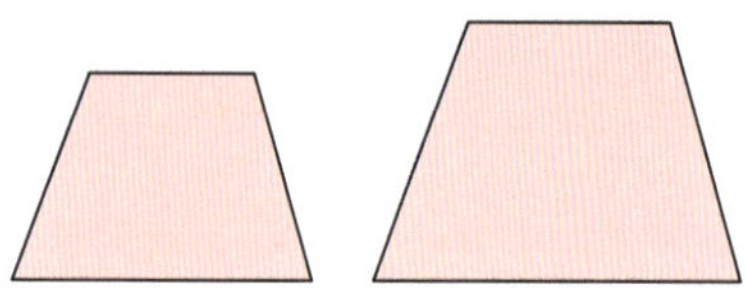

b

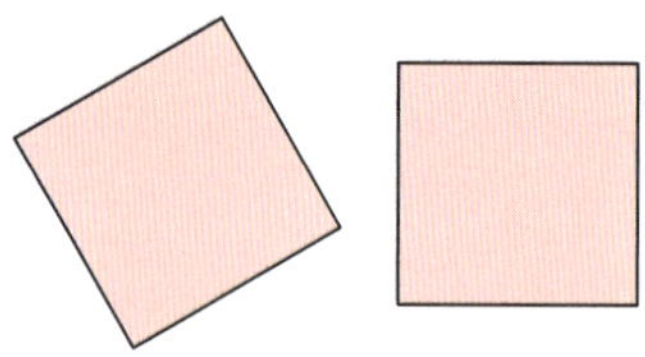

c

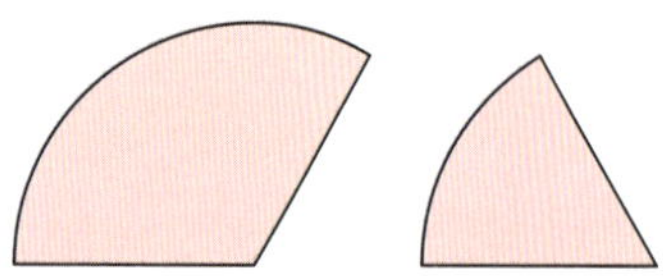

d

2 Which of the following figures are congruent?

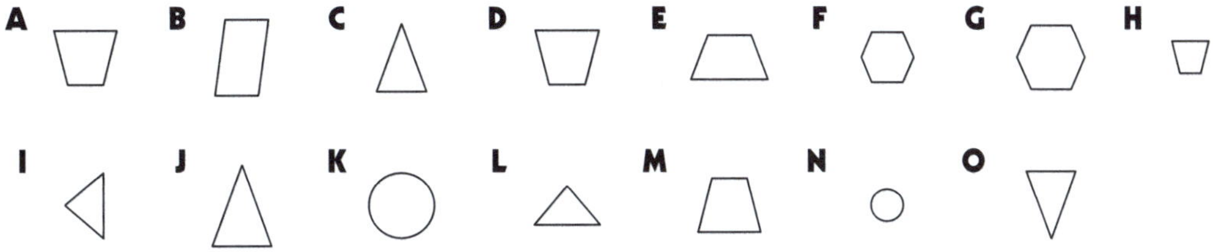

3 Quadrilaterals ABCD and PQRS are congruent.

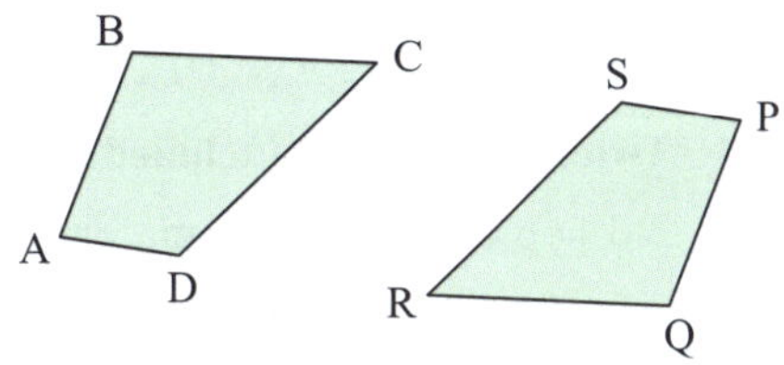

a Name another side which has the same length as:

i [BC] **ii** [RS]

b Name another angle which has the same measure as:

i $A\widehat{B}C$ **ii** $R\widehat{S}P$

4 Quadrilaterals ABCD and PQRS are congruent. Determine the:

a length of [QR]

b perimeter of PQRS.

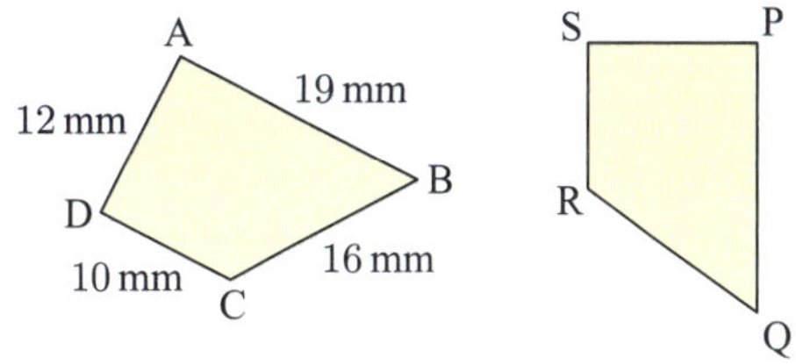

5 ABC $\cong$ LMN.

a Determine the length of [BC].

b Determine the size of:

i $B\widehat{A}C$ **ii** $L\widehat{M}N$ **iii** $L\widehat{N}M$

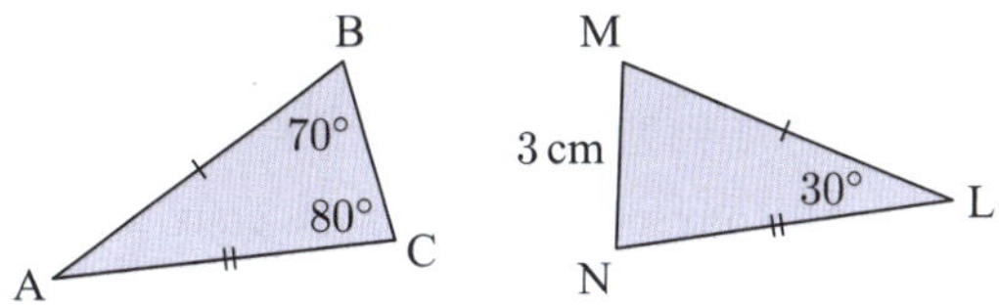

B CONGRUENT TRIANGLES

INVESTIGATION CONSTRUCTING TRIANGLES

If several people are asked to accurately draw a triangle ABC in which AB = 3 cm and BC = 2 cm, many differently shaped triangles will probably be drawn.

For example, these three triangles all satisfy the given information.

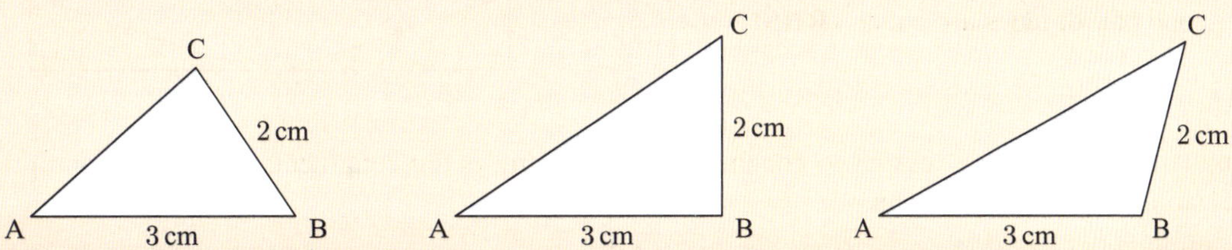

The information given is insufficient to draw a triangle of one particular shape.

PRINTABLE WORKSHEET

In this Investigation we consider what information is sufficient to *uniquely* describe the size and shape of a triangle. All triangles drawn using this information will be congruent.

What to do:

For each of the following cases:

- **a** Draw a triangle ABC from the information given.
- **b** Decide whether *all* triangles drawn using the given information will be congruent.

1 **Three sides**
AB = 4 cm, BC = 5 cm, and AC = 6 cm.

2 **Three angles**
$A\widehat{B}C = 50°$, $B\widehat{C}A = 60°$, and $C\widehat{A}B = 70°$.

3 **Two sides and the included angle between them**
AB = 6 cm, BC = 5 cm, and $A\widehat{B}C = 45°$.

4 **Two sides and a non-included angle**
AB = 6 cm, BC = 5 cm, and $C\widehat{A}B = 45°$.

5 **Two angles and a side**
$A\widehat{B}C = 70°$, $B\widehat{C}A = 30°$, and BC = 7 cm.

6 **A right angle, a hypotenuse, and a side**
$A\widehat{B}C = 90°$, AC = 10 cm, and AB = 8 cm.

From the **Investigation** you should have made the following discoveries:

Two triangles are **congruent** if any one of the following is true:

- All corresponding sides are equal in length. **(SSS)**

- Two sides and the **included angle** are equal. **(SAS)**

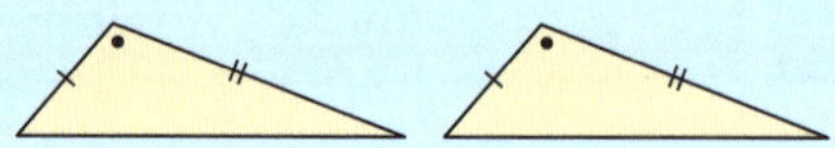

- Two angles and a pair of **corresponding sides** are equal. **(AAcorS)**

- For right angled triangles, the hypotenuses and one pair of sides are equal. **(RHS)**

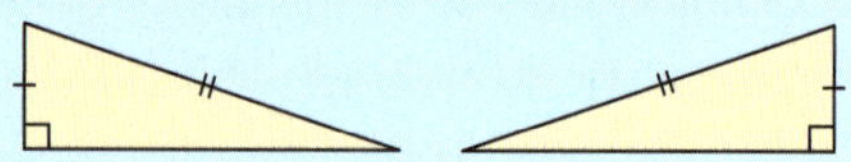

When we use one of these reasons to conclude that two triangles are congruent, we use the appropriate abbreviation to record our reason.

From the **Investigation** you should also have found that:

- If we know two side lengths and a non-included angle, there may be two ways to construct the triangle. We therefore *cannot* conclude that the two triangles are congruent.
- If we know all angles of a triangle, the triangle may still vary in size. We therefore *cannot* conclude that the two triangles are congruent.

Example 1 **Self Tutor**

Determine whether each pair of triangles is congruent, giving reasons for your answers.

a

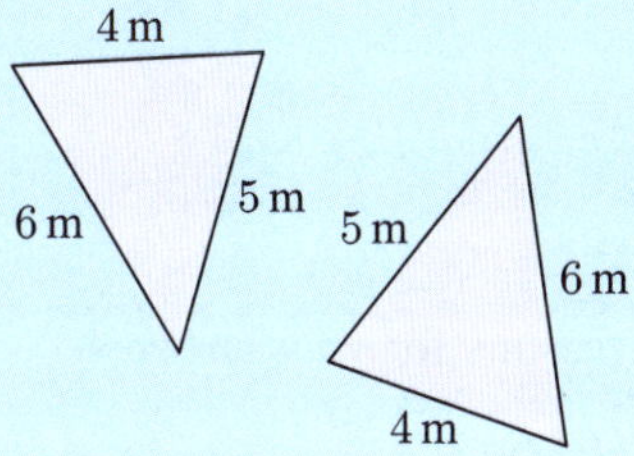

b

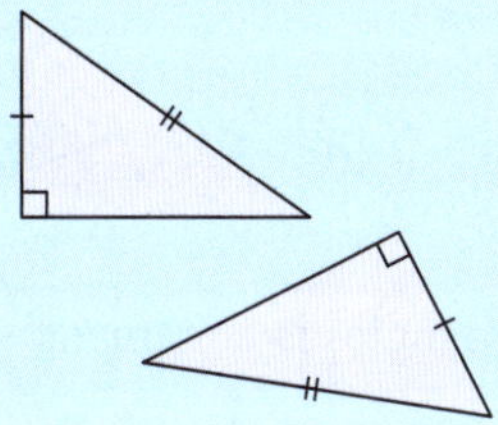

c

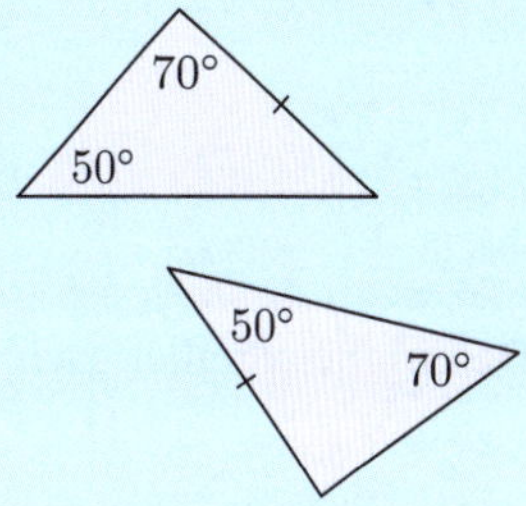

a The triangles are congruent. {SSS}

b The triangles are congruent. {RHS}

c The third angle of each triangle is $180° - 70° - 50° = 60°$. {angle sum of a triangle}

$\therefore$ the triangles are not isosceles.

This is *not* AAcorS since the equal sides are not in corresponding positions. One is opposite the 50° angle, the other is opposite the 70° angle.

$\therefore$ the triangles are not congruent.

Once we have established that two triangles are congruent, we can deduce that the remaining corresponding sides and angles of the triangle are equal.

Example 2 **Self Tutor**

a Show that these triangles are congruent.

b What can be deduced from this congruence?

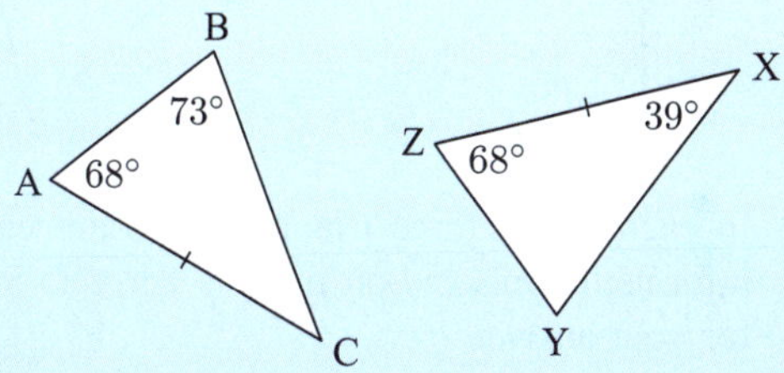

a $A\widehat{C}B = 180° - 68° - 73°$ {angle sum of a triangle}

$= 39°$

In $\triangle$s ABC and XYZ:

- AC = ZX
- $B\widehat{A}C = Y\widehat{Z}X$
- $A\widehat{C}B = Z\widehat{X}Y$

$\therefore$ $\triangle ABC \cong \triangle ZYX$ {AAcorS}

b We can deduce that:

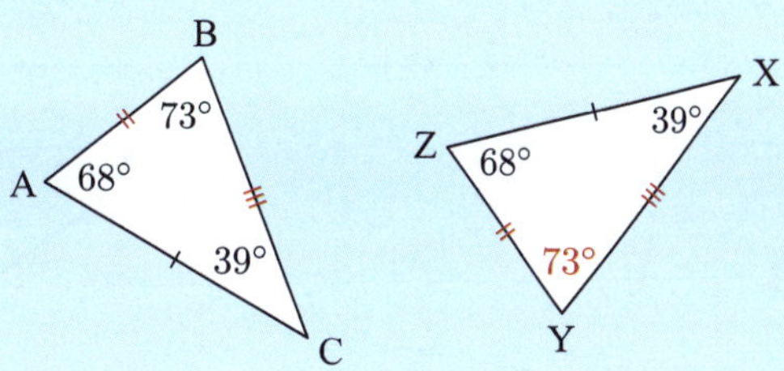

- $X\widehat{Y}Z = 73°$
- AB = ZY
- BC = YX

EXERCISE 22B

1 Explain why we *cannot* conclude that these triangles are congruent:

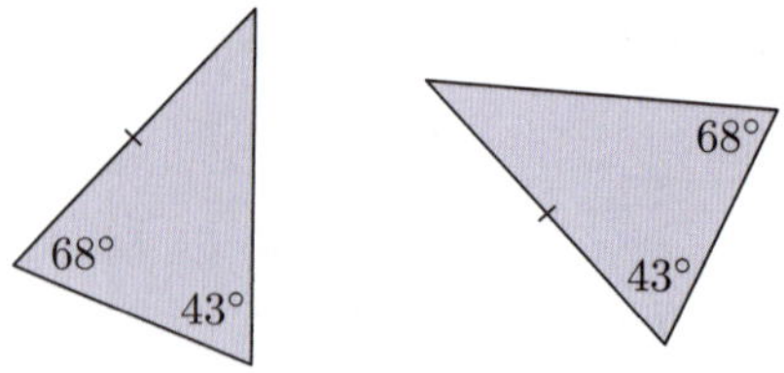

2 Determine whether each pair of triangles is congruent, giving reasons for your answers.

a

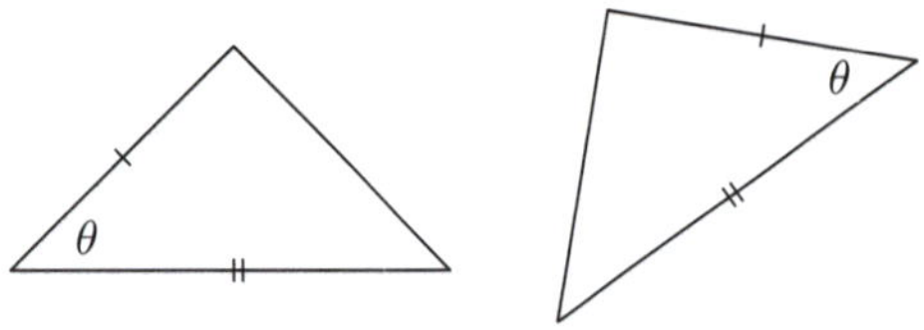

b

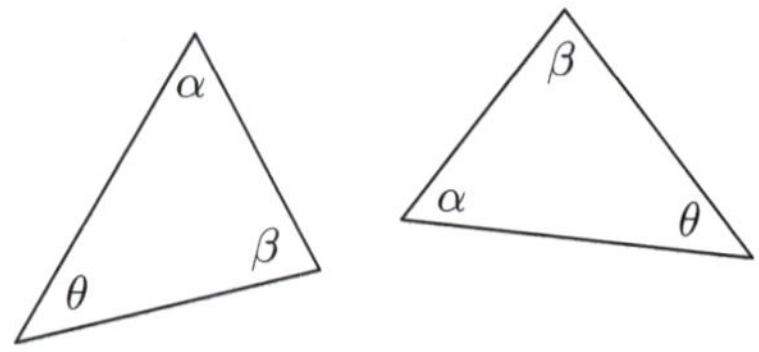

c

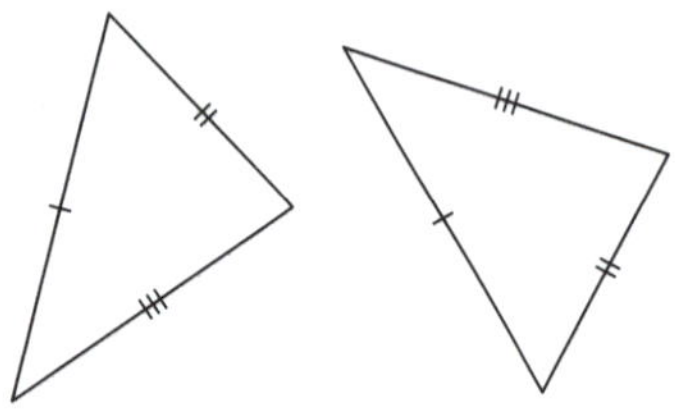

d

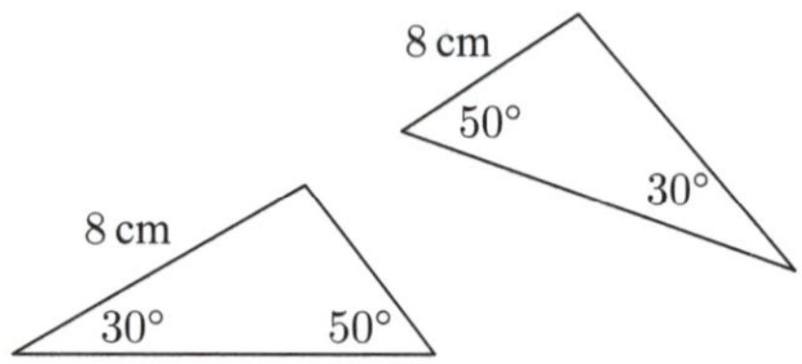

e

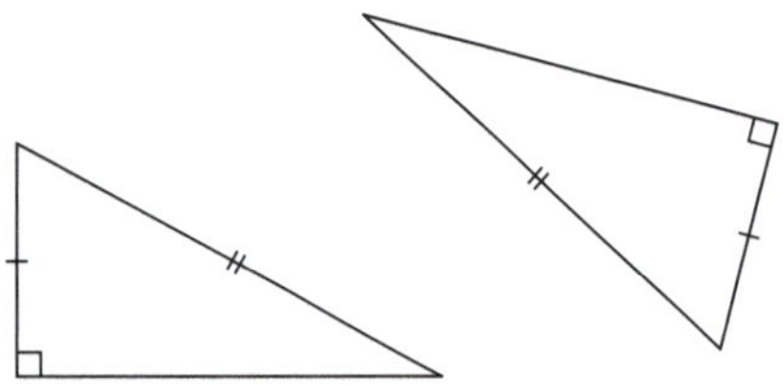

f

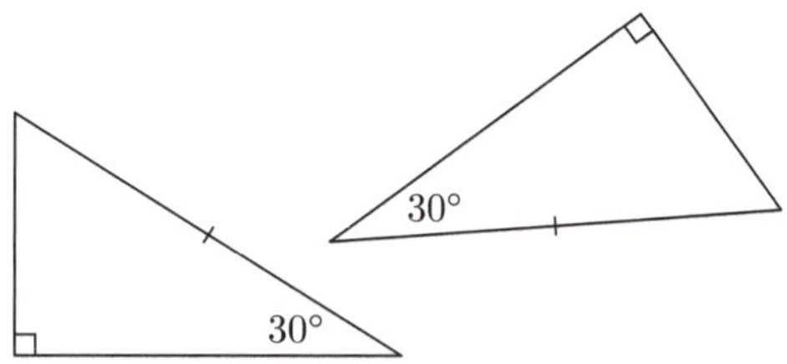

3 In each set of three triangles, two are congruent. The diagrams are *not* drawn to scale, but the information marked on them is correct. State which pair of triangles is congruent, giving a reason for your answer.

a

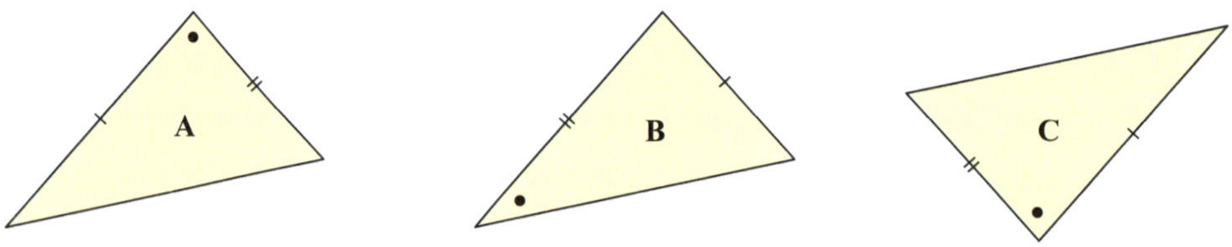

b

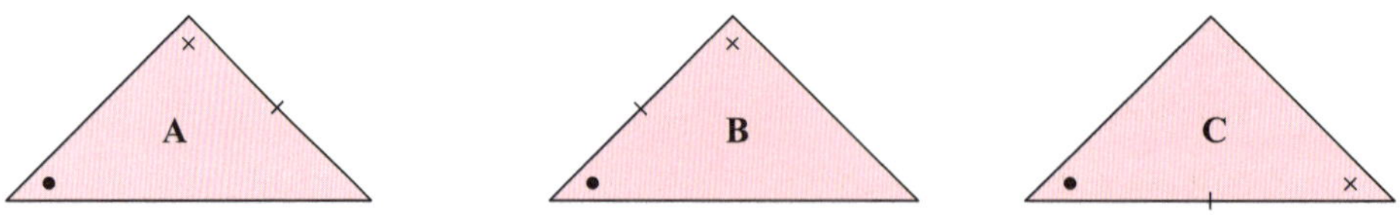

c

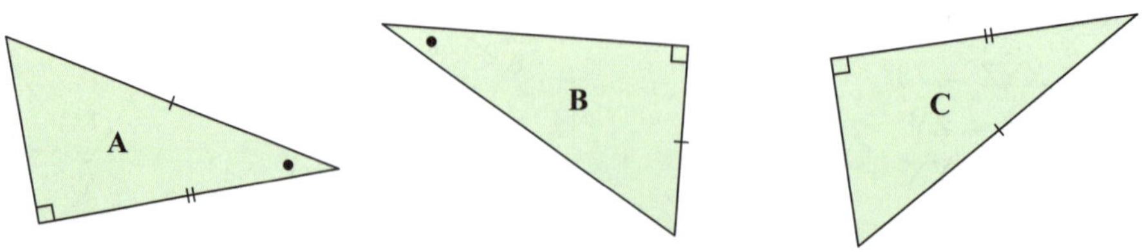

4 The following pairs of triangles are not drawn to scale, but the information on them is correct.

i Determine whether the triangles are congruent.

ii If the triangles are congruent, state what else we can deduce about them.

a

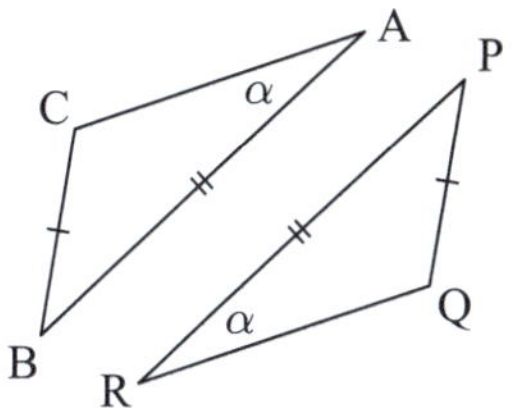

b

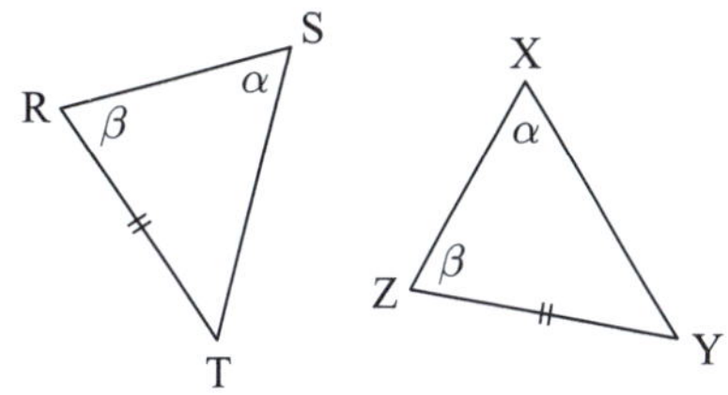

c

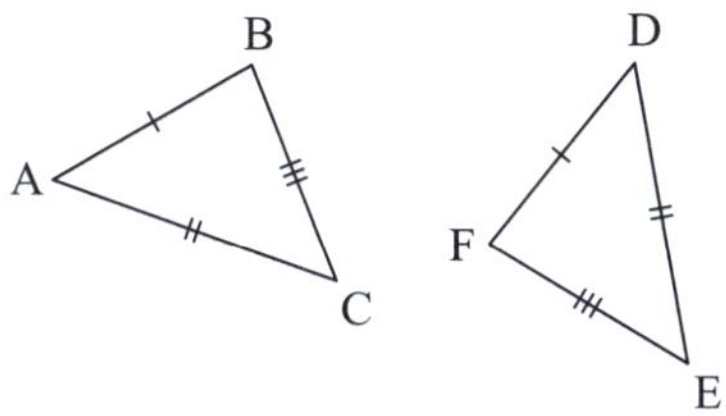

d

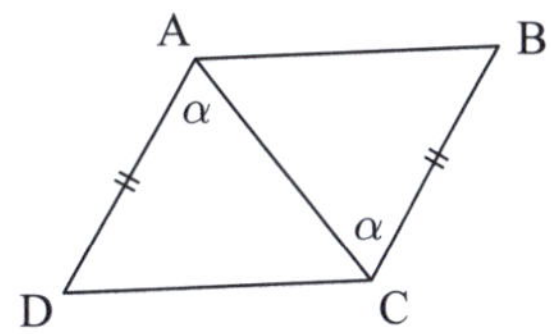

e

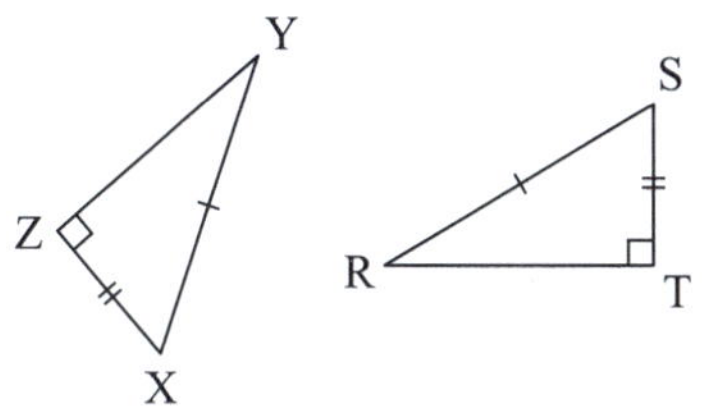

f

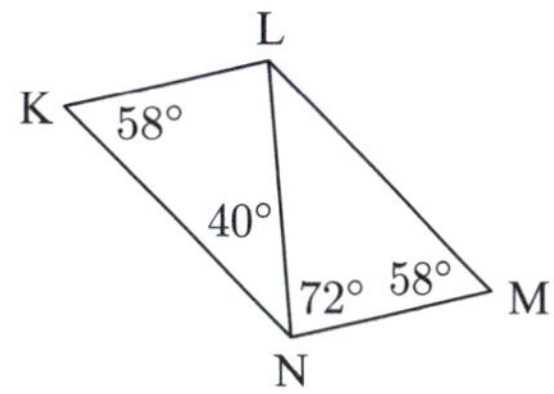

g

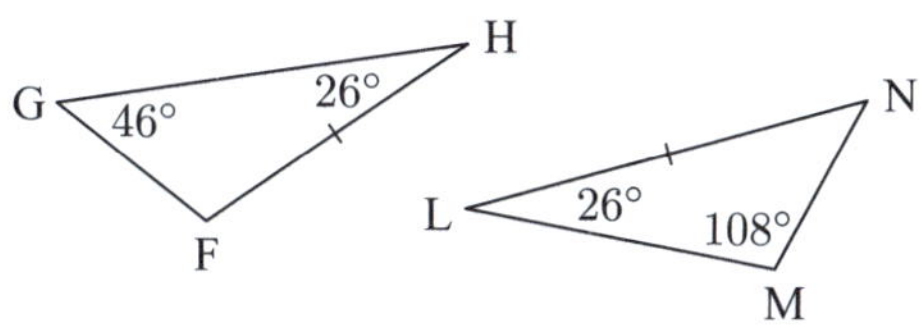

h

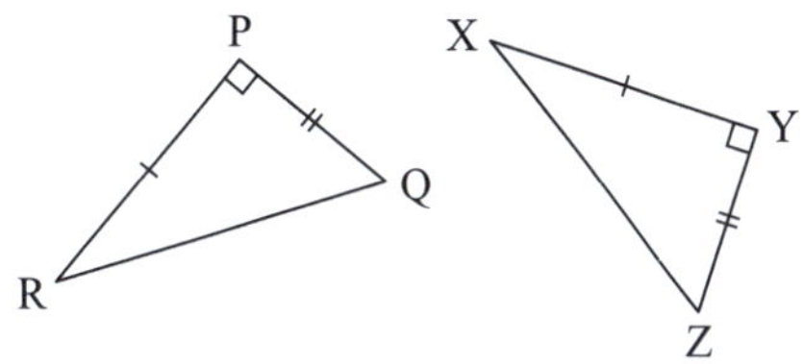

Example 3 **Self Tutor**

Triangle ABC is isosceles with AC = BC.

[BC] and [AC] are produced to E and D respectively so that CE = CD.

Prove that AE = BD.

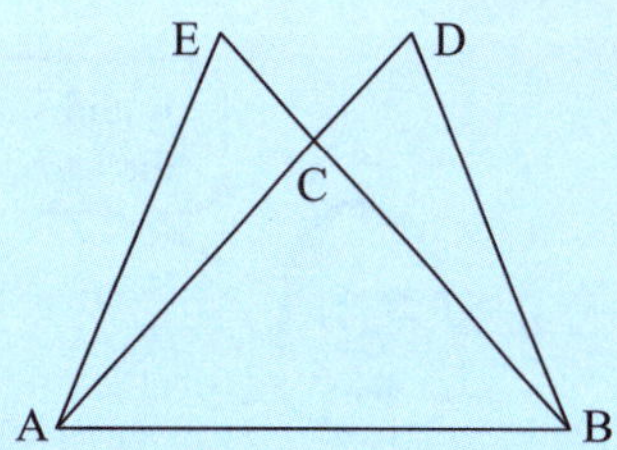

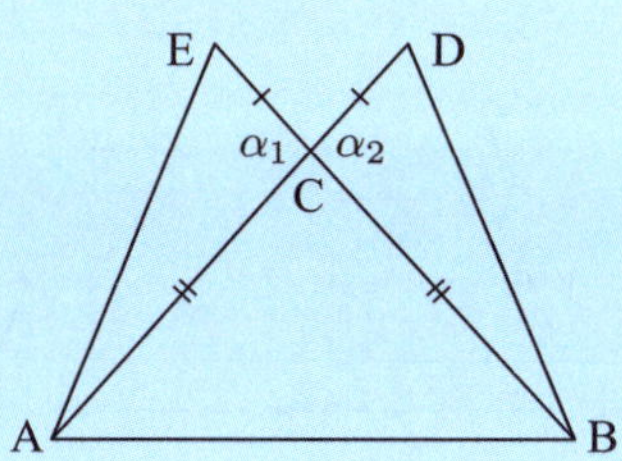

In $\triangle$s ACE and BCD:

- AC = BC {given}
- $\alpha_1 = \alpha_2$ {vertically opposite angles}
- CE = CD {given}

$\therefore$ $\triangle ACE \cong \triangle BCD$ {SAS}

$\therefore$ AE = BD

5 If AC = 5 cm, find the length of [EC].

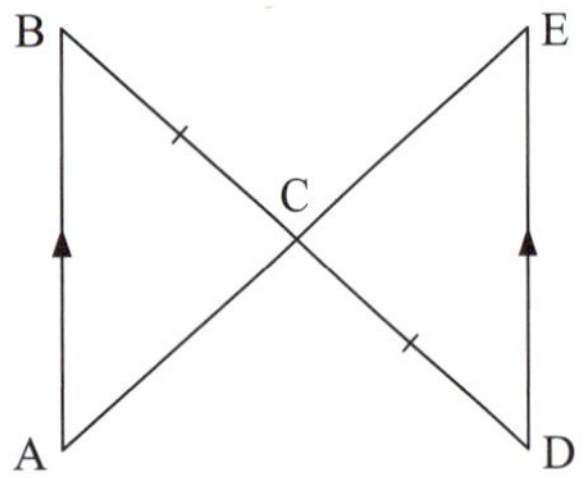

6 [AB] and [CD] are diameters of the circle.
Prove that AC = BD.

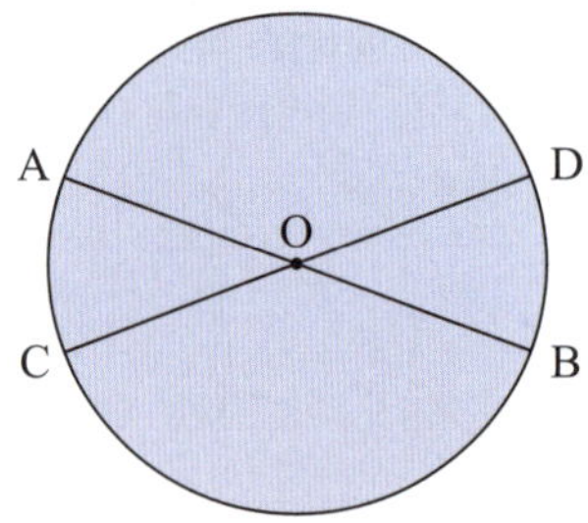

7 Two diagonals are drawn from vertex P of a regular pentagon. Prove that these diagonals are equal in length.

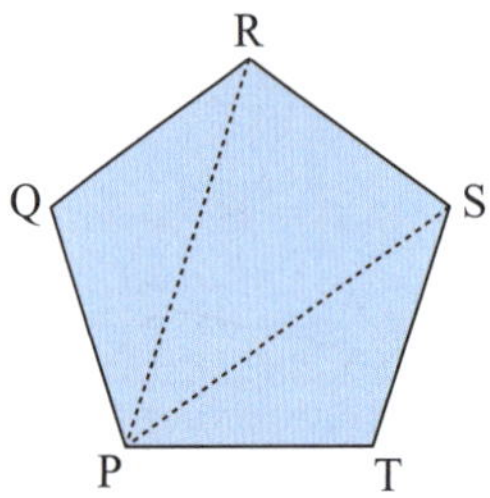

8 In △ABC, [BM] is drawn perpendicular to [AC], and [CN] is drawn perpendicular to [AB].
Suppose these perpendiculars are equal in length.

- **a** Prove that triangles BCM and CBN are congruent.
- **b** Hence prove that △ABC is isosceles.

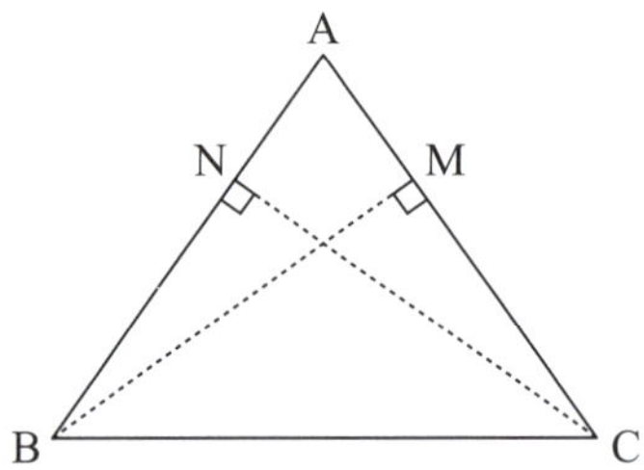

9 Use congruent triangles to show that any point on the perpendicular bisector of [AB] is equidistant from A and B.

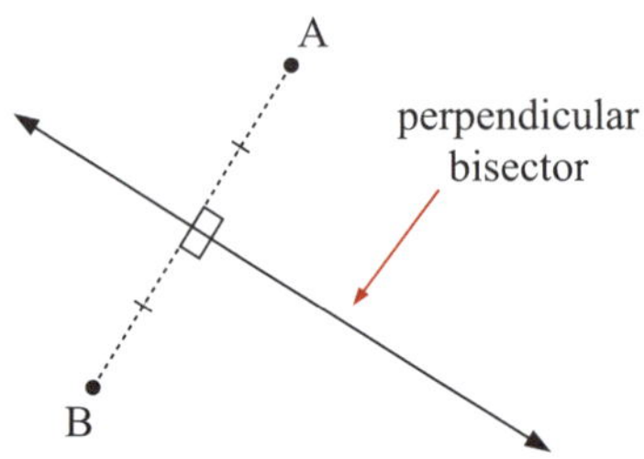

10 An **isosceles trapezium** is a trapezium where the non-parallel sides are equal in length.

- **a** By drawing two lines perpendicular to [DC] through A and B respectively, show that $A\widehat{D}C = B\widehat{C}D$.
- **b** Hence prove that the diagonals of an isosceles trapezium are equal in length.

DISCUSSION

If we are not able to conclude that a pair of triangles *are* congruent, can we necessarily conclude that the triangles *are not* congruent?

For each pair of triangles below, answer these questions:

- Is there sufficient information to conclude that the triangles are congruent?
- Can we say with certainty that the triangles are not congruent?

- 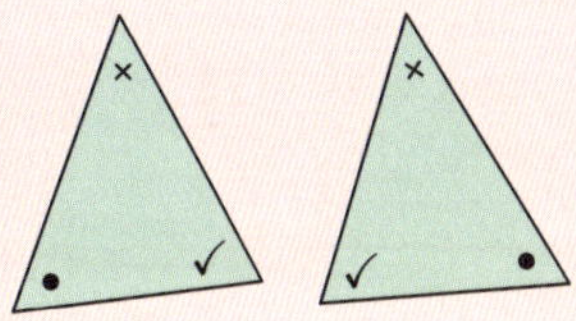
- 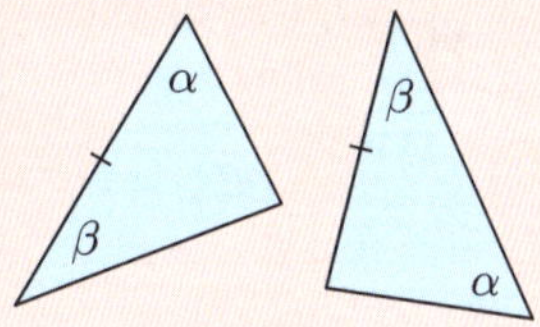

- 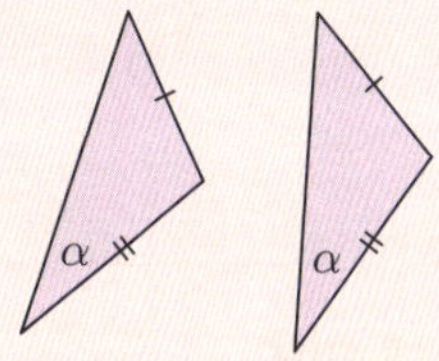

PUZZLE ARE ALL TRIANGLES ISOSCELES?

Charles Dodgson (1832 - 1898), better known by his pen name **Lewis Carroll**, was both a mathematician and writer of children's books including *Alice in Wonderland.*

In this online puzzle you will investigate Dodgson's "proof" that all triangles are isosceles!

PUZZLE

C SIMILARITY

Similar figures have the same *shape* but different *size.*

Two figures are similar if one can be placed exactly on top of the other using an enlargement or reduction, and if necessary a combination of a translation, a rotation, and a reflection.

If two figures are similar then their corresponding side lengths are in the **same ratio**. This ratio is the scale factor of the enlargement from one figure to the other.

For example, in the figures alongside, A′B′C′D′ is an enlargement of ABCD with scale factor $k = 1.5$. The two figures are therefore *similar.*

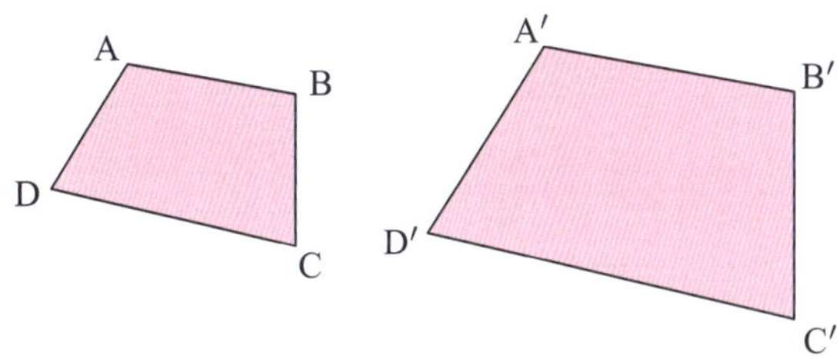

Since $k = 1.5$, $\frac{A'B'}{AB} = \frac{B'C'}{BC} = \frac{C'D'}{CD} = \frac{D'A'}{DA} = \frac{B'D'}{BD} = \frac{A'C'}{AC} = ... = 1.5$.

When a figure is enlarged or reduced, its angles do not change. We say that the figures are **equiangular**.

Two figures are **similar** if:

- the figures are **equiangular** *and*
- the corresponding side lengths are in the **same ratio**.

Example 4 | **Self Tutor**

Find the value of x given that the figures are similar:

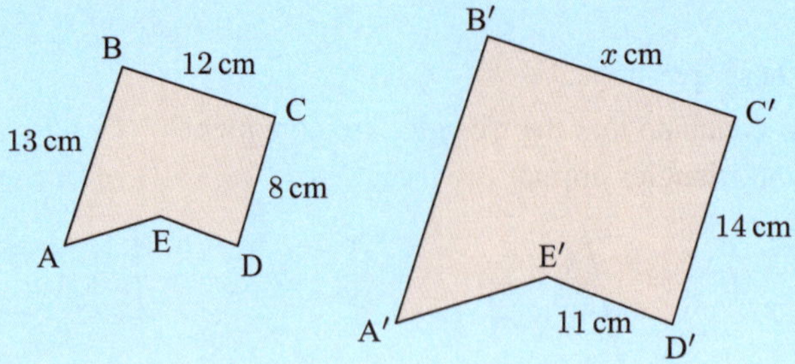

Since the figures are similar, their corresponding sides are in the same ratio.

$\therefore \frac{B'C'}{BC} = \frac{C'D'}{CD}$

$\therefore \frac{x}{12} = \frac{14}{8}$

$\therefore x = \frac{14}{8} \times 12$

$\therefore x = 21$

EXERCISE 22C

1 Determine whether each pair of figures is similar:

a

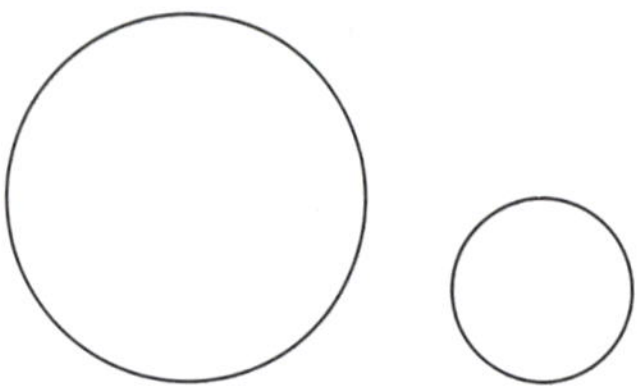

b

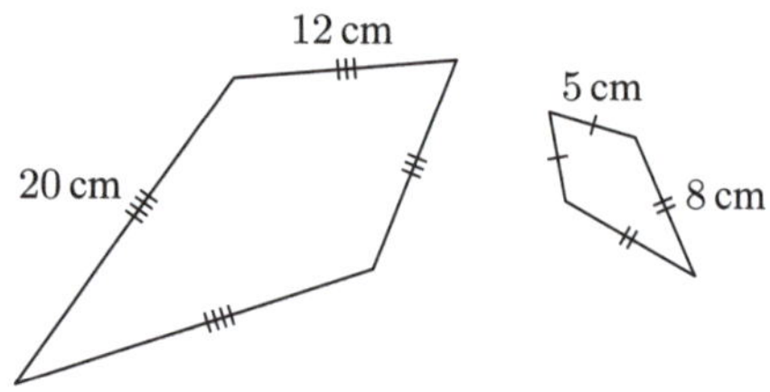

c

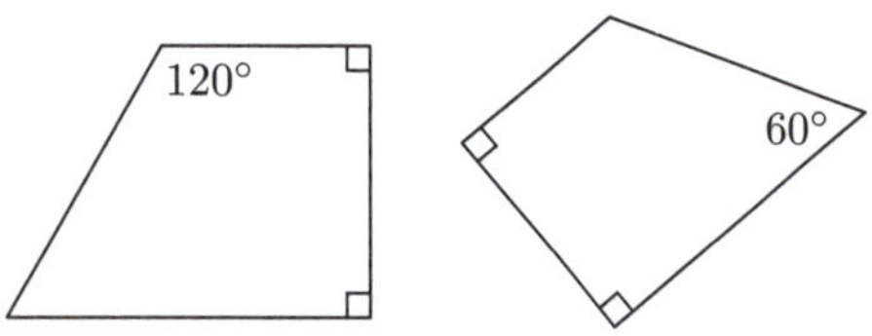

d

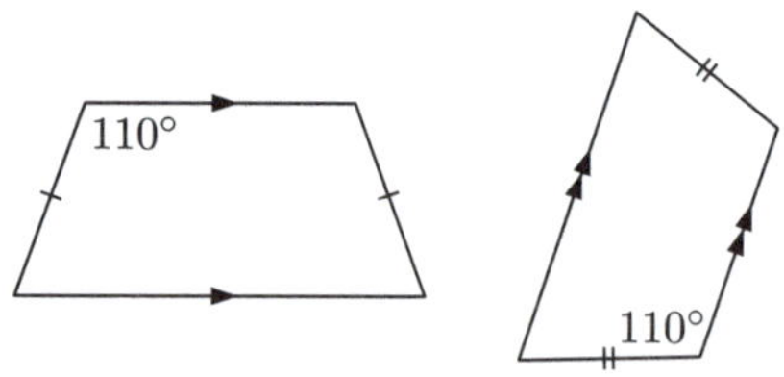

2 Find the value of x given that the figures are similar:

a

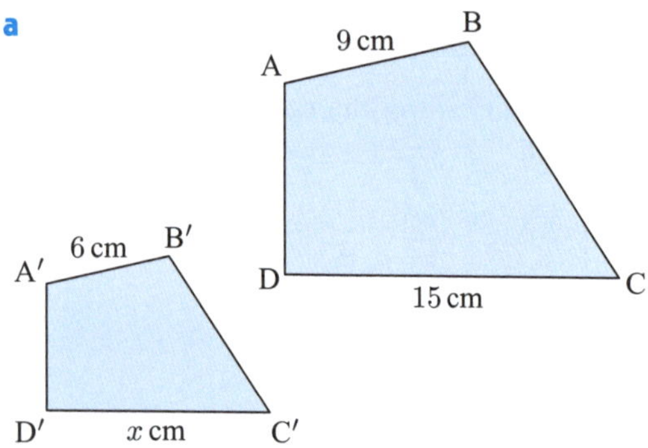

b

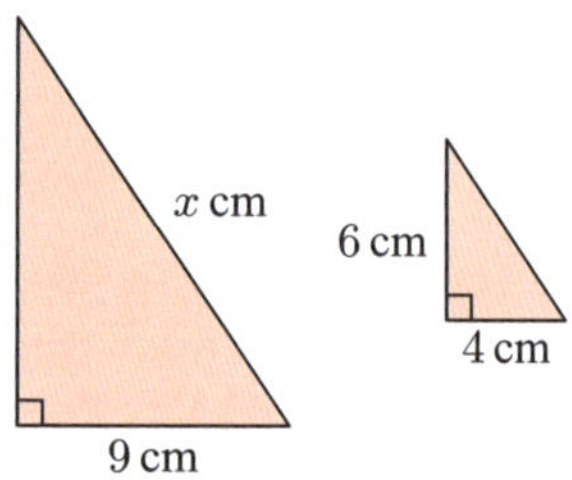

c

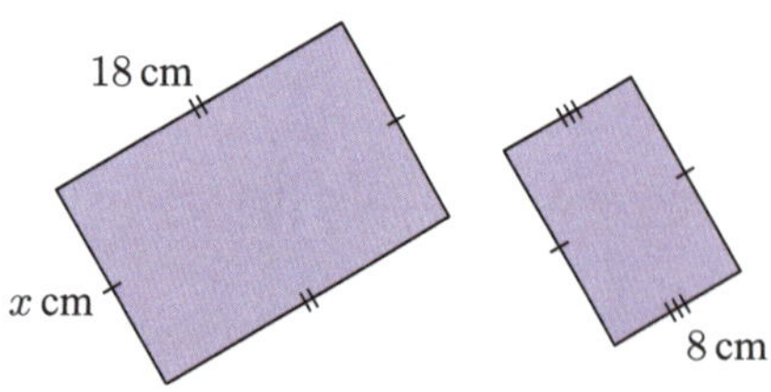

d

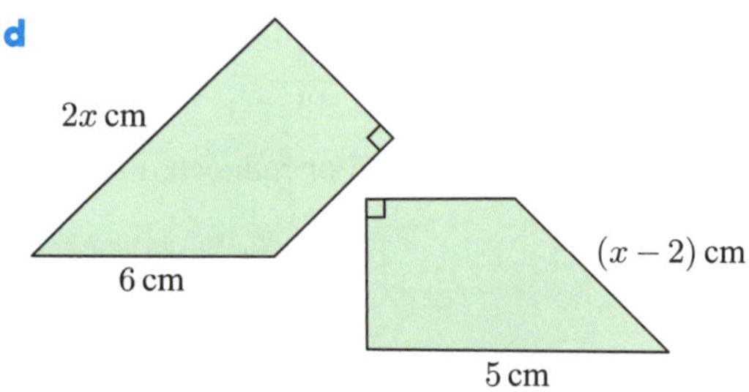

3 Suppose figure **A** is similar to figure **B**, and figure **B** is similar to figure **C**. Is figure **A** similar to figure **C**? Explain your answer.

4 **a** Use a ruler to verify that the side lengths marked equal are indeed equal.

b Measure the angles of each quadrilateral. Hence explain why the quadrilaterals are *not* similar.

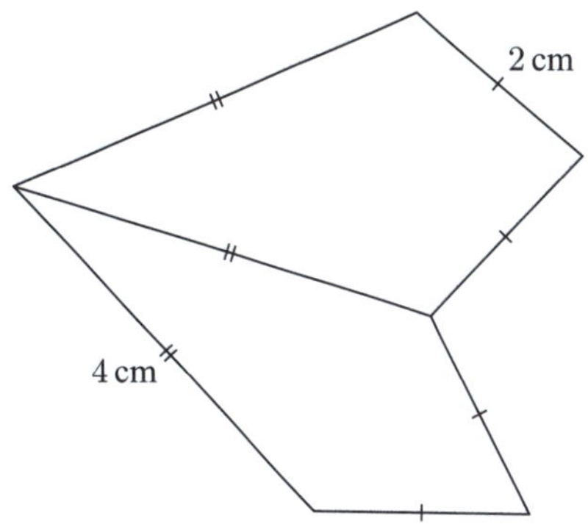

5 Consider the quadrilaterals ABEF and ACDF.

a Show that the quadrilaterals are equiangular.

b Explain why the quadrilaterals are *not* similar.

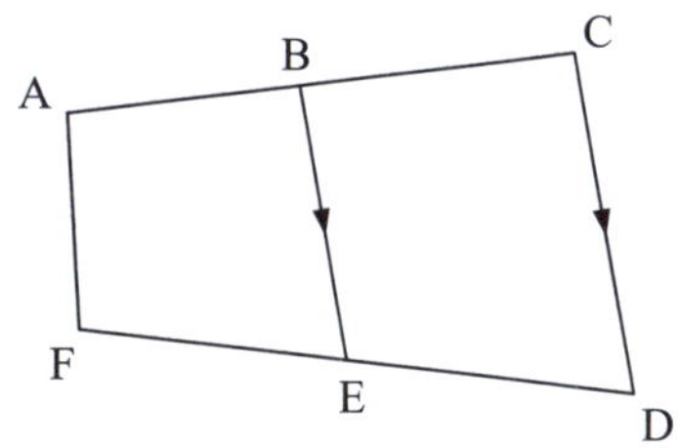

D SIMILAR TRIANGLES

In the previous Exercise we saw that quadrilaterals which are equiangular are not necessarily similar, and quadrilaterals with sides in proportion are not necessarily similar. However, if *triangles* are equiangular then their corresponding sides *must* be in the same ratio, and vice versa. So, to show that two triangles are similar, we only need to show that *one* of these properties is true.

TESTS FOR TRIANGLE SIMILARITY

Two triangles are **similar** if *either*:
- they are **equiangular** *or*
- their side lengths are in the **same ratio**.

Notice that since the angles of any triangle sum to 180°, if two angles of one triangle are equal to two angles of another triangle, then the remaining angles of the triangles must also be equal.

EXERCISE 22D.1

1 Explain why these triangles are similar:

a

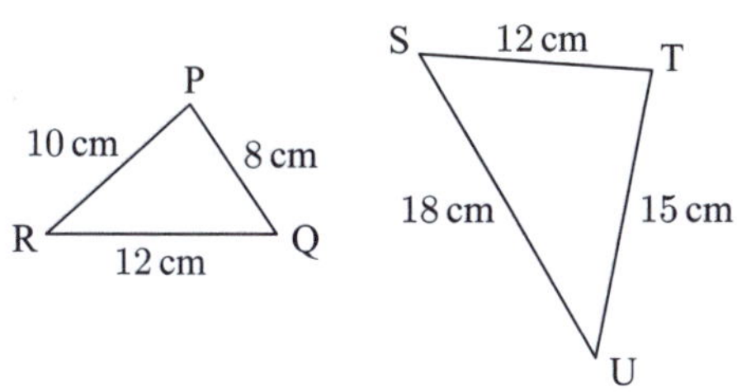

b

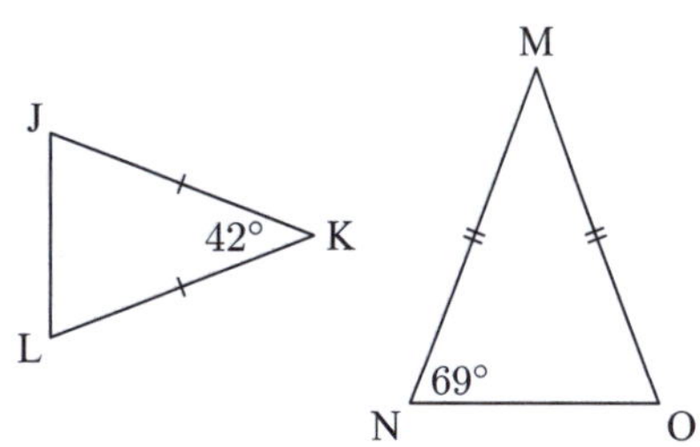

Example 5 — Self Tutor

Show that these figures possess similar triangles:

a

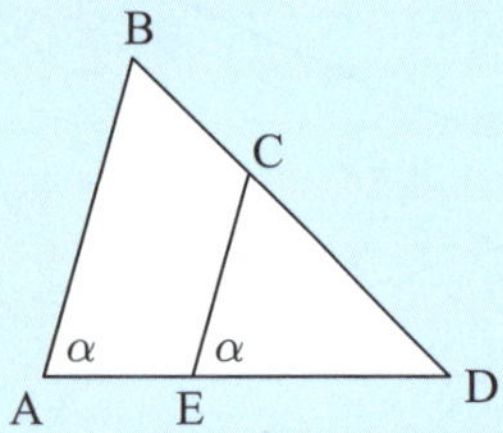

b

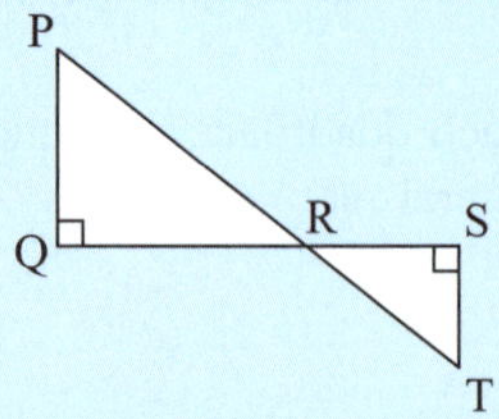

a

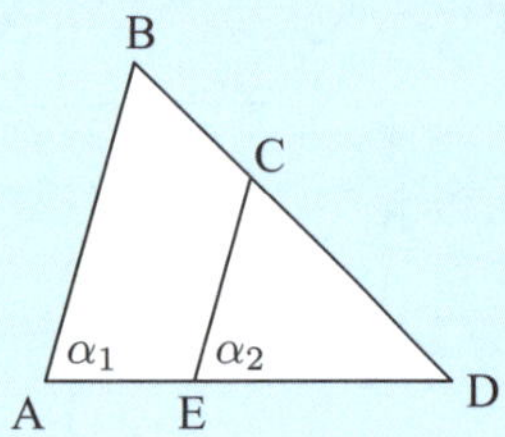

$\triangle$s ABD and ECD are equiangular as:

- $\alpha_1 = \alpha_2$ {given}
- the angle at D is common to both triangles

$\therefore$ the triangles are similar.

b

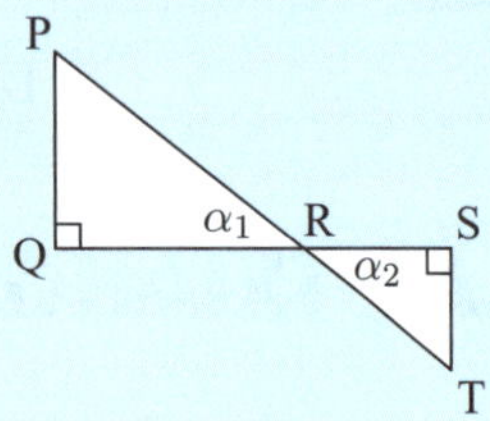

$\triangle$s PQR and TSR are equiangular as:

- $P\widehat{Q}R = T\widehat{S}R = 90^\circ$ {given}
- $\alpha_1 = \alpha_2$ {vertically opposite angles}

$\therefore$ the triangles are similar.

If two triangles are similar, we list corresponding vertices in the same order.

2 Show that these figures possess similar triangles:

a

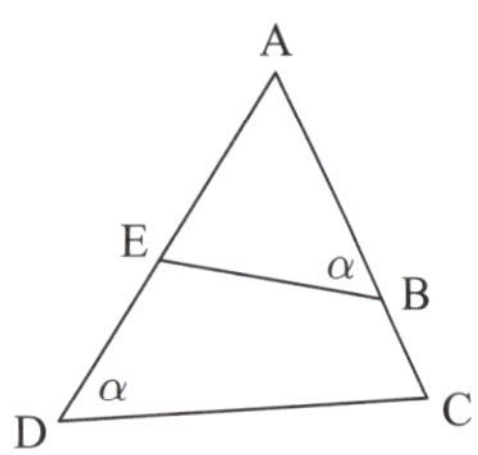

b

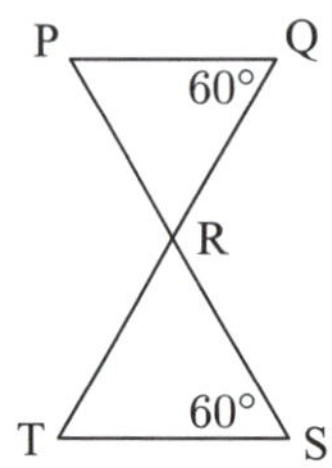

c

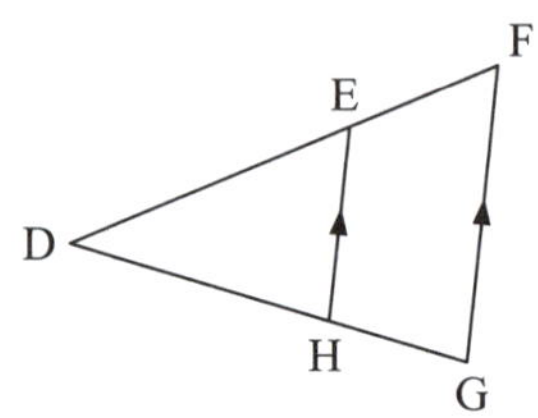

d

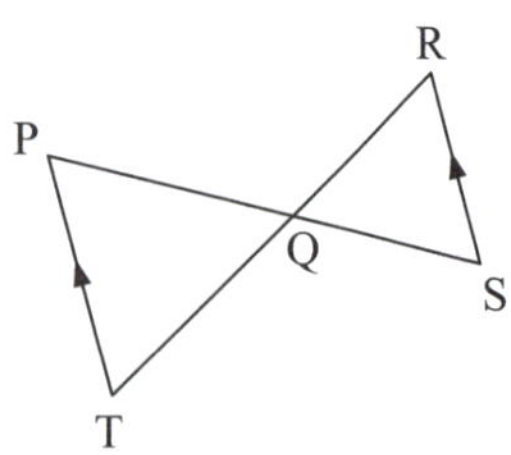

e

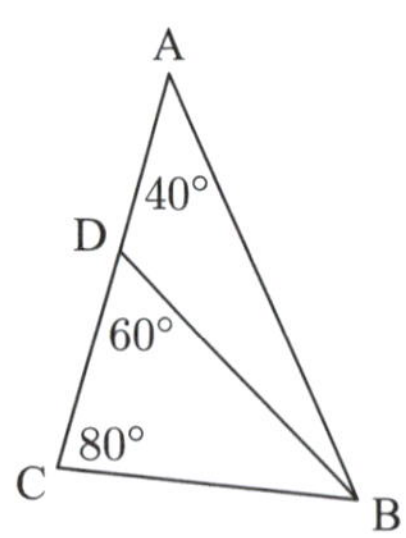

f

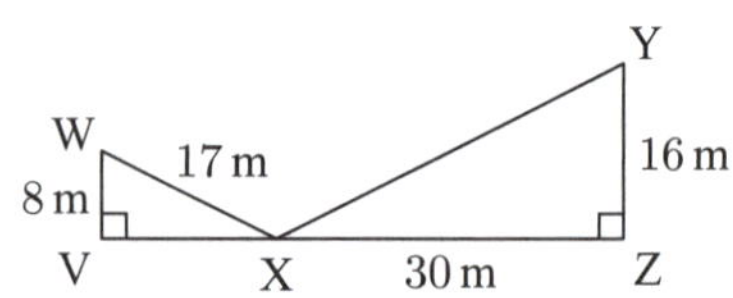

3 Congruent triangles ABC and CDE are placed next to each other so that ACE is a straight line.
Show that the shaded triangles are similar.

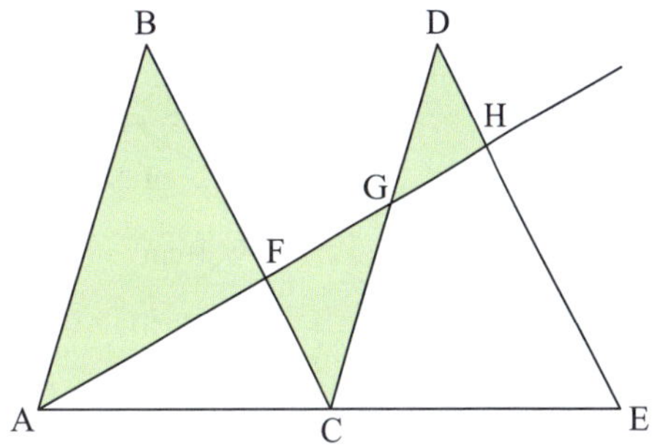

FINDING SIDE LENGTHS

Once we have shown that two triangles are similar, we can use the fact that corresponding sides are in the same ratio to find unknown lengths.

Example 6 **Self Tutor**

In the figure alongside, BD = 20 cm.

Show that a pair of triangles is similar, and hence find the value of x.

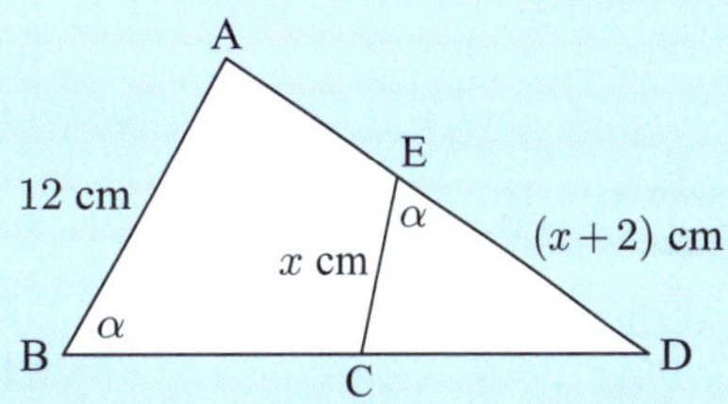

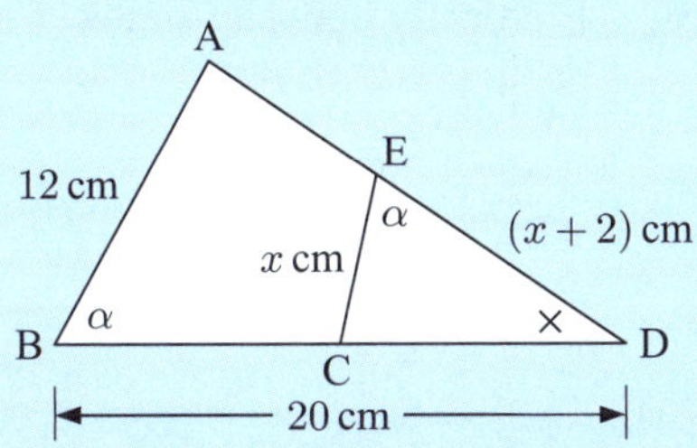

$\triangle$s CED and ABD are equiangular since:

- $\widehat{CED} = \widehat{ABD}$ {given}
- the angle at D is common

$\therefore$ the triangles are similar.

$$\therefore \frac{ED}{BD} = \frac{CE}{AB} \quad \text{\{same ratio\}}$$

$$\therefore \frac{x+2}{20} = \frac{x}{12}$$

$$\therefore 12(x+2) = 20x$$

$$\therefore 12x + 24 = 20x$$

$$\therefore 24 = 8x$$

$$\therefore x = 3$$

EXERCISE 22D.2

1 In each figure, establish that a pair of triangles is similar. Hence find the value of x.

a

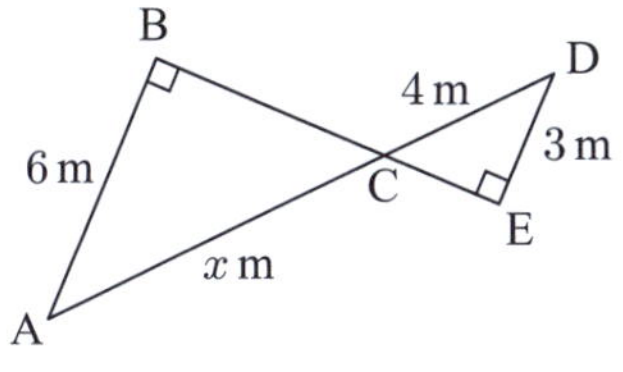

b

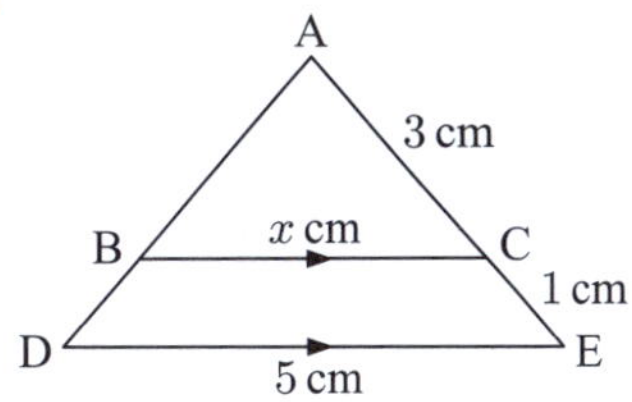

c

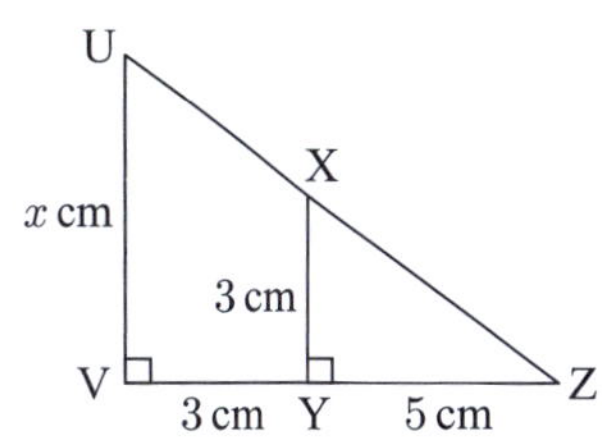

d

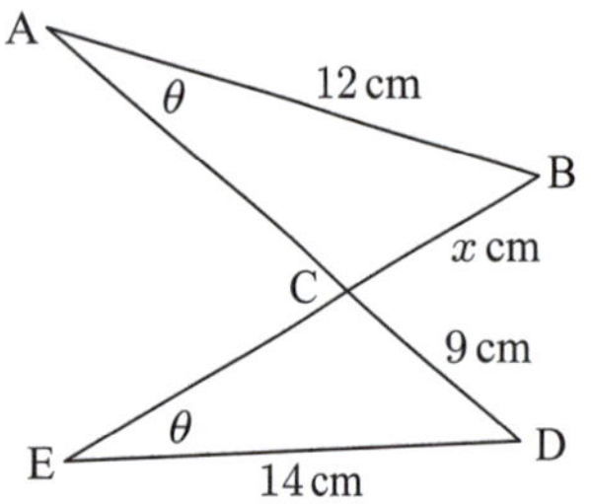

e

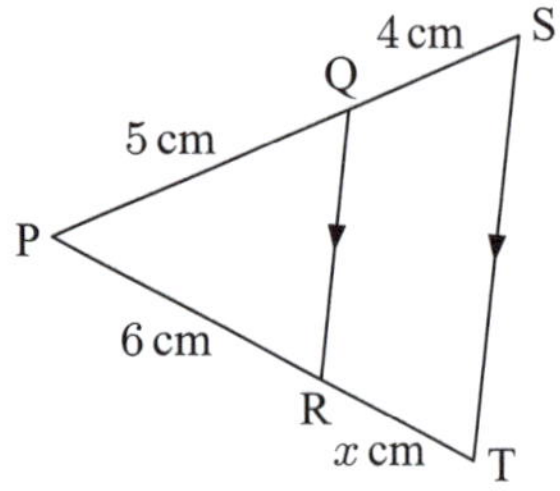

f

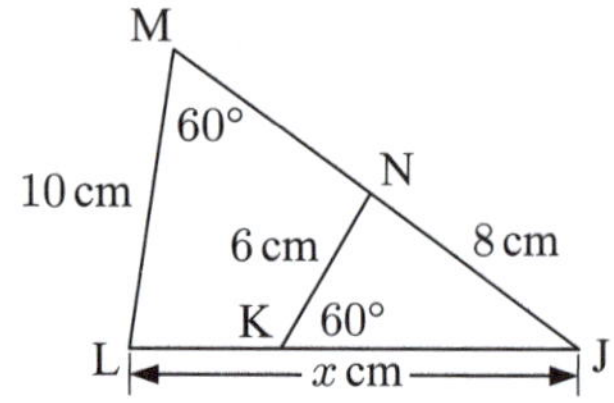

2 Find the value of x:

a

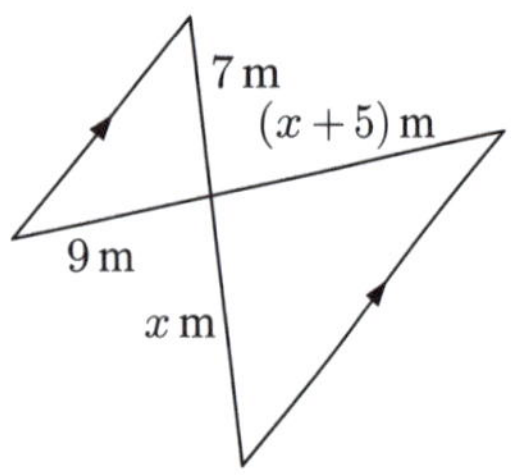

b

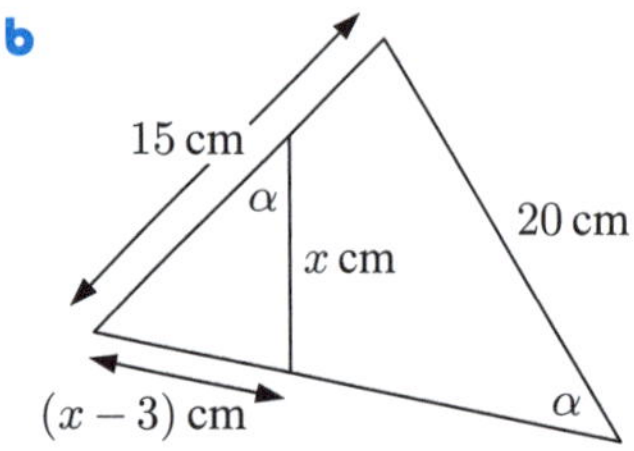

c

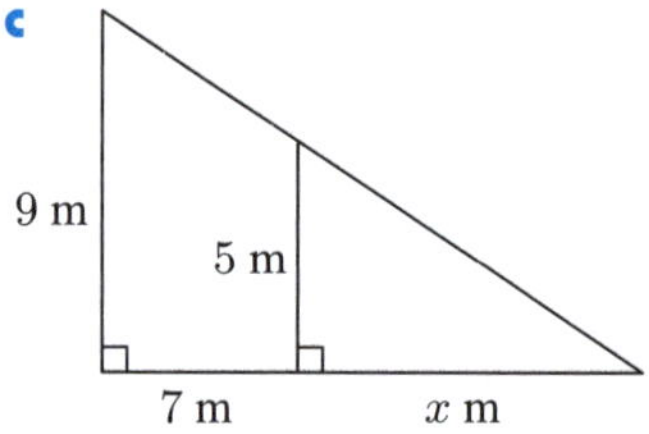

Example 7 **Self Tutor**

Find the value of x:

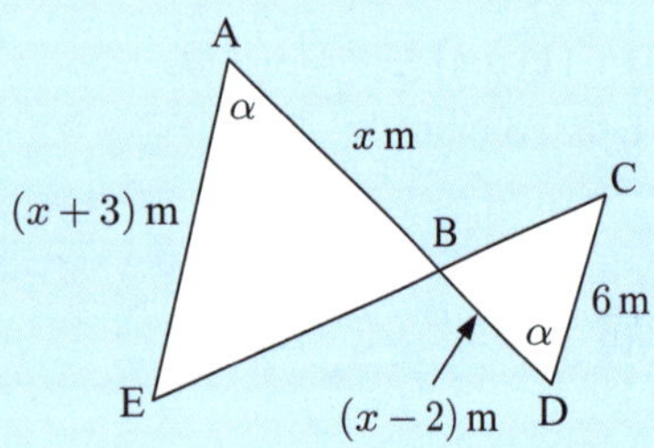

$\triangle$s ABE and DBC are equiangular as:

- $B\widehat{A}E = B\widehat{D}C$ {given}
- $A\widehat{B}E = D\widehat{B}C$ {vertically opposite angles}

$\therefore$ the triangles are similar.

$$\therefore \frac{AE}{DC} = \frac{AB}{DB} \quad \{\text{same ratio}\}$$

$$\therefore \frac{x+3}{6} = \frac{x}{x-2}$$

$$\therefore (x+3)(x-2) = 6x$$

$$\therefore x^2 + x - 6 = 6x$$

$$\therefore x^2 - 5x - 6 = 0 \quad \{\text{making RHS} = 0\}$$

$$\therefore (x+1)(x-6) = 0 \quad \{\text{factorising}\}$$

$$\therefore x + 1 = 0 \quad \text{or} \quad x - 6 = 0$$

$$\therefore x = -1 \text{ or } 6$$

But $x > 0$, so $x = 6$.

3 Find the value of x:

a

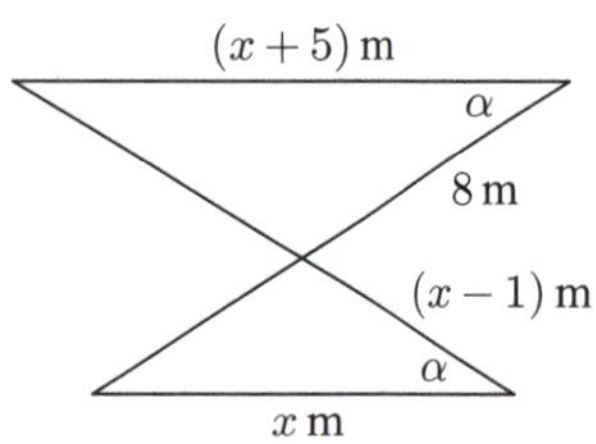

b

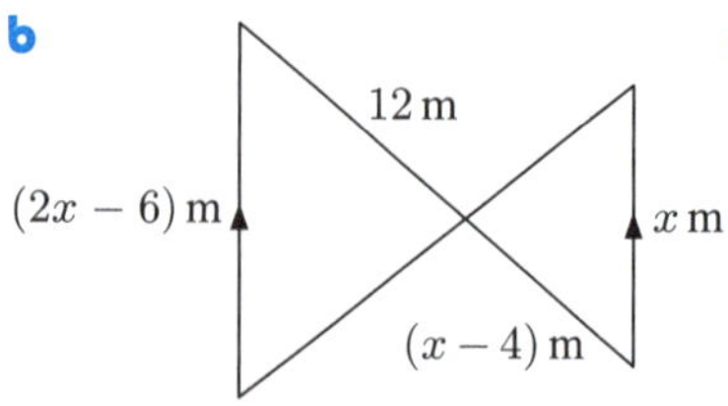

c

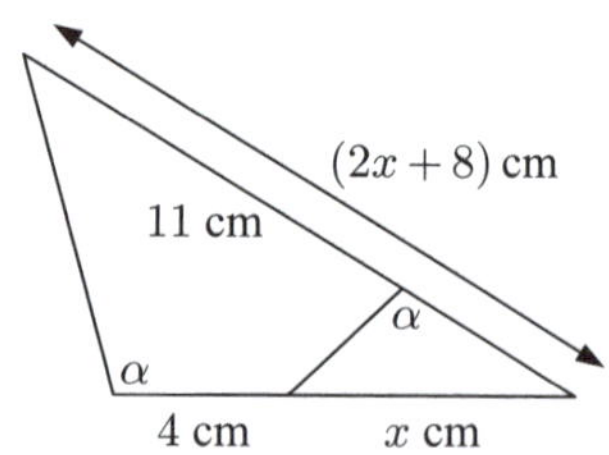

4 Find the values of x and y:

a

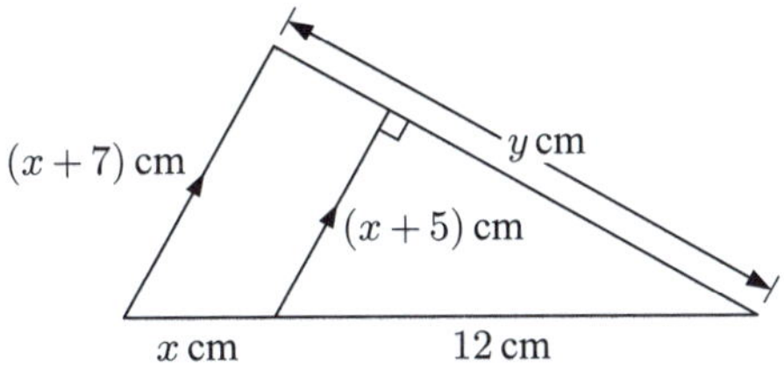

b

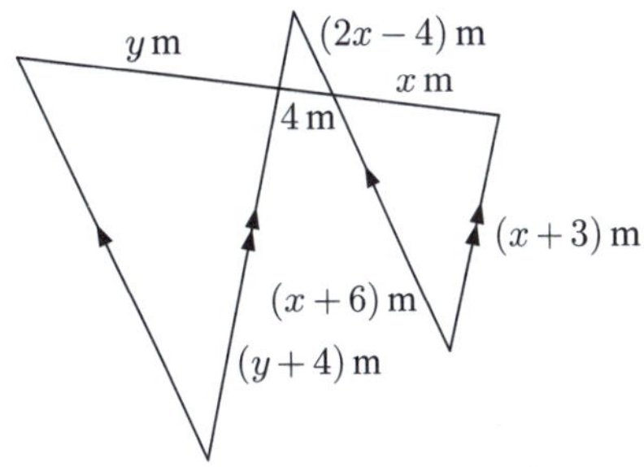

Example 8 **Self Tutor**

A 1.5 m tall boy stands 8 m from the base of a 6.3 m light pole.

Find the length of the boy's shadow.

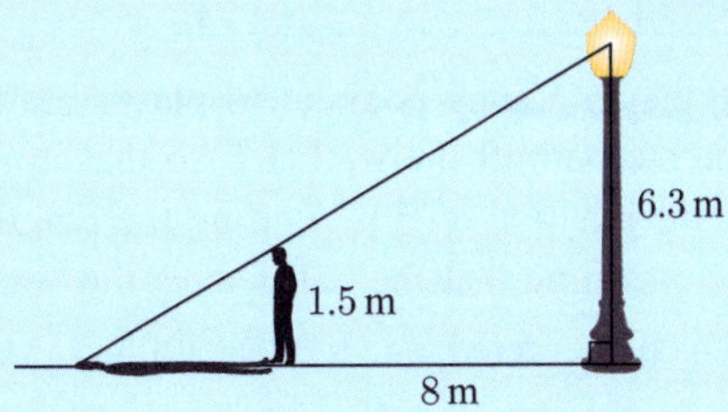

Let the length of the shadow be x m.

$\triangle$s ABE and ACD are equiangular as:

- $A\widehat{E}B = A\widehat{D}C = 90°$
- the angle at A is common

$\therefore$ the triangles are similar.

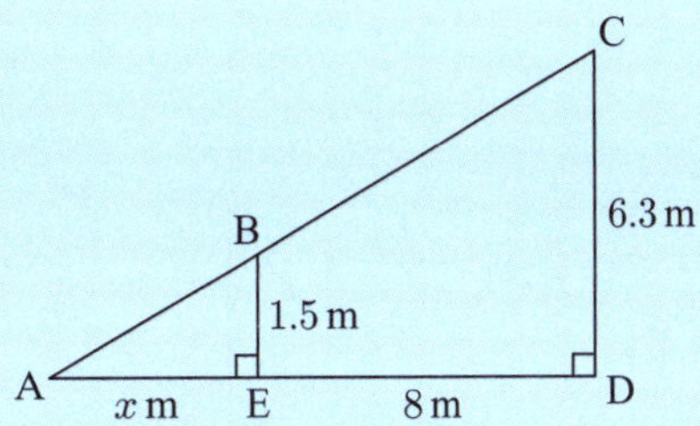

$\therefore \dfrac{AE}{AD} = \dfrac{BE}{CD}$ {same ratio}

$\therefore \dfrac{x}{x+8} = \dfrac{1.5}{6.3}$

$\therefore 6.3x = 1.5(x+8)$

$\therefore 6.3x = 1.5x + 12$

$\therefore 4.8x = 12$

$\therefore x = 2.5$ The boy's shadow is 2.5 m long.

5 A road has a constant slope of 2 in 11. This means that if you travel 11 m horizontally along the road, you will rise 2 m.

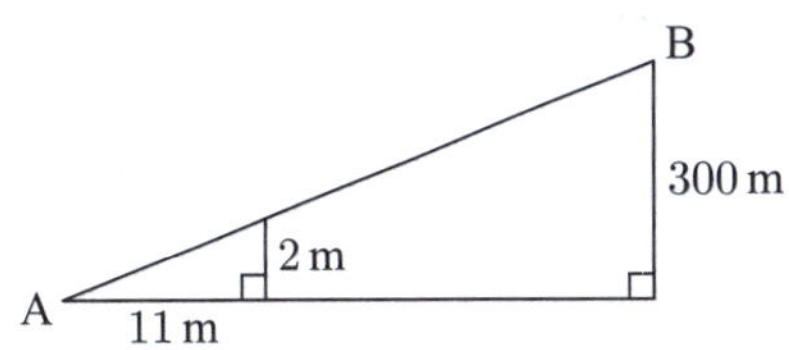

a Over what horizontal distance will the road rise 300 m?

b Find the length of the road [AB] to the nearest metre.

6 Enrico is standing at point A on one side of a river, directly opposite a tree T. Enrico walks 10 m to P and then a further 6 m to C. He then walks 8 m directly away from the river until he reaches D, which is in line with P and T.

Find the width of the river at A.

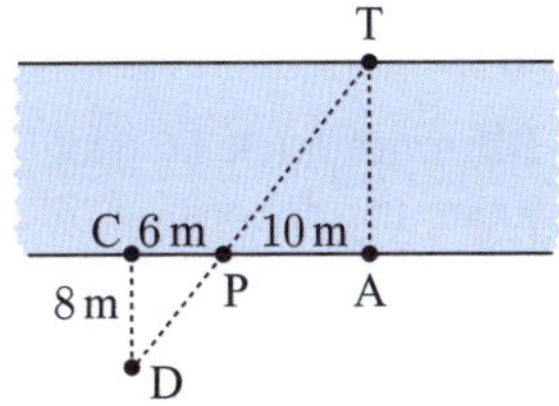

7 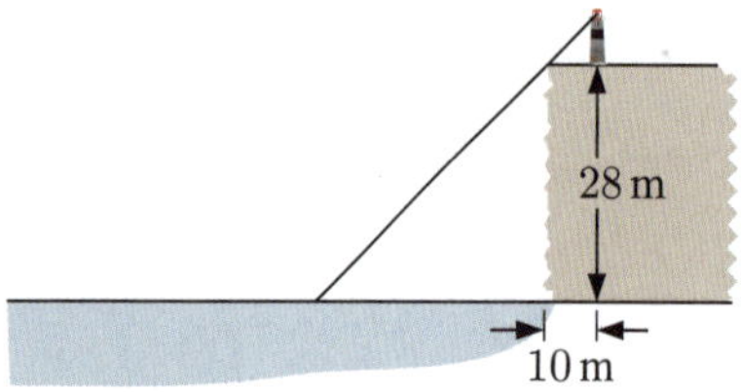

A lighthouse is situated 10 m back from the edge of a 28 m high cliff. The shadow extends 35 m from the cliff. Find the height of the lighthouse.

8 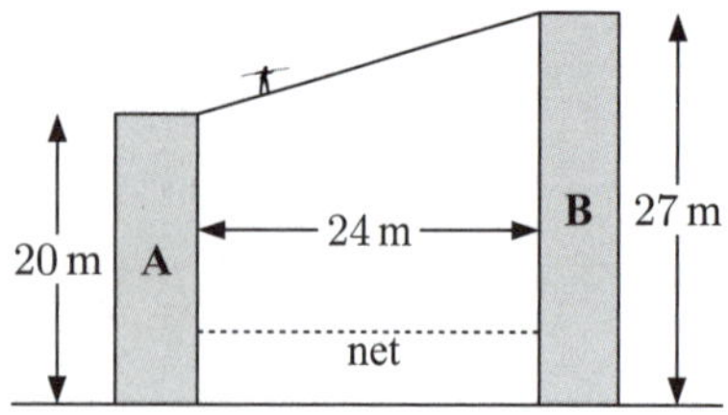

A tightrope walker is walking between the roofs of two buildings A and B. The buildings are 24 m apart. There is a safety net 5 m above ground level.

a Find the length of the tightrope.

b After travelling 15 m, the tightrope walker falls off. How far will he fall before landing in the net?

9 Squash player Kenny is about to play a shot at K. He wants to rebound the ball off the wall at P so that it hits the corner at Q. Locate the position of P by finding the length of [PR].
You may assume that the ball leaves the wall at the same angle with which it entered, so $\text{K}\widehat{\text{P}}\text{R} = \text{Q}\widehat{\text{P}}\text{S}$.

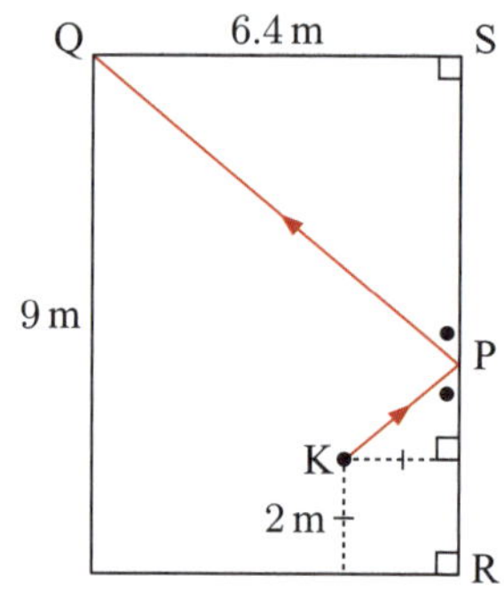

10 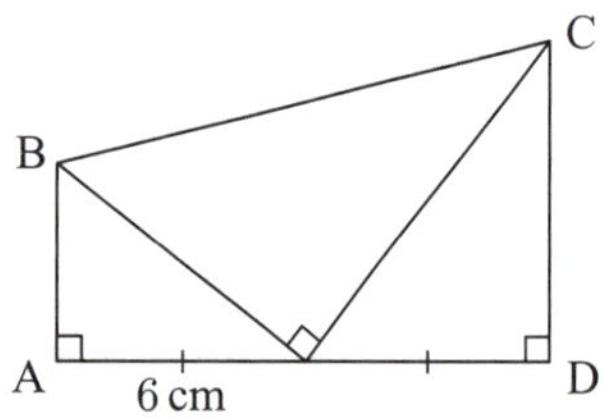

Trapezium ABCD can be divided into three right angled triangles. Given that [DC] is 5 cm longer than [AB], find the area of the trapezium.

11 Alison and Brianna are in the basement of a department store. Alison steps onto an escalator to the ground floor, and Brianna steps onto an escalator to the first floor. The escalators' paths cross at P. When Alison reaches P, she has completed $\frac{13}{20}$ of her escalator trip.

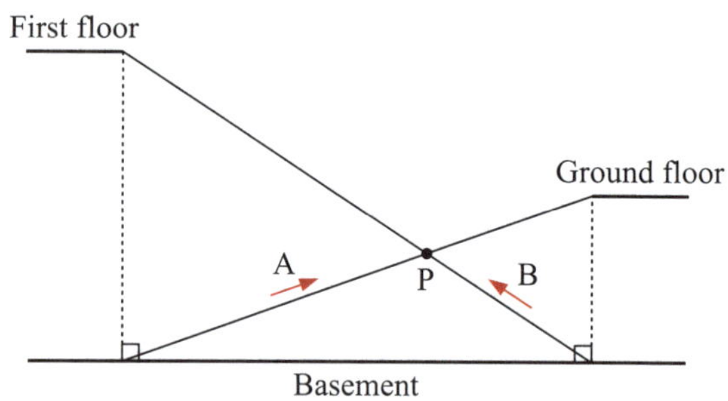

a What fraction of Brianna's escalator trip has been completed when she reaches P?

b Find the ratio of the height from the basement to the ground floor, compared with the height from the ground floor to the first floor.

ACTIVITY — THE BILTMORE STICK

A **Biltmore stick** is a tool invented in the late 19th century which is used to estimate the dimensions of a tree.

Click on the icon to obtain this practical Activity.

THE BILTMORE STICK

E AREAS OF SIMILAR FIGURES

We have seen that if two objects are similar, then there is a relationship between the lengths of their sides. The sides are *in proportion.*

For 2-dimensional figures, there is also a relationship between their *areas*.

For example:

- Circle **B** is an enlargement of circle **A** with scale factor k.

  Area of **B** $= \pi(kr)^2$
  $= k^2 \times \pi r^2$
  $= k^2 \times$ area of **A**

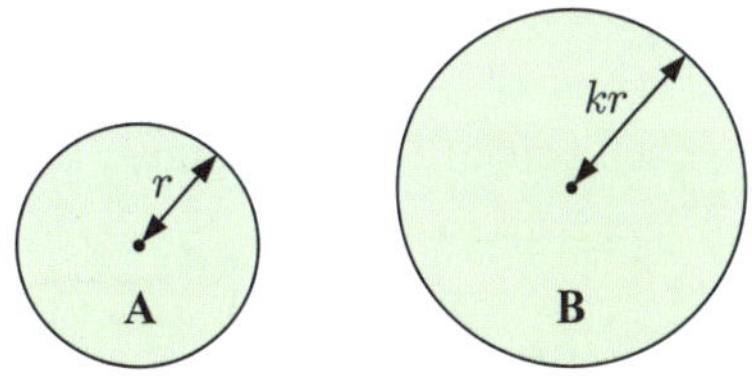

- Rectangle **B** is an enlargement of rectangle **A** with scale factor k.

  Area of **B** $= ka \times kb$
  $= k^2 \times ab$
  $= k^2 \times$ area of **A**

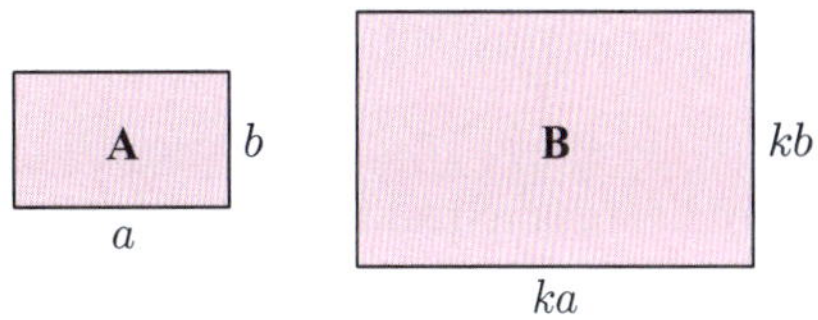

Using examples like this we can conclude that:

> If a figure is enlarged with scale factor k to produce a similar figure, then the new area $= k^2 \times$ the old area.

Example 9 Self Tutor

For each pair of similar figures, find the value of x:

a

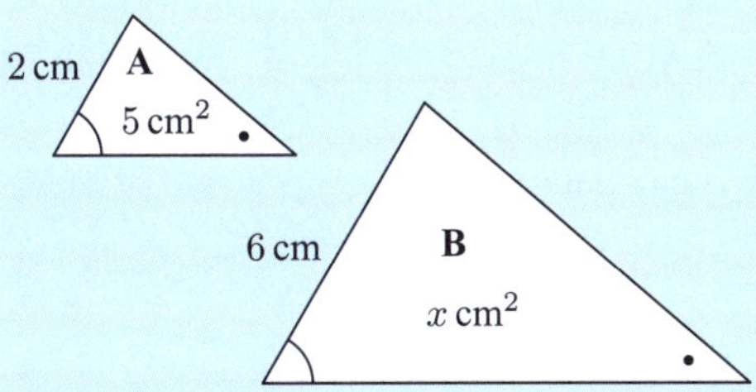

b

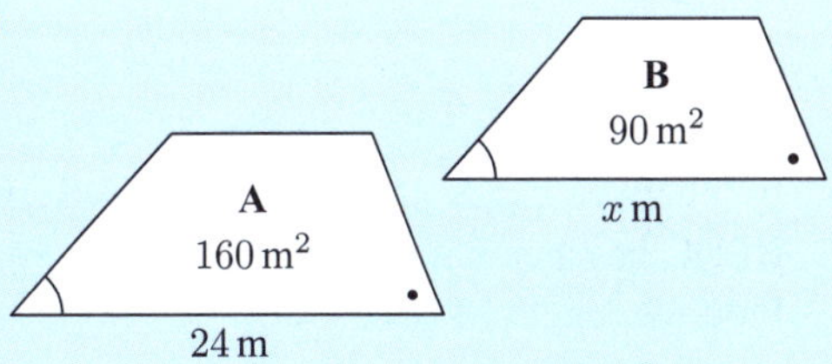

a Suppose figure **A** is enlarged with scale factor k to give figure **B**.
Using the known corresponding side lengths, $k = \frac{6}{2} = 3$.

Area of **B** $= k^2 \times$ area of **A**
$\therefore\ x = 3^2 \times 5$
$= 45$

b Suppose figure **A** is reduced with scale factor k to give figure **B**.

Area of **B** $= k^2 \times$ area of **A**
$\therefore\ 90 = k^2 \times 160$
$\therefore\ \frac{9}{16} = k^2$
$\therefore\ k = \frac{3}{4}$ {as $k > 0$}
Now $x = \frac{3}{4} \times 24$ {corresponding sides}
$\therefore\ x = 18$

EXERCISE 22E

1 For each pair of similar figures, find the unknown area:

a

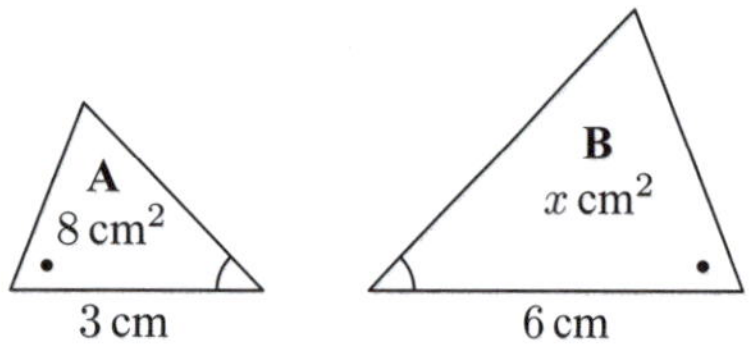

b

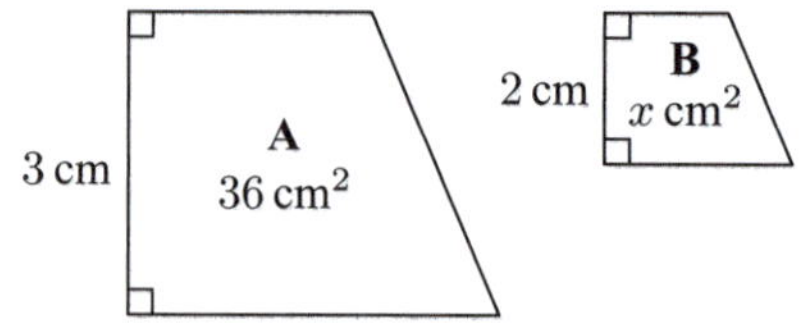

2 For each pair of similar figures, find the unknown length:

a

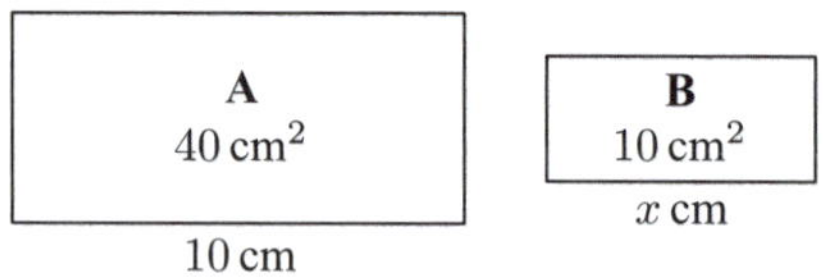

b

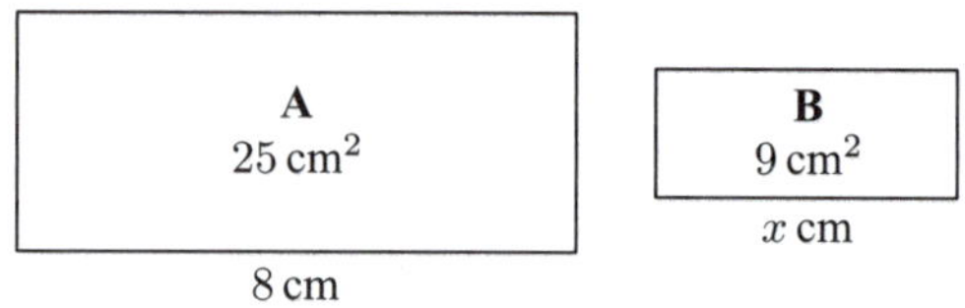

3 The side lengths of triangle A are 5 cm, 6 cm, and 7 cm. The longest side of a similar triangle B is 28 cm.

- **a** Find the scale factor k when enlarging triangle A to give triangle B.
- **b** Find the lengths of the other two sides of triangle B.
- **c** How many times larger is the area of triangle B than the area of triangle A?

4

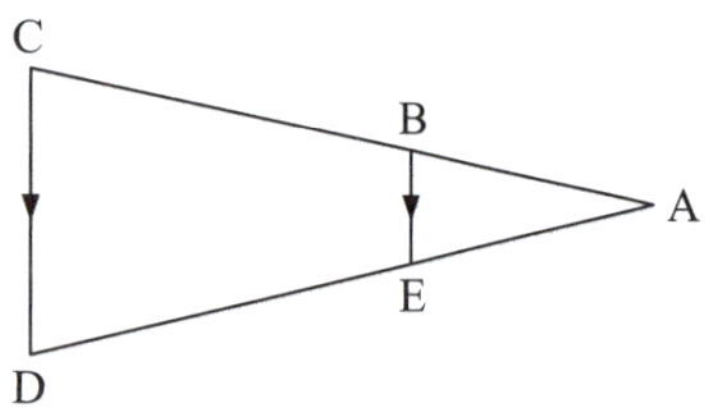

In the given figure, DE = 6 cm, AE = 4 cm, and $\triangle$ABE has area 16 cm^2.

- **a** Show that $\triangle$ABE and $\triangle$ACD are similar.
- **b** Find the scale factor to enlarge $\triangle$ABE into $\triangle$ACD.
- **c** Find the area of:
  - **i** $\triangle$ACD
  - **ii** trapezium BCDE.

5 In the given figure, AE = 3 cm, $\triangle$ABE has area 5 cm^2, and BEDC has area 6 cm^2.
Find the length of [ED].

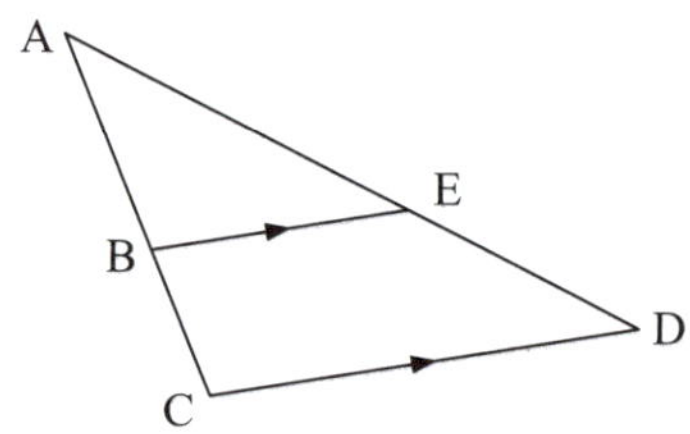

6 Figures **A** and **B** are similar. The dots shown are equally spaced at intervals of 1 unit.
If figure **A** has area 1.6 units^2, find the area of figure **B**.

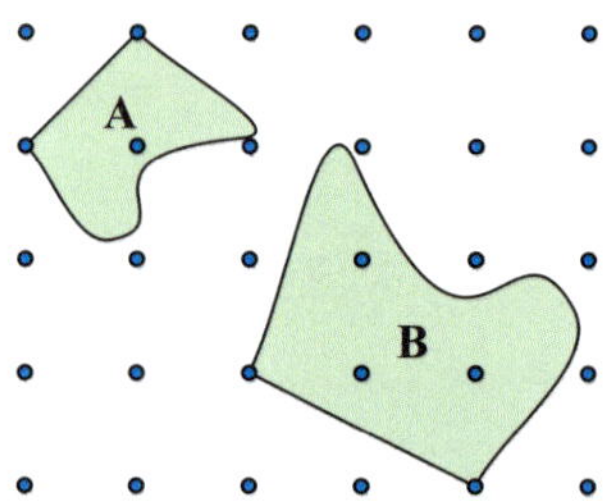

7

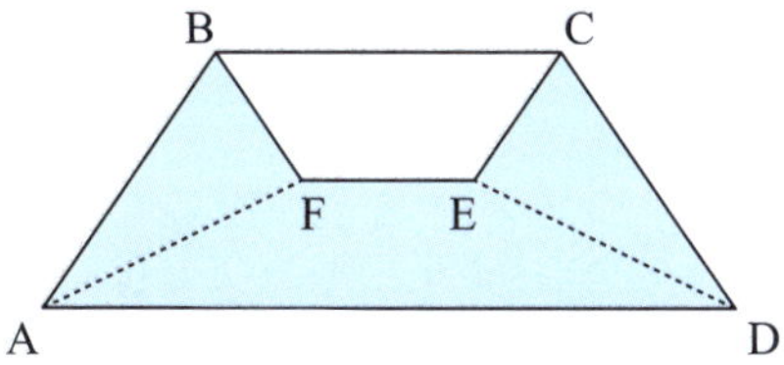

Trapezium ABCD is similar to trapezium CEFB. The ratio of the shaded to unshaded area is $3 : 1$. Find the ratio of the areas of CEFB and AFED.

F VOLUMES OF SIMILAR SOLIDS

For 3-dimensional solids which are similar, there is also a relationship between their *volumes*.

For example, suppose box **B** is an enlargement of box **A** with scale factor k.

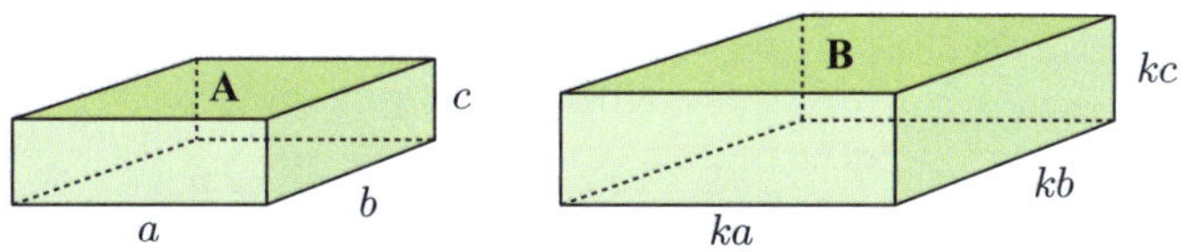

Volume of box **B**
$= ka \times kb \times kc$
$= k^3 \times abc$
$= k^3 \times$ volume of box **A**

If a 3-dimensional solid is enlarged with scale factor k to produce a similar solid, then the new volume $= k^3 \times$ the old volume.

Example 10 Self Tutor

For each pair of similar solids, find the unknown value:

a

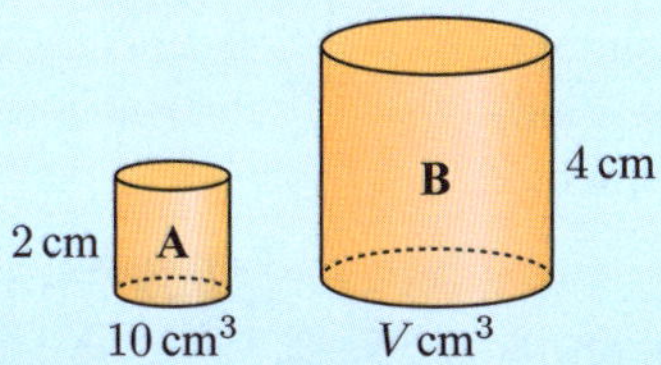

b

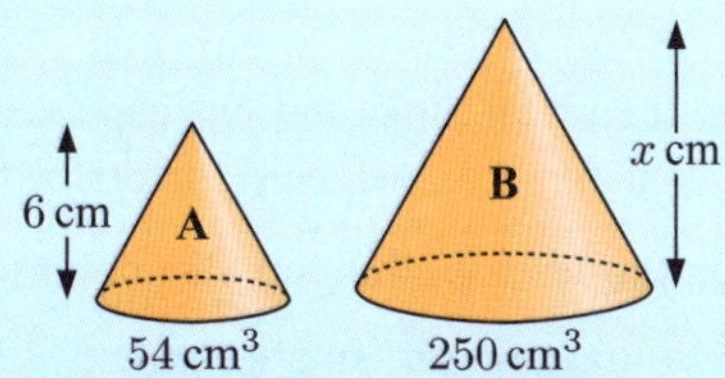

a Suppose cylinder **A** is enlarged with scale factor k to give cylinder **B**.

Using the heights, $k = \frac{4}{2} = 2$

Volume of **B** $= k^3 \times$ volume of **A**

$\therefore\ V = 8 \times 10$

$= 80$

b Suppose cone **A** is enlarged with scale factor k to give cone **B**.

Volume of cone **B** $= k^3 \times$ volume of cone **A**

$\therefore\ 250 = k^3 \times 54$

$\therefore\ \frac{125}{27} = k^3$

$\therefore\ k = \frac{5}{3}$

So, the height of cone **B** $= \frac{5}{3} \times$ the height of cone **A**

$\therefore\ x = \frac{5}{3} \times 6$

$= 10$

EXERCISE 22F

1 For each pair of similar solids, find the unknown volume:

a

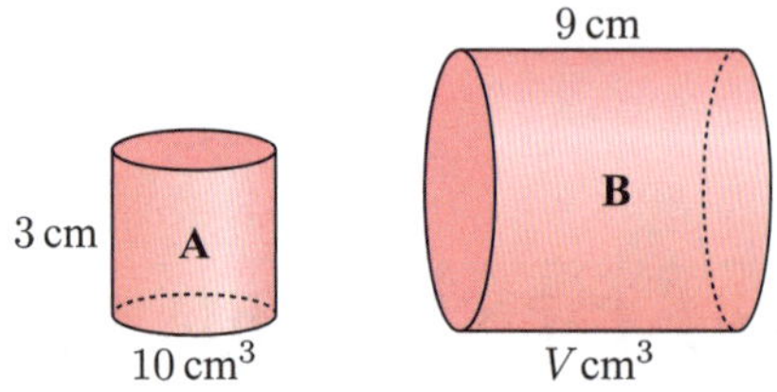

b

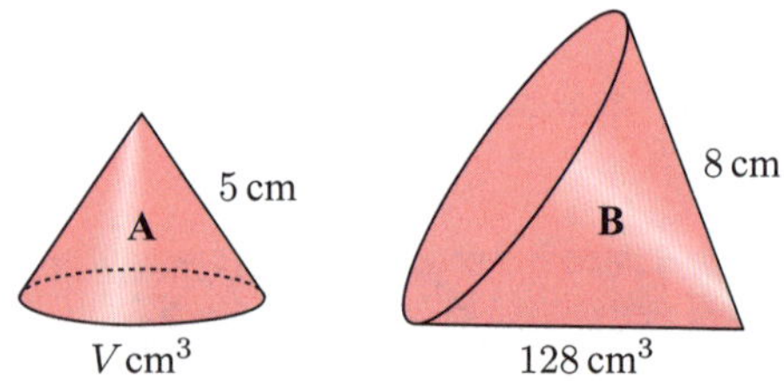

2 For each pair of similar solids, find the unknown length:

a

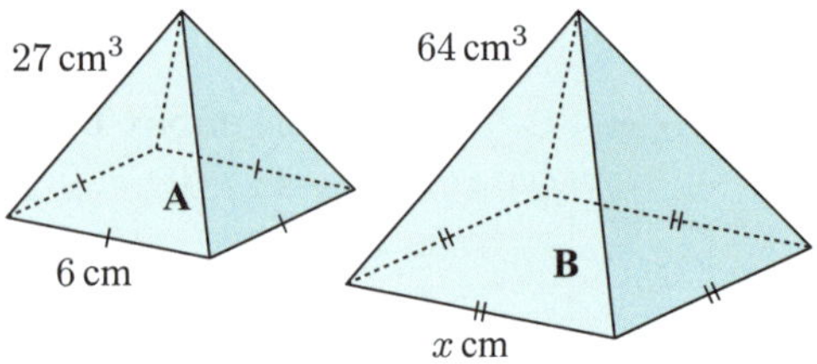

b

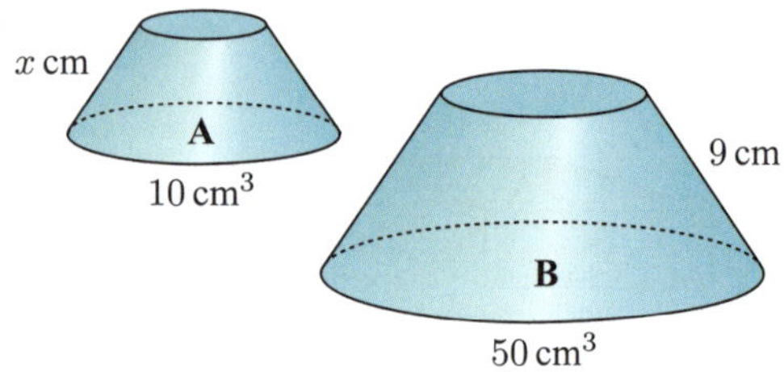

3 Two solid wooden spheres have radii a cm and $3a$ cm respectively. If the smaller sphere has volume 250 cm^3, find the volume of the larger sphere.

4 Two similar pyramids have volumes 7500 cm^3 and 480 cm^3. If the larger pyramid is 30 cm high, find the height of the smaller pyramid.

5 A pet store sells small and large fish tanks. The small fish tank is 30 cm by 60 cm by 50 cm high. The large fish tank is similar to the small tank, and has capacity 150 litres. Find, to 3 significant figures, the scale factor for the enlargement.

6 Two similar statues have heights 3.5 m and 4.9 m.

a If the volume of the smaller statue is 1.5 m^3, find the volume of the larger statue.

b If the surface area of the larger statue is 7.35 m^2, find the surface area of the smaller statue.

7 Two cylindrical containers have capacities 875 mL and 2240 mL.
Thin straws are placed in the cylinders as shown.
Are the cylinders similar? Explain your answer.

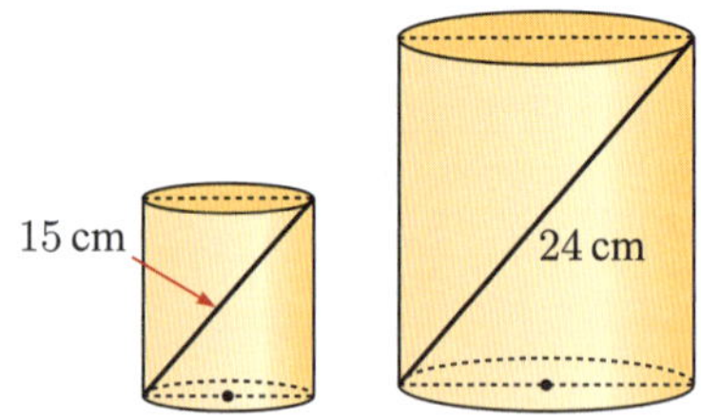

8 Answer the **Opening Problem** on page **348**.

9 The ratio of the surface areas of two similar solids is 25 : 9.
Find the ratio of their volumes.

10 A store sells trophies in two sizes which are similar. The mass of a large trophy is 72.8% greater than the mass of a small trophy. By what percentage is the base area of a large trophy greater than the base area of a small trophy?

GLOBAL CONTEXT — TESSELLATIONS

Global context: Personal and cultural expression

Statement of inquiry: An understanding of geometric transformations helps us design and appreciate art.

Criterion: Communicating

GLOBAL CONTEXT

QUICK QUIZ

MULTIPLE CHOICE QUIZ

REVIEW SET 22A

1 Quadrilaterals ABCD and GHEF are congruent.
Determine the:

a length of [BC]

b perimeter of GHEF.

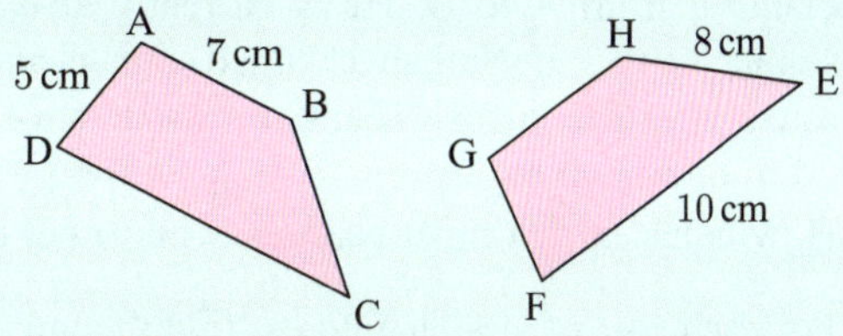

2 Determine whether each pair of triangles is congruent, giving reasons for your answers.

a

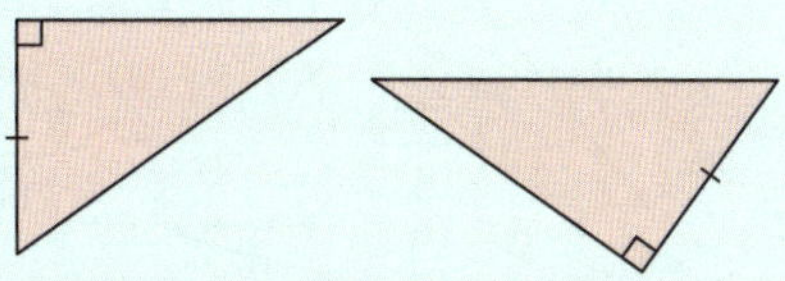

b

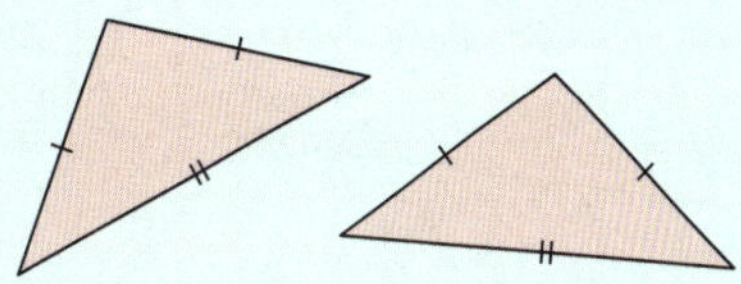

3

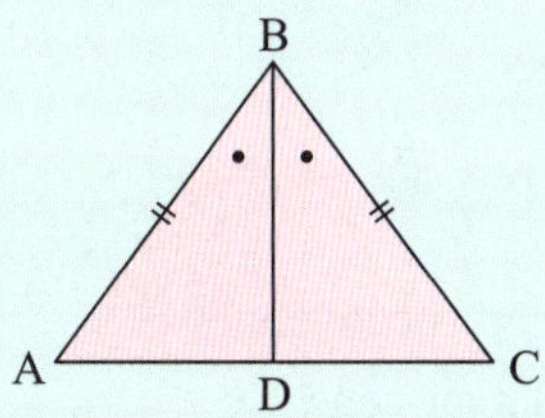

Triangle ABC is isosceles with $BA = BC$.

[BD] is the angle bisector of $A\widehat{B}C$.

a Prove that triangles DAB and DCB are congruent.

b Hence prove that [BD] is perpendicular to [AC].

4 Find the values of x and y given that the figures are similar:

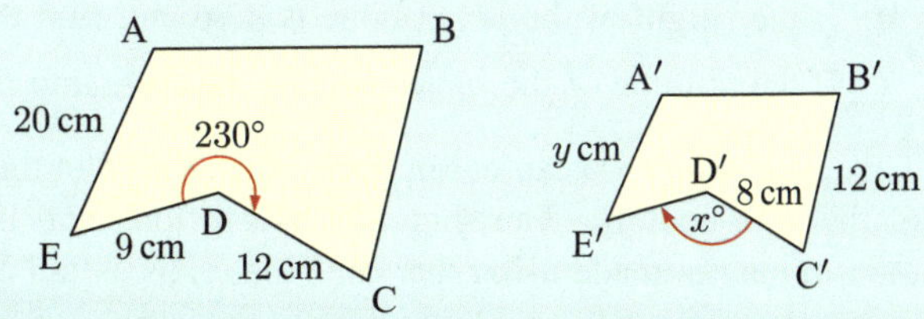

5 Determine whether each pair of triangles is similar:

a

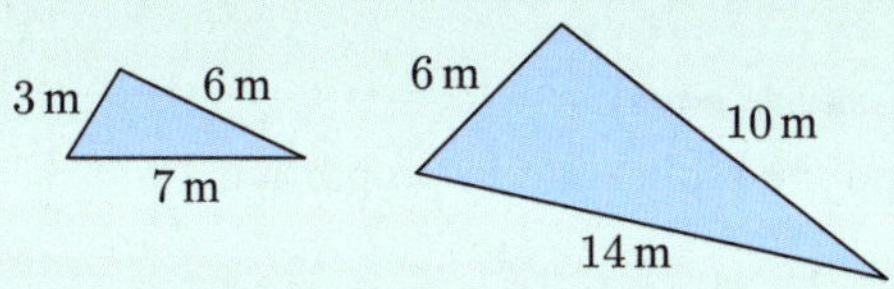

b

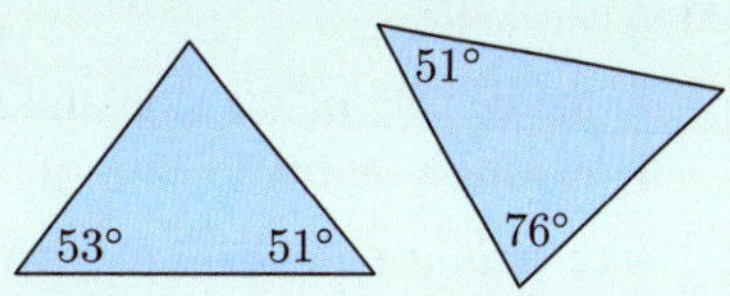

6 In each figure, establish that a pair of triangles is similar. Hence find the value of x.

a
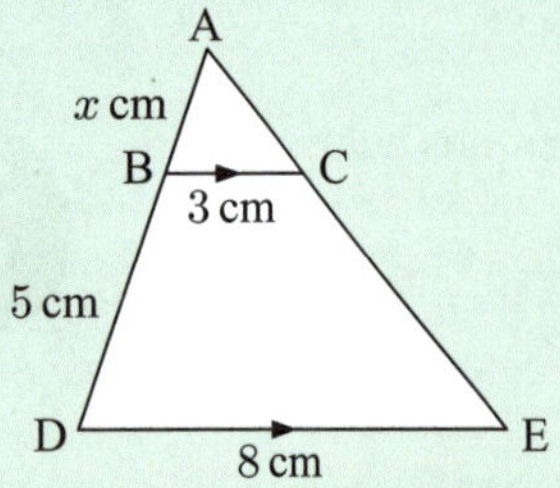

b
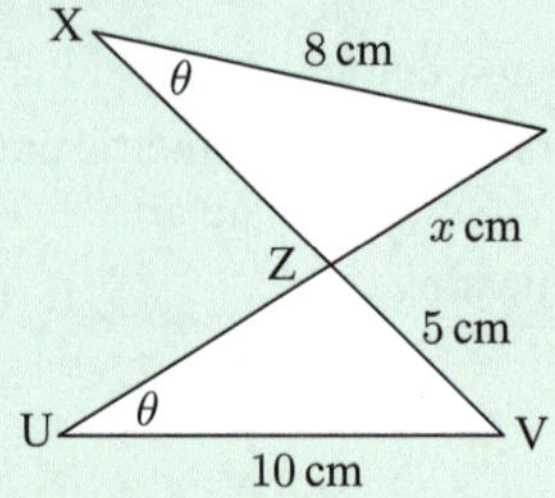

c
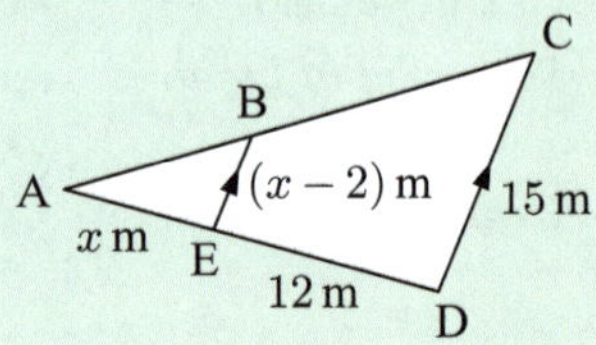

7 Figure ABCD is a rectangle. Show that the three triangles in this figure are all similar to each other.

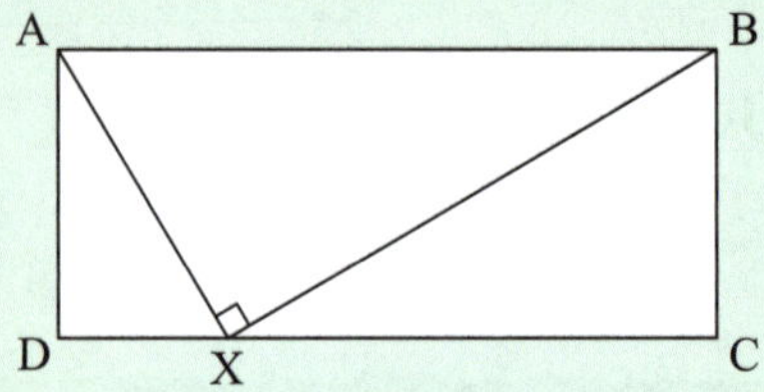

8 A river is 48 m wide. Nathan is swimming across the diagonal from A to B. He is currently at N, having swum 30 m. If Nathan changed course and swam directly for the opposite bank, he would save himself 20 m of swimming.
How much farther must Nathan swim to get to B?

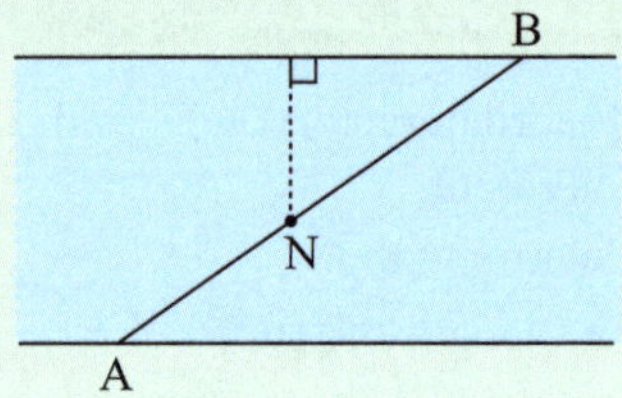

9 For each pair of similar figures, find the unknowns:

a
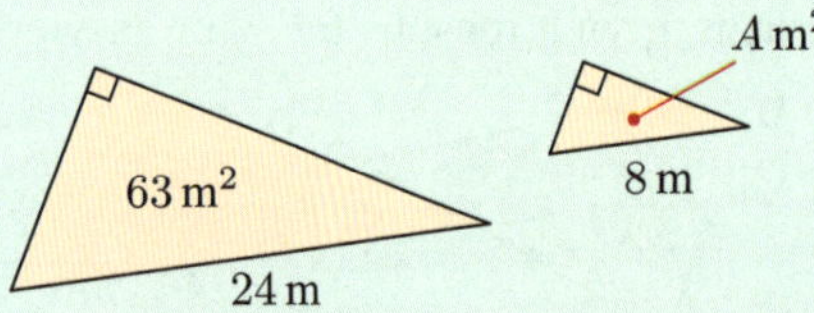

b
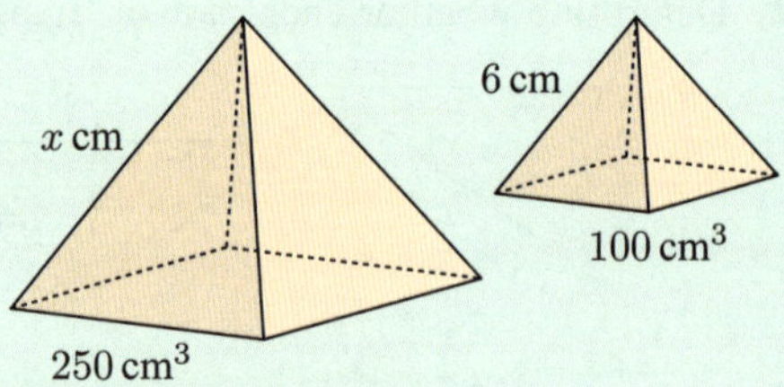

c
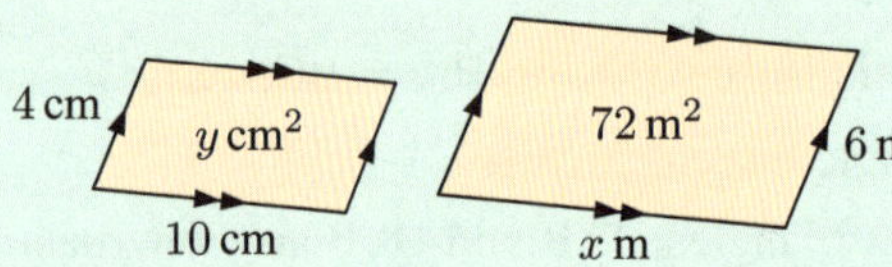

10 The surface areas of two similar cones are 4.2 cm^2 and 67.2 cm^2 respectively.

a Find the scale factor k to enlarge the smaller cone into the larger cone.

b If the height of the larger cone is 3.96 cm, find the height of the smaller cone.

c If the volume of the larger cone is 32 cm^3, find the volume of the smaller cone.

11 The *Queen Mary* was an ocean liner which sailed in the North Atlantic Ocean during the middle of the 20th century. The ship was 310 m long and had mass 80 000 tonnes. A scale replica 1.55 m long is made using the same materials as the original.

a Find the scale factor for this reduction.

b Explain why:

i the density of the ship and the replica are equal

ii the mass and volume of the ship and the replica are in proportion.

c Hence estimate the mass of the replica.

REVIEW SET 22B

1 Determine whether each pair of triangles is congruent, giving reasons for your answers.

a

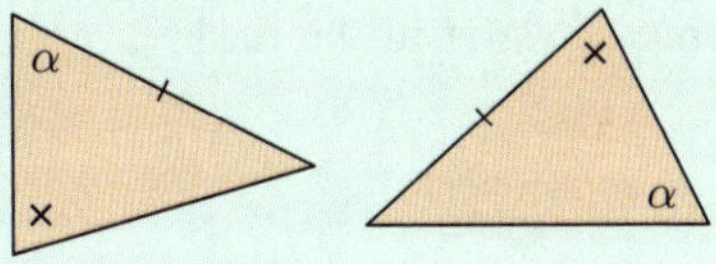

b

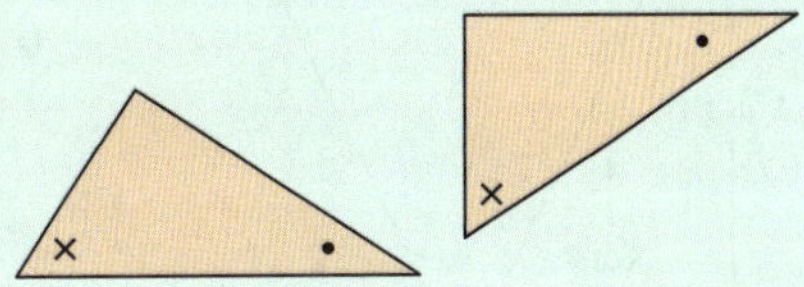

2 Two of these triangles are congruent. State which pair is congruent, and give a reason for your answer.

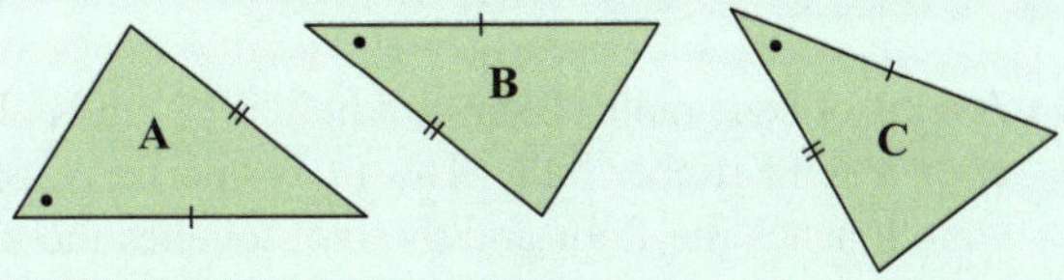

3

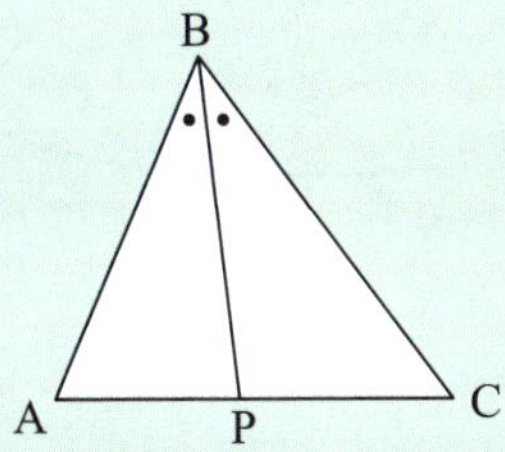

In triangle ABC alongside, the angle bisector of $A\widehat{B}C$ meets [AC] at P.

Prove that P is equidistant from [AB] and [CB].

Hint: Draw perpendicular lines from P to [AB] and [CB].

4 Determine whether each pair of figures is similar:

a

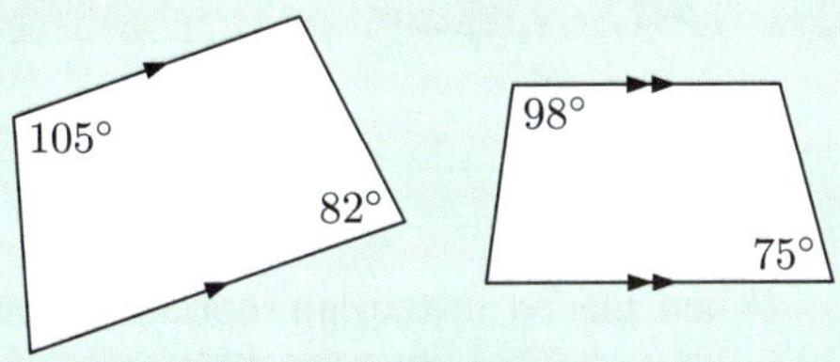

b

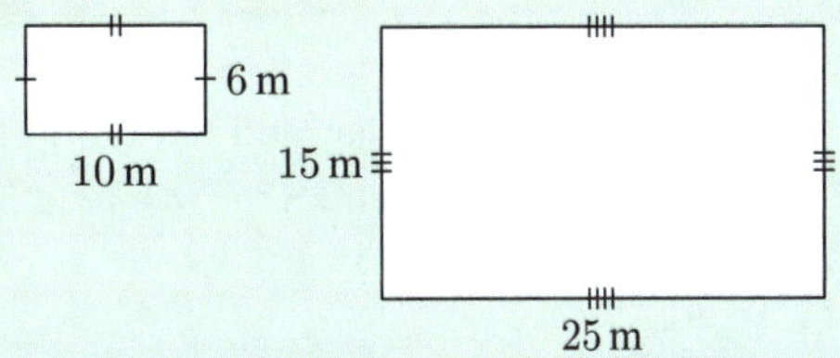

5 Show that these figures possess similar triangles:

a

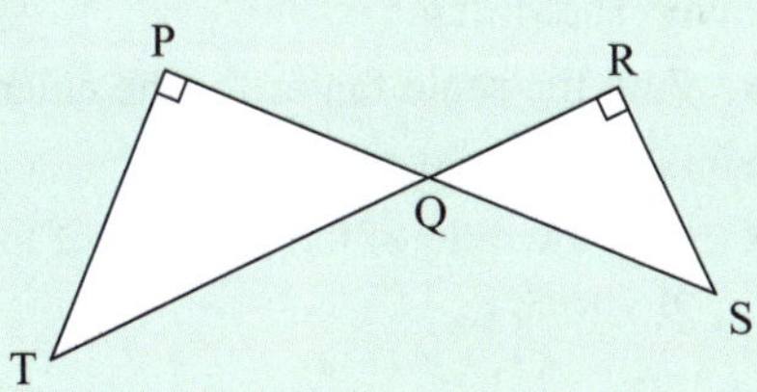

b

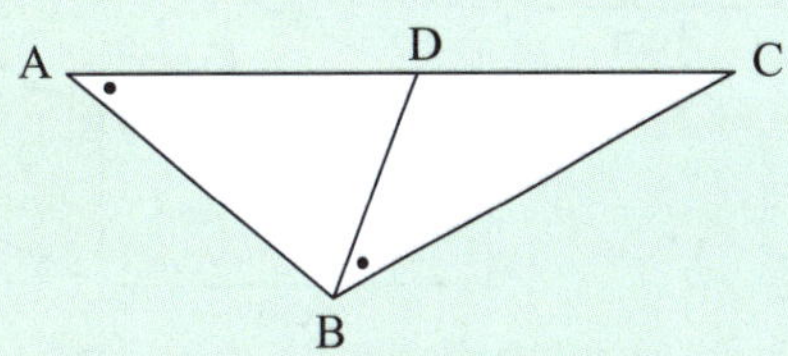

6 In each figure, establish that a pair of triangles is similar. Hence find the value of x.

a

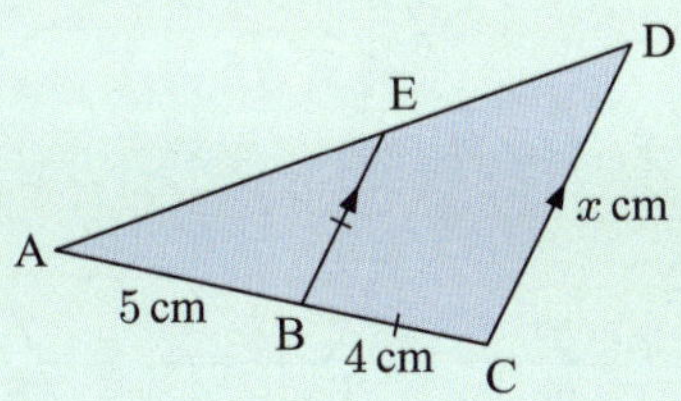

b

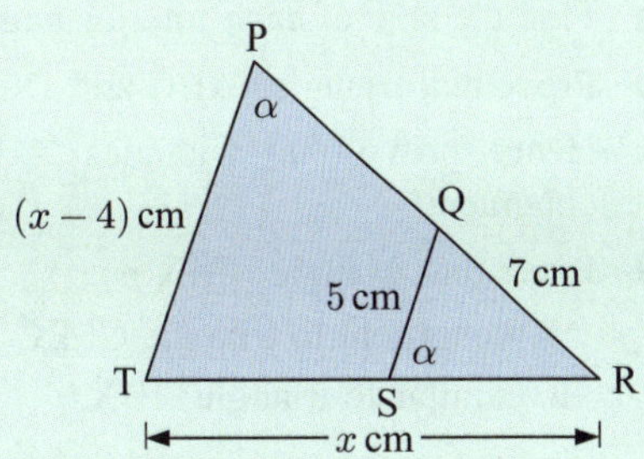

7 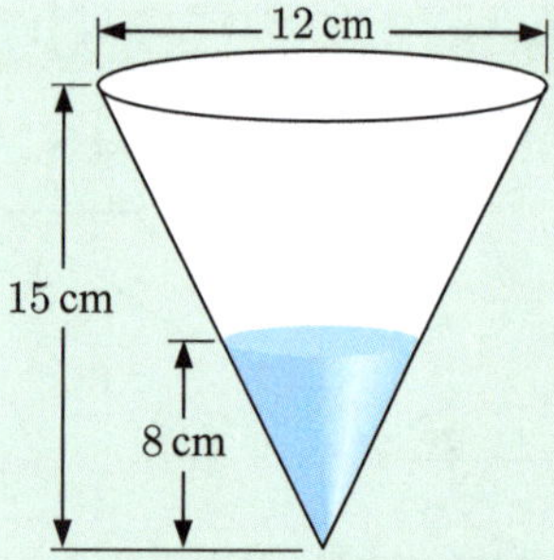

A conical flask has height 15 cm and base diameter 12 cm. Water is poured into the flask to a depth of 8 cm.

a Find the diameter of the surface of the water.

b Find the volume of water in the flask.

8 A cone of base radius 5 cm and height 12 cm is placed over a solid rubber ball. The fit is "perfect", so the cone touches the floor and the ball touches the cone. Find the radius of the ball.

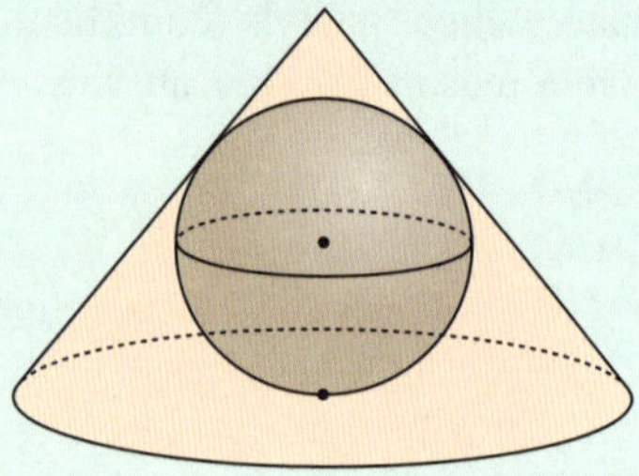

9 Find the value of x given that each pair of figures is similar:

a 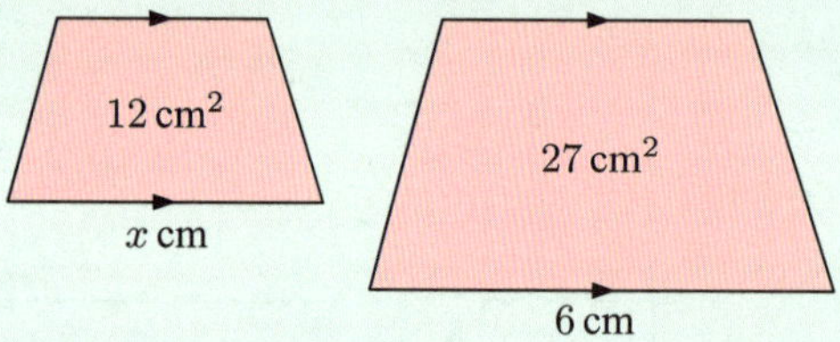

b 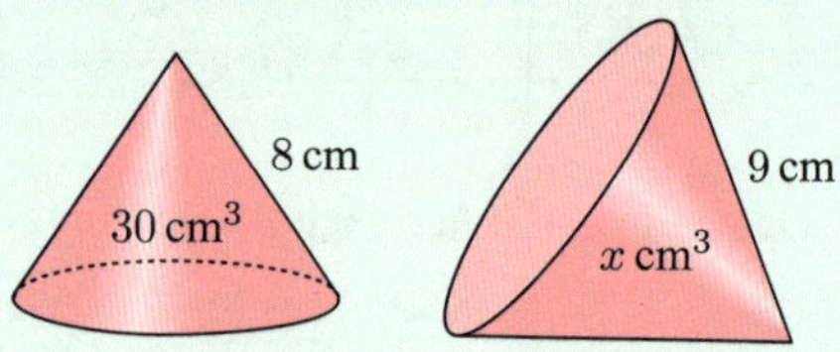

10 Two solids are similar, and the ratio of their volumes is $343 : 125$.
Find the ratio of their: **a** lengths **b** surface areas.

11 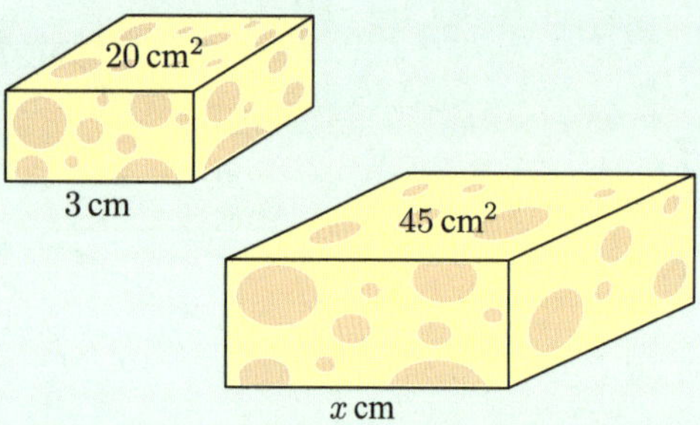

When placed in the appropriate orientation, it is clear that two blocks of cheese are similar. The surface areas which are uppermost are 20 cm^2 and 45 cm^2.

a Find the scale factor for the enlargement.

b Find the value of x.

c Find the ratio of the volumes of the blocks of cheese.

12 ABCDEF is a regular hexagon. M is the midpoint of [BC].

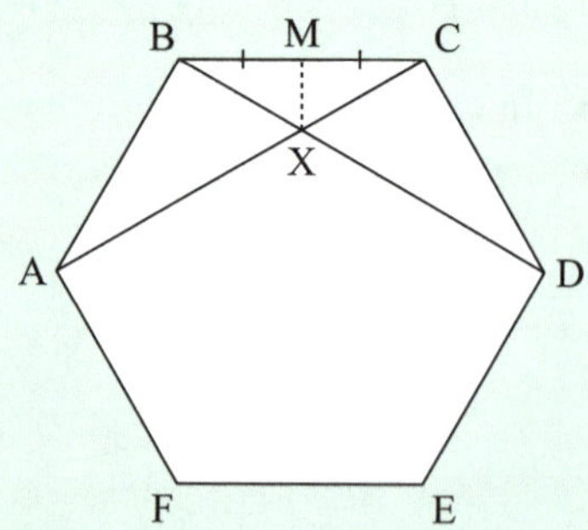

a Find the size of each interior angle of the hexagon.

b Prove that triangles ABC and DCB are congruent.

c Hence prove that triangles ABX and DCX are congruent.

d Prove that triangle BMX is:

i congruent to triangle CMX

ii similar to triangle ABX.

e Find the ratio of the area of triangle BMX to the area of triangle ABX.

Chapter 23

Trigonometry

Contents:

OPENING PROBLEM

Sometimes it is difficult or even impossible to measure angles, heights, and distances directly.

Suppose you are in a boat at sea. You look up at a mountain in front of you and wonder how high it is.

You have no way of measuring the height directly, but what you *can* do is measure the angle from horizontal to the top of the mountain.

Suppose you make two measurements from points A and B which are 300 m apart in a direct line with the mountain.

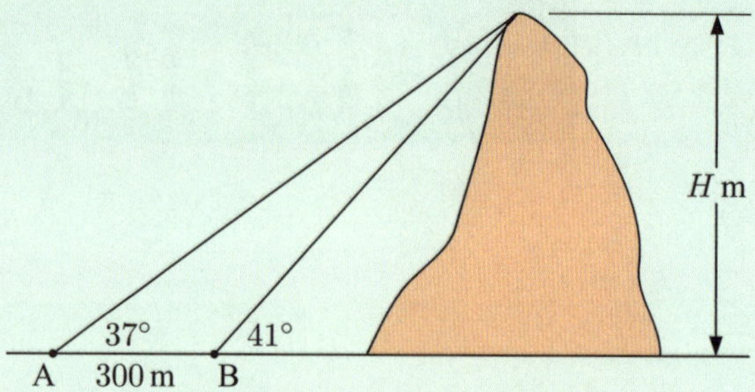

Things to think about:

a Could you use these measurements to draw a scale diagram and hence *estimate* the mountain's height?

b Would your scale diagram be *similar* to the real world?

c Is there a mathematical method for calculating the height of the mountain *without* using a scale diagram?

Trigonometry is the study of the relationship between the side lengths and angles of triangles.

HISTORICAL NOTE

The word *trigonometry* is derived from the Greek word *trigonometria*, meaning "triangle measuring".

This subject was first studied in ancient Egypt and Babylon in the 2nd millennium BC. The Egyptians used trigonometry in the construction of the pyramids, and the scribe **Ahmes** included a problem of trigonometry in the *Rhind Mathematical Papyrus*. The Babylonian astronomers, meanwhile, studied the relationships between angles and distances in their astronomy.

Hipparchus

The Greek astronomer **Hipparchus of Nicaea** (180 - 125 BC) is credited with being the founder of trigonometry. To aid his astronomical calculations, he produced a table of numbers in which the lengths of chords of a circle were related to the length of the radius.

His work was extended by **Menelaus of Alexandria**, who recorded Menelaus' theorem in his book *Sphaerica*, and **Claudius Ptolemy**, who expanded Hipparchus' table of values in his book *Almagest.*

Today, we use trigonometry alongside algebra, arithmetic, and geometry to find unknown lengths and angles of triangles. This has important applications in engineering, architecture, construction, navigation, and surveying.

A SCALE DIAGRAMS IN GEOMETRY

Scale diagrams can be used to estimate side lengths and angles in geometrical figures.

Example 1 Self Tutor

Jake has been contracted to install lights to shine on the front of a hotel. He wants to install the lights at ground level at a point A on the near side of a drain. The drain is 36.5 m from the base of the hotel. From this point, the angle up to the top of the hotel is 50°.

How high is the hotel?

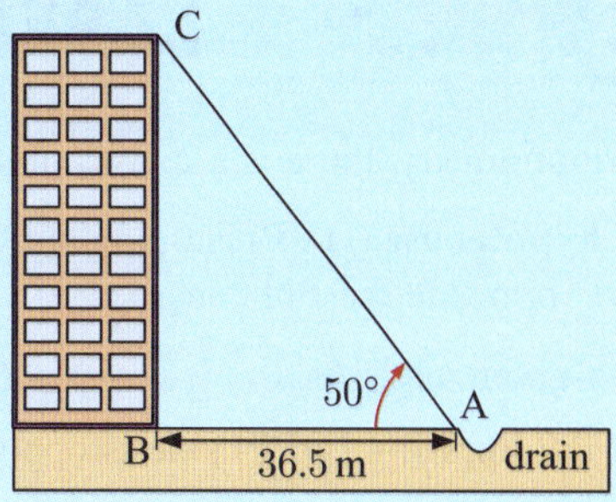

We choose a suitable *scale*, in this case 1 mm ≡ 1 m.

We draw a horizontal line [BA] 36.5 mm long.

From B we draw a vertical line.

We use a protractor to measure a 50° angle at A. We draw a line at this angle to meet the vertical at C.

Using a ruler, we find that BC ≈ 43.5 mm

∴ the building is approximately 43.5 metres high.

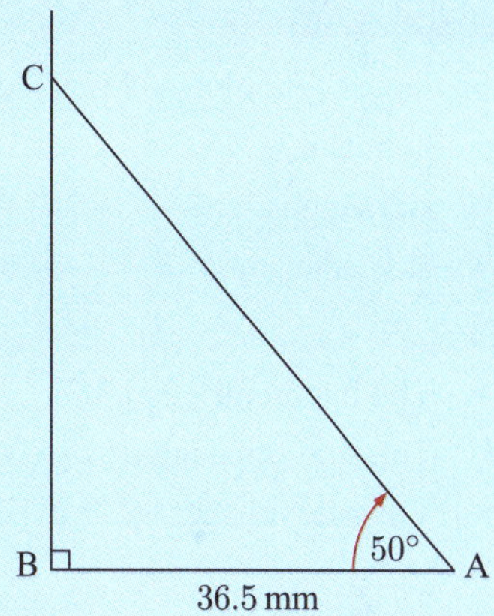

EXERCISE 23A

1 **a** Draw an accurate scale diagram using the measurements given. Use the scale 1 cm ≡ 1 m.

b Use your scale diagram to estimate the actual length of:

i [BC] **ii** [AC]

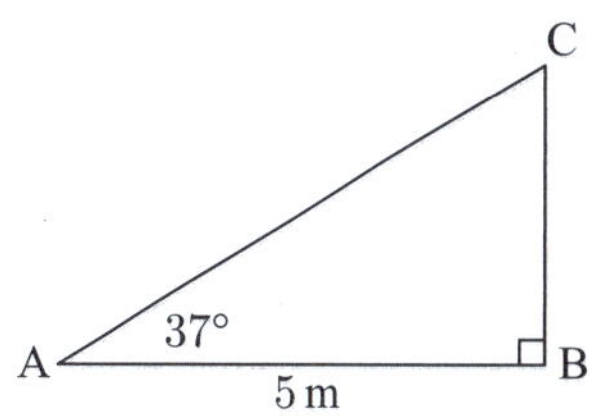

2 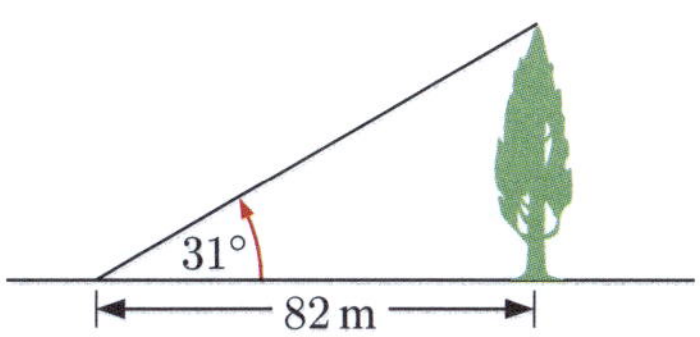

Use a scale diagram with scale 1 cm ≡ 10 m to estimate the height of the tree.

3 Use a scale diagram to estimate the height of the mountain in the **Opening Problem**.

DISCUSSION

What are the most likely causes of error when using scale diagrams?

How accurate are the answers when using scale diagrams?

B LABELLING RIGHT ANGLED TRIANGLES

In trigonometry, there is a convention for labelling the sides of right angled triangles.

The **hypotenuse (HYP)** is the longest side of a right angled triangle, and is opposite the right angle.

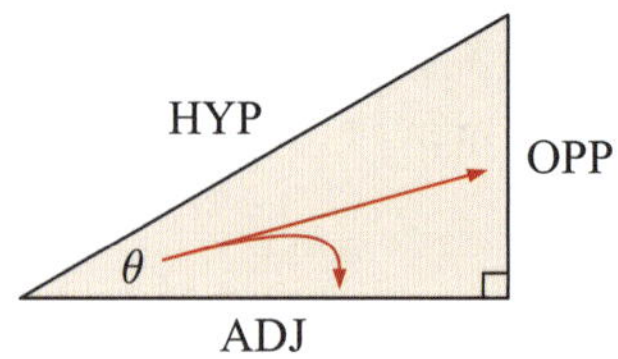

For a given angle θ:

- the **opposite (OPP)** side is opposite θ
- the **adjacent (ADJ)** side is adjacent to θ.

Example 2 — Self Tutor

For the triangle alongside, name the:

- **a** hypotenuse
- **b** side opposite θ
- **c** side adjacent to θ.

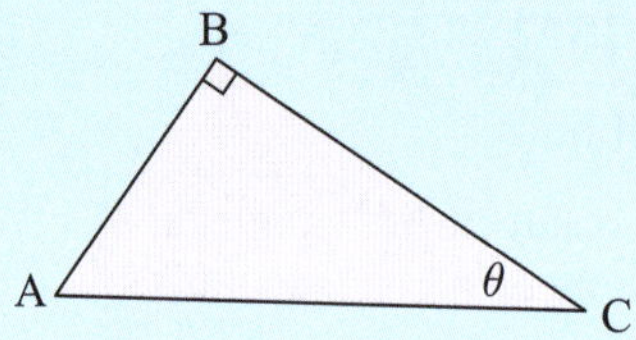

- **a** The hypotenuse is [AC].
- **b** The side opposite θ is [AB].
- **c** The side adjacent to θ is [BC].

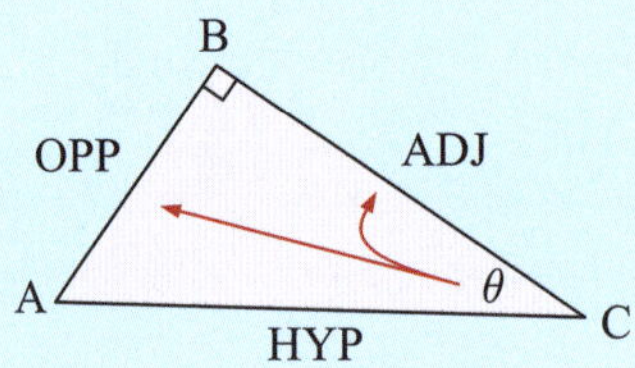

EXERCISE 23B

1 For each triangle, name the:

i hypotenuse **ii** side opposite θ **iii** side adjacent to θ.

a

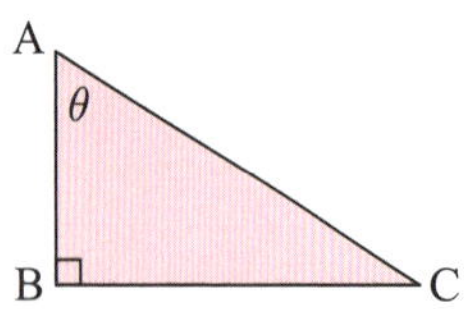

b

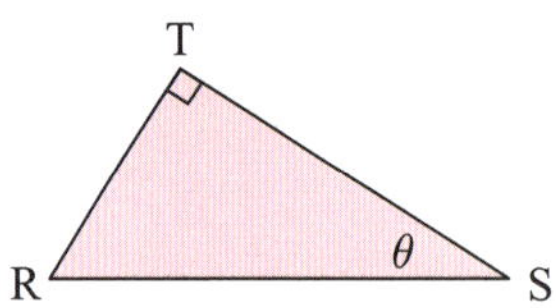

c

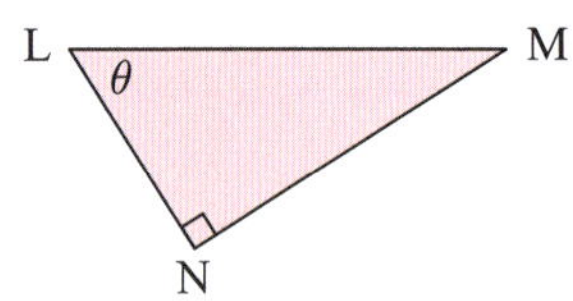

2 For the given right angled triangle, state the length of the side:

- **a** opposite θ
- **b** adjacent to θ
- **c** opposite ϕ
- **d** adjacent to ϕ.

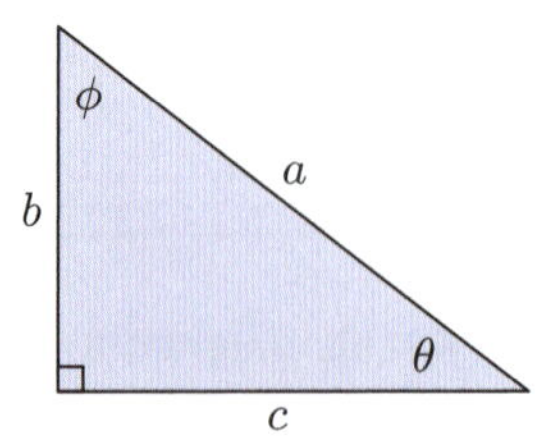

C THE TRIGONOMETRIC RATIOS

We have seen that similar triangles are equiangular, and that the lengths of their sides are in the same ratio.

INVESTIGATION RATIO OF SIDES OF RIGHT ANGLED TRIANGLES

In this Investigation we consider the ratios $\frac{\text{OPP}}{\text{HYP}}$, $\frac{\text{ADJ}}{\text{HYP}}$, and $\frac{\text{OPP}}{\text{ADJ}}$ in a series of *similar* right angled triangles.

What to do:

1 For each set of four similar triangles below, complete a table like the one shown. Measure each side length to the nearest millimetre, and round each of the ratios to 2 decimal places.

Triangle	HYP	OPP	ADJ	$\frac{\text{OPP}}{\text{HYP}}$	$\frac{\text{ADJ}}{\text{HYP}}$	$\frac{\text{OPP}}{\text{ADJ}}$
1						
2						
3						
4						

PRINTABLE WORKSHEET

a

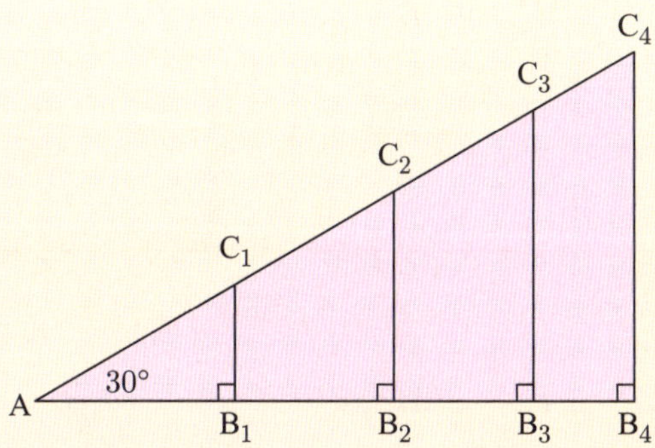

b

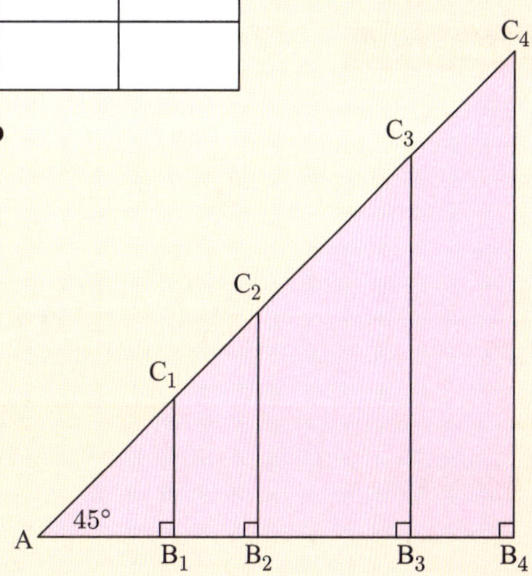

2 Use your calculator to evaluate $\frac{\sqrt{3}}{2}$, $\frac{1}{\sqrt{3}}$, and $\frac{1}{\sqrt{2}}$.

3 Summarise your results from **1** and **2**.

From the **Investigation**, you should have found that for any set of *similar* right angled triangles, the ratios $\frac{\text{OPP}}{\text{HYP}}$, $\frac{\text{ADJ}}{\text{HYP}}$, and $\frac{\text{OPP}}{\text{ADJ}}$ are constant.

These ratios have the traditional names **sine**, **cosine**, and **tangent** respectively. We abbreviate them to **sin**, **cos**, and **tan**.

In any right angled triangle with one angle θ:

$$\sin\theta = \frac{\textbf{OPP}}{\textbf{HYP}}, \quad \cos\theta = \frac{\textbf{ADJ}}{\textbf{HYP}}, \quad \tan\theta = \frac{\textbf{OPP}}{\textbf{ADJ}}$$

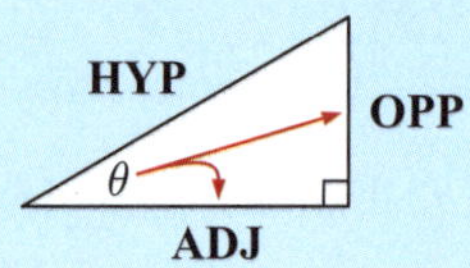

Notice that $\dfrac{\sin\theta}{\cos\theta} = \dfrac{\frac{\text{OPP}}{\text{HYP}}}{\frac{\text{ADJ}}{\text{HYP}}} = \dfrac{\text{OPP}}{\text{ADJ}} = \tan\theta$. So, $\boldsymbol{\tan\theta = \dfrac{\sin\theta}{\cos\theta}}$

From the **Investigation**, you should also have found some particular values of the trigonometric ratios:

$$\sin 30^\circ = 0.5 \qquad \sin 45^\circ = \frac{1}{\sqrt{2}} \approx 0.71$$

$$\cos 30^\circ = \frac{\sqrt{3}}{2} \approx 0.87 \qquad \cos 45^\circ = \frac{1}{\sqrt{2}} \approx 0.71$$

$$\tan 30^\circ = \frac{1}{\sqrt{3}} \approx 0.58 \qquad \tan 45^\circ = 1$$

Trigonometric ratios for any angle can be found using a calculator. You should first check that your calculator is in "degree" mode.

GRAPHICS CALCULATOR INSTRUCTIONS

Example 3 — Self Tutor

For the given triangle, find $\sin\theta$, $\cos\theta$, and $\tan\theta$.

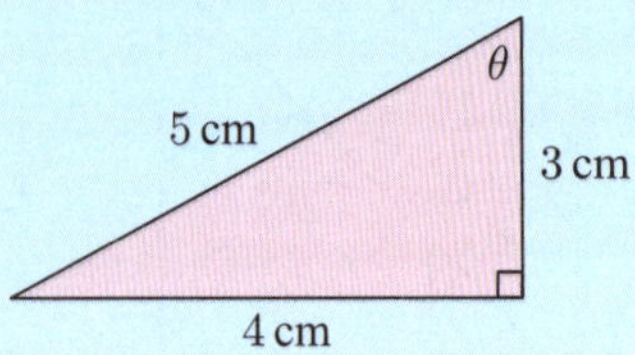

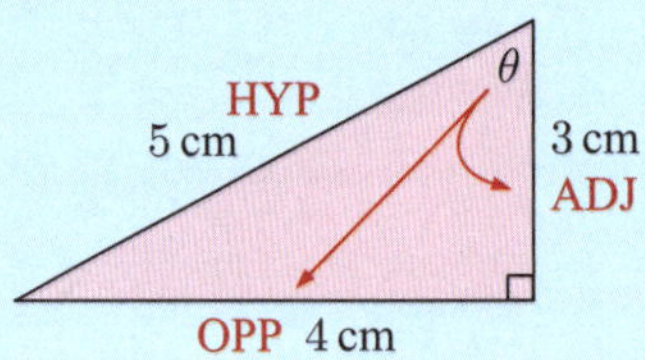

$\sin\theta = \dfrac{\text{OPP}}{\text{HYP}} = \frac{4}{5}$

$\cos\theta = \dfrac{\text{ADJ}}{\text{HYP}} = \frac{3}{5}$

$\tan\theta = \dfrac{\text{OPP}}{\text{ADJ}} = \frac{4}{3}$

EXERCISE 23C

1 For each triangle, find: **i** $\sin\theta$ **ii** $\cos\theta$ **iii** $\tan\theta$

a

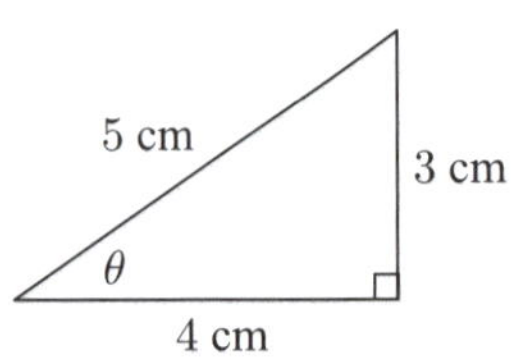

b

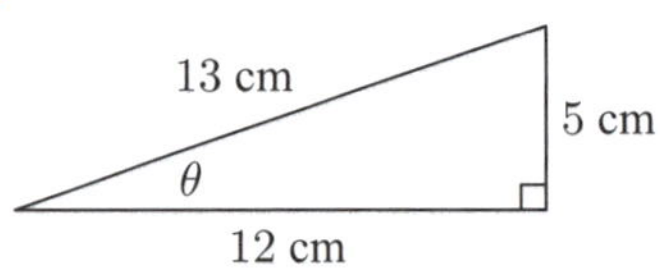

c

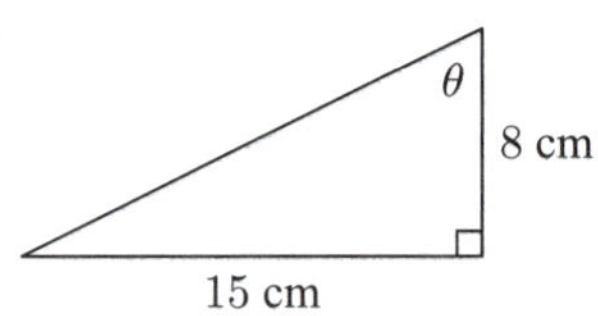

d

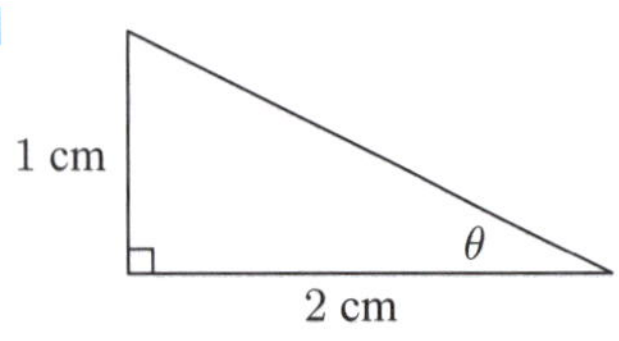

e

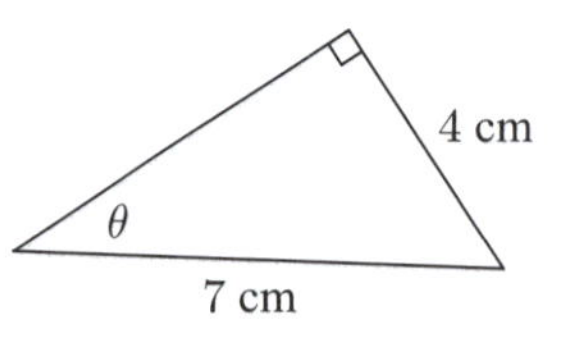

f

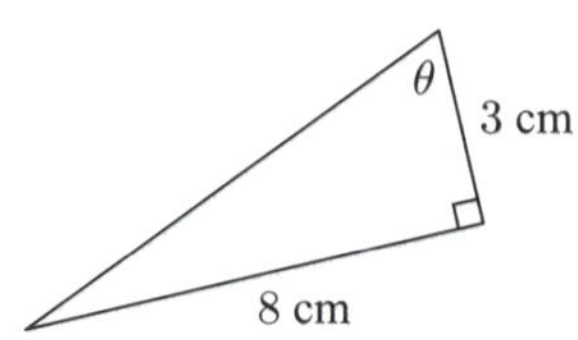

2 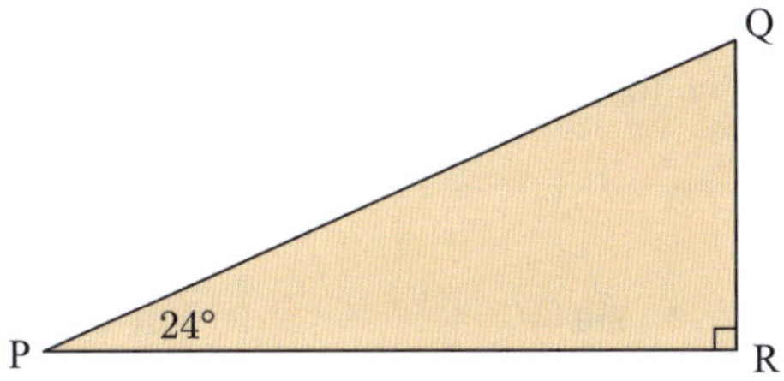

a Use a ruler to measure the length of each side of the triangle, to the nearest millimetre.

b Hence estimate the value of:

i $\sin 24°$ **ii** $\cos 24°$ **iii** $\tan 24°$

c Check your answers using a calculator.

3 The side lengths in the figure alongside have been rounded to the nearest millimetre.

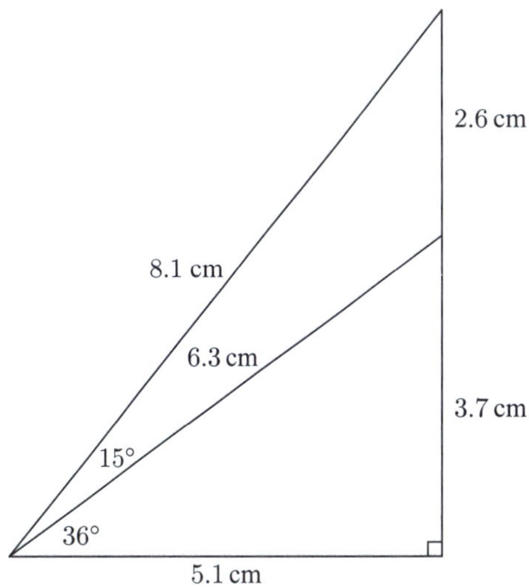

a Use the diagram to estimate:

i $\sin 36°$ **ii** $\cos 51°$
iii $\tan 54°$ **iv** $\cos 39°$

b Use your calculator to evaluate, to 3 decimal places:

i $\sin 36°$ **ii** $\cos 51°$
iii $\tan 54°$ **iv** $\cos 39°$

4 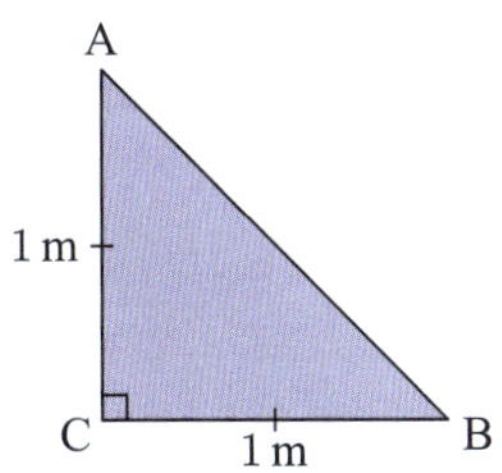

ABC is a right angled isosceles triangle with equal sides of length 1 m.

a Explain why $A\widehat{B}C = 45°$.

b Find the exact length of the hypotenuse [AB].

c Hence find the exact value of:

i $\sin 45°$ **ii** $\cos 45°$ **iii** $\tan 45°$

5 $\cos^2\theta$ is used to denote $(\cos\theta)^2$ or $\cos\theta \times \cos\theta$.

a Use your calculator to find, correct to 4 decimal places:

i $\cos^2 16°$ **ii** $\sin^2 16°$ **iii** $\cos^2 16° + \sin^2 16°$
iv $\cos^2 65°$ **v** $\sin^2 65°$ **vi** $\cos^2 65° + \sin^2 65°$

b For any value of θ, what do you suspect is the value of $\cos^2\theta + \sin^2\theta$?

c Prove your answer to **b** using this figure:

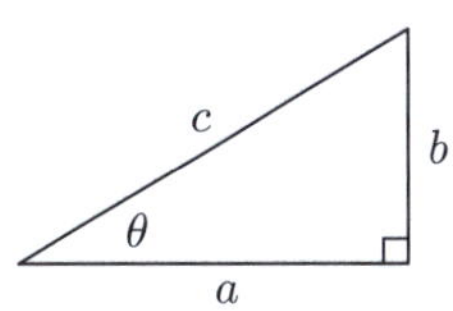

6 **a** Use your calculator to find:

i $\tan 70° \times \tan 20°$ **ii** $\tan 63° \times \tan 27°$ **iii** $\tan 49° \times \tan 41°$

b What do you suspect is the relationship between $\tan\theta$ and $\tan(90° - \theta)$?

c Prove your answer to **b** using this figure:

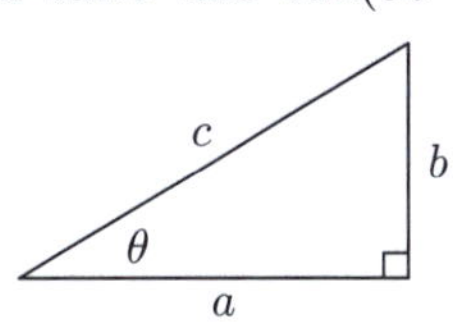

7 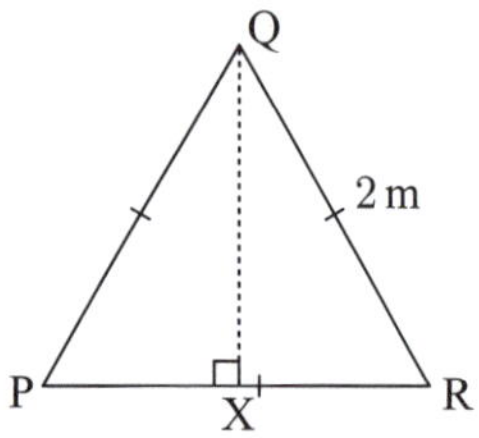

PQR is an equilateral triangle with side lengths 2 m.
Use the diagram to prove that:

a $\cos 60° = \frac{1}{2}$ **b** $\sin 60° = \frac{\sqrt{3}}{2}$ **c** $\tan 60° = \sqrt{3}$

d $\cos 30° = \frac{\sqrt{3}}{2}$ **e** $\sin 30° = \frac{1}{2}$ **f** $\tan 30° = \frac{1}{\sqrt{3}}$

DISCUSSION

Suppose θ is an acute angle of a right angled triangle.

What is the range of possible values of:

- $\sin\theta$
- $\cos\theta$
- $\tan\theta$?

D FINDING SIDE LENGTHS

Suppose we are given the angles of a right angled triangle, and the length of a side. We can use the appropriate trigonometric ratio to construct equations and hence find the other side lengths.

Example 4 **Self Tutor**

Find the value of x, rounding your answer to 2 decimal places:

a 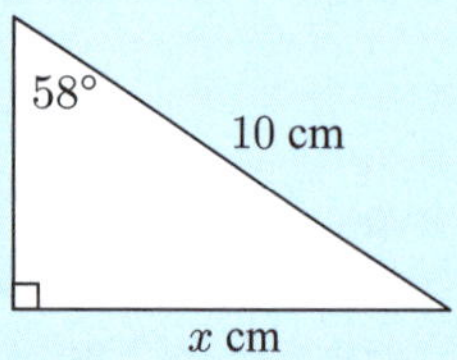

b 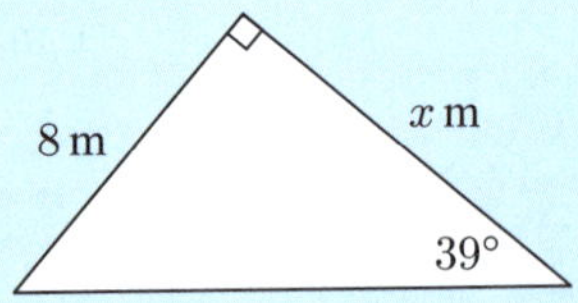

a The relevant sides are OPP and HYP, so we use the *sine* ratio.

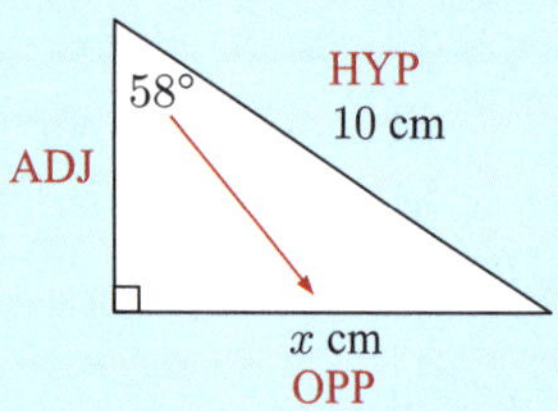

$$\sin\theta = \frac{\text{OPP}}{\text{HYP}}$$

$$\therefore \sin 58° = \frac{x}{10}$$

$$\therefore \sin 58° \times 10 = x \quad \{\times 10 \text{ both sides}\}$$

$$\therefore x \approx 8.48 \quad \{\text{calculator}\}$$

b The relevant sides are OPP and ADJ, so we use the *tangent* ratio.

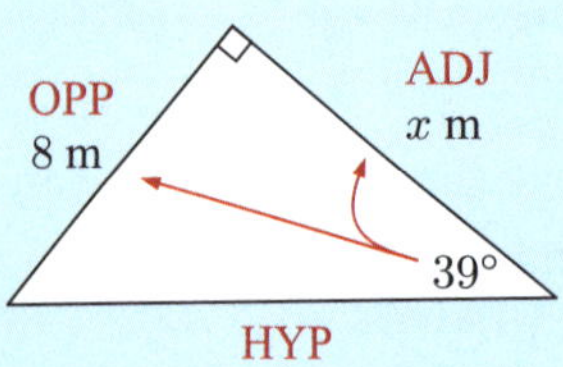

$$\tan\theta = \frac{\text{OPP}}{\text{ADJ}}$$

$$\therefore \tan 39° = \frac{8}{x}$$

$$\therefore x \times \tan 39° = 8 \quad \{\times x \text{ both sides}\}$$

$$\therefore x = \frac{8}{\tan 39°} \quad \{\div \tan 39° \text{ both sides}\}$$

$$\therefore x \approx 9.88 \quad \{\text{calculator}\}$$

EXERCISE 23D

1 Write down a trigonometric equation connecting the angle and the sides given:

a 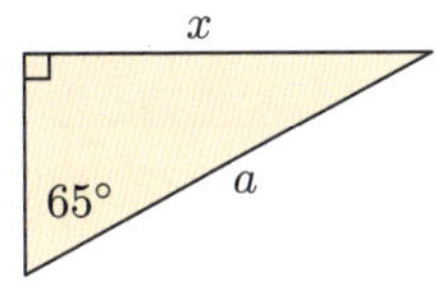

b 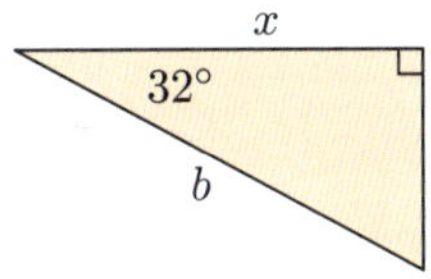

c 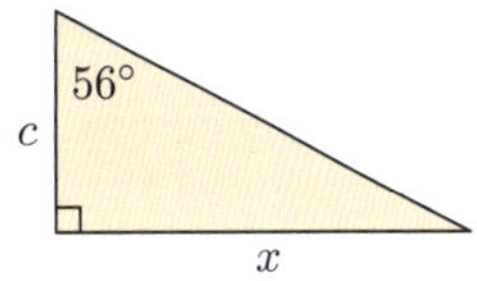

d 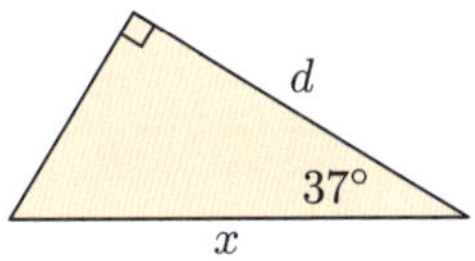

e 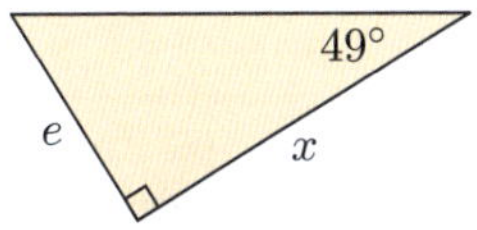

f 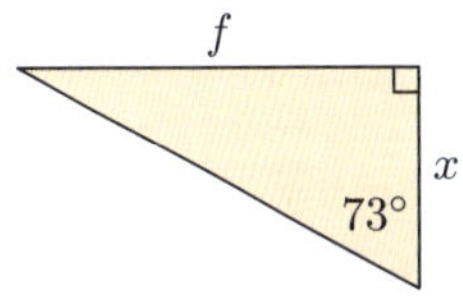

2 Find the value of x, rounding your answer to 2 decimal places:

a 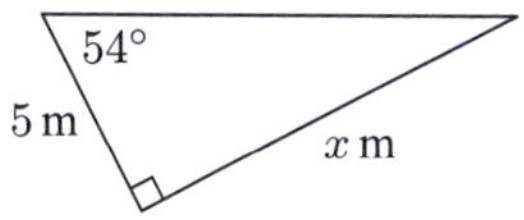

b 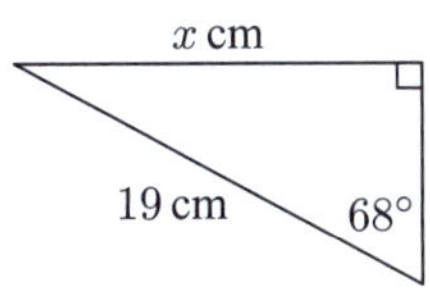

c 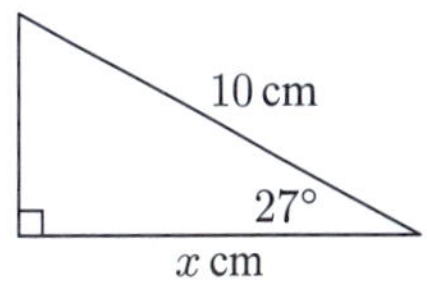

d 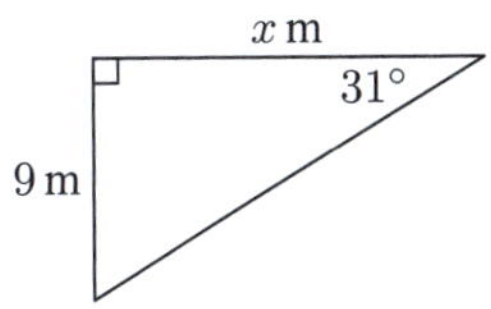

e 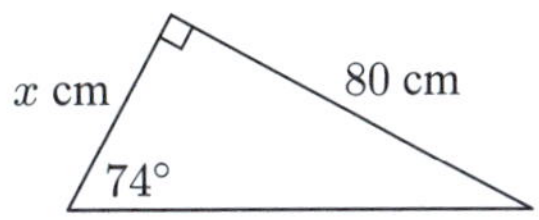

f 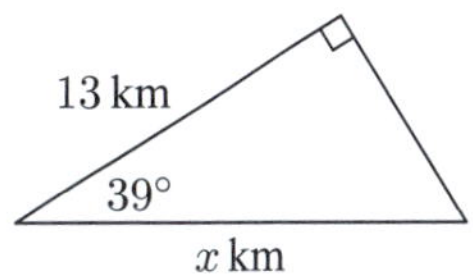

g 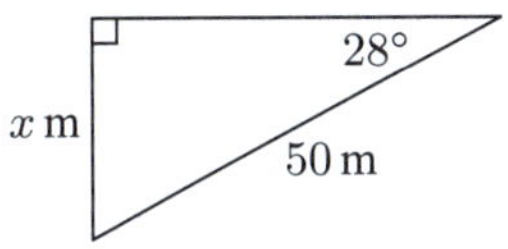

h 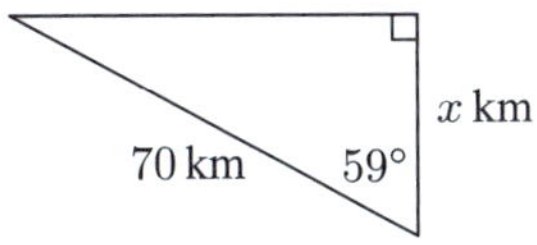

i 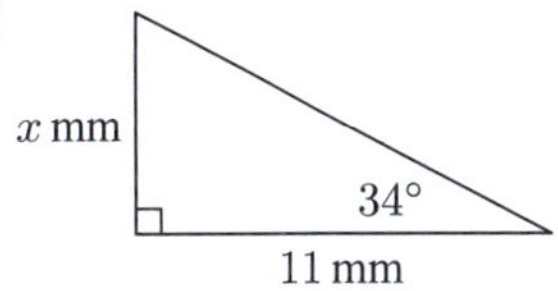

j 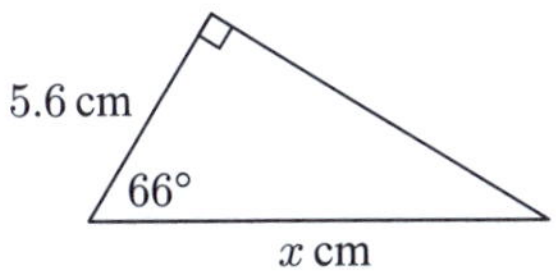

k 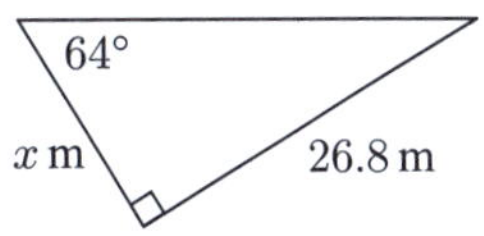

l 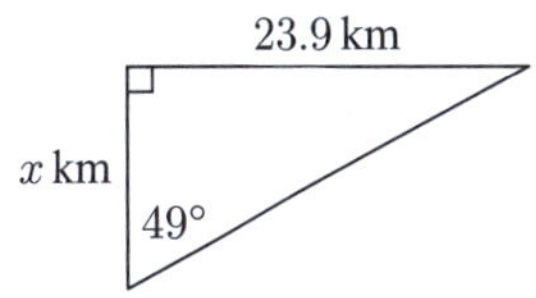

3 Find, to 1 decimal place, *all* unknown angles and sides:

a 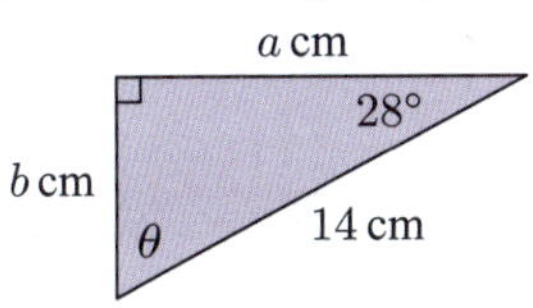

b 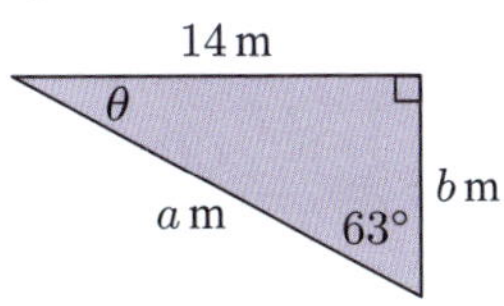

c 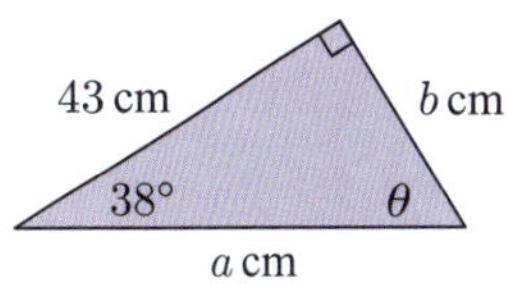

4 Find the perimeter of triangle ABC.

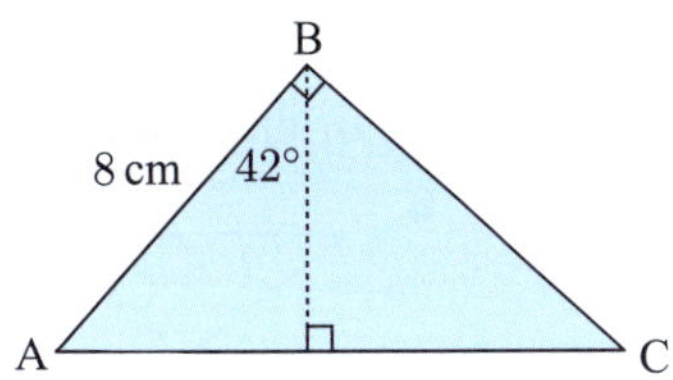

E FINDING ANGLES

If we know two side lengths of a right angled triangle, we can use trigonometry to find the acute angles.

In the right angled triangle shown, $\cos\theta = \frac{3}{5}$.

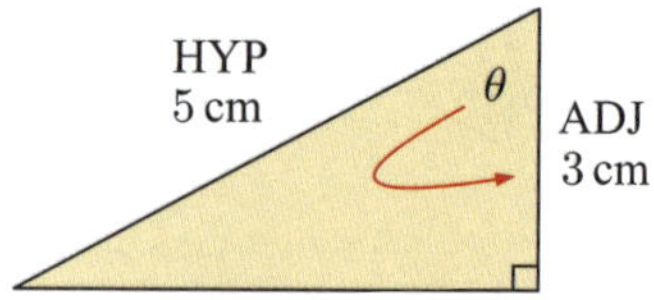

To find θ, we need an angle whose cosine is $\frac{3}{5}$.

We say that θ is the **inverse cosine** of $\frac{3}{5}$, and write $\theta = \cos^{-1}(\frac{3}{5})$.

We can use a calculator to evaluate inverse cosines. Click on the icon for instructions.

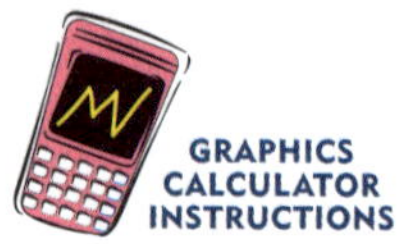

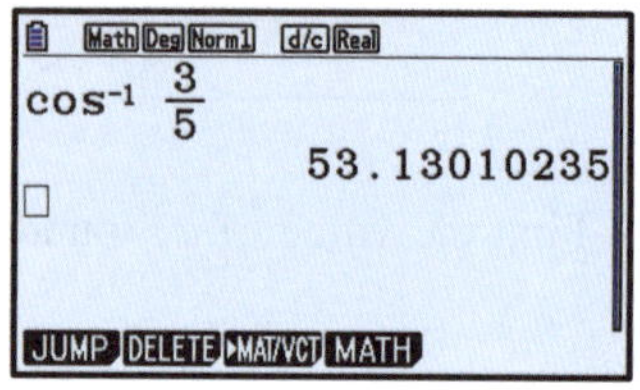

For the right angled triangle above, we find that $\theta \approx 53.1°$.

We define **inverse sine** and **inverse tangent** in a similar way.

Example 5 — Self Tutor

Find, to 1 decimal place, the measure of the angle marked θ.

a

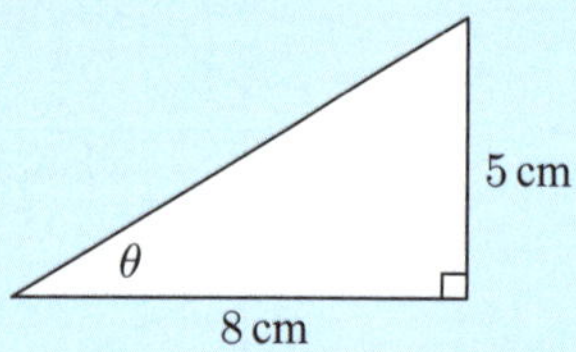

b

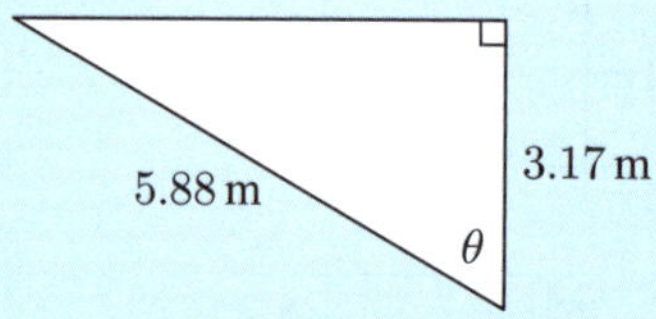

a

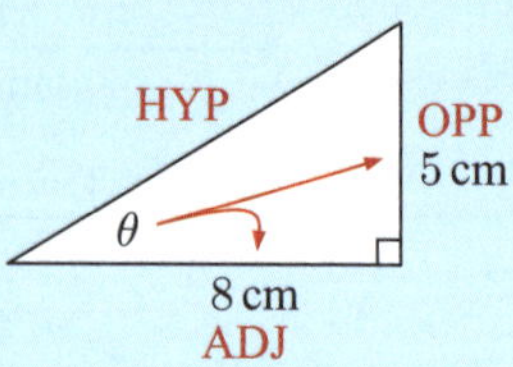

$\tan\theta = \dfrac{\text{OPP}}{\text{ADJ}} = \frac{5}{8}$

$\therefore\ \theta = \tan^{-1}(\frac{5}{8})$

$\therefore\ \theta \approx 32.0°$

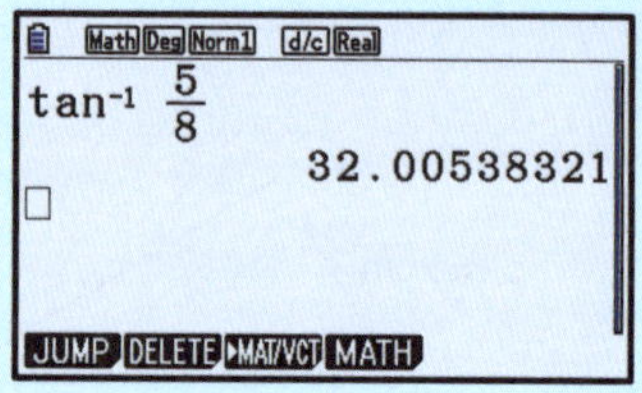

b

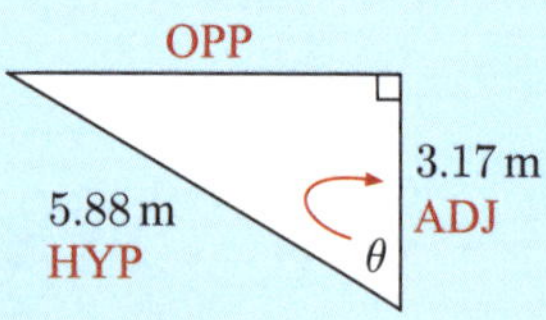

$\cos\theta = \dfrac{\text{ADJ}}{\text{HYP}} = \dfrac{3.17}{5.88}$

$\therefore\ \theta = \cos^{-1}\left(\dfrac{3.17}{5.88}\right)$

$\therefore\ \theta \approx 57.4°$

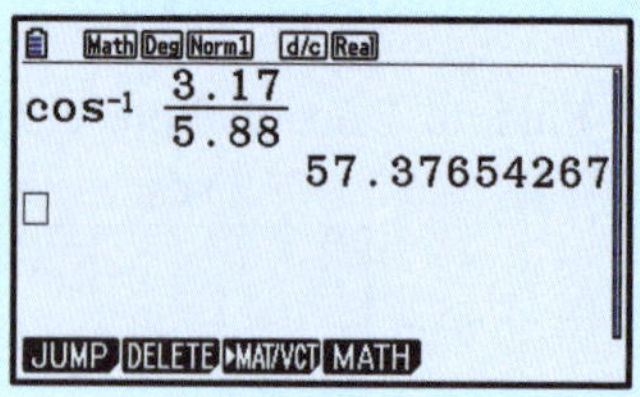

EXERCISE 23E

1 Find, to 1 decimal place, the measure of the angle marked θ.

a

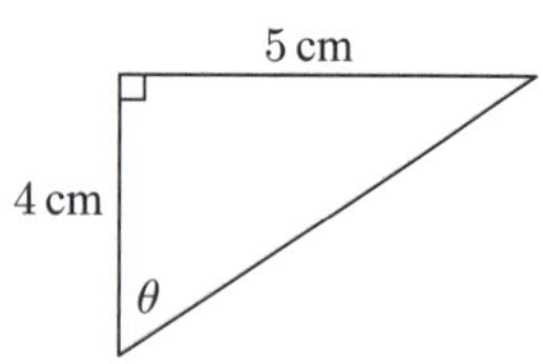

b

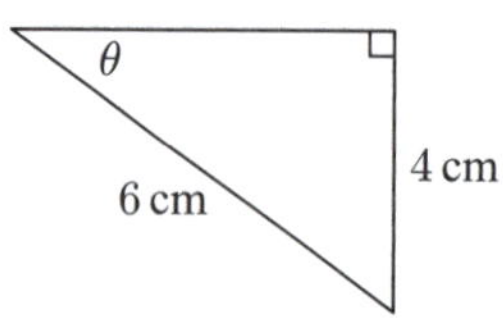

c

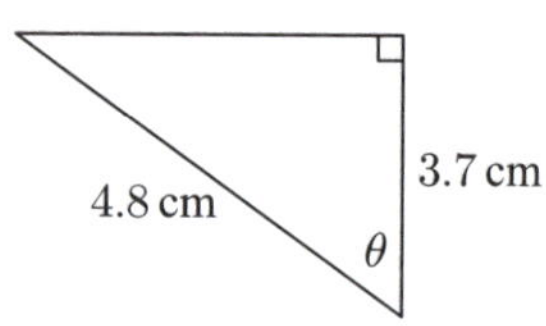

d
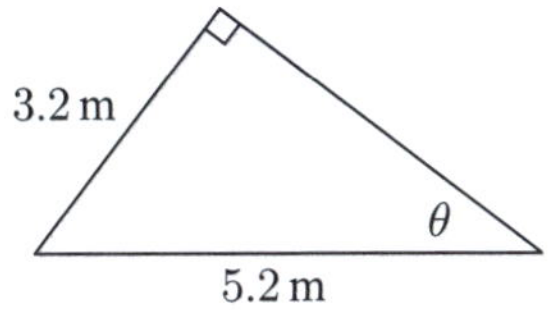

e
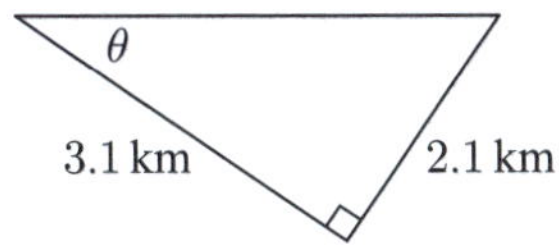

f
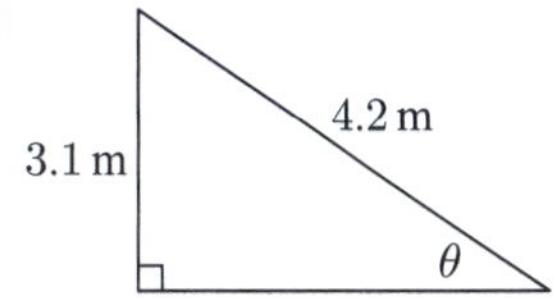

g
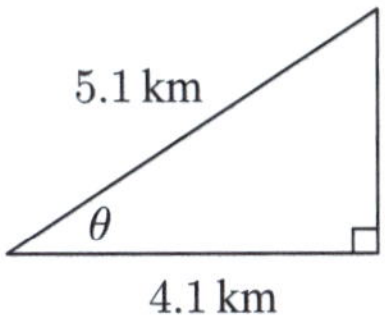

h
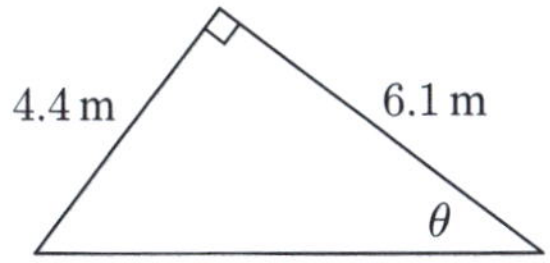

i
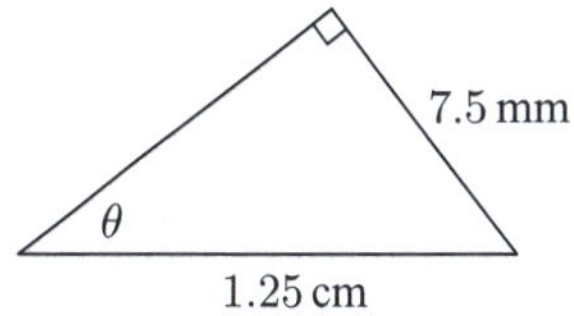

2 Find, to 1 decimal place, *all* unknown sides and angles:

a
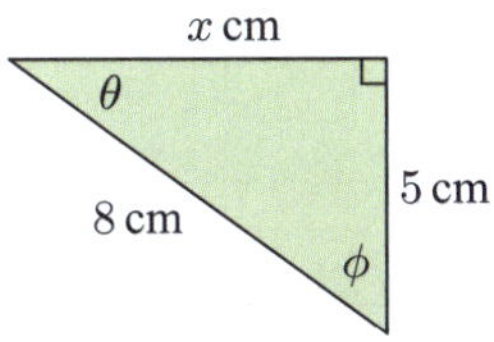

b
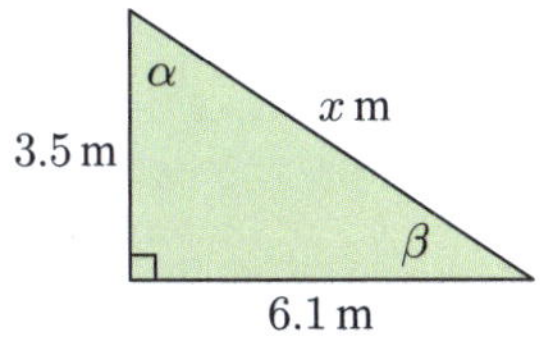

c
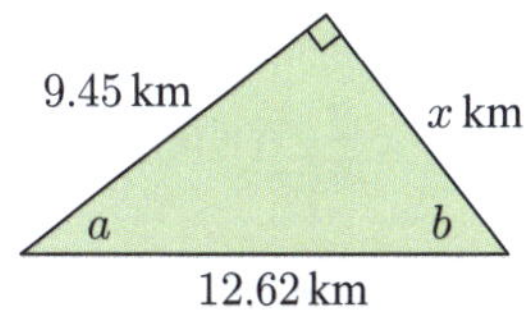

3 **a** State the exact value of:

i $\sin\theta$ **ii** $\cos\theta$ **iii** $\tan\theta$

b Use each of these values to find θ. Check that you get the same answer each time.

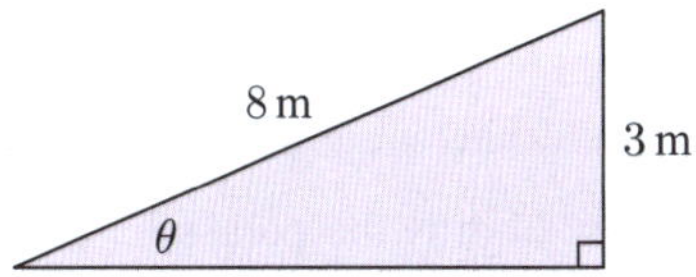

4 Try to find θ using the information given. Explain your results.

a
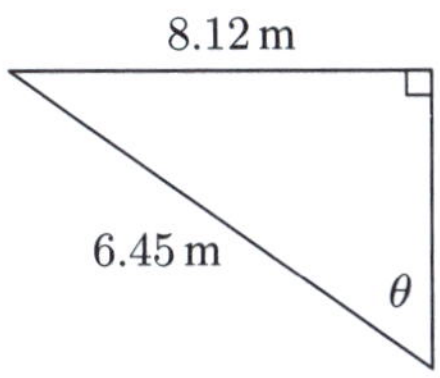

b
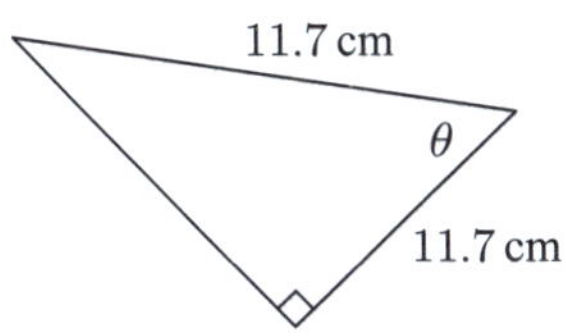

PUZZLE THE SPIRAL OF THEODORUS

The **spiral of Theodorus** is a series of right angled triangles.

The initial triangle ABC is isosceles with legs of length 1.

For each subsequent triangle, one leg is the hypotenuse of the previous triangle, and the other leg has length 1.

How many triangles can be drawn before they start to overlap?

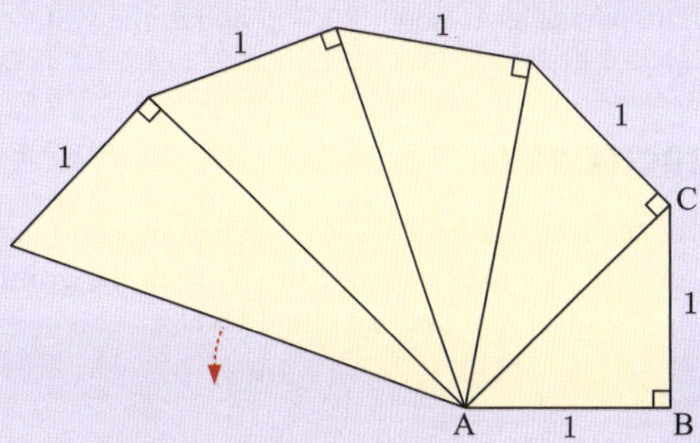

F PROBLEM SOLVING

The trigonometric ratios can be used to solve problems involving right angled triangles.

When solving these problems, you should follow these steps:

Step 1: Draw a **diagram** to illustrate the situation.

Step 2: Mark on the diagram the **unknown** angle or side that needs to be calculated. We often use x for a length and θ for an angle.

Step 3: Identify a **right angled triangle** in your diagram.

Step 4: Write an **equation** connecting an angle and two sides of the triangle using an appropriate trigonometric ratio.

Step 5: **Solve** the equation to find the unknown.

Step 6: **Write** your answer in a sentence.

Example 6 **Self Tutor**

A parasailer is towed behind a jet boat. The towing cable is 40 m long and is attached to the boat at water level. The cable makes an angle of 40° to the water.

How high is the parasailer above the water?

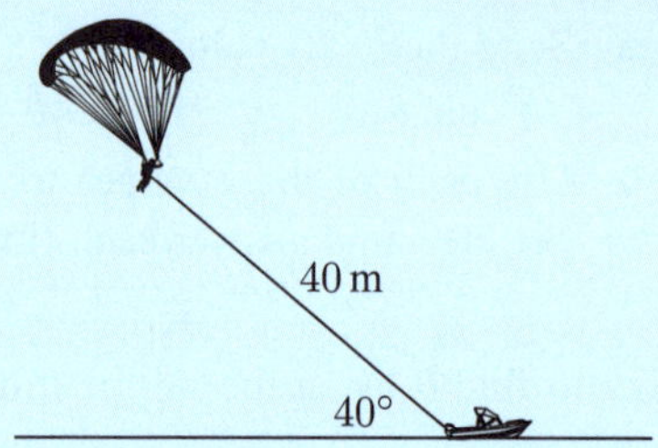

Let the height be x m.

$$\sin\theta = \frac{\text{OPP}}{\text{HYP}}$$

$$\therefore \sin 40° = \frac{x}{40}$$

$$\therefore x = 40\sin 40°$$

$$\therefore x \approx 25.7$$

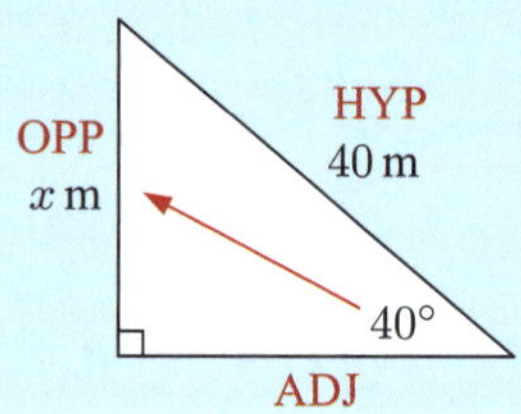

The parasailer is about 25.7 m above the water.

EXERCISE 23F

1 A see-saw has length 5.2 m. When one end is resting on the ground, it makes an angle of 23° with the ground. How high is the other end above ground level?

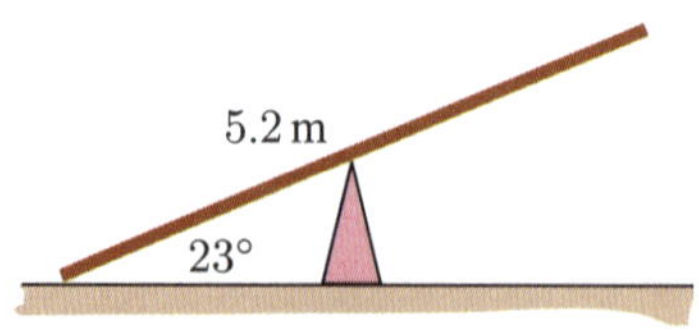

2 A driver travels 2 km up a long steady incline angled 8° to the horizontal. How far has the driver moved horizontally?

3 The shadow of a tree is 40 m long. The angle from the end of the shadow to the top of the tree is $33°$.
Find the height of the tree.

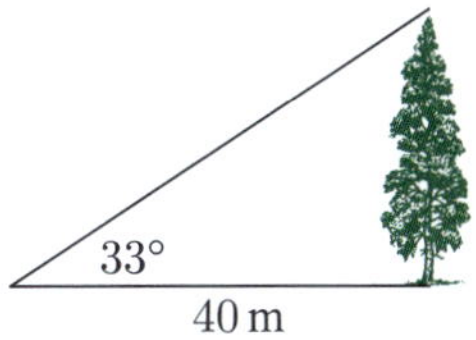

4 When a ladder makes an angle of $75°$ with the ground, it rests 2.4 m up a wall.
Find the length of the ladder.

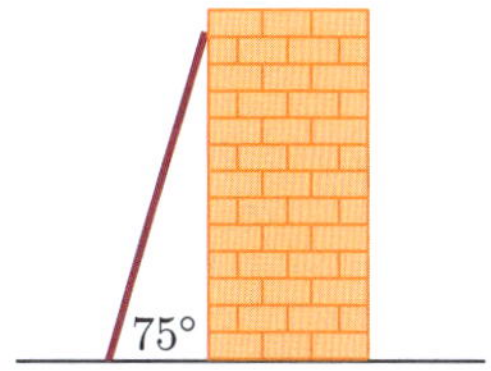

5 A boat has an anchor chain 55 m long. The boat drifts with the current so the chain makes an angle of $63°$ with the surface of the water. Find the depth of the water at the point where the anchor lies on the bottom.

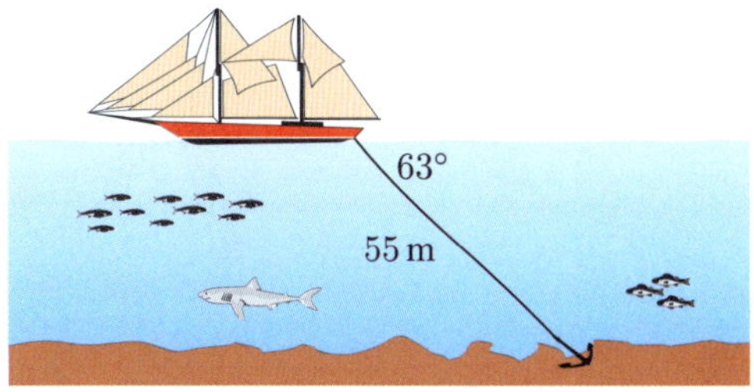

Example 7

Self Tutor

A staircase 3.6 m long leads to a tree house platform 2.9 m above the ground.

Find, using trigonometry:

a the angle the staircase makes with the ground

b the distance from the foot of the staircase to the post supporting the tree house.

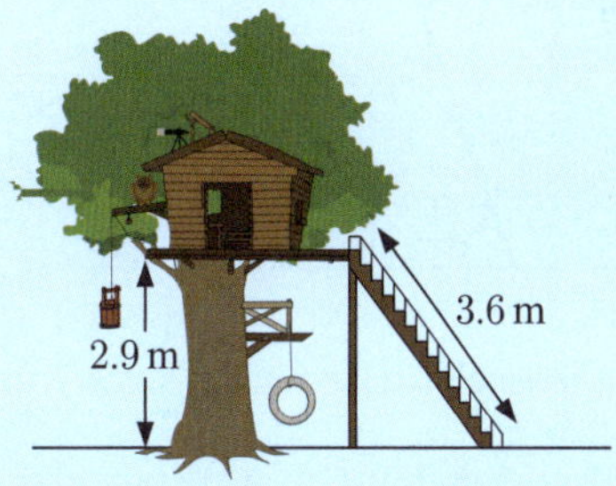

a

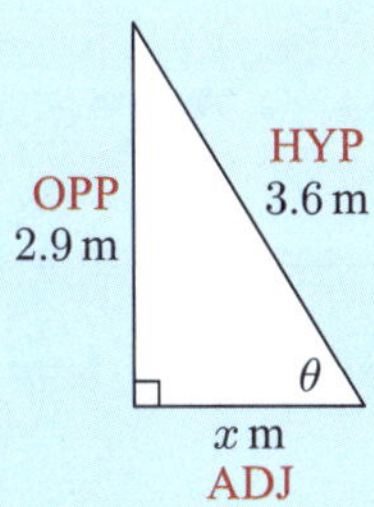

$$\sin\theta = \frac{\text{OPP}}{\text{HYP}} = \frac{2.9}{3.6}$$

$$\therefore \ \theta = \sin^{-1}\left(\frac{2.9}{3.6}\right)$$

$$\therefore \ \theta \approx 53.7°$$

The staircase makes an angle of about $53.7°$ with the ground.

b

$$\cos\theta = \frac{\text{ADJ}}{\text{HYP}}$$

$$\therefore \ \cos 53.7° \approx \frac{x}{3.6}$$

$$\therefore \ 3.6 \times \cos 53.7° \approx x$$

$$\therefore \ x \approx 2.13$$

The foot of the staircase is about 2.13 m from the post.

Check the answer using Pythagoras' theorem.

6 A kite is attached to a 30 m long string. The other end of the string is secured to the ground using a tent peg. If the kite is flying 21 m above ground level, find the angle θ that the string makes with the ground.

7 An advertising billboard by a highway is supported by two 5 m long beams fixed to the ground 2 m behind the billboard.
Find the angle that the beams make with the ground.

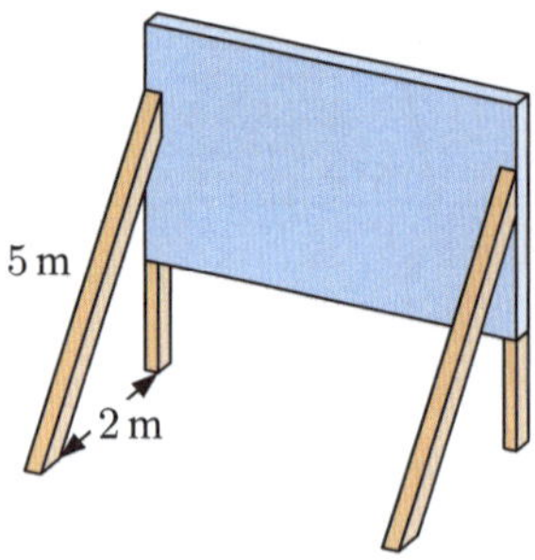

8 A flagpole is supported by four metal cables, each of which makes an angle of $55°$ with the ground and meets the pole 3.2 m above ground level.
Find the total length of the cables.

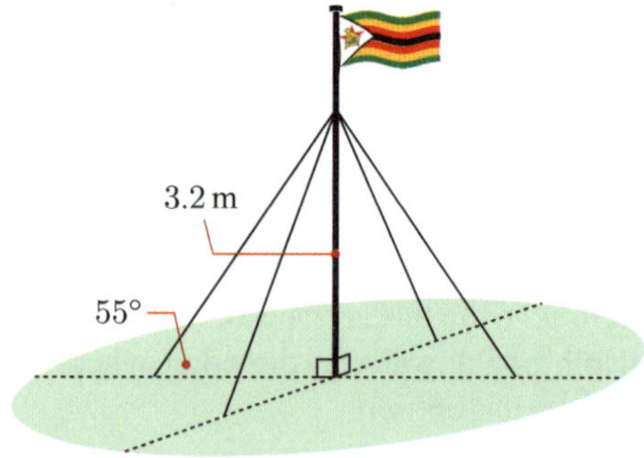

9 An A-frame house has the shape of an isosceles triangle with base angles $67°$. The walls are 13.2 m long. Find:

a the width of the building at ground level

b the height of the apex above the ground.

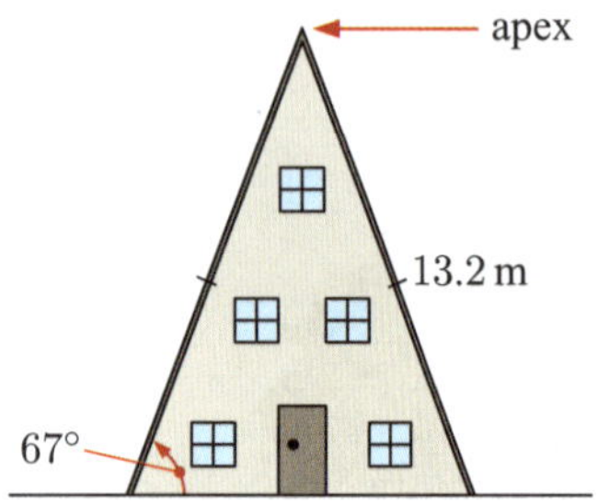

10 An isosceles triangle has sides 7 cm, 7 cm, and 8 cm in length. Find the measure of the base angles.

11 The metal frame of a rectangular gate has height 1.4 m and length 3.2 m. A diagonal strut is welded in place to support the frame. Find the measure of the angle θ between the strut and the top of the frame.

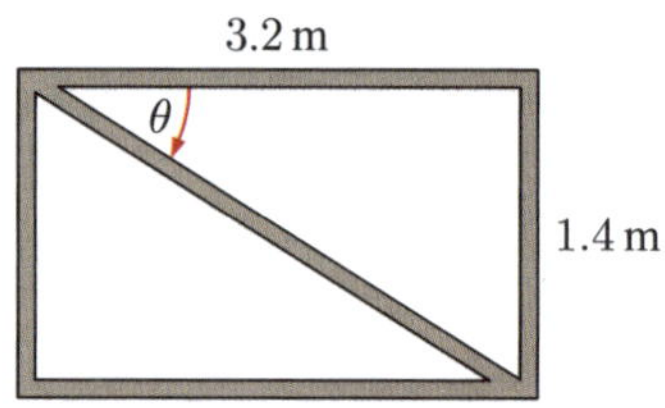

12 The diagram shows the cross-section of a farm shed. Find:

a the height of the apex A above ground level

b the area of this cross-section.

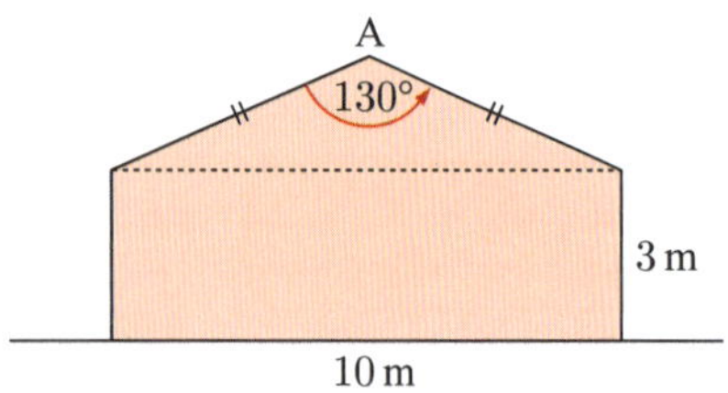

13 Find the measure of $Q\widehat{T}S$.

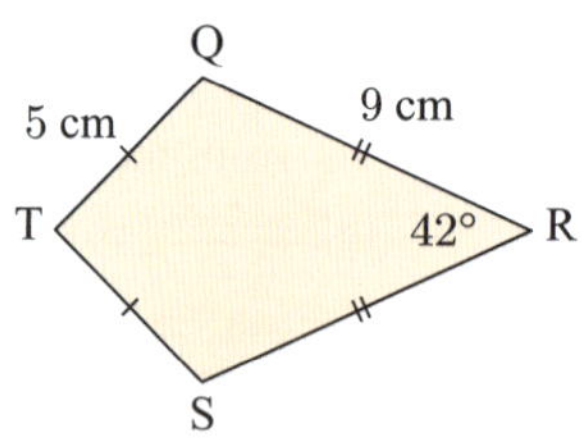

14 In a rhombus ABCD, AC = 4.8 m and BD = 6.6 m. Find the measure of $B\widehat{D}C$.

15 

An aeroplane takes off at an angle of 18°. Its average speed in the first 20 seconds of flight is 240 km/h. Find the altitude of the plane at this time.

16 A rectangle is 16 cm long and 10 cm wide. Find the acute angle at which the diagonals of the rectangle intersect.

17 A square tiling pattern is created from four congruent triangles and a square as shown.
Find:

- **a** the perimeter of each triangular tile
- **b** the area of each triangular tile
- **c** the side length of the square tile
- **d** the percentage of the pattern occupied by the square tile.

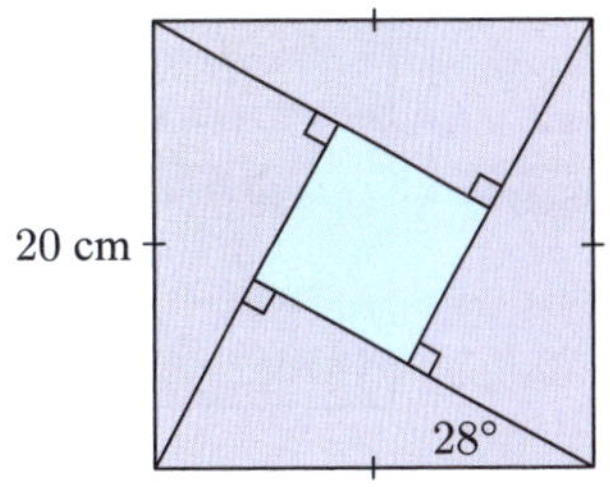

18 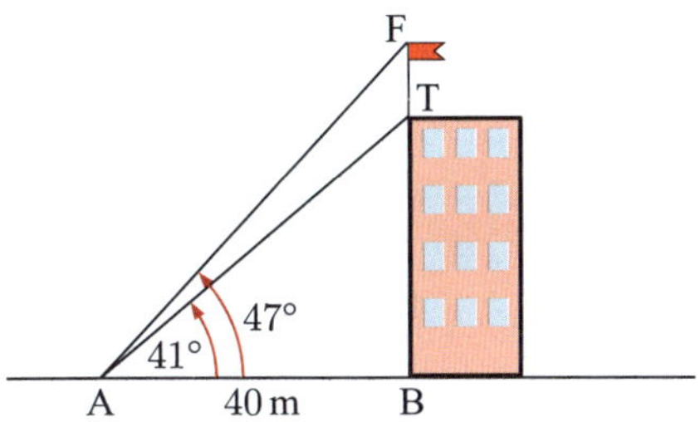

A surveyor takes the measurements shown. Find:

- **a** the height of the building BT
- **b** the height of the flagpole FT.

19 Use trigonometry to find the height of the mountain in the **Opening Problem** on page **372**.

20 Julie, Tina, and Brenton have baked a 12 cm by 12 cm square cake. They want to share the cake equally between them.

- **a** Brenton suggests cutting the cake at 30° angles from one corner.
  - **i** Find the area of each slice.
  - **ii** Would each person receive the same amount of cake?

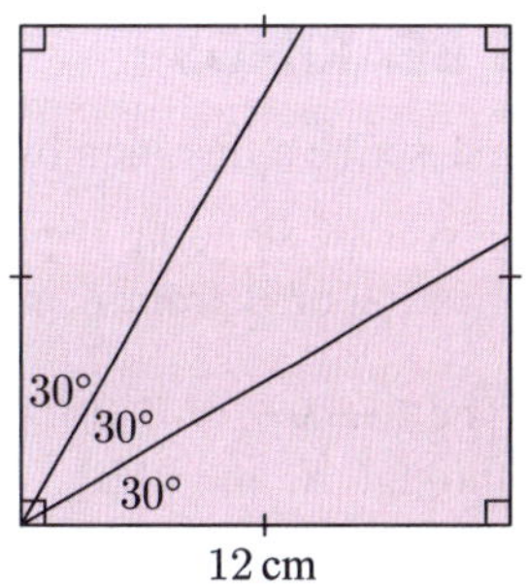

- **b** Find the angles $x°$, $y°$, and $z°$ needed to cut the cake into three slices of equal size.

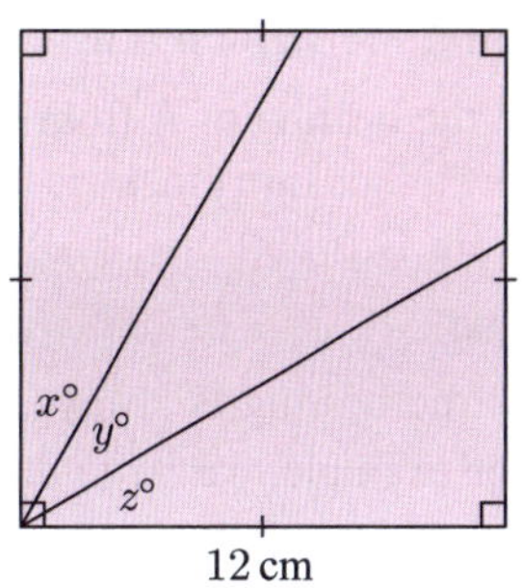

21 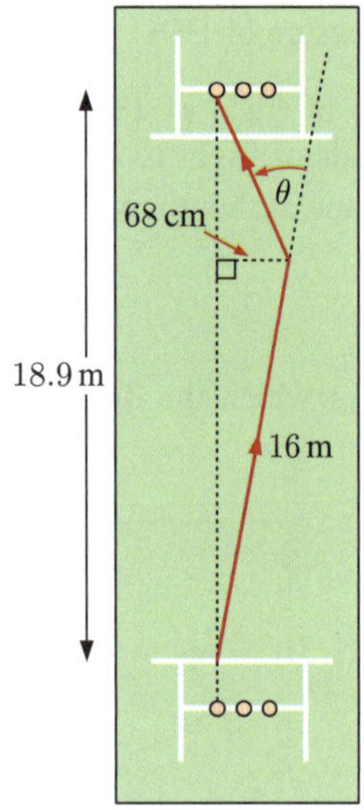

During a test match in England, a leg spin bowler delivers the ball from a point 18.9 m from the stumps, and level with the off stump.

The ball travels 16 m before it bounces, and lands 68 cm to the side as shown.

Find the angle θ through which the ball spins in order to hit the off stump as shown.

G BEARINGS

COMPASS BEARINGS

In previous years we have studied the bearings on a compass.

We commonly use the **cardinal directions** north, south, east, and west, and the **ordinal directions** northeast, southeast, southwest, and northwest.

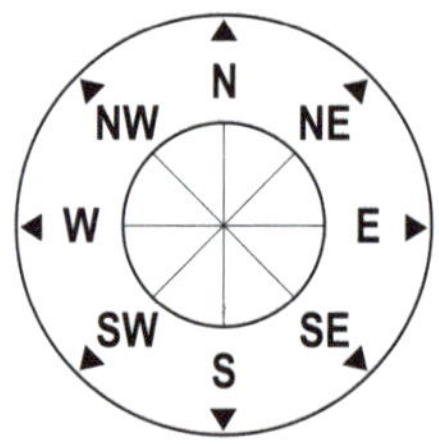

TRUE BEARINGS

We can describe a direction by comparing it with the **true north direction**. We call this a **true bearing**.

Imagine you are standing at point A, facing north. If you turn clockwise through an angle until you face B, the bearing of B from A is the angle through which you have turned.

The **bearing of B from A** is the clockwise measure of the angle between the "north" line through A, and [AB].

In the diagram alongside, the bearing of B from A is 120° from true north. We write this as 120°.

To find the bearing of A from B, we place ourselves at point B, face north, then turn clockwise until we face A. The true bearing of A from B is 300°.

Note that:

- A true bearing is always written using three digits. For example, we write 065° rather than 65°.
- The bearing of A from B always differs from the bearing of B from A by 180°.

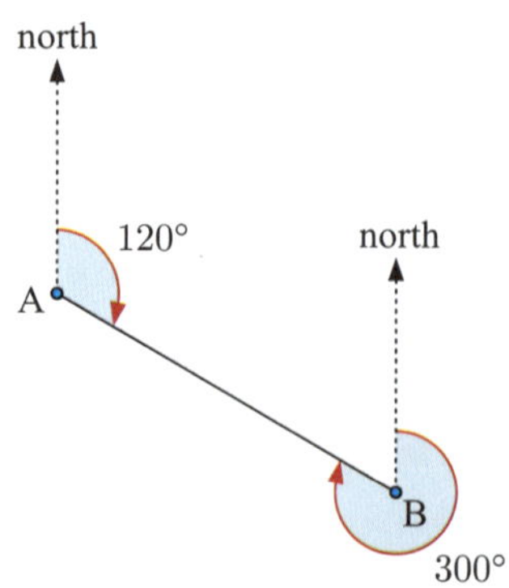

EXERCISE 23G

1 Draw diagrams to represent bearings from O of:

a 055° **b** 140° **c** 330° **d** 255°

2 Find the bearing of Q from P if the bearing of P from Q is:

a 124° **b** 068° **c** 244° **d** 321°

3 In the centre of a city is the square shown.
State the compass bearing and true bearing of:

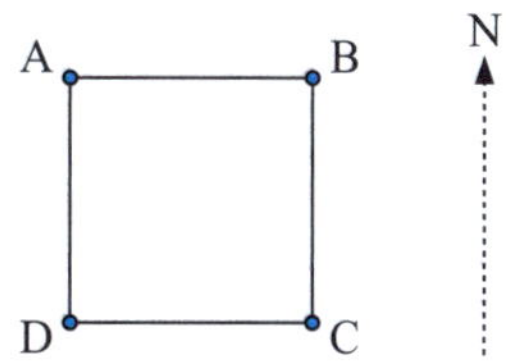

a B from A **b** C from A **c** D from A
d A from B **e** D from B **f** A from C
g B from C **h** B from D.

4 Write down the bearing of B from A:

a
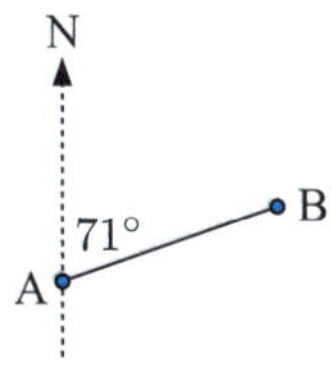

b
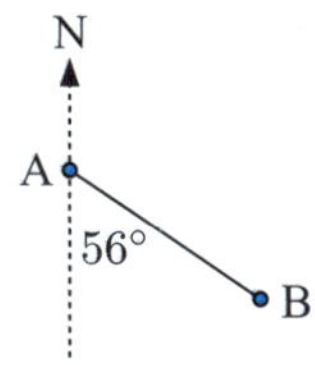

c
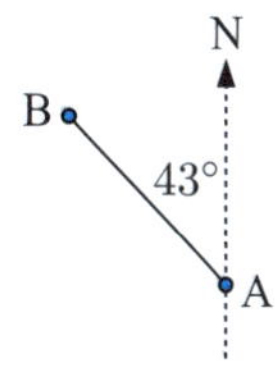

d
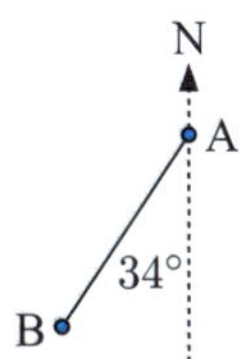

e
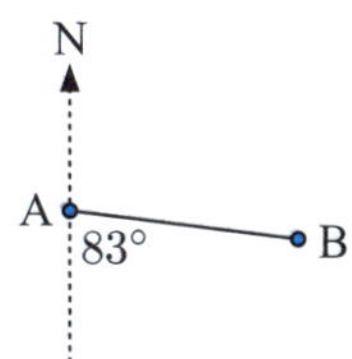

f
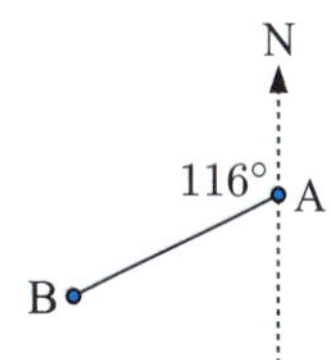

5 A, B, and C are checkpoints in an orienteering course.
Find the bearing of:

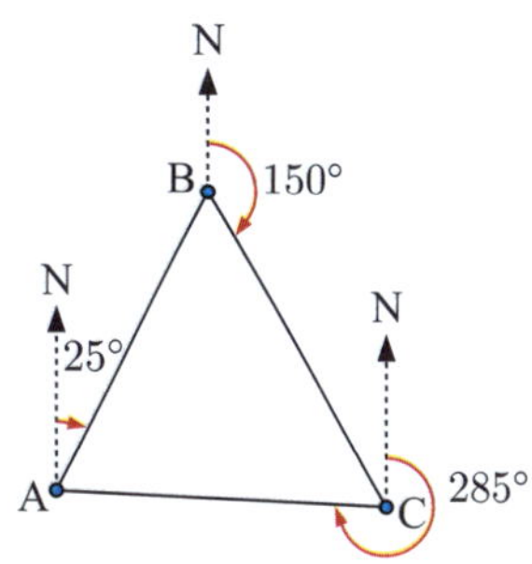

a B from A **b** C from B **c** B from C
d C from A **e** A from B **f** A from C.

6 Find the bearing of:

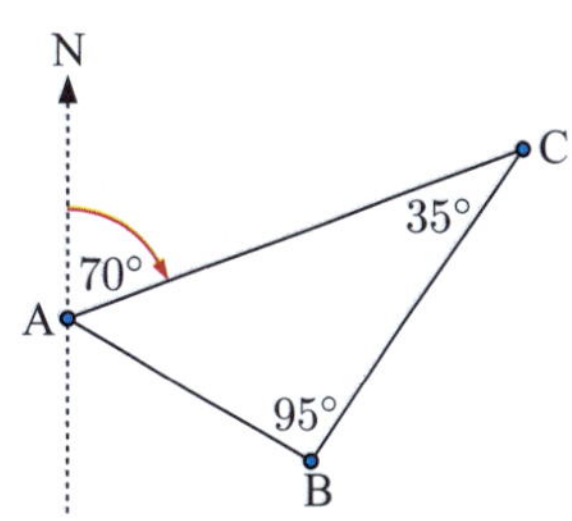

a B from A **b** A from B **c** C from A
d A from C **e** C from B **f** B from C.

Example 8 **Self Tutor**

A cyclist rides to a destination 21.3 km west and 13.8 km north of her starting point. Find the bearing of the destination from the starting point, to the nearest degree.

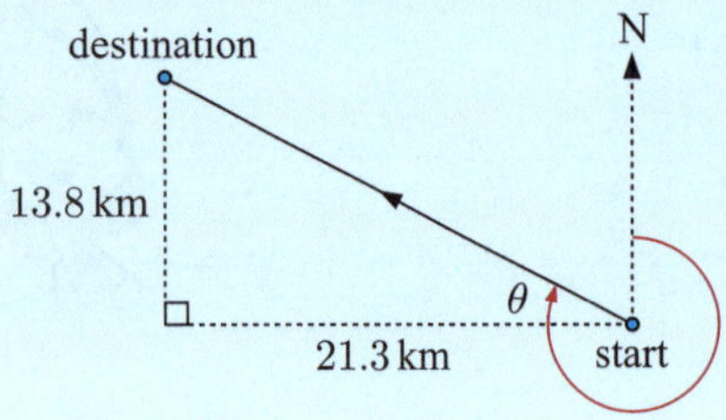

$\tan\theta = \dfrac{\text{OPP}}{\text{ADJ}}$

$\therefore\ \tan\theta = \dfrac{13.8}{21.3}$

$\therefore\ \theta = \tan^{-1}\left(\dfrac{13.8}{21.3}\right) \approx 32.9°$

$\therefore\ 270° + \theta \approx 303°$

So, the bearing of the destination from the starting point is about 303°.

7 An orienteer is studying the map for a competition.
Each grid unit represents 1 km. Find the distance and true bearing of:

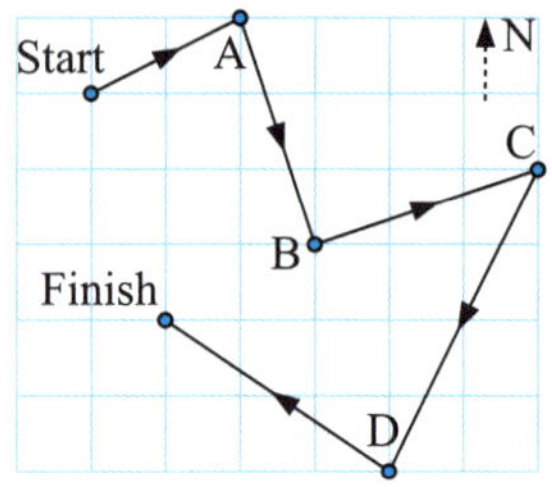

- **a** A from the start
- **b** B from A
- **c** C from B
- **d** D from C
- **e** the finish from D.

8 A small plane flies between two country airfields. Its destination is 18.4 km south and 27.3 km west of its origin. Find, to the nearest degree, the bearing on which the plane flies.

Example 9 **Self Tutor**

A ship sails for 18 km on the bearing 238°. How far west is the ship from its starting point?

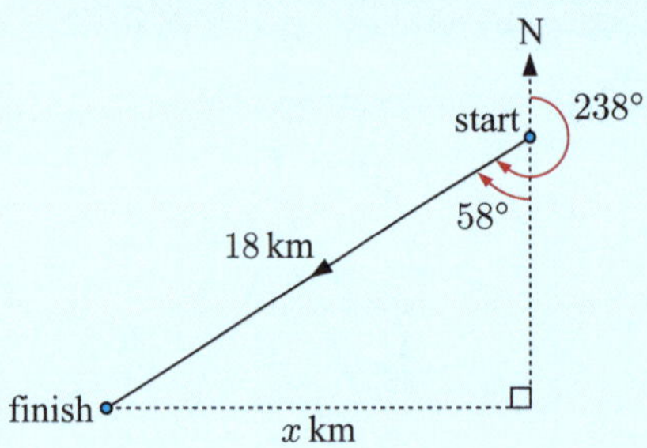

$\sin\theta = \dfrac{\text{OPP}}{\text{HYP}}$

$\therefore\ \sin 58° = \dfrac{x}{18}$

$\therefore\ x = 18 \times \sin 58°$

$\therefore\ x \approx 15.26$

The ship is about 15.3 km west of its starting point.

9 At the beach, a swimmer swims for 68 m on the bearing 036°.

- **a** How far north is the swimmer from her starting point?
- **b** How far east is the swimmer from her starting point?

10 A helicopter pilot flies in the direction 147°. He lands when he is 12 km south of his starting point. How far did he fly?

11 An aeroplane flies on the bearing 315° until it is 650 km west of its starting point. How far north is the aeroplane from its starting point?

QUICK QUIZ

MULTIPLE CHOICE QUIZ

REVIEW SET 23A

1 **a** Draw an accurate scale diagram using the measurements given. Use the scale $1 \text{ cm} \equiv 1 \text{ m}$.

b Use your scale diagram to estimate the actual length of [PQ].

c Use trigonometry to find the length of [PQ].

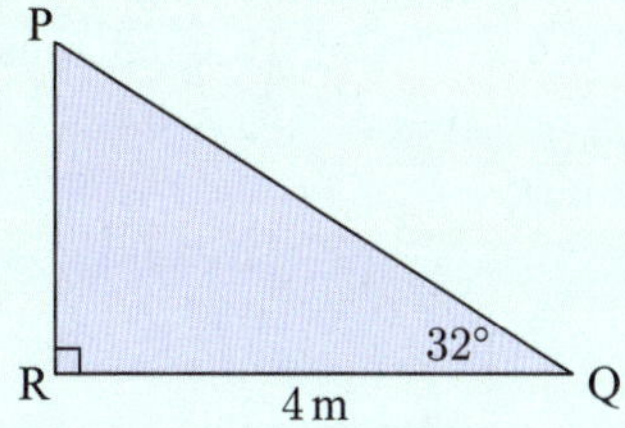

2 Find $\sin\theta$, $\cos\theta$, and $\tan\theta$ for this triangle.

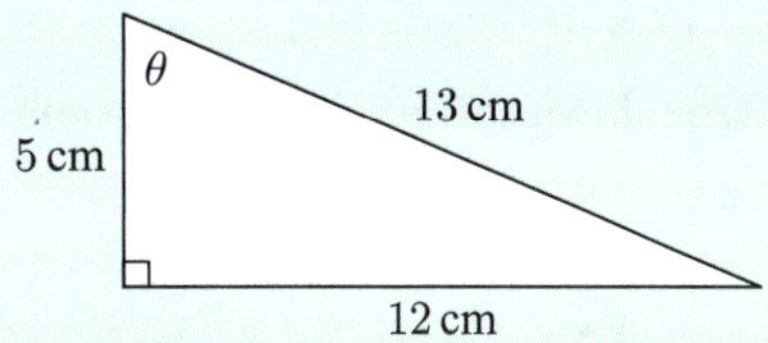

3

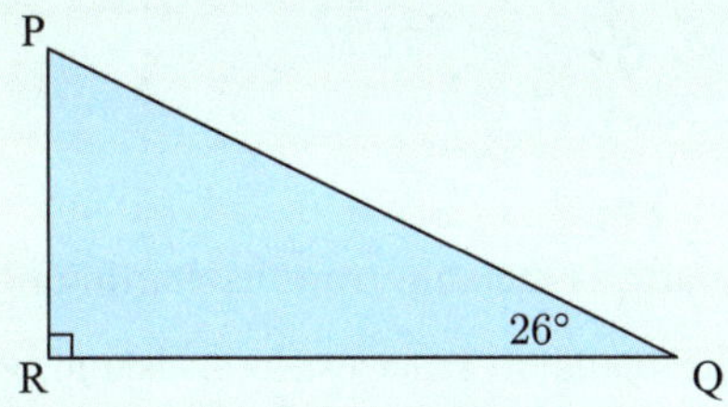

a Use a ruler to measure the length of each side of the triangle, to the nearest millimetre.

b Hence estimate the value of:

i $\sin 26°$ **ii** $\cos 26°$ **iii** $\tan 26°$

c Check your answers using a calculator.

4 Find the value of x, rounding your answer to 2 decimal places:

a

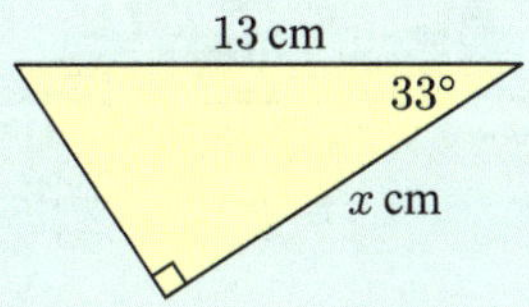

b

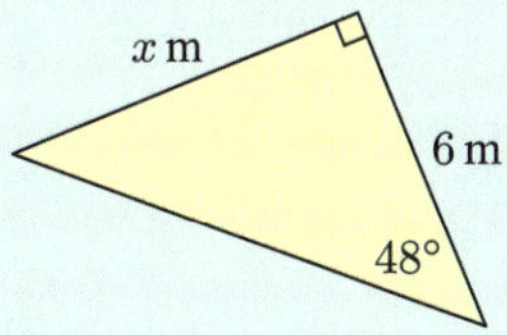

c

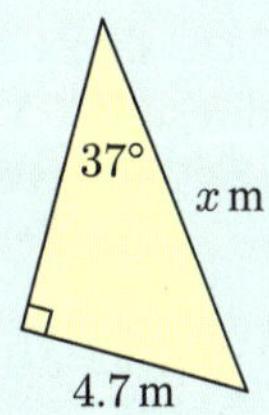

5 Find, to 1 decimal place, the measure of the angle marked θ:

a

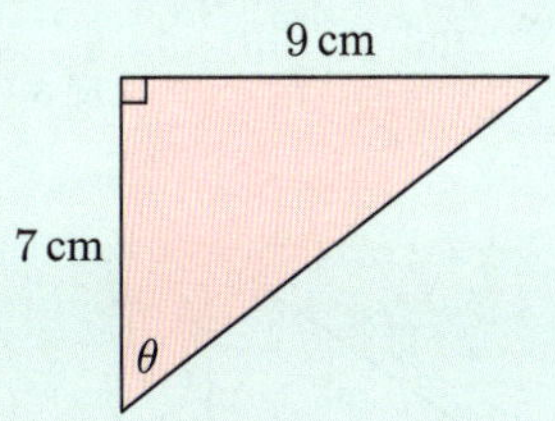

b

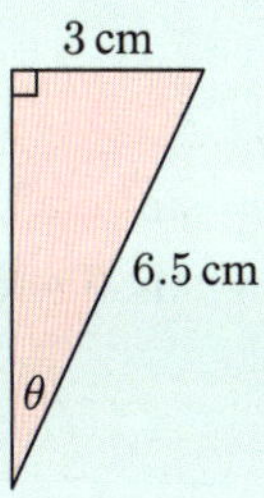

c

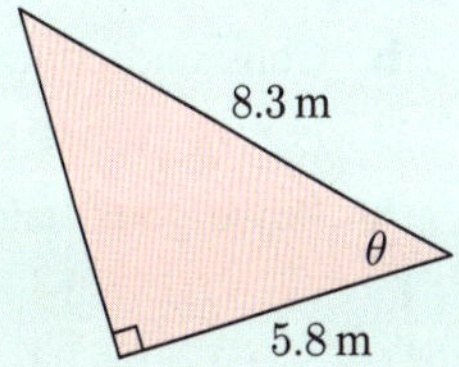

6 A cyclist travels for 5 km up a steady incline. In that time she climbs a vertical distance of 800 m. Find the angle of the incline.

7 Find, to 1 decimal place, *all* unknown sides and angles:

a

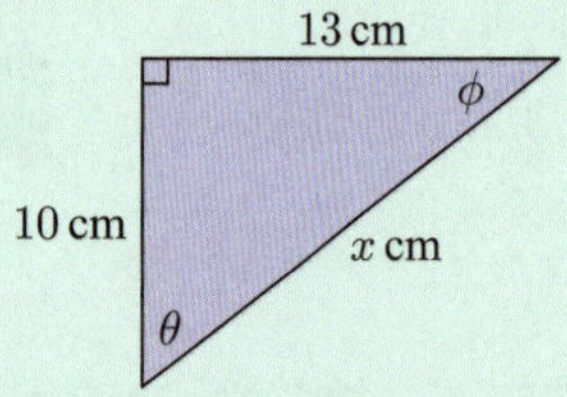

b

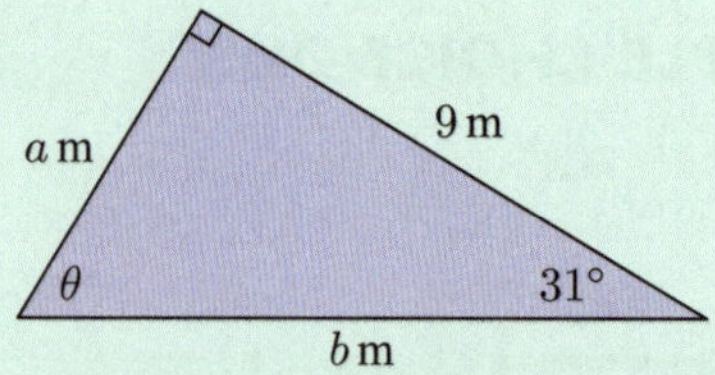

8 During a storm, a flagpole falls and breaks over a fence as shown. How tall was the flagpole?

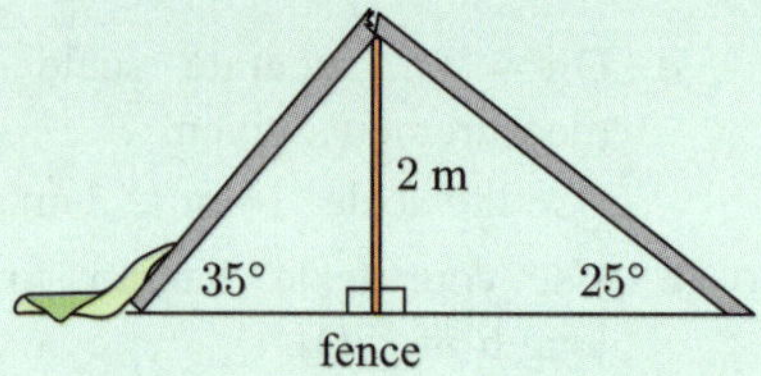

9 An isosceles triangle has equal sides of length 12 cm, and base angles 58°. Find the length of the base of the triangle.

10 Write down the bearing of Q from P:

a

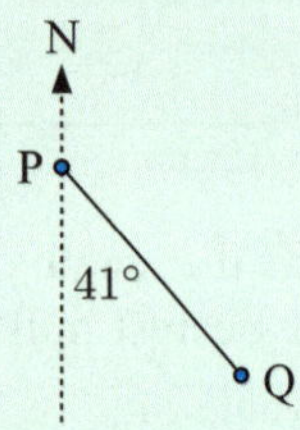

b

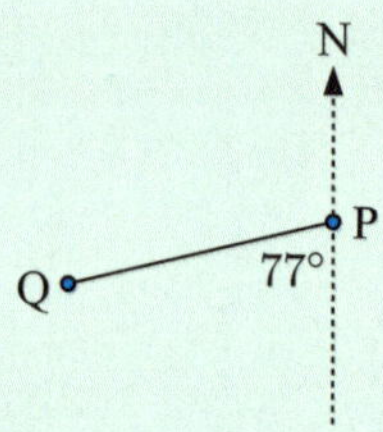

11 A boat sails 50 km on the bearing 054°. How far north is the ship from its starting point?

12 A ship leaves port A and travels 60 km south. It then sails 40 km west to an island port B.

a Draw a diagram of the situation.

b How far is B from A?

c The ship's captain wishes to sail back directly to A from B. On what bearing does the ship need to sail?

13 **a** Use your calculator to find, correct to 3 decimal places:

i $\cos 20°$ and $\sin 20°$ **ii** $\cos 70°$ and $\sin 70°$

iii $\cos 35°$ and $\sin 35°$ **iv** $\cos 55°$ and $\sin 55°$

v $\cos 41°$ and $\sin 41°$ **vi** $\cos 49°$ and $\sin 49°$

b Copy and complete:

For any acute angle θ, $\cos\theta = \sin(......)$.

c Prove your statement in **b** using this figure:

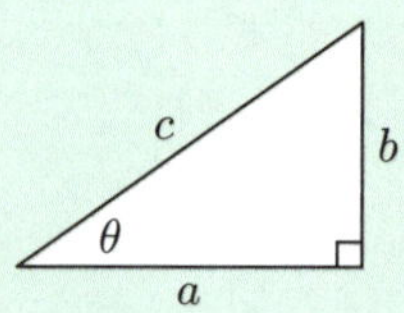

REVIEW SET 23B

1 Use a scale diagram with scale $1\text{ cm} \equiv 40\text{ m}$ to estimate the height of this volcano.

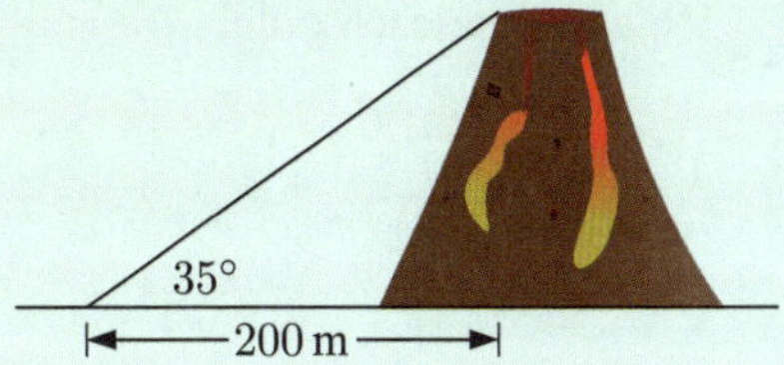

2 Use this figure to find the exact value of:

a $\cos 30°$ **b** $\sin 60°$
c $\cos 60°$ **d** $\sin 30°$

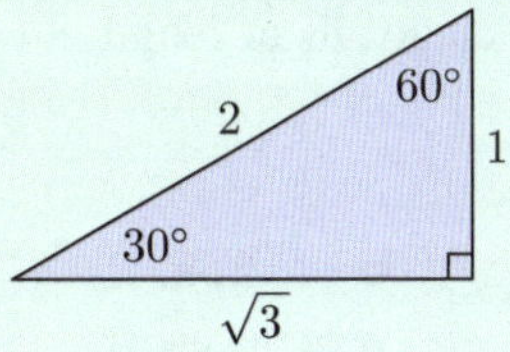

3 Find the value of x, rounding your answer to 1 decimal place:

a

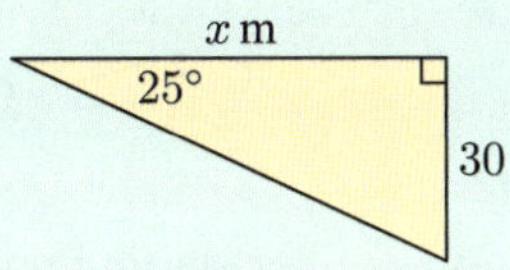

b

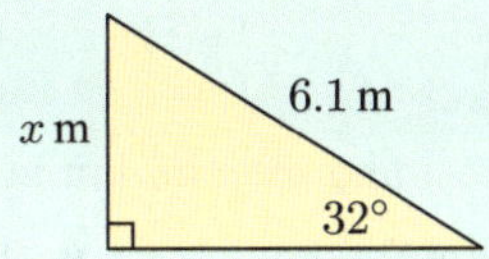

c

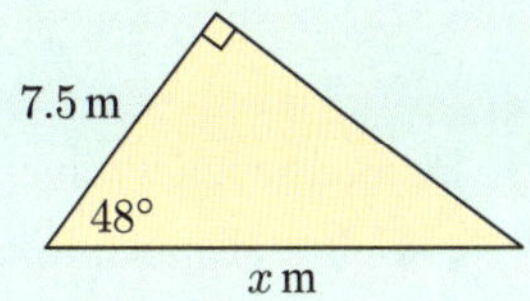

4 Find, to 1 decimal place, the measure of the angle marked θ.

a

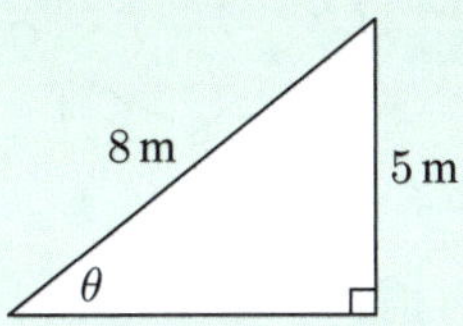

b

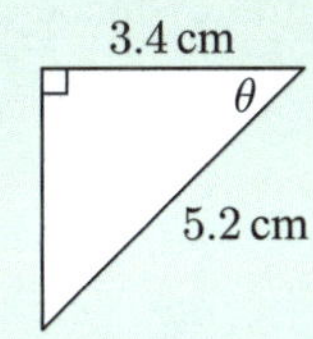

c

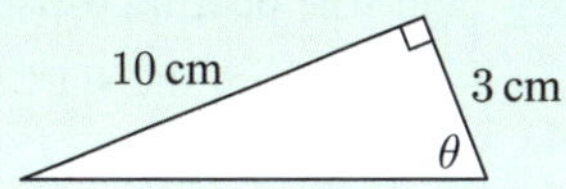

5 Find, to 1 decimal place, *all* unknown sides and angles.

a

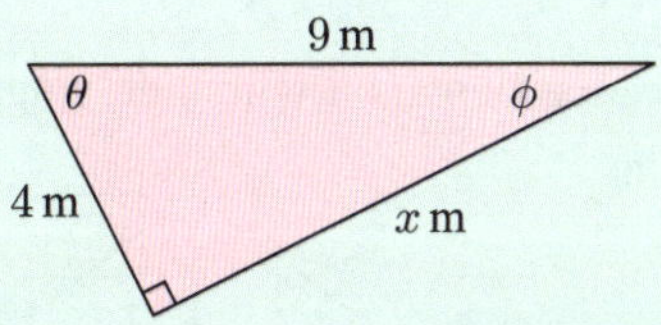

b

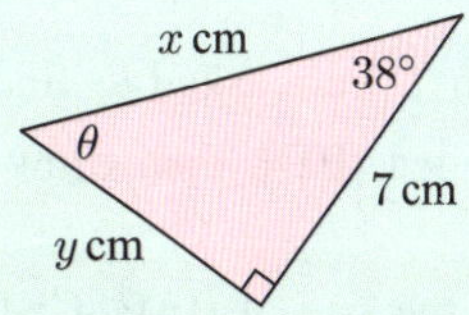

6 Find the perimeter of triangle PQR.

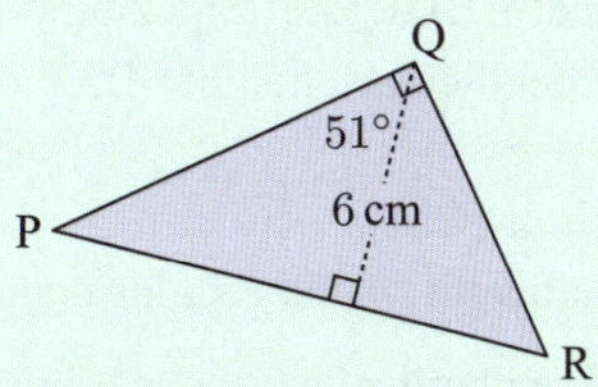

7

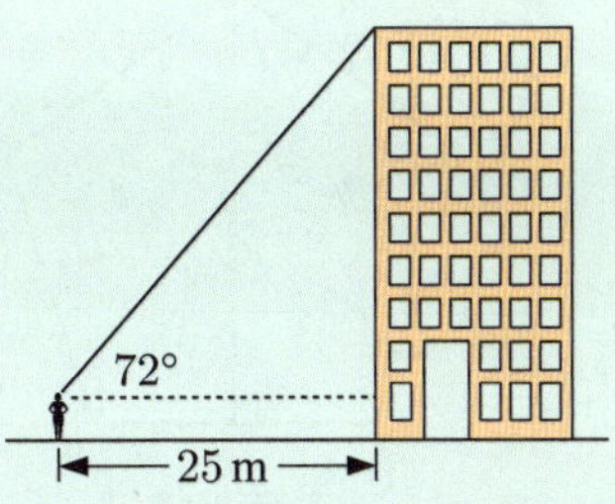

A young girl's eyes are one metre above ground level. She stands 25 m from the base of a tall building. She looks up to the top of the building at an angle of 72°. Find the height of the building.

8 A boundary fence is reinforced with a series of metal poles as shown.
How long is each pole?

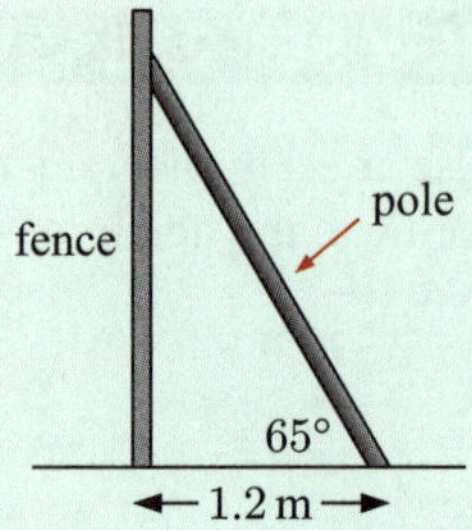

9 A rhombus has diagonals of length 8 cm and 11 cm. Find the measure of the smaller angles of the rhombus.

10 Find the bearing of:

a B from A **b** A from B **c** C from A
d A from C **e** C from B **f** B from C.

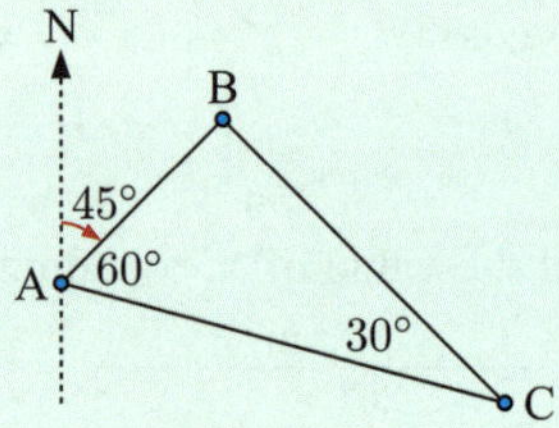

11 A motorcyclist travels 87 km west and then 63 km north. Find:
a her distance from her starting point **b** her bearing from her starting point.

12 The course for a yachting race is shown alongside. Each grid unit represents 1 km. Find the distance and true bearing of:

a A from the start
b B from A
c C from B
d D from C
e the finish from D.

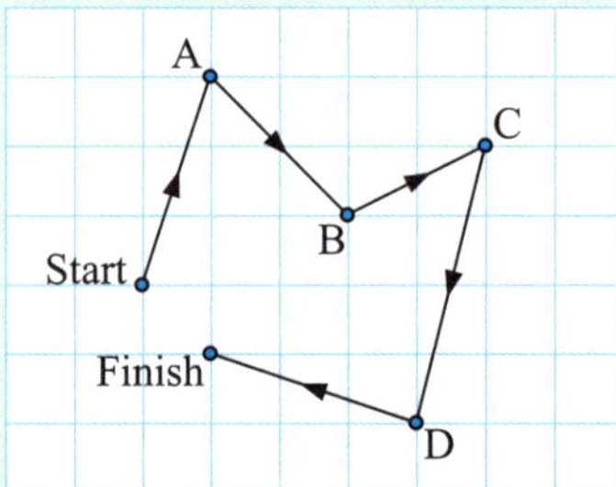

13 Simon and Sonya were asked to find the height of a regular pentagon with side length 10 cm.

a Simon joined [BD], then drew [CM] perpendicular to [BD].

i Find the size of $B\widehat{A}E$, $C\widehat{B}D$, and $A\widehat{B}D$.
ii Hence show that ABDE is a trapezium.
iii Find the perpendicular height BX of the trapezium.
iv Find the length of [CM].
v Hence find the height of the regular pentagon.

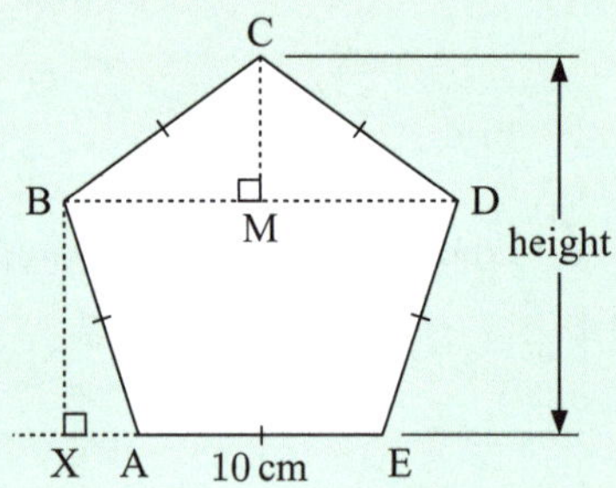

b Sonya contructed five identical isosceles triangles by joining each vertex to the centre of the pentagon O, then drew [OM] perpendicular to [AE].

i Find the size of $O\widehat{A}E$.
ii Hence find the length of [OA].
iii Find the length of [OM].
iv Hence find the height of the regular pentagon.

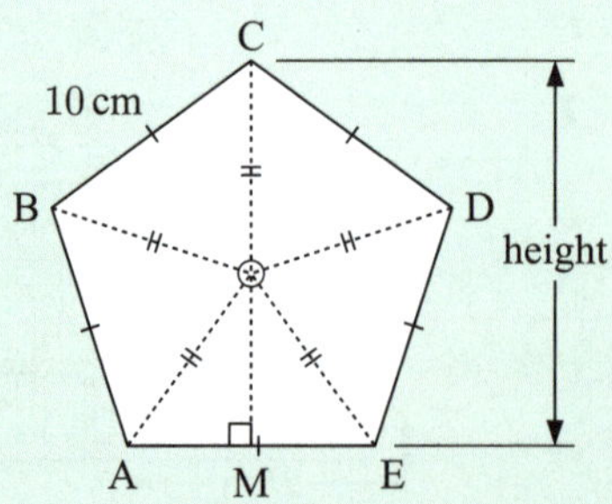

Chapter 24

Deductive geometry

Contents:

OPENING PROBLEM

GEOMETRY PACKAGE

Kelly draws many different "stars" like the one shown. The stars vary in shape and size.

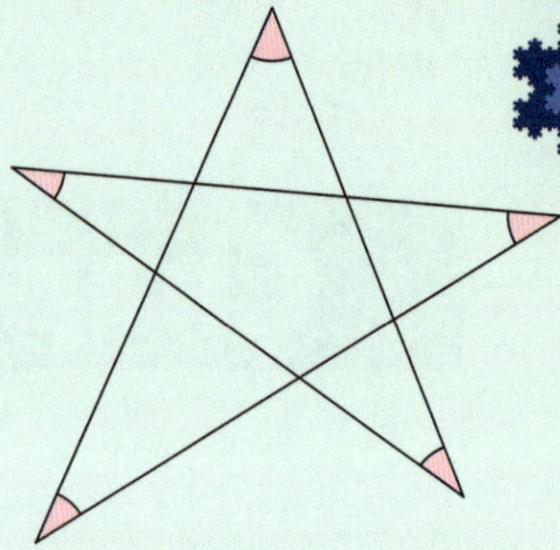

For each star she draws, Kelly measures the five interior angles with a protractor. She adds them together and always gets an answer of approximately 180°.

Kelly therefore *conjectures* that they actually do add to 180°.

Things to think about:

- **a** What do we mean by a *conjecture*?
- **b** Can you prove that Kelly's conjecture is true?

HISTORICAL NOTE

Euclid was one of the great mathematical thinkers of ancient times. He founded a school in Alexandria during the reign of Ptolemy I, which lasted from 323 BC until 284 BC.

Euclid

Euclid's most famous mathematical writing is the set of 13 books called *Elements*. It is the most complete study of geometry ever written, and has been a major source of information for the study of geometric techniques, logic, and reasoning.

In *Elements*, Euclid suggested a set of **axioms**, or unproveable statements he assumed to be true, as a foundation for geometry. Based on these axioms, he wrote definitions, theorems, and proofs.

Euclidean geometry is the study of geometrical figures based on the axioms of Euclid.

From previous years you should be familiar with the following geometric facts. These results are **theorems** which are always true.

Title	*Theorem*	*Example*
Angles at a point	Angles at a point add to 360°.	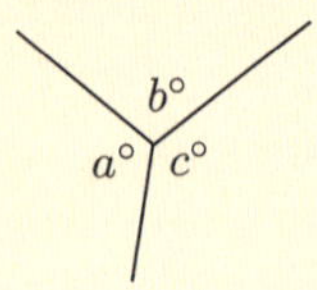 $a+b+c=360$
Angles on a straight line	Angles on a line are supplementary, which means their sum is 180°.	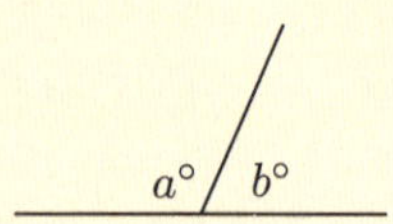 $a+b=180$

Title	*Theorem*	*Example*
Angles in a right angle	Angles in a right angle are complementary, which means their sum is 90°.	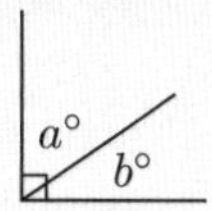 $a+b=90$
Vertically opposite angles	Vertically opposite angles are equal.	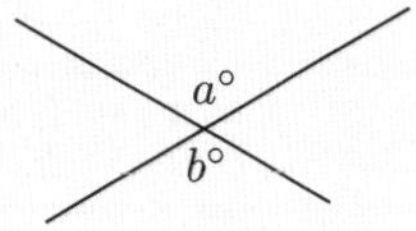 $a=b$
Corresponding angles	When two parallel lines are cut by a transversal, angles in corresponding positions are equal.	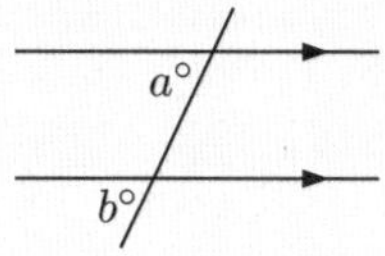 $a=b$
Alternate angles	When two parallel lines are cut by a transversal, angles in alternate positions are equal.	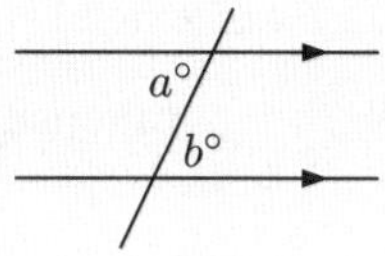 $a=b$
Co-interior or allied angles	When two parallel lines are cut by a transversal, angles in co-interior positions are supplementary.	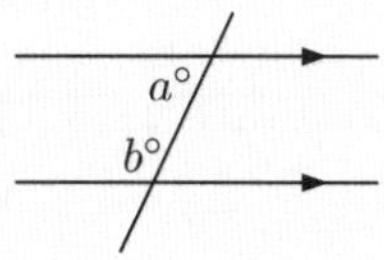 $a+b=180$
Angle sum of a triangle	The sum of the interior angles of a triangle is 180°.	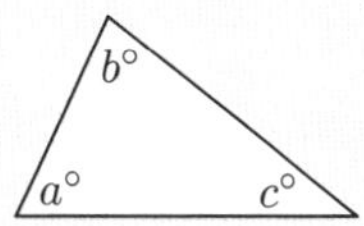 $a+b+c=180$
Exterior angle of a triangle	Any exterior angle of a triangle is equal to the sum of the opposite interior angles.	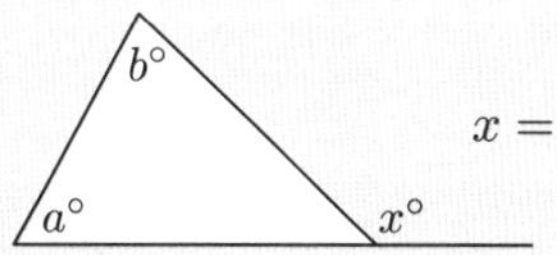 $x=a+b$
Isosceles triangle theorem	In any isosceles triangle: • the base angles are equal • the line joining the apex to the midpoint of the base bisects the vertical angle and meets the base at right angles.	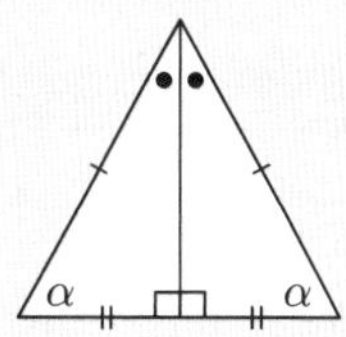
Angle sum of an n-sided polygon	The sum of the interior angles of an n-sided polygon is $(n-2)\times 180°$.	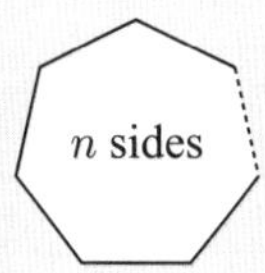 sum of angles $=(n-2)\times 180°$

A DEDUCTIVE GEOMETRY

In **deductive geometry** we apply definitions and theorems to *deduce* further results about a figure.

When using a theorem, we should state the theorem's name as a reason to justify a particular step in our argument.

Example 1 Self Tutor

Find the value of a:

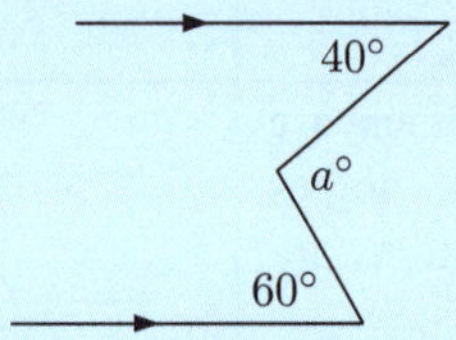

We label the figure and extend [DC] to meet [AB] at X.

$B\widehat{X}D = X\widehat{D}E$ {alternate angles}

$\therefore$ $B\widehat{X}D = 60°$

$\therefore$ $a = 60 + 40$ {exterior angle of a triangle}

$\therefore$ $a = 100$

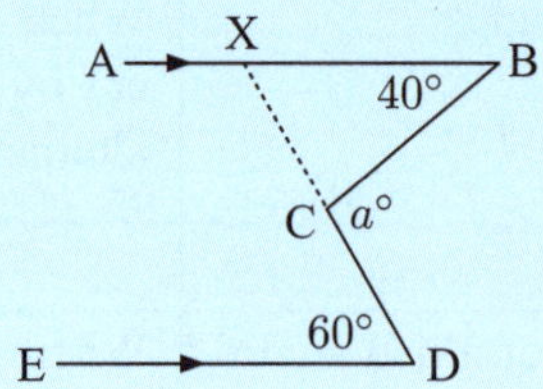

EXERCISE 24A

1 Find the values of the unknowns:

a
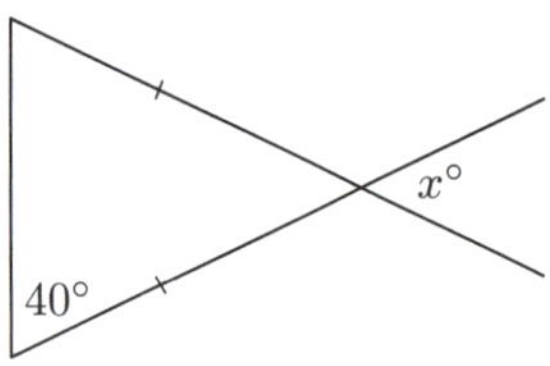

b
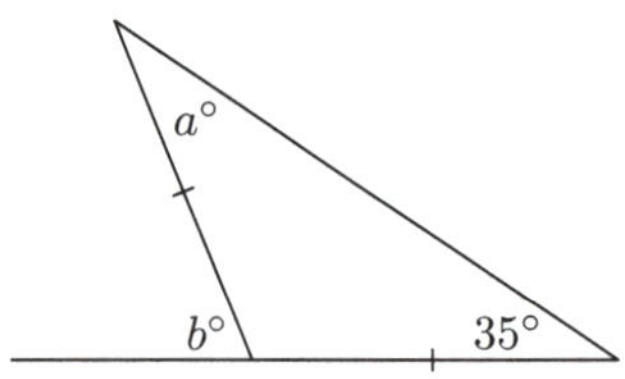

c
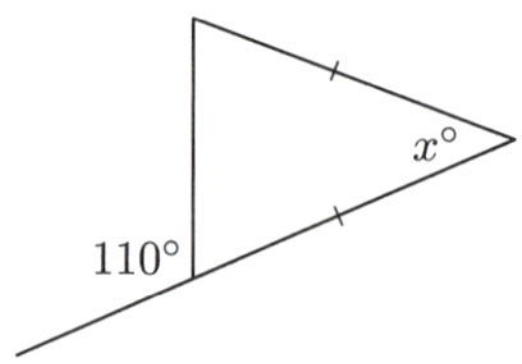

d
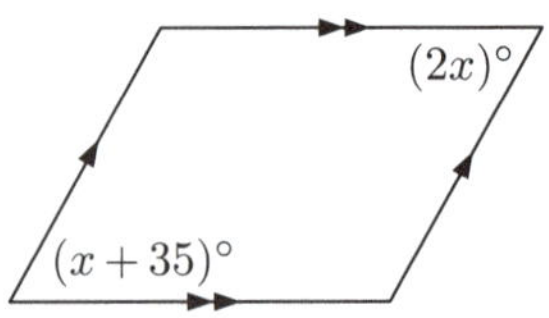

e
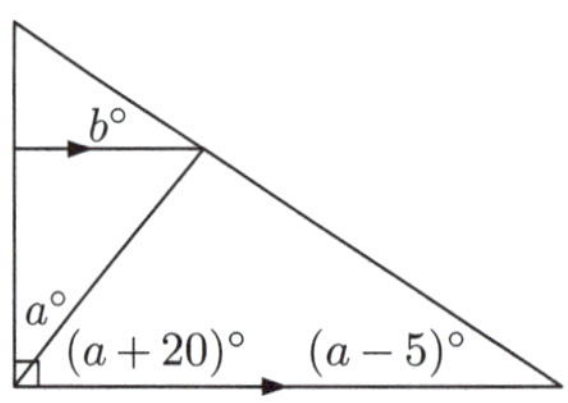

f
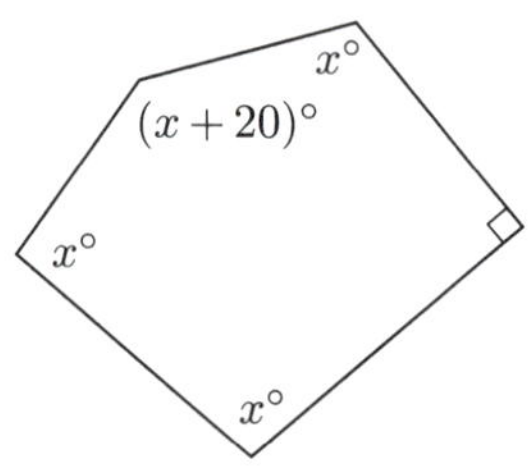

2 By adding a construction line, find the value of a:

a
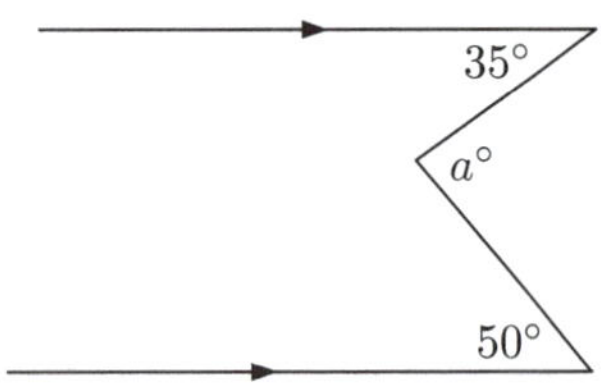

b
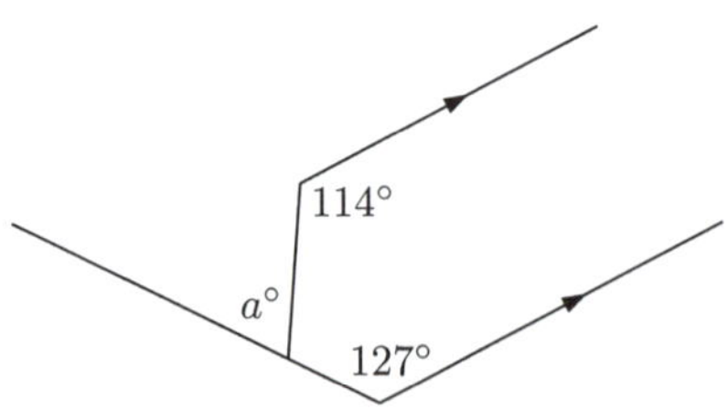

Example 2 Self Tutor

Prove that the angle bisectors of $\widehat{ABD}$ and $\widehat{CBD}$ form a right angle.

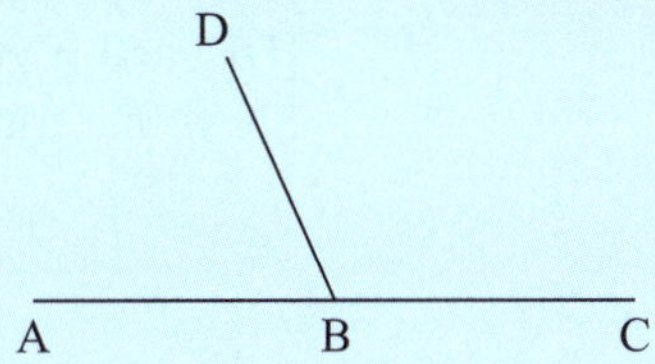

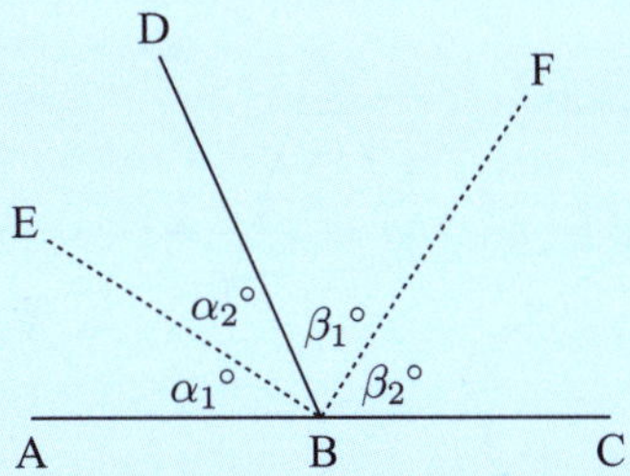

We draw angle bisectors [BE] and [BF].

$\therefore\ \alpha_1 = \alpha_2$ and $\beta_1 = \beta_2$

But $2\alpha + 2\beta = 180$ {angles on a line}

$\therefore\ \alpha + \beta = 90$

$\therefore\ \widehat{EBF} = 90°$

3 ABC is an isosceles triangle in which AB = AC. [BA] is extended to D, and [AE] is drawn parallel to [BC]. Let $\widehat{ABC} = x°$.

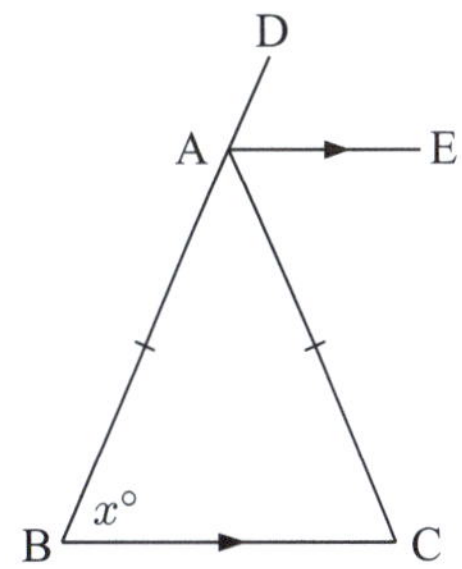

a Find, in terms of x:

i $\widehat{ACB}$ **ii** $\widehat{CAE}$ **iii** $\widehat{DAE}$

b Copy and complete:

"The line segment [AE] ...... $\widehat{CAD}$."

4 The opposite angles of this quadrilateral are equal in size. Let $\widehat{ABC} = \alpha°$ and $\widehat{BCD} = \beta°$.

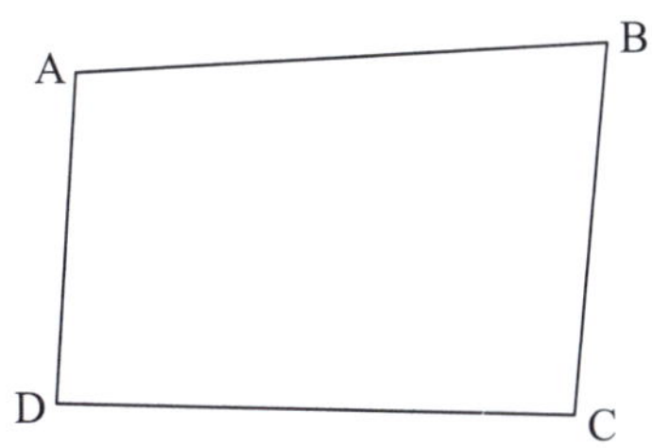

a Find the value of $\alpha + \beta$.

b Explain why [AB] is parallel to [DC], and [AD] is parallel to [BC].

c Copy and complete: "If the opposite angles of a quadrilateral are equal, then the quadrilateral is a ......"

5 ABCD is a rhombus. M is the point on [AD] such that [CM] bisects $\widehat{ACD}$. Show that $\widehat{DMC}$ is 3 times larger than $\widehat{ACM}$.

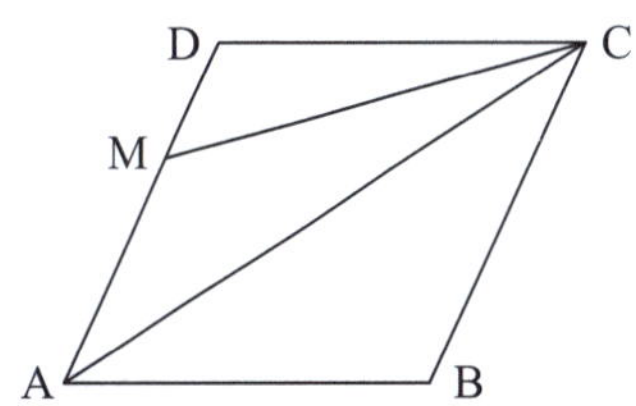

6 Triangle ABC is isosceles with AB = AC. [CB] is extended to D, and [AB] is extended to E such that BE = DE. Show that [DE] and [AC] are parallel.

7 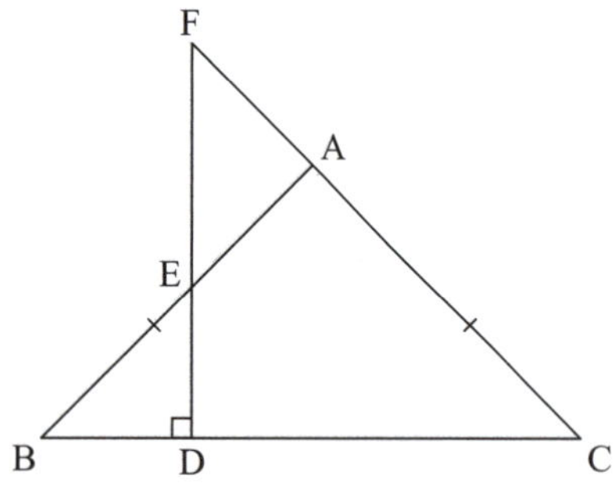

ABC is an isosceles triangle in which AB = AC.
[DE] is perpendicular to [BC].
[DE] and [CA] are both extended to meet at F.
Show that triangle AEF is also isosceles.

8 Consider the **Opening Problem** on page **394**.
Prove that the five interior angles of the star add to $180°$.

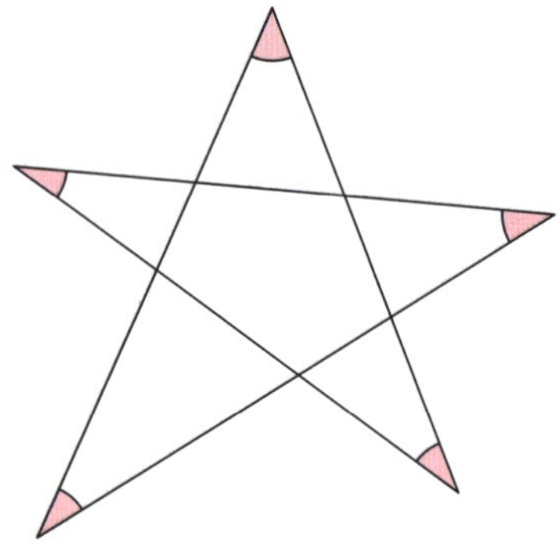

ACTIVITY — PAPPUS' THEOREM

In this Activity you will discover Pappus' theorem, named after **Pappus of Alexandria** (290 - 350).

What to do:

1. Draw two long straight lines on a sheet of paper. They do not have to be parallel, but do not let them intersect on the page.

   GEOMETRY PACKAGE

2. Mark any three points A, B, and C on one line, and three points D, E, and F on the other line.

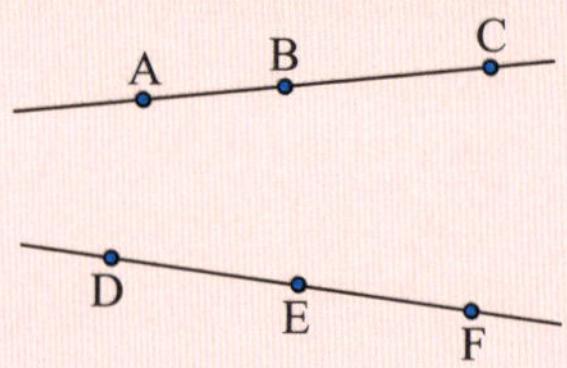

3. **a** Draw lines (AE) and (BD), and mark their intersection point X.
   b Draw lines (AF) and (CD), and mark their intersection point Y.
   c Draw lines (BF) and (CE), and mark their intersection point Z.
4. What do you notice about the points X, Y, and Z?
5. Repeat these steps with another pair of straight lines. Comment on your observations. Does it matter if your chosen lines intersect?

B MIDPOINT THEOREM

INVESTIGATION 1 MIDPOINT THEOREM

What to do:

1. Construct a large triangle ABC.
2. Mark the midpoints P and Q of [AB] and [AC] respectively.
3. Join [PQ].
4. Measure the lengths of [PQ] and [BC]. What do you notice?
5. Measure $A\widehat{P}Q$ and $A\widehat{B}C$. What conclusion can be drawn about [PQ] and [BC]?
6. Repeat the procedure with different sized triangles.
7. Copy and complete: *The line joining the midpoints of two sides of a triangle is ...... to the third side and ...... its length.*

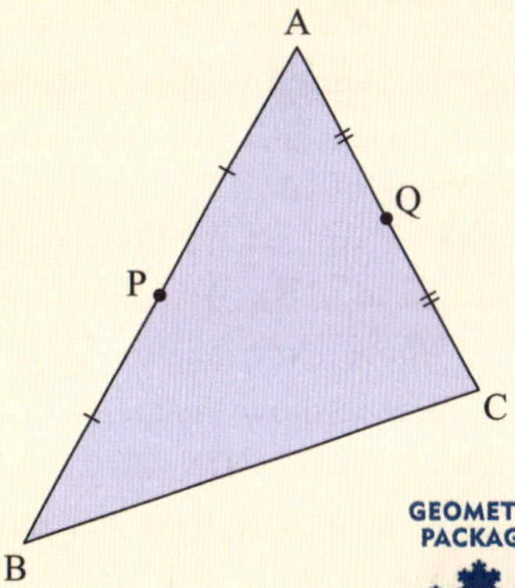

GEOMETRY PACKAGE

The **midpoint theorem** states that:

The line joining the midpoints of two sides of a triangle is parallel to the third side and half its length.

- [MN] is parallel to [BC]
- $MN = \frac{1}{2}BC$

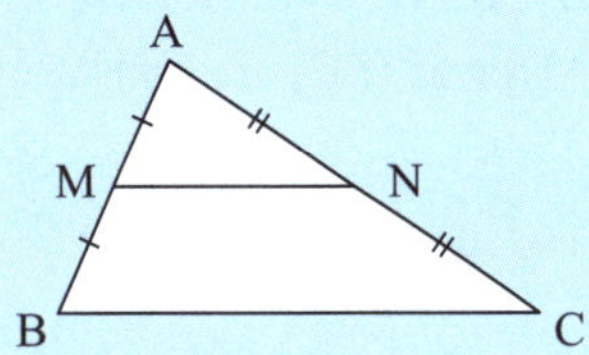

The **converse of the midpoint theorem** states that:

The line drawn from the midpoint of one side of a triangle, parallel to a second side, bisects the third side.

$$AN = NC$$

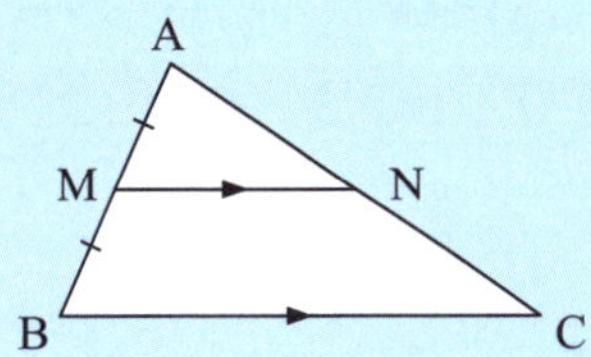

EXERCISE 24B

1. Find the values of the unknowns, giving reasons for your answers.

a

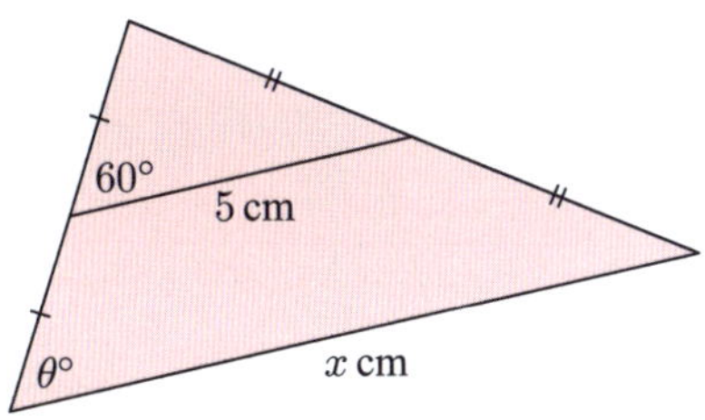

b

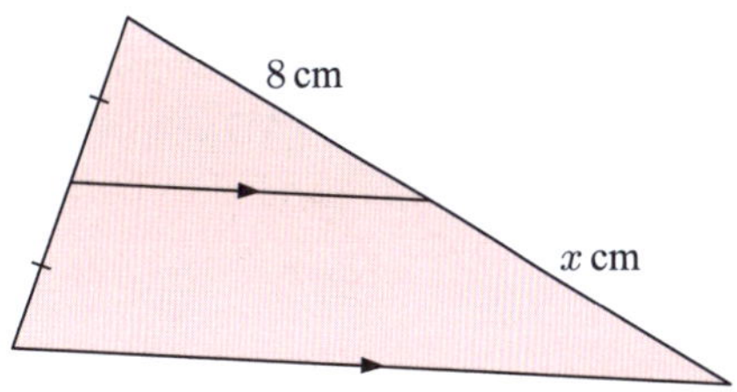

2 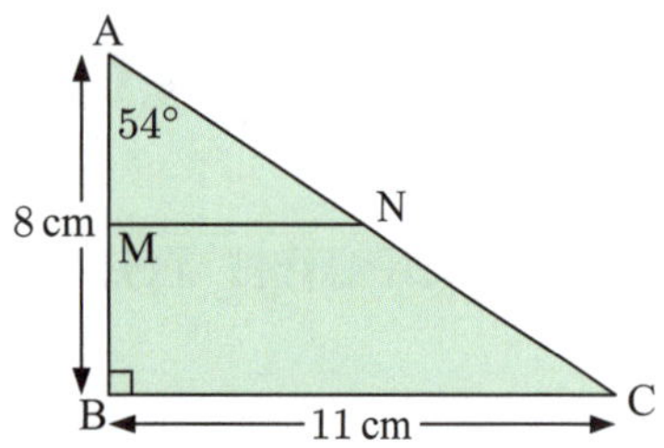

M and N are the midpoints of [AB] and [AC] respectively. Find:

- **a** the length of [MN]
- **b** the size of $\widehat{ANM}$
- **c** the area of triangle ANM.

3 In this question we prove the *midpoint* theorem.

In triangle ABC, M is the midpoint of [AB] and N is the midpoint of [AC].

Suppose we extend [MN], and draw a line through C parallel to [AB]. We suppose these lines intersect at D.

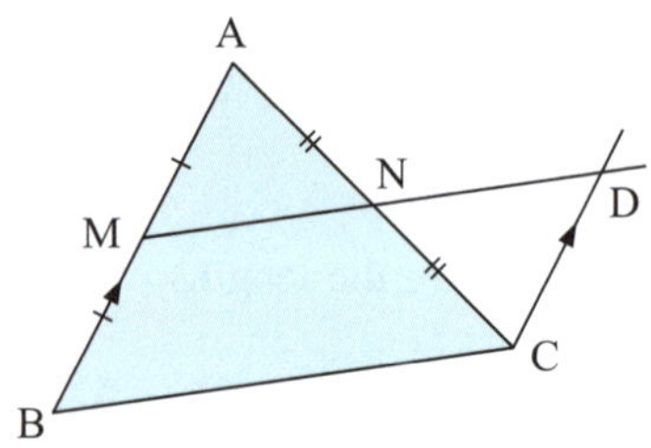

- **a** Show that triangles AMN and CDN are congruent.
- **b** Hence show that:
  - **i** MN = DN **ii** BM = CD
- **c** Show that BCDM is a parallelogram.
- **d** Hence show that:
  - **i** [MN] is parallel to [BC] **ii** $MN = \frac{1}{2}BC$

4 Use similar triangles to prove the *converse* of the midpoint theorem.

Example 3 **Self Tutor**

ABCD is a parallelogram. [AB] is extended to E such that AB = BE. [AD] and [EC] are both extended to meet at F. Prove that EC = CF.

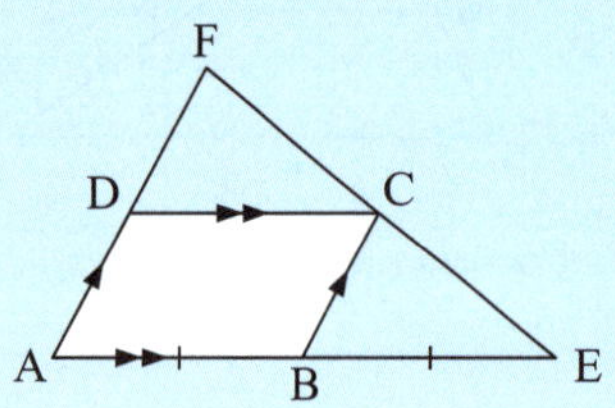

Since ABCD is a parallelogram, [BC] is parallel to [AF].

Now B is the midpoint of [AE],
so C is the midpoint of [EF].
{converse of midpoint theorem}

$\therefore$ EC = CF.

5 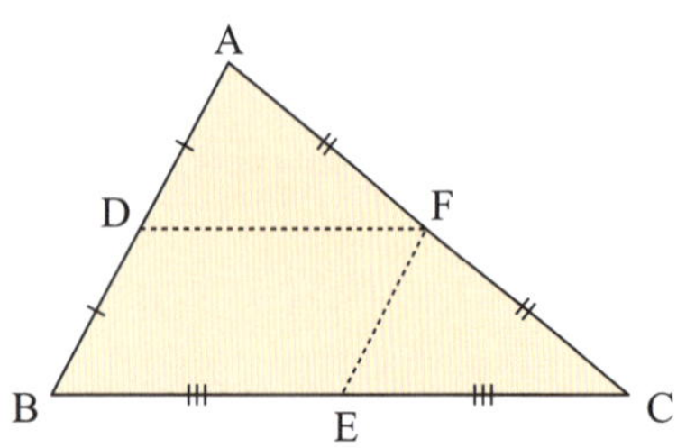

ABC is a triangle. D, E, and F are the midpoints of its sides as shown.

- **a** Show that DFEB is a parallelogram.
- **b** Name two other parallelograms which can be drawn using the given points.

6 ABCD is a parallelogram whose diagonals meet at E. M is the midpoint of [AD].

Show that [ME] is parallel to [AB] and half its length.

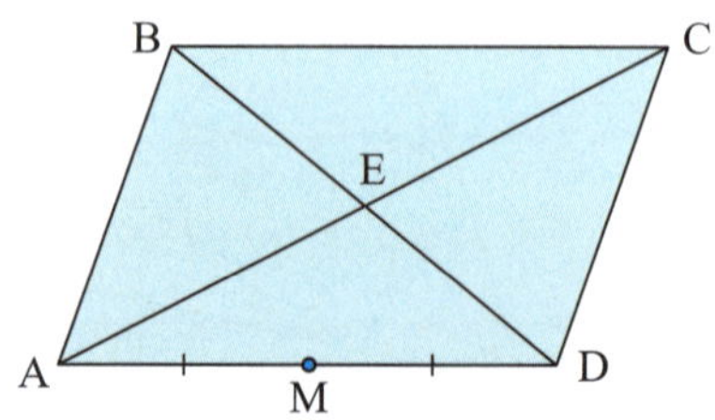

7 A and B are the midpoints of sides [PQ] and [PR]. Y is any point on [QR]. Prove that X is the midpoint of [PY].

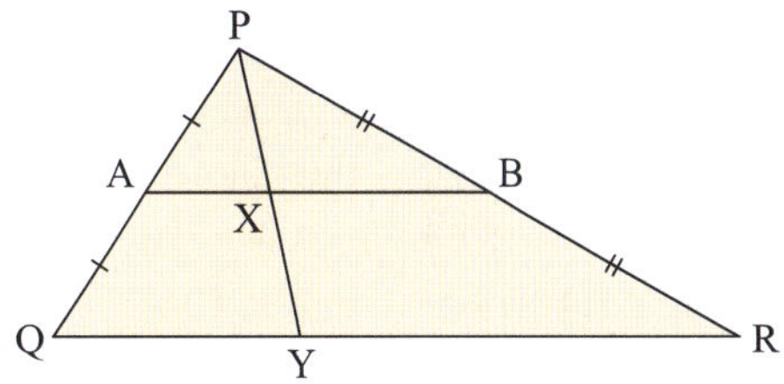

8 For any quadrilateral ABCD, let W, X, Y, and Z be the midpoints of [AB], [BC], [CD], and [DA] respectively. Prove that WXYZ is a parallelogram.

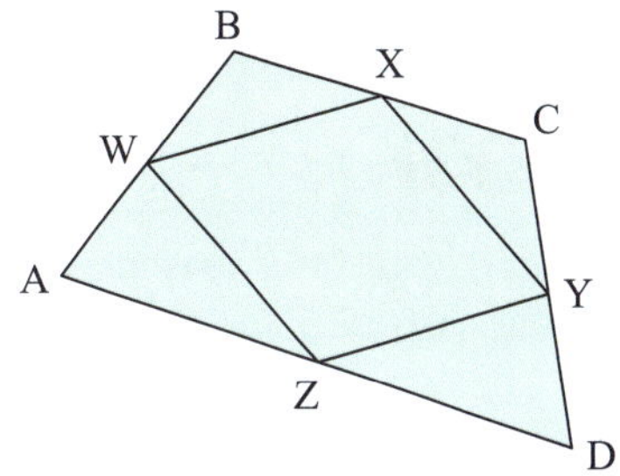

C ANGLE IN A SEMI-CIRCLE THEOREM

INVESTIGATION 2 ANGLE IN A SEMI-CIRCLE THEOREM

What to do:

1. Draw a circle and construct a diameter. Label it as shown.
2. Mark any point P not at A or B on the circle. Draw [AP] and [PB].
3. Measure $A\widehat{P}B$.
4. Repeat for different positions of P and for different circles. What do you notice?
5. Copy and complete: *The angle in a semi-circle is ......*

GEOMETRY PACKAGE

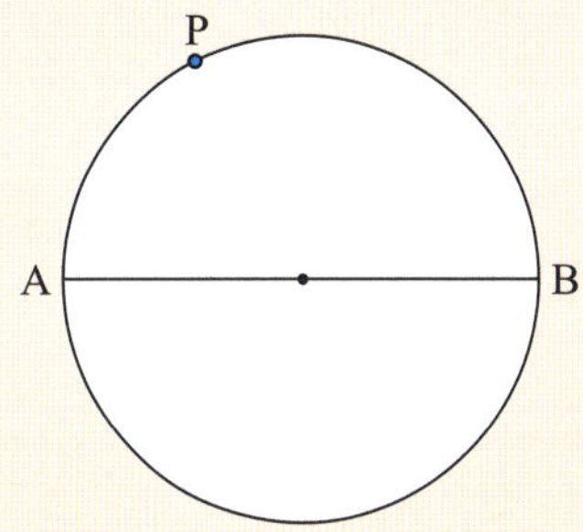

The **angle in a semi-circle theorem** states that:

The angle in a semi-circle is a right angle.

$$A\widehat{B}C = 90°$$

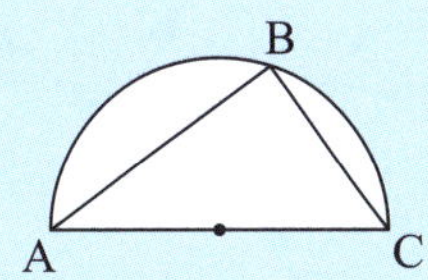

The **converse of the angle in a semi-circle theorem** states that:

If line segment [AB] subtends a right angle at C, then the circle through A, B, and C has diameter [AB].

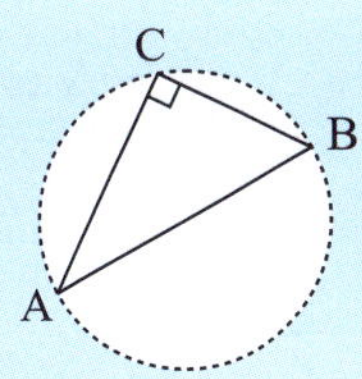

EXERCISE 24C

1 In this question we prove the *angle in a semi-circle* theorem.
Suppose [AC] is the diameter of a circle with centre O, and B is any point on the semi-circle.

Let $\widehat{OAB} = x°$ and $\widehat{OCB} = y°$.

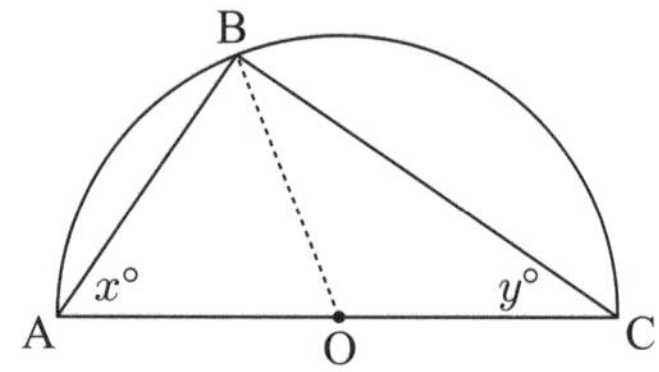

a Find $\widehat{OBA}$ and $\widehat{OBC}$ in terms of x and y.

b Hence prove that $\widehat{ABC} = 90°$.

Example 4 — **Self Tutor**

Find x, giving brief reasons for your answer.

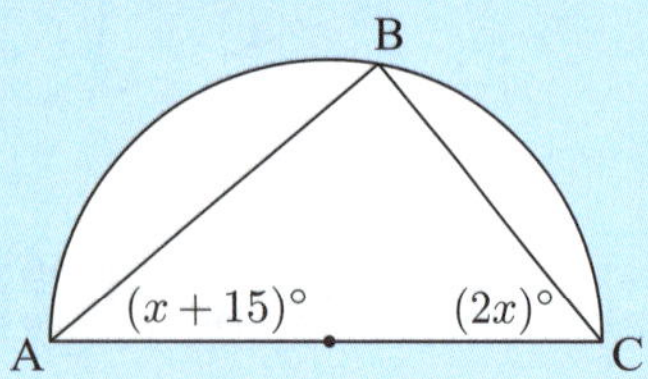

$\widehat{ABC}$ measures 90° {angle in a semi-circle}

$\therefore\ (x+15)+2x+90 = 180$ {angle sum of a triangle}

$\therefore\ 3x+105 = 180$

$\therefore\ 3x = 75$

$\therefore\ x = 25$

2 Find the value of x:

a

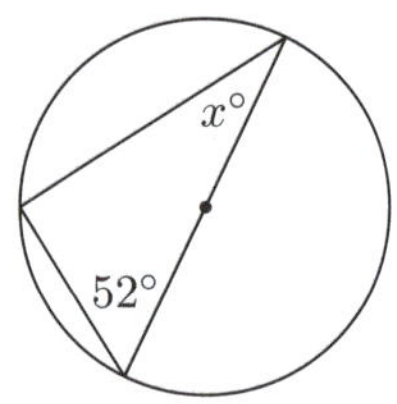

b

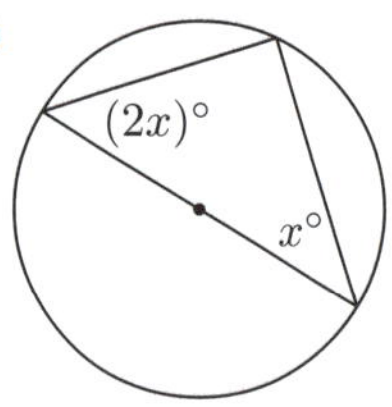

c

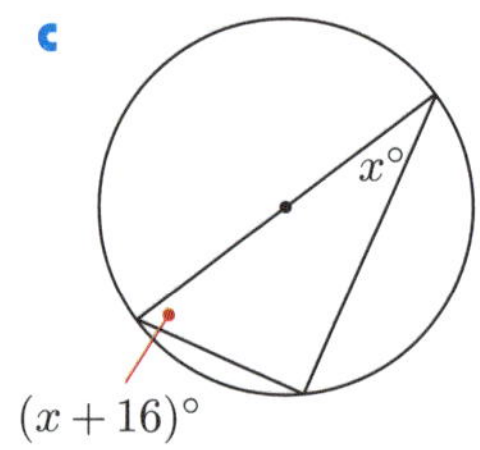

d

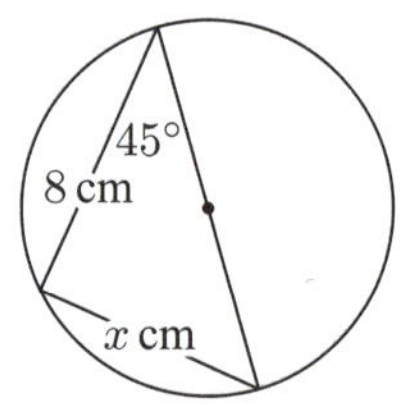

3

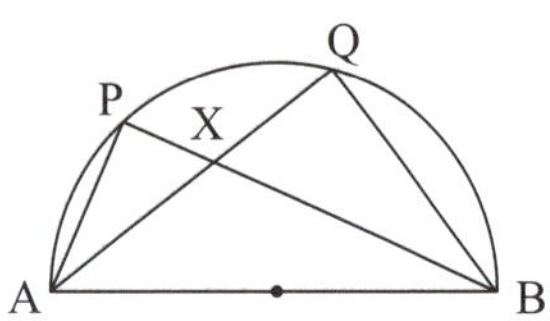

P and Q are any two points on a semi-circle with diameter [AB]. Show that triangles APX and BQX are similar.

4 Find the exact value of x:

a

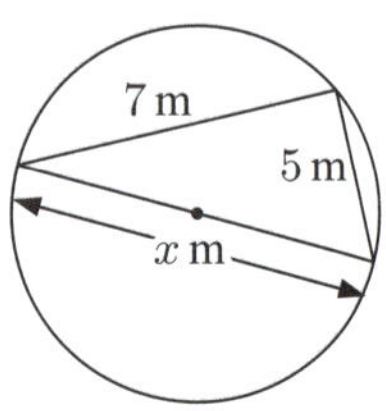

b

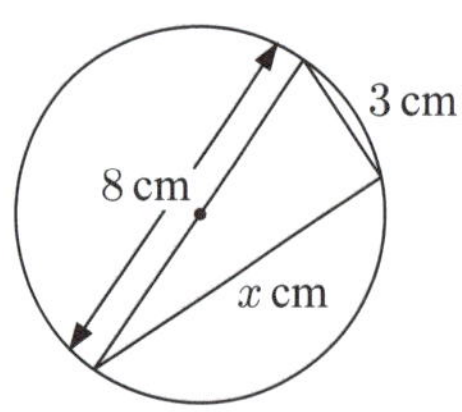

5

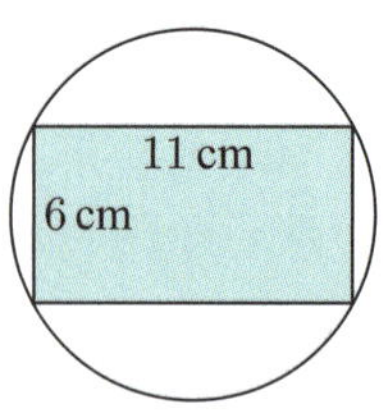

A rectangle with side lengths 11 cm and 6 cm is inscribed in a circle. Find the radius of the circle.

6 Use trigonometry to find the value of x:

a

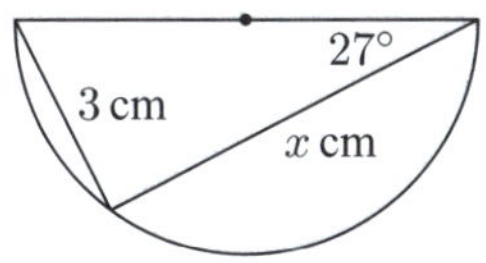

b

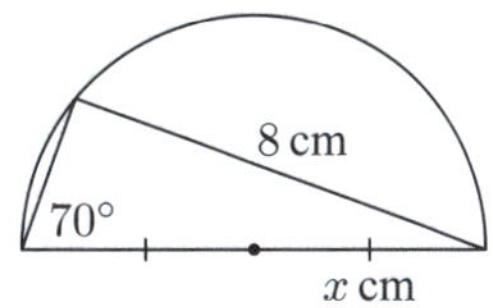

7 A circle has diameter [AB] of length 10 cm. C is a point on the circle such that [AC] is 8 cm. Find the length of [BC].

8 Suppose [AB] is a diameter of a circle. C is another point on the circle such that [AC] is half the length of [AB], and [BC] is 12 cm long. Find the radius of the circle.

9

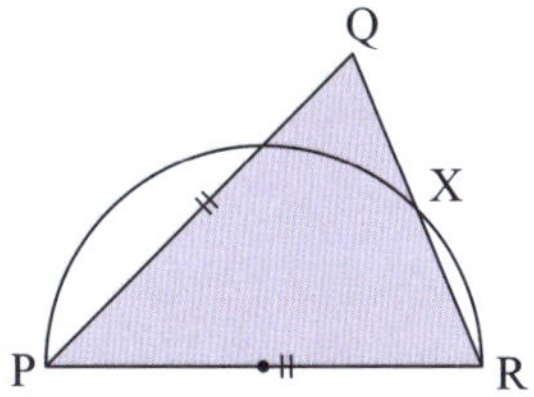

Triangle PQR is isosceles with PQ = PR.
A semi-circle with diameter [PR] is drawn to cut [QR] at X.
Prove that X is the midpoint of [QR].

D CHORDS OF A CIRCLE THEOREM

INVESTIGATION 3 CHORDS OF A CIRCLE THEOREM

What to do:

GEOMETRY PACKAGE

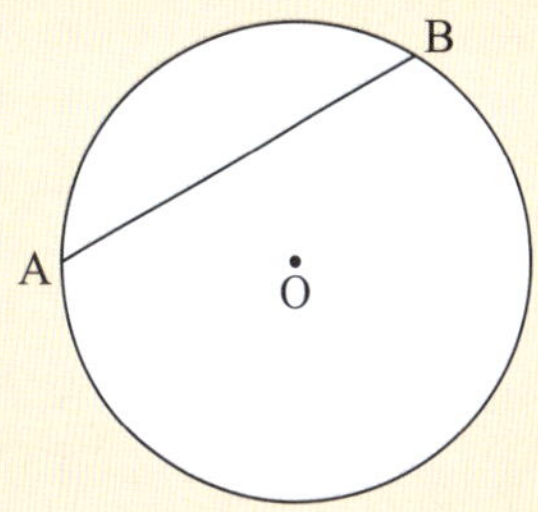

1. Draw a circle with centre O. Construct any chord [AB].
2. Construct the perpendicular from O to [AB] which cuts the chord at M.
3. Measure the lengths of [AM] and [BM]. What do you notice?
4. Repeat this procedure with another circle and chord.
5. Copy and complete: *The perpendicular from the centre of a circle to a chord ......*

The **chords of a circle theorem** states that:

The perpendicular from the centre of a circle to a chord bisects the chord.

AM = BM

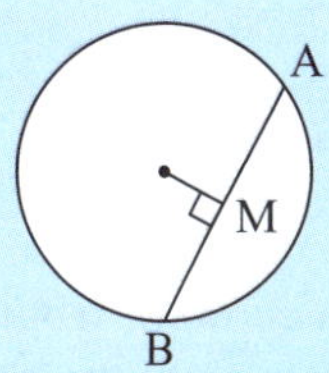

The **converse of the chords of a circle theorem** states that:

The perpendicular bisector of a chord of a circle passes through its centre.

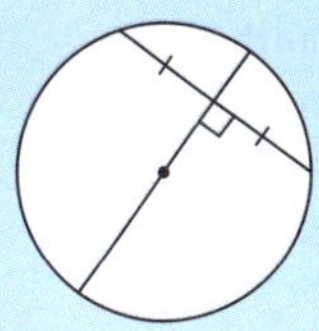

EXERCISE 24D

1 In this question we prove the *chords of a circle* theorem.
Suppose [AB] is a chord of a circle with centre O, and [OM] is perpendicular to [AB].

  a Show that triangles OAM and OBM are congruent.

  b Hence prove that $AM = BM$.

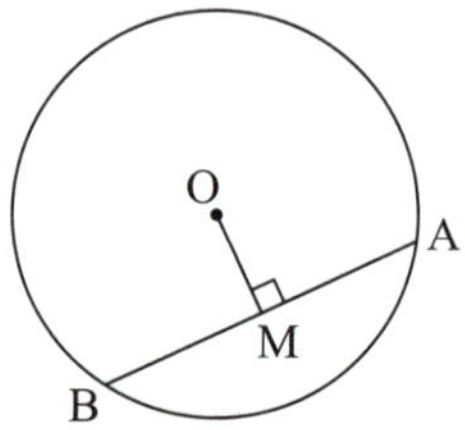

Example 5 — Self Tutor

A circle has a chord of length 8 cm. The shortest distance from the circle's centre to the chord is 2 cm. Find, to the nearest mm, the circle's radius.

Suppose the radius is r cm.

The shortest distance from the centre to the chord is the perpendicular distance.

Let M be the foot of the perpendicular from the centre O to the chord [AB].

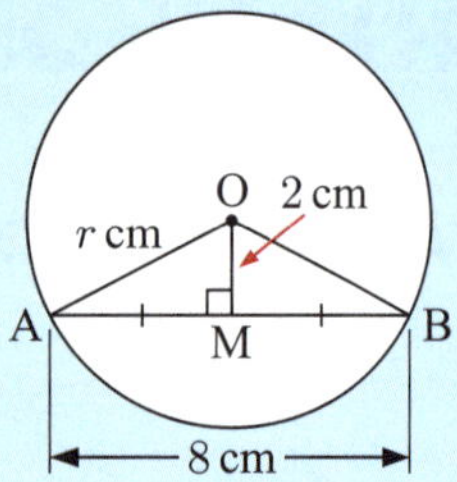

$\therefore$ M is the midpoint of [AB] {chords of a circle}

$\therefore$ AM = 4 cm

In triangle OAM, $r^2 = 2^2 + 4^2$ {Pythagoras}

$\therefore\ r^2 = 20$

$\therefore\ r = \sqrt{20} \approx 4.5$ {as $r > 0$}

So, the radius is about 4.5 cm long.

2 Find the value of x, rounding your answer to 2 decimal places:

a

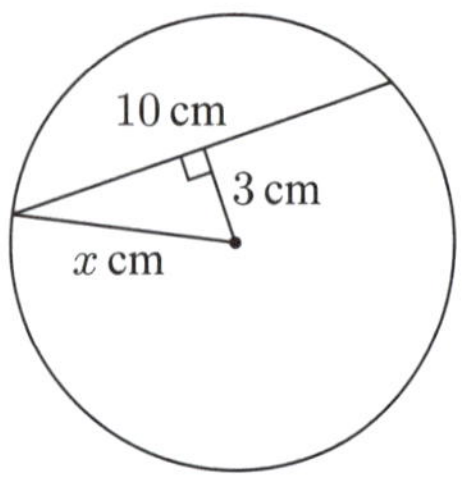

b

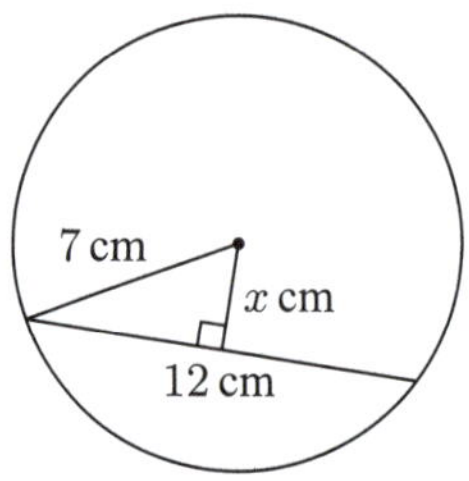

c

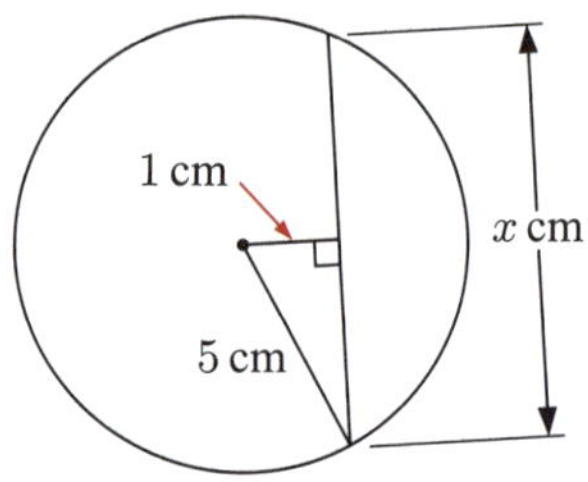

3 A circle has a chord of length 12 cm. The shortest distance from the chord to the circle's centre is 2 cm. Find, to the nearest mm, the circle's radius.

4 A circle has a chord of length 15 cm. The radius of the circle is 9 cm. Find, to the nearest mm, the shortest distance from the chord to the circle's centre.

5 A circle has radius 5 cm. The shortest distance from the circle's centre to a chord is 2 cm. Find, to the nearest mm, the length of this chord.

6 A, B, and C lie on a circle. The perpendicular bisectors of [AB] and [AC] meet at X.

a Explain why X must be the centre of the circle.

b Hence find the shaded area.

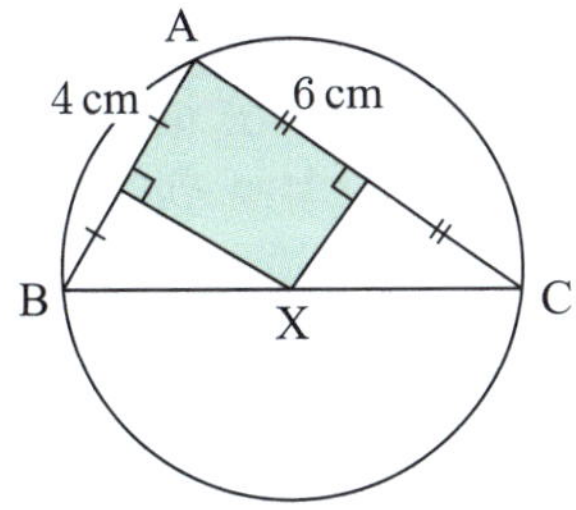

E RADIUS-TANGENT THEOREM

INVESTIGATION 4 RADIUS-TANGENT THEOREM

What to do:

1 Draw a circle with centre O, and mark on it a point A.

2 At A, draw as accurately as possible a tangent [TA].

3 Draw the radius [OA].

4 Measure $O\widehat{A}T$.

5 Repeat this procedure with another circle and tangent.

6 Copy and complete: *The tangent to a circle is ...... to the radius at the point ......*

GEOMETRY PACKAGE

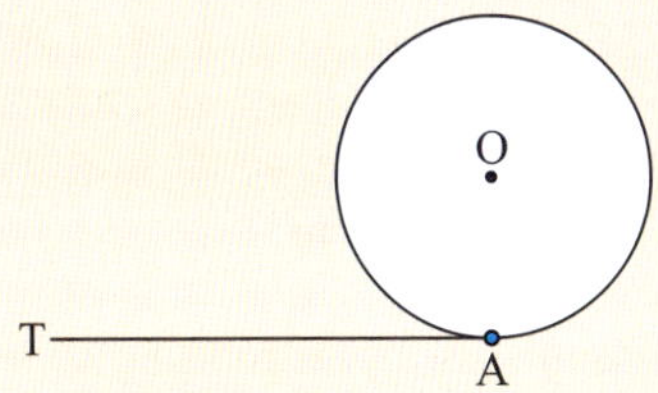

The **radius-tangent theorem** states that:

The tangent to a circle is perpendicular to the radius at the point of contact.

$$O\widehat{A}T = 90°$$

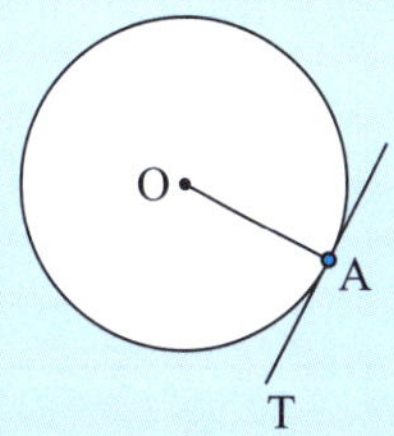

Example 6

Self Tutor

A circle has radius 6 cm. A tangent is drawn at point T on the circle. The tangent is 10 cm long, and ends at point P. Find the distance from P to the centre of the circle.

Let the distance from P to the centre of the circle be x cm.

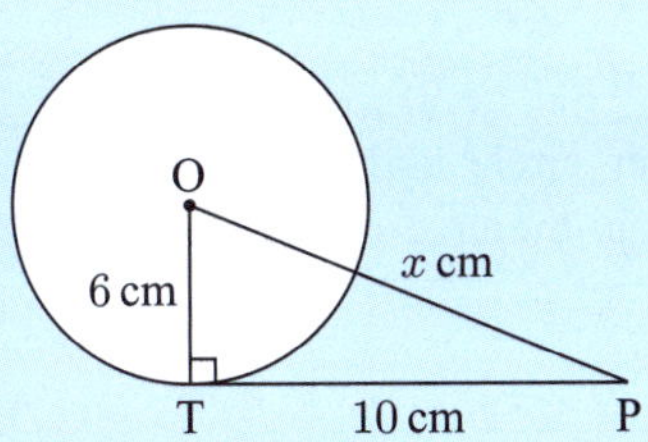

$O\widehat{T}P = 90°$ {radius-tangent}

$\therefore\ x^2 = 6^2 + 10^2$ {Pythagoras}

$\therefore\ x^2 = 136$

$\therefore\ x = \sqrt{136} \approx 11.7$ {as $OP > 0$}

So, P is about 11.7 cm from the centre of the circle.

EXERCISE 24E

1 In the following diagrams, O is the centre of the circle, and [TP] is a tangent. Find the value of x, rounding your answer to 1 decimal place where necessary.

a

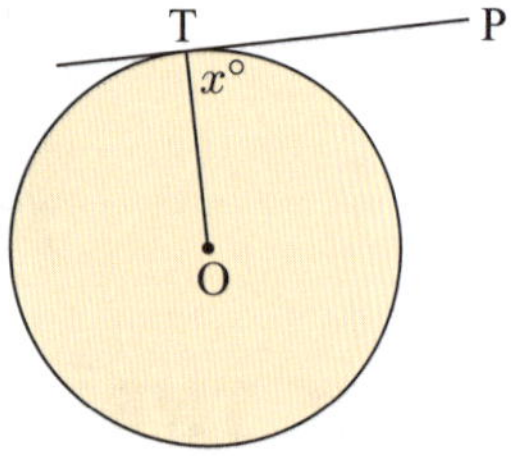

b

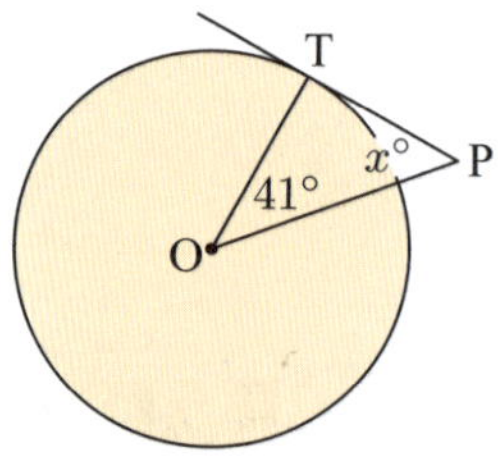

c

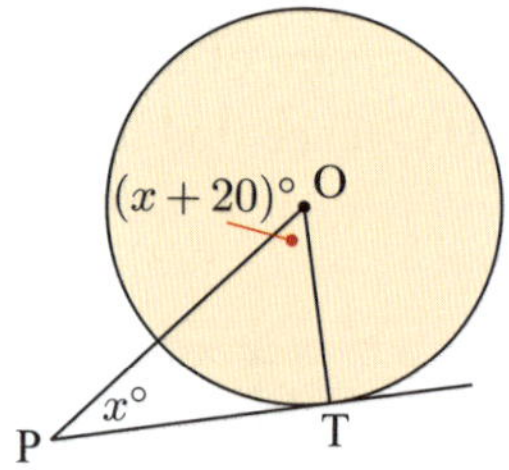

d

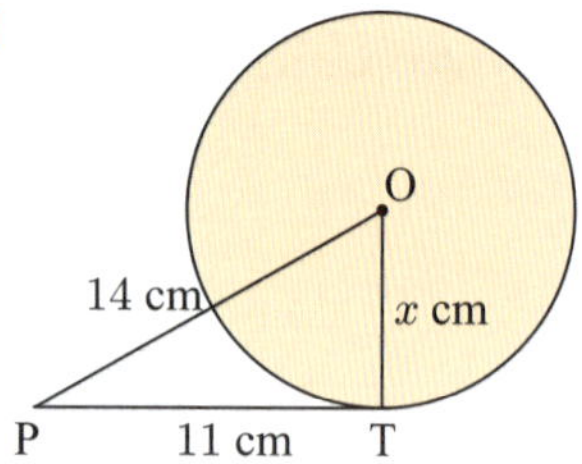

e

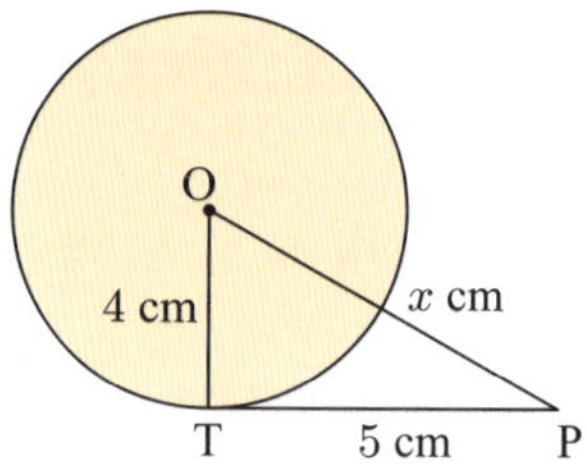

f

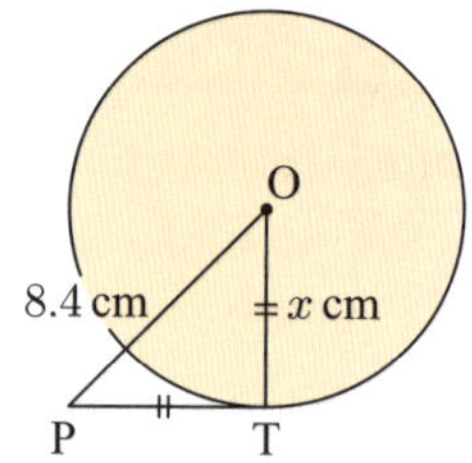

2 Point P is 9 cm from the centre of a circle. A tangent from P to the circle is 5 cm long. Find the radius of the circle.

3 Point Q is 12 cm from the centre of a circle with radius 7 cm. A tangent is drawn from Q to touch the circle at P. Find the length of [PQ].

4 In the figure alongside, the tangent [TP] has the same length as the radius of the circle. What percentage of triangle OTP is inside the circle?

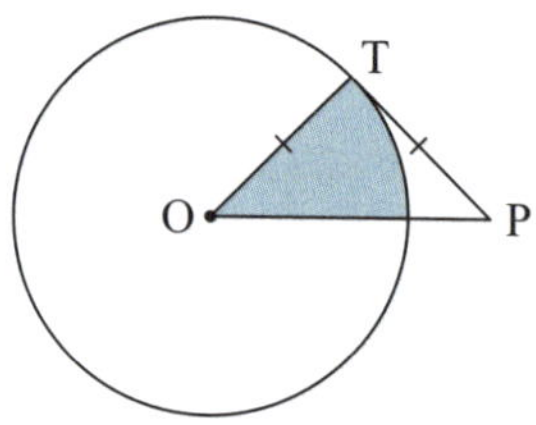

5

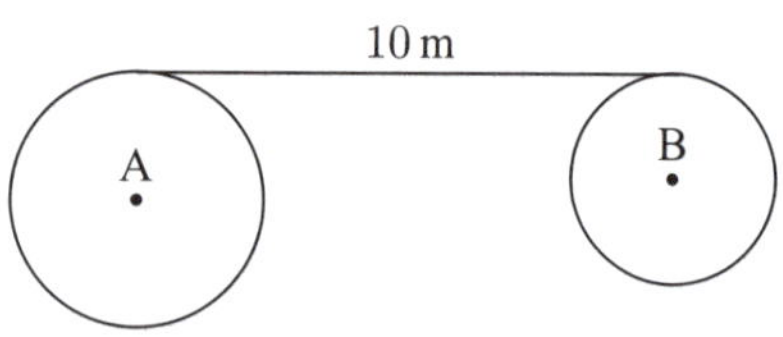

A and B are the centres of two circles with radii 4 m and 3 m respectively. They have a common tangent with length 10 m. Find the distance between the centres, correct to 2 decimal places.

6 Two circles are drawn so they do not intersect. The larger circle has radius 6 cm. A common tangent is 10 cm long and the centres are 11 cm apart. Find the radius of the smaller circle, correct to 3 significant figures.

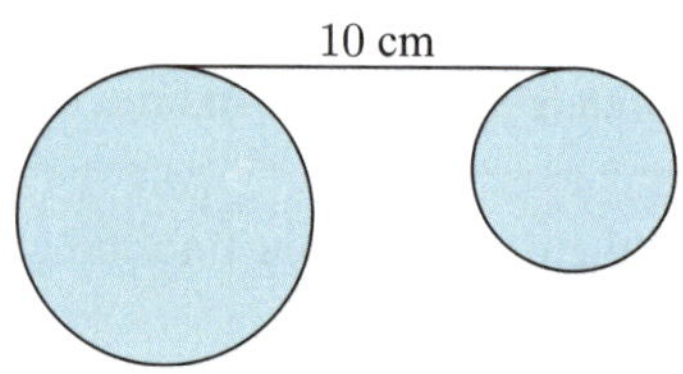

7

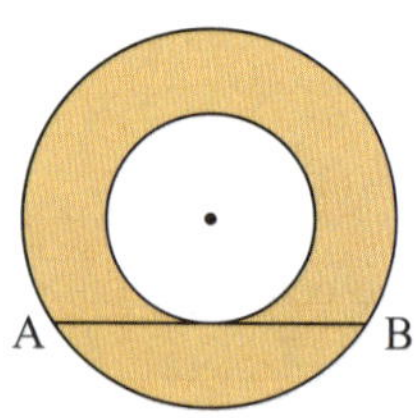

[AB] is a tangent to the inner circle and is a chord of the larger circle. If [AB] is 6 m long, find the area of the shaded annulus.

F TANGENTS FROM AN EXTERNAL POINT THEOREM

INVESTIGATION 5 TANGENTS FROM AN EXTERNAL POINT THEOREM

What to do:

1. Draw a circle with centre O.
2. From an external point P, draw the two tangents to the circle. Let the tangents meet the circle at A and B.
3. Measure [AP] and [BP].
4. Repeat this procedure with a circle of a different size.
5. Copy and complete: *Tangents from an external point to a circle are ......*

GEOMETRY PACKAGE

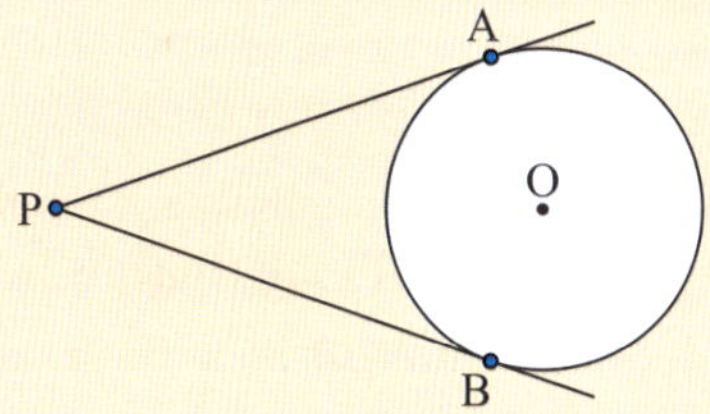

The **tangents from an external point theorem** states that:

Tangents from an external point are equal in length.

$$AP = BP$$

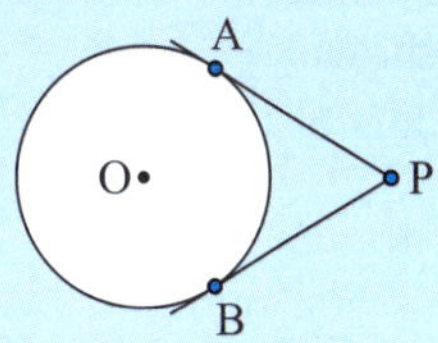

EXERCISE 24F

1. In this question we prove the *tangents from an external point* theorem.
   We join [OA], [OB], and [OP].
   a. Show that triangles AOP and BOP are congruent.
   b. Hence show that AP = BP.

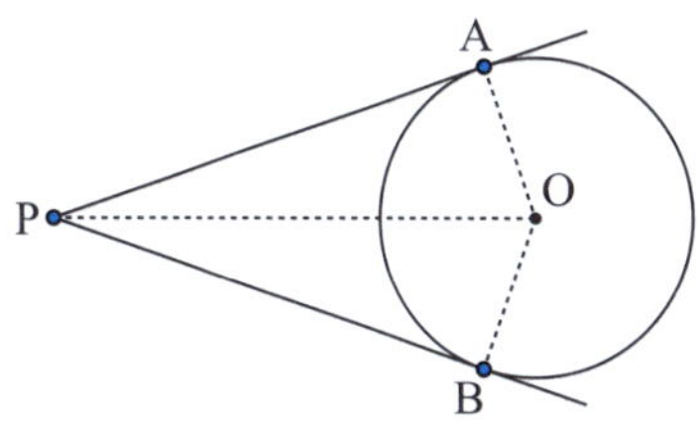

2. In each diagram, O is the centre of the circle, and [TP] and [SP] are tangents. Find the values of the unknowns.

a 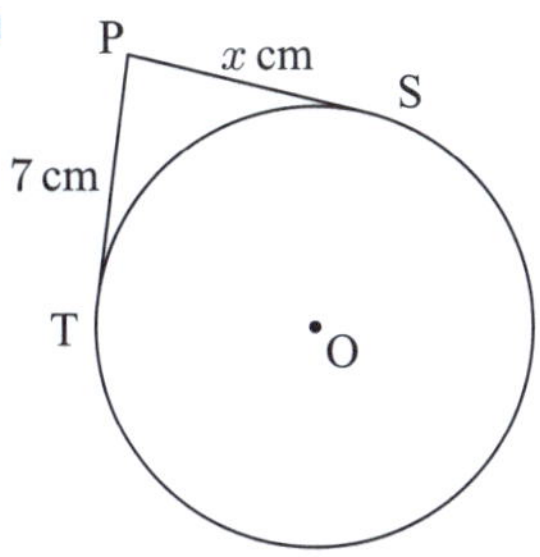

b 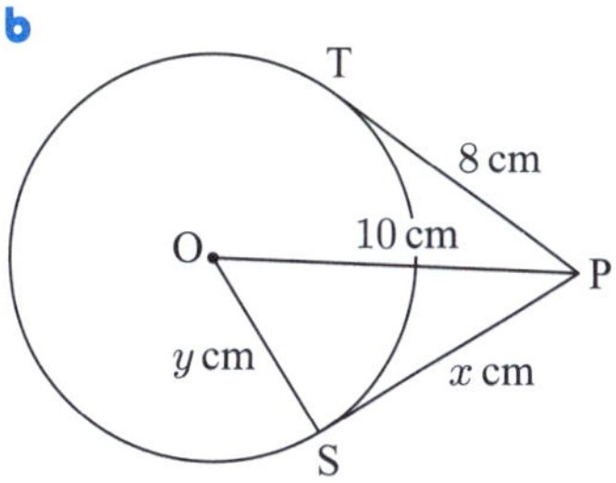

c 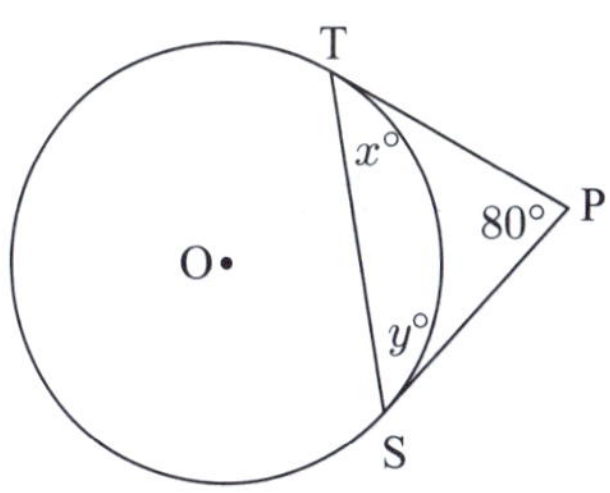

d

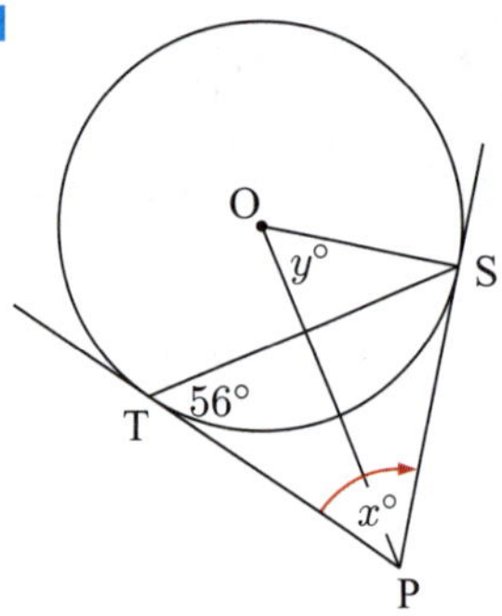

e

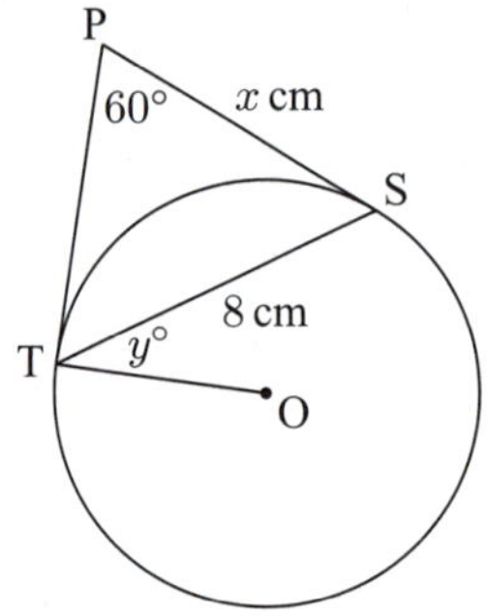

f

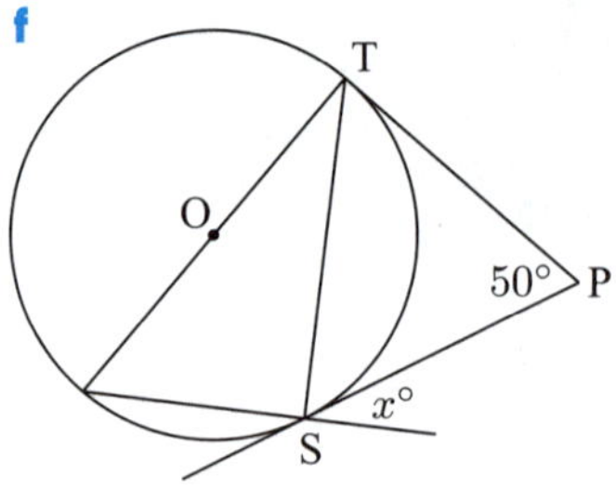

3 In the figure alongside, O is the centre of the circle, and [AP] and [BP] are tangents to the circle.

Show that $A\widehat{O}B$ is twice the size of $B\widehat{A}P$.

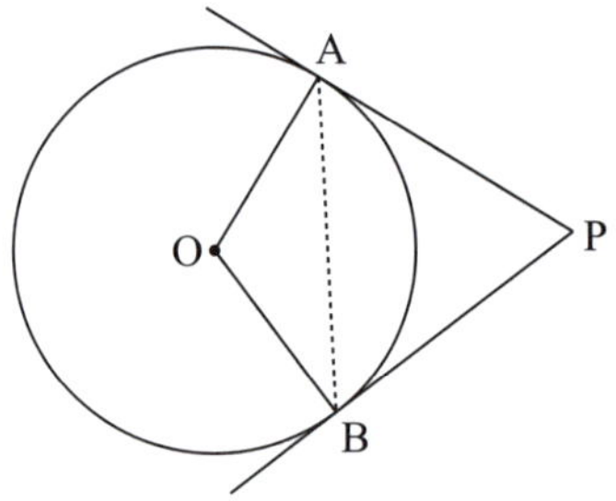

4 A circle touches the three sides of a triangle as shown.

Find the radius of the circle.

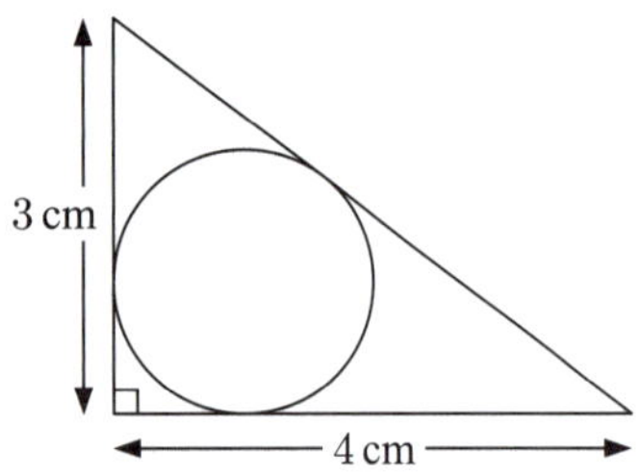

5 Find the length of [AD].

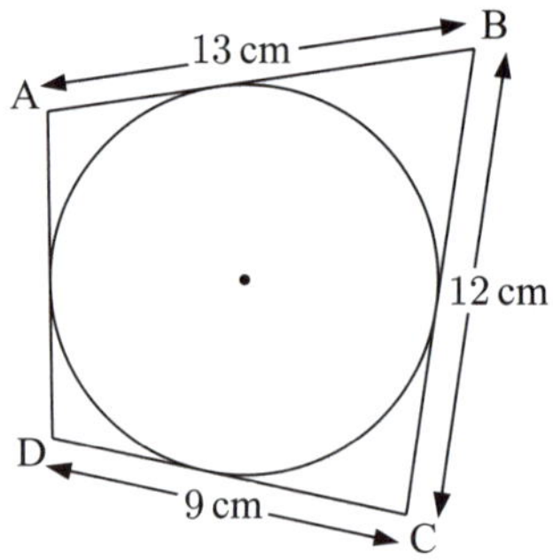

6 A circle with radius 3 cm fits exactly in triangle ABC. Find the perimeter of the triangle.

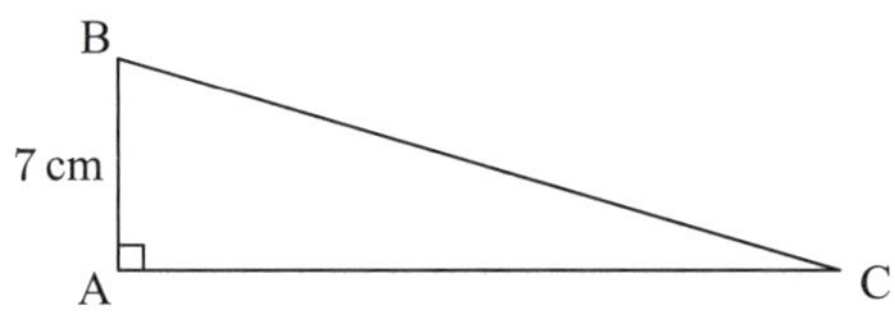

QUICK QUIZ

MULTIPLE CHOICE QUIZ

REVIEW SET 24A

1 Find the value of x:

a

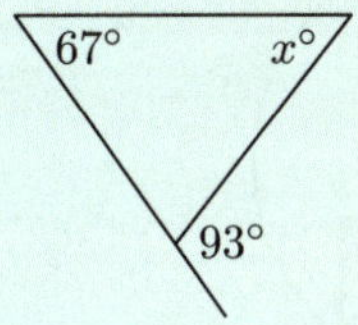

b

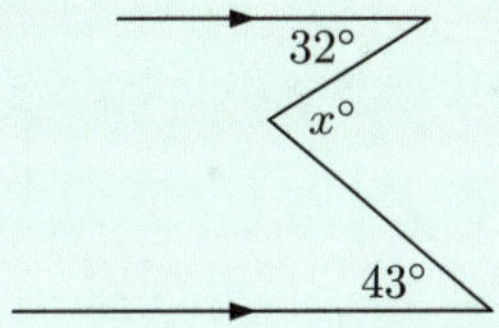

c

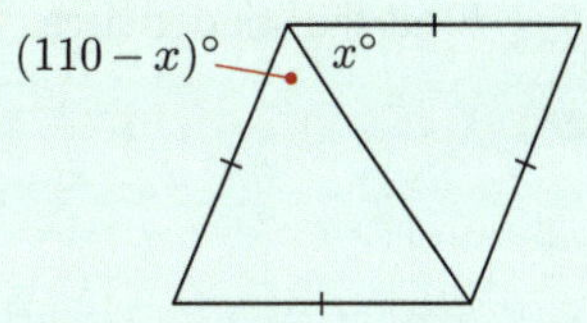

2

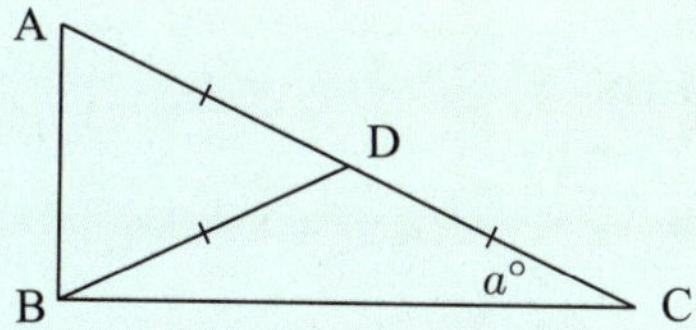

a Find, in terms of a:

i $D\widehat{B}C$ **ii** $A\widehat{D}B$ **iii** $D\widehat{B}A$

b Show that $A\widehat{B}C$ is a right angle.

3

In the given figure $D\widehat{A}B = C\widehat{B}A$ and $A\widehat{D}C = B\widehat{C}D$. Show that [AB] is parallel to [DC].

4

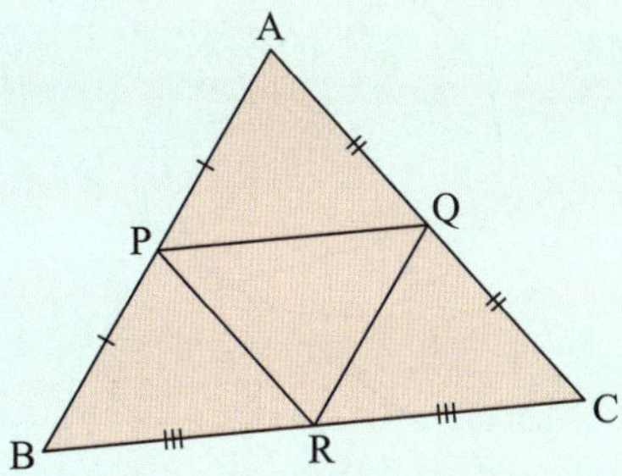

In triangle ABC, let P, Q, and R be the midpoints of [AB], [AC], and [BC] respectively. Prove that the four small triangles are all congruent.

5 Find x:

a

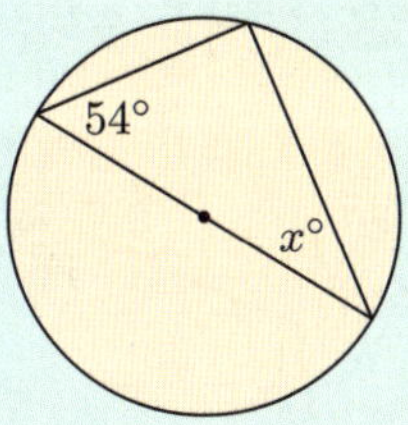

b

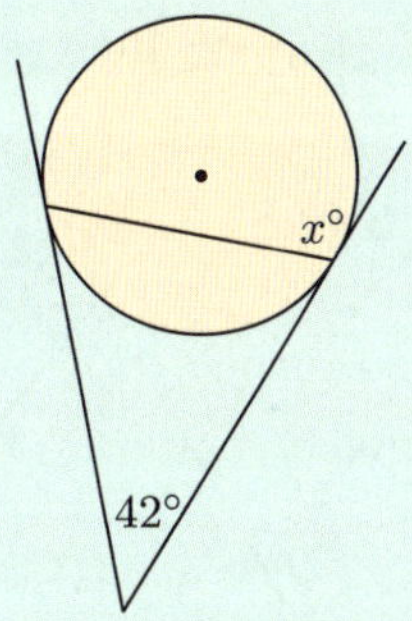

c

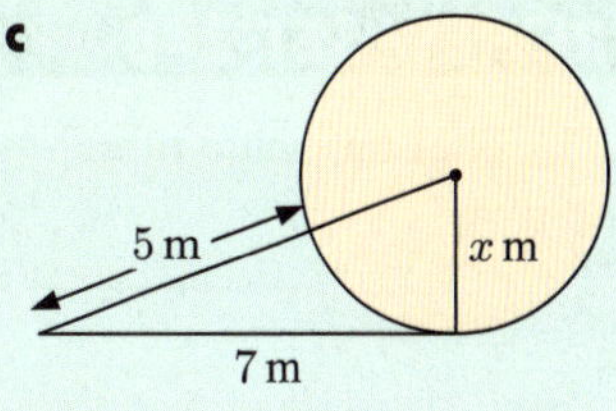

6 In this question we prove the *tangents from an external point* theorem.

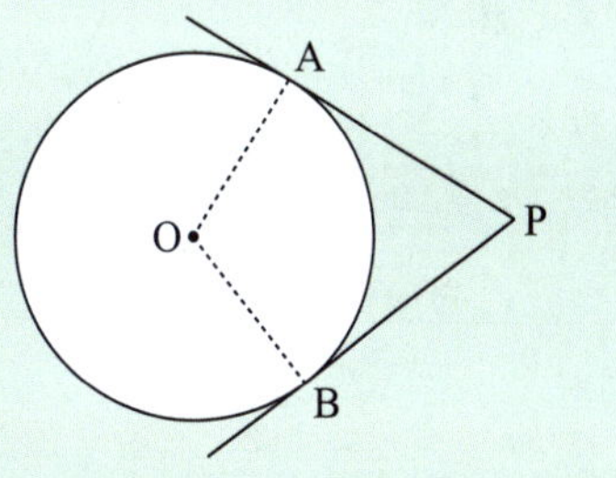

a Show that:

i $AP^2 = OP^2 - OA^2$

ii $BP^2 = OP^2 - OB^2$.

b Hence prove that $AP = BP$.

7 A circle has a chord of length 12 cm. The shortest distance from the circle's centre to the chord is 5 cm. Find the circle's radius, to the nearest mm.

8 This circle has radius 5 cm, and AB $= 6$ cm.
Find the shaded area.

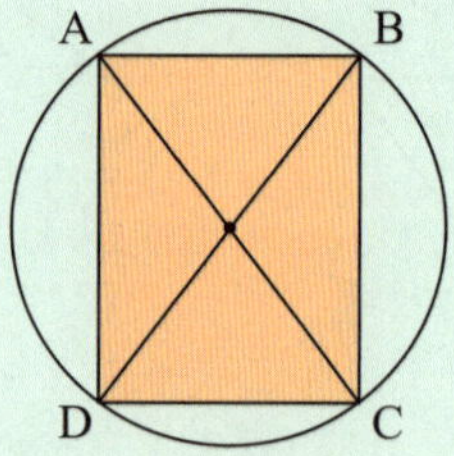

9 A circle with diameter 50 cm has a chord of length 14 cm.
Find the shortest distance from the chord to the circle's centre.

10 [TP] is a tangent such that triangle OTP has the same area as the circle.
Show that the length of the tangent is equal to the circumference of the circle.

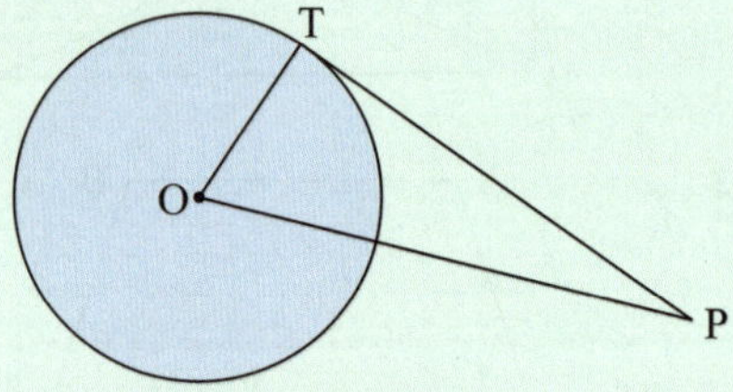

11 The circles shown share a common tangent at P. XPY is a straight line.
Show that the tangents at X and Y are parallel.

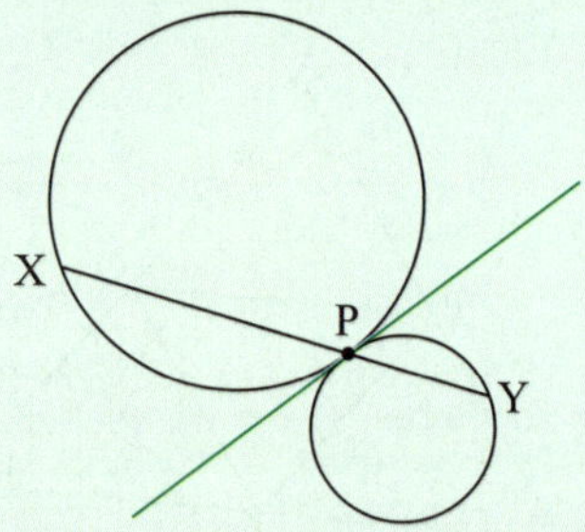

REVIEW SET 24B

1 Find the value of x:

a

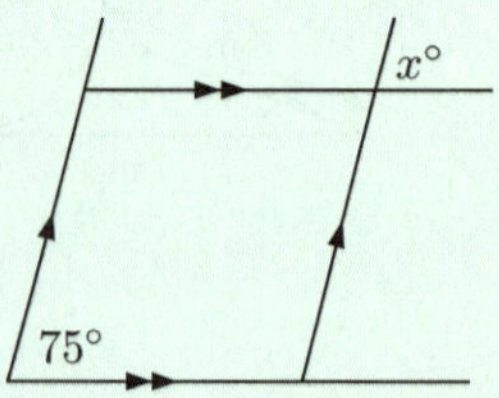

b

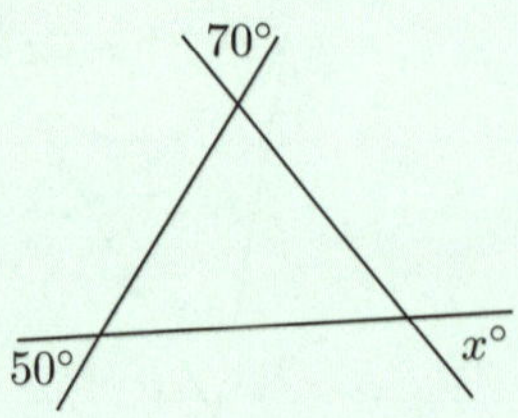

c

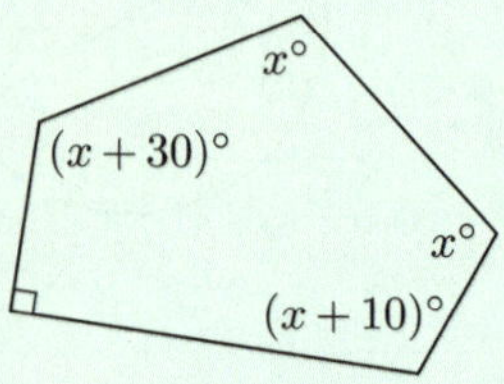

d

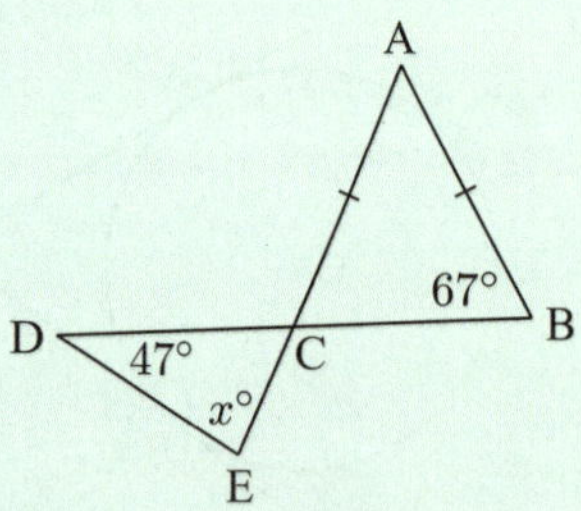

e

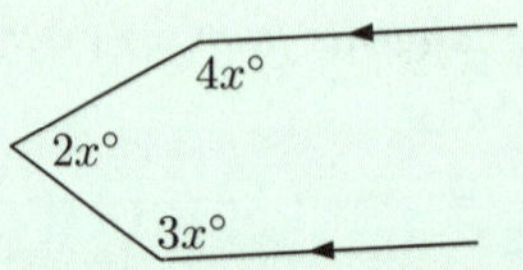

2 ABC is an isosceles triangle with $AB = AC$.
From a point X on the base [BC], perpendiculars to [AB] and [AC] are drawn, meeting these line segments at Y and Z respectively.
Show that $B\widehat{A}C$ is twice the size of $C\widehat{X}Z$.

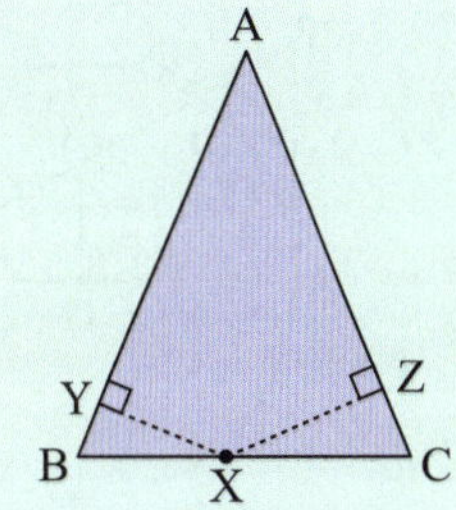

3 ABCD is a trapezium in which [AB] is parallel to [DC], and $AD = BC$.
Prove that $A\widehat{D}C = B\widehat{C}D$.

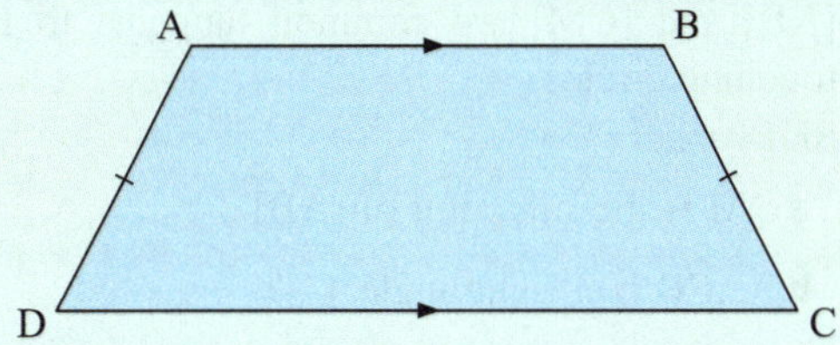

4

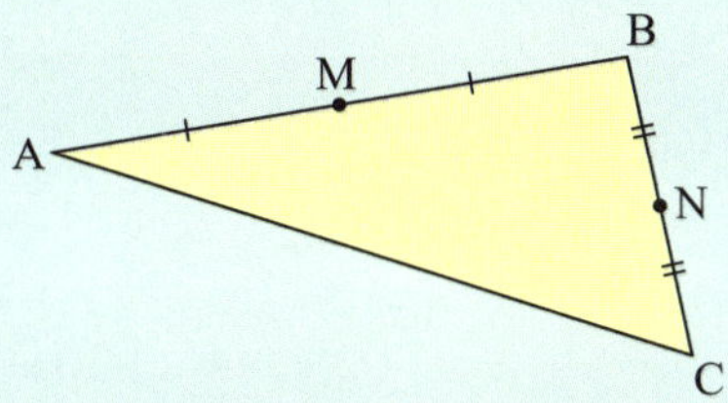

a Show that triangles BAC and BMN are similar.
b Given that triangle BAC has area 100 cm^2, find the area of:
  i triangle BMN
  ii quadrilateral AMNC.

5 Find the values of the unknowns:

a

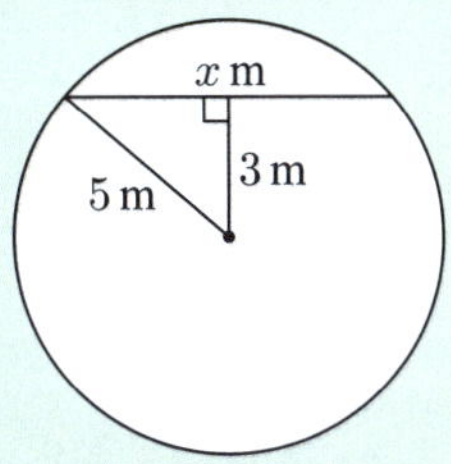

b

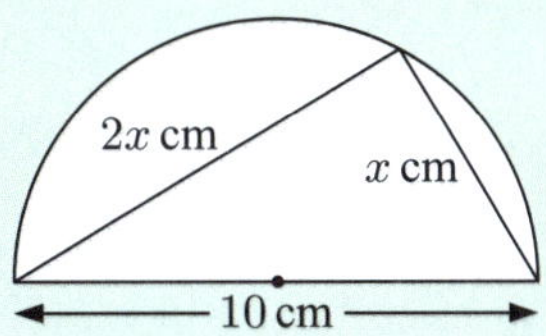

c

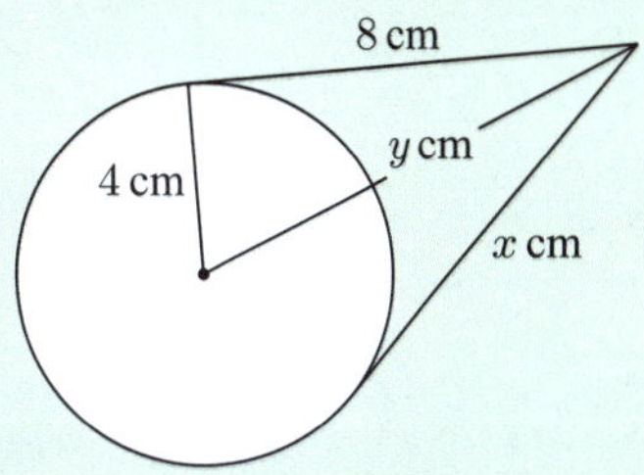

6 A circle has radius 8 cm. The shortest distance from the circle's centre to a chord is 5 cm. Find the length of this chord, to the nearest mm.

7

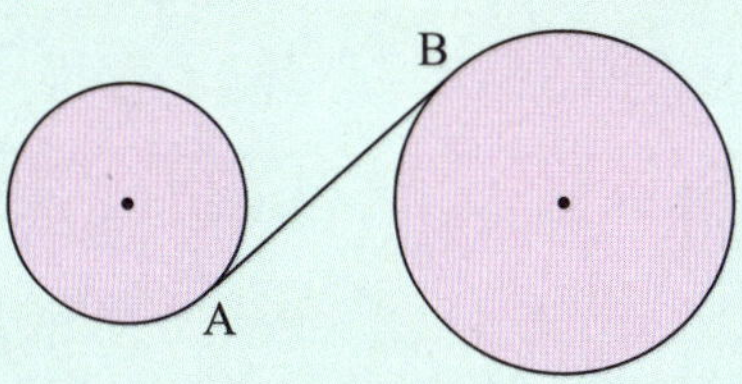

The circles illustrated have radii of length 5 cm and 7 cm respectively.
Their centres are 18 cm apart. Find the length of the common tangent [AB].

8 [AC] is a radius of the large circle, and a diameter of the small circle. A line through A cuts the small circle at X and the large circle at B.
Prove that X is the midpoint of [AB].

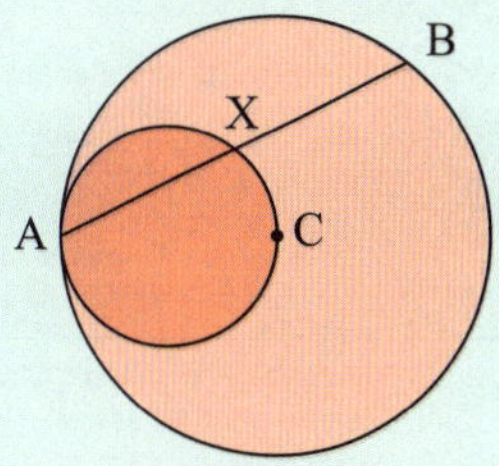

9 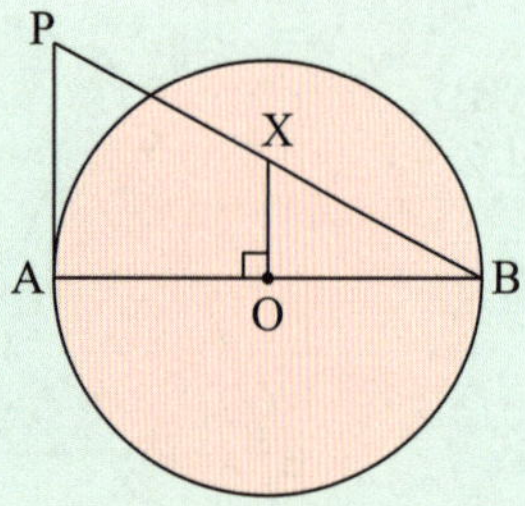

[AB] is a diameter of the circle shown, and [AP] is a tangent.

A line is drawn through the centre of the circle and perpendicular to [AB], meeting [PB] at X.

Show that X is the midpoint of [PB].

10 [AB] and [CM] are common tangents to two touching circles.

Show that:

a M is the midpoint of [AB]

b $A\widehat{C}B$ is a right angle.

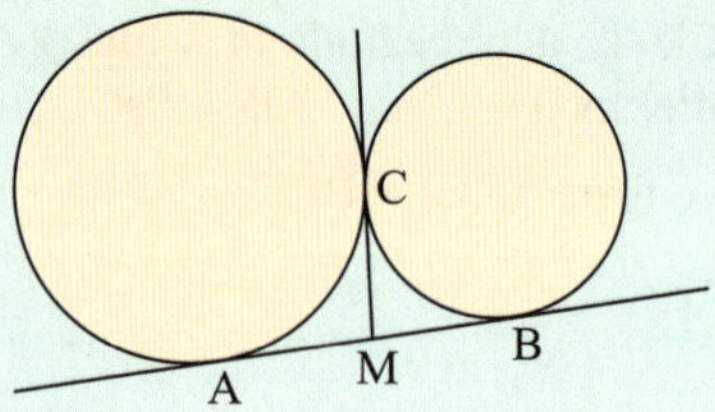

Chapter 25

Proportion

Contents:

OPENING PROBLEM

A stall sells soft drink bottles for \$4 each. Suppose we buy n bottles and the total cost is $\$C$.

To study the relationship between the *number of bottles* and the *total cost*, we can use a table of values or a graph.

n	0	1	2	3	4	5
C	0	4	8	12	16	20

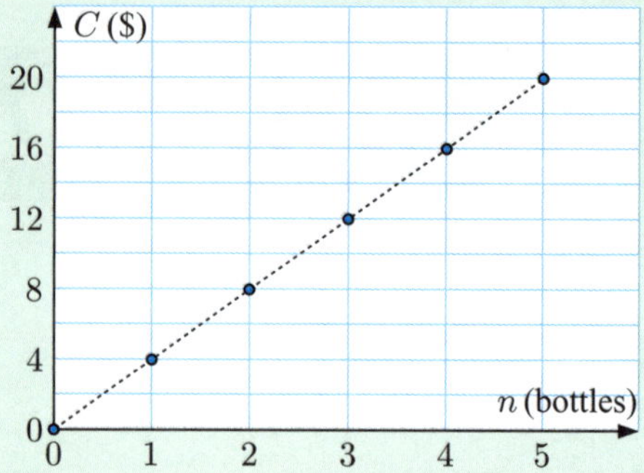

The total cost *depends* on the number of bottles bought.

The number of bottles is the independent variable, so it is placed on the horizontal axis.

The total cost is the dependent variable, so it is placed on the vertical axis.

Since we can only buy a whole number of bottles, the graph of C against n consists of discrete points. However, an imagined line passing through these points would also pass through the origin.

Things to think about:

a Are these statements true or false?

i Doubling the number of bottles doubles the total cost.

ii Halving the number of bottles halves the total cost.

iii Increasing the number of bottles by 30% increases the total cost by 30%.

b How can we *describe* the relationship between n and C?

A DIRECT PROPORTION

The rectangles alongside are **similar**. Their sides are in the same **ratio**.

The ratios of the side lengths can be written as the *proportion* $1 : 3 = 1.5 : 4.5$.

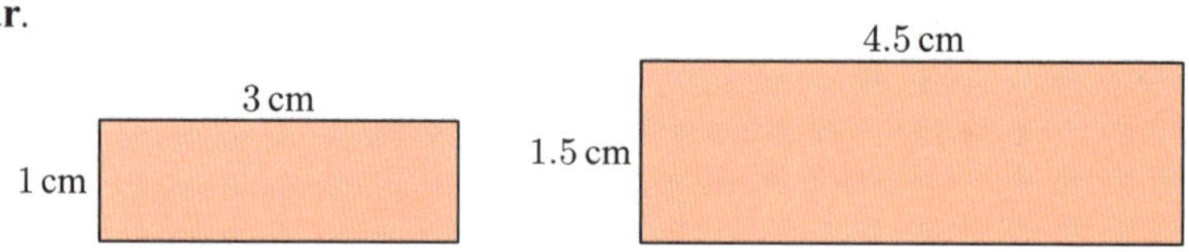

A rectangle with sides x cm and y cm will be similar to both of the rectangles above if $x : y = 1 : 3$. For any given value of x, this occurs when $y = 3x$.

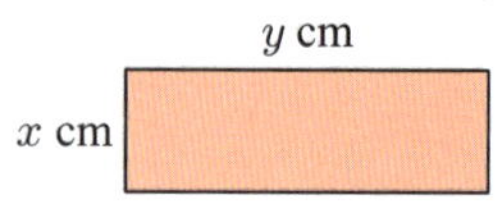

Since the sides are in proportion, we say that y is *directly proportional* to x.

Two variables are **directly proportional** if multiplying one of them by a number results in the other one being multiplied by the same number.

In the **Opening Problem**, C *is directly proportional to* n.

- When n is tripled from 1 to 3, C is tripled from 4 to 12.
- If we plot C against n, the points lie in a straight line with gradient 4 which passes through the origin. The variables are connected by the formula $C = 4n$.

If two quantities x and y are **directly proportional**, we write $y \propto x$.

$\propto$ reads "*is directly proportional to*".

If $y \propto x$ then $y = kx$ where k is a constant called the **proportionality constant**.

When y is graphed against x, the graph is a straight line with gradient k which passes through the origin.

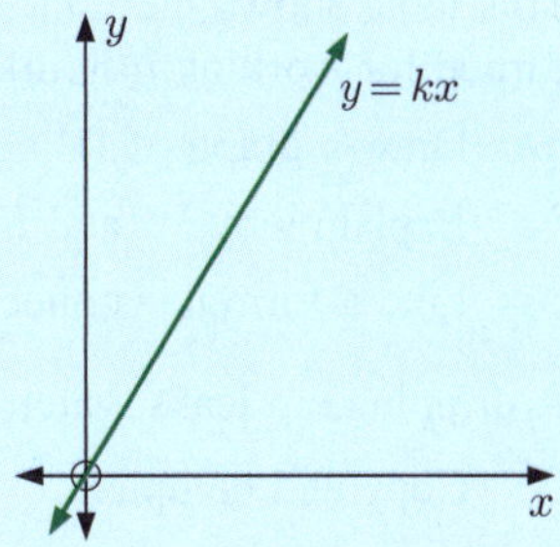

Example 1 — Self Tutor

At the village bakery, fruit buns are sold for 60 cents each.

a Use a graph to show that $\$y$, the total price paid, is directly proportional to x, the number of fruit buns bought.

b Find a formula connecting x and y.

a

x	0	1	2	3	4	5
y	0	0.60	1.20	1.80	2.40	3.00

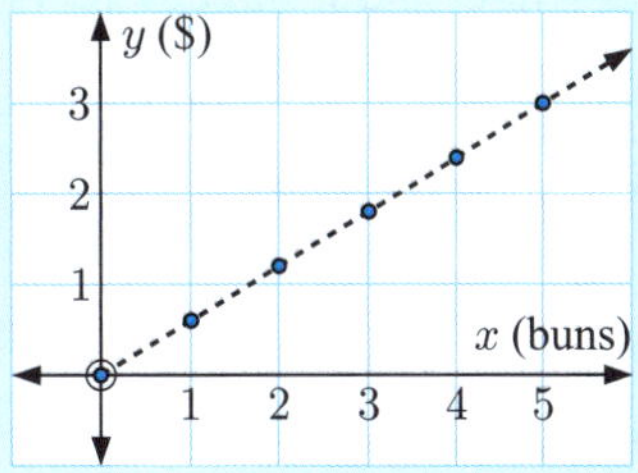

The points lie in a straight line passing through $(0, 0)$, so $y \propto x$.

b The gradient of the line $= \dfrac{0.60 - 0}{1 - 0} = 0.6$

$\therefore$ the proportionality constant k is 0.6.

$\therefore$ the formula connecting x and y is $y = 0.6x$.

EXERCISE 25A

1 Which graph indicates that y is directly proportional to x?

A

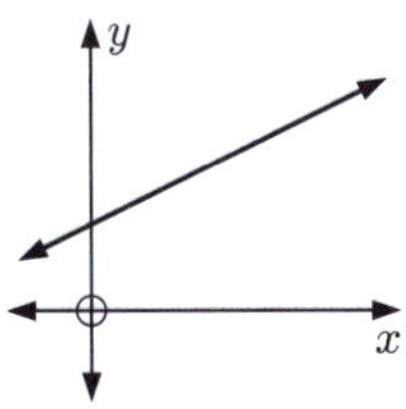

B

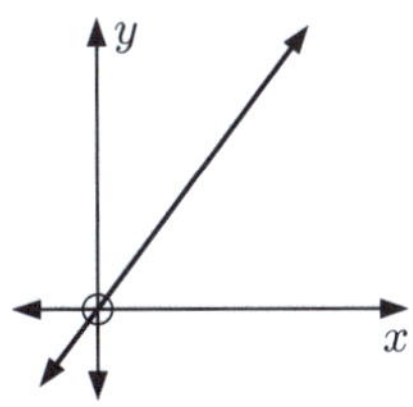

C

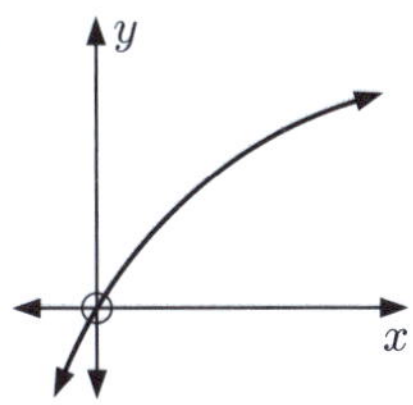

D

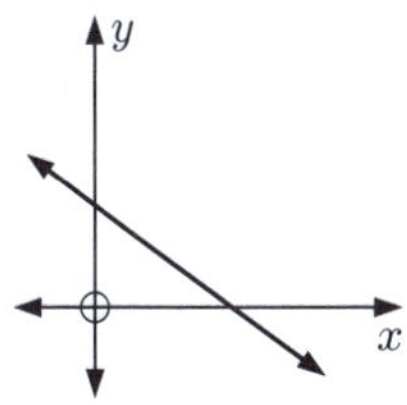

2 For each table of values:

i Plot the graph of y against x.

ii Determine whether y is directly proportional to x. If it is, find the proportionality constant.

a

x	0	1	2	3	4
y	0	3	6	9	12

b

x	0	1	2	3	4
y	3	5	7	9	11

c

x	0	1	3	5	6
y	0	4	8	12	16

d

x	2	3	5	6	8
y	5	7.5	12.5	15	20

3 This table shows the total wages $\$W$ earned for working h hours.

h	0	1	2	3	4	5
W	0	22.50	45.00	67.50	90.00	112.50

a Draw a graph of W against h.

b Explain why W and h are directly proportional.

c Find a formula connecting W and h.

4 A dripping tap leaks water at a rate of 10 mL per minute.

a Copy and complete:

Time (t minutes)	0	2	4	6	8	10
Amount of water leaked (A mL)						

b Plot the graph of A against t.

c Are A and t directly proportional? If so, find the proportionality constant.

5 A dry cleaner charges \$3.20 each for cleaning the first three items, then \$2.20 for each extra item.

a Copy and complete:

Number of items (n)	0	1	2	3	4	5
Cost ($\$C$)						

b Plot C against n.

c Are C and n directly proportional? If so, find the proportionality constant.

6 Suppose y is directly proportional to x. State what happens to:

a y if x is trebled

b x if y is doubled

c y if x is halved

d x if y is decreased by 30%

e y if 2 is added to x

f y if 3 is subtracted from x.

7 The law connecting the circumference C and radius r of a circle is $C = 2\pi r$.

a Explain why $C \propto r$ and state the proportionality constant.

b State what happens to:

i C if r is doubled

ii r if C is increased by 50%.

Example 2 **Self Tutor**

Suppose $y \propto n$ and that $y = 40$ when $n = 3$. Find n when $y = 137$.

Method 1:

Since $y \propto n$, $y = kn$ where k is the proportionality constant.

When $n = 3$, $y = 40$

$\therefore\ 40 = k \times 3$

$\therefore\ \frac{40}{3} = k$

$\therefore\ y = \frac{40}{3}n$

So, when $y = 137$,

$\frac{40}{3}n = 137$

$\therefore\ n = 137 \times \frac{3}{40}$

$\therefore\ n \approx 10.3$

Method 2:

n	3	?
y	40	137

$\times \frac{137}{40}$

To change y from 40 to 137, we multiply by $\frac{137}{40}$.

Since $y \propto n$, we must also multiply n by $\frac{137}{40}$.

$\therefore\ n = 3 \times \frac{137}{40} \approx 10.3$

8 Suppose $y \propto x$ and that $y = 35$ when $x = 7$. Find:

a y when $x = 28$ **b** x when $y = 5$.

9 Suppose $y \propto x$ and that $y = 20$ when $x = 6$. Find:

a y when $x = 18$ **b** x when $y = 70$.

10 Complete this table given that $y \propto x$:

x	5	15	
y	20		120

11 The amount of petrol Jill's car uses each day is directly proportional to the distance she travels. Yesterday she travelled 81 km and used 6 litres of petrol.

a If Jill travelled 54 km in one day, how much petrol would she use?

b Jill used 9 litres of petrol in one day. How far did she travel?

12 The resistance R ohms to the flow of electricity in a wire varies in direct proportion to the length l cm of the wire. A 10 cm length of wire has resistance 0.06 ohms.

a Find:

i the law connecting R and l

ii the resistance when the wire is 50 cm long

iii the length of wire which has resistance 3 ohms.

b Plot the graph of R against l.

13 A 4 litre can of paint will cover 18 m^2 of wall. Find:

a the area of wall which can be covered using 10 litres of paint

b the amount of paint required to paint a room with wall area 40.5 m^2.

14 Rachael eats l lollies every 5 days. Find, in terms of l:

a the number of lollies Rachael eats in 20 days

b the number of days it will take Rachael to eat 30 lollies.

15 The speed of a falling object is directly proportional to the time for which it falls. The speed of an object which has fallen for 5 seconds is 49 m/s.

a Find the speed of an object which falls for 8 seconds.

b How long will it take a falling object to reach a speed of 100 m/s?

16 The dimensions of a model car are directly proportional to the dimensions of the actual car it was designed from. The actual car is 4 m long, and the model car is 30 cm long.

a The actual car is 1.4 m wide. Find the width of the model car.

b The model car is 9 cm high. Find the height of the actual car.

DISCUSSION

What variables do you commonly use, which are in direct proportion?

B POWERS IN DIRECT PROPORTION

The formula for the area A of a circle of radius r is $A = \pi r^2$.

We can construct the table of values:

r	0	1	2	3	4
A	0	3.14	12.57	28.27	50.27

The graph of A against r is shown alongside. The graph is not a straight line, so A is *not* directly proportional to r.

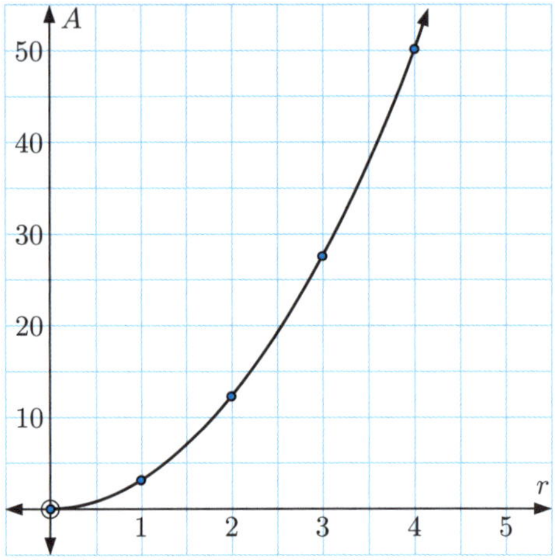

However, suppose we include a row in the table for the values of r^2:

r	0	1	2	3	4
r^2	0	1	4	9	16
A	0	3.14	12.57	28.27	50.27

The graph of A against r^2 is a straight line through the origin O. A is directly proportional to r^2, and we write $A \propto r^2$. In this case we know from the formula that the proportionality constant is $k = \pi$.

Notice from the table that if r is doubled from 1 to 2, both r^2 and A are multiplied by 4. If r is trebled from 1 to 3, both r^2 and A are multiplied by 9.

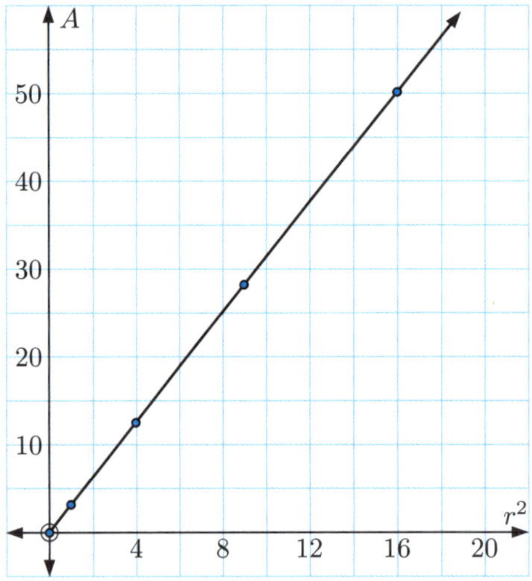

Example 3 — Self Tutor

Consider $y = \dfrac{x^4}{5}$.

State which two variables are directly proportional, and determine the proportionality constant k.

Since $y = \frac{1}{5}x^4$, we conclude that $y \propto x^4$ and $k = \frac{1}{5}$.

EXERCISE 25B

1 State which two variables are directly proportional, and determine the proportionality constant k:

a $A = 4.9t^2$ **b** $K = 2x^3$ **c** $T = \dfrac{\sqrt{l}}{5}$

d $V = 500t^4$ **e** $P = 4\sqrt[3]{x}$ **f** $V = \frac{4}{3}\pi r^3$

Example 4 — Self Tutor

Consider the table of values shown.

x	2	4	6
y	6	24	54

a By plotting y against x^2, establish that $y \propto x^2$.

b Hence write down the rule connecting y and x.

a

x	2	4	6
x^2	4	16	36
y	6	24	54

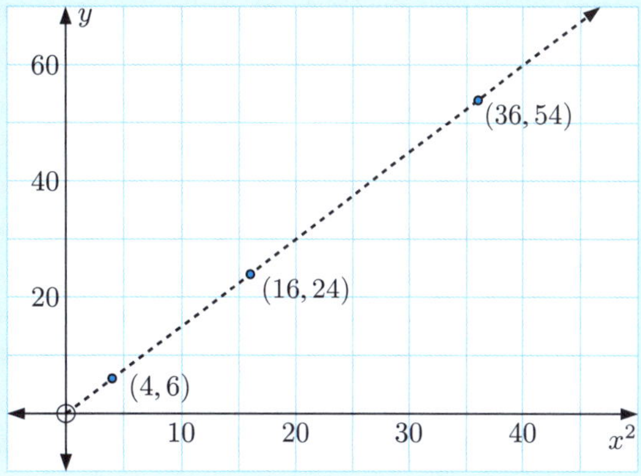

The graph of y against x^2 is a straight line through the origin, so $y \propto x^2$.

b The line has gradient $= \dfrac{24-6}{16-4}$

$= \frac{18}{12}$

$= 1.5$

$\therefore \quad y = 1.5x^2$

2 By plotting y against x^2, establish that $y \propto x^2$. Hence write down the rule connecting y and x.

a

x	2	3	5
y	16	36	100

b

x	1	3	4	5
y	3	27	48	75

3 By plotting y against x^3, establish that $y \propto x^3$ and find the proportionality constant k.

a

x	1	2	3
y	5	40	135

b

x	2	3	5
y	4	13.5	62.5

Example 5 — Self Tutor

Suppose T is directly proportional to d^2, where $d > 0$. When $d = 2$, $T = 100$. Find:

a T when $d = 3$

b d when $T = 200$.

a

$\times \frac{3}{2}$

d	2	3
T	100	

d is multiplied by $\frac{3}{2}$

$\therefore \quad d^2$ is multiplied by $\left(\frac{3}{2}\right)^2$

$\therefore \quad T$ is multiplied by $\left(\frac{3}{2}\right)^2$ $\{$as $T \propto d^2\}$

$\therefore \quad T = 100 \times \left(\frac{3}{2}\right)^2 = 225$

b

d	2	
T	100	200

$\times 2$

T is multiplied by 2

$\therefore \quad d^2$ is multiplied by 2 $\{$as $T \propto d^2\}$

$\therefore \quad d$ is multiplied by $\sqrt{2}$ $\{$as $d > 0\}$

$\therefore \quad d = 2 \times \sqrt{2} = 2\sqrt{2}$

4 Suppose P is directly proportional to a^2, where $a > 0$. When $a = 5$, $P = 300$. Find:

a P when $a = 2$

b a when $P = 2700$.

5 Suppose M is directly proportional to the cube of x, and that when $x = 2$, $M = 24$. Find:

a M when $x = 3$ **b** x when $M = 120$.

6 The power required to drive a motor boat is directly proportional to the cube of its speed. If an 80 hp motor drives a boat at 20 km/h, what power is required to drive the same size boat at 30 km/h?

7 The value of a gemstone is directly proportional to the square of its weight. If a 4 carat stone is valued at \$200, find the value of a 5 carat stone of the same quality.

8 The surface area of a sphere is directly proportional to the square of its radius. A sphere of radius 6 cm has a surface area of about 452 cm^2. Find, to the nearest mm, the radius of a sphere with surface area 1000 cm^2.

Example 6 **Self Tutor**

The *period* or time for one complete swing of a pendulum is directly proportional to the square root of its length. When the length is 25 cm, the period is 1.00 seconds.

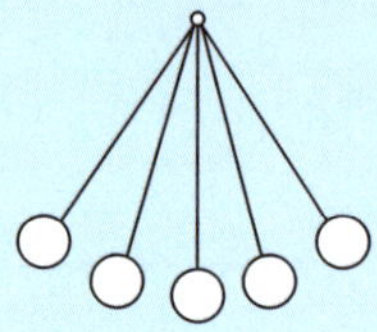

a Find the period of a pendulum with length 70 cm. Give your answer to 2 decimal places.

b Find the length of a pendulum with period 2 seconds.

If T is the period and l is the length, then $T \propto \sqrt{l}$.

a $\times \frac{70}{25}$

l	25	70
T	1.00	

l is multiplied by $\frac{70}{25}$

$\therefore$ $\sqrt{l}$ is multiplied by $\sqrt{\frac{70}{25}}$

$\therefore$ T is multiplied by $\sqrt{\frac{70}{25}}$ {as $T \propto \sqrt{l}$}

$\therefore$ $T = 1.00 \times \sqrt{\frac{70}{25}} \approx 1.67$

So, the period would be 1.67 seconds.

b

l	25	
T	1.00	2

$\times 2$

T is multiplied by 2

$\therefore$ $\sqrt{l}$ is multiplied by 2 {as $T \propto \sqrt{l}$}

$\therefore$ l is multiplied by 2^2

$\therefore$ $l = 25 \times 2^2 = 100$

So, the length would be 100 cm.

9 When a stone falls freely, the time taken for it to hit the ground is directly proportional to the square root of the distance fallen. If it takes a stone 4 seconds to fall 78.4 m, how long will it take for a stone to fall 500 m down a mine shaft?

10 At sea, the distance in km of the visible horizon is directly proportional to the square root of the height in metres of the observer's eye above sea level. So, $D \propto \sqrt{h}$.
The horizon is 9 km away when my eye is 5.4 m above sea level.

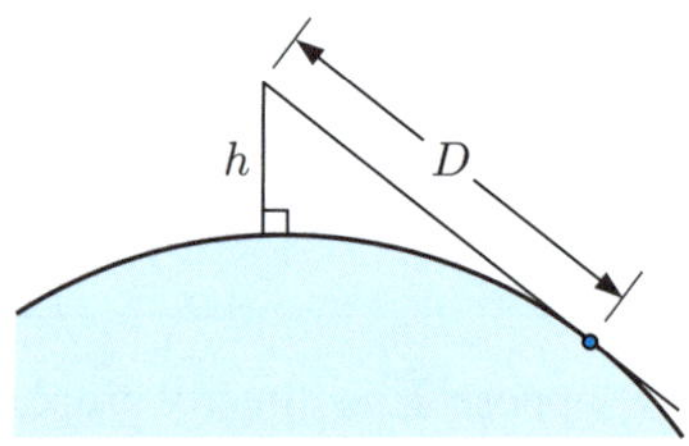

a How far can I see from a height of 10 m?

b From what height can I see a distance of 6 km?

11 The volume of blood flowing through a blood vessel is directly proportional to the square of the internal diameter.

a If the diameter is multiplied by 1.1, what happens to the volume?

b If the diameter is reduced by 20%, what percentage reduction in blood flow will occur?

12 The volume V of a sphere with radius r is given by $V = \frac{4}{3}\pi r^3$.

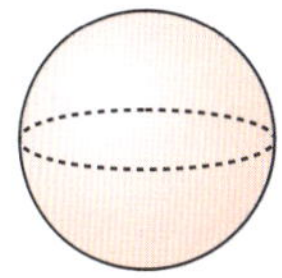

a State which two variables are directly proportional.

b Find the percentage change in volume if the radius is reduced by 10%.

c What percentage change in radius is necessary to double the volume?

13 The volume of a cone is given by $V = \frac{1}{3}\pi r^2 h$, where r is the radius and h is the height.

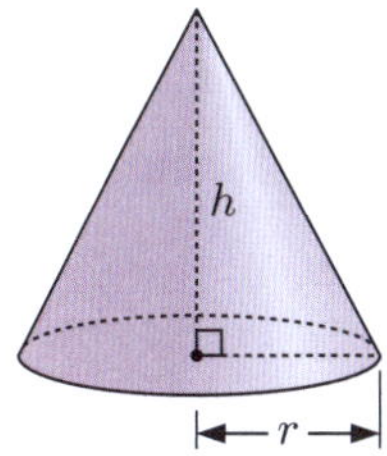

a If we consider cones of fixed radius but variable height, what proportionality exists between V and h?

b If we consider cones of fixed height but variable radius, what proportionality exists between V and r?

c What happens to the volume of a cone if the base radius is increased by 20% and the height is decreased by 15%?

14 For a cylinder with height $2r$ and radius r, the surface area is given by $A = 6\pi r^2$, and the volume is given by $V = 2\pi r^3$.

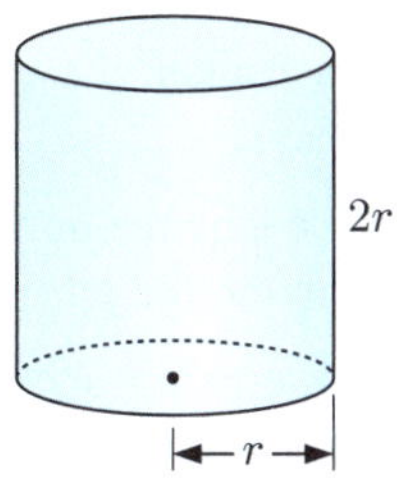

a Write r in terms of A.

b Hence write V in terms of A.

c Hence find the proportionality that exists between V and A.

DISCUSSION

In the world around us, there are many proportional relationships which we can anticipate using our knowledge of measurements and similarity.

A souvenir shop in New York sells different size scale models of the Statue of Liberty. They are all solid and made with the same material.

1 Predict the relationship between the *height* of a model and its:

- *volume*
- *mass*
- *surface area.*

2 If we sampled 100 models at random from the souvenir shop and took measurements of them, will the data match *exactly* with the relationships you predicted in **1**? What factors might cause the data to not match the relationships exactly?

3 Now suppose the models are made with different materials. Which of the pairs of variables in **1** do you expect to now vary differently? Will there necessarily still be a proportional relationship between them?

C INVERSE PROPORTION

DISCUSSION

If two painters can paint a house in three days, how long would it take four painters to paint the house, each working at the same rate?

The two variables in this example are the *number of painters* and the *time taken.*

The four painters will be able to do twice as much work in the same amount of time. It will therefore take them a half of the time, or $1\frac{1}{2}$ days, to complete the job.

In a case like this where doubling one variable halves the other, we have an **inverse proportion**.

Two variables are **inversely proportional** if, when one is *multiplied* by a constant, the other is *divided* by the same constant.

Dividing by k is the same as multiplying by $\frac{1}{k}$, so if one of the variables is multiplied by 2, the other must be multiplied by $\frac{1}{2}$.

Consider again the example of two painters completing a job in three days.

Suppose x is the number of painters and y is the number of days to complete the job.

The table below shows some possible combinations of x and y.

x	1	2	3	4	6
y	6	3	2	$1\frac{1}{2}$	1

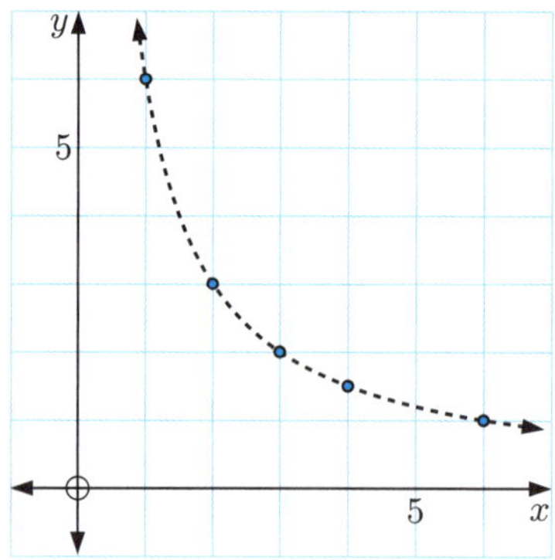

When these points are plotted, they lie in a curve called a **hyperbola**.

Suppose we include a row in the table for the values of $\frac{1}{x}$.

x	1	2	3	4	6
$\frac{1}{x}$	1	$\frac{1}{2}$	$\frac{1}{3}$	$\frac{1}{4}$	$\frac{1}{6}$
y	6	3	2	$1\frac{1}{2}$	1

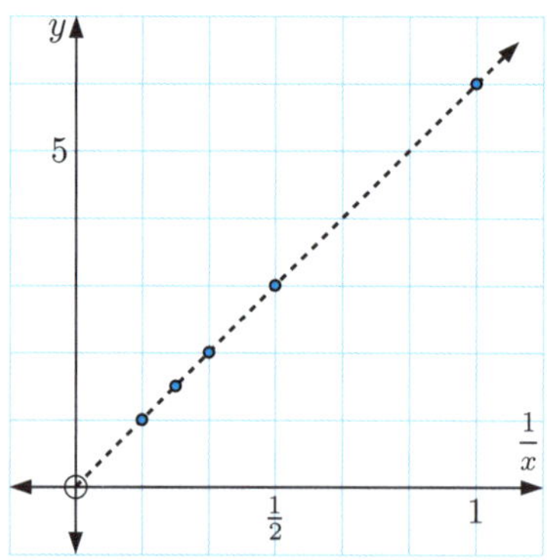

The points on the graph of y against $\frac{1}{x}$ form a straight line through the origin, so y is directly proportional to $\frac{1}{x}$.

If y is **inversely proportional** to x, then y is **directly proportional** to $\frac{1}{x}$.

If the proportionality constant is k, then $\boldsymbol{y = \frac{k}{x}}$ or $\boldsymbol{xy = k}$.

In the painters example, notice that for all points in the table, $xy = 6$.

Example 7 — Self Tutor

Suppose M is inversely proportional to t, and that $M = 200$ when $t = 3$. Find:

a M when $t = 5$ **b** t when $M = 746$.

Since M is inversely proportional to t, $M \propto \frac{1}{t}$.

a

$\times \frac{5}{3}$

t	3	5
M	200	

t is multiplied by $\frac{5}{3}$

$\therefore$ M is multiplied by $\frac{3}{5}$ $\{$as $M \propto \frac{1}{t}\}$

$\therefore$ $M = 200 \times \frac{3}{5} = 120$

b

t	3	
M	200	746

$\times \frac{746}{200}$

M is multiplied by $\frac{746}{200}$

$\therefore$ t is multiplied by $\frac{200}{746}$ $\{$as $M \propto \frac{1}{t}\}$

$\therefore$ $t = 3 \times \frac{200}{746} \approx 0.804$

EXERCISE 25C

1 **a** Copy and complete the table alongside.

b Plot the graph of y against $\frac{1}{x}$.

c Explain why y is inversely proportional to x.

d Calculate xy for each point in the table.

x	1	2	4	8
$\frac{1}{x}$				
y	40	20	10	5

2 For each table of values, calculate the value of xy for each point and hence decide whether x and y are inversely proportional. If an inverse proportionality exists, determine the law connecting the variables, and draw the graph of y against x.

a

x	1	2	3	6
y	12	6	4	2

b

x	2	3	6	9
y	12	8	4	3

c

x	7	6	12	1
y	12	14	7	84

3 Which graph(s) could indicate that y is inversely proportional to x?

A

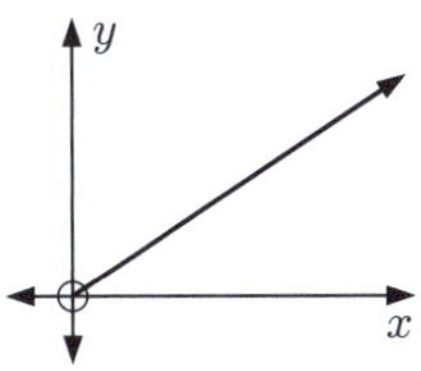

B

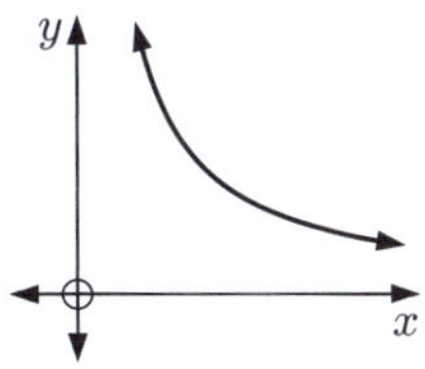

C

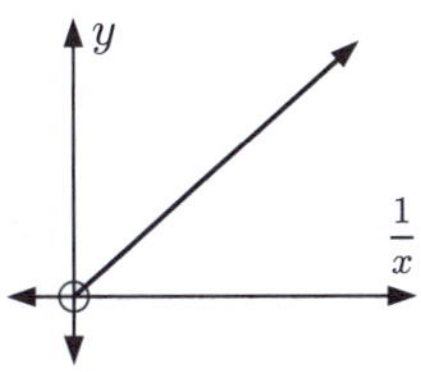

D

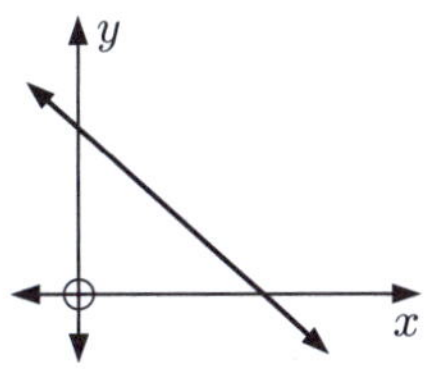

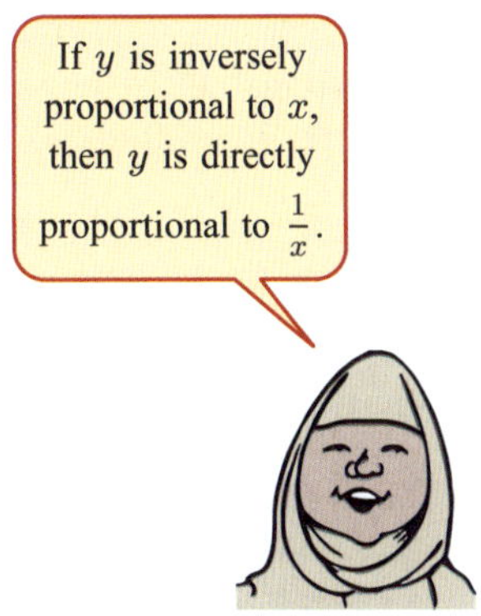

4 The formula connecting average speed s, distance travelled d, and time taken t, is $s = \frac{d}{t}$.
Complete the following statements by inserting "directly" or "inversely":

a If d is fixed, s is ...... proportional to t. **b** If t is fixed, s is ...... proportional to d.

c If s is fixed, d is ...... proportional to t.

5 Suppose y is inversely proportional to x. Describe what happens to:

a y if x is multiplied by 4 **b** y if x is multiplied by $\frac{7}{9}$ **c** x if y is multiplied by $\frac{8}{3}$.

6 Suppose y is inversely proportional to x, and that $y = 20$ when $x = 2$. Find:

a y when $x = 5$ **b** x when $y = 100$.

7 Suppose p is inversely proportional to n, and that $p = 24$ when $n = 6$. Find:

a p when $n = 8$ **b** n when $p = 12$.

8 Copy and complete this table given that $y \propto \frac{1}{x}$:

x	6	10	
y	25		3

9 The time taken to complete a certain job is inversely proportional to the number of workers doing the task. If 20 workers could do the job in 6 days, how long would it take 15 workers to do the job?

10 Four people are hired to set up 5000 seats for a concert. Working together, they could complete the job in 12 hours. However, one person is unwell, and is unable to help set up the chairs.

a How many chairs will the remaining three people be able to set up in 12 hours?

b How long will it take them to complete the job?

11 A rectangle with dimensions x m by y m has area 60 m^2.

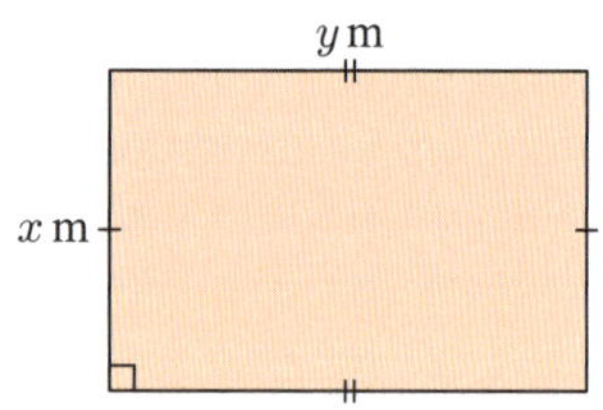

a Explain why y is inversely proportional to x.

b Copy and complete:

x	5		15
y		10	

c Plot the graph of y against x.

12 The chamber of an airtight cylinder is filled with the yellow gas fluorine. For a given mass of gas kept at a constant temperature, the volume is inversely proportional to the pressure. When $V = 10$ cm^3, the pressure $P = 40$ units.

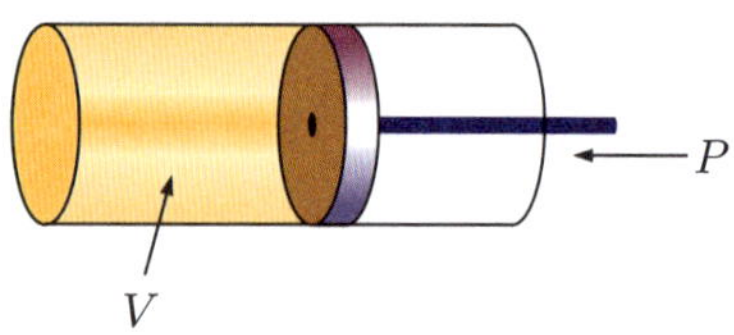

a Find the pressure when the volume is 20 cm^3.

b Find the volume when the pressure is 100 units.

c Find the formula connecting P and V.

d Plot the graph of P against V.

13 The average court time for each player in a netball team is inversely proportional to the number of players the team has. If the team has 10 players, the average court time for each player is 28 minutes.

a If the team has 12 players, what is the average court time for each player?

b If the average court time is 35 minutes, how many players does the team have?

c Given that 7 players are allowed on court at one time, how long does each game last?

D POWERS IN INVERSE PROPORTION

In this Section we consider the inverse proportion between one variable and the power of another. We use the same techniques as for direct proportion.

Example 8 **Self Tutor**

Suppose y is inversely proportional to the square of x, and that $x > 0$. When $x = 5$, $y = 36$.

a Find y when $x = 15$. **b** Find x when $y = 49$.

a

$\times 3$

x	5	15
y	36	

x is multiplied by 3

$\therefore$ x^2 is multiplied by $3^2 = 9$

$\therefore$ $\frac{1}{x^2}$ is multiplied by $\frac{1}{9}$

$\therefore$ y is multiplied by $\frac{1}{9}$ $\{$as $y \propto \frac{1}{x^2}\}$

$\therefore$ $y = 36 \times \frac{1}{9} = 4$

b

x	5	
y	36	49

$\times \frac{49}{36}$

y is multiplied by $\frac{49}{36}$

$\therefore$ $\frac{1}{x^2}$ is multiplied by $\frac{49}{36}$ $\{$as $y \propto \frac{1}{x^2}\}$

$\therefore$ x^2 is multiplied by $\frac{36}{49}$

$\therefore$ x is multiplied by $\sqrt{\frac{36}{49}} = \frac{6}{7}$ $\{$as $x > 0\}$

$\therefore$ $x = 5 \times \frac{6}{7} = \frac{30}{7}$

EXERCISE 25D

1 Suppose y is inversely proportional to the square of x, and that $x > 0$. When $x = 10$, $y = 4$.

a Find y when $x = 5$. **b** Find x when $y = 25$.

2 Suppose F is inversely proportional to the square root of c. When $c = 11$, $F = 20$.

a Find F when $c = 33$. **b** Find c when $F = 30$.

3 A heavy ball is suspended from a spring, and its weight is initially supported. The ball is then dropped so it will bounce up and down. The time between successive bounces is called the *period* of the motion, and is inversely proportional to the square root of the stiffness of the spring. The period of a spring of stiffness 100 units is 0.2 seconds. Find:

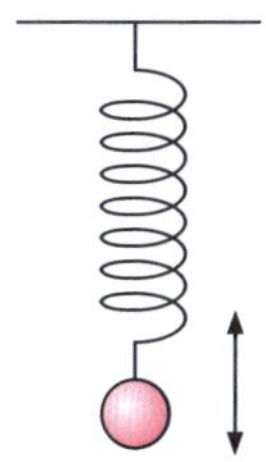

a the period of a spring of stiffness 300 units

b the stiffness required for a 0.1 second period.

4

The radiant heat received by a body is inversely proportional to the square of the distance of the body from the heat source. If the distance from the source is decreased by 20%, what effect does this have on the heat received?

5 The children in a class are each given the same amount of plasticine, and asked to build a solid cylinder. The radius of the cylinder produced is inversely proportional to the square root of the height. Xian's cylinder has radius 2.7 cm and height 13.1 cm.

a Beth's cylinder has radius 5.4 cm. Find its height.

b Roger's cylinder is 8 cm high. Find its radius.

c What volume of plasticine was each child given?

6 The intensity of light on a screen is inversely proportional to the square of the distance between the screen and the light source.
If a screen is illuminated by a light source 20 m away, the intensity is one fifth of what is required. Where should the light be placed?

RESEARCH

Indigenous Australians use fires to manage the land they live on. They deploy low-intensity fires to reduce the fuel load, which helps create firebreaks in the landscape.

To manage these fires, Indigenous Australians take into account several factors which affect the rate at which a fire spreads.

What to do:

Research the relationship between each of the factors below and the rate at which a fire spreads. Discuss whether the relationship involves a direct proportion, an inverse proportion, or neither.

- wind speed
- ambient temperature
- moisture content of the fuel
- humidity
- the steepness of the land
- density of the fuel

MULTIPLE CHOICE QUIZ

QUICK QUIZ

REVIEW SET 25A

1 Which graph indicates that y is directly proportional to x?

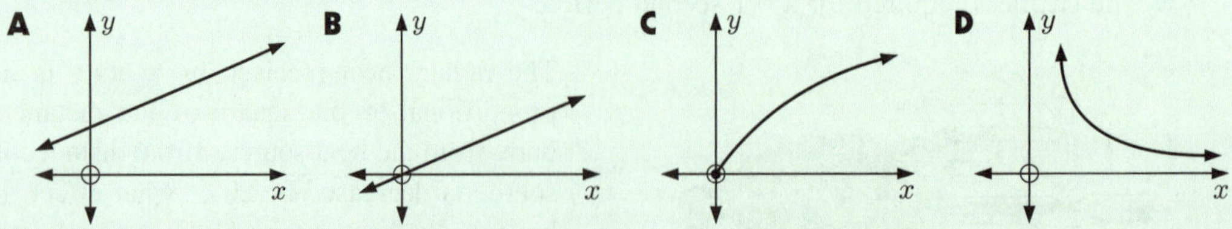

2 Consider the table of values alongside.

x	2	3	6	11
y	6	9	18	33

a Plot the graph of y against x.

b Explain why y is directly proportional to x.

c Find a formula connecting x and y.

3 Copy and complete this table given that $y \propto x$:

x	12	60	
y	3		20

4 Suppose x is inversely proportional to y, and that $y = 65$ when $x = 10$. Find:

a y when $x = 12$ **b** x when $y = 150$.

5 A school's quiz night raises money in direct proportion to the number of people attending. Last year 85 people attended the quiz night, and the event raised \$1275. This year the entry fees were the same, but the school only raised \$1080. How many people attended this year?

6 Jamela thinks that if y is inversely proportional to x, then a 25% increase in y will produce a 25% decrease in x. Explain why Jamela is incorrect.

7 If 3 workers can build a house in 60 days, how long would it take 5 workers to build the house?

8 Suppose M is directly proportional to $\sqrt{t}$, and that $M = 20$ when $t = 4$. Find:

a M when $t = 36$ **b** t when $M = 30$.

9 The volume of a cylinder of fixed length is directly proportional to the square of its radius.

a Find the change in volume produced by doubling the radius.

b Find the change in radius needed to produce a 60% increase in volume.

10 The kinetic energy E of a solid with mass m and speed v, is given by $E = \frac{1}{2}mv^2$.

a State the proportional relationship that exists between:

i E and m **ii** E and v **iii** m and v.

b Describe the change in the kinetic energy of an object if its speed is reduced by 12%.

11 This table shows the diameter and price of three sizes of frypan.

Diameter (x cm)	16	25	36
Price (\$$P$)	22	27.50	33

a Sketch the graph of P against $\sqrt{x}$.

b Hence write a formula for P in terms of x.

c What would you expect the price of a frypan with diameter 30 cm to be?

d Describe the proportionality relationship between the price P and the area A of the frypan.

REVIEW SET 25B

1 The variables p and q in the table are directly proportional. Find a and b.

p	4	7	a
q	12	b	42

2 If 7 litres of petrol are needed to drive 100 km, how far could you travel on 16 litres of petrol?

3 Suppose $y \propto x$ and that $y = 77.9$ when $x = 4.1$.

a Find the proportionality constant. **b** Find the value of y when $x = 6.3$.

4 A person's height in centimetres is directly proportional to their height in inches.
Luke is 127 cm or 50 inches tall. His sister Melanie is 160 cm tall. Find Melanie's height in inches.

5 Suppose k is directly proportional to the cube of n. What happens to k when n is doubled?

6 For the table alongside, show that x and y are inversely proportional, and find the law connecting x and y.

x	3	6	9
y	12	6	4

7 6 pumps can empty a lake in 45 minutes. How long would 10 pumps take to empty the lake?

8 Suppose y is inversely proportional to the square of x, and that $x > 0$. When $y = 50$, $x = 7$.

a Find y when $x = 5$. **b** Find x when $y = 75$.

9 When a train travels around a bend, the safe speed of the train is directly proportional to the square root of the radius of the curve. For a bend with radius 100 m, the safe speed is 25 m/s.

a Find the safe speed for a bend with radius 150 m.

b Find the bend radius for which the safe speed is 45 m/s.

10 The surface area of a sphere with radius r is given by $A = 4\pi r^2$.

a State which two variables are directly proportional.

b Find the percentage change in the radius if the surface area is increased by 20%.

11 Temperatures in **Kelvin** T_k are related to temperatures in degrees Celsius T_c using the formula $T_k = T_c + 273.15$.

a Assuming constant pressure, the density of dry air ρ is inversely proportional to T_k.

i Is ρ inversely proportional to T_c? Explain your answer.

ii At 20°C, the density of dry air is 1.20 kg/m^3. Find the density of dry air at 40°C.

b The temperature T_k of a gas is directly proportional to its volume V and pressure P.
If air at 30°C is compressed in a piston to $\frac{1}{10}$ of its volume and 20 times its pressure, what will its new temperature be?

Chapter 26

Probability

Contents:

OPENING PROBLEM

Players A, B, C, and D are the final four left in a knock-out tennis tournament. In the semi-finals, A will play B, and C will play D. The winners of these matches will play in the final.

The table alongside shows the probabilities of each player beating the others. For example, the yellow shaded square indicates that A has a 40% chance of beating B.

winner \ loser	A	B	C	D
A		40%	55%	47%
B	60%		42%	54%
C	45%	?		35%
D	53%	46%	65%	

Things to think about:

- **a** What is the probability that B will beat A?
- **b** What is the sum of the probability that C will beat D and the probability that D will beat C? Can you explain why this must be true?
- **c** Can you complete the table with the probability that C will beat B?
- **d** How can we illustrate the possible outcomes of the finals series?
- **e** Which player is most likely to win the final?

The **probability** of an event is the chance or likelihood of it occurring.

An **impossible** event has 0% chance of happening, and is assigned the probability 0.

A **certain** event has 100% chance of happening, and is assigned the probability 1.

All other events have probability between 0 and 1.

Probabilities may be given as decimals, proper fractions, or percentages.

This number line shows how we could interpret different probabilities:

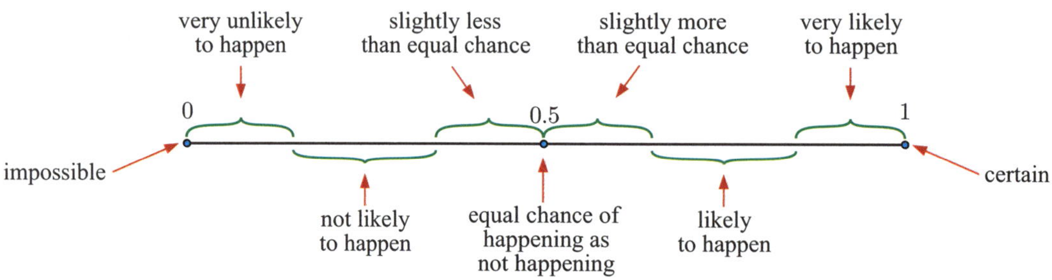

We can determine probabilities based on:

- what we theoretically expect to happen
- the results of an experiment.

The study of chance has important applications in physical and biological sciences, economics, politics, sport, life insurance, quality control, production planning, and many other areas.

Probability theory can be applied to card and dice games to try to increase our chances of success. It may therefore appear that an understanding of probability encourages gambling. However, a better knowledge of probability theory actually helps us to understand why most habitual gamblers lose in the long term.

HISTORICAL NOTE

Chevalier de Méré (1607 - 1684) was a French aristocrat and gambler in the 17th century. He wanted to know the answer to this question:

> "Should I bet even money on the occurrence of at least one 'double six' when rolling a pair of dice 25 times?"

De Méré's experience of playing dice games convinced him that the answer was yes, but he did not know how to prove it. He therefore asked his friend, the French mathematician **Blaise Pascal** (1623 - 1662), for help.

In a series of letters between Pascal and fellow mathematician **Pierre de Fermat** (1607 - 1665), the problem was solved. In the process, they became interested in solving other questions of this kind, and together they laid the foundations of a new branch of mathematics called **theoretical probability**.

Pierre de Fermat

A SAMPLE SPACE AND EVENTS

SAMPLE SPACE

The **sample space** of an experiment is the set of its possible outcomes.
We usually use U to represent the sample space, as it is the universal set of possible outcomes.

We can represent sample spaces using:

- lists
- 2-dimensional grids
- tree diagrams.

Suppose we toss a coin and spin the spinner shown.

From the coin we could obtain a head (H) or a tail (T).

From the spinner we could obtain A, B, or C.

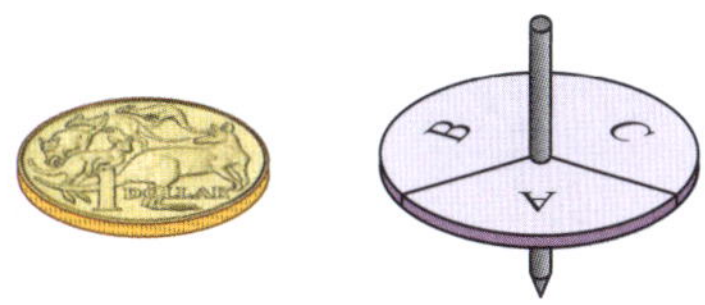

LISTING OUTCOMES

The simplest way to represent a sample space is to **list** the possible outcomes using set notation.

{HA, HB, HC, TA, TB, TC}

2-DIMENSIONAL GRIDS

If an experiment involves *two* operations, we can use a **2-dimensional grid**.

The two possibilities for tossing the coin are listed on the horizontal axis.

The three possibilities for spinning the spinner are listed on the vertical axis.

Each point on the grid represents one of the six possible outcomes of the experiment.

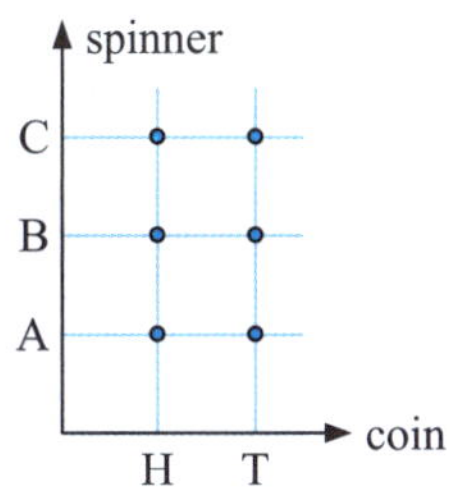

TREE DIAGRAMS

If an experiment involves *two or more* operations, we can use a **tree diagram**.

The two possibilities for tossing the coin are shown on the first set of branches.

The three possibilities for spinning the spinner are shown on the second set of branches.

Each path on the tree diagram represents one of the six possible outcomes of the experiment.

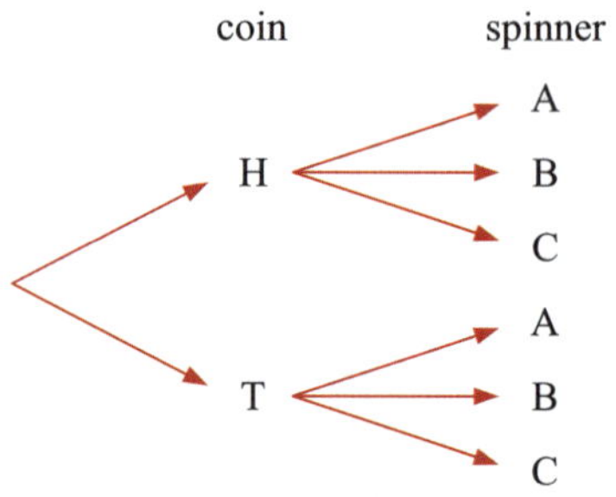

EVENTS

An **event** is a set of outcomes in the sample space that have a particular property.

Example 1 — Self Tutor

Suppose we toss a coin and spin the spinner shown.
Highlight the outcomes corresponding to the event "a tail and a consonant" on:

a a list of possible outcomes
b a 2-dimensional grid
c a tree diagram.

Let H represent a "head" and T represent a "tail".

a {HA, HB, HC, TA, TB, TC}

b

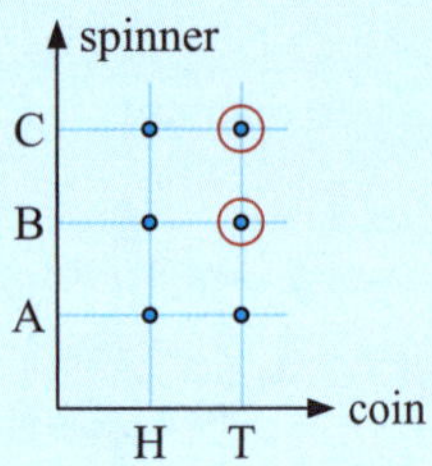

c

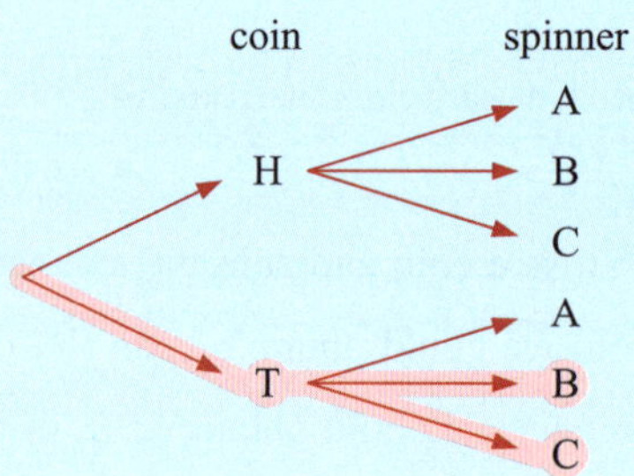

EXERCISE 26A

1 List, in set notation, the sample space for:

a selecting a ticket from the container shown
b the genders of a 2-child family
c the order in which 3 men A, B, and C can be lined up
d tossing 4 coins simultaneously.

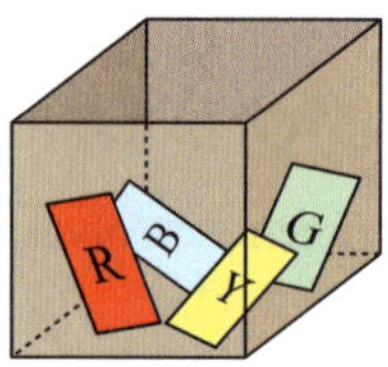

2 Lucy randomly selects one letter from the alphabet.

a Write down the sample space U.
b List the outcomes in the event:

i A that Lucy selects a consonant
ii B that Lucy selects a letter in the word EVENT.

3 Draw a 2-dimensional grid to illustrate the sample space for:

a rolling a die and tossing a coin simultaneously

b rolling a pair of dice

c twirling a square spinner marked A, B, C, D and a triangular spinner marked 1, 2, 3.

4 Draw a 2-dimensional grid to illustrate the sample space when this spinner is spun twice.
Circle the outcomes corresponding to the event "spinning a red on either spin".

5 Draw a 2-dimensional grid to illustrate the sample space when selecting one even number and one odd number from the integers 1 to 10.
Circle the outcomes corresponding to the event "the sum of the numbers is 13".

6 Use a tree diagram to illustrate the sample space for:

a tossing a 10 cent coin and a 20 cent coin simultaneously

b rolling a 6-sided die and tossing a coin simultaneously.

7 Bag A contains red and white marbles, and bag B contains blue and yellow marbles. A bag is selected and one marble is taken from it. Draw a tree diagram to illustrate the sample space. Highlight the outcome corresponding to the event "a blue marble from bag B".

8 A coin is tossed and a spinner with sectors A, B, and C is spun. Draw a tree diagram to illustrate the sample space. Highlight the outcomes corresponding to the event "head *or* A".

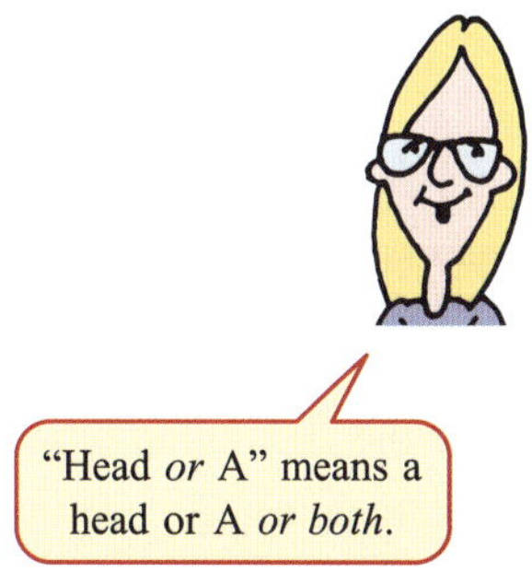

Example 2 **Self Tutor**

Illustrate, using a tree diagram, the sample space when drawing three counters from a bag containing many red counters and blue counters.
Highlight the outcomes corresponding to the event "drawing two blue counters".

Let R represent a red counter, and B represent a blue counter.

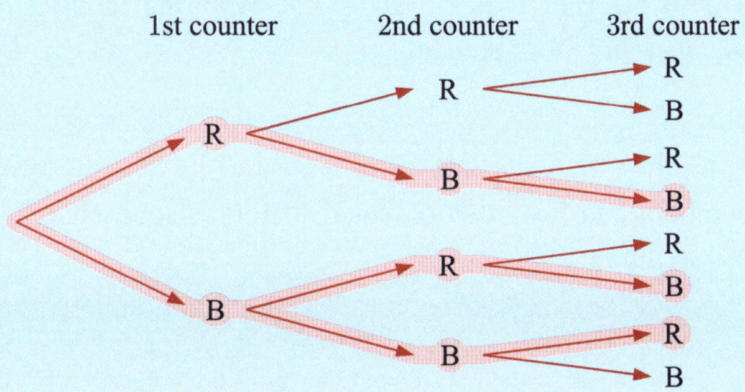

9 Use a tree diagram to illustrate the sample space for the genders of a 3-child family. Highlight the outcomes corresponding to the event "two girls and a boy".

10 Hats A, B, and C each contain pink and white tickets. A hat is selected and then two tickets are taken from it. Draw a tree diagram to illustrate the sample space. Highlight the outcomes corresponding to the event "two pink tickets".

11 Use a tree diagram to illustrate the sample space for the order in which 4 different rowing teams A, B, C, and D could finish a race.

B THEORETICAL PROBABILITY

The sample space for rolling this die is $\{1, 2, 3, 4, 5, 6\}$. There are 6 possible outcomes. Since the die is a cube and hence symmetrical, the outcomes are *equally likely* to occur.

If the possible outcomes of an experiment are equally likely, the theoretical probability of an event E is given by

$$\text{P}(E) = \frac{\textbf{number of outcomes in } E}{\textbf{total number of possible outcomes}}.$$

Example 3 — Self Tutor

A wooden shape is randomly selected from a box containing 6 squares, 4 triangles, and 5 pentagons. Find the probability of selecting a:

a triangle **b** triangle or square **c** circle **d** polygon.

There are $6 + 4 + 5 = 15$ shapes which could be selected with equal probability.

a $\text{P(triangle)} = \frac{4}{15}$

b $\text{P(triangle or square)} = \frac{6+4}{15} = \frac{10}{15} = \frac{2}{3}$

c $\text{P(circle)} = \frac{0}{15} = 0$

d All of the shapes are polygons.

$\therefore \text{P(polygon)} = \frac{15}{15} = 1$

The **complement** of event E is E', and is the event that E does *not* occur.

For any event E, $\text{P}(E') = 1 - \text{P}(E)$.

EXERCISE 26B

1 A spinner with the numbers 1 to 5 written on equal sectors is spun once. Find the probability of spinning:

a a 4 **b** a 1 or a 2 **c** an odd number.

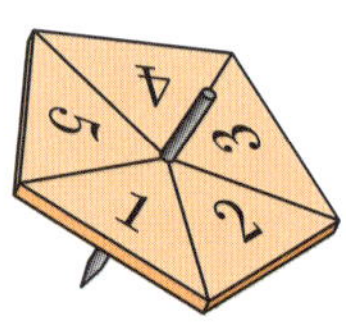

2 Candice has baked 6 blueberry muffins and 12 raspberry muffins. She chooses one to eat at random. Let E be the event that the muffin is blueberry.

a Explain the meaning of the event E'. **b** Find $\text{P}(E)$ and $\text{P}(E')$.

3 A symmetrical octahedral die has the numbers 1 to 8 marked on its faces. If it is rolled once, find the probability that the number rolled is:

a 4 **b** not 4 **c** less than 5 **d** greater than 8.

4 A bag contains 4 red and 3 green buttons. One button is randomly selected from the bag. Find the probability that the button is:

a red **b** green **c** red or green.

5 Amy is invited to select a treat from a bag containing 6 strawberry lollies, 3 mints, and 5 chocolates. If Amy selects a treat at random, find the probability that she selects:

a a mint **b** a mint or a chocolate.

6 A 52 card pack is well shuffled, then one card is dealt from the top of the pack. Find the probability that the card is:

a a Jack **b** a black card

c a diamond **d** a diamond or an ace.

7 Find the probability that a person randomly selected in the street has his or her birthday in:

a May **b** February.

8 37 people applied for a job at a toy store.
11 were selected for an interview, and from these people, 3 were offered a position at the store.

a Find the probability that a randomly selected applicant was:

i offered a position **ii** interviewed, but not offered a position

iii not selected for an interview.

b Find the sum of the probabilities in **a**. Explain your answer.

9 There are 5 different pairs of socks in Vanessa's sock drawer. The light in her room is not working, so she removes 2 socks from her drawer at random. Find the probability that Vanessa has removed a matching pair.

Example 4 **Self Tutor**

Two square spinners, each with 1, 2, 3, and 4 on their edges, are twirled simultaneously.

a Draw a 2-dimensional grid to illustrate the sample space.

b Hence find the probability of getting:

i a 3 with each spinner **ii** a 3 and a 1 **iii** an even result with each spinner.

a

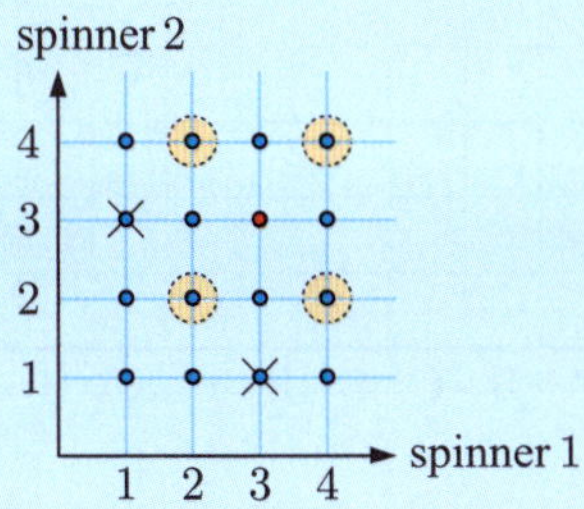

b There are 16 possible outcomes, each of which is equally likely.

i P(a 3 with each spinner) $= \frac{1}{16}$ {•}

ii P(a 3 and a 1) $= \frac{2}{16} = \frac{1}{8}$ {crossed points}

iii P(an even result with each spinner)

$= \frac{4}{16} = \frac{1}{4}$ {circled points}

10 A coin and a die are tossed simultaneously.

a Draw a 2-dimensional grid to illustrate the sample space.

b Hence find the probability of getting:

i a tail and a 6 **ii** a tail or a 6

iii neither a 2 nor a 6 **iv** neither a tail nor a 5

v a head and an odd number **vi** a head or an odd number.

11 An ordinary die is tossed and a triangular spinner labelled A, B, and C is spun simultaneously.

a Draw a 2-dimensional grid to illustrate the sample space.

b Hence find the probability of getting:

i B and 5 **ii** A and a prime number **iii** A or C, and a multiple of 3.

12 The 36 different possible outcomes when a pair of dice is rolled is illustrated on the 2-dimensional grid.
Use the grid to find the probability of getting:

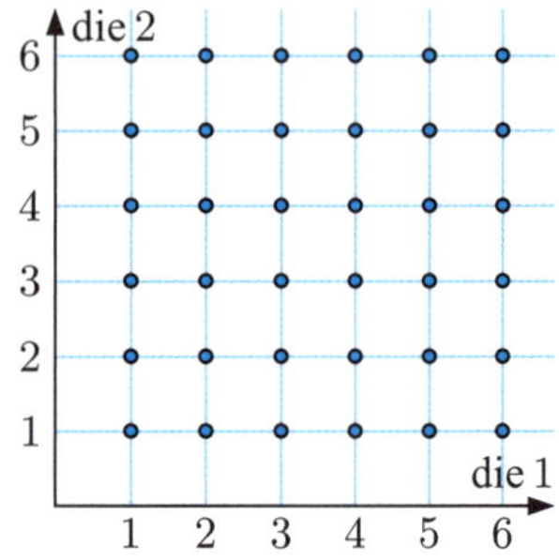

a two 3s **b** a 5 and a 6

c a 5 or a 6 **d** at least one 6

e exactly one 6 **f** no 6s

g a sum of 7 **h** a sum of 7 or 11

i a sum greater than 8 **j** a sum of no more than 8.

13 Two students each select a day of the week at random. Find the probability that:

a a Tuesday and a Friday are selected

b at least one student selected a weekend day

c the selected days start with the same letter

d between them, the selected days contain at least 15 letters.

14 In a role-playing game, Melanie and Neil duel by each rolling an eight-sided die. Neil's character has a higher level, so to win the duel Melanie must roll a number at least 2 greater than Neil's roll. Find the probability that Melanie will win the duel.

PUZZLE

Ralph begins a game of "Battleship" by placing an aircraft carrier on a 10×10 grid. The aircraft carrier must be placed horizontally or vertically, occupying 5 squares. One possible position is illustrated.

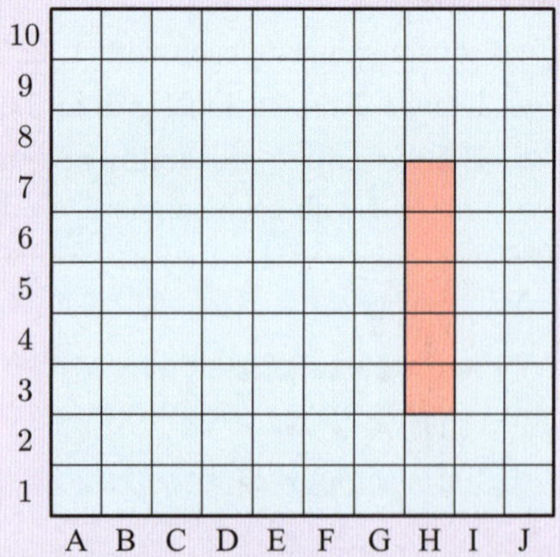

1 In how many different positions can the aircraft carrier be placed?

2 Suppose Ralph selects one of these positions at random. Find the probability that the aircraft carrier:

a is placed horizontally

b occupies at least one edge or corner square

c occupies the square B3

d occupies the square E5.

GAME

Click on the icon to play a game involving 2-dimensional grids.

GAME

C PROBABILITIES FROM VENN DIAGRAMS

In **Chapter 7** we studied **Venn diagrams**. We saw that they consist of a rectangle which represents the universal set, and circles within it which represent subsets. In probability questions, the circles in a Venn diagram are used to represent events.

Example 5 — Self Tutor

A wheel with sectors numbered 1 to 20 is spun once. Let A be the event of spinning an odd number, and B be the event of spinning a number greater than 13.

a Represent A and B on a Venn diagram.

b Find the probability that the number spun is:

  i odd *and* greater than 13

  ii odd *or* greater than 13.

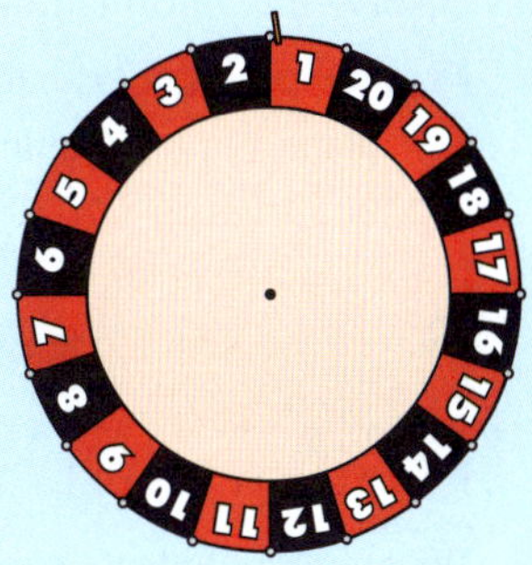

a $A = \{1, 3, 5, 7, 9, 11, 13, 15, 17, 19\}$

$B = \{14, 15, 16, 17, 18, 19, 20\}$

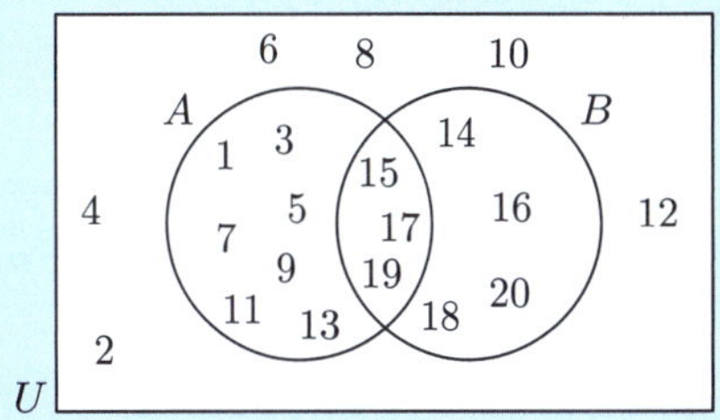

b **i** 3 of the 20 elements are in the region $A \cap B$.

$\therefore$ P(odd *and* greater than 13) $= \frac{3}{20}$

ii 14 of the 20 outcomes are in the region $A \cup B$.

$\therefore$ P(odd *or* greater than 13) $= \frac{14}{20} = \frac{7}{10}$

EXERCISE 26C

1 The octagonal spinner alongside is spun once. Let A be the event that the spinner lands on a pink sector, and B be the event that an even number is spun.

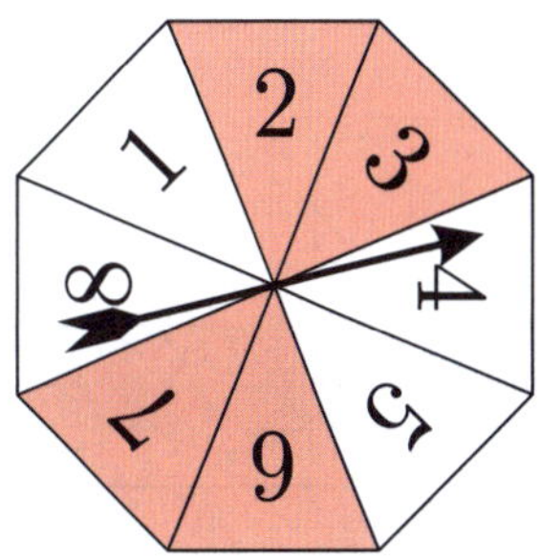

a Copy and complete this Venn diagram.

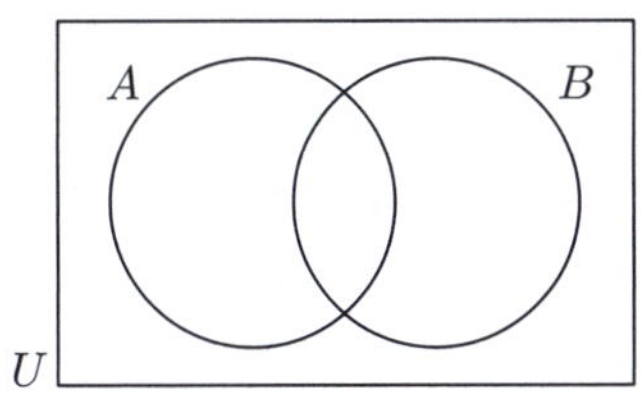

b Find the probability that the spinner lands on:

i a pink even number

ii a pink sector *or* an even number.

2 The numbers from 1 to 25 are placed in a bag, and one number is selected at random. Let A represent selecting a multiple of 3, and B represent selecting a multiple of 4.

  a Represent A and B on a Venn diagram.

  b Find the probability of selecting:

   i a multiple of 3
   ii a multiple of 3 *and* 4
   iii a multiple of 4 but *not* a multiple of 3
   iv a multiple of 3 or 4, but not both.

3 A letter of the alphabet is chosen at random. Use a Venn diagram to find the probability that the letter is in:

  a both STATISTICS and PROBABILITY
  b STATISTICS but not PROBABILITY
  c STATISTICS or PROBABILITY
  d neither word.

Example 6 — Self Tutor

In a music class of 24 students, 13 can play the piano, 9 can play the guitar, and 4 can play neither.

  a Display this information on a Venn diagram.

  b Find the probability that a randomly selected student plays:

   i piano but not guitar
   ii either instrument.

a Let P represent students who can play piano, and G represent students who can play guitar.

Let $n(P \cap G) = x$

$\therefore \ n(P \cap G') = 13 - x$ and $n(P' \cap G) = 9 - x$

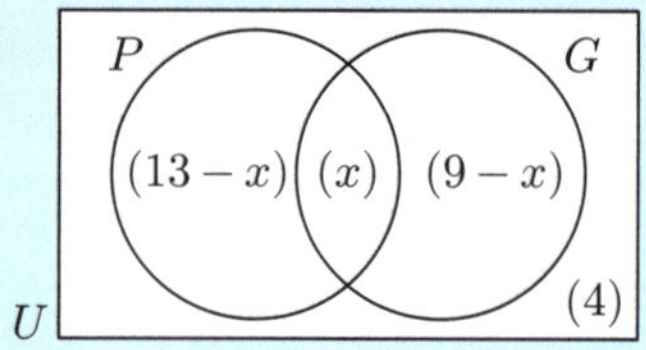

We know 4 students can play neither instrument.

Since there are 24 students in the class,

$(13 - x) + x + (9 - x) + 4 = 24$

$\therefore \ 22 - x = 20$

$\therefore \ x = 2$

$\therefore \ n(P \cap G') = 11$ and $n(P' \cap G) = 7$

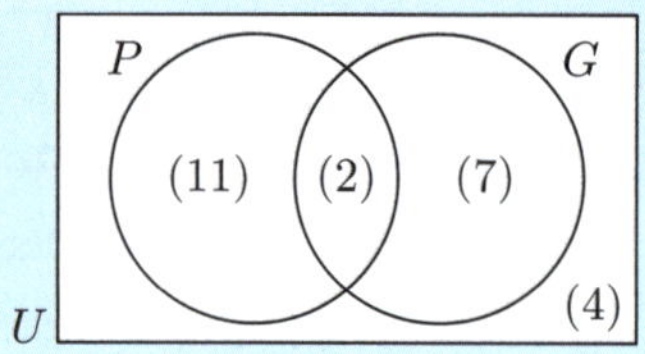

b **i** P(piano but not guitar) $= \text{P}(P \cap G')$

$= \frac{11}{24}$

ii P(either instrument) $= \text{P}(P \text{ or } G \text{ or both})$

$= \text{P}(P \cup G)$

$= \dfrac{11 + 2 + 7}{24}$

$= \frac{20}{24} = \frac{5}{6}$

4

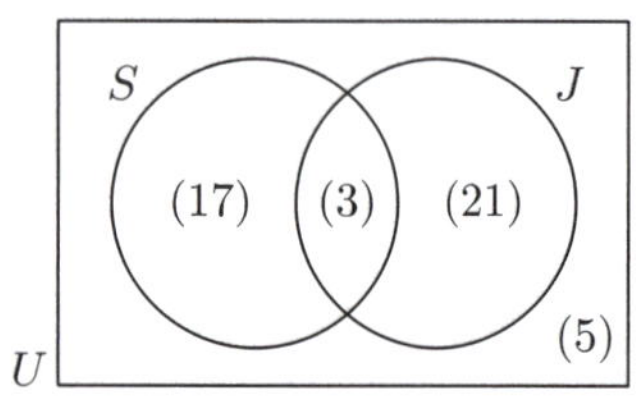

A group of university students were asked whether they study full time (S) and if they have a job (J).

The results are shown in the Venn diagram.

  a How many students participated in the survey?

  b Find the probability that a randomly selected student:

   i studies full time *and* has a job
   ii studies full time
   iii has a job but does not study full time.

5 Out of the 47 guests staying at an island resort, 21 signed up for snorkelling, 15 signed up for waterskiing, and 4 signed up for both.

- **a** Display this information on a Venn diagram.
- **b** Find the probability that a randomly selected guest signed up for:
  - **i** neither activity
  - **ii** exactly one activity.

6 In a group of 65 international travellers, 49 are vaccinated against hepatitis A and 32 are vaccinated against typhoid. All of the travellers have received at least one of the vaccinations.

- **a** Display this information on a Venn diagram.
- **b** Find the probability that a randomly selected traveller is vaccinated against:
  - **i** both diseases
  - **ii** hepatitis A but not typhoid.

7 In a class of 40 students, 19 play tennis, 20 play netball, and 8 play neither of these sports. A student is randomly chosen from the class. Determine the probability that the student:

- **a** plays tennis
- **b** does not play netball
- **c** plays at least one of the two sports
- **d** plays exactly one of the sports.

8 The medical records for a class of 30 children show that 24 have had measles, 12 have had measles *and* mumps, and 26 have had measles *or* mumps. If a child from the class is selected at random, determine the probability that he or she has had:

- **a** mumps
- **b** mumps but not measles
- **c** neither mumps nor measles.

9 Neville checked his letter box each day for 80 days. He received letters on 40 days, bills on 14 days, and junk mail on 41 days. He received letters *and* bills on 9 days, and bills *and* junk mail on 7 days. He received all three items on 4 days, and no items on 14 days.

- **a** Display this information on a Venn diagram.
- **b** Find the probability that, on a randomly chosen day, Neville received:
  - **i** junk mail only
  - **ii** letters or junk mail, but not both
  - **iii** bills and letters, but not junk mail.

D INDEPENDENT EVENTS

When we calculate the probability of two or more events occurring, we are calculating the probability of **compound events** or **combined events**.

Consider tossing a coin and rolling a die simultaneously.

We can illustrate the sample space on a 2-dimensional grid, and use this grid to calculate probabilities.

For example, only 1 of the 12 outcomes is a head and a 5, so

P(a head *and* a 5) $= \frac{1}{12}$.

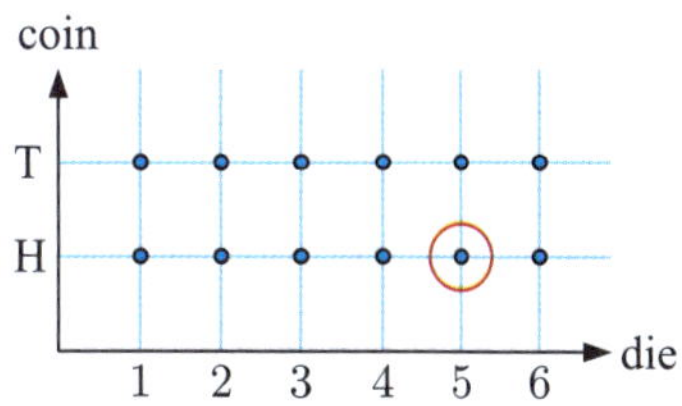

For a compound event like this, we can also calculate the probability *without* using a grid.

The outcome of the coin toss does not affect the outcome of the dice roll, and vice versa. We say the events are *independent*.

Notice that P(a head) $= \frac{1}{2}$ and P(a 5) $= \frac{1}{6}$, so P(a head) $\times$ P(a 5) $= \frac{1}{12}$.

So, P(a head **and** a 5) = P(a head) $\times$ P(a 5).

Two events are **independent** if the occurrence of each event does not affect the occurrence of the other.

If two events A and B are **independent** then $\text{P}(A \textbf{ and } B) = \text{P}(A) \times \text{P}(B)$.

Example 7 **Self Tutor**

Sunil has probability $\frac{4}{5}$ of hitting a target and Monika has probability $\frac{5}{6}$.

If they both fire simultaneously at the target, determine the probability that:

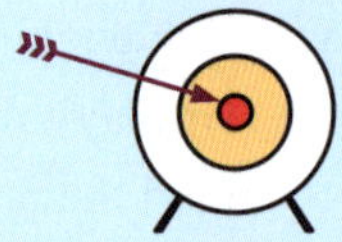

a they both hit **b** they both miss.

Let S be the event of Sunil hitting the target and M be the event of Monika hitting it.

a $\text{P(both hit)} = \text{P}(S \text{ and } M)$
$= \text{P}(S) \times \text{P}(M)$
$= \frac{4}{5} \times \frac{5}{6}$
$= \frac{2}{3}$

b $\text{P(both miss)} = \text{P}(S' \text{ and } M')$
$= \text{P}(S') \times \text{P}(M')$
$= \frac{1}{5} \times \frac{1}{6}$
$= \frac{1}{30}$

EXERCISE 26D

1 The spinners alongside are each spun once. Find the probability that both spinners will finish on green.

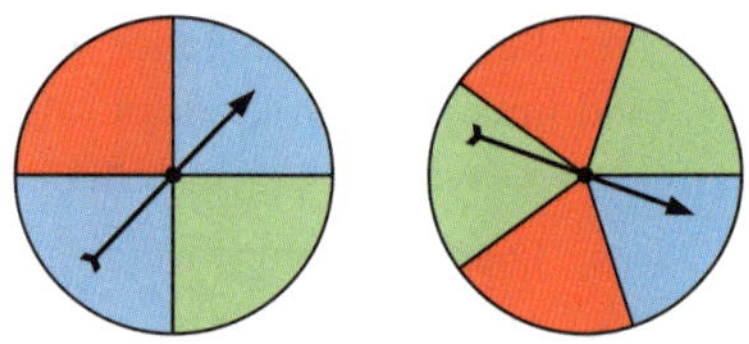

2 A coin and a 6-sided die are tossed simultaneously. Find the probability of getting:

a a head and a 5 **b** a tail and an odd number

c a head and a number greater than 2.

3 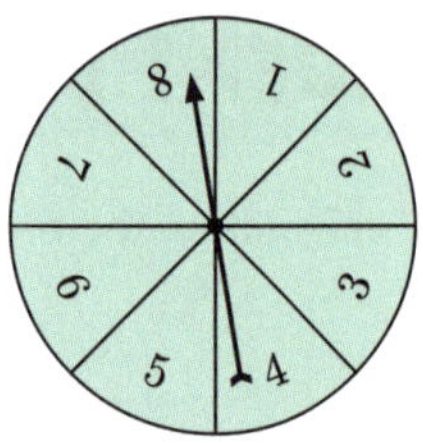

Tony and Ben each spin this spinner. Find the probability that Tony spins a prime number, and Ben spins a multiple of 3.

4 Golfers George and Brett each have a putt 5 metres from the hole. George has probability 0.6 of holing the putt, and Brett has probability 0.3 of holing the putt. Find the probability that:

a they both hole the putt **b** they both miss the putt.

5 On any given day at the office, there is probability 0.7 that Brian will wear purple, probability 0.4 that Hannah will wear jeans, and probability 0.8 that Patrick will wear a T-shirt.

Find the probability that on a particular day:

a Brian wears purple and Patrick wears a T-shirt

b Hannah wears jeans and Brian does *not* wear purple

c Brian wears purple, Hannah wears jeans, and Patrick does *not* wear a T-shirt.

6 Each time Morris the dog is thrown a frisbee, he has probability 0.7 of catching it. If a frisbee is thrown to Morris three times, find the probability that Morris will:

a catch it all three times
b miss the first throw but catch the next two
c catch the first throw but miss the next two.

E DEPENDENT EVENTS

A bag contains 5 blue discs and 3 yellow discs. One disc is selected at random, and put to one side. A second disc is then selected from the bag.

If the first disc was blue,

P(second disc is blue) $= \frac{4}{7}$ ← 4 blue discs remaining, ← 7 discs to choose from

If the first disc was yellow,

P(second disc is blue) $= \frac{5}{7}$ ← 5 blue discs remaining, ← 7 discs to choose from

So, the probability of the second disc being blue *depends* on what colour the first disc was.

> Two events are **dependent** if they are **not independent**.
> The occurrence of one of the events *does* affect the occurrence of the other event.

The rule for finding compound event probabilities for dependent events is different from the rule for independent events.

> If A and B are dependent events, then P(A and B) = P(A) × P(B given that A has occurred).

Example 8 Self Tutor

A bag contains 3 orange and 4 green balls. Two balls are randomly selected one after the other from the bag, the first being removed before the second ball is chosen. Find the probability that:

a both balls are orange
b the first ball is orange and the second ball is green.

a P(both balls are orange)
= P(first ball is orange and the second ball is orange)
= P(first ball is orange) × P(second ball is orange given that the first ball is orange)
$= \frac{3}{7} \times \frac{2}{6}$ ← 2 orange balls remaining, ← 6 balls to choose from
$= \frac{6}{42}$
$= \frac{1}{7}$

b P(first ball is orange and the second ball is green)
= P(first ball is orange) × P(second ball is green given that the first ball is orange)
$= \frac{3}{7} \times \frac{4}{6}$ ← 4 green balls remaining, ← 6 balls to choose from
$= \frac{12}{42}$
$= \frac{2}{7}$

EXERCISE 26E

1 A box contains 5 red and 2 purple balls. Two balls are randomly selected from the box. The first is *not* replaced in the box before the second is selected. Determine the probability that:

a both balls are red **b** the first ball is purple and the second is red.

2 Amelie has a bag containing 4 red apples and 6 green ones. She selects one apple at random, eats it, and then takes another, also at random.

a Determine the probability that:

i both apples were red **ii** both apples were green

iii the first was red and the second was green

iv the first was green and the second was red.

b Find the sum of the probabilities in **a**. Explain your answer.

3 Marion has a carton containing 12 cans of soup. 4 cans are tomato and the rest are pumpkin. She selects two cans at random without looking at the labels.

a Find the probability that both cans are:

i tomato soup **ii** pumpkin soup.

b Hence find the probability that Marion selects one can of each flavour.

4 A raffle has 150 tickets which are placed in a barrel. Two tickets are drawn at random from the barrel to decide the prizes. John has 3 tickets in the raffle. Determine his probability of winning:

a first and second prize **b** second prize but not first prize

c none of the prizes.

5 When Jamie plays football, he kicks with his right foot 80% of the time, and his left foot 20% of the time. His right foot kicks are accurate 70% of the time, and his left foot kicks are accurate 40% of the time. Find the probability that Jamie's next kick will be:

a with his right foot and accurate **b** with his left foot and inaccurate.

Example 9 **Self Tutor**

A hat contains 15 tickets with the numbers 1 to 15 printed on them.
If three tickets are drawn from the hat without replacement, determine the probability that they are all multiples of 3.

The multiples of 3 are $\{3, 6, 9, 12, 15\}$.

$\therefore$ P(three multiples of 3)

= P(1st is a multiple *and* 2nd is a multiple *and* 3rd is a multiple)

$= \frac{5}{15}$ ← 5 multiples out of 15 numbers

$\times \frac{4}{14}$ ← 4 multiples out of 14 numbers after a successful first draw

$\times \frac{3}{13}$ ← 3 multiples out of 13 numbers after two successful draws

≈ 0.0220

6 Reyn has a carton containing a dozen eggs, 5 of which have a double-yolk. He randomly selects 3 eggs from the carton to bake a cake. Find the probability that:

a all 3 have a double-yolk **b** none of them have a double-yolk.

7 A bag contains three white and five red marbles. Three marbles are selected from the bag without replacement. Determine the probability that:

a they are all red **b** they are all white **c** the first is red and the others are white.

8 A box contains 15 tickets with the numbers 1 to 15 printed on them. If three tickets are drawn from the box without replacement, find the probability that:

a they are all odd **b** they are all even **c** the first two are odd and the third is even.

F PROBABILITIES FROM TREE DIAGRAMS

Tree diagrams can be particularly useful for helping us calculate the probabilities of compound events.

For the archers Sunil and Monika in **Example 7**, let S be the event that Sunil hits the target, and M be the event that Monika hits the target.

The tree diagram below shows the possible outcomes. The probability for each person's possible results are marked on the branches. We calculate the probability of each outcome by **multiplying** the probabilities along its path.

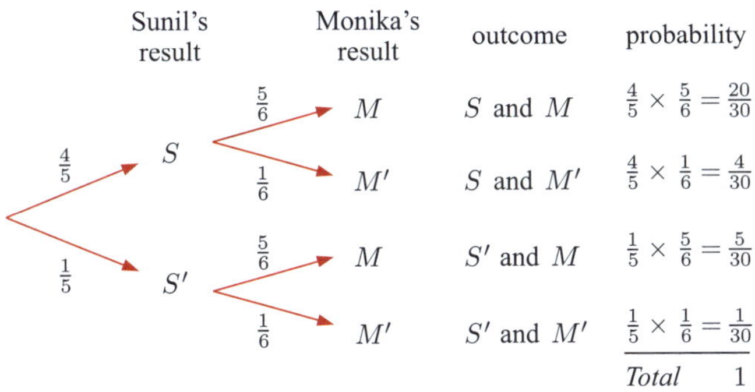

Example 10 Self Tutor

At an airport there are two machines which compare people's faces with their passport photos. 60% of passengers use machine A which recognises 98% of faces successfully.

The remaining passengers use machine B which recognises 97% of faces successfully.

a Draw a tree diagram to illustrate this situation.

b Find the probability that a passenger uses:

i machine A successfully **ii** machine B but is unsuccessful.

a Let A be the event that a passenger uses machine A and S be the event that a passenger's face is successfully recognised.

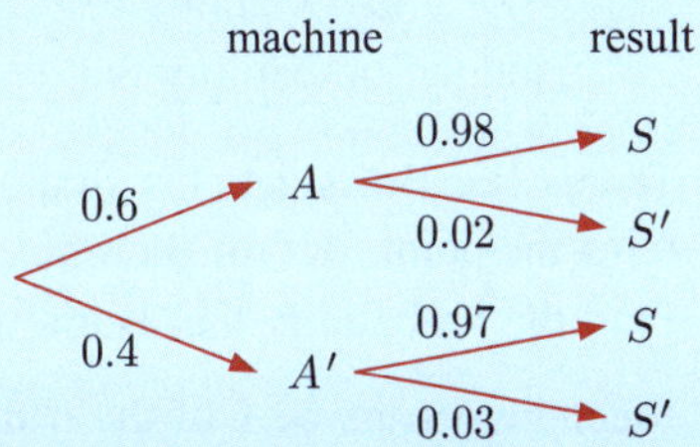

b **i** $\text{P}(A \text{ and } S)$
$= \text{P}(A) \times \text{P}(S)$
$= 0.6 \times 0.98$
$= 0.588$

ii $\text{P}(A' \text{ and } S')$
$= \text{P}(A') \times \text{P}(S')$
$= 0.4 \times 0.03$
$= 0.012$

If an event contains more than one outcome, we need to **add** the probabilities for each outcome in the event.

Example 11 — Self Tutor

Bag A contains 4 red jelly beans and 1 yellow jelly bean. Bag B contains 2 red and 3 yellow jelly beans. A bag is randomly selected by tossing a coin, and one jelly bean is removed from it. Find the probability that the jelly bean is yellow.

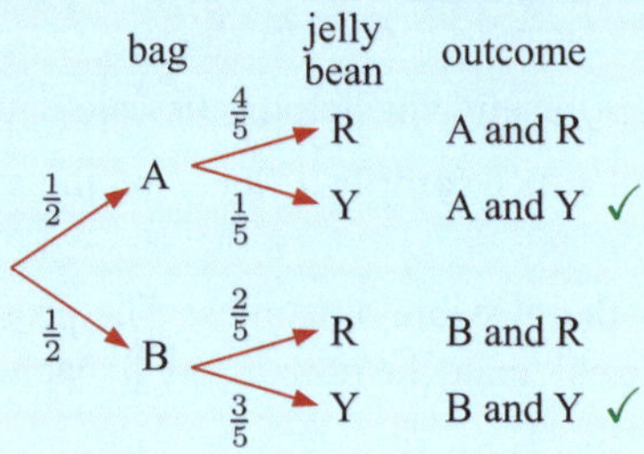

$$\begin{aligned} \text{P(yellow)} &= \text{P(A and Y)} + \text{P(B and Y)} \\ &= \tfrac{1}{2} \times \tfrac{1}{5} + \tfrac{1}{2} \times \tfrac{3}{5} \quad \{\text{branches marked } \checkmark\} \\ &= \tfrac{4}{10} \\ &= \tfrac{2}{5} \end{aligned}$$

EXERCISE 26F

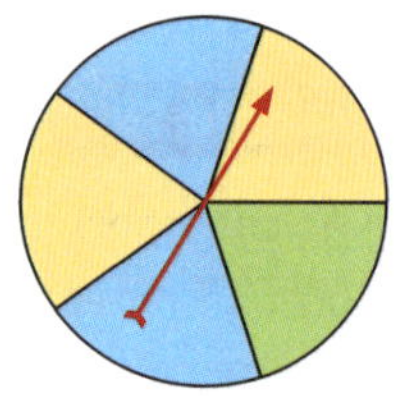

1. Suppose this spinner is spun twice.
   a. Draw a tree diagram to illustrate the sample space.
   b. Find the probability that:
      i. blue appears on both spins
      ii. green appears on both spins
      iii. different colours appear on the two spins
      iv. blue appears on *either* spin.

2. The probability that it will rain tomorrow is $\frac{2}{5}$. If it rains, the probability that Natalie will go for a jog is $\frac{1}{15}$. If it is fine, the probability that she will go for a jog is $\frac{7}{10}$.
   a. Draw a tree diagram to illustrate the sample space.
   b. Find the probability that Natalie will go for a jog tomorrow.

3. Tennis star Boris gets his first serve in play 72% of the time. If he gets his first serve in play, he wins the point 85% of the time. If not, he only wins the point 50% of the time.
   Find the probability that Boris will win the next point he serves.

4. Machine A cans 60% of the fruit at a factory. Machine B cans the rest. Machine A spoils 3% of its produce, while Machine B spoils 4%. Find the probability that the next can inspected at this factory will be spoiled.

5. Box A contains 2 black and 3 white blocks. Box B contains 5 black and 1 white block. A box is chosen by tossing a coin, and one block is taken at random from that box. Find the probability that the block is white.

6 A sporting club runs a raffle with 200 tickets. Two winning tickets are drawn at random, in succession, without replacement.

Adam bought 8 tickets in the raffle. Determine the probability that he:

a wins first prize

b does not win first prize

c wins both prizes

d wins neither prize.

7 Consider the **Opening Problem** on page **430**.

a Copy and complete the tree diagram alongside.

b Find the probability that player A will win the final.

c Who is most likely to win the final?

d Suppose that instead, A played C and B played D in the semi-finals. Who would now be most likely to win the final?

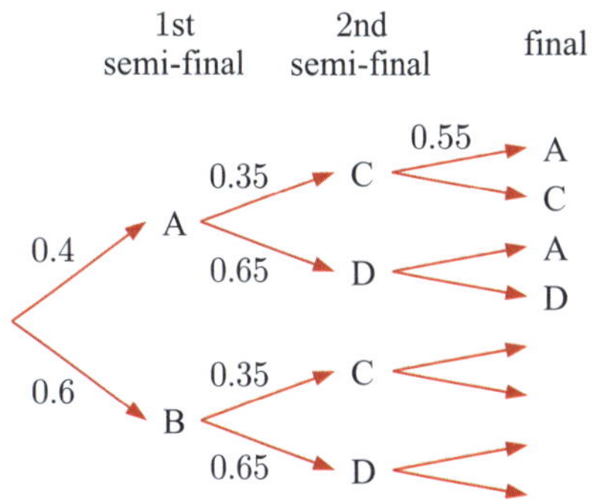

8 In the manufacture of transformers for electric lights, there is a 3% chance that a transformer is made faulty. Benny leaves for a job to install 2 new lights. He carries 3 transformers in his van.

a Draw a tree diagram to illustrate the sample space.

b Find the probability that Benny will not be able to finish the job with the transformers in his van.

ACTIVITY **THE FINALS SERIES**

In many sports competitions, the final is played in a best of 3, 5, or 7 game series, rather than a single match.

In this Activity we consider the probability of one team defeating the other.

ACTIVITY

G EXPERIMENTAL PROBABILITY

In situations where we cannot calculate theoretical probabilities, we can *estimate* the probability of an event by performing many trials of the experiment. This estimate is known as an **experimental probability**.

In a probability experiment:

- the **frequency** of a particular event is the number of times that this event is observed
- the **relative frequency** of an event is the frequency of that event divided by the total number of trials
- the relative frequency is the **experimental probability** of the event.

For example, when tossing a tin can in the air 250 times, it comes to rest on an end 37 times. We say that:

side

end

- the number of trials is 250
- the possible outcomes are *end* and *side*
- the frequency of *end* is 37
- the frequency of *side* is 213
- the relative frequency of *end* $= \frac{37}{250} = 0.148$
- the relative frequency of *side* $= \frac{213}{250} = 0.852$.

In the absence of any further data, the relative frequency of each outcome is our best estimate of the probability of it occurring.

We write $\text{P}(end) \approx 0.148$, $\text{P}(side) \approx 0.852$.

> The accuracy of an experimental probability is improved by increasing the number of trials of the experiment.

EXERCISE 26G

1 A coin falls heads 47 times in 100 tosses. Find the experimental probability that the coin falls:

a heads **b** tails.

2 Clem fired 200 arrows at a target and hit the target 168 times. Estimate the probability that Clem's next shot will:

a hit the target **b** miss the target.

3 Jackson leaves for work at the same time each day. Over a period of 227 working days he had to wait for a train at the railway crossing 58 times. Estimate the probability that on a given day, Jackson will have to wait for a train on his way to work.

4 A goalkeeper saved 49 penalty kicks and conceded 105. Estimate the probability that she will save the next penalty kick.

5 A circular spinner has 3 sectors marked P, Q, and R.
Ravi twirled the spinner 417 times and obtained a Q on 98 occasions.
Shiva twirled the spinner 280 times and obtained a Q on 59 occasions.

a Find the experimental probability that a Q will result when the spinner is spun using:

i Ravi's results **ii** Shiva's results **iii** the combined results.

b Which of the experimental probabilities you have calculated do you expect to be most accurate? Explain your answer.

6 A drawing pin was tossed into the air 600 times. It landed on its back  243 times and on its side  for the remainder.

a Estimate the probability that when tossed once, this drawing pin will land on its:

i back **ii** side.

b Suppose the drawing pin is tossed twice. Estimate the probability that it will land:

i on its back both times **ii** on its side both times.

H PROBABILITIES FROM TABLED DATA

If data is given in a table, we can use relative frequencies to estimate the probabilities of various events.

Example 12 **Self Tutor**

George keeps a record of the number of apples he eats each day.

Estimate the probability that on a randomly selected day, George will eat:

a exactly 1 apple **b** at least 3 apples.

Number of apples	*Frequency*
0	10
1	17
2	12
3	3
4	1

George has recorded data for $10 + 17 + 12 + 3 + 1 = 43$ days.

a $\text{P(1 apple)} \approx \frac{\text{frequency}}{\text{number of trials}}$

$\approx \frac{17}{43}$

≈ 0.395

b $\text{P(at least 3 apples)} \approx \frac{\text{frequency}}{\text{number of trials}}$

$\approx \frac{3+1}{43}$

≈ 0.0930

EXERCISE 26H

1 Kelly recorded the number of times she went jogging in each of the last 25 weeks.
Estimate the probability that next week Kelly will:

a not go jogging
b go jogging 4 times
c go jogging at least twice.

Number of times	*Frequency*
0	4
1	6
2	5
3	4
4	3
5	1
6	2
Total	25

2 Paula recorded the number of phone calls she received over a period of consecutive days.

a For how many days did Paula's survey last?
b How many calls did Paula receive over this period?
c Estimate the chance that on a particular day, Paula will receive:
 i no phone calls
 ii 5 or more phone calls
 iii less than 3 phone calls.

Number of calls	*Frequency*
0	2
1	7
2	11
3	8
4	7
5	4
6	3
7	0
8	1

3 José recorded the lengths of TV commercials in seconds. His results are summarised in the table.
Estimate the probability that a randomly chosen TV commerical will last:

a between 20 and 40 seconds
b at least a minute
c between 20 seconds and a minute.

Length (t seconds)	*Frequency*
$0 \leqslant t < 20$	17
$20 \leqslant t < 40$	38
$40 \leqslant t < 60$	19
$t \geqslant 60$	4

4 This table shows how long Nathan has slept each night recently.

Estimate the probability that tonight he will sleep for:

a between 6 and 7 hours
b at least 7 hours
c between 5 and 8 hours.

Sleep (h hours)	*Frequency*
$5 \leqslant h < 6$	7
$6 \leqslant h < 7$	29
$7 \leqslant h < 8$	46
$8 \leqslant h < 9$	39

Example 13 — Self Tutor

To investigate the breakfast habits of teenagers, a survey was conducted amongst some students from a high school. The results are shown alongside.

Use this table to estimate the probability that a randomly selected student from the school:

a is male
b is male *and* regularly eats breakfast
c is female *or* regularly eats breakfast.

	Eats breakfast Regularly	Irregularly
Male	87	68
Female	53	92

We extend the table to include totals:

	Eats breakfast Regularly	Irregularly	*Total*
Male	87	68	155
Female	53	92	145
Total	140	160	300

a There are 155 males amongst the 300 students surveyed.

$\therefore$ P(male) $\approx \frac{155}{300} \approx 0.517$

b 87 of the 300 students are male and regularly eat breakfast.

$\therefore$ P(male *and* regularly eats breakfast) $\approx \frac{87}{300} \approx 0.29$

c $53 + 92 + 87 = 232$ out of the 300 are female or regularly eat breakfast.

$\therefore$ P(female *or* regularly eats breakfast) $\approx \frac{232}{300} \approx 0.773$

5 310 students at a high school in Wales were surveyed on the question "Do you like watching rugby on TV?". The results are shown in the table.

	Like	Dislike
Junior students	87	38
Senior students	129	56

a Copy and complete the table to include totals.

b Estimate the probability that a randomly selected student:

i likes watching rugby on TV and is a junior student
ii likes watching rugby on TV and is a senior student
iii dislikes watching rugby on TV.

6 A random selection of students in a youth club were asked whether they wore glasses, contact lenses, or neither. The results were further categorised by gender.

	Glasses	Contact lenses	Neither
Male	15	6	26
Female	14	8	31

a How many students were surveyed?

b Estimate the probability that a randomly chosen student in the club:

i wears glasses
ii is female and wears contact lenses
iii is male and wears neither
iv is female or wears glasses.

7 The table alongside describes the types of cars advertised for sale in a newspaper.
Estimate the probability that a randomly selected car for sale:

	Manual	Automatic
Hatchback	26	27
Sedan	30	39
4WD	9	16

a is a sedan **b** is a manual hatchback
c is automatic, but not a sedan.

8 50 people were asked whether they like eating eggs and whether they like eating bacon.
The results were:

- 41 people said that they like eggs
- of the people who like eggs, 26 also like bacon
- of the people who *dislike* eggs, 6 like bacon.

a Construct a two-way table to display this information.
b Estimate the probability that a randomly selected person will:
  i like bacon **ii** like eggs but not bacon **iii** dislike both eggs and bacon.

9 A popular restaurant requires patrons to make reservations for lunch and dinner on weekends.
Over the past month the restaurant received:

- 198 weekend reservations in total
- 102 reservations for Saturday of which 35 were for lunch
- 63 reservations for dinner on Sunday.

a Construct a two-way table to display this information.
b Estimate the probability that the next reservation will be for a:
  i Sunday **ii** Saturday dinner **iii** Sunday lunch.

I EXPECTATION

The probability of an event can be used to predict the number of times the event will occur in a number of trials.

For example, from the golf tournaments last year, it was found that if Rory has a putt from less than 2 metres, the experimental probability that he will hit the ball in the hole is 0.861. If Rory has 62 putts from less than 2 metres in his next tournament, we would *expect* him to make $62 \times 0.861 \approx 53$ of them.

If there are n trials of an experiment, and the probability of an event occurring in each trial is p, then the **expectation** of the occurrence of that event is np.

Example 14 Self Tutor

In one week, 79 out of 511 trains were late to the station at Keswick. In the next month, 2369 trains are scheduled to pass through the station. How many of these would you expect to be late?

The experimental probability of a train being late to Keswick is $p = \frac{79}{511}$.

We expect $2369 \times \frac{79}{511} \approx 366$ of the trains in the next month to be late.

EXERCISE 26I

1 In a particular region in Africa, the probability that it will rain on any one day is 0.177. On how many days of the year would you expect it to rain?

2 At football practice, Tony kicked 53 out of 74 goals from the penalty goal spot. If he performs as well throughout the season and has 18 attempts to kick penalty goals, how many would you expect him to score?

3 When a paper plane was thrown 50 times, it landed the right way up 38 times. If the plane is thrown 140 more times, how many times would you expect it to land the right way up?

4 A hat contains three yellow discs and four green discs. A disc is drawn from the hat, its colour recorded, then the disc is returned to the hat. The procedure is performed 350 times. How many times would you expect a green disc to be drawn?

5 The table alongside shows the sales for a coffee truck in one morning. In total, 80 coffees were sold.

Size	*Frequency*	*Relative frequency*
Small	34	
Medium	28	
Large		

a Copy and complete the table.

b The truck sold 720 coffees in total for the week. How many of these coffees would you expect to be large?

6 The pair of spinners alongside are spun 600 times. How many times would you expect the sum of the spins to equal 5?

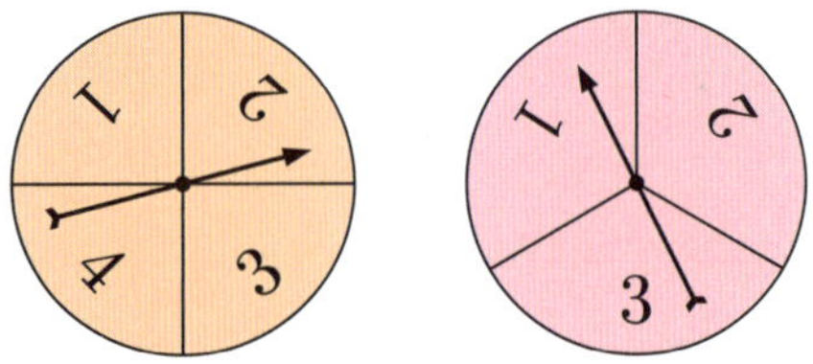

7 200 people are each given a bag containing 2 red balls and 3 green balls. Each person draws two balls from their bag without replacement. How many people would you expect to draw:

a two red balls **b** two green balls?

8 When Stephanie runs the hurdles, she has a 30% chance of knocking at least one hurdle. If she knocks a hurdle, she has a 20% chance of breaking 14 seconds. If she does not knock a hurdle, she has a 65% chance of breaking 14 seconds. If Stephanie runs the hurdles 28 times in a season, estimate the number of times she will break 14 seconds.

9 Each hockey game, the probabilities of Gemma, Tiffany, and Melanie scoring a goal are 0.4, 0.15, and 0.2 respectively. Over a 40 game season, in how many games would you expect:

a Gemma to score a goal **b** Melanie, but not Tiffany, to score a goal

c none of them to score a goal **d** exactly one of them to score a goal?

GLOBAL CONTEXT GENETICS

GLOBAL CONTEXT

Global context: Identities and relationships

Statement of inquiry: Knowledge of probability can help us understand the natural world.

Criterion: Communicating

MULTIPLE CHOICE QUIZ

QUICK QUIZ

REVIEW SET 26A

1 Bags A, B, and C contain green and yellow tickets. A bag is selected and then a ticket is taken from it. Draw a tree diagram to illustrate the sample space. Highlight the outcome corresponding to the event "a green ticket from bag C".

2 A farmer fences his rectangular property into 9 rectangular paddocks as shown.

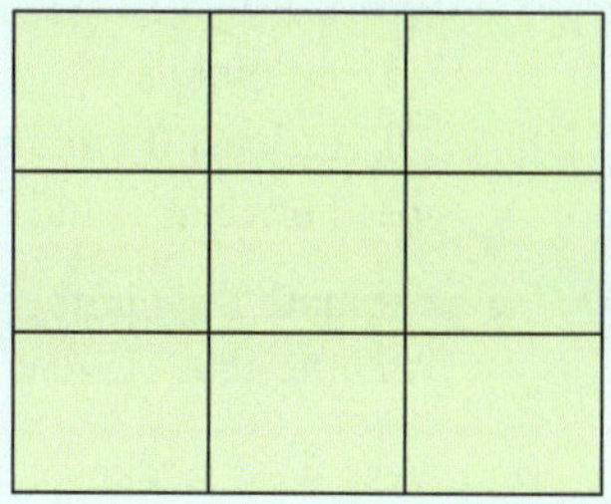

A paddock is selected at random. Find the probability that it has:

a no fences on the boundary of the property

b one fence on the boundary of the property

c two fences on the boundary of the property.

3 Two dice are rolled simultaneously.

a Draw a 2-dimensional grid to illustrate the sample space.

b Hence find the probability of getting:

i two 5s

ii at least one 4

iii a sum greater than 9

iv a sum of 5 or 6.

4 In an office of 15 employees, 10 drive a car to work, 6 use public transport, and 3 use both.

a Display this information on a Venn diagram.

b Find the probability that a randomly selected employee uses neither a car nor public transport to get to work.

5 On a given day, there is probability 0.8 that Kim will walk her dog, and probability 0.6 that Tam will walk her dog. Find the probability that on a particular day:

a both girls walk their dog

b neither girl walks their dog.

6 70% of the spectators at a football match are adults. 80% of the adult spectators support the home team, and 90% of the children spectators support the home team.

Find the probability that a randomly chosen spectator:

a is an adult who does not support the home team

b is a child who supports the home team.

7 Bag X contains 10 mints and 6 chocolate caramels. Bag Y contains 8 mints and 8 chocolate caramels. A bag is randomly chosen by tossing a coin, and a lolly is then taken from it.

a Draw a tree diagram to illustrate this situation.

b Find the probability of selecting:

i a chocolate caramel

ii a mint.

8 A survey of forty five 18 year olds was conducted. It was found that 19 enjoyed camping.

a Estimate the probability that a randomly selected 18 year old likes camping.

b How can the accuracy of this estimate be improved?

9 At his local squash club, Pietro has won 18 out of his last 39 matches.

a Estimate the probability that Pietro will win his next match at the club.

b How many matches out of the next 60 would you expect Pietro to win?

10 Over a 35 day period, Lorna recorded the number of emails she received each day. The data is shown in the table alongside.

Number of emails	*Frequency*
0	2
1	5
2	8
3	12
4	5
5	3

a Estimate the probability that on a randomly selected day, Lorna will receive:

i no emails

ii at least 3 emails

iii between 1 and 5 emails inclusive.

b Estimate the number of days in the next year on which Lorna will receive at least 1 email. Assume that $1 \text{ year} = 365$ days.

11 In a Year 9/10 hockey team, a different player is chosen to captain the team each week. The current captain chooses the captain for the next week.

Year 9 players choose another Year 9 captain 60% of the time, and Year 10 players choose another Year 10 captain 70% of the time.

The current captain is in Year 9. Find the probability that the captain in three weeks' time will be in:

a Year 9 **b** Year 10.

12 The probability of a delayed flight on a foggy day is $\frac{9}{10}$. When it is not foggy the probability of a delayed flight is $\frac{1}{12}$. The probability of a foggy day is $\frac{1}{20}$.

a Draw a tree diagram to show this information.

b Hence find the probability of:

i a foggy day and a delayed flight **ii** a delayed flight

iii a flight which is not delayed.

c Comment on your answers to **b ii** and **iii**.

REVIEW SET 26B

1 Draw a 2-dimensional grid to illustrate the sample space when this spinner is spun twice.

Circle the outcomes corresponding to the event "spinning at least one odd number".

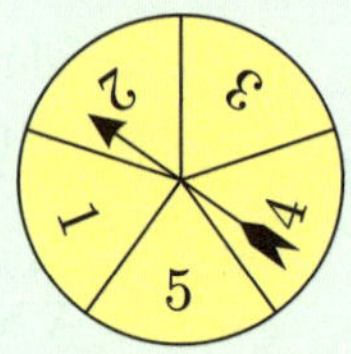

2 A letter of the alphabet is chosen at random. Find the probability that the letter is:

a M or N **b** contained in the word CHOCOLATE

c after the letter S in the alphabet.

3 Marisa and Caleb each roll a die. Find the probability that Marisa rolls a number greater than 1, and Caleb rolls a 3 or a 4.

4 A hat contains 12 tickets with the numbers 1 to 12 printed on them. If two tickets are drawn from the hat without replacement, find the probability that they are both prime numbers.

5 A chess piece is placed on a random square of an 8×8 chess board. A second piece is then placed at random on one of the unoccupied squares.
Find the probability that:

a the pieces lie on the same row

b the pieces lie on the same row or column.

6 Of the 200 visitors to a museum one day, 154 saw the dinosaur exhibit, and 127 saw the Egyptian exhibit. 31 visitors did not see either exhibit.

a Display this information on a Venn diagram.

b Find the probability that a randomly selected visitor saw both exhibits.

7 Bag X contains 3 white and 2 red marbles. Bag Y contains 1 white and 3 red marbles.
A bag is randomly chosen and two marbles are drawn from it.

a Draw a tree diagram to illustrate this situation.

b Find the probability that the two marbles drawn are different colours.

8 Matthew is taking a Mathematics test. There is a 2% chance that his calculator will not work. If his calculator works, Matthew has a 70% probability of passing the test. If his calculator does not work, Matthew has a 55% probability of passing the test.
Find the probability that Matthew passes the test.

9 A sample of teenagers were surveyed about how they most like to spend their spare time. The results of the survey are shown alongside.

Activity	*Frequency*	*Relative frequency*
Playing sport	32	
Social media	51	
Playing games	36	
Watching movies	17	
Total		

a How many teenagers were surveyed?

b Copy and complete the table.

c Estimate the probability that a randomly selected teenager most likes spending their spare time:

i watching movies **ii** playing sport or games.

10 In a test there are 3 multiple choice questions, each with 4 choices. A student randomly guesses the answers to each question.

a Draw a tree diagram to show all possible ways the student can answer the questions, and the probability for the result of each question.

b Find the probability that the student gets:

i the first question right, but the second and third questions wrong

ii exactly 1 of the 3 multiple choice questions correct.

11 The table alongside shows the results from asking the question "Do you like the school uniform?".
If a student is randomly selected from these year groups, estimate the probability that the student:

	Likes	Dislikes
Year 8	129	21
Year 9	108	42
Year 10	81	69

- **a** likes the school uniform
- **b** dislikes the school uniform
- **c** is in Year 8 and dislikes the uniform.

12 Last year, Karen borrowed 57 books from the library. She returned 42 of the books on time.
- **a** Estimate the probability that Karen will return her next library book on time.
- **b** This year Karen borrowed 85 books. How many of the books would you expect her to return on time?

13 Two dice are rolled simultaneously. If the dice are rolled 900 times, how many times would you expect the sum of the rolls to be prime?

Chapter 27

Statistics

Contents:

OPENING PROBLEM

The heights of all students in Year 9 and Year 12 were measured at a particular school. Graphs for the heights of the Year 9 students, the Year 12 students, and both year groups combined are shown below.

A

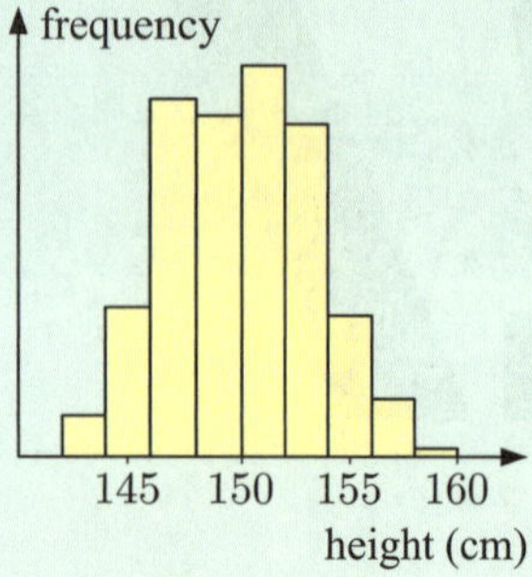

B

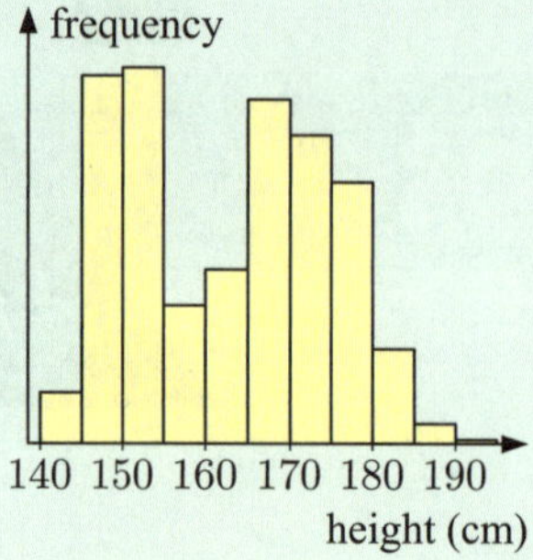

C

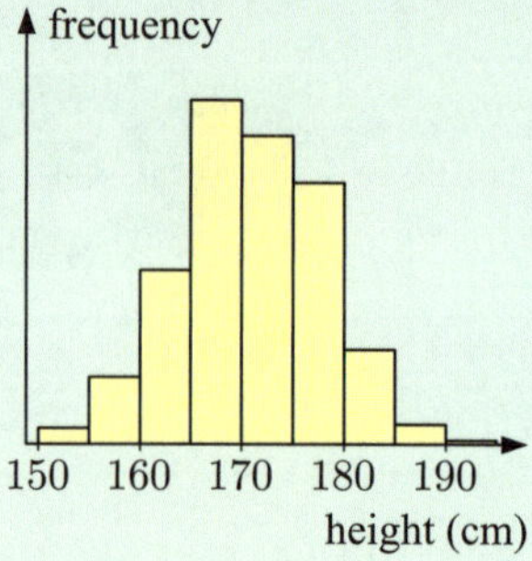

Things to think about:

a Which graph corresponds to the heights of:

i the Year 9 students **ii** the Year 12 students **iii** both year groups combined?

b What words would you use to describe the *shape* of each distribution?

c With this type of graph, do we know the exact height of any particular students?

d How could we estimate the *average* height of the students?

Every day, vast quantities of information are recorded. We record the populations of countries and how they are distributed, the number of children in families, the number of unemployed people, how much food and manufactured goods are produced, daily temperatures, and much more.

Such pieces of information are called **data**. The singular of data is one *datum*.

Statistics is the study of collecting and analysing data.

Government departments, businesses, and scientific researchers all use statistics to help them discover facts, make decisions, and predict outcomes.

In statistics:

- The target **population** is the group of things or people we are interested in finding information about. A **sample** is a subset of the population.
- A **census** is the process of collecting data from *every* member of the population.
  A census is the most accurate way to investigate a population of interest. However, in most situations, it is impractical and expensive to do this.
- A **survey** is the process of collecting data from a sample.
  Conclusions based on data from samples always involves some error. However, we can use the properties of the sample to **estimate** the properties of the population. We can make our estimate more reliable by choosing a sample that is **unbiased** and **sufficiently large**.
  A sample is **biased** if the data is unfairly influenced by the collection process, and is therefore not truly representative of the whole population.

A TYPES OF DATA

When we collect data, we measure or observe a particular feature or **variable** associated with the population. The variables we observe are described as either **categorical** or **numerical**.

CATEGORICAL DATA

A **categorical variable** describes a particular quality or characteristic.
The data is divided into **categories**, and the information collected is called **categorical data**.

Some examples of categorical variables are:

- *continent of birth*: The categories are Europe, Asia, Africa, ...
- *occupation of workers at a hospital*: The categories are doctor, nurse, secretary, cleaner, ...

We have seen in previous years that categorical data can be displayed using a **column graph**, a **bar graph**, or a **pie chart**.

NUMERICAL DATA

A **numerical variable** describes a quantity which takes a value which is a number.
The information collected is called **numerical data**.

Numerical variables can either be **discrete** or **continuous**.

A **discrete numerical variable** takes exact numerical values. It is usually a result of **counting**.

Some examples of discrete numerical variables are:

- *the number of goals scored*: The variable could take the values 0, 1, 2, 3, ...
- *the number rolled on a die*: The variable could take the values 1, 2, 3, 4, 5, or 6.

A **continuous numerical variable** can take any numerical value within a certain range.
It is usually a result of **measuring**.

Some examples of continuous numerical variables are:

- *the weight of pumpkins*: The variable would likely be between 1 kg and 15 kg.
- *the width of a child's hand*: The variable could take values from 2 cm to 10 cm.

EXERCISE 27A

1 Classify each variable as either categorical or numerical:

a the brand of shoes a person wears
b the number of cousins a person has
c voting intention at the next election
d the number of cars in a household
e the temperature of coffee in a mug
f a person's favourite type of apple
g the town or city where a person was born
h the cost of houses on a street.

2 Write down the possible categories for each categorical variable:

a hair colour
b favourite musical instrument
c language spoken at home
d employment status.

3 Classify each numerical variable as discrete or continuous:

a the number of clocks in a house
b the weights of students in a class
c the number of kittens in a litter
d the temperature each day of the week
e the number of leaves on a rose stem
f the amount of soup in a can
g the number of road deaths in a year
h the rainfall each day of the month
i the distances thrown in the javelin
j the number of cars in a car park.

4 Determine whether each numerical variable is discrete or continuous.
If the variable is discrete, list the values the variable is likely to take.
If the variable is continuous, write down the *range* of values the variable is likely to take.

a the number of days in a month
b the length of a person's thumb
c the number of pins knocked down in one frame of ten pin bowling
d the time it takes to run 100 metres
e the number of students in a class
f the electricity used by a person in a year

B DISCRETE NUMERICAL DATA

If there is only a small number of different data values, discrete numerical data can be organised in a **tally and frequency table** and displayed in a **column graph** or **dot plot**.

The most frequently occurring value in a data set is called the **mode**.

Example 1 Self Tutor

The scores of 30 contestants in a trivia quiz are shown alongside.

5 4 5 7 4 5 3 6 5 4
4 5 6 4 6 5 5 6 4 5
5 4 5 5 3 4 6 5 5 4

a Illustrate the data using a:
  i dot plot
  ii column graph.

b State the mode of the data and explain what it means.

a We first organise the data using a tally and frequency table.

Score	*Tally*	*Frequency*
3	||	2
4	卌 ||||	9
5	卌 卌 |||	13
6	卌	5
7	|	1

i **Quiz scores**

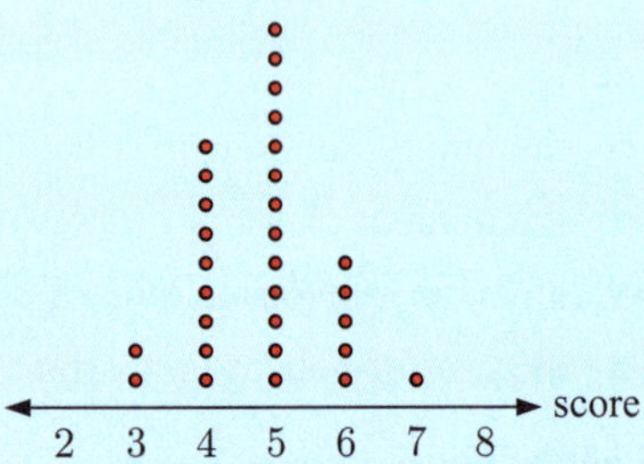

ii **Quiz scores**

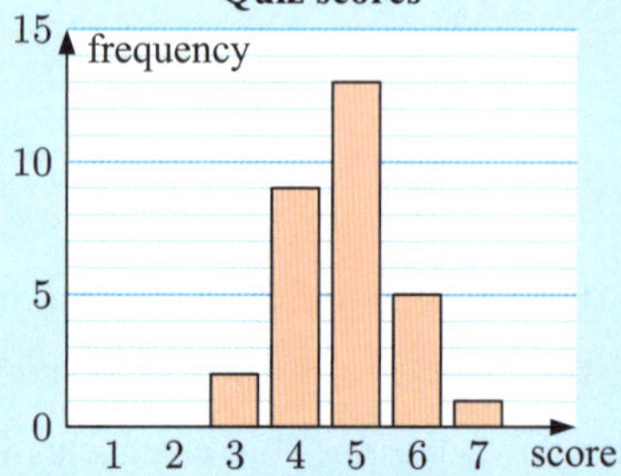

b The mode of the data is 5. This means that the most common score in the trivia quiz was 5.

You can use the **statistics package** or your **calculator** to display discrete numerical data.

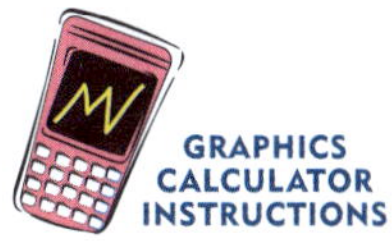

EXERCISE 27B.1

1 When Aimee walked through a shopping mall, she recorded the number of customers in each shop. The data are displayed in a dot plot.

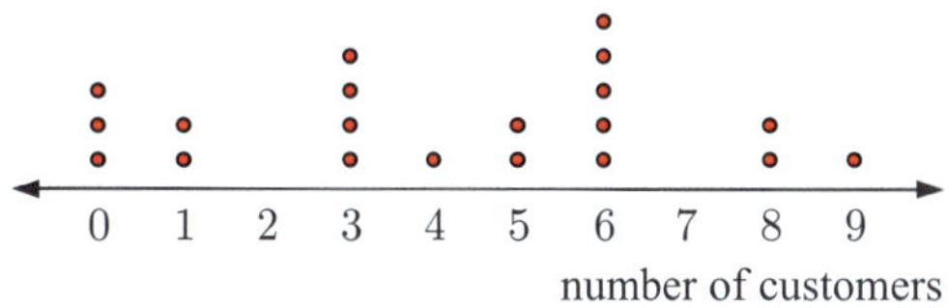

- **a** How many shops did Aimee collect data from?
- **b** How many shops had 3 customers?
- **c** What percentage of shops had:
  - **i** 5 customers
  - **ii** at least 7 customers?
- **d** State the mode of the data and explain what it means.

2 This column graph shows the ages of students in an Irish dancing class.

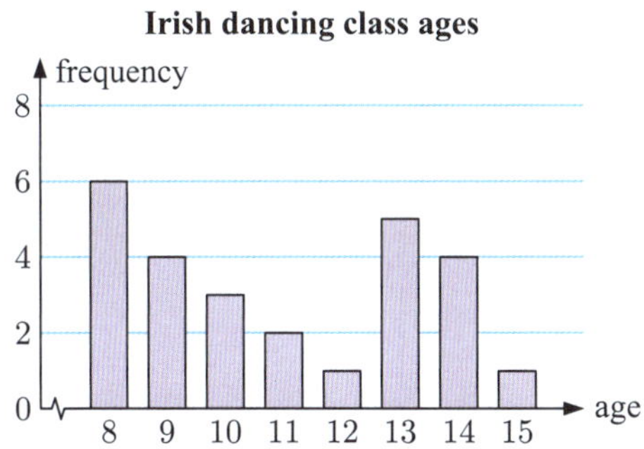

- **a** How many students are in the class?
- **b** State the mode of the data and explain what it means.
- **c** What fraction of the students are younger than 12 years?

3 Stanley works for a ridesharing company. He records the number of trips he completes each day for 28 days:

8 7 8 8 10 4 9 8 10 7 9 7 8 9
5 10 8 8 6 5 8 10 8 9 4 8 7 8

- **a** Draw a dot plot to display the data.
- **b** State the mode of the data and explain what it means.
- **c** On how many days did Stanley complete 10 trips?

4 An ice hockey player scored the following numbers of goals in his last 30 matches:

1 1 3 2 0 0 4 2 2 4 3 1 0 1 0
2 2 5 1 3 7 2 3 1 1 2 1 4 0 3

- **a** Organise the data in a tally and frequency table.
- **b** Construct a column graph to display the data.
- **c** State the mode of the data and explain what it means.
- **d** What was the highest number of goals scored in any game?
- **e** In what fraction of games did the player score at least 3 goals?

GROUPED DISCRETE DATA

If there are a large number of different data values, it may be appropriate to *group* the data into **class intervals**. We can display grouped data using a **column graph** as before, except the columns correspond to the class intervals, rather than individual values.

The **modal class** is the class interval with the highest frequency.

On a column graph, the modal class has the highest column.

Example 2 Self Tutor

A local hardware store is studying the number of people visiting the store at lunch time. Over 30 consecutive weekdays they recorded the data:

37 30 17 13 46 23 40 28 38 24 23 22 18 29 16
35 24 18 24 44 32 54 31 39 32 38 41 38 24 32

a Organise the data in a tally and frequency table using class intervals of width 10.

b Display the data on a column graph.

c State the modal class for the data.

a

Number of people	*Tally*	*Frequency*
10 to 19	𝍸	5
20 to 29	𝍸 𝍷𝍷𝍷𝍷	9
30 to 39	𝍸 𝍸 𝍷	11
40 to 49	𝍷𝍷𝍷𝍷	4
50 to 59	𝍷	1
	Total	30

b **Lunch-time customers**

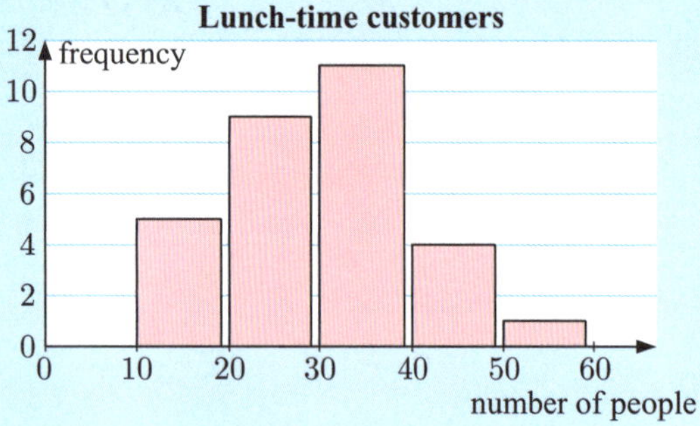

c The modal class is 30 to 39 customers.

EXERCISE 27B.2

1 A furniture production company recorded the number of chairs they completed each day for 26 days:

38 27 29 33 18 22 42 27 36 30 16 23 34
22 19 37 28 44 37 25 33 40 22 16 39 25

a Construct a tally and frequency table for this data using class intervals of width 10.

b Draw a column graph to display the data.

c On how many days were at least 30 chairs made?

d Find the modal class for the data.

2 The values alongside are the test scores (out of 100) for a Science test for 50 students.

92	29	78	67	68	58	80	89	92	69
66	56	88	81	70	73	63	55	67	64
62	74	56	75	90	56	47	59	64	89
39	51	87	89	76	59	72	80	95	68
80	64	53	43	61	71	38	44	88	62

a Construct a tally and frequency table for this data using class intervals of width 10.

b Display the data on a column graph.

c What percentage of the students scored 80 or more for the test?

d What percentage of students scored less than 50 for the test?

e Copy and complete the following:
More students had a test score in the interval ...... than in any other interval.

3 Over a 6 week period, a museum keeps a record of the number of visitors it receives each day. The results are shown alongside.

515	432	674	237	445	510	585	411	605	332	196
432	537	421	318	570	640	298	554	611	458	322
584	232	674	519	174	377	543	630	490	501	139
549	322	612	222	625	582	381	459	609		

a Construct a tally and frequency table for this data using class intervals of width 100.

b Draw a column graph to display the data.

c On how many days did the museum receive at least 500 visitors?

d What is the modal class for the data?

DISCUSSION

How many class intervals should we have?

How should we choose the class interval width?

Do we always need to choose class intervals of equal width?

STEM-AND-LEAF PLOTS

A **stem-and-leaf plot** is a method of writing data in groups without losing the actual data values.

For each data value, the last digit is used as a **leaf**, and the digits before it determine the **stem** on which the leaf is placed.

We include a **scale** to tell us the place value of each leaf.

STATISTICS PACKAGE

Once a stem-and-leaf plot has been constructed, we can write the leaves in ascending order to form an **ordered stem-and-leaf plot**.

HISTORICAL NOTE

Stem-and-leaf plots were invented in 1972 by American mathematician **John Tukey** (1915 - 2000), based on earlier work by **Sir Arthur Bowley** (1869 - 1957) in the early 1900s.

During the 1970s, statistical graphs were not very common because most computers lacked the computing power to generate them.

However, due to the popularity of monospaced (or typewriter) fonts at the time, stem-and-leaf plots were relatively easy to produce.

```
0 | 1
1 | 5 7 8
2 | 1 2 2 3 5
3 | 3 6 7 9
4 | 2
```

Example 3 — Self Tutor

The scores for a test out of 50 were recorded for 36 students.

25	36	38	49	23	46	47	15	28	38	34	9
30	24	27	27	42	16	28	31	24	46	25	31
37	35	32	39	43	40	50	47	29	36	35	33

a Display the data using a stem-and-leaf plot.

b What percentage of students scored 40 or more marks?

a The stems will be 0, 1, 2, 3, 4, 5 to account for scores from 9 to 50.

Unordered stem-and-leaf plot

Stem	Leaf
0	9
1	5 6
2	5 3 8 4 7 7 8 4 5 9
3	6 8 8 4 0 1 1 7 5 2 9 6 5 3
4	9 6 7 2 6 3 0 7
5	0

Scale: 2 | 4 means 24 marks

Ordered stem-and-leaf plot

Stem	Leaf
0	9
1	5 6
2	3 4 4 5 5 7 7 8 8 9
3	0 1 1 2 3 4 5 5 6 6 7 8 8 9
4	0 2 3 6 6 7 7 9
5	0

Scale: 2 | 4 means 24 marks

b 9 students scored 40 or more marks.

$\therefore$ the percentage of students scoring 40 or more marks $= \frac{9}{36} \times 100\% = 25\%$.

EXERCISE 27B.3

1 The stem-and-leaf plot alongside shows the ten pin bowling scores Ron has obtained this year.

Stem	Leaf
7	6 8
8	2 5 8
9	1 1 2 5 9
10	1 2 4
11	0
12	6
13	2 7

Scale: 8 | 2 means 82

a How many games has Ron played?

b Find Ron's:

i lowest score **ii** highest score.

c How many times has Ron scored less than 100?

d What percentage of Ron's scores have been 130 or higher?

2 The following marks were scored for a test out of 50 marks:

47 32 32 29 36 39 40 46 43 39 44 18 38 45 35 46 7 44 27 48

a Construct a stem-and-leaf plot to display the data.

b How many students scored 40 or more marks?

c What percentage of the students scored less than 30 marks?

d A score of 25 or more is required to pass the test. What percentage of the students passed?

3 The total scores for a basketball team in the matches during a season were:

84	108	79	86	91	84	100	92	87	82
95	80	93	102	92	99	76	88	97	93

a Construct a stem-and-leaf plot to display the data.

b In what fraction of matches did the team score more than 90 points?

c In what percentage of games did the team score less than 85 points?

C CONTINUOUS NUMERICAL DATA

When we consider a *continuous* numerical variable, no two data values will be *exactly* the same.

For example, even though we may record the heights of two people as 1.68 m, they are not *exactly* 1.68 m tall, nor are they *exactly* the same height.

Since the data values are all different, the data is organised using **class intervals**.

We commonly use a **frequency histogram** or **histogram** to display continuous data. A histogram is similar to a column graph but because the data is continuous, the columns are joined together.

Column graph

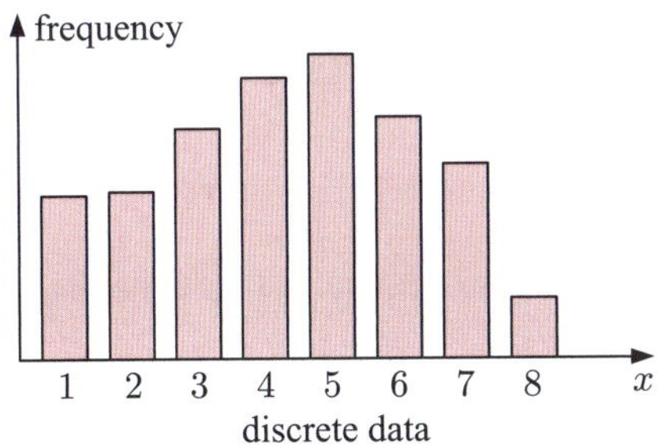

Frequency histogram

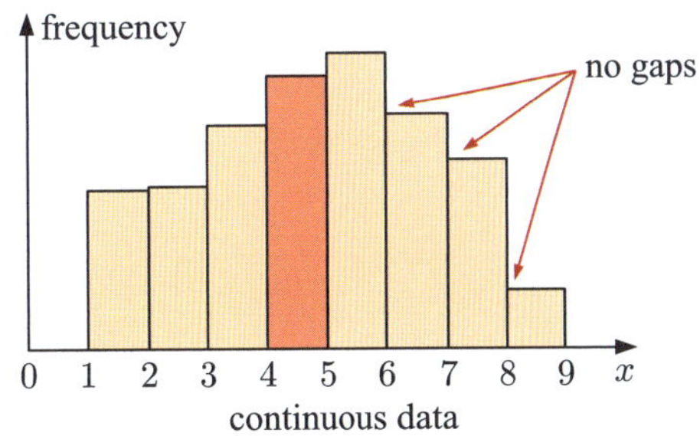

The *sides* of the columns are labelled on the horizontal axis to show the end points of each class interval. For example, in the frequency histogram above, the red shaded column corresponds to data values in the class interval $4 \leqslant x < 5$.

The **modal class** is the class interval with the highest frequency. On a frequency histogram, the modal class has the highest column.

Example 4 Self Tutor

The weights w of 17 pineapples were recorded in kilograms. Each measurement was rounded *down* to the nearest 0.1 kg.

2.1, 3.0, 0.6, 1.5, 1.9, 2.4, 3.2, 4.2, 2.6, 3.1, 1.8, 1.7, 3.9, 2.4, 0.3, 1.5, 1.2

a Organise the data into class intervals.

b Draw a frequency histogram to display the data.

c Find the modal class.

a The lowest weight recorded was 0.3 kg and the highest was 4.2 kg, so we use class intervals of width 1 kg.

Weight (w kg)	*Frequency*
$0 \leqslant w < 1$	2
$1 \leqslant w < 2$	6
$2 \leqslant w < 3$	4
$3 \leqslant w < 4$	4
$4 \leqslant w < 5$	1

b **Weight of pineapples**

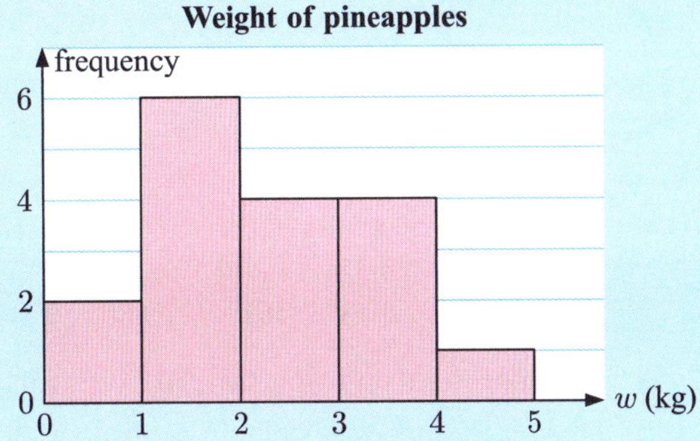

c The modal class is $1 \leqslant w < 2$ kg.

EXERCISE 27C

1 The speeds v of vehicles travelling along a section of highway were recorded and displayed in the frequency histogram alongside.

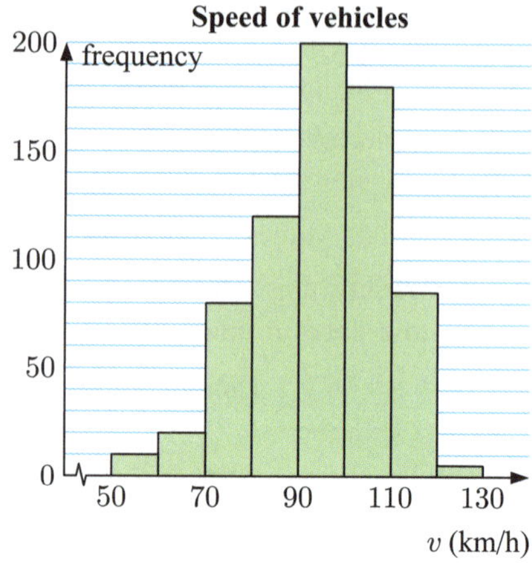

- **a** State the modal class.
- **b** How many vehicles were included in this survey?
- **c** What percentage of vehicles were travelling at speeds between 100 and 110 km/h?
- **d** What percentage of vehicles were travelling at speeds less than 80 km/h?
- **e** The owners of the vehicles travelling at 110 km/h or more were fined $252 each. What amount was collected in fines?

2

Height (h cm)	*Frequency*
$170 \leqslant h < 175$	1
$175 \leqslant h < 180$	8
$180 \leqslant h < 185$	9
$185 \leqslant h < 190$	11
$190 \leqslant h < 195$	9
$195 \leqslant h < 200$	3
$200 \leqslant h < 205$	3

This table shows the heights of players in a volleyball squad.

- **a** Explain why *height* is a continuous variable.
- **b** Draw a frequency histogram for the data.
- **c** What is the modal class? Explain what this means.
- **d** What percentage of players are shorter than 180 cm?

3 The masses m of lambs in a paddock were recorded in kilograms. Each measurement was rounded *down* to the nearest 0.1 kg.

8.8	12.4	14.6	9.7	11.3	18.2	14.3	20.6	17.9	12.5
13.7	16.6	15.8	14.7	14.5	16.0	17.3	19.7	14.5	16.2
18.1	9.8	11.5	13.6	14.1	12.8	18.2	13.9	16.4	19.8

- **a** Organise the data using the class intervals $8 \leqslant m < 10$, $10 \leqslant m < 12$, ..., $20 \leqslant m < 22$.
- **b** Draw a frequency histogram to display the data.
- **c** State the modal class.
- **d** What percentage of lambs were:
  - **i** less than 12 kg
  - **ii** at least 16 kg?

4 The average monthly water usage w of each apartment in a building was recorded in kilolitres. Each measurement was rounded *down* to the nearest 0.01 kL.

9.37	8.22	10.14	8.78	11.13	10.03	9.63	7.87	12.46	10.55
8.62	13.09	9.66	7.46	9.14	8.12	8.77	10.43	9.93	11.58
10.11	9.38	11.36	8.53	12.71	10.82	7.92	9.45	8.80	9.62
7.71	8.11	9.70	10.30	7.01	9.12	11.59	8.62		

- **a** Organise the data using the class intervals $7 \leqslant w < 8$, $8 \leqslant w < 9$, ..., $13 \leqslant w < 14$.
- **b** Draw a frequency histogram to display the data.
- **c** State the modal class.
- **d** What percentage of apartments use an average of at least 10 kL of water per month?

DISCUSSION

You might have noticed in the previous Exercise how measurements are rounded *down*.

Suppose the heights (h cm) of students in a class were measured and sorted into the class intervals $150 \leqslant h < 155$, $155 \leqslant h < 160$, $160 \leqslant h < 165$, and $165 \leqslant h < 170$.

Sam is 164.9 cm tall. When his height was measured, it was rounded to 165 cm because this was the nearest whole number.

- Will Sam's height be placed in the correct class interval?
- Is this problem avoided if measurements are always rounded *down*?

D DESCRIBING THE DISTRIBUTION OF DATA

When we display *categorical* data on a graph, the shape of the graph is not particularly meaningful. We can change the shape by simply changing the order in which the categories are listed.

When we display *numerical* data on a graph, the axis which shows the data values is a *number line*. The *shape* of the graph therefore gives us information about the **distribution** of the data.

Placing a curve over the data can help us to describe its distribution:

- This data is **symmetric** about the mode.

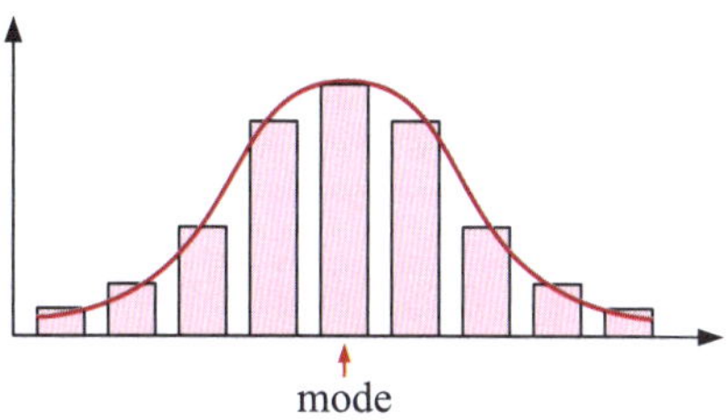

- This data is "stretched" on the right or positive side of the mode. We say the data is **positively skewed**.

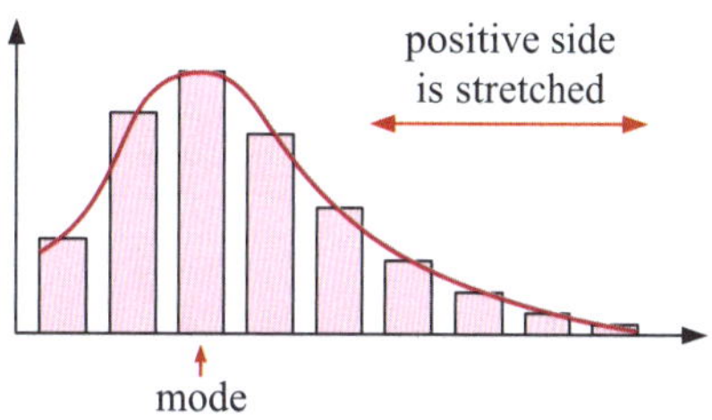

- This data is "stretched" on the left or negative side of the mode. We say the data is **negatively skewed**.

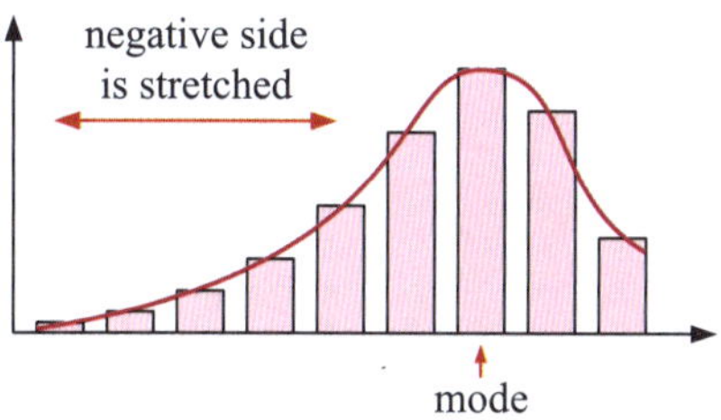

- This data contains two distinct peaks. We say the data is **bimodal**.

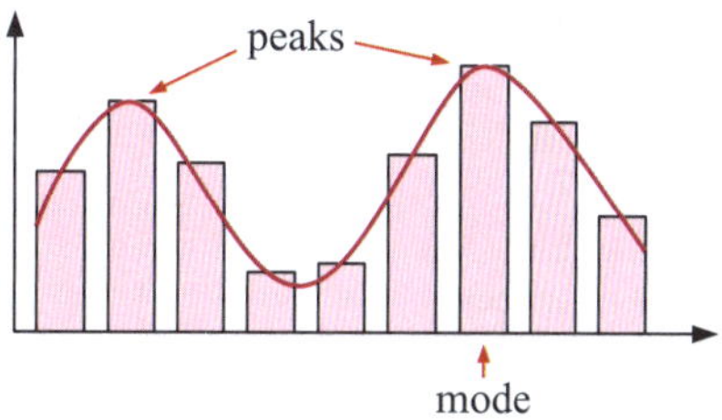

OUTLIERS

Outliers are data values that are either much larger or much smaller than the general body of data.

On a graph, an outlier appears separated from the rest of the data.

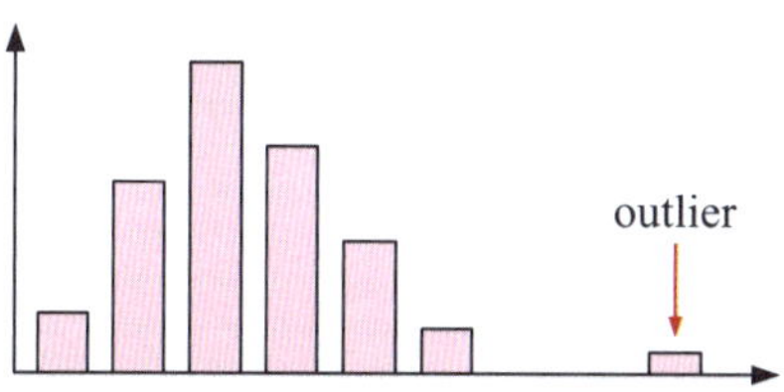

Example 5 **Self Tutor**

A class of students was asked how many houses they had lived in. The results are shown in the table alongside.

a Draw a column graph to display the data.

b Find the mode of the data.

c Describe the distribution of the data.

Number of houses	*Frequency*
1	6
2	8
3	5
4	3
5	2
6	0
7	0
8	1

a

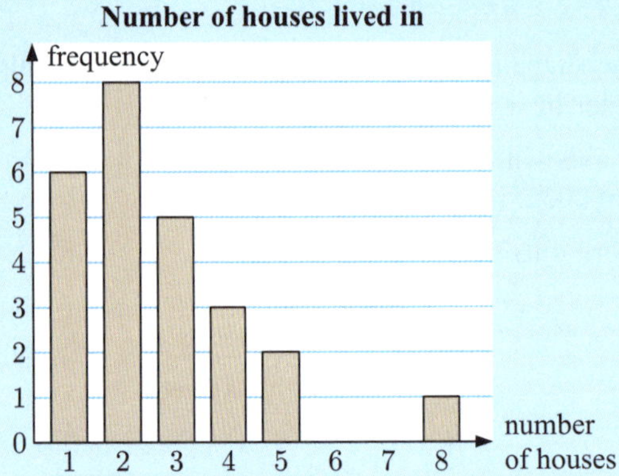

b The mode of the data is 2 houses.

c The data is "stretched" to the positive side, so the data is positively skewed.
The data value "8" is an outlier.

EXERCISE 27D

1 For each graph, describe the distribution of the data:

a

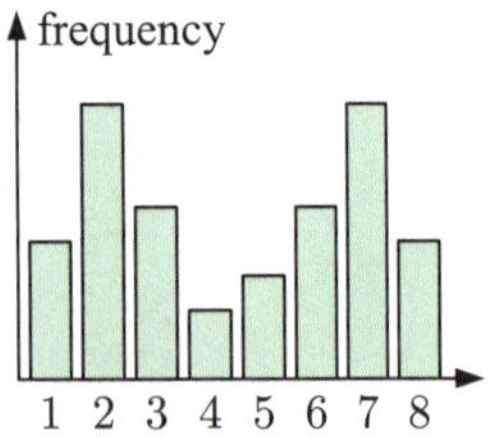

b

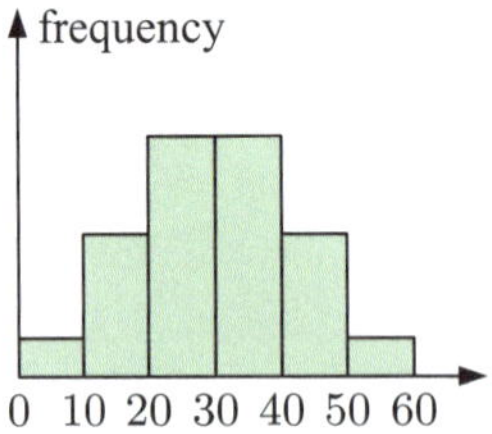

c

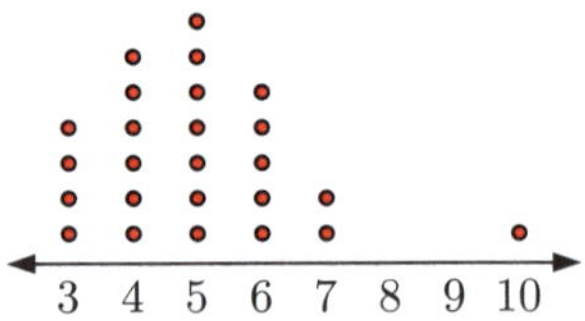

d

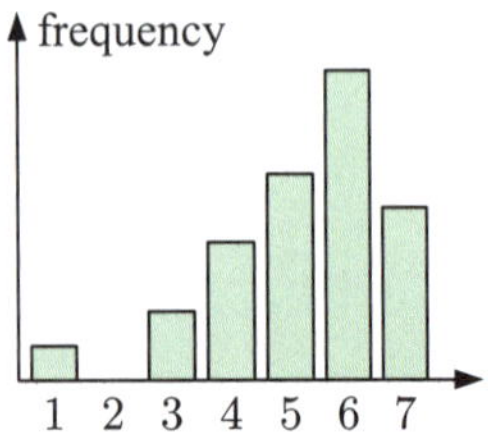

e

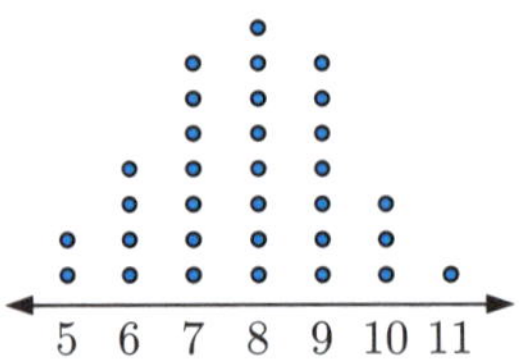

f

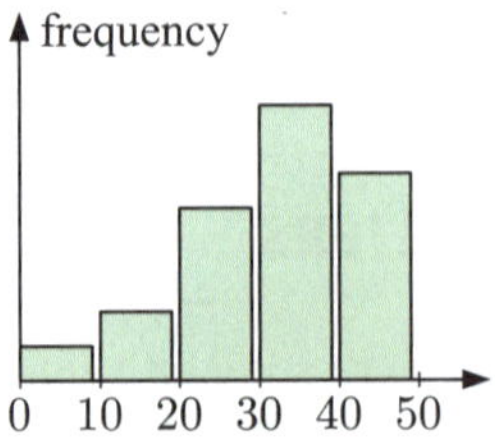

2 For each stem-and-leaf plot, describe the distribution of the data:

a

Stem	Leaf
2	7 8
3	0 1 3
4	2 3 5 5 7
5	1 1 3 7 9 9
6	3 4 6 7
7	0 2 8
8	1

Scale: 2 | 7 means 27

b

Stem	Leaf
6	3 6
7	2 2 6 8 9
8	0 5 7 7 8 9 9
9	1 3 4 6 7
10	2 4 5 8
11	1 4 7
12	4
13	
14	2

Scale: 6 | 3 means 6.3

c

Stem	Leaf
3	4 7 8
4	0 1 1 5 7
5	2 3 6 6 7 8 8
6	1 4 7 9
7	2 7
8	1 4 4 6 9
9	0 2 5 5 6 7 7
10	3 3 4 5 8
11	0 2 3

Scale: 3 | 4 means 34

3 50 boxes of sweets were selected, and the sweets in each box were counted.

a Draw a column graph to display the data.

b State the mode of the data.

c Describe the distribution of the data.

Number of sweets	*Frequency*
25	1
26	0
27	4
28	7
29	10
30	16
31	12

4 A test score out of 60 marks is recorded for a group of 45 students:

34	37	44	51	53	39	33	58	40	42	43	43	47	37	35
41	43	48	50	55	44	44	52	54	59	39	31	29	44	57
45	34	29	27	22	49	41	42	37	42	43	43	45	34	51

a Organise the data in a tally and frequency table using class intervals of width 5.

b Draw a column graph to display the data.

c Describe the distribution of the data.

d An A is awarded to students who scored 50 or more in the test. What percentage of students scored an A?

5 The number of peanuts in a jar varies from jar to jar. A manufacturer samples sixty jars of peanuts and records the number of peanuts in each. The results are:

201	184	213	224	221	193	194	195	178	185	196	210
187	203	207	190	204	192	188	205	207	188	215	185
209	217	189	191	194	194	198	195	183	192	213	224
227	185	198	186	188	213	216	241	182	193	194	203
200	206	210	228	181	196	186	197	199	208	204	189

a Construct a stem-and-leaf plot for the data. Split the stems by using the stem 17 for numbers from 170 to 174, 17* for numbers from 175 to 179, and so on.

b What fraction of the jars had 200 peanuts or more?

c Describe the distribution of the data.

d The manufacturer would like at least 95% of his jars to have within 20 peanuts of the stated number 200. Is this the case for this sample?

6 Lewis is a physical education teacher. He recorded the times each of his students took to sprint 100 m. He drew two histograms to display his results, one for each gender.

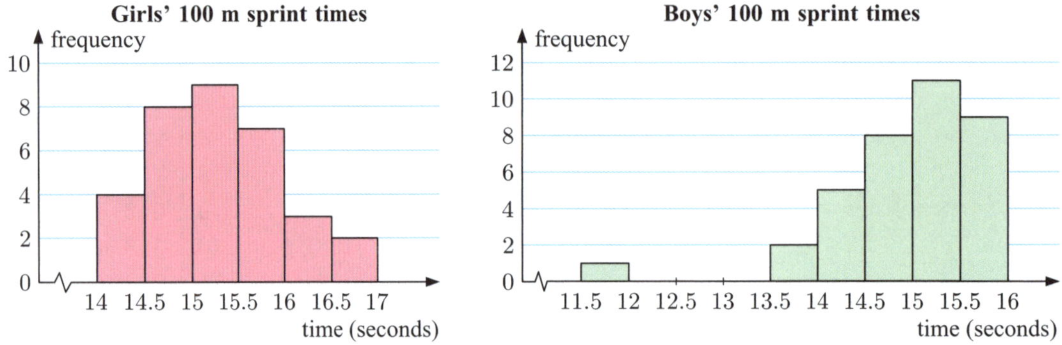

a State the modal class of each distribution.

b Describe the distribution of each data set.

c Which gender had more students with times in the interval $15 \leqslant t < 16$ seconds?

d Which gender generally had faster times for the 100 m sprint?

7 Students in a cooking class were instructed to bake a cake. The table alongside shows the times taken for the students to complete the task.

a How many students were in the class?

b Draw a frequency histogram to display the data.

c Find the modal class.

d Describe the distribution of the data.

e Students who took 2 or more hours to complete the task received a penalty. What percentage of students received a penalty?

Time (t min)	*Frequency*
$50 \leqslant t < 60$	2
$60 \leqslant t < 70$	9
$70 \leqslant t < 80$	11
$80 \leqslant t < 90$	7
$90 \leqslant t < 100$	4
$100 \leqslant t < 110$	8
$110 \leqslant t < 120$	13
$120 \leqslant t < 130$	10
$130 \leqslant t < 140$	5

E MEASURES OF CENTRE

We can gain a better understanding of a data set by locating the **middle** or **centre** of the data, and by measuring its **spread**. Knowing one of these without the other is often of little use.

There are three statistics that are used to measure the **centre** of a data set:

- **mean**
- **median**
- **mode**.

THE MEAN

The **mean** $\overline{x}$ of a data set is the *arithmetic average* of the data values.

$$\textbf{mean} = \frac{\textbf{sum of data values}}{\textbf{number of data values}}$$

The mean is not necessarily a member of the data set.

THE MEDIAN

The **median** of a data set is the *middle* value when the set is written in *order*.

The median splits the data in two halves. Half of the data are less than or equal to the median, and half are greater than or equal to the median.

If there is an *odd* number of data values, there is one middle value which is the median.

If there is an *even* number of data values, there are two middle values. The median is the average of these values.

If there are n data values, the median is the $\left(\frac{n+1}{2}\right)$th data value.

For example:

- When $n = 13$, $\frac{n+1}{2} = 7$, so the median is the 7th ordered data value.
- When $n = 18$, $\frac{n+1}{2} = 9.5$, so the median is the average of the 9th and 10th ordered data values.

MEDIAN

THE MODE

We have seen that for ungrouped discrete numerical data:

The **mode** is the most frequently occurring value in the data set.

If there are two data values which are the most frequently occurring, both values are modes and we say the data is **bimodal**.

If there are more than two data values which are the most frequently occurring, we say the mode is **undefined** and do not use it.

DISCUSSION

In the previous Section, we used "bimodal" to describe a distribution of data with two distinct "peaks".

Discuss the similarities and differences between the use of "bimodal" for *frequency* and the use of "bimodal" for *shape*.

In which way(s) are these data sets "bimodal"?

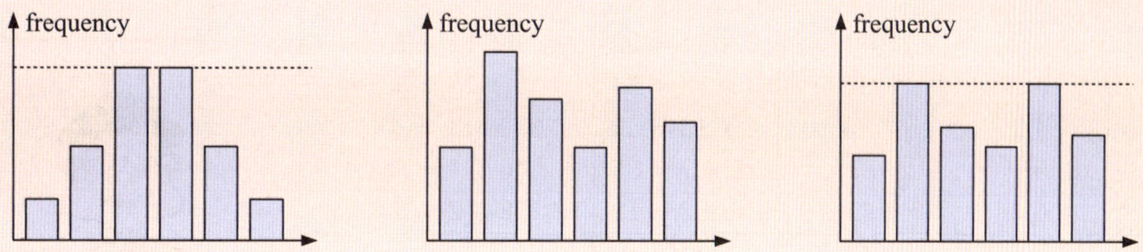

Example 6 — Self Tutor

12 people were shown a series of television commericals, and then were asked to write down the commericals that they remembered. The data below shows how many commercials each person remembered:

7 10 12 10 8 13 10 6 5 11 6 10

For this data set, find the:

a mean **b** median **c** mode.

---

a mean $\overline{x} = \dfrac{7+10+12+10+8+13+10+6+5+11+6+10}{12}$ ← sum of data values / ← number of data values

$= \dfrac{108}{12}$

$= 9$ commercials

b The ordered data set is: ~~5 6 6 7 8~~ 10 10 ~~10 10 11 12 13~~ $\{n = 12, \ \dfrac{n+1}{2} = 6.5\}$

$\therefore$ the median $= 10$ commercials

c 10 is the value which occurs most often.

$\therefore$ the mode $= 10$ commercials

Given a frequency table, we can add:

- a *product* column to record the sum of scores with each particular value, and help us calculate the mean
- a *cumulative frequency* column to record the number of data values less than or equal to that particular score, and help us calculate the median.

Example 7 — Self Tutor

The table alongside shows the number of aces served by tennis players in their first set of a tournament.

For this data set, find the:

a mode **b** median **c** mean.

Number of aces	*Frequency*
1	4
2	11
3	18
4	13
5	7
6	2

Number of aces	*Frequency*	*Product*	*Cumulative frequency*
1	4	$1 \times 4 = 4$	4
2	11	$2 \times 11 = 22$	$4 + 11 = 15$
3	18	$3 \times 18 = 54$	$15 + 18 = 33$
4	13	$4 \times 13 = 52$	$33 + 13 = 46$
5	7	$5 \times 7 = 35$	$46 + 7 = 53$
6	2	$6 \times 2 = 12$	$53 + 2 = 55$
Total	55	179	

a The most common number of aces was 3, so the mode $= 3$ aces.

b $n = 55$, so $\dfrac{n+1}{2} = \dfrac{56}{2} = 28$

The 28th data value is 3, so the median = 3 aces.

c mean $\overline{x} = \dfrac{\text{sum of data values}}{\text{number of data values}}$

$= \dfrac{179}{55}$

≈ 3.25 aces

Click on the icon for instructions on finding the measures of centre using your **calculator**.

Alternatively, you can use the **statistics package**.

GRAPHICS CALCULATOR INSTRUCTIONS

STATISTICS PACKAGE

EXERCISE 27E.1

1 Find the mean, median, and mode of each data set:

a 3, 9, 8, 6, 8, 4, 5, 5

b 17, 23, 14, 37, 25, 30, 22, 27, 31, 36

c 105, 106, 107, 107, 107, 109, 113, 120, 124, 132

d 173, 146, 128, 132, 116, 129, 141, 163, 187, 153, 162, 184

2 The selling prices of the last 10 houses sold in Everton Hills were:

$692 800 $655 200 $822 000 $785 000 $912 800
$664 800 $696 000 $659 000 $662 800 $673 000

a Calculate the mean and median selling prices of these houses, and comment on the results.

b Which measure would you use if you were:

i a vendor wanting to sell your house **ii** looking to buy a house in the district?

3 The weights of Peter's 15 racing pigeons are, in grams:

291.5 331.0 344.1 319.5 350.1 328.5 312.4 296.5
290.8 346.4 311.5 296.6 310.0 282.0 260.4

a Find the mean weight of the pigeons. **b** Find the median weight of the pigeons.

c Explain why the mode is not an appropriate measure of centre for this data.

4 25 children were asked how many birthday presents they received last year. The results are shown alongside.
For this data, find the:

a mode **b** median **c** mean.

Number of presents	*Frequency*
3	2
4	3
5	6
6	8
7	4
8	2

5 Packets of chocolate almonds were opened and their contents counted. The results are shown alongside.
For this data, find the:

a mode **b** median **c** mean.

Number in packet	*Frequency*
32	6
33	8
34	9
35	13
36	10
37	3
38	2

6 Find the mean and median of each data set:

a

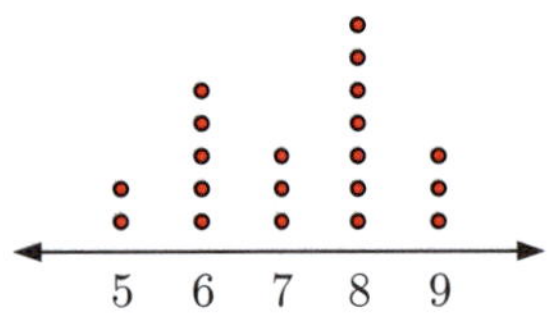

b

Stem	Leaf
2	5 8 9
3	1 3 3 6 8
4	0 5 6
5	1 3
6	0 7

Scale: 2 | 5 means 25

7 In a golf tournament, the scores for each player on a par 5 hole are recorded in the column graph.

a Construct a frequency table from the graph.

b How many players were in the tournament?

c Describe the distribution of the data.

d For this data, find the:

i mean **ii** median **iii** mode.

Golf scores

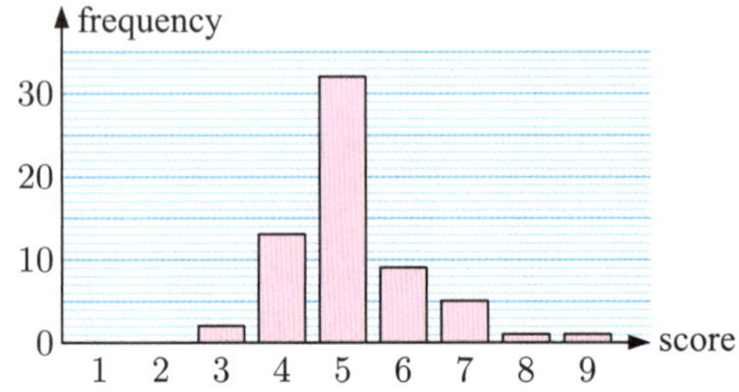

8 Consider these column graphs:

A

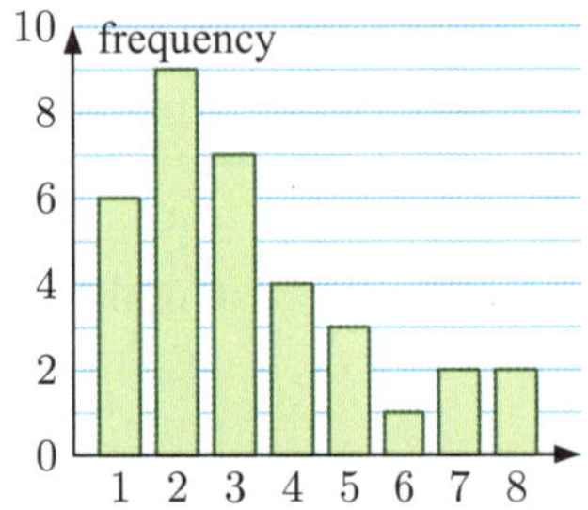

B

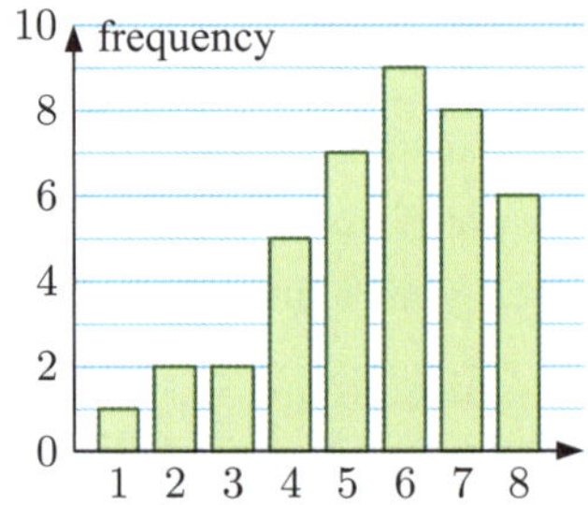

C

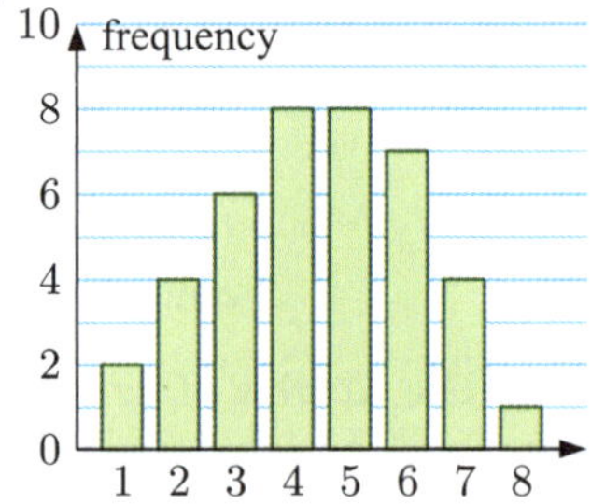

a Describe the distribution of each data set.

b Find the mean, median, and mode of each data set.

c Copy and complete:

i For positively skewed data, the mean is ...... than the median.

ii For negatively skewed data, the mean is ...... than the median.

iii For symmetric data, the mean and median are approximately ......

Example 8 **Self Tutor**

The mean of six scores is 78.5. Find the sum of the scores.

$$\frac{\text{sum}}{6} = 78.5$$

$$\therefore \text{ sum} = 78.5 \times 6$$

$$= 471$$

$\therefore$ the sum of the scores is 471.

9 The mean of 12 scores is 8.8. Find the sum of the scores.

10 While on a camping holiday, Lachlan hiked an average of 12.4 km per day for 8 days. How far did Lachlan hike in total?

11 The mean monthly sales for a department store were \$216 000. Calculate the total sales for the store for the year.

12 A set of data values have sum 1297.7 kg and mean 68.3 kg. How many data values are in the set?

Example 9 **Self Tutor**

Find x if 10, 7, 3, 6, and x have a mean of 8.

There are 5 scores.

$$\therefore \quad \frac{10+7+3+6+x}{5} = 8$$

$$\therefore \quad \frac{26+x}{5} = 8$$

$$\therefore \quad 26+x = 40$$

$$\therefore \quad x = 14$$

13 Find x if:

a 2, 2, 6, 4, 7, and x have a mean of 5

b 9, x, 14, 18, x, x, 8, 10, and 4 have a mean of 11.

14 Over a semester, Jamie did 8 science tests. Each was marked out of 30, and Jamie averaged 25. However, when checking his files, he could only find 7 of the 8 tests. For these he scored 29, 26, 18, 20, 27, 24, and 29. How many marks did he score for the eighth test?

15 Towards the end of the season, a netballer had played 14 matches and scored an average of 16.5 goals per match. In the final two matches of the season the netballer threw 21 goals and 24 goals. Find the netballer's new average.

16 A sample of 12 measurements has mean 16.5, and a sample of 15 measurements has mean 18.6. Find the mean of all 27 measurements.

INVESTIGATION 1 MEASURING THE CENTRE OF A POPULATION FROM A SAMPLE

At a particular school, there are 200 Year 9 students. The data below shows the grade each student received for mathematics in the previous year, converted to a numerical value (A $\equiv$ 5, B $\equiv$ 4, ...).

2	2	4	3	4	3	4	3	3	2	3	5	5	3	3	3	2	3	2	4	3	3	4	4	4
3	2	2	3	3	4	2	4	4	1	3	4	3	2	3	3	4	4	2	3	1	4	3	3	4
3	3	3	3	4	3	3	3	3	3	2	3	3	1	2	3	3	4	4	2	3	2	4	2	4
4	3	2	2	3	2	3	3	3	4	5	3	2	3	2	3	4	3	3	4	3	4	5	2	3
2	4	2	4	3	2	3	3	4	4	3	2	4	4	4	3	4	3	4	2	4	3	2	5	2
4	3	3	3	4	2	2	4	2	4	2	1	3	3	3	2	3	4	2	3	3	3	3	3	4
5	4	1	4	1	3	3	4	4	3	2	4	3	3	2	3	2	3	4	2	2	4	3	5	3
3	5	3	2	3	3	4	3	2	3	3	2	2	3	4	2	3	4	4	4	2	3	4	3	4

What to do:

1 **a** Construct a tally and frequency table for the data.

b For the *population*, find the:

i mean **ii** median **iii** mode.

2 Marianne is a Year 9 student. She wants to investigate the mathematics grades of her peers.

a Explain why Marianne would want to collect data from a *sample* of students, rather than the entire population.

b Click on the icon to run a simulation which randomly selects 10 students from the population. Generate 20 samples and record the mean, median, and mode for each sample in a table like this:

SIMULATION

Sample number	*Mean*	*Median*	*Mode*
1			
2			
⋮			

c Explain why the measures of centre of a sample are *estimates* of the population measures.

3 Marianne is unsure whether collecting data from a sample of 10 students is sufficient to give a good estimate of the population statistics.
Click on the icon to investigate the effect of changing the sample size.

DEMO

a Copy and complete the table alongside.

b Comment on how each sample measure of centre compares to the corresponding population measure as the sample size increases.

c How large do you think Marianne's sample should be? Explain your answer.

Sample size	*Mean*	*Median*	*Mode*
10			
20			
50			
75			
100			

ESTIMATING THE MEAN OF GROUPED DATA

When data is presented in **class intervals**, either for grouped discrete data or for continuous data, the actual data values are not known. This makes it impossible to calculate the exact mean of the data set.

To *estimate* the mean of the data, we use the **midpoint** of an interval to represent all of the data values within the interval. For example:

- For the discrete data class interval 20 to 29, the interval midpoint is $\dfrac{20+29}{2} = 24.5$.
- For the continuous data interval $20 \leqslant m < 30$, the interval midpoint is $\dfrac{20+30}{2} = 25$.

Example 10 **Self Tutor**

The speeds of 129 cars were measured at a particular point on a country road. They are summarised in the table alongside.

Estimate the mean speed of the cars.

Speed (v km/h)	*Frequency*
$40 \leqslant v < 50$	1
$50 \leqslant v < 60$	3
$60 \leqslant v < 70$	17
$70 \leqslant v < 80$	39
$80 \leqslant v < 90$	48
$90 \leqslant v < 100$	17
$100 \leqslant v < 110$	4

Speed (v km/h)	*Frequency*	*Interval midpoint*	*Product*
$40 \leqslant v < 50$	1	45	45
$50 \leqslant v < 60$	3	55	165
$60 \leqslant v < 70$	17	65	1105
$70 \leqslant v < 80$	39	75	2925
$80 \leqslant v < 90$	48	85	4080
$90 \leqslant v < 100$	17	95	1615
$100 \leqslant v < 110$	4	105	420
Total	129		10 355

$$\therefore \quad \text{mean} = \frac{\text{sum of data values}}{\text{number of data values}} \approx \frac{10\,355}{129} \approx 80.3 \text{ km/h}$$

EXERCISE 27E.2

1 Fifty people were asked: "How many times have you been to a restaurant in the last three months?" The results are given in the table.

Number of times	*Frequency*
0 to 4	3
5 to 9	10
10 to 14	16
15 to 19	14
20 to 24	7

   a Extend the table to include an *interval midpoint* and a *product* column.

   b Estimate the mean of the data.

2 50 adults each kicked a football. The table alongside shows the distances the ball travelled in the air before it bounced.

Distance (d m)	*Frequency*
$10 \leqslant d < 20$	10
$20 \leqslant d < 30$	19
$30 \leqslant d < 40$	14
$40 \leqslant d < 50$	7

   a Extend the table to include an *interval midpoint* and a *product* column.

   b Estimate the mean distance the ball was kicked.

3

Area (A m^2)	*Frequency*
$400 \leqslant A < 500$	5
$500 \leqslant A < 600$	11
$600 \leqslant A < 700$	12
$700 \leqslant A < 800$	10
$800 \leqslant A < 900$	8

The table alongside shows the areas of the blocks of land along a street.

   a Find the modal class.

   b Estimate the mean land block size.

4 The frequency histogram shows the weights of cats at a cat show.

   a Describe the distribution of the data.

   b Construct a frequency table for the data.

   c Estimate the mean weight of the cats at the show.

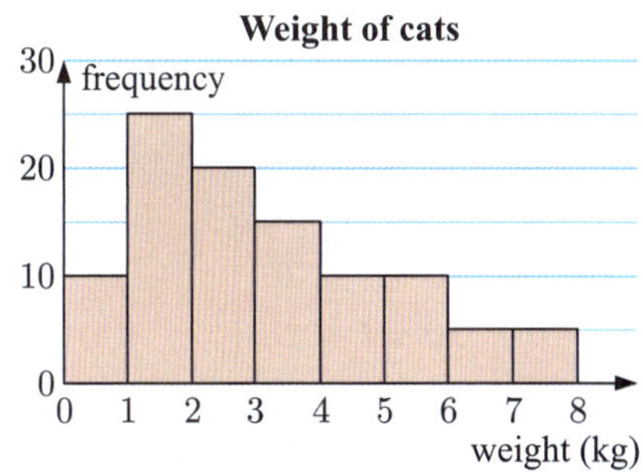

5 The times t that visitors spent in an art gallery were, in minutes:

21.6	35.7	26.2	32.3	22.2	38.0	17.8	31.5	37.5	13.0	29.8	39.1
34.0	12.8	25.4	24.1	33.3	23.3	36.0	14.7	20.1	27.2	32.9	21.6

a Find the exact mean of the data.

b Group the data into the class intervals $10 \leqslant t < 15$, $15 \leqslant t < 20$, ..., $35 \leqslant t < 40$.

c Use your grouped data to *estimate* the mean time spent in the art gallery. Compare your answer with the exact mean.

6 40 students received these marks out of 100 for a Chemistry examination:

70	65	50	40	58	72	39	85	90	65	53	75	83	92	66	78	82	88	56	68
43	90	80	85	78	72	59	83	71	75	54	68	75	89	92	81	77	59	63	80

a Find the exact mean of the data.

b Group the data into class intervals of width 10.

c Hence *estimate* the mean of the data. Compare your answer with the exact mean.

F CUMULATIVE FREQUENCY GRAPHS

When we calculated the median from a frequency table, we used a **cumulative frequency** column to record the number of data values less than or equal to a particular value.

A **cumulative frequency graph** shows the cumulative frequency plotted against the data values.

We can use a cumulative frequency graph to estimate:

- the number of data values less than or equal to a particular value
- the value corresponding to a particular cumulative frequency
- the median of the data set.

Example 11 Self Tutor

The table summarises the times 50 people spent on an exercise bike at a gym in one day.

a Add a cumulative frequency column to the table.

b Draw a cumulative frequency graph for the data.

c Estimate how many people used the bike for:

  i less than 12 minutes **ii** more than 23 minutes.

d Estimate the median time spent on the bike.

Time (t min)	*Frequency*
$0 \leqslant t < 5$	5
$5 \leqslant t < 10$	8
$10 \leqslant t < 15$	10
$15 \leqslant t < 20$	12
$20 \leqslant t < 25$	9
$25 \leqslant t < 30$	6

a

Time (t min)	*Frequency*	*Cumulative frequency*
$0 \leqslant t < 5$	5	5
$5 \leqslant t < 10$	8	$5 + 8 = 13$
$10 \leqslant t < 15$	10	$13 + 10 = 23$
$15 \leqslant t < 20$	12	$23 + 12 = 35$
$20 \leqslant t < 25$	9	$35 + 9 = 44$
$25 \leqslant t < 30$	6	$44 + 6 = 50$

← 23 people used the bike for less than 15 minutes.

b

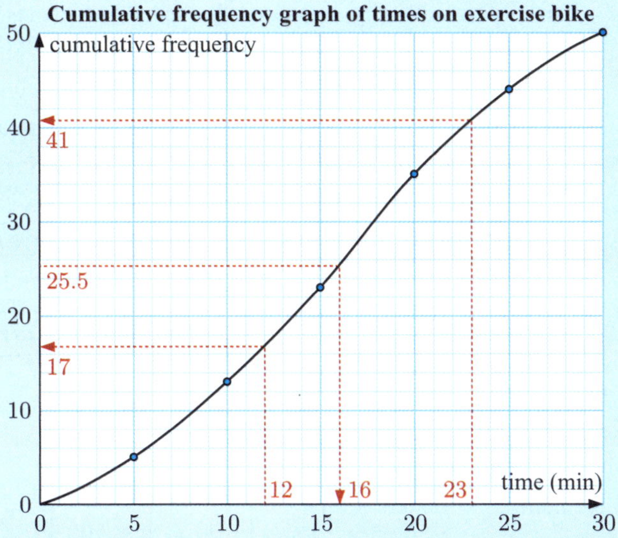

c Reading from the graph, we estimate that:

- **i** 17 people used the bike for less than 12 minutes
- **ii** $50 - 41 = 9$ people used the bike for more than 23 minutes.

d The median is the average of the 25th and 26th times.
Reading from the graph, the median $\approx$ 16 minutes.

EXERCISE 27F

1 This cumulative frequency graph shows the distances jumped by students in a long jump competition.

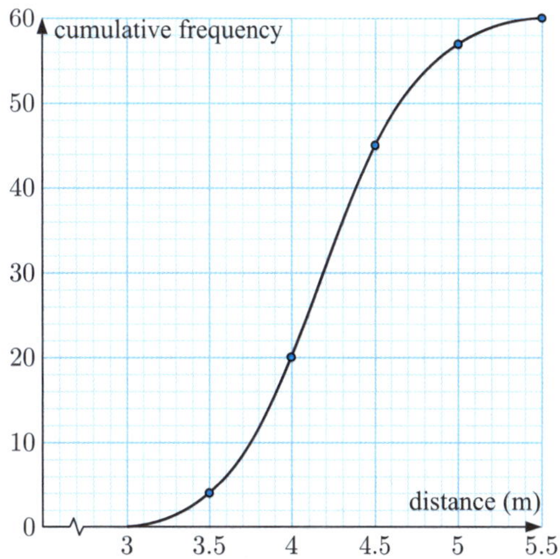

- **a** How many students participated in the competition?
- **b** What can you say about the:
  - **i** shortest jump
  - **ii** longest jump?
- **c** How many students jumped:
  - **i** less than 4 m
  - **ii** more than 4.5 m?
- **d** Estimate the median jump.

2 This table shows the heights of dancers in a ballet class.

Height (h cm)	*Frequency*
$120 \leqslant h < 130$	5
$130 \leqslant h < 140$	7
$140 \leqslant h < 150$	12
$150 \leqslant h < 160$	10
$160 \leqslant h < 170$	6

- **a** Add a cumulative frequency column to the table.
- **b** How many dancers are in the class?
- **c** Draw a cumulative frequency graph for the data.
- **d** Estimate the number of dancers who are:
  - **i** shorter than 155 cm
  - **ii** taller than 145 cm.
- **e** Estimate the median height of the dancers.

3

Speed (v km/h)	*Frequency*
$200 \leqslant v < 220$	12
$220 \leqslant v < 240$	16
$240 \leqslant v < 260$	21
$260 \leqslant v < 280$	18
$280 \leqslant v < 300$	13

An airport authority measured the speeds of planes as they touched down on the runway during a particular day.

- **a** Add a cumulative frequency column to the table.
- **b** Draw a cumulative frequency graph for the data.
- **c** Estimate the number of planes travelling faster than 230 km/h when they touched down.
- **d** Estimate the median speed at which the planes touched down.

4 The table alongside shows the weights of competitors in a boxing tournament.

- **a** Estimate the mean weight of the boxers.
- **b** Draw a cumulative frequency graph for the data.
- **c** Boxers weighing between 79 kg and 91 kg are in the cruiserweight class. Estimate the number of boxers in this class.
- **d** Estimate the median weight of the boxers.

Weight (w kg)	*Frequency*
$50 \leqslant w < 60$	14
$60 \leqslant w < 70$	19
$70 \leqslant w < 80$	28
$80 \leqslant w < 90$	23
$90 \leqslant w < 100$	16
$100 \leqslant w < 110$	18
$110 \leqslant w < 120$	14

DISCUSSION

Which of these cumulative frequency graphs indicates:

- **a** a symmetric distribution
- **b** a positively skewed distribution
- **c** a negative skewed distribution?

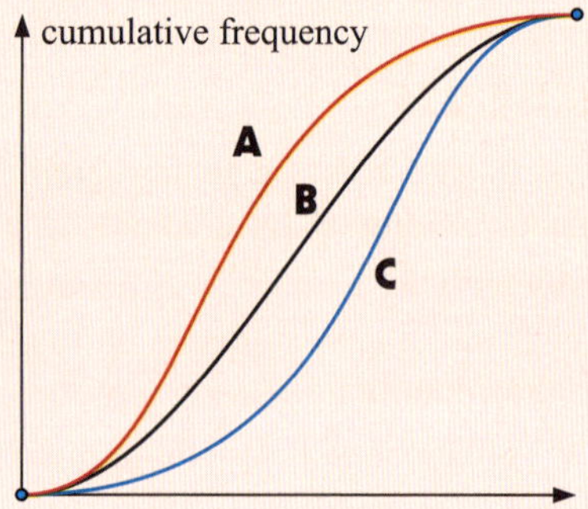

G MEASURES OF SPREAD

Knowing the middle of a data set can be quite useful, but for a more complete understanding of the data we also need to know its **spread**.

To measure the **spread** of a data set, we can calculate its:

- **range**
- **interquartile range**.

THE RANGE

The **range** of a data set is the difference between the **maximum** or largest data value, and the **minimum** or smallest data value.

$$\textbf{range} = \textbf{maximum value} - \textbf{minimum value}$$

Example 12 — Self Tutor

The number of bedrooms in the houses on a street are listed below:

2 3 1 3 2 2 4 3 2 2 5 1 2 1 3 1

Find the range of this data set.

The minimum value is 1, and the maximum value is 5.

$\therefore$ the range $= 5 - 1 = 4$.

THE INTERQUARTILE RANGE

We have seen how the median divides an ordered data set into two halves. By finding the middle value of each half, we divide the original data into *quarters*. These middle values are therefore called **quartiles**.

The middle value of the *lower* half is called the **lower quartile** or $\mathbf{Q_1}$. One quarter or 25% of the data values are less than or equal to Q_1.

The middle value of the *upper* half is called the **upper quartile** or $\mathbf{Q_3}$. Three quarters or 75% of the data values are less than or equal to Q_3.

The **interquartile range (IQR)** is the range of the middle half or 50% of the data.

interquartile range = upper quartile − lower quartile

or $\mathbf{IQR = Q_3 - Q_1}$

Example 13 — Self Tutor

For the data set 7, 3, 4, 2, 5, 6, 7, 5, 5, 9, 3, 8, 3, 5, 6, find the:

a median **b** lower and upper quartiles **c** interquartile range.

The ordered data set is: 2 3 3 3 4 5 5 5 5 6 6 7 7 8 9 {15 data values}

a Since $n = 15$, $\dfrac{n+1}{2} = 8$ $\therefore$ the median $= 5$

b As the median is a data value, we now ignore it and split the remaining data into two:

lower		upper
2 3 3 **3** 4 5 5	5	5 6 6 **7** 7 8 9
Q_1		Q_3

Q_1 = median of lower half = 3

Q_3 = median of upper half = 7

c $\text{IQR} = Q_3 - Q_1 = 7 - 3 = 4$

You can use the **statistics package** or your **calculator** to help you calculate the range and interquartile range of a data set.

Be careful when using other computer packages as they may calculate quartiles differently.

Example 14 **Self Tutor**

For the data set 6, 10, 7, 8, 13, 7, 10, 8, 1, 7, 5, 4, 9, 4, 2, 5, 9, 6, 3, 2, find the:

a median **b** lower and upper quartiles **c** interquartile range.

The ordered data set is:

~~1 2 2 3 4 4 5 5 6~~ 6 7 ~~7 7 8 8 9 9 10 10 13~~ {20 data values}

a Since $n = 20$, $\dfrac{n+1}{2} = 10.5$

$\therefore$ the median $= \dfrac{6+7}{2} = 6.5$

b As the median is not a data value, we split the data into two halves:

lower: ~~1 2 2 3~~ 4 4 ~~5 5 6 6~~ upper: ~~7 7 7 8~~ 8 9 ~~9 10 10 13~~

$Q_1 = \dfrac{4+4}{2} = 4$ $Q_3 = \dfrac{8+9}{2} = 8.5$

c $\text{IQR} = Q_3 - Q_1$
$= 8.5 - 4$
$= 4.5$

EXERCISE 27G

1 Kimmi runs a weekly painting class. The numbers of people attending the class over the last 10 weeks have been:

17 14 21 18 19 26 20 22 16 25

Find the range of the data.

2 For each data set, order the data if necessary, then find the:

i range **ii** median **iii** lower and upper quartiles **iv** interquartile range.

a 4, 5, 5, 7, 8, 10, 10, 12, 15, 17, 18, 21, 23 **b** 5, 7, 7, 8, 10, 11, 13, 14, 14, 15, 18

c 39, 28, 50, 63, 37, 46, 61, 33, 25, 47, 29 **d** 7.5, 13.3, 8.6, 6.5, 9.9, 11.5, 13.4, 6.6, 8.2

3 The following amounts of money were withdrawn from an ATM on a particular day:

$100 $60 $120 $90 $130 $150 $200 $120 $180
$70 $140 $100 $50 $200 $120 $80 $100 $150

Find the:

a median **b** lower quartile **c** upper quartile **d** interquartile range.

4 This stem-and-leaf plot shows the heights of bean seedlings two weeks after the seeds germinated. Find the:

a minimum value **b** maximum value
c range **d** median
e lower quartile **f** upper quartile
g interquartile range.

Stem	Leaf
6	0 3 8
7	0 1 5 6 7
8	1 1 2 4 4 8 9 9
9	0 4 7 9
10	1

Scale: 7 | 5 means 7.5 cm

5 The times spent (in minutes) by 24 people in a queue at a supermarket were:

1.4 5.2 2.4 2.8 3.4 3.8 2.2 1.5 0.8 0.8 3.9 2.3
4.5 1.4 0.5 0.1 1.6 4.8 1.9 0.2 3.6 5.2 2.7 3.0

a Find the:
  i range **ii** median **iii** lower quartile
  iv upper quartile **v** interquartile range.

b Copy and complete:
  i 50% of the waiting times were greater than ...... minutes.
  ii 75% of the waiting times were less than ...... minutes.
  iii The minimum waiting time was ...... minutes and the maximum waiting time was ...... minutes. The waiting times were spread over ...... minutes.

6 The table below shows the number of occupants in the cars passing through a drive-through fast food outlet one night.

Number of occupants	1	2	3	4	5	6
Frequency	32	29	15	17	11	5

a How many cars passed through the drive-through?
b How many people passed through the drive-through?
c Find the range and interquartile range of the data.

7 The intelligence quotient of the members of a book club are:

118 115 123 108 119 98 111 120 104 109 125 116 101 119 121

a Find the range and interquartile range of the data.
b A person with an intelligence quotient of 136 joins the club. Find the new range and interquartile range of the data.
c A person with an intelligence quotient of 89 also joins the club. Find the new range and interquartile range of the data.
d Which measure of spread is most affected by the introduction of outliers?

DISCUSSION

Compare the advantages and disadvantages of the range and IQR. Consider:

- How easy is it to calculate each statistic?
- How *accurately* does each statistic measure the spread of a data set?
- How much do outliers affect the value of each statistic?

H BOX PLOTS

THE FIVE-NUMBER SUMMARY OF A DATA SET

The **five-number summary** of a data set consists of its:

- minimum value (min)
- lower quartile (Q_1)
- median (Q_2)
- upper quartile (Q_3)
- maximum value (max)

BOX PLOTS

A **box plot** or **box-and-whisker plot** is a visual display of the five-number summary of a data set.

In a box plot:

- The rectangular box represents the middle half of the data set. Its length is the interquartile range.
- The lower whisker represents the 25% of the data with the lowest values.
- The upper whisker represents the 25% of the data with the highest values.
- The distance between the ends of the whiskers is the range.

For example, this box plot is for a data set with:

- min $= 1$
- $Q_1 = 4$
- median $= 6.5$
- $Q_3 = 8.5$
- max $= 13$
- range $= 13 - 1 = 12$
- IQR $= 8.5 - 4 = 4.5$

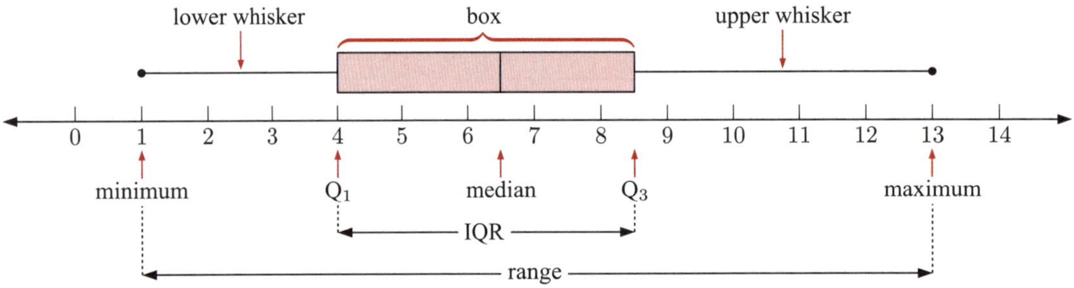

Example 15 — Self Tutor

Consider the data set: 5 6 7 6 2 8 9 8 4 6 7 4 5 4 3 6 6

a Find the five-number summary for the data.

b Find the range and IQR.

c Draw a box plot to display the data.

---

a The ordered data set is:

2 3 4 4 4 5 5 6 6 6 6 6 7 7 8 8 9 {17 data values}

min $= 2$ $Q_1 = 4$ median $= 6$ $Q_3 = 7$ max $= 9$

So, the five-number summary is:

min $= 2$, $Q_1 = 4$, median $= 6$, $Q_3 = 7$, max $= 9$

b **i** Range $=$ max $-$ min $= 9 - 2 = 7$ **ii** IQR $= Q_3 - Q_1 = 7 - 4 = 3$

c

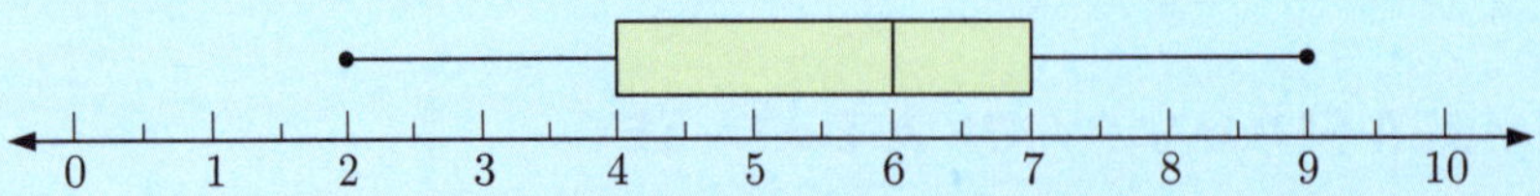

You can use the **statistics package** or your **calculator** to draw box plots.

EXERCISE 27H

1 The box plot summarises the ages of children at a family gathering. State the value of:

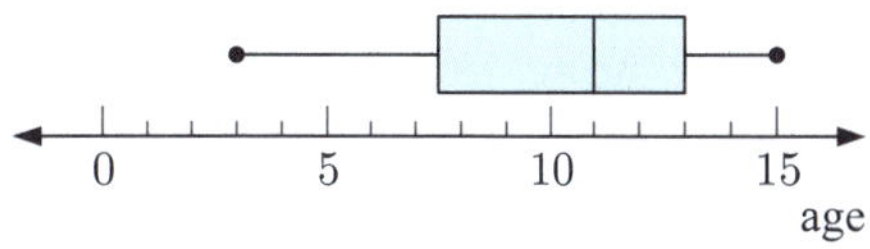

- **a** min
- **b** Q_1
- **c** the median
- **d** Q_3
- **e** max
- **f** the range
- **g** the IQR.

2 This box plot shows the distributions of marks for a class in a test out of 100.

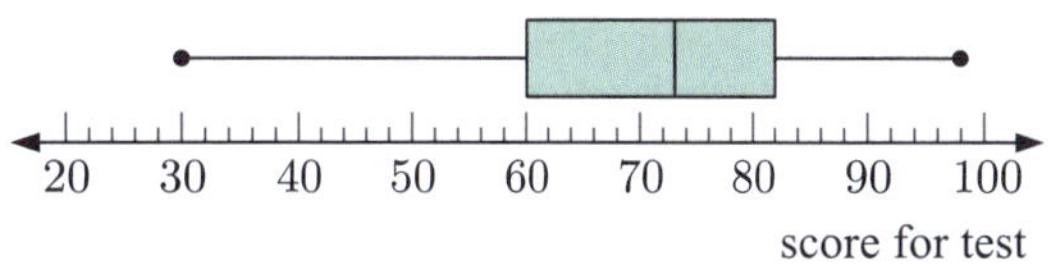

- **a** Find the:
  - **i** highest mark
  - **ii** lowest mark.
- **b** What was the median test score for the class?
- **c** Find the range of marks scored for the test.
- **d** What percentage of students scored 60 or more for the test?
- **e** What was the interquartile range of the test scores?
- **f** If you scored 70 for the test, would you be in the top 50% of students in the class?

3 Consider the data set: 7 13 16 9 12 17 13 10 11

- **a** Find the five-number summary for the data.
- **b** Find the range and IQR.
- **c** Draw a box plot to display the data.

4 Julie examines a new variety of bean. She counts the number of beans in 33 pods. Her results are:

5 8 10 4 2 12 6 5 7 7 5 5 5 13 9 3 4
4 7 8 9 5 5 4 3 6 6 6 6 9 8 7 6

- **a** Find the five-number summary for the data.
- **b** Find the range and IQR.
- **c** Draw a box plot for the data.

5 The data set below shows the number of parking tickets handed out by an inspector each day for a month.

21 18 27 25 16 22 23 19 22 24
15 21 22 26 14 18 17 19 21 14
13 19 24 28 23 25 16 15 20 25

- **a** Find the five-number summary for the data.
- **b** Find the range and IQR.
- **c** Draw a box plot to display the data.

I COMPARING NUMERICAL DATA

Analysing a single set of data can be useful, but to fully understand a situation we often need to **compare** two or more data sets. For example:

- To test whether a medical treatment is effective, we compare the outcomes for patients using the treatment with the outcomes for patients who are not.
- We often want to compare the performances of different groups of individuals.

We can compare data sets:

- **numerically** by calculating measures of:
  - **centre**: mean, median, and mode or modal class
  - **spread**: range and interquartile range.
- **graphically** by displaying multiple data sets on the same set of axes:
  - **back-to-back bar chart**

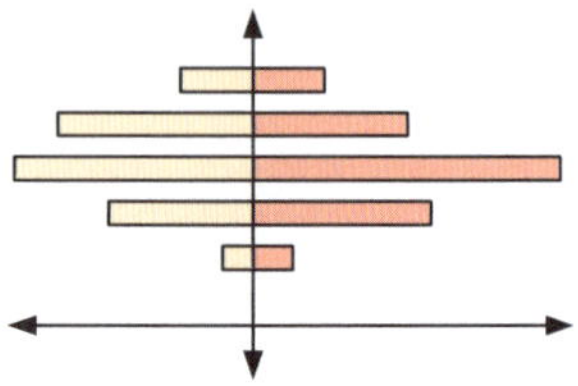

  - **side-by-side column graph**

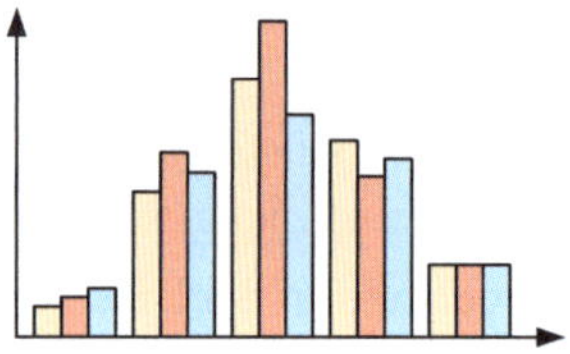

  - **back-to-back histogram**

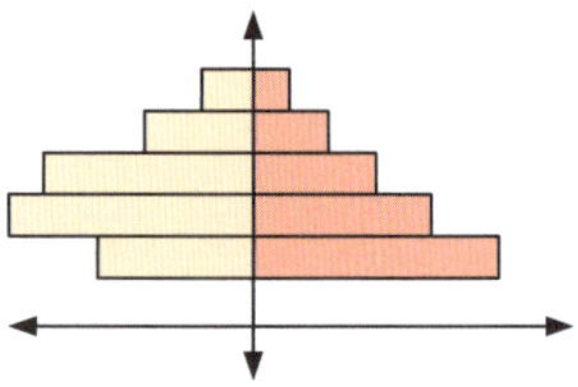

  - **back-to-back stem-and-leaf plot**

	Stem	
	6	1
2 0	7	0 5 6
7 5 3 2 1	8	2 2 4 5 8
9 8 7 6 6 5 3 3 2 0	9	0 3 4 5 5 7 9
7 1	10	6 7 8 8
4	11	2 3 7
	12	6

Scale: 6 | 1 means 6.1

  - **parallel box plot**

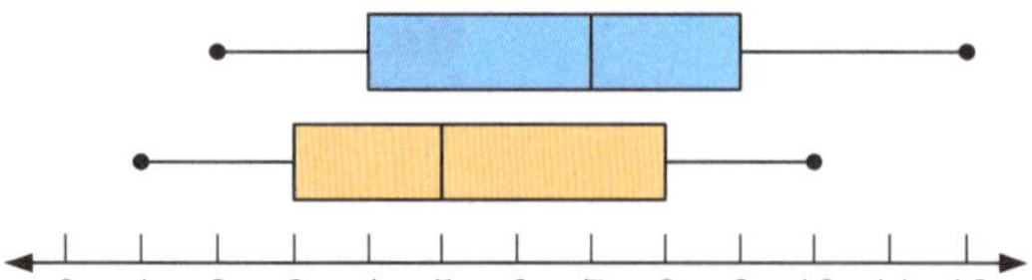

STATISTICS PACKAGE

GRAPHICS CALCULATOR INSTRUCTIONS

DISCUSSION

The parallel box plot is particularly useful for comparing data sets which have different amounts of data. Why is it more useful than other graph types in these cases?

Example 16 **Self Tutor**

Sam and Kate are each given a weekly spelling test out of 20 marks. The results over 35 weeks are displayed in a side-by-side column graph.

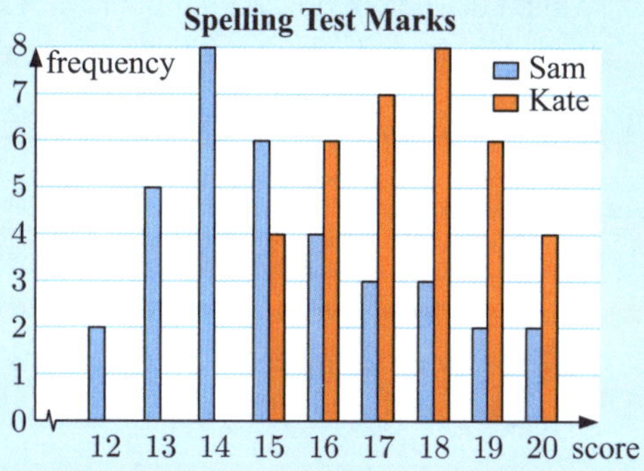

a How many times did each student score full marks?

b Describe the distribution of each data set.

c Find the range of the marks for each student.

d Who generally scored higher marks?

e Who had the greater spread of marks?

---

a Sam scored full marks 2 times. Kate scored full marks 4 times.

b Sam's data is positively skewed. Kate's data is approximately symmetrical.

c The range of Sam's marks is $20 - 12 = 8$ marks.
The range of Kate's marks is $20 - 15 = 5$ marks.

d Kate's scores are generally to the right of Sam's scores on the number line, so Kate generally scored higher.

e Comparing the ranges, Sam had the greater spread of marks.

Example 17 **Self Tutor**

Pedro operates a Mexican restaurant. He is about to launch a new advertising campaign, and he wants to be able to measure its effect. He records the number of customers in his restaurant for thirty days before he starts advertising, and then for thirty days after he starts advertising. He collects the following data:

Before advertising:

74	82	63	77	89	91	74	71	57	62	83	79	75	62	59
84	93	78	68	94	73	77	84	66	72	61	56	76	88	64

After advertising:

93	98	108	84	89	76	92	69	93	85	101	78	67	99	90
88	70	93	104	108	95	80	103	79	92	98	84	106	98	101

a Construct a back-to-back stem-and-leaf plot for this data.

b Without doing any calculations, state which period generally had a higher number of customers.

c Calculate the mean of each data set. Do your results agree with your observations in **b**?

d Discuss the effect of the advertising.

---

a

Before advertising		*After advertising*
9 7 6	5	
8 6 4 3 2 2 1	6	7 9
9 8 7 7 6 5 4 4 3 2 1	7	0 6 8 9
9 8 4 4 3 2	8	0 4 4 5 8 9
4 3 1	9	0 2 2 3 3 3 5 8 8 8 9
	10	1 1 3 4 6 8 8

Scale: 6 | 7 means 67 customers

b From the back-to-back stem-and-leaf plot, the *after advertising* data values are generally higher than the *before advertising* data values.
So, the period after advertising generally had more customers.

c *Before advertising*: $\text{mean} = \dfrac{74 + 82 + 63 + ... + 88 + 64}{30} = 74.4$ customers

After advertising: $\text{mean} = \dfrac{93 + 98 + 108 + ... + 98 + 101}{30} = 90.7$ customers

Since $90.7 > 74.4$, these calculations agree with our observations in **b**.

d The average number of customers has *increased* by approximately 16 customers per night as a result of the advertising.

Example 18 **Self Tutor**

Mrs Freeman has a Year 8 and a Year 12 music class. She asks each student in her classes how many times they have practised their musical instrument in the last month. The responses are:

Year 8 class:	5	2	10	8	3	16	6	10	9	9	3	11	5	7	6
Year 12 class:	13	19	15	10	14	11	15	13	14	16	15	18	12		

a Draw a parallel box plot for the data sets.
b Which class has generally practised their musical instruments more often?
c Which class is more varied in their number of practices?

a We first construct the five-number summary for each data set.

Year 8 class:	min $= 2$	$Q_1 = 5$	median $= 7$	$Q_3 = 10$	max $= 16$
Year 12 class:	min $= 10$	$Q_1 = 12.5$	median $= 14$	$Q_3 = 15.5$	max $= 19$

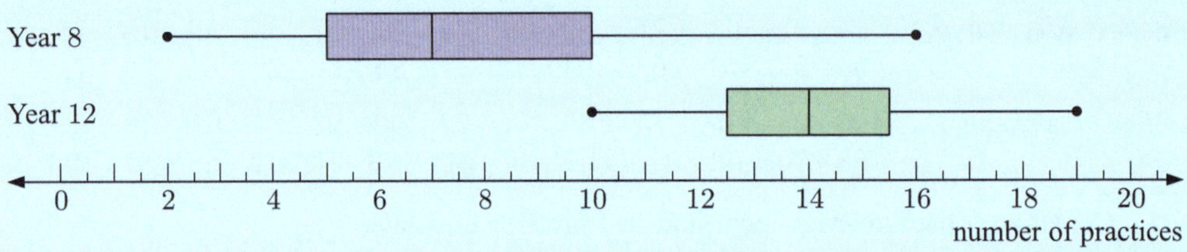

b Comparing each statistic in the five-number summaries, the Year 12 class has generally practised more times than the Year 8 class.
On the parallel box plot, the box for the Year 12 class is much further to the right.

c The IQR for the Year 12 class is 3, compared with 5 for the Year 8 class.
The range for the Year 12 class is 9, compared with 14 for the Year 8 class.
So, the Year 8 class is more varied in their number of practices.
On the parallel box plot, the box and whiskers for Year 8 are longer than those for Year 12.

EXERCISE 27I

1 Olivia randomly selected 50 Grade 9 classes from California and Oregon, and recorded the number of students in each class. The results are displayed in the back-to-back bar chart below.

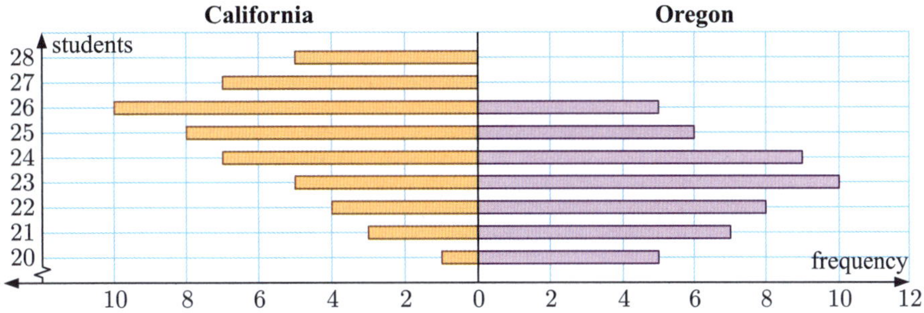

a How many of the selected classes from each state contain at least 25 students?

b Describe the distribution of each data set.

c Copy and complete: "The selected classes from ...... generally have larger class sizes."

2 This back-to-back stem-and-leaf plot shows the cricket scores for a batsman in the 2020 and 2021 seasons.

a In which season was the batsman more consistent?

b In which season did the batsman generally score more runs?

2020 season		*2021 season*
4 3 0	0	
8 8 6 1	1	4
9 6 5 5	2	0 6
9 7 6 2 0	3	1 3 3 6 9
7 4	4	2 2 4 5 8 9
7	5	2 3 6 7
2	6	1 1

Scale: 2 | 0 means 20 runs

3 This parallel box plot shows the numbers of students on school buses A and B over a one month period.

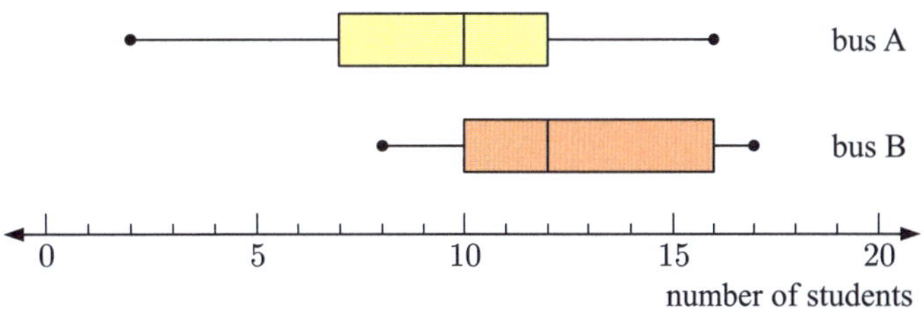

a On what percentage of trips were there 10 or more students on:

i bus A **ii** bus B?

b Which bus generally has the higher number of students?

c On which bus is there a greater spread in the number of students? Explain your answer.

4 The chess ratings for members of two chess clubs are:

The Rookies:	1176	1208	1322	1105	1098	1301	1111	1225	1076
	1099	1233	1176	1153	1258	1358	1236	1083	1184
Best Mates:	1208	1053	1133	1181	1313	1214	1102	1107	1057
	1157	1410	1030	972	1095	1020	1046	1152	1216

a Calculate the mean and median rating for the members of each club.

b Calculate the range and interquartile range for each club.

c Which club generally has higher rated players?

d Which club has a greater spread of player ratings?

5 Maria and Sophie play in the same softball team. They are fierce but friendly rivals when it comes to scoring the most runs. The parallel box plot below shows the numbers of runs scored by each player during a season.

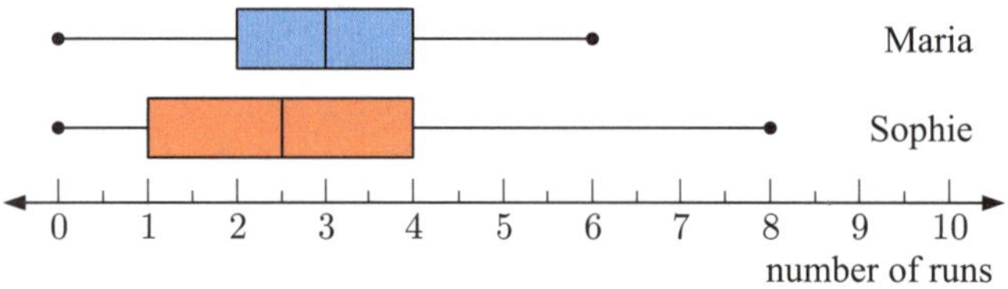

a Find the five-number summary for each player.

b For each player, find the:

  i range **ii** interquartile range.

c Compare the performances of the players.

6 A psychologist wants to know whether single men or women own more pets. She surveys 50 men and 50 women, each of whom live by themselves, and records how many pets they own. The results are shown alongside.

Pets	*Single men*	*Single women*
0	15	7
1	20	10
2	10	15
3	3	10
4	2	5
5	0	3

a Display the data on a back-to-back bar chart.

b Calculate the mean and median of each data set.

c What conclusion could the psychologist make?

7 Vasilios owns a vending machine. He records the number of drinks and snacks sold each week.

Drinks

32	57	44	36	40	28	23	31
22	32	27	33	24	42	25	64
39	60	37	48	30	61	72	67
47	36	52	49	41	52	75	48
39	56	81	55	74	50	84	68

Snacks

44	58	42	50	53	67	49	51
37	28	35	47	60	62	73	56
66	42	23	31	26	37	39	43
70	73	52	74	48	67	53	46
61	77	68	40	31	38	26	50

a Construct a back-to-back stem-and-leaf plot to display the data.

b Describe the distribution of each data set.

c Find the mean and median of each data set.

d Find the range and IQR of each data set.

e Does Vasilios generally sell more drinks or snacks?

f Is the range of sales greater for drinks or snacks?

8 The data below shows the number of test matches played by players from the West Indies and the India cricket teams.

West Indies:	56	9	8	19	53	123	36	38	14	17	13	68	4
India:	19	43	34	2	6	99	2	43	28	25	12	7	61

a Find the five-number summary for each data set.

b Draw a parallel box plot for the data sets.

c Compare the five-number summary statistics for the data sets.

d Compare the spread of the data sets.

9 A group of randomly selected 14 year olds from Australia and Japan were asked how many subjects they studied at school. The results were:

Australia: 6 7 6 5 6 7 7 5 6 7 8 7 6 7 6 8 5 7 6 5 7 6 6
5 7 4 6 7 6 5 8 7 6 7 7 7 6 5 7 4 5 7 8 6 7

Japan: 7 6 6 8 5 7 6 8 7 5 6 6 7 9 6 7 5 8 7 6 5 8 7
5 8 7 6 8 9 8 7 6 8 5 7 8 6 5 7 6 9 8 5 7 7

a Draw a side-by-side column graph to illustrate the data.

b Find the mean, median, and range for each data set.

c Copy and complete: "The selected students from ...... generally studied more subjects."

10 The ages of members from two youth groups are displayed in this side-by-side column graph.

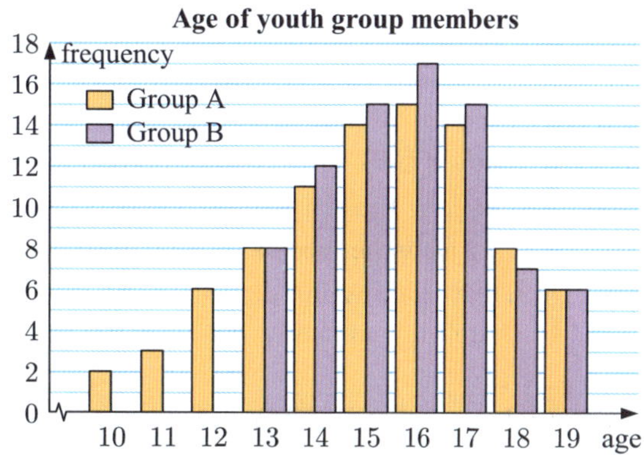

a How many members from each group are aged 14 years or younger?

b Describe the distribution of each data set.

c Find the mean and median for each youth group.

d Find the range and IQR for each youth group.

e Which youth group has the greater range in the ages of its members?

f For which youth group are the members generally older?

11 The marks for boys and girls in a science test are given in the table.

Marks	*Girls*	*Boys*
21 to 30	2	0
31 to 40	6	5
41 to 50	10	8
51 to 60	8	11
61 to 70	4	5
71 to 80	5	6
81 to 90	3	4
91 to 100	1	0

a Draw a back-to-back bar chart to illustrate the data.

b Compare the distributions of the two data sets.

c Estimate the mean mark for each gender, and compare your results.

12 At a local factory there are two machines which place screw caps on plastic bottles. The quality control officer Hugh suspects that machine A causes fewer faulty screw caps than machine B. Each day 1000 bottles are randomly selected from the machines, and their caps are tested for faults. Hugh records the number of faulty caps each day for 50 days:

Machine A: 4 7 5 6 6 5 6 6 7 6 6 5 5 8 6 6 5 7 4 8 9 6 7 5 6
6 8 6 5 9 6 8 7 5 6 6 4 6 5 5 7 5 6 4 6 5 6 6 5 6

Machine B: 5 9 9 7 3 9 8 10 8 6 7 7 9 8 5 9 9 4 11 9 6 9 7 9 8
9 7 9 8 8 9 4 10 8 6 9 7 3 8 7 5 9 8 7 6 8 4 9 10 9

a Clearly state the problem that Hugh wants to solve.

b How has Hugh tried to make a fair comparison?

c Find the five-number summary for each data set.

d Draw a parallel box plot to display the data sets.

e Find the mean for each data set.

f Which machine is more varied in the faults it produces? Explain your answer.

g What conclusion should Hugh make? Justify your answer.

13 The sets of data below show the weights, in kilograms, of 32 adults before and after a lengthy diet and exercise program.

Before program

95	82	96	87	102	78	76	86
108	94	96	88	72	111	105	89
76	80	106	115	92	94	90	103
95	78	83	124	106	83	96	84

After program

94	78	91	80	95	76	71	79
97	87	98	81	67	104	94	90
72	77	101	107	83	89	91	96
97	72	79	111	102	78	91	78

a Organise each data set into a frequency table using the class intervals $60 \leqslant w < 70$, $70 \leqslant w < 80$, ..., $120 \leqslant w < 130$.

b Draw a back-to-back histogram to display the data.

c Describe the distribution of each data set.

d Calculate the mean and median for each data set.

e Do you think the diet and exercise program has been effective in reducing the participants' weights? Explain your answer.

GLOBAL CONTEXT — THE PRINTING PRESS

Global context: Orientation in space and time

Statement of inquiry: Analysing data helps us to identify important events in history.

Criterion: Communicating

GLOBAL CONTEXT

INVESTIGATION 2 — EXPLORING DATA

Click on the icon to obtain a spreadsheet of data collected from students at a high school. The spreadsheet contains both numerical and categorical data.

SPREADSHEET

What to do:

1 Use the data to answer the following questions. You can use the filters at the top of the **Data** sheet to help you select subsets of the data.

a What is the median age of the Year 9 boys?

b What percentage of Year 10 girls chose *red* as their favourite colour?

c What is the mean length of time that Year 11 students spend each day watching television?

d What is the most common method of travelling to school used by Year 12 boys?

e Do Year 8 girls or Year 8 boys generally have more sisters?

f Do Year 12 students or Year 7 students generally send more text messages?

2 The **Pivot table** sheet contains an empty pivot table and chart for the data. Pivot tables can be used to quickly summarise and explore the data's characteristics. Use the pivot table to help you generate:

- **a** a column graph showing the number of students in each year level
- **b** a pie chart showing the favourite colours of the students
- **c** a side-by-side column graph showing the students' favourite colours, separated by gender
- **d** a side-by-side column graph showing the number of sisters the students have, separated by year level.

3 Experiment with different combinations of variables, calculations, and graph types to see what else you can learn about the data.

QUICK QUIZ

MULTIPLE CHOICE QUIZ

REVIEW SET 27A

1 Classify each numerical variable as discrete or continuous:

- **a** the lengths of the hedgehogs in a pet shop
- **b** the number of books sold by a book store each week.

2 A survey is conducted in an airport in Beijing to find out the most popular holiday destinations for Chinese people.

- **a** What type of variable is *favourite holiday destination*?
- **b** Discuss ways in which this sample may be biased.

3 Marion counted the number of cats she saw on the way to school each day for 20 days:

3 2 5 1 1 4 4 0 8 2 0 1 3 7 10 2 4 3 0 3

- **a** Draw a dot plot to display the data.
- **b** State the mode of the data and explain what it means.
- **c** Find the mean and median of the data.

4 A randomly selected sample of small businesses were asked "How many full-time employees are there in your business?" This column graph shows the results.

- **a** How many small businesses were surveyed?
- **b** How many of the businesses had only one or two full-time employees?
- **c** What percentage of the businesses had five or more full-time employees?
- **d** Describe the distribution of the data.
- **e** Find the mean of the data.

Full-time employees

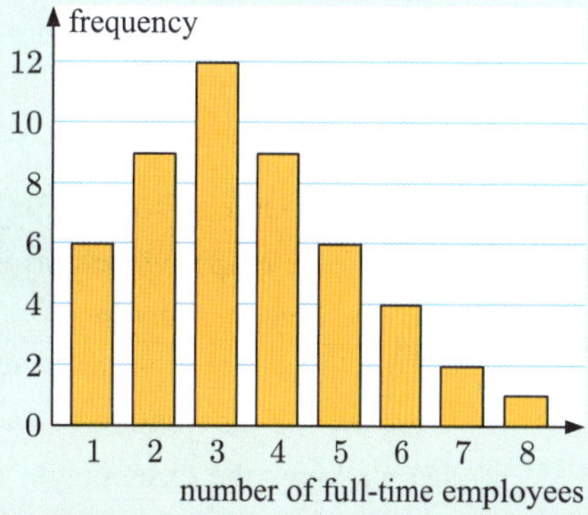

5 The stem-and-leaf plot alongside shows the weights of calves in kilograms.

Stem	Leaf
2	8
3	1 4 9
4	0 3 3 6 7
5	1 4 4 4 5 5 8
6	3 5

Scale: 6 | 3 means 63 kg

a Explain what 5 | 1 means.

b Find the minimum calf weight.

c Find the range of the calf weights.

d What fraction of calves weigh less than 45 kg?

e Describe the distribution of the data.

6 Sixty people were asked: "How many times have you been to the cinema in the last year?". The results are given in the table alongside.

Number of times	*Frequency*
0 to 2	19
3 to 5	24
6 to 8	10
9 to 11	5
12 to 14	2

a Draw a column graph to display the data.

b Can you tell how many people did not go to the cinema at all? Explain your answer.

c What percentage of people went to the cinema between 3 and 8 times?

d Describe the distribution of the data.

7 A transport authority recorded the number of vehicles travelling along a stretch of road each day for 40 days. The data is displayed in this column graph.

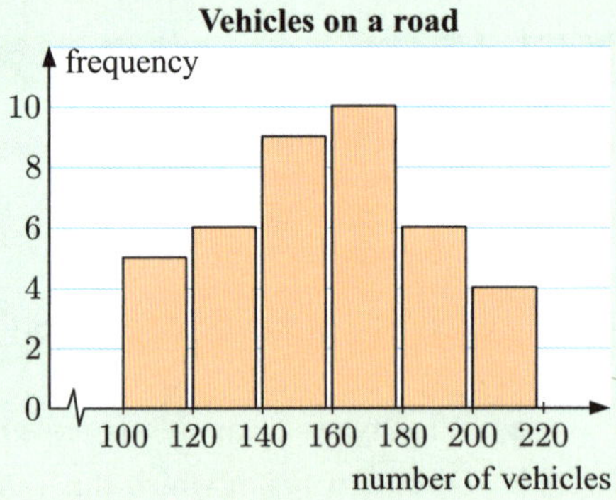

a On how many days were there at least 180 vehicles?

b On what percentage of days were there less than 160 vehicles?

c Find the modal class.

d Estimate the mean number of vehicles on the stretch of road each day.

8 Six scores have mean 8. A seventh score is added, which increases the mean by 1. Find the seventh score.

9 This table summarises the best 50 metre freestyle times of 100 swimmers.

Time (t s)	*Frequency*
$25 \leqslant t < 30$	5
$30 \leqslant t < 35$	17
$35 \leqslant t < 40$	34
$40 \leqslant t < 45$	29
$45 \leqslant t < 50$	15

a Draw a frequency histogram of the data.

b Find the modal class.

c What percentage of swimmers have a best time less than 35 seconds?

d Estimate the mean best time of the swimmers.

10 Consider the **Opening Problem** on page **456**.

a Identify the graph which corresponds to the heights of:

i the Year 9 students **ii** the Year 12 students

iii both year groups combined.

b Describe the distribution of each graph.

c Do we know the exact height of any particular students? Explain your answer.

11 The cumulative frequency graph shows the results for a test out of 60 marks.

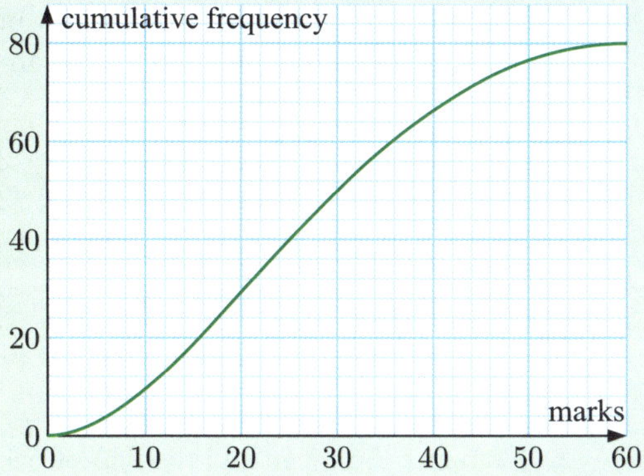

a How many students did the test?

b Estimate the number of students who scored:

i less than 15 marks

ii more than 40 marks.

c Estimate the median test score.

12 Jenny's golf scores for her last 20 rounds were:

90	106	84	103	112	100	105	81	104	98
107	95	104	108	99	101	106	102	98	101

a Find the five-number summary for the data set.

b Find the interquartile range of the data set and explain its meaning.

c Draw a box plot to display the data.

13 32 Year 9 students and 32 Year 10 students were selected at random. They were given a general knowledge test containing 20 questions worth one mark each. The results were:

Year 9:	18	10	11	13	12	14	7	15	14	13	16	12	16	17	15	13
	18	14	12	15	17	11	17	16	15	9	17	12	15	14	13	14
Year 10:	8	14	16	19	12	14	15	13	10	16	15	15	14	18	17	17
	11	14	14	15	17	15	18	16	14	9	13	12	15	13	11	13

a Draw a side-by-side column graph for this data.

b For each data set, calculate the:

i mean **ii** median **iii** range **iv** interquartile range.

c Which group had the greater spread of test scores?

d Which group generally performed better in the test?

14 The parallel box plot below compares times taken by athletes to run 1000 m at the start and at the end of the athletic season. The times are shown in seconds.

a For each set of data, find the:

i fastest time

ii slowest time

iii median time

iv range

v IQR.

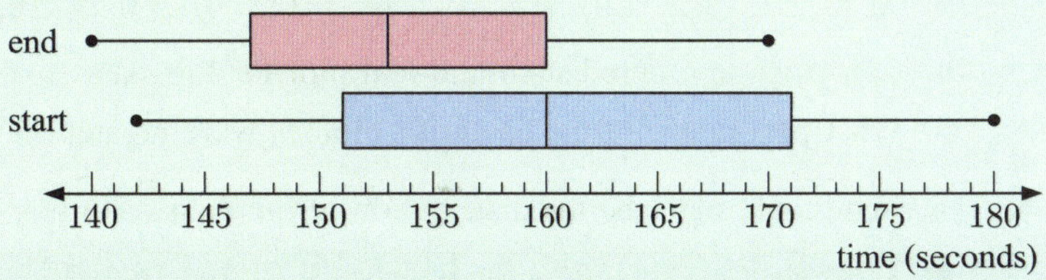

b At which recording time were the times more varied? Explain your answer.

c What percentage of athletes took 160 seconds or more to run the 1000 m:

i at the start of the season **ii** at the end of the season?

d Discuss whether there was an improvement in times during the season.

15 To find out whether doctors or lawyers work longer hours, 20 doctors and 20 lawyers were each asked how many hours they worked in an average month.
The following data were collected:

Doctors:	212	203	228	215	200	194	216	222	210	232
	211	225	209	217	217	199	213	235	219	207
Lawyers:	227	216	238	221	206	228	233	214	229	232
	220	210	197	238	224	226	219	171	225	236

a Draw a back-to-back stem-and-leaf plot to compare the two data sets.

b Are there any outliers in the data?

c Compare the mean and median of each data set.

d Copy and complete:
"From the samples above, it appears that ...... generally work longer hours than ...."

REVIEW SET 27B

1 Classify each variable as categorical, discrete numerical, or continuous numerical:

a the number of kits in a litter of ferrets

b the flavours of milkshakes sold in a café

c the weights of pumpkins at a supermarket.

2 A squad of netballers each had 20 shots at goal. The number of goals scored by each netballer is shown in the table alongside.

a Draw a column graph of the data.

b Find the mean and median of the data.

Goals scored	*Frequency*
12	2
13	1
14	3
15	6
16	5
17	4
18	0
19	2
20	1

3 A microbiologist measured the diameter of a number of bacteria colonies 12 hours after seeding. The results, in cm, were:

0.4, 2.1, 3.4, 3.9, 1.7, 3.7, 0.8, 3.6, 4.1, 0.9, 2.5, 3.1, 1.5, 2.6, 1.3, 3.5

a Construct an ordered stem-and-leaf plot for this data.

b What percentage of the bacteria colonies were greater than 2 cm in diameter?

4 For each graph, describe the distribution of the data:

a

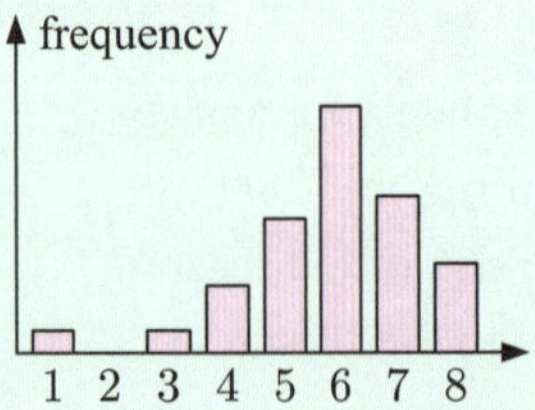

b

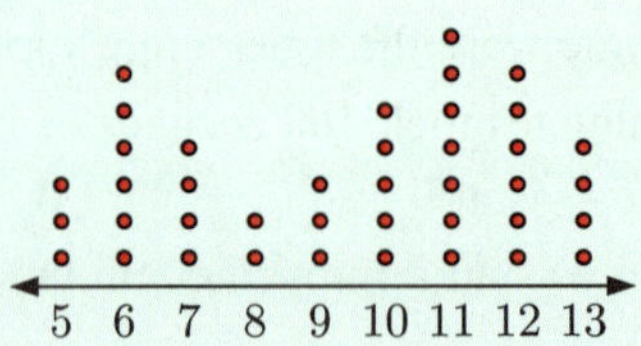

c

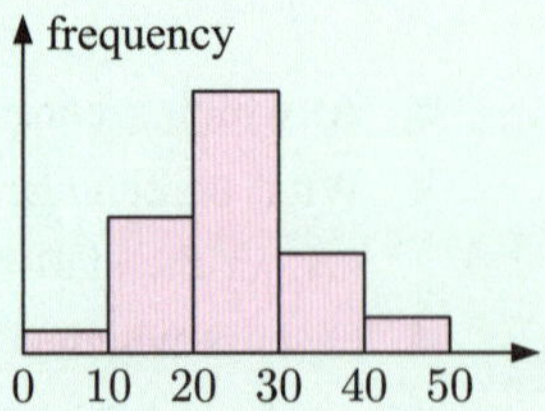

5 For the data set in the stem-and-leaf plot alongside, find the:

1	2 4 4 6 8
2	2 3 7 7 7 9 9
3	0 0 2 4 5 8
4	0 2 3

Scale: 4 | 0 means 40

a minimum value **b** maximum value

c median **d** lower quartile

e upper quartile **f** range

g interquartile range.

6 The number of call-outs each day for a city fire department over a period of 25 days were:

29 14 23 32 28 30 9 24 31 30 18 22 27
15 32 26 22 19 16 23 38 12 8 22 31

a Construct a tally and frequency table for the data using class intervals of width 10.

b Find the modal class for the data.

c Display the data on a column graph.

d On how many days were there at least 20 call-outs?

e On what percentage of days were there less than 10 call-outs?

f Describe the distribution of the data.

7 The numbers of potatoes growing on each of 100 potato plants were recorded and summarised in the table below.

Number of potatoes	0 to 2	3 to 5	6 to 8	9 to 11	12 to 14	15 to 17
Frequency	7	11	25	40	15	2

Estimate the mean number of potatoes per plant.

8 Eight scores have an average of 6. Scores of 15 and x increase the average to 7. Find x.

9 12 of 29 measurements are below 20 cm, and 13 measurements are above 21 cm. Find the median if the other 4 measurements are 20.1 cm, 20.4 cm, 20.7 cm, and 20.9 cm.

10 The table alongside summarises the masses of 50 domestic cats chosen at random.

Mass (m kg)	*Frequency*
$0 \leqslant m < 2$	5
$2 \leqslant m < 4$	18
$4 \leqslant m < 6$	12
$6 \leqslant m < 8$	9
$8 \leqslant m < 10$	5
$10 \leqslant m < 12$	1

a What is the width of each class interval?

b Draw a frequency histogram of the data.

c What is the modal class?

d Estimate the mean mass of a domestic cat.

e Find the percentage of domestic cats with mass 8 kg or more.

11 The times taken for trampers to climb Ben Nevis were recorded. The results are shown alongside.

Time (t min)	*Frequency*
$175 \leqslant t < 190$	11
$190 \leqslant t < 205$	35
$205 \leqslant t < 220$	74
$220 \leqslant t < 235$	32
$235 \leqslant t < 250$	8

a Add a cumulative frequency column to the table.

b Draw a cumulative frequency graph for the data.

c Estimate the median time for the walk.

d Guides on the walk say that anyone who completes the walk in 3 hours 15 min or less is extremely fit. Estimate the number of extremely fit trampers in this group.

12 The data set below shows the ages of children at a scouts camp:

10	11	13	12	11	9	10	10	11	12	12	9	11	10
13	12	11	9	11	10	11	14	8	11	12	14	10	13

a Draw a dot plot to display the data.

b Describe the distribution of the data.

c Find the five-number summary.

d Draw a box plot of the data.

13 Over the course of a 35 game season, basketballers Rhys and Evan kept a record of the number of rebounds they made each game. The results were:

Rhys

8	6	4	9	6	10	9	9	8
7	5	7	8	7	4	7	8	5
8	9	6	3	9	10	8	7	10
9	7	8	8	10	9	6	8	

Evan

7	7	6	8	6	7	6	6	8
6	5	7	7	8	5	6	6	5
7	9	7	6	6	5	8	7	7
8	5	7	7	6	8	7	8	

a Construct a back-to-back bar chart for the data.

b Which player generally collected more rebounds?

14 Frank's company makes batteries. He claims that his brand lasts *twice as long* as those manufactured by a competitor. Thirty of each brand of battery were randomly selected and tested. The results, to the nearest hour, were:

Frank:	135	139	141	145	150	163	135	128	145	143	140	149	136	146	129
	139	142	132	146	146	135	137	144	146	145	128	151	136	154	152
Competitor:	87	89	76	88	86	90	88	91	84	94	90	97	92	93	87
	89	87	92	94	87	88	93	83	89	88	90	85	95	87	96

a Find the five-number summary for each data set.

b Draw a parallel box plot to display the data sets.

c Find the mean of each data set.

d Does Frank's claim appear to be justified? Explain your answer.

15 The heights of players from two rugby squads are displayed on a back-to-back histogram.

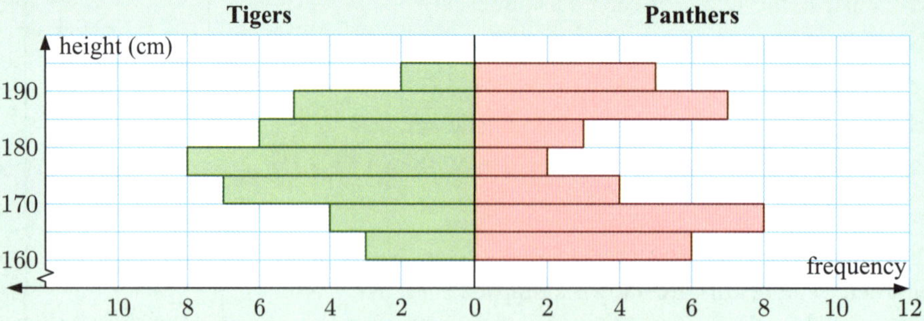

a How many players from each squad are at least 180 cm tall?

b Find the modal class for each squad.

c Describe the distribution of each data set.

d Estimate the mean of each data set.

e Which team do you think has the greater mean height? Discuss the reliability of your answer.

Chapter 28

Networks

Contents:

CHAPTER

Click on the icon to access this chapter

Chapter 29

Non-right angled triangle trigonometry

Contents:

CHAPTER

Click on the icon to access this chapter

Chapter 30

Mathematical logic

Contents:

CHAPTER

Click on the icon to access this chapter

ANSWERS

EXERCISE 1A

1 a 8 b 16 c 27 d 50 e 100 f 135 g 315 h 1188 i 360 j 13 475 k 17 500 l 2100

2 a $3^3 \times 5$ b $2^2 \times 3^2$ c $2 \times 3 \times 5^3$

3 a -9 b 9 c -1 d -1 e 25 f -25 g 16 h -16 i -27 j -27 k 27 l 64

4 a -72 b 108 c -648

5 a 4096 b $-78\,125$ c 32 768 d $-59\,049$ e 5.0625 f ≈ 8.91 g ≈ -0.653 h ≈ -5.87

6 a 2^4 b 2^6 c 2^8 d 2^9 e 2^{10}

7 a 3^2 b 3^4 c 3^5 d 3^7 e 3^{10}

8 2 9 a 9 b 9 c 7

EXERCISE 1B

1 a 2^3 b 3^3 c 5^3 d 2^7 e 7^3 f 3^6 g 19^2 h 11^3

2 a $54 = 2 \times 3^3$ b $108 = 2^2 \times 3^3$
c $360 = 2^3 \times 3^2 \times 5$ d $228 = 2^2 \times 3 \times 19$
e $196 = 2^2 \times 7^2$ f $756 = 2^2 \times 3^3 \times 7$
g $936 = 2^3 \times 3^2 \times 13$ h $1225 = 5^2 \times 7^2$
i $588 = 2^2 \times 3 \times 7^2$ j $945 = 3^3 \times 5 \times 7$
k $910 = 2 \times 5 \times 7 \times 13$ l $1274 = 2 \times 7^2 \times 13$

3 a 2 b 24 c 15 d 2 e 5 f 87

4 $a = 3,\ b = 4$ or $a = 4,\ b = 3$

5 a For each factor of 90, the exponent of:
- 2 can be 0 or 1
- 3 can be 0, 1, or 2
- 5 can be 0 or 1.

There are $2 \times 3 \times 2 = 12$ combinations of these exponents.
$\therefore$ there are 12 factors of 90.

b i $540 = 2^2 \times 3^3 \times 5$ ii 24 factors

c $(a+1) \times (b+1) \times (c+1)$ factors

EXERCISE 1C

1 a -2 b 1 c 24 d -4 e 6 f -10 g -3 h 18 i 18 j 3 k 19 l -3 m 12 n 31 o 44

2 a $6 + 32 \div 8 + 12 = 22$ b $6 - 32 + 8 \times 12 = 70$
c $6 \times 32 - 8 \times 12 = 96$
d $15 \times 3 - 12 \times 4 = -3$ *or* $15 \div 3 - 12 + 4 = -3$
e $15 + 3 \times 12 \times 4 = 159$ f $15 \times 3 + 12 \div 4 = 48$

3 a 3 b 6 c 6 d 14 e 2 f 21 g 30 h -15 i -6

4 a -2 b 24 c -2 d 21 e 9 f 12 g 2 h -62 i 7

5 a 5 b 3 c $\frac{1}{2}$ d 5 e 7 f 7 g 18 h 1 i $-\frac{1}{4}$

6 a 2 b -3 c 2

EXERCISE 1D

1 a 6 b 1 c 9 d 11 e 13 f 4 g 23 h 0

2 a $|7| > |5|$ b $|-4| < |-9|$ c $|-3| = |3|$ d $|9| < |-10|$

3 a 7 b 11 c 24 d 45 e 6 f 8

4 a i 6 ii 0 iii 3 iv 5
b $|x - 3|$ is the distance between x and 3 on the number line.

5 a 22 b 20 c 2 d 38 e 41 f 7

6 a 30 b 14 c 12 d 3 e 10 f 5

7 a 6 b 11 c 3

REVIEW SET 1A

1 a 125 b 140 c -16

2 a 19 683 b 65 536 c 64.363 43

3 4 4 a 2^6 b 13^2 c 5^4

5 a $60 = 2^2 \times 3 \times 5$ b $84 = 2^2 \times 3 \times 7$
c $750 = 2 \times 3 \times 5^3$

6 264 7 a 3 b -40 c 12

8 a $|8| < |-10|$ b $|-6| = |6|$ c $|-11| > |-7|$

9 a 11 b 10 c 48

10 a 13 b $36 \div 3 - (1 \times 2 + 6 \div 2)$
c $(36 \div 3 - 1 \times (2 + 6)) \div 2$

REVIEW SET 1B

1 $3^2 \times 5 \times 7^3$ 2 a 24 b -50

3 $32 = 2^5$ 4 $a = 4,\ b = 2$

5 a $135 = 3^3 \times 5$ b $1050 = 2 \times 3 \times 5^2 \times 7$

6 a -18 b 18 c -4

7 a 18 b 11 c $\frac{30}{7}$ 8 a -8 b -18

9 a i $9 = 3^2$ ii $20 = 2^2 \times 5$
iii $40 = 2^3 \times 5$ iv $100 = 2^2 \times 5^2$
v $108 = 2^2 \times 3^3$ vi $144 = 2^4 \times 3^2$
b 9, 100, and 144
c The exponent of each prime factor is even.

10 a Lift A: $2 - 6$, Lift B: $3 - 5 + 10$, Lift C: $3 \times 2 - 8$, Lift D: $-2 + 4 \times 3$
b Lift A: -4, Lift B: 8, Lift C: -2, Lift D: 10
c i The distance a lift is from ground level is the absolute value of its position.
ii Lift C (2 floors from ground level)

EXERCISE 2A

1 a $9x$ b $5f$ c pq d $6cd$ e $4ab$ f $2yz$ g stu h $3lmn$ i $ab + c$ j $d - xy$ k $2a + 3b$ l $4 + 2m + p$

2 a $4k$ b $2a + 3b$ c $3t - 2u$
d $5 + 3m$ e $8 - 2c$ f $9 + g + 3h$
g $2a - 2b$ h $ac + 2b$ i $2x - 3y$

3 a $8(a - b)$ b $6(c + d)$ c $xy(z + 5)$

4 a y^2 b b^4 c $4x^2$ d $2a^2$ e $7y^3$ f $3x^2y$ g m^2n^2 h $2c^3d$ i $6p^2q^3$

5 a $y \times y \times y$ b $4 \times x \times x \times x \times x \times x$ c $3 \times a \times a \times b$
d $6 \times p \times 6 \times p$ e $2 \times g \times g + h \times h \times h$
f $m \times m \times m \times m \times m - 7 \times n \times n$

6 **a** $9a-4ab^2$ **b** t^2+s^2t **c** $2xy^2z-3z^2$ **d** a^3-2a **e** $5m^2+6n^3$ **f** m^3n^2-2m **g** $3x-8y^2$ **h** $3d-6cd$ **i** $2s+5st^3$ **j** $4g^3-3g$ **k** $2x+5-xy^2$ **l** $3ab^2+ab+b$ **m** $10v^3w-6vw^2+7w$

EXERCISE 2B

1 **a** double a **b** the product of p and q **c** the square root of m **d** the square of a **e** 3 less than a **f** the sum of b and c **g** the sum of double x and c **h** the square of twice a **i** twice the square of a **j** c less than the square of a **k** a minus the square of b **l** the square of all of a minus b

2 **a** $a+c$ **b** $p+q+r$ **c** ab **d** $r+s^2$ **e** $(r+s)^2$ **f** r^2+s^2 **g** $2a+b$ **h** p^3+5 **i** b^2-a **j** $\frac{a+b}{2}$ **k** $a+\frac{b}{4}$ **l** $\sqrt{m+n}$ **m** $x+\frac{1}{x}$ **n** $\frac{a+b}{4}$ **o** $\sqrt{x^2+y^2}$

3 $\sqrt{x^2-13}$

EXERCISE 2C

1 **a** 12 **b** 45 **c** 30 **d** -30 **e** -9 **f** -2 **g** -11 **h** -58

2 **a** $-\frac{4}{3}$ **b** -2 **c** $\frac{1}{2}$ **d** -2 **e** $\frac{3}{2}$ **f** -2 **g** $\frac{3}{2}$ **h** 3

3 **a** 1 **b** 64 **c** 26 **d** 36 **e** -126 **f** 45 **g** -1 **h** 50

4 **a** 1 **b** $\sqrt{3}\approx 1.73$ **c** $\sqrt{7}\approx 2.65$ **d** 1 **e** 5 **f** 2 **g** 3 **h** 4

5 **a** $\frac{3}{16}$ **b** $\frac{1}{5}$ **6** **a** 1 **b** 4

7 **a** **i** $2h$ m **ii** $(2h-3)$ m **b** **i** 10 m **ii** 7 m

8 **a** \$3129 **b** **i** $2x$ shares **ii** $\frac{x}{2}$ shares **iii** $\$\left(3.40\times x+2.19\times 2x+5.30\times\frac{x}{2}\right)$ **c** When $x=300$, the expression in **b iii** gives \$3129 ✓

9 **a** $\frac{24-x}{2}$ chocolates **b** **i** 10 chocolates **ii** 8 chocolates **iii** 5 chocolates

10 **a** $\sqrt{x(2x+1)}$ **b** 6 **11** **a** $\frac{x+y+z}{3}$ **b** 5.4

EXERCISE 2D

1 **a** equation **b** expression **c** equation **d** expression **e** equation **f** expression

2 **a** 3 terms **b** 2 terms **c** 4 terms

3 **a** 6 **b** -3 **c** 2 **d** -1 **e** 1 **f** $-\frac{1}{2}$ **g** 5 **h** -6

4 **a** 3 **b** -1 **c** -3 **d** $\frac{2}{3}$

5 **a** 2 **b** -5 **c** $\frac{1}{2}$ **d** 5 and -1

6 **a** expression **b** 5 terms **c** **i** 2 **ii** -4 **d** 7 **e** $-x$ and $5x$

7 **a** $3x$ and $-x$ **b** $-2xy$ and xy **c** $2x^2$ and x^2 **d** -2 and 7 **e** $2xy$ and $-xy$

EXERCISE 2E

1 **a** $a+8$ **b** $2b+5$ **c** $c+3$ **d** $2x-3$ **e** $5m$ **f** $3m$ **g** $-2t$ **h** $2t$ **i** cannot be simplified **j** $2x^2+2x$ **k** cannot be simplified **l** $-2a^2+3$ **m** $7x-4$ **n** $3x$ **o** $4ab$ **p** $8x^2y$

2 **a** 0 **b** $5x$ **c** cannot be simplified **d** $3cd$ **e** $-2pq$ **f** $2m^2$ **g** $9b^2-6$ **h** $4a+6$ **i** $-2a+2$ **j** $c+7d$ **k** $2a+6$ **l** $3x^2+4x$ **m** $3xy-1$ **n** $-4b-3$ **o** $3a^2+4a$ **p** cannot be simplified **q** cannot be simplified **r** $2x^2-11x$

3 **a** $6pq$ **b** $3m^2-1$ **c** $4g-8$ **d** $-6a-2b$ **e** $2ab-2a+4b$ **f** cannot be simplified **g** $3x^3+5x^2$ **h** cannot be simplified **i** $4x^3+x^2-3x$

EXERCISE 2F

1 **a** $10x$ **b** $12x$ **c** $-12x$ **d** $-21x$ **e** $3x^3$ **f** $18x^2$ **g** $-4d^2$ **h** $18x^3$ **i** $-2x^4$ **j** $-15x^3$ **k** a^2 **l** $3k^4$ **m** $6x^4$ **n** $-10t^4$ **o** $8x^5$ **p** $27x^3$

2 **a** $14x$ **b** $30a-14b^2$ **c** $16d^2$ **d** $10xy$ **e** x^2 **f** $-7xy$ **g** $15ab$ **h** $-4xy$ **i** $-7cd$

3 **a** $5xy-3x^2y$ **b** $12a^2b^3-a^3b^2$ **c** $2m^2n^2$

EXERCISE 2G

1 **a** $3x$ **b** $\frac{2}{y}$ **c** x^2 **d** $\frac{5}{9}$ **e** $5x$ **f** $\frac{1}{a^2}$ **g** $2y^2$ **h** $\frac{2}{5t}$ **i** 1 **j** $\frac{1}{7c^2}$ **k** $\frac{x^3}{4}$ **l** $\frac{4}{3z^2}$

2 **a** $\frac{1}{y}$ **b** n **c** $\frac{3a}{b}$ **d** $\frac{1}{2y^2}$ **e** $\frac{1}{4}$ **f** $\frac{3p}{q}$ **g** $\frac{y^2}{5}$ **h** $\frac{3}{2mn}$ **i** $\frac{7x}{y^3}$ **j** $\frac{b}{a}$ **k** $\frac{2x^2}{y^3z}$ **l** $\frac{3q^2}{5pr^2}$

EXERCISE 2H

1 **a** 3 **b** a **c** 3 **d** 1 **e** x **f** $6y$ **g** $5x$ **h** $4x^2$ **i** $4m$

2 **a** 2 **b** $3x$ **c** x **d** b **e** y **f** $7x$ **g** $6xy$ **h** $3xz$ **i** pr **j** $3a$ **k** $2xz$ **l** $2pqr^2$

3 **a** x **b** $x+1$ **c** $x(x+1)$ **d** $x-1$ **e** $2(x-1)$ **f** $x+1$ **g** $3(x+2)$ **h** $x+3$ **i** $2x(x+1)$

REVIEW SET 2A

1 **a** $8t$ **b** $7ab$ **c** $3y+2z$

2 **a** half of the sum of a and b **b** the sum of a and half b **c** 3 less than the square root of a **d** the square root of all of a minus 3

3 **a** $a+\sqrt{b}$ **b** $\sqrt{a+b}$ **4** **a** 7 **b** 4 **c** -2

5 **a** 2 **b** -3 **c** $\frac{3}{4}$

6 **a** $\left(\frac{100-x}{4}\right)$ dollars **b** **i** \$20 **ii** \$16 **iii** \$10

7 **a** expression **b** 4 terms **c** -2 **d** $3x$ and $-4x$

8 **a** $x+8$ **b** $5x-6y$ **c** $2a^2+4a$ **d** cannot be simplified **e** $5x^3y+2xy^3$ **f** x^3+2x^2

9 a $-28x$ b $45x^2$ c $9d^2$

10 a $\frac{5}{d^3}$ b $\frac{x}{3}$ c $\frac{4p}{q}$

11 a $2m$ b $3ab$ c $x(x-2)$

12 a $(x-5)^2$ b i 25 ii 1 iii 81
c No, as the expression is a perfect square which cannot be negative.

REVIEW SET 2B

1 a $5k^3$ b $2y-3y^4$ c a^3b-2b d x^2+4xyz^2

2 a x^2+3 b $(3+x)^2$ **3** 3

4 a 4 b $\frac{1}{2}$ c 2

5 a 4 terms b 3 c -1 **6** a 2 b -2

7 a $7-x^2+6x$ b $3t^3+t^2+t$

8 a $-6x^3$ b $2xy$ c x^3+8x^2

9 a $\frac{2a}{5b^3}$ b $\frac{x^2y}{5z}$ **10** a $2q$ b $2(x+1)$

11 a ab chocolates b $2ab$ chocolates c $3ab$ chocolates
d 720 chocolates

EXERCISE 3A

1 a 5^6 b 6^{11} c 3^7 d $(-2)^7$
e x^5 f a^6 g a^{10} h b^8
i a^{2+n} j b^{m+9} k x^{a+b} l p^7

2 a 3^4 b 7^6 c 5^3 d 4^4
e a^5 f x^2 g b^7 h p^{n-2}
i y^{5-b} j x^{a-b} k x^3 l b^{2n-1}

3 a 2^6 b 3^{21} c 5^{28} d 2^{ab}
e a^{12} f m^8 g 3^{2m} h q^9
i d^{6n} j x^{8y} k x^{pq} l g^{10a}

4 a b^3 b x^9 c a^{40} d x^{y-3}
e z^{3n} f t^{8+k} g x^{2m+n} h c^{3x-4y}

5 a $(x^3)^4=x^{12}$ b $n^3\times n^2=n^5$ c $(2^5)^3=2^{15}$
d $k^8\div k^5=k^3$ e $p^3\times p^4=p^7$ f $\frac{r^9}{r^1}=r^8$

6 a $15c^5$ b $28x^6y^6$ c a^2b^3 d $6d^5$
e $7p^2q^3$ f $36s^4t^3$ g $5a^2b^{16}$ h $2x^5y$

7 a 2^4 b 3^3 c 5^3 d 2^6
e 3^6 f 2^{a+2} g 7^{t-1} h 2^{4k}
i 2^{5-x} j 5^3 k $7^{4(a-1)}$ l 3^2
m 3^{y-2x} n 2^{2y-3x} o 11^{2x} p 5^3

8 a p^3q^3 b a^5b^5 c x^4y^4 d $p^5q^5r^5$
e $27a^3$ f $25b^2$ g $32n^5$ h $64a^3b^3$

9 a $\frac{x^4}{y^4}$ b $\frac{a^7}{b^7}$ c $\frac{m^3}{8}$ d $\frac{25}{p^2}$
e $\frac{16x^2}{y^2}$ f $\frac{a^3}{27b^3}$ g $\frac{c^6}{64d^6}$ h $\frac{10\,000s^4}{t^4}$

10 a $\frac{36}{49}$ b $\frac{125}{27}$ c $\frac{81}{256}$ d $\frac{1}{128}$

11 a $9a^4$ b x^3y^9 c $p^{20}q^5$ d $32x^{20}$
e $16a^8b^4$ f $27c^3d^{12}$ g $m^{18}n^{30}$ h $625x^8y^{20}$

12 a $\frac{x^4}{49}$ b $\frac{x^4}{y^8}$ c $\frac{w^4}{4v^2}$ d $\frac{8r^3}{27s^6}$
e $\frac{16}{p^8q^4}$ f $\frac{a^8}{9n^{14}}$ g $\frac{8x^6}{y^{18}}$ h $\frac{81x^8y^{12}}{256z^4}$

EXERCISE 3B

1 a 1 b 1 c 1 d 5 e 1 $\{a\neq 0\}$
f 1 g 3 $\{y\neq 0\}$ h 1 $\{y\neq 0\}$ i 12
j d^2 $\{d\neq 0\}$ k 3 l 7

2 a 1 b a^2 c x d $\frac{a}{2}$

3 a $\frac{1}{2}$ b $\frac{1}{3}$ c $\frac{1}{6}$ d $\frac{1}{8}$ e $\frac{1}{9}$ f $\frac{1}{8}$
g $\frac{1}{25}$ h $\frac{1}{100}$ i $\frac{2}{3}$ j $1\frac{1}{7}$ k $3\frac{1}{2}$ l $5\frac{1}{25}$

4 a $\frac{1}{t}$ b $\frac{1}{x^2}$ c $\frac{1}{y^4}$ d $\frac{1}{z^5}$

5 a 5^{-2} b 3^{-4} c 2^{-5} d 7^{-8}

6 a 3 b 5 c $\frac{7}{3}$ d $\frac{2}{5}$ e 10 f $\frac{36}{25}$
g $\frac{16}{81}$ h $\frac{27}{8}$

7 a $\frac{1}{2x}$ b $\frac{2}{x}$ c $\frac{3}{b}$ d $\frac{1}{3b}$ e $\frac{1}{4b^2}$ f $\frac{b^2}{4}$
g n^2 h $\frac{n^2}{3}$ i $\frac{1}{ab}$ j $\frac{a}{b}$ k $\frac{a}{b^2}$ l $\frac{1}{a^2b^2}$
m $\frac{1}{2ab}$ n $\frac{2}{ab}$ o $\frac{2a}{b}$ p a^2b^3

8 a 2^4 b 2^{-4} c 5^2 d 5^{-2} e 3^4
f 3^{-4} g 2^{-10} h 2^{-a} i $3^2\times 5^{-2}$
j $2^5\times 3^{-4}$ k $2^2\times 3\times 5^{-1}$ l $3\times 5^2\times 2^{-3}$
m 2^{-3x} n 3^{-4k} o $2^{-2m}\times 3^{-m}$

EXERCISE 3C.1

1 C

2 a $376=3.76\times 10^2$ b $8000=8\times 10^3$
c $0.04=4\times 10^{-2}$ d $0.005\,07=5.07\times 10^{-3}$
e $9\,040\,000=9.04\times 10^6$ f $0.000\,000\,23=2.3\times 10^{-7}$

3 a 4.25×10^2 b 4.25×10^5 c 4.25×10^0
d 4.25×10^{-1} e 2.01×10^1 f 2.01×10^4
g 2.01×10^{-3} h 2.01×10^6 i 3.87×10^3
j 3.87×10^{-2} k 3.87×10^7 l 3.87×10^{-4}

4 a 4.0075×10^4 km b 4×10^{-4} mm
c 4×10^7 bacteria d 1.4162×10^{-7}
e 1×10^{-2} mm f 1.5×10^7 °C

5 a 50 000 b 3000 c 18 000 000 d 810
e 650 000 f 11 g 275 000 000 h 8 000 000

6 a 0.03 b 0.000 09 c 0.007
d 0.000 41 e 0.000 008 2 f 0.761
g 0.000 000 325 h 0.000 000 02

7 a 0.000 000 475 m b 7 530 000 000 people
c 0.000 012 m d 12 700 km

8 a 1.817×10^7 b 9.34×10^{10} c 4.1×10^{-4}

9 a 2.2×10^8 b 3.5×10^{-5} c 4.9×10^3
d 6.41×10^8 e 1.01×10^{-6} f 9×10^{-8}

10 a 6×10^7 b 2.4×10^7 c 4×10^{12}
d 8.1×10^{19} e 9×10^{-5} f 4.9×10^{-5}
g 4×10^{-1} h 2×10^{11} i 2.5×10^3

11 a China b 1.5×10^{10} kg c 12 times more

EXERCISE 3C.2

1 a 6.00×10^{-10} b $\approx 2.39\times 10^{10}$ c 2.19×10^{11}
d $\approx 4.01\times 10^{10}$ e $\approx 6.24\times 10^{-10}$ f $\approx 1.03\times 10^{-7}$

2 **a** $\approx 9.55 \times 10^{9}$ **b** $\approx 9.62 \times 10^{-7}$ **c** 4.90×10^{-14}
d $\approx 6.50 \times 10^{-7}$ **e** 4.00×10^{-7} **f** $\approx 4.37 \times 10^{14}$

3 **a** **i** 9.11×10^{-28} g **ii** 1.67×10^{-24} g
b ≈ 1830 times heavier

4 **a** $\approx 4.80 \times 10^{8}$ mm **b** $\approx 3.15 \times 10^{7}$ seconds
c $\approx 5.26 \times 10^{8}$ minutes **d** 5×10^{-7} kg

5 **a** 8.64×10^{4} km **b** $\approx 6.05 \times 10^{5}$ km
c $\approx 6.31 \times 10^{7}$ km

6 **a** 602 000 000 000 000 000 000 000
b 6.02×10^{29} atoms **c** ≈ 1.66 tonnes

7 $\approx 1.02 \times 10^{20}$ electrons

EXERCISE 3D

1 **a** 6×10^{-6} m **b** 4.3×10^{-9} m **c** 1.2×10^{-3} m
d 2.7×10^{-8} m **e** 1.68×10^{-4} m **f** 7.6×10^{-4} m

2 **a** 3.4 μm **b** 2.7 nm **c** 69 μm **d** 850 nm

3 **a** 9×10^{-3} s **b** 4×10^{-6} s **c** 2×10^{-9} s
d 3.8×10^{-5} s **e** 4.1×10^{-1} s **f** 2.85×10^{-8} s

4 **a** 8 ms **b** 78 μs **c** 12.6 ns **d** 450 μs

5 **a** 8×10^{6} L **b** 5.2×10^{9} L **c** 3.14×10^{4} L
d 1.72×10^{-4} L **e** 4×10^{-4} L **f** 1.49×10^{5} L

6 **a** 4.8 kL **b** 1.64 mL **c** 329 ML **d** 68.5 μL

7 **a** 10^{3} times larger **b** 10^{6} times larger
c 10^{6} times larger **d** 10^{9} times larger

8 **a** DNA: 2.5×10^{-9} m, paper: 1×10^{-4} m
b 4×10^{4} times thicker

9 **a** $\approx 2.59 \times 10^{9}$ MWh **b** $\approx 2.59 \times 10^{6}$ GWh

10 **a** 5×10^{-9} s **b** 5×10^{-3} μs **c** 5 ns

11 **a** 710 ML more **b** 12 000 times more

12 **a** 15 μg/L **b** 3000 ng **c** yes (1.25×10^{-5} g/L)

REVIEW SET 3A

1 **a** q^{7} **b** a^{2} **c** t^{4x} **d** x^{8k-3}

2 **a** $10d^{9}$ **b** $4x^{6}$ **c** $3a^{21}b^{2}$

3 **a** $a^{3}b^{3}$ **b** $\dfrac{32}{z^{5}}$ **c** $125p^{9}q^{6}$ **d** $\dfrac{16a^{8}}{d^{12}}$

4 **a** 1 **b** 5 **c** $5b$

5 **a** $\frac{1}{16}$ **b** $\frac{1}{27}$ **c** $\frac{7}{8}$ **d** $8\frac{1}{16}$

6 **a** $\dfrac{1}{5c}$ **b** $\dfrac{7}{k^{2}}$ **c** $\dfrac{1}{64d^{6}}$ **d** $\dfrac{1}{p^{2}q^{2}}$

7 **a** 2^{-3} **b** 3^{-2t} **c** $3^{3} \times 5^{-2}$

8 **a** 1.7×10^{3} **b** 4.682×10^{5} **c** 5.8×10^{-2}
d 6.06×10^{-4}

9 **a** 0.007 71 **b** 40 000 **c** 0.000 017 02 **d** 58 000 000

10 1.398×10^{5} km

11 **a** 2.8×10^{10} **b** 9×10^{-8} **c** 5×10^{-6}

12 **a** $\approx 2.57 \times 10^{6}$ km **b** $\approx 1.80 \times 10^{7}$ km
c $\approx 9.37 \times 10^{8}$ km

13 **a** 7×10^{-6} s **b** 8.52×10^{9} L **c** 2.3×10^{-8} m
d 5.04×10^{-4} g

14 **a** 6.7 μm **b** 9.42 kL **c** 38 GW **d** 150 ns

15 $\approx 5.71 \times 10^{9}$ grains

16 **a** $\approx 4.13 \times 10^{13}$ km
b **i** 1.496×10^{8} km **ii** $\approx 2.76 \times 10^{5}$ times farther

REVIEW SET 3B

1 **a** $(-5)^{10}$ **b** x^{4n-3} **c** c^{5k}

2 **a** $t^{9} \times t^{3} = t^{12}$ **b** $\dfrac{k^{8}}{k^{4}} = k^{4}$

3 **a** 5^{6} **b** 2^{6k} **c** 3^{2t-2}

4 **a** $81p^{4}$ **b** $\frac{8}{27}$ **c** $\dfrac{a^{10}}{b^{5}}$

5 **a** $\frac{1}{4}$ **b** $10\frac{1}{3}$ **c** $\frac{999}{1000}$ **6** **a** $\dfrac{1}{k^{3}}$ **b** 7^{-5}

7 **a** $\frac{5}{4}$ **b** $\frac{49}{4}$ **c** $\frac{9}{100}$ **8** **a** $\dfrac{b}{a^{2}}$ **b** $\dfrac{25}{a^{4}}$ **c** $\dfrac{p^{3}}{5}$

9 **a** 2.6357×10^{2} **b** 5.11×10^{-4} **c** 8.634×10^{8}

10 **a** 2.78 **b** 39 900 000 **c** 0.002 081

11 30 times larger

12 **a** $\approx 3.06 \times 10^{-4}$ **b** $\approx 1.69 \times 10^{9}$ **c** $\approx 4.75 \times 10^{8}$

13 7.4×10^{7} L **14** **a** 5.3 μs **b** 457 nm **15** 8.76 GJ

16 **a** 3.58×10^{-10} m **b** 1.405×10^{10} years
c **i** $\approx 3.85 \times 10^{-22}$ g **ii** $\approx 2.59 \times 10^{22}$ atoms

EXERCISE 4A

1 **a** $3x+3$ **b** $10-2x$ **c** $5x+10$
d $12-4x$ **e** $5x-5y$ **f** $63+9x$
g $x^{2}+3x$ **h** $7m-m^{2}$ **i** $a^{2}+ab$

2 **a** $10x+6$ **b** $4a+8b$ **c** $6x-3y$
d $6m-24n$ **e** $5x^{2}+20x$ **f** $-6x^{2}+6y^{2}$
g $4a^{2}+7a$ **h** $2x^{2}-10x$ **i** $15p^{2}-3pq$

3 **a** $-2x-8$ **b** $-3x-6$ **c** $-4x+12$
d $-6x+3$ **e** $-7+2x$ **f** $-2x+2y$
g $-a^{2}+ab$ **h** $-3x^{2}-8x$ **i** $-4x+2x^{2}$

4 **a** $4x+20$ **b** $5x-35$ **c** $6x-2$
d $x^{2}-6x$ **e** $12-9x$ **f** $-4x-8$
g $2x^{2}-x$ **h** $-5a^{2}-6a$ **i** $4x^{2}+4x$
j $-2c^{2}+cd$ **k** $12m^{2}+3mn$ **l** $-6x+42x^{2}$

5 **Hint:** The overall area is equal to the sum of the areas of the three small rectangles.

6 **a** $3x-3y+3$ **b** $-x^{2}-x+4$
c $2a-6b-2ab$ **d** $6x^{2}+3xy-3x$
e $2x^{3}-2x^{2}-4x$ **f** $-3x^{2}-3x+3x^{3}$

7 **a** $2x+5$ **b** $1-4x$ **c** $3x-1$ **d** $2-4x$
e x^{2} **f** $6x-1$ **g** $2x-2$ **h** $x+10$
i $6x-x^{2}$ **j** $15x-8$ **k** $7x-3x^{2}$ **l** $2x^{2}-x$
m $-2x^{2}+2x$ **n** $3a-2b$ **o** $a+2b$
p $2ab+a^{2}$ **q** $2x^{2}-10x$ **r** $x^{2}-x$

8 **a** $5x-2$ **b** $15-3y$ **c** $-6y-10$
d $5-3x$ **e** $8x-8$ **f** $a+5b$
g $x^{2}+6x-6$ **h** $x^{2}+2x+6$ **i** $x^{2}-y^{2}$
j $-a-b$ **k** $3ab-a^{2}$ **l** $6x^{2}-22x$

EXERCISE 4B

1 **a** ac **b** ad **c** bc **d** bd **e** $(a+b)(c+d)$
$(a+b)(c+d) = ac+ad+bc+bd$

2 **a** $x^{2}+3x+2$ **b** $x^{2}+7x+12$ **c** $x^{2}+4x-5$
d $x^{2}-2x-8$ **e** $x^{2}+4x-21$ **f** $x^{2}-7x+12$
g $x^{2}+10x+21$ **h** $x^{2}+x-20$ **i** $x^{2}+3x-18$
j $x^{2}-7x+10$ **k** $x^{2}-5x-24$ **l** $6x^{2}+11x+4$

m $1+2x-8x^2$ **n** $12+5x-2x^2$ **o** $6x^2-x-2$
p $25-10x-3x^2$ **q** $-4x^2+29x-7$ **r** $3-7x-6x^2$

3 **a** x^2-4 **b** a^2-25 **c** $16-x^2$ **d** b^2-16
e x^2-81 **f** $36-y^2$ **g** $4x^2-1$ **h** $25a^2-9$
i $16-9a^2$

4 **a** x^2+6x+9 **b** x^2-4x+4 **c** $9x^2-12x+4$
d $25+20x+4x^2$ **e** $1-6x+9x^2$ **f** $x^2+2xy+y^2$
g $9-24x+16x^2$ **h** $25x^2-10xy+y^2$
i $a^2-2ab+b^2$ **j** $16x^2+16x+4$
k $4x^2+4xy+y^2$ **l** x^4+2x^2+1

5 **a** x^2+8x+1 **b** $x^2+7x-10$ **c** $3a^2-30$
d $5x^2-2x-15$ **e** $3x^2+8x-24$ **f** $-y-31$
g $-k^2+2k-7$ **h** $2x^2-40$

EXERCISE 4C

1 **a** x^2-4 **b** x^2-4 **c** $4-x^2$ **d** $4-x^2$
e x^2-1 **f** $1-x^2$ **g** x^2-49 **h** c^2-64
i d^2-25 **j** x^2-y^2 **k** $16-d^2$ **l** $25-e^2$

2 **a** $4x^2-1$ **b** $9x^2-4$ **c** $16y^2-25$ **d** $4y^2-25$
e $9x^2-1$ **f** $1-9x^2$ **g** $4-25y^2$ **h** $9-16a^2$
i $16-9a^2$ **j** $4a^2-b^2$ **k** a^2-4b^2 **l** $16x^2-y^2$
m $16x^2-25y^2$ **n** $4x^2-9y^2$ **o** $49x^2-4y^2$

3 **a** **i** $43\times 37 = (40+3)(40-3) = 40^2-3^2$
ii $24\times 26 = (25-1)(25+1) = 25^2-1^2$

b **i** 396 **ii** 2499 **iii** 9991

EXERCISE 4D

1 **a** a^2 **b** ab **c** ab **d** b^2 **e** $(a+b)^2$
$(a+b)^2=a^2+2ab+b^2$

2
$$\begin{aligned}(5+3)^2 &= 5^2+2(5)(3)+3^2\\ &= 5^2+30+3^2\\ &= 5^2+3^2+30\end{aligned}$$
$\therefore\ (5+3)^2\neq 5^2+3^2$

3 **a** $x^2+10x+25$ **b** $x^2+8x+16$ **c** $x^2+14x+49$
d a^2+4a+4 **e** $9+6c+c^2$ **f** $4+4x+x^2$

4 **a** x^2-6x+9 **b** x^2-4x+4 **c** $y^2-16y+64$
d $a^2-14a+49$ **e** $25-10x+x^2$ **f** $16-8y+y^2$

5 **a** $9x^2+24x+16$ **b** $4a^2-12a+9$ **c** $9y^2+6y+1$
d $4x^2-20x+25$ **e** $9y^2-30y+25$ **f** $49+28a+4a^2$
g $1+10x+25x^2$ **h** $49-42y+9y^2$ **i** $9+24a+16a^2$

6 **a** x^4+4x^2+4 **b** y^4-6y^2+9 **c** $9a^4+24a^2+16$
d $1-4x^2+4x^4$ **e** $x^4+2x^2y^2+y^4$ **f** $x^4-2a^2x^2+a^4$

7 **a** $-x^2-3x-8$ **b** x^2+x+2 **c** $2x^2+6x+5$
d $-6x-25$ **e** $11-13x+3x^2$ **f** $10x^2-7x-5$
g $3x^2-2x-10$ **h** $3x^2-x-10$ **i** $2x^2+2x+5$
j $3x^2-10x-8$

8 **Hint:** Let the middle integer be n, and show that
$$\frac{(n-1)^2+(n+1)^2}{2}=n^2+1.$$

EXERCISE 4E

1 **a** x^3+3x^2+3x+2 **b** x^3+4x^2+x-6
c x^3+3x^2-2x+8 **d** x^3+5x^2-7x-6
e $3x^3+4x^2+16x+5$ **f** $5x^3-7x^2+22x-8$
g $2x^3+7x^2+x-6$ **h** $6x^3-5x^2+9x-4$
i $6x^3+7x^2-7x-6$

2 **a** x^3+3x^2+3x+1 **b** $x^3+9x^2+27x+27$
c x^3-3x^2+3x-1 **d** $x^3-6x^2+12x-8$
e $8x^3+12x^2+6x+1$ **f** $8x^3-36x^2+54x-27$

3 **a** x^3+6x^2+8x **b** x^3-2x^2-15x
c x^3-5x^2+4x **d** $2x^3+2x^2-12x$
e $-2x^3+12x^2+14x$ **f** $-x^3+3x^2+18x$
g $-10x^3-25x^2+15x$ **h** $4x^3-13x^2+3x$
i $3x^4+12x^3+12x^2$

4 **a** $x^3+7x^2+14x+8$ **b** x^3-4x^2+x+6
c $x^3-4x^2-7x+10$ **d** $2x^3-11x^2+13x-4$
e $3x^3+16x^2+23x+6$ **f** $16x^3+64x^2-x-4$
g $3x^3+x^2-8x+4$ **h** $-5x^3+28x^2-13x-10$
i $-3x^3-14x^2-3x+36$

5 **a** **i** x^3-1 **ii** x^4-1 **iii** x^7-1 **b** $x^{n+1}-1$

REVIEW SET 4A

1 **a** $6+3x$ **b** $-4x+2$ **c** $3x^2+9x$

2 **a** $4x-1$ **b** $7x-8$ **c** $-x^2-9x+12$

3 **a** $x^2+11x+28$ **b** $-5x^2-x+6$ **c** $20x^2-11x-4$

4 **a** $x^2-20x+100$ **b** $9-24x+16x^2$
c $x^2+4xy+4y^2$

5 **a** x^2-81 **b** $9k^2-4$ **c** $1-49x^2$

6 **a** x^2+8x-6 **b** $-10x-49$ **c** $8x^2-8x$

7 **a** $x^3+5x^2-3x-28$ **b** $4x^3-7x^2-17x+15$
c $4x^3+7x+4$ **d** $8x^3-12x^2+6x-1$
e x^3-x^2-30x **f** $4x^3-5x^2-7x+2$

8 **a** shaded area = area of larger square − area of smaller square $=a^2-b^2$

b **i** $a-b$ **ii** $\dfrac{(a+b)(a-b)}{2}$

c shaded area $=2\times$ area of trapezium $=(a+b)(a-b)$
$\therefore$ using **a**, $(a+b)(a-b)=a^2-b^2$

REVIEW SET 4B

1 **a** $-3x-12$ **b** $xy+3x$ **c** $2a^2-10a$

2 **a** $10x-11$ **b** x^2-3x+8 **c** x^2+4x-3

3 **a** $x^2+5x-36$ **b** $2x^2+5x-42$ **c** $2a^2-11a+12$

4 **a** y^2+6y+9 **b** $9x^2+12x+4$ **c** $16a^2-8ab+b^2$

5 **a** $25-x^2$ **b** $9y^2-16$ **c** $1-16x^2$

6 **a** $-x^2-6x-9$ **b** x^4-10x^2+25
c $n-9$ **d** $y^2+5y-40$

7 **a** $-x^3+5x^2-10x+8$ **b** x^3+6x^2-32
c $4x^3-12x^2-x+3$ **d** $x^3+12x^2+48x+64$
e $x^3-15x^2+75x-125$

8 **a** $4x^2+12x+9$, sum of coefficients = 25
b $8x^3+36x^2+54x+27$, sum of coefficients = 125
c **i** 625 **ii** 5^n

EXERCISE 5A

1 **a** **i** $C=\{\text{black, yellow}\}$ **ii** $D=\{3, 7, 9\}$
iii $L=\{\text{B, C, S, T}\}$
b **i** $n(C)=2$ **ii** $n(D)=3$ **iii** $n(L)=4$

2 **a** $P=\{1, 3, 5, 9, 15, 45\}$, $n(P)=6$

b $Q = \{\text{A, N, T, R, C, I}\}$, $n(Q) = 6$

3 **a** $M = \{4, 8, 12, 16, 20, 24, 28\}$,
$C = \{4, 6, 8, 9, 10, 12, 14, 15, 16, 18, 20, 21, 22, 24, 25, 26, 27, 28\}$

b **i** false **ii** true
c **i** $n(M) = 7$ **ii** $n(C) = 18$ **d** yes

4 **a** yes **b** no **c** no **d** yes

5 **a** $\varnothing$, $\{\square\}$ **b** $\varnothing$, $\{a\}$, $\{b\}$, $\{a, b\}$
c $\varnothing$, $\{-1\}$, $\{0\}$, $\{2\}$, $\{-1, 0\}$, $\{-1, 2\}$, $\{0, 2\}$, $\{-1, 0, 2\}$

6 **a** $x = 2, 4, 6,$ or 8 **b** $x = 4$ **c** $x = -4, -2, 0,$ or 4

7 $B = \{2, 3, 4, 5, 6, 8, 9, 10, 12\}$

EXERCISE 5B

1 **a** $A' = \{2, 3, 6, 8, 9, 10\}$ **b** $B' = \{1, 3, 4, 6, 7, 9, 10\}$
c $C' = \{5, 6, 7, 8, 9, 10\}$ **d** $D' = \{1, 3, 5, 7, 9\}$
e $E' = \{6\}$

2 **a** $M' = \{\text{Mercury, Venus}\}$
This set represents the planets without moons.
b $R' = \{\text{Mercury, Venus, Earth, Mars}\}$
This set represents the planets without rings.

3 **a** **i** $F = \{1, 2, 3, 4, 6, 8, 12, 24\}$
ii $F' = \{5, 7, 9, 10, 11, 13, 14, 15, 16, 17, 18, 19, 20, 21, 22, 23\}$
iii $M = \{4, 8, 12, 16, 20, 24\}$
iv $M' = \{1, 2, 3, 5, 6, 7, 9, 10, 11, 13, 14, 15, 17, 18, 19, 21, 22, 23\}$
v $P = \{2, 3, 5, 7, 11, 13, 17, 19, 23\}$
vi $P' = \{1, 4, 6, 8, 9, 10, 12, 14, 15, 16, 18, 20, 21, 22, 24\}$

b **i** $n(F) = 8$ **ii** $n(F') = 16$ **iii** $n(M) = 6$
iv $n(M') = 18$ **v** $n(P) = 9$ **vi** $n(P') = 15$

4 **a** Every element of U is either in A or in A'.
$\therefore \ n(A) + n(A') = n(U)$
b **i** $n(A') = 13$ **ii** $n(B') = 5$

5 **a** **i** true **ii** true **iii** false **iv** true **b** D and E

6 **a** $\varnothing$ **b** U **c** A

7 Since $A \subseteq B$, each element which is *not* in B is also *not* in A.
$\therefore \ B' \subseteq A'$

EXERCISE 5C

1 **a** $A \cap B = \{b, d\}$, $A \cup B = \{a, b, c, d, f, h\}$
b $A \cap B = \{4, 5\}$, $A \cup B = \{1, 2, 3, 4, 5, 6, 7, 8\}$
c $A \cap B = \{\text{vanilla, chocolate, strawberry}\}$,
$A \cup B = \{\text{vanilla, lemon, chocolate, strawberry, coffee}\}$

2 **a** $P \cap Q = \varnothing$ **b** yes

3 **a** $A = \{3, 6, 9, 12, 15, 18, 21, 24, 27\}$,
$B = \{1, 2, 3, 5, 6, 10, 15, 30\}$
b **i** $A \cap B = \{3, 6, 15\}$ **ii** $n(A \cap B) = 3$
iii $A \cup B = \{1, 2, 3, 5, 6, 9, 10, 12, 15, 18, 21, 24, 27, 30\}$
iv $n(A \cup B) = 14$

4 A and B, A and C, C and D

5 **a** $J = \{\text{January, June, July}\}$,
$K = \{\text{January, March, May, July, August, October, December}\}$
b **i** $J \cap K = \{\text{January, July}\}$
This set represents the months which begin with a "J" and contain 31 days.
ii $J \cup K = \{\text{January, March, May, June, July, August, October, December}\}$
This set represents the months which begin with a "J" or contain 31 days.
iii $(J \cup K)' = \{\text{February, April, September, November}\}$
This set represents the months which do not begin with a "J" nor contain 31 days.

6 **a** Each element in U is in either A or not in A.
$\therefore \ A \cup A' = U$
b An element cannot be both in A and not in A.
$\therefore \ A \cap A' = \varnothing$
c If an element is in A and B then it must be in A.
$\therefore \ A \cap B \subseteq A$
d If an element is in A and B then it must be in A or B.
$\therefore \ A \cap B \subseteq A \cup B$

7 **a** true; an element cannot be both in a set and its complement.
b false; there may be elements which do not belong to either set.
c false; for example if $A = C \neq \varnothing$ then A and C are not disjoint.

8 **a** **i** $A = \{12, 14, 16, 18\}$ **ii** $B = \{12, 15, 18\}$
iii $A' = \{13, 15, 17\}$ **iv** $B' = \{13, 14, 16, 17\}$
v $A \cup B = \{12, 14, 15, 16, 18\}$
vi $A' \cap B' = \{13, 17\}$
b $(A \cup B)' = \{13, 17\} = A' \cap B'$ ✓

9 **a** **i** $P = \{2, 3, 5, 7\}$ **ii** $Q = \{1, 3, 5, 7, 9\}$
iii $P' = \{1, 4, 6, 8, 9, 10\}$ **iv** $Q' = \{2, 4, 6, 8, 10\}$
v $P \cap Q = \{3, 5, 7\}$ **vi** $P \cup Q = \{1, 2, 3, 5, 7, 9\}$
b $(P \cap Q)' = \{1, 2, 4, 6, 8, 9, 10\}$
$P' \cup Q' = \{1, 2, 4, 6, 8, 9, 10\}$
$= (P \cap Q)'$ ✓
c If an element is not in both P and Q then it is either not in P or not in Q.

EXERCISE 5D

1

Number	$\mathbb{N}$	$\mathbb{Z}$	$\mathbb{Q}$	$\mathbb{R}$
19	✓	✓	✓	✓
$-\frac{2}{3}$	✗	✗	✓	✓
$\sqrt{7}$	✗	✗	✗	✓
5.6389	✗	✗	✓	✓
$\sqrt{16}$	✓	✓	✓	✓
2π	✗	✗	✗	✓
-11	✗	✓	✓	✓
$\frac{6}{0}$	✗	✗	✗	✗
$\sqrt{-2}$	✗	✗	✗	✗

2 **a** true **b** false **c** true **d** true
e false **f** true **g** true **h** false

3 **a** true **b** true **c** true **d** false

4 **a** $\mathbb{N}$ and $\mathbb{Z}^-$ **b** $\mathbb{Q}$ and $\mathbb{Q}'$

5 **a** $\varnothing$ **b** $\mathbb{Z}$ **c** $\mathbb{Q}'$ **d** $\mathbb{R}$ **6** $(\mathbb{Z}^- \cup \mathbb{Z}^+)' = \{0\}$

EXERCISE 5E

1 **a**
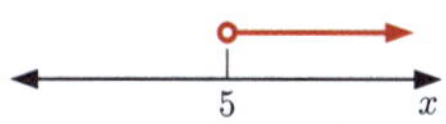

b
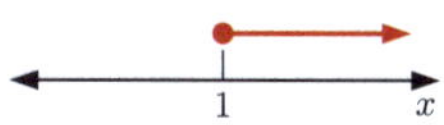

c 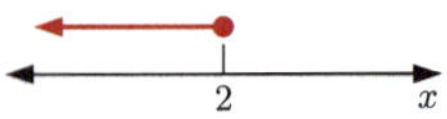 **d** 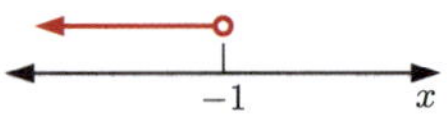

e  **f** 

g  **h** 

i 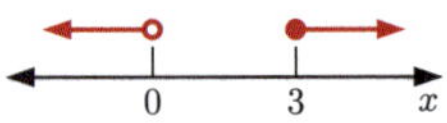 **j** 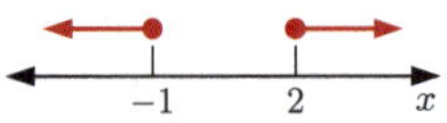

k 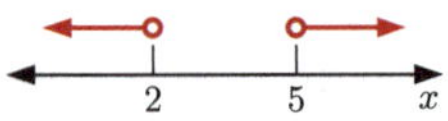 **l** 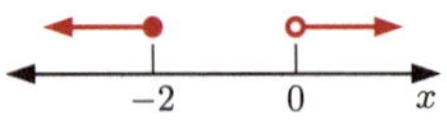

2 **a** $x > 0$ **b** $x \leqslant -2$ **c** $x \geqslant 100$
d $x < 0.2$ **e** $-1 \leqslant x \leqslant 4$ **f** $3 \leqslant x < 10$
g $-5 < x < 5$ **h** $7 < x \leqslant 20$
i $x < 2$ or $x \geqslant 7$ **j** $x < -3$ or $x > 0$
k $x \leqslant 4$ or $x \geqslant 17$ **l** $x \leqslant 21$ or $x > 33$

3 **a** The set of real numbers greater than 4.
b The set of natural numbers less than or equal to 5.
c The set of integers greater than 0 and less than 8.
d The set of real numbers greater than or equal to 1 and less than or equal to 4.
e The set of positive integers greater than 2 and less than 7.
f The set of real numbers less than or equal to 3 or greater than 6.

4 **a** $\{2, 3, 4, 5, 6\}$ **b** $\{...., -2, -1, 0, 1, 2, 3, 4, 5\}$
c $\{3, 4, 5, 6, ...\}$ **d** $\{0, 1, 2, 3, 4, 5\}$
e $\{-4, -3, -2, -1\}$ **f** $\{7, 8, 9, 10, ...\}$

5 **a** $\{x \in \mathbb{Z} \mid -5 \leqslant x \leqslant -1\}$
b $\{x \in \mathbb{N} \mid x \leqslant 5\}$ or $\{x \in \mathbb{Z} \mid 0 \leqslant x \leqslant 5\}$
c $\{x \in \mathbb{N} \mid x \geqslant 4\}$ or $\{x \in \mathbb{Z} \mid x \geqslant 4\}$
d $\{x \in \mathbb{Z} \mid x \leqslant 1\}$ **e** $\{x \in \mathbb{Z} \mid -5 \leqslant x \leqslant 1\}$
f $\{x \in \mathbb{Z} \mid x \leqslant 44\}$

6 **a** $\{x \in \mathbb{Z} \mid -1 \leqslant x \leqslant 4\}$
b $\{x \in \mathbb{N} \mid x \leqslant 6\}$ or $\{x \in \mathbb{Z} \mid 0 \leqslant x \leqslant 6\}$

7 **a** 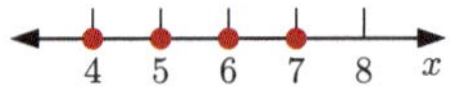 **b** 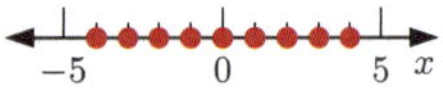

c 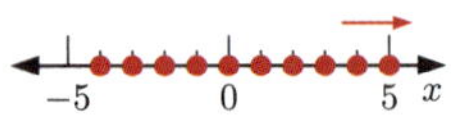

8 **a** finite **b** infinite **c** infinite **d** finite
e finite **f** infinite

9 **a** **i** $P \cap Q = \{x \in \mathbb{Z} \mid 12 \leqslant x \leqslant 18\}$
ii $P \cup Q = \{x \in \mathbb{Z} \mid 1 \leqslant x \leqslant 25\}$
iii $P' = \{x \in \mathbb{Z} \mid 19 \leqslant x \leqslant 30\}$
iv $Q' = \{x \in \mathbb{Z} \mid 1 \leqslant x \leqslant 11$ or $26 \leqslant x \leqslant 30\}$
b **i** $n(P) = 18$ **ii** $n(Q) = 14$
iii $n(P \cap Q) = 7$ **iv** $n(P \cup Q) = 25$

10 **a** true **b** true **c** true **d** false

11 Between any two rational numbers we can find an irrational number, and between any two irrational numbers we can find a rational number. So, on any given interval we would need to place an infinite number of dots with gaps between them.

REVIEW SET 5A

1 **a** $C = \{\text{red, white, green, yellow, black, blue}\}$
b $n(C) = 6$ **c** **i** true **ii** false

2 **a** yes **b** no **c** yes

3 **a** $P' = \{1, 2, 4, 6, 9\}$ **b** $Q' = \{2, 3, 5, 6, 7, 8\}$
c $R' = \{1, 2, 3, 5, 7\}$

4 **a** $A \cup B = \{a, b, c, d, e, i, o, u\}$ **b** $A \cap B = \{a, e\}$

5 **a** $M = \{5, 10, 15\}$ **b** $F = \{1, 2, 5, 10\}$
c $M \cup F = \{1, 2, 5, 10, 15\}$ **d** $M \cap F = \{5, 10\}$
e $(M \cap F)' = \{1, 2, 3, 4, 6, 7, 8, 9, 11, 12, 13, 14, 15\}$
f $M \cap F' = \{15\}$

6 **a** false **b** false **c** true **7** **a** $\mathbb{N}$ **b** $\mathbb{R}$

8 **a** The set of negative integers greater than or equal to -8.
b $\{-8, -7, -6, -5, -4, -3, -2, -1\}$
c 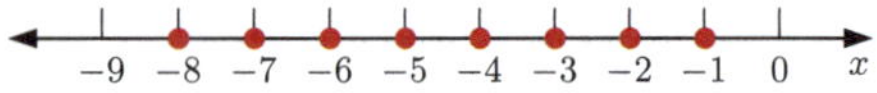

9 **a** $\{x \mid x > 3\}$ **b** $\{x \mid -5 \leqslant x < 4\}$
c $\{x \mid x < -11$ or $x > -7\}$

10 **a** $U = \{x \in \mathbb{N} \mid x \leqslant 30\}$
b no; the elements 5, 15, and 25 are in A but not in B.
c $A \cap B' = \{5, 15, 25\}$
This represents the numbers less than or equal to 30 which are multiples of 5 but not 10.

11 **a** $A \cap B = \{x \in \mathbb{Z}^+ \mid 7 < x < 16\}$
b $A \cup B = \{x \in \mathbb{Z}^+ \mid x \geqslant 4\}$ **c** $A' = \{x \in \mathbb{Z}^+ \mid x \leqslant 7\}$
d $B' = \{x \in \mathbb{Z}^+ \mid x < 4$ or $x \geqslant 16\}$

12 Let A be the ages of people involved in Trial A, and B be the ages of people involved in Trial B.
a $A = \{25, 26, 27, ..., 39, 40\}$
b no; ages are given as natural numbers only.
c $A = \{x \in \mathbb{N} \mid 25 \leqslant x \leqslant 40\}$
d **i** $B = \{x \in \mathbb{N} \mid x \geqslant 30\}$
ii $A \cap B = \{x \in \mathbb{N} \mid 30 \leqslant x \leqslant 40\}$
iii $(A \cup B)' = \{x \in \mathbb{N} \mid x < 25\}$

REVIEW SET 5B

1 $P = \{23, 29, 31, 37\}$

2 $\varnothing, \{2\}, \{5\}, \{7\}, \{2, 5\}, \{2, 7\}, \{5, 7\}, \{2, 5, 7\}$

3 $n(A') = 5$

4 **a** $F \cap M = \{4, 8, 16\}$
b $F \cup M = \{1, 2, 4, 8, 12, 16, 20, 24, 28, 32\}$

5 **a** $A \cap C = \{1, 3, 5, 7, 9, 11, 13, 15, 17, 19\}$
b no; neither set contains 20

6 All natural numbers can be expressed as fractions of integers, so $\mathbb{N} \subseteq \mathbb{Q}$. For example, $7 = \frac{7}{1}$.

7 **a** false **b** true

8 **a** $\{3, 4, 5, ..., 12, 13\}$ **b** $\{1, 2, 3, ..., 9, 10\}$

9 **a** $\{x \in \mathbb{Z} \mid -4 \leqslant x \leqslant 1\}$ **b** $\{x \mid x < 1\}$

10 

11 **a** $n(P) = 6, \quad n(G) = 5$
b **i** $P \cap G = \{\text{thyme, basil}\}$
This set represents the herbs for sale at both "Pete's Plants" and "Garden World".
ii $P \cup G = \{\text{lavender, rosemary, parsley, thyme, mint, basil, sage, lemongrass, chives}\}$
This set represents the herbs for sale at either "Pete's Plants" or "Garden World" or both.

iii $P' \cap G = \{\text{sage, lemongrass, chives}\}$
This set represents the herbs for sale at "Garden World" which are not for sale at "Pete's Plants".

c no; $P \cap G = \{\text{thyme, basil}\} \neq \varnothing$

12 a infinite
b i A is the set of real numbers greater than or equal to -3 and less than 7.
ii $A = \{x \mid -3 \leqslant x < 7\}$
c $A \cap \mathbb{N} = \{x \in \mathbb{N} \mid x < 7\}$
d i A' is the set of real numbers less than -3 or greater than or equal to 7.
ii $A' = \{x \mid x < -3 \text{ or } x \geqslant 7\}$
iii 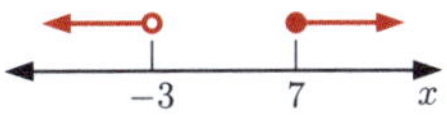

EXERCISE 6A

1 a $x=-4$ b $x=7$ c $x=5$ d $x=-11$ e $x=100$ f $x=-12$ g $x=10$ h $x=-32$

2 a $x=3$ b $x=3$ c $x=2$ d $x=-7$ e $x=12$ f $x=-11$ g $x=6$ h $x=2$ i $x=-3$ j $x=-2$ k $x=\frac{17}{8}$ l $x=\frac{1}{3}$

3 a $x=-20$ b $x=-17$ c $x=13$ d $x=-\frac{2}{3}$ e $x=5$ f $x=\frac{19}{2}$ g $x=-\frac{8}{3}$ h $x=18$ i $x=\frac{19}{5}$

4 a $x=3$ b $x=5$ c $x=-1$ d $x=13$

5 a $x=3$ b $x=4$ c $x=-5$ d $x=-\frac{3}{2}$ e $x=-1$ f $x=-3$ g $x=4$ h $x=1$ i $x=-6$ j $x=\frac{1}{2}$

6 a $x=5$ b $x=2$ c $x=\frac{3}{5}$ d $x=-\frac{9}{5}$ e $x=\frac{3}{7}$ f $x=\frac{3}{2}$

7 a $x=-\frac{5}{4}$ b $x=-\frac{1}{8}$ c $x=1$ d $x=2$ e $x=-7$ f $x=5$ g $x=\frac{26}{5}$ h $x=-\frac{2}{17}$

8 a The equation reduces to $-5=-5$. This equation is an identity, which is true for all x.
b There is no solution for x as the equation reduces to $-12=3$.

EXERCISE 6B

1 a $x=\frac{10}{9}$ b $x=2$ c $x=-3$ d $x=\frac{13}{11}$ e $x=\frac{24}{19}$ f $x=\frac{7}{8}$ g $x=-4$ h $x=3$ i $x=-2$

2 a $x=12$ b $x=-\frac{24}{5}$ c $x=-\frac{22}{3}$ d $x=-\frac{9}{7}$ e $x=4$ f $x=\frac{3}{4}$

3 a $x=\frac{36}{11}$ b $x=1$ c $x=-44$ d $x=-\frac{7}{5}$ e $x=-25$ f $x=\frac{4}{17}$

4 a $x=\frac{1}{3}$ b $x=-\frac{1}{4}$ c $x=\frac{3}{2}$ d $x=-\frac{1}{6}$ e $x=\frac{21}{2}$ f $x=\frac{10}{3}$ g $x=\frac{7}{6}$ h $x=-28$

5 $\dfrac{9}{6x}=\dfrac{2}{6x}$ $\therefore$ $9=2$, which is absurd.
The equation has no solutions.

6 a $x=-2$ b $x=-\frac{1}{2}$ c $x=-1$ d $x=\frac{17}{2}$ e $x=-12$ f $x=-1$

g $x=-\frac{13}{10}$ h $x=-\frac{1}{4}$ i $x=\frac{2}{3}$

7 a $x=2$ b $x=\frac{1}{2}$ c $x=\frac{13}{4}$

EXERCISE 6C.1

1 a The number is 9. b The number is 5. c The number is 7. d The number is 91. e The number is -4. f The number is 12.

2 7 packs **3** 6 shirts **4** There are 7 passengers in each row.

5 Paige ate 5 sweets. **6** Carol drank 580 mL of water.

7 The number is 11. **8** The fraction is $\frac{8}{12}$.

9 The fraction is $\frac{-2}{-6}$. **10** There were 4 people at Table A.

EXERCISE 6C.2

1 Isaac buys 6 shirts. **2** eight 50 cent coins **3** 6 apples

4 Ellie's son is 10 years old now **5** Adrian is 7 years old now

6 Lana has 11 20 cent coins and 15 50 cent coins.

7 2700 \$20 tickets, 900 \$35 tickets, and 1300 \$50 tickets were sold.

8 Olivia used 9 kg of dried fruit and 21 kg of nuts.

9 Tom invested \$4500 in Canadian Airways shares.

EXERCISE 6C.3

1 $1\frac{1}{2}$ L **2** 2 L **3** $1\frac{1}{3}$ L

4 Dominic should mix 10 L of 4% saline mixture and 5 L of 7% saline mixture together.

EXERCISE 6D

1 a $S \leqslant 40$ b $A \geqslant 18$ c $a > 3$ d $b \leqslant -3$ e $d < 5$ f $-20 \geqslant x$ g $4 < y$ h $z \geqslant 0$

2 a $x > 2$ b $b < 5$ c $c \geqslant 2\frac{1}{2}$ d $d \leqslant -7$ e $a < -19$ f $p > -3$

3 a 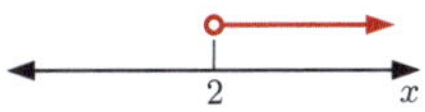

b 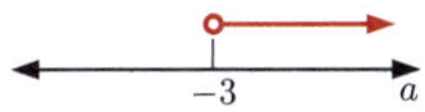

c 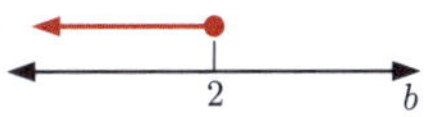

d 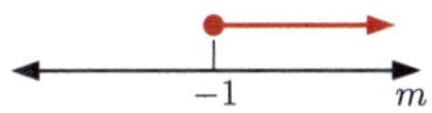

e 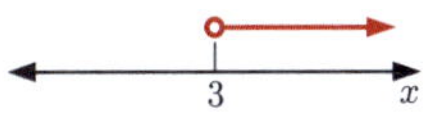

f 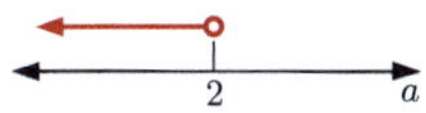

g 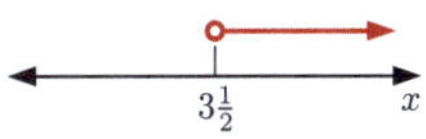

h 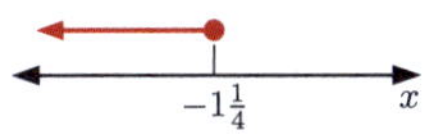

4 a yes b no c no d yes

5 a 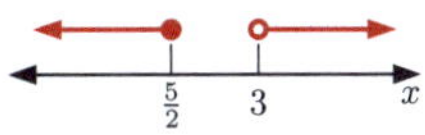

b $x=3$

6 16 **7** 55 **8** 3 **9** -30

EXERCISE 6E

1 a $a > 2$ 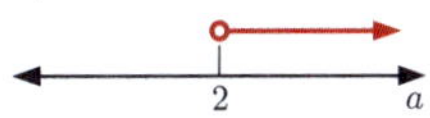

b $b \leqslant -3$ 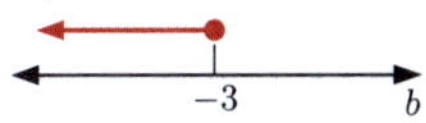

c $s < 6$ 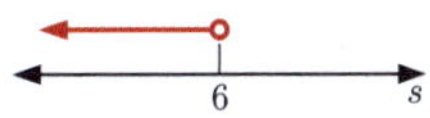

d $c < 10$ 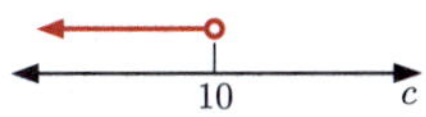

e $x \geqslant 5$ **f** $b < -4$

g $t > -5$ **h** $m \leqslant -60$

2 **a** $x > 3$ **b** $a \geqslant \frac{10}{3}$

c $a < -2$ **d** $b > 5$

e $s > -2$ **f** $b \geqslant \frac{1}{3}$

g $n < -2$ **h** $x < 13$

i $b \geqslant \frac{7}{4}$

3 **a** $x > 9$ **b** $x \leqslant -5$

c $x \geqslant -\frac{14}{3}$ **d** $x > -6$

e $x > \frac{5}{2}$ **f** $x < \frac{28}{3}$

g $x < 15$ **h** $x \leqslant -2$

i $x \geqslant \frac{17}{3}$

4 **a** $x < 1$ **b** $c \geqslant -2$ **c** $b < -2$ **d** $a \geqslant 21$
e $d > 14$ **f** $p \leqslant 0$

5 **a** $x > 2$ **b** $x \leqslant -3$

c $x < -1$ **d** $x < \frac{4}{5}$

e $x \leqslant -3$ **f** $x \geqslant -2$

g $x \leqslant \frac{3}{2}$ **h** $x < -\frac{4}{5}$

i $x < \frac{7}{17}$

6 **a** $x < -\frac{17}{3}$ **b** $x \geqslant 13$ **c** $x \leqslant \frac{7}{6}$
d $x > 3$ **e** $x \geqslant 2$ **f** $x > -\frac{23}{14}$

REVIEW SET 6A

1 **a** $x = -7$ **b** $x = -2$ **c** $x = -2$
2 **a** $x = \frac{47}{5}$ **b** $x = -9$ **3** The number is 6.
4 **a** $x = \frac{26}{3}$ **b** $x = -\frac{9}{11}$ **c** $x = \frac{14}{15}$
5 **a** $x = \frac{13}{12}$ **b** $x = -\frac{21}{5}$ **6** The number is $\frac{13}{5}$.
7 Evan is 7 years old now.
8 Elaine has 36 5 cent coins and 52 10 cent coins.
9 Simon invested \$2400 in technology shares and \$4800 in petroleum shares.
10 **a** **b** **c**
11 **a** $x < 10$ **b** $x \leqslant \frac{7}{3}$
c $x < -\frac{9}{2}$
12 **a** $x < -\frac{1}{3}$ **b** $x \leqslant -\frac{4}{7}$
13 **a** $7\frac{1}{2}$ L **b** **i** 0.2 L **ii** 0.32 L

REVIEW SET 6B

1 **a** $x = 7$ **b** $x = -28$
2 **a** $x = \frac{1}{9}$ **b** $x = \frac{19}{15}$ **3** Dara went on 4 rides.
4 **a** $x = -5$ **b** $x = \frac{39}{2}$ **c** $x = -\frac{15}{2}$
5 The fraction is $\frac{20}{24}$. **6** Amanda is 9 years old now.
7 40 90 g tins and 72 200 g tins were sold.
8 $2\frac{1}{2}$ L **9** **a** $k < -2$ **b** $7 \geqslant y$
10 **a** $x \leqslant -4$ **b** $x > -11$
c $x \geqslant 0$
11 **a** **i** 12 **ii** 17 **b** **i** 7 **ii** 26
12 **a** $x > -10$ **b** $x \leqslant \frac{1}{3}$
c $-10 < x \leqslant \frac{1}{3}$

EXERCISE 7A

1 **a**

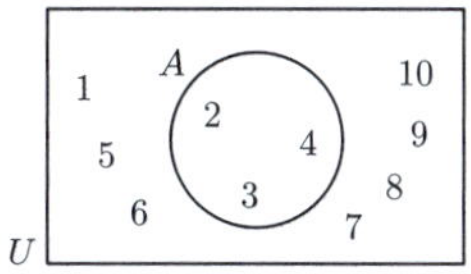

b

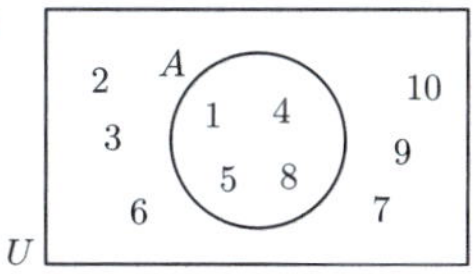

c

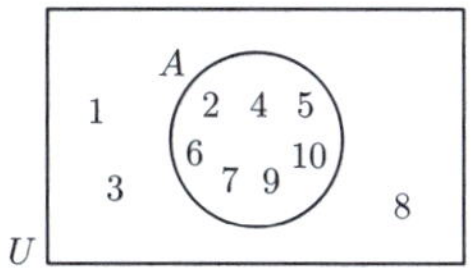

2 **a** $A = \{a, c, f, h, i\}$ **b** $B = \{c, d, g, i\}$
c $A' = \{b, d, e, g\}$ **d** $B' = \{a, b, e, f, h\}$
e $A \cap B = \{c, i\}$ **f** $A \cup B = \{a, c, d, f, g, h, i\}$
g $U = \{a, b, c, d, e, f, g, h, i\}$

3 **a**

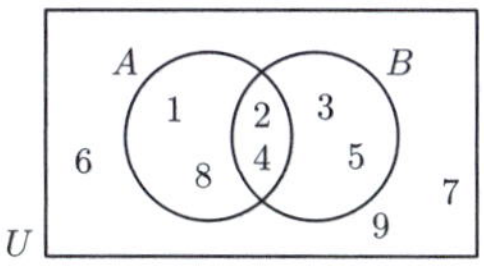

b

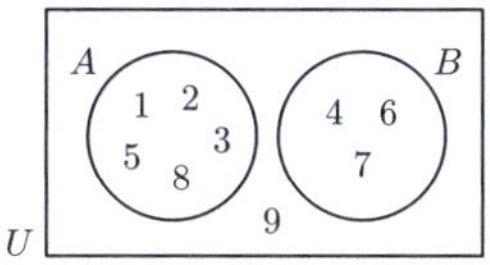

c

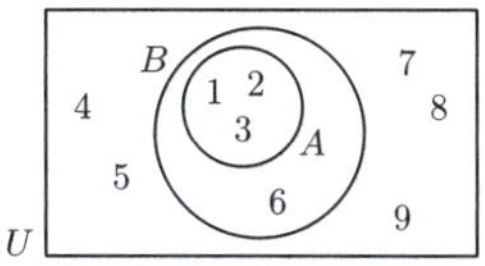

d

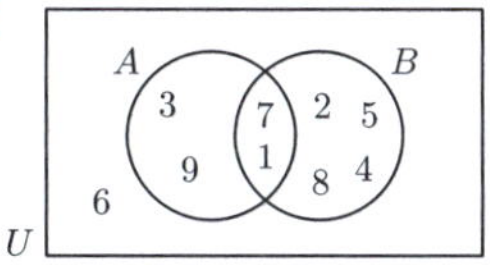

4 **a** **i** $A = \{1, 2, 3, 4, 6, 12\}$ **ii** $B = \{2, 3, 5, 7, 11\}$
iii $A \cap B = \{2, 3\}$
iv $A \cup B = \{1, 2, 3, 4, 5, 6, 7, 11, 12\}$

b

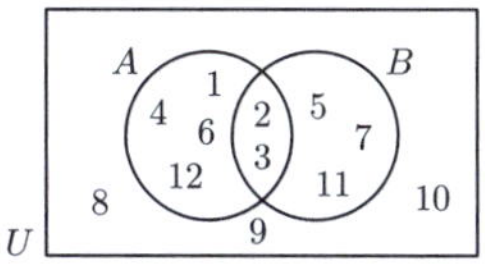

5 **a** **i** $R = \{2, 3, 5, 7, 11, 13, 17, 19\}$
ii $S = \{4, 6, 8, 9, 10, 12, 14, 15, 16, 18\}$
iii $R \cap S = \varnothing$
iv $R \cup S = \{2, 3, 4, 5, 6, 7, 8, 9, 10, 11, 12, 13, 14, 15, 16, 17, 18, 19\}$

b

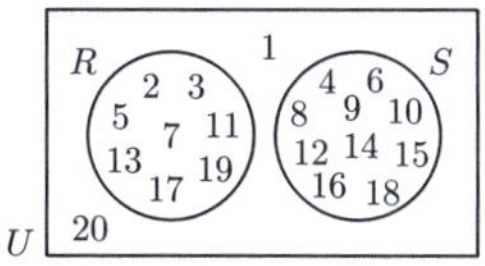

6 **a** $M =$ {Shaun, Ken, Ling, Don, Vivian},
$H =$ {Paul, Ken, Pam, Ling, Stefan, Reg, Don}

b

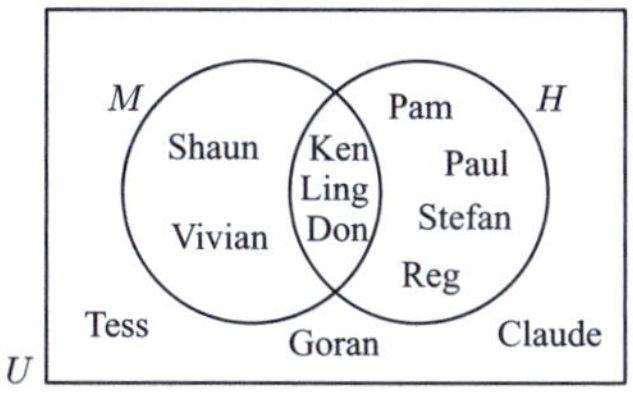

c $M \cup H =$ {Paul, Shaun, Ken, Pam, Ling, Stefan, Reg, Don, Vivian}
This set represents the students who scored 15 or more for at least one of the tests.

d $M \cap H' =$ {Shaun, Vivian}
This set represents the students who scored 15 or more for the Mathematics test, but less than 15 for the History test.

7 **a** $Z \subseteq Q$ **b**

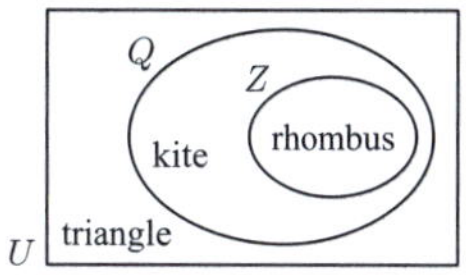

8 **a**

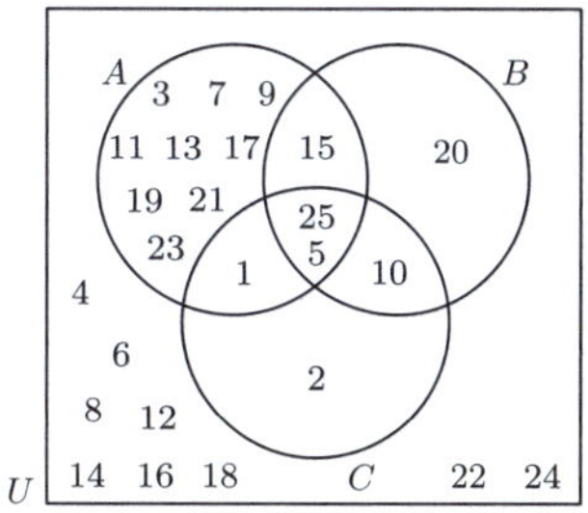

b **i** {5, 15, 25}
ii {3, 7, 9, 11, 13, 15, 17, 19, 21, 23} **iii** {10}

EXERCISE 7B

1 **a**

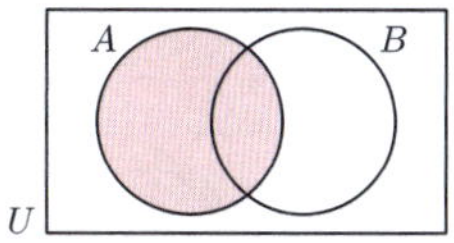

b

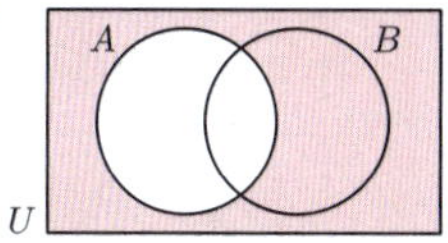

c

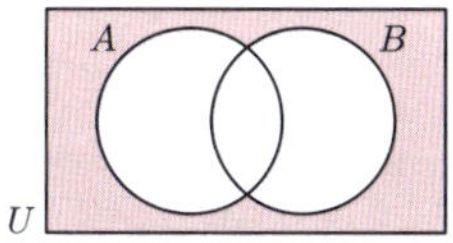

d

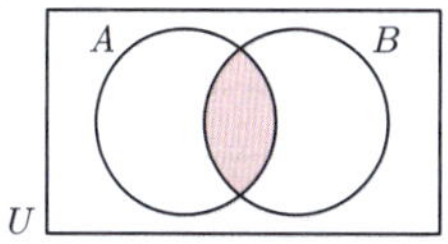

e

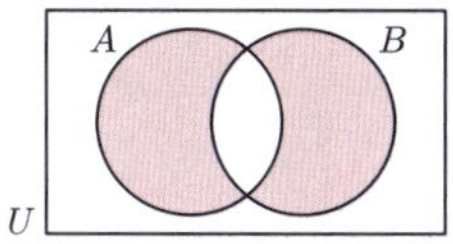

2 **a**

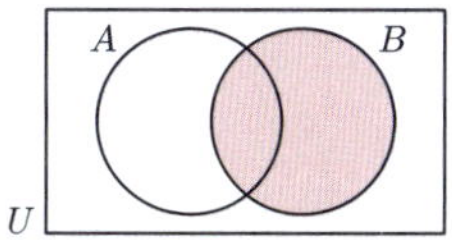

b

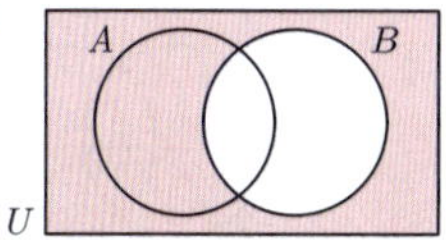

c

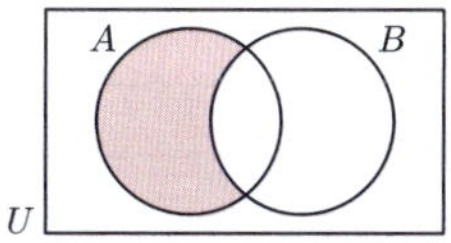

d

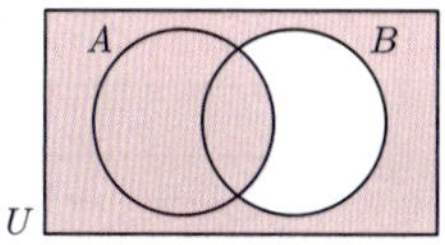

e

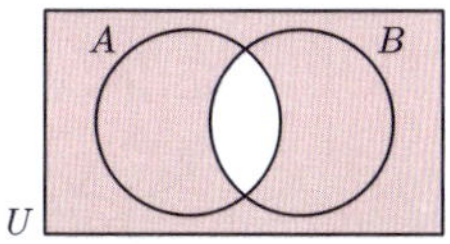

f

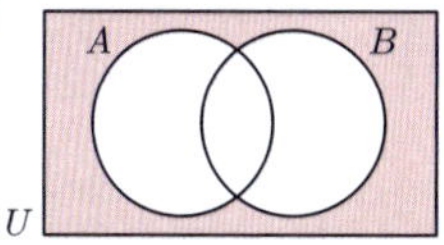

3 **a** in X but not in Y
b in both X and Y or in neither X nor Y
c in neither X nor Y

4 **a**
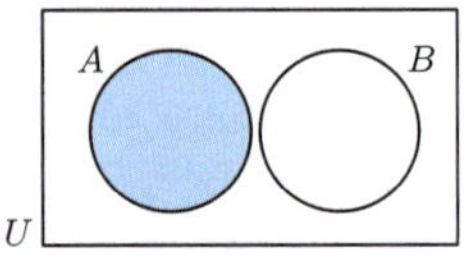

b
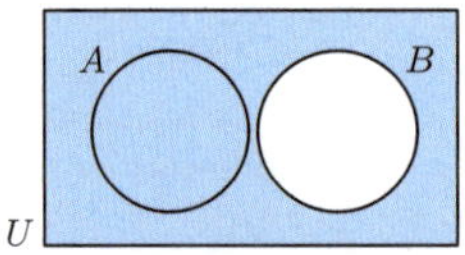

c
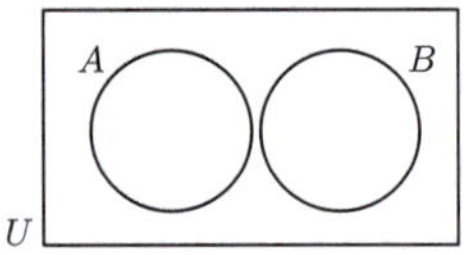

d
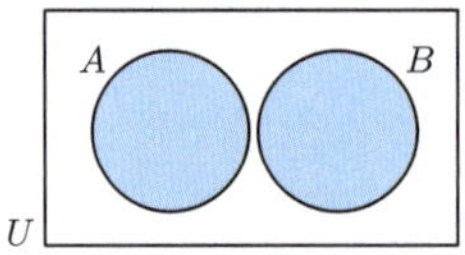

e
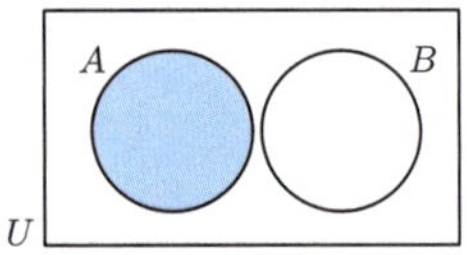

f
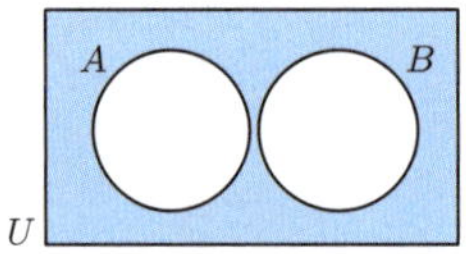

5 **a**
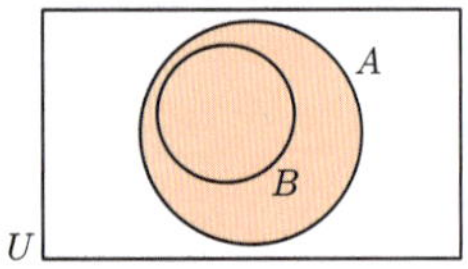

b
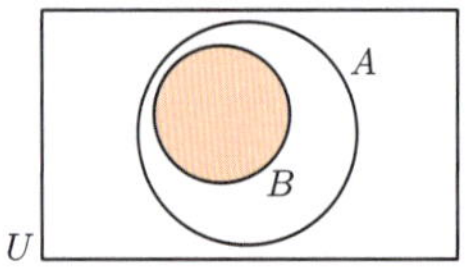

c
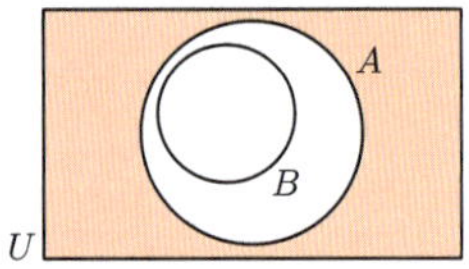

d
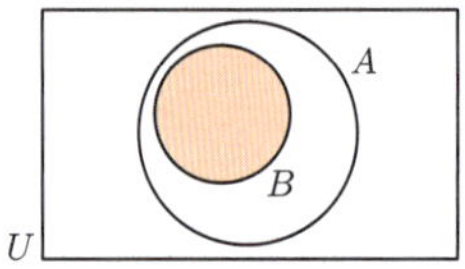

e
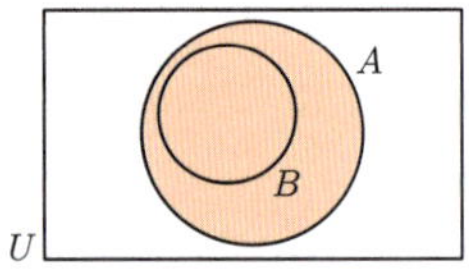

f
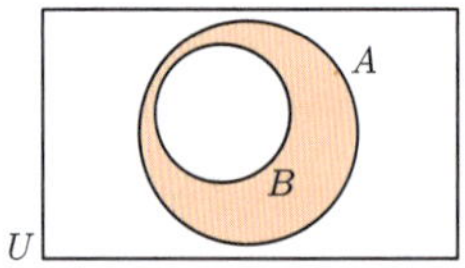

6 **a**
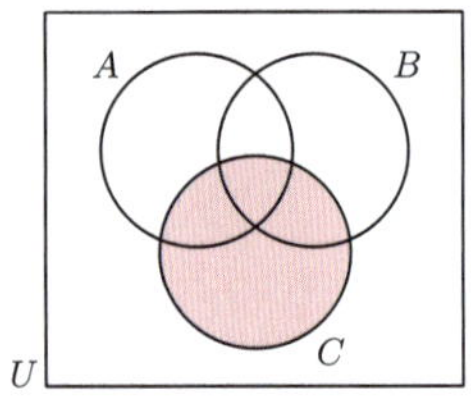

b
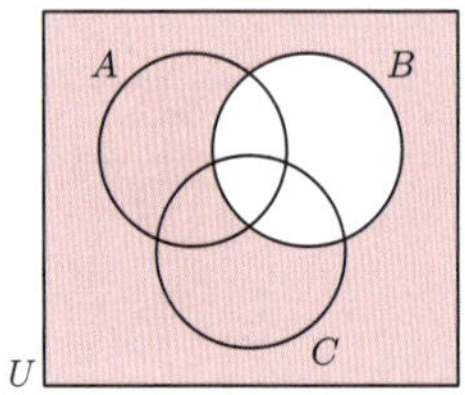

c
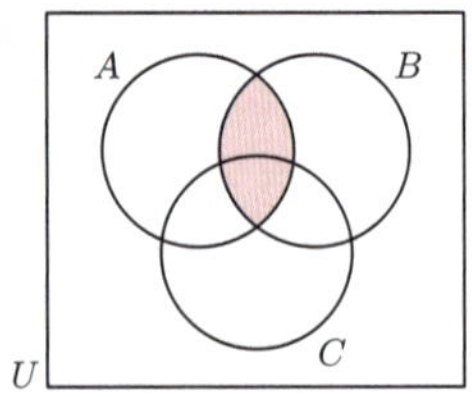

d
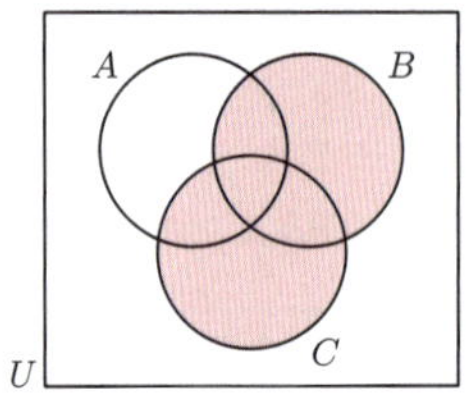

e
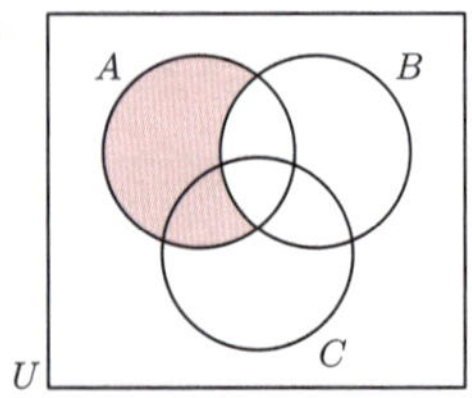

f
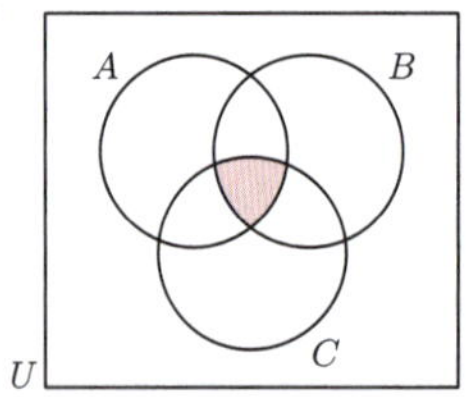

EXERCISE 7C

1 **a** 7 **b** 9 **c** 15 **d** 8 **e** 5 **f** 19

2 **a** $b+c$ **b** $c+d$ **c** $a+c$ **d** b **e** $a+c+d$ **f** d

3 **a** 12 **b** 16 **c** 42 **d** 18 **e** 8 **f** 4

4 **a** **i** $n(Q) = 2a - 4$ **ii** $n(P \cup Q) = 5a - 4$
iii $n(Q') = 4a + 5$ **iv** $n(U) = 6a + 1$
b $a = 7$

5 **a** $n(A) + n(B) - n(A \cap B) = a + b + b + c - b$
$= a + b + c$
$= n(A \cup B)$ ✓

b If A and B are disjoint sets, then $n(A \cap B) = 0$.
So, $n(A \cup B) = n(A) + n(B) - n(A \cap B)$
$= n(A) + n(B)$

6 **a**
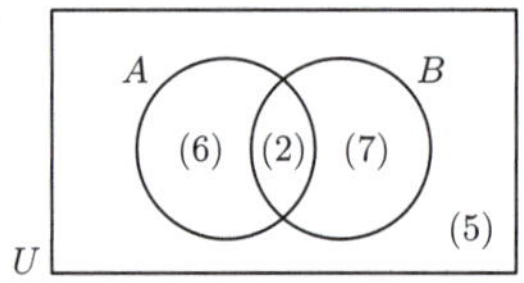

b **i** $n(A \cup B) = 15$
ii $n(B \cap A') = 7$

7 **a**
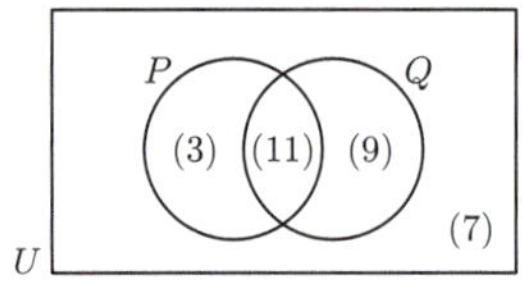

b **i** 3
ii 12

8 **a** $n(R) = 19$ **b** $n((N \cup R)') = 4$
c $n(N' \cap R) = 15$

9 **a**
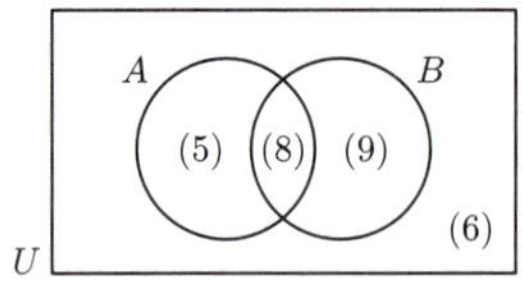

b **i** 8
ii 5

10 **a** 15 **b** 31 **11** **a** 17 **b** 18

EXERCISE 7D

1 **a** 48 members **b** 27 members **c** 23 members
d 16 members **e** 30 members

2 **a**
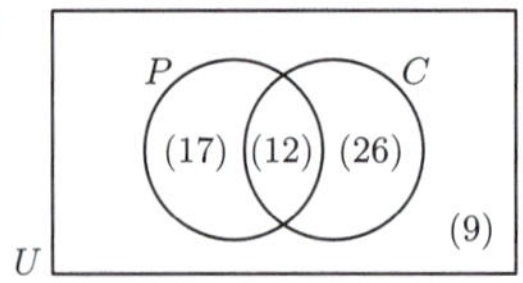

b **i** 17 entrants
ii 26 entrants
iii 9 entrants

3 **a**
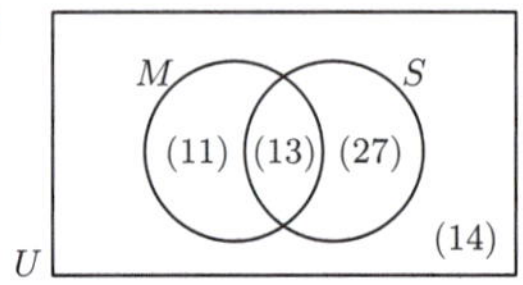

b **i** 38 students
ii 14 students

4 **a**
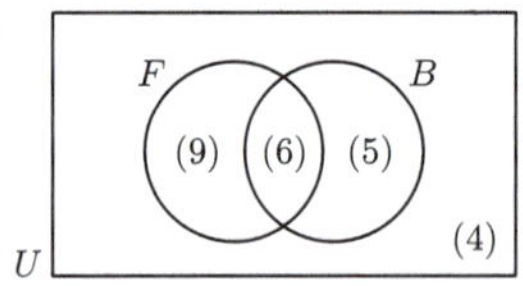

b **i** 5 swimmers
ii 20 swimmers
iii 9 swimmers
iv 4 swimmers
v 14 swimmers

5 **a**

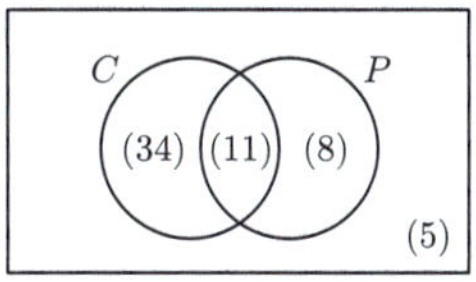

b **i** 13 households **ii** 53 households **iii** 42 households **iv** 8 households **v** 34 households

6 **a** 12 books **b** 14 books

7 **a**

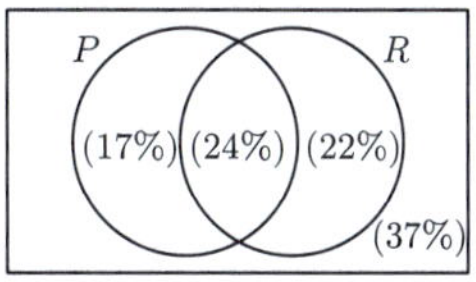

b **i** 24% **ii** 63% **iii** 17% **iv** 39%

8 **a** 150 people **b** 150 people

9 **a** 7 shops **b** 12 shops **c** 18 shops

10 39 guests **11** 1 worker

12 **a** $n(P \cap E \cap D) = 8$

The shapes in $P \cap E \cap D$ are quadrilaterals with at least one pair of parallel sides, at least two equal sides, and diagonals which intersect at right angles.

b

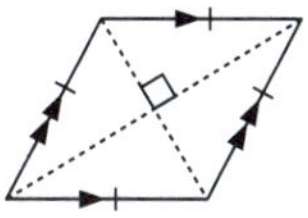

$P \cap E \cap D$

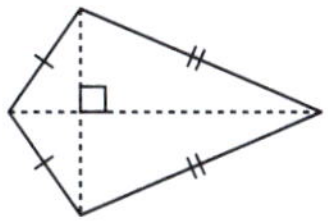

$P' \cap E \cap D$

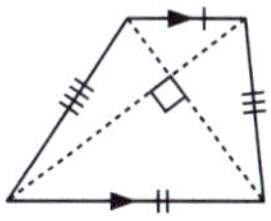

$P \cap E' \cap D$

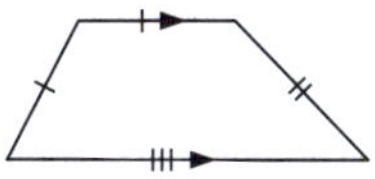

$P \cap E \cap D'$

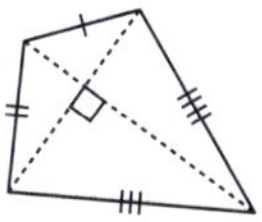

$P' \cap E' \cap D$

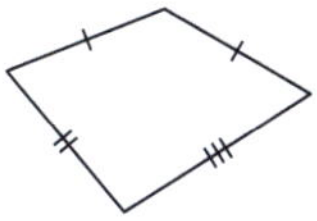

$P' \cap E \cap D'$

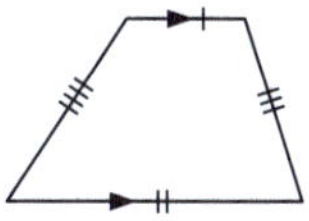

$P \cap E' \cap D'$

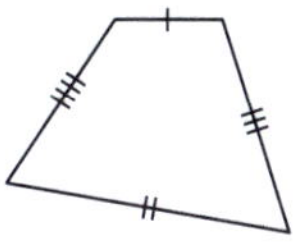

$P' \cap E' \cap D'$

REVIEW SET 7A

1 **a**

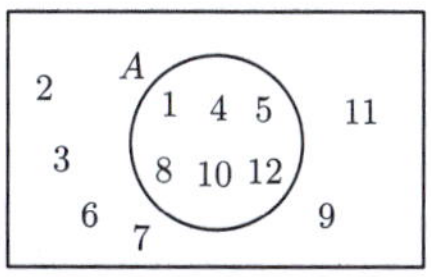

b

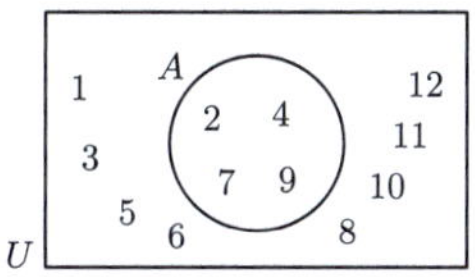

2 **a** **i** $A = \{1, 2, 3, 4, 5\}$ **ii** $B = \{1, 2, 7\}$

iii $U = \{1, 2, 3, 4, 5, 6, 7\}$

iv $A \cup B = \{1, 2, 3, 4, 5, 7\}$ **v** $A \cap B = \{1, 2\}$

b **i** $n(A) = 5$ **ii** $n(B) = 3$ **iii** $n(A \cup B) = 6$

3 **a** **i** $A' = \{c, f, g, h\}$ **ii** $A \cap B = \{a, b\}$

iii $A \cup B = \{a, b, d, e, g\}$

b

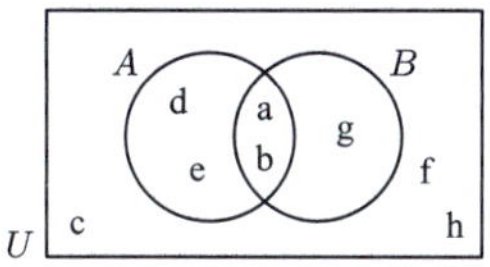

4 **a** not in X **b** in X or Y, but not both

c in X or not in Y

5 **a**

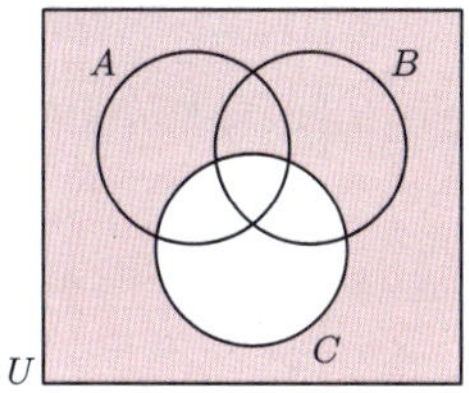

b

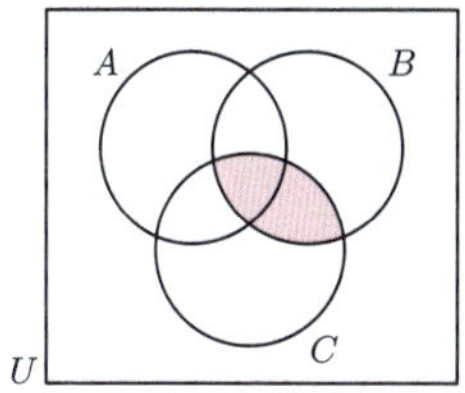

c

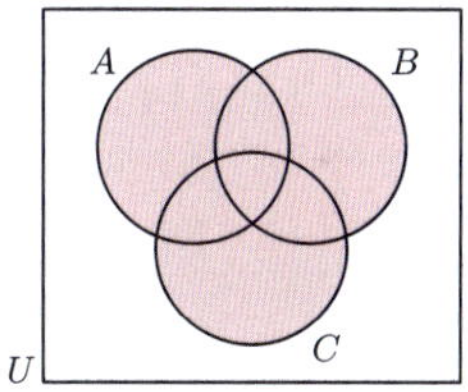

6 **a** 10 **b** 7 **c** 11

7 **a**

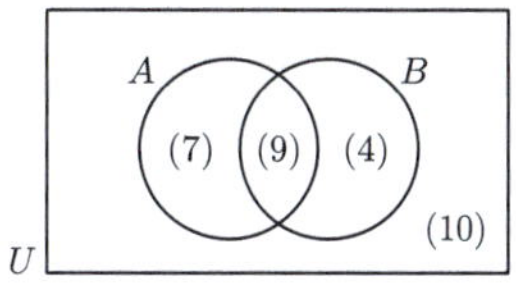

b **i** 4 **ii** 10

8 **a**

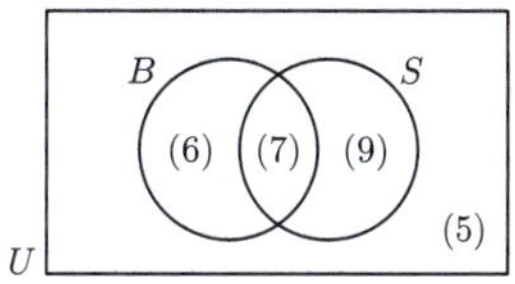

b **i** 5 students **ii** 9 students

9 **a**

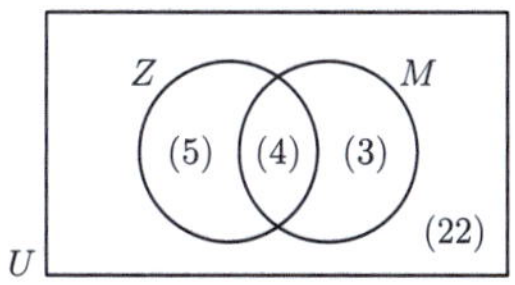

b **i** 5 times **ii** 3 times

10 **a**

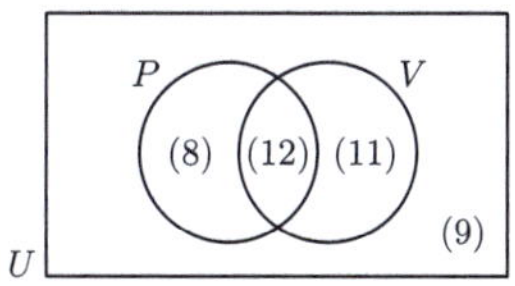

b 12 days

11 **a**

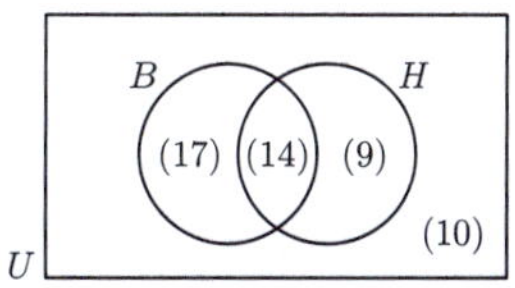

b **i** 17 tourists **ii** 14 tourists

12 **a**

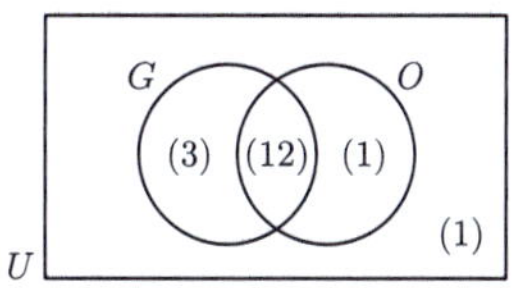

b **i** 3 dishes **ii** 1 dish **iii** 12 dishes

REVIEW SET 7B

1

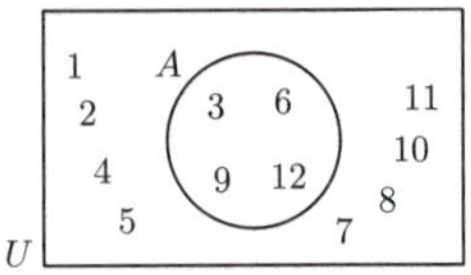

2 a $P = \{c, d, f, h, i\}$ b $Q' = \{a, b, d, e, g, h\}$

3 a

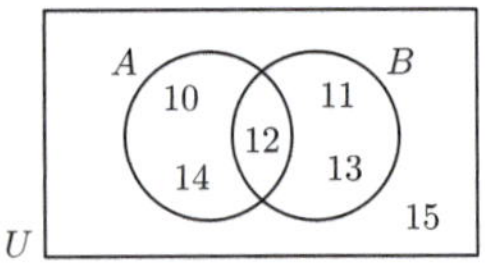

b

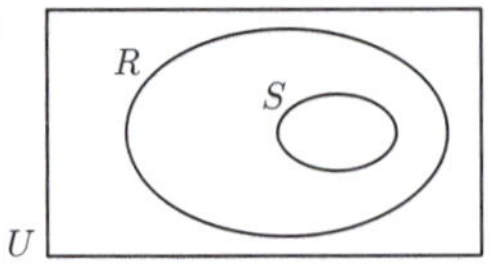

4

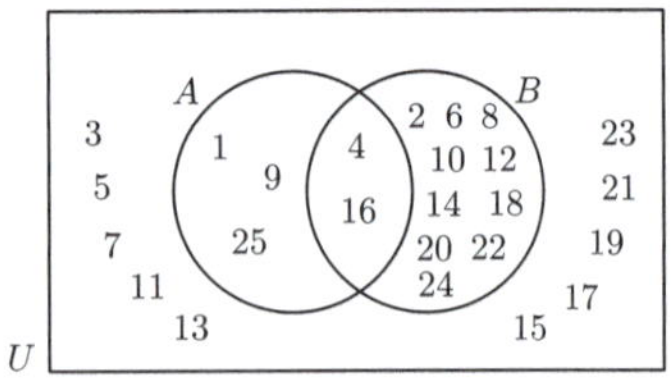

5 a

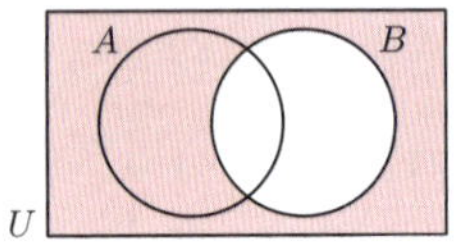

b

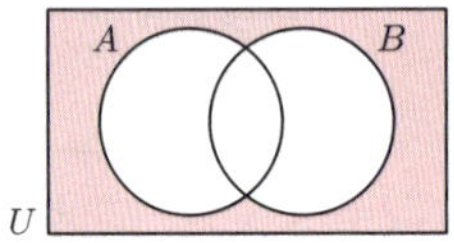

6 a 5 b 28 c 36 7 a 11 b 8

8 a $n(Y \cup Z) = 3x + 1$ b $x = 6$

9 a

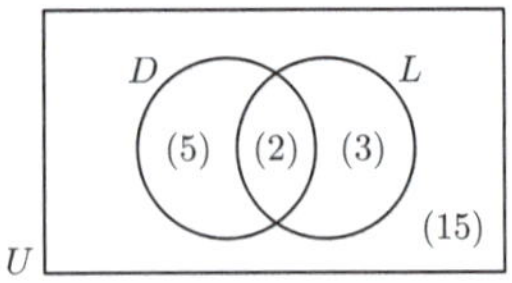

b 15 students

10 a 19 apartments b 25 apartments

11 a

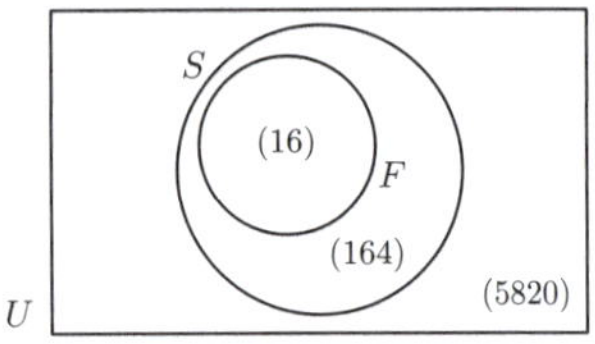

b i 164 people
ii 5820 people

c $\approx 0.267\%$

12 a, c

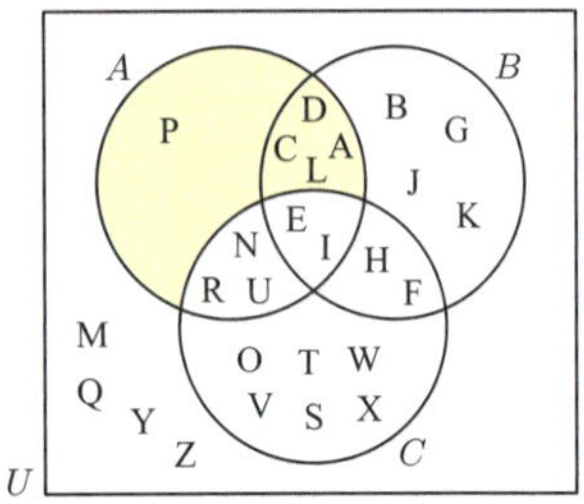

b $n(A \cap B) = 6$

c $A \cap C'$ represents the letters in the word PERPENDICULAR which are not used when writing the first six counting numbers.

d $(A \cup B \cup C)' = \{M, Q, Y, Z\}$

EXERCISE 8A

1 a 2 b 3 c 1 d 6 e 0 f 11
g 9 h 12

2 a $\frac{1}{3}$ b $\frac{1}{5}$ c $\frac{1}{8}$ d $\frac{2}{3}$

3 a ≈ 1.732 b ≈ 3.742 c ≈ 4.690 d ≈ 5.568
e ≈ 8.485 f ≈ 11.402 g ≈ 12.083 h ≈ 14.142
i ≈ 1.476 j 0.3 k ≈ 0.791 l ≈ 2.343

4 a surd, ≈ 3.162 b not a surd, 10 c surd, ≈ 8.062
d surd, ≈ 2.646 e surd, ≈ 5.196 f not a surd, 16
g not a surd, $\frac{1}{2}$ h not a surd, $\frac{3}{4}$ i surd, ≈ 0.480
j not a surd, 0.2 k surd, ≈ 1.039 l not a surd, 1.5

5 a 1 and 2 b 4 and 5 c 2 and 3 d 4 and 5
e 7 and 8 f 10 and 11 g 9 and 10 h 8 and 9

6 a 5 b 13 c 19 d 20 e 27 f 80
g 72 h 24 i 50 j 48 k 44 l 63
m 5 n 3 o 8 p 45

7 a 6 b 14 c 12 d 66 e -10 f -6
g 16 h 72

8 a 7 b 49 c $\frac{1}{7}$ d $\frac{9}{7}$ e 7 f $\frac{1}{7}$
g $\frac{7}{4}$ h $\frac{1}{49}$

EXERCISE 8B

1 a $\sqrt{10}$ b $\sqrt{6}$ c $\sqrt{35}$ d $\sqrt{66}$ e $\sqrt{230}$
f $\sqrt{51}$ g $\sqrt{65}$ h $\sqrt{70}$ i $\sqrt{30}$

2 a $\sqrt{4} \times \sqrt{9} = 2 \times 3 = 6$ b $\sqrt{4} \times \sqrt{9} = \sqrt{36} = 6$

3 a $2\sqrt{6}$ b $5\sqrt{6}$ c $6\sqrt{10}$ d $-2\sqrt{15}$
e $6\sqrt{21}$ f $-12\sqrt{10}$

4 a $\sqrt{2}$ b $\sqrt{2}$ c $\sqrt{13}$ d $\sqrt{7}$ e $\sqrt{34}$ f $\sqrt{21}$

5 a $\frac{3}{4}$ b $\frac{6}{5}$ c $\frac{4}{9}$ d $\frac{3}{8}$ e $\frac{10}{7}$ f $\frac{11}{9}$

6 a $\frac{3}{2}$ b $\frac{2}{3}$ c $\frac{2}{5}$ d $\frac{3}{5}$ e $\frac{5}{4}$ f $\frac{10}{3}$

7 a $2\sqrt{2}$ b $3\sqrt{2}$ c $5\sqrt{2}$ d $7\sqrt{2}$
e $10\sqrt{2}$ f $12\sqrt{2}$ g $100\sqrt{2}$ h $\frac{1}{2}\sqrt{2}$

8 a $2\sqrt{3}$ b $3\sqrt{3}$ c $5\sqrt{3}$ d $\frac{1}{3}\sqrt{3}$

9 a $2\sqrt{5}$ b $3\sqrt{5}$ c $5\sqrt{5}$ d $\frac{1}{5}\sqrt{5}$

EXERCISE 8C

1 a $2\sqrt{2}$ b $2\sqrt{5}$ c $3\sqrt{2}$ d $4\sqrt{2}$
e $3\sqrt{3}$ f $2\sqrt{13}$ g $4\sqrt{3}$ h $5\sqrt{2}$
i $6\sqrt{3}$ j $5\sqrt{3}$ k $3\sqrt{5}$ l $3\sqrt{11}$
m $2\sqrt{15}$ n $2\sqrt{22}$ o $2\sqrt{35}$ p $5\sqrt{11}$

2 a $\frac{\sqrt{96}}{\sqrt{6}} = \sqrt{\frac{96}{6}} = \sqrt{16} = 4$ b $\frac{\sqrt{96}}{\sqrt{6}} = \frac{4\sqrt{6}}{\sqrt{6}} = 4$

3 a $2028 = 2^2 \times 3 \times 13^2$ b $26\sqrt{3}$

EXERCISE 8D

1 a 1 b -1 c 2 d -2 e 0
f -4 g 4 h -6 i -10 j 7

2 a 3 b 2 c 5 d 3 e 2

3 a ≈ 1.587 b ≈ -2.520 c ≈ 4.642
d ≈ 4.606 e ≈ -14.422

4 a real b not a real number c real d real
e not a real number

5 a $\frac{1}{5}$ b $-\frac{2}{3}$ c $\frac{7}{4}$ d $-\frac{5}{3}$

6 a $216 = 2^3 \times 3^3$ b 6

EXERCISE 8E

1 **a** $x=\pm 3$ **b** $x=\pm 7$ **c** $x=\pm 6$ **d** $x=0$ **e** $x=\pm 1$ **f** $x=\pm\sqrt{17}$ **g** $x=\pm\sqrt{23}$ **h** $x=\pm 10$ **i** no real solutions **j** no real solutions **k** $x=\pm\sqrt{27}$ **l** no real solutions

2 **a** $x=2$ **b** $x=3$ **c** $x=0$ **d** $x=4$ **e** $x=-1$ **f** $x=-5$ **g** $x=\frac{1}{2}$ **h** $x=\sqrt[3]{36}$ (≈ 3.30)

3 **a** $x=\pm 2$ **b** $x=\pm 3$ **c** $x=\pm 5$ **d** $x=\pm 4$ **e** $x=\pm 6$ **f** $x=\pm 5$ **g** $x=\pm\sqrt{50}$ **h** $x=\pm\sqrt{44}$ **i** $x=\pm\sqrt{12}$

4 **a** $x=1$ **b** $x=-2$ **c** $x=6$ **d** $x=100$ **e** $x=\sqrt[3]{16}$ (≈ 2.52) **f** $x=\sqrt[3]{-800}$ (≈ -9.28)

5 **a** $x=\pm 3$ **b** $x=\pm 4$ **c** $x=\pm 2$ **d** $x=\pm 4$ **e** $x=\pm\sqrt{30}$ **f** $x=\pm\sqrt{20}$ **g** $x=\pm 3$ **h** $x=\pm\sqrt{7}$ **i** $x=\pm\sqrt{\frac{1}{5}}$

EXERCISE 8F

1 **a** $2\sqrt{2}$ **b** 0 **c** $\sqrt{2}$ **d** $\sqrt{3}$ **e** $7\sqrt{7}$ **f** $-3\sqrt{5}$ **g** $-3\sqrt{2}$ **h** $6\sqrt{5}$ **i** $6\sqrt{2}$ **j** $-3\sqrt{2}$ **k** $4\sqrt{3}$ **l** $4\sqrt{5}-10$

2 **a** $2\sqrt{2}+7\sqrt{3}$ **b** $5\sqrt{2}-\sqrt{3}$ **c** $-7\sqrt{2}+4\sqrt{3}$ **d** $11\sqrt{5}-5\sqrt{2}$ **e** $2\sqrt{2}-10\sqrt{7}$ **f** $2\sqrt{2}+3\sqrt{11}$ **g** $1-3\sqrt{2}$ **h** $9\sqrt{3}-13$

3 **a** $6\sqrt{2}$ **b** $8\sqrt{3}$ **c** $4\sqrt{5}$

4 **a** $\sqrt{12}+\sqrt{27}=2\sqrt{3}+3\sqrt{3}=5\sqrt{3}=\sqrt{75}$ **b** $\sqrt{80}-\sqrt{5}=4\sqrt{5}-\sqrt{5}=3\sqrt{5}=\sqrt{45}$

5 **a** $12+4\sqrt{2}$ **b** $3\sqrt{2}+3\sqrt{3}$ **c** $20-5\sqrt{7}$ **d** $6\sqrt{11}-24$ **e** $\sqrt{2}+2$ **f** $2-5\sqrt{2}$ **g** $2\sqrt{3}+6$ **h** $3-\sqrt{6}$ **i** $6\sqrt{5}-5$ **j** $10-\sqrt{5}$ **k** $10+\sqrt{15}$ **l** $2\sqrt{7}+7+\sqrt{14}$

6 **a** $-4\sqrt{2}-2$ **b** $-3\sqrt{2}+2$ **c** $-2+\sqrt{14}$ **d** $-3\sqrt{3}-3$ **e** $-5\sqrt{3}+3$ **f** $-6-\sqrt{15}$ **g** $-2\sqrt{10}+\sqrt{15}$ **h** $-4-2\sqrt{6}$ **i** $-20\sqrt{2}+20$ **j** $-14-4\sqrt{7}$ **k** $-2\sqrt{11}+11$ **l** $-2\sqrt{3}+4\sqrt{6}$

7 **a** $7+6\sqrt{2}$ **b** $11+6\sqrt{2}$ **c** $\sqrt{2}$ **d** $9+\sqrt{3}$ **e** 1 **f** $1-\sqrt{7}$ **g** 9 **h** $2\sqrt{2}+1$ **i** $8+2\sqrt{2}$

8 **a** $4+2\sqrt{3}$ **b** $27+10\sqrt{2}$ **c** $11-6\sqrt{2}$ **d** $8+2\sqrt{7}$ **e** $5-2\sqrt{6}$ **f** $21-8\sqrt{5}$ **g** $8+2\sqrt{15}$ **h** $15-6\sqrt{6}$ **i** $9-6\sqrt{2}$ **j** $17+12\sqrt{2}$ **k** $17-12\sqrt{2}$ **l** $59-30\sqrt{2}$

9 **a** 7 **b** 2 **c** 22 **d** -13 **e** -2 **f** 2 **g** -18 **h** -16 **i** 14 **j** 3 **k** -4 **l** 7

EXERCISE 8G

1 **a** $\frac{\sqrt{3}}{3}$ **b** $\sqrt{3}$ **c** $3\sqrt{3}$ **d** $\frac{11\sqrt{3}}{3}$ **e** $\sqrt{2}$ **f** $3\sqrt{2}$ **g** $6\sqrt{2}$ **h** $\frac{\sqrt{6}}{2}$ **i** $\sqrt{5}$ **j** $3\sqrt{5}$ **k** $-\frac{3\sqrt{5}}{5}$ **l** $40\sqrt{5}$ **m** $\sqrt{7}$ **n** $3\sqrt{7}$ **o** $\frac{2\sqrt{11}}{11}$ **p** $2\sqrt{13}$

2 **a** $\frac{\sqrt{15}}{3}$ **b** $\frac{\sqrt{14}}{2}$ **c** $\frac{\sqrt{30}}{10}$ **d** $\frac{\sqrt{14}}{4}$

3 **a** $\frac{\sqrt{2}}{8}$ **b** $\frac{\sqrt{5}}{15}$ **c** $\frac{\sqrt{2}}{3}$ **d** $\frac{5\sqrt{2}}{7}$ **e** $\frac{\sqrt{6}}{9}$ **f** $\frac{\sqrt{15}}{10}$ **g** $\frac{\sqrt{3}}{12}$ **h** $\frac{5\sqrt{6}}{6}$

REVIEW SET 8A

1 **a** surd, ≈ 4.899 **b** not a surd, 4 **c** surd, ≈ 1.225

2 **a** 4 and 5 **b** 10 and 11

3 **a** 7 **b** 45 **c** 40

4 **a** $\sqrt{15}$ **b** $\sqrt{7}$ **c** $\sqrt{22}$ **5** $\frac{10}{7}$

6 **a** $2\sqrt{10}$ **b** $6\sqrt{5}$ **7** **a** 3 **b** 10 **c** $\frac{2}{5}$

8 **a** $x=\pm 8$ **b** $x=\sqrt[3]{81}$ (≈ 4.33) **c** $x=\sqrt[3]{-10}$ (≈ -2.15)

9 **a** $x=\pm\sqrt{35}$ **b** $x=\pm 3$

10 **a** $12\,672$ **b** **i** $8\sqrt{2}$ **ii** $3\sqrt{11}$ **c** $24\sqrt{22}$

11 **a** $3\sqrt{11}$ **b** $6\sqrt{7}-3\sqrt{3}$ **c** $10\sqrt{2}$

12 **a** $2\sqrt{3}+2$ **b** $3\sqrt{2}-2$ **c** $8+2\sqrt{7}$ **d** $49-12\sqrt{5}$ **e** 19 **f** $1-2\sqrt{2}$

13 **a** $2\sqrt{5}$ **b** $\frac{\sqrt{6}}{3}$ **c** $\frac{4\sqrt{2}}{3}$

REVIEW SET 8B

1 ≈ 8.544 **2** **a** 8 and 9 **b** 12 and 13

3 **a** 63 **b** 30 **c** -21 **4** **a** $\sqrt{55}$ **b** 2

5 **a** $\frac{2}{5}$ **b** $\frac{9}{4}$ **c** $\frac{8}{11}$

6 **a** $5\sqrt{5}$ **b** $3\sqrt{10}$ **c** $10\sqrt{3}$ **7** $-\frac{2}{7}$

8 **a** ≈ 2.224 **b** ≈ -3.684 **c** ≈ 4.162

9 **a** no real solutions **b** $x=\sqrt[3]{50}$ (≈ 3.68) **c** $x=\sqrt[3]{-70}$ (≈ -4.12) **d** $x=\pm\sqrt{5}$

10 **a** 3 and 4 **b** no **c** perimeter $=4\sqrt{12}$ cm $=\sqrt{192}$ cm

11 **a** irrational **b** **i** 75 **ii** $375\sqrt{3}$ **iii** 5625 **c** n even **d** $x=-5\sqrt{3}$

12 **a** $6\sqrt{2}+2\sqrt{5}$ **b** $3\sqrt{3}$

13 **a** $\sqrt{14}-\sqrt{7}$ **b** $-6-\sqrt{15}$ **c** $4-\sqrt{3}$ **d** $5+2\sqrt{6}$ **e** -1 **f** $2-3\sqrt{2}$

14 **a** $\frac{8\sqrt{3}}{3}$ **b** $\frac{\sqrt{91}}{13}$ **c** $\frac{\sqrt{10}}{7}$

EXERCISE 9A

1 **a** 13 cm **b** ≈ 5.83 m **c** ≈ 5.02 cm

2 **a** $\sqrt{41}$ cm **b** $3\sqrt{5}$ cm **c** $5\sqrt{5}$ cm

3 **a** $2\sqrt{6}$ cm **b** 12 cm **c** $2\sqrt{53}$ m

4 **a** ≈ 5.29 m **b** ≈ 5.90 cm **c** ≈ 5.39 cm

5 **a** $\sqrt{58}$ cm **b** $2\sqrt{5}$ cm **c** $2\sqrt{14}$ m **d** $\sqrt{137}$ cm **e** $2\sqrt{34}$ m **f** $3\sqrt{15}$ m

6 **a** $x=\sqrt{3}$ **b** $x=\frac{\sqrt{5}}{2}$ **c** $x=\sqrt{6}$ **d** $x=\frac{\sqrt{10}}{2}$ **e** $x=\frac{1}{2}$ **f** $x=\sqrt{6}$

7 **a** $x=3\sqrt{3}$ **b** $x=2\sqrt{13}$ **c** $x=2$

8 **a** $x=\sqrt{5},\ y=\sqrt{6}$ **b** $x=5,\ y=\sqrt{26}$ **c** $x=2\sqrt{51}$ **d** $x=\sqrt{37}$ **e** $x=\sqrt{13},\ y=\sqrt{29}$ **f** $x=\sqrt{2},\ y=\sqrt{3},\ z=2$

EXERCISE 9B

1 a $5^2+12^2=25+144=169=13^2$ ✓
b $7^2+24^2=49+576=625=25^2$ ✓
c $8^2+15^2=64+225=289=17^2$ ✓

2 a yes b yes c no d yes e no f yes

3 a $k=15$ b $k=10$ c $k=21$ d $k=25$ e $k=24$ f $k=60$

4 a $(3k)^2+(4k)^2=9k^2+16k^2$
$=25k^2$
$=(5k)^2$
$\therefore$ $\{3k, 4k, 5k\}$ is a Pythagorean triple for any $k\in\mathbb{Z}^+$.
b i $\{6, 8, 10\}$ ii $\{9, 12, 15\}$ iii $\{12, 16, 20\}$ iv $\{15, 20, 25\}$

5 a If $a^2+b^2=c^2$ then $(ka)^2+(kb)^2=k^2a^2+k^2b^2$
$=k^2(a^2+b^2)$
$=k^2c^2$
$=(kc)^2$
$\therefore$ the triangle with side lengths ka, kb, and kc is right angled.
b If $\{a, b, c\}$ is a Pythagorean triple then $a, b, c\in\mathbb{Z}^+$.
$\therefore$ for any $k\in\mathbb{Z}^+$, $ka, kb, kc\in\mathbb{Z}^+$ and $(ka)^2+(kb)^2=(kc)^2$
$\therefore$ $\{ka, kb, kc\}$ is a Pythagorean triple.
c If a, b, and c have a common factor greater than 1, then there will be other Pythagorean triples with numbers in the ratio $a:b:c$.

6 a $(2m)^2+(m^2-1)^2=4m^2+m^4-2m^2+1$
$=m^4+2m^2+1$
$=(m^2+1)^2$
$\therefore$ $\{2m, m^2-1, m^2+1\}$ is a Pythagorean triple for any $m\in\mathbb{Z}^+$, $m>1$.
b i $\{6, 8, 10\}$ ii $\{12, 35, 37\}$ iii $\{14, 48, 50\}$
c no

EXERCISE 9C

1 ≈ 2.33 m 2 ≈ 4.72 m 3 $4\sqrt{2}$ m

4 a $5\sqrt{10}$ km b $10\sqrt{5}$ km c $10\sqrt{5}$ km d $5\sqrt{26}$ km e $5\sqrt{34}$ km f $10\sqrt{10}$ km

5 Yes; the ladder will reach ≈ 3.22 m up the wall.

6 ≈ 0.894 m 7 $\frac{5}{2}\sqrt{3}$ cm 8 ≈ 19.8 m

9 ≈ 0.619 km ≈ 619 m 10 ≈ 27.0 m 11 ≈ 91.8 m

12 ≈ 10.4 m 13 ≈ 3.63 km 14 ≈ 7.48 cm

15 40 cm 16 $\sqrt{3}$ cm 17 ≈ 5.83 m

18 130 cm 19 ≈ 7.07 m 20 ≈ 6.32 m

REVIEW SET 9A

1 a 17 cm b $2\sqrt{13}$ m c $3\sqrt{10}$ cm

2 $x\approx 9.80$, $y\approx 5.66$

3 a $x=\sqrt{26}$ b $x=2\sqrt{2}$ c $x=2\sqrt{5}$

4 a no b yes c no 5 ≈ 7.81 km

6 ≈ 4.82 m 7 ≈ 5.77 cm 8 ≈ 12.7 cm

9 a $x=40$ b i ≈ 2.26 m ii ≈ 9.05 m c ≈ 22.6 m

REVIEW SET 9B

1 a ≈ 10.3 cm b ≈ 3.72 km c ≈ 5.03 m

2 $5\sqrt{2}$ cm 3 $x=\sqrt{7}$, $y=4$ 4 $k=61$

5 72 cm by 54 cm 6 32 cm 7 ≈ 13.1 m

8 ≈ 71.6 m

9 a i The square of an even number is even.
ii The square of an odd number is odd.
iii The sum of two odd numbers is even.
iv The sum of two even numbers is even.
v The sum of an odd number and an even number is odd.
b If all three numbers in a Pythagorean triple were odd, then we would have:
$$\underbrace{(\text{odd})^2+(\text{odd})^2}_{\text{even}}=\underbrace{(\text{odd})^2}_{\text{odd}}$$
which is impossible.
c $\{6, 8, 10\}$
d i If a and b are both even, c must be even.
ii If one of a and b is even and the other is odd, then c must be odd.

EXERCISE 10A

1 a $F=30+60\times 2$ b $F=30+60h$

2 a $A=250\times 6$ b $A=250m$ c $A=Dm$

3 a $F=50-40$ b $F=50-p$ c $F=n-p$

4 a $M=20+12\times 5$ b $M=20+5b$ c $M=20+bn$ d $M=w+bn$

5 a $A=100-6\times 12$ b $A=100-12b$ c $A=100-bp$ d $A=d-bp$

6 a $H=2\times 5+4\times 8$
b $H=5r+(7-r)\times 8=56-3r$
c $H=5r+(d-r)\times 8=8d-3r$

7 a $G=\dfrac{6-3\times 1.2}{3+1}$ b $G=\dfrac{6-3w}{3+1}$ c $G=\dfrac{d-3w}{3+1}$ d $G=\dfrac{d-nw}{n+1}$

EXERCISE 10B

1 a $D=7$ b $n=10$

2 a $F=17$ b $g=7$ c $h=\frac{12}{5}=2.4$

3 a i \$3.80 ii \$8 b 64 minutes

4 a 38 wheels b $m=7$ c $c=14$

5 a i 22 500 kg m/s ii 1.5 kg m/s b 40 m/s

6 a i $T=14$ ii $T=77$ iii $T=50$
b i $a=\pm 1$ ii $a=\pm 3$ iii $a=\pm 8$

7 a i $S=32$ ii $S=864$
b i $n=3$ ii $n=5$

8 a €7.50 b €2.87 c ≈ 19.9 cm

9 a ≈ 20.3 m² b $2\sqrt{2}$ cm² ≈ 2.83 cm²

10 a $C=n^3$ b i $C=27$ ii $n=6$
c i The block has 6 faces, and each face has $(n-2)^2$ cubes with one exposed face.
$\therefore$ $F=6(n-2)^2$
ii $F=54$
d The block has 12 edges, and each edge has $(n-2)$ cubes with two exposed faces.
$\therefore$ $G=12(n-2)$
e i $n=7$ ii $F=150$ iii $C=343$
f i $H=(n-2)^3$ ii $H=512$

EXERCISE 10C

1 a $x = \frac{y}{5}$ b $x = y + 3$ c $x = \frac{y-2}{3}$
d $x = 2y + 2$ e $x = 3 - y$ f $x = 2 - 3y$

2 a $x = \frac{B-n}{7}$ b $x = \frac{T}{3y}$ c $x = G + 5p$
d $x = \frac{l-K}{9}$ e $x = \frac{P-yz}{w}$ f $x = \frac{C+z}{4y^2}$

3 a $y = \frac{9-2x}{3}$ b $y = 3x - 4$ c $y = \frac{3-4x}{2}$

4 $T = \frac{PV}{nR}$

5 a $a = xy$ b $a = \frac{c}{K}$ c $a = \frac{Mn}{bc}$
d $a = pT - s$ e $a = \frac{b+d}{2N}$ f $a = \frac{2Fg+d}{3}$

6 a $M = DV$ b $V = \frac{M}{D}$

7 a $p = \frac{10}{r}$ b $p = \frac{qs}{r}$ c $p = \frac{rs}{q}$
d $p = \frac{qs}{r}$ e $p = \frac{r^2}{s}$ f $p = \frac{r^3}{2s}$

8 a $x = \sqrt{D-1}$ b $x = \sqrt{\frac{K}{5}}$ c $x = \sqrt{Fy}$
d $x = \sqrt{\frac{7z}{N}}$ e $x = \sqrt{\frac{C-y^2}{2}}$ f $x = \sqrt{\frac{T+w}{3z}}$

9 $t = \sqrt{\frac{2s}{g}}$ 10 a $M = BH^2$ b $H = \sqrt{\frac{M}{B}}$

11 a $x = \frac{3}{2-y}$ b $x = \frac{2y}{y-3}$ c $x = 5 - \frac{9}{y^2}$
d $x = -\frac{19y}{2}$ e $x = \sqrt{\frac{3}{y^2-1}}$ f $x = \pm\frac{y}{\sqrt{2}},\ y \neq 0$

12 $u = \frac{fv}{v-f}$

EXERCISE 10D

1 a $t = \frac{Ms^3}{4}$ b i $t = 81$ ii $t = 30$ iii $t = 62.5$

2 a $d = st$
i 180 km ii 120 km iii $\approx$ 127 km
b $t = \frac{d}{s}$
i 3 h ii 1 h 40 min iii 2 h 12 min

3 a $f = \frac{E}{h}$
b i $\approx 6.00 \times 10^{14}$ hertz ii $\approx 7.50 \times 10^{11}$ hertz
iii $\approx 1.50 \times 10^{16}$ hertz

4 a $a = \sqrt{c^2 - b^2}$
b i $a = 6$ ii $a = 2\sqrt{6} \approx 4.90$ iii $a \approx 2.24$

5 a 1080 newtons b $v = \sqrt{\frac{Fr}{m}}$
c i $\approx$ 21.2 m/s ii $\approx$ 24.5 m/s iii 30 m/s

REVIEW SET 10A

1 a $C = 40 + 20n$ b $C = F + an$

2 a 50 m/s b 4.8 s

3 a i $P = 23$ ii $P = 121$
b i $k = 4$ ii $k = 10$

4 a $x = \frac{y-c}{m}$ b $x = \frac{2}{a}$ c $x = \frac{5-Np}{2}$

5 a $y = \frac{5x-8}{3}$ b $y = \frac{3b}{a}$ c $y = \sqrt{\frac{K-x^2}{3}}$

6 a $d \approx \sqrt{0.08\frac{P}{I}}$ b i $\approx$ 89.4 m ii $\approx$ 283 m

7 a $400 b i $P = \frac{12Rn}{in+1}$ ii $10 714.29
c i $i = \frac{12R}{P} - \frac{1}{n}$ ii 3.5% p.a.

REVIEW SET 10B

1 a $B = 20 - 0.6 \times 8 - 1.4 \times 5$ b $B = l - 0.6b - 1.4c$

2 $\approx$ 1.72 seconds 3 a $A = 9$ b $b = 3$

4 a $x = \frac{3q-y}{2}$ b $x = \frac{4}{3y}$

5 a $b = \frac{c-2a}{3}$ b $b = \frac{4}{a-c}$ c $b = \frac{4c}{a}$

6 a $p = \frac{2a}{r}$ b 36 cm

7 a i 4.3875 kWh ii 3.2625 kWh
b i $H = \frac{E}{s^2 \times c}$ ii ≈ 7.14 kWh/m^2
c i $s = \sqrt{\frac{E}{H \times c}}$ ii 3.2 m

EXERCISE 11A

1 a 1.25 b 0.9 c 0.81 d 1.082 e 0.925 f 2.5

2 a 20% increase b 30% decrease c 45% increase
d 55% decrease e 13% increase f 4% decrease
g 111% increase h 6.5% decrease

3 a $52.50 b $28 c £624 d $78.20
e $1888 f €4162.50

4 a 25% increase b 20% decrease
c $\approx$ 3.53% decrease d 12.5% increase
e $\approx$ 14.7% increase f $\approx$ 31.4% decrease

5 $24.09 6 $26 7 $\approx$ 11.6% decrease

8 $242.55 per week

EXERCISE 11B

1 $24 2 $225 3 $6480 4 $504

5 a $49.30 b £120.90 c €697 6 $1.40

7 80% profit 8 $\approx$ 25.3% loss 9 a 8% b 22.5%

10 a $17.55
b loss; the selling price is less than the cost price
c 2.5% loss

11 16% 12 $31.38 per kg 13 $1.05 14 $80

15 $34.50 per kg 16 a €1080 b €180

17 a $4950 b $450 18 a €181 b €45.25

EXERCISE 11C

1 \$533.03 **2** \$2426.76 **3** \$3400
4 **a** £1842.38 **b** 5 years **5** €1655.36
6 **a** \$1440 **b** \$1600 **7** **a** \$271.05 **b** 27.65%
8 $\approx$ 1.70% depreciation **9** $\approx$ 0.499% depreciation
10 **a** €900.05 **b** €379.60 **11** **a** 30% **b** $\approx$ 9.14%

EXERCISE 11D

1 **a** \$660 **b** \$2062.50 **c** \$15 500 **d** \$3881.25
2 \$12 960 **3** **a** \$63 **b** \$170 **c** \$288 750 **d** \$1625
4 **a** \$15 820 **b** \$8303.87
5 **a** \$7912.50 **b** \$23 115 **6** **a** \$31 000 **b** \$645.84
7 \$529.17 **8** \$118
9 **a** $P = \dfrac{I}{in}$ **b** $i = \dfrac{I}{Pn}$ **c** $n = \dfrac{I}{Pi}$ **10** \$4500
11 \$3750 **12** \$67 741.94 **13** **a** \$3500 **b** \$71.95
14 **a** 7% p.a. **b** $\approx$ 6.82% p.a. **15** $\approx$ 11.1% p.a.
16 9 years **17** 6 years 3 months
18 **a** 6 years 6 months **b** \$1512.31 **19** 3 years 6 months

EXERCISE 11E.1

1 **a**

Year	*Initial amount*	*Interest*	*Final amount*
1	\$16 000	6% of \$16 000 = \$960	\$16 960
2	\$16 960	6% of \$16 960 = \$1017.60	\$17 977.60
3	\$17 977.60	6% of \$17 977.60 = \$1078.66	\$19 056.26
4	\$19 056.26	6% of \$19 056.26 = \$1143.38	\$20 199.64

b **i** \$20 199.64 **ii** \$4199.64
2 **a** **i** \$1350 **ii** \$1411.66
b compound interest; in this method, interest can be earned on interest
3 **a** \$5324 **b** \$23 217.71 **4** \$2178.15

EXERCISE 11E.2

1 \$4410 **2** **a** \$7498.58 **b** \$498.58
3 **a** \$7311.62 **b** \$811.62 **4** \$1270.50
5 **a** Option 1: \$10 160, Option 2: \$10 077.70
Option 1 should be chosen as it would be worth \$82.30 more than option 2 after 3 years.
b Yes. Option 1: \$11 600, Option 2: \$11 754.62
After 5 years, option 2 would be worth \$154.62 more than option 1.
6 **a** \$2407.26 **b** Yes, by about \$23.75.
7 **a** $P = \dfrac{F}{(1+i)^n}$ **b** **i** \$7500 **ii** \$95 000
8 \$15 372.52 **9** \$6000 **10** $\approx$ 5.84% p.a.

REVIEW SET 11A

1 **a** 1.37 **b** 0.905
2 **a** 15% decrease **b** 18.75% increase
3 \$290 **4** $\approx$ 11.1% loss **5** \$80 **6** \$41.13
7 \$497 353.92 **8** **a** \$562.41 **b** 10.24%
9 **a** €23 506.89 **b** €21 156.20
10 **a** \$6700 **b** \$139.59
11 **a** \$19 546.74 **b** \$7546.74
12 $\approx$ 5.88% p.a. **13** \$6200
14 **a** **i** €5600 **ii** €5532
b Bank A. After 5 years, Carlos' investment will be worth €68 more if he uses bank A instead of bank B.
c Yes. Bank A: €6240, Bank B: €6298.12
After 7 years, Carlos' investment will be worth €58.12 more if he uses bank B instead of bank A.

REVIEW SET 11B

1 **a** 70% increase **b** 29% decrease **c** 145% increase
2 **a** \$78.40 **b** £1600 **3** \$357.50 **4** 16%
5 \$384.75 **6** **a** \$230 **b** \$23 **7** \$53.30
8 $\approx$ 38.6% **9** \$420 **10** 2 years 5 months
11 **A**. Option **A** would be worth \$87.62 more than option **B** after 4 years.
12 **a** £34 333.36 **b** 7 years
13 **a** 10.25% **b** €4410 **c** **i** €4210 **ii** €5613.33
14 **a** \$10 099.82 **b** \$2099.82 **c** $\approx$ 6.56% p.a.

EXERCISE 12A

1 **a** 52 000 m **b** 4.8 m **c** 1150 mm
d 5.4 cm **e** 165 cm **f** 5.28 km
g 4210 m **h** 0.425 m **i** 2 m
j 6300 mm **k** 7 km **l** 250 000 mm
2 5.3 km **3** 12.375 km
4 256 666 lengths **5** 2125 steps
6 **a** 0.375 mm **b** 1.17 m **c** 320 cards

EXERCISE 12B

1 **a** 21 cm **b** 18 cm **c** 12 m
d 144 mm **e** 12 m **f** 29 cm
g 22 cm **h** $\approx$ 19.1 cm **i** $\approx$ 15.6 m
2 **a** $\approx$ 18.8 m **b** $\approx$ 20.6 cm **c** $\approx$ 23.0 m
d $\approx$ 60.1 cm
3 **a** 9 m **b** 19 mm **c** $\approx$ 26.4 cm
d $\approx$ 5.92 cm **e** $\approx$ 3.08 m
4 **a** 36 mm **b** 22 cm **c** $\approx$ 32.9 m
d $\approx$ 451 m **e** $\approx$ 82.3 cm **f** $\approx$ 77.1 mm
g $\approx$ 94.2 cm **h** 38 m **i** $\approx$ 37.7 cm
5 38.4 m
6 **a** $P = 2x + y$ **b** $P = 4x + 6$ **c** $P = 2\pi r + 2d$
7 **a** 24.4 m **b** 97.6 m **c** \$3855.20 **8** 14.5 m
9 **a** **i** $\approx$ 565.5 m **ii** $\approx$ 502.7 m **b** $\approx$ 62.8 m
10 $\approx$ 133 times
11 **a** $\approx$ 20.9 cm **b** $\approx$ 36.9 cm **c** $\approx$ 3.33 cm
12 $\approx$ 61 cm **13** $\approx$ 64.6 minutes

EXERCISE 12C

1 **a** 0.23 cm^2 **b** 36 000 m^2 **c** 0.0726 m^2
d 7 600 000 mm^2 **e** 0.853 ha **f** 35 400 000 cm^2
g 1354 mm^2 **h** 4 320 000 cm^2 **i** 4820 mm^2
2 **a** 3 000 000 m^2 **b** 70 ha **c** 6 600 000 m^2
d 6.6 km^2 **e** 500 cm^2 **f** 0.0025 m^2
g 0.052 cm^2 **h** 720 000 000 000 mm^2
3 100 rectangles **4** 2100 chickens

EXERCISE 12D

1 **a** 49 cm^2 **b** 152 mm^2 **c** 7.5 mm^2
d 13 cm^2 **e** 186.34 mm^2 **f** 5.46 cm^2
g 1.02 m^2 **h** 24 cm^2 **i** 3.68 cm^2
2 **a** 30 cm^2 **b** $\approx$ 31.2 cm^2 **c** $\approx$ 26.9 cm^2
3 **a** $h = 6$ **b** $h = 3.6$ **c** $h = 5$
4 **a** $\approx$ 6.09 m^2 **b** $\approx$ 7.31 m^2 **c** \$207.68

5 The square has greater area (306.25 cm^2).
Area of rectangle is 300 cm^2.

6 5.09 m^2 **7** 375 tiles **8** 48 cm^2
9 ≈ 5.66 cm **10** $\approx 25.5\%$ **11** 144 cm^2
12 80 cm^2 **13** $\approx 9.98 \text{ cm}^2$ **14** ≈ 27.4 mm

15 **a** $A = \frac{a^2\sqrt{3}}{4}$ **b** $A = \frac{b\sqrt{4a^2 - b^2}}{4}$ **c** $A = a\sqrt{b^2 - c^2}$

16 **a** 112 m^2 **b** 74 m^2 **c** 84 cm^2
d 31.5 cm^2 **e** 189 cm^2 **f** 36 cm^2

17 area of hexagon : area of triangle = 3 : 2

EXERCISE 12E

1 **a** $\approx 154 \text{ cm}^2$ **b** $\approx 1.42 \text{ m}^2$ **c** $\approx 26.2 \text{ m}^2$
d $\approx 30.2 \text{ cm}^2$

2 $9\pi \text{ cm}^2$ **3** $\approx 73.90 \text{ cm}^2$ **4** $\approx 139 \text{ cm}^2$

5 **a** $\frac{4}{3}\pi$ cm **b** $16 + \frac{4}{3}\pi$ cm **c** $\frac{16}{3}\pi \text{ cm}^2$

6 **a** 3 m **b** ≈ 2.52 cm **7** $\approx 23.9°$ **8** $\approx 154 \text{ m}^2$

9 **a** $\approx 6.85 \text{ cm}^2$ **b** $\approx 30.6 \text{ cm}^2$ **c** $\approx 30.9 \text{ cm}^2$
d $\approx 6430 \text{ m}^2$ **e** $\approx 113 \text{ cm}^2$ **f** $\approx 36.9 \text{ cm}^2$

10 $\approx 2.85 \text{ m}^2$ **11** $\approx 46.2\%$ **12** $\approx 21.5\%$

13 **a** $\approx 9.42 \text{ m}^2$ **b** $74.46

14 **a** ≈ 44.0 m **b** 13 m^2 **c** $\approx 35.0\%$

15 **a** $A = 2ab + \pi a^2$ **b** $A = \pi(R^2 - r^2)$
c $A = (x + 2a)^2 - \pi a^2$

16 **Hint:** Area of B = area of quarter circle
$-$ 2 × area of semi-circles + area of A

REVIEW SET 12A

1 **a** 6700 m **b** 0.93 cm **c** $25\,900 \text{ m}^2$

2 **a** ≈ 158 cm **b** 48 m **c** ≈ 215 mm

3 5.4 km **4** ≈ 565 m **5** 5.75 times larger

6 **a** 40 cm^2 **b** $\approx 4.51 \text{ m}^2$ **c** 40 m^2

7 **a** $\approx 105 \text{ cm}^2$ **b** 30 cm^2 **c** $\approx 177 \text{ cm}^2$ **8** ≈ 10.0 m

9 **a** $A = \frac{1}{2}a\sqrt{h^2 - a^2}$ **b** $A = (8 + \frac{\pi}{2})x^2$
c $A = (4 - \pi)r^2$

11 **a** $14\,000 \text{ m}^2$ **b** $1 260 000 **c** **i** 112 m **ii** 474 m

REVIEW SET 12B

1 **a** 7.26 km **b** $34\,000 \text{ mm}^2$ **c** 0.0382 m^2

2 3.36 m **3** ≈ 31.9 cm **4** $\approx 262°$

5 **a** $P = a + 2b + c + \sqrt{(a-b)^2 + (c-b)^2}$
b $P = 2x(1 + \sqrt{2} + \frac{\pi}{4})$

6 **a** $\approx 34.6 \text{ cm}^2$ **b** $\approx 2.28 \text{ m}^2$ **c** $\approx 1.45 \text{ cm}^2$

7 112 ha **8** 3300 tiles **9** **a** 5.96 m^2 **b** $187.74

10 ≈ 13.6 cm **11** $\theta = 90$

12 **a** 140 cm = 1.4 m **b** 1200 cm^2
c **i** 50 cm **ii** 48 cm
d **i** 24 cm **ii** 18 cm **iii** 32 cm
e **i**

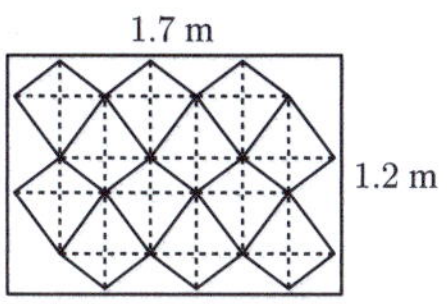

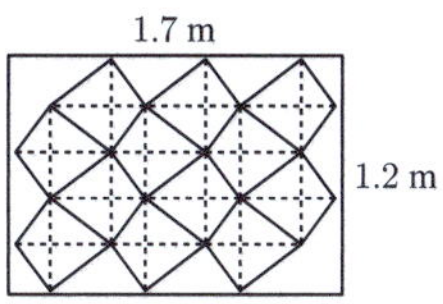

ii $\approx 29.4\%$

EXERCISE 13A

1 **a** 150 cm^2 **b** 541.5 mm^2

2 **a** 276 cm^2 **b** 6880 mm^2 **c** 8802 m^2

3 **a** $\approx 198 \text{ m}^2$ **b** $\approx 496 \text{ cm}^2$ **c** $\approx 148 \text{ cm}^2$

4 1160 cm^2 **5** **a** 360 cm^2 **b** 340 m^2 **c** $\approx 9840 \text{ cm}^2$

6 **a** 576 cm^2 **b** 216 m^2 **c** $\approx 823 \text{ m}^2$
d $\approx 1.73 \text{ m}^2$ **e** $\approx 63\,500 \text{ cm}^2$

7 **a** $A = 2(ab + ac + bc)$ **b** $A = 8ab + 6a^2$
c $A = a\sqrt{4b^2 - a^2} + a^2$

8 **a** $\approx 1010 \text{ m}^2$ **b** $\approx$ $25 300

9 $3717 **10** 26.52 m^2

EXERCISE 13B

1 **a** $\approx 207 \text{ cm}^2$ **b** $\approx 339 \text{ cm}^2$ **c** $\approx 196 \text{ cm}^2$
d $\approx 56.7 \text{ m}^2$ **e** $\approx 124 \text{ m}^2$ **f** $\approx 79.5 \text{ cm}^2$

2 $\approx 199 \text{ cm}^2$ **3** $\approx 1.30 \text{ m}^2$

4 **a** $h = \frac{A}{2\pi r} - r$ **b** **i** ≈ 7.89 cm **ii** ≈ 1.26 m

5 **a** $\approx 0.704 \text{ m}^2$ **b** $\approx 0.784 \text{ m}^2$ **6** $\approx 1370 \text{ cm}^2$

EXERCISE 13C

1 **a** $\approx 204 \text{ cm}^2$ **b** $\approx 56.5 \text{ m}^2$ **c** $\approx 298 \text{ cm}^2$
d $\approx 49.5 \text{ mm}^2$ **e** $\approx 302 \text{ cm}^2$ **f** $\approx 95.1 \text{ m}^2$

2 $\approx 1230 \text{ cm}^2$ **3** $\approx 63.5 \text{ cm}^2$

4 **a** $s = \frac{A}{\pi r} - r$ **b** ≈ 5.70 cm **c** ≈ 3.59 cm

EXERCISE 13D

1 **a** $\approx 201 \text{ cm}^2$ **b** $\approx 616 \text{ mm}^2$ **c** $\approx 763 \text{ cm}^2$

2 $\approx 84.9 \text{ cm}^2$ **3** **a** $9\pi \text{ m}^2$ **b** $192\pi \text{ cm}^2$

4 28 foam balls **5** ≈ 4.64 cm **6** ≈ 10.4 cm

7 **a** $\approx 179.1 \text{ cm}^2$ **b** $\approx 238.8 \text{ cm}^2$ **8** 8 wedges

EXERCISE 13E

1 **a** 250 mm^3 **b** $83\,000 \text{ cm}^3$ **c** 0.598 cm^3
d $0.009\,81 \text{ m}^3$ **e** 0.6351 m^3 **f** $81\,500 \text{ mm}^3$

2 **a** 21.25% **b** $\approx 4.29\%$ **3** $\approx 2610 \text{ cm}^3$ **4** 120 cm^3

5 $27\,200 \text{ cm}^3$ **6** **a** $27\,700 \text{ m}^3$ **b** 1.108×10^6 boxes

EXERCISE 13F

1 **a** 1200 cm^3 **b** 124.416 cm^3 **c** 0.912 m^3

2 **a** $\approx 352 \text{ cm}^3$ **b** $\approx 35.3 \text{ m}^3$ **c** $\approx 1.54 \text{ m}^3$

3 **a** 45 cm^3 **b** 320 cm^3 **c** 704 cm^3
d 432 cm^3 **e** 288 cm^3 **f** $\approx 2320 \text{ cm}^3$

4 **a** 2000 cm^3 **b** $\approx 0.982 \text{ cm}^3$ **c** $\approx 183 \text{ cm}^3$
d 2196 cm^3 **e** $\approx 2380 \text{ cm}^3$

5 1.4 cm **6** ≈ 0.813 cm **7** ≈ 16.0 mm

8 ≈ 1.50 cm **9** **a** 32.5 m^2 **b** 195 m^3

10 **a** $V = \frac{1}{4}\pi d^2 h$ **b** $V = \frac{1}{2}abc$ **c** $V = c(a^2 - b^2)$

12 **a** **i** $\approx 3.93 \text{ m}^3$ **ii** $\approx 9.05 \text{ m}^3$ **b** $\approx 13.0 \text{ m}^3$
c $1846 (buy 13 m^3)

13 $\approx 170 \text{ cm}^3$

EXERCISE 13G

1 **a** 32 cm^3 **b** $\approx 134 \text{ cm}^3$ **c** 120 cm^3

2 **a** $\approx 314 \text{ cm}^3$ **b** $\approx 23.4 \text{ cm}^3$ **c** $\approx 1140 \text{ cm}^3$

3 **a** $r = \sqrt{\frac{3V}{\pi h}}$ **b** **i** ≈ 3.99 cm **ii** ≈ 4.37 cm

4 $\approx 258 \text{ cm}^3$ **5** $\approx 599 \text{ cm}^2$

EXERCISE 13H

1 **a** ≈ 113 cm^3 **b** ≈ 905 mm^3 **c** ≈ 262 cm^3
2 $\approx 14\,100$ cm^3 **3** **a** ≈ 408 m^3 **b** ≈ 688 cm^3
4 19 862 chocolate spheres **5** ≈ 3.85 cm^3
6 **a** $r = \sqrt[3]{\dfrac{3V}{4\pi}}$
b **i** ≈ 2.12 cm **ii** ≈ 5.76 cm **iii** ≈ 62.0 cm
7 **a** 143 spheres **b** ≈ 1.00 cm^3 **c** ≈ 6.21 mm

EXERCISE 13I

1 **a** 5600 mL **b** 524 000 L **c** 0.28 L
d 6.3 ML **e** 3500 kL **f** 0.083 kL
g 310 mL **h** 0.743 ML **i** 840 000 mL
2 5.475 kL **3** $\approx 9.04\%$
4 **a** 3 L **b** ≈ 198 mL **c** 240 L
5 **a** ≈ 53.8 L **b** ≈ 618 km
6 ≈ 7.55 cm **7** 37 tubs
8 **a** ≈ 154 cm^2 **b** ≈ 15.0 cm^3 **c** ≈ 165 mL
9 ≈ 71.4 cm^2 **10** ≈ 87.2 L

REVIEW SET 13A

1 **a** 216 m^2 **b** ≈ 66.5 cm^2 **c** ≈ 2630 mm^2
2 **a** 200 mm^3 **b** 0.057 cm^3 **c** 9 040 000 cm^3
3 **a** surface area: 182.88 cm^2, volume: 154.224 cm^3
b surface area: ≈ 534 cm^2, volume: ≈ 942 cm^3
c surface area: ≈ 66.6 m^2, volume: 32 m^3
4 **a** ≈ 170 cm^3 **b** ≈ 46.0 m^3
5 3 m **6** ≈ 5.24 cm **7** 7.29 ML **8** 2842 m^3
9 **a** $A = 6\pi x^2$ **b** $A = (\sqrt{35}+1)x^2$
c $A = \pi s\sqrt{s^2-h^2} + 2\pi(s^2-h^2)$
10 **a** $V = \dfrac{\pi h^3}{3}$ m^3 **b** $h = \sqrt[3]{\dfrac{3V}{\pi}}$ **c** ≈ 1.68 m
11 **a** 9.5 m^2 **b** 1.5 kL
12 $S = (4+n)\pi r^2$ **13** 5.76 kL
14 **a** 6750 mm^3 **b** 6.75 mL **c** 222 figurines
d ≈ 29.8 cm
15 **a** ≈ 33.5 kL **b** ≈ 1 h 52 min

REVIEW SET 13B

1 15% **2** **a** 0.82 L **b** 3070 kL **c** 720 000 mL
3 **a** ≈ 94.2 cm^2 **b** ≈ 62.8 m^2 **c** ≈ 350 cm^2
4 **a** surface area: ≈ 692 cm^2, volume: 864 cm^3
b surface area: 300 cm^2, volume: 300 cm^3
c surface area: ≈ 61.3 cm^2, volume: ≈ 28.2 cm^3
5 1600 cubes **6** ≈ 7.98 cm
7 **a** ≈ 4.14 m^2 **b** ≈ 0.353 m^3 **8** 227 labels
9 **a** $A = \dfrac{\theta}{360}(2\pi rh + 2\pi r^2) + 2hr$ **b** $V = \dfrac{\theta}{360}\pi r^2 h$
10 **a** $\approx 10\,900$ km **b** $\approx 37\,900\,000$ km^2
c $\approx 22\,000\,000\,000$ km^3
11 **a** $V = \dfrac{\pi d^3}{12}$ cm^3 **b** $V = 100\pi x^3$ cm^3
c $V = \frac{4\sqrt{7}}{3}x^3$ cm^3
12 **a** ≈ 905 cm^2 **b** ≈ 11.3 cm **13** ≈ 14.0 m^3
14 **a** 375 kL **b** 350 kL
15 **a** $\approx 2\,460\,000$ m^3 **b** $\approx 6.56 \times 10^6$ t **c** ≈ 327 kL

EXERCISE 14A

1 **a** $3x+6 = 3(x+2)$ **b** $5x-15 = 5(x-3)$
c $6a+12 = 6(a+2)$ **d** $20-5m = 5(4-m)$
e $4x+8x^2 = 4x(1+2x)$ **f** $3x^2-6x = 3x(x-2)$
g $3p+9 = 3(p+3)$ **h** $6x-24 = 6(x-4)$
i $4m+8mn = 4m(1+2n)$ **j** $15a-5a^2 = 5a(3-a)$
k $cd+de = d(c+e)$ **l** $4ab-6bc = 2b(2a-3c)$
2 **a** $2(x+3)$ **b** $2(x-3)$ **c** $5(a+2)$ **d** $5(a-2)$
e $2(2+3x)$ **f** $2(2-3x)$ **g** $a(1+b)$ **h** $a(1-b)$
3 **a** $3(x-2)$ **b** $8(d+5)$ **c** $10(x-3)$ **d** $2(2+x)$
e $5(p+q)$ **f** $a(b+d)$ **g** $7(6-5x)$ **h** $h(g+1)$
i $y(1+z)$ **j** $p(1-q)$ **k** $e(3+f)$ **l** $x(9-y)$
m $k(j-3h)$ **n** $s(r-t)$ **o** $y(3x-2z)$ **p** $m(4n+l)$
4 **a** $x(x+5)$ **b** $x(5-2x)$ **c** $4x(x+2)$ **d** $7x(2-x)$
e $6x(x+2)$ **f** $x^2(x+9)$ **g** $xy(x+y)$
h $2x^2(2x-3)$ **i** $9x(x^2-2y)$ **j** $3xy(2y+3)$
5 **a** $a(a^2+a+1)$ **b** $2(a^2+2a+4)$ **c** $3a(a^2-2a+3)$
6 **a** $x(6-2x)$ is not fully factorised since $(6-2x)$ still has the common factor 2.
b $6(2x^2-3xy)$ is not fully factorised since $(2x^2-3xy)$ still has the common factor x.
7 **a** $9(b-a)$ **b** $3(2b-1)$ **c** $4(b-2a)$
d $c(d-7)$ **e** $a(b-1)$ **f** $6x(2-x)$
g $5x(3x-1)$ **h** $2b(2a-b)$
8 **a** $-6(a+b)$ **b** $-4(1+2x)$ **c** $-3(y+2z)$
d $-c(9+d)$ **e** $-x(1+y)$ **f** $-5x(x+4)$
g $-3y(4+y)$ **h** $-9a(2a+b)$
9 **a** $(x-7)(2+x)$ **b** $(x+3)(a+b)$ **c** $(x+2)(4-x)$
d $(x+9)(x+1)$ **e** $(b+4)(a-1)$ **f** $(b+c)(a+d)$
g $(m+n)(a-b)$ **h** $(x+3)(x-1)$ **i** $(x-2)(2x-1)$
10 **a** $(x+3)(x-1)$ **b** $(x-7)(x+7)$ **c** $(x+6)(x-4)$
d $(x-2)(x-8)$ **e** $x+2$ **f** $(a+b)(4-a)$
g $3(a-2)(a-4)$ **h** $(x+4)(4x+1)$
i $5(x-1)(6-x)$ **j** $-(x+5)(4x+17)$
11 **a** $2x^2$ units^2 **b** $2x(x-2)$ units^2 **c** $4x(x-1)$ units^2

EXERCISE 14B

1 **a** $(d+e)(d-e)$ **b** $(p+q)(p-q)$ **c** $(q+p)(q-p)$
d $(x+y)(x-y)$ **e** $(x+3)(x-3)$ **f** $(x+10)(x-10)$
g $(5+x)(5-x)$ **h** $(y+7)(y-7)$ **i** $(9+a)(9-a)$
j $(2b+1)(2b-1)$ **k** $(1+2x)(1-2x)$
l $(4x+3)(4x-3)$ **m** $(6y+7)(6y-7)$
n $(9+d)(9-d)$ **o** $(11+3y)(11-3y)$
2 **a** $3(x+5)(x-5)$ **b** $8(x+2)(x-2)$
c $5(a+2)(a-2)$ **d** $5(x+1)(x-1)$
e $9(b+10)(b-10)$ **f** $5(4+t)(4-t)$
g $6(k+2)(k-2)$ **h** $15(1+y)(1-y)$
i $r(r+7)(r-7)$ **j** $z(z+1)(z-1)$
k $x^3(x+1)(x-1)$ **l** $xy(x+y)(x-y)$
3 **a** $(5a+b)(5a-b)$ **b** $(x+9y)(x-9y)$
c $(4x+7y)(4x-7y)$ **d** $(5p+6q)(5p-6q)$
e $(9j+k)(9j-k)$ **f** $(st+3)(st-3)$
g $(7x+yz)(7x-yz)$ **h** $(2d+5e)(2d-5e)$

4 **a** $2(2x+1)(2x-1)$ **b** $3(2+3x)(2-3x)$
c $2(y+2x)(y-2x)$ **d** $a(3+4a)(3-4a)$

5 **a** $x(x+4)$ **b** $(x+3)(x-9)$ **c** $(x-1)(21-x)$
d $(x+1)(17-x)$ **e** $(x+8)(x+6)$ **f** $-(x+7)(x+1)$

6 **a** **i** Area of garden and path $= (x+2)^2$ m^2
Area of path $= (x+2)^2 -$ area of garden
$= [(x+2)^2 - x^2]$ m^2
ii Area of path $= 4(x+1) \times 1$ m^2
$= (4x+4)$ m^2
b $(x+2)^2 - x^2 = [(x+2)+x][(x+2)-x]$
$= (2x+2)(2)$
$= 2(2x+2)$

7 **a** $8(x+1)$ **b** $3x(x-2)$ **c** $5x(x-2)$

EXERCISE 14C

1 **a** perfect square, $(x+4)^2$ **b** not a perfect square
c perfect square, $(3+a)^2$ **d** not a perfect square
e perfect square, $(t-1)^2$ **f** not a perfect square

2 **a** $(x-1)^2$ **b** $(x+5)^2$ **c** $(x+2)^2$ **d** $(x-7)^2$
e $(x+11)^2$ **f** $(x-10)^2$ **g** $(x+6)^2$ **h** $(x-3)^2$
i $(x+12)^2$ **j** $(7+x)^2$ **k** $(x-9)^2$ **l** $(x-y)^2$

3 **a** perfect square, $(3x+1)^2$ **b** not a perfect square
c not a perfect square

4 **a** $(2x-1)^2$ **b** $(5x+2)^2$ **c** $(8x-3)^2$
d $(3x+7)^2$ **e** $(6x-5)^2$ **f** $(4x-5)^2$
g $(3x-4)^2$ **h** $(7x+1)^2$ **i** $(2x-3y)^2$

5 **a** $2(x+3)^2$ **b** $2(x-1)^2$ **c** $3(x+2)^2$
d $4(x-4)^2$ **e** $5(x+5)^2$ **f** $-(x-6)^2$
g $-(x+8)^2$ **h** $-2(x-10)^2$ **i** $-3(x-5)^2$

6 961, $\sqrt{961} = 31$

7 **a** $x(x+1)^2$ **b** $(x^2+1)^2$ **c** $(x+1)^2(x-1)^2$

EXERCISE 14D

1 **a** 3 and 4 **b** 2 and 5 **c** 2 and 8 **d** 2 and 9
e -3 and 7 **f** -7 and 3 **g** -6 and 2 **h** -2 and 15

2 **a** $(x+1)(x+3)$ **b** $(x+2)(x+12)$ **c** $(x+3)(x+7)$
d $(x+6)(x+9)$ **e** $(x+4)(x+5)$ **f** $(x+2)(x+6)$
g $(x+4)(x+6)$ **h** $(x+2)(x+7)$ **i** $(x+2)(x+4)$

3 **a** $(x-1)(x-2)$ **b** $(x-1)(x-3)$ **c** $(x-2)(x-3)$
d $(x-3)(x-11)$ **e** $(x-3)(x-13)$ **f** $(x-3)(x-16)$
g $(x-4)(x-7)$ **h** $(x-2)(x-12)$ **i** $(x-2)(x-18)$

4 **a** $(x-8)(x+1)$ **b** $(x+7)(x-3)$ **c** $(x-2)(x+1)$
d $(x-4)(x+2)$ **e** $(x+8)(x-3)$ **f** $(x-5)(x+2)$
g $(x+9)(x-6)$ **h** $(x+9)(x-8)$ **i** $(x-7)(x+3)$
j $(x-3)(x+2)$ **k** $(x-12)(x+5)$ **l** $(x+12)(x-5)$
m $(x+6)(x-3)$ **n** $(x+2)(x-9)$ **o** $(x-15)(x+3)$

5 **a** No two integers have a sum of 3 and a product of -2.
b No two integers have a sum of 0 and a product of 2.
c No two integers have a sum of -8 and a product of 4.

6 **a** $(x+6)(x+1)$ **b** $(x-9)(x+7)$ **c** $(x-2)(x-9)$
d $(x+8)(x-2)$ **e** $(x-1)(x-4)$ **f** $(x+7)(x+5)$
g $(x-5)(x+4)$ **h** $(x-11)(x+2)$ **i** $(x+12)(x-4)$
j $(x-7)(x+4)$ **k** $x(x+13)$ **l** $(x-7)^2$

7 **a** $2(x-4)(x+1)$ **b** $3(x+4)(x-1)$
c $5(x+3)(x-1)$ **d** $4(x+5)(x-4)$
e $2(x-5)(x+3)$ **f** $3(x+7)(x-3)$
g $-2(x-5)(x+4)$ **h** $-3(x-2)^2$
i $-7(x+4)(x-1)$ **j** $-(x+2)(x+1)$
k $-(x-3)(x-2)$ **l** $-(x-3)(x-6)$
m $5(x+5)(x-2)$ **n** $-2(x+7)(x-3)$
o $-(x-8)(x+4)$

EXERCISE 14E

1 **a** $x(x+49)$ **b** not possible **c** $-x(x+49)$
d $(7-x)(7+x)$ **e** $4a(a+4)$ **f** $4a(a-4)$
g $4(a^2+4)$ **h** $4(a-2)(a+2)$ **i** $st(t-2)$
j $st(s-1)$ **k** $(st-1)(st+1)$ **l** $s(t+2)$

2 **a** $3(b^2+5)$ **b** $h^2(1+2h)$ **c** $-2(x^2+9)$
d $(r+7)(r-6)$ **e** $(x+1)(x-1)$
f $(9y-7x)(9y+7x)$ **g** $(x-5)(2-x)$
h $(y-3)(y-4)$ **i** $(s+t)(s-t)$

3 **a** $(x-4)(x-8)$ **b** $(x-9)(x+6)$ **c** $(x-6)(x+3)$
d $(x-13)(x+13)$ **e** $(x-5)^2$
f $(x+5)(x+6)$ **g** $(x+9)(x-2)$
h $(x+6)^2$ **i** $(x-8)(x+3)$
j $5(x-5)(x+5)$ **k** $3(x+6)(x-4)$
l $4(x-5)(x+3)$ **m** $3(x-11)(x-3)$
n $9(x-2)(x+1)$ **o** $-(x-7)(x-2)$
p $6(x-3)^2$ **q** $-(x+4)(x+9)$
r $2(x-8)^2$ **s** $-2(x+9)(x-2)$
t $5x(2-x)(2+x)$ **u** $x(x+2)^2$

EXERCISE 14F

1 **a** $(b+c)(x+y)$ **b** $(p+q)(x+3)$
c $(2a+b)(3x+1)$ **d** $(a+b)(m-n)$
e $(r-s)(3d+1)$ **f** $(a+b)(2c-5)$

2 **a** $(x+5)(x+7)$ **b** $(x-2)(x-6)$ **c** $(x+3)^2$
d $(x+1)(x+8)$ **e** $(x+1)(2x+3)$
f $(x-1)(3x+1)$ **g** $(x+5)(2x+3)$
h $(3x+2)(2x-1)$ **i** $(x+2)(4x+1)$
j $(2x+1)(3x+5)$ **k** $(2x+3)(3x-2)$
l $(x-2)(4x+1)$ **m** $(x+11)(3x+4)$
n $(3x-2)(6x+1)$ **o** $(2x-7)(5x+2)$

EXERCISE 14G

1 **a** $(x+4)(2x+1)$ **b** $(2x+1)(x+4)$

2 **a** $(2x+3)(x+2)$ **b** $(2x+5)(x+5)$
c $(2x+5)(x+3)$ **d** $(3x+1)(x+2)$
e $(3x+5)(x+3)$ **f** $(3x+2)(2x+5)$

3 **a** $(2x+5)(2x-3)$ **b** $(2x-3)(2x+5)$

4 **a** $(2x-1)(x+3)$ **b** $(2x+3)(x-1)$
c $(3x-2)(x+2)$ **d** $(3x+1)(x-3)$
e $(5x+2)(x-2)$ **f** $(2x+3)(3x-1)$

5 **a** $(2x-1)(x-5)$ **b** $(2x-3)(x-4)$
c $(2x-5)(x-2)$ **d** $(3x-2)(x-4)$
e $(5x-2)(2x-1)$ **f** $(3x-1)(3x-4)$

6 **a** $p, q > 0$ **b** $p, q < 0$

c p and q have opposite signs
d p and q have opposite signs

7 **a** $(3x+4)(4x-1)$ **b** $(3x-1)(3x+4)$ **c** $(4x+3)(2x-1)$ **d** $(3x-2)(2x-5)$ **e** $(5x+1)(2x-3)$ **f** $(4x+3)(5x+4)$

8 **a** $2(2x-3)(x+2)$ **b** $3(3x+2)(2x+1)$ **c** $2(2x-1)(2x-3)$

REVIEW SET 14A

1 **a** $3(x-6)$ **b** $2a(b+5c)$ **c** $y(x-6z)$ **d** $-2x(x+16)$

2 **a** $(t+2)(d-4)$ **b** $(x-1)(x-2)$ **c** $(x+3)(2x-5)$

3 **a** $(x+5)(x-5)$ **b** $(10+k)(10-k)$ **c** $(3+4x)(3-4x)$ **d** $(3a+2b)(3a-2b)$ **e** $6(x+2)(x-2)$ **f** $4(2m+5n)(2m-5n)$

4 **a** $A_1=b^2,\ A_2=b(a-b),\ A_3=b(a-b),\ A_4=(a-b)^2$

b Total area $=a^2$ and

$$\begin{aligned}\text{Total area} &= A_1+A_2+A_3+A_4\\ &= b^2+2b(a-b)+(a-b)^2\\ &= b^2+2ab-2b^2+(a-b)^2\\ &= (a-b)^2+2ab-b^2\end{aligned}$$

c Rearrange result from **b** to give $a^2-2ab+b^2=(a-b)^2$.

5 **a** not a perfect square **b** perfect square, $(4x+3)^2$

6 **a** $(x-4)^2$ **b** $(x+7)^2$ **c** $(2x+5)^2$

7 No two integers have a sum of -2 and a product of 4.

8 **a** $(x+3)(x+9)$ **b** $(x-12)(x-4)$ **c** $(x+11)(x-2)$ **d** $(x-6)(x-4)$ **e** $(x-4)(x-9)$ **f** $-3(x+4)(x-2)$

9 **a** $(x+2)(a+b)$ **b** $(x-8)(x-3)$ **c** $(2x+3)(3x-1)$

10 **a** $(2x+1)(x+8)$ **b** $(2x-1)(x+8)$ **c** $(x-5)(3x-2)$ **d** $(2x-5)(3x+2)$ **e** $(3x+2)(4x-1)$ **f** $2(2x+1)(3x-2)$

REVIEW SET 14B

1 **a** $x(x-3)$ **b** $3n(m+2n)$ **c** $ax^2(x+2)$ **d** $-3x(x-6)$

2 **a** $(x+1)(8+x)$ **b** $(a+5)(a-8)$ **c** $3(g+1)(g-2)$

3 **a** $(3+x)(3-x)$ **b** $3(x+4)(x-4)$ **c** $y(5y+6)(5y-6)$

4 **a** $(x+10)(x-4)$ **b** $(x-3)(7-x)$ **c** $(3x+1)(x+5)$

5 **a** $(x+3)^2$ **b** $(x-14)^2$ **c** $(3x-1)^2$

6 **a** The middle term is not $\pm 2(2x)(3)=\pm 12x$.
$\therefore\ 4x^2-20x+9$ is not a perfect square.
b $k=25$ **c** $(2x-5)^2$

7 **a** $5(x-2)^2$ **b** $-3(x-3)^2$ **c** $6(x-10)^2$

8 **a** $(x+3)(x+5)$ **b** $(x+5)(x-3)$ **c** $(x-5)(x+3)$ **d** $4(x+3)(x-1)$ **e** $2(x-6)(x+5)$ **f** $3(x-8)(x-2)$

9 **a** $(m+n)(x-y)$ **b** $(2x+3)(x+4)$ **c** $(5x+1)(x-2)$

10 **a** $(3x+1)(x-6)$ **b** $(3x-1)(x-6)$ **c** $(4x-1)(2x+5)$ **d** $(4x+1)(3x+1)$ **e** $(x-2)(12x+1)$ **f** $3(2x+7)(x-2)$

EXERCISE 15A

1 **a** 3 **b** 2 **c** $\frac{1}{2}$ **d** -6 **e** $\frac{9}{2}$ **f** 1 **g** $\frac{9}{2}$ **h** 12 **i** 2 **j** 6

2 **a** -2 **b** $-\frac{1}{2}$ **c** $\frac{1}{3}$ **d** 8 **e** 4 **f** -1 **g** $\frac{1}{2}$ **h** -3 **i** $-\frac{2}{3}$ **j** -4

3 **a** -5 **b** $\frac{16}{7}$ **c** 3 **d** -1 **e** $\frac{1}{4}$ **f** 8 **g** $\frac{1}{3}$ **h** 1

EXERCISE 15B.1

1 **a** $\dfrac{2}{x}$ **b** $3x$ **c** 3 **d** $\frac{1}{2}$ **e** c **f** $\dfrac{1}{b}$ **g** $\dfrac{y}{z}$ **h** a **i** $\dfrac{1}{a}$ **j** 1 **k** $\dfrac{1}{2p}$ **l** $\dfrac{x}{2}$ **m** $\dfrac{2}{3y^2}$ **n** $\dfrac{1}{2x}$ **o** $\dfrac{4y^2}{x}$

2 **a** $-\dfrac{a}{2}$ **b** $-\dfrac{1}{2a}$ **c** $\dfrac{a^2}{4}$ **d** -4 **e** $-\dfrac{1}{3x^2}$ **f** $\dfrac{2}{t}$ **g** $-2d$ **h** $-\dfrac{b}{2}$ **i** $-\dfrac{2b}{3a}$ **j** $\dfrac{2x}{y^2}$

3 **a** t **b** cannot be simplified **c** y **d** cannot be simplified **e** $\dfrac{a}{b}$ **f** cannot be simplified **g** cannot be simplified **h** cannot be simplified **i** $\dfrac{7c}{4d}$ **j** cannot be simplified

4 **a** $25a$ **b** $5a^2$ **c** $\dfrac{5}{a}$ **d** $-4a$ **e** $\dfrac{3b^2}{2}$ **f** $2c^2$ **g** 9 **h** e^3

5 **a** $a+3$ **b** $2(x+2)$ **c** $4(b+5)$ **d** $\dfrac{d-3}{3}$ **e** $\dfrac{1}{2(z+3)}$ **f** $\dfrac{2}{x+1}$ **g** $\dfrac{3}{2-x}$ **h** $\dfrac{3}{2(1+x)}$

6 **a** 1 **b** cannot be simplified **c** 1

7 **a** 5 **b** $\frac{1}{3}$ **c** 2 **d** $\frac{3}{4}$ **e** $\dfrac{x+y}{3}$ **f** $\dfrac{y}{3}$ **g** $\dfrac{y+2}{3}$ **h** $\dfrac{3(p+q)}{2}$ **i** $\dfrac{x+1}{3}$

EXERCISE 15B.2

1 **a** $x+5$ **b** $\dfrac{6(x+2)}{5}$ **c** $\dfrac{2(x-3)}{3}$ **d** $x-3$ **e** $6a+2$ **f** $\dfrac{5(3a+1)}{3}$ **g** $\dfrac{5(b-2)}{2}$ **h** $b-2$ **i** $x-5$ **j** $\dfrac{3(x-5)}{5}$ **k** $\dfrac{4(2y+1)}{3}$ **l** $2y+1$ **m** $2x+8$ **n** $x+4$ **o** $\dfrac{x+4}{2}$ **p** $x+2$

2 **a** $\frac{2}{3}$ **b** $\frac{2}{3}$ **c** 2 **d** $\frac{5}{3}$ **e** 7 **f** $\frac{1}{2}$ **g** x **h** 3

3 **a** $a+c$ **b** b **c** $\dfrac{1}{a}$ **d** $a+c+1$

4 **a** $x+3$ **b** $x-2$ **c** $3x$ **d** $5x$ **e** 4 **f** $\frac{7}{2}$ **g** $2x$ **h** x **i** x^2-1 **j** x **k** b^2 **l** $\dfrac{y}{2}$

5 **a** -1 **b** $\dfrac{a+b}{-(a-b)}$ **c** -2 **d** $-\frac{4}{3}$

e $-\frac{1}{3}$ **f** $-\frac{1}{4}$ **g** $-\frac{1}{r}$ **h** $-\frac{y}{2}$

6 **a** $\frac{1}{x+1}$ **b** $a-1$ **c** $\frac{1}{2-m}$ **d** $x-y$

e $\frac{x-1}{3}$ **f** $-(x+4)$ **g** $\frac{a+b}{b}$ **h** $-\frac{x}{y}$

7 **a** $x-2$ **b** $x+7$ **c** $\frac{1}{x-5}$ **d** $\frac{1}{2-x}$

e $\frac{x}{x-4}$ **f** $\frac{x+6}{x+4}$ **g** $\frac{x+2}{x-4}$ **h** $\frac{x-3}{x}$

i $\frac{1-x}{x+3}$ **j** $\frac{2(x-3)}{x+2}$ **k** $\frac{x+4}{x-4}$ **l** $\frac{x+5}{5-x}$

EXERCISE 15C

1 **a** $\frac{x}{4}$ **b** $\frac{b}{2}$ **c** $\frac{2}{a}$ **d** $\frac{x^2}{6}$ **e** $\frac{1}{2}$ **f** $3x$

g $\frac{12}{x}$ **h** $4n$ **i** $\frac{m}{7}$ **j** $\frac{3a}{4b}$ **k** 1 **l** 2

m $\frac{ab}{5}$ **n** $\frac{x^2}{9}$ **o** 1 **p** $2x$ **q** $\frac{m^2}{4}$ **r** $\frac{4}{c}$

s $\frac{8}{x}$ **t** $\frac{2}{b^2}$ **u** $\frac{a^2}{8}$ **v** $\frac{1}{m^3}$ **w** $\frac{2}{3}$ **x** 1

2 **a** $\frac{3}{2y}$ **b** $\frac{1}{3xy}$ **c** $\frac{x^2}{2}$ **d** $\frac{xy}{4}$ **e** 2 **f** $\frac{8x}{3y^2}$

3 **a** $\frac{x-1}{x+1}$ **b** 1 **c** $\frac{x-1}{2x}$

EXERCISE 15D

1 **a** a **b** $\frac{a}{b^2}$ **c** $\frac{b^2}{a}$ **d** a **e** $\frac{4}{3}$ **f** $\frac{c^2}{12}$

g $\frac{3}{4}$ **h** $\frac{12}{c^2}$ **i** a^2 **j** $\frac{1}{3x}$ **k** $5x$ **l** $\frac{1}{b^2}$

2 **a** 8 **b** 6 **c** $\frac{1}{4}$ **d** $3a$ **e** $6a$ **f** $\frac{5}{2x}$

g $\frac{c}{b^2}$ **h** $\frac{16}{ab}$

EXERCISE 15E

1 **a** $\frac{2b}{3}$ **b** $\frac{a-4}{5}$ **c** $\frac{2x+1}{3}$ **d** $\frac{3-x}{4}$

2 **a** $\frac{5b}{6}$ **b** $\frac{a}{6}$ **c** $\frac{23a}{24}$ **d** $\frac{3a-4b}{12}$

e $\frac{17n}{10}$ **f** $-\frac{11p}{21}$ **g** $\frac{17k}{12}$ **h** $-\frac{s}{8}$

3 **a** $\frac{n+3}{3}$ **b** $\frac{20+y}{4}$ **c** $\frac{14-d}{2}$ **d** $\frac{4n}{3}$

e $\frac{3x}{2}$ **f** $\frac{m-6}{3}$ **g** $\frac{3-2x}{3}$ **h** $\frac{9x}{4}$

4 **a** $\frac{5}{x}$ **b** $\frac{5}{2x}$ **c** $\frac{x+2}{2x}$ **d** $\frac{2-x}{2x}$

e $\frac{5q+3p}{pq}$ **f** $\frac{7s+2r}{rs}$ **g** $\frac{4q+3p}{pq}$ **h** $\frac{3t}{2s}$

i $\frac{ac+2b}{bc}$ **j** $\frac{9}{2b}$ **k** $\frac{5a+3b}{5b}$ **l** $\frac{3a+c}{3b}$

5 **a** $\frac{3+a}{a}$ **b** $\frac{b-2}{b}$ **c** $\frac{c^2+1}{c}$ **d** $\frac{2-3d}{d}$

e $\frac{4p+3}{p}$ **f** $\frac{7q-2}{q}$ **g** $\frac{24r+1}{3r}$ **h** $\frac{8s-3}{2s}$

6 **a** $\frac{4}{p}$ **b** $\frac{13}{6q}$ **c** $\frac{2y+5x}{xy}$ **d** $-\frac{2}{3x}$

e $\frac{19a}{6}$ **f** $\frac{5x}{2}$ **g** $\frac{3a+2}{a^2}$ **h** $\frac{5+b}{b^2}$

i $\frac{6d-1}{d^2}$ **j** $\frac{5-4x}{x^2}$ **k** $\frac{x^2+1}{x^2}$ **l** $\frac{x^3+4}{x^2}$

m $\frac{20-a^2}{4}$ **n** $\frac{6b^2-2}{b^2}$ **o** $\frac{2x+x^2}{2}$ **p** $\frac{x^3-1}{x^2}$

q $\frac{4x^2+2}{x}$ **r** $\frac{6x^2-1}{x}$ **s** $\frac{3x^2-4}{4x}$ **t** $\frac{1+2x^2}{2x}$

7 **a** $\frac{a}{2}+2$ **b** $x+\frac{3}{2}$ **c** $2x+\frac{3}{2}$ **d** $3-\frac{3x}{2}$

e $\frac{x^2}{2}+x$ **f** $1-\frac{6}{x}$ **g** $\frac{x^2}{6}-x$ **h** $\frac{5}{x}-3x$

8 **a** $\frac{1}{r}$ **b** **i** $\frac{t-r}{rt}$ **ii** $\frac{r}{t}$

9 **a** **i** $2x$ dollars **ii** $\frac{2x}{3}$ dollars **b** $\frac{11x}{3}$ dollars

c \$55 each week

REVIEW SET 15A

1 **a** $-\frac{3}{4}$ **b** 6 **c** $\frac{1}{4}$ **d** $-\frac{3}{8}$

2 **a** $\frac{a^2}{2}$ **b** $\frac{x^2}{2y}$ **c** 4

3 The numerator and denominator have no common factors.

4 **a** $\frac{4}{x-3}$ **b** $\frac{a-b}{5a}$

5 **a** $\frac{y}{5}$ **b** $\frac{x+2}{x-2}$ **c** $\frac{x-5}{x-3}$

6 **a** $\frac{15}{x}$ **b** 12 **c** $\frac{4y}{3x}$

7 **a** $\frac{3a}{b^2}$ **b** $6x$ **c** $\frac{2c^3}{d^2}$

8 **a** $\frac{19a}{28}$ **b** $\frac{9-2n}{3}$ **c** $-\frac{x}{12}$

9 **a** $\frac{15}{2x}$ **b** $\frac{21x}{4}$ **c** $\frac{2x^2-3}{3x}$

10 **a** $\frac{c}{2}+3$ **b** $3x-\frac{1}{3}$ **c** $\frac{x^2}{2}-x$

11 **a** $\frac{3a+2b}{12}$ **b** **i** $\frac{2n+3m}{mn}$ **ii** $\frac{11}{20}$

REVIEW SET 15B

1 **a** $-\frac{7}{3}$ **b** $\frac{25}{7}$ **c** -10

2 **a** $\frac{b}{c}$ **b** $-\frac{x}{5y^2}$ **c** $-16t$

3 **a** $\frac{a-4}{3}$ **b** $\frac{5}{3}$ **c** $\frac{2}{5(k+3)}$

4 **a** $x+5$ **b** $3y$ **c** -1

5 **a** $\frac{5x}{2}$ **b** $x-5$ **c** $\frac{x+5}{2x}$

6 **a** $\frac{1}{2}$ **b** $\dfrac{5k}{2}$ **c** $\dfrac{25}{p^2}$ **d** $\dfrac{xy}{4}$

7 **a** $\frac{1}{4}$ **b** $\dfrac{3}{t^2}$ **c** $\frac{6}{5}$ **d** $\dfrac{a^3}{6}$

8 **a** $\dfrac{4x}{5}$ **b** $\dfrac{7x}{10}$ **c** $\dfrac{12+n}{2}$

9 **a** $\dfrac{10}{3x}$ **b** $\dfrac{5b-4a}{ab}$ **c** $\dfrac{x^3+4}{x}$ **10** $4-\dfrac{12}{x}$

11 **a** $\dfrac{x+6}{3}$ **b** $\dfrac{6-x}{3}$ **c** $\dfrac{6}{a}$ **d** $\dfrac{3a^3}{50}$
e $\dfrac{1}{2x}$ **f** $\dfrac{5x+4}{2x^2}$

12 **a** **i** $\dfrac{a^2-9}{a}$ **ii** $\dfrac{3-a}{3}$ **b** $-\dfrac{3(a+3)}{a}$
c **i** -12 **ii** undefined **iii** $-\frac{24}{5}$

EXERCISE 16A.1

1 **a** 5 units **b** 4 units **c** 3 units
d $\sqrt{41}$ units **e** 5 units **f** $\sqrt{13}$ units

2 **a** $\sqrt{34}$ units **b** $\sqrt{29}$ units **c** $\sqrt{34}$ units
d $\sqrt{26}$ units **e** $\sqrt{17}$ units **f** $\sqrt{13}$ units

3 **a**

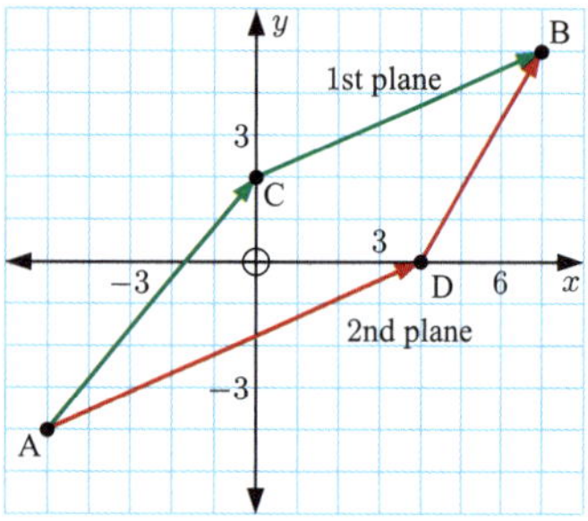

b 150 km
c **i** ≈ 154.3 km
ii ≈ 156.8 km
d the first plane

4 **a**

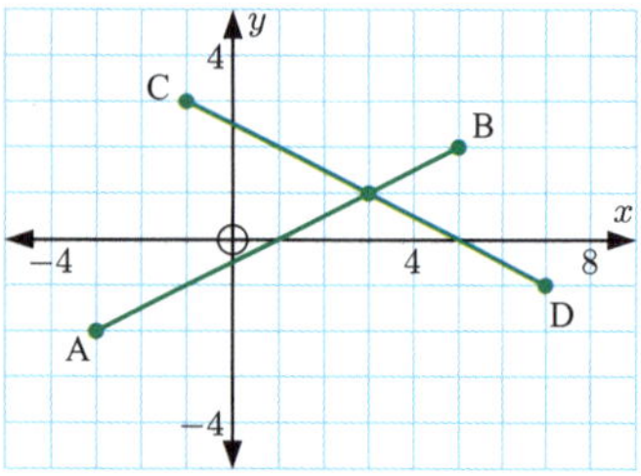

b X(3, 1) **c** B is closest and is $\sqrt{5}$ units from X.

EXERCISE 16A.2

1 **a** $\sqrt{10}$ units **b** $\sqrt{13}$ units **c** $4\sqrt{2}$ units
d $\sqrt{41}$ units **e** $5\sqrt{2}$ units **f** $\sqrt{13}$ units
g $2\sqrt{41}$ units **h** $2\sqrt{6}$ units

2 $d=\sqrt{a^2+b^2}$ units

3 **a** isosceles **b** scalene **c** isosceles **d** equilateral

4 **a** right angled at A **b** not right angled
c not right angled **d** right angled at B

5 ≈ 13.7 units

6 **a** LM = MN = $\sqrt{20}$ units $\therefore$ $\triangle$LMN is isosceles.
b $LM^2 + MN^2 = 40 = LN^2$
$\therefore$ $\triangle$LMN is right angled at M.
c $\widehat{MLN} = 45°$ **d** 10 units^2

7 **a** WX = YZ = $\sqrt{5}$ units, XY = WZ = $3\sqrt{5}$ units
b $5\sqrt{2}$ units
c $WX^2 + XY^2 = 50 = WY^2$
$\therefore$ $\triangle$WXY is right angled at X.
$\therefore$ $\widehat{WXY}$ is a right angle.
d a rectangle **e**
f 15 units^2

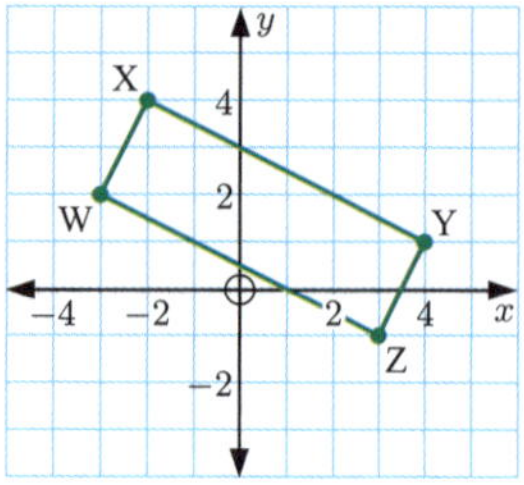

EXERCISE 16B

1 **a** M(−1, 2) **b** AM = BM = $\sqrt{5}$ units ✓

2 **a** M(3, 5) **b** M(4, 0) **c** M(−1, 1) **d** M(3, $1\frac{1}{2}$)
e M($-1\frac{1}{2}$, 2) **f** M(1, 1) **g** M($2\frac{1}{2}$, 4)
h M($-3\frac{1}{2}$, $3\frac{1}{2}$) **i** M($\frac{a+2}{2}$, 2) **j** M(0, 2b)

3 **a** C(3, 2) **b** B(5, $\frac{1}{2}$) **c** D(1, $3\frac{1}{2}$) **4** N(4, $3\frac{1}{4}$)

5 **a** Q(−1, 11) **b** Q(1, −10) **c** Q(6, 0) **d** Q($-1\frac{1}{2}$, 6)

6 **a** A(10, −9) **b** A(−6, −8)

7 **a** B(0, 5) **b** D(−2, 9) **8** **a** A(−2, 7) **b** 10 units

9 **a** X($3\frac{1}{2}$, $4\frac{1}{2}$) **b** C(6, 7)

EXERCISE 16C.1

1 **a** $\frac{1}{2}$ **b** $\frac{1}{5}$ **c** 2 **d** undefined
e 0 **f** $-\frac{3}{2}$ **g** $-\frac{1}{2}$ **h** −3 **i** $-\frac{2}{5}$

2 **a** **A** and **D** **b** **B**

3

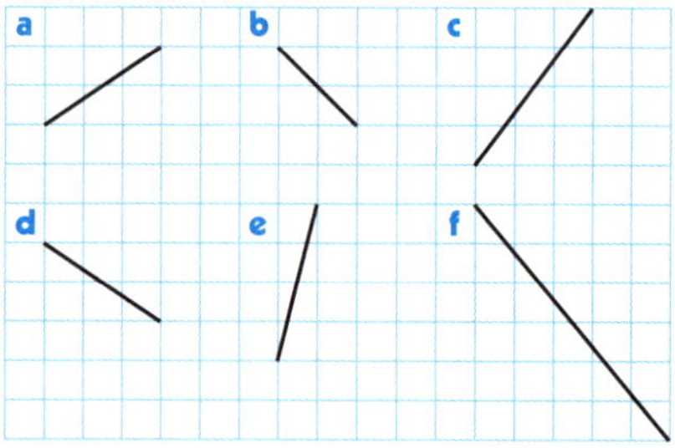

4 **a**

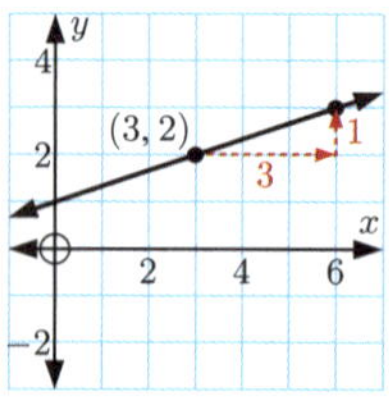

b

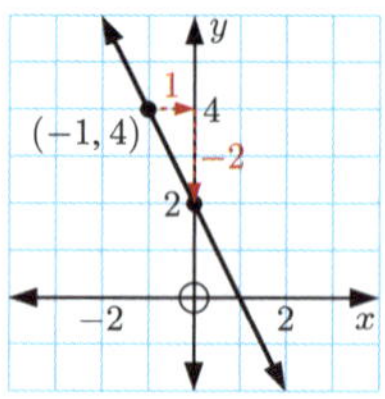

c

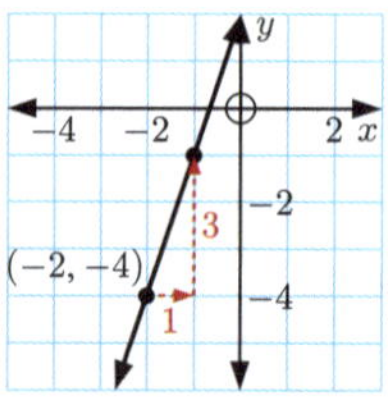

d

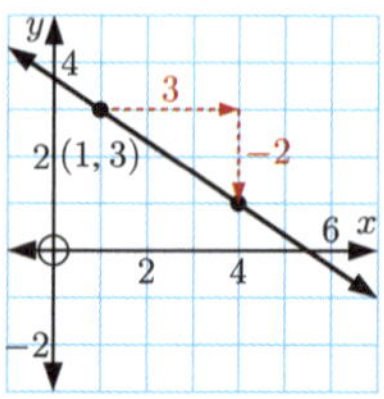

e

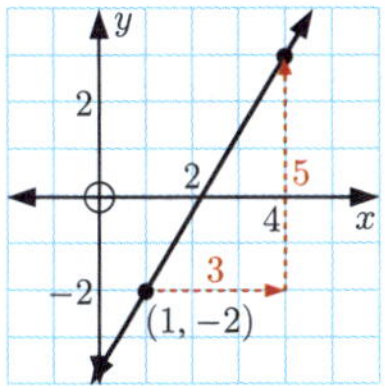

f

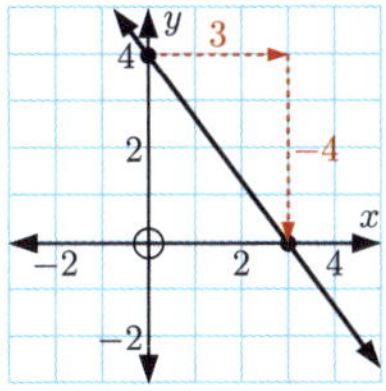

5

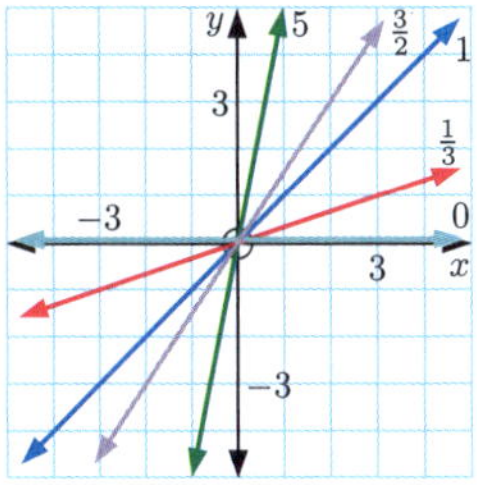

6

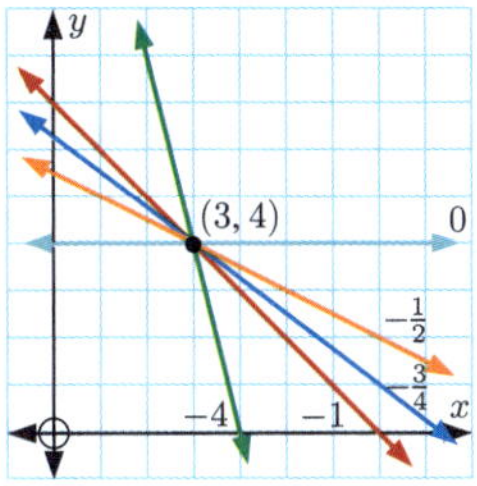

7

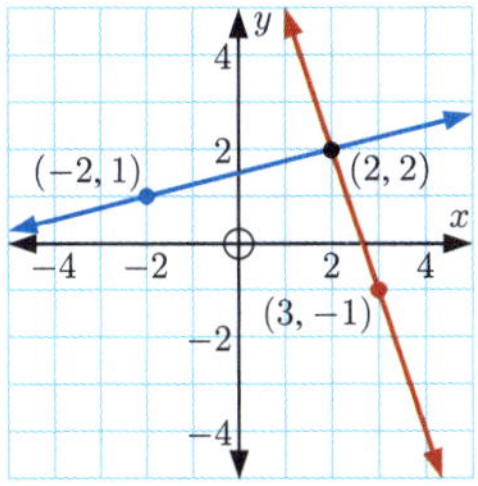

EXERCISE 16C.2

1 **a** $\frac{5}{2}$ **b** -4 **c** $\dfrac{b}{a}$ **2** **a** $\frac{2}{7}$ **b** -4 **c** $-\frac{2}{3}$

3 **a** 5 **b** 4 **c** $\frac{1}{4}$ **d** 1 **e** undefined

f 0 **g** $\frac{7}{2}$ **h** $-\frac{7}{2}$ **i** 2

4 **a** $\frac{4}{5}$ **b** $-\frac{9}{2}$ **c** $-\frac{6}{19}$

5 **a** **i** $\frac{2}{3}$ **ii** $\frac{2}{3}$

b $\dfrac{y_2 - y_1}{x_2 - x_1} = \dfrac{-(y_1 - y_2)}{-(x_1 - x_2)} = \dfrac{y_1 - y_2}{x_1 - x_2}$

$\therefore$ gradient of [AB] = gradient of [BA]

6 **a** gradient of [AB] $= \frac{1}{2}$, gradient of [BC] $= -\frac{1}{3}$,
gradient of [CD] $= \frac{8}{3}$, gradient of [AD] $= -1$

b **i** 2 **ii** $\frac{5}{6}$

7 **a** $a = 4$ **b** $a = 3$ **c** $a = \frac{10}{3}$ **d** $a = -\frac{7}{4}$ **e** $a = 1$

EXERCISE 16D

1 **a** $\frac{3}{5}$ **b** $-\frac{5}{3}$ **c** 6 **d** $-\frac{1}{6}$

2 **a** $-\frac{5}{4}$ **b** -3 **c** $\frac{7}{3}$ **d** $-\frac{1}{2}$ **e** $\frac{1}{11}$

f undefined **g** 0 **h** $\frac{2}{3}$

3 **a** **i** $\frac{2}{3}$ **ii** -1 **iii** $\frac{2}{3}$ **iv** $\frac{11}{9}$ **v** $\frac{1}{9}$ **vi** 1

b **i** [AB] and [CD] **ii** [BC] and [BD]

4 **a** **i** 5 **ii** $-\frac{1}{5}$ **b** right angled at B

5 **a** $a = 6$ **b** $a = 0$ **6** **a** $t = -3$ **b** $t = 9$

EXERCISE 16E

1 **a** AB = AC = $\sqrt{20}$ units $\therefore$ $\triangle$ABC is isosceles.

b M(2, 3)

c gradient of [AM] = 1, gradient of [BC] = -1
Gradients are negative reciprocals so [AM] $\perp$ [BC].

d

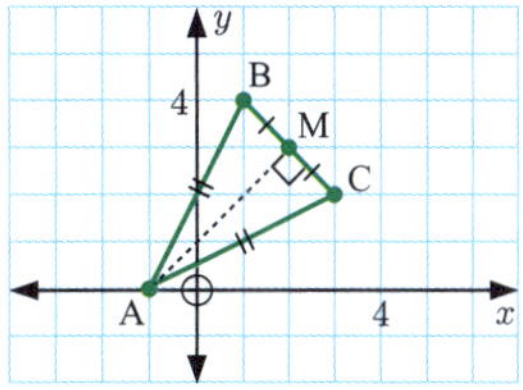

2 **a** gradient of [MN] = -3, gradient of [AC] = -3
$\therefore$ [MN] $\parallel$ [AC]

b MN = $\sqrt{10}$ units, AC = $\sqrt{40} = 2\sqrt{10}$ units

3 **a** **i** gradient of [AB] = 0, gradient of [CD] = 0
$\therefore$ [AB] $\parallel$ [CD]

ii gradient of [AD] $= \frac{3}{5}$, gradient of [BC] $= \frac{3}{5}$
$\therefore$ [AD] $\parallel$ [BC]

b a parallelogram

4 **a** 5 units **b** $a = 4$ **c** **i** $\frac{1}{2}$ **ii** -2

d Gradients are negative reciprocals so [AP] $\perp$ [BP]
$\therefore$ $\widehat{APB}$ is a right angle.

REVIEW SET 16A

1 **a** $6\sqrt{5}$ units **b** (1, 0)

2 **a** isosceles **b** right angled at P **3** D(1, 2)

4 **a**

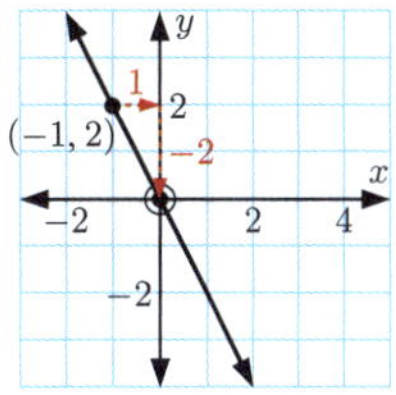

b

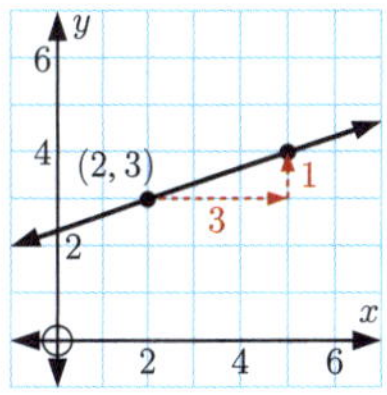

5 **a** $\frac{5}{2}$ **b** undefined **6** **a** $-\frac{1}{4}$ **b** 4

7 **a** $a = 5$ **b** $a = \frac{5}{4}$

8 **a** **i** A($\frac{1}{2}$, $1\frac{1}{2}$) **ii** S(2, 0)

b gradient of [QS] = -1, gradient of [PR] = 1
Gradients are negative reciprocals so [QS] $\perp$ [PR].

9 **a** **i** $2\sqrt{13}$ units **ii** $2\sqrt{13}$ units **iii** $2\sqrt{13}$ units
iv $2\sqrt{13}$ units

b AC = BD = $2\sqrt{26}$ units

c midpoint of [AC] : (4, −3), midpoint of [BD] : (4, −3)

d a square

REVIEW SET 16B

1 **a** [BC] **b** $(1, -\frac{1}{2})$

2 Q(12, 2) **3**

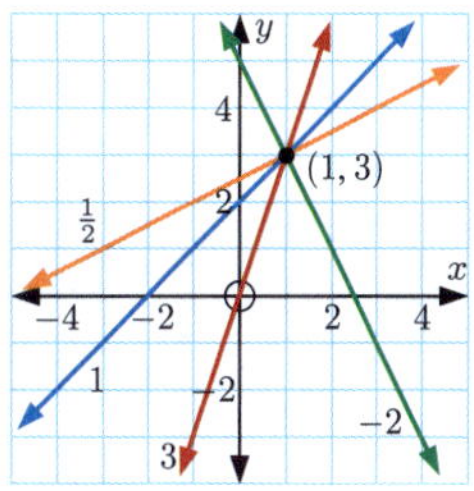

4 **a** $\frac{3}{2}$ **b** $-\frac{3}{4}$ **5** **a** $t = 10$ **b** $t = \frac{5}{2}$

6 **a** $\frac{1}{2}$ **b** $-\frac{4}{3}$ **c** $\frac{3}{10}$

7 **a** gradient of [PQ] $= \frac{3}{4}$, gradient of [SR] $= \frac{3}{4}$
$\therefore$ [PQ] $\parallel$ [SR]

b gradient of [PQ] $= \frac{3}{4}$, gradient of [PS] $= -\frac{4}{3}$
Gradients are negative reciprocals so [PQ] $\perp$ [PS].

8 **a** M(0, −1) **b** $OM^2 + MA^2 = 4 = OA^2$
$\therefore$ $\triangle$OMA is right angled at M.

c **i** 2 units
ii 2 units
iii 2 units

d equilateral

e

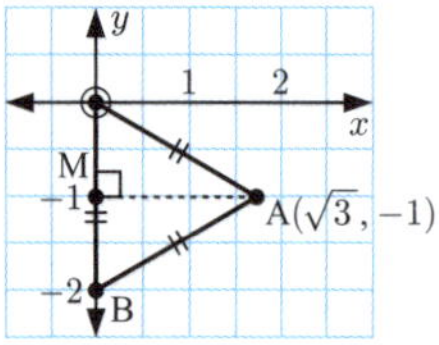

9 **a** **i** P(−1, 4) **ii** Q(5, $-\frac{1}{2}$) **iii** R(2, −7)
iv S(−4, $-2\frac{1}{2}$)

b **i** gradient of [PQ] $= -\frac{3}{4}$, gradient of [SR] $= -\frac{3}{4}$
$\therefore$ [PQ] $\parallel$ [SR]
ii gradient of [QR] $= \frac{13}{6}$, gradient of [PS] $= \frac{13}{6}$
$\therefore$ [QR] $\parallel$ [PS]

c a parallelogram

EXERCISE 17A

1 **a**

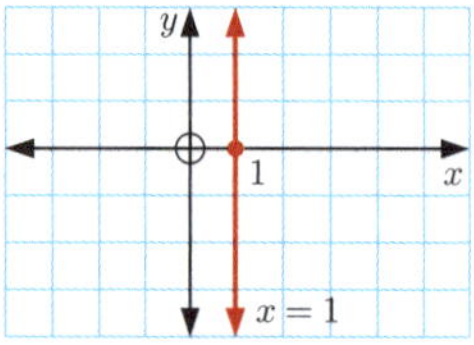

gradient is undefined

b

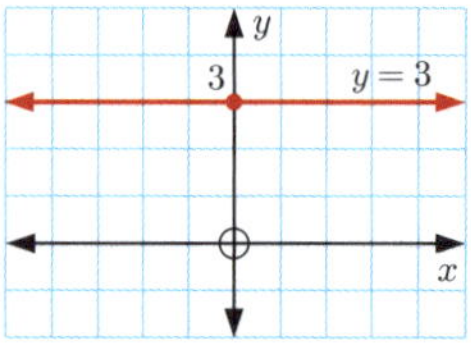

gradient is 0

c

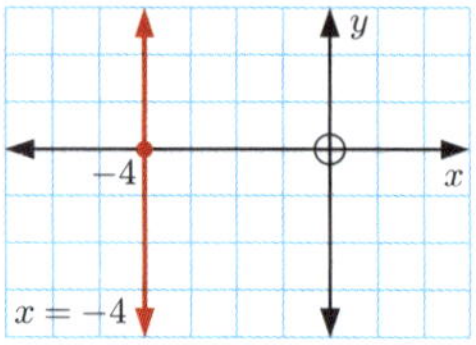

gradient is undefined

d

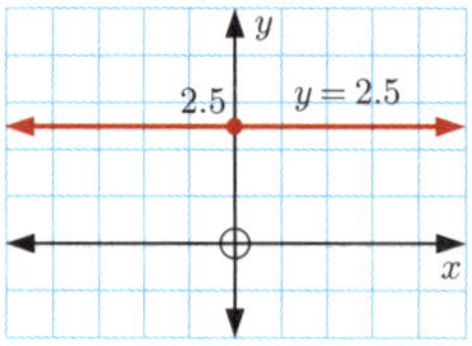

gradient is 0

e

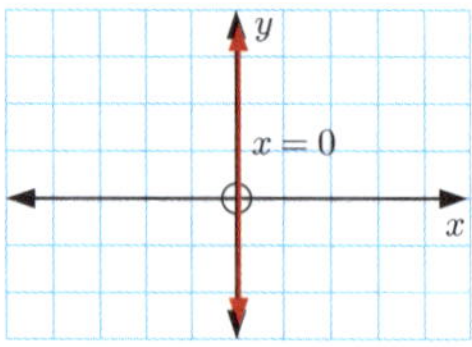

gradient is undefined

2 **a** $y = 0$ **b** $x = 0$ **c** $y = -3$ **d** $x = 4$

3 **a** $y = 3$ **b** $x = 4$

4 **a**

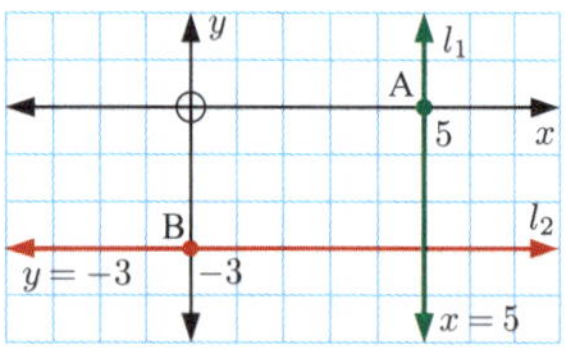

b A(5, 0), B(0, −3) **c** $\frac{3}{5}$

5 **a**

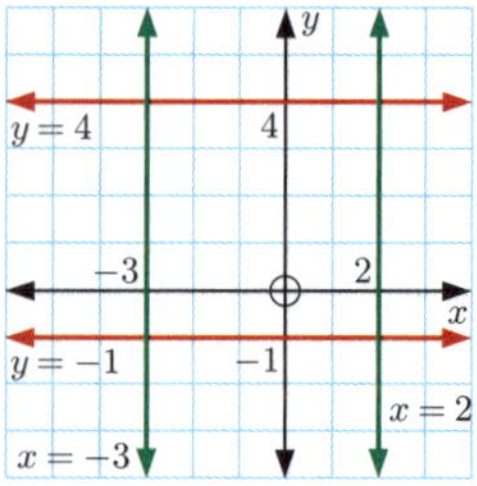

b **i** a square
ii 25 units^2
iii 1 and −1

EXERCISE 17B

1 **a** no **b** yes **c** yes **d** no **e** yes **f** no

2 **a** $a = 2$ **b** $a = 13$ **c** $a = 5$ **d** $a = 4$
e $a = 3$ **f** $a = -2$ **g** $a = 2$ **h** $a = -2$

3 above

EXERCISE 17C

1 **a** x-intercept is −3, y-intercept is 4, gradient is $\frac{4}{3}$
b x-intercept is −4, y-intercept is −2, gradient is $-\frac{1}{2}$
c x-intercept is 1, y-intercept is 4, gradient is −4

2 **a**

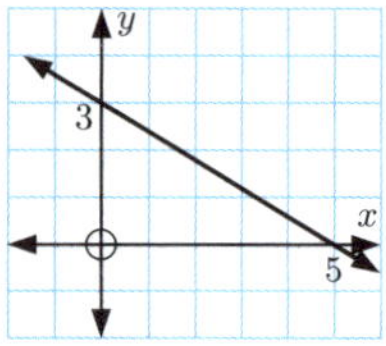

gradient is $-\frac{3}{5}$

b

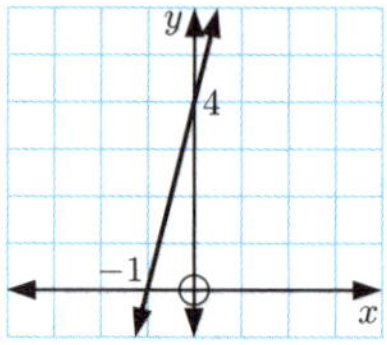

gradient is 4

c

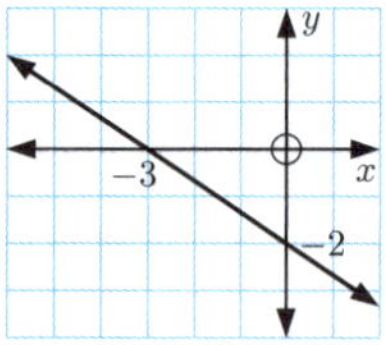

gradient is $-\frac{2}{3}$

d

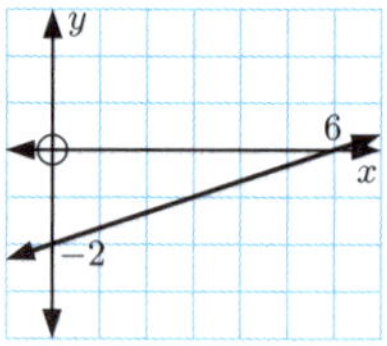

gradient is $\frac{1}{3}$

3 **a** 5 **b** −3 **c** 0 **d** 5 **e** 1 **f** 2
g −4 **h** 6

4 **a** 4 **b** −2 **c** 1 **d** −3 **e** 4 **f** 3
g 6 **h** −4

5 **a** c **b** $-\dfrac{c}{m}$ **c** m

EXERCISE 17D

1 **a** **i**

x	−3	−2	−1	0	1	2	3
y	−3	−2	−1	0	1	2	3

ii

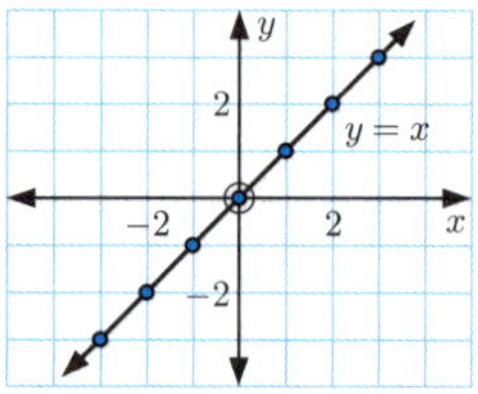

iii gradient is 1,
x-intercept is 0,
y-intercept is 0

b **i**

x	−3	−2	−1	0	1	2	3
y	3	2	1	0	−1	−2	−3

ii

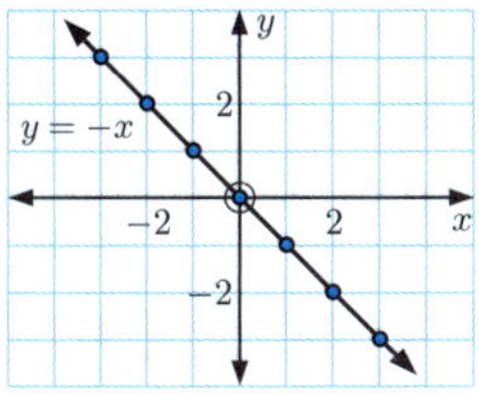

iii gradient is -1, x-intercept is 0, y-intercept is 0

c **i**

x	-3	-2	-1	0	1	2	3
y	-6	-4	-2	0	2	4	6

ii

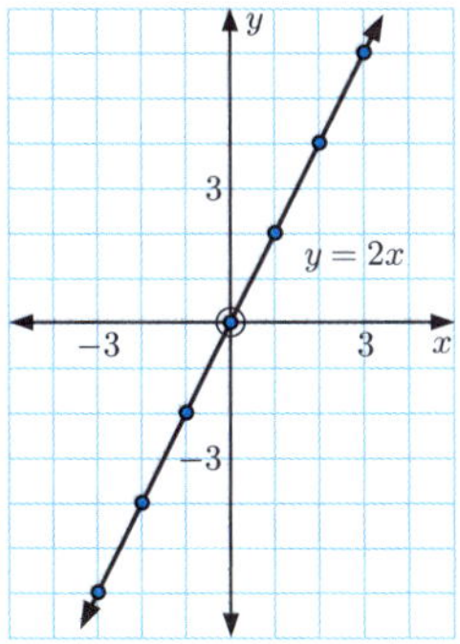

iii gradient is 2, x-intercept is 0, y-intercept is 0

d **i**

x	-3	-2	-1	0	1	2	3
y	6	4	2	0	-2	-4	-6

ii

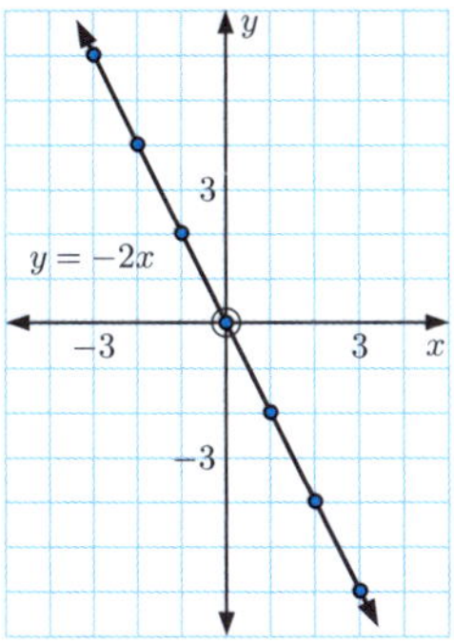

iii gradient is -2, x-intercept is 0, y-intercept is 0

e **i**

x	-3	-2	-1	0	1	2	3
y	-7	-5	-3	-1	1	3	5

ii

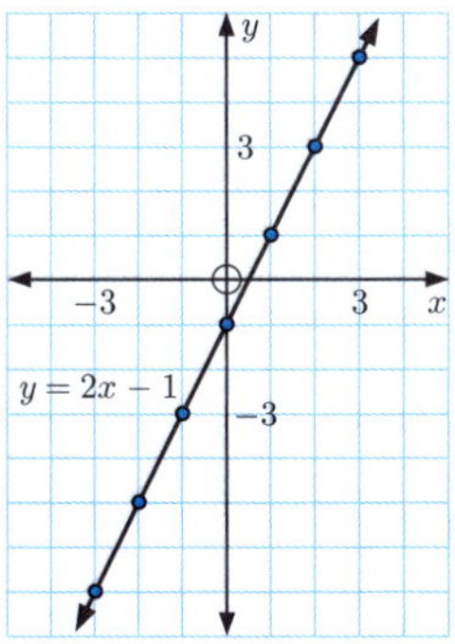

iii gradient is 2, x-intercept is $\frac{1}{2}$, y-intercept is -1

f **i**

x	-3	-2	-1	0	1	2	3
y	$\frac{1}{2}$	1	$1\frac{1}{2}$	2	$2\frac{1}{2}$	3	$3\frac{1}{2}$

ii

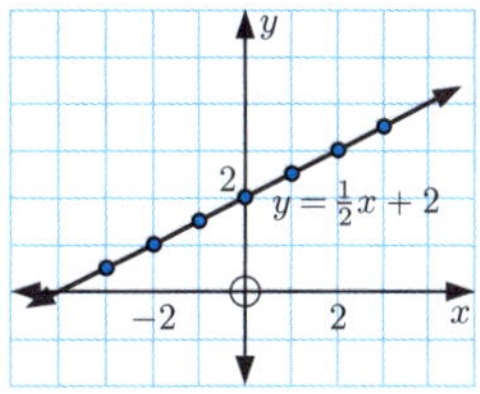

iii gradient is $\frac{1}{2}$, x-intercept is -4, y-intercept is 2

g **i**

x	-3	-2	-1	0	1	2	3
y	$3\frac{1}{2}$	3	$2\frac{1}{2}$	2	$1\frac{1}{2}$	1	$\frac{1}{2}$

ii

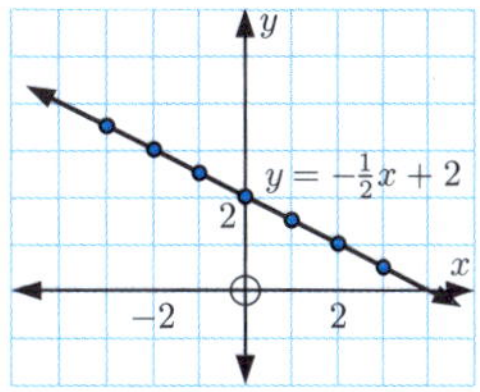

iii gradient is $-\frac{1}{2}$, x-intercept is 4, y-intercept is 2

h **i**

x	-3	-2	-1	0	1	2	3
y	5	4	3	2	1	0	-1

ii

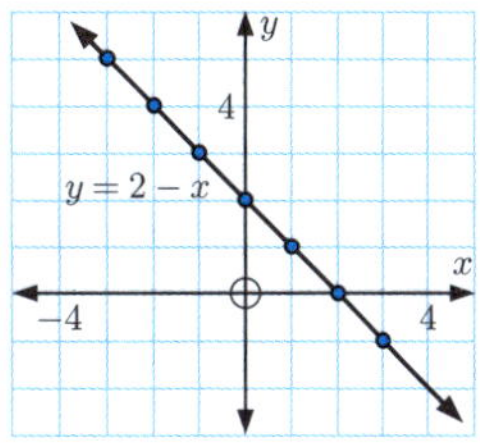

iii gradient is -1, x-intercept is 2, y-intercept is 2

i **i**

x	-3	-2	-1	0	1	2	3
y	$3\frac{1}{2}$	2	$\frac{1}{2}$	-1	$-2\frac{1}{2}$	-4	$-5\frac{1}{2}$

ii

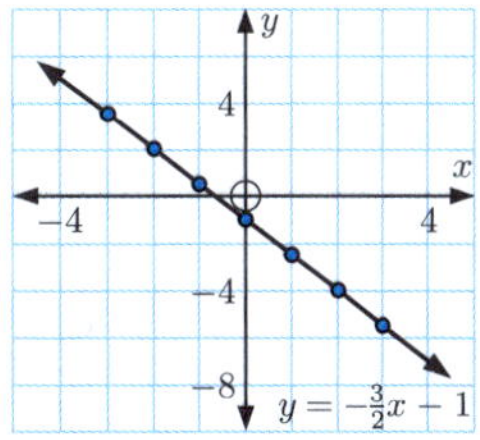

iii gradient is $-\frac{3}{2}$, x-intercept is $-\frac{2}{3}$, y-intercept is -1

2 **a**

x	-2	0	2	$-\dfrac{c}{m}$
y	$-2m+c$	c	$2m+c$	0

b x-intercept is $-\dfrac{c}{m}$, y-intercept is c **c** gradient is m

d A straight line with equation $y = mx + c$ has gradient m and y-intercept c.

EXERCISE 17E

1 **a** gradient is 2, y-intercept is -4
b gradient is -4, y-intercept is 1
c gradient is $\frac{2}{3}$, y-intercept is -3
d gradient is $-\frac{1}{3}$, y-intercept is 7
e gradient is $\frac{3}{4}$, y-intercept is $\frac{1}{2}$
f gradient is $-\frac{3}{2}$, y-intercept is $-\frac{1}{4}$
g gradient is $\frac{1}{2}$, y-intercept is $-\frac{3}{2}$

h gradient is $\frac{5}{4}$, y-intercept is $\frac{1}{4}$

i gradient is $-\frac{2}{3}$, y-intercept is $\frac{5}{3}$

2 **a**
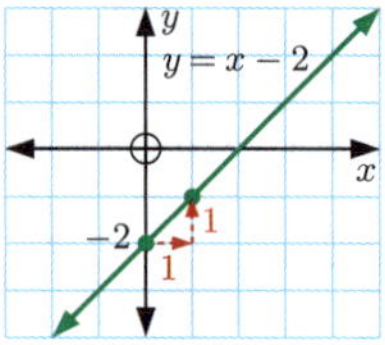

b
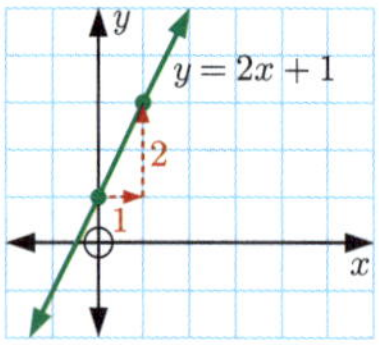

c
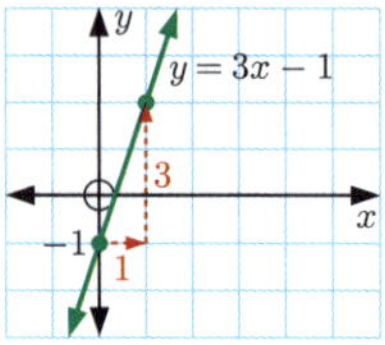

d
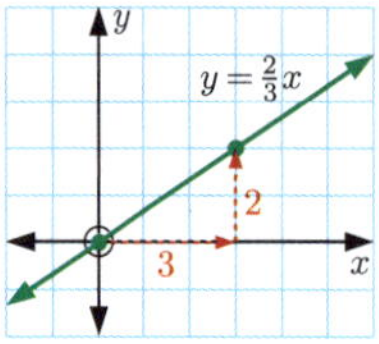

e
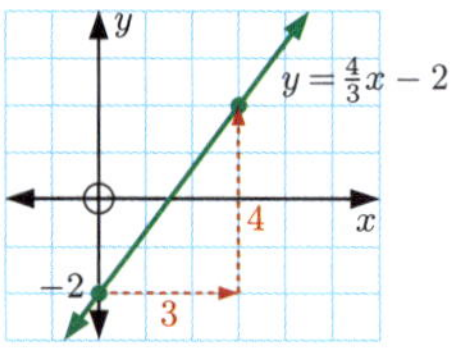

f
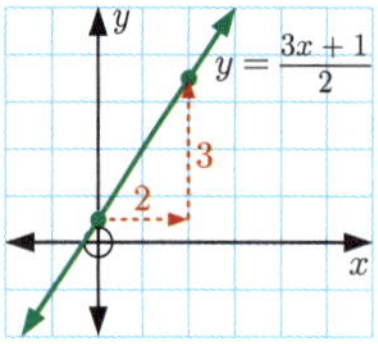

3 **a**
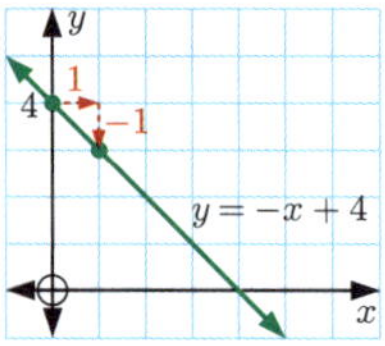

b
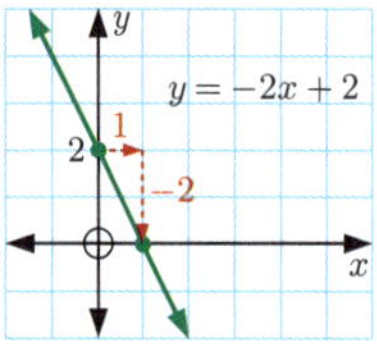

c
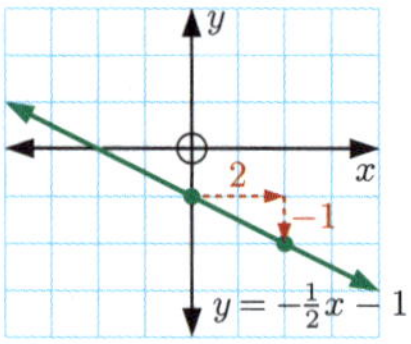

d
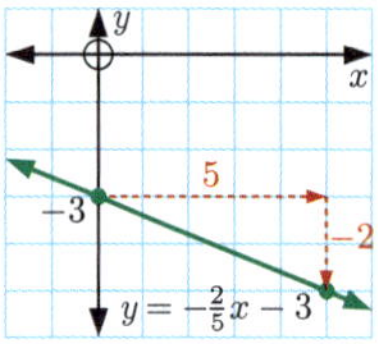

e
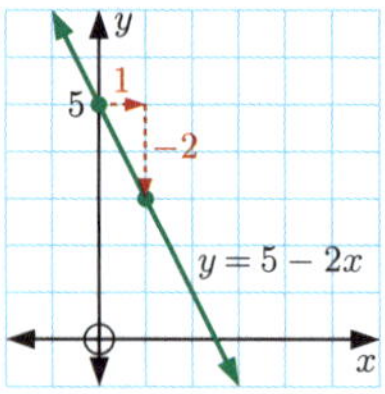

f
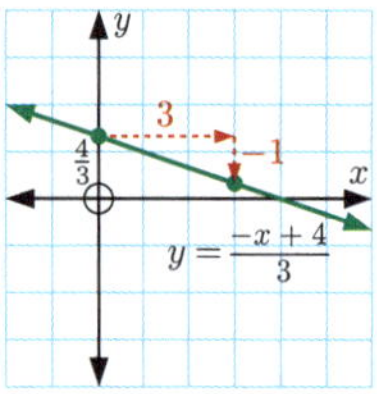

4 **a** **i** 3
ii -5
iii $\frac{5}{3}$

b
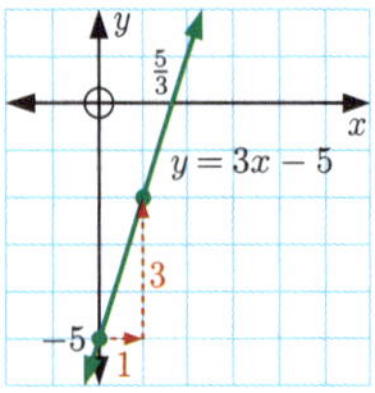

c $\frac{25}{6}$ units^2

5 **a** **i** $-\frac{2}{3}$
ii 2
iii 3

b
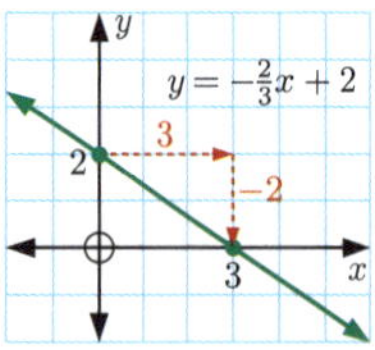

c yes

6
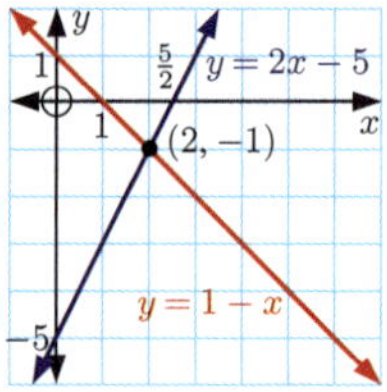

7 **a**
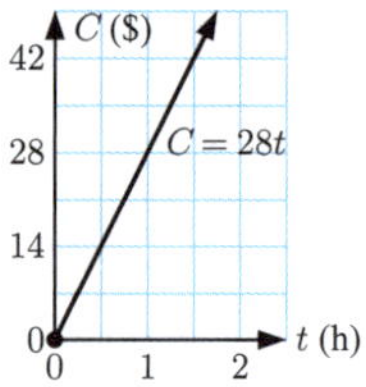

b 28; the cost of hiring a squash court is \$28 per hour.

8 **a**
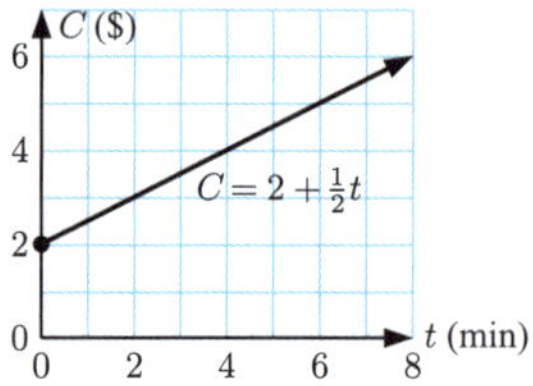

b $\frac{1}{2}$; the cost of hiring a scooter is 50 cents per minute.

c 2; the base cost of hiring a scooter is \$2.

EXERCISE 17F

1 **a** $2x+y=11$ **b** $3x-y=4$ **c** $x+5y=6$
d $6x+7y=5$ **e** $5x-8y=1$ **f** $4x-9y=-18$

2 **a** **i** $y=-3x+5$ **ii** -3
b **i** $y=-\frac{2}{5}x+2$ **ii** $-\frac{2}{5}$
c **i** $y=-\frac{7}{4}x-\frac{9}{4}$ **ii** $-\frac{7}{4}$
d **i** $y=6x-1$ **ii** 6
e **i** $y=\frac{5}{11}x-\frac{2}{11}$ **ii** $\frac{5}{11}$
f **i** $y=\frac{9}{2}x+\frac{5}{2}$ **ii** $\frac{9}{2}$

3 **a**
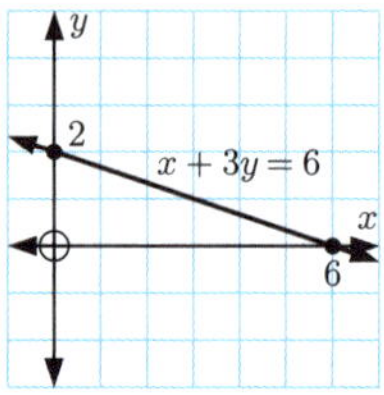

b
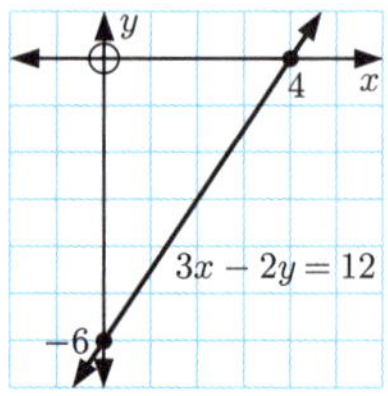

c
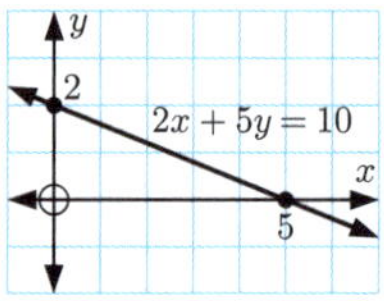

d
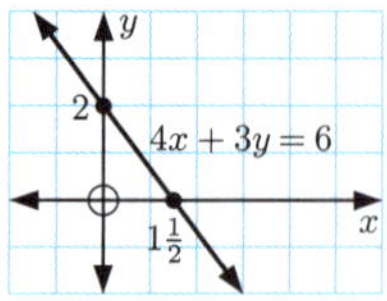

e
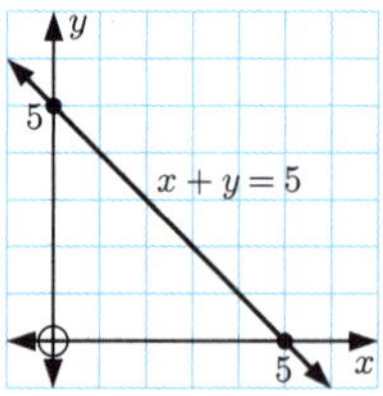

f
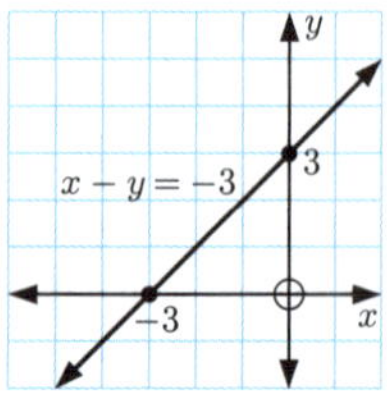

g 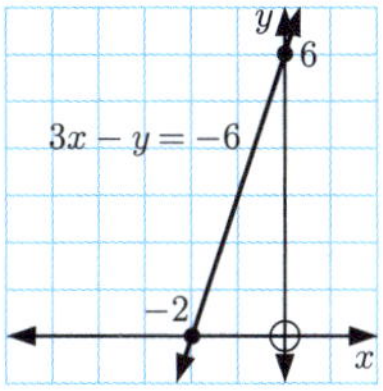

h 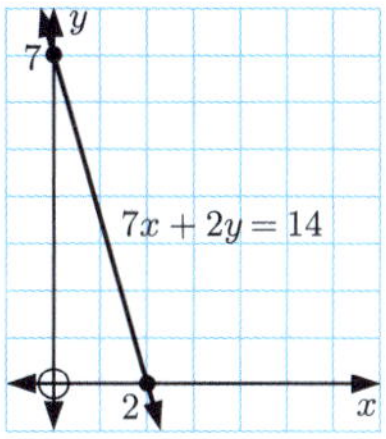

i 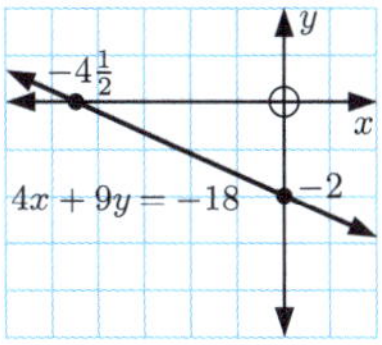

4 **a** x-intercept is 5, y-intercept is -3

b **i** yes **ii** no

c 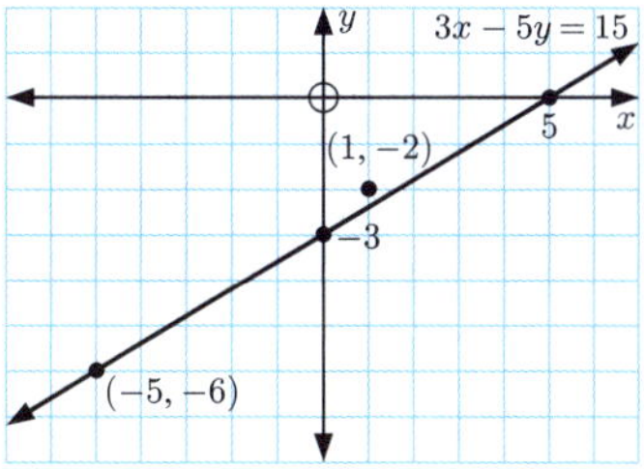

5 **a** 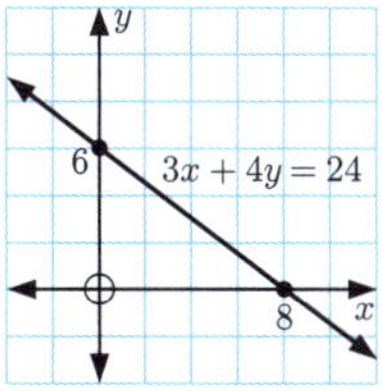

b D **c** A

6 **a** The total cost of x carrots at \$0.20 each, and y potatoes at \$0.50 each is \$5.

$\therefore \ 0.2x + 0.5y = 5$

or $2x + 5y = 50$

b 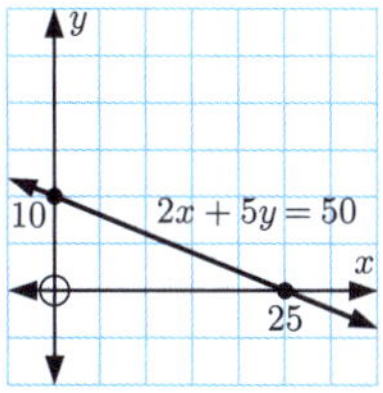

c **i** x-intercept is 25; this is the number of carrots bought if no potatoes were bought.

ii y-intercept is 10; this is the number of potatoes bought if no carrots were bought.

7 **a** $-\frac{a}{b}$ **b** $b \neq 0$

EXERCISE 17G.1

1 **a** $y = 3x + 4$ **b** $y = 7x - 1$ **c** $y = -x + \frac{1}{3}$

d $y = \frac{1}{2}x$ **e** $y = 5$ **f** $y = -\frac{5}{2}x - \frac{2}{3}$

2 **a** $y = \frac{1}{4}x - 6$ **b** 24

3 **a** The tank loses 8 litres per minute.

$\therefore$ the gradient must be -8.

b 480; there is 480 L of water in the tank initially.

c $V = 480 - 8t$

d 60; the tank will be empty after 60 minutes or 1 hour.

EXERCISE 17G.2

1 **a** $y = 2x + 6$ **b** $y = -3x - 4$ **c** $y = 5x + 14$

d $y = \frac{3}{4}x + 6$ **e** $y = 4$ **f** $y = -\frac{1}{2}x - \frac{13}{2}$

g $y = -\frac{1}{3}x + \frac{13}{3}$ **h** $y = \frac{2}{3}x + \frac{19}{3}$

2 **a** $y = 3x - 2$ **b** $y = -\frac{1}{4}x - 1$ **c** $y = 2x - 1$

3 **a** $x - 2y = 2$ **b** $3x - 4y = 15$ **c** $3x - y = 11$

d $3x + 4y = -6$ **e** $2x + y = 4$ **f** $3x + y = 4$

4 $x - 3y = 7$

EXERCISE 17G.3

1 **a** $y = 2x + 2$ **b** $y = \frac{2}{5}x + 3$ **c** $y = -x - 1$

d $y = -2x + 4$ **e** $y = \frac{1}{2}x - \frac{1}{2}$ **f** $y = -\frac{2}{5}x$

2 **a** $y = \frac{4}{3}x + 4$ **b** $y = 4x - 5$ **c** $y = -3x + 7$

d $y = \frac{3}{8}x - 1$

3 **a** $y = x - 1$ **b** $y = \frac{1}{2}x + 1$ **c** $y = -\frac{3}{4}x + 3$

d $y = -\frac{1}{3}x + 3$ **e** $y = -2x - 4$ **f** $y = 3$

4 **a** $y = 2x + 3$ **b** $y = -2x + 12$ **c** $y = \frac{2}{3}x - 2$

d $y = -3$ **e** $y = \frac{3}{2}x - 5$ **f** $y = -\frac{4}{3}x + \frac{1}{3}$

5 **a** $y = 3x - 2$ **b** $y = -\frac{1}{3}x + 4$ **c** $y = 7$

d $y = -\frac{2}{5}x - 2$ **e** $y = \frac{5}{4}x + \frac{5}{2}$ **f** $x = -4$

6 **a** $3x + y = 4$ **b** $y = 2$ **c** $x = 3$

d $5x + 3y = 12$ **e** $5x - 4y = -14$ **f** $x - 5y = 9$

7 **a** $C = \frac{1}{2}n + 2$ **b** $V = 5 - t$ **c** $N = \frac{1}{2}t + 2$

d $H = 2 - g$ **e** $F = \frac{3}{10}x + 5$ **f** $P = -\frac{1}{3}t - 2$

8 **a** $C(0, 2\sqrt{3})$ **b** $M(1, \sqrt{3})$, $N(-1, \sqrt{3})$

c $x = 0$, $\sqrt{3}x - 3y = -2\sqrt{3}$, $\sqrt{3}x + 3y = 2\sqrt{3}$

REVIEW SET 17A

1 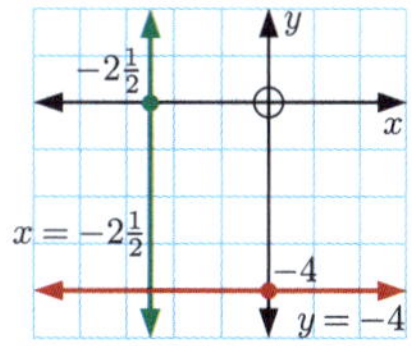

2 **a** yes **b** no

3 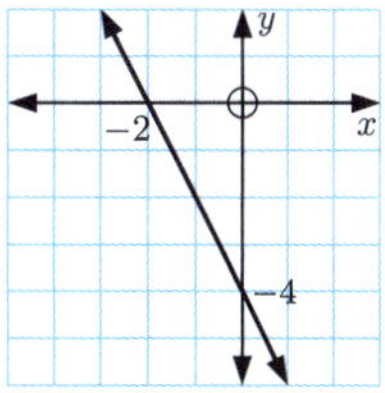

gradient is -2

4 **a**

x	-3	-2	-1	0	1	2	3
y	$\frac{1}{2}$	0	$-\frac{1}{2}$	-1	$-1\frac{1}{2}$	-2	$-2\frac{1}{2}$

b 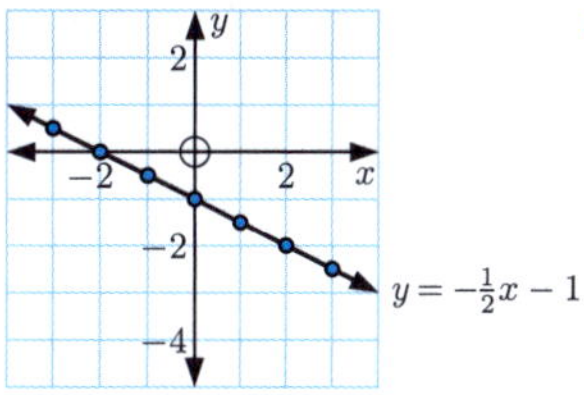

c gradient is $-\frac{1}{2}$, x-intercept is -2, y-intercept is -1

5 **a** gradient is 4, y-intercept is -3
b gradient is -1, y-intercept is 2
c gradient is $\frac{3}{2}$, y-intercept is $-\frac{1}{2}$

6 **a** $3x - 2y = -8$ **b** $y = \frac{5}{4}x - 10$

7 x-intercept is 3, y-intercept is 2, gradient is $-\frac{2}{3}$

8 **a**

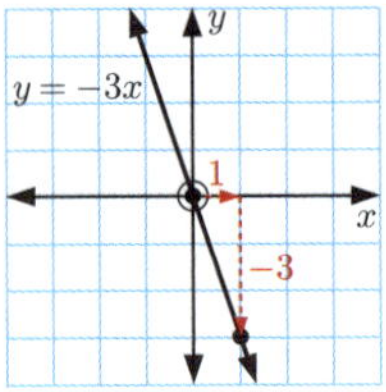

b

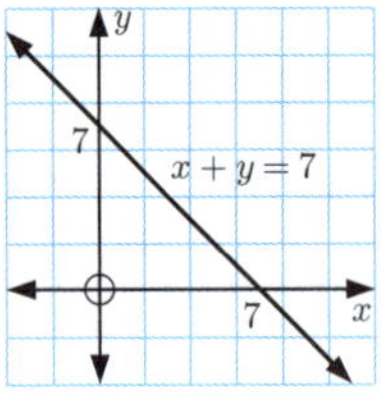

c

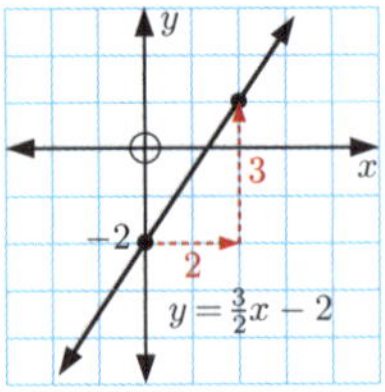

d

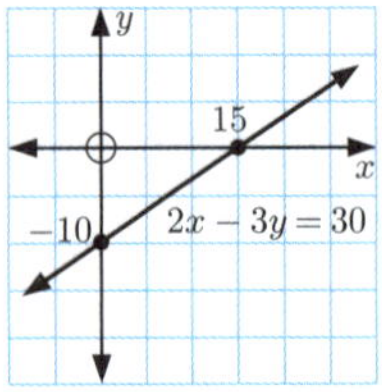

e

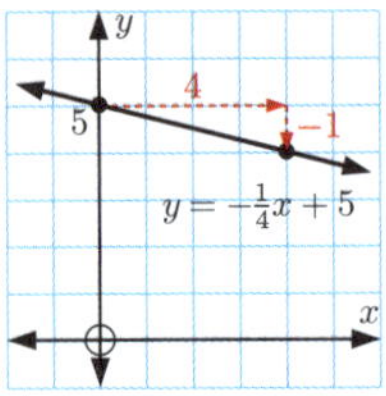

f

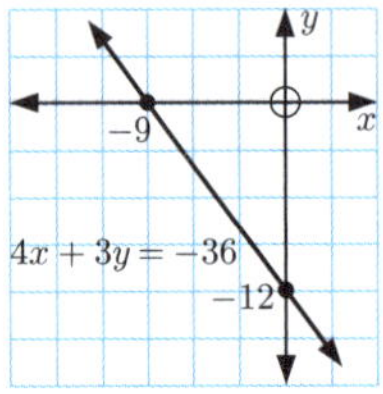

9 **a** $y = 2x - 3$ **b** $y = 2 - x$ **c** $y = \frac{3}{2}x - 5$

10 **a** The total cost of x bats at \$10 each, and y dozen balls at \$5 per dozen is \$140.
$\therefore\ 10x + 5y = 140$

b

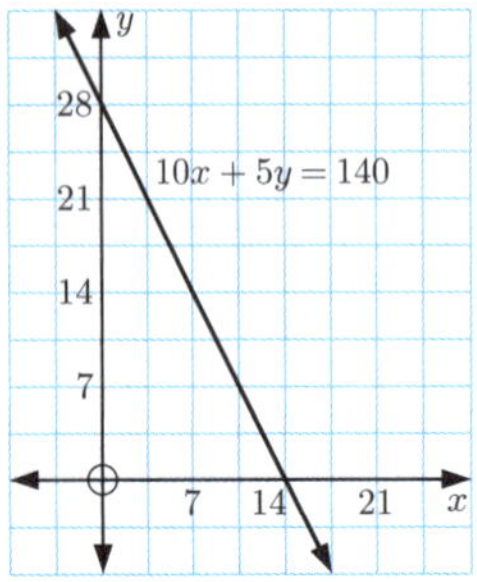

c **i** x-intercept is 14; this is the number of bats bought if no balls were bought.
ii y-intercept is 28; this is the number of dozens of balls bought if no bats were bought.

11 **a** $5x + 2y = -10$ **b** $8x - 3y = -7$ **c** $2x - 3y = 18$

12 **a** $y = x - 3$ **b** $y = -\frac{2}{3}x + \frac{2}{3}$
c line 1: $y = \frac{6}{5}x + \frac{8}{5}$, line 2: $y = -\frac{5}{6}x + \frac{29}{6}$

13 **a** 150; the cost of hiring a hall is \$150 for each hour.
b 100; the initial hiring fee is \$100.
c $C = 150t + 100$ **d** \$1300

14 **a** **i** $y = \frac{1}{3}x + 2$
ii When $x = 6$, $y = \frac{1}{3}(6) + 2 = 4$ ✓
$\therefore$ line 1 passes through Q(6, 4).
b **i** $y = -3x + 22$ **ii** $k = 1$ **c** $y = 1$

d

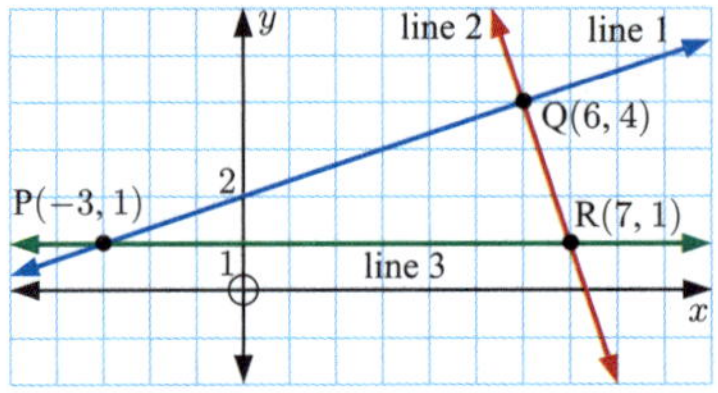

e 15 units^2

REVIEW SET 17B

1 **a** $y = -3$ **b** $x = 5$ **2** **a** $k = -1$ **b** $k = 4$

3 **a** -6 **b** -35

4 **a**

x	-3	-2	-1	0	1	2	3
y	$-1\frac{1}{2}$	0	$1\frac{1}{2}$	3	$4\frac{1}{2}$	6	$7\frac{1}{2}$

b

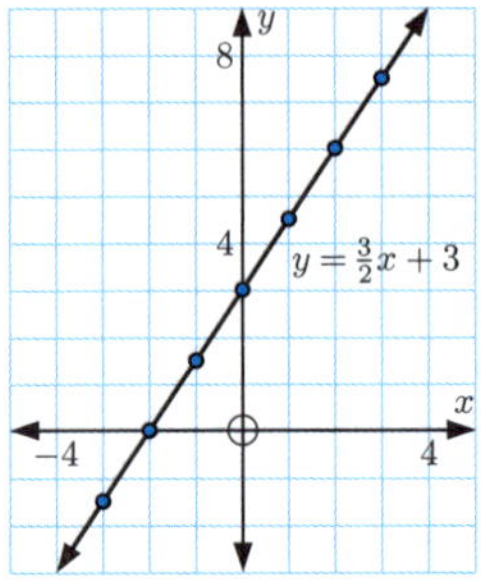

c gradient is $\frac{3}{2}$,
x-intercept is -2,
y-intercept is 3

5 **a** gradient is -2, y-intercept is 5
b gradient is $\frac{7}{2}$, y-intercept is -3
c gradient is $\frac{4}{3}$, y-intercept is $\frac{1}{3}$

6 **a** $y = \frac{4}{3}x - 8$ **b** $\frac{4}{3}$

7 **a** x-intercept is 5,
y-intercept is -2

b

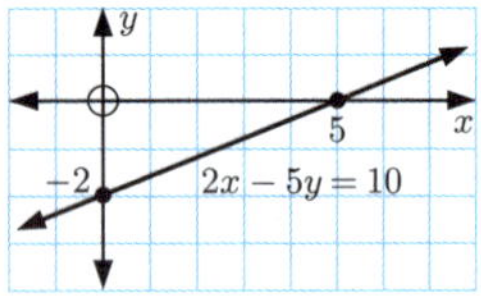

8 **a**

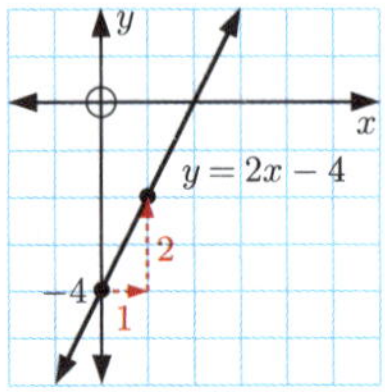

b

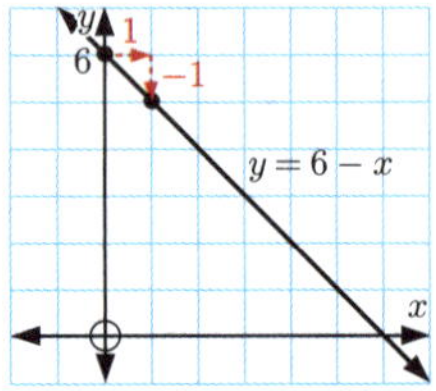

c

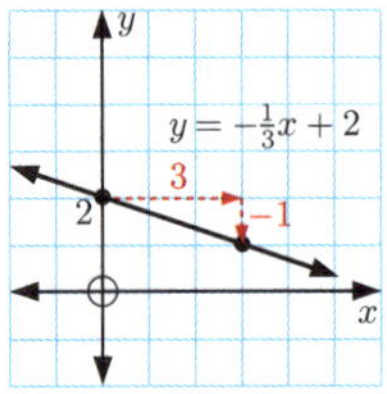

d

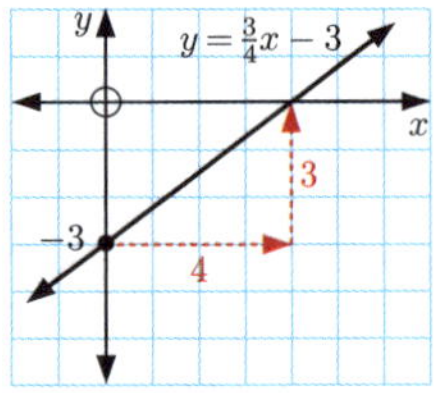

e

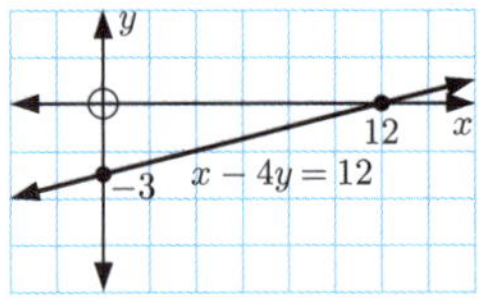

f

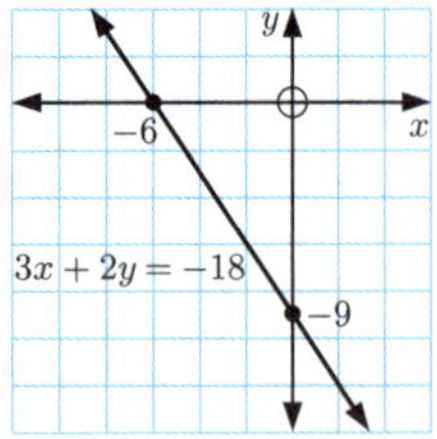

9 **a**

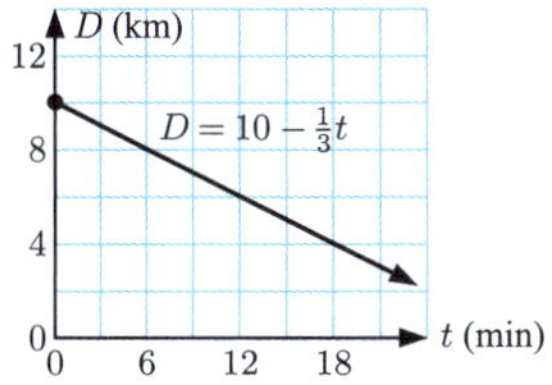

b 10; the school is 10 km from Tina's house.

c $-\frac{1}{3}$; Tina cycles home at a constant speed of $\frac{1}{3}$ km/min.

d 30; Tina arrives at home after 30 minutes.

10 **a** $y = -2x + 3$ **b** $y = \frac{2}{3}x + 6$ **c** $y = 3x - 5$

11 **a** $4x + 7y = 6$ **b** $x + 2y = 6$

12 **a** $R = -\frac{3}{4}t + 3$ **b** $P = r + 4$ **c** $M = -\frac{2}{5}x - \frac{4}{5}$

13 **a** **i** 4 **ii** 10

b **i** yes **ii** no

c $b = \frac{15}{2}$

d

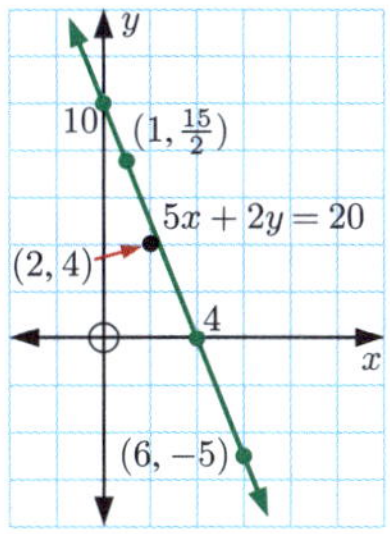

14 **a** 5 units **b** B(8, 4)

c **i** $y = \frac{1}{2}x$ **ii** $y = -2x + 10$

d Gradients of (OB) and (AC) are negative reciprocals so (OB) ⊥ (AC).

e M(4, 2)

f When $x = 4$, $y = -2(4) + 10 = 2$ ✓

$\therefore$ M lies on diagonal [AC].

EXERCISE 18A

1 **a** no **b** yes **c** yes **d** yes

2 **a** $x = 3,\ y = 1$ **b** $x = 7,\ y = 4$ **c** $x = 3,\ y = 5$
d $x = 3,\ y = 9$ **e** $x = 3,\ y = 1$ **f** $x = 4,\ y = 2$
g $a = 1,\ b = 0$ **h** $p = 5,\ q = 2$ **i** $x = -5,\ y = 16$

EXERCISE 18B

1 **a** $x = 1,\ y = 1$ **b** $x = -1,\ y = 2$
c $x = -3,\ y = -7$ **d** $x = -2,\ y = 7$
e $x = 1,\ y = -1$ **f** $x = -3,\ y = -5$

2 **a** $x = 3,\ y = 4$ **b** $x = -1,\ y = 3$
c $x = 2,\ y = -3$ **d** $x = -2,\ y = -4$
e $x = 3,\ y = 3$ **f** $x = 4,\ y = -3$

3 **a**

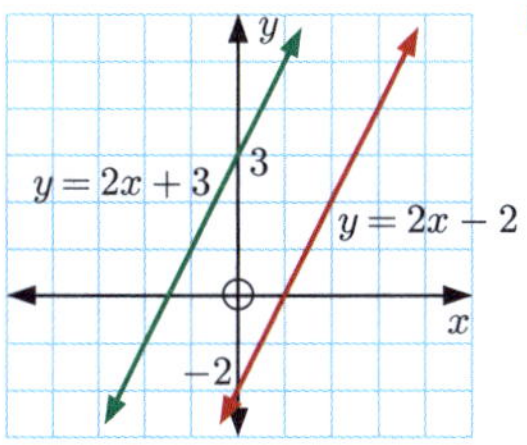

The lines are parallel.
$\therefore$ there are no solutions.

b

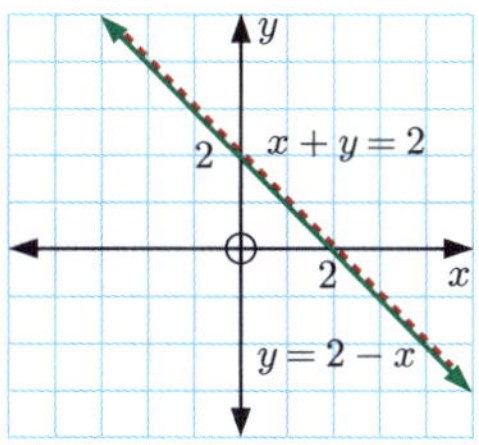

The lines are coincident.
$\therefore$ there are infinitely many solutions.

c

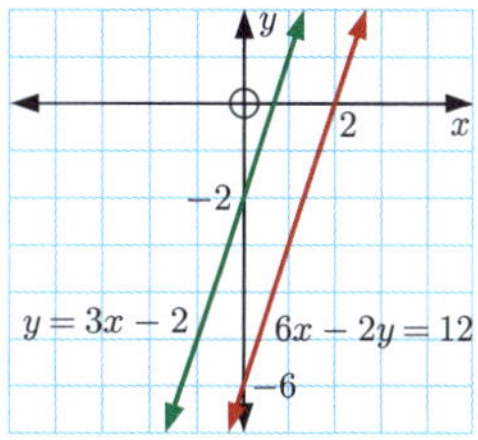

The lines are parallel.
$\therefore$ there are no solutions.

EXERCISE 18C

1 **a** $x = 4,\ y = 5$ **b** $x = -3,\ y = 1$ **c** $x = 1,\ y = 3$
d $x = 0,\ y = -4$ **e** $x = 4,\ y = 0$ **f** $x = -5,\ y = 3$

2 **a** $(4, -2)$ **b** $(\frac{5}{3}, 0)$ **c** $(-\frac{13}{4}, -\frac{11}{4})$ **d** $(\frac{1}{2}, 1)$

3 **a** We obtain $4 = -2$ which is absurd.

b

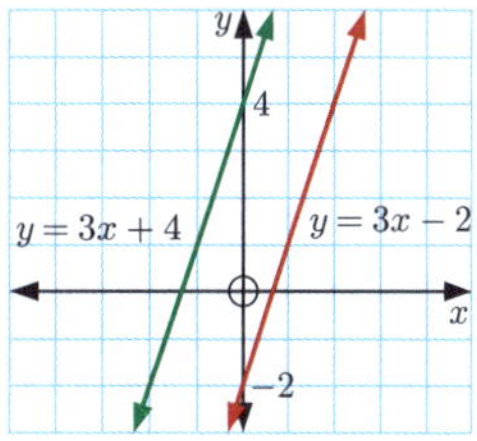

c The lines are parallel.
$\therefore$ there are no solutions.

EXERCISE 18D

1 **a** $x = 3,\ y = 5$ **b** $x = -2,\ y = -3$
c $x = 1,\ y = -6$ **d** $x = -7,\ y = -3$
e $x = 0,\ y = 3$ **f** $x = -\frac{1}{2},\ y = \frac{1}{2}$

2 **a** $x = \frac{2}{5},\ y = -\frac{7}{5}$ **b** $x = \frac{5}{4},\ y = 3$
c $x = \frac{25}{17},\ y = \frac{10}{17}$

3 **a** $(2, 5)$ **b** $(3, 1)$ **c** $(\frac{5}{3}, \frac{10}{3})$ **d** $(0, 4)$

4 **a** We obtain $20 = 7$ which is absurd.

b

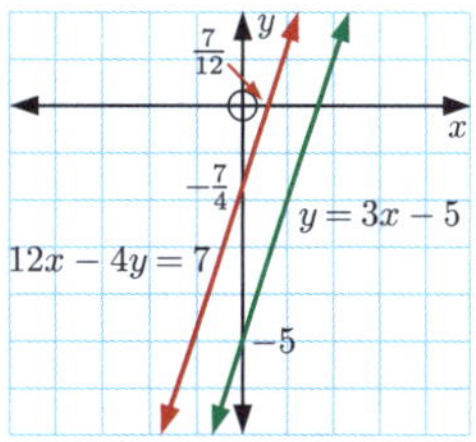

c The lines are parallel.
$\therefore$ there are no solutions.

5 **a** We obtain $15 = 15$ which is always true.

b

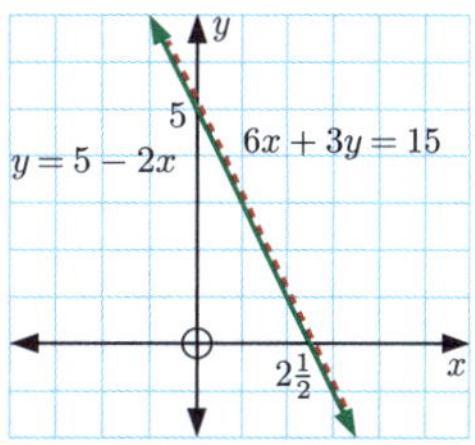

c The lines are coincident.
$\therefore$ there are infinitely many solutions.

EXERCISE 18E.1

1 **a** $6x = 6$ **b** $-y = 8$ **c** $5x = 7$
d $-6x = -30$ **e** $8y = 4$ **f** $-2y = -16$

2 **a** $x = 2,\ y = -1$ **b** $x = -2,\ y = 5$
c $x = 3,\ y = 2$ **d** $x = -2,\ y = -1$
e $x = 5,\ y = -3$ **f** $x = 4,\ y = -3$

EXERCISE 18E.2

1 a $9x + 12y = 6$ b $-2x + 8y = -14$
c $25x - 5y = -15$ d $-21x - 9y = 12$
e $8x + 20y = -4$ f $-3x + y = 1$

2 a $x = -3, \ y = 1$ b $x = 3, \ y = 2$
c $x = 8, \ y = 7$ d $x = -2, \ y = 3$
e $x = 5, \ y = -1$ f $x = -\frac{5}{2}, \ y = 0$
g $x = 3, \ y = 1$ h $x = 4, \ y = -\frac{3}{2}$

3 a $x = 2, \ y = 1$ b $x = 12, \ y = 4$
c $x = 3, \ y = 2$ d $x = 5, \ y = 2$
e $x = 4, \ y = 1$ f $x = -1, \ y = -3$
g $x = 2, \ y = -5$ h $x = 19, \ y = -17$

4 a Yes, dividing an equation by a non-zero constant is equivalent to multiplying the equation by the reciprocal of the constant.
b Dividing $5x - 10y = 15$ by 5, we obtain $x - 2y = 3$.
c $x = 5, \ y = 1$

5 a There are infinitely many solutions as the lines are coincident.
b There are no solutions as the lines are parallel.

EXERCISE 18F

1 The numbers are 97 and 181.
2 The numbers are $81\frac{1}{2}$ and $118\frac{1}{2}$.
3 The numbers are 9 and 17.
4 hammer: \$14, screwdriver: \$6
5 adult's ticket: \$28, child's ticket: \$12
6 nectarine: 56 cents, peach: 22 cents
7 19 five cent coins and 14 twenty cent coins
8 14 rabbits and 21 pheasants
9 74 one litre cartons and 23 two litre cartons
10 a $x = 3, \ y = 5$ b 15 cm^2
11 a $x = 39, \ y = 55$ b $94°$
12 a $x = \frac{3}{5}, \ y = 1$ b 36 cm
13 a $a = \frac{5}{9}, \ b = -\frac{160}{9}$ b $25°$C 14 134 m 15 $2\sqrt{2}$ cm

REVIEW SET 18A

1 $x = 2, \ y = 5$ 2 $x = -1, \ y = -3$
3 a $x = -5, \ y = -7$ b $x = -\frac{2}{5}, \ y = \frac{9}{5}$
4 a $x = 8, \ y = -13$ b $x = \frac{1}{3}, \ y = -\frac{2}{3}$
5 a $x = 4, \ y = -1$ b $x = 3, \ y = -4$
c $x = -5, \ y = 2$
6 a $x = 4, \ y = -2$ b $x = -3, \ y = 5$
7 pencil: 24 cents, ruler: 50 cents
8 9 of 5 kg packets, 11 of 2 kg packets
9 a $y = -4x + 29$ b $x = 8, \ y = -3$
10 a $x = 5, \ y = 11$ b 55 cm^2
11 a i 3 minutes ii 5 minutes b 38 minutes

REVIEW SET 18B

1 $x = 2, \ y = -6$ 2 $x = 3, \ y = 7$
3 a $x = 4, \ y = 4$ b no solutions
4 a $x = 2, \ y = 1$ b $x = 3, \ y = \frac{1}{3}$
5 a $x = 2, \ y = -1$ b $x = -1, \ y = -7$
c $x = 2, \ y = -1$
6 a $x = \frac{24}{5}, \ y = -\frac{23}{5}$ b $x = 2, \ y = -1$
7 The numbers are 37 and 48.
8 13 ten cent coins and 8 fifty cent coins

9 small bus: 23 people, large bus: 34 people
10 a $x = 7, \ y = 5$ b 9 cm
11 a 2 b $\frac{k}{3}$
c $k \neq 6$; when $k \neq 6$, the lines are not parallel and hence intersect at a single point.
d For $k = 6$, the second equation is $6x - 3y = 3$ or $y = 2x - 1$. The lines are parallel, so there are no solutions.

EXERCISE 19A

1 a 5 units right b 3 units down
c 2 units right, 2 units down d 4 units left, 1 unit up
e 3 units right, 3 units down f 3 units left, 3 units down

2 a
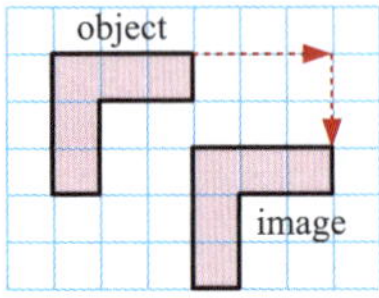

b
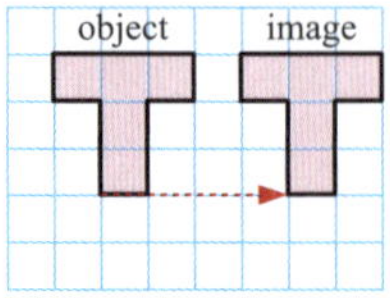

c
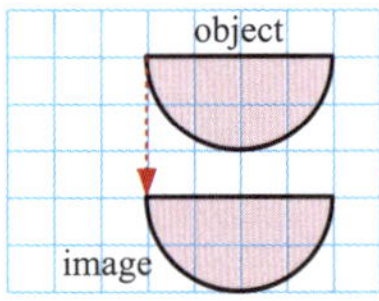

d
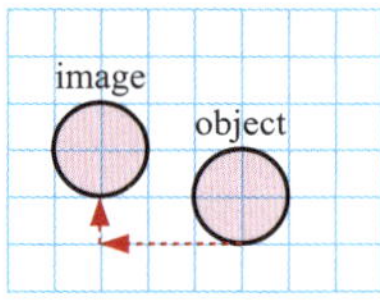

e
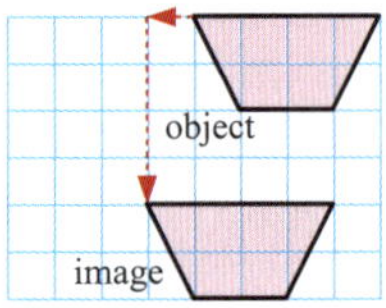

f
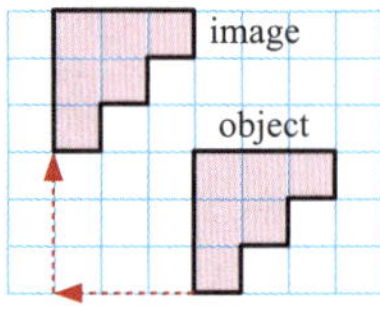

3 a no b yes (2 units right, 3 units down) c no
4 a I (5 units left, 6 units down)
b D (7 units right, 5 units up) c H
5 a 5 units left, 3 units up b 2 units right, 7 units down
c 3 units right, 4 units up d 3 units left, 4 units down

6 a i
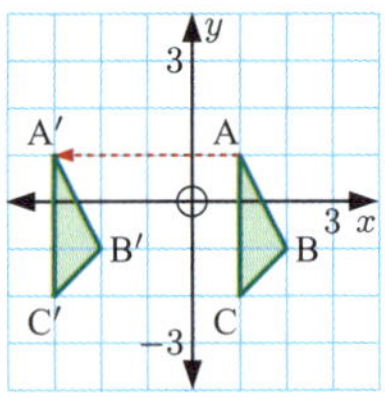

ii A′(−3, 1)
B′(−2, −1)
C′(−3, −2)

b i
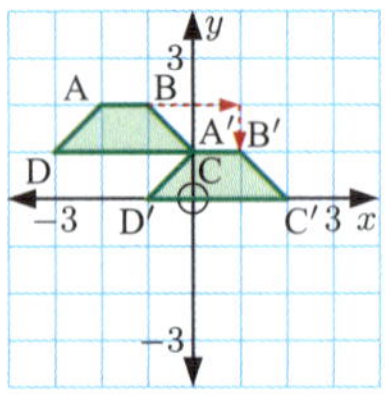

ii A′(0, 1)
B′(1, 1)
C′(2, 0)
D′(−1, 0)

c i
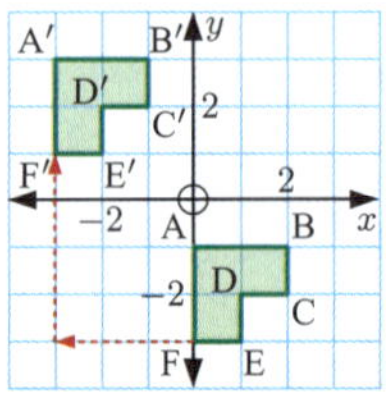

ii A′(−3, 3)
B′(−1, 3)
C′(−1, 2)
D′(−2, 2)
E′(−2, 1)
F′(−3, 1)

7 **a** P(−2, −2), Q(−1, 3), R(1, −1)
b R′(2, 1) or R′(3, −4)

8 **a**

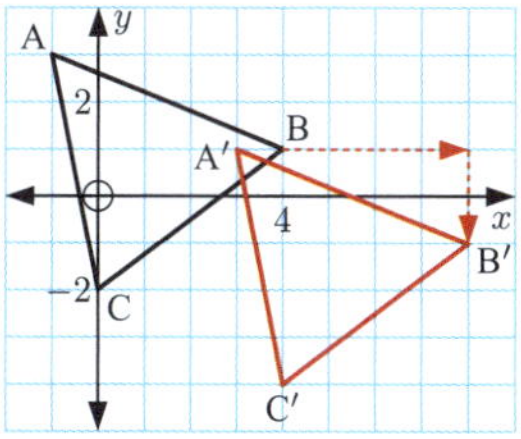

b A′(3, 1)
B′(8, −1)
C′(4, −4)

c [AA′]: $-\frac{1}{2}$, [BB′]: $-\frac{1}{2}$, [CC′]: $-\frac{1}{2}$
The line segments connecting corresponding object and image vertices all have the same gradient.

d AA′ = $2\sqrt{5}$ units, BB′ = $2\sqrt{5}$ units, CC′ = $2\sqrt{5}$ units
The image vertices are all the same distance from the corresponding object vertices.

9 **a** A′$(x + h, y + k)$ **b** $\sqrt{h^2 + k^2}$ units
c A translation h units to the left and k units down.

10 **a** (3, 6) **b** (−6, 2)

EXERCISE 19B

1 **a**

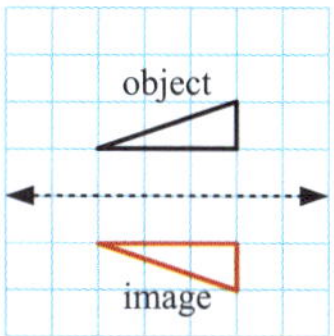

b

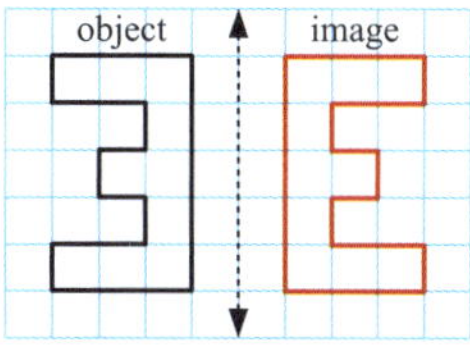

c

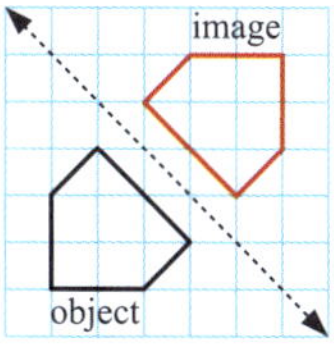

d

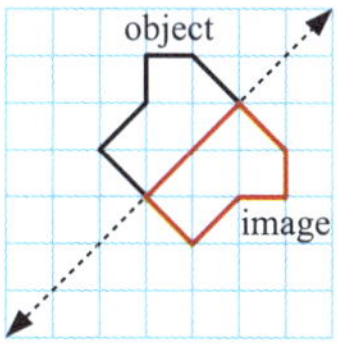

e

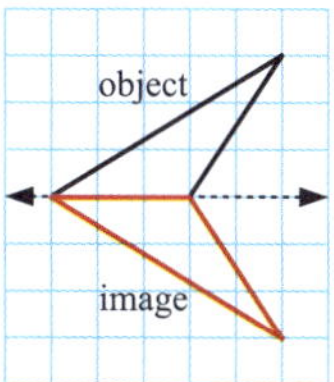

f

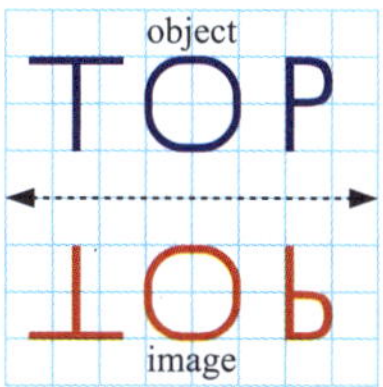

2 **a** no

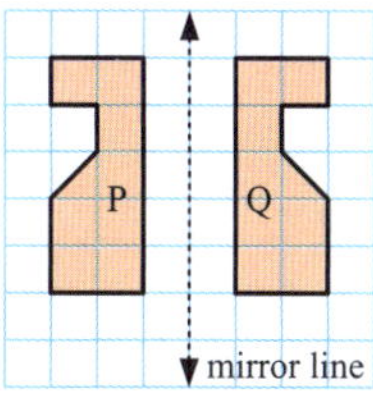

b yes

c no

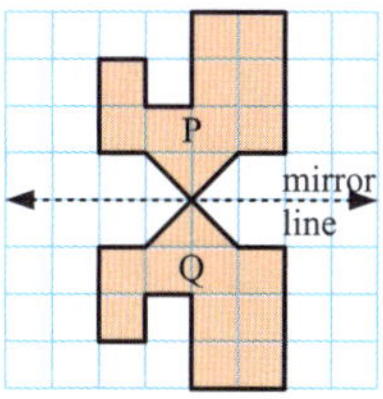

3 **a**

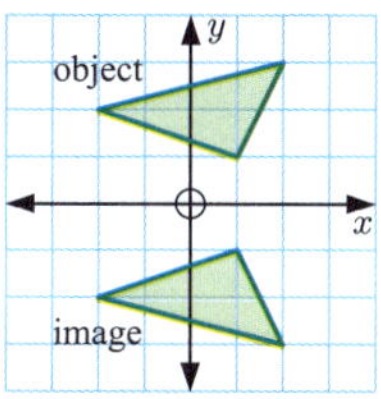

b

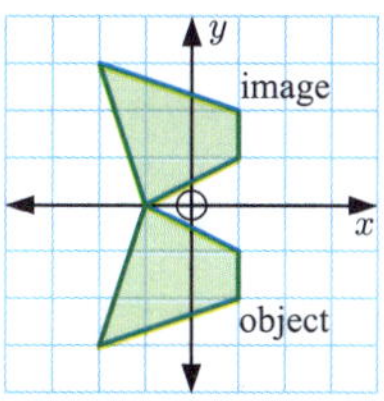

c

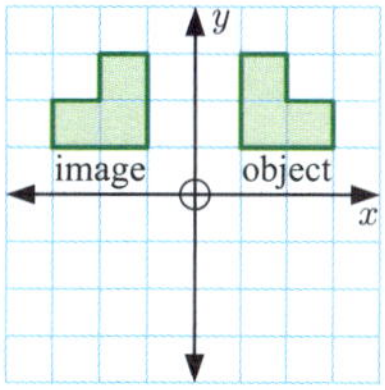

d

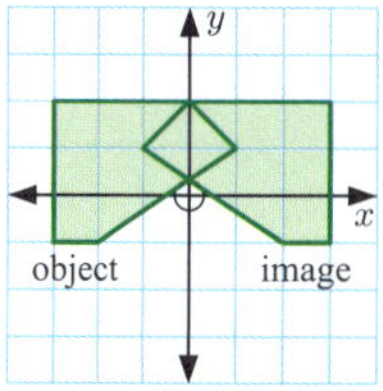

e

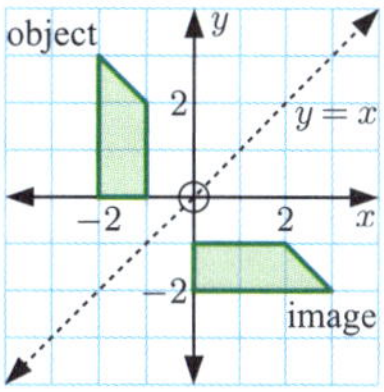

f

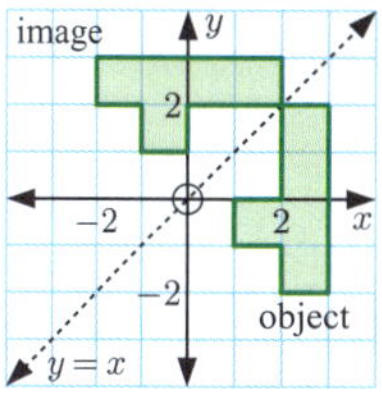

g

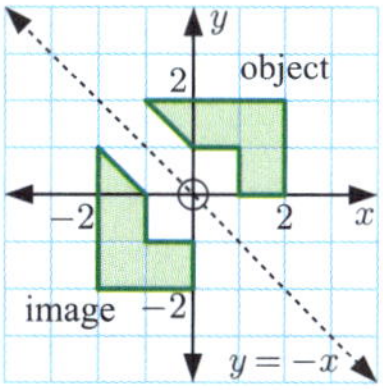

h

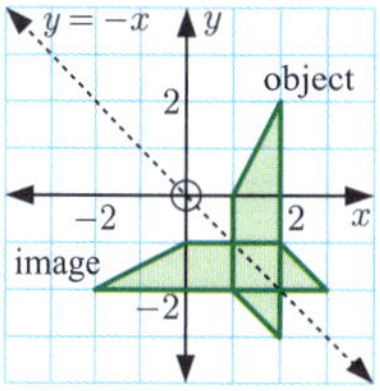

4 **a** A and C, x-axis
b A and D; 6 units right, 7 units down (from A to D)

5 **a** **i**

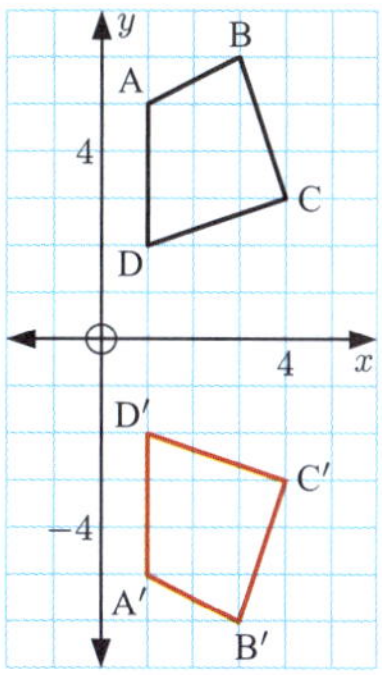

A′(1, −5)
B′(3, −6)
C′(4, −3)
D′(1, −2)

ii

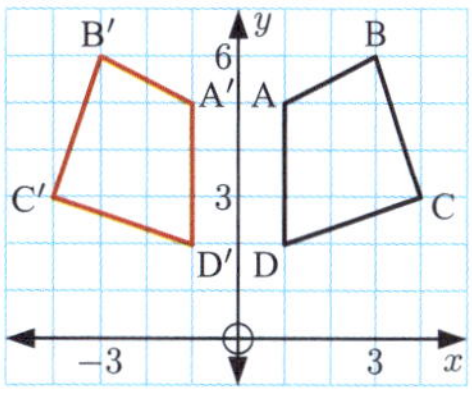

A′(−1, 5), B′(−3, 6), C′(−4, 3), D′(−1, 2)

iii

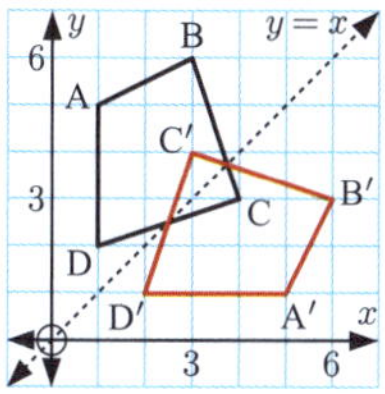

A′(5, 1), B′(6, 3), C′(3, 4), D′(2, 1)

iv

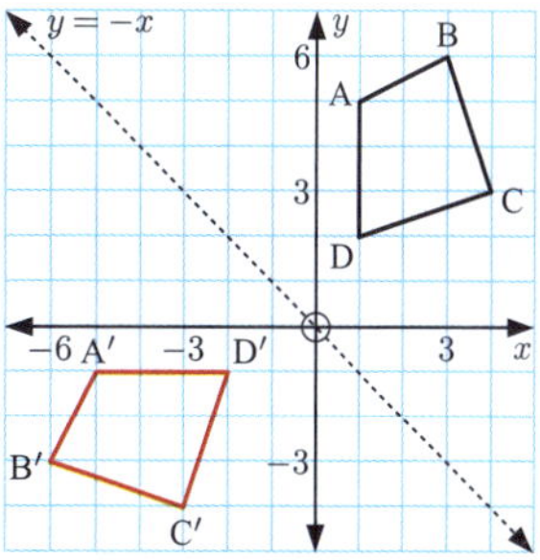

A′(−5, −1)
B′(−6, −3)
C′(−3, −4)
D′(−2, −1)

b **i** $P'(x, -y)$ **ii** $P'(-x, y)$ **iii** $P'(y, x)$
iv $P'(-y, -x)$

6 **a**

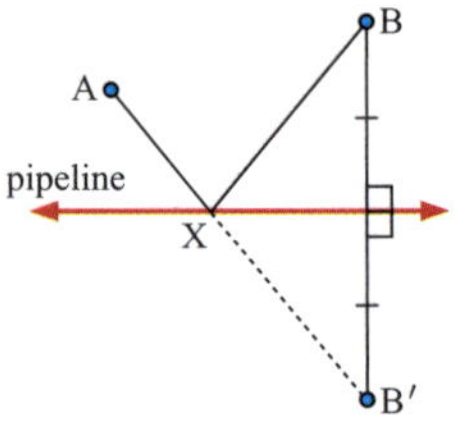

b $\triangle BXB'$ is isosceles since the pipeline perpendicularly bisects $[BB']$.

c If Y is any other point on the pipeline, then
$AY + YB' > AX + XB'$ {A, X, and B′ are collinear}
$\therefore$ $AY + YB > AX + XB$ $\{YB' = YB,\ XB' = XB\}$
So, $AX + XB$ is the shortest distance.

7 The mirror line is a line of symmetry of the object.

EXERCISE 19C

1 **a** D **b** B **c** C

2 Rotate 75° anticlockwise about O. **3** Rotate 180° about O.

4 **a**

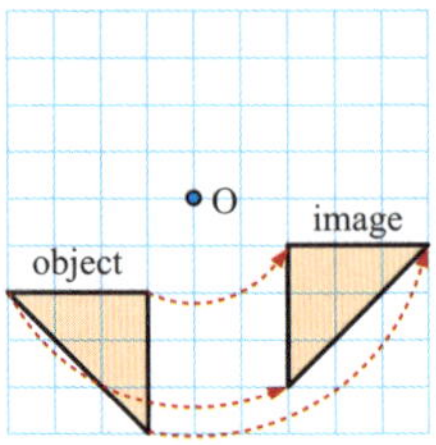

b

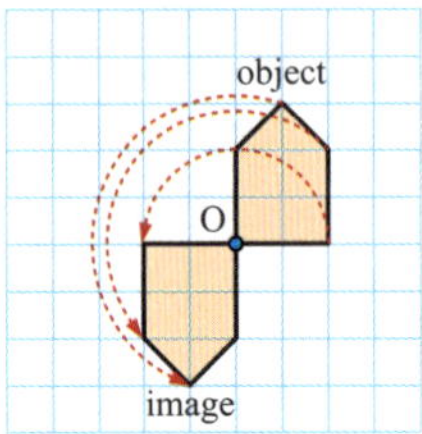

c

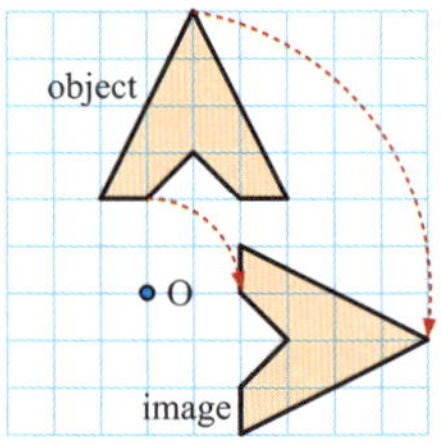

d

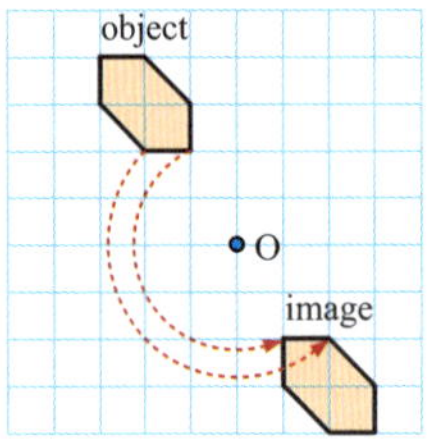

e

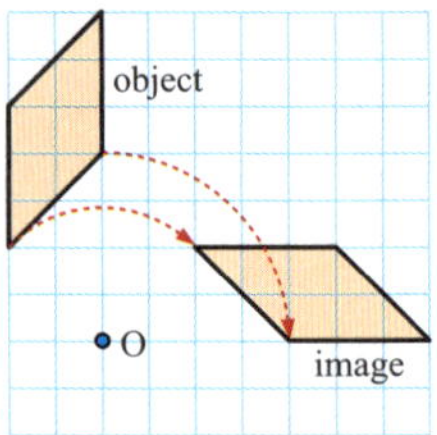

f

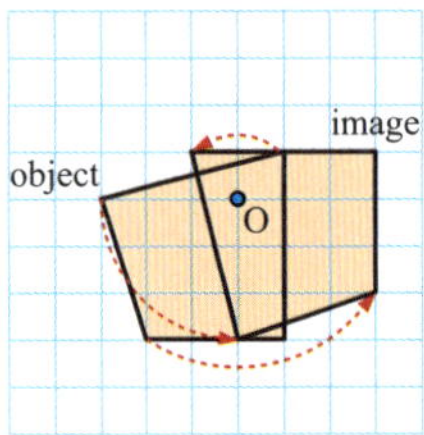

5 **a** C **b** 180°

6 **a** $A(-2, 4)$
$B(-1, 4)$
$C(-2, 1)$

b

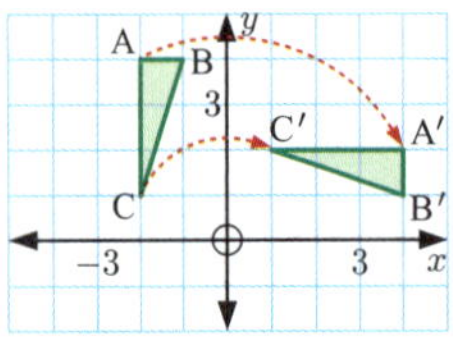

c $A'(4, 2)$
$B'(4, 1)$
$C'(1, 2)$

7 **a** $P(0, -2)$
$Q(2, -1)$
$R(2, -4)$

b

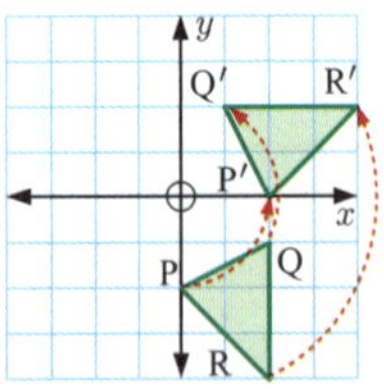

c $P'(2, 0)$
$Q'(1, 2)$
$R'(4, 2)$

8 **a** **i**

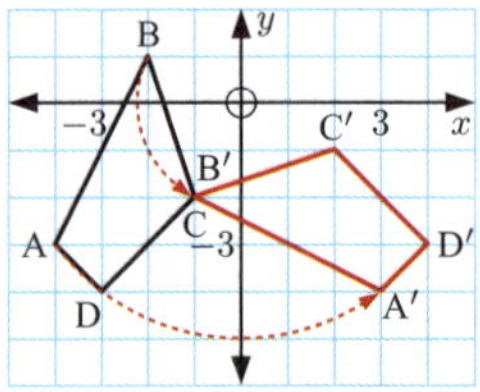

$A'(3, -4)$
$B'(-1, -2)$
$C'(2, -1)$
$D'(4, -3)$

ii

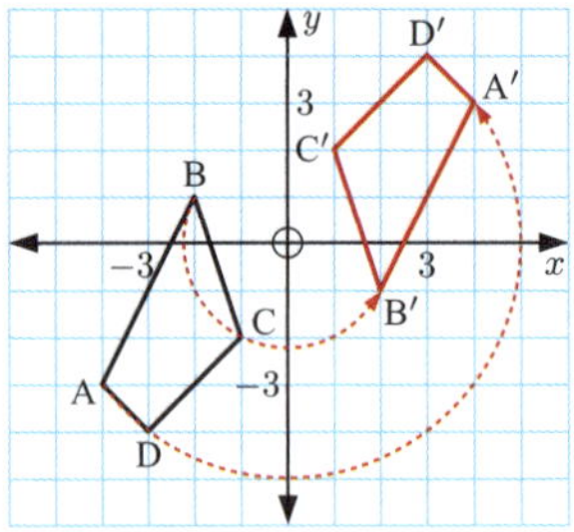

$A'(4, 3)$
$B'(2, -1)$
$C'(1, 2)$
$D'(3, 4)$

iii

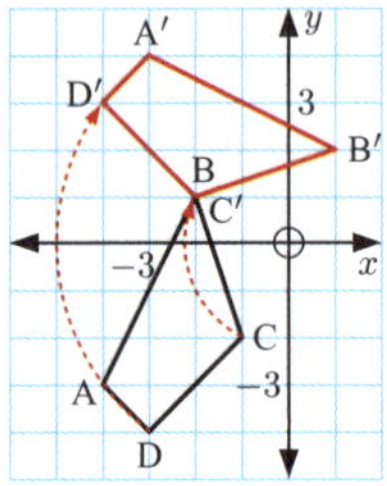

$A'(-3, 4)$
$B'(1, 2)$
$C'(-2, 1)$
$D'(-4, 3)$

b **i** $P'(-y, x)$ **ii** $P'(-x, -y)$ **iii** $P'(y, -x)$

9 **a**

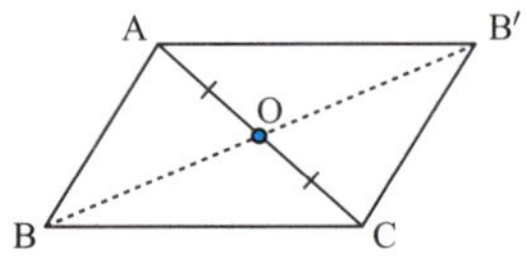

b $ABCB'$ is a parallelogram.

c opposite sides are equal and parallel, diagonals bisect each other, opposite angles are equal

d **i** rectangle **ii** rhombus **iii** square

10 The object has rotational symmetry of an order that is a multiple of 4 and its centre of rotational symmetry is at O, or the object is a circle with centre O.

EXERCISE 19D.1

1 **a** enlargement, scale factor 2 **b** reduction, scale factor $\frac{1}{2}$
c reduction, scale factor $\frac{1}{6}$ **d** enlargement, scale factor 3
e enlargement, scale factor $\frac{3}{2}$ **f** reduction, scale factor $\frac{3}{4}$

2 **a** $\frac{3}{2}$ **b** $\frac{2}{3}$ **c** 2 **d** $\frac{1}{2}$ **e** $\frac{4}{3}$ **f** $\frac{3}{4}$

EXERCISE 19D.2

1 **a**

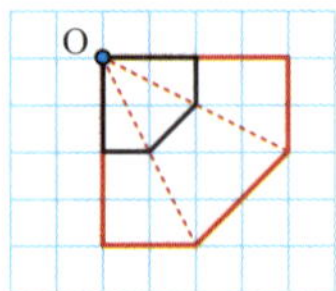

b

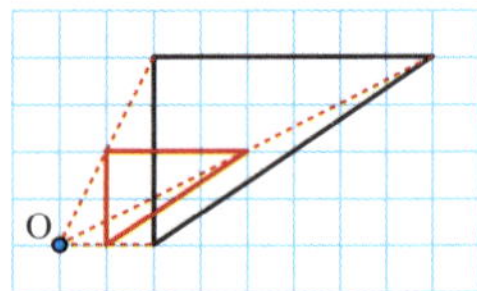

c

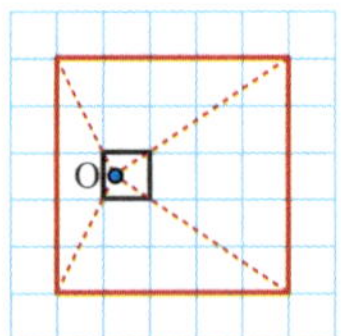

2 **a** 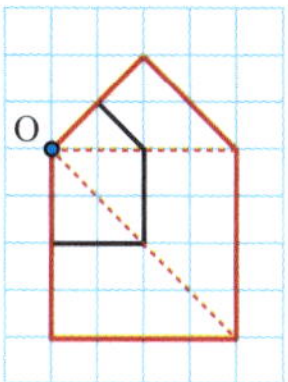

b 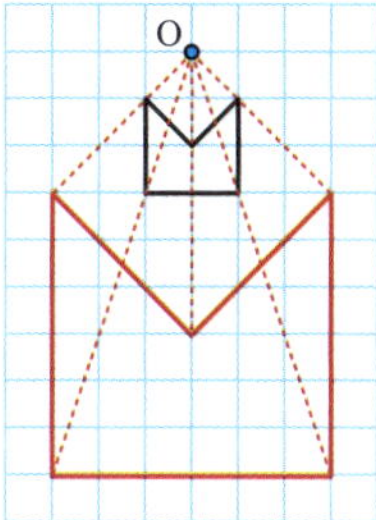

c 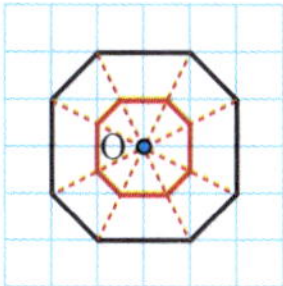

d 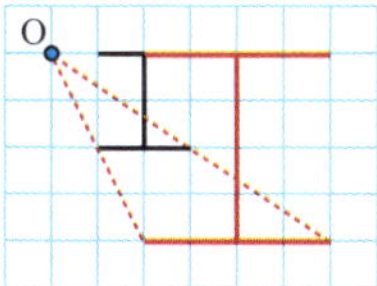

e 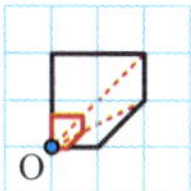

f 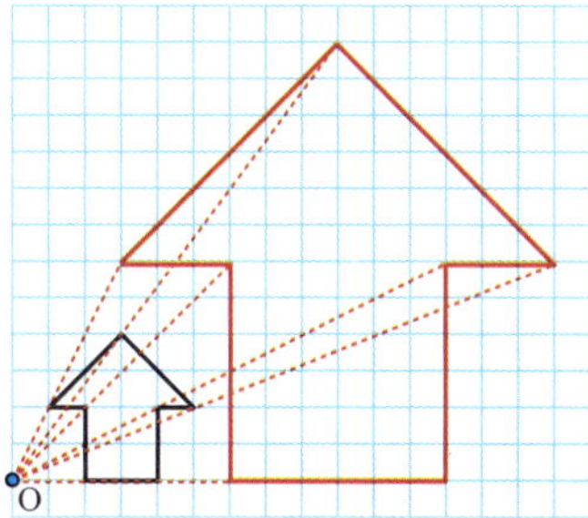

3 **a** **i** 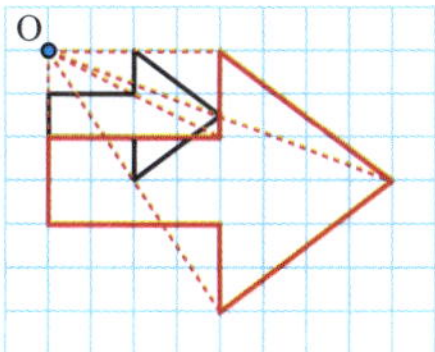

ii 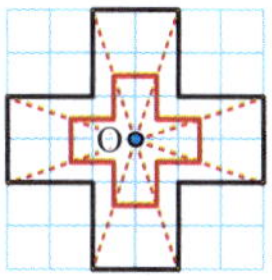

iii 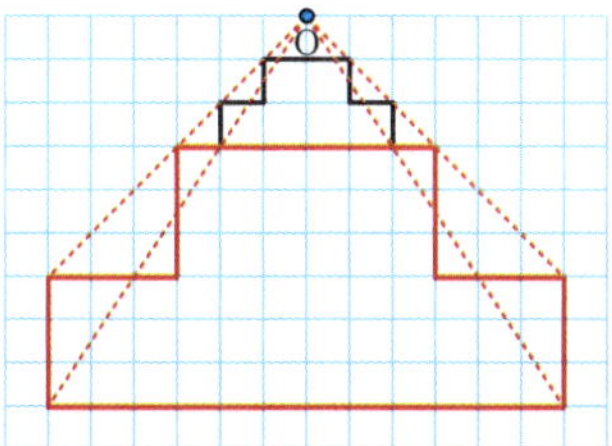

b

	Area of object	*Area of image*	$\dfrac{\textit{Area of image}}{\textit{Area of object}}$	k	k^2
i	5 units^2	20 units^2	4	2	4
ii	20 units^2	5 units^2	$\frac{1}{4}$	$\frac{1}{2}$	$\frac{1}{4}$
iii	6 units^2	54 units^2	9	3	9

c $\dfrac{\text{area of image}}{\text{area of object}} = k^2$ or area of image $= k^2 \times$ area of object

4 **a** **i** 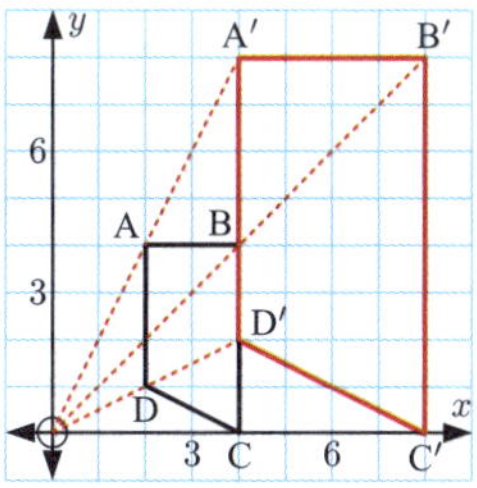

$A'(4, 8)$, $B'(8, 8)$, $C'(8, 0)$, $D'(4, 2)$

ii 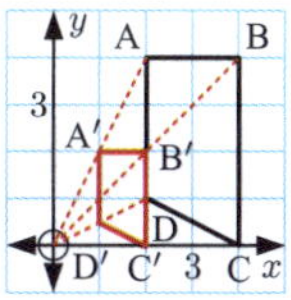

$A'(1, 2)$, $B'(2, 2)$, $C'(2, 0)$, $D'(1, \frac{1}{2})$

b $P'(kx, ky)$ **c** **i** $(5, 7\frac{1}{2})$ **ii** $(-4, 6)$

EXERCISE 19E

1 **a** horizontal stretch with scale factor 2
b vertical stretch with scale factor 3
c vertical stretch with scale factor $\frac{3}{4}$
d horizontal stretch with scale factor $\frac{1}{4}$

2 **a** 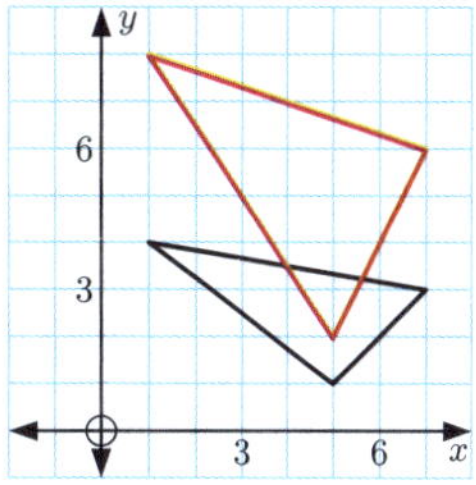

b 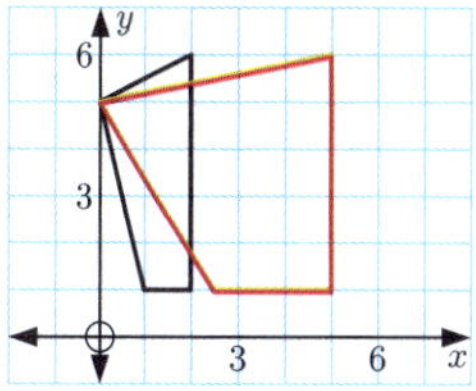

c 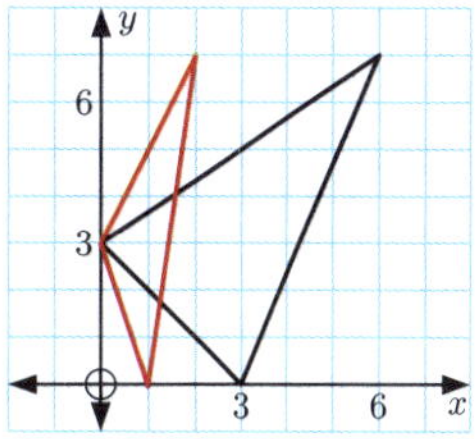

d 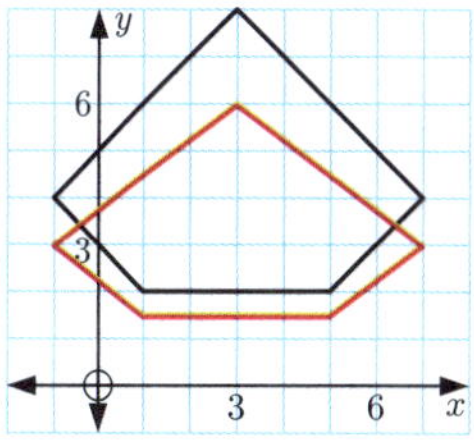

3 **a** vertical stretch with scale factor 3
b horizontal stretch with scale factor 3
c enlargement with scale factor 3 and centre O
d horizontal stretch with scale factor 3
e vertical stretch with scale factor $\frac{1}{3}$
f reduction with scale factor $\frac{1}{3}$ and centre O

4 **a** 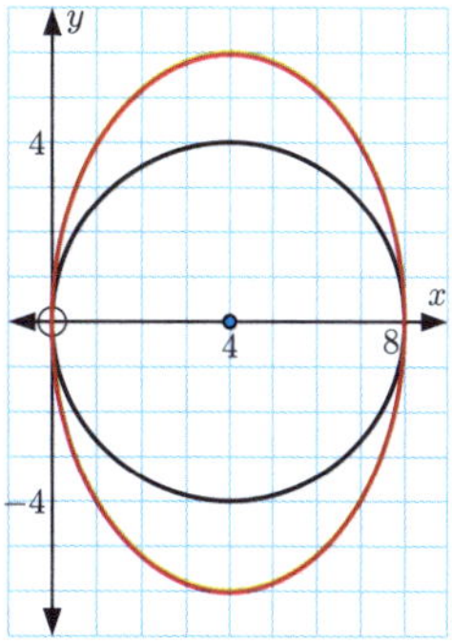

b 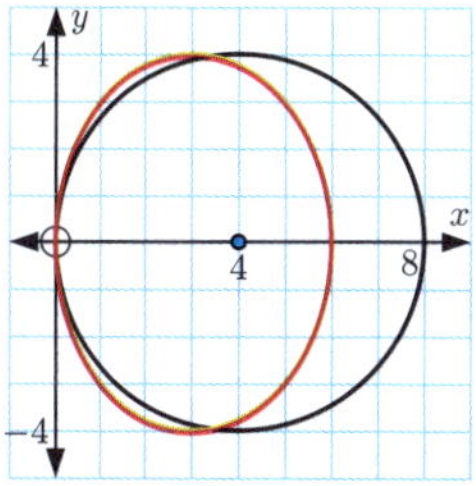

5 **a** $k = 4$ **b** $k = \sqrt{2}$

EXERCISE 19F

1 **a** 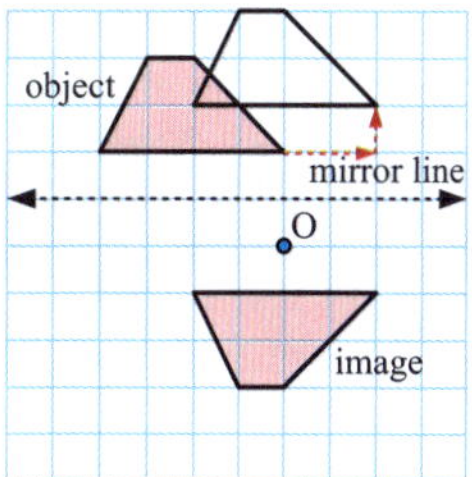

b 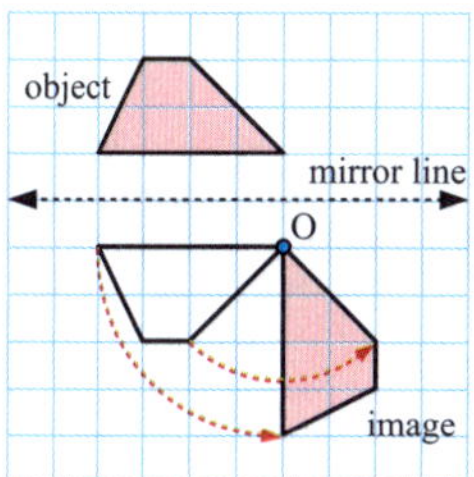

2 **a**

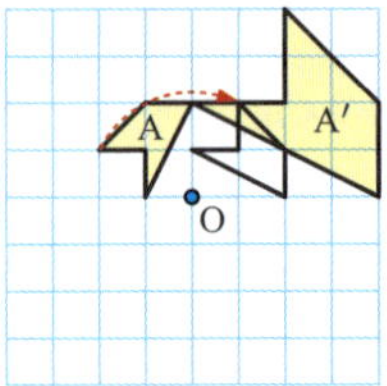

b A reduction with scale factor $\frac{1}{2}$ and centre O, then a rotation 90° anticlockwise about O.

3 **a** A reduction with scale factor $\frac{2}{3}$ and centre O, followed by a translation 4 units to the right and 2 units down.

b A translation 6 units to the right and 3 units down, followed by a reduction with scale factor $\frac{2}{3}$ and centre O.

4

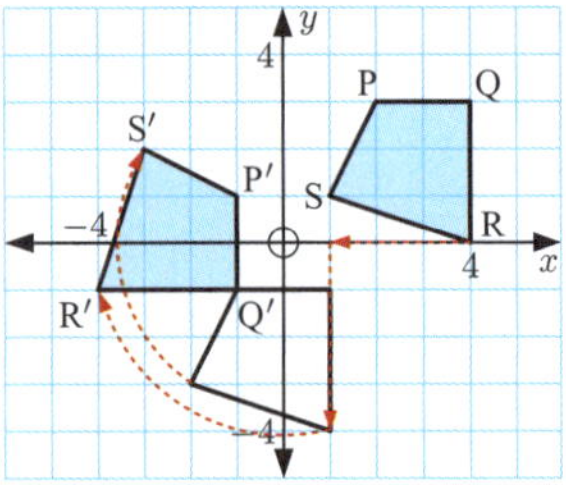

P′(−1, 1)
Q′(−1, −1)
R′(−4, −1)
S′(−3, 2)

5 **a**

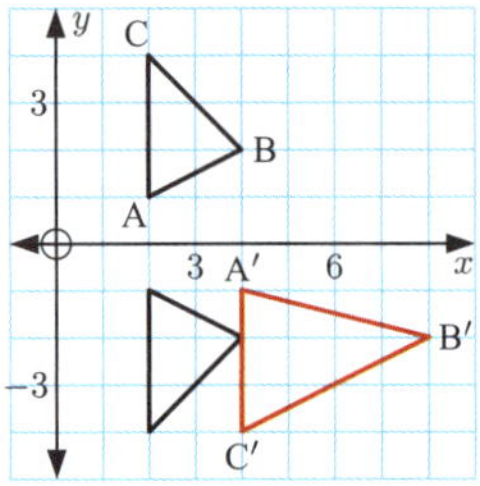

A′(4, −1)
B′(8, −2)
C′(4, −4)

b P′(2x, −y)

REVIEW SET 19A

1 **a**

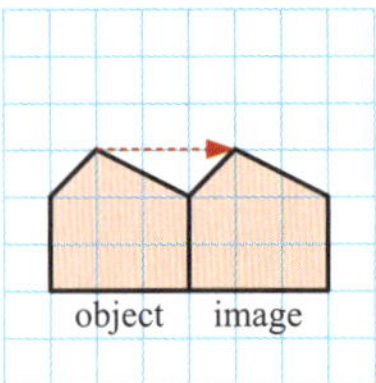

b

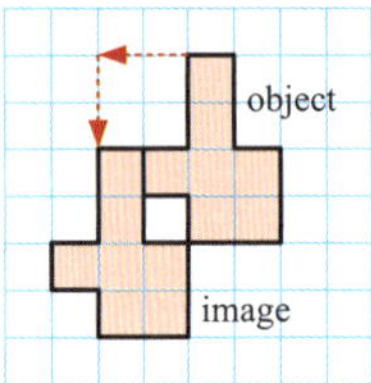

c

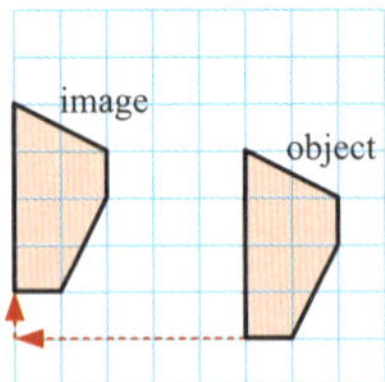

2 **a** (3, 3) **b** (−8, −4)

3 **a**

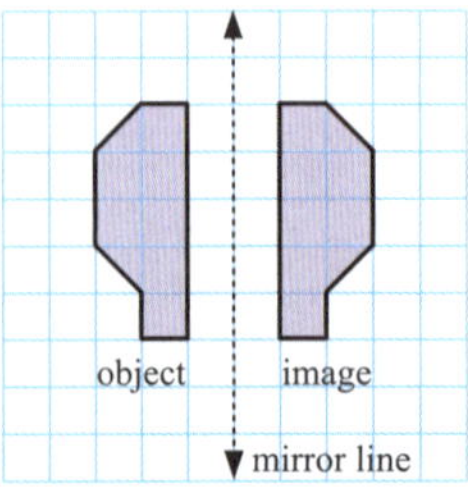

b

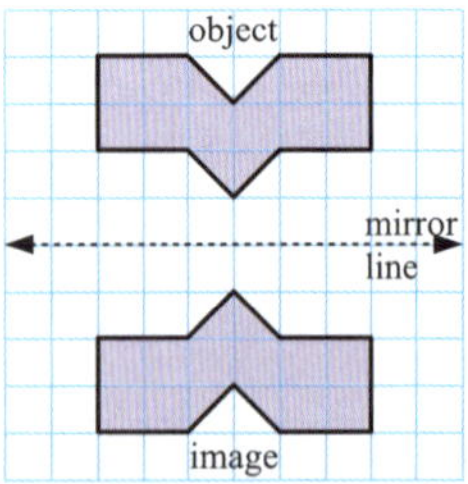

c

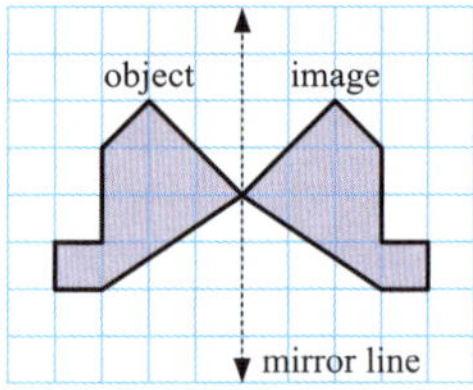

4 **a**

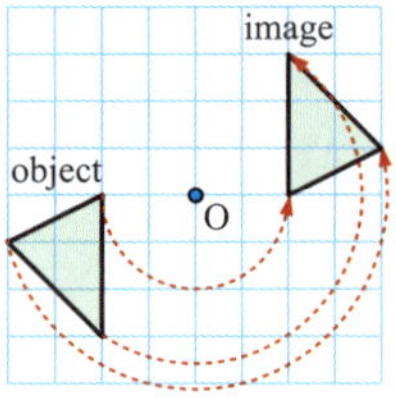

b

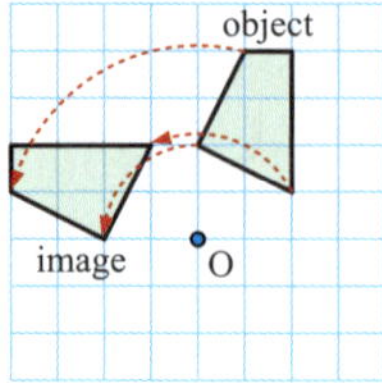

c

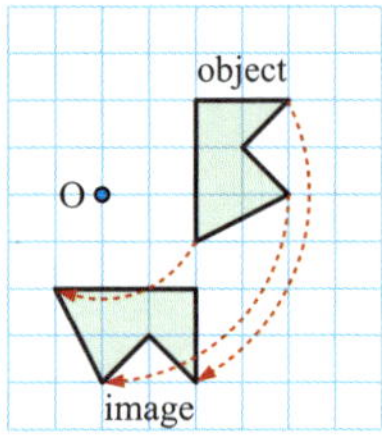

5 **a** P(−4, −3)
Q(−2, 0)
R(−1, −2)
S(−1, −3)

b

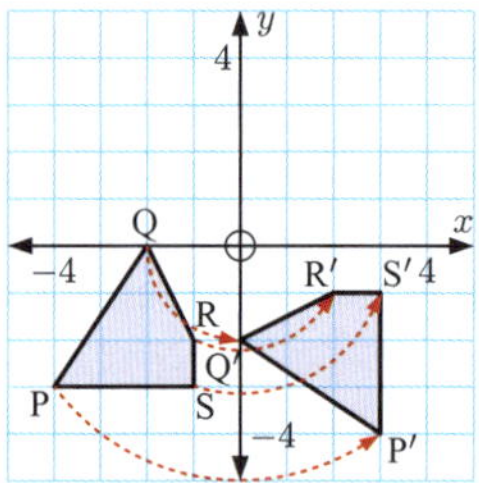

c P′(3, −4)
Q′(0, −2)
R′(2, −1)
S′(3, −1)

6 **a** enlargement, scale factor 3 **b** reduction, scale factor $\frac{1}{2}$

c enlargement, scale factor $\frac{5}{3}$

7 **a** (12, −20) **b** $(4, -\frac{4}{5})$

8 **a** vertical stretch with scale factor $\frac{5}{3}$

b horizontal stretch with scale factor 3

9 **a** $k = 5$ **b** $k = \frac{\sqrt{3}}{2}$

10

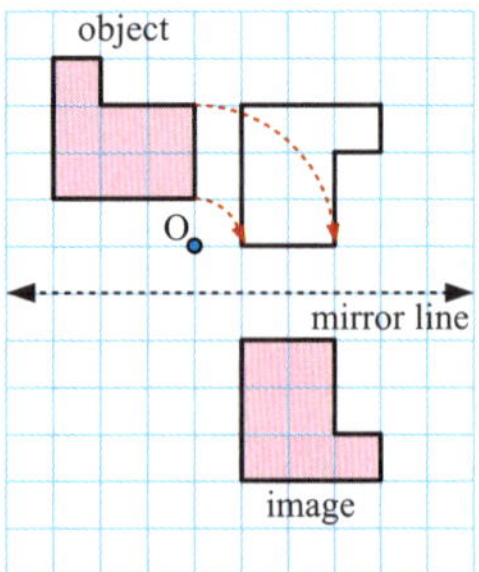

11

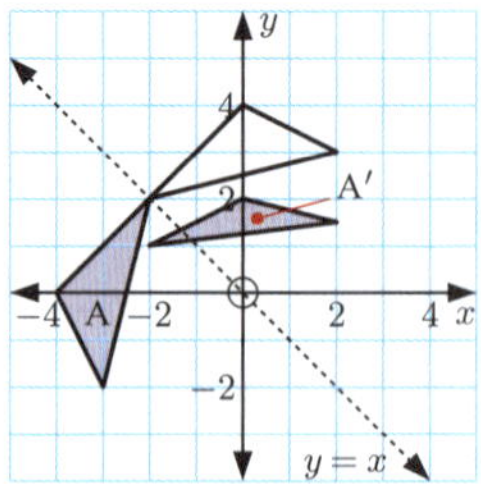

$(-2, 1)$, $(0, 2)$, $(2, 1\frac{1}{2})$

REVIEW SET 19B

1 **a** 4 units right, 3 units up **b** 4 units right, 1 unit down

c 3 units left, 1 unit down

2 **a**

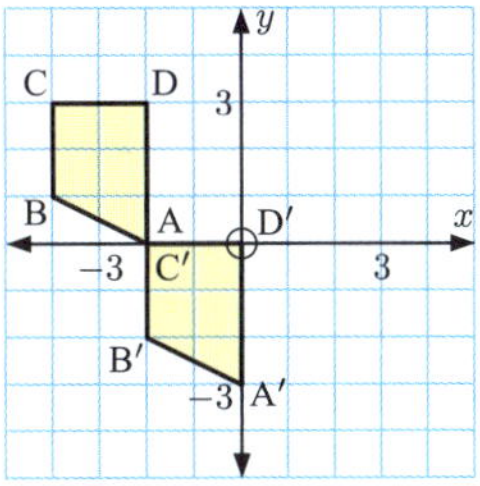

b A′(0, −3)
B′(−2, −2)
C′(−2, 0)
D′(0, 0)

3 **a**

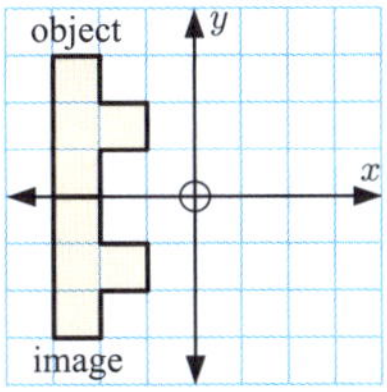

b

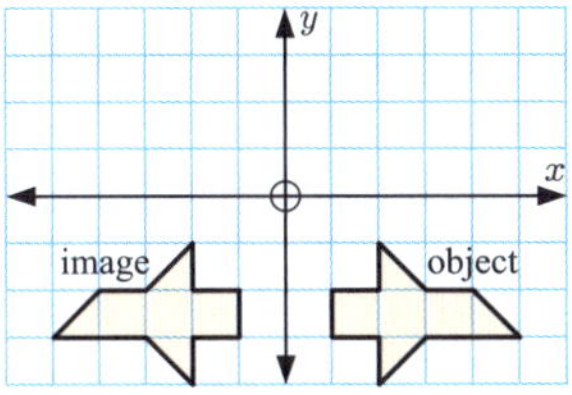

c

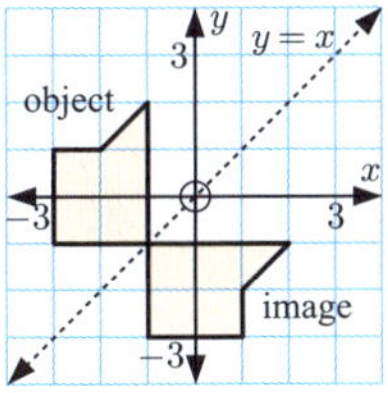

4

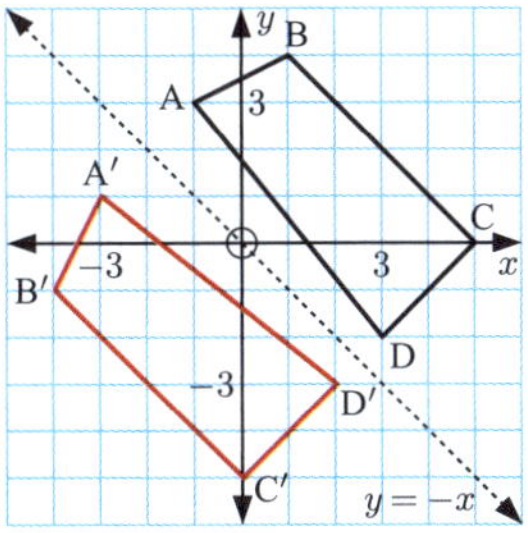

A′(−3, 1)
B′(−4, −1)
C′(0, −5)
D′(2, −3)

5 **a** D **b** 90° clockwise

6 **a, b**

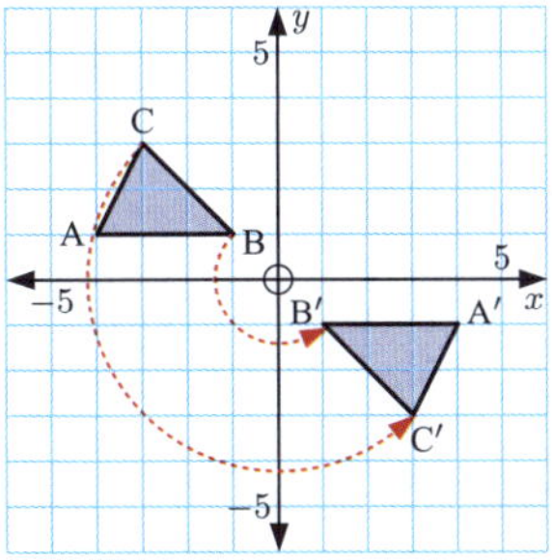

c A′(4, −1)
B′(1, −1)
C′(3, −3)

d A rotation 180° about O.

7

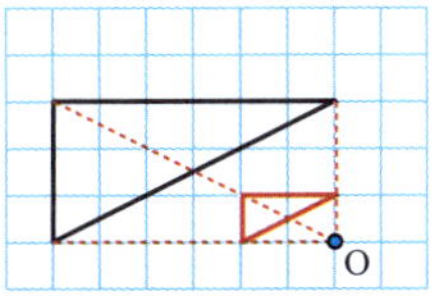

8 **a**

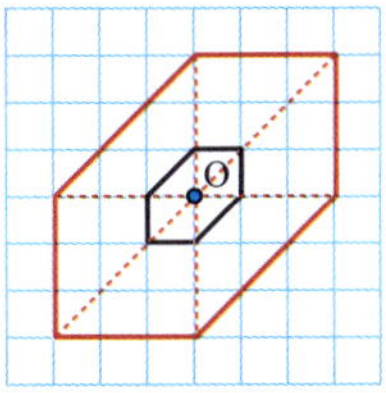

b

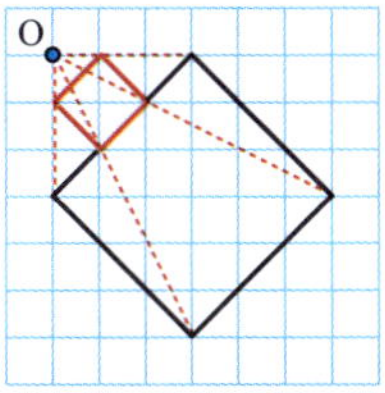

c

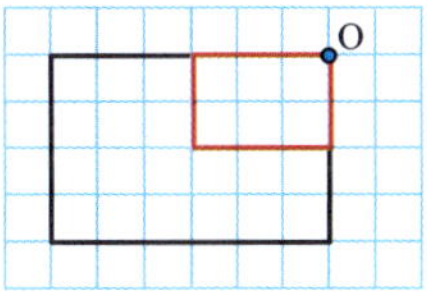

9 **a** horizontal stretch with scale factor 4
b vertical stretch with scale factor 4
c enlargement with scale factor 4 and centre O
d horizontal stretch with scale factor 4
e vertical stretch with scale factor $\frac{1}{4}$
f reduction with scale factor $\frac{1}{4}$ and centre O

10 **a**

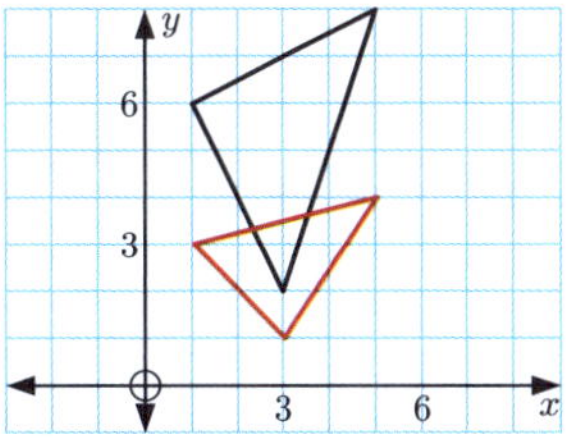

b

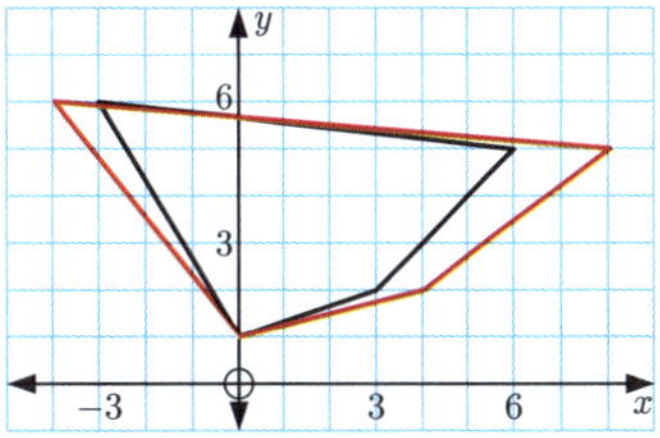

11 **a**

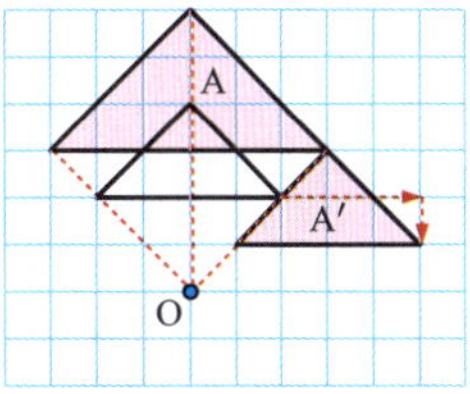

b A translation 3 units left and 1 unit up, followed by an enlargement with scale factor $\frac{3}{2}$ and centre O.

12 **a**

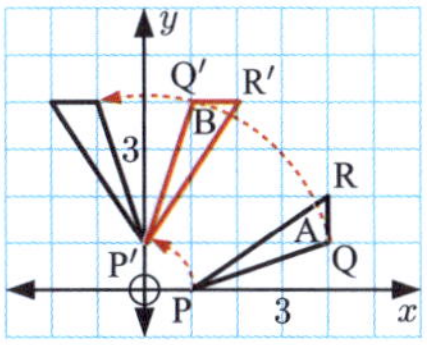

P′(0, 1), Q′(1, 4), R′(2, 4)

b A reflection in the line $y = x$.

EXERCISE 20A

1 **a** quadratic equation **b** not quadratic equation
c not quadratic equation **d** quadratic equation
e not quadratic equation **f** quadratic equation

2 **a** 1 and 6 **b** −4 and 2 **c** 0 and 5 **d** 3 **e** −2 and $\frac{1}{2}$

3 If $x^2 + 5 = 0$, then $x^2 = -5$.
But x^2 is always $\geqslant 0$. $\therefore$ there are no real solutions.

EXERCISE 20B

1 **a** $x = \pm 4$ **b** $x = \pm\sqrt{19}$ **c** $x = \pm 3$
d $x = \pm\sqrt{6}$ **e** no real solutions **f** $x = 0$
g $x = \pm 5$ **h** no real solutions **i** $x = \pm\sqrt{3}$

2 **a** $x = 5$ or -1 **b** $x = 1$ or -9 **c** no real solutions
d $x = 4 \pm \sqrt{2}$ **e** no real solutions **f** $x = 2$
g $x = -5 \pm \sqrt{3}$ **h** $x = 6 \pm \sqrt{2}$ **i** $x = -8 \pm \sqrt{13}$
j $x = -\frac{5}{3}$ **k** $x = \frac{1}{5}$ or $-\frac{3}{5}$ **l** $x = \frac{1 \pm \sqrt{2}}{2}$

EXERCISE 20C

1 **a** $a = 0$ or $c = 0$ **b** $x = 0$ or $y = 0$
c $a = 0,\ b = 0,$ or $c = 0$ **d** $x = 0$
e $x = 0$ or $x - 3 = 0$ **f** $x = 0$
g $x = 0$ or $x + 6 = 0$ **h** $x = 0$ or $y = 0$

2 **a** $x = 0$ or 1 **b** $x = 0$ or -5 **c** $x = 0$ or -2
d $x = 1$ **e** $x = 0$ or 4 **f** $x = 0$ or -3
g $x = 0$ or $-\frac{1}{2}$ **h** $x = 0$ or $\frac{3}{4}$ **i** $x = 0$ or $-\frac{5}{3}$

3 **a** $x = 1$ or 5 **b** $x = -2$ or 4 **c** $x = -3$ or -7
d $x = -7$ or 11 **e** $x = 0$ or 8 **f** $x = -12$ or 5
g $x = 0$ or -7 **h** $x = -\frac{1}{2}$ or 3 **i** $x = -6$ or $\frac{1}{3}$
j $x = -\frac{1}{2}$ or -6 **k** $x = 3$ **l** $x = 31$ or -11
m $x = -4$ or $\frac{1}{4}$ **n** $x = 0$ or $-\frac{3}{7}$ **o** $x = 2$ or $-\frac{4}{3}$

EXERCISE 20D.1

1 **a** $x = 0$ or 1 **b** $x = 0$ or 13 **c** $x = 0$ or -8
d $x = 0$ or -3 **e** $x = 0$ or -2 **f** $x = 0$ or 5
g $x = 0$ or 12 **h** $x = 0$ or -7 **i** $x = 0$ or 4
j $x = 0$ or $\frac{7}{2}$ **k** $x = 0$ or 5 **l** $x = 0$ or -4

2 **a** $x = 0$ or 3 **b** $x = 0$ or 10 **c** $x = 0$ or -6
d $x = 0$ or $-\frac{7}{2}$ **e** $x = 0$ or 5 **f** $x = 0$ or 8
g $x = 0$ or 6 **h** $x = 0$ or $\frac{6}{5}$ **i** $x = 0$ or -6
j $x = 0$ or 2 **k** $x = 0$ or 1 **l** $x = 0$ or $\frac{1}{3}$

3 **a** $x = 2$ or 3 **b** $x = -4$ or 5 **c** $x = 6$ or -1

EXERCISE 20D.2

1 **a** $x = \pm 4$ **b** $x = \pm 7$ **c** $x = \pm 12$ **d** $x = \pm 2$
e $x = \pm 3$ **f** $x = \pm 2$ **g** $x = \pm 2$ **h** $x = \pm 2$
i no real solutions

2 **a** $x = \pm\frac{2}{3}$ **b** $x = \pm\frac{1}{3}$ **c** $x = \pm\frac{2}{5}$ **d** $x = \pm\frac{1}{2}$
e $x = \pm\frac{3}{2}$ **f** $x = \pm\frac{5}{4}$

3 **a** $x = 1$ **b** $x = -2$ **c** $x = \frac{2}{3}$ or -4 **d** $x = \frac{1}{2}$ or 3

EXERCISE 20D.3

1 **a** $x = 2$ or 5 **b** $x = -2$ or -4 **c** $x = -1$ or -10
d $x = 2$ or 6 **e** $x = 1$ or 4 **f** $x = 3$ or 8
g $x = -5$ **h** $x = -3$ or 6 **i** $x = 2$ or -9
j $x = 11$ **k** $x = 3$ **l** $x = -1$ or 6
m $x = 4$ or -15 **n** $x = 3$ or -21 **o** $x = -4$ or 16
p $x = 5$ or 14

2 **a** $x = -1$ or 15 **b** $x = 1$ or 2 **c** $x = 7$ or -4
d $x = 5$ or -4 **e** $x = 1$ or -8 **f** $x = 8$ or -3
g $x = 9$ or -5 **h** $x = 1$ **i** $x = -1$ or 20

3 **a** $x = 3$ or -5 **b** $x = -5$ or 12 **c** $x = 1$ or 6
d $x = -2$ or 3 **e** $x = 1$ **f** $x = 2$ or -4

4 **a** $x = 3$ or -5 **b** $x = 6$ **c** $x = -5$ or 4
d $x = 3$ or 4 **e** $x = 7$ or -6 **f** $x = -2$ or -6

5 **a** $x = -4$ or 3 **b** $x = 2$ or 8 **c** $x = -4$ or 8

EXERCISE 20E

1 The number is 6 or -7. **2** The number is 0 or 5.
3 The number is 8 or -7. **4** The numbers are 12 and -3.
5 The integers are 12 and 13 or -13 and -12.
6 The numbers are 11 and 13 or -11 and -13.
7 The numbers are 13 and 17 or -13 and -17.
8 The width is 8 cm.
9 **a** **i** $x(x+5)$ cm^2 **ii** $(4x + 10)$ cm
b The area is 22.91 cm^2. **c** The perimeter is 54 cm.
10 The base is 11 cm long. **11** The perimeter is 50 cm.
12 The outer radius is 7 cm.
13 **a** $x = 3$ **b** $x = 7$ **c** $x = 8$ **d** $x = 8$ **e** $x = 4$ or 9
14 **a** The perimeter is 36 cm. **b** The area is 54 cm^2.
15 [BX] is 2 m long.
16 **a** $b = 3,\ c = -10$ **b** $b = 3,\ c = \frac{9}{4}$
17 **a** **i**

$$\begin{aligned} x^2 + (x+1)^2 + (x+2)^2 &= 110 \\ \therefore\ x^2 + x^2 + 2x + 1 + x^2 + 4x + 4 &= 110 \\ \therefore\ 3x^2 + 6x - 105 &= 0 \\ \therefore\ 3(x^2 + 2x - 35) &= 0 \\ \therefore\ 3(x+7)(x-5) &= 0 \\ \therefore\ x &= -7 \text{ or } 5 \end{aligned}$$

The integers are $-7, -6,$ and -5 or $5, 6,$ and 7.

ii

$$\begin{aligned} (x-1)^2 + x^2 + (x+1)^2 &= 110 \\ \therefore\ x^2 - \cancel{2x} + 1 + x^2 + x^2 + \cancel{2x} + 1 &= 110 \\ \therefore\ 3x^2 &= 108 \\ \therefore\ x^2 &= 36 \\ \therefore\ x &= \pm 6 \end{aligned}$$

The integers are $-7, -6,$ and -5 or $5, 6,$ and 7.

b The method in **a ii** is easier as the equation reduces to a power equation.

18 The numbers are 9, 11, and 13 or $-13, -11,$ and -9.
19 **a** rectangular garden bed: 4 m by 16 m, square garden bed: 8 m by 8 m **b** 72 m
20 The radius is 5 cm.
21 The larger perfect cube is $9^3 = 729$.

EXERCISE 20F

1 **a** **i** 1^2 **ii** $(x+1)^2 = 3$
b **i** 1^2 **ii** $(x-1)^2 = 5$
c **i** 3^2 **ii** $(x+3)^2 = 5$
d **i** 3^2 **ii** $(x-3)^2 = 16$
e **i** 5^2 **ii** $(x+5)^2 = 2$
f **i** 4^2 **ii** $(x+4)^2 = 3$
g **i** 3^2 **ii** $(x+3)^2 = 11$
h **i** $(\frac{3}{2})^2$ **ii** $(x + \frac{3}{2})^2 = \frac{5}{4}$
i **i** $(\frac{7}{2})^2$ **ii** $(x - \frac{7}{2})^2 = \frac{69}{4}$

2 **a** $x = 2 \pm \sqrt{6}$ **b** $x = -1 \pm \sqrt{7}$ **c** $x = 1 \pm \sqrt{3}$
d $x = -2 \pm \sqrt{5}$ **e** $x = -3 \pm 2\sqrt{5}$ **f** $x = 3 \pm 2\sqrt{2}$
g $x = -4 \pm \sqrt{11}$ **h** $x = -\frac{1}{2} \pm \frac{\sqrt{13}}{2}$ **i** $x = \frac{3}{2} \pm \frac{\sqrt{29}}{2}$

3 **a** no real solutions **b** $x = -2 \pm \sqrt{3}$ **c** no real solutions
d $x = -3 \pm \sqrt{11}$ **e** $x = \frac{1}{2} \pm \frac{\sqrt{17}}{2}$ **f** no real solutions

4 $x = -\frac{3}{2} \pm \frac{\sqrt{11}}{2}$

5 **a** $x = -2 \pm \sqrt{\frac{7}{2}}$ **b** $x = 1 \pm \frac{1}{\sqrt{3}}$ **c** $x = -1 \pm \frac{2}{\sqrt{5}}$

REVIEW SET 20A

1 **a** quadratic equation **b** not quadratic equation
c not quadratic equation

2 -2 and 3 **3** **a** $x = \pm 4$ **b** $x = \pm 5$

4 **a** $x = 5$ or -7 **b** no real solutions **c** $x = \frac{1 \pm \sqrt{5}}{3}$

5 **a** $m = 0$ or $n = 0$ **b** $p = 0$ **c** $a = 0$ or $b = 0$

6 **a** $x = 0$ or $-\frac{2}{3}$ **b** $x = -2$ or 8 **c** $x = \frac{1}{2}$

7 **a** $x = 0$ or -2 **b** $x = -3$ or 4 **c** $x = \pm\frac{9}{2}$

8 **a** $x = -\frac{4}{3}$ or -2 **b** $x = 8$ or -3 **c** $x = 9$ or -2
d $x = 5$ or -2 **e** $x = 6$ or -2 **f** $x = 1$ or 4
g $x = 5$ or -2 **h** $x = 3$ or 11

9 The number is -10 or 11.

10 The numbers are -1 and 2 or -2 and 1.

11 square: 16.8 cm by 16.8 cm, rectangle: 11.2 cm by 12.6 cm

12 $x = 7$; the perimeter is 70 m.

13 **a** $0 < x < 15$
b Area of L-shape $= 15 \times x + x \times (20 - x)$
$= [15x + x(20 - x)]$ cm^2
c Area of L-shape $= \frac{1}{2} \times 15 \times 20 = 150$ cm^2
$\therefore\ 15x + x(20 - x) = 150$
$\therefore\ 35x - x^2 = 150$
d 2 solutions; $x = 5$ or 30 **e** $x = 5$ $\{0 < x < 15\}$

14 **a** $x = 1 \pm \sqrt{11}$ **b** $x = -\frac{5}{2} \pm \frac{\sqrt{33}}{2}$

15 **a** $x = -\frac{3}{2} \pm \frac{\sqrt{13}}{2}$
b Completing the square we get $(x + \frac{3}{2})^2 = -\frac{7}{4}$ which has no real solutions.
c **i** $k > -\frac{9}{4}$ **ii** $k = -\frac{9}{4}$ **iii** $k < -\frac{9}{4}$

REVIEW SET 20B

1 $p = 0$, $q = 0$, or $r = 0$ **2** $-\frac{3}{5}$ and 1

3 If $x^2 + 9 = 0$, then $x^2 = -9$.
But x^2 is always $\geqslant 0$. $\therefore$ there are no real solutions.

4 **a** $x = \pm 5$ **b** $x = \frac{-3 \pm \sqrt{3}}{2}$ **c** no real solutions

5 **a** $x = 5$ or -3 **b** $x = 0$ or $-\frac{3}{4}$ **c** $x = 5$ or $\frac{1}{2}$

6 **a** $x = 0$ or $-\frac{1}{2}$ **b** $x = 0$ or 2 **c** $x = 2$ or $-\frac{5}{3}$

7 **a** $x = \pm 4$ **b** $x = \pm\frac{3}{5}$

8 **a** $x = 3$ or 11 **b** $x = -4$

9 **a** $x = 4$ or -3 **b** $x = 1$ or 3

10 **a** $x = -1$ or 5 **b** $x = -6$ or 8 **c** $x = -9$ or 5

11 The numbers are $8, 10, 12$ or $-12, -10, -8$.

12 The area is 160 m^2.

13 The numbers are -2 and 3 or -3 and 2.

14 The radii are 4 m and 6 m.

15 **a** $x = 5 \pm 3\sqrt{3}$ **b** no real solutions

EXERCISE 21A

1 **a** quadratic function; $a = 2$, $b = 1$, $c = 4$
b not a quadratic function; no x^2 term
c quadratic function; $a = -1$, $b = -1$, $c = -1$
d quadratic function; $a = 1$, $b = 0$, $c = 0$
e quadratic function; $a = \frac{1}{2}$, $b = 0$, $c = -6$
f not a quadratic function; contains x^3 term
g quadratic function; $a = -3$, $b = 0$, $c = 0$
h not quadratic function; contains x^{-2} term
i quadratic function; $a = -2$, $b = 0$, $c = -\frac{5}{2}$

2 **a** $y = 10$ **b** $y = -5$ **c** $y = -5$

3 **a** $y = 4$ **b** $y = 14$ **c** $y = 34$

4 **a** $y = 1$ **b** $y = 1$ **c** $y = \frac{9}{2}$

5 **a** $k = 3$ **b** $k = -39$ **c** $k = -7$ **d** $k = 4$

6 **a** yes **b** no **c** yes **7** **a** no **b** no **c** yes

8 **a** $x = 0$ or -2 **b** $x = -1$ **c** $x = 3$ or -5

9 **a** $x = -1$ or 5 **b** $x = 0$ or 4 **c** $x = 2$

10 **a** $x = 0$ or -5 **b** $x = -2$ or -3 **c** $x = 2$ or -7

11 **a** $k = 2$ or -5 **b** $k = 1$ or -11

12 **a** **i** 55 m **ii** 135 m
b **i** 2 s and 10 s **ii** 4 s and 8 s **c** 12 s

13 2 seconds

14 **a** **i** \$3200 loss **ii** \$6400 profit **iii** \$8800 profit
b 3 cabinets or 13 cabinets

EXERCISE 21B

1 **a**

x	-3	-2	-1	0	1	2	3
y	-1	-3	-3	-1	3	9	17

b

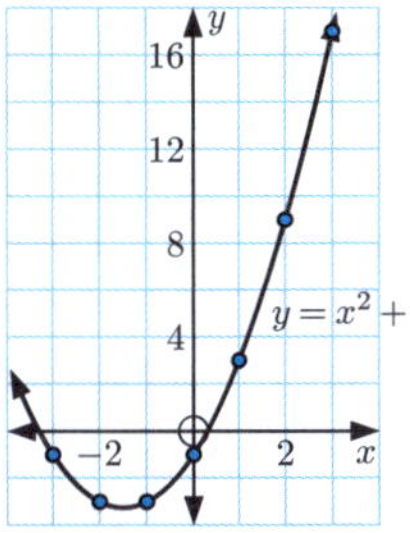

2 **a**

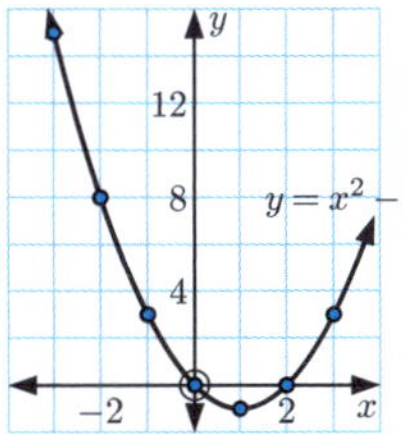

b

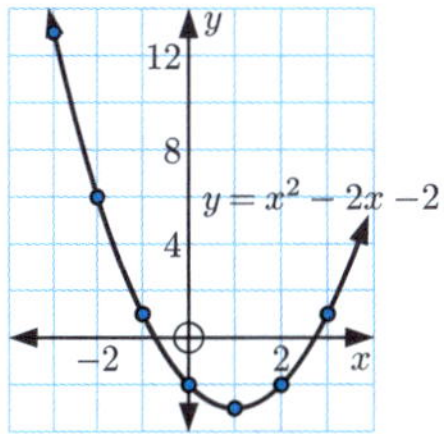

c

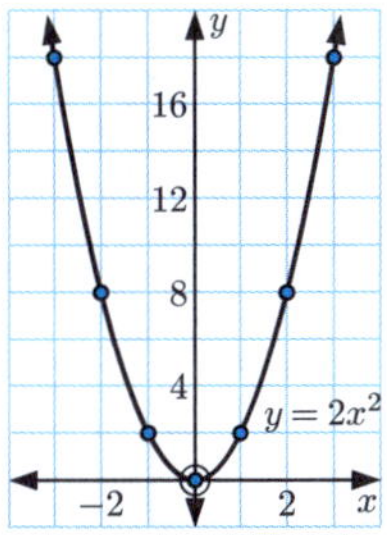

d

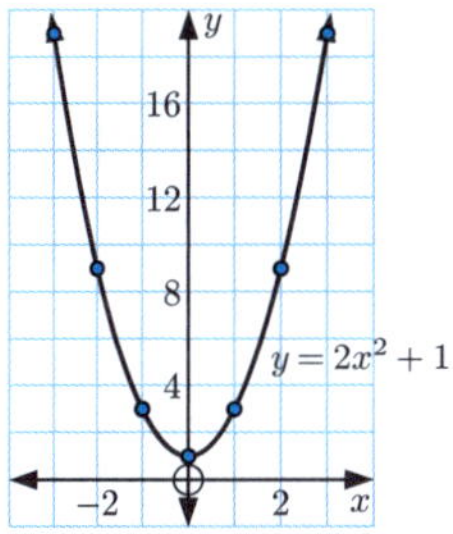

e

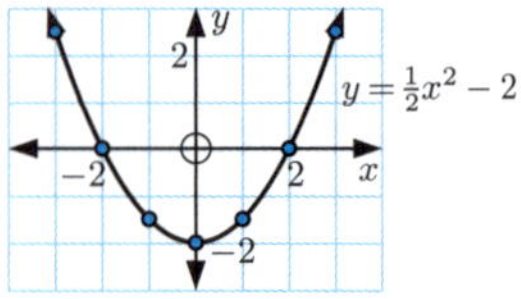

f

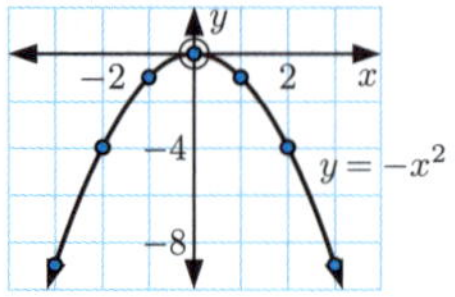

g

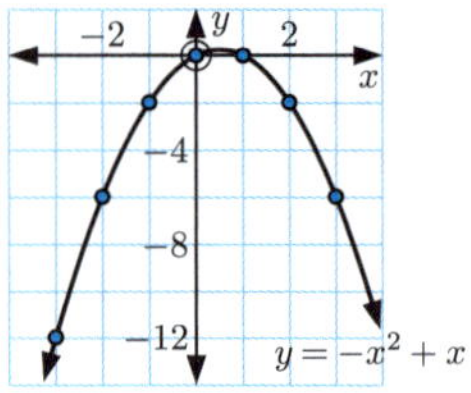

h

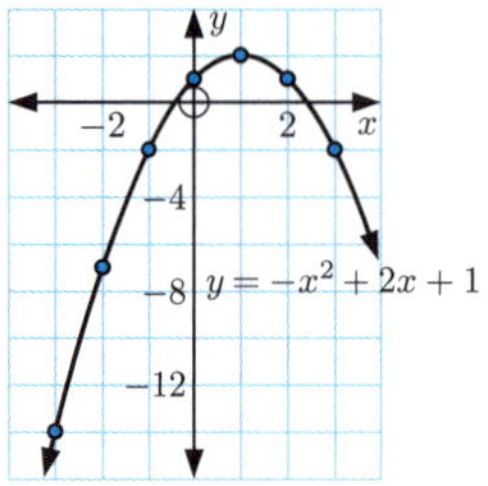

i

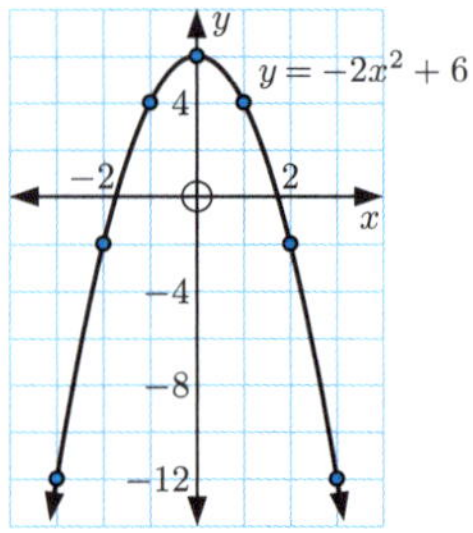

3 **a**

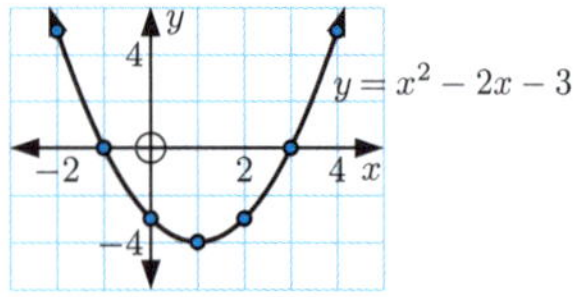

b $x = 1$

c $(1, -4)$

d **i** two points

ii one point

iii no points

e **i** $k > -4$ **ii** $k = -4$ **iii** $k < -4$

4 **a**

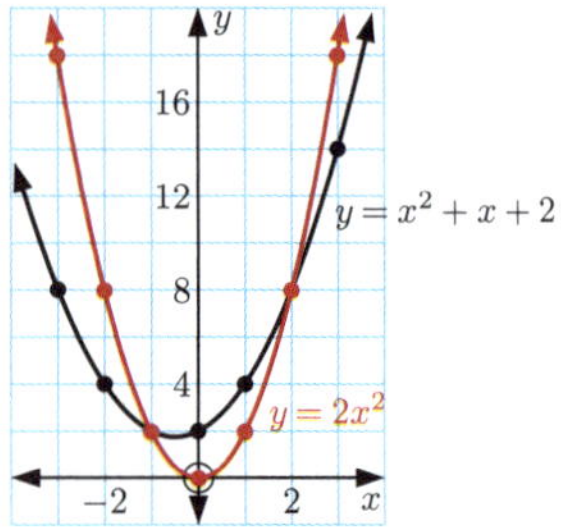

b, c $x = -1$ or 2

EXERCISE 21C

1 **a**

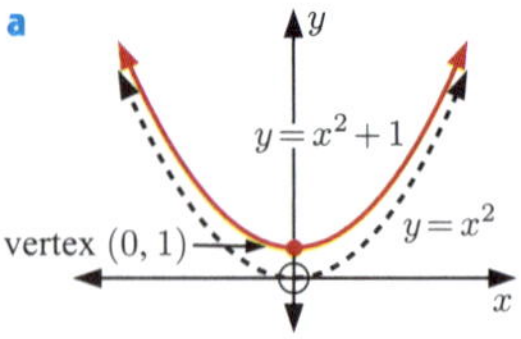

b

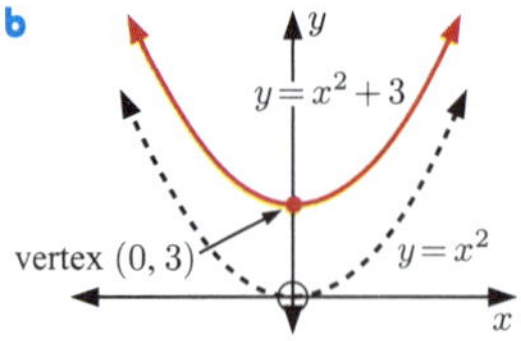

c

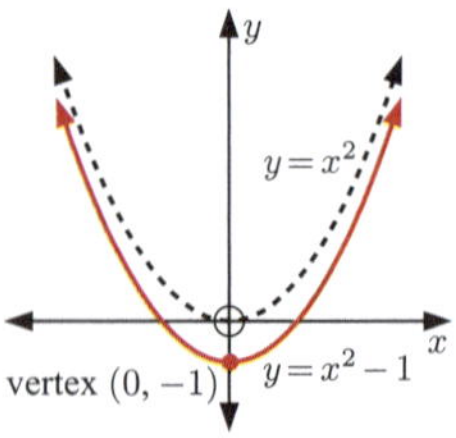

d

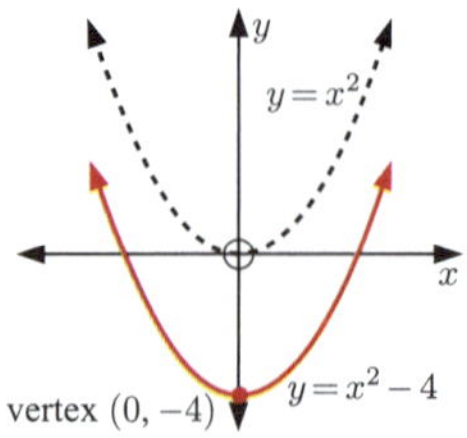

e

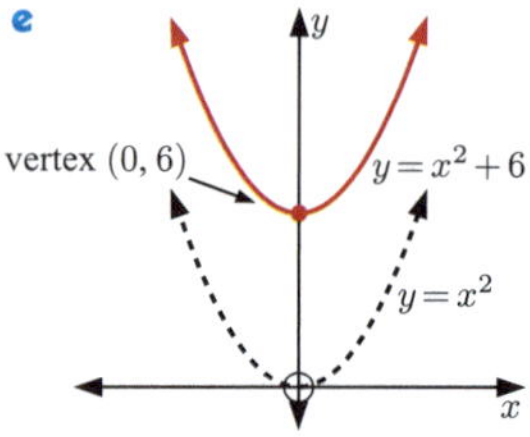

f

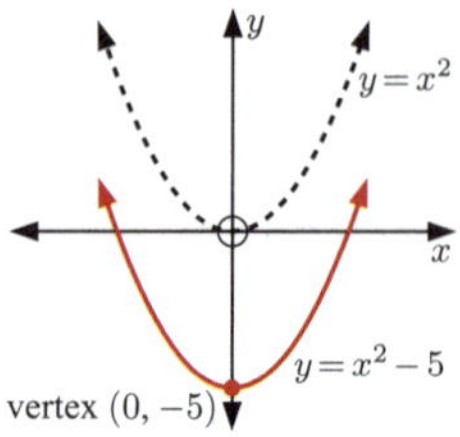

2 **a**

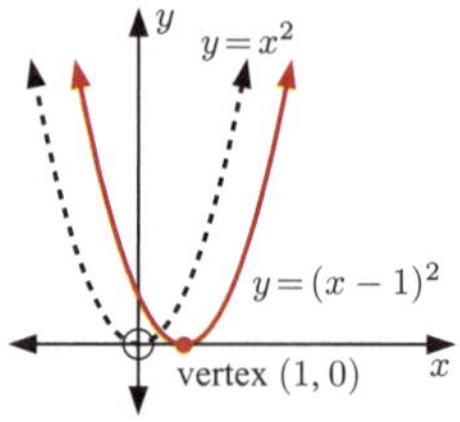

b

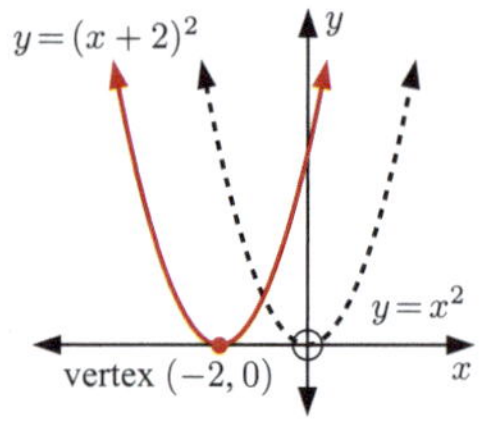

c

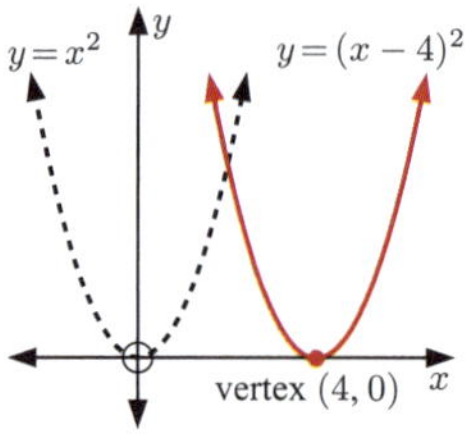

d

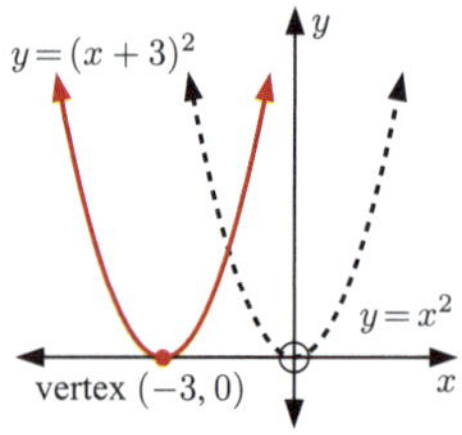

e

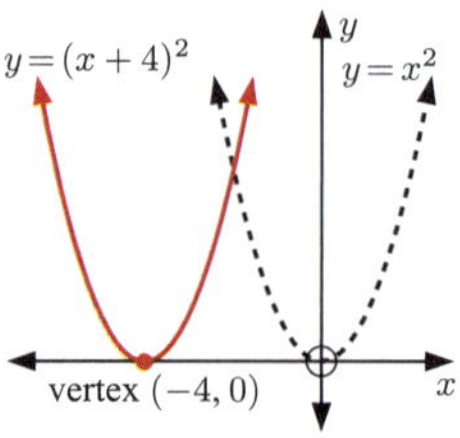

f

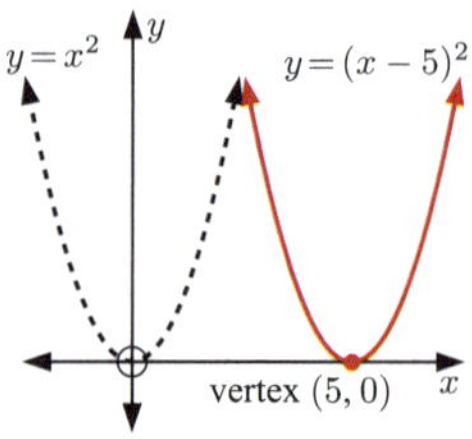

3 **a** $(0, 2)$ **b** $(-1, 0)$ **c** $(0, -3)$ **d** $(6, 0)$

4 **a** $y = (x-2)^2 + 1$

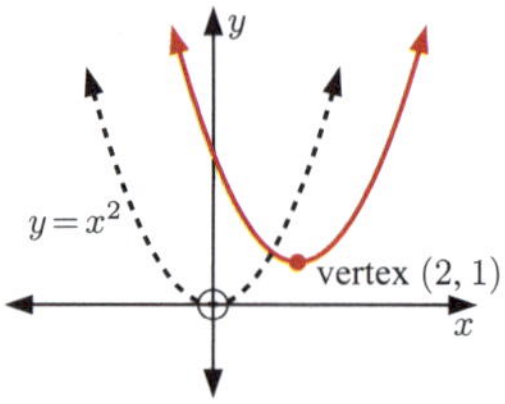

b $y = (x+3)^2 + 4$

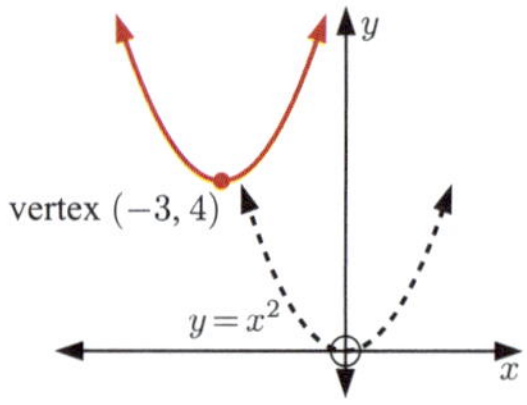

c $y = (x-1)^2 - 3$

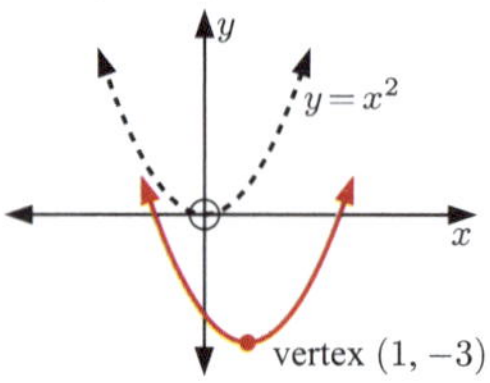

d $y = (x+4)^2 - 2$

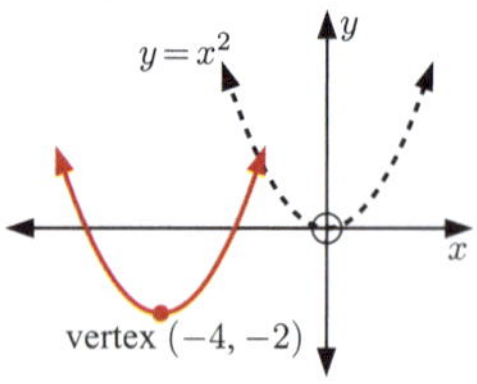

e $y = (x-4)^2 - 5$

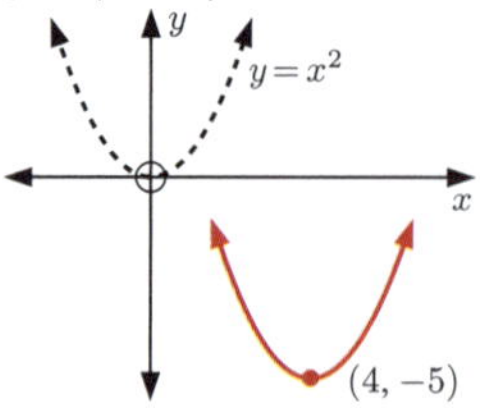

f $y = (x+5)^2 + 1$

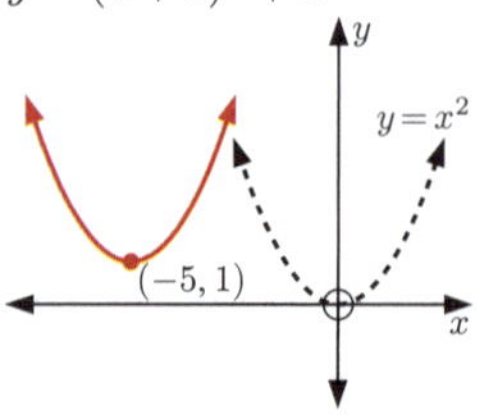

5 **a** $(1, 5)$
b $(3, -1)$
c $(-3, -6)$

6 **b**

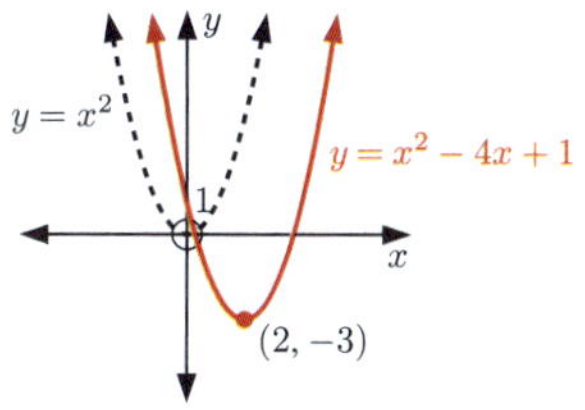

7

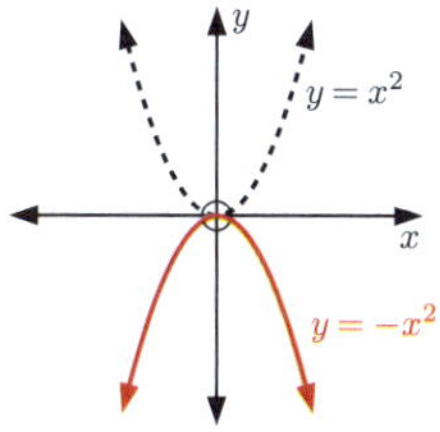

8 **a**

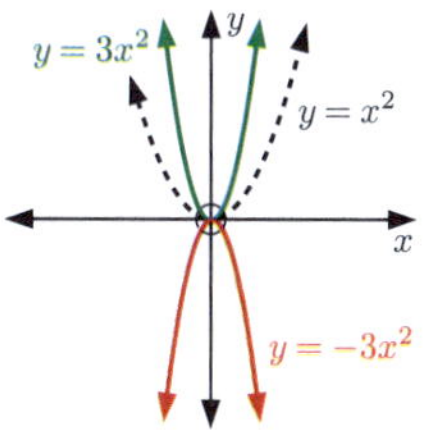

b

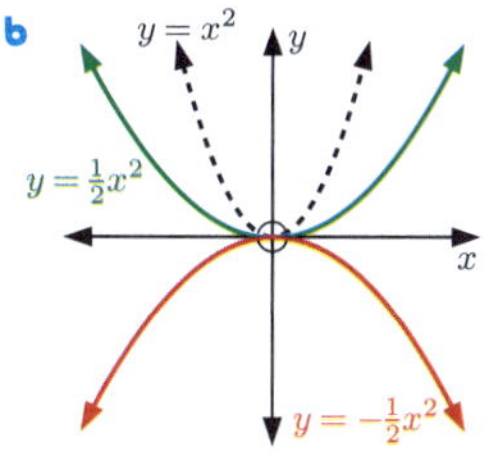

9 **a** C **b** D **c** E **d** B **e** F **f** A

10 **a** **i** $y = (x + 2)^2 - 6$

ii

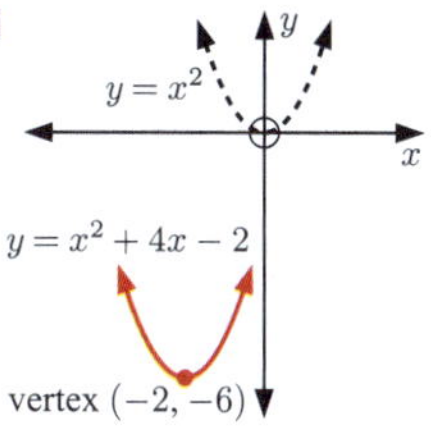

b **i** $y = (x - 4)^2 + 3$

ii

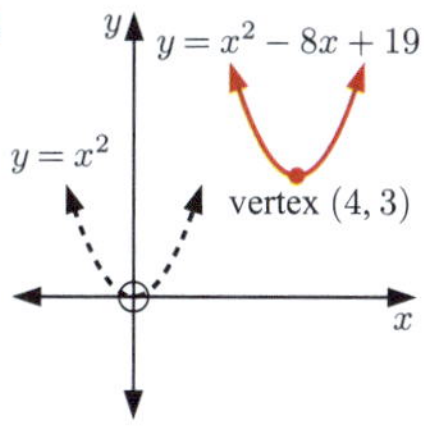

c **i** $y = (x + \frac{3}{2})^2 + \frac{3}{4}$

ii

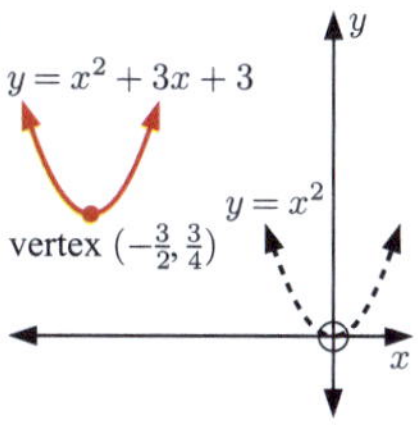

11 **a**

$$\begin{aligned} (x-3)^2 &\geqslant 0 \quad \text{for all } x \\ \therefore\ -(x-3)^2 &\leqslant 0 \quad \text{for all } x \\ \therefore\ -(x-3)^2 + 1 &\leqslant 1 \quad \text{for all } x \\ \therefore\ y &\leqslant 1 \quad \text{for all } x \end{aligned}$$

b 2 and 4
c 3 units right, 1 unit up
d

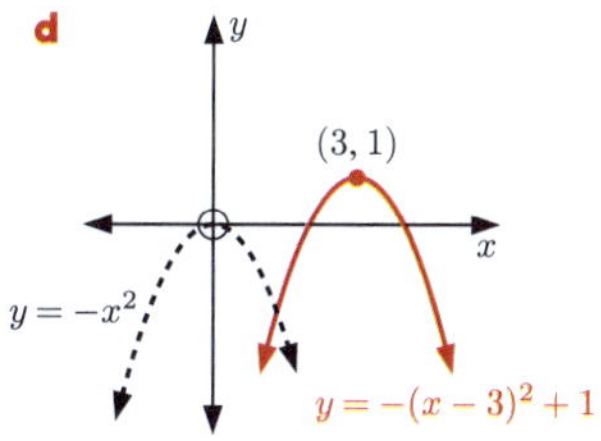

12 **a** The graph of $y = ax^2$ is translated h units horizontally and k units vertically to give the graph of $y = a(x - h)^2 + k$.
The vertex of $y = ax^2$ is $(0, 0)$.
$\therefore$ the vertex of $y = a(x - h)^2 + k$ is (h, k).

b **i** $(4, 2)$ **ii** $(-3, -4)$ **iii** $(-4, -1)$

EXERCISE 21D

1 **a** 1 **b** 3 **c** -2 **d** 4 **e** 6 **f** 1
g 4 **h** 3 **i** 7

2 **a** 1 and -3 **b** -2 and 3 **c** 3 and -7
d 2 and 5 **e** -5 **f** 7

4 **a** -2 **b** -3 **c** 2 **d** 0 **e** 4 **f** -20

5 **a** $a\alpha\beta$ **b** $a\alpha^2$

6 **a** 2 and -2 **b** 3 and -3 **c** 3 and 4 **d** 3 and -2
e 0 and 3 **f** 2 and 5 **g** 3 **h** -1
i 1 and -6 **j** 1 and -5 **k** -2 and 8 **l** 3 and -7

7 **a** y-intercept: 2, x-intercept: -2
b y-intercept: -6, x-intercepts: -2 and 1
c y-intercept: 24, x-intercepts: -3 and 8

8 **a** $-\frac{3}{2} \pm \frac{\sqrt{29}}{2}$ **b** $\frac{1}{2}$ and -4 **c** 2 and $-\frac{1}{3}$
d $2 \pm 2\sqrt{3}$ **e** $\frac{5}{2}$ and $-\frac{1}{2}$ **f** $1 \pm \frac{3}{\sqrt{2}}$

EXERCISE 21E

1 **a**

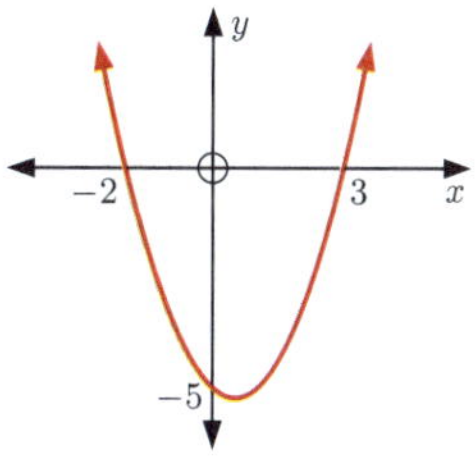

b

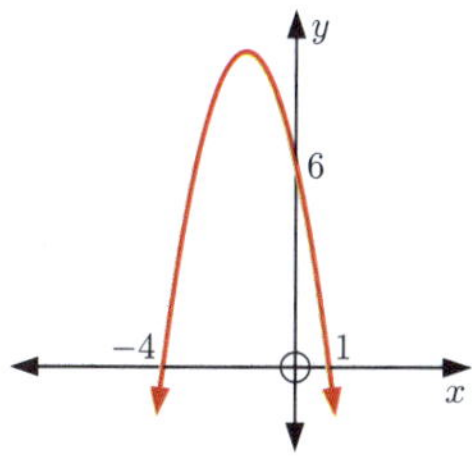

c

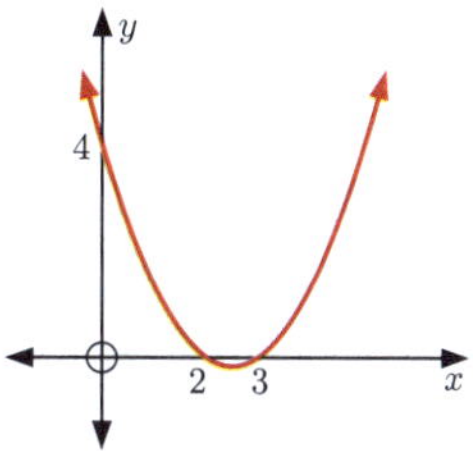

d

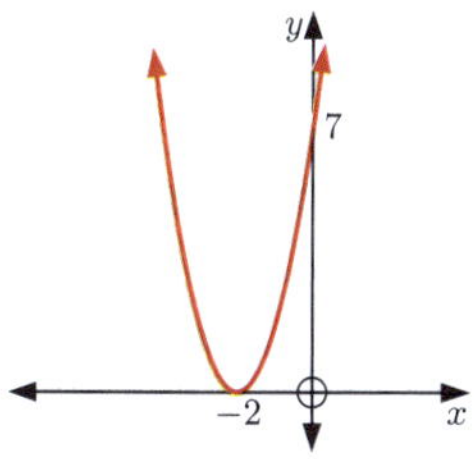

e

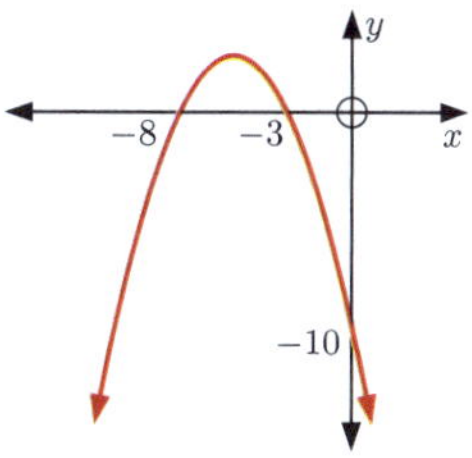

f

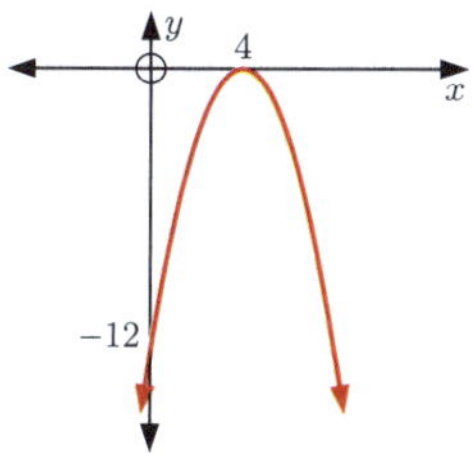

2 **a**

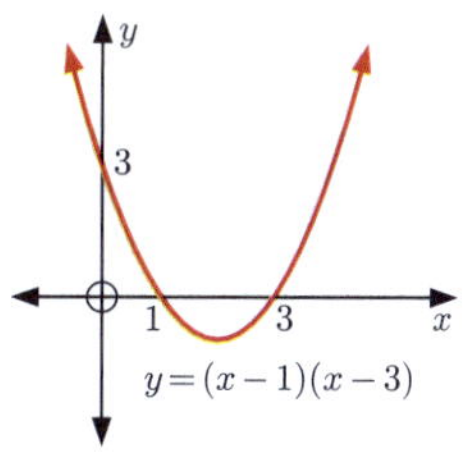

b

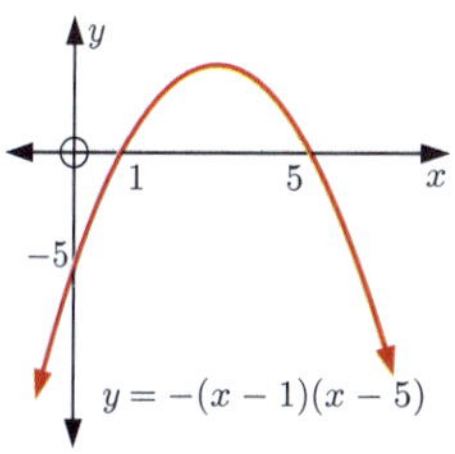

c 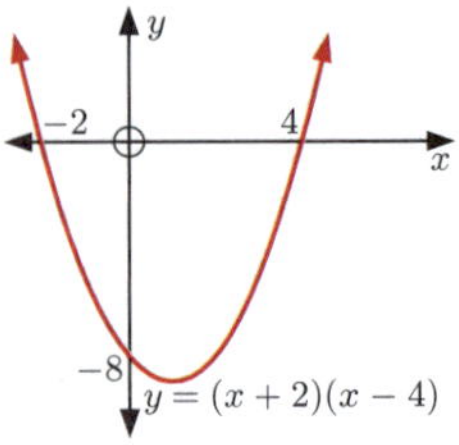

d 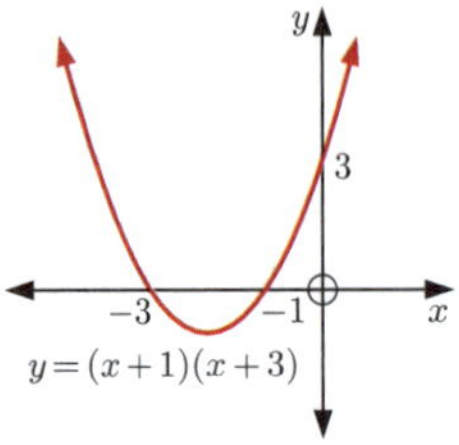

e 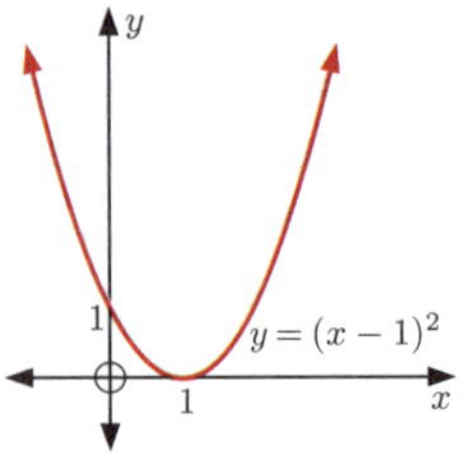

f 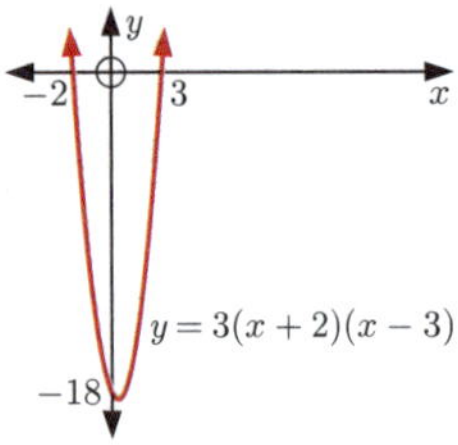

g 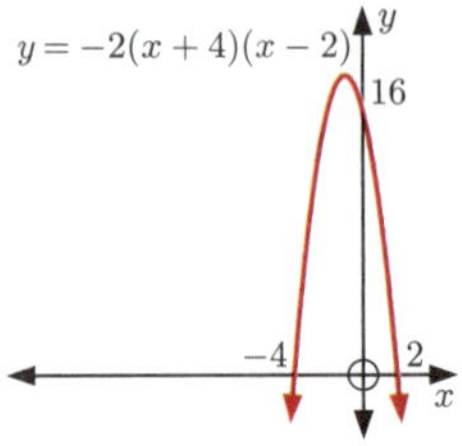

h 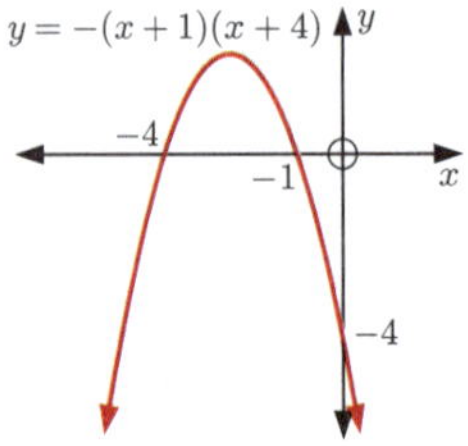

i 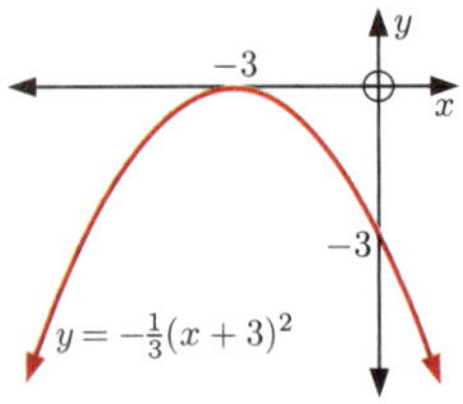

3 **a** 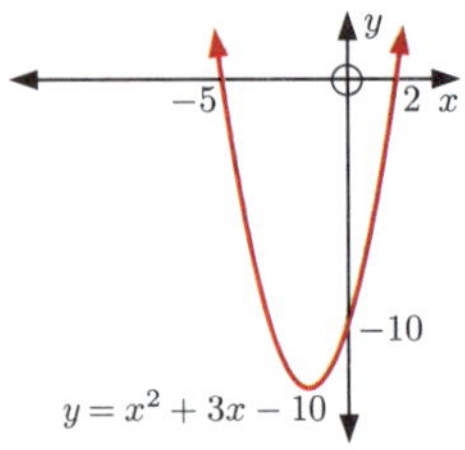

b 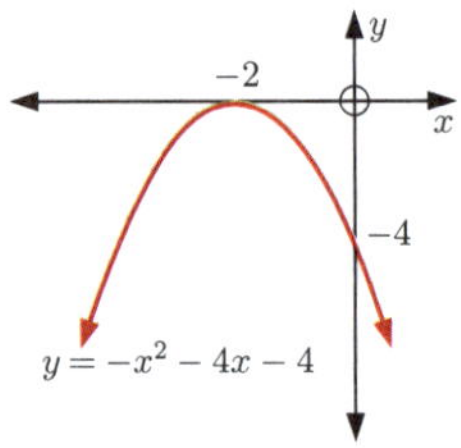

c 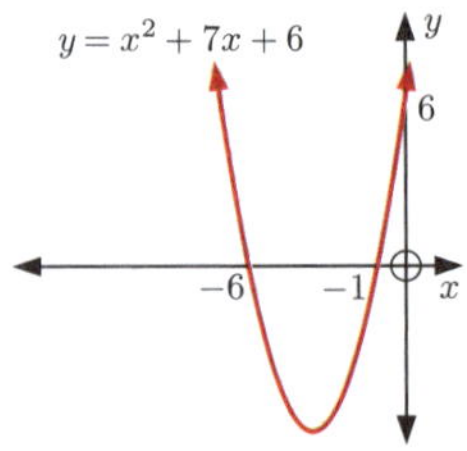

d 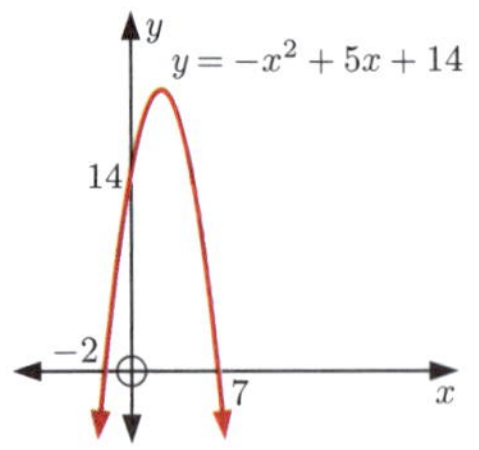

e 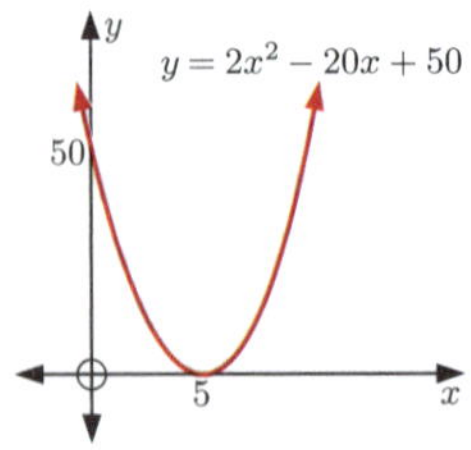

f 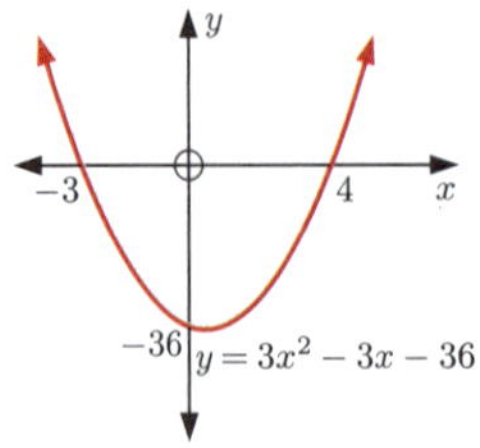

4 **a** 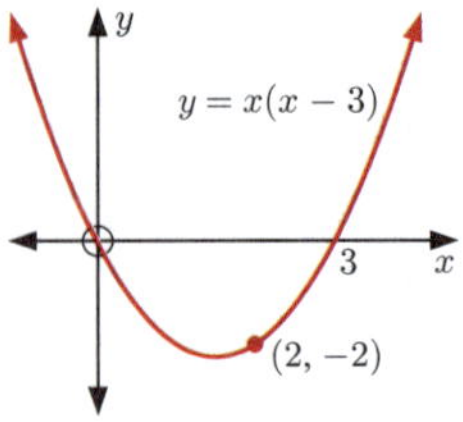

b 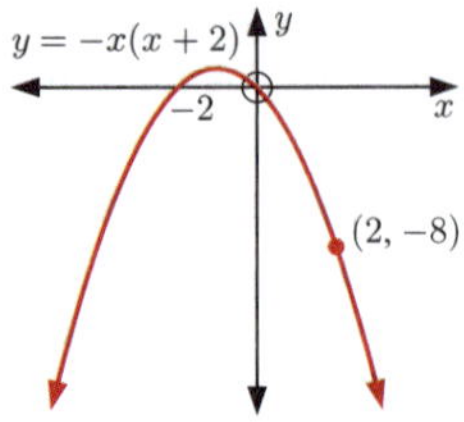

c 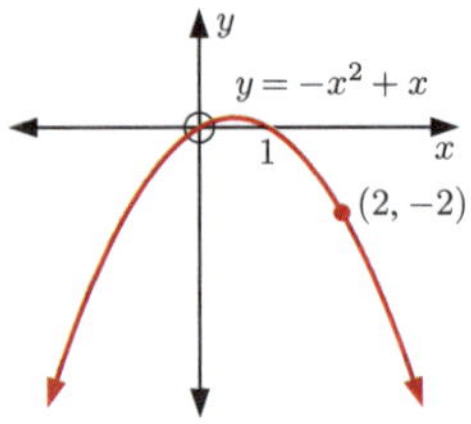

5 **a** $(4, -4)$

b **i** $y = x^2 - 8x + 12$

ii $y = (x-2)(x-6)$

c y-intercept: 12,
x-intercepts: 2 and 6

d 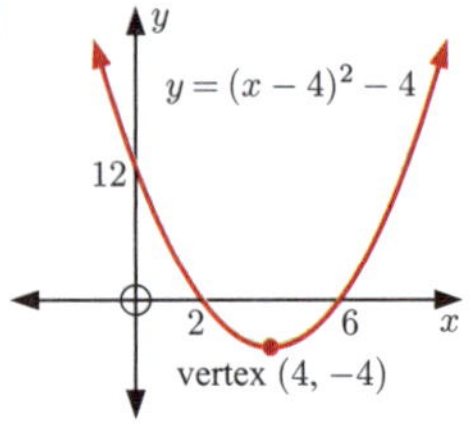

6 **a** **B** **b** **C** **c** **E** **d** **A** **e** **F** **f** **D**

7 **a** 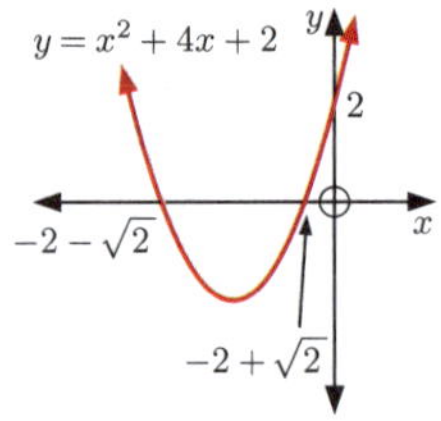

b 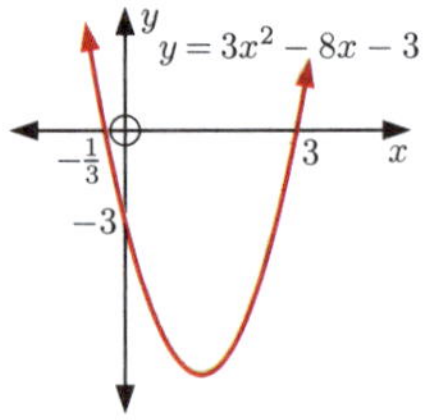

c 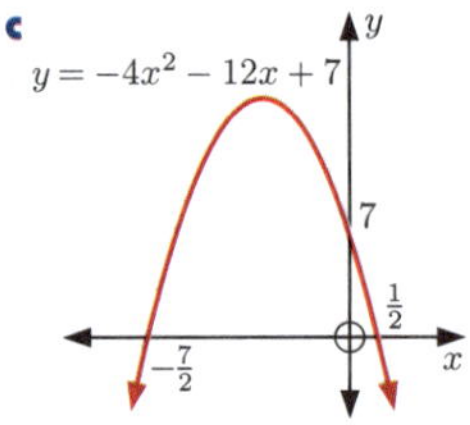

d 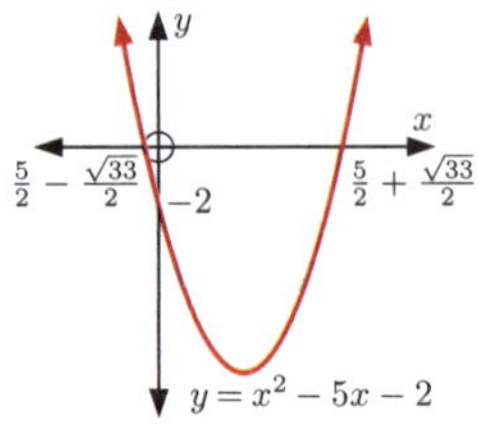

e 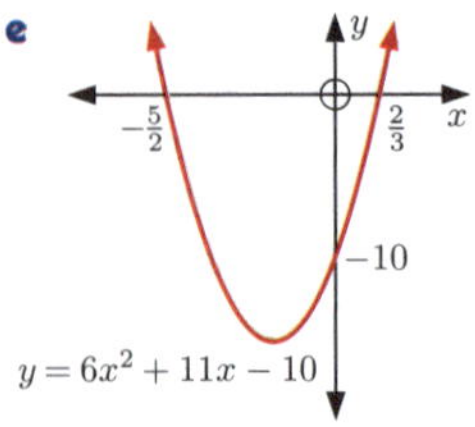

f 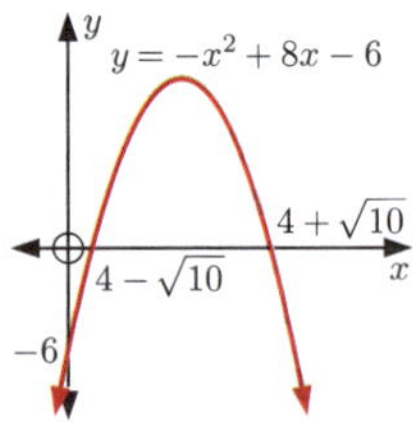

EXERCISE 21F

1 **a** 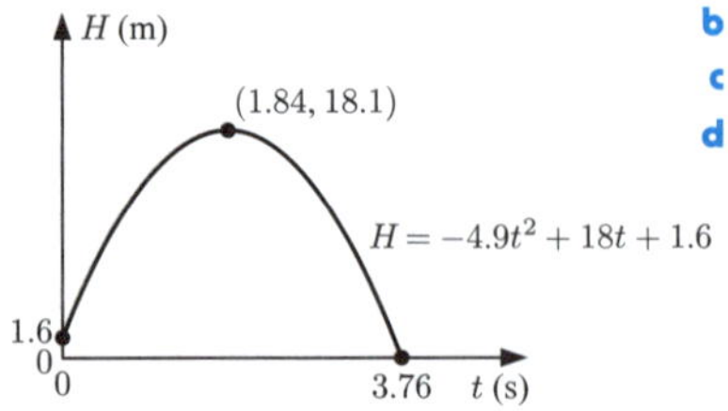

b ≈ 18.1 m

c ≈ 1.84 s

d ≈ 3.76 s

2 **a**
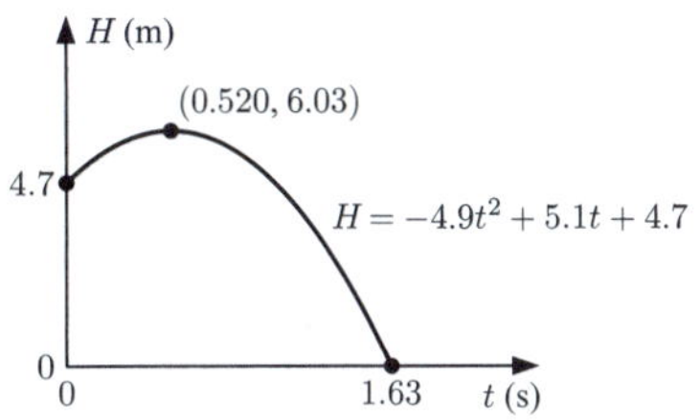

b ≈ 6.03 m
c ≈ 1.63 s

3 **a**
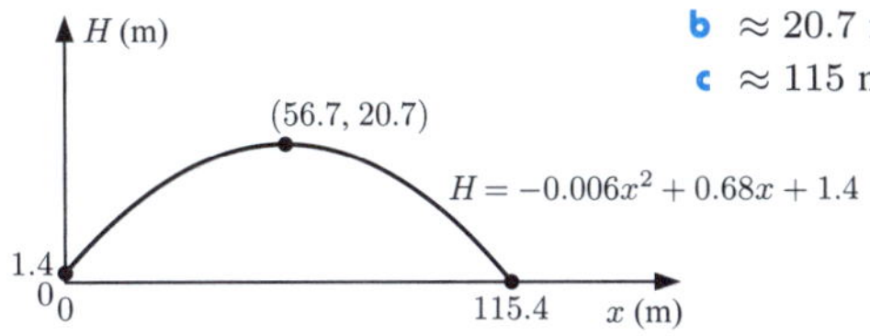

b ≈ 20.7 m
c ≈ 115 m

4 **a** a quadratic relationship **b** **i** 6 m **ii** 28.4 m
c yes **d** ≈ 33.8 m

REVIEW SET 21A

1 **a** $y = 24$
b $y = 3$
c $y = 29$

2 **a** $t = 11$
b $t = 0$ or 3

3
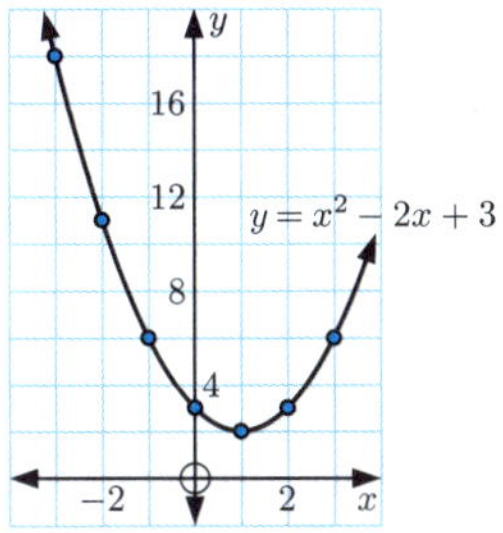

4 **a**
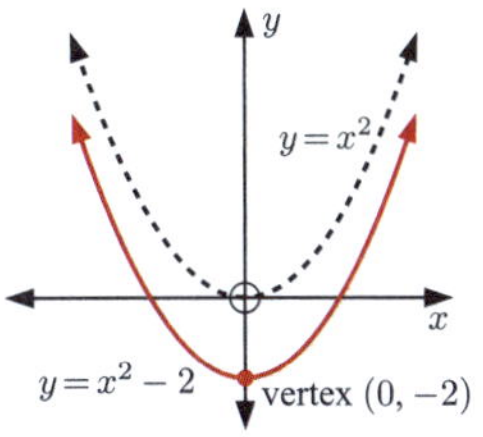

b
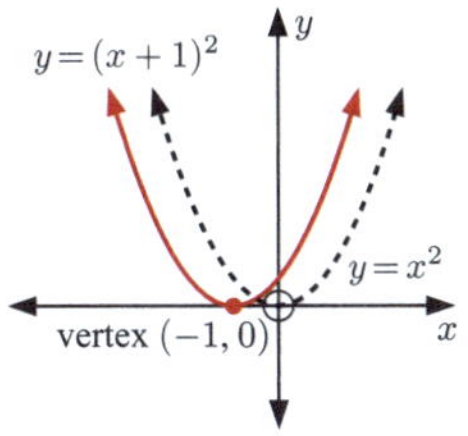

5 **a**
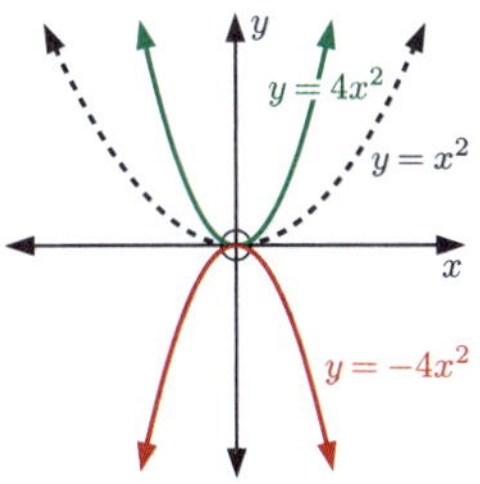

b
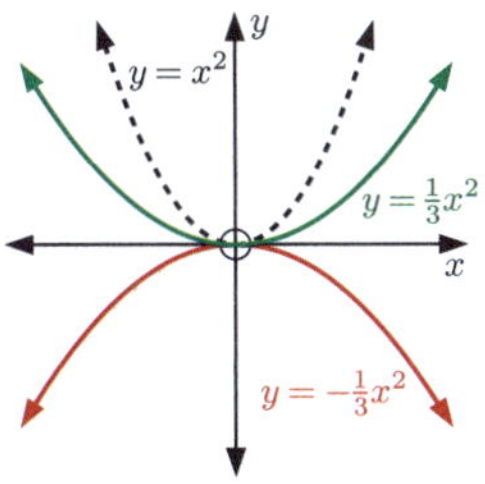

6 **a** $y = (x - 3)^2 - 7$
b
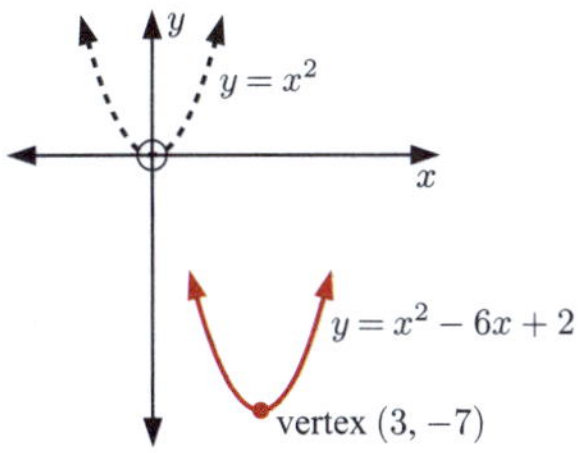

7 **a** -4 **b** 10 **c** 9

8 **a** -2 and 3 **b** -3 and 4 **c** -4

9 **a**
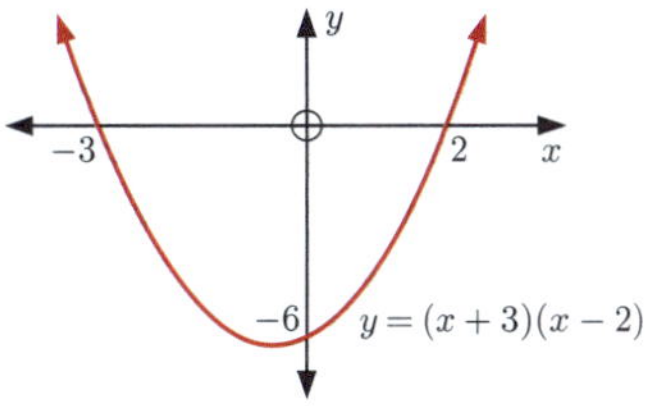

b
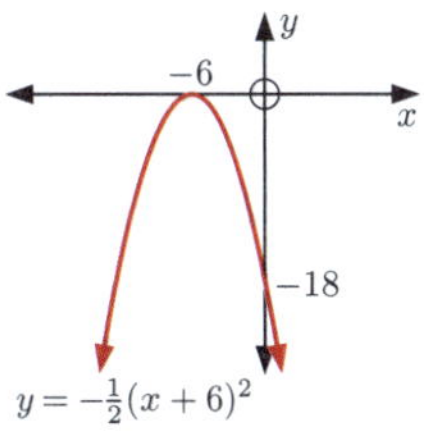

c
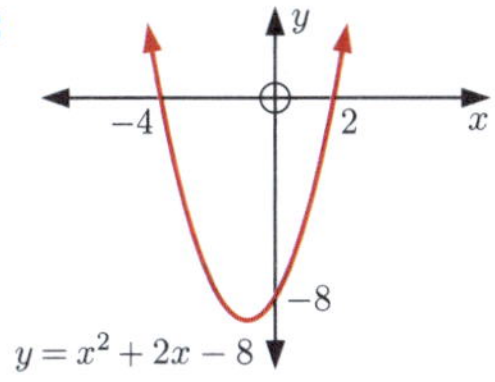

10
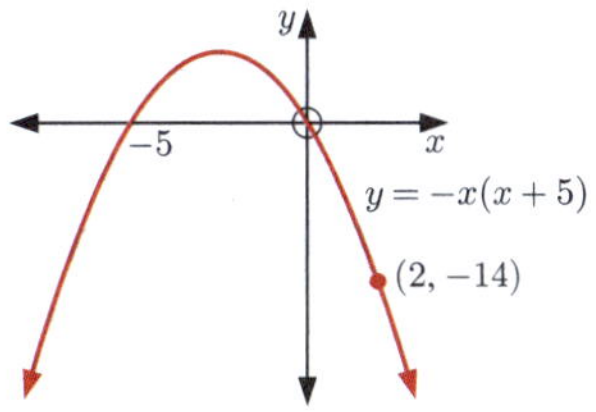

11 **a** $(3, -16)$
b $y = x^2 - 6x - 7$
c y-intercept: -7,
x-intercepts: -1 and 7

d
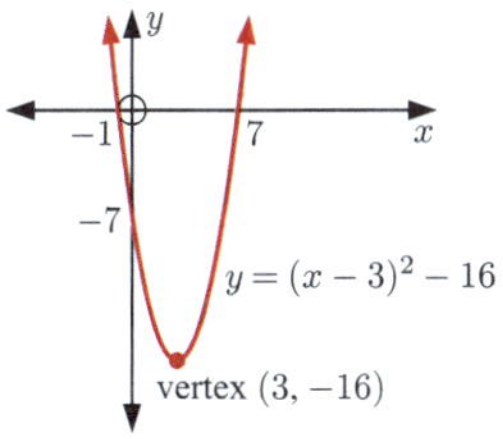

12 **a**
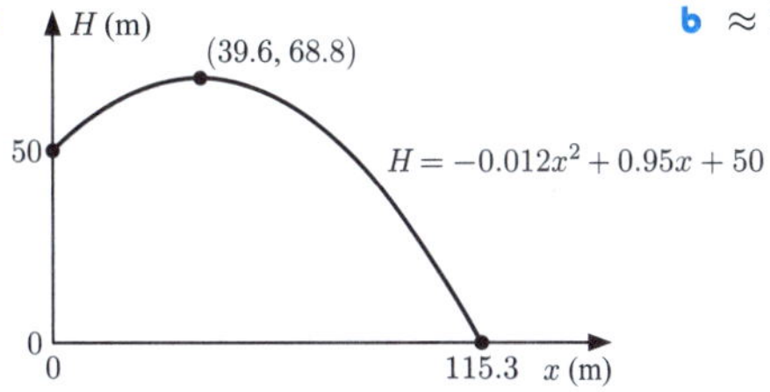

b ≈ 115 m

13 **a**
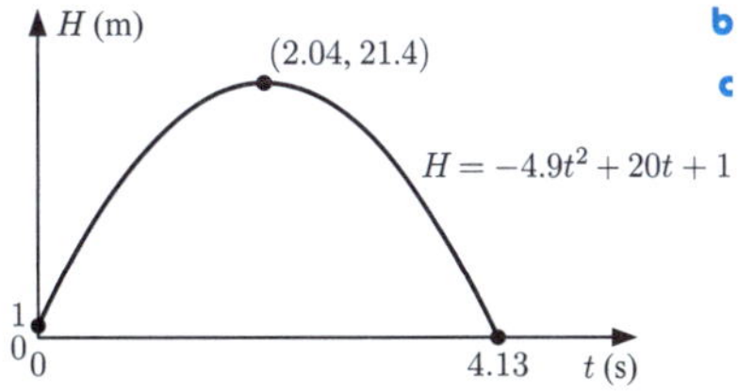

b ≈ 21.4 m
c ≈ 4.13 s

REVIEW SET 21B

1 **a** $y = 11$
b $x = -1$ or 5

2 $a = -1$ or -2

3
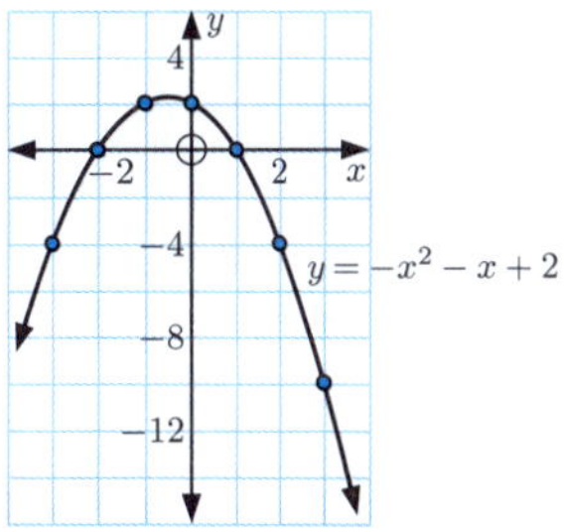

4 **a** 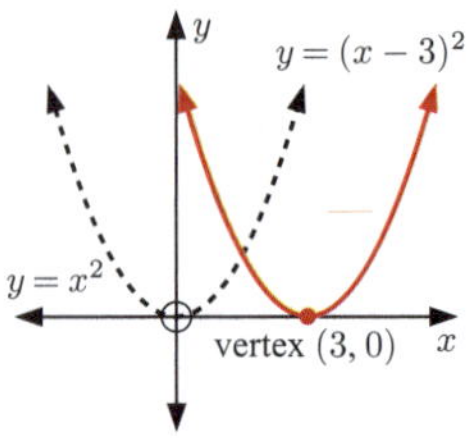

b 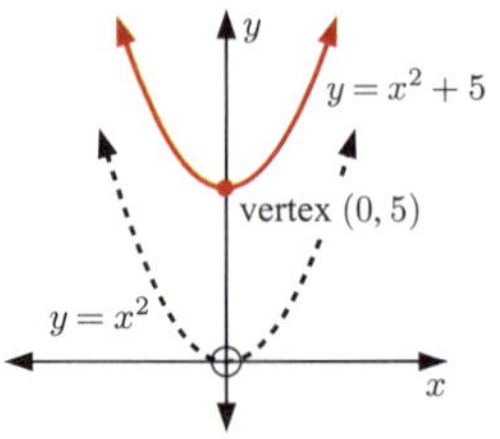

5 **a** $(-4, 0)$
b $(0, -5)$
c $(6, 3)$

6 **b** 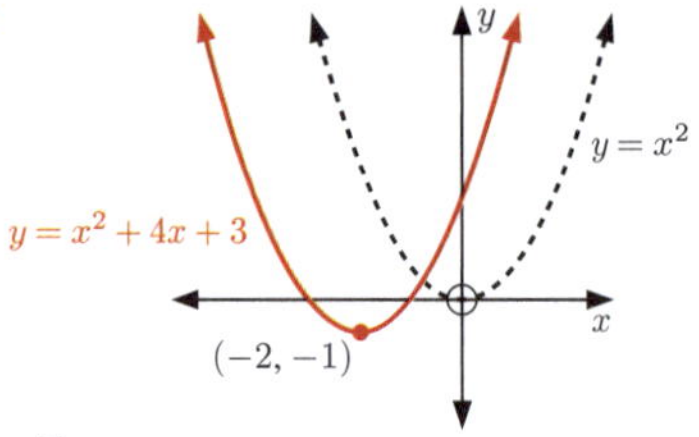

7 **a** 0
b 12
c -21

8 **a** -3 and 6 **b** $\pm 4\sqrt{2}$ **c** -3 and 2

9 **a** 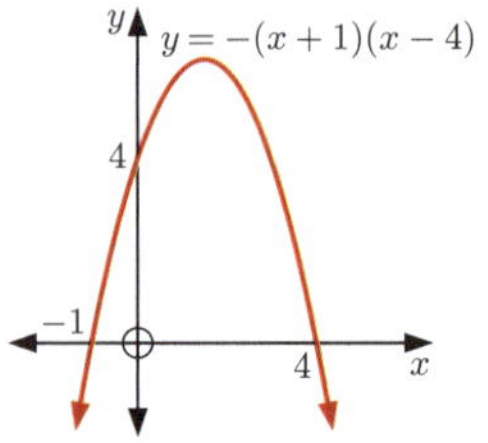

b 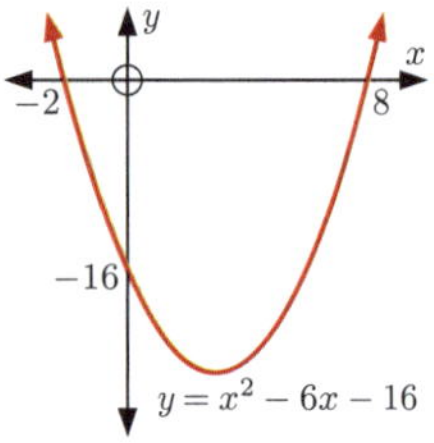

c 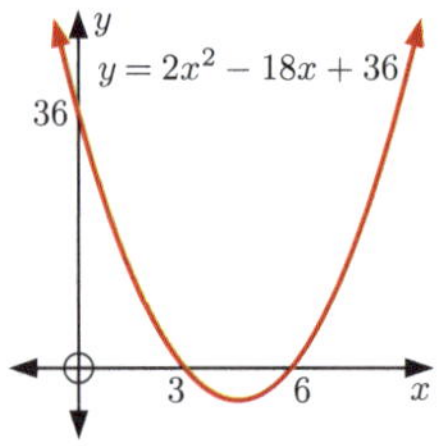

10 **a** 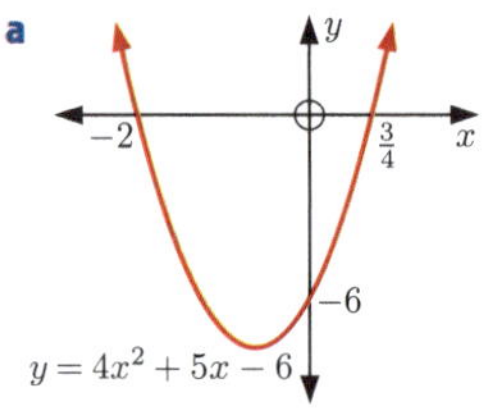

b 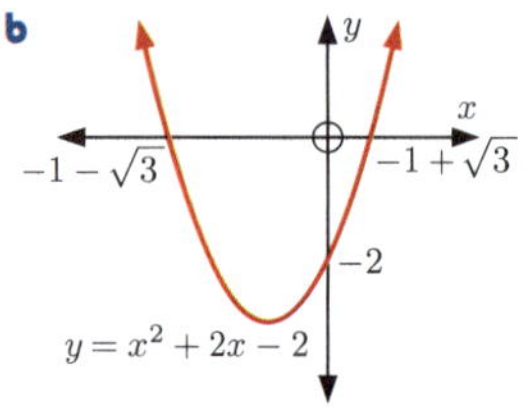

c 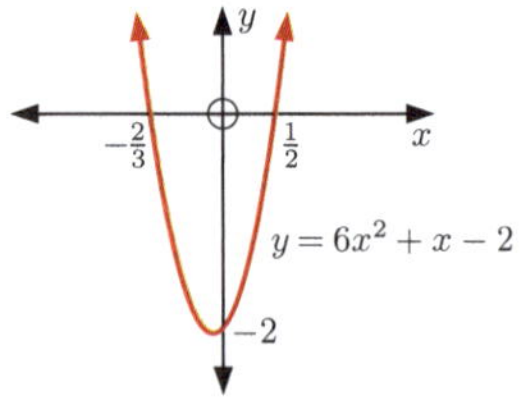

11 **a** 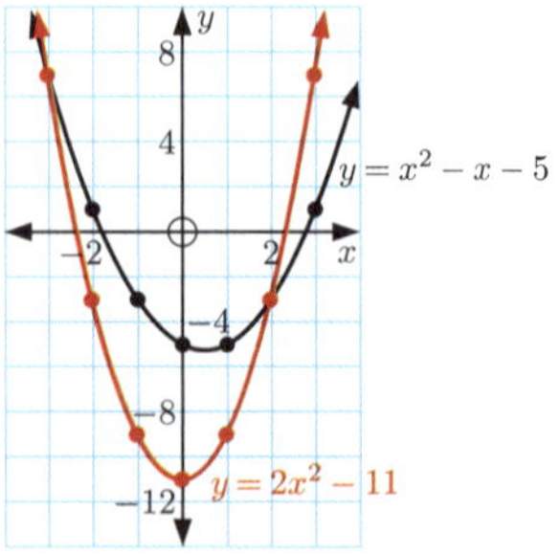

b, c $x=-3$ or 2

12 **a** $(-4, -1)$
b **i** $y=x^2+8x+15$ **ii** $y=(x+3)(x+5)$
c y-intercept: 15, x-intercepts: -5 and -3
d 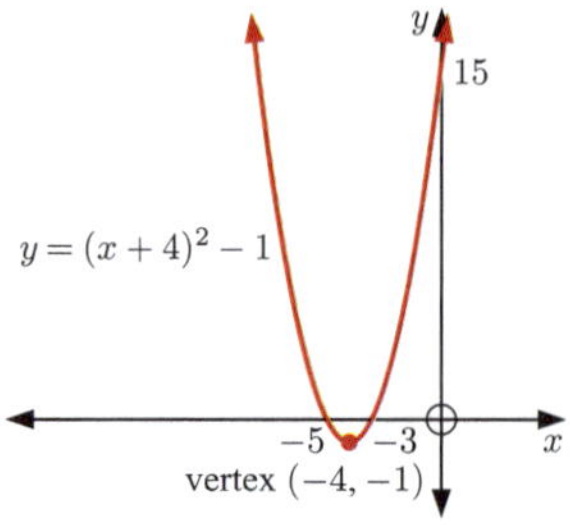

13 **a** 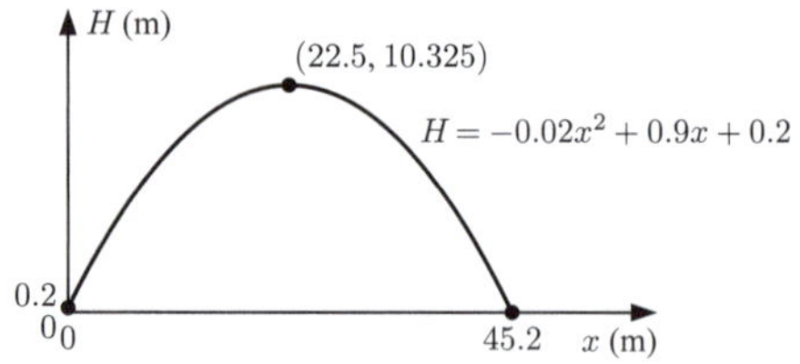

b 10.325 m **c** ≈ 45.2 m
d Yes, when $x=40$, $H=4.2>3$.

EXERCISE 22A

1 **a** no **b** yes **c** no **d** yes
2 **A**, **D**, and **M**; **C** and **O**; **I** and **L**
3 **a** **i** [QR] **ii** [CD] **b** **i** $\widehat{PQR}$ **ii** $\widehat{CDA}$
4 **a** 16 mm **b** 57 mm
5 **a** 3 cm **b** **i** 30° **ii** 70° **iii** 80°

EXERCISE 22B

1 The equal sides are not in corresponding positions. One is opposite the 43° angle, the other is opposite the 68° angle.
2 **a** yes {SAS} **b** No, they could be different sizes.
c yes {SSS}
d No, the equal sides are not in corresponding positions.
e yes {RHS} **f** yes {AAcorS}
3 **a** A and C {SAS} **b** B and C {AAcorS}
c A and C {RHS}
4 **a** **i** not congruent
b **i** $\triangle RST \cong \triangle ZXY$ {AAcorS}
ii $\widehat{RTS}=\widehat{ZYX}$, $RS=ZX$, and $ST=XY$
c **i** $\triangle ABC \cong \triangle DFE$ {SSS}
ii $\widehat{ABC}=\widehat{DFE}$, $\widehat{BCA}=\widehat{FED}$, and $\widehat{CAB}=\widehat{EDF}$
d **i** $\triangle ABC \cong \triangle CDA$ {SAS}
ii $AB=CD$, $\widehat{CAB}=\widehat{ACD}$, and $\widehat{ABC}=\widehat{CDA}$
e **i** $\triangle XYZ \cong \triangle SRT$ {RHS}
ii $YZ=RT$, $\widehat{ZYX}=\widehat{TRS}$, and $\widehat{ZXY}=\widehat{TSR}$
f **i** not congruent **g** **i** not congruent
h **i** $\triangle PQR \cong \triangle YZX$ {SAS}
ii $RQ=XZ$, $\widehat{PQR}=\widehat{YZX}$, and $\widehat{PRQ}=\widehat{YXZ}$
5 5 cm **6** **Hint:** Show that $\triangle AOC \cong \triangle BOD$.
7 **Hint:** Show that $\triangle PQR \cong \triangle PTS$.
8 **a** **Hint:** Use RHS.
9 **Hint:** Let P be any point on the perpendicular bisector, and M be the midpoint of [AB].
Show that $\triangle AMP \cong \triangle BMP$.
Consider also the case where $P=M$.

10 **a** **Hint:** Show that $\triangle AED \cong \triangle BFC$.
b **Hint:** Show that $\triangle ACD \cong \triangle BDC$.

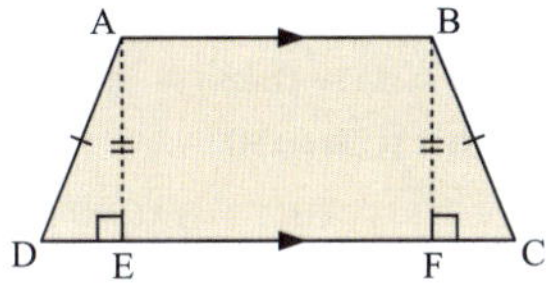

EXERCISE 22C

1 **a** yes **b** no **c** not enough information
d not enough information

2 **a** $x = 10$ **b** $x = \frac{9\sqrt{13}}{2} \approx 16.2$ **c** $x = 12$
d $x = 5$

3 The angles of **A** are equal to the angles of **B**, and the angles of **B** are equal to the angles of **C**. **A** and **B** have corresponding sides in the same ratio, and **B** and **C** have corresponding sides in the same ratio.
$\therefore$ **A** and **C** are equiangular and have corresponding sides in the same ratio. $\therefore$ **A** is similar to **C**.

4 **b**

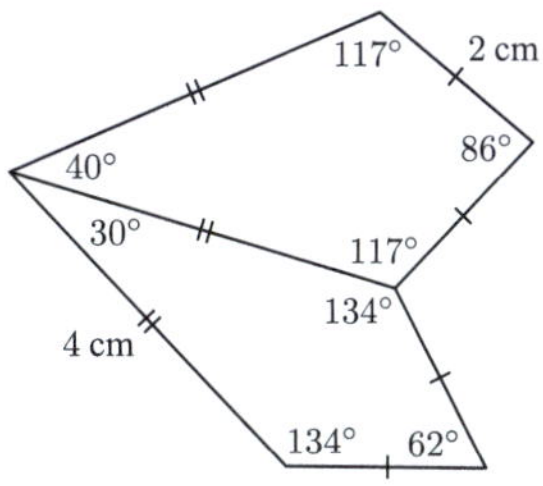

The quadrilaterals are not equiangular.
$\therefore$ they are not similar.

5 **a** $\widehat{ABE} = \widehat{ACD}$ and $\widehat{BEF} = \widehat{CDF}$ {equal corresponding angles}
and the angles at A and F are common.
$\therefore$ ABEF and ACDF are equiangular.

b $\frac{AC}{AB} \neq \frac{AF}{AF} = 1$
$\therefore$ the corresponding side lengths are not in the same ratio.
$\therefore$ the quadrilaterals are not similar.

EXERCISE 22D.1

1 **a** $PQ = \frac{2}{3}TS$, $QR = \frac{2}{3}SU$, $RP = \frac{2}{3}UT$
b $\widehat{KJL} = \widehat{KLJ} = \widehat{MNO} = \widehat{MON} = 69°$ {base angles of isosceles triangle}
$\widehat{NMO} = 42°$ {angle sum of a triangle}
$= \widehat{JKL}$

2 **a** $\widehat{ABE} = \widehat{ADC}$ {given}, angle A is common
b $\widehat{PQR} = \widehat{TSR}$ {given}
$\widehat{PRQ} = \widehat{TRS}$ {vertically opposite angles}
c angle D is common,
$\widehat{DEH} = \widehat{DFG}$ {equal corresponding angles}
d $\widehat{PTQ} = \widehat{SRQ}$ {equal corresponding angles}
$\widehat{PQT} = \widehat{SQR}$ {vertically opposite angles}
e angle C is common,
$\widehat{CBD} = 40°$ {angle sum of a triangle}
$= \widehat{CAB}$
f $VX = 15$ m, $XY = 34$ m {Pythagoras}
$\therefore$ $YZ = 2WV$, $XY = 2XW$, $XZ = 2XV$

3 **Hint:** Show that [AB] is parallel to [CD].

EXERCISE 22D.2

1 **a** $\widehat{ABC} = \widehat{DEC}$ {given}
$\widehat{ACB} = \widehat{DCE}$ {vertically opposite angles}; $x = 8$
b angle A is common,
$\widehat{ABC} = \widehat{ADE}$ {equal corresponding angles}; $x = 3.75$
c $\widehat{UVZ} = \widehat{XYZ}$ {given}, angle Z is common; $x = 4.8$
d $\widehat{BAC} = \widehat{DEC}$ {given}
$\widehat{ACB} = \widehat{ECD}$ {vertically opposite angles};
$x = \frac{54}{7} \approx 7.71$
e angle P is common,
$\widehat{PQR} = \widehat{PST}$ {equal corresponding angles}; $x = 4.8$
f $\widehat{JKN} = \widehat{JML}$ {given}, angle J is common;
$x = \frac{40}{3} \approx 13.3$

2 **a** $x = 17.5$ **b** $x = 12$ **c** $x = 8.75$
3 **a** $x = 5$ **b** $x = 12$ **c** $x = 2$
4 **a** $x = 3$, $y \approx 11.2$ **b** $x = 6$, $y = 8$
5 **a** 1650 m **b** ≈ 1677 m
6 ≈ 13.3 m **7** 8 m **8** **a** 25 m **b** 19.2 m
9 PR ≈ 3.67 m **10** 78 cm^2 **11** **a** $\frac{7}{20}$ **b** 7 : 6

EXERCISE 22E

1 **a** $x = 32$ **b** $x = 16$ **2** **a** $x = 5$ **b** $x = 4.8$
3 **a** $k = 4$ **b** 20 cm and 24 cm **c** 16 times
4 **a** angle A is common,
$\widehat{ABE} = \widehat{ACD}$ {equal corresponding angles}
b $k = \frac{5}{2}$ **c** **i** 100 cm^2 **ii** 84 cm^2
5 ≈ 1.45 cm **6** 4 units^2 **7** 3 : 5

EXERCISE 22F

1 **a** $V = 270$ **b** $V = 31.25$ **2** **a** $x = 8$ **b** $x \approx 5.26$
3 6750 cm^3 **4** 12 cm **5** ≈ 1.19
6 **a** 4.116 m^3 **b** 3.75 m^2
7 No, comparing lengths scale factor $= 1.6$,
comparing capacities scale factor ≈ 1.37.
8 **a** The objects are similar. **b** **i** 2 times **ii** 56 kL
9 125 : 27 **10** 44% greater

REVIEW SET 22A

1 **a** 8 cm **b** 30 cm
2 **a** No, need another pair of equal angles or sides.
b yes {SSS}
3 **a** **Hint:** Use SAS. **4** $x = 130$, $y = \frac{40}{3} \approx 13.3$
5 **a** no **b** yes
6 **a** angle A is common,
$\widehat{ABC} = \widehat{ADE}$ {equal corresponding angles}; $x = 3$
b $\widehat{YXZ} = \widehat{VUZ}$ {given}
$\widehat{XZY} = \widehat{UZV}$ {vertically opposite angles}; $x = 4$
c angle A is common,
$\widehat{ABE} = \widehat{ACD}$ {equal corresponding angles}; $x = 8$
7 **Hint:** Let $\widehat{DAX} = \alpha$, and find other angles in terms of α.
8 50 m **9** **a** $A = 7$ **b** $x \approx 8.14$ **c** $x = 15$, $y = 32$
10 **a** $k = 4$ **b** 0.99 cm **c** 0.5 cm^3
11 **a** $\frac{1}{200}$

b **i** The replica is made using the same materials as the original ship.

ii Since the density of the ship and replica are equal and density $= \frac{\text{mass}}{\text{volume}}$, the mass and volume of the ship and the replica are in proportion.

c 10 kg

REVIEW SET 22B

1 **a** No, the equal sides are not in corresponding positions.
b No, the triangles may be different sizes.

2 B and C {SAS}

4 **a** not enough information **b** yes

5 **a** $\widehat{QPT} = \widehat{QRS}$ {given}
$\widehat{PQT} = \widehat{RQS}$ {vertically opposite angles}
b $\widehat{BAC} = \widehat{DBC}$ {given}, angle C is common

6 **a** angle A is common,
$\widehat{AEB} = \widehat{ADC}$ {equal corresponding angles}; $x = 7.2$
b angle R is common, $\widehat{RSQ} = \widehat{RPT}$ {given}; $x = 14$

7 **a** 6.4 cm **b** ≈ 85.8 cm^3 **8** ≈ 3.33 cm

9 **a** $x = 4$ **b** $x \approx 42.7$ **10** **a** 7 : 5 **b** 49 : 25

11 **a** $\frac{3}{2}$ **b** $x = 4.5$ **c** 8 : 27

12 **a** 120° **b** **Hint:** Use SAS. **c** **Hint:** Use AAcorS.
d **i** **Hint:** Use SAS. **ii** **Hint:** $\triangle$ABC is isosceles.
e 1 : 4

EXERCISE 23A

1 **a** 1 cm ≡ 1 m

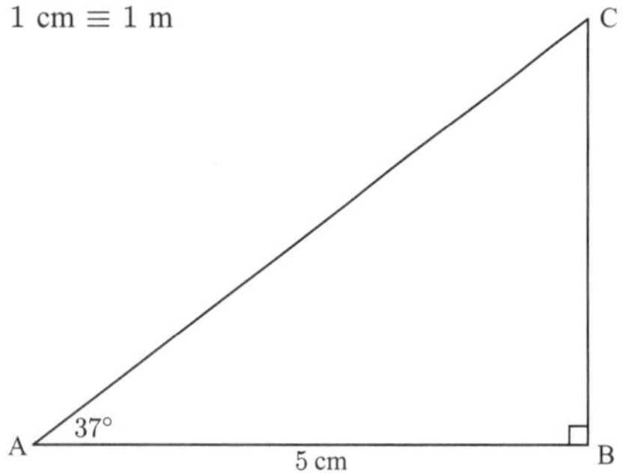

b **i** ≈ 3.8 m **ii** ≈ 6.3 m

2 ≈ 49 m **3** ≈ 1700 m

EXERCISE 23B

1 **a** **i** [AC] **ii** [BC] **iii** [AB]
b **i** [RS] **ii** [RT] **iii** [ST]
c **i** [LM] **ii** [MN] **iii** [LN]

2 **a** b units **b** c units **c** c units **d** b units

EXERCISE 23C

1 **a** **i** $\frac{3}{5}$ **ii** $\frac{4}{5}$ **iii** $\frac{3}{4}$ **b** **i** $\frac{5}{13}$ **ii** $\frac{12}{13}$ **iii** $\frac{5}{12}$
c **i** $\frac{15}{17}$ **ii** $\frac{8}{17}$ **iii** $\frac{15}{8}$ **d** **i** $\frac{1}{\sqrt{5}}$ **ii** $\frac{2}{\sqrt{5}}$ **iii** $\frac{1}{2}$
e **i** $\frac{4}{7}$ **ii** $\frac{\sqrt{33}}{7}$ **iii** $\frac{4}{\sqrt{33}}$
f **i** $\frac{8}{\sqrt{73}}$ **ii** $\frac{3}{\sqrt{73}}$ **iii** $\frac{8}{3}$

2 **a** PQ ≈ 59 mm, PR ≈ 54 mm, QR ≈ 24 mm
b **i** ≈ 0.407 **ii** ≈ 0.915 **iii** ≈ 0.444
c **i** ≈ 0.407 **ii** ≈ 0.914 **iii** ≈ 0.445

3 **a** **i** ≈ 0.587 **ii** ≈ 0.630 **iii** ≈ 1.378 **iv** ≈ 0.778
b **i** ≈ 0.588 **ii** ≈ 0.629 **iii** ≈ 1.376 **iv** ≈ 0.777

4 **a** $\widehat{ACB} = 90°$
$\therefore$ $\widehat{ABC} + \widehat{BAC} = 90°$ {angles in a triangle sum to 180°}
$\therefore$ $\widehat{ABC} = \widehat{BAC} = 45°$ {isosceles}
b $\sqrt{2}$ m **c** **i** $\frac{1}{\sqrt{2}}$ **ii** $\frac{1}{\sqrt{2}}$ **iii** 1

5 **a** **i** ≈ 0.9240 **ii** ≈ 0.0760 **iii** 1
iv ≈ 0.1786 **v** ≈ 0.8214 **vi** 1
b $\cos^2\theta + \sin^2\theta = 1$

6 **a** **i** 1 **ii** 1 **iii** 1 **b** $\tan\theta \times \tan(90° - \theta) = 1$

EXERCISE 23D

1 **a** $\sin 65° = \frac{x}{a}$ **b** $\cos 32° = \frac{x}{b}$ **c** $\tan 56° = \frac{x}{c}$
d $\cos 37° = \frac{d}{x}$ **e** $\tan 49° = \frac{e}{x}$ **f** $\tan 73° = \frac{f}{x}$

2 **a** $x \approx 6.88$ **b** $x \approx 17.62$ **c** $x \approx 8.91$ **d** $x \approx 14.98$
e $x \approx 22.94$ **f** $x \approx 16.73$ **g** $x \approx 23.47$ **h** $x \approx 36.05$
i $x \approx 7.42$ **j** $x \approx 13.77$ **k** $x \approx 13.07$ **l** $x \approx 20.78$

3 **a** $\theta = 62°$, $a \approx 12.4$, $b \approx 6.6$
b $\theta = 27°$, $a \approx 15.7$, $b \approx 7.1$
c $\theta = 52°$, $a \approx 54.6$, $b \approx 33.6$

4 ≈ 28.8 cm

EXERCISE 23E

1 **a** $\theta \approx 51.3°$ **b** $\theta \approx 41.8°$ **c** $\theta \approx 39.6°$
d $\theta \approx 38.0°$ **e** $\theta \approx 34.1°$ **f** $\theta \approx 47.6°$
g $\theta \approx 36.5°$ **h** $\theta \approx 35.8°$ **i** $\theta \approx 36.9°$

2 **a** $\theta \approx 38.7°$, $\phi \approx 51.3°$, $x \approx 6.2$
b $\alpha \approx 60.2°$, $\beta \approx 29.8°$, $x \approx 7.0$
c $a \approx 41.5°$, $b \approx 48.5°$, $x \approx 8.4$

3 **a** **i** $\frac{3}{8}$ **ii** $\frac{\sqrt{55}}{8}$ **iii** $\frac{3}{\sqrt{55}}$
b In each case, $\theta \approx 22.0°$.

4 These triangles do not exist. The hypotenuse must be longer than the other sides.

EXERCISE 23F

1 ≈ 2.03 m **2** ≈ 1.98 km **3** ≈ 26.0 m **4** ≈ 2.48 m
5 ≈ 49.0 m **6** $\theta \approx 44.4°$ **7** $\approx 66.4°$ **8** ≈ 15.6 m
9 **a** ≈ 10.3 m **b** ≈ 12.2 m **10** $\approx 55.2°$
11 $\theta \approx 23.6°$ **12** **a** ≈ 5.33 m **b** ≈ 41.7 m^2
13 $\approx 80.3°$ **14** $\approx 36.0°$ **15** ≈ 412 m **16** $\approx 64.0°$
17 **a** ≈ 47.0 cm **b** ≈ 82.9 cm^2 **c** ≈ 8.27 cm
d $\approx 17.1\%$
18 **a** ≈ 34.8 m **b** ≈ 8.12 m **19** ≈ 1698 m
20 **a** **i** triangular slice ≈ 41.6 cm^2, kite slice ≈ 60.9 cm^2
ii no
b $x° \approx 33.7°$, $y° \approx 22.6°$, $z° \approx 33.7°$
21 $\theta \approx 15.6°$

EXERCISE 23G

1 **a**

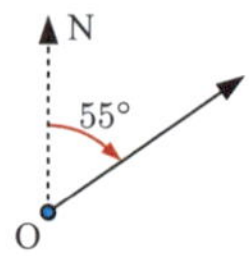

b

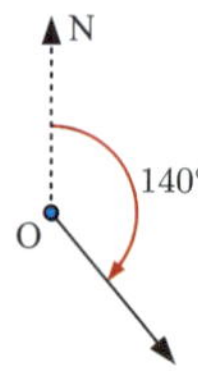

c 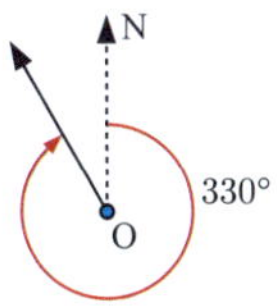

d 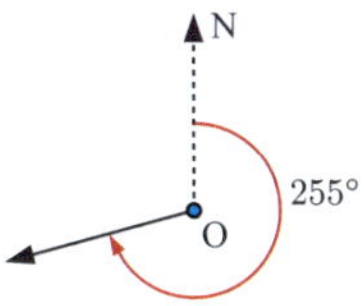

2 **a** 304° **b** 248° **c** 064° **d** 141°

3 **a** east, 090° **b** southeast, 135° **c** south, 180° **d** west, 270° **e** southwest, 225° **f** northwest, 315° **g** north, 000° **h** northeast, 045°

4 **a** 071° **b** 124° **c** 317° **d** 214° **e** 097° **f** 244°

5 **a** 025° **b** 150° **c** 330° **d** 105° **e** 205° **f** 285°

6 **a** 120° **b** 300° **c** 070° **d** 250° **e** 035° **f** 215°

7 **a** ≈ 2.24 km, $\approx 063°$ **b** ≈ 3.16 km, $\approx 162°$ **c** ≈ 3.16 km, $\approx 072°$ **d** ≈ 4.47 km, $\approx 207°$ **e** ≈ 3.61 km, $\approx 304°$

8 $\approx 236°$ **9** **a** ≈ 55.0 m **b** ≈ 40.0 m

10 ≈ 14.3 km **11** 650 km

REVIEW SET 23A

1 **a** 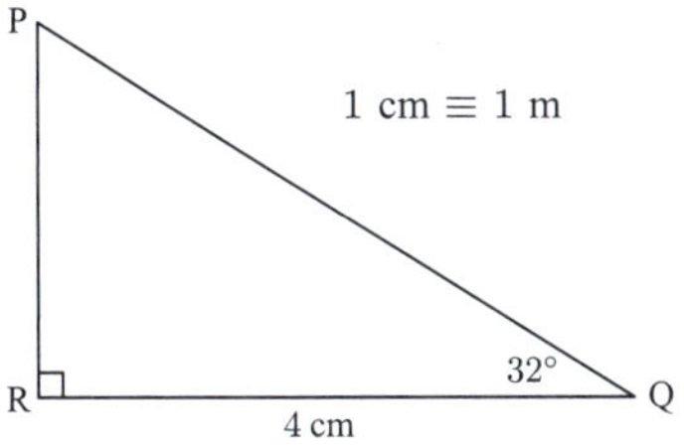

b ≈ 4.7 m **c** ≈ 4.72 m

2 $\sin\theta = \frac{12}{13}$, $\cos\theta = \frac{5}{13}$, $\tan\theta = \frac{12}{5}$

3 **a** PR ≈ 2.2 cm, QR ≈ 4.5 cm, PQ ≈ 5.0 cm

b **i** ≈ 0.44 **ii** ≈ 0.9 **iii** ≈ 0.489

c **i** ≈ 0.438 **ii** ≈ 0.899 **iii** ≈ 0.488

4 **a** $x \approx 10.90$ **b** $x \approx 6.66$ **c** $x \approx 7.81$

5 **a** $\theta \approx 52.1°$ **b** $\theta \approx 27.5°$ **c** $\theta \approx 45.7°$

6 $\approx 9.21°$

7 **a** $x \approx 16.4$, $\theta \approx 52.4°$, $\phi \approx 37.6°$

b $a \approx 5.4$, $b \approx 10.5$, $\theta = 59°$

8 ≈ 8.22 m **9** ≈ 12.7 cm

10 **a** 139° **b** 257° **11** ≈ 29.4 km

12 **a** 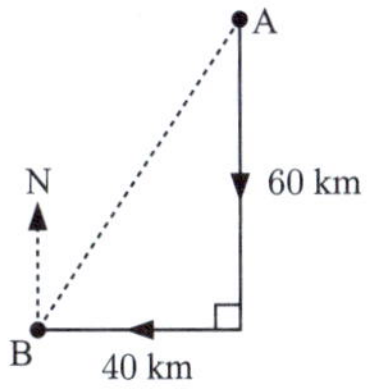

b ≈ 72.1 km **c** $\approx 034°$

13 **a** **i** $\cos 20° \approx 0.940$, $\sin 20° \approx 0.342$

ii $\cos 70° \approx 0.342$, $\sin 70° \approx 0.940$

iii $\cos 35° \approx 0.819$, $\sin 35° \approx 0.574$

iv $\cos 55° \approx 0.574$, $\sin 55° \approx 0.819$

v $\cos 41° \approx 0.755$, $\sin 41° \approx 0.656$

vi $\cos 49° \approx 0.656$, $\sin 49° \approx 0.755$

b For any acute angle θ, $\cos\theta = \sin(90° - \theta)$

REVIEW SET 23B

1 ≈ 140 m **2** **a** $\frac{\sqrt{3}}{2}$ **b** $\frac{\sqrt{3}}{2}$ **c** $\frac{1}{2}$ **d** $\frac{1}{2}$

3 **a** $x \approx 64.3$ **b** $x \approx 3.2$ **c** $x \approx 11.2$

4 **a** $\theta \approx 38.7°$ **b** $\theta \approx 49.2°$ **c** $\theta \approx 73.3°$

5 **a** $\theta \approx 63.6°$, $\phi \approx 26.4°$, $x \approx 8.1$

b $\theta = 52°$, $x \approx 8.9$, $y \approx 5.5$

6 ≈ 29.5 cm **7** ≈ 77.9 m **8** ≈ 2.84 m **9** $\approx 72.1°$

10 **a** 045° **b** 225° **c** 105° **d** 285° **e** 135° **f** 315°

11 **a** ≈ 107 km **b** $\approx 306°$

12 **a** ≈ 3.16 km, $\approx 018°$ **b** ≈ 2.83 km, 135° **c** ≈ 2.24 km, $\approx 063°$ **d** ≈ 4.12 km, $\approx 194°$ **e** ≈ 3.16 km, $\approx 288°$

13 **a** **i** $B\widehat{A}E = 108°$, $C\widehat{B}D = 36°$, $A\widehat{B}D = 72°$

ii $B\widehat{A}E + A\widehat{B}D = 108° + 72° = 180°$

$\therefore$ [BD] $\parallel$ [AE] {supplementary co-interior angles}

$\therefore$ ABDE is a trapezium.

iii BX ≈ 9.51 cm **iv** ≈ 5.88 cm **v** ≈ 15.4 cm

b **i** 54° **ii** ≈ 8.51 cm **iii** ≈ 6.88 cm **iv** ≈ 15.4 cm

EXERCISE 24A

1 **a** $x = 100$ **b** $a = 35$, $b = 70$ **c** $x = 40$ **d** $x = 35$ **e** $a = 35$, $b = 30$ **f** $x = 107\frac{1}{2}$

2 **a** $a = 85$ **b** $a = 61$

3 **a** **i** $A\widehat{C}B = x°$ **ii** $C\widehat{A}E = x°$ **iii** $D\widehat{A}E = x°$

b The line segment [AE] *bisects* $C\widehat{A}D$.

4 **a** $\alpha + \beta = 180°$ **b** co-interior angles are supplementary

c the quadrilateral is a parallelogram.

5 **Hint:** $\triangle$ADC is isosceles.

6 **Hint:** Show that $B\widehat{D}E = B\widehat{C}A$.

7 **Hint:** Let $A\widehat{B}C = \alpha$. Find $A\widehat{E}F$ and $A\widehat{F}E$ in terms of α.

8 **Hint:** Use the exterior angle in a triangle theorem to place all marked angles within the same triangle.

EXERCISE 24B

1 **a** $x = 10$, $\theta = 60$ {midpoint theorem, corresponding angles}

b $x = 8$ {converse of midpoint theorem}

2 **a** 5.5 cm **b** $A\widehat{N}M = 36°$ **c** 11 cm^2

3 **a** $\triangle$AMN $\cong$ $\triangle$CDN {AAcorS}

c **Hint:** A pair of opposite sides are parallel and equal in length.

d **i** **Hint:** Opposite sides of a parallelogram are parallel.

ii **Hint:** Opposite sides of a parallelogram are equal in length.

4 **Hint:** Show that the triangles are equiangular and the ratio of their side lengths is 2 : 1.

5 **b** ADEF, DECF

6 **Hint:** The diagonals of a parallelogram bisect each other.

7 **Hint:** Use the midpoint theorem and the converse.

8 **Hint:** Show that [WX] $\parallel$ [ZY] and [XY] $\parallel$ [WZ].

EXERCISE 24C

1 **a** $O\widehat{B}A = x°$, $O\widehat{B}C = y°$

b $2x + 2y = 180$ {angles in a triangle}

$\therefore$ $x + y = 90$

and $A\widehat{B}C = (x + y)° = 90°$

2 **a** $x = 38$ **b** $x = 30$ **c** $x = 37$ **d** $x = 8$

3 $\widehat{APX} = \widehat{BQX} = 90°$ {angles in a semi-circle}
$\widehat{AXP} = \widehat{BXQ}$ {vertically opposite angles}

4 **a** $x = \sqrt{74}$ **b** $x = \sqrt{55}$ **5** ≈ 6.26 cm

6 **a** $x \approx 5.89$ **b** $x \approx 4.26$ **7** 6 cm

8 ≈ 6.93 cm **9** **Hint:** Show that $\widehat{PXR} = 90°$.

EXERCISE 24D

1 **a** **Hint:** Use RHS.

2 **a** $x \approx 5.83$ **b** $x \approx 3.61$ **c** $x \approx 9.80$

3 ≈ 6.3 cm **4** ≈ 5.0 cm **5** ≈ 9.2 cm

6 **a** By the converse of the chords of a circle theorem, the perpendicular bisectors of [AB] and [AC] each pass through the centre of the circle.
$\therefore$ their intersection is the centre of the circle.
$\therefore$ X is the centre of the circle.

b 24 cm^2

EXERCISE 24E

1 **a** $x = 90$ **b** $x = 49$ **c** $x = 35$ **d** $x \approx 8.7$
e $x \approx 6.4$ **f** $x \approx 5.9$

2 ≈ 7.48 cm **3** ≈ 9.75 cm **4** $\approx 78.5\%$

5 ≈ 10.05 m **6** ≈ 1.42 cm **7** $9\pi \approx 28.3$ m^2

EXERCISE 24F

1 **a** **Hint:** Use RHS.

2 **a** $x = 7$ **b** $x = 8,\ y = 6$ **c** $x = y = 50$
d $x = 68,\ y = 56$ **e** $x = 8,\ y = 30$ **f** $x = 25$

3 **Hint:** Show that $\triangle$ABP is isosceles.

4 1 cm **5** 10 cm **6** 56 cm

REVIEW SET 24A

1 **a** $x = 26$ **b** $x = 75$ **c** $x = 55$

2 **a** **i** $\widehat{DBC} = a°$ **ii** $\widehat{ADB} = (2a)°$ **iii** $\widehat{DBA} = (90 - a)°$

3 **Hint:** Show that co-interior angles are supplementary.

4 **Hint:** Use SSS.

5 **a** $x = 36$ **b** $x = 111$ **c** $x = 2.4$

6 **a** **Hint:** $\widehat{OAP} = \widehat{OBP} = 90°$ **7** ≈ 7.8 cm

8 48 cm^2 **9** 24 cm **10** **Hint:** $\widehat{OTP} = 90°$

11 **Hint:** Use the *tangents from an external point* theorem.

REVIEW SET 24B

1 **a** $x = 75$ **b** $x = 60$ **c** $x = 102\frac{1}{2}$
d $x = 66$ **e** $x = 40$

2 **Hint:** $\widehat{ABC} = \widehat{ACB}$

3 **Hint:** Draw a line from B to [DC], parallel to [AD].

4 **a** **Hint:** Use the midpoint theorem.
b **i** 25 cm^2 **ii** 75 cm^2

5 **a** $x = 8$ **b** $x \approx 4.47$ **c** $x = 8,\ y \approx 8.94$

6 ≈ 12.5 cm **7** ≈ 13.4 cm **8** **Hint:** Join [CX].

9 **Hint:** Show that [AP] and [OX] are parallel.

10 **a** **Hint:** Use the tangents from an external point theorem.
b **Hint:** Show that $\triangle$s ACM and BCM are isosceles.

EXERCISE 25A

1 **B**

2 **a** **i**

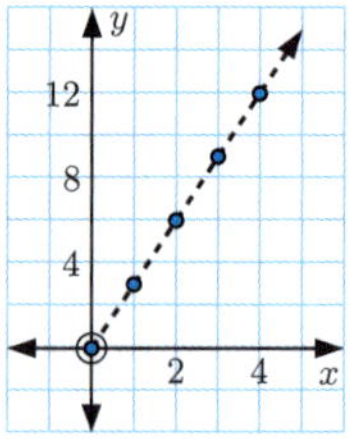

ii directly proportional, $k = 3$

b **i**

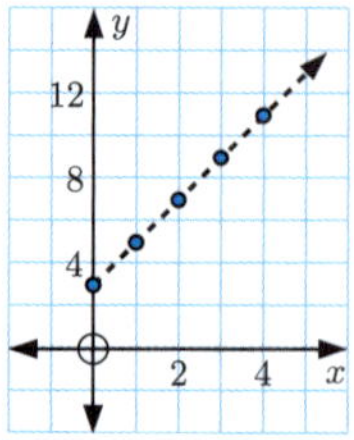

ii not directly proportional

c **i**

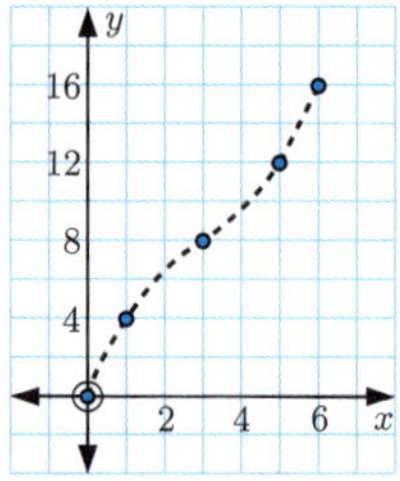

ii not directly proportional

d **i**

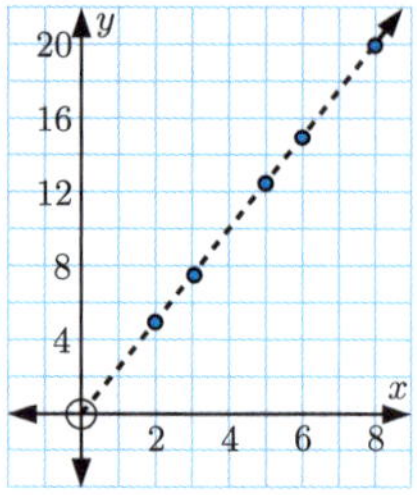

ii directly proportional, $k = 2.5$

3 **a**

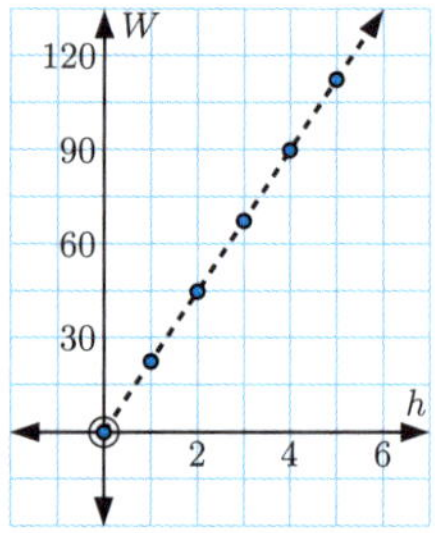

b The graph of W against h is a straight line through the origin.

c $W = 22.5h$

4 **a**

Time (t minutes)	0	2	4	6	8	10
Amount of water leaked (A mL)	0	20	40	60	80	100

b

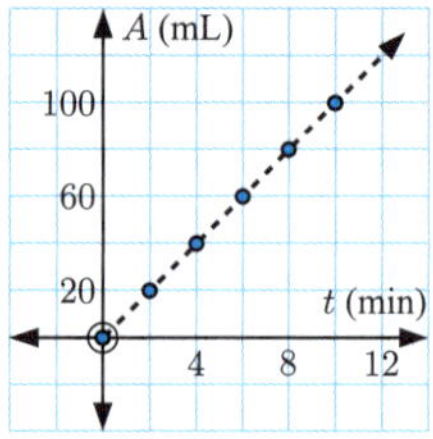

c yes, $k = 10$

5 **a**

Number of items (n)	0	1	2	3	4	5
Cost (\$$C$)	0	3.20	6.40	9.60	11.80	14

b

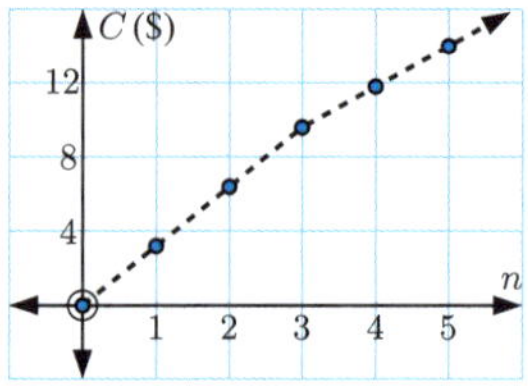

c no

6 **a** y is trebled **b** x is doubled
c y is halved **d** x is decreased by 30%
e cannot be determined **f** cannot be determined

7 **a** 2π is a constant so the graph of C against r is a straight line through the origin. The proportionality constant is $k = 2\pi$.

b **i** C is doubled **ii** r is increased by 50%

8 **a** $y = 140$ **b** $x = 1$ **9** **a** $y = 60$ **b** $x = 21$

10

x	5	15	30
y	20	60	120

11 **a** 4 litres **b** 121.5 km

12 **a** **i** $R = 0.006l$ **ii** 0.3 ohms **iii** 500 cm

b

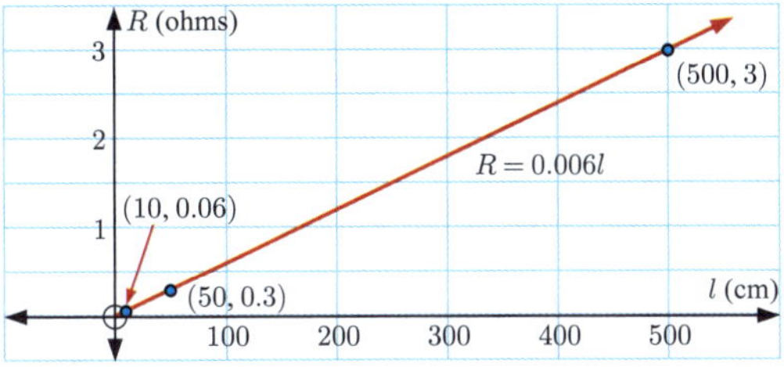

13 **a** 45 m^2 **b** 9 litres **14** **a** $4l$ lollies **b** $\frac{150}{l}$ days

15 **a** 78.4 m/s **b** ≈ 10.2 s **16** **a** 10.5 cm **b** 1.2 m

EXERCISE 25B

1 **a** $A \propto t^2$, $k = 4.9$ **b** $K \propto x^3$, $k = 2$
c $T \propto \sqrt{l}$, $k = \frac{1}{5}$ **d** $V \propto t^4$, $k = 500$
e $P \propto \sqrt[3]{x}$, $k = 4$ **f** $V \propto r^3$, $k = \frac{4}{3}\pi$

2 **a**

x^2	4	9	25
y	16	36	100

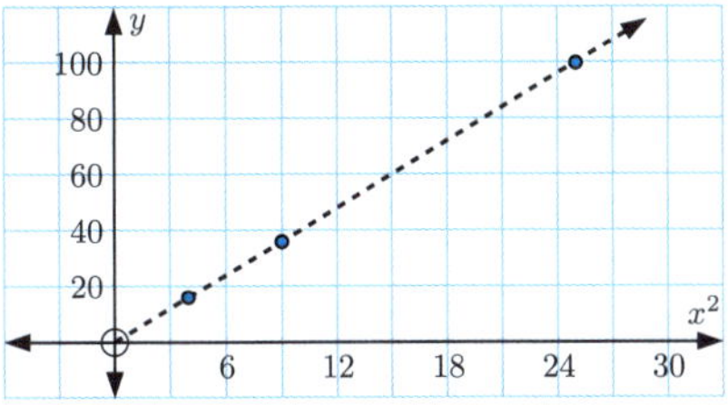

The graph is a straight line through the origin, so $y \propto x^2$.
The rule is $y = 4x^2$.

b

x^2	1	9	16	25
y	3	27	48	75

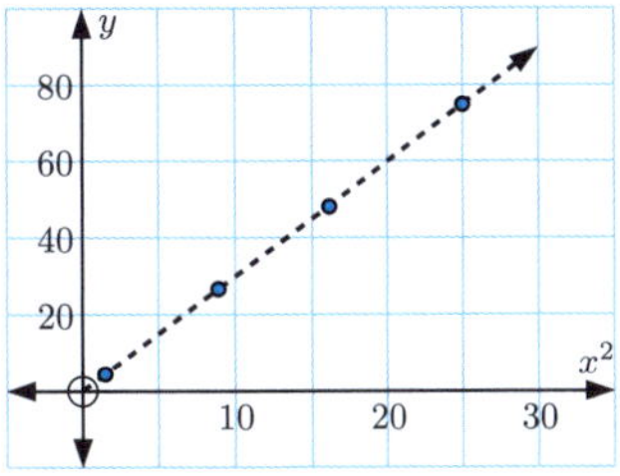

The graph is a straight line through the origin, so $y \propto x^2$. The rule is $y = 3x^2$.

3 **a**

x^3	1	8	27
y	5	40	135

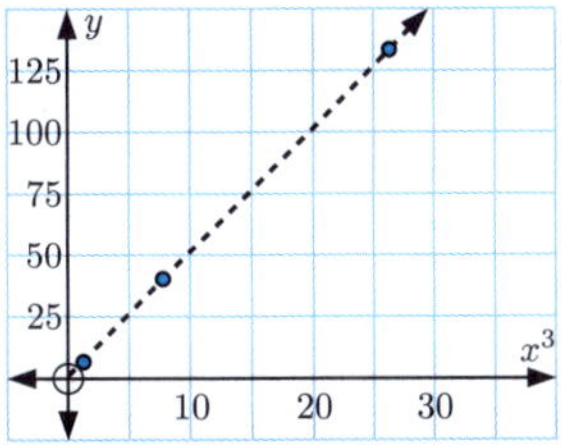

The graph is a straight line through the origin, so $y \propto x^3$ and $k = 5$.

b

x^3	8	27	125
y	4	13.5	62.5

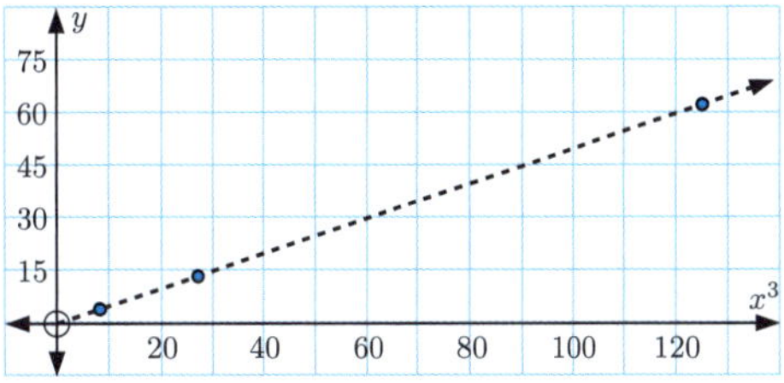

The graph is a straight line through the origin, so $y \propto x^3$ and $k = \frac{1}{2}$.

4 **a** $P = 48$ **b** $a = 15$
5 **a** $M = 81$ **b** $x = 2\sqrt[3]{5} \approx 3.42$
6 270 hp **7** \$312.50 **8** ≈ 8.9 cm
9 ≈ 10.1 seconds **10** **a** ≈ 12.2 km **b** 2.4 m
11 **a** The volume is multiplied by 1.21.
b The volume is reduced by 36%.
12 **a** $V \propto r^3$ **b** 27.1% reduction **c** $\approx 26.0\%$ increase
13 **a** $V \propto h$ **b** $V \propto r^2$
c The volume is increased by 22.4%.
14 **a** $r = \sqrt{\frac{A}{6\pi}}$ **b** $V = \frac{A^{\frac{3}{2}}}{3\sqrt{6\pi}}$ **c** $V \propto A^{\frac{3}{2}}$

EXERCISE 25C

1 **a, d**

x	1	2	4	8
$\frac{1}{x}$	1	$\frac{1}{2}$	$\frac{1}{4}$	$\frac{1}{8}$
y	40	20	10	5
xy	40	40	40	40

b

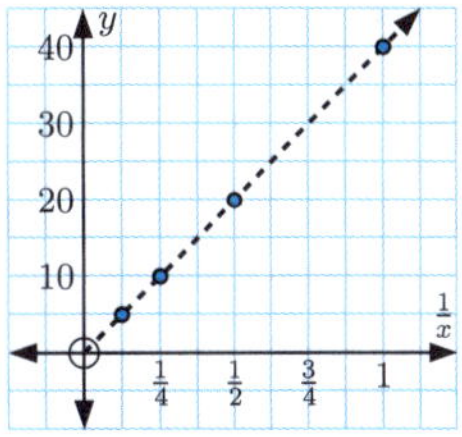

c The points of y against $\frac{1}{x}$ form a straight line through the origin,
$\therefore$ y is directly proportional to $\frac{1}{x}$
$\therefore$ y is inversely proportional to x.

2 **a**

x	1	2	3	6
y	12	6	4	2
xy	12	12	12	12

x and y are inversely proportional, and $xy = 12$.

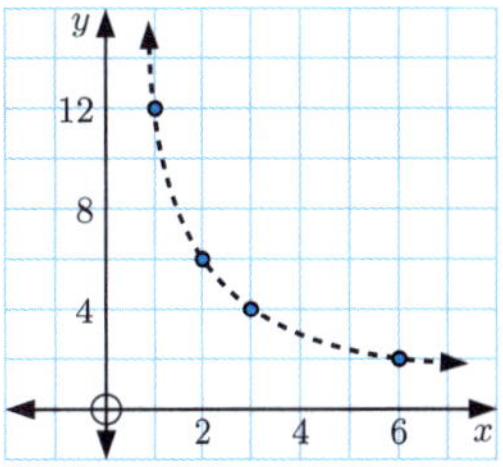

b

x	2	3	6	9
y	12	8	4	3
xy	24	24	24	27

x and y are not inversely proportional.

c

x	7	6	12	1
y	12	14	7	84
xy	84	84	84	84

x and y are inversely proportional, and $xy = 84$.

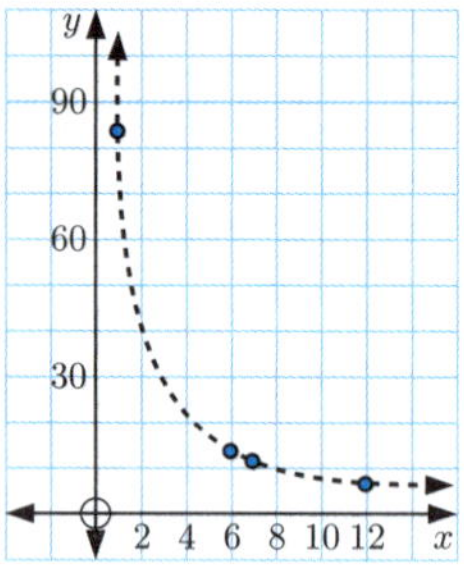

3 **B** and **C**

4 **a** If d is fixed, s is **inversely** proportional to t.
b If t is fixed, s is **directly** proportional to d.
c If s is fixed, d is **directly** proportional to t.

5 **a** y is multiplied by $\frac{1}{4}$ **b** y is multiplied by $\frac{9}{7}$
c x is multiplied by $\frac{3}{8}$

6 **a** $y = 8$ **b** $x = \frac{2}{5}$ **7** **a** $p = 18$ **b** $n = 12$

8

x	6	10	50
y	25	15	3

9 8 days **10** **a** 3750 chairs **b** 16 hours

11 **a** The area of a rectangle is fixed.
$\therefore$ $xy = 60$, so x and y are inversely proportional.

b

x	5	6	15
y	12	10	4

c

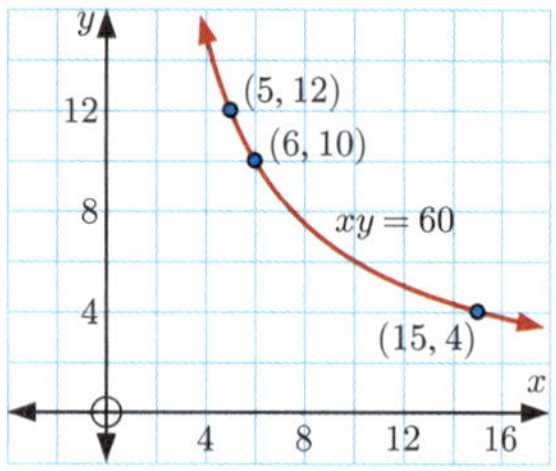

12 **a** 20 units **b** 4 cm^3 **c** $P = \dfrac{400}{V}$

d

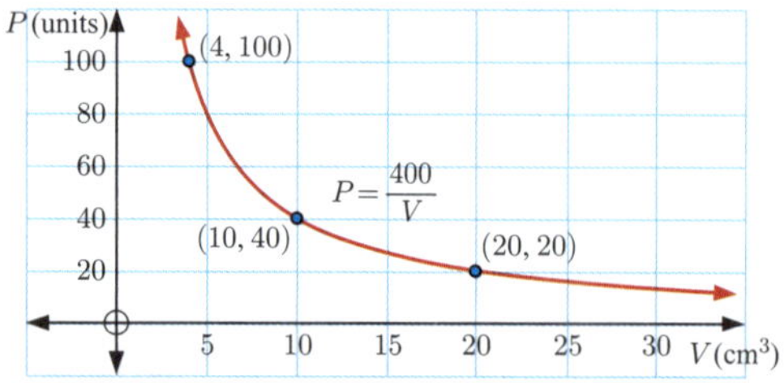

13 **a** 23 min 20 s **b** 8 players **c** 40 min

EXERCISE 25D

1 **a** $y = 16$ **b** $x = 4$

2 **a** $F = \frac{20}{\sqrt{3}} \approx 11.5$ **b** $c = \frac{44}{9} \approx 4.89$

3 **a** ≈ 0.115 seconds **b** 400 units **4** 56.25% increase

5 **a** 3.275 cm **b** ≈ 3.46 cm **c** ≈ 300 cm^3

6 ≈ 8.94 m away

REVIEW SET 25A

1 **B**

2 **a**

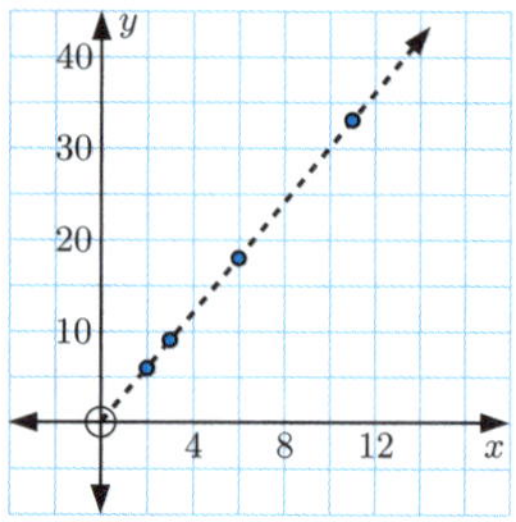

b The graph of y against x is a straight line through the origin.
c $y = 3x$

3

x	12	60	80
y	3	15	20

4 **a** $y \approx 54.2$ **b** $x \approx 4.33$ **5** 72 people

6 If y is increased by 25%, then y is multiplied by 1.25.

$\therefore$ $\dfrac{1}{x}$ is multiplied by 1.25.

$\therefore$ x is multiplied by $\frac{1}{1.25} = \frac{4}{5}$

So, x is decreased by 20%.

7 36 days **8** **a** $M = 60$ **b** $t = 9$

9 **a** The volume is multiplied by 4.
b The radius needs to increase by $\approx 26.5\%$.

10 **a** **i** $E \propto m$ if v is constant.
ii $E \propto v^2$ if m is constant.
iii $m \propto \dfrac{1}{v^2}$ if E is constant.

b If m is constant, $E \propto v^2$.
$\therefore$ E is reduced by $\approx 22.6\%$.

11 **a**

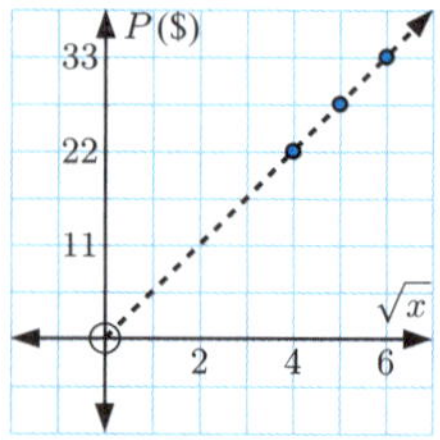

b $P = 5.5\sqrt{x}$
c $\approx \$30.12$
d $A \propto P^4$

REVIEW SET 25B

1 $a = 14$, $b = 21$ **2** ≈ 229 km

3 **a** $k = 19$ **b** $y = 119.7$

4 ≈ 63.0 inches **5** k is multiplied by 8

6 $xy = 36$ for all points
$\therefore$ x and y are inversely proportional, and $y = \dfrac{36}{x}$.

7 27 minutes **8** **a** $y = 98$ **b** $x = 7\sqrt{\frac{2}{3}} \approx 5.72$

9 **a** ≈ 30.6 m/s **b** 324 m

10 **a** $A \propto r^2$ **b** The radius is increased by $\approx 9.54\%$.

11 **a** **i** The line $T_k = T_c + 273.15$ does not pass through the origin.

$\therefore$ T_c is not directly proportional to T_k.

$\therefore$ T_c is not inversely proportional to ρ. $\{$as $\rho \propto \dfrac{1}{T_k}\}$

ii ≈ 1.12 kg/m^3

b $333.15°$C

EXERCISE 26A

1 **a** {R, B, Y, G} **b** {BB, BG, GB, GG}

c {ABC, ACB, BAC, BCA, CAB, CBA}

d {HHHH, HHHT, HHTH, HTHH, THHH, HHTT, HTHT, THHT, HTTH, THTH, TTHH, HTTT, THTT, TTHT, TTTH, TTTT}

2 **a** $U =$ {A, B, C, D, E, F, G, H, I, J, K, L, M, N, O, P, Q, R, S, T, U, V, W, X, Y, Z}

b **i** $A =$ {B, C, D, F, G, H, J, K, L, M, N, P, Q, R, S, T, V, W, X, Y, Z}

ii $B =$ {E, N, T, V}

3 **a**

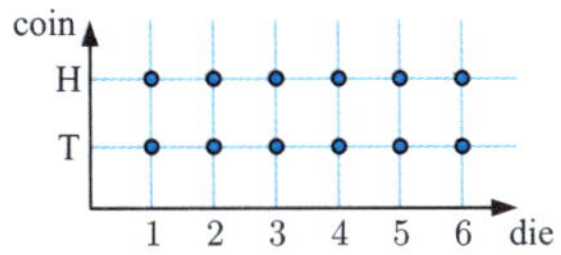

b

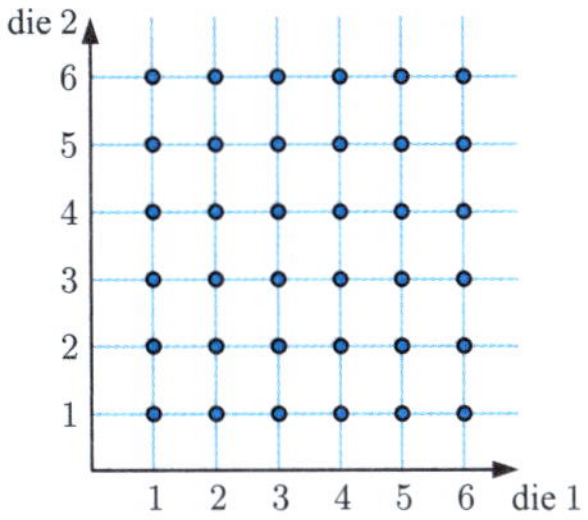

c

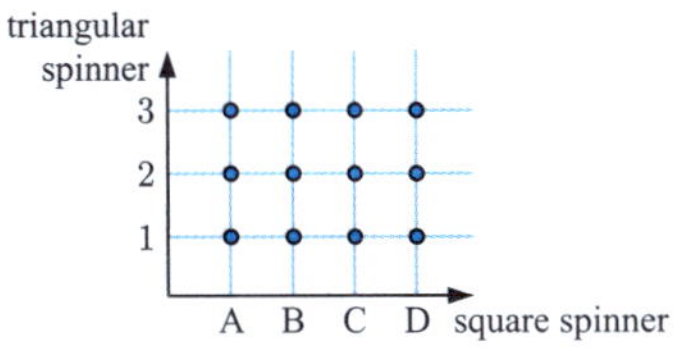

4

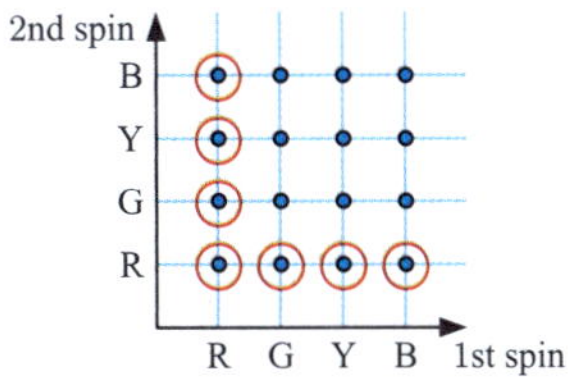

5

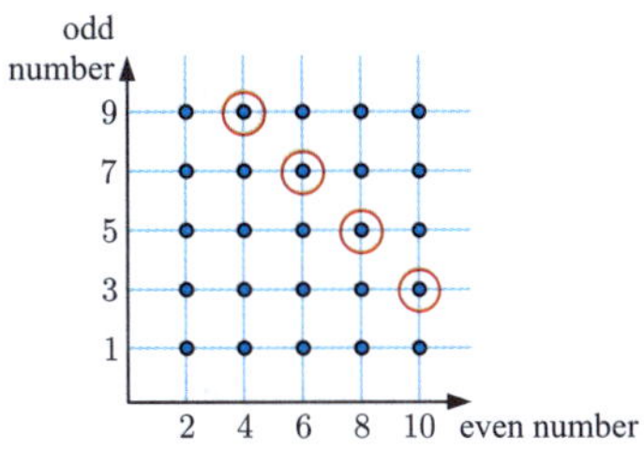

6 **a**

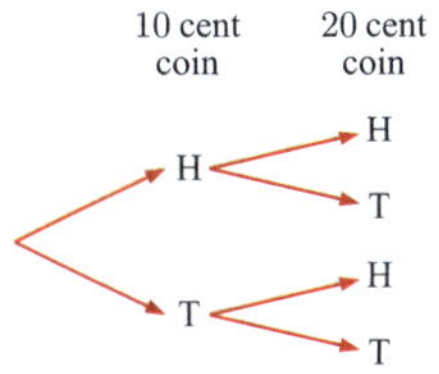

b

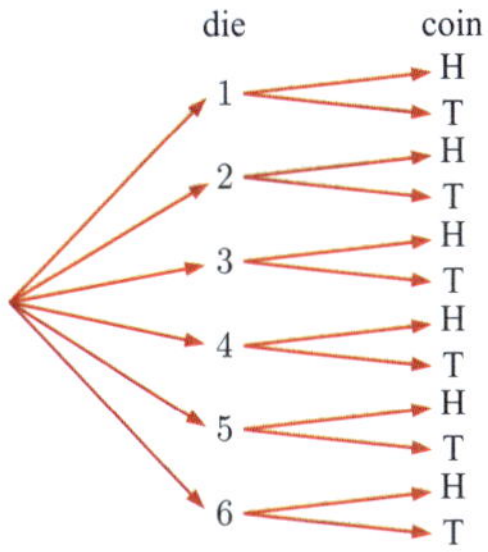

7

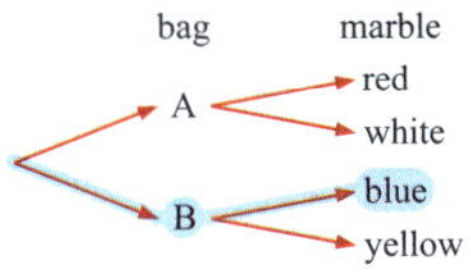

8

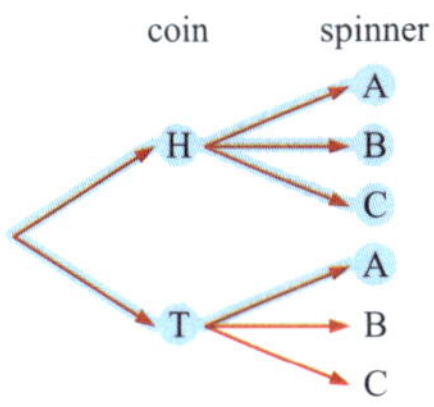

9

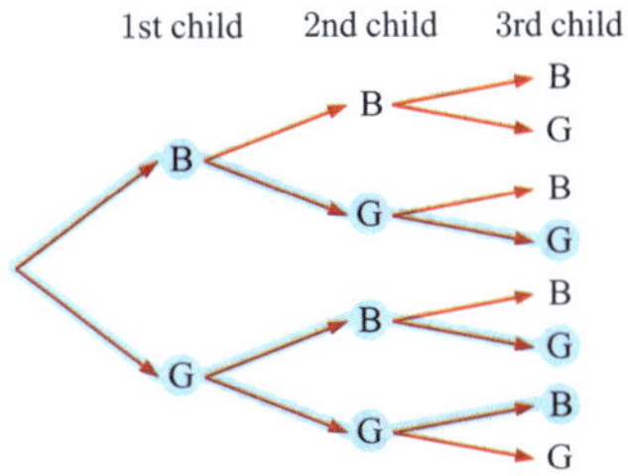

10

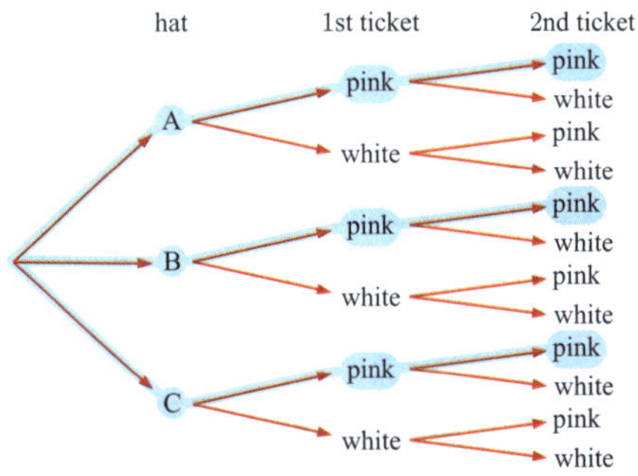

11

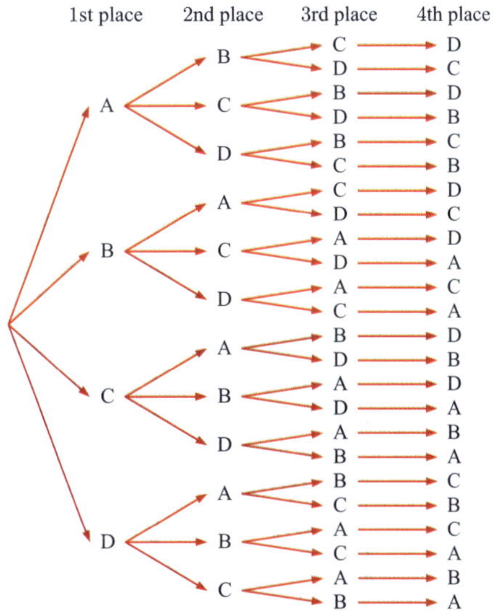

EXERCISE 26B

1 **a** $\frac{1}{5}$ **b** $\frac{2}{5}$ **c** $\frac{3}{5}$

2 **a** E' is the event that the muffin is not blueberry
$\therefore$ E' is the event that the muffin is raspberry.
b $P(E) = \frac{1}{3}$, $P(E') = \frac{2}{3}$

3 **a** $\frac{1}{8}$ **b** $\frac{7}{8}$ **c** $\frac{1}{2}$ **d** 0

4 **a** $\frac{4}{7}$ **b** $\frac{3}{7}$ **c** 1 **5** **a** $\frac{3}{14}$ **b** $\frac{4}{7}$

6 **a** $\frac{1}{13}$ **b** $\frac{1}{2}$ **c** $\frac{1}{4}$ **d** $\frac{4}{13}$

7 For a 4-year cycle: **a** $\frac{124}{1461}$ **b** $\frac{113}{1461}$

8 **a** **i** $\frac{3}{37}$ **ii** $\frac{8}{37}$ **iii** $\frac{26}{37}$
b 1. This is the sum of all possible outcomes.

9 $\frac{1}{9}$

10 **a**
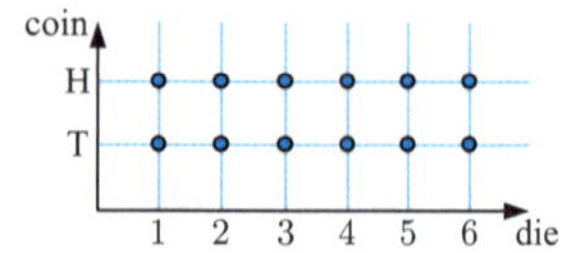

b **i** $\frac{1}{12}$ **ii** $\frac{7}{12}$ **iii** $\frac{2}{3}$ **iv** $\frac{5}{12}$ **v** $\frac{1}{4}$ **vi** $\frac{3}{4}$

11 **a**
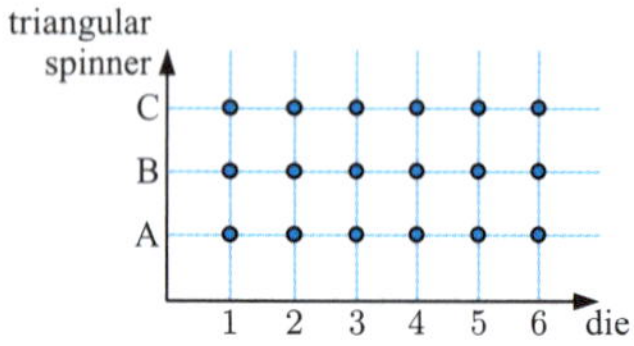

b **i** $\frac{1}{18}$ **ii** $\frac{1}{6}$ **iii** $\frac{2}{9}$

12 **a** $\frac{1}{36}$ **b** $\frac{1}{18}$ **c** $\frac{5}{9}$ **d** $\frac{11}{36}$ **e** $\frac{5}{18}$ **f** $\frac{25}{36}$
g $\frac{1}{6}$ **h** $\frac{2}{9}$ **i** $\frac{5}{18}$ **j** $\frac{13}{18}$

13 **a** $\frac{2}{49}$ **b** $\frac{24}{49}$ **c** $\frac{11}{49}$ **d** $\frac{3}{7}$ **14** $\frac{21}{64}$

EXERCISE 26C

1 **a**
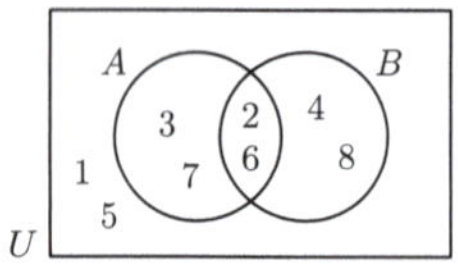

b **i** $\frac{1}{4}$ **ii** $\frac{3}{4}$

2 **a**
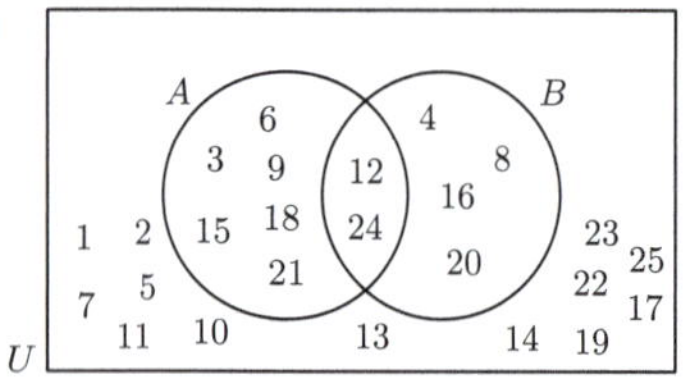

b **i** $\frac{8}{25}$ **ii** $\frac{2}{25}$ **iii** $\frac{4}{25}$ **iv** $\frac{2}{5}$

3 **a** $\frac{3}{26}$ **b** $\frac{1}{13}$ **c** $\frac{11}{26}$ **d** $\frac{15}{26}$

4 **a** 46 students **b** **i** $\frac{3}{46}$ **ii** $\frac{10}{23}$ **iii** $\frac{21}{46}$

5 **a**
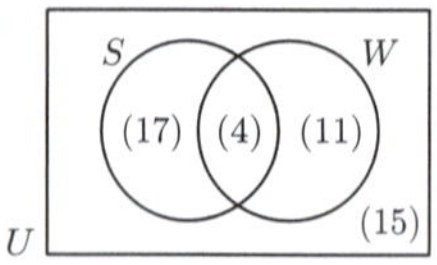

b **i** $\frac{15}{47}$ **ii** $\frac{28}{47}$

6 **a**
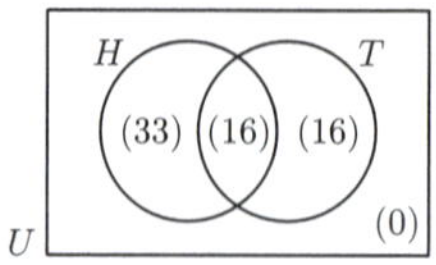

b **i** $\frac{16}{65}$ **ii** $\frac{33}{65}$

7 **a** $\frac{19}{40}$ **b** $\frac{1}{2}$ **c** $\frac{4}{5}$ **d** $\frac{5}{8}$

8 **a** $\frac{7}{15}$ **b** $\frac{1}{15}$ **c** $\frac{2}{15}$

9 **a**
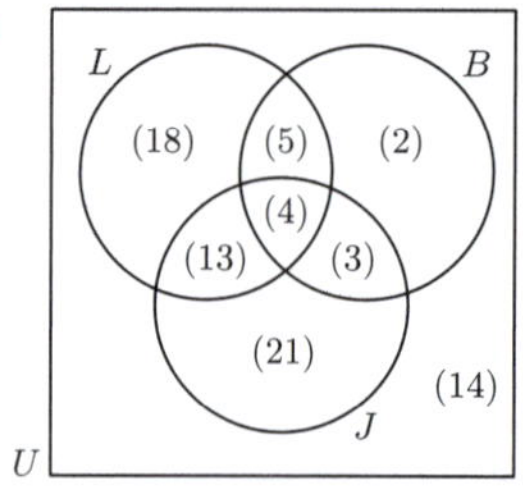

b **i** $\frac{21}{80}$ **ii** $\frac{47}{80}$ **iii** $\frac{1}{16}$

EXERCISE 26D

1 $\frac{1}{10}$ **2** **a** $\frac{1}{12}$ **b** $\frac{1}{4}$ **c** $\frac{1}{3}$ **3** $\frac{1}{8}$

4 **a** 0.18 **b** 0.28 **5** **a** 0.56 **b** 0.12 **c** 0.056

6 **a** 0.343 **b** 0.147 **c** 0.063

EXERCISE 26E

1 **a** $\frac{10}{21}$ **b** $\frac{5}{21}$

2 **a** **i** $\frac{2}{15}$ **ii** $\frac{1}{3}$ **iii** $\frac{4}{15}$ **iv** $\frac{4}{15}$
b 1. This is the sum of all possible outcomes.

3 **a** **i** $\frac{1}{11}$ **ii** $\frac{14}{33}$ **b** $\frac{16}{33}$

4 **a** $\frac{1}{3725}$ **b** $\frac{147}{7450}$ **c** $\frac{3577}{3725}$

5 **a** 0.56 **b** 0.12 **6** **a** $\frac{1}{22}$ **b** $\frac{7}{44}$

7 **a** $\frac{5}{28}$ **b** $\frac{1}{56}$ **c** $\frac{5}{56}$ **8** **a** $\frac{8}{65}$ **b** $\frac{1}{13}$ **c** $\frac{28}{195}$

EXERCISE 26F

1 **a**
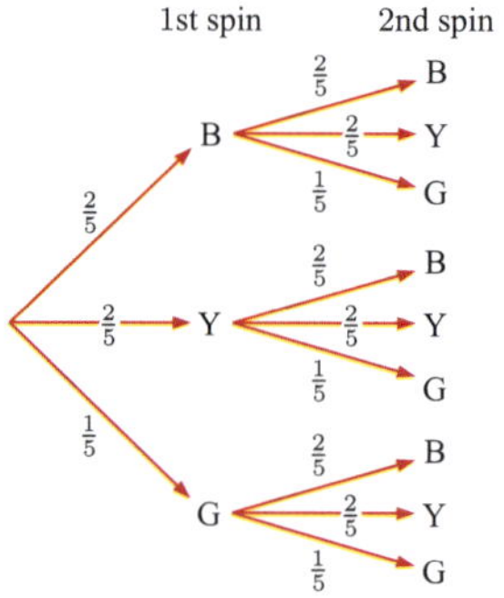

b **i** $\frac{4}{25}$ **ii** $\frac{1}{25}$ **iii** $\frac{16}{25}$ **iv** $\frac{16}{25}$

2 **a**
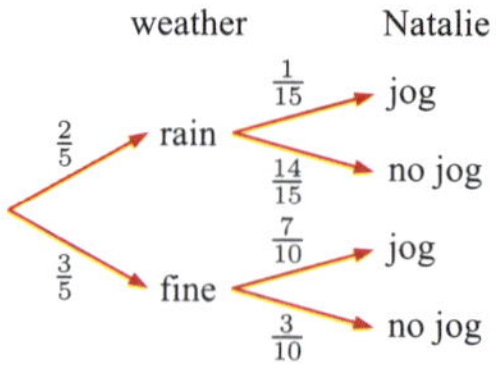

b $\frac{67}{150}$

3 0.752 **4** 0.034 **5** $\frac{23}{60}$

6 **a** $\frac{1}{25}$ **b** $\frac{24}{25}$ **c** $\frac{7}{4975}$ **d** $\frac{4584}{4975}$

7 **a**
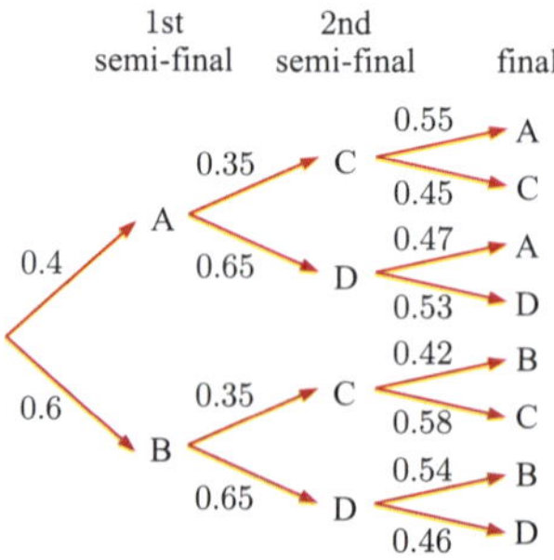

b 0.1992
c player D
d player B

8 **a**

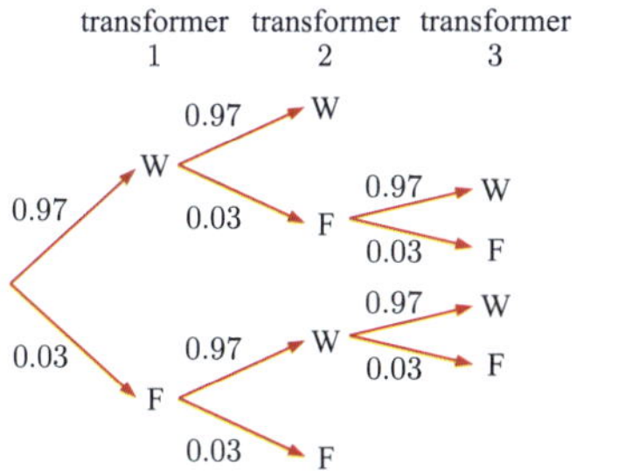

b 0.002 646

EXERCISE 26G

1 **a** 0.47 **b** 0.53 **2** **a** ≈ 0.84 **b** ≈ 0.16

3 ≈ 0.256 **4** ≈ 0.318

5 **a** **i** ≈ 0.235 **ii** ≈ 0.211 **iii** ≈ 0.225

b We expect **a iii** to be the most accurate because it uses the largest sample size.

6 **a** **i** ≈ 0.405 **ii** ≈ 0.595

b **i** ≈ 0.164 **ii** ≈ 0.354

EXERCISE 26H

1 **a** ≈ 0.16 **b** ≈ 0.12 **c** ≈ 0.6

2 **a** 43 days **b** 127 calls

c **i** ≈ 0.0465 **ii** ≈ 0.186 **iii** ≈ 0.465

3 **a** ≈ 0.487 **b** ≈ 0.0513 **c** ≈ 0.731

4 **a** ≈ 0.240 **b** ≈ 0.702 **c** ≈ 0.678

5 **a**

	Like	Dislike	*Total*
Junior students	87	38	125
Senior students	129	56	185
Total	216	94	310

b **i** ≈ 0.281 **ii** ≈ 0.416 **iii** ≈ 0.303

6 **a** 100 students

b **i** ≈ 0.29 **ii** ≈ 0.08 **iii** ≈ 0.26 **iv** ≈ 0.68

7 **a** ≈ 0.469 **b** ≈ 0.177 **c** ≈ 0.293

8 **a**

		Bacon		
		Like	Dislike	*Total*
Eggs	Like	26	15	41
	Dislike	6	3	9
	Total	32	18	50

b **i** ≈ 0.64 **ii** ≈ 0.3 **iii** ≈ 0.06

9 **a**

	Lunch	Dinner	*Total*
Saturday	35	67	102
Sunday	33	63	96
Total	68	130	198

b **i** ≈ 0.485 **ii** ≈ 0.338 **iii** ≈ 0.167

EXERCISE 26I

1 65 days **2** 13 of them **3** 106 times **4** 200 times

5 **a**

Size	*Frequency*	*Relative frequency*
Small	34	0.425
Medium	28	0.35
Large	18	0.225

b 162 coffees

6 150 times **7** **a** 20 people **b** 60 people

8 ≈ 14 times

9 **a** 16 games **b** ≈ 7 games **c** ≈ 16 games

d ≈ 18 games

REVIEW SET 26A

1

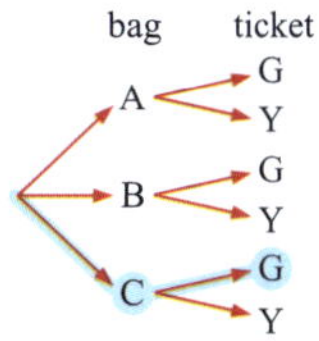

2 **a** $\frac{1}{9}$ **b** $\frac{4}{9}$ **c** $\frac{4}{9}$

3 **a**

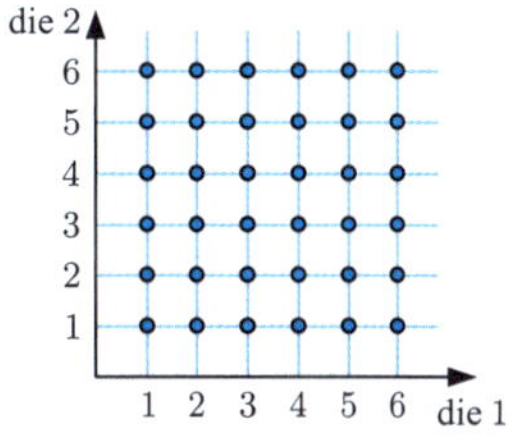

b **i** $\frac{1}{36}$ **ii** $\frac{11}{36}$ **iii** $\frac{1}{6}$ **iv** $\frac{1}{4}$

4 **a**

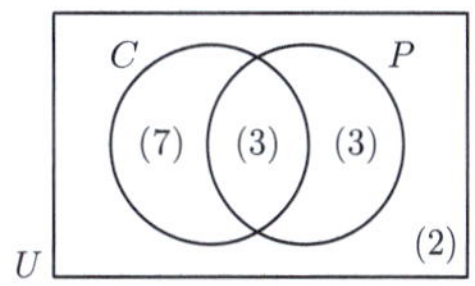

b $\frac{2}{15}$

5 **a** 0.48 **b** 0.08 **6** **a** 0.14 **b** 0.27

7 **a**

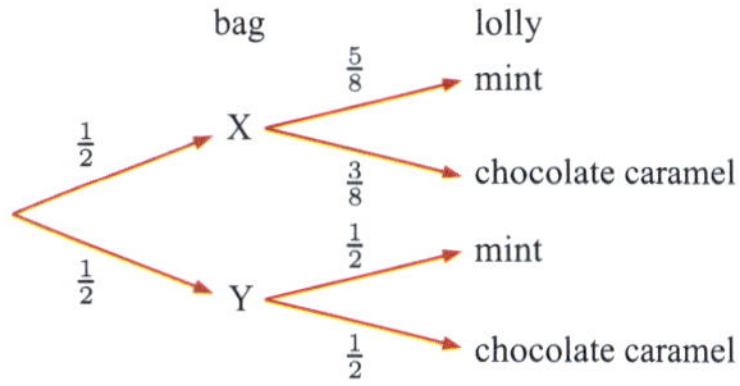

b **i** $\frac{7}{16}$ **ii** $\frac{9}{16}$

8 **a** ≈ 0.422

b The accuracy of the estimate can be improved by using a larger sample size.

9 **a** ≈ 0.462 **b** 28 matches

10 **a** **i** ≈ 0.0571 **ii** ≈ 0.571 **iii** ≈ 0.943

b ≈ 344 days

11 **a** 0.444 **b** 0.556

12 **a**

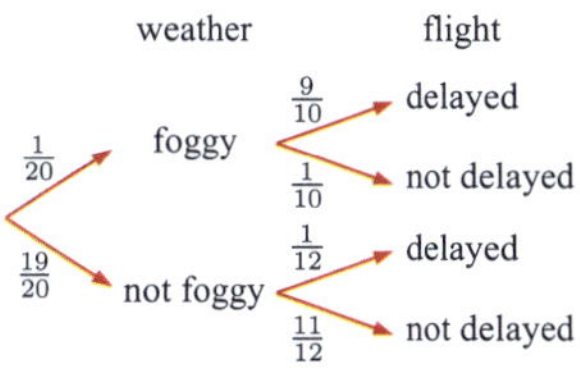

b **i** $\frac{9}{200}$ **ii** $\frac{149}{1200}$ **iii** $\frac{1051}{1200}$

c A flight is either delayed or not delayed, so these cases cover all possible outcomes and the probabilities sum to 1.

REVIEW SET 26B

1

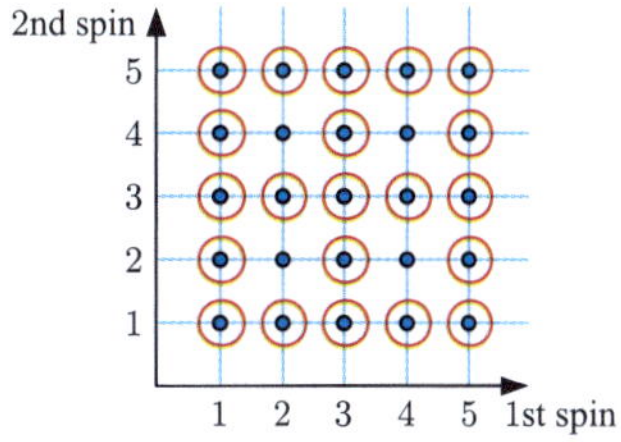

2 **a** $\frac{1}{13}$ **b** $\frac{7}{26}$ **c** $\frac{7}{26}$

3 $\frac{5}{18}$ **4** $\frac{5}{33}$ **5** **a** $\frac{1}{9}$ **b** $\frac{2}{9}$

6 **a**

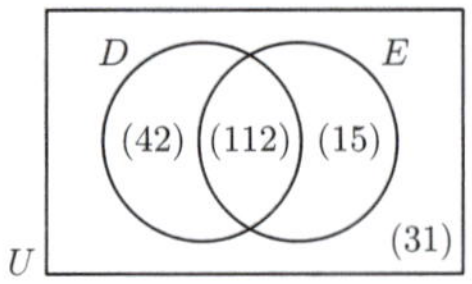

b $\frac{14}{25}$

7 **a**

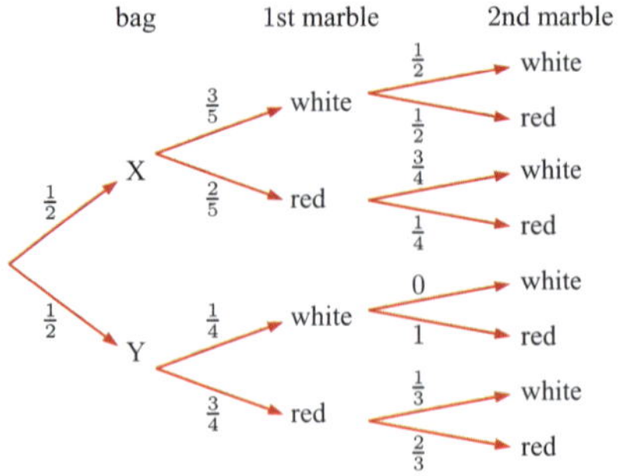

b $\frac{11}{20}$

8 0.697

9 **a** 136 teenagers

b

Activity	Freq.	Rel. freq.
Playing sport	32	≈ 0.235
Social media	51	0.375
Playing games	36	≈ 0.265
Watching movies	17	0.125
Total	136	1

c **i** ≈ 0.125 **ii** ≈ 0.5

10 **a**

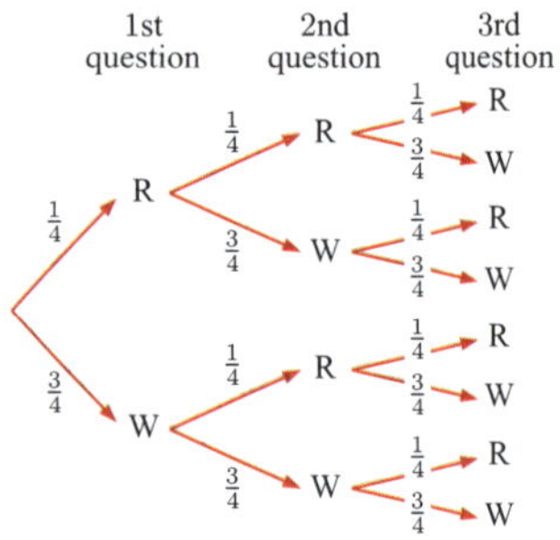

b **i** $\frac{9}{64}$ **ii** $\frac{27}{64}$

11 **a** ≈ 0.707 **b** ≈ 0.293 **c** ≈ 0.0467

12 **a** ≈ 0.737 **b** 63 books **13** 375 times

EXERCISE 27A

1 **a** categorical **b** numerical **c** categorical **d** numerical
e numerical **f** categorical **g** categorical **h** numerical

2 **a** black, blond, brown, grey, red
b piano, guitar, trumpet, violin, clarinet
c English, Mandarin, Arabic, Italian, Vietnamese
d full-time, part-time, unemployed.

3 **a** discrete **b** continuous **c** discrete
d continuous **e** discrete **f** continuous
g discrete **h** continuous **i** continuous **j** discrete

4 **a** discrete; 28, 29, 30, 31 **b** continuous; 2 cm - 8 cm
c discrete; 0, 1, 2, 3, 4, 5, 6, 7, 8, 9, 10
d continuous; 9 seconds - 25 seconds
e discrete; 1, 2, 3, 4, ..., 50
f continuous; 2000 kWh - 15 000 kWh

EXERCISE 27B.1

1 **a** 20 shops **b** 4 shops **c** **i** 10% **ii** 15%
d 6 customers, it is the most common number of customers in a shop.

2 **a** 26 students **b** 8 years, it is the most common age. **c** $\frac{15}{26}$

3 **a**

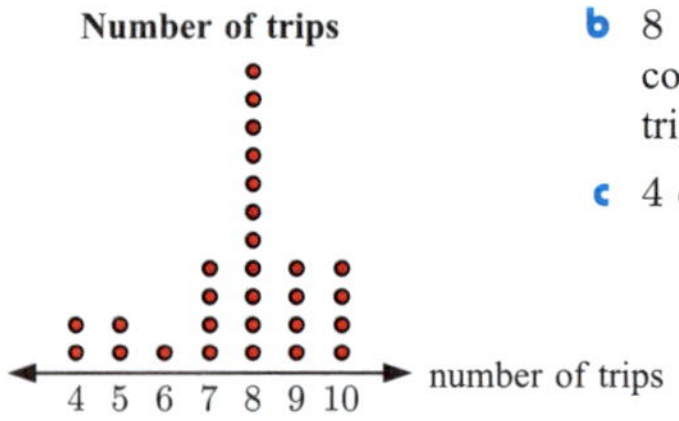

b 8 trips, it is the most common number of trips made in one day.
c 4 days

4 **a**

Number of goals	Tally	Frequency
0	𝍸	5
1	𝍸 |||	8
2	𝍸 ||	7
3	𝍸	5
4	|||	3
5	|	1
6		0
7	|	1
	Total	30

b

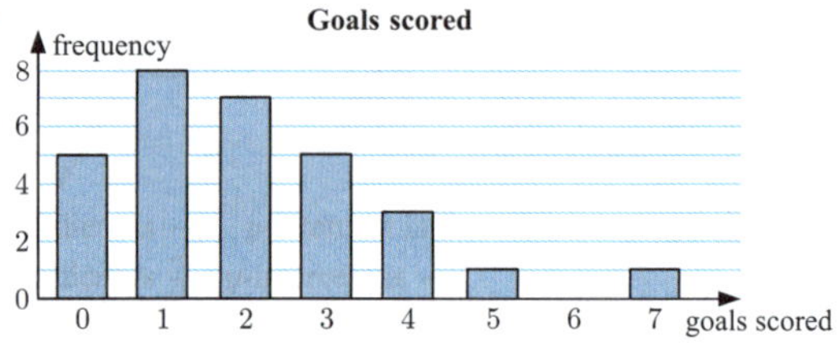

c 1 goal, it is the most common number of goals scored in a match.
d 7 goals **e** $\frac{1}{3}$

EXERCISE 27B.2

1 **a**

Chairs completed	Tally	Frequency
10 to 19	||||	4
20 to 29	𝍸 𝍸	10
30 to 39	𝍸 ||||	9
40 to 49	|||	3
	Total	26

b

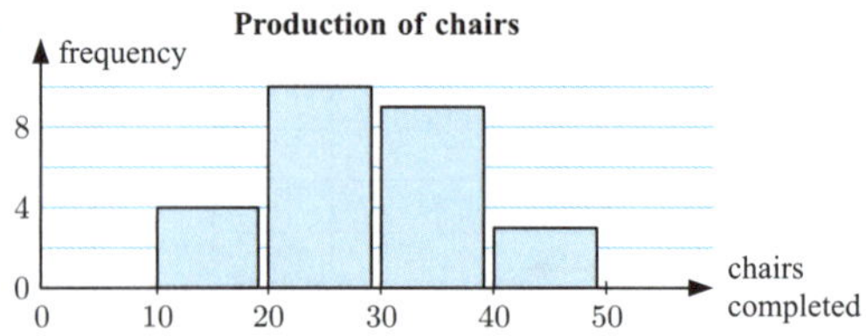

c 12 days **d** 20 to 29 chairs

2 **a**

Test score	Tally	Frequency
20 to 29	|	1
30 to 39	||	2
40 to 49	|||	3
50 to 59	𝍸 ||||	9
60 to 69	𝍸 𝍸 |||	13
70 to 79	𝍸 |||	8
80 to 89	𝍸 𝍸	10
90 to 99	||||	4
	Total	50

b Science test scores

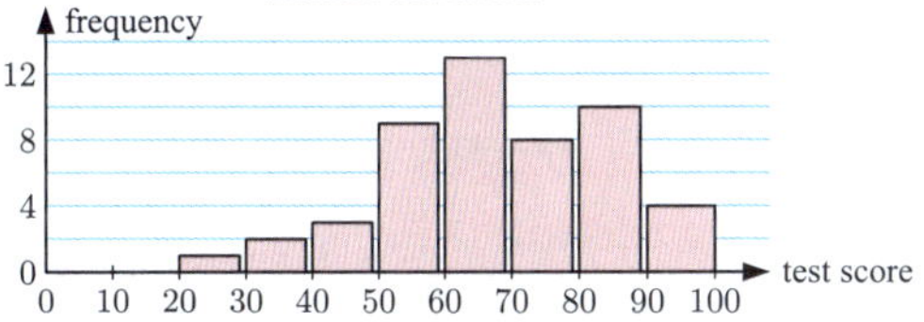

c 28% **d** 12%

e More students had a test score in the interval 60 to 69 than in any other interval.

3 **a**

Visitors	*Tally*	*Frequency*
100 to 199	ǀǀǀ	3
200 to 299	ǀǀǀǀ	4
300 to 399	卌 ǀ	6
400 to 499	卌 ǀǀǀ	8
500 to 599	卌 卌 ǀǀ	12
600 to 699	卌 ǀǀǀǀ	9
	Total	42

b Museum visitors

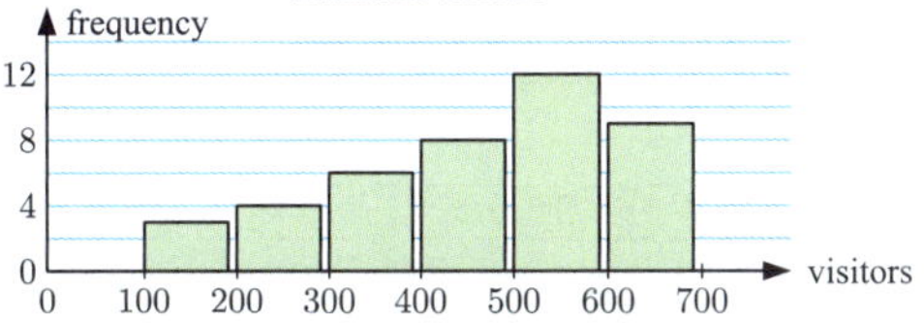

c 21 days **d** 500 to 599 visitors

EXERCISE 27B.3

1 **a** 17 games **b** **i** 76 points **ii** 137 points
c 10 times **d** $\approx 11.8\%$

2 **a**

Stem	Leaf
0	7
1	8
2	7 9
3	2 2 5 6 8 9 9
4	0 3 4 4 5 6 6 7 8

Scale: 1 | 8 means 18 marks

b 9 students **c** 20% **d** 90%

3 **a**

Stem	Leaf
7	6 9
8	0 2 4 4 6 7 8
9	1 2 2 3 3 5 7 9
10	0 2 8

Scale: 7 | 6 means 76 points

b $\frac{11}{20}$ **c** 30%

EXERCISE 27C

1 **a** $90 \leqslant v < 100$ km/h **b** 700 vehicles
c $\approx 25.7\%$ **d** $\approx 15.7\%$ **e** \$22 680

2 **a** *Height* can take any value within each class interval.

b Height of volleyball players

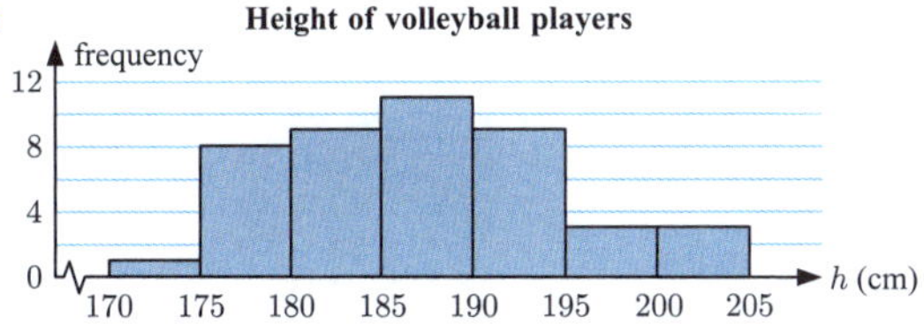

c $185 \leqslant h < 190$ cm. This class interval has the highest frequency.

d $\approx 20.5\%$

3 **a**

Mass (m kg)	*Frequency*
$8 \leqslant m < 10$	3
$10 \leqslant m < 12$	2
$12 \leqslant m < 14$	6
$14 \leqslant m < 16$	7
$16 \leqslant m < 18$	6
$18 \leqslant m < 20$	5
$20 \leqslant m < 22$	1
Total	30

b Mass of lambs

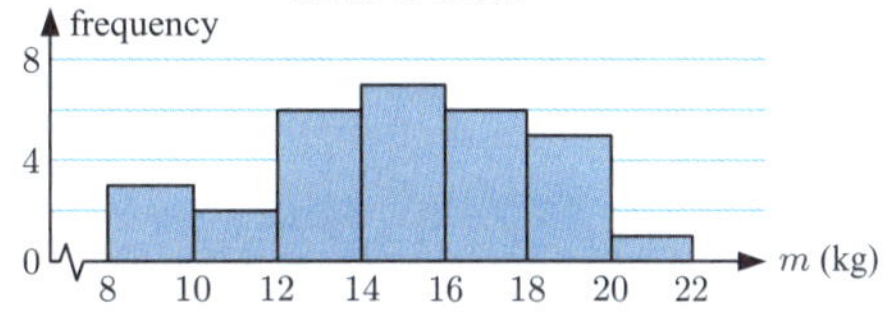

c $14 \leqslant m < 16$ kg **d** **i** $\approx 16.7\%$ **ii** 40%

4 **a**

Water usage (w kL)	*Frequency*
$7 \leqslant w < 8$	5
$8 \leqslant w < 9$	9
$9 \leqslant w < 10$	10
$10 \leqslant w < 11$	7
$11 \leqslant w < 12$	4
$12 \leqslant w < 13$	2
$13 \leqslant w < 14$	1
Total	38

b Monthly water usage

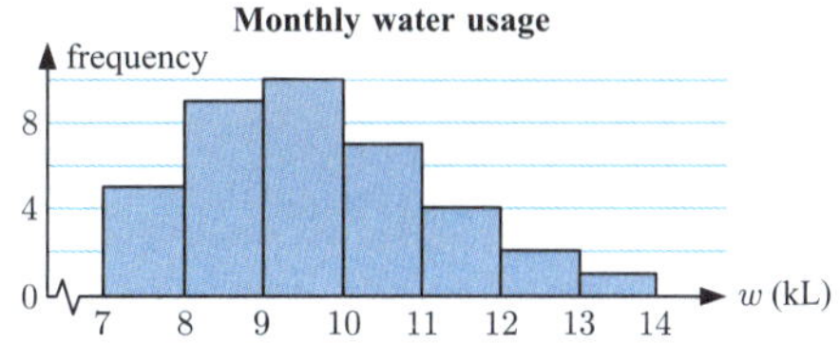

c $9 \leqslant w < 10$ kL **d** $\approx 36.8\%$

EXERCISE 27D

1 **a** bimodal **b** symmetric
c positively skewed, "10" is an outlier
d negatively skewed, "1" is an outlier **e** symmetric
f negatively skewed

2 **a** symmetric **b** positively skewed, "14.2" is an outlier
c bimodal

3 **a** Sweets per box

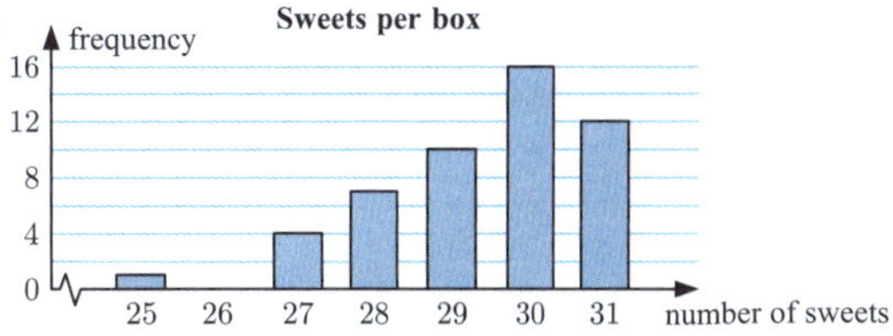

b 30 sweets **c** negatively skewed, "25" is an outlier

4 **a**

Test score	*Tally*	*Frequency*
20 to 24	ǀ	1
25 to 29	ǀǀǀ	3
30 to 34	卌	5
35 to 39	卌 ǀ	6
40 to 44	卌 卌 卌	15
45 to 49	卌	5
50 to 54	卌 ǀ	6
55 to 59	ǀǀǀǀ	4
	Total	45

b 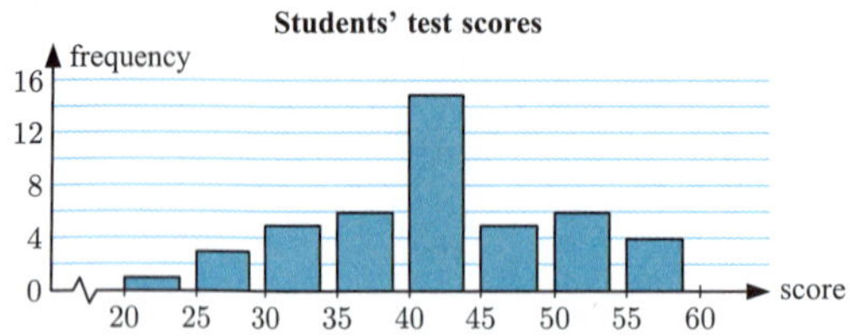

c negatively skewed **d** $\approx 22.2\%$

5 **a**

Stem	Leaf
17*	8
18	1 2 3 4
18*	5 5 5 6 6 7 8 8 8 9 9
19	0 1 2 2 3 3 4 4 4 4
19*	5 5 6 6 7 8 8 9
20	0 1 3 3 4 4
20*	5 6 7 7 8 9
21	0 0 3 3 3
21*	5 6 7
22	1 4 4
22*	7 8
23	
23*	
24	1

Scale: 18 | 1 means 181 peanuts
18* | 5 means 185 peanuts

b $\frac{13}{30}$ **c** positively skewed, "241" is an outlier

d no, only $\approx 88.3\%$ are within 20 peanuts of 200.

6 **a** girls: $15 \leqslant t < 15.5$ s, boys: $15 \leqslant t < 15.5$ s

b girls: approximately symmetric,
boys: negatively skewed, "$11.5 \leqslant t < 12$" is an outlier

c boys **d** boys

7 **a** 69 students

b 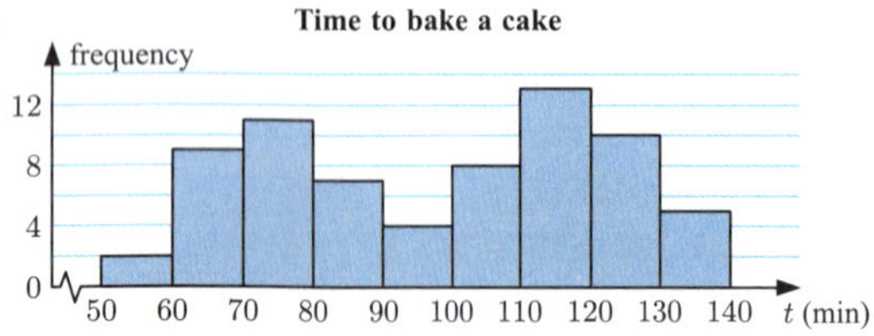

c $110 \leqslant t < 120$ min **d** bimodal **e** $\approx 21.7\%$

EXERCISE 27E.1

1 **a** mean $= 6$, median $= 5.5$, mode $= 5$ and 8
b mean $= 26.2$, median $= 26$, no mode
c mean $= 113$, median $= 108$, mode $= 107$
d mean ≈ 151, median $= 149.5$, no mode

2 **a** mean $= \$722\,340$, median $= \$682\,900$
The mean has been affected by the extreme values.
b **i** mean selling price **ii** median selling price

3 **a** 311.42 g **b** 311.5 g
c The data is continuous and each value is different.

4 **a** 6 presents **b** 6 presents **c** 5.6 presents

5 **a** 35 chocolate almonds **b** 35 chocolate almonds
c ≈ 34.6 chocolate almonds

6 **a** mean $= 7.2$, median $= 7.5$ **b** mean $= 41$, median $= 38$

7 **a**

Score	*Frequency*
3	2
4	13
5	32
6	9
7	5
8	1
9	1

b 63 players
c positively skewed
d **i** ≈ 5.14
ii 5
iii 5

8 **a** **A**: positively skewed, **B**: negatively skewed,
C: symmetric

b **A**: mean ≈ 3.29, median $= 3$, mode $= 2$
B: mean $= 5.6$, median $= 6$, mode $= 6$
C: mean $= 4.45$, median $= 4.5$, mode $= 4$ and 5

c **i** For positively skewed data, the mean is *greater* than the median.
ii For negatively skewed data, the mean is *less* than the median.
iii For symmetric data, the mean and median are approximately *equal*.

9 105.6 **10** 99.2 km **11** $\$2\,592\,000$
12 19 data values **13** **a** $x = 9$ **b** $x = 12$
14 27 marks **15** 17.25 goals per match **16** ≈ 17.7

EXERCISE 27E.2

1 **a**

Number of times	*Frequency*	*Interval midpoint*	*Product*
0 to 4	3	2	6
5 to 9	10	7	70
10 to 14	16	12	192
15 to 19	14	17	238
20 to 24	7	22	154
Total	50		660

b ≈ 13.2 times

2 **a**

Distance (d m)	*Frequency*	*Interval midpoint*	*Product*
$10 \leqslant d < 20$	10	15	150
$20 \leqslant d < 30$	19	25	475
$30 \leqslant d < 40$	14	35	490
$40 \leqslant d < 50$	7	45	315
Total	50		1430

b ≈ 28.6 m

3 **a** $600 \leqslant A < 700$ m^2 **b** ≈ 661 m^2

4 **a** positively skewed
b

Weight (w kg)	*Frequency*
$0 \leqslant w < 1$	10
$1 \leqslant w < 2$	25
$2 \leqslant w < 3$	20
$3 \leqslant w < 4$	15
$4 \leqslant w < 5$	10
$5 \leqslant w < 6$	10
$6 \leqslant w < 7$	5
$7 \leqslant w < 8$	5
Total	100

c ≈ 3.15 kg

5 **a** 27.0875 min
b

Time (t min)	*Frequency*
$10 \leqslant t < 15$	3
$15 \leqslant t < 20$	1
$20 \leqslant t < 25$	6
$25 \leqslant t < 30$	4
$30 \leqslant t < 35$	5
$35 \leqslant t < 40$	5
Total	24

c ≈ 27.1 min, the estimated mean and exact mean are equal to 1 decimal place.

6 **a** 71.225 marks
b

Chemistry exam mark	*Frequency*
30 to 39	1
40 to 49	2
50 to 59	7
60 to 69	6
70 to 79	10
80 to 89	10
90 to 99	4
Total	40

c ≈ 71.5 marks, which is fairly close to the exact mean.

EXERCISE 27F

1 **a** 60 students
b **i** The shortest jump was between 3 m and 3.5 m.
ii The longest jump was between 5 m and 5.5 m.
c **i** 20 students **ii** 15 students **d** ≈ 4.2 m

2 **a**

Height (h cm)	*Frequency*	*Cumulative frequency*
$120 \leqslant h < 130$	5	5
$130 \leqslant h < 140$	7	12
$140 \leqslant h < 150$	12	24
$150 \leqslant h < 160$	10	34
$160 \leqslant h < 170$	6	40

b 40 dancers
c
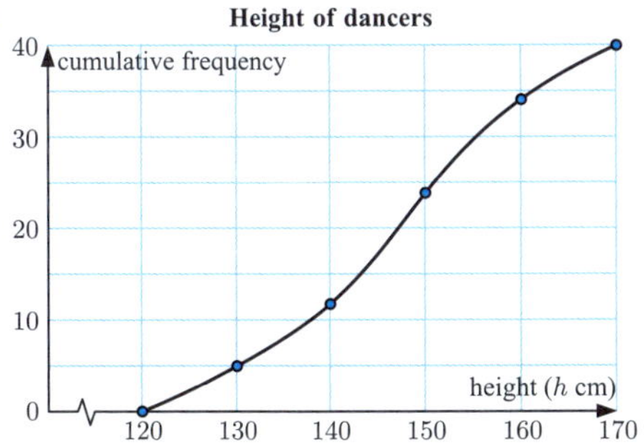

d **i** ≈ 30 dancers **ii** ≈ 23 dancers **e** ≈ 147 cm

3 **a**

Speed (v km/h)	*Frequency*	*Cumulative frequency*
$200 \leqslant v < 220$	12	12
$220 \leqslant v < 240$	16	28
$240 \leqslant v < 260$	21	49
$260 \leqslant v < 280$	18	67
$280 \leqslant v < 300$	13	80

b
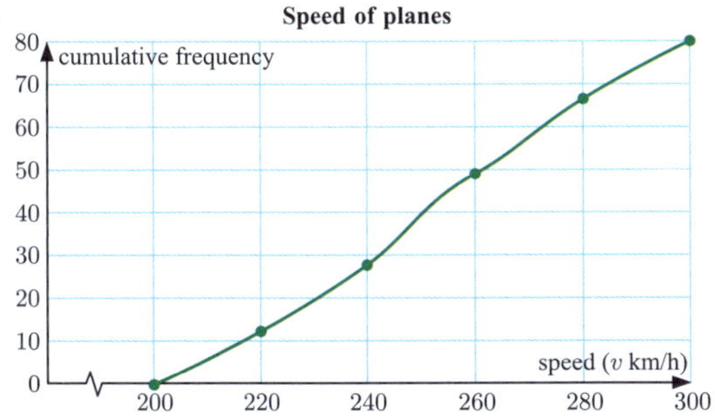

c ≈ 60 planes **d** ≈ 250 km/h

4 **a** ≈ 83.9 kg
b
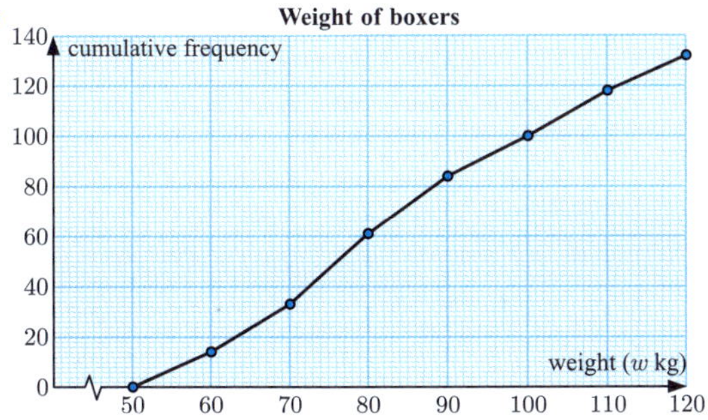

c ≈ 24 boxers **d** ≈ 83 kg

EXERCISE 27G

1 12

2 **a** **i** 19 **ii** 10 **iii** $Q_1 = 6$, $Q_3 = 17.5$ **iv** 11.5
b **i** 13 **ii** 11 **iii** $Q_1 = 7$, $Q_3 = 14$ **iv** 7
c **i** 38 **ii** 39 **iii** $Q_1 = 29$, $Q_3 = 50$ **iv** 21
d **i** 6.9 **ii** 8.6 **iii** $Q_1 = 7.05$, $Q_3 = 12.4$ **iv** 5.35

3 **a** \$120 **b** \$90 **c** \$150 **d** \$60

4 **a** 6.0 cm **b** 10.1 cm **c** 4.1 cm **d** 8.2 cm
e 7.3 cm **f** 8.95 cm **g** 1.65 cm

5 **a** **i** 5.1 min **ii** 2.35 min **iii** 1.4 min
iv 3.7 min **v** 2.3 min
b **i** 50% of the waiting times were greater than 2.35 minutes.
ii 75% of the waiting times were less than 3.7 minutes.
iii The minimum waiting time was 0.1 minutes and the maximum waiting time was 5.2 minutes. The waiting times were spread over 5.1 minutes.

6 **a** 109 cars **b** 288 people **c** range = 5, IQR = 3

7 **a** range = 27, IQR = 12 **b** range = 38, IQR = 12
c range = 47, IQR = 14.5 **d** range

EXERCISE 27H

1 **a** 3 **b** 7.5 **c** 11 **d** 13 **e** 15 **f** 12 **g** 5.5

2 **a** **i** 98 marks **ii** 30 marks **b** 73 marks
c 68 marks **d** 75% **e** 22 marks **f** no

3 **a** min = 7, $Q_1 = 9.5$, median = 12, $Q_3 = 14.5$, max = 17
b range = 10, IQR = 5
c
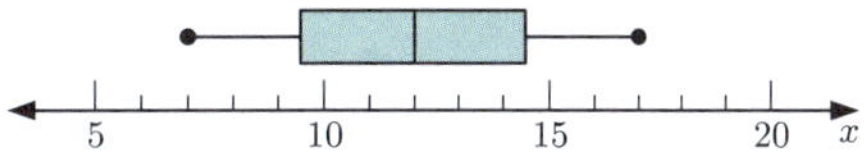

4 **a** min = 2 beans, $Q_1 = 5$ beans, median = 6 beans, $Q_3 = 8$ beans, max = 13 beans
b range = 11 beans, IQR = 3 beans
c
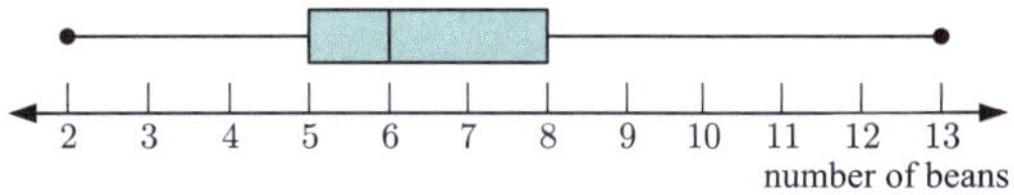

5 **a** min = 13 tickets, $Q_1 = 17$ tickets, median = 21 tickets, $Q_3 = 24$ tickets, max = 28 tickets
b range = 15 tickets, IQR = 7 tickets
c
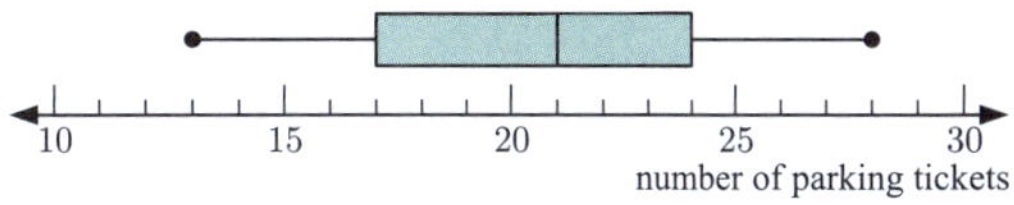

EXERCISE 27I

1 **a** California: 30 classes; Oregon: 11 classes
b California: negatively skewed;
Oregon: approximately symmetric
c The selected classes from *California* generally have larger class sizes.

2 **a** 2021 season **b** 2021 season

3 **a** **i** 50% **ii** 75% **b** bus B
c bus A; its range is greater.

4 **a** *The Rookies*: mean = 1189, median = 1180;
Best Mates: mean = 1137, median = 1120
b *The Rookies*: range = 282, IQR = 131;
Best Mates: range = 438, IQR = 155
c *The Rookies* **d** *Best Mates*

5 **a**

	Maria	*Sophie*
min.	0	0
Q_1	2	1
median	3	2.5
Q_3	4	4
max.	6	8

b **i, ii** Maria: range = 6 runs, IQR = 2 runs; Sophie: range = 8 runs, IQR = 3 runs

c Maria was more consistent (her scores had less spread) and had a higher median than Sophie. Sophie had a higher maximum but her scores were positively skewed.

6 **a**

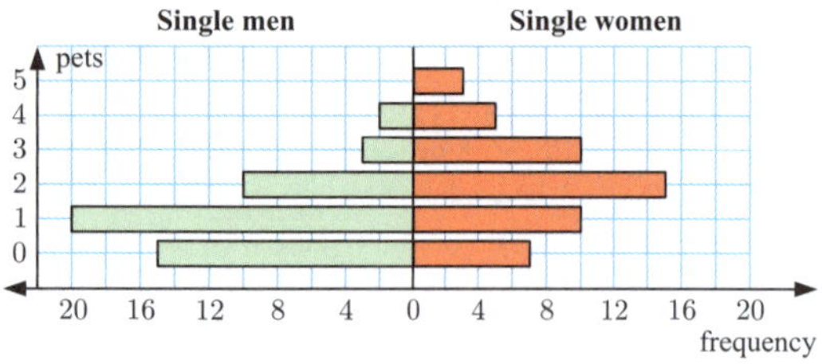

b Single men: mean = 1.14, median = 1;
Single women: mean = 2.1, median = 2

c In general, single women own more pets than single men.

7 **a**

Drinks		*Snacks*
8 7 5 4 3 2	2	3 6 6 8
9 9 7 6 6 3 2 2 1 0	3	1 1 5 7 7 8 9
9 8 8 7 4 2 1 0	4	0 2 2 3 4 6 7 8 9
7 6 5 2 2 0	5	0 0 1 2 3 3 6 8
8 7 4 1 0	6	0 1 2 6 7 7 8
5 4 2	7	0 3 3 4 7
4 1	8	*Scale*: 2 | 3 means 23

b Drinks: positively skewed;
Snacks: approximately symmetric

c Drinks: mean ≈ 47.0, median = 45.5;
Snacks: mean ≈ 49.8, median = 49.5

d Drinks: range = 62, IQR = 26;
Snacks: range = 54, IQR = 23

e snacks **f** drinks

8 **a**

	West Indies	*India*
min.	4	2
Q_1	11	6.5
median	19	25
Q_3	54.5	43
max.	123	99

b

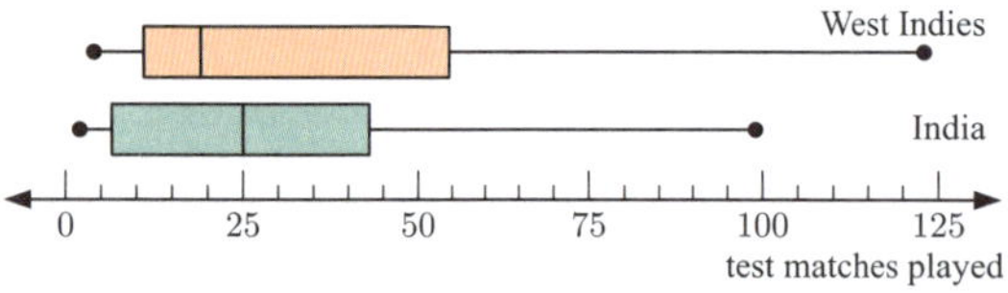

c The median number of matches played is higher for India, but all of the other five-number summary statistics are higher for the *West Indies*.

d West Indies has a greater spread than India.

9 **a**

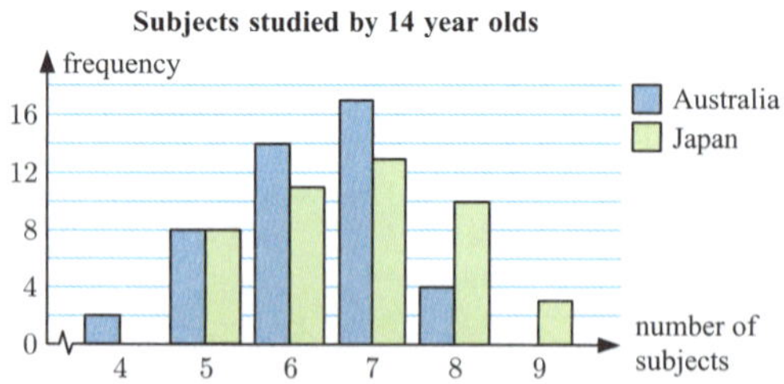

b Australia: mean ≈ 6.29 subjects, median = 6 subjects, range = 4;
Japan: mean ≈ 6.76 subjects, median = 7 subjects, range = 4

c The selected students from *Japan* generally studied more subjects.

10 **a** Group A: 30 members; Group B: 20 members

b Group A: negatively skewed;
Group B: approximately symmetric

c Group A: mean ≈ 15.3 years, median = 15 years;
Group B: mean = 15.8 years, median = 16 years

d Group A: range = 9 years, IQR = 3 years;
Group B: range = 6 years, IQR = 2.5 years

e Group A **f** Group B

11 **a**

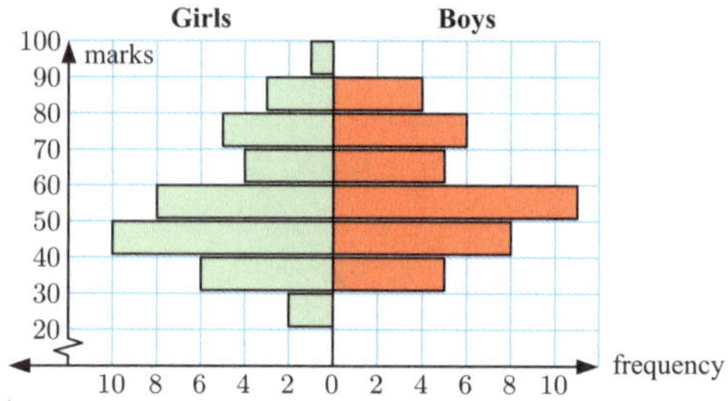

b The distribution of the girls' marks are more positively skewed and have a larger range compared to the distribution of the boys' marks.

c girls: ≈ 55.2 marks; boys: ≈ 58.3 marks
The boys generally performed better on the test than the girls.

12 **a** Hugh wants to determine which machine causes fewer faulty screw caps.

b Each day, the same number of bottles are randomly selected from each machine.

c

	Machine A	*Machine B*
min.	4	3
Q_1	5	7
median	6	8
Q_3	6	9
max.	9	11

d

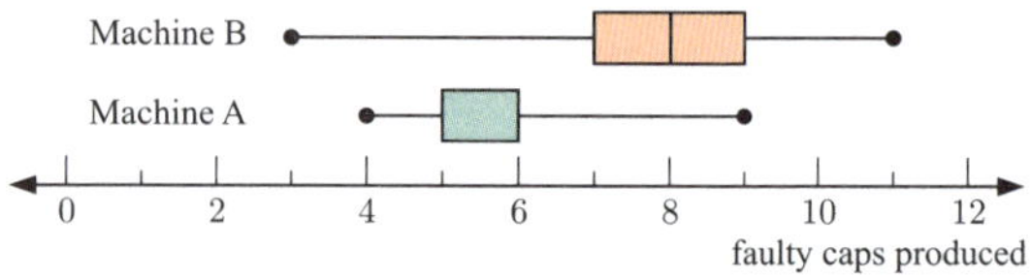

e Machine A: 5.98 caps; Machine B: 7.56 caps

f Machine B; on the parallel box plot, the box and whiskers for machine B are longer than those for machine A.

g Machine A generally causes fewer faulty caps because it has a lower mean and median. On the parallel box plot, the box for machine A is also further to the left.

13 **a**

Weight (w kg)	*Before*	*After*
$60 \leqslant w < 70$	0	1
$70 \leqslant w < 80$	5	10
$80 \leqslant w < 90$	9	5
$90 \leqslant w < 100$	9	11
$100 \leqslant w < 110$	6	4
$110 \leqslant w < 120$	2	1
$120 \leqslant w < 130$	1	0

b

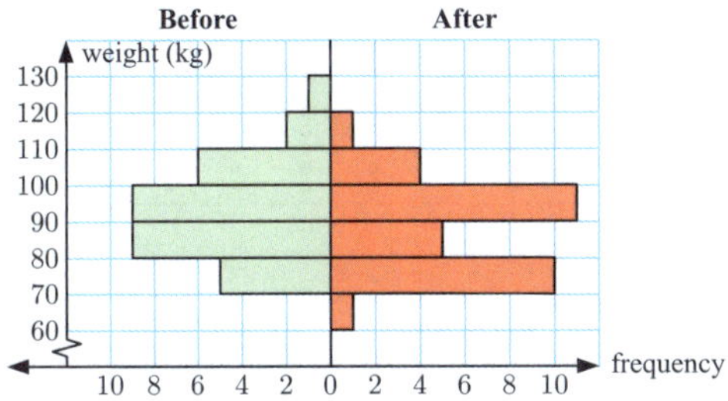

c before: positively skewed; after: bimodal

d before: mean ≈ 92.8 kg, median = 93 kg;
after: mean ≈ 87.7 kg, median = 89.5 kg

e Yes, the mean and median weights were both lower after the diet and exercise program.

REVIEW SET 27A

1 **a** continuous **b** discrete

2 **a** categorical

b
- The sample is only of Chinese people already at the airport, it is not representative of the whole population.
- The responses could be influenced by the time of year.

3 **a**

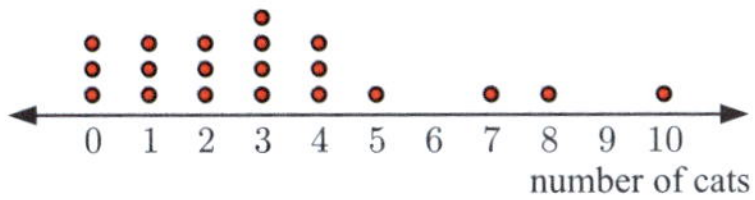

b 3 cats, it is the most common number of cats Marion saw.

c mean = 3.15 cats, median = 3 cats

4 **a** 49 businesses **b** 15 businesses **c** ≈ 26.5%

d positively skewed **e** ≈ 3.51 employees

5 **a** 5 | 1 means that a calf weighs 51 kg. **b** 28 kg

c 37 kg **d** $\frac{7}{18}$ **e** negatively skewed

6 **a**

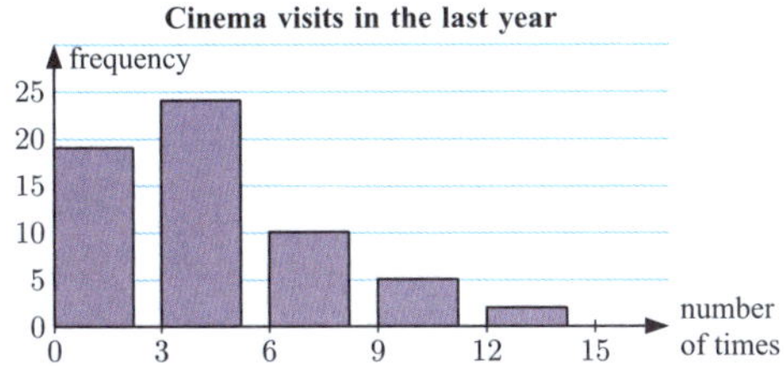

b No, we do not know the exact number of times people in the "0 to 2" class interval visited.

c ≈ 56.7% **d** positively skewed

7 **a** 10 days **b** 50% **c** 160 to 179 vehicles

d ≈ 159 vehicles

8 15

9 **a**

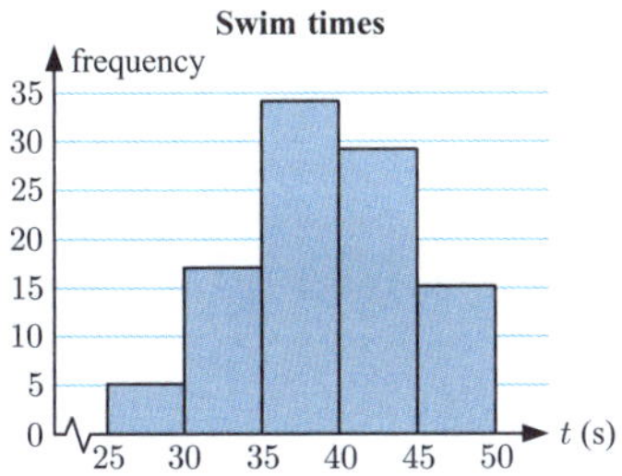

b $35 \leqslant t < 40$ s

c 22%

d ≈ 39.1 s

10 **a** **i** **A** **ii** **C** **iii** **B**

b **A**: approximately symmetric; **B**: bimodal;
C: approximately symmetric

c No, the graphs show the number of students in each class interval, not the height of any particular students.

11 **a** 80 students **b** **i** ≈ 19 students **ii** ≈ 13 students

c ≈ 25 marks

12 **a** min = 81, $Q_1 = 98$, median = 101.5, $Q_3 = 105.5$, max = 112

b IQR = 7.5. The middle 50% of the data lies within an interval of length 7.5.

c

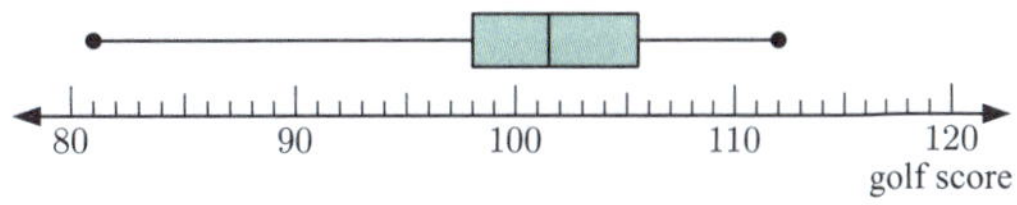

13 **a**

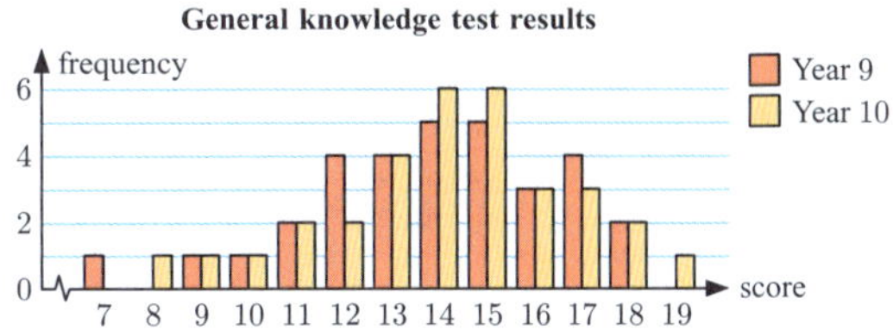

b **i** Year 9: ≈ 13.9 marks; Year 10: ≈ 14.2 marks

ii Year 9: 14 marks; Year 10: 14 marks

iii Year 9: 11, Year 10: 11 **iv** Year 9: 4, Year 10: 3

c Year 9 **d** Year 10

14 **a**

	start	*end*
i	142 s	140 s
ii	180 s	170 s
iii	160 s	153 s
iv	38 s	30 s
v	20 s	13 s

b The "start of the season" times show greater variability as indicated by greater range and IQR values.

c **i** 50% **ii** 25%

d Yes, there was an improvement. The median time improved by 7 seconds. The slowest and fastest times improved also.

15 **a**

Doctors		*Lawyers*
	17	1
	18	
9 4	19	7
9 7 3 0	20	6
9 7 7 6 5 3 2 1 0	21	0 4 6 9
8 5 2	22	0 1 4 5 6 7 8 9
5 2	23	2 3 6 8 8

Scale: 20 | 6 means 206 hours

b Yes, "171" is an outlier.

c

	Mean (hours)	*Median* (hours)
Doctors	214.2	214
Lawyers	220.5	224.5

The mean and median time is higher for lawyers.

d From the samples above, it appears that *lawyers* generally work longer hours than *doctors*.

REVIEW SET 27B

1 **a** discrete numerical **b** categorical

c continuous numerical

2 **a**

b mean = 15.625 goals, median = 15.5 goals

3 **a**

0	4 8 9
1	3 5 7
2	1 5 6
3	1 4 5 6 7 9
4	1

Scale: 1 | 3 means 1.3 cm

b 62.5%

4 **a** negatively skewed, "1" is an outlier **b** bimodal

c approximately symmetric

5 **a** 12 **b** 43 **c** 29 **d** 20 **e** 34.5 **f** 31 **g** 14.5

6 **a**

Number of call-outs	*Tally*	*Frequency*
0 to 9	||	2
10 to 19	~~||||~~ |	6
20 to 29	~~||||~~ ~~||||~~	10
30 to 39	~~||||~~ ||	7
	Total	25

b 20 to 29 call-outs

c **Fire department call-outs**

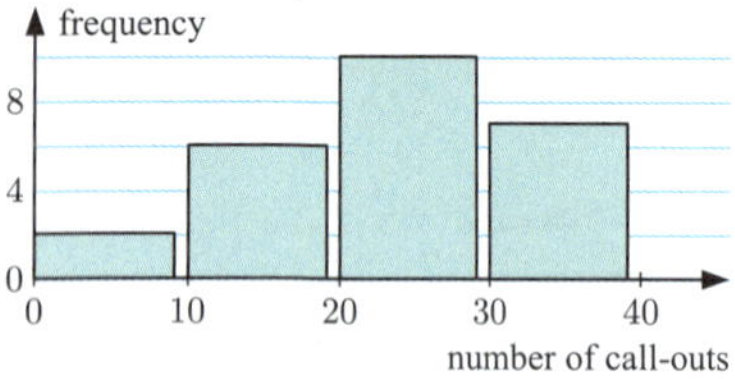

d 17 days **e** 8% **f** negatively skewed

7 ≈ 8.53 potatoes per plant **8** $x = 7$ **9** 20.7 cm

10 **a** 2 kg **b** **Mass of domestic cats**

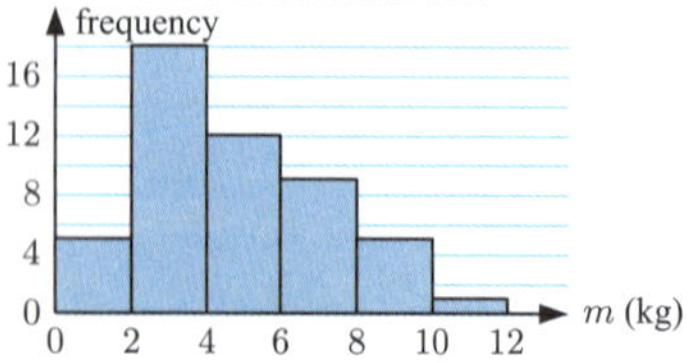

c $2 \leqslant m < 4$ kg **d** ≈ 4.76 kg **e** 12%

11 **a**

Time (t *min*)	*Frequency*	*Cumulative frequency*
$175 \leqslant t < 190$	11	11
$190 \leqslant t < 205$	35	46
$205 \leqslant t < 220$	74	120
$220 \leqslant t < 235$	32	152
$235 \leqslant t < 250$	8	160

b **Cumulative frequency graph of Ben Nevis climb data**

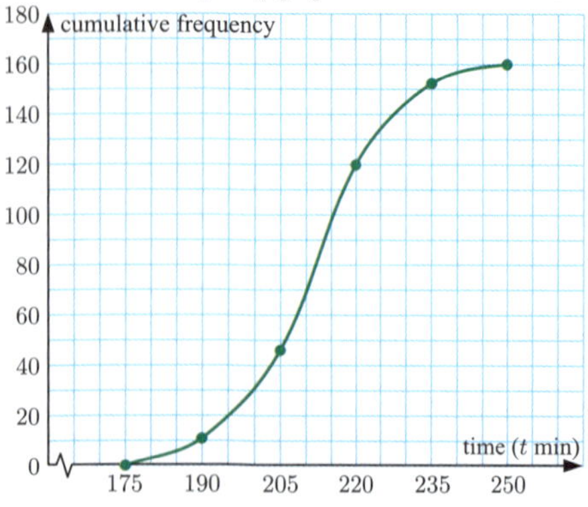

c ≈ 212 min **d** ≈ 20 trampers

12 **a** **Ages of children at a scouts camp**

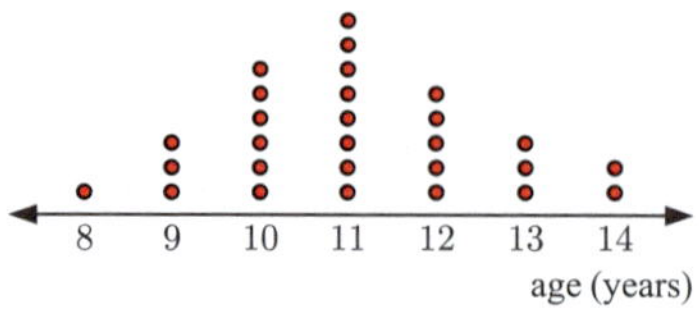

b symmetric

c min = 8 years, Q_1 = 10 years, median = 11 years, Q_3 = 12 years, max = 14 years

d

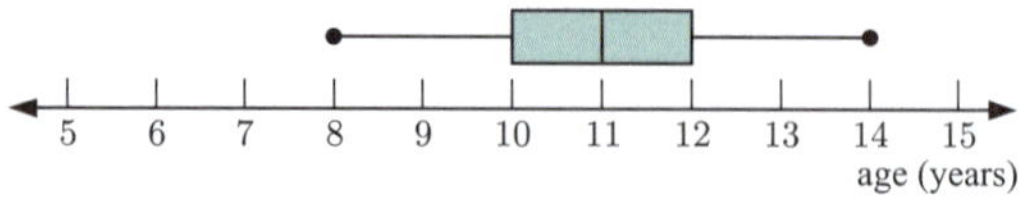

13 **a**

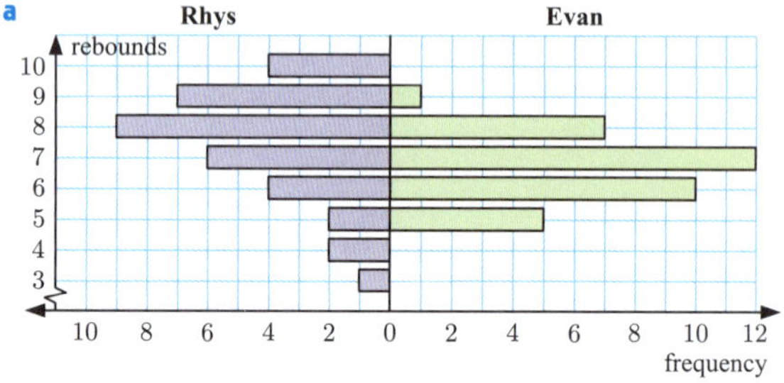

b Rhys

14 **a**

	Frank	*Competitor*
min.	128	76
Q_1	136	87
median	142.5	89
Q_3	146	92
max.	163	97

b

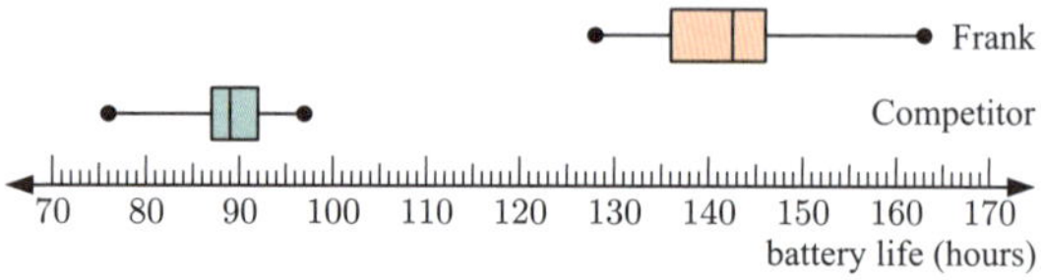

c Frank: ≈ 141.9 hours; Competitor: ≈ 89.2 hours

d No. The values in the five-number summary for Frank's batteries are not double their corresponding value for the competitor's batteries. The mean for Frank's batteries is not double the mean of the competitor's batteries.

15 **a** Tigers: 13 players; Panthers: 15 players

b Tigers: $175 \leqslant h < 180$ cm; Panthers: $165 \leqslant h < 170$ cm

c Tigers: approximately symmetric; Panthers: bimodal

d Tigers: ≈ 177.2 cm; Panthers: ≈ 176.6 cm

e The Tigers appear to have a slightly greater average height. This conclusion may be unreliable as we can only estimate the means, and the estimates of the means are quite close to each other.

INDEX